D1

Harry's Cosmeticology

Authors and revisers include the following:

P. Alexander, BSc, CChem, FRSC (Unilever Ltd)
S. F. Bloomfield, BPharm, PhD, MPS (Chelsea College)
C. Bouillon, Docteur ès-Sciences (L'Oreal)
R. J. Clarkson, LRSC, MInstPkg (Roure Bertrand Dupont)
F. J. G. Ebling, DSc, PhD, FIBiol (University of Sheffield)
R. A. Gunn-Smith, MInstPkg (Metal Box Ltd)
B. Jayasekara, BPharm (Ceyl), DipBact, MRSH (Yardley of London Ltd)
G. Kalopissis, Docteur ès-Sciences (L'Oreal)
R. E. Leak, BPharm, MPS (Chelsea College)
J. R. Nixon, BPharm, PhD, MPS (Chelsea College)
W. E. Parish, MA, PhD, BVSc, FRCPath (Unilever Research)
R. P. Reeves, CChem, FRSC (Consultant)
P. M. Scott, MInstPkg (Avon Cosmetics Ltd)
H. Smith, BTech, PhD (formerly of Elida Gibbs Ltd)
J. Thompson, BSc, PhD (formerly of Wilkinson-Match Ltd)
K. Tomlinson, BA, MSc, CChem, FRSC (formerly of Colgate-Palmolive Ltd)
D. A. Wheeler, BSc, MSc, CChem, MRSC (Elite Assembly Co.)

Patents

In this book, the patent literature has been treated as a source of information. Certain formulae and processes have been included in the interests of science, notwithstanding the existence of actual or potential patent rights.

Mention of a patent does not necessarily indicate that the patent is currently in force, but in so far as materials and processes are protected by letters-patent, their inclusion neither conveys nor implies licence to manufacture. Each manufacturer should ascertain for himself the patent position existing in his own country at that time.

Legislation

Legislation concerned with permitted materials, limitations on use and methods of sale of toilet preparations is in a state of continual change, notably in the USA and the European Economic Community. While every effort has been made to take count of the latest position, inclusion of a particular ingredient in any one illustrative formula cannot be taken as indicating that this formula will be within the limits of legal permission in any one country at the time when it may be under consideration. As with patents (above) every manufacturer must ascertain for himself the legal position existing in his country or that to which he exports at that time.

Harry's Cosmeticology

Seventh edition

Edited by
J. B. Wilkinson, MA, BSc, CChem, FRSC
R. J. Moore, BSc, CChem, MRSC, MIInfSc

George Godwin London

First published in New York 1940 by Leonard Hill Books
Second edition (London) 1944
Third edition (New York) 1947
Fourth edition (London) 1955
Fifth edition (London) 1962
Sixth edition (London) 1973
Seventh edition published in Great Britain 1982 by
George Godwin
Longman House
Burnt Mill, Harlow, Essex, UK
A division of Longman Group Ltd, London

British Library Cataloguing in Publication Data

Harry, Ralph G.
Harry's cosmeticology—7th ed.
1. Cosmetics
I. Title II. Wilkinson, J. B. III. Moore, R. J.
668'.55 TP983

ISBN 0-7114-5679-8

Published in the USA by Chemical Publishing Company, Inc., New York

Printed and bound by Mansell Bookbinders Ltd, Witham, Essex

Contents

Preface xvii

PART ONE: THE SKIN AND SKIN PRODUCTS 1

1 **The Skin** 3
Introduction 3
Epidermis and keratinizing system 5
Pigmentary system 7
Langerhans cells 9
Dermis 10
Nerves and sense organs 11
Blood vessels 12
Eccrine sweat glands 13
Hair follicles 14
Sebaceous glands 15
Apocrine glands 16
Common disorders of the skin 16

2 **Irritation and Sensitization of the Skin** 27
Introduction 27
Irritants and inflammation 28
Hypersensitivity and allergy 33
Tests to predict irritation or sensitization 38

3 **Nutrition and Hormonal Control of the Skin** 42
Nutrition of the skin 42
Skin conditions related to nutritional deficiency 43
Percutaneous absorption 45
Hormones 45

4 **Skin Creams** 50
Introduction 50
Classification of skin creams 51
Cleansing creams 53
Night and massage creams 60
Moisturizing, vanishing and foundation creams 62
Pigmented foundation creams 67
Hand creams and hand-and-body creams 69
All-purpose creams 70

5 **Astringents and Skin Tonics** 74
Introduction 74

Types of astringent 74
Astringent products 75

6 Protective Creams and Hand Cleansers 82
Introduction 82
Barrier materials—protective creams and gels 83
Hand cleansers 88

7 Bath Preparations 92
Foam baths 92
Introduction 92
Formulation of foam baths 93
Types of product 98
Product assessment 100
Bath salts 101
Ingredients and formulation 101
Bath oils 103
Introduction 103
Floating or spreading oils 104
Dispersible or blooming oils 106
Soluble oils 107
Foaming oils 107
After-bath products 108
Body or dusting powders 108
After-bath emollients 109

8 Skin Products for Babies 111
Introduction 111
Skin problems in babies 111
Functional requirements of baby products 112
Safety of baby products 113
Example formulations 114
Cleansing of nappies (diapers) 117

9 Skin Products for Young People 119
Introduction 119
Adolescent skin problems 119
Products for oily skins 120
Specific treatments for acne 121

10 Antiperspirants and Deodorants 124
Introduction 124
Perspiration and its control 124
Antiperspirant ingredients 127
Evaluation of antiperspirants 130
Mechanism of deodorants and deodorant ingredients 132
Assessment of deodorants 133

Product formulation—antiperspirants 134
Product formulation—deodorants 139

11 Depilatories 142
Introduction 142
Epilation 142
Electrolysis 144
Chemical depilation 144
Facial depilatories for black skin 150
The 'ideal' depilatory 151
Evaluation of depilatory efficacy 152

12 Shaving Preparations 156
Wet shaving preparations 156
Introduction 156
Beard softening cream 157
Lather shaving cream 159
Lather shaving stick 161
Aerosol shaving foams 161
Brushless or non-lathering cream 171
Brushless shaving stick 173
Novel compositions for wet shaving 174
Dry shaving preparations 175
Introduction 175
Pre-electric shave lotion 176
Collapsible foam pre-electric shave lotion 178
Pre-electric shave gel stick 178
Pre-electric shave talc stick 178
Pre-electric shave powder 179
After-shave preparations 180
After-shave lotion 181
Quick-break foam after-shave 184
Crackling foam aerosol after-shave lotion 185
After-shave gel 185
After-shave cream and balm 186
After-shave powder 187

13 Foot Preparations 190
Introduction 190
Influence of footwear 190
Foot malodours 191
Foot ailments 191
Foot infections 192
Foot care and hygiene 193
Bathing the feet 193
Foot powders 196
Foot sprays 197
Foot creams 198
Corn and callus preparations 200
Chilblain preparations 201

Athlete's foot preparations 202
Other developments 204

14 Insect Repellents 206
Introduction 206
Repellent materials 206
Formulation 213

15 Sunscreen, Suntan and Anti-sunburn Preparations 222
Sunlight and the human body 222
Introduction 222
Tanning 222
Beneficial effects of sunlight 223
Adverse effects of sunlight 223
Solar radiation and its effect on skin 225
Protective mechanism of the skin 229
Sunscreens and suntan preparations 231
Introduction 231
Sunscreen agents 231
Evaluation of sunscreen preparations 242
Formulation of sunscreens 251
Palliative preparations 256
Artificial suntan preparations 258

16 Skin Lighteners or Bleaches 264
Colour of the skin 264
Chemistry of melanin 265
Mechanism of depigmentation 265
Skin lightening agents and formulations 266

17 Face Packs and Masks 276
Introduction 276
Wax-based systems 276
Rubber-based systems 277
Vinyl-based systems 277
Hydrocolloid-based systems 278
Earth-based systems (argillaceous masks) 280
Anti-wrinkle preparations 282

18 Face Powders and Make-up 285
Face powder 285
Function and properties 285
Covering power 285
Absorbency 289
Slip 292
Adhesion 293
Bloom 294
Colour 295

Perfume 296
Formulation 297
Manufacture 299
Compact powder 301
Cake make-up 304
Make-up cream 307
Liquid powder 307
Liquid make-up 310
Stick make-up 311

19 Coloured Make-up Preparations 314
Lipstick 314
Introduction 314
Ingredients of lipsticks 314
Example formulations 323
Manufacture of lipsticks 327
Transparent lipsticks 330
Lip salves 330
Liquid lipsticks 332
Rouge 333
Introduction 333
Dry rouge (compact rouge) 334
Wax-based rouge 336
Cream rouge 336
Liquid rouge 340
Eye make-up 341
Introduction 341
Mascara (eyelash cosmetic) 341
Eyeshadow 347
Eyeliner 351
Eyebrow pencils 352

20 The Application of Cosmetics 355
Introduction 355
Care and cleansing of the skin 355
Cosmetic application 357

PART TWO: THE NAILS AND NAIL PRODUCTS 361

21 The Nails 363
Biology of the nails 363
Pathology of the nails 365

22 Manicure Preparations 369
Cuticle remover 369
Nail bleach 371

Nail cream 372
Nail strengthener 372
Nail white 373
Nail polish 374
Nail lacquer (nail varnish) 375
Introduction 375
Ingredients of nail lacquer 376
Formulation 383
Manufacture of nail lacquer 385
Base coats and top coats 385
Enamel remover 386
Nail drier 389
Plastic fingernails and elongators 389
Nail mending compositions 391

PART THREE: THE HAIR AND HAIR PRODUCTS 395

23 The Hair 396
Introduction 396
The hair follicle 396
Hormonal influences 401
Nutritional influences 402
Chemistry of hair 403
Hair colour 413
Hair disorders 416
Dandruff 419

24 Shampoos 427
Introduction 427
Detergency 428
Evaluation of detergents as shampoo bases 429
Raw materials for shampoos 431
Principal and auxiliary surfactants 432
Additives 443
Formulation of shampoos 448
Clear liquid shampoos 449
Liquid cream or lotion shampoos 450
Solid cream and gel shampoos 451
Oil shampoos 453
Powder shampoos 453
Aerosol shampoos 454
Dry shampoos 454
Conditioning shampoos 455
Baby shampoos 457
Anti-dandruff and medicated shampoos 458
Acid-balanced shampoos 459
Safety of shampoos 460

25 Hair Setting Lotions, Sprays and Dressings 470
Use and purpose of hair dressings 470
Women's hair dressings 470
Setting lotions 470
Heated curlers and blow drying 473
Hair sprays 474
Men's hair dressings 483
Formulation 484
Brilliantines 484
Non-oily fixatives 489
Aerosols 490
Emulsions 491
Gels 494

26 Hair Tonics and Conditioners 498
Introduction 498
Formulation of medicated hair tonics 498
Conditioners 506
Evaluation of conditioning 512
Hair thickeners 512
Rinses 513

27 Hair Colorants 521
Introduction 521
Hair colouring systems 521
Characteristics of an ideal hair colorant 522
The process of hair colouring 524
Temporary hair dyes 526
Dyestuffs 526
Types of commercial temporary product and their formulation 528
Semi-permanent colorants 528
Dyestuffs 528
Commercial semi-permanent products and their formulation 532
Permanent hair dyes 533
Bases 534
Couplers or modifiers 535
Formation of colours in the hair 535
Toxicity and dangers of para dyes 538
Formulation of permanent hair dyes 540
Other dyes for hair 544
Aromatic polyhydroxy compounds 544
Vegetable hair dyes 545
Metallic hair dyes 546
Hair dye removers 547
Bleaching and lightening 547

28 Permanent Waving and Hair Strengtheners 555
Introduction 555
Chemistry of hair waving 556
Evaluation of permanent waving 563

Hot waving processes 567
Cold waving processes 569
Tepid 'warm air' wave 574
Roller and pin permanent waves 574
Instant permanent waves 575
Perfuming of thioglycollate lotions 575
Toxicity 575
Hair strengthening preparations 575

29 Hair Straighteners 581
Introduction 581
Hot comb method 581
Caustic preparations 581
Chemical hair reducing agents 583

PART FOUR: THE TEETH AND DENTAL PRODUCTS 587

30 The Tooth and Oral Health 588
Introduction 588
The tooth and its surroundings 588
Structure of the tooth 588
Saliva 590
Acquired integuments of the tooth 590
Major problems of oral health 593
Magnitude of the problem 593
Dental caries 594
Periodontal disease 599
Use of prophylactic toothpastes 600
Active ingredients 600

31 Dentifrices 608
Basic requirements of a dentifrice 608
Toothpastes 609
Basic structure 609
Ingredients 609
Formulation of toothpastes 615
Manufacture of toothpastes 616
Toothpowders 617
Manufacture of toothpowders 618
Solid dentifrice 618
Performance tests 618
Abrasive action 619
Lustre (gloss or polish) 620
The toothbrush and toothbrushing 621
Denture cleansers 622

32 Mouthwashes 626
Introduction 626
Choice of antibacterial agent 627

Flavouring of mouthwashes 628
Aerosol mouth fresheners 629

PART FIVE: PRODUCT INGREDIENTS AND MANUFACTURE 631

33 Surface-active Agents 632
Introduction 632
Classification of surfactants 632
Properties of surface-active agents 633
Selection and use of surface-active agents 637
Biological properties of surface-active agents 639

34 Humectants 641
Introduction 641
Drying out 641
Types of humectant 642
Hygroscopicity 643
Stability of emulsions 650
Safety 651
Skin moisturizing 651

35 Antiseptics 653
Introduction 653
Microbial flora of the body 654
Effects of antibacterial agents on body flora 655
Antibacterial soap bars and other skin degerming preparations 657
Antimicrobial agents commonly used in antiseptic products 659

36 Preservatives 675
Introduction 675
Microbial metabolism 676
Clinical significance of contamination 678
Origins of contamination 681
Microbial growth in products 683
Preservative requirements 686
Factors influencing the effectiveness of preservatives 686
Selection of a preservative 696
Safety aspects 699
Tests for preservative effectiveness 701
Current UK regulations 703

37 Antioxidants 707
Introduction 707
General autoxidative theory 707
Antioxidants 714
Measurement of oxidation and the assessment of antioxidant efficiency 717
Choice of antioxidant 721

Phenolic antioxidants 722
Non-phenolic antioxidants 724
Photo-deterioration 725

38 Emulsions 729
Introduction 729
Basic principles 729
Stabilization of cosmetic emulsions 733
Other factors affecting the stability of emulsions 742
Practical aspects of emulsifier choice 745
Assessment of emulsion stability 749
Characteristics of emulsions 750
Determination of emulsion type 754
Quality control and emulsion analysis 754

39 The Manufacture of Cosmetics 757
Introduction 757
Mixing—and the manufacture of bulk cosmetic products 759
Solid–solid mixing 760
Manufacture of pigmented powder products 762
Mixing processes involving fluids 767
General principles of fluid mixing 767
Mixing equipment for fluids 769
Solid–liquid mixing 788
Suspension of solids in agitated tanks 792
Liquid–liquid mixing 793
Miscible liquids 793
Immiscible liquids 793

40 Aerosols 800
Introduction 800
The aerosol 800
Containers 801
Valves 804
Propellants 808
Filling of aerosols 817
Types of dispensed aerosol product 821
Two-phase systems 821
Three-phase systems 825
Personal care products with alternative propellants 831
Corrosion in aerosol containers 833
Alternative systems 839
Propellant-free dispensing pumps 843

41 Packaging 849
Introduction 849
Principles of packaging 849
Marketing and packaging 849

Technology and components 850
Plastics 850
Metals 852
Laminates 853
Glass 854
Paper and board 855
Printing and decorations 858
Package development and design 858
Technical aspects of design 859
Closures 860
Package testing and compatibility 861

42 The Use of Water in the Cosmetics Industry 864
Properties and cosmetic uses of water 864
Composition of mains water 864
Water purity requirements for cosmetics 865
Further purification of mains water 866
Distribution systems 872
Good housekeeping 876

43 Cleanliness, Hygiene and Microbiological Control in Manufacture 877
Introduction 877
Sources of contamination 878
Cleaning and disinfection 880
Cleaning staff 880
Equipment cleaning 881
Equipment disinfection 884
Parameters of cleaning, disinfection and rinsing 888
Control of contamination 891
Hazards from personnel 891
Washrooms and toilets 892
Raw materials 893
Storage areas 894
Product packaging 894
Microbiological standards 895
Conclusion 896

Appendix Proprietary Materials Cited in this Book 899

Index 915

PART ONE

The Skin and Skin Products

Chapter One

The Skin

Introduction

The skin[1–3] is not simply a protective wrap for the body; it is a busy frontier which mediates between the organism and the environment. It not only controls the loss of valuable fluid, prevents the penetration of noxious foreign materials and radiation and cushions against mechanical shock, but also regulates heat loss and transduces incoming stimuli. Moreover, by its colour, texture and odour it transmits sexual and social signals which may possibly be physiologically enhanceable by cosmetic science but certainly are culturally enhanceable by cosmetic art. For cosmeticians, whether they are concerned with the improvement of the skin by pharmacology or the prevention of damage as a result of artifice, an understanding of skin structure and function is essential.

The total area of the skin ranges from about 2500 cm^2 at birth to 18 000 cm^2 in the adult, when it weighs about 4·8 kg in men and 3·2 kg in women.

There are two main kinds of human skin: hairy and glabrous. Over most of the body the skin possesses hair follicles with their associated sebaceous glands. However, the amount of hair varies greatly; at the extreme, the scalp, with its large hair follicles, may be contrasted with the female face, which has large sebaceous glands associated with very small follicles which produce fine, short vellus hairs. The skin of the palms and soles lacks hair follicles and sebaceous glands, and is grooved on its surface by continuously alternating ridges and sulci which form patterns of whorls, loops or arches, unique to each individual, known as dermatoglyphics (Figure 1.1). Glabrous skin is also characterized by its thick epidermis and by the presence of encapsulated sense organs within the dermis.

The barriers to permeability are situated in the several layers of closely packed cells which form the overlying epidermis; mechanical protection is provided by the thicker underlying dermis which is composed mainly of connective tissue, that is, material secreted by cells and lying outside of them. Isolated epidermis is as impermeable as whole skin, whereas once the epidermis is removed the dermis is completely permeable. If the epidermal layers are progressively stripped by adhesive tape, the permeability of the skin increases, and there is little doubt that the bonded, interlocked, horny cells of the stratum corneum constitute the barrier. It is unlikely that emulsified fat on the skin surface greatly affects permeability, or that the sweat glands and hair follicles are more permeable than the surface epithelium, though material may possibly reach the sebaceous glands by the follicular route.

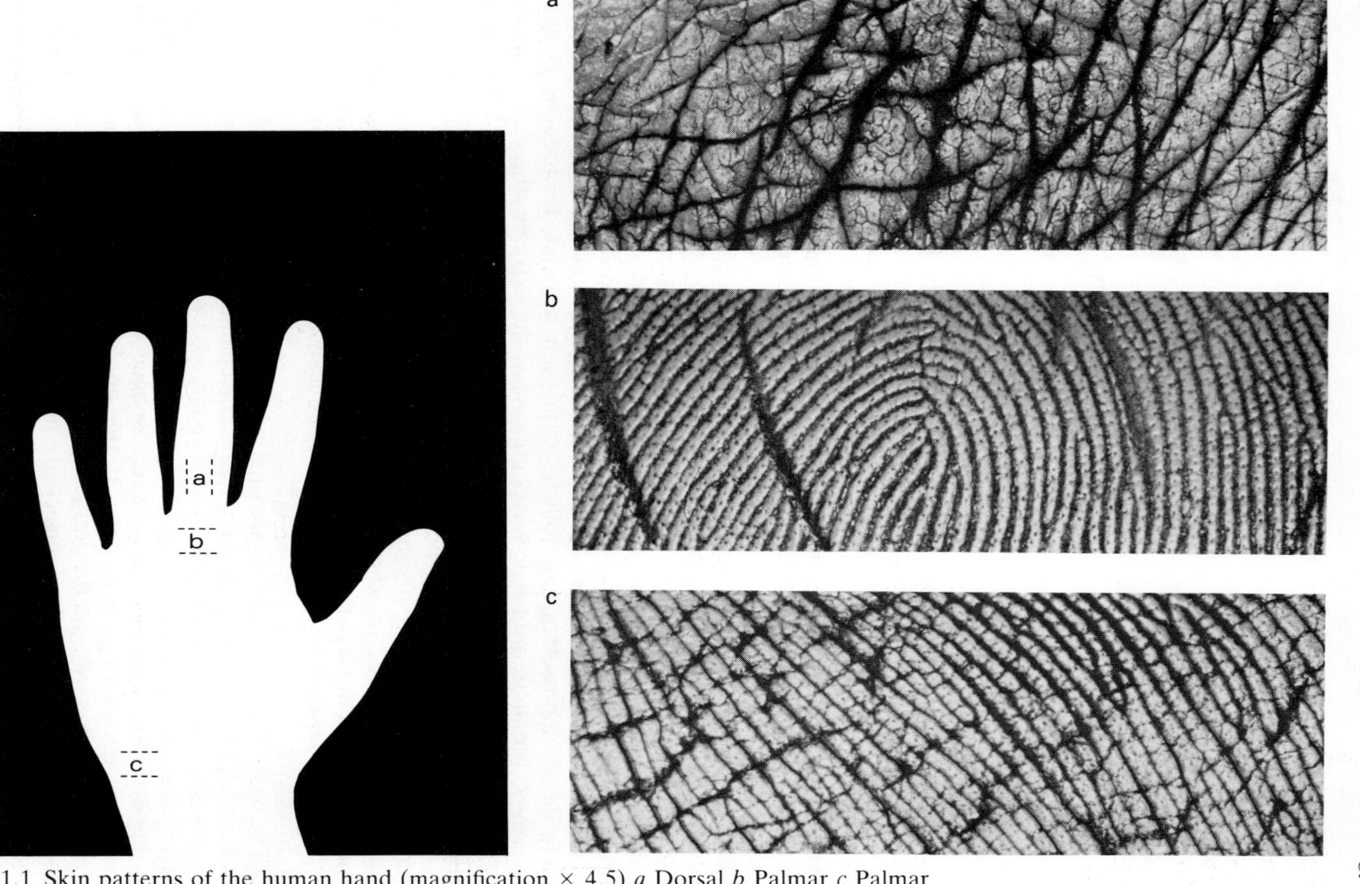

Figure 1.1 Skin patterns of the human hand (magnification × 4.5) *a* Dorsal *b* Palmar *c* Palmar

Epidermis and the Keratinizing System

The epidermis consists of a number of layers. The stratification is the result of changes in the *keratinocytes* as they move outwards from the basal layer, in which they are continuously formed by mitosis, to the skin surface where they are lost.[4–7] Three other cell types are present: *melanocytes* or pigment cells, *Langerhans cells,* which are colourless and dendritic in form, and *Merkel cells,* which are concerned with sensation (see below).

Dermo–Epidermal Junction

The dermo–epidermal junction is undulating in section; so-called rete pegs or epidermal ridges project from the epidermis into the dermis. At the junction is a basement membrane, which under the electron microscope is seen as a convoluted plasma membrane studded with *semi-* or *junctional desmosomes,* separated from the underlying electron-dense *basal lamina* by a clear *lamina lucida.* The basal lamina is anchored in the dermis below by fibrils and bundles of fine filaments.[7,8]

Stratum Basale

The *stratum basale* or *stratum germinativum* is a continuous layer which gives rise to all the keratinocytes. It is usually described as one cell thick,[7] but in thick normal or pathological epidermis it appears that mitosis may not be confined to cells in contact with the basement membrane.[9–11] Do cells destined for differentiation arise as daughters of progenitors permanently committed to cell division? One view is that both daughters equally retain the capacity to divide for a time, but that for each division a basal cell moves into the stratum spinosum, either at random[12,13] or by precedence of age.[14] However, Potten[15] supports the traditional concept of permanent stem-cells, though he concedes that daughters may remain capable of a few 'amplification' divisions before differentiating.

Between one division and the next the cell undergoes a cycle.[16–19] Immediately following mitosis (M) is a growth phase (G_1), which is succeeded by a period of active nuclear DNA synthesis (S) and a short premitotic growth phase (G_2). Each period has a *transit time*; for the complete cycle the term 'cell cycle time' should be used. The expression '*turnover time*', and its synonyms '*regeneration time*', or '*replacement time*', refer to the time for complete replacement of a cell population. Although frequently stated to be equivalent to the cell cycle time, this would only be true if all the cells were continually cycling. In fact, it is likely that there is a substantial compartment (G_0) of non-cycling cells. It is, moreover, important to distinguish the turnover time of the stratum corneum from that of the viable epidermis.

The average duration of the cell cycle has been variously estimated for normal human epidermis as 163 hours,[19] 308 hours,[20] 457 hours[21] and 213 hours,[22] and for psoriatic epidermis as 37 hours.[23] However, these measurements have assumed that in normal epidermis all cells cycle continuously. An alternative explanation is that psoriatic epidermis differs from normal not because of a shorter cell cycle, but because it has a much higher proportion of cycling cells.

The replacement time for the whole viable epidermis is probably about 42 days[24] and for the stratum corneum about 14 days,[25,26] and it is generally agreed that the times are considerably less in psoriatic skin.[27–29]

Cells of the stratum basale have large nuclei; under the electron microscope their cytoplasm reveals many ribosomes, mitochondria and, sometimes, smooth membranes. In particular, they contain numerous fine *tonofilaments,* about 5 nm in diameter, which occur mainly in loose bundles, the *tonofibrils.*

Stratum Spinosum

The *stratum spinosum* or prickle cell layer is so called because the cells are given a spiny appearance by the numerous *desmosomes* or attachment plaques at their surfaces. These were once believed to be intercellular bridges through which the tonofibrils maintained the tonus of the epidermis. Ultrastructural studies reveal that they are laminated structures. In the upper region of the stratum spinosum, *membrane-coating granules,*[30,31] also known as lamellated or Odland bodies,[32] make their appearance. These are ovoid bodies about 100–500 nm long. In the stratum intermedium they ultimately migrate towards the periphery of the cell and appear to increase in numbers in the intercellular spaces. Their function is unknown, though they appear to contain mucopolysaccharides and it has been suggested that they may constitute the intercellular cement.[31]

Stratum Granulosum

The stratum spinosum is succeeded by the *stratum intermedium*, or *stratum granulosum,* which contains basophil granules of a material called *keratohyalin.*[33]

Stratum Lucidum

The *stratum lucidum*, unstainable by the usual histological methods, can be recognized only in palmar and plantar skin.

Stratum Corneum

In the *stratum corneum*[6,7,34,35] the keratinocytes have lost their nuclei and virtually all of their cytoplasmic organelles and contents, including the keratohyalin granules. The cells are flattened and completely filled with keratin, in the form of bundles of filaments embedded in an opaque interfilamentous material. At the transition between the stratum intermedium and stratum corneum, transition cells or T-cells[7,36] are recognizable. The cornified cells in their epidermis, though not those of glabrous skin, can be shown to be arranged in regular vertical stacks, which must reflect the underlying dynamic organization.[37–42] Most authors now believe that both the filamentous structures of the lower epidermal layers and the keratohyalin of the stratum intermedium contribute to the formation of keratin.[43,44] Some, however, have held that the fibrils contribute nothing;[45–48] others have questioned the contribution of keratohyalin.[49] The most attractive, if unproven, hypothesis is that the fibrillar material, with helically arranged peptide chains, is transformed in the stratum intermedium by a sulphur-rich matrix which makes possible cystine links.[7] Various attempts to characterize chemically pure 'pre-keratin' have proposed units with molecular weights of 640 000,[50,51] 100 000–200 000[52] or 50 000.[53]

Horny cells are continuously shed from the skin surface. If skin sites are protected by cups for long periods, exfoliated material is trapped, but the thickness of the coherent stratum corneum remains unchanged.[54,55] It seems, therefore, that the horny layer desquamates at a final level which is not much influenced by external forces.

Pigmentary System

Although skin owes some of its colour[56] to red haemoglobin in the blood vessels and yellow carotenoids in the hypodermal fat, the major determinant is a dark pigment, *melanin,* which is the product of special cells known as *melanocytes.* The skin colour of human subjects can be measured by reflectance spectrophotometry.[57]

Melanocytes are derived from the neural crest in the embryo[58,59] and migrate to many tissues of the body, including the basal layers of the epidermis and the hair bulb. They differ from other cells of the stratum basale by the possession of dendritic (that is, finger-like) processes (Figure 1.2), by which they transfer

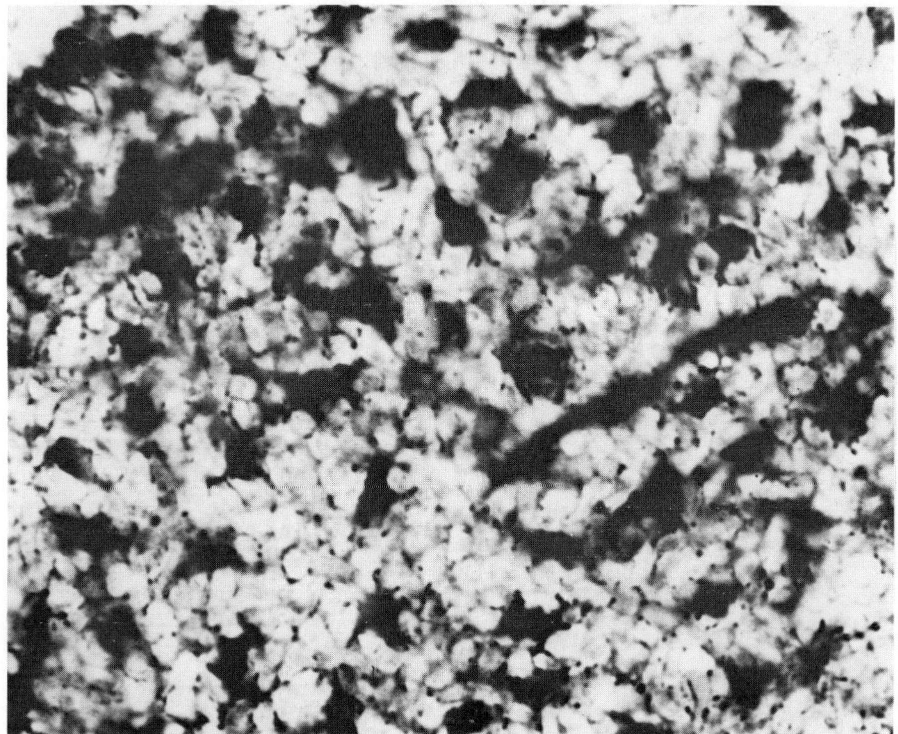

Figure 1.2 Melanocytes on underside of epidermis (magnification × 500): finger-like processes extend from the centres of the melanocytes

pigment to a group of keratinocytes, the whole forming an 'epidermal melanin unit'.[60] They have no desmosomes.

The characteristic feature of melanocytes is a special cytoplasmic organelle known as a *melanosome* (Figure 1.3) on which the melanin is formed by the action of the enzyme tyrosinase. The melanosomes arise as spherical, membrane-bounded vesicles in the zone of the Golgi apparatus. Filaments are at first

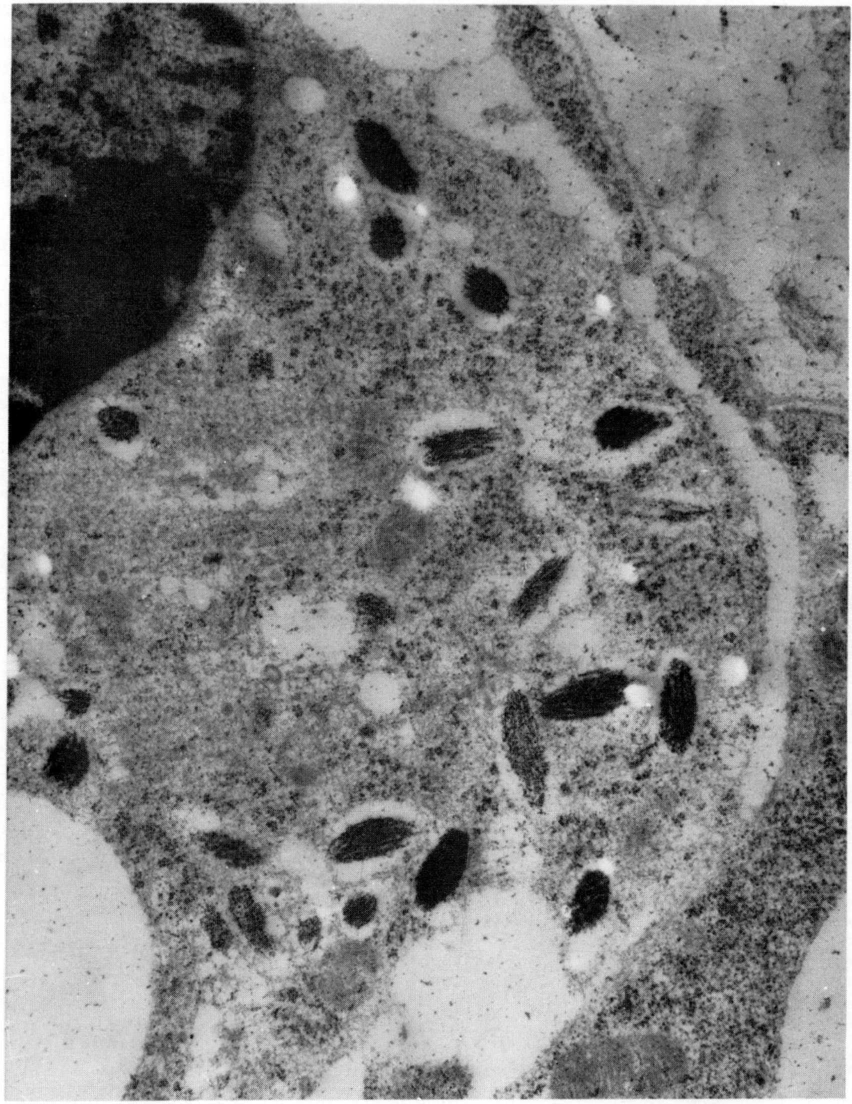

Figure 1.3 Melanosomes in various stages of melanization (magnification × 29 000): incompletely melanized granules have a striped appearance

visible in them, but the build-up of melanin ultimately results in a dense structure.[61]

Melanins are quinoid polymers of two kinds. *Phaeomelanins,* which are yellow or red in colour, differ from the brown or black *eumelanins* in being soluble in dilute alkali. Both are formed by the same initial steps which involve oxidation of tyrosine to 3,4-dihydroxyphenylalanine (dopa) and its dehydrogenation to dopa quinone.[62] The formation of eumelanins then involves several further steps to produce indole-5,6-quinone, which polymerizes and becomes linked to protein. It is now believed that eumelanin is not a homopolymer composed solely of indole-5,6-quinone units, but a poikilopolymer which includes several intermediates. Phaeomelanins are formed by a different route. The dopa quinone interacts with cysteine to form 5-S- and 2-S-cysteinyldopa, and these isomers are further oxidized to a series of intermediates which then polymerize.[63,64]

Skin colour has a *constitutive*—that is, genetic—component and a *facultative*—that is, environmental—component. Thus various degrees of pigmentation occur in different ethnic groups; the differences are in the amount of melanin produced, not in the numbers of melanocytes present. Pigmentation can be enhanced by exposure to sun, or by endocrine factors, for example in pregnancy. Melanogenesis is influenced by certain polypeptide hormones of the pituitary[65–67] and to some extent by steroid hormones. From hog pituitary, two melanocyte-stimulating hormones α-MSH and β-MSH, containing respectively 13 and 18 amino acid residues have been isolated.[68,69] The human pituitary lacks α-MSH, but produces a β-MSH with 22 residues. However, it seems likely that the active sequence is actually part of two larger molecules, β-lipoprotein with 91 amino acids and γ-lipotropin with 58.[70]

There are long-standing reports that testosterone increases skin pigmentation in castrated men[71,72] and in women.[73] The same may be true for certain specialized areas of skin in some animals, but experimental studies on the guinea pig failed to reveal any effect of androgens,[74,75] though oestrogens clearly increased skin pigmentation in a number of areas.[76,77]

The major function of melanin is undoubtedly protection against solar radiation.[78–80] In general, pigment is geographically distributed in relation to the solar intensity experienced by the various ethnic groups, being greatest in the tropics, reduced in temperate zones, and partly reappearing in areas of snow-glare.[81] There are exceptions: for example, American Indians do not noticeably differ in colour throughout the continent. The damaging effects of ultraviolet light are well illustrated by the high incidence of epidermal carcinoma in Europeans exposed to the tropical sun. Melanin pigmentation may be useful in two ways. As well as providing direct protection from radiation, it may be activated to a free radical state by incident light and thus could possibly eliminate genetically damaged cells by a phototoxic mechanism.

Langerhans Cells

Langerhans cells are dendritic cells similar in form to melanocytes but free from pigment and unable to form it when they are incubated with dihydroxyphenylalanine (that is, they are dopa-negative). They were first demonstrated in human skin by the use of gold chloride[82] and can be stained with ATPase.[83] Under the electron microscope they resemble melanocytes in having a lobulated nucleus, but differ in lacking melanosomes, having instead characteristic granules which are rod- or racquet-shaped.[84–88]

The origin and affinities of Langerhans cells have been much debated, and their function remains undecided. The view that they are effete melanocytes is discarded.[89–91] It is currently believed that Langerhans cells are of mesenchymal origin and equivalent or closely related to dermal histiocytes,[92] in which identical granules have been described.[93–96] Various possible functions have been ascribed to them. For example, opinion is divided about whether they may[97] or may not[98,99] control proliferation of keratinocytes and the pattern of epidermal cell columns. Another suggested role might be the loosening of intercellular connections.[100,101] Langerhans cells are capable of limited phagocytosis, but they should not be regarded as functional macrophages.[102,103] Recently attention has become focussed on the possibility that they are concerned with immune functions.

Dermis

The *dermis*[1,104,105] is a tough and resilient tissue which cushions the body against mechanical injury and provides nutriment to the epidermis and cutaneous appendages. It consists of an association of protein fibres with an amorphous ground substance containing mucopolysaccharide. There are few cells in this matrix; most of them are *fibroblasts* which secrete the dermal constituents; others are *mast cells*, histiocytes or macrophages, lymphocytes and other leucocytes, and melanocytes. The dermis also houses blood, lymphatic and nervous systems, and surrounds the invaginated epidermal appendages, namely the hair follicles, with its associated glands and the eccrine sweat glands.

Collagen

The major fibrous constituent of the dermis, accounting for 75 per cent of the dry weight and 18–30 per cent of the volume, is *collagen*.[105–111] Under the light microscope collagen fibres appear as colourless, branching wavy bands about 15 μm in width. The electron microscope reveals that each fibre is composed of unbranched fibrils about 100 nm (1000 Å) wide and is characteristically cross-striated with a periodicity of 60–70 nm. Collagen fibres can be disintegrated by 0·01 per cent acetic acid, forming molecules with a molecular weight of 300 000–360 000, about 180 nm long. When these acid solutions of *tropocollagen* are neutralized, the 64 nm periodicity reappears, which may be explained on the hypothesis that native collagen is composed of molecules of tropocollagen associated side by side with a regular overlap of a quarter of their length.[112]

Skin collagen is characterized by a high content of glycine, which forms a third of all the residues, and of proline and hydroxyproline, which together make up a further fifth. Tropocollagen molecules[107] consist of three polypeptide chains each containing about 1000 amino acids. The fibroblasts produce a precursor known as *procollagen* which has 300–400 additional amino acids in each of its chains; these extensions are removed after secretion.[113,114]

Elastin and Reticulin

Elastic fibres[115–124] make up 4 per cent of the dry weight and 1 per cent of the volume of the dermis. They are delicate, straight, freely branching fibres which can be stretched by 100 per cent or more but return to their original length when the stress is removed. Elastin differs from collagen in having only about a quarter or a third the amount of basic and acid amino acids, only one tenth the amount of hydroxyproline, a relatively large amount of valine, and an amino acid known as desmosene[125] which appears to be unique to it and to be concerned with cross-linkage.

Not all fibrous constituents can be clearly identified as collagen or elastin on the basis of their tinctorial properties. In addition to true elastin, two other similar fibres have been distinguished and given the names of oxytalan and elaunin.[126] Moreover, about 0·4 per cent of the dry weight of the dermis is made up of fine branching fibres which, unlike collagen, stain black with silver nitrate, and are known as *reticulin*. Their axial periodicity is identical with that of collagen.[116]

Ground Substance

The amorphous ground substance[127–135] in which the fibres and cells lie contains a variety of carbohydrates, proteins and lipids, of which the most important are the acid mucopolysaccharides. These are macromolecules made up of two different saccharide units which alternate regularly. In dermis the major forms are *hyaluronic acid*, in which D-glucosamine, with an acetylated amino group, alternates with D-glucuronic acid, and *dermatan sulphate*, in which L-iduronic acid alternates with D-galactosamine.

Fibroblasts

The term *fibroblast*[105,136,137] should, strictly, designate a cell at any early stage and *fibrocyte* one which is fully differentiated,[138] but most authors use fibroblast to describe an actively secreting cell and fibrocyte for an inactive one.[105] Fibroblasts are derived from the mesenchyme. It is not doubted that fibroblasts secrete collagen.[139] It is probable that they are the source of elastin[140] and, though Asboe-Hansen[141] has implicated the mast cell, also of mucopolysaccharides.[142]

Mast Cells

Mast cells[143–146] also originate from wandering cells of the mesenchyme. They are characterized by a cytoplasm filled with granules which stain metachromatically with basic aniline dyes—purple with methylene blue. They contain, and can release, heparin and histamine. Rupture of the cells, with release of the granules, is observed in many types of skin damage, and histamine is responsible for many of the events associated with inflammation, irritation and other skin disorders. This subject is dealt with at greater length in the next chapter.

Nerves and Sense Organs

The skin is innervated with about one million afferent nerve fibres; most terminate in the face and extremities; relatively few supply the back.

Sensory endings fall into two major groups: corpuscular, which incorporate non-nervous elements, and free, which do not.[147–152] Corpuscular endings, in turn, are subdivided into encapsulated receptors, of which there is a range in the dermis, and non-encapsulated, such as the epidermal Merkel 'touch spot'.[153,154]

The largest encapsulated receptors are the elaborate *Pacinian corpuscles*[155,156] which are ovoid bodies about 1 mm in length and lamellated in cross-section like an onion. Others are the *Golgi-Mazzoni corpuscles* found in the subcutaneous tissue of the finger, the *Krause end-bulbs* in the superficial layers of the dermis, and the *Meissner corpuscles*[147,151,152,157] in the papillary ridges of the glabrous skin. Of somewhat different structure are the branching terminals of *Ruffini.*[151]

Free nerve endings occur both in the dermis and in the epidermis. Hair follicles have nerve terminals of varying degrees of complexity.

The way these miscellaneous receptors function has been much debated. As it is easy to map separate sensory spots for the several kinds of stimuli, the classical view was that receptors were specific for the qualities of touch (Meissner's corpuscles), warmth (Ruffini end-organs), cold (Krause end-bulbs) and pain (free nerve endings). The hypothesis came under attack on the grounds that it did not explain why hairy skin could also distinguish between the stimuli, even though it lacked the encapsulated structures.[158]

The existence of functionally specific afferent units has in recent years been reaffirmed by electrophysiological experiments. Two major categories of units have been established: mechanoreceptors and thermoreceptors,[150,151] and a third category, pain receptors, respond only to high threshold stimulation, mechanical, thermal or chemical. Mechanoreceptors have been further classified into 'slowly adapting', as exemplified by the Ruffini endings and Merkel cells, and 'rapidly adapting', namely the hair follicle receptors, Meissner corpuscles, and the laminated Pacinian and Golgi-Mazzoni corpuscles.[159]

The autonomic nervous system supplies both adrenergic and cholinergic fibres to the arrector pili muscles and the blood vessels. Stimulation of the arrector pili muscle by its associated nerve causes the hair shaft to rise to a more perpendicular position with respect to the skin surface. This slows down the passage of air over the skin and consequently reduces the rate of heat loss. This phenomenon is the cause of 'goose flesh'. Regulation of the amount of blood flowing through the superficial layers of the dermis also influences heat loss (see next section).

Eccrine sweat glands are also richly supplied with nerves.[160] Anticholinergic substances are able to inhibit sweating, and most of the nerves appear to be cholinergic, though a few adrenergic fibres can be demonstrated. It seems likely that the glands of the palms and soles, which secrete sweat to increase the grip of the skin, are influenced by adrenergic fibres, whereas those of the general body surface, which regulate body heat, are under cholinergic control.[161]

Blood Vessels

The arteries entering the skin form a deep plexus, from which a network arises which gives branches to the cutaneous appendages and to a subpapillary plexus, which in turn sends loops into the papillary layer just below the dermo–epidermal frontier. From these capillaries the blood is drained by veins which descend into the intermediate plexuses.[162–164]

All the nutriment for the epidermal cells has to pass through the dermo–epidermal junction; no blood vessels enter the epidermis. The vasculature is much more elaborate than would be necessary solely for nutrition; indeed, the metabolic rate in skin is lower than in many less well-perfused organs. Temperature control thus appears to be a most important function. When the superficial loops are fully dilated, the skin appears flushed and heat loss is at a maximum. However, shunts are provided between arterial and venous systems deeper in the dermis which can carry all or most of the blood when heat loss must be kept to a minimum. In these circumstances the superficial capillary loops are found to be almost completely closed.

The regulation of the total blood volume in the skin as opposed to its distribution is mediated by constriction and dilation of the cutaneous circulation, and allows a large reservoir of blood to be made rapidly available for vital central functions in times of stress. The mechanism of constriction of the lumen of a blood vessel in the dermis can be either by a general activation of contractile myoepithelial cells in the wall of the capillary, or by activation of '*glomerae*' which are small contractible cuffs around the vessel, and which effectively strangle the vessel and cut the blood flow. The operations of constriction and dilation are mediated via the local secretion of chemicals (for example acetylcholine) from nerves, hormones (for example adrenalin) and, in cases of skin damage, histamine from the mast cells in the dermis.

As distinct from this widely held view, Ryan[165] has stressed the oxygenating function of the vasculature in a tissue which is exposed to many kinds of injury. Finally, it must be remembered that the blood supply carries all the materials for making the products of the hair follicle and its associated glands, as well as the hormones which influence their manufacture, and the substances which are excreted by the sweat glands.

Eccrine Sweat Glands

Eccrine sweat glands[166] are the most numerous skin appendages and occur over the majority of the body surface. In some areas they number as many as 600 cm^{-2}. They have a cylindrical spiral duct lined with epidermal cells extending from their visible opening in the epidermis down into the deep dermis where the duct becomes coiled and convoluted into a ball (Figure 1.4). Part of the tangled duct is secretory and manufactures the odourless sweat which rises up the duct to be released on the skin surface. It is thought that the duct of the gland has the ability to modify the sweat as it flows upward, by removing salts or water.[167] The analogy with the nephron of the kidney is frequently drawn. Though the lining of the duct is said to be epidermal, it is not highly pigmented even in people with pigmented skin.

The sweat glands of the general body surface are concerned with both control of body temperature and excretion. The evaporation of sweat has a cooling effect. The glands thus respond to environmental temperature, but also to other stimuli, such as ultraviolet light, emotional stress and rises in body temperature due to fevers. On the palms and soles, however, the secretion from the glands serves to increase surface friction. In both areas, sweating is under nervous control, though different types of fibre may be involved (see previous section).

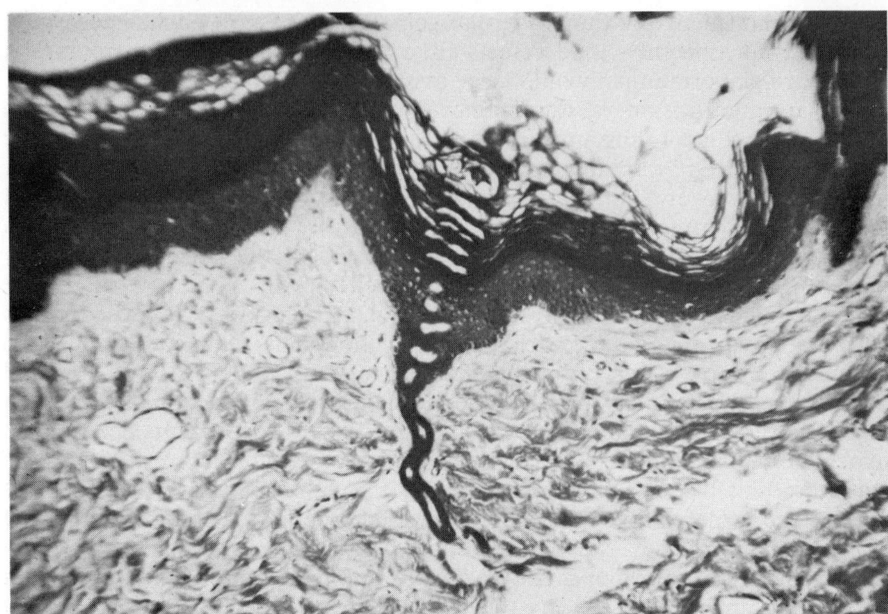

Figure 1.4 Section of human toe skin showing a spiralling sweat duct (magnification × 140): invagination of the epidermal tissue along the length of the duct can be clearly seen

Sweating appears to involve activation of myoepithelial cells which line the ducts of the glands. Although sweating is considered to be a continuous process, it seems that sweat is ejected in small bursts, perhaps 6–7 per minute, suggesting a peristaltic action by the ducts.[168] The composition of eccrine sweat is variable but consists of electrolyte ions, urea, amino acids, small quantities of sugars and possibly some lipid. The normal range of sodium chloride concentration in eccrine sweat is stated to be between 10 and 100 milli-equivalents per litre.[169]

Hair Follicles

Hair follicles are tubular inpushings of the epidermis. The hair is produced by keratinization of cells formed by division in the matrix at the base of the follicle.[170–174] This epidermal matrix surrounds a small dermal papilla which becomes invaginated into its base.

There are about 120 000 follicles on the human scalp. Each one undergoes a cycle of activity[174] in which an active phase (anagen), which lasts for 1 to 3 years or even longer, is followed by a short transition phase (catagen) and a resting phase (telogen) (see Figure 23.2). This process involves a cessation of mitosis in the matrix and the keratinization of the expanded base of the hair to form a 'club', which is retained until the follicle again becomes active, when it is shed (Figure 23.3). Thus about 100 hairs are normally lost from the scalp each day.[175,176]

Such cyclic activity of hair follicle may be considered as a remnant of the moult in other mammals. In contrast to the human scalp, where the activity of each follicle appears to be independent of its neighbours, some animals, such as rats and mice, exhibit wavelike patterns of new hair growth and moulting, which start in the mid-venter and spread over the flanks to the back.[177] These have proved interesting models for experimentation on the factors controlling hair growth, but it should not be supposed that this has any direct relevance to human baldness. It appears that hair follicles have an intrinsic rhythm, of which the mechanism remains undiscovered, but that this can be greatly modified by circulating hormones and thus, in turn, by environmental factors acting through the hypothalamus and the pituitary.[177,178] Thus moulting, like reproductive activity, is seasonally controlled. Perhaps even the human scalp retains a reflection of the seasonal moult, with increased shed of club hairs in the autumn.[179]

In the axillary and pubic regions of both sexes, and on the face of the male, coarse *terminal* hair—as distinct from fine *vellus*—develops at puberty, and continues to increase in amount for several years.[180] The growth of this hair is initiated by and dependent upon androgens (male steroid hormones) which are secreted by the testicles of the male and by the adrenal glands and the ovaries in the female. Male-type body hair is also androgen-dependent, though its amount and distribution vary greatly between individuals. Unacceptable amounts of facial and body hair in women, known as *hirsutism*, may result from abnormal high androgen production, but individual variations in the sensitivity of the target hair follicles is also important. Compounds which block the action of androgen, known as *anti-androgens*, offer possibilities for the alleviation of female hirsutism.[181]

Male pattern alopecia, a condition in which vigorously growing terminal hair is gradually replaced by miserably small and cosmetically useless fibres over areas of the scalp, appears to be hereditary, but requires the presence of male hormone. Hence eunuchs, even if genetically disposed, do not go bald, unless treated with testosterone,[182] and women rarely develop conspicuous bald patches, though they frequently suffer diffuse hair loss which may be the female equivalent. Why male hormones should promote hair growth on the face and body and ruin it on the vertex of the scalp, so far eludes any consistent explanation.

The structure and growth of hair is further considered in Chapter 23.

Sebaceous Glands

Sebaceous glands[183,184] secrete sebum, which forms the majority of the lipid which covers the skin and hair. They occur throughout most of the body and are normally, though not invariably, associated with hair follicles. The greatest concentrations (400–900 cm^{-2}) are found on the scalp, face and upper chest and shoulders, and there are none on the palms and soles.

The glands are *holocrine*, that is to say the cells of the gland pass through a development and maturation stage, during which they accumulate lipid, becoming several times their original size, and subsequently disintegrate completely, releasing their contents into the lumen of the gland. New cells are formed continually from the lining of the gland by cell division to replace those lost.

Sebaceous gland activity is under hormonal control. It is stimulated by androgens. In human males, the glands are minute during the prepubertal period, but undergo vast enlargement at puberty when the output increases more than fivefold.[185] Eunuchs secrete about half as much sebum as normal males, but substantially more than boys; it seems that the secretion is dependent on adrenal androgens. Adult women secrete only a little less than men; their sebaceous activity appears to be maintained by androgens from the ovary as well as from the adrenal cortex.

Circumstantial evidence from man, and experimental evidence from animals, indicates that pituitary hormones may also influence sebaceous secretion. Sebum secretion is abnormally high in acromegalics.[186] The response of the rat sebaceous glands to testosterone is greatly diminished when the pituitary is removed. Bovine growth hormone[187] and synthetic α-MSH[188,189] have each been shown to have some direct effect on sebaceous secretion, and to facilitate the response of the glands to testosterone.

Oestrogens, or anti-androgens such as cyproterone acetate, will inhibit sebaceous secretion in man[181] as well as in rats.[190]

Human sebum[191] is composed of glycerides and free fatty acids (57·5 per cent), wax esters (26·0 per cent), squalene (12·0 per cent), cholesterol esters (3·0 per cent) and cholesterol (1·5 per cent). Lipid produced from the superficial epidermis differs in lacking wax esters and squalene, and having much higher proportions of cholesterol esters and cholesterol. Skin lipids appear to differ greatly between species.

Apocrine Glands

The so-called *apocrine glands*[192] are tubular glands attached to the hair follicle and, like the sebaceous glands, developed in association with it. Though rudiments are formed throughout the body in the foetus, the glands become canalized and functional almost exclusively in the axillary, anal and genital regions and in the areola of the nipple; few are found elsewhere. The axillary glands only become functional at puberty and it seems probable that, like similar derivatives in other animals, for example the rabbit,[193] they are androgen-sensitive.

The secretion of human apocrine glands is milky, viscous and at first without noticeable odour, which is said to develop through bacterial action. Secretory activity is controlled by adrenergic nerves.

The function of the glands in the human species has been much debated. In many other mammals they constitute or contribute to scent glands. Odour is undoubtedly important in human communication,[194,195] though little information has been recorded since Havelock Ellis wrote down his entertaining if anecdotal evidence.[196]

Common Disorders of the Skin

The cosmetic chemist is concerned not with serious clinical disorders of the skin but with lesser, if often chronic, conditions that affect large numbers of the population and which are only presented to the clinician when extremely severe. Discussion in this chapter is confined to a few which appear to come within the purview of the cosmetic scientist. For detailed accounts of these and of other disorders the reader is referred to textbooks of dermatology.[197,198]

Pigmentary Disorders

Ephelides, Lentigens and Moles. It is not easy to discover a consistent classification for the small hyperpigmented areas which occur on the skin of most Caucasians. It is generally agreed that freckles (ephelides) are pale, variably coloured, not usually raised, and harmless. Their pigmentation is due to an increased local synthesis of melanin in the epidermis. The predispositions for these are apparently genetically determined. They are found predominantly on the exposed areas of fair or red-haired people and are stimulated by exposure to UV or X-irradiation. Children do not usually have freckles until after their sixth year of life.

It is usually considerations of degree which differentiate between freckles and the more pronounced lentigens which are generally associated with age, and moles (junctional naevi). These latter are usually more heavily pigmented, fewer in number and are associated with a thickening of the epidermis. They are rarely present at birth, and in women become considerably darker during pregnancy, as do other areas.[200] It should be mentioned that in the most severe cases these naevi can become malignant but this will not be considered further in this book.

Vitiligo. Apart from hyperpigmented disorders there are considerable cosmetic problems associated with hypopigmentation diseases, the most common of which is vitiligo, a patchy depigmentation of the skin afflicting a considerable number of non-Caucasians. Although it does occur in Caucasians, it is not usually cosmetically troublesome. This condition has been referred to by a former Prime Minister as 'India's national disease'. It is the more distressing on account of its resemblance to the early stages of leprosy, when depigmentation also occurs, and therefore it can carry a social stigma without any foundation.

Vitiligo is usually associated with an absence not only of melanin but of melanocytes in the affected areas. The aetiology is unknown. It frequently exhibits a degree of bilateral symmetry and is also seen to follow superficial nerve trunks, but there is little support for the hypothesis that it is linked with nerve function.[201] An autoimmune hypothesis is based on its clinical association with a number of other supposedly autoimmune disorders.[202]

Vitiligo has been treated by systemic psoralens (photosensitizing compounds obtained from certain umbelliferous plants) followed by exposure to the sun or UV radiation,[203] or by topical corticosteroid preparations.[204] Treatments are usually not very satisfactory, and cosmetic camouflage is often the best recourse.

Disorders of the Sebaceous and Sweat Glands

Acne vulgaris[205–207] is a chronic disorder of the pilosebaceous follicles which is so frequent among Caucasoids as to be regarded as physiological in adolescents.

The lesions, which may include papules, pustules, and even cysts and severe scarring, are so well known as to need no detailed description. Comedones (blackheads) may be present, but these do not always progress to pustules.

The condition involves inflammation of the pilosebaceous apparatus. It seems to develop by hyperkeratinization of the neck of the follicle, a build-up of sebum within the gland, and a rupture into the dermis. Acne sufferers have, on average, a higher rate of sebum production than normal subjects.[208] The bacteria *Corynebacterium acnes* and *Staphylococcus epidermidis* are almost always present in the pustular contents.[209,210]

The prime cause of acne has been much debated. Undoubtedly it requires the presence of androgens, for prepubertal children and eunuchs do not normally develop it.[211] But, in males at least, the mean levels of androgen among sufferers from acne are not greater than those in normal subjects.[212] Genetic factors are undoubtedly important.

The factors in acne thus appear to be predisposition, the presence of androgens together with an abnormal sensitivity to them of the sebaceous gland and its duct, and infection by bacteria. It is possible that the bacteria cause the release of free fatty acids from the sebum in the occluded glands and that this produces the inflammation,[213] but which is the prime mover of all these distressing events remains an open question.

Like other intractable conditions, acne has been attacked by many forms of therapy. Such treatments have usually aimed at reducing sebum secretion or controlling bacterial growth. Physiological doses of oestrogens given systemically[214] will reduce sebum production, and so will anti-androgens such as cyproterone acetate,[181] but neither is suitable for males, since possible consequences such as gynaecomastica and loss of libido might prove less acceptable than the condition. Broad-spectrum antibiotics, such as tetracycline, have proved safe and fairly effective.[215]

Various topical medicaments are aimed either at bacteriostasis or at inducing exfoliation. Retinoic acid has recently achieved popularity.[216]

Miliaria. This name is given to several disorders in which the sweat duct becomes to some extent obstructed. The most common is *Miliaria rubra* or *prickly heat*.[217,218]

The lesions of prickly heat are uniform minute reddish papules, which are associated with an unbearable prickling sensation. They occur especially in areas of friction with clothing and in flexures. Infants are especially susceptible and often have lesions on the face as well as the neck, groins, axillae and elsewhere.

Prickly heat is most common in hot, humid conditions, though it may occur in deserts, and can affect up to 30 per cent of people exposed to these climates. It almost invariably accompanies profuse sweating and can be produced experimentally by occluding the skin under polythene for a few days. Hölzle and Kligman[217] have postulated that the condition results from an increase in the density of aerobic bacteria, notably cocci. These, in turn, secrete a toxin which injures the luminal cells and precipitates a cast in the lumen; infiltration of leucocytes then completes the obstruction.

There is no satisfactory medication for prickly heat. Topical application of antibiotics or other antibacterial preparations has achieved little success, though calamine lotion, followed by bland emollients, may relieve the discomfort. Oral vitamin C has been reported as helpful,[219] and systemic antibiotics may be useful as a prophylaxis.[217] The only effective treatment is to limit sweating. To withdraw the sufferer from the environment into an air-conditioned room for a few hours a day may suffice.

Skin Scaling Disorders

Psoriasis. This condition is the province of the dermatologist, not, at present, of the cosmetic chemist. It is considered briefly here because it is very widespread, affecting nearly 2 per cent of the population of North-West Europe, including the United Kingdom.[220]

The lesions are well-defined pink or dull-red plaques surmounted by characteristic silvery scales which, on removal, often show a small bleeding point. There seems little doubt that a genetic factor is involved, but the clinical manifestation of the condition is sometimes delayed until late in life, and various metabolic, infective, environmental and even psychogenic factors may precipitate it.

The plaques result from a greatly increased rate of epidermal proliferation, coupled with a much accelerated turnout of the cells through the layers of the epidermis. The cells retain their nuclei even in the stratum corneum, which is thus described as *parakeratotic*. According to several authors, the increase in cell production is achieved by a shortening of the average period from one division to the next in a population in which all cells are cycling.[221] The alternative explanation that in normal epidermis only a minor proportion of cells is cycling, whereas in psoriasis almost all become mitotically active, appears more probable.[222,223]

The treatments proposed for psoriasis have been manifold. Generally speaking they are directed against the division of epidermal cells and include topical application of coal tar, dithranol and corticosteroids.

Dandruff. Sometimes known as *pityriasis capitis*, this condition is characterized by the massive desquamation of small flakes of stratum corneum from the otherwise normal scalp (Figure 1.5). The scales may be dry or trapped in a film of sebum. Dandruff is uncommon in infancy and early childhood, but by puberty about half of all males and females become affected and in many it persists throughout life. It must therefore be considered as a physiological state rather than a disease and, as such, falls very much in the cosmetic rather than the clinical field.

The causation of dandruff is still debatable. Perhaps constitution or, as in acne, stimulation by androgens or other physiological factors plays a part. Micro-organisms may well be involved;[224] both *Pityrosporum ovale*[225] and *Pityrosporum orbiculare* are more abundant in affected than in non-affected persons.[226] Other suggestions are that the condition is caused by an allergen in sweat,[277] or is a physiological error in the normal process of desquamation.[228]

Dandruff has been treated with ointments containing 2 per cent salicyclic acid. Shampoos containing selenium disulphide or zinc pyrithione are currently favoured, and appear to work by reducing epidermal turnover.[229] Other preparations are based on supposed ability to reduce the yeast flora.

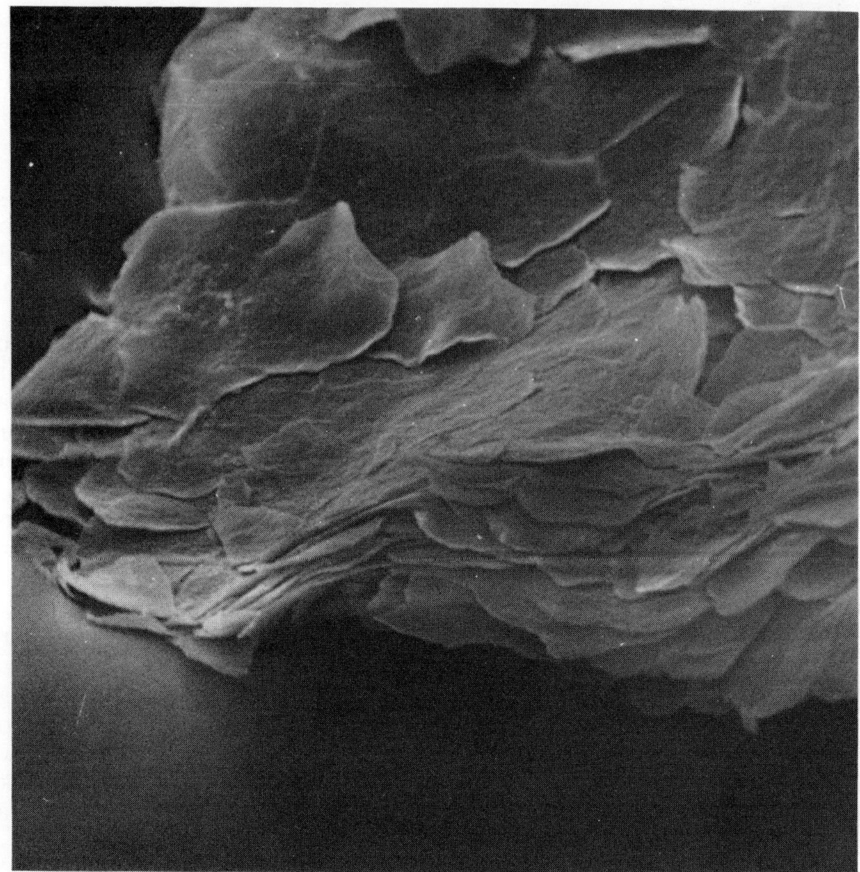

Figure 1.5 Dandruff flake (magnification × 660)

REFERENCES

1. Montagna, W., and Parakkal, P. F., *The Structure and Function of Skin*, 3rd edn, New York, Academic Press, 1974.
2. Rothman, S., *Physiology and Biochemistry of Skin*, Chicago, University Press, 1954.
3. Tregear, R. T., *Theoretical and Experimental Biology*, Vol. 5, *Physical Functions of Skin*, London, Academic Press, 1966.
4. Breathnach, A. S., *An Atlas of the Ultrastructure of Human Skin*, London, Churchill, 1971.
5. Breathnach, A. S., *J. invest. Dermatol.*, 1975, **65,** 2.
6. Brody, I., *The Epidermis*, ed. Montagna, W., and Lobitz, W. J., Jnr, New York, Academic Press, 1964, p. 251.
7. Brody, I., *J. Jadassohn's Handbuch der Haut und Geschlechtskrankheiten*, Vol. 1, part 1, ed. Gans, O., and Steigleder, G. K., Berlin, Springer, 1968, p. 1.

8. Briggaman, R. A., and Wheeler, C. E., Jnr, *J. invest. Dermatol.*, 1975, **65,** 71.
9. Pinkus, H., *Physiology and Biochemistry of the Skin*, ed. Rothman, S., Chicago, University Press, 1954, p. 584.
10. Pinkus, H., *Br. J. Dermatol.*, 1970, **94,** 351.
11. Pinkus, H., and Hunter, R., *Arch. Dermatol.*, 1966, **94,** 351.
12. Greulich, R. C., *The Epidermis*, ed. Montagna, W. and Lobitz, W. C., Jnr, New York, Academic Press, 1964, p. 117.
13. Leblond, C. P., Greulich, R. C., and Pereira, J. P. M., *Advances in Biology of Skin*, Vol. 5, *Wound Healing*, ed. Montagna, W. and Billingham, R. E., New York, Academic Press, 1964, p. 39.
14. Iverson, O. H., *Cell Tissue Kinet.*, 1968, **1,** 351.
15. Potten, C. S., *J. invest. Dermatol.*, 1975, **65,** 488.
16. Baserga, R., *Cell Tissue Kinet.*, 1968, **1,** 167.
17. Bullough, W. S., *Cancer Res.*, 1965, **25,** 1683.
18. Epifanova, O. I., and Terskikh, V. V., *Cell Tissue Kinet.*, 1965, **2,** 75.
19. Weinstein, G. D. and Frost, P., *J. invest. Dermatol.*, 1968, **50,** 254.
20. Weinstein, G. D. and Frost, P., *J. nat. Cancer Inst.*, 1969, **30,** 225.
21. Weinstein, G. D. and Frost, P., *Arch. Dermatol.*, 1971, **103,** 33.
22. Heenan, M. and Galand, P., *J. invest. Dermatol.*, 1971, **56,** 425.
23. Weinstein, G. D., *Br. J. Dermatol.*, 1975, **92,** 229.
24. Halprin, K. M., *Br. J. Dermatol.*, 1972, **86,** 14.
25. Baker, H. and Kligman, A., *Arch. Dermatol.*, 1967, **95,** 408.
26. Porter, D. and Shuster, S., *J. invest. Dermatol.*, 1967, **49,** 251.
27. Porter, D. and Shuster, S., *Arch. Dermatol.*, 1968, **98,** 339.
28. Rothberg, S., Crounse, R. G. and Lee, J. L., *J. invest. Dermatol.*, 1961, **37,** 497.
29. Rowe, L. and Dixon, W. J., *J. invest. Dermatol.*, 1972, **58,** 16.
30. Matoltsy, A. G., *J. ultrastruct. Res.*, 1966, **15,** 510.
31. Matoltsy, A. G. and Parakkal, P. F., *J. Cell Biol.*, 1965, **24,** 297.
32. Odland, G. F., *J. invest. Dermatol.*, 1960, **34,** 11.
33. Matoltsy, A. G. and Matoltsy, M.N., *J. Cell Biol.*, 1970, **47,** 593.
34. Brody, I., *Z. mikrosk. anat. Forsch.*, Leipzig, 1972, **86,** S.305.
35. Brody, I., *Dermatology*, 1977, **16,** 245.
36. Roth, S. I. and Clark, W. H., Jnr, *The Epidermis*, ed. Montagna, W. and Lobitz, W. C., Jnr, New York, Academic Press, 1964, p. 303.
37. Christophers, E., *J. invest. Dermatol.*, 1971, **56,** 165.
38. Christophers, E., Wolf, H. H. and Laurence, E. B., *J. invest. Dermatol.*, 1974, **62,** 555.
39. Christophers, E., and Laurence, E. B., *Curr. Probl. Dermatol.*, 1976, **6,** 87.
40. Menton, D. N. and Eisen, A. Z., *J. ultrastruct. Res.*, 1971, **35,** 247.
41. Mackenzie, I. C. and Linder, J. E., *J. invest. Dermatol.*, 1973, **61,** 245.
42. Mackenzie, I. C. *J. invest. Dermatol.*, 1975, **65,** 45.
43. Mercer, E. H., *Keratin and Keratinisation*, Oxford, Pergamon, 1961.
44. Odland, G. F., *The Epidermis*, ed. Montagna, W. and Lobitz, W. C., Jnr, New York, Academic Press, 1964, p. 237.
45. Jarrett, A., *Symp. zool. Soc. London*, 1964, **12,** 55.
46. Jarrett, A., *The Physiology and Pathophysiology of the Skin*, Vol. 1, ed. Jarrett, A., London, Academic Press, 1973, p. 161.
47. Jarrett, A. and Spearman, R. I. C., *Histochemistry of the Skin: Psoriasis*, London, English Universities Press, 1964.
48. Snell, R. S., *Z. Zellforsch. mikrosk. Anat.*, 1965, **65,** 829.
49. Braun-Falco, O., Kint, A. and Vogell, W., *Arch. Klin. exp. Dermatol.*, 1963, **217,** 627.
50. Matoltsy, A. G., *Nature* (*London*), 1964, **201,** 1130.

51. Matoltsy, A. G., *Biology of the Skin and Hair Growth*, ed. Lyne, A. G. and Short, B. F., Sydney, Angus and Robertson, 1965, p. 291.
52. Baden, H. P., *J. invest. Dermatol.*, 1970, **55,** 185.
53. Crounse, R. G., *Nature* (*London*), 1966, **211,** 1301.
54. Goldschmidt. H. and Kligman, A. M. *Arch. Dermatol.*, 1964, **88,** 709.
55. Kligman, A. M., *The Epidermis*, ed. Montagna, W. and Lobitz, W. C., Jnr, New York, Academic Press, 1964, p. 387.
56. Edwards, E. A. and Duntley, S. Q., *Am. J. Anat.*, 1939, **65,** 1.
57. Gibson, I. M., *J. Soc. cosmet. Chem.*, 1971, **22,** 725.
58. Boyd, J. D., *Progress in the Biological Sciences in Relation to Dermatology*, ed. Rook, A. J., Cambridge, University Press, 1960, p. 3.
59. Rawles, M. E., *Physiol. Rev.*, 1947, **28,** 383.
60. Fitzpatrick, T. B. and Breathnach, A. S., *Dermatol. Wochenschr.*, 1963, **147,** 481.
61. Jimbow, K. and Kukita, A., *Biology of Normal and Abnormal Melanocytes*, ed. Kawamura, T., Fitzpatrick, T. B. and Seiji, M., Tokyo, University of Tokyo Press, 1971, p. 171.
62. Mason, H. S., *Advances in Biology of Skin*, Vol. 8, *The Pigmentary System*, ed. Montagna, W. and Hu, F., Oxford, Pergamon, 1967, p. 293.
63. Prota, G. and Nicolaus, R. A., *Advances in Biology of Skin*, Vol. 8, *The Pigmentary System*, ed. Montagna, W. and Hu, F., Oxford, Pergamon, 1967, p. 323.
64. Prota, G. and Thomson, R. H., *Endeavour*, 1976, **35,** 32.
65. Lee, T. H. and Lerner, A. B., *Pigment Cell Biology*, ed. Gordon, M., New York, Academic Press, 1959, pp. 435–444.
66. Lerner, A. B. and McGuire, J. S., *Nature* (*London*)., 1961, **189,** 176.
67. McGuire, J. S. and Lerner, A. B., *Ann. N. Y. Acad. Sci.*, 1963, **100,** 622.
68. Lerner, A. B. and Lee, T. H., *J. Am. chem. Soc.*, 1955, **77,** 1066.
69. Lee, T. H. and Lerner, A. B. *J. biol. Chem.*, 1956, **221,** 943.
70. Scott, A. P. and Lowry, P. J., *Biochem. J.*, 1974, **139,** 593.
71. Edwards, E. A., Hamilton, J. B., Duntley, S. Q. and Hubert, G., *Endocrinology*, 1941, **28,** 119.
72. Hamilton, J. B. and Hubert G., *Sciences* (*NY*), 1938, **88,** 481.
73. Hamilton, J. B., *Proc. Soc. exp. Biol. Med.*, 1939, **40,** 502.
74. Bischitz, P. G. and Snell, R. S., *J. invest. Dermatol.*, 1959, **33,** 299.
75. Snell, R. S. and Bischitz, P. G., *Z. Zellforsch. mikrosk. Anat.*, 1959, **50,** 825.
76. Bischitz, P. G. and Snell, R. S., *J. Endocrinol*, 1960, **20,** 312.
77. Snell, R. S. and Bischitz, P. G., *J. invest. Dermatol.*, 1960, **35,** 73.
78. Blum, H. F., *Physiol. Rev.*, 1945, **25,** 483.
79. Blum, H. F., *Q. Rev. Biol.*, 1961, **36,** 50.
80. Wasserman, H. P., *Ethnic Pigmentation*, Amsterdam, Exerpta Medica, 1974.
81. Fleure, H. J. *Geogr. Rev.*, 1945, **35,** 580.
82. Langerhans, P., *Virchows Arch. pathol Anat. Physiol.*, 1868, **44,** 325.
83. Jarrett, A. and Riley, P. A., *Br. J. Dermatol.*, 1963, **75,** 79.
84. Birbeck, M. S., Breathnach, A. S. and Everall, J. D., *J. invest. Dermatol.*, 1961, **37,** 51.
85. Breathnach, A.S. and Wyllie, L. M. *Advances in Biology of Skin*, Vol. 8, ed. Montagna, W. and Hu, F., Oxford, Pergamon, 1967, p. 97.
86. Hashimoto, K., *Arch. Dermatol.*, 1970, **102,** 280.
87. Hashimoto, K., *Arch. Dermatol.*, 1971, **104,** 148.
88. Zelickson, A. S., *J. invest. Dermatol.*, 1965, **56,** 10.
89. Billingham, R. E., *J. Anat.*, 1949, **83,** 109.
90. Billingham, R. E. and Medawar, P. B., *Philos. Trans. R. Soc. London Ser. B.*, 1953, **237,** 151.

91. Masson, P., *Cancer*, 1951, **4,** 9.
92. Riley, P. A., *The Physiology and Pathophysiology of the Skin*, Vol. 3, *The Dermis and the Dendrocytes*, ed. Jarrett, A., London, Academic Press, 1974, p. 1101.
93. Basset, F. and Nezelof, C., *Bull. Mém. Soc. Méd. Hôp. Paris*, 1966, **117,** 413.
94. Breathnach, A. S., Gross, M., Basset, F. and Nezelof, C., *Br. J. Dermatol.*, 1973, **89,** 571.
95. Gianotti, F. and Caputo, R., *Arch. Dermatol.*, 1969, **100,** 342.
96. Kobayasi, T. and Asboe-Hansen, G., *Acta Derm. Vernereol.*, 1972, **52,** 257.
97. Potten, C. S. and Allen, T. D., *Differentiation*, 1976, **5,** 43.
98. Mackenzie, I. C., *J. invest. Dermatol.*, 1975, **65,** 45.
99. Mackenzie, I. C., Zimmerman, K. L. and Wheelock, D. A., *Am. J. Anat.*, 1975, **144,** 461.
100. Breathnach, A. S., *Br. J. Dermatol.*, 1968, **80,** 688.
101. Prunieras, M., *J. invest. Dermatol.*, 1969, **52,** 1.
102. Wolf, K. and Schreiner, E., *J. invest. Dermatol.*, 1970, **54,** 37.
103. Sagebiel, R. W., *J. invest. Dermatol.*, 1972, **58,** 47.
104. Montagna, W., Bentley, J. P. and Dobson, R. L., eds, *Advances in Biology of Skin*, Vol. 10, *The Dermis*, New York, Appleton–Century–Crofts, 1970.
105. Jarrett, A., ed., *The Physiology and Pathophysiology of the Skin*, Vol. 3, *The Dermis and Dendrocytes*, London, Academic Press, 1974.
106. Bornstein, P., *Annu. Rev. Biochem.*, 1974, **43,** 567.
107. Gross, J., *Harvey Lect.*, 1974, **68,** 351.
108. Harkness, R. D., *Biol. Rev.*, 1961, **36,** 399.
109. Harkness, R. D., *Progress in the Biological Sciences in Relation to Dermatology*, ed. Rook, A. and Champion, R. H., Cambridge, University Press, 1964, p. 3.
110. Jackson, D. S., *Advances in Biology of Skin*, Vol. 10, *The Dermis*, ed. Montagna, W., Bentley, J. P. and Dobson, R. L., New York, Appleton–Century–Crofts, 1970, p. 39.
111. Ramachandran, G. N. and Gould, B. S., eds, *Treatise on Collagen*, Vols. 1 and 2, London, Academic Press, 1967, 1968.
112. Cassel, J. M., *Biophysical Properties of the Skin*, ed. Elden, H. R., New York, Wiley Interscience, 1971, p. 63.
113. Martin, G. R., Byers, P. H. and Piez, K. A., *Adv. Enzymol. relat. Areas mol. Biol.*, 1975, **42,** 167.
114. Schofield, J. D. and Prockop, D. J., *Clin. Orthop.*, 1973, **97,** 175.
115. Piez, K. A., Miller, E. J. and Martin, G. R., *Advances in Biology of Skin*, Vol. 6, *Ageing*, ed. Montagna, W., Oxford, Pergamon, 1965, p. 245.
116. Smith, J. G., Jnr, Sams, W. M., Jnr and Finlayson, G. R., *Modern Trends in Dermatology*, Vol. 3, ed. McKenna, R. M. B., London, Butterworth, 1966, p. 110.
117. Banga, I., *Structure and Function of Elastin and Collagen*, Budapest, Akademiai Kiado, 1966.
118. Ross, R. and Bornstein, P., *J. cell Biol.*, 1969, **40,** 366.
119. Partridge, S. M., *Advances in Biology of Skin*, Vol. 10, *The Dermis*, ed. Montagna, W., Bentley, J. P. and Dobson, R. L., New York, Appleton–Century–Crofts, 1970, p. 69.
120. Hall, D. A., *Biophysical Properties of the Skin*, ed. Elden, H. R., New York, Wiley Interscience, 1971, p. 187.
121. Bodley, H. D. and Wood, R. L., *Anat. Rec.*, 1972, **172,** 71.
122. Gotte, L., Mammi, M. and Pezzin, G., *Connect. Tissue Res.*, 1972, **1,** 61.
123. Gotte, L., Giro, M. G., Volpin, D. and Horne, R. W., *J. ultrastruct. Res.*, 1974, **46,** 23.
124. Jarrett, A., *The Physiology and Pathophysiology of the Skin*, Vol. 3, ed. Jarrett, A., London, Academic Press, 1974, p. 847.

125. Thomas, J., Elsden, D. F. and Partridge, S. M., *Nature* (*London*), 1963, **200,** 651.
126. Colta-Pereira, G., Guerra Rodrigo, F. and Bittencourt-Sampaio, S., *J. invest. Dermatol.*, 1976, **66,** 143.
127. Dorfman, A. *J. Histochem. Cytochem.*, 1963, **11,** 2.
128. Muir, H., *The Biochemistry of Mucopolysaccharides of Connective Tissue*, ed. Clark, F. and Grant, J. K., Cambridge, University Press, 1961, p. 4.
129. Muir, H., *Progress in the Biological Sciences in Relation to Dermatology*, Vol. 2, ed. Rook, A. and Champion, R. H., Cambridge, University Press, 1964, p. 25.
130. Muir, H. *International Review of Connective Tissue Research*, Vol. 2, ed. Hall, D. A., New York, Academic Press, 1964, p. 101.
131. Pearce, R. H. and Grimmer, B. J., *Advances in Biology of Skin*, Vol. 10, *The Dermis*, ed. Montagna, W., Bentley, J. P. and Dobson, R. L., New York, Appleton–Century–Crofts, 1970, p. 89.
132. Rogers, H. J., *The Biochemistry of Mucopolysaccharides of Connective Tissue*, ed. Clark, F. and Grant, J. K., Cambridge, University Press, 1961, p. 51.
133. Davidson, E. A., *Advances in Biology of Skin*, Vol. 6, *Ageing*, ed. Montagna, W., Oxford, Pergamon, 1965, p. 255.
134. Schubert, M., *Biophys. J.*, 1964, **4** (Supplement), 119.
135. Smith, J. G., Jnr, Davidson, E. A. and Taylor, R. W., *Advances in Biology of Skin* Vol. 6, *Ageing*, ed. Montagna, W., Oxford, Pergamon, 1965, p. 211.
136. Branwood, A. W., *International Review of Connective Tissue Research*, Vol. 1, ed. Hall, D. A. New York, Academic Press, 1963, p. 1.
137. Porter, K. R., *Biophys. J.*, 1964, **4** (Supplement), 167.
138. Szirmai, J. A., *Advances in Biology of Skin*, Vol. 10, *The Dermis*, ed. Montagna, W., Bentley, J. P. and Dobson, R. L., New York, Appleton–Century–Crofts, 1970, p. 1.
139. Bellamy, G. and Bornstein, P., *Proc. natl. Acad. Sci. USA*, 1971, **68,** 1138.
140. Ayer, J. P., *International Review of Connective Tissue Research*, Vol. 2, ed. Hall, D. A., New York, Academic Press, 1964, p. 33.
141. Asboe-Hansen, G., *Connective Tissue*, ed. Tunbridge, R. E., Oxford, Blackwell, 1957, p. 30.
142. Gersh, L. and Catchpole, H. R., *Am. J. Anat.*, 1949, **85,** 457.
143. Smith, D. E., *Ann. N. Y. Acad. Sci.*, 1963, **103,** 40.
144. Selye, H., *The Mast Cells*, Washington, Butterworth, 1965.
145. Comaish, J. S., *Br. J. Dermatol.*, 1965, **77,** 92.
146. Lagunoff, D., *J. invest. Dermatol.*, 1972, **58,** 296.
147. Weddell, G., Palmer, E. and Pallie, W., *Biol. Rev.*, 1955, **30,** 159.
148. Allenby, C. F., *An Introduction to the Biology of Skin*, ed. Champion, R. H., Gilman, T., Rook, A. J. and Sims, R. T., Oxford, Blackwell, 1970, p. 124.
149. Kenshalo, D. R., ed., *The Skin Senses*, Springfield, Charles C. Thomas, 1968.
150. Iggo, A., ed., *Handbook of Sensory Physiology*, *Somatosensory System*, Berlin, Springer, 1973.
151. Iggo, A., *The Peripheral Nervous System*, ed. Hubbard, J. I., New York, Plenum, 1974, p. 347.
152. Sinclair, D. C., *The Physiology and Pathophysiology of the Skin*, Vol. 2, *The Nerves and Blood Vessels*, ed. Jarrett, A., London, Academic Press, 1973, p. 347.
153. Hashimoto, K., *J. Anat,* 1972, **111,** 99.
154. Winkelmann, R. K. and Breathnach, A. S., *J. invest. Dermatol.*, 1973, **60,** 2.
155. Hunt. C. C., *The Peripheral Nervous System*, ed. Hubbard, J. I., New York, Plenum, 1974, p. 405.
156. Winkelmann, R. K., *Advances in Biology of Skin*, Vol. 1, *Cutaneous Innervation*, ed. Montagna, W., Oxford, Pergamon, 1960, p. 48.

157. Hashimoto, K., *J. invest. Dermatol.*, 1973, **60,** 20.
158. Weddell, G., *Advances in Biology of Skin*, Vol. 1, *Cutaneous Innervation*, ed. Montagna, W., Oxford, Pergamon, 1960, p. 112.
159. Iggo, A. and Gottischaldt, K.-M, *Rheinisch-Westfälische Acad. Wissenschaften*, 1976, **53,** 153.
160. Uno, H, *J. invest. Dermatol.*, 1977, **69,** 112.
161. Robertshaw, D., *J. invest. Dermatol.*, 1977, **69,** 121.
162. Montagna, W. and Ellis, R. A., ed., *Advances in Biology of Skin*, Vol. 2, *Blood Vessels and Circulation*, Oxford, Pergamon, 1961.
163. Moretti, G., *Jadassohn's Handbuch der Haut und Geschlechtskrankheiten*, Vol. 1, Part 1, Berlin, Springer, 1968, p. 491.
164. Ryan, T. J., *The Physiology and Pathophysiology of the Skin*, Vol. 2, ed. Jarrett, A., London, Academic Press, 1973, p. 557.
165. Ryan, T. J., *J. invest. Dermatol.*, 1976, **67,** 110.
166. Montagna, W., Ellis, R. A. and Silver, A. F., *Advances in Biology of Skin*, Vol. 3, *Eccrine Sweat Glands and Eccrine Sweating*, Oxford, Pergamon, 1962.
167. Gordon, R. S., and Cage, G. W., *Lancet*, 1966, **i,** 1246.
168. Randall, W. C. and McChine, W., *Am. J. Physiol.*, 1948, **155,** 462.
169. Cage, G. W. and Dobson, R. L., *J. clin. Invest.*, 1965, **44,** 1270.
170. Hamilton, J. B. and Light, A. E., eds, *Ann. N. Y. Acad. Sci.*, 1950, **53,** 461.
171. Montagna, W. and Ellis, R. A., eds, *The Biology of Hair Growth*, New York, Academic Press, 1958.
172. Lubowe, I. I., *Ann. N. Y. Acad. Sci.,* 1959, **83,** 539.
173. Montagna W. and Dobson, R. L., eds, *Advances in Biology of Skin*, Vol. 9, *Hair Growth*, Oxford, Pergamon, 1969.
174. Jarrett, A., ed., *The Physiology and Pathophysiology of the Skin*, Vol. 4, London, Academic Press, 1977.
175. Kligman, A. M., *J. invest. Dermatol.*, 1959, **33,** 307.
176. Kligman, A. M., *Arch. Dermatol.*, 1961, **83,** 175.
177. Ebling, F. J. and Hale, P. A., *Mem. Soc. Endocrinol.*, 1970, **18,** 215.
178. Ebling, F. J., *J. invest. Dermatol.*, 1976, **67,** 98.
179. Orentreich, N., *Advances in Biology of Skin*, Vol. 9, *Hair Growth*, ed. Montagna, W. and Dobson, R. L., Oxford, Pergamon, 1969, p. 99.
180. Hamilton, J. B., *The Biology of Hair Growth*, ed. Montagna, W. and Ellis, R. A., New York, Academic Press, 1958, p. 399.
181. Ebling, F. J., Thomas, A. K., Cooke, I. D., Randall, V. A., Skinner, J. and Cawood, M., *Br. J. Dermatol.*, 1977, **97,** 371.
182. Hamilton, J. B. *Am. J. Anat.*, 1942, **71,** 451.
183. Montagna, W., Ellis, R. A. and Silver, A. F., ed., *Advances in Biology of Skin*, Vol. 4, *Sebaceous Glands*, Oxford, Pergamon, 1963.
184. Ebling, F. J., *Dermatotoxicology and Pharmacology*, ed. Marzulli, F. N. and Maibach, H. I., Washington, Hemisphere, 1977, p. 55.
185. Strauss, J. S. and Pochi, P. E., *Recent Prog. Horm, Res.*, 1963, **19,** 385.
186. Burton, J. L., Libman, L. J., Cunliffe, W. J., Hall, R. and Shuster, S., *Br. med. J.*, 1972, **139,** 406.
187. Ebling, F. J., Ebling, E., Randall, V. A. and Skinner, J., *Br. J. Dermatol*, 1975, **92,** 235.
188. Thody, A. J. and Shuster, S., *J. Endocrinol.*, 1975, **64,** 503.
189. Ebling, F. J., Ebling, E., Randall, V. A. and Skinner, J., *J. Endocrinol.*, 1975, **66,** 407.
190. Ebling, F. J., *Acta Endocrinol.*, 1973, **72,** 361.
191. Greene, R. S., Downing, D. T., Pochi, P. E. and Strauss, J. S., *J. invest. Dermatol.*, 1970, **54,** 240.

192. Hurley, M. J. and Shelley, W. B., *The Human Apocrine Sweat Gland in Health and Disease*, Springfield, Charles C. Thomas, 1960.
193. Wales, N. A. M. and Ebling, F. J., *J. Endocrinol.*, 1971, **51,** 763.
194. Le Magnen, J., *Arch. Sci. Physiol.*, 1952, **6,** 125.
195. Russell, M. J., *Nature* (*London*), 1976, **260,** 520.
196. Ellis, H. *Studies in the Psychology of Sex*, 4, *Sexual Selection in Man*, Philadelphia, F. A. Davis, 1905.
197. Rook, A., Wilkinson, D. S. and Ebling, F. J., *Textbook of Dermatology*, 3rd edn, Oxford, Blackwell, 1979.
198. Fitzpatrick, T. B., Eisen, A. Z., Wolff, K., Freedberg, I. M. and Austin, K. F., *Dermatology in General Medicine*, 2nd edn, New York, McGraw–Hill, 1979.
199. Brues, A. M., *Am. J. hum. Genet.*, 1950, **2,** 215.
200. Ferrara, R. J., *Cutis*, 1960, **2,** 561.
201. Lerner, A. B., Snell, R. S., Chanco-Turner, M. L. and McGuire, J. S., *Arch. Dermatol.*, 1966, **94,** 269.
202. Cunliffe, W. J., Hall, R., Newell, D. J. and Stevenson, C. J., *Br. J. Dermatol.*, 1968, **80,** 135.
203. El Mofty, A. M., *Vitiligo and Psoralens*, Oxford, Pergamon, 1968.
204. Bleehan, S. S., *Br. J. Dermatol.*, 1976, **94** (Supplement 12), 43.
205. Cunliffe, W. J. and Cotterill, J. A., *The Acnes*, London, Saunders, 1975.
206. Plewig, G. and Kligman, A. M., *Acne*: *Morphogenesis and Treatment*, Berlin, Springer, 1975.
207. Frank, S. B., ed., *Acne: Update for the Practitioner*, New York, Yorke Medical, 1979.
208. Pochi, P. E. and Strauss, J. S., *J. invest. Dermatol.*, 1974, **62,** 191.
209. Shehadeh, N. D. and Kligman, A. M., *Arch. Dermatol.*, 1963, **88,** 829.
210. Marples, R. R., *J. invest. Dermatol.*, 1974, **62,** 326.
211. Hamilton, J. B., *J. clin. Endocrinol. Metab.*, 1941, **1,** 570.
212. Forstrom, L., Mustakallio, K. K., Dessypris, A., Uggeldahl, P. E. and Adlersceutz, H., *Acta Derm. Venereol.*, 1974, **54,** 369.
213. Strauss, J. S. and Pochi, P. E., *Arch. Dermatol.*, 1965, **92,** 443.
214. Strauss, J. S. and Pochi, P. E., *Arch. Dermatol.*, 1962, **86,** 757.
215. Akers, W. A., Allen, A. M., Burnett, J. W., *et. al.*, *Arch. Dermatol.*, 1975, **111,** 1630.
216. Hersh, K., *Dermatologica*, 1972, **145,** 187.
217. Hölzle, E. and Kligman, A. M., *Br. J. Dermatol.*, 1978, **99,** 117.
218. Lobitz, W. C., *Dermatoses due to Environmental and Physical Factors*, ed. Rees, P. B., Springfield, Charles C. Thomas, 1962.
219. Hindson, T. C. and Worsley, D. E., *Br. J. Dermatol.*, 1969, **81,** 226.
220. Kidd, C. B. and Meenan, J. C., *Br. J. Dermatol.*, 1961, **73,** 129.
221. Weinstein, G. D. and Frost, P., *J. invest. Dermatol.*, 1968, **50,** 254.
222. Gelfant, S., *Br. J. Dermatol.*, 1976, **95,** 577.
223. Duffill, M. B., Appleton, D. R., Dyson, P., Shuster, S. and Wright, N. A., *Br. J. Dermatol*, 1977, **96,** 493.
224. Van Abbe, N. J., *J. Soc. cosmet. Chem.*, 1964, **15,** 609.
225. Weary, P. E., *Arch. Dermatol.*, 1968, **98,** 408.
226. Roia, F. C. and Vanderwyck, R. W., *J. Soc. cosmet. Chem.*, 1969, **20,** 113.
227. Berrens, L. and Young, E., *Dermatologica*, 1964, **128,** 3.
228. Brody, I., *J. ultrastruct. Res.*, 1963, **8,** 580.
229. Plewig, G. and Kligman, A. M., *Arch. Klin. exp. Dermatol.*, 1970, **236,** 406.

Chapter Two

Irritation and Sensitization of the Skin

Introduction

The manufacturers of cosmetics, toilet preparations and similar products have a moral obligation, which is becoming enforced by increasingly stringent legal requirements, not to sell substances harmful to the user. Such substances applied to the skin may induce several harmful effects, the most likely being irritation and allergic sensitization. Less frequently encountered responses are contact urticaria resulting from a cytotoxic release of histamine, 'stinging', phototoxicity and photoallergy (Table 2.1). Much care is necessary to assess the potential adverse effects of substances to be applied to the skin, and where appropriate to do the biological tests to ensure safety in use so that the potential for adverse reactions is reduced to the minimum.

Considerations of safety apply not only to the users of products but also to those preparing the products, who are likely to handle the ingredients in large amounts.

Table 2.1 Inflammatory and Allergic Responses that may be Induced in Skin by Topical Application of Substances

Irritation (irritant dermatitis) (a) Acute or primary irritation (b) Repeated exposure or secondary irritation	
Contact urticaria	A transient oedematous response mediated by pharmacological mediators secreted by mast cells, or cytotoxic release from mast cells, induced by the applied substance
Stinging	A transient sensation distinct from irritation and allergy, but which may be considered to be an irritability of sensory nerve endings
Allergic urticaria	A response similar in appearance to contact urticaria above, but induced by antigen in the applied substance reacting with specific antibody initiating the release or generation of pharmacological mediators from mast cells
Allergic contact dermatitis (contact eczema)	
Phototoxic dermatitis	
Photoallergic dermatitis	

Irritants and Inflammation

There are several definitions of irritants, the simplest being that they are substances that induce inflammation or, in more detail, non-corrosive substances and preparations which, through immediate, prolonged or repeated contact with the skin or mucous membranes, cause inflammation.

A primary irritant induces an inflammatory response on first contact with the skin, though the contact may be of several hours' duration. A secondary irritant is a substance that is outwardly harmless on first contact, but of which repeated applications induce inflammation which becomes progressively more severe. Other definitions of irritants relate to the intensity of reactions in a proportion of rabbits used in predictive tests. Such definitions are legal expediencies.

'Inflammation' is the term for all the changes occurring in living tissue when it is injured, provided that the injury is not so severe as to immediately kill the cells or destroy the tissue structure.

Inflammation

It is not possible in a short chapter to do more than indicate the nature of the changes occurring in irritated (inflamed) skin, and the complex interactions between the epidermis, the infiltrating leucocytes and the pharmacologically active substances released or generated. For an account of inflammation particularly in relation to skin, the reader is referred to Parish and Ryan[1] and, for more detailed reviews, to Zweifach *et al.*[2] and Lepow and Ward.[3]

The clinical signs of inflammation are redness, swelling, heat and pain. Not all lesions show all four features. Mild irritations, such as occur following use of some cosmetic preparations, may result in redness and mild itching or sting, with inappreciable swelling or heat. Such lesions usually progress to an accumulation of dry scales, fine surface fissures and slight thickening of the skin. The redness is a manifestation of the increased flow of blood through the superficial dilated vessels, and hence the greater number of red cells in the tissue. The subsequent desquamation of scales and thickening of the skin results from the shedding of the surface corneal squames, which may have been damaged, and the slight excess of new squames formed as part of the reactive hyperplasia. The epidermis also has several additional layers of keratinocytes, and the dermis is likely to be infiltrated by leucocytes and plasma contributing to the increased thickness of the skin. Within one to two weeks, depending upon the severity of the initial inflammation, the appearance of the skin returns to normal, though there will still be histological evidence of the episode.

Inflammation progresses to regeneration or repair. Regeneration is healing with the same tissue elements in almost the same form as were present before the damage, though regenerated skin, if the inflammation was severe, seldom returns exactly to the pre-damage state. Repair is healing resulting in distortion of the original tissues and deposition of scar tissue. This is a very unlikely consequence of the application of a cosmetic preparation.

Changes in inflamed stratum corneum and keratinocytes are summarized in Tables 2.2 and 2.3; inflammation induced in hypersusceptible persons by cosmetics which have previously been examined for safety is mild, and significant coagulative degeneration and necrosis are unlikely to occur. However, an overview of inflammation is incomplete without their consideration.

Table 2.2 Changes in Stratum Corneum after Application of Irritants

Removal of lipids
Removal of soluble cellular substances and water
Denaturation and unfolding of proteins
Vacuolation
Maceration
Desquamation
Changes in detectable enzyme content
Hyperkeratosis and parakeratosis

The changes in irritated skin are induced by the physical and chemical toxic actions of the irritant, and by the pharmacological mediators released or activated in the inflammatory response. Thus solvents may extract lipids from the stratum corneum, macerate the cells, impair the water barrier function, and damage or kill some of the underlying keratinocytes. These changes are a direct effect of the applied substance. During the ensuing inflammatory response, lysosomal proteolytic and other enzymes from infiltrating leucocytes and damaged epidermal cells degrade tissue elements and activate other pharmacologically active systems, for example complement and the kinins. These mediators attract more leucocytes and also release other active substances, for example histamine and proteolytic chymases from mast cells. The complex cascade of inflammatory events results in more tissue change than that induced directly by the toxic substance.

Among the changes induced in the stratum corneum by applied substances are removal of lipids, soluble proteins and other cellular substances, denaturation of soluble proteins and unfolding of fibrillar proteins such as keratin (Table 2.2). These result in impairment of physiological function, for example loss of the water barrier or water retention properties, impaired resistance to penetration by micro-organisms or environmental substances, and loss of plasticity or

Table 2.3 Summary of Features of Epidermal Inflammation, Omitting the Participation by Leucocytes (after Parish and Ryan)[1]

Stimulation	Hydropic degeneration (vacuolation)	Coagulative degeneration
Metabolism	Cell enzyme activation	Condensation of cytoplasm and nucleus
Cell migration	Chromatin aggregation	↓
Mitosis	Cell swelling	Disappearance of enzymes and lysosomes
↓	↓	
Hyperplasia	Nuclear pyknosis	↓
	Perinuclear vacuolation	Persistence of shrunken cell, possibly retaining some of its superficial form (necrosis)
	Swelling	
	Complete autolysis or disruption	

elasticity which may lead to fine ruptures and desquamation. There are also histological changes, an altered affinity for histological stains, deformation of the cells, changes in the detectable enzymes which are either unmasked within the cells or have permeated into the squames from the underlying epidermis and dermis. Eventually, proliferation of the underlying epidermal cells (hyperplasia) leads to a transitory increase in numbers of corneal squames (hyperkeratosis) to replace those damaged, some of which may retain condensed nuclear material (parakeratosis). As the increased numbers of corneal squames are shed, the thickness of this surface stratum returns to normal.

Thus it will be appreciated that the dry flaking skin commonly observed in mild irritation may result from the direct effect of the irritant or from the subsequent transient hyperactive response to the damage.

The possible changes in the basal layer and keratinocytes are much more complex, and these cells are more subject to stimuli from infiltrating leucocytes and substances permeating from the dermis and blood. The summary of possible changes, not comprehensive, and omitting the effects induced by leucocytes and non-epidermal mediators (Table 2.3), reflects the direct responses to irritants. Each group of changes represents the dominant histological response which may be observed at a particular time, as there is a shift in the nature of the response and its intensity with time and all responses end with a phase of stimulation of metabolism and hyperplasia, with epidermal cell migration from the edges to cover the area if the original damage resulted in cell death and slough.

A common epidermal response to irritants is vacuolation resulting from poisoning of the osmotic regulatory process within the cell, so that excessive amounts of fluid are absorbed; this is enhanced by the release of the lysosomal enzymes which autolyse the cytoplasm, releasing more fluid.

The lysosomal enzymes, acidic and neutral proteases, phosphatases and nucleases, when released from living or dying cells, contribute to the degradation of surrounding tissues and activation of other substances, for example complement in plasma, that attract leucocytes.

Inflammatory changes in the dermis (Table 2.4) resemble those found in many tissues, and vary much according to the severity and duration of the injury. The immediate response in small blood vessels to mild irritation is erythema (increased blood flow), increased permeability leading to oedema, and stickiness

Table 2.4 Features of Inflammation in the Dermis (common to many tissues)

Erythema
Oedema
Leucocyte adhesion and infiltration
Polymorphonuclear leucocytes
Mononuclear cells
Fibrin deposition and thrombosis
Degradation of tissues
Granulomata
Capillary proliferation during resolution
Fibrosis and scar

of the endothelium so that within minutes leucocytes adhere to the surface and some emigrate from the vessel, particularly neutrophils. These are the chief changes following mild irritation of short duration.

In more severe or prolonged irritation, in addition to dense accumulations of neutrophils there is also infiltration by macrophages which may be obscured at first by the neutrophils. In the later stages of resolution of the damage, the macrophages ingest and remove dead cells and tissue debris, release enzymes to degrade damaged tissue and release other substances that stimulate cells, promoting repair. At this stage fibroblast activity is intensified while new connective tissue elements are laid down. Other changes (Table 2.4) usually occur after more severe irritation.

Skin that has recently recovered from an inflammatory episode tends to be more susceptible to further damage for several days; dividing cells are more susceptible to toxic change, newly formed capillaries and venules are hypersusceptible to many stimuli, and the residua of plasma substances, for example fibrin, potentiate the further activation of pharmacological mediators.

Changes observed in inflammation are mediated by substances derived from the plasma, from cells of the damaged tissue and from infiltrating leucocytes. The effects observed are the resultant of the stimuli promoting change, and the many inhibitors that modify or prevent the change. For detailed reviews see Lepow and Ward,[3] Cochrane[4] and Wasserman;[5] Parish and Ryan[1] provide a summary of activities.

Mediators of inflammatory change generated or activated in the plasma are bradykinin and complement. These substances together mediate vasodilatation in capillaries, increase vascular permeability leading to oedema, attract leucocytes, and a complement component releases other mediators from connective tissue mast cells. Many inflammation-promoting substances are released from cells. Platelets on aggregation release histamine, clotting promotors and neutral pH proteases. The infiltrating neutrophils and macrophages release a wide range of lysosomal degradative enzymes; macrophages also synthesize one or more complement components, enzymes activating complement, and prostaglandins. Prostaglandins, also generated by damaged cells, for example epidermis, contribute further stimuli to increased vascular leakage and modulate, or influence, the release of mediators from other types of cell. Connective tissue mast cells are a source of potent inflammatory mediators including histamine, a substance which increases blood flow in capillaries, increases leakage from venules leading to oedema, and constricts smooth muscle. Mast cells also release substances that attract or 'arrest' (that is, stop further movement of) eosinophils and neutrophils, leading to the accumulation of these cells and augmenting the concentration of inflammation-promoting mediators. Thus one type of cell enhances the activities of others, stimulating a cyclic sequence of cell and mediator interaction until the initial effects of the inducing agent are eliminated.

All these interacting changes occur, with varying intensity, even in mild inflammation.

Contact Urticaria

Urticaria is a transient erythematous eruption, with oedema mainly in the dermis. The spontaneous clinical disorder in man that may extend over much of the body has many presumed causes, the majority of which are not relevant to cosmetics. Contact urticaria refers to the local oedema and erythema at the site of application of a substance.

The oedema results from the release of mast cell histamine which increases the permeability of the cutaneous vessels, augmented by the activation of kinins with similar activity.

It is regrettable that there is a current tendency to designate as contact urticaria all immediate transient, erythematous, oedematous reactions at the site of contact without consideration of the cause. This type of response in allergic anaphylactic reactions to specific antigen is considered in the section on allergy. The designation 'contact urticaria' is best reserved for the non-allergic induced secretion, or cytotoxic release from mast cells, of histamine. Incidentally, nettle rash from contact with the nettle (urtica), an often cited example of a local urticarial response, mainly results from histamine of the plant penetrating with the fine hairs of the leaf into the skin, and not from histamine released from the mast cells.

Non-allergic contact urticaria is induced by some substances used in cosmetics. Chronic and generalized urticaria and asthma occur in some persons, for example in bakers and hairdressers, following occupational exposure to persulphates, or in persons using hair bleach preparations containing these salts. There is good evidence that dipersulphate ions (of potassium and ammonium salts) have a selective ability to release histamine from mast cells without morphological cytotoxic membrane damage.[6]

However, though the release of histamine from mast cells by persulphates is not an allergic response, as can be shown in tests on isolated normal mast cells *in vitro*, it is possible that those persons who are susceptible to this immediate action of persulphates, whether urticarial or asthmatic, may have a concomitant delayed hypersensitivity (lymphocyte-mediated) to the persulphate ion.[6] Cinnamic aldehyde is another substance inducing immediate urticarial reactions and is also a contact sensitizing allergen.[7]

With the increasing interest in the phenomenon, a similar activity is likely to be reported for several substances affecting only a very small proportion of those exposed.

Stinging

There is an ill-defined response of the skin to some topically applied substances that is generally referred to as stinging, though other descriptions of the sensation are itch, tingle, burn or sore. The response starts within minutes of application of the substance, intensifies over the next 5 to 10 minutes and then wanes. The response is peculiar to the face, particularly the nasolabial folds, and to a lesser degree the cheeks.[8] Not all persons are susceptible; fair-skinned females who 'blush easily' appear to be the most susceptible.[8]

The phenomenon is distinct from irritation, and does not lead to inflammatory change. Irritants may not sting, whereas non-irritants may do so. A wide range of substances, acids and alkalis, but not strictly pH-dependent, have this property.[8] However, in a test with creams containing urea, the preparation with an acidic pH caused stinging in 13 out of 60 persons, but that with approximately neutral pH had no such effect.[9]

A few animal tests have been reported to predict stinging activity for man, but as the effect on man is a transient discomfort with no residual inflammation, the author considers that man himself offers the more appropriate test system.

Frosch and Kligman[8] have reported a procedure to identify 'stingers' and to use them to test the activity of substances. A simpler procedure based on their method could be devised.

Variations in Susceptibility to Irritants

There is little significant variation in response to irritants inducing moderate to severe inflammation, but there are variations in susceptibility to irritants that induce very mild inflammation, manifested as a slight redness followed by a dry skin with surface scales. There are changes in susceptibility of normal skin, changes with age, and with the oestral cycle in women. Reference has already been made to the increased susceptibility of skin already recovering from inflammation. Another stimulus is change in 'accommodation' of the skin. Skin adapts to the repeated application of products. A change to another similar product may bring about transient, mild changes indicative of stimulation while the skin adapts.

Environmental conditions, temperature, relative humidity, and exposure to sunlight also influence the susceptibility to irritation.

Hypersensitivity and Allergy

There are many descriptions of immunological (allergic) responses. Parish[10] gives an account of immunology with special reference to skin; Champion and Parish[11] summarize skin allergic disorders; Coombs and Gell[12] and Fudenberg *et al.*[13] describe allergies and immunological mediators. All these authors give references to further reading.

Hypersensitivity is an immunological responsiveness of any type that is greater than that normally occurring in response to antigens of the environment. It applies to all immunological responses to induced antigenic stimulation, for example responses to vaccines. The term 'allergy' designates an antigen-specific, altered reactivity of the tissues to substances compared with the response of the first exposure to the same substances. In the strict sense of the terms, hypersensitivity and allergy are synonymous, but in practice 'allergy' is restricted to clinically observable altered reactivity. Immunological sensitivity or hypersensitivity is a state of immunological responsiveness. Allergy is the clinical change or condition resulting from the exposure of the hypersensitive person to the antigen.

An antigen is a substance which stimulates the formation of antibody, or alters the reactivity of certain cells (the cell-mediated response) and when mixed with that antibody *in vitro*, or applied to the tissues as in skin tests, reacts specifically to induce an observable reaction. Antigens that induce allergy are frequently known as allergens.

Some substances, for instance low-molecular-weight chemicals, need to combine with protein before they become antigenic. These are known as haptens and when combined with the protein induce formation of antibody, or delayed hypersensitivity (cell-mediated) responses specific for the hapten. Haptens are important sensitizing agents in contact dermatitis.

An important property of antigens is specificity. Antibodies or sensitized lymphocytes react with the antigen inducing the sensitization. However, some cross-reactions with other antigens may occur if they share some of the determinants conferring specificity. Thus a person sensitized to one antigen may react to another of similar chemical form, though the cross-reaction is usually weaker.

Antigenic sensitization results from a complex series of events in which antigens (for example from a cosmetic preparation which may penetrate through the skin, mucous membranes of the mouth or respiratory tract) enter the body. The antigens are modified by macrophages, or by Langerhans cells of the epidermis, and the stimulus to specificity to the particular chemical structure (determinants) of the antigen is transferred to lymphoid cells. The lymphoid cells with the acquired specific responsiveness to that antigen undergo numerous cell divisions to form clones, which give rise to circulating T lymphocytes. The T lymphocytes are the effector cells mediating the changes of delayed hypersensitivity, or contact dermatitis reactions. These lymphocytes have several other activities, one of which is to act as 'helper' cells contributing to the antigenic activation of another group of lymphocytes, the B lymphocytes.

B lymphocytes synthesize immunoglobulins (antibodies) which remain bound to the cell membrane. They are the precursors of the plasma cells which synthesize the antibody released into the blood. A few antigens can stimulate B lymphocyte activation without T lymphocyte cooperation.

Man forms five classes of immunoglobulin, each having special physical and chemical properties and biological activities. Although four (and probably all five) classes of antibody participate in various allergic reactions, the antibody of greatest importance in allergy to cosmetics is IgE (Ig designates immunoglobulin). This used to be known as the reagin and is the anaphylactic antibody of man, mediating the immediate-type allergic responses, for example hay fever, asthma and some allergic gastro-enteritis. In the skin it mediates erythema (reddening), allergic urticaria (reddening with oedematous swellings) and allergic angio-oedema (dermal and subcutaneous oedematous swelling).

Of the four types of allergic response, only two are of importance in responses to cosmetics: *delayed hypersensitivity*, which is manifested by contact dermatitis or eczema, and *anaphylactic sensitivity*, which is manifested in the various forms of erythema and oedema and by 'contact allergic urticaria'.

Delayed Hypersensitivity

Delayed hypersensitivity is also designated as the cell-mediated immune response, because the clinical response is mediated by the T lymphocytes in the absence of specific antibody, though antibodies may be formed concomitantly with the activated lymphocytes to the same antigen.

The sequence of changes in a delayed hypersensitivity response (Figure 2.1) is as follows. The allergen penetrating the skin must be bound in the tissue for about two hours. When a lymphocyte, specifically sensitized to the antigen and randomly moving through the tissue, encounters the antigen, the cell binds to it through the specific receptors in its membrane. This reaction causes the

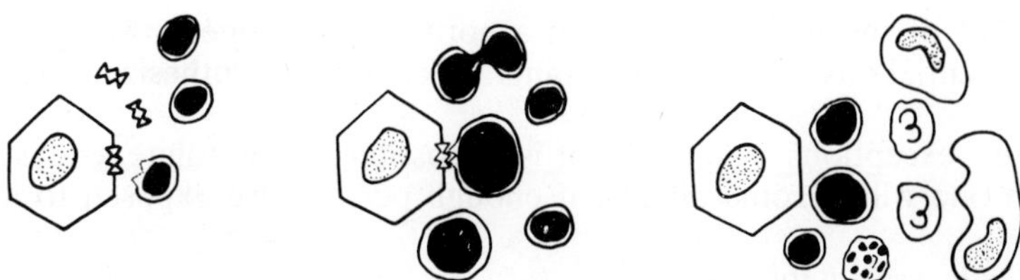

Figure 2.1 Diagrammatic representation of the delayed hypersensitivity response. Antigen (allergen: sensitizer) is bound in the tissue for a few hours. T (effector) lymphocytes specifically sensitized to the antigen react with it, resulting in enlargement of the cell (transformation) preliminary to several cell divisions. At the same time other lymphocytes, not sensitized to the antigen, are attracted and activated. Other substances are generated that attract neutrophils, sometimes basophils, and also activate macrophages. (Modified from Parish[10])

lymphocyte to transform (enlarge with increased synthesis of DNA) and subsequently it divides, and several further cell divisions may follow. At the same time, the transforming lymphocyte synthesizes several other substances, known as products of activated lymphocytes or lymphokines. These substances attract other lymphocytes, not sensitized to the antigen, which also transform and synthesize more lymphocyte products, thus augmenting the effect of the few specifically sensitized lymphocytes initiating the reaction. The substances also attract neutrophils and attract and activate macrophages. In some responses there is an early infiltration by large numbers of basophils. The neutrophils disappear within 24 hours, and at 48 hours the predominant change at the site is the infiltration of mononuclear cells. Other inflammation-promoting substances are synthesized and contribute to the damage of the epidermal cells in the area, or damage to other tissues according to the site of reaction.

The diagnostic skin test procedure is the patch test in which antigen is applied on special absorbent discs to the skin for 24 or 48 hours. The response usually reaches its maximum intensity in 48 hours from first application, and appears as a raised, red, firm area, which may have minute papules or vesicles which in strong reactions may develop into larger bullae.

In general, persons do not have a predisposition to delayed contact sensitivity similar to the predisposition to anaphylactic sensitivity in atopic persons. There is however some individual constitutional susceptibility to contact dermatitis. Strong allergens sensitize the majority of persons; but weak allergens sensitize only a small proportion of those exposed, and a person sensitized to one substance is not necessarily susceptible to sensitization by another.

Laboratory predictive tests to identify substances likely to induce contact dermatitis, enabling their elimination from formulations, ensure that the great majority of cosmetic preparations do not contain such allergens, generally known as sensitizers. However, there are always some susceptible persons who become sensitized to substances harmless to the great majority of the population. Perfume ingredients appear to be the commonest source of adverse reactions of this type. Consistent with this are the numerous persons developing contact allergy to common garden flowers, as vegetable allergens are among the most potent of those encountered in a normal environment. Many perfumes contain essential oils derived from plants, or chemicals synthesized to resemble them.

With few exceptions, like water, it is impossible to formulate a substance to which no one will become allergic if enough persons are exposed to it.

Anaphylactic Sensitivity

At one time anaphylactic sensitivity was known as the immediate allergic reaction because the signs appear within minutes of exposure to the antigen. This response is mediated by IgE antibodies which have the special property of binding to mast cells and to basophil leucocytes. These are the only cells with receptors for the anaphylactic antibodies, and they contain pharmacological substances, for example histamine, which mediate the anaphylactic changes.

The sensitized person has IgE antibodies bound to receptors of mast cells, for example in the dermis, and on basophil leucocytes (Figure 2.2). The antibodies are bound by the Fc portion, leaving the antigen-binding portion free. IgE antibodies mediate most anaphylactic reactions in man, but in a small proportion of persons there is an IgG short-term sensitizing (S-TS) antibody that apparently binds to the same receptors as the IgE.[10]

Antigen penetrating through the skin reacts with the cell-bound antibodies. If two antibody molecules are bridged by antigen, a series of changes in the cell membrane and cytoplasm results in the secretion of preformed mediators, for example histamine, and eosinophil chemotactic factor, or generation of others, for example slow reacting substance of anaphylaxis. These pharmacologically active substances mediate the changes observed in the skin, for example capillary dilatation (reddening) and increased vascular permeability (oedema and swelling).

Reference has already been made to the contact allergic urticaria and other signs observed in anaphylactic responses in the skin. The reactions are usually apparent within minutes and, unless severe or unless the antigen persists at the

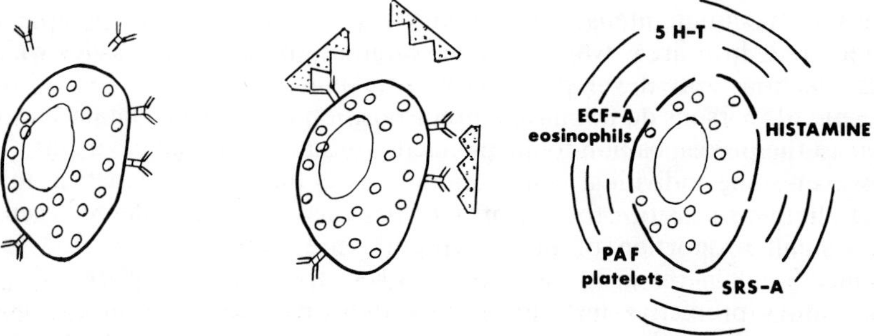

Figure 2.2 Diagrammatic representation of the anaphylactic response. Anaphylactic antibody F_c binds to the specific receptors on mast cell or basophil membranes. Antigen (serrated blocks) reacts with free F_{ab} (antigen binding) sites to bridge or link two antibody molecules. Antigen reacting with one antibody molecule does not initiate the reaction. Formation of the complex on the cell membrane initiates the release or generation of the pharmacological agents—histamine, 5-hydroxytryptamine in rodents, slow reacting substance of anaphylaxis (SRS-A), platelet activating factor (PAF). (Parish[10])

site, the response reaches a maximum intensity in fifteen minutes and subsides in one to two hours. The diagnostic test for soluble antigens is the prick test, in which a drop of the solution of suitable concentration is applied to the skin and a small prick with a fine needle is made into the surface to facilitate penetration. The time course of the skin test reaction is described above.

Individuals vary in their susceptibility to anaphylactic sensitization. About 15–20 per cent of persons have a genetically determined liability to form IgE reaginic antibodies to common antigens of their environment, and have an increased susceptibility to anaphylactic disorders. Such persons, known as atopic subjects, may respond to antigenic products applied to their skin by urticarial eruptions and itching. Itching is one of the sensations induced by histamine and is felt in the early stages of a prick test response.

Laboratory predictive tests are not yet well developed to detect the potential of substances to induce formation of IgE antibodies in man, and most information is obtained from history of use in the market place.

Allergic Contact Urticaria. This clinical condition differs from the contact urticaria discussed above, in that it is initiated by antigen reacting with anaphylactic antibody to release the histamine and other substances from mast cells, whereas in non-allergic contact urticaria the inducing substance acts directly on the mast cells. Thereafter, both responses are similar in that the same mediators are released or activated.

The recent interest in urticarial responses following soon after application of substances to skin in patch tests is a reflection of good clinical observation, and not an indication of a new type of response.

Phototoxic and Photo-allergic Reaction

Some substances which are innocuous and well tolerated become harmful when activated by light. The effects induced by the light-activated substance or its metabolites may be phototoxic (that is, a light-induced inflammation), photo-allergic, where the antigenic stimulus is activated by light, or photo-cancerous after frequent and prolonged exposure. Penetration of the substances inducing reactions may follow percutaneous application, ingestion or even inhalation. The substance (or its metabolites) is activated by light to give rise to molecules in an electronically excited state which are tissue damaging. There are natural de-activation processes in the tissues which limit the intensity of such reactions.

The energy to activate the light-sensitive substance is derived from the radiations in the 300 to 800 nm wavelengths of the ultraviolet and visible light absorbed by the system. It is good practice to examine substances under consideration for use in cosmetics for their ability to absorb light at wavelengths greater than 290 nm. Any substance doing so should be considered suspect until tests show that it is free of photobiological activity or that such activity is not relevant to its intended use.

There are several substances which are, or used to be, used in cosmetics which induce phototoxicity or photo-allergy. Eosin, once used in lipsticks, is a much quoted example of a phototoxic substance. Other examples are *p*-aminobenzoic acid derivatives and digalloyl trioleate in sunscreen preparations, bithionol and hexachlorophene in toilet soaps and deodorants, and the halogenated salicylanilides, for example tetrachlorosalicylanilide (T4CS) as a bacteriostatic substance. Perfumes compounded with essential oils from some plants, for example bergamot and particularly the psoralens contained in this oil, are a potent source of phototoxins.

Phototoxic reactions are inflammatory changes induced by wavelengths of light which would be well tolerated if the skin had not been made susceptible by the photo-activated chemical. The histological changes do not differ significantly from other acute inflammatory responses, for example to mild chemical irritants.

Photo-allergies are immunological reactions mediated by lymphocytes, as in delayed hypersensitivity contact dermatitis, or by antibodies mediating anaphylactic urticarial changes. There is no reason to believe that these allergies differ from their counterpart allergic responses which are not dependent upon light. It is believed that the photo-energy modifies the susceptible chemical, transforming it into an allergen which first stimulates and later elicits the typical allergic susceptibility and response.

Diagnosis of phototoxicity and photo-allergy is complicated by the need for lamps emitting light of the appropriate wavelengths and for control test skin sites to determine the susceptibility of the patient to light alone.

Tests to Predict the Potential of Substances to Induce Irritation or Sensitization

There are numerous descriptions of methods to predict the potential activity of substances to induce irritation or sensitization in man. Most predictive procedures are on animals, though in some laboratories tests are made on man. Details of procedures and relevant references are given by Marzulli and Maibach,[14] Drill and Lazar[15] and The National Academy of Sciences.[16]

Predictive tests, properly made, have been very effective for many years in detecting substances or products likely to be harmful to man, enabling the rejection of harmful products, or appropriate warning on the label if the product has weak activity for a particular tissue, for example the eyes. It should be understood that, despite all care in examination of products, such is the diversity of human susceptibility that a few persons will show adverse reactions if sufficient numbers are exposed to the products. Nevertheless, predictive tests for safety in use have ensured that products are harmless to the very great majority.

Primary Irritation

The test most frequently used to detect potential primary irritants is that of Draize[17] or slight modifications of it. Albino rabbits are clipped and the test substance is applied to intact skin and to abraded or lightly scarified skin, and covered with a closed patch for 24 hours. The sites of application are then examined at intervals, and the changes seen are assessed in severity according to a scale of numerical values for various features.

The skin of the rabbit is more susceptible to irritation than that of man, so that it is possible to identify any substance likely to have an effect on man. However, the method of assessing results may lead to false positives and rejection of materials harmless to man. There are also laboratory variations in technique and in recording results. It is preferable to compare the effect of the test substance with that of a similar substance known to be harmless to users, rather than to use the score system incorporated in the test procedure of the USA or France. Furthermore, in Europe proposals are being debated that the period of application of the test substance should be four hours or less, which is just as effective for inducing irritation and is a milder treatment for the animals. The necessity to test on abraded skin is also being questioned.

Several procedures have been proposed to examine the effects of repeated application[14] but these tend to be tests particular to each laboratory, with no accepted standard protocol.

Having obtained the results of tests on animals, and those of other tests to determine the safety of the substance or product, the product may be tested for its irritation potential for man by several techniques including repeated application to skin, patch tests, arm immersion tests and simulated in-use tests. Such tests on man confirm the results of the animal tests, albeit on a small number of persons who cannot be considered truly representative of all members of a large population. The results of patch tests also indicate concentrations of ingredients that are suitable as patch test reagents should they be required by dermatologists to examine any adverse effect, allergy or irritation occurring in an individual user of a product.

Delayed Hypersensitivity (Contact Allergic Dermatitis)

There are several techniques designed to detect the potential of substances to induce contact allergic dermatitis in man. Techniques vary much in the regimen and form of application of the test substance, and in the use of adjuvants to potentiate antigenic activity. They also vary in ability to detect allergens (sensitizers). All techniques detect the more potent allergens, but those in which adjuvants are used detect a larger number of weak allergens. Comparisons between the discriminating abilities of several methods are given by Fahr *et al.*[18] and Magnusson,[19] and by Klecak in *Dermatotoxicology and Pharmacology*.[14]

The most widely used technique not requiring an adjuvant is the method of Draize (1959).[17] Guinea pigs are injected with the test substance on ten occasions during three weeks, and subsequently challenged by injection on the 35th day. Evidence for allergic responses is sought during the sensitizing and after the challenge dose. Marzulli and Maibach[14] comment on the technique.

The most discriminating procedure is probably that of Magnusson and Kligman,[19,20] generally known as the maximization test. The test substance is injected into guinea pigs with Freund's adjuvant, together as an emulsion or in separate sites. The treated animals are subsequently challenged by topical application patch tests.

No one method is suitable for tests on all substances and there has been criticism that injection tests do not truly reflect human exposure resulting from topical application. Klecak[14] proposes the use of his Open Epicutaneous Test (OET), in which the test substance is painted onto the intact skin. This method, however, requires three to four weeks of daily, or 5-days-a-week, treatment.

It has been advocated that sensitizing potential should be examined on man instead of guinea pigs. There is increasing strong criticism of human tests on the grounds that it is not ethical wilfully to sensitize man, predisposing the subjects to adverse reactions, which could be severe, on subsequent chance contact with the antigen or cross-reacting antigen. Furthermore, a test on 10 or 20 persons is no more effective than some tests on guinea pigs in predicting the allergenic potential of a substance exposed to several thousand persons, a few of whom will have a special ability to respond to the determinants of the substance. In detection of allergens the guinea pig is as effective as limited tests on man.[19]

Predictive procedures establish only the allergenic potential or sensitizing activity of a substance, and not the actual risk of sensitization. Risk can only be assessed by considering the results of predictive tests in relation to concentrations of the substances in the product, nature of use, frequency of exposure and many other considerations.

The proper performance of predictive tests and careful consideration of the use of substances has done much to ensure that cosmetic preparations are safe for use by millions of people.

Requirements to Test for Irritation and Sensitization Potential

At a time when there are many impending changes in legislation controlling the safety of cosmetic products, it is sufficient to indicate that tests to examine the potential of products to induce irritation and allergic sensitization (delayed hypersensitivity) will become obligatory.

Statutory documents, as in the Council Directive of the EEC[21] and the UK Consumer Protection Cosmetic Products Regulations,[22] have stated no specific requirements for tests, but the need for testing for safety was covered by a general requirement that '. . . cosmetic products must not be harmful under normal or foreseeable conditions of use; whereas in particular it is necessary to take into account the possibility of danger to zones of the body that are contiguous to the area of application'. Similarly, in the USA cosmetics are controlled primarily under the authority of the Federal Food, Drug and Cosmetic (FD and C) Act of 1938, modified in 1960, that a cosmetic is considered to be adulterated if it contains a poisonous or deleterious substance which may cause it to be injurious to users under customary conditions of use. The Draize technique was later defined in detail and stipulated to be the reference procedure for tests for irritancy.[23]

More specific requirements to test for primary irritation on cosmetics and beauty products were stated in French Cosmetic Laws of 1971, modified in 1973.[24] Dossiers on safety data of products should include the results of irritation tests on intact and scarified skin, based on the Draize technique.

Among the proposals for requirements or legislation in the future are those of the draft model data sheets of the Comité de Liaison des Syndicats Européennes de I'Industrie, de la Parfumerie et des Cosmetiques (COLIPA) which include provision for data on skin irritation and sensitization.

REFERENCES

1. Parish, W. E. and Ryan, T. J., *Textbook of Dermatology*, 3rd edn, ed. Rook, A., Wilkinson, D. S. and Ebling, F. J. G., Oxford, Blackwell, 1979, p. 231.
2. Zweifach, B. W., Grant, L. and McCluskey, R. T., eds, *The Inflammatory Process*, 2nd edn, 3 Vols, New York/London, Academic Press, 1974.

3. Lepow, I. H. and Ward, P., eds, *Inflammation Mechanisms and Control*, New York/London, Academic Press, 1972
4. Cochrane, C. G., *The Role of Immunological Factors in Infectious Allergic and Auto-immune Processes*, ed. Beers, R. F. and Basset, E., New York, Raven Press, 1976, p. 237; also Orange, R. P., on p. 223.
5. Wasserman, S. I., *J. invest. Dermatol.*, 1976, **67,** 620.
6. Parsons, J. F., Goodwin, B. F. J. and Safford, R. J., *Fd. Cosmet. Toxicol.*, 1979, **17,** 129.
7. Nater, J. P., de Jong, M. C. J. M., Boar, A. J. M. and Bleumink, E., *Contact Dermatitis*, 1977, **3,** 151.
8. Frosch, P. J. and Kligman, A. M., *J. Soc. cosmet. Chem.*, 1977, **28,** 197.
9. Fredriksson, T. and Gip, I., *Int. J. Dermatol.*, 1975, **14,** 442.
10. Parish, W. E., *Textbook of Dermatology*, 3rd edn, ed. Rook, A., Wilkinson, D. S. and Ebling, F. J. G., Oxford, Blackwell, 1979, p. 249.
11. Champion, R. H. and Parish, W. E., in *Clinical Aspects of Immunology*, 3rd edn, ed. Gell, P. G. H., Coombs, R. R. A. and Lachman, P. J., Oxford, Blackwell, 1975.
12. Coombs, R. R. A. and Gell, P. G. H., *Clinical Aspects of Immunology*, 3rd edn, ed. Gell, P. G. H., Coombs, R. R. A. and Lachman, P. J., Oxford, Blackwell, 1975, p. 761.
13. Fudenberg, H. H., Stiles, D. P., Caldwell, J. L. and Wells, J. V., eds, *Basic and Clinical Immunology*, Los Altos, Calif., Lange Medical, 1976.
14. Marzulli, F. N. and Maibach, H. I., eds, *Dermatotoxicology and Pharmacology, Advances in Modern Toxicology*, Vol. 4, New York/London, Wiley, 1977.
15. Drill, V. A. and Lazar, P., eds, *Cutaneous Toxicity*, New York/London, Academic Press, 1977.
16. *Principles and Procedures for Evaluating the Toxicity of Household Substances*, Report, National Academy of Sciences, Washington, 1977.
17. Draize, J. H., Woodward, G. and Calvery, H. O., *J. Pharmacol. exp. Therap.*, 1944, **82,** 377; also Draize, J. H., in *Appraisal of the Safety of Chemicals in Foods, Drugs and Cosmetics*, Austin, Texas, Association of Food and Drug Officials of the United States, 1959.
18. Fahr, H., Noster, U. and Schultz, K. H., *Contact Dermatitis*, 1976, **2,** 335.
19. Magnusson, B., *Contact Dermatitis*, 1980, **6,** 46.
20. Magnusson, B. and Kligman, A. M., *Allergic Contact Dermatitis in the Guinea-pig. Identification of Contact Allergens*, Springfield, Charles C. Thomas, 1970.
21. Council Directive of 27th July 1976 (76/768/EEC), *Official Journal of the European Communities*, L 262/169.
22. Consumer Protection, SI No. 1354, *The Cosmetic Products Regulations*, London, HMSO, 1978.
23. Code of Federal Regulations Title 49 Part 173, 210 (1975) and incl. Title 16 Part 1500.41 (1976).
24. *J. Officiel de la République Française*, 1973, 5 June.

Chapter Three

Nutrition and Hormonal Control of the Skin

Nutrition of the Skin

In common with all the other tissues of the body, the skin requires materials for the maintenance of its structure and its metabolic activity. Its needs are clearly considerable. For example, the constant production and loss of keratinized cells in the superficial epidermis and the hair follicle demands supplies of amino acids, and the secretion of the sebaceous glands requires the components for the synthesis of lipids. These materials are brought from within the body by the circulation. The blood also brings other important materials, such as hormones, which may profoundly affect the function of the skin structures.

Important questions are how far the condition of the skin can be affected by lack of such major materials, whether there are any special requirements that are peculiar to skin, and whether any deficiencies can be remedied by internal medication. A further problem of particular interest to the cosmetician is the extent to which skin can be affected by materials, either nutrients or hormones, applied to its exterior.

Substances entering skin cells thus suffer one of two fates: either they are broken down to produce energy, or they are synthesized into large molecules which may be of structural significance or act as energy stores.

Carbohydrates

Carbohydrates[1–4] are a major source of energy needed to maintain the cutaneous cells and for the synthesis of their products. However, they also contribute to the structural components, for example mucopolysaccharides, and they may not be the only source of energy.

Glucose, on entering the cell, first becomes phosphorylated to glucose-6-phosphate. In catabolism, two major pathways are open: the glycolytic sequence (Embden-Meyerhof pathway) followed by complete oxidation in the tricarboxylic acid (Krebs) cycle, or the hexose-monophosphate (Warburg-Dickens) shunt.

The glycolytic sequence involves a series of some 20 transformations in which the glucose-6-phosphate is broken down to pyruvic acid with a low yield of energy in the form of adenosine triphosphate. Under anaerobic conditions the pyruvic acid is converted to lactic acid; when oxygen is present it enters the tricarboxylic acid cycle as acetyl co-enzyme A, condensing with oxaloacetic acid to form citric acid, which then undergoes a series of transformations yielding a large amount of energy-rich phosphate bonds. The final 4-C oxaloacetic acid then combines with more acetyl co-enzyme A.

In the hexose-monophosphate pathway, glucose-6-phosphate is initially dehydrogenated to 6-phosphogluconate and thence to a series of 3-C, 5-C and 7-C sugar phosphates. The 5-C residue may enter into nucleic acid synthesis.

The two synthetic pathways open to glucose-6-phosphate both involve a coupling to uridine diphosphate. This can then either be taken through 1,4-glucosile to glycogen or can be dehydrogenated to uronic acid and hence to mucopolysaccharides.

Lipids

Lipids are synthesized in the skin both by the sebaceous glands and in the epidermis. The sebaceous gland lipids are secreted as sebum but the epidermal lipids are believed to have a structural role in the preservation of the barrier function and structural integrity of the stratum corneum. Skin surface lipids differ from those of other tissues in their content of branched chain and odd-numbered fatty acids, two types of diester waxes, and intermediates in the pathway of cholesterol synthesis, ranging from squalene to lathosterol.[5] By incubating skin slices with radioactive materials it can be shown that a wide range of precursors, including acetate, propionate and butyrate, intermediates of carbohydrate metabolism, and a number of amino acids can be incorporated into lipids.[5] It is not known which are the preferred substrates *in vivo*.

Lipids may also be catabolized by skin, feeding into the tricarboxylic acid cycle by way of co-enzyme A. Skin can respire *in vitro* for several hours in the absence of any added substances, and it seems that under such conditions utilization of carbohydrate and protein accounts for less than half the endogenous respiration. This suggests that fatty acids may provide an important substrate, albeit that glucose normally takes precedence.[1]

Amino Acids

The epidermis and hair (see Chapter 23) contain most of the twenty-two amino acids which normally occur in living tissues, though certain proteins may contain exceptionally large amounts of individual acids—for example, histidine.[6] Protein is thought to be synthesized in the epidermis and hair follicle in ways similar to those employed in other tissues. Amino acids are assembled by attachment to ribonucleic acid into chains of appropriate constitution and are then joined together in specialized particles in the cytoplasm of cells called ribosomes, being released from these as protein molecules.

The amino acid tyrosine is the substrate for the manufacture of eumelanin and phaeomelanin in the melanocytes. However, there is little detectable difference in tyrosine content between heavily and lightly pigmented skins.

Skin Conditions Related to Nutritional Deficiency

Nutritional deficiencies frequently cause widespread disorders throughout the body, of which skin changes are only one symptom. Here will be considered only those conditions in which skin changes are a major feature. Vitamin deficiencies and severe protein malnutrition are the most important.

Vitamin A

Deficiency of vitamin A causes dryness and follicular keratosis.[7] Vitamin A appears to slow down the differentiation of epidermal cells[8] and it was shown in a classic study that addition of large amounts to normal embryonic chick skin *in vitro* would cause the epidermis to be transformed into a mucus-secreting epithelium.[9]

Riboflavine (Vitamin B_2)

Riboflavine deficiency causes a number of symptoms, of which one may be a scaly dermatitis around the nose, eyes and ears.[10]

Nicotinic Acid

Deficiency of nicotinic acid causes *pellagra*,[11] of which the symptoms are diarrhoea, dementia, and a dermatitis. The skin becomes red and scaly, especially in areas exposed to friction, pressure, sunlight, or heat, and the tongue atrophies and may become purple.

Vitamin C

Deficiency of vitamin C (ascorbic acid) results in scurvy, a disease which used to be associated with long sea voyages and other occasions when green vegetables and fruit were scarce. Treatment with ascorbic acid will alleviate the symptoms, which are usually described as haemorrhage, lengthening of wound healing times, and bleeding of gums, though other symptoms not associated with skin are seen in the body. One feature of scurvy is deficiency of collagen fibres, resulting from a fault in the hydroxylation of proline.[12]

Protein

Severe protein malnutrition is the cause of the disease *kwashiorkor*.[13] In consequence, this is one of the commonest and most widespread disorders, especially amongst children, in countries where the sparse diet consists of corn, rice or beans with little animal protein. The skin manifestations can be useful in diagnosis; they consist of purplish patches associated with crinkling and flaking, like paint. Normally dark hair may become pale brown or red, become prematurely grey, or have a banded appearance. The linear growth may be decreased by as much as one half.[14] The percentage of hairs in anagen may be very much reduced, and the follicles will be much diminished in size.[15]

Essential Fatty Acids

When rats are deprived of certain polyunsaturated fatty acids with long carbon chains, the skin becomes scaly and hair is lost.[16] There is also a greatly increased loss of water. It seems that linoleic acid, which must be derived entirely from dietary sources, and arachidonic acid, which can be made from it, are essential components of the phospholipids within the lipoproteins of the cell membranes and are involved in the integrity of the water barrier.[17,18]

Human beings who are deficient in essential fatty acids because of disease or surgery also show symptoms of scaliness and increased water loss. Such changes can be reversed by topical application of sunflower seed oil.[19] Dermatitis of the scalp, some alopecia and lightening of the colour of the hair have been reported in a patient on a fat-free diet for a long period.[20]

Percutaneous Absorption

Since one of the main functions of skin is to prevent the penetration into the body of noxious materials, including water, it is not surprising that passage of most applied materials is usually negligible or very slow. That the barrier consists of the whole stratum corneum is demonstrated by the fact that sequential stripping of epidermal layers of adhesive tape progressively increases permeability. Pilosebaceous units and sweat glands may provide routes of penetration. Movement of materials along hair follicles can be observed by various techniques, and they may reach the sebaceous glands. On the other hand, sweat glands are probably not important, since palmar skin, rich in eccrine glands, is extremely impermeable.

Human skin is slightly permeable to water, but relatively impermeable to ions in aqueous solution.[21] Permeability to many covalent substances, for example glucose and urea, is slight but to others, for example some aliphatic alcohols, it is relatively high. Solutes in organic liquids show a similar permeability to the solvents themselves. Some solids, for example corticosteroids, will continue to penetrate long after evaporation of a volatile solvent.[22]

The integrity of the skin barrier depends upon the degree of hydration of the stratum corneum. Absorption of materials depends on the vehicle used. If the material is soluble in one of the phases of a two-phase (oil-in-water or water-in-oil) vehicle, it will penetrate better if it is in the continuous phase. Occlusion by dressings or polyethylene sheeting or by soft paraffin will increase penetration.[23] Salicylic acid, which has a keratolylic action, is sometimes incorporated into ointments; it is doubtful if it impairs the barrier in normal skin, though it may have a greater effect on the parakeratotic epidermis in eczema and psoriasis.

Certain solvents, of which dimethylsulphoxide (DMSO) is the most potent, not only penetrate rapidly but greatly enhance the penetration of substances dissolved in them.[24,25] It seems they may act by temporarily over-hydrating the stratum corneum or dissolving lipid in cell walls, but any such damage does not appear to be permanent.

Hormones

The skin is affected by a range of steroid hormones, including oestrogens, androgens and corticosteroids.[26] In addition, protein or polypeptide hormones of the pituitary may affect the skin structures, either directly or by enhancing their response to steroid hormones. Finally, there is a deal of circumstantial and experimental evidence of the existence of local hormones, stimulators or inhibitors, though such substances remain to be characterized.

Androgens

Androgens are responsible for the development of the secondary sexual characters in males. They are secreted by the testes and adrenals, and the levels rise steeply at about the time of puberty when these organs are activated by trophic hormones of the pituitary. Females also produce androgens from the adrenals and, to some extent, from the ovaries.

Androgens induce the hair follicles in the pubic and axillary regions, and the male face, to produce coarse terminal hair instead of fine vellus. Paradoxically, they are also a prerequisite for the development of male pattern baldness in persons who are constitutionally disposed.

Male hormones also markedly stimulate sebaceous secretion, probably by increasing cell division in the glands as well as the lipid synthesized within each cell.[27–30] They probably also stimulate the apocrine glands, and cell division in the epidermis.[31]

Oestrogens

Oestrogens are secreted by the ovary in the female. They affect the growth of the female reproductive system and their cyclic production is responsible for the changes during the menstrual cycle.

The circumstantial evidence that appearance, texture and tone of skin are important sexual features of the female and that regressive changes occur after the menopause, favours the view that oestrogens stimulate the cutaneous tissues. Evidence that topical application of oestrogenic ointments to the backs of senile human subjects locally increased the size of the epidermal cells and accentuated the waviness of the basal layer[32] appeared to be reinforced by the claim that oestrogens stimulate epidermal cell division in mice.[33] Punnonen and Rauramo[34] revived the issue by the claim that, in women, epidermal thickness and thymidine-labelling index were decreased by ovariectomy, but could be restored by oral treatment with oestriol succinate or oestradiol valerate. Against this must be set the failure to demonstrate any mitotic effects of systemically administered oestradiol in adult rats,[35,36] or any noticeable clinical improvement when oestrogen creams were applied to women's faces.[37]

In pharmacological doses, at least, oestrogens suppress sebaceous activity in both man[28] and experimental animals.[27,29,36] There is evidence to suggest that the effect is directly on the target organ, but some central action, for example to reduce androgen levels, cannot be ruled out.

Anti-androgens

Anti-androgens are synthetic steroids or other compounds that antagonize the action of androgens. The anti-androgenic steroids 17α-methyl-B-nortestosterone[38] and cyproterone acetate[39] are each effective inhibitors of testosterone-stimulated sebaceous secretion when administered systemically in the rat, requiring a dose 10 times or more that of the androgen. They act, at least in part, by inhibiting cell division, in contrast to oestrogens, which appear to act, at very much lower doseage, mainly by inhibiting intracellular lipid synthesis.

Cyproterone acetate, given orally, has been successfully used for the treatment of female hirsutism and acne,[40,41] and may have a potentiality for the treatment of some diffuse alopecias in women. The treatment does reduce levels of androgens in the blood plasma, but the evidence suggests that antagonism at the target site is the mechanism of paramount importance.

The above-mentioned anti-androgens are believed to compete with testosterone in binding to the intracellular receptor proteins. However, other steroidal antagonists exist which act by inhibition of the enzyme 5α-reductase, which is necessary for conversion of testosterone to an active metabolite, 5α-dihydrotestosterone.

Corticosteroids
Preparations of adrenocortical hormones, such as cortisone, or their synthetic homologues, many of which are fluorinated, are widely used as topical medicaments and are of considerable value in dermatology. They alleviate inflammation and allergic sensitization and are useful in many forms of eczema and in psoriasis. However, prolonged use of the more potent compounds may cause atrophic changes in the epidermis and dermis, and systemic absorption, particular in infants, may suppress the production of adrenocorticotrophic hormone and the natural function of the adrenal gland.

Whatever can be measured, be it capillary blood flow in inflammation,[42] epidermal mitosis,[43] epidermal thickness[44] or skin thickness,[45] is decreased by corticosteroids.

Pituitary Hormones
A number of pituitary hormones have a direct effect on the skin. Several polypeptides have been shown to influence pigmentation and are thus known as melanocyte-stimulating hormones. Two such peptides were first isolated from hog pituitary; one known as α-MSH had 13 amino acid residues; the other, β-MSH, had 18.[46] Man does not produce α-MSH, and it is uncertain whether the active melanotrophic hormone is β-MSH or a larger molecule, β-lipotropin with 91 amino acids, but containing the β-MSH sequence.[47]

α-MSH also appears to act upon sebaceous secretion. When given to rats, it not only has a direct effect, but also enhances the response to testosterone.[48,49]

Similar effects have been demonstrated for some other much larger molecules secreted by the pituitary. In hypophysectomized rats, the response of the sebaceous glands to testosterone is very much diminished.[27,50] The existence of a separate sebotrophic hormone was proposed,[51] but such a material has not been characterized. On the other hand, a pig growth hormone and a sheep prolactin have each been shown to restore the response,[52] as was a bovine growth hormone which also exerted a significant independent action.[53]

Local Hormones
Several events provide circumstantial evidence for the existence of locally produced substances which influence the activity of various skin cells. Wounding of skin is followed after about 40 hours by a burst of mitotic activity. Some authors (see Montagna and Billingham[54]) believe that a wound hormone is liberated by the wounded cells; others favour the view that cell division is normally kept in check by inhibitors or 'chalones', and that mitosis is initiated when such substances are dispersed.[55–57] A chalone extracted from the epidermis of pigs and cod[58] was stated to be an antigenic protein or glycoprotein with a molecular weight of 30 000–40 000, but chalones have not, as yet, been further characterized. They have been prepared from other tissues, such as the sebaceous glands and the melanocytes, and appear to be highly tissue specific but not species specific.

Roles for inhibitors[59] or wound hormones[60] have also been postulated in the control of the hair follicle cycle.

REFERENCES

1. Cruikshank, C. N. D., *Symp. zool, Soc. London*, 1964, **12,** 25.
2. Freinkel, R., *The Epidermis*, ed. Montagna, W. and Lobitz, W. C. Jnr, New York, Academic Press, 1964, p. 485.
3. Weber, G., *The Epidermis*, ed. Montagna, W. and Lobitz, W. C. Jnr, New York, Academic Press, 1964, p. 453.
4. Jarrett, A., *The Physiology and Pathophysiology of the Skin*, Vol. 1, *The Epidermis*, ed. Jarrett, A., London, Academic Press, 1973, p. 45.
5. Wheatley, V. R., *J. invest. Dermatol.*, 1974, **62,** 245.
6. Gimucio, J., *J. invest. Dermatol.*, 1967, **49,** 545.
7. Moore, T., *Vitamin A*, Amsterdam, Elsevier, 1957.
8. Hunter, R. and Pinkus, H., *J. invest. Dermatol.*, 1961, **37,** 459.
9. Fell, H. B. and Mellanby, E., *J. Physiol. London*, 1953, **119,** 470.
10. Sherman, H., *Vitam. Horm. NY*, 1950, **8,** 55.
11. Stratigos, J. D. and Katsambas, A., *Br. J. Dermatol.*, 1977, **96,** 99.
12. Robertson, W. van B., *Biophys. J.*, 1964, **4** (Supplement), 93.
13. Gillman, T., *Comparative Physiology and Pathology of the Skin*, ed. Rook, A. and Walton, G. S., Oxford, Blackwell, 1965, p. 355.
14. Sims, R. T., *Arch. Dis. Child.*, 1967, **42,** 397.
15. Bradfield, R. B., *Protein Deficiencies and Calorie Deficiencies*, ed. McCance, R. and Widdowson, E. M., London, Churchill, 1968, p. 213.
16. Kingery, F. A. J. and Kellum, R. E., *Arch. Dermatol.*, 1965, **91,** 272.
17. Hartop, P. J. and Prottey, C., *Br. J. Dermatol.*, 1976, **92,** 255.
18. Prottey, C., *Br. J. Dermatol.*, 1976, **94,** 579.
19. Prottey, C., Hartop, P. J. and Press, M., *J. invest. Dermatol.*, 1975, **64,** 228.
20. Skolnik, P., Eaglstein, W. H. and Ziboh, V. A., *Arch. Dermatol.*, 1977, **113,** 939.
21. Tregear, R. T., *Physical Functions of Skin*, London, Academic Press, 1966.
22. Scheuplein, R. J. and Ross, L. W., *J. invest. Dermatol.*, 1974, **62,** 353.
23. Wahlberg, J. E. *Curr. Probl. Dermatol.*, 1973, **5,** 1.
24. Stoughton, R. B., *Arch. Dermatol.*, 1965, **91,** 657.
25. Baker, H., *J. invest. Dermatol.*, 1968, **50,** 283.
26. Ebling, F. J., *Biochem. Soc. Trans.*, 1976, **4,** 597.
27. Ebling, F. J., *J. Endocrinol.*, 1957, **15,** 297.
28. Strauss, J. S. and Pochi, P. E., *Recent Prog. Horm. Res.*, 1963, **19,** 385.
29. Ebling, F. J., *J. invest. Dermatol.*, 1974, **62,** 161.
30. Shuster, S. and Thody, A. J., *J. invest. Dermatol.*, 1974, **62,** 172.
31. Ebling, F. J., *Proc. r. Soc. Med.*, 1964, **57,** 523.
32. Eller, J. J. and Eller, W. D., Arch. Dermatol. Syphilol., 1949, **59,** 449.
33. Bullough, W. S., *Ciba Found. Colloq. Endocrinol.*, 1953, **6,** 278.
34. Punnonen, R. and Rauramo, L., *Acta Obstet. Gynecol. Scand.*, 1974, **53,** 267.
35. Carter, S. B., *J. Endocrinol.*, 1953, **9,** 19.
36. Ebling, F. J., *J. Endocrinol.*, 1954, **10,** 147.
37. Behrman, H. T., *J. Am. med. Assoc.*, 1954, **155,** 119.
38. Ebling, F. J., *J. Endocrinol.*, 1967, **38,** 181.
39. Ebling, F. J., *Acta Endocrinol.* (*Copenhagen*), 1973, **72,** 361.
40. Hammerstein, J., Meckies, J., Leo-Rossberg, I., Matzl, L. and Zielske, F., *J. Steroid Biochem.*, 1975, **6,** 827.

41. Ebling, F. J., Thomas, A. K., Cooke, I. D., Randall, V., Skinner, J. and Cawood, M., *Br. J. Dermatol.*, 1977, **97,** 371.
42. Wilson, L., *Br. J. Dermatol.*, 1976, **94** (Supplement 12), 33.
43. Goodwin, P., *Br. J. Dermatol.*, 1976, **94** (Supplement 12), 95.
44. Winter, G. D. and Burton, J. L., *Br. J. Dermatol.*, 1976, **94** (Supplement 12), 107.
45. Kirby, J. D. and Munro, D. D., *Br. J. Dermatol.*, 1976, **94** (Supplement 12), 111.
46. Lerner, A. B. and Lee, T. H., *J. Am. chem. Soc.*, 1955, **77,** 1066.
47. Gilkes, J. J. H., Bloomfield, G. A., Scott, A. P. and Rees, L. H., *Proc. r. Soc. Med.*, 1974, **67,** 876.
48. Ebling, F. J., Ebling, E., Randall, V. and Skinner, J., *J. Endocrinol.*, 1975, **66,** 407.
49. Thody, A. J. and Shuster, S., *J. Endocrinol.*, 1975, **64,** 503.
50. Lasher, N., Lorincz, A. L. and Rothman, S., *J. invest. Dermatol.*, 1954, **22,** 25.
51. Lorincz, A. L., *Advances in Biology of Skin*, Vol. 4, *The Sebaceous Glands*, ed. Montagna, W., Ellis, R. A. and Silver, A. F., Oxford, Pergamon, 1963, p. 188.
52. Ebling F. J., Ebling E. and Skinner, J., *J. Endocrinol.*, 1969, **45,** 245.
53. Ebling, F. J., Ebling, E., Randall, V. and Skinner, J., *Br. J. Dermatol.*, 1975, **92,** 325.
54. Montagna, W. and Billingham, R. E., *Advances in Biology of Skin*, Vol. 5, *Wound Healing*, Oxford, Pergamon, 1964.
55. Bullough, W. S. *Biol. Rev.*, 1962, **37,** 307.
56. Bullough, W. S. and Laurence, E. B., *Proc. r. Soc. London Ser. B*, **151,** 517.
57. Bullough, W. S., *Natl. Cancer Inst. Monogr.*, 1973, **38,** 99.
58. Bullough, W. S., Hewett, C. L. and Laurence, E. B., *Exp. Cell Res.*, 1964, **36,** 192.
59. Chase, H. B., *Physiol Rev.*, 1954, **34,** 113.
60. Argyris, T. S., *Advances in Biology of Skin*, Vol. 9, *Hair Growth*, ed. Montagna, W. and Dobson, R. L., Oxford, Pergamon, 1969, p. 339.

Chapter Four

Skin Creams

Introduction

The word 'cream' is in such common use that definition is almost superfluous. Indeed, 'creamy' is often used to describe the texture or appearance of objects or products that cannot themselves claim to be creams.

In the context of cosmetics, the term 'cream' usually signifies a solid or semi-solid emulsion, although it may equally well be applied to non-aqueous products such as wax–solvent-based mascaras, liquid eyeshadows and ointments. If an emulsion is of sufficiently low viscosity to be pourable—that is, it can be made to flow under the influence of gravity alone—then it may no longer be referred to as a cream, but as a 'lotion'. For the purposes of the ensuing discussion, however, creams and lotions are both dealt with under the general heading of skin creams. The theoretical basis of the preparation of such emulsions is dealt with in Chapter 38.

Such are the number and variety of raw materials that are available to the cosmetic scientist for the formulation of skin creams and lotions, that no general textbook could spare the space needed to list them all. The good, stable formulae to which these ingredients can give rise are too numerous for a complete catalogue even to be contemplated. Further, new materials—emulsifiers, emollients, moisturizers, healing materials—are being produced and made available by suppliers continuously, so that any catalogue would be out of date even before it was printed.

The materials and formulae given in this chapter are therefore to be regarded generally as illustrating only well-tried, traditional types of product which, while still being of great value, are starting-points only and capable of benefiting from new technology as it becomes available. The literature and advice of established suppliers of good quality raw materials should be sought and listened to; many such suppliers spend considerable effort in developing good cream formulae that illustrate the best use of their own products. The skill of the cosmetic scientist is not to follow these formulae slavishly, but to study them and adapt them to his own needs, incorporating, perhaps, the best features of a number of published formulae into a series of experiments of his own.

Raw materials are given here under their official CTFA adopted names, to avoid the difficulties and ambiguities that sometimes arise over the precise chemical nature of some of these substances.[1]

Classification of Skin Creams

Traditionally, cosmetic creams have been marketed and sold on the basis of their 'function'—that is, the broad claims which are made for them by advertising—and on the packaging which contains them. Thus customers have come to learn what type of emulsion they can expect from a jar marked 'cold cream' or 'night cream'. It is undoubtedly true, however, that this is not a particularly precise means of classification, since the number of variations in appearance, texture, subjective feel, ease of spreading and speed of 'rubbing in' far outstrip the number of functional categories and there is, perforce, a considerable amount of overlap. The customer, therefore, is likely to make her own judgment on the basis of these subjective features, using the manufacturers' functional labels as a guide to end-use and quality.

Table 4.1 Characteristics of Skin Creams

Functional	Physicochemical	Subjective
Cleansing creams Cold creams Massage creams Night creams	Medium-to-high oil content Oil-in-water *or* water-in-oil Low slip-point oil phase Neutral pH May contain surfactants to improve penetration and suspension properties	Oily Difficult to 'rub in' May be stiff and 'rich' Also popular as lotions
Moisturizing creams Foundation creams Vanishing creams	Low oil content Usually oil-in-water Low slip-point oil phase Neutral to slightly acidic pH May contain emollients and special moisturizing ingredients	Easily spreadable and 'rub in' quickly Available as creams or lotions
Hand and body protective	Low-to-medium oil content Usually oil-in-water Medium slip-point oil phase May have slightly alkaline or acidic pH May contain 'protective factors', especially silicones and lanolin	Easily spreadable but do not 'rub in' with the ease of vanishing creams Very popular in lotion form
All-purpose creams	Medium oil content Oil-in-water *or* water-in-oil	Very often slightly oily but should be easy to spread

The cosmetic scientist, however, may view the problem somewhat differently. He is concerned with such physicochemical features as the volume ratio of oil to water, the nature of the continuous phase, the pH of the emulsion, the type of emollients used, the slip-point of the oil phase and so on.

To a certain extent, these three methods of classification (functional, subjective, physicochemical) can be correlated and Table 4.1 is an attempt to do this. A purely objective correlation between the percentage oil phase and the claimed functionality of some 236 skin creams is given in the histograms (Figures 4.1 to 4.6). These have the value of presenting, in an easily digestible form, the range of values which by common consent are used for commercially viable products; it will be observed that, apart from singling out two groups having a particularly high average oil content (cleansing creams and night creams), they do not represent an easy method of identifying function by a purely chemical means.

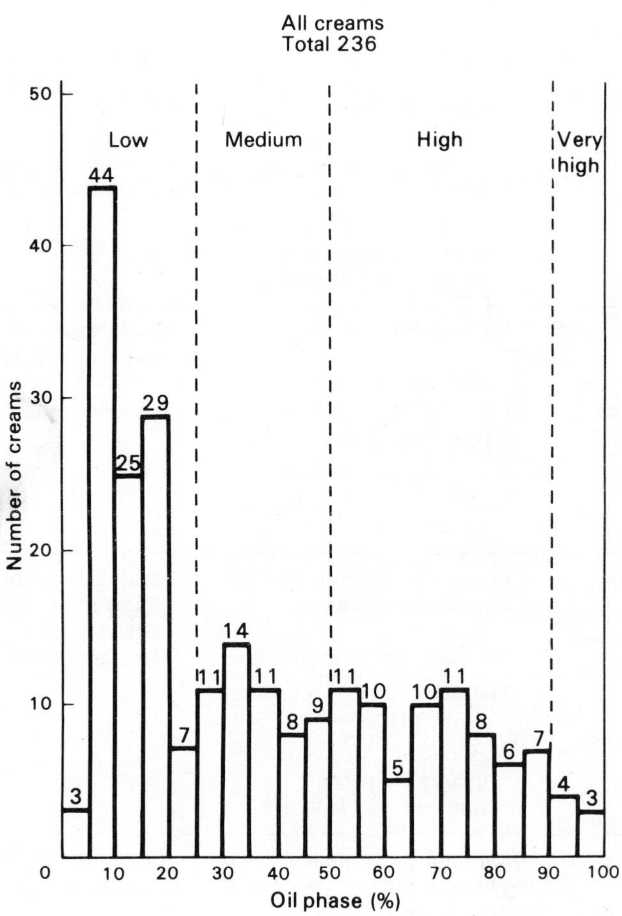

Figure 4.1 Classification of 236 skin creams according to content of oil phase

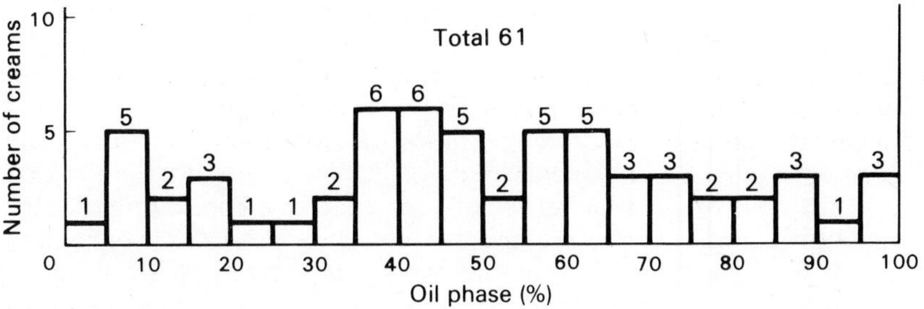

Figure 4.2 Classification of cleansing (cold) creams according to content of oil phase

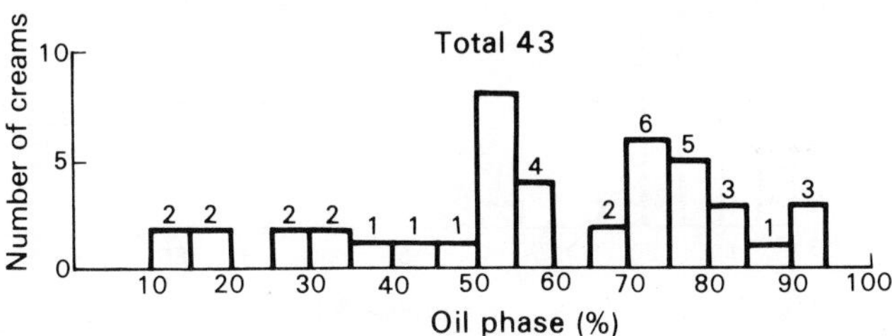

Figure 4.3 Classification of night creams according to content of oil phase

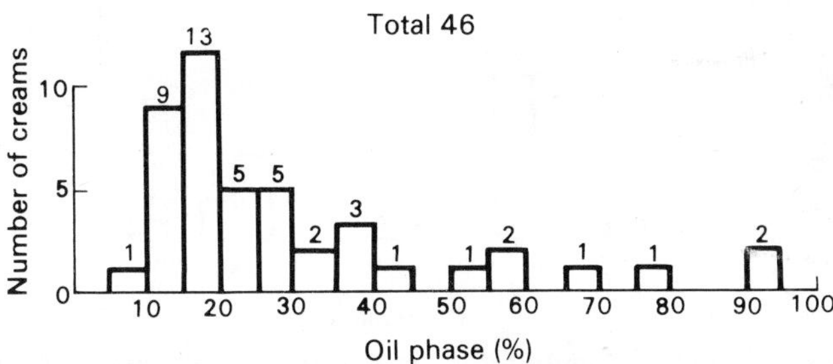

Figure 4.4 Classification of vanishing creams according to content of oil phase

Cleansing Creams

To remain healthy and of good appearance the skin surface requires frequent cleansing to remove grime, sebum and other secretions, dead cells, crusts and applied make-up. Water is a very cheap and effective cleansing agent for certain types of facial soil but is ineffective on its own against oils. By the process of emulsification, soaps and other detergents are able to improve the cleansing properties of water dramatically. However, this combination suffers from disadvantages: it is inconvenient to use outside the bathroom, and it may remove too much oil from the surface, leaving it feeling dry and rough—a feature that is made worse by the alkalinity of soap, which may cause the outermost cells to lift and separate from their neighbours.

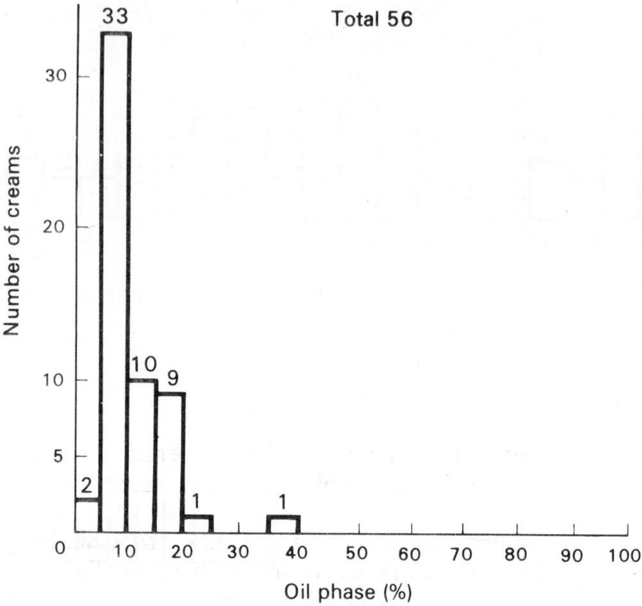

Figure 4.5 Classification of hand-and-body creams according to content of oil phase

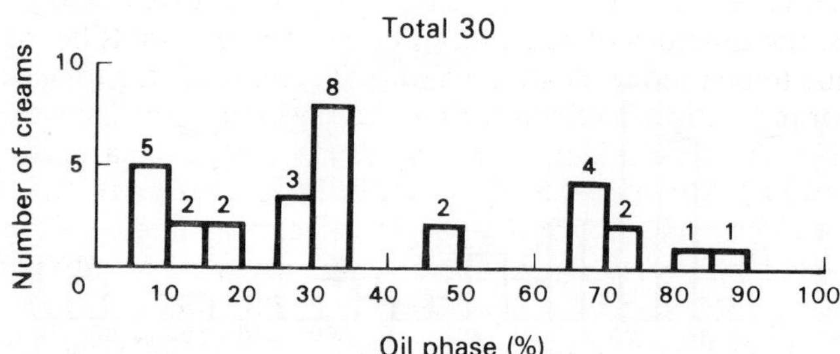

Figure 4.6 Classification of all-purpose creams according to content of oil phase

Cleansing creams and lotions can, by a combination of water and the solvent action of oils, effect the cleansing of the skin surface efficiently and pleasantly. Moreover, if they are properly formulated they can accomplish this without completely degreasing the skin (indeed, by leaving behind a very thin emollient layer of clean oil, they may give a healthy supple feel to the skin surface) and this is combined with a much more convenient method of usage than soap and water.[2,3]

In use, a cleansing cream or lotion is spread onto the skin, using the finger-tips, and massaged onto the surface. This action serves to loosen and suspend the grime and soil in the emulsion. A subsequent wipe with a tissue or cotton wool pad removes the majority of the applied cleansing emulsion, and with it the skin soil, grime or make-up. It follows, therefore, that the emulsion should have a medium-to-high percentage oil phase, should be easily spreadable, should not 'rub in' and should not irritate the skin. If, in addition, it can also leave a residual emollient film on the skin, so much the better.

Related to cleansing creams is a group of emulsions known collectively as 'cold creams' (the name stems from the cooling effect of such products on the skin). Cold creams are of particular historical interest since they are among the first cosmetic emulsions to be described in the literature.[4] These ancient (second century) emulsions were composed primarily of natural waxes and vegetable oils (traditionally beeswax and olive oil). At the turn of the twentieth century, mineral oil was substituted for the less stable unsaturated vegetable oil and the basis of modern cold creams was established. The inclusion of borax into the formulae imparted increased stability since, by reaction with the free fatty acids in the natural waxes, it was able to form sodium soaps, thus producing an emulsifier *in situ*.

Today, beeswax–borax emulsions are still popular and commercially viable, although the development of secondary or alternate emulsifiers has enabled the formulator to produce a wider range of emulsions around the beeswax–borax theme.

Beeswax itself suffers from two disadvantages as an ingredient in skin creams. The first of these is that it has a distinctive smell which usually has to be masked in the final product; the beeswax odour is not unpleasant, but it is not compatible with the sophisticated skin-care image which many manufacturers nowadays build into their products. Secondly, being a natural product, the quality and price of beeswax tends to vary according to region of origin and time of year. Nevertheless, the qualities of neutralized beeswax as an emulsifier are such that it will continue to be used in cleansing and cold creams for the forseeable future.

When a borax solution is mixed with molten beeswax, the sodium salts of the wax acids will be formed at the oil–water interface. It is usual to use rather less than the theoretical quantity of borax since this gives a more stable, textured cream. Usually, 5–6 per cent of the weight of beeswax is used. The amount of borax-neutralized beeswax in a cold cream can vary from 5 to 16 per cent. The lower levels produce softer creams which can be stiffened (if required) by incorporating other waxes. Examples 1, 2 and 3 are all water-in-oil emulsions.

	(1)	(2)	(3)
	per cent	*per cent*	*per cent*
Beeswax	5·0	16·0	12·0
Mineral oil	45·0	50·0	—
Borax	0·2	0·8	0·5
Microcrystalline wax	7·0	—	—
Spermaceti	—	—	12·5
Sesame oil	—	—	40·0
Water	32·8	33·2	35·0
Perfume, preservative	*q.s.*	*q.s.*	*q.s.*
Paraffin wax	10·0	—	—

Alternatives to waxes as thickeners for a continuous oil phase are bentones (quaternary hectorites and related chemical species); Polon[5] describes the mechanisms of thickening by this type of inorganic agent. Example 4 is a product of this type:

	(4)
	per cent
Beeswax	12·0
Mineral oil	53·0
Quaternium-18 hectorite	0·7
Borax	0·7
Water	33·2
Isopropanol	0·4

Procedure: Mill the bentone with the isopropanol and some of the mineral oil. Heat the resultant gel with the beeswax and the remainder of the oil to 75°C. Dissolve the borax in the water, bring to 70°C and slowly pour into the oil phase with stirring. Continue stirring to 45°C, adding perfume at a late stage.

It is a peculiarity of the beeswax–borax system that both water-in-oil and oil-in-water creams may be produced without the aid of secondary emulsifiers. Factors which influence the type of emulsion include the ratio of oil to water, the proportion of the beeswax that is saponified, the constituents of the cream (which will affect the HLB requirement) and the temperature. Salisbury *et al.*[6] studied a simple three-component system of mineral oil, water and beeswax fully neutralized with borax and found that under these conditions 45 per cent was a critical level for the water phase; below this level, the creams were water-in-oil, above it they were oil-in-water. Such a critical level probably exists for all beeswax formulations although this may often be well below 45 per cent. Pickthall,[7] commenting on Salisbury's paper, points out that preparation at high temperature tends to produce cold creams of the water-in-oil type. It is also possible that phase inversion may occur during processing. Almost certainly, phase inversion occurs on the skin when an oil-in-water emulsion is spread onto the skin surface and the water phase begins to evaporate.

Nonionic emulsifiers can be used to supplement beeswax–borax emulsions, adding increased flexibility and stability to the emulsion. By far the most popular as co-emulsifiers are sorbitan fatty acid esters. Examples 5 and 6 illustrate the point; example 5 is a water-in-oil emulsion whereas 6 is oil-in-water.

	(5)	(6)
	per cent	*per cent*
Beeswax	10·0	10·0
Mineral oil	50·0	20·0
Lanolin	3·1	3·0
Borax	0·7	0·7
Hydrogenated vegetable oil	—	25·0
Antioxidant	—	0·5
Sorbitan sesquioleate	1·0	—
Sorbitan stearate	—	5·0
Polysorbate 60	—	2·0
Water	35·2	33·8
Perfume, preservative	*q.s.*	*q.s.*

Moving away from beeswax–borax as the primary emulsifier system, these same nonionic emulsifiers can be used on their own—although beeswax itself is sometimes retained. Examples 7 and 8 are water-in-oil and examples 9 and 10 are oil-in-water.

	(7) *per cent*	(8) *per cent*	(9) *per cent*	(10) *per cent*
Petrolatum	31·0	35·0	—	—
Mineral oil	20·0	15·0	50·0	23·0
Paraffin wax	7·0	5·0	—	—
Lanolin	3·0	3·0	0·5	4·0
Ceresin	—	5·0	—	—
Sorbitan sesquioleate	4·0	4·0	0.5	—
Sorbitol solution 70%	2·5	2·5	—	12·2
Magnesium sulphate	0·2	0·2	—	—
Water	32·3	30·3	23·5	41·8
Beeswax	—	—	15·0	2·0
Spermaceti	—	—	2·0	—
PEG-40 Sorbitan lanolate	—	—	4·0	—
Sorbitan stearate	—	—	4·0	—
Magnesium aluminium silicate	—	—	0·5	—
Stearic acid	—	—	—	15·0
Sorbitan trioleate	—	—	—	1·0
Polysorbate-85	—	—	—	1·0
Perfume, preservative	*q.s.*	*q.s.*	*q.s.*	*q.s.*

If required, the external phase of oil-in-water cleansing creams may be thickened by the use of cellulose derivative alginates and other hydrocolloids (example 11).

	(11) *per cent*
Beeswax	8·0
Mineral oil	49·0
Paraffin wax	7·0
Cetyl alcohol	1·0
PEG-15 Cocamine	1·0
Borax	0·4
Carbomer-934	0·2
Water	33·4
Perfume, preservative	*q.s.*

A number of beeswax derivatives are manufactured with modified emulsifier properties. There is, for example, a range of ethoxylated beeswax derivatives available with HLB values ranging from 5 to 9. Although they still have the beeswax odour, it is claimed that creams made from them are softer, liquefy readily, allow the incorporation of larger amounts of water, are neutral and are stable at 50°C. Examples 12 and 13 are both oil-in-water creams.

	(12)	(13)
	per cent	*per cent*
Mineral oil	50·0	50·0
Beeswax	—	7·0
PEG-8 Sorbitan beeswax	12·0	—
PEG-20 Sorbitan beeswax	3·0	8·0
Polysorbate-40	—	2·0
Water	35·0	33·0
Perfume, preservative	*q.s.*	*q.s.*

Lighter creams of the oil-in-water type with a medium oil content can function as cleansing creams—indeed they are preferred by the consumer in some sections of the market. Many successful products have been built around conventional triethanolamine stearate emulsifier systems or self-emulsifying glyceryl stearate. Of the following examples, 14–16 are creams, 17 and 18 are lotions.

	(14)	(15)	(16)	(17)	(18)
	per cent	*per cent*	*per cent*	*per cent*	*per cent*
Mineral oil	30·0	29·0	18·0	10·0	—
Stearic acid	10·0	13·5	—	3·0	4·0
Triethanolamine	2·0	1·8	—	1·8	1·0
Glyceryl stearate (SE)	—	—	15·0	—	—
Carbomer 934	0·5	—	—	—	—
Water	57·5	51·9	55·0	84·7	66·0
Glycerin	—	2·0	5·0	—	1·0
Sodium alginate	—	1·8	—	—	—
Cetyl alcohol	—	—	2·0	0·5	—
Spermaceti	—	—	5·0	—	—
Squalane	—	—	—	—	28·0
Perfume, preservative	*q.s.*	*q.s.*	*q.s.*	*q.s.*	*q.s.*

Since the discovery that the pH of skin is acidic[8–10] and that buffered acidic cleansing creams allow a more rapid return to normal skin pH than their more alkaline counterparts, some interest in acidic cleansing emulsions has been generated. There are fewer emulsifiers available for acidic formulations, the most popular being glyceryl stearates, cetyl or stearyl alcohols, phosphated fatty alcohols and fatty alcohol sulphates. Example 19 utilizes lemon juice as the acidic ingredient.

	(19)
	per cent
Sorbitan sesquioleate	4·0
Ozokerite	8·0
Petrolatum	30·0
Mineral oil	10·0
Lanolin	12·0
Water	30·0
Lemon juice	6·0
Perfume, preservative	*q.s.*

Traditional beeswax–borate emulsions are difficult to remove from the skin with water alone. Many of the more 'sophisticated' formulae given thus far show rather better washability, a feature which is appreciated by a large number of women who use soap and water as part of their facial cleaning regime (examples 14 and 15 show particularly good 'wash-off' properties).

However, washability can be further improved by the use of detergents as part of the emulsifier system. Examples 20 and 21 utilize sodium cetyl sulphate. This is preferred to the lauryl sulphate since it is a better emulsifier, foams less and is less irritant to the skin.

	(20) *per cent*	(21) *per cent*
Mineral oil	40·0	52·0
Ozokerite	3·0	—
Cetyl alcohol	2·0	3·0
Sodium cetyl sulphate	1·0	3·0
Water	54·0	23·0
Beeswax	—	5·6
Paraffin wax	—	5·0
Petrolatum	—	8·4
Perfume, preservative	*q.s.*	*q.s.*

More recently, the 'wash-off' principle has been extended to emulsions that will actually lather on the skin during use, particularly if a little extra water is added. Examples 22 and 23 are products of this type, formula 23 being the subject of a patent.[11]

	(22) *per cent*	(23) *per cent*
Stearic acid	10·0	12·5
Mineral oil	5·0	—
Petrolatum	2·0	—
Cetostearyl alcohol	1·5	2·0
Isopropyl myristate	3·0	5·0
Sorbitan monolaurate	2·0	—
Glycerin	6·5	—
Sodium laureth sulphate	5·0	—
Triethanolamine	1·5	—
Polyoxyethylene sorbitan monolaurate	2·0	—
Water	61·5	68·1
Lanolin	—	0·4
Coco-sodium isethionate	—	12·0
Perfume, preservative	*q.s.*	*q.s.*

There are cleansing preparations available other than emulsions, the most popular being detergent gels based on neutralized carbomers or cellulose derivatives (example 24) or non-emulsified lotions, the latter usually being simple aqueous or aqueous–alcoholic solutions of mild detergents with or without a humectant (examples 25–27).

	(24) *per cent*
Sodium magnesium silicate	4·0
Sodium lauroyl sarcosinate	15·0
PPG-12-PEG-50 Lanolin	5·0
Hydroxyethylcellulose	0·3
Water	75·7
Perfume, perservative	*q.s.*

	(25) *per cent*	(26) *per cent*	(27) *per cent*
TEA laureth sulphate	5·0	—	5·0
Sulphonated olive oil	—	10·0	—
Propylene glycol	—	—	10·0
Water	95·0	90·0	85·0
Perfume, preservative	*q.s.*	*q.s.*	*q.s.*

Night and Massage Creams

Traditionally, products described as night or massage creams are designed to be left on the skin for several hours or to remain mobile on the skin even after vigorous rubbing. Evidently, therefore, they must be composed with a substantial oil phase which will spread easily without disappearing but also without rubbing off onto clothing or bed linen in use. Such creams tend to be high-oil-content, water-in-oil, soft solid or viscous liquid creams.

The benefits to be expected from the use of night or 'overnight creams' have undoubtedly been overstated in the past. There is no doubt that the occlusive layer they provide for the skin surface slows the rate of transepidermal water loss and can therefore claim to have a 'moisturizing' effect. Certainly they will, like most creams, make the skin surface feel smooth by the action of lubricating the surface and allowing any 'saw tooth' cells in the outer layer of the stratum corneum to be smoothed down. From time to time, however, formulators have been tempted to add the term 'nutritive' to their description of such products: this is a term which can hardly be justified, irrespective of the constituents of such creams, since the stratum corneum is completely dead and any materials (such as hormones) which penetrated this layer would, by current definition, alter the status of such a product from cosmetic to pharmaceutical.

Massage, however, has a valuable part to play in skin care since it is well known that vigorous rubbing of the skin helps to prevent the build-up of excessive numbers of dead surface cells and keeps the epidermal blood supply in good condition.

The term 'moisturizing', which has already been referred to, has also been applied to water-in-oil creams of this type. With the advances in scientific research on skin care which have taken place in recent years, the concept of moisturizing has broadened from the simple occlusive skin barrier principle; many modern moisturizing creams are comparatively light and easy to rub in compared with those of the overnight and massage types, although there still remains a market for the heavier moisturizing creams.

Apart from constituents which can be shown to have a moisturizing effect or a UV-filtering effect, claims are made, from time to time, that materials have been discovered which have a more obscure beneficial effect on the skin and these often find their way into night or massage creams. While it is true that many such products have been a commercial success, few have stood the test of careful scientific investigation. Pre-eminent among these are 'natural' products—particularly vitamins. Some of the evidence advanced for the use of vitamins in skin creams is given below—the reader will be in a position draw his or her own conclusions.

Vitamins in Skin Creams

De Ritter *et al.*[12] consider that local vitamin deficiencies can be alleviated by topical application of vitamins in sufficient quantities to give high local concentrations. According to Lorincz[13] severe vitamin deficiencies are rare, and are best remedied systemically via oral administration. There is evidence that the increased epidermal thickness and the decreased rate of keratinization noted after topical applications on rats do not occur with human epidermis. Jarrett[14] suggests that the vitamin must be in a water-solubilized form in the cream and that oil-soluble preparations are of little value.

Ellot,[15] however, claims that fat-soluble as well as water-soluble vitamins are capable of being taken up through the skin, which justifies their use in cosmetic preparations for external application provided that they contain sufficient content of suitably stabilized vitamins. Several formulae are given.

The claims for royal (bee) jelly[16] and pollen,[17] which have been stated to have almost magical properties, chiefly by ingestion but also by topical application, must be based largely on the vitamin B content.

Pantothenic acid, a part of the water-soluble vitamin B complex, its precursor and the related materials panthenol,[18] pantethine[19] and pangamic acid[20] have all been quoted as having a beneficial action on the skin and being useful in skin and/or hair preparations. Although there is no certain proof that they penetrate the skin and reach the location where they might exert an influence, vitamin B complex, panthenol and vitamin B_6-pyridoxin are used in some cosmetics.

Ascorbic acid (vitamin C) and its isomer iso-ascorbic acid are added to some toilet preparations, but this is not usually for their effect on skin. An improvement in skin condition during the curing of scurvy demands ingestion of vitamin C.

Vitamin D, like vitamin A, is oil-soluble and is essential for skin health, but deficiencies are best corrected by oral administration to achieve a systemic effect. However, vitamins D_2 and D_3 (calciferol) are used, sometimes in conjunction with vitamin A. A mixture of vitamins A, E and D_3 has been claimed to be synergistic.[21]

Vitamin E is said to enhance percutaneous resorption, and vitamin H is claimed to help fat and cholesterol synthesis.

Other vitamins which have been mentioned in the literature as having some use in topical preparations include the so-called vitamin F, now known as essential (unsaturated) fatty acids (EFA), which can undoubtedly heal the skin symptoms characteristic of rats brought to a chronic state of EFA deficiency. The relevant point here is that it would be virtually impossible to bring the human to a corresponding EFA-deficient state.

Oil Phase Constituents
The predominant oil phase constituents in massage and night creams are petrolatum, mineral oil, lanolin and low-melting-point waxes such as beeswax and low-melting mineral waxes (ceresins and paraffin). Esters such as isopropyl palmitate, isopropyl myristate and purcellin oil are reserved for lighter 'vanishing cream' types of product. Suppliers' literature abounds with formulae for night creams, each illustrating the virtues of the company's particular products. Alternatively, any of the examples 1, 2, 3, 5, 7, 8 and 11 already given will serve as a starting formula.

Moisturizing, Vanishing and Foundation Creams

As the term 'vanishing' implies, creams and lotions falling within this category are designed to spread easily and to seem to disappear rapidly when they are rubbed into the skin.

Moisturizers
Of all the beneficial properties claimed for cosmetic creams, 'moisturizing' is possibly the most widely used. The term stems from the discovery that water is the only material which will plasticize the outer dead layers of the epidermis to give the much desired attribute we call 'soft, smooth skin'.

If water is lost more rapidly from the stratum corneum than it is received from the lower layers of the epidermis, the skin becomes dehydrated and loses its flexibility. Blank[22] has shown that oil alone will not restore the flexibility.

There are two basic types of dry skin. The first is due to prolonged exposure to low humidity and air movement, which modifies the normal hydration gradient of the stratum corneum. The second is due to physical or chemical changes in the skin due to processes such as aging, continual degreasing, etc.

Changes due to aging are held to be largely due to the influence of ultraviolet light, which seems justified when one considers the difference between the skin on parts of the body which are habitually covered and on those not covered.[23–25]

The approach to restoring water to dry skin has taken three different routes—occlusion, humectancy, and restoration of deficient materials—which may be (and often are) combined.

Occlusion consists in reducing the rate of transepidermal water loss through old or damaged skin or in protecting otherwise healthy skin from the effect of a severely drying environment.

It has been demonstrated[26] that the occlusion of skin in this way results in an immediate decrease in the rate of water loss through the epidermis. This has the desired effect of causing the stratum corneum to become more hydrated, making it softer and more supple; however, the eventual effect of this extra hydration is to increase the diffusion coefficient of water across the epidermis, so that within three hours of the application of, for example, petroleum jelly to healthy skin the rate of water loss actually increases to a value higher than the pre-treatment value. (This, of course, in no way detracts from the usefulness of this approach to moisturization, since it achieves the desired hydration of the stratum corneum.)

To this end, many occlusive, non-water-permeable substances can be used, among them mineral and vegetable oils, lanolin and silicones. These simple materials have occasionally been augmented by the use of mixtures of lipids and other fatty chemicals which have been designed to imitate the composition of the skin's natural oily secretions. (Such secretions, it has been argued, may have ceased to occur in dry skin, this being at least partly the cause of the dryness.) The use of such artificial skin lipid mixtures has not been a noted commercial or even scientific success, largely because they are difficult to formulate into an emulsion that can be preserved microbiologically, and because they have not been shown to be an improvement over simpler, less expensive and readily available oils such as those mentioned above. As would be expected, the measured chemical composition of skin secretions varies greatly with skin site, age and time.

Alternatively, the use of simpler film-forming materials which only approximate in composition or are of 'allied' composition to natural skin secretions can be considered. In this category can be included albumin,[27] mucopolysaccharide,[28,29] a mixture of twenty amino acids such as occur in skin keratin,[30] gelatin[31] and hydrolysed protein.[32] If fruit and vegetable extracts have any value, it is possibly by virtue of the polyuronic acids, sugars, amines and amino acids such as are claimed in a patent[33] for the use of bamboo extracts to hydrate and protect the skin. Cactus extract is claimed in another patent,[34] while Massera and Fayaud[35] extolled the use of various fruit juices as a supplement to naturistic diets. Some of these materials are said to be successfully used by beauty consultants, but they are not generally used in mass market products.

More recently, skin substantive barrier materials (mainly based on quaternary ammonium complexes) have become available which seem to be able to influence the rate of transepidermal water loss without putting an occlusive or greasy barrier layer on the surface. These materials can be shown to be substantive to skin (and hair) and act not only as moisturizers, but as emollients and skin conditioning agents.[36] Examples are Quaternium-19, a hydroxyethylcellulose derivative, and Quaternium-22, which is based upon gluconic acid. The dry skin lotion, example 28, is taken directly from supplier's literature.

	(28) *per cent*
Isopropyl linoleate	2·0
Glyceryl stearate	3·0
Diisopropyl adipate	2·0
Myristyl myristate	1·0
PEG-40 stearate	1·0
Cetyl alcohol	1·5
Ceteareth-20	0·5
Quaternium-22	2·0
Hydroxyethylcellulose (2% aqueous)	25·0
Propylene glycol	3·0
Water	59·0
Perfume, preservative	*q.s.*

A second approach to the moisturizing problem is the use of humectants to attract water from the atmosphere, so supplementing the skin's water content.

Although popular in use, such a concept is, to say the least, somewhat doubtful from the physiological viewpoint. It is, after all, easy to demonstrate that externally applied water will not increase the flexibility of the stratum corneum—in fact, it can have precisely the opposite effect.

The humectants most frequently used as moisturizers are glycerol, ethylene glycol, propylene glycol and sorbitol which can be used alone or in admixture at various levels. Whether or not they penetrate the skin surface is a moot point, but at least they will attract moisture to the skin. Fox *et al.*[37] measured the water uptake of mixtures of some of these materials and another humectant, sodium lactate, with dried callus and found it to be strictly additive. None of the humectants had any effect in increasing the uptake of moisture by callus, except sodium lactate which is one of the water-extractable materials in skin.

Osipow[38] claims that sodium lactate acid acts as a buffer as well as a humectant and moisture loss is reasonably independent of pH. The lactates are compatible with most cosmetic ingredients and cream formulae containing sodium lactate solution are given.

The third and perhaps the most valuable approach to moisturization of skin is to determine the precise mechanism of the natural moisturization process, to assess what has gone wrong with it in the case of dry skin and to replace any materials in which such research has shown damaged skin to be deficient.

During the last twenty-five years many workers, including Jacobi, Blank, Shapiro and Rothman, have amply demonstrated that there is a natural moisturizing factor (NMF) in the skin which can be removed by means of water, other polar solvents and detergent solutions. This material has been shown to have an amino–lipid nature. According to Curri[28] it may be a mucoprotein complex or a lipomucopolysaccharide complex.

Working on the basis that the material should have both polar and non-polar moieties, or should in some way resemble a material found naturally in skin and associated with the natural moisturizing factor, there are several approaches that should be recorded. Working with the water-soluble materials, Laden and Spitzer[39] identified sodium 2-pyrrolidone-5-carboxylate as a naturally occurring humectant. This has been shown to be useful in moisturizing dry flaky skin.[40] At an earlier date, Ciocca, Rovesti and Rocchegani[41] had synthesized a material, furyl glycine, and an intermediate, furyl hydantoin, which combine amino and carboxylic functions with a lipophilic nucleus. As a result of limited tests they claimed that the glycine compound did have favourable skin properties and recommended its use in cosmetics.

Apparently, then, the facile addition of water to skin does not suffice to plasticize it; in the skin it is bound up in protein–lipid complexes (possibly within the dead cells themselves) and only in this form is it effective in keeping the skin soft. Unfortunately, simple applications of water-soluble components of NMF (for example sodium pyrrolidone-carboxylate and sodium lactate) display little affinity for the outermost layers of the epidermis, either as solutions or as oil-in-water emulsions. It has been suggested that they would be better applied locked in lipid lamellae as, for example, aqueous dispersions of nonionic lipids which have been called 'niosomes'.[42] These certainly show some promise, judging by the experimental results reported.

Emollients

'Emollience' is another ill-defined term often used in connection with skin creams. The general understanding of the word is the imparting of a smoothness and general sense of well-being to the skin, as determined by touch. (In a sense, therefore, water is an emollient.) Further, it has been shown that traditional emollients may also cause flattening of the surface contours of the skin, plumping of individual corneocytes and general smoothing and diminishing of facial lines.[43–45] The precise cause of these effects is not discussed in detail, although it may be simply due to hydration caused by the occlusive effect of the emollients. Certainly, however, the effect of various liquids in lubricating the skin surface (and diminishing the rough feel associated with 'saw tooth' dead cells in the outermost skin layer) is well established.

The list of emollients is almost endless, since virtually every liquid, semi-solid or low-melting-point solid of a bland nature and cosmetic quality has been used for this purpose. Among the most popular water-soluble emollients are glycerin, sorbitol, propylene glycol, and various ethoxylated derivatives of lipids. Oil-soluble emollients include hydrocarbon oils and waxes, silicone oils, vegetable oils[46,47] and fats, alkyi esters, fatty acids and alcohols,[48–50] together with ethers of fatty alcohols (including polyhydric alcohols).[51] The choice is determined by personal preference, data on potential skin irritation, the degree of 'greasiness' and apparent residual film on the skin, cheapness and availability.

Mineral oils and silicone oils do not 'disappear' from the skin very readily when used in any quantity and are therefore useful, as has already been noted, in cleansing and night creams. Propylene glycol is widely used and is an efficient preservative against certain micro-organisms at concentrations of more than 8 per cent, but it is a potential sensitizer.

The alkyl esters represent a range of interesting emollients ranging, as they do, through lactates, oleates, myristates, adipates, linoleates with the possibility of straight-chained, branch-chained,[52–55] unsaturated, or saturated precursors. Some are almost water-thin liquids which rub quickly into the skin (decyl and isodecyl oleates, isopropyl myristate) and others are waxy solids which melt near body temperature and give 'body' to creams.

Lanolin was considered once to be an extremely desirable emollient and the claim 'contains lanolin' was felt to be a product 'plus'. Presently, this same declaration is required by European law to warn consumers of the possible risk that it constitutes a primary sensitizer.[56] However, much work is currently underway to show that lanolin and its derivatives are not sensitizers to healthy skin and still have great value in skin creams as emollients.[57]

Vanishing and Foundation Creams

It will be clear from the foregoing that many of the creams hitherto described as night creams, massage creams and creams of high oil-content can also legitimately be described as 'moisturizing' and 'emollient'. The modern trend is, however, towards moisturizers and emollient creams that are close in composition and in-use characteristics to vanishing and foundation creams.

In order to achieve their rapid 'rub-in' effect, vanishing creams are composed, in the oil phase, of emollient esters which leave little apparent film on the skin; for this reason also, a low percentage oil phase is usually chosen.

Foundation creams possess many of the same properties; these creams are for day-time use to protect and 'condition' the cleansed skin. They must therefore leave the skin surface non-greasy and preferably matt so that other make-up can easily be applied over it. Modern foundation creams are of excellent appearance and stability; they contain not only emollients and moisturizers but also (in many cases) sun-screen agents which help to protect the consumer's skin from the harmful, aging effects of short-wave solar radiation.[58–60]

The traditional formula for a vanishing cream is based on high quality stearic acid as the oil phase. This provides an oil phase which melts above body temperature and crystallizes in a suitable form so as to be invisible in use and give a non-greasy film; it can, moreover, impart a very attractive appearance to the product. The emulsifier is soap, which is frequently formed *in situ* by adding sufficient alkali or base to neutralize a portion, usually 20–30 per cent, of the free fatty acids. Ristic[61] defined these creams as suspensions of stearic acid in a gel of stearate soap (hydrogel suspension).

A typical 'simple' vanishing cream can be made to the following formula:

	(29) *per cent*
Stearic acid	15·0
Potassium hydroxide	0·7
Glycerin	8·0
Water	76·3
Perfume, preservative	*q.s.*

Although this formula appears simple it is, in fact, not so for two reasons. Firstly, commercial stearin, although frequently referred to as stearic acid, is not a single entity but a mixture of palmitic and stearic acids together with a small quantity of oleic acid. Secondly, although in the past vanishing creams of this type were assumed to be alkaline because they contained soap, they usually have pH values of between 6·0 and 6·9. This can be explained by the existence of 'acid soap'. Ryer[62] prepared a number of stearic acid soaps of well defined composition corresponding to the following formulae:

(1) R—COONa·R—COOH
(2) 2R—COONa·3R—COOH
(3) R—COONa·2R—COOH

but found no evidence of

$$2R\text{—}COONa\cdot R\text{—}COOH \ (R = C_{17}H_{35})$$

It is therefore likely that a vanishing cream to the formula given above will contain not only normal soaps and free fatty acids of each of the constituent acids of the stearin but also acid soaps of the three acids. By using mixed bases for the neutralization it can be envisaged that the number of possible constituents is greatly increased and there is considerable scope for varying the appearance and properties of the cream.

This selection of formulae represents a by no means exhaustive range of good, pearly white, shiny vanishing creams, day creams and moisturizing creams which are attainable with the vast selection of modern raw materials. They do, however, cover a number of very successful basic products into which extra moisturizing factors, UV-absorbers and other desirable additives may be introduced, as already discussed.

Pigmented Foundation Creams

Pigmented foundation creams can contain from 3 to 25 per cent of pigments. Those with between 3 and 10 per cent can form a suitable substrate for the subsequent use of powder, whereas those with higher pigment concentrations can be used as complete make-up and are often termed powder creams. They can be water-continuous or oil-continuous systems in liquid or solid form. There are difficulties in the manufacture of these preparations, in particular (a) the high specific surface of the pigment which may preferentially absorb the emulsifying agents and, in some cases, cause inversion of the emulsion, and (b) the obtaining of adequate dispersion of the pigments to give reproducible colours.

Pigments can be suspended by the use of cellulose derivatives or inorganic silicates such as bentonite or hydrated magnesium silicate.

Jacobi[54] states that some branched-chain organic compounds can give porosity to film-building materials and hence do not interfere with the insensible respiration of the skin. These materials are called porositones and are ideal for incorporation into a liquid make-up. The formula cited in example 30 is said to give a more natural look and not to interfere with the physiological function of the skin. It allows the passage of 96 per cent of the amount of water vapour transmitted by uncovered skin, compared with 50 to 70 per cent by most other make-up creams on the market.

	(30) *per cent*
Monoglyceride of polyunsaturated acids	0·5
Isopropyl myristate	2·0
Glyceryl stearate	2·5
TEA-stearate	2·5
Propylene glycol	5·0
Talc	4·0
Titanium dioxide	5·0
Iron oxide pigments	*q.s.* to shade
Cellulose gum	0·8
Hydrated magnesium silicate	0·8
Isopropyl lanolate	2·5
Branched-chain fatty acids and esters	5·0
Allantoin	0·2
Hexachlorophene	0·5
Water, perfume, preservative, etc.	to 100·0

Some of the basic types of tinted foundation cream are illustrated by the following examples.

Water-in-oil creams		(31) Solid *per cent*	(32) Liquid *per cent*
Light mineral oil		4·0	30·0
Isopropyl myristate		8·0	—
Lanolin		—	8·0
Ceresin		19·2	—
Micro-crystalline wax		—	1·0
Sorbitan sesquioleate		2·8	2·3
Polysorbate-60		—	0·1
Powder base		*q.s.*	8·0
Titanium dioxide		3·0	—
Glycerin		—	5·0
Water, perfume, preservative	to	100.0	100·0

Oil-in-water solid creams		(33) *per cent*	(34) *per cent*	(35) *per cent*
Mineral oil		30·0	—	—
Stearic acid		3·0	8·0	12·0
Isopropyl palmitate		—	—	1·0
Glyceryl stearate		3·0	—	—
Sorbitan stearate		—	—	2·0
Polysorbate-60		—	—	1·0
Cetyl alcohol		2·0	—	—
Triethanolamine		1·0	—	—
Glycerin		—	10·0	—
Sorbitol		—	—	2·5
Propylene glycol		—	—	12·0
Lanette wax		—	8·0	—
Pigment and powder base		5·0	10·0	11·0
Water, perfume, preservative, etc.	to	100·0	100·0	100·0

Oil-in-water liquid make-up		(36) *per cent*
Mineral oil		20·0
Cetyl alcohol		1·0
Spermaceti		1·0
Sodium lauryl sulphate		0·5
Glyceryl stearate		1·0
Bentonite		2·5
Powder base and colour		8·0
Water, perfume, preservative, etc.	to	100·0

There are some foundation preparations which do not contain water, for example:

	(37)
	per cent
Sesame oil	64
Zinc oxide	11
Oxycholesterol	2
Triglyceryl stearate (polyglyceryl-3 stearate)	1
Perfume and colouring	6
Titanium dioxide	16
Preservative	*q.s.*

Hand Creams and Hand-and-body Creams

The hands obviously represent the main unprotected area of the body other than the face. In some ways, the hands are even more vulnerable to effects of environment than is the face and it is just as important that the skin which covers them should remain soft and smooth. Perhaps the most damaging environment of all is hot detergent solution since this has the effect of solubilizing lipids and can be shown to damage cell walls. Skin can then be deprived of its natural moisturizing factor and its natural protective secretions and become dry, scurfy and flaky—a condition which has been dubbed 'dish-pan hands'. Hand creams may be expected to provide some sort of remedy for this condition by softening and moisturizing the damaged skin. The main features of good hand creams or lotions are therefore that they should be easy and quick to apply without leaving a tacky film, and that they should soften the hands and perhaps help them to heal without interfering with normal hand perspiration. They are usually coloured and are lightly perfumed to make their use pleasant.

The distinction between hand creams and hand-and-body creams is not at all clear since these same criteria are to be applied to the latter also. However, it may be said that for covering large areas of the body easily and quickly, lotions are generally to be preferred to solid creams.

It follows that many of the formulae already given for vanishing and moisturizing creams will also fulfil the function of hand creams and hand-and-body creams. Since the hands are particularly vulnerable to cracking and splitting of skin, however, it has become usual to add to the emollients and moisturizing agents materials which have been shown to assist in soothing and healing broken skin, and also antiseptics.

Among the healing agents, the most popular is allantoin (2,5-dioxo-4-imidazolidinyl-urea) whose cell-proliferating, cell-cleaning and soothing properties were already well known in the 1930s and 1940s.[63–69] Allantoin is still in popular use today, as are some of its weak metal complexes such as aluminium dihydroxy allantoinate.[70] Additionally, some of the newer, skin-substantive quaternary salts have also been shown to have healing and soothing effects—for example, Quaternium-19, a hydroxyethylcellulose derivative.

Further examples of formulations suitable for the starting-point of a hand lotion or cream are given below.

	(38)
	per cent
Glyceryl stearate SE	2·7
Cetyl alcohol	1·5
Dimethicone	1·5
Lanolin oil	2·0
Squalane	3·0
Sodium lauryl sulphate	0·3
Water, perfume, preservative	*q.s.*

	(39)
	per cent
DEA-oleth-3 phosphate	0·5
Lanolin alcohol	1·0
Mineral oil	4·0
Stearic acid	1·0
Glycerin	3·0
Triethanolamine	0·5
Carbomer 941	0·1
Dimethicone	1·0
Water, perfume, preservative	*q.s.*

	(40)
	per cent
Stearic acid	7·0
Lanolin	0·5
Sorbitan oleate	0·5
Polysorbate-60	0·5
Sorbitol	10·0
Water, perfume, preservative	*q.s.*

	(41)
	per cent
Cetrimonium bromide	1·5
Isopropyl myristate	3·0
Cetyl alcohol	2·5
Lanolin	2·0
Glycerin	8·0
Water, perfume, preservative	*q.s.*

All-purpose Creams

The title 'all-purpose', although popular, is something of a misnomer since really to serve all purposes such a preparation should comply with the following requirements:

(1) As a foundation cream for general use it must provide a satisfactory foundation base for make-up without being too greasy.
(2) As a cleaning cream it should liquefy readily, be of an oily nature but should be free from 'drag'. It should not be readily absorbed by the skin.
(3) As a hand cream it should be emollient yet not leave a greasy or sticky film on the skin.

(4) As a protective cream and as an emollient cream it should leave a continuous but non-occlusive oil film on the skin.

All-purpose creams are also sometimes referred to as 'sports creams'. This term arose in Europe where creams based on lanolin esters or extracts, including one leading cream sold as an all-purpose cream, were popular for skiing and other outdoor activities where the elements could damage the skin. Obviously no single preparation can satisfy all the conflicting requirements mentioned above, and any preparation attempting to do so must be a compromise which performs each of the expected functions satisfactorily without excelling in any, or indeed performing specific functions as well as functionally specialized creams.

However, there appears to be a market for an all-purpose cream and some of the possible sales outlets for such a product are:

(a) the unsophisticated user who lacks space and/or money and who therefore buys one cream to do as much as possible;
(b) the slightly more sophisticated user who buys a speciality cream for one particular function and relies on an all-purpose cream for all other functions;
(c) the user who finds the all-purpose cream ideally suited to her particular skin for a particular function and uses it as a speciality cream;
(d) the user who normally fragments her skin creams but resorts to an all-purpose cream when travelling or on holiday;
(e) for general family use and protection against the elements.

Some suggested starting formulae for all-purpose creams are given in examples 42–44.

	(42) *per cent*
Trioleyl phosphate	3·0
Petrolatum	18·0
Glyceryl stearate	5·0
Isopropyl palmitate	4·0
Cetyl alcohol	2·0
Stearyl heptanoate	0·5
Cetearyl octanoate	0·5
Sorbitol	5·0
Water, perfume, preservative	*q.s.*

	(43) *per cent*	(44) *per cent*
Stearic acid	15·0	15·0
Lanolin	4·0	2·0
Beeswax	2·0	2·0
Mineral oil	23·0	24·0
Polysorbate-85	1·0	—
Sorbitan trioleate	1·0	—
PEG-40 stearate	—	5·0
Sorbitol	12·0	10·0
Water, perfume, preservative	*q.s.*	*q.s.*

REFERENCES

1. *CTFA Cosmetic Ingredient Dictionary*, CTFA, Washington, 1977.
2. Frazier, C. N. and Blank, J. H., *A Formulatory for External Therapy of the Skin*, Springfield, Charles C. Thomas, 1956.
3. Latuen, A. R., *Am. Perfum.*, 1958, **72,** 29.
4. Martin, E. W. and Cook, E. F., eds, *Remington's Practice of Pharmacy*, Mack Publishing, Easton, Penna, 1956, p. 8, 624.
5. Polon, J. A., *J. Soc. cosmet. Chem.*, 1970, **21,** 347.
6. Salisbury, R., Leuallen, E-E. and Charkin, L. T., *J. Am. Pharm. Assoc. Sci. Ed.*, 1954, **43,** 117.
7. Pickthall, J., *Soap Perfum. Cosmet.*, 1954, **27,** 1270.
8. Schade, H. and Marchionini, A., *Klin. Wochenschr.*, 1928, **7,** 12.
9. Rothman, S., *Physiology and Biochemistry of the Skin*, Chicago, University Press, 1954, pp. 224–226.
10. Jacobi, O. and Heinrich, H., *Proc. sci. Sect. Toilet Goods Assoc.*, 1954, (21), 6.
11. French Patent 1 467 447, Unilever NV, 1966.
12. De Ritter, E., Mafid, L. and Sleezer, P. E., *Am. Perfum. Aromat.*, 1959, **73** (5), 54.
13. Lorincz, A. L., *Drug Cosmet. Ind.*, 1961, **88** (4), 442.
14. Jarrett, A., *J. Pharm. Pharmacol.*, 1961, **13,** 35T.
15. Ellot, B., *Kosmet. Parfüm. Drogen Rundsch.*, 1970, **17** (3–4), 33, 36,
16. Willson, R. B., *Drug Cosmet. Ind.*, 1957, **81,** 452.
17. Entrich, M., *Am. Perfum. Aromat.*, 1956, **68** (5), 25.
18. Walker, G. T., *Seifen Öle Fette Wachse*, 1963, **89,** 203.
19. US Patent 3 285 818, Dai-ichi Seivaku Co., 1964.
20. Bigi, B., *Parfums Cosmet. Savons*, 1966, **9,** 565.
21. Eller, J. J. and Eller, W. D., *Arch. Dermatol. Syphilol.*, 1949, **59,** 449.
22. Blank, J. H., *Proc. sci. Sect. Toilet Goods Assoc.*, 1955, (23), 19.
23. Smith, J. G. and Findlayson, C. R., *J. Soc. cosmet. Chem.*, 1965, **16,** 527.
24. Nix, T. E., Nordquist, R. E. and Everett, M. A., *J. ultrastruct. Res.*, 1965, **12,** 547.
25. Ippen, M. and Ippen, H., *J. Soc. cosmet. Chem.*, 1965, **16,** 305.
26. Cooper, E. R. and Van Duzee, B. F., *J. Soc. cosmet. Chem.*, 1976, **27,** 555.
27. British Patent 1 038 415, Helene Curtis, 1966.
28. Curri, S. B., *Soap Perfum. Cosmet.*, 1967, **40,** 109.
29. Pichler, E., *Kosmet. Monatschr.*, 1967, **16** (8), 10.
30. Belgian Patent 669 090, Nestle SA, 1964.
31. US Patent 3 016 334, Lewis, J. T., 1962.
32. S. African Patent 67/0945, Colgate-Palmolive Co., 1966.
33. Netherlands Patent 6 508 251, Roger and Gallet, 1964.
34. US Patent 3 227 616, Warner Lambert, 1962.
35. Massera, A. and Fayaud, A., *Perfum. essent. Oil Rec.*, 1958, **49,** 812.
36. Faucher, J. A. and Goddard, E. D., *J. Soc. cosmet. Chem.*, 1976, **27,** 543.
37. Fox, C., Tassoff, J. A., Reiger, M. M. and Deem, D. E., *J. Soc. cosmet. Chem.*, 1962, **13,** 263.
38. Osipow, L. I., *Drug Cosmet. Ind.*, 1961, **88,** 438.
39. Laden, K. and Spitzer, R., *J. Soc. cosmet. Chem.*, 1967, **18,** 351.
40. Middleton, J. D.and Roberts, M. E., *J. Soc. cosmet. Chem.*, 1978, **29,** 201.
41. Ciocca, B., Rovesti, P. and Rocchegani, G., *J. Soc. cosmet. Chem.*, 1959, **10,** 77.
42. Handjani-Vila, R. M., Ribier, A., Rondot, B. and Vanlerberghie, G., *Int. J. cosmet. Sci.*, 1979, **1** (5), 303.
43. Nicholls, S., King, C. S. and Marks, R., *J. Soc. cosmet. Chem.*, 1978, **29,** 617.
44. Packman, E. W. and Gans, E. H., *J. Soc. cosmet. Chem.*, 1978, **29,** 79.

45. Packman, E. W. and Gans, E. H., *J. Soc. cosmet. Chem.*, 1978, **29,** 91.
46. Rutkowski, A. *et al.*, *Tluszcze, Srodki Piorace Kosmet.*, 1969, **13** (2), 44.
47. Hutterer, G., *Riechst. Aromen, Koerperpflegem.*, 1970, **20** (6), 222.
48. Weitzel, G. and Fretzdorff, A. M., *Riechst. Kosmet. Seifen*, 1965, **67** (1), 26.
49. Coppersmith, M. and Rutkowski, A. J., *Drug Cosmet. Ind.*, 1965, **96,** 630.
50. Murray, H. E., *Aust. J. Chem.*, 1967, **18,** 149.
51. US Patent 3 098 795, Van Dyk and Co., 1958.
52. Weitzel, G. and Fretzdorff, A. M., *Fette Seifen Anstrichm.*, 1961, **63,** 171.
53. Jacobi, O. K., *J. Soc. cosmet. Chem.*, 1967, **18,** 149.
54. Jacobi, O. K. and Maruszewski, A., *Am. Perfum. Cosmet.*, 1967, **82** (10), 83.
55. British Patent 1 004 774, Kolmar AG, 1961.
56. EEC Cosmetics Directive, 27 July 1976, 76/768/EEC.
57. Guillot, J. P., Giauffret, J. Y., Martin, M. C., Gonnet, J. F. and Soule, G., *Int. J. cosmet. Sci.*, 1980, **2,** 1.
58. Meybeck, A., *Int. J. cosmet. Sci.*, 1979, **1** (4), 199.
59. De Rio, G., Chan, J. T., Black, H. S., Rudolph, A. H. and Knox, J. M., *J. invest. Dermatol.*, 1978, **70,** 123.
60. Volden, G., *Br. J. Dermatol.*, 1978, **99,** 53.
61. Ristic, N., *Arh. Farm.*, 1969, **19** (1), 28, (Croat.) (*Chem. Abs.*, **71,** 116476).
62. Ryer, F. V., *Oil, Soap* (*Chicago*), 1946, **23,** 310.
63. Mecca, S. B., *Phila. Med.*, 1946, **41,** 1109.
64. Comunale, A. R., *J. med. Soc. N.J.*, 1937, **34,** 619.
65. Kaplan, T., *J. Am. med. Assoc.*, 1937, **108,** 968.
66. Robinson, W., *Smithsonian Institution Annual Report*, 1937, p. 451.
67. Robinson, W., *J. Bone Jt. Surg.*, 1935, **17,** 267.
68. Gordon, I., *Int. J. orthod. oral Surg.*, 1937, **23,** 840.
69. Rice, E. C., *J.natl. Assoc. Chirop.*, 1936, **26,** 5.
70. Mecca, S. B., *Soap Perfum. Cosmet.*, 1964, **37,** 132.

Chapter Five

Astringents and Skin Tonics

Introduction

Astringents are a class of materials which are identified by their local effect on skin when applied topically. These effects may include all or some of the following (not all astringents are equally active): the erection of hairs, the tightening of skin (or at least the sensation of tightening), the temporary reduction of pore size, antiperspirancy, the mitigation of 'oily skin', the rapid coagulation of blood from a fresh wound, skin healing, the promotion of tissue growth and other, more subjective sensations such as a refreshing or invigorating feeling. While these beneficial properties ensure that astringents are regarded as important and valuable cosmetic materials, not all the claims made for them can be substantiated by careful experiment.

Chemically, the known astringents can be placed, with few exceptions, into one of three categories: metal salts of organic or inorganic acids, low-molecular-weight organic acids and the lower alcohols. It is not without significance that all three categories have in common the ability to precipitate proteins from their suspensions.[1]

Types of Astringent

Metal Salt Astringents

The astringent effects of many metal ions have long been known. The list of active metals includes iron, chromium, aluminium, zinc, lead, mercury, tin, copper, silver and zirconium, although they vary in their degree of astringency.[2] Not surprisingly, some of these salts are unsuitable for cosmetic use because of their toxicity or because of the discoloration or irritation they cause. The only accepted practical choice is, therefore, between the salts of zinc, aluminium and zirconium.

The effect of the anion in metal salt astringents is not entirely passive. It can be shown, for example, that the astringency of a metal ion is partially dependent upon the identity of the anion. Furthermore, the anion will also help to determine the solubility of the salt in various cosmetic media. Whereas, for example, aluminium chlorhydrate is extensively used in aqueous antiperspirant preparations, the satisfactory development of aerosol antiperspirants containing the active ingredient in solution had to await the invention of an alcohol-soluble variation, aluminium chlorhydrex,[3] in which basic aluminium chlorhydrate is complexed with a glycol. The list of possible anions is extremely long, including acetate, chloride, sulphate, chlorhydroxide, phenolsulphonate, lactate, glycollate, citrate, tartrate, salicylate, formate, gluconate, benzoate and alums.

Acetates are rarely used because of their disagreeable odour, or formates because of their tendency to cause irritation.

Organic Acids
Low-molecular-weight organic acids with an ionizable proton show astringent properties, although formic and acetic acids are to be avoided because of their ability to cause skin damage. Most commonly encountered are lactic and citric acids.

Alcohols
Both ethanol and (less frequently) isopropanol are used as astringents, usually as aqueous solutions of strength up to 60 per cent w/v. Solutions of ethanol greater in strength than 20 per cent may cause stinging when first applied although this may be regarded as beneficial in certain types of product.

Alcohols are sometimes included in astringent products as tinctures or as 'witch hazel distillate'. ('Witch hazel' itself—a material obtained from the plant *Hamamelis virginiana*—is a powerful astringent probably because of the presence of tannin and tannic acid. The distillate owes its astringency to the alcohol subsequently added to prevent the decomposition of the solution.)

Auxiliary Additives
Apart from the materials that constitute the vehicle for the astringent, other materials are often added to astringent preparations to assist or enhance their effect. Menthol included to produce a cooling effect in an after-shave, an antibacterial in a styptic stick, and rose water as a refreshing adjunct to a skin toner are various agents added to tonics to promote soothing and the healing of damaged skin.

Astringent Products

Many product types contain raw materials with astringent properties. It is possible to categorize these various products according to the particular astringent feature which they utilize. Antiperspirants, for example, exploit the property which some zinc, aluminium and zirconium compounds have of causing 'anhidrosis' or decrease in activity of the sweat glands.

Astringent lotions, another product type, may be subdivided into a number of related products. These include pre-shave and after-shave lotions, skin tonics, toners, colognes and fresheners.

Astringent emulsion products include antiperspirant creams as well as plain astringent creams and milks together with emulsion colognes and after-shaves.

Stick astringents are generally formed from sodium stearate–alcohol gels. Among the products presented in this form are antiperspirants, deodorants, colognes, after-shaves and styptic pencils.

Some astringents are made in gel form, including some cleansing products and face masks.

Because they are dealt with more fully in other chapters, antiperspirant products will not be discussed further here and shaving products only briefly.

Aqueous and Aqueous–Alcoholic Lotions

Toners. Toners have become an accepted part of the facial skin treatment regime. They are normally recommended for use after cleansing and before the application of a moisturizing cream. The primary purpose of such a product is to tighten the skin, reduce the pore size and reduce any tendency of areas of the face and neck to oiliness. Some manufacturers also claim that the toner will help to remove any residual cleansing cream. Especially 'mild' toners may contain no alcohol at all, as in examples 1 to 3. Alcohol (usually denatured ethanol) may be added in quantities up to 60 per cent. In some product ranges, toners are offered in a series of increasing alcoholic strength for dry, normal and oily skins (examples 4, 5 and 6).

Skin toners—aqueous	(1)	(2)	(3)
	per cent	*per cent*	*per cent*
Potassium alum	1·0	—	4·0
Zinc sulphate	0·3	1·0	—
Glycerin	5·0	—	6·0
Zinc phenolsulphonate	—	2·0	—
Rose water	50·0	57·0	35·0
Orange flower water	—	—	35·0
Water	43·7	40·0	20·0
Perfume, preservative	*q.s.*	*q.s.*	*q.s.*

There is a variety of water-soluble or alcohol-soluble emollients that can be used to offset the drying effect of ethanol. In examples 4, 5 and 6 use is made of an ethoxylated lanolin derivative, together with propylene glycol.

Skin toners—aqueous–alcoholic	(4)	(5)	(6)
	per cent	*per cent*	*per cent*
Denatured ethanol	20·0	35·0	50·0
Water	72·0	58·0	42·0
Propylene glycol	5·0	5·0	2·0
Laneth-10 acetate	3·0	3·0	1·0
Perfume, preservative	*q.s.*	*q.s.*	*q.s*

Skin Tonics. From toners, the transition to skin tonics can be made by the addition of auxiliary additives and perhaps menthol or camphor to produce a slightly 'medicated' fragrance. Possibly the most commonly used skin healing and soothing additive is allantoin, a chemical of the purine group. Allantoin has been shown to have regenerative, healing, softening, soothing and keratolytic properties.[4] More recently, two allantoin combination compounds, chlorohydroxyaluminium allantoinate and dihydroxyaluminium allantoinate have become available. These possess mild astringent qualities in addition to the healing and soothing properties of allantoin itself and their use has been specifically recommended in astringent lotions of the 'tonic type'.[5,6]

Azulene and its derivatives, particularly guaiazulene, have long been accredited with healing and soothing properties.[7,8]

Newer soothing ingredients to become available include a cationic cellulose polymer, which has been given the CTFA adopted name of Quaternium-19.[9,10]

No doubt other skin-substantive cationic polymers will be found to have similar desirable properties.

The various benefits to be derived from the use of such additives make them logical ingredients of pre-shave and after-shave products. Pre-shave lotions—especially those designed to be used before electric shaving—make use of the ability of astringents to make facial hairs stand erect. After-shave lotions may be designed to soothe and cool the skin, to tighten it, to close the pores, to sterilize and stem the flow of blood from any inadvertent nicks or cuts.

Skin tonic/after-shave	(7)	(8)
	per cent	*per cent*
Alcohol	40·00	20·00
Water	55·30	77·75
Allantoin	0·20	—
Polysorbate-80	1·50	—
Sorbitol	1·50	—
Glycerin	1·50	2·00
Quaternium-19	—	0·25
Perfume, colour, preservative	*q.s.*	*q.s.*

The normal pH of skin is slightly acid; body lotions and shaving preparations often make use of acids to 'restore' this acidic condition as well as for reasons of astringency, as in examples 9 and 10.

Skin tonic/shaving lotions	(9)	(10)
	per cent	*per cent*
Alcohol	10·00	45·00
Zinc sulphate	0·50	—
Citric acid	1·00	—
Sorbitol	6·00	5·00
Water	82·50	48·90
Lactic acid	—	1·00
Menthol	—	0·10
Perfume, colour, preservative	*q.s.*	*q.s.*

Astringent lotions such as these may be thickened with cellulose ethers (examples 11, 12) or carbomer (examples 13, 14) or any other suitable agent, such as an alginate.

Body lotion	(11)
	per cent
Alcohol	43·00
Aluminium chlorhydroxyallantoinate	0·20
Propylene glycol	3·00
Menthol	0·05
Aluminium chlorhydrate (50%)	5·00
Hydroxypropylmethylcellulose (3%)	47·75
Mica (and) titanium dioxide	1·00
Perfume, colour, preservative	*q.s.*

Body lotion	(12)
	per cent
Magnesium aluminium silicate	1·50
Hydroxypropylmethylcellulose	0·75
Water	80·05
Acetylated monoglyceride	3·00
Di-isopropyl adipate	5·00
Camphor	0·40
Menthol	0·80
Methyl salicylate	6·50
Zinc phenolsulphonate	1·00
Mica (and) titanium dioxide	1·00
Perfume, colour, preservative	*q.s.*

Examples 11 and 12 contain a small quantity of titania coated micas to give a 'pearly' effect, which is slightly unusual in products of this kind.

Skin freshener	(13)
	per cent
Carbomer 940	2·00
Alcohol	73·00
Propylene glycol dicaprate	1·00
PEG-25 castor oil	5·00
Di-(2-ethyl hexyl)amine	2·00
Water	17·00
Perfume, colour, preservative	*q.s.*

Astringent gel	(14)
	per cent
Alcohol	50·25
Water	47·26
Carbomer 940	0·70
Menthol	0·04
Di-isopropylamine	0·50
Octoxynol-9	1·25
Perfume, colour, preservative	*q.s.*

Another form of astringent in gel form is the face mask. Example 16 is a powder blend to which water is added (1 part to 3 parts powder) just before use to form a spreadable paste.

Acid face mask	(15)	(16)
	per cent	*per cent*
Magnesium aluminium silicate	6·00	20·00
Water	83·00	—
Alcohol	4·00	—
Glycerin	4·00	—
Sulphated castor oil	3·00	—
Talc	—	55·00
Citric acid	—	9·90
Kaolin	—	15·00
Captan	—	0·10
Perfume, colour, preservative	*q.s.*	*q.s.*
Buffer	to pH 5·5	

Clear gel face mask	(17) *per cent*
Sodium magnesium silicate	8·00
PEG-75	1·00
Alcohol	5·00
Carbomer 940 (2% aqueous)	to pH 7·5
Water	to 100·00
Perfume, colour, preservative	*q.s.*

Astringent Emulsions

Astringent creams or lotions (of the emulsion variety) serve many purposes in the cosmetic field, particularly as antiperspirants, shaving products and perfumed creams. In practice, there is little difficulty in incorporating astringents into the water phase of an emulsion provided only that the emulsifier system is chosen to be compatible with them. For this reason, many astringent emulsions are formed with nonionic or cationic emulsifiers (the free cations associated with many astringent materials in solution are less likely to survive intact with an anionic emulsifier). The following examples illustrate a few of the ways in which astringents can be used in emulsions. Example 20 is nonionic and 21 is anionic.

Alcoholic astringent cream (anionic)	(18) *per cent*
Sodium magnesium silicate	2·00
Isopropyl myristate	5·00
Triethanolamine	0·80
Glycerin	2·00
Stearic acid	3·00
Cetyl alcohol	0·50
Triclosan	0·10
Alcohol	30·00
Water	56·60
Colour, perfume	*q.s.*

Alcoholic aftershave lotion (nonionic)	(19) *per cent*
Laneth-40	0·5
Oleth-10	0·5
Mineral oil	1·0
Alcohol	15·0
Water	72·5
Glycerin	1·0
Di-isopropylamine (10% aqueous)	1·5
Carbomer 941 (1% dispersion)	8·0
Perfume, colour, preservative	*q.s.*

Astringent/antiperspirant creams	(20) *per cent*	(21) *per cent*
Stearic acid	14·00	10·60
Mineral oil	1·00	1·00
Beeswax	2·00	1·00
Sorbitan stearate	5·00	—
Polysorbate 60	5·00	—
Aluminium chlorhydrate (50% aqueous)	40·00	32·00
Water	33·00	42·70
Glyceryl stearate	—	6·40
Propylene glycol	—	5·00
Sodium lauryl sulphate	—	1·30
Colour, perfume, preservative	*q.s.*	*q.s.*

Astringent lotion nonionic/cationic	(22) *per cent*
Glyceryl stearate	5·00
Quaternium-7	5·00
Aluminium chlorhydrate	15·00
Water	67·00
PEG-20 stearate	3·00
PEG-8	5·00
Colour, perfume, preservative	*q.s.*

Astringent cream with witch hazel	(23) *per cent*
Water	69·00
Witch hazel extract	10·00
Cetyl alcohol	3·00
Propylene glycol stearate	12·00
Sorbitol	5·00
Isopropyl myristate	1·00
Colour, perfume, preservative	*q.s.*

Astringent sticks

A popular product form for an antiperspirant, deodorant or perfume/cologne is the alcohol–sodium stearate soap gel stick, the basic formula of which is given below:

Alcohol–sodium stearate stick	(24)
Water or aqueous solution	to 100·00 g
Alcohol	12–45 ml
Sodium stearate	6·0 g
Propylene glycol or sorbitol or glycerin	3·0 g
Perfume and colour	*q.s.*

The water or alcohol may carry any soluble astringents such as (in antiperspirant sticks) sodium aluminium chlorhydroxy lactate, the total concentration of which is usually 5–15 per cent.

The gel may be manufactured by forming sodium stearate *in situ* from stearic acid and sodium hydroxide by reflux.

Styptic sticks can be made from traditional alcohol–soap gels, but these are not satisfactory because of the tendency of small sticks to dry out too quickly. They are more usually manufactured from potassium alum crystals by melting them, incorporating a filler—usually talc—and a humectant. Finally the molten product is poured into heated moulds and allowed to solidify, the final product being polished with a moist cloth.

REFERENCES

1. Govett, T. and de Navarre, M. G., *Am. Perfum.*, 1947, **49,** 365.
2. Sollmann, T., *A Manual of Pharmacology*, 8th edn, Philadelphia, W. B. Saunders, 1957.
3. *Cosmetic Ingredient Dictionary*, 2nd edn, Washington, CTFA, 1977, p. 16.
4. Van Abbe, N. J., *Chem. Prod.* 1956, **19,** 3.
5. Fabre, R., *Press Méd.*, 1962, **70,** 1042.
6. Mecca, S. B., *Soap Perfum. Cosmet.*, 1976, **49,** 10.
7. Nöcker, I. and Schleusing, G., *Münch. med. Wochenschr.*, 1958, **100,** 495.
8. Ritschel, W., *Pharmacol-Acta. Helv.*, 1959, **34,** 162.
9. *Cosmetic Ingredient Dictionary*, 2nd edn, Washington, CTFA, 1977, p. 279.
10. Faucher, J. A., Goddard, E. D., Hannan, R. B. and Kligman, A. M., *Cosmet. Toiletries*, 1979, **92,** 39.

Chapter Six

Protective Creams and Hand Cleansers

Introduction

One of the main functions of skin is to protect the body from the hazards of its environment—to keep dirt out and water in. During its long evolution, however, the developing skin tissue probably never encountered the chemical and physical hazards to which modern technology gives rise in the home and the workplace; consequently this same technology has had to provide some extra protection to prevent the skin itself from becoming damaged.

The chemicals that are in everyday use and are capable of inflicting damage on unprotected skin are far too numerous to allow a definitive list to be drawn up, but include domestic materials such as detergents, floor and metal polishes, bleaches, oven cleaners, paints and varnishes. Industrial hazards include acids, alkalis, organic solvents, resins, dyestuffs, weedkillers, insecticides, lubricants and many more.

The obvious and complete way to afford protection against such a vast array of potential irritants is the use of protective clothing made of suitable material—particularly the wearing of gloves. However, such clothing may be deemed uncomfortable, unglamorous and restrictive for constant use, even in the workplace. Thus the way is open for the cosmetic formulator to provide skin protective products of a more subtle nature, allowing the hands and body to be used naturally and unencumbered while still affording a sufficient degree of protection. The influence of comfort and aesthetic appeal in determining the extent to which protective skin products are used should not be underestimated. Experience has shown that workers will not apply an unattractive product, no matter how effective it may be.

For convenience, the types of hazard from which the skin is to be protected can be placed in a number of categories—for example, the following:

(1) Dry solids, dust and dirt.
(2) Aqueous solutions or suspensions.
(3) Non-aqueous materials, including oils, fats and solvents.
(4) Emulsions.
(5) Physical hazards, such as heat, cold, UV radiation and abrasion.

It seems unlikely that any one product would effectively protect the wearer against such a range of hazards; some degree of product specialization is therefore to be expected (although 'all-purpose' barrier creams have been marketed).

The damage caused by topically applied materials can be thought of either as direct damage to the skin surface (as with strong acids, alkalis, oxidizing agents, abrasives) or as physiological damage (as with the absorption of hormones and other chemicals interfering with normal metabolism or with primary irritants and sensitizers and allergenic agents).

The aim of the formulator should be to develop a product which, besides forming a continuous, impervious, flexible barrier to the appropriate hazard, is easy to apply and to remove when required, is of pleasing consistency, odour and appearance, and is itself non-irritating.

Barrier Materials—Protective Creams and Gels

Of all the barrier materials available for use in protective creams and gels, those giving protection against waterborne hazards are the most numerous. There are many hydrophobic substances that can be spread upon the skin in a continuous film to form a water-repellent occlusive film. These include petrolatum, paraffin, waxes, vegetable oils, lanolin, silicones and occlusive esters. Additionally other water-repellent materials can be included which are not themselves capable of forming a continuous film, but which are capable of modifying the film-forming agent to improve its aesthetic or functional qualities; these include alumina, zinc oxide, zinc stearate, talc, titanium dioxide, kaolin and stearic acid. Oil-repellent films can be formed from water-swellable polymers such as alginates, cellulose derivatives, bentonites and natural clays. Combined water-repellent and oil-repellent materials can form the basis of a general purpose barrier cream.

Example Formulae—Creams and Lotions

Example 1 is an unsophisticated cream utilizing the barrier properties of lanolin, petrolatum and kaolin. It also contains a small amount of sodium stearate to facilitate easy wash-off. Such a cream might best be used to protect the skin against dust and dry powders.

	(1) *per cent*
Stearic acid	6·00
Cetyl alcohol	3·00
Lanolin	3·00
Petrolatum	2·00
Sodium hydroxide	0·65
Water	67·35
Kaolin	18·00
Colour, preservative, perfume	*q.s.*

Examples 2 and 3 are water-in-oil emulsions using nonionic emulsifiers which are thickened by the presence of stearic acid. Example 2 utilizes the barrier properties of silicone oil whereas example 3 illustrates use of the water-repellent properties of zinc stearate together with a film-forming material, methylcellulose. The presence of sorbitol in each case prevents 'rolling' on application and provides an additional emollient effect. (Example 2 can be used as an aerosol if packaged in a nitrogen-pressurized container.)

Protective hand creams	(2) *per cent*	(3) *per cent*
Stearic acid	20·00	15·00
Dimethicone	5·00	—
Zinc stearate	—	5·00
Isopropyl myristate	2·00	—
Sorbitan stearate	1·50	1·50
Polysorbate 60	3·50	2·00
Sorbitol	20·00	6·00
Methylcellulose (4% aqueous)	—	25·00
Water	48·00	45·50
Perfume, colour, preservative	*q.s.*	*q.s.*

Examples 4 and 5 also utilize the barrier properties of silicones; the emulsifier system consists of a mixture of polyethylene glycol esters of cetyl alcohol and the viscosity is controlled by the presence of another film-forming agent, Carbomer 934 (neutralized with triethanolamine).

Protective hand lotion	(4) *per cent*	(5) *per cent*
Dimethicone	10·0	10·0
Ceteth-2	2·6	4·0
Ceteth-10	4·9	—
Ceteth-20	—	3·6
Triethanolamine	0·2	0·2
Carbomer 934	0·2	0·2
Water	82·1	82·0
Perfume, colour, preservative	*q.s.*	*q.s.*

Example 6 is a more complex oil-in-water protective hand lotion utilizing lanolin as well as silicone and a magnesium aluminium silicate film as barrier material.

Protective lotion	(6) *per cent*
DEA-oleth-3 phosphate	2·0
Cetyl alcohol	0·5
Lanolin	0·5
Dimethicone	2·0
Isopropyl myristate	2·0
Stearic acid	3·0
Triethanolamine	0·5
Propylene glycol	5·0
Magnesium aluminium silicate	0·5
Water	84·0
Perfume, colour, preservative	*q.s.*

Procedure: Disperse the magnesium aluminium silicate as a 20 per cent dispersion in some of the water before adding it to the emulsified product at 60°C.

A simpler cream containing silicone, mineral oil and sodium magnesium sulphate is represented by example 7.

Barrier cream	(7)
	per cent
Dimethicone	19·0
Mineral oil	19·0
Glyceryl oleate	2·0
Sodium magnesium silicate	2·0
Water	58·0
Colour, preservative, perfume	*q.s.*

Film-forming materials such as cellulose ethers, PVP and silicates have been shown to present some barrier properties, although they may be adversely affected by water and are thus more appropriate for protection against organic solvents and other non-aqueous irritants. In formulations containing these materials, protection against waterborne irritants can be given, as has been demonstrated, by water-repellent constituents. There have been attempts, however, to utilize film-forming materials of a type completely impervious to and unaffected by water. A method of using such polymeric films is to apply the material in a water-soluble form and then to convert it, *in situ*, into an insoluble analogue. The use, for example, of acid polymers that are soluble as alkaline salts but insoluble in the free acid form could be considered. If the substance were to be presented as an ammonium salt, for example, the volatile alkaline end of the molecule would detach itself leaving the polymer behind in the free acid form. Similarly, soluble sodium salts of certain film-formers (for example, alginates) can be converted into an insoluble form by the subsequent application of calcium or other suitable ions.[1]

Non-aqueous Barrier Products

Some formulators have proposed that protection against water-soluble irritants can best be given by film-forming materials applied from a non-aqueous medium. The use of nitrocellulose films (in conjunction with silicones) has, for example, been patented.[2] Similarly, compositions containing fluorocarbon polymers[3] and acrylates of acrylamides and other polymers have been advocated from time to time. Such compositions have two disadvantages, however. Firstly, the solvents that may be required in the formulation to dissolve or disperse these polymers may themselves be irritant or harmful to skin or unpleasant to use. Secondly, such impervious films may seriously impair the normal functioning of the skin surface, particularly the transport of water across the epidermis, leading to maceration. Moreover, the eventual removal of such films may require the application of further solvents which may irritate or damage the skin.

Perhaps the most widely used non-aqueous protective product of all is the zinc and castor oil ointment which has been smeared onto the more vulnerable skin surfaces of countless babies and young children. Less sticky, anhydrous, modern equivalents are given in the chapter on baby products—these are based upon the protective action of zinc oxide, silica and paraffin hydrocarbons. The following anhydrous formulation is suggested as protection against flash burns.[4]

Protective base	(8) *per cent*
Dimethicone	50·0
Titanium dioxide	30·0
Magnesium stearate	18·0
Iron oxide	2·0

A much more pleasant and sophisticated product is the gelled oil formulation given in example 9.[5]

Protective gel	(9) *per cent*
Lanolin oil	12·00
PEG-75 Lanolin wax	2·50
Mineral oil	50·48
Olive oil BP	20·00
Sorbitan oleate	5·00
BHT	0·02
Silica	10·00

Barrier Properties of Cationic Polymers

More recently, it has become possible to show that certain water-soluble cationic resins seem to be able to afford protection to the skin against irritants and allergens and to reduce the damage done to the barrier function of skin caused by prolonged exposure to water. It is claimed that the measure of protection given by such polymers is greater than might be expected from the simple mechanical barrier of uncharged polymer films.[6] Evidence is offered that such polymers may penetrate into the stratum corneum by reason of their cationic charge, thus modifying the bulk and surface properties of the stratum corneum so as to decrease its sensitivity to soaps and detergents, to alkalis present in depilatory creams and to certain allergens.[7–11] Example 10 illustrates the use of such a polymer in a protective hand lotion.[12]

Protective hand lotion	(10) *per cent*
Mineral oil	2·40
Isopropyl myristate	2·40
Stearic acid	2·90
Lanolin	0·50
Cetyl alcohol	0·40
Glyceryl stearate	1·00
Triethanolamine	0·95
Propylene glycol	4·80
Quaternium-19	0·20
Water	84·45
Colour, perfume, preservative	*q.s.*

Testing of Protective Preparations

Although laboratory screening tests to determine the relative efficacy of protective creams have been described, none of them in its present form presents a real criterion by which the behaviour of the protective cream in use can actually be predicted, though some of them give some useful indications, particularly in the negative aspect, in that they can help to eliminate preparations which are obviously unsuitable, before the selected preparations are submitted to actual practical user trials.

In opening a paper on occupational dermatitis, which concerns itself among other things with test methods for assessing barrier substances, Porter[13] called attention to the deficiencies in the test methods so far proposed.

Perhaps the simplest of the tests carried out is that in which a film of the barrier or protective cream is applied to a series of clean glass microscope slides which, after being allowed to dry for a prescribed period, are immersed in various solvents (such as water, alcohol, acetone, and various oils and other substances) against which the resistance of the product is to be assessed. The slide may be subjected to standardized conditions of agitation, and the integrity of the barrier film after such treatment examined, preferably against a control preparation which has been found to be of promise.

In another test the preparation is applied to a porous supporting membrane (which can be filter or other paper, animal skin or other desired substance) and after drying, various solutions—the resistance of the barrier preparation to which it is desired to determine—are applied. The length of time which the test solution takes to penetrate the membrane containing the protective film is taken as a measure of the resistance of the preparation to that solution. Various modifications of this procedure, in which the pH of the test solution is made to change the colour of an indicator solution spotted on the reverse side of the membrane, in which the penetration of the aqueous phase renders a suitable indicator fluorescent under ultraviolet light, or in which the test solution merely contains a soluble dye which, on penetration, stains the far side of the membrane, have been employed. One of the main difficulties has been to obtain the protective substance on the test membrane in a layer of standardized thickness and area.

In order to overcome this difficulty, Schwartz, Mason and Albritton[14] controlled the thickness of the film by placing it on the test paper by means of a specially shaped sheet of metal of standardized thickness which, after filling with the preparation to be tested, was subjected to a pressure of 5000 psi (35 MPa) by which the thickness of the film of protective cream became that of the standard metal shim.

An apparatus for measuring the permeability of films of barriers cream was devised by Marriott and Sadler[15] in which the passage of the test solution through the barrier could be measured by reading the meniscus from time to time in a glass tube of fine bore, graduated in hundredths of a millilitre.

All the above tests, while they may have value as sorting tests in that they serve to reject those creams which are unsatisfactory, omit to take into account the flexibility of the barrier film, and also its resistance under conditions of use where conditions of high humidity or electrolyte content may be encountered on one side of the film only.

Porter[13] proposed the use of a technique based on that used for testing the water-penetration of leather, which flexes a film of cream held on the filter paper folded in a boat-shape and containing a liquid against which the cream's performance is to be judged.

When it comes to the final choice between a number of different experimental samples, all of which have shown promise in laboratory tests, the choice must be based on practical trials.

It has been found in practice that by applying the protective preparation to the hands, allowing to dry, flexing the fingers and then immersing the hands in various solutions (against which it is desired to determine the resistance of the product, such solutions being highly coloured with an innocuous but soluble dye) a good idea of both the flexibility and general behaviour of the preparations in use may be obtained. It is of interest to note that in the experiments carried out by Schwartz, Mason and Albritton, mentioned above, anhydrous lanolin and petroleum jelly fell in the same class as the best commercial protectives, not only against alkali and acid, as might have been expected, but also against oil. It was also noteworthy that a far higher proportion of the preparations, stated by their manufacturers to be protective against oils or solvents, actually fell into this category under laboratory tests than those which claimed to be protective against water.

Barrier creams will usually remain effective for at least 4 hours, and they are usually applied twice a day. However, when dealing with particularly corrosive substances or whenever the cream is likely to be rubbed off more regularly, more frequent applications of the barrier cream are indicated in order to provide full protection for the hands.

All protective creams should be properly labelled, and the substances against which protection will be conferred should be clearly indicated, so that the user may avoid contact with irritants against which a particular cream will not protect him.

Hand Cleansers

Skin cleansers of various types are now considered to be a valuable part of the skin care regime along with toners and moisturizers. Such cleansers, however, are formulated to remove everyday grime, secretions and make-up and they may not be equally effective against the heavy stains, greases, resins, adhesives, oils, paint, tar, and dyestuffs with which the skin (particularly the hands) may get covered in the modern home, garage or workplace. Heavy-duty cleansers offer the possibility of removing many of these problem contaminants with little risk of permanent damage to the skin.

Historically, the first heavy-duty skin cleaners other than soap and water were sulphonated oils—these being particularly valuable in the removal of oils and solvents where the habitual and frequent use of soap had caused skin irritation. Sulphonated oils, however, have now been largely superseded by the so-called waterless hand cleansers. The term 'waterless' is misleading because many of them contain water in the formulation, and 'waterless' refers to the fact that they can be used without the use of *additional* water (although a final rinse-off in water is often recommended by the manufacturer).

Waterless hand cleansers can be formulated as pastes, creams, gels, lotions or clear liquid and consist of a cleansing agent, a thickener, an emulsifier and (usually) water.

Since most of the contaminants with which heavy-duty cleansers have to deal are not water-soluble, the cleansing agents most commonly employed in these products are aliphatic solvents, these being reasonably effective, cheap, innocuous and readily available. Thus the majority of examples given below contain odourless kerosene, mineral spirits or mineral oils.

Any agent which will thicken either the water phase or oil phase of the product may be used as a thickener. In the examples given, use is made of magnesium aluminium silicate (example 17), sodium magnesium silicate (example 16) a methyl vinyl ether–maleic anhydride copolymer (example 18) and methylcellulose (example 17). The soap or nonionic detergents which are often used for emulsification and for their detergent action also cause thickening by the formation of a gel. Ancillary emulsifiers may also be employed to increase the cleansing power of the product (as with the amphoteric in example 12, the sulphonate in example 13 and the cocamide in example 11).

The soaps used as emulsifiers to produce gels may be the sodium, triethanolamide or monoethanolamide salts of stearic or oleic acids or a mixture of both.

Emollients are added to improve the application properties and to prevent the defatting of the skin. The choice of emollient is wide: in the following formulae, lanolin (example 11), ethoxylated lanolin (example 12), myristyl myristate (example 19) and propylene glycol (example 14) have been used.

Waterless hand gels	(11) *per cent*	(12) *per cent*
Deodorized kerosene	35·00	25·00
Lanolin	10·00	—
Cocamide DEA	4·00	—
Stearic acid	2·43	6·00
Oleic acid	3·64	8·00
Sodium hydroxide	0·38	0·80
Amphoteric-2	—	2·00
PEG-75 Lanolin	—	0·50
Water	44·55	57·70
Perfume, colour, preservative	*q.s.*	*q.s.*

Waterless hand gel	(13) *per cent*
Deodorized kerosene	20·00
Alkylaryl sulphonate (amine neutralized)	5·00
Cocamide DEA	2·00
Oleic acid	8·00
Monoethanolamide (20%)	8·00
Water	57·00
Perfume, colour, preservative	*q.s.*

Waterless hand cleansers	(14) *per cent*	(15) *per cent*
Deodorized kerosene	20·00	55·00
Mineral oil	20·00	—
Glyceryl stearate	3·00	—

Waterless hand cleansers (cont.)

	per cent	*per cent*
Stearic acid	5·00	—
Oleic acid	—	4·50
Stearamide-MEA stearate	—	6·00
Propylene glycol	5·00	—
Triethanolamine	1·50	1·50
PEG-8 Cocoate	—	3·00
Water	45·50	30·00
Perfume, colour, preservative	*q.s.*	*q.s.*

Skin cleansing gel	(16)
	per cent
Mineral oil	15·50
Stearic acid	4·40
Triclosan	0·10
Sodium magnesium silicate	2·00
Triethanolamine	1·60
Water	76·40
Perfume, colour, preservative	*q.s.*

Example 16 illustrates the possibility of using antiseptic agents (such as triclosan) and other additives to provide additional benefits to the product. If the product is designed to remove particularly tenacious grime (such as the ink from carbon paper or typewriter ribbons) the formulator may choose to include a mild abrasive such as finely ground pumice. Such a product would not be recommended, however, for use on sensitive skin.

Example 17 is soapless, utilizing nonionic emulsifiers to produce a cream. Such a formulation might be favoured by users who find that alkaline products are irritating to their skin. Example 18 is a much milder form of cleanser in simple liquid form:

Waterless hand cleansing cream	(17)
	per cent
Magnesium aluminium silicate	2·50
Sorbitan stearate	2·00
Polysorbate 60	8·00
Deodorized kerosene	35·00
Methylcellulose	0·50
Water	52·00
Perfume, colour, preservative	*q.s.*

Liquid hand cleanser	(18)
	per cent
PVM/MA polymer	0·40
PEG-6-32	5·00
Octoxynol-9	5·00
Water	89·60
Potassium hydroxide	to pH 7
Perfume, colour, preservative	*q.s.*

The final example is somewhat unusual since it comprises a base which may be turned into a number of different products of varying consistency by the addition of solvent together with variable amounts of oleic acid:

Waterless hand cleanser (19)

A Base (smooth paste)

	per cent
Cocamidopropylamine oxide	18·00
Cocamide DEA	38·00
Dioctyl sulphosuccinate	15·00
Myristyl myristate	15·00
Oleic acid	14·00

Mix until smooth at 45°–55°C.

B Finished products	Gel	Cream	Lotion
	per cent	*per cent*	*per cent*
Base	15·00	13·90	12·95
Odourless kerosene	30·00	27·70	25·90
Oleic acid	1·50	1·40	1·40
Water	53·50	57·00	59·75
Perfume, colour, preservative	*q.s.*	*q.s.*	*q.s.*

Procedure: Heat and stir all the ingredients except the water until clear at 55°–65°C. Add water slowly at 55°C with high-speed stirring until the product is homogeneous and smooth.

REFERENCES

1. British Patent 1 122 796, Givardière, G., 1968.
2. British Patent 754 844, Morgulis, S., 1954.
3. British Patent 797 992, British Oxygen Co., 1956.
4. Cook, M. K., *Drug Cosmet. Ind.*, 1959, **84,** 32.
5. Silverman, H. I. *et al.*, *Drug Cosmet. Ind.*, 1974, **114,** 30.
6. Union Carbide Corporation, *Polymer JR for Skin Care*, 1977.
7. Faucher, J. A. and Goddard, E. D., *J. Soc. cosmet. Chem.*, 1976, **27,** 543.
8. Goddard, E. D., Hannan, R. B. and Faucher, J. A., *The Absorption of Charged and Uncharged Cellulose Ethers*, paper presented at the International Congress on Detergency, Moscow, September, 1976.
9. Goddard, E. D., Phillips, T. S. and Hannan, R. B., *J. Soc. cosmet. Chem.*, 1975, **26,** 461.
10. Goddard, E. D. and Hannan, R. B., *J. Colloid Interface Sci.*, 1976, **55,** 73.
11. Faucher, J. A., Goddard, E. D., Hannan, R. B. and Kligman, A. M., *Cosmet. Toiletries*, 1977, **92,** 39.
12. Goddard, E. D. and Lueng, P. S., *Cosmet. Toiletries*, 1980, **95,** 67.
13. Porter, R., *Br. J. Dermatol.*, 1959, **71,** 22.
14. Schwartz, L., Mason, H. S. and Albritton, H. R., *Occup. Med.*, 1946, **1,** 376.
15. Marriott, R. H. and Sadler, C. G. A., *Br. med. J.*, 1946, **2,** 769.

Chapter Seven

Bath Preparations

The bath products market has undergone considerable change in recent years in terms both of volume and of the range of products available. In particular, bath salts, tablets and crystals, which previously dominated the market, are now much less popular and have been replaced to a large extent by bubble bath products. The range of bath preparations nowadays includes bath oils, shower gels, after-bath body lotions and the even newer hydroalcoholic products sometimes referred to as bath satins.

FOAM BATHS

Introduction

Foam baths are undoubtedly the most popular bath preparations currently on the market and they have enjoyed a very healthy growth in recent years. Generally, they are available in liquid, gel and powder forms. Body cleansing, the primary function of a bath, is performed very well by a bubble bath, which in addition offers the opportunity to apply many desirable health and beauty ingredients to the skin, although not necessarily so efficiently as when these aids are used individually. Functioning as a body cleanser, a bubble bath soaks off and suspends dirt, grime and body oils and prevents the formation of the 'bath-tub ring' that usually results from the use of soap. In this respect a bubble bath is superior to the traditional soap and water bath. Additionally, a well formulated bubble bath will condition the skin, deodorize, perfume the body and the bathroom, stimulate the senses, yet promote relaxation.

A good foam bath should exhibit the following characteristics:

(1) It should provide copious foam at minimal detergent concentration.
(2) The foam should be stable, particularly in the presence of soap and soil and within wide limits of temperature. Simultaneous attributes of stable foam and ready removal from the bath are unattainable in practice and a compromise must be aimed at to provide reasonable but not excessive foam stability. By delaying the use of soap it is, of course, possible to prevent premature breakdown of the foam, thereby satisfying the aesthetic requirements of the bather and facilitating the subsequent removal of the dirty water from the bath.
(3) It should prevent bath-tub ring formation.
(4) It must be non-irrritant to the eyes, skin and mucous membranes. Bubble baths have been claimed to produce symptoms of irritation in the lower urinary tract and it is essential to check the irritation potential of all these products before marketing.

(5) It should have adequate detergent power so that it will cleanse the body efficiently. In order to counteract any excessive harshness to the skin it is advisable to include a low level of skin emollient.

Formulation of Foam Baths

As already mentioned, foam baths are available in several physical forms and clearly the choice of raw materials is highly dependent on the final product form. Before discussing the product types in detail, it is of interest to examine the raw materials available.

Foaming Agents

Obviously, the foaming agent is the most important ingredient in all foaming bath products and great care should be taken in its selection. When selecting a surfactant it is important to bear in mind the properties that good foam baths should exhibit. Of the many surfactants currently available, anionics are the most widely used; both nonionics and amphoterics are also of considerable interest. Cationics, however, because of their incompatibility with soaps and other anionics and their much greater eye irritancy, are seldom if ever used in bubble bath formulations.

Among the most popular anionic surfactants used are the sodium, ammonium and alkanolamine salts of fatty alcohol sulphates, fatty alcohol ether sulphates and (sometimes) alkyl benzene sulphonates. Soaps are not suitable surfactants for use in foam baths, since in hard water calcium and magnesium soaps precipitate as a dirty hard scum.

The fatty alcohol sulphates, principally the lauryl sulphates, were the first anionics of any importance to be used as the primary foaming agents in bubble baths. Although they give less flash foam than the even more widely used fatty alcohol ether sulphates, their foam is often judged to be much creamier; this attribute, together with relatively low irritation potential, gentle 'feel' on the skin and inhibition of bath-tub ring formation, has contributed to the widespread use of lauryl sulphates. The versatility of the various salts available offers formulators great freedom in creating products in many forms. Highly concentrated sodium lauryl sulphate, for example, can be dry blended into powder products. Sodium lauryl sulphate, however, is not particularly recommended for liquid products because of its relative insolubility and hence high cloud point values—for instance, a 30 per cent solution of sodium lauryl sulphate has a cloud point of about 20°C. Ammonium lauryl sulphate is considered by some to be the best foamer and cleanser and it has the advantage over the sodium salt of being less prone to hydrolysis at low pH values. One problem, however, is that the pH must be kept acidic in order to prevent the release of ammonia. The various alkanolamine lauryl sulphates, by virtue of their greater solubility and lower viscosity, permit the formulation of more highly concentrated liquid products but tend to produce less foam.

Perhaps the most popular surfactants used in foam baths are the fatty alcohol ether sulphates, especially the sodium salts and, in particular, those based on lauryl–myristyl alcohols and containing 2–3 moles of ethylene oxide per mole of the alkyl ether sulphate. Copious foaming independent of water hardness, reasonable foam stability in the presence of soap, fragrances, oil additives and body debris, together with good skin compatibility, make these materials an obvious choice for bath products. A high degree of proficiency as a lime soap dispersant prevents bath-tub ring even in very hard water. Other advantages of these surfactants are good colour, which permits the use of very delicate pastel colours if required, a good viscosity response to electrolyte and unusual solubilizing powers for perfumes.

Primarily for reasons of cost, alkyl benzene sulphonates (mainly branched-chain dodecyl benzene sulphonates) soon found their way from household detergent uses into toiletry items such as bubble baths. Of all the synthetic surfactants, the sodium alkyl benzene sulphonates are probably the most suitable for spray drying and until recently most of the commercial products on the market in bead form consisted of this material and various salt builders and extenders. When biodegradability considerations became critical, the importance of linear alkyl benzene sulphonates grew. However, in 1970 an increasing number of complaints involving irritation and infection, mainly in young girls, were reported by the US Food and Drug Administration. Linear alkyl benzene sulphonates were suspected and, in order to avoid threatened regulatory actions,[1] the use of both the sodium salt in powder bubble baths and the more soluble triethanolamine salt in liquid foam baths declined.

Other anionic surfactants worthy of mention are the alpha-olefin sulphonates, lauryl sulphoacetates, the half ester sulphosuccinates and the paraffin sulphonates.

The alpha-olefin sulphonates are commercially attractive and have been used as alternatives to linear alkyl benzene sulphonates in powder foam baths. Their toxicological properties, however, have yet to be clarified, while in liquid products viscosity regulation with electrolytes is said to be a problem.

Lauryl sulphoacetates are sometimes used in high-priced powder or granular bubble baths but limited solubility has greatly restricted their use in liquid products.

The sulphosuccinates, particularly the monoesters such as the di-sodium lauryl alcohol polyglycol ether sulphosuccinate, Rewopol SBFA30, are considered to be very mild detergents with good foaming properties and free from any tendency to irritate the skin and mucous membranes. Furthermore, they are also claimed to increase the tolerance of the skin to other detergents such as fatty alcohol ether sulphates and alkyl benzene sulphonates.

The paraffin sulphonates produced by the sulphoxidation of *n*-paraffins are relatively cheap and, therefore, are of undoubted interest. In particular, the secondary alkane sulphonates known commercially as Hostapur SAS are interesting. This material, in addition to being biodegradable and exhibiting good physiological properties, is also a good foamer with high solubility in water. The use of Hostapur SAS on its own, however, normally results in excessive degreasing of the skin and moreover the finished product is difficult to thicken. When it is combined with other surfactants such as alkyl ether sulphates, these problems can be overcome.

Nonionics do not foam particularly well and consequently these surfactants are not used as primary foamers. Instead, the nonionic components of bubble baths are used to stabilize foam, to enhance the viscosity of the product, or to solubilize skin care ingredients and fragrances. Among the nonionic surfactants used in bubble baths, the alkanolamides and amine oxides are possibly the most widely used; in addition, ethoxylated derivatives such as ethoxylated fatty alcohols, ethoxylated fatty acids, alkyl phenol ethoxylates, ethoxylated alkanolamides, ethoxylated propylene oxide condensates (Pluronics) and ethoxylated sorbitan fatty acid condensates are occasionally employed.

The alkanolamides comprise a very broad class of surfactants and there are many types commercially available. They are derived by condensing various cuts of fatty acids *ex*-coconut oil with alkanolamines. The diethanolamides tend to be used in liquid foam baths because of their greater water solubility. Both the Kritchevsky amides (2 moles of diethanolamine reacted with 1 mole of fatty acid) and the super amides (1 mole of diethanolamine reacted with 1 mole of fatty acid methyl ester) perform well in this application and are hard to surpass in terms of economy for foam boosting, foam stabilization and viscosity building in conventional foam baths based on alkyl sulphate and alkyl ether sulphate. The water-insoluble monoethanolamides and isopropanolamides tend to be used mainly in dry products. The fatty acid portions of these amides are generally 12 to 18 carbons in length, with the purest C-12 fatty acid amide usually giving the greatest foam stability and compatibility with soap. The level of alkanolamide used will vary depending on which primary detergent is used, but is usually less than 3–4 per cent. The presence of high levels of an emollient, perfume or pearlescing agent may occasionally necessitate increasing this level.

The ethoxylated alkanolamides contribute to the overall formulation in much the same way as the alkanolamides but are more soluble as expected and in most cases do not thicken the finished product to the same extent.

Amine oxides are claimed to have superior foam boosting properties to alkanolamides but this is open to question. Perhaps the most widely used is the C-12 dimethylamine oxide, since cloud point problems can be experienced with the C-14 dimethylamine oxide.

The ethoxylated nonionics previously mentioned tend to be used as emulsifiers, solubilizers and emollients, rather than foam boosters. The Pluronic block polymers, for example, have been recommended as useful perfume solubilizers. Water-soluble gums have been claimed to act as foam stabilizers when used at low concentrations and they are believed to act by supporting the walls of foam bubbles and thereby increasing their resistance to collapse.

Amphoteric surfactants whose charge can vary according to the pH of the system appear to be one of the faster growing speciality surfactant groups. Much interest has recently been focussed on the alkyl imidazoline betaines and, in particular, two alternative coconut imidazoline betaines, Empigen CDR10 and Empigen CDR30, manufactured by Albright and Wilson. These two materials are very similar but differ considerably in their effect on viscosity. Both products are claimed to exhibit outstanding mildness and, in addition, excellent foaming properties comparable to those of conventional anionic surfactants; the foam stability is also said to be good, even in the presence of soil and soap. Nevertheless, for improved cleaning, viscosity control and cost performance it is recommended that these materials be combined with conventional anionic surfactants. Other amphoterics that have been used in foam baths are the alkyl amido betaines, which are claimed to be good foam producers in addition to having foam stabilizing properties. In particular, these materials are claimed to be exceptionally good at promoting foam stability in the presence of oil and sebum. They are available commercially from a number of sources, for example Tego-Betaine L7 *ex* Goldschmidt, Steinapon AM-B13 *ex* Rewo and Empigen BT *ex* Albright and Wilson. The alkyl dimethyl betaines, for example Empigen BB *ex* Albright and Wilson, on the other hand, tend to be used only as foam stabilizers.

Emollients
In an attempt to overcome any possible harsh effects of foam baths on the skin, special ingredients known as emollients are often added to the formulation to help achieve and maintain a healthy and attractive skin. The value of many of these conditioners has been challenged but they continue to be used with apparent success. Although it has been argued that the bather stays in the bath for too short a time to receive any real benefit, equally strong opinions have been expressed that, even in this short time, the skin absorbs active substances. Many such ingredients are available. The following list, though not exhaustive, gives some of the most popular in common use: branched chain esters, for example isopropyl myristate; decyl oleate; ethoxylated partial glyceride fatty acid esters (Softigen 767); protein derivatives; lanolin derivatives; and fatty alcohol ethoxylates; etc. Polymer JR from Union Carbide, which is a cationic cellulose ether derivative, has been recommended for use in foam baths and, because of its cationic nature, is said to be more substantive to skin than many other emollients.

Two further interesting and relatively new materials are Aethoxal and Cetiol HE *ex* Henkel. In addition to emolliency, these oils exhibit mild surface activity and do not interfere with the foaming of other surfactants. Both are more soluble in cold water than in hot and consequently when they are added to warm bath water an instant bloom occurs.

It is worth while to remember that careful formulation is necessary when using some of these ingredients in order to prevent foam loss and to protect against product instability.

Perfumes
There is no question that the perfume used in a foam bath is extremely important: it is perhaps equal in importance to the foaming agent. Most of the larger companies marketing foam baths spend a great deal of time and money in selecting the perfumes for their products. A good perfume must, of course, convey the marketing image of the brand and it should also fulfil the following requirements:

(1) It should be acceptable when sniffed in the bottle.
(2) In use it should be fresh and have sufficient volatility to give a strong impact.
(3) It should linger on the skin to give a feeling of freshness and wellbeing.
(4) It must have an acceptable shelf life in the product.

The level of perfume used will vary between 1 and 5 per cent, depending on the cost limitations. The nature of the perfume ingredients may require the use of additional solubilizers; the most commonly used are nonionics such as ethoxylated fatty alcohols, ethoxylated fatty acid esters, ethoxylated sorbitan fatty acid esters and ethoxylated propylene oxide condensates (for example, Pluronics). A recent interesting publication by Blakeway *et al.*[2] on perfume stabilization merits attention.

Because of their complex nature, perfumes often cause problems of product instability. They not only affect odour stability, but can cause discoloration, upset preservation systems and cause instability in clear, opacified and emulsion products. The need to test the shelf life of all new products adequately cannot be overstressed. The ingredients used in perfumes should also be adequately safety tested before use.

Herbal extracts, while not strictly to be classed as perfumery ingredients, are used in bath preparations, usually to help convey the brand image and to justify therapeutic claims with respect to minor skin disorders. Herbal extracts of most living plants can be obtained if required, but their medicinal nature is open to speculation.

Viscosity Controllers

The problem of achieving the required viscosity of liquid products is not simple since this depends on many factors such as the choice and level of surfactant and foam booster. Even certain perfumes have been found to have significant effect on viscosity. Generally, inorganic salts such as sodium and potassium chloride are used to thicken the product where possible, whereas alcohol, hexylene glycol, propylene glycol and polyethylene glycols are used to lower the viscosity. Certain thickening problems, however, can only be solved by the use of natural gums such as tragacanth and gum acacia, or synthetic gums such as methyl cellulose and hydroxyethyl cellulose.

Colour

Colour of foam baths is clearly important in marketing and care should be taken to select colours that are stable in the chosen product. Obviously, if the products are marketed in clear bottles adequate light testing should be carried out. Interesting colour effects in the bath can be achieved by the use of indicator colours and fluorescein. Before finally selecting a colour system the legislative requirements of the country in which the products are to be sold should be carefully checked.

Preservatives

Except in the case of dry bubble bath preparations and those with a very high detergent concentration, foam baths should contain an adequate amount of preservative to prevent attack by moulds and bacteria, particularly the Pseudomonas species. Bacterial attack can produce opacity in products that are intended to be clear, separation in emulsified and pearlescent products and can cause changes in both perfume and colour system. Prevention is a matter of selecting a suitable preservative in accordance with the legislative requirements in the country of sale. Suitable preservatives include ethanol; methyl, propyl and butyl hydroxy benzoate; phenylmercuric nitrate; formaldehyde; Bronopol; and many others. The best preservative for a particular foam bath can only be determined by properly designed microbiological testing. Good housekeeping in the manufacturing unit is, however, just as important as choice of preservative if product contamination is to be avoided.

Opacifying Agents
When an opaque liquid foam bath is required, an opacifier is needed. The following are the most commonly used: higher alcohols such as stearyl or cetyl alcohols; ethylene glycol mono- and distearates; glyceryl and propylene glycol stearates and palmitates; and the magnesium, calcium and zinc salts of stearic acid.

Clearly, viscosity is an important factor in the stability of such systems but also the manufacturing technique used is vitally important in achieving maximum stability. Ideally, when using the opacifiers mentioned above, all the ingredients (except perfume) should be heated to 65°–70°C and allowed to cool slowly to ambient temperature with gentle mixing, when the pearl will develop. Rapid cooling will produce less pearly products that may be unstable. It is possible, however, to obtain blended opacifier–detergent concentrates of some of these opacifiers which eliminate the need to heat the batch. Other opacifiers, for example polymeric materials such as the Antara range *ex* GAF and the Morton Williams E Series opacifiers, can also be incorporated without heat, but these tend to be less pearly in appearance.

Types of Product

Liquids
Liquids can be further divided into clear, translucent, opaque, pearlescent and multi-layer products. The formulation possibilities for a medium-priced clear liquid foam bath are endless; a typical and very simple product formulation would be as follows:

	(1) *per cent*
Sodium lauryl ether sulphate (28% active)	50
Coconut diethanolamide	3
Perfume	1–2
Citric acid	*q.s.* to pH 7
Colour, preservative, emollients, solubilizer	*q.s.*
Sodium chloride	*q.s.* to required viscosity
Water	to 100

More expensive but milder formulations can be achieved by replacing part of the sodium lauryl ether sulphate by a coconut imidazoline betaine, for example Empigen CDR10 (Albright and Wilson), or the disodium salt of lauryl alcohol polyglycol ether sulphosuccinate (Rewopol SBFA30—Rewo). In addition, an alkyl amido betaine, for example Empigen BT (Albright and Wilson), could be introduced to improve foam stability, particularly in the presence of soap:

	(2) *per cent*	(3) *per cent*
Sodium lauryl ether sulphate (28% active)	25	30
Empigen CDR10	25	—
Rewopol SBFA30	—	40
Empigen BT	4	—
Coconut diethanolamide	—	3
Perfume	1–2	1–2
Citric acid	*q.s.* to pH 7	*q.s.* to pH 7
Colour, preservative, emollients	*q.s.*	*q.s.*
Sodium chloride		*q.s.* to required viscosity
Water	to 100	100

Translucent and pearlescent products can be created by the addition of insoluble stearates, as already discussed. These are readily available from all major surfactant manufacturers and the depth of opacity is governed by the level of incorporation, which is usually between 1 and 5 per cent. Opaque non-pearlescent products are achieved by the use of polymeric materials such as Antara 430 (GAF).

Multi-layer products can be achieved and a typical formulation taken from a British Patent[3] is as follows:

	(4) *per cent*
Sodium lauryl ether sulphate (28%)	50·0
Coconut diethanolamide	9·0
Hexylene glycol	14·0
Neutral monoethanolamine citrate	13·0
Citric acid	3·0
Perfume, colour, preservative, water	to 100·0

Gels

Basically gels are very similar to liquid products, except that they have a much higher viscosity. This is achieved by increasing the level of detergent, foam stabilizer or electrolyte content, depending on the particular formulation. Shower gels may also be considered here since they are virtually identical in formulation to the highly viscous liquid bath products; however, because they are sold for direct application to the body they must be very mild, and it is usual to adopt for shower gels the types of formulation, already mentioned, that have a milder action on the skin and eyes.

Dry Bubble Baths

Dry bubble baths, of considerably less importance than liquid bubble baths, are nevertheless worthy of discussion. Basically they consist of a mixture of one or more foaming agents, fillers and water softeners to add bulk or act as carriers, perfume, colour and free-flow agents.

The major surfactant ingredients are usually dry products, sometimes fortified with liquid surfactants, for example sodium lauryl ether sulphate, to give good flash foam. The principal surfactants employed in the past were linear alkyl benzene sulphonates; because of sporadic incidents of alleged urinary tract irritation, these tend to have been replaced by alpha-olefin sulphonates. Other surfactants that have been used include sodium lauryl sulphate, sodium lauryl sulphoacetate and isethionate derivatives.

Inorganic fillers such as sodium chloride and sodium sulphate are often used in the more inexpensive products. These can, however, be replaced by functional fillers which, in addition to acting as fillers, also have water-softening properties. These include sodium hexametaphosphate, sodium sesquicarbonate and tetrasodium pyrophosphate. One of the main problems in formulating a powdered bubble bath is to keep it free flowing and prevent it from caking. This can be achieved by adding tricalcium phosphate, calcium silicate or sodium silica aluminate. Bentonite or starch is usually employed to absorb the perfume, in order to disperse it throughout the product, while colour is usually incorporated by premixing with one of the fillers. Example formulations are as follows:

	(5) *per cent*
Alpha-olefin sulphonate (40% active spray-dried beads)	20
Lauric isopropanolamide	3
Sodium sesquicarbonate (low density)	60
Sodium chloride	14
Perfume	3
Colour	*q.s.*

	(6) *per cent*
Sodium lauryl sulphate	30
Sodium lauryl sulphoacetate	10
Lauric isopropanolamide	3
Sodium sesquicarbonate (low density)	50
Calcium silicate	4
Perfume	3
Colour	*q.s.*

Product Assessment

Complete assessment of foam bath products is difficult, since interpretation of the initial and residual feel on skin, efficacy of special ingredients, fragrance and the amount and texture of foam is subjective and requires large consumer panels to get significant results. It is normal, therefore, to limit laboratory testing to the assessment of foam volume and stability in the presence of, perhaps, soil and soap. This is not unreasonable, since foaming power is probably the single most important property as far as the consumer is concerned. A convenient technique for measuring foam is detailed in a paper by Beh and James.[4] Using a relatively simple method, it was demonstrated that the addition of a small quantity of calcium ions actually increased foam volume, although in the presence of soap foam stability was worse when calcium ions were present. It was also demonstrated that the effect of calcium ions could be destroyed by the incorporation of

EDTA. A number of foam stabilizers were also examined and from this study an alkylamido betaine was shown to be the best stabilizing agent against soap, with the N-alkyldimethyl betaine second. Some apparently anomalous results in this paper highlight the importance of carrying out foaming measurements on all finished products in order to detect the possible inclusion of foam destabilizers.

BATH SALTS

Bath salts, also known as bath crystals, were among the first bath additives to be used but today they form a relatively unimportant part of the total bath market. They consist of soluble inorganic salts attractively perfumed and coloured and they are designed to give fragrance, colour and, in most cases, water-softening properties to the bath. Some are effervescent in action while others include oily additives for emolliency. Fragrance is unquestionably the most important property of bath salts and this should be refreshing and relaxing, with sufficient strength to pervade the bathing area.

Size, colour and attractiveness of the crystals are clearly also important; in addition the crystals must be free flowing and easily dispersed and must dissolve rapidly in the bath water. Low alkalinity and mildness to the skin are also very important and, whether water-softening or not, bath salts should never be deleterious to soap lather or detergency—nor should they contribute to a bath-tub ring.

Ingredients and Formulation

Salts

Sodium sesquicarbonate ($Na_2CO_3{\cdot}NaHCO_3{\cdot}2H_2O$), which is a mixed salt, is probably the most popular material used in the preparation of bath salts. It is available in uniformly sized, elongated, attractive translucent crystals which are extremely stable, non-caking and free-flowing. It dissolves rapidly and completely in water, is easy to colour and perfume, is an excellent water softener and is quite mild to the skin, having a pH value of about 9·8 for a 1 per cent solution.

Other carbonates that have been used are sodium carbonate decahydrate, $Na_2CO_3{\cdot}10H_2O$, and the monohydrate, $Na_2CO_3{\cdot}H_2O$. The decahydrate, otherwise known as washing soda, is a good water softener and consists of large attractive crystals that are readily soluble in warm water. Unfortunately, it has several serious disadvantages including a low melting point of 35°C, at which temperature it dissolves in its own water of crystallization. This clearly precludes its use in hot climates. Even under ordinary storage conditions it tends to effloresce to become powdery and unsightly, although this can be overcome by coating the crystals with a film of humectant such as glycerin. It is also more highly alkaline than the sesquicarbonate. The monohydrate is the most stable form of sodium carbonate and is available in attractive crystal agglomerates with excellent stability. Its main disadvantages are its slow rate of dissolution and the fact that it is more alkaline than the sesquicarbonate.

Phosphates are often included in bath salts to improve the water-softening properties, the most commonly used being sodium hexametaphosphate

(Calgon), tetrasodium pyrophosphate and sodium tripolyphosphate. Trisodium phosphate is a good water softener but because of its highly alkaline nature it cannot be used without buffering. However, it can be used in combination with either sodium sesquicarbonate or borax, both of which buffer it quite effectively.

Borax ($Na_2B_4O_7 \cdot 10H_2O$) is less alkaline than the carbonates and possesses a mild detergent action, although it is less effective as a water softener than the carbonates and is slow to dissolve.

Rock salt, NaCl, is used in bath salts of the type that provide fragrance only; it is very stable and its large crystals are attractive and easily coloured. It is of course non-alkaline and mild to the skin. Its disadvantage are that it has no water-softening properties and, if used in quantity, it tends to interfere with soap lathering; furthermore, the large crystals do not dissolve easily.

For effervescent systems, sodium bicarbonate and tartaric or citric acid are included.

Fragrance

Instability of fragrance can be a problem in bath salts, since in these products the perfume is distributed in a thin film on the surface of an inorganic salt which can be quite alkaline. Consequently, the perfume must have good fixative properties in addition to adequate stability to alkalis, light and oxidation. Oil-absorbing powders such as calcium silicate or fumed silica are sometimes used to aid deposition, retention and stability of the fragrance and they may also improve the flow characteristics of the product.

Colour

As with fragrance, the choice of colour is limited by stability to both alkalis and light. Colour stability may also be affected by perfume. Insoluble colours have been recommended because of their better stability to alkalis and light. These are best dispersed at a concentration of about 1 per cent in a suitable medium such as a glycol or liquid nonionic surfactant.

Example Formulations

Bath salts—fragrance only	(7)
	per cent
Sodium chloride	95–99
Perfume	1–5
Colour	*q.s.*

Water-softening bath salts	(8)
	per cent
Sodium sesquicarbonate	95–99
Perfume	1–5
Colour	*q.s.*

Effervescent bath salts	(9)
	per cent
Sodium sesquicarbonate	25
Sodium bicarbonate	50
Tartaric acid	20
Perfume, colour	5

The manufacture of bath salts is a straightforward process and can be carried out in most types of dry powder blender. Colouring is done either by spraying with colour solution, mixing and drying or by immersing the salts in the colour solution, followed by drying. Colour solutions should be hydroalcoholic or, if possible, alcoholic. Alcohol, in addition to reducing the drying time, helps to prevent solution of the bath salts. Fragrance is added either by spraying an alcoholic solution or by dispersing the perfume in the colour solution. Perfume can also be mixed with oil-absorbent powders. Clearly, in colouring effervescent salts no water can be used.

Bath Cubes and Tablets
Bath cubes and tablets employ the same materials as bath salts, but in powder form. The powdered bath salts are first granulated with starch and a small amount of alcohol-soluble gum binder is normally added. The granules are then compressed into tablets or cubes. Prior to compressing, a suitable disintegrating material such as sodium lauryl sulphate or starch should be included to aid the dissolving process in the bath.

BATH OILS

Introduction

The primary function of a bath is to cleanse the body. It is essential, therefore, to point out straightaway that the function of bath oils is not one of cleansing but of skin lubrication; in addition, they are sometimes used to impart fragrance to the body.

The oil bath has emerged as the simplest and most effective method of lubrication for generalized skin dryness. Dry skin affects both young and old and, in its mildest form, appears as a slight roughening and scaling of the skin. Severe dryness can result in disturbing itching and both the incidence and severity of itching becomes progressively worse as the individual becomes older. With age, atrophic changes in the cutaneous and subcutaneous layers cause thinning of the skin and, because the sweat and sebaceous glands slow down, the skin surface becomes dry, flaky and tends to form fissures more easily. This is a result of the horny layer losing water to the environment more rapidly than it is receiving moisture from the lower epidermal and dermal layers. In the winter months, particularly in heated homes where the relative humidity falls as low as 10 per cent, skin dryness can be considerably exacerbated. Clinical efforts to overcome skin dryness are based on the concept that a surface oil or lipid film on the skin retards water loss through evaporation—hence the importance of oil baths in combating dry skin.

A number of workers have carried out investigations aiming to quantify the adsorption of different oils by skin. Taylor[5,6] made one of the first objective attempts to quantify oil deposition onto skin, using the technique of arm immersion in a bath of oil. He concluded that products based on mineral oil adhere to skin better than vegetable oil formulations. Further, an oilated oatmeal preparation (colloidal oatmeal combined with mineral oil and lanolin) was found to be adsorbed very poorly by the skin, the likely explanation for this being that oatmeal is a better adsorptive substrate for oils than is the skin. Taylor also observed that adsorption increased as the temperature of the bath was raised and as the concentration of oil increased. Soaking for longer than 20 minutes, however, did not cause a significant increase in oil adsorption. Knox and Ogiva,[7] employing a modification of the Taylor technique but using ground stratum corneum, obtained results in close agreement with those reported by Taylor. The incorporation of surfactants into bath oils may, of course, be expected to change the adsorption characteristics and this has in fact been shown by Knox and Ogiva.[8]

Bath oils can be classified into four main categories: floating or spreading bath oils which are water-immiscible; a dispersible or blooming type which turns milky on addition to water; a soluble type which forms a clear dispersion in water; and a foaming type similar to the foam bath.

Floating or Spreading Oils

Floating or spreading bath oils are hydrophobic in nature. By virtue of their lower specific gravity they float on top of the bath water, covering the bather's skin with an oily film on emergence from the water. In addition to providing the bather with a luxurious emollient coat, this type of product is ideal for augmenting the aesthetic nature of the bath by providing a pleasant fragrance to the bathroom, since the oil layer on top of the hot bath water allows the fragrance to distil readily into the atmosphere. One problem with this type of product is the occurrence of an unsightly 'ring' around the bath caused by the oil deposit. This deposit is compounded by a soap scum if the bather also uses a true soap in conjunction with hard water. Furthermore, since oils are natural foam depressants, the floating oil layer is likely to impede the lathering performance of the soap. Ideally, a floating bath oil should fully cover the surface of the water and be deposited on the skin in a very thin film, covering as much of the skin surface as possible. A lubricating bath oil should not be deposited in a heavy greasy layer, which is unattractive to the user, nor should it leave in the bath a heavy oily film that is difficult to remove.

In order to formulate a spreading oil system it is necessary to understand the physicochemical principles involved. If a drop of bath oil is placed on the surface of water in which it is insoluble it will either spread into a film or remain as a lens-like blob. Clearly, for use as a bath oil a non-spreading oil would be unsatisfactory since it would provide only a patchy deposit on the skin. Whether the drop spreads or remains intact depends on a balance of two surface forces. The first force is the work of cohesion, W_c, which is that component of surface free energy that causes any drop of liquid to take on the shape of minimal surface area. The second force is the work of adhesion, W_a, and this is the component of surface free energy that maximizes the interface between two immiscible liquids. The difference between these two forces, known as the spreading coefficient, S, determines whether the drop spreads or not. This relationship can be expressed as follows:

$$S = W_a - W_c = \gamma_w - \gamma_o - \gamma_{ow}$$

where γ_w is the surface tension of the aqueous phase, γ_o is the surface tension of the oil phase and γ_{ow} is the interfacial tension between the two phases. The derivation of this equation can be obtained from any standard textbook on physical chemistry, or by reference to a specific article on floating bath oils by Becher and Courtney.[9]

This equation predicts that spreading will occur when $S>0$ and non-spreading when $S<0$. Table 7.1 lists some typical spreading coefficients for oils commonly used in bath products, based on the surface tension value for water of 72 dyn cm^{-1} at 25 °C. From these data it is obvious that only the light mineral oil will not spread on water. Spreading, however, can be achieved by the addition of a suitable surfactant, of which the effect on the surface tension of the oil is usually small compared with the dramatic reduction in the interfacial tension between the oil and the water. In fact, with a high-performance surfactant such as polyoxyethylene polyol fatty acid ester (Arlatone T), the interfacial tension of mineral oil–water can be reduced almost to zero with the addition of 1 per cent, and this results in a spreading coefficient of approximately +40, which is optimum.

A high positive spreading coefficient is desirable from several considerations. Firstly, the area covered by the oil is proportional to the spreading coefficient. Secondly, a rapidly spreading oil is important for aesthetic reasons because it is more appealing to the user, and since the velocity of spreading is directly proportional to the spreading coefficient divided by the viscosity of the liquid on which it is spread, the merits of a high spreading coefficient are obvious. The tendency for spreading coefficients to decrease with increasing temperature is noteworthy, since in some cases a reversal from spreading to non-spreading can occur with a temperature change of about 20°C. It is important, therefore, to measure the performance at bath water temperatures (40°–50°C). This can be done relatively easily by the following simple method; a pan about 25×25 cm is filled with water at 50°C and dusted with starch; a small measured quantity of oil is then dropped into the centre; the oil will move the starch away from the centre to the edge of the pan, showing both the spread and velocity of spread quite clearly.

Although it has been shown that the HLB value (hydrophilic–lipophilic balance) of the surfactant is directly related to the spreading coefficient (that is, spreading coefficient increases in magnitude with increasing HLB value), the surfactant with the highest HLB is not necessarily the best choice. Clearly, for an acceptable product the surfactant must be soluble in the oil and in the case of high HLB surfactants this might not be the case. As already mentioned,

Table 7.1 Spreading Coefficients at 25°C

Oil	**Surface tension, γ_o** (dyn cm^{-1})	**Interfacial tension, γ_{ow}** (dyn cm^{-1})	**Spreading coefficient, S** (dyn cm^{-1})
Hexadecyl alcohol	28·6	22·6	20·8
Hexadecyl stearate	30·6	23·2	18·2
Isopropyl myristate	29·1	25·2	17·7
Light mineral oil	29·0	50·6	−7·6

Arlatone T, which has an HLB of 9·0, has been found to give both good solubility and spreadability in many systems.

Possibly the most widely used oil in such formulae is mineral oil, because of its economy, safety, availability and, of course, emolliency. It is often used, however, in association with isopropyl myristate, which helps to overcome the greasiness of mineral oil and is a much better perfume solubilizer. Many other emollients have been used and these include vegetable oils such as olive oil, cottonseed oil, peanut oil, safflower and castor oil, etc. Lanolin and lanolin derivatives, as well as fatty acids, fatty alcohols and their esters, have also been used to give better emollient effects such as better skin feel. One of the newer emollients is a fatty acid propoxylate (Arlamol E) and this is said to be particularly suitable for bath oils because of its distinctive feel on the skin and its exceptional solvency for perfumes. Indeed it is claimed to be more efficient than isopropyl myristate for carrying perfume into mineral oil.

Perfume is obviously an important ingredient and its level of incorporation is very much dependent on the cost requirements. Other ingredients sometimes included are antioxidants, colours and sunscreens.

Manufacture of these products is relatively straightforward and usually involves simple mixing. Sometimes filtration or even chilling before filtering is necessary in order to produce a perfectly clear product.

Typical formulae for floating bath oils are given in examples 10 and 11.

	(10) *per cent*
Arlamol E	49
Arlatone T	1
Light mineral oil	45
Perfume	5

	(11) *per cent*
Light mineral oil	46
Isopropyl myristate	48
Arlatone T	1
Perfume	5

Dispersible or Blooming Oils

Dispersible bath oils consist of emollient oils and perfume oils and contain a surfactant selected to emulsify the oils in the water instead of making them spread on the surface. When poured into the bath water they bloom into a milky cloud. They are sometimes preferred to the floating type because the oils are dispersed uniformly throughout the bath water, providing thorough contact with the body during the bath. When properly formulated they leave very little oily stickness or ring in the tub after the water is drained. The emollient oils used tend to be similar to the ones used in floating bath oils, with perfume levels between 5 and 10 per cent.

One of the most commonly employed surfactants in the formulation of blooming bath oils is Brij 93 (polyoxyethylene (2) oleyl ether). It has a low HLB (4·9) which is indicative of its good solubility in oil; however, it is sufficiently hydrophilic to disperse the oils in bath water. Since it is nonionic, it is effective in either hard or soft water.

A typical formula for a dispersible bath oil would be as follows.

	(12) *per cent*
Mineral oil	65
Isopropyl myristate	20
Brij 93	10
Perfume	5

Mineral oil is used as the principal emollient because of its low cost. Isopropyl myristate also adds to the emolliency and helps to dissolve the perfume oils. The amount of Brij 93 will vary with the emulsion requirements of the emollient and perfume selected. For example, the bloom can be increased by reducing the isopropyl myristate and increasing the mineral oil, or by increasing the Brij 93 content.

Soluble Oils

Soluble bath oils contain large quantities of surfactants to solubilize the high-fragance oil concentrations and to dissolve or disperse these oils readily in the bath water. They leave no residue in the bath and have no emollient effect on the skin. Soluble bath oils are either anhydrous concentrates consisting of perfume and surfactant or solubilized products consisting of perfume, surfactant and water. They contain between 5 and 20 per cent perfume oil, which can usually be solubilized quite readily by means of a hydrophilic surfactant. Tween and Brij surfactants in the HLB range of 12 to 18 are widely used in perfume oil solubilization and a typical formulation is shown in example 13.

	(13) *per cent*
Perfume	5
Tween 20	5–25
Preservative	*q.s.*
Water	to 100

The quantity of Tween 20 (polyoxyethylene sorbitan monolaurate) is clearly dependent on the type of perfume used. A higher-viscosity product can be achieved by the use of Tween 80 (polyoxyethylene sorbitan mono oleate).

Foaming Oils

Foaming bath oils can be considered either as bubble baths with high fragrance levels or as soluble bath oils with foaming agents and stabilizers added. These products provide both fragrance and foaming action and also serve to eliminate the bath-tub ring. Like the soluble bath oils they generally have no emolliency properties. The foaming agents and stabilizers discussed under foam baths are used, while Tween 20 is often used to solubilize the fragrance in the foaming bath oil. Thickeners such as carboxymethyl cellulose, methyl cellulose and other gums are commonly added, as are sometimes sequestrants. A typical formulation is shown in example 14.

		(14) *per cent*
Perfume		5
Tween 20		20
Sodium lauryl ether sulphate (28% active)		40
Coconut diethanolamide		2
Preservative		*q.s.*
Water	to	100

Very occasionally emollient oils are included in these products at significant levels but most of these have such a foam-depressing action that the final product can barely be considered to be a foam bath.

AFTER-BATH PRODUCTS

After-bath products include body or dusting powders and the various lotions for use after a bath or shower.

Body or Dusting Powders

Body or dusting powders are also known as body talcs or talcum powders and have a wide appeal because of the smooth feeling and cooling effect which they impart while they temporarily absorb moisture. The cooling effect is due to the extra heat loss from the large surface area of the talc particles.

Talc is the major ingredient in these formulations, which should have good slip characteristics, covering power, body adhesion and absorbency. The slip and texture properties are essentially based on the talc. It is essential, therefore, that grit-free, alkali-free high quality cosmetic talc is used. Talc should, of course, be free of bacteria and sterilized grades should always be used. In order to improve adhesion properties, metallic stearates such as zinc or magnesium stearate and kaolin are incorporated, while magnesium carbonate, starch, kaolin and precipitated chalk all improve absorbency. Zinc and titanium oxides at low levels along with earth colours can be incorporated for tinting purposes when required. Perfume oils are easily incorporated and should be sufficiently powerful to cover the base odour yet not interfere with other perfumes that may be used. Other ingredients sometimes included are boric acid to act as a skin buffering agent and fumed silica to give a powder of lower density.

A typical formulation is as follows:

	(15) *per cent*
Talc	75
Kaolin	10
Fumed silica	2
Magnesium carbonate	6
Zinc stearate	6
Perfume	1

After-bath Emollients

After-bath emollients or moisturizers are applied to the body with the object of replacing natural skin liquids removed during bathing. Their function, therefore, is to prevent the occurrence of dry skin. Basically four product types exist: anhydrous oil-based systems; oil-in-water emulsions; water-in-oil emulsions; and hydroalcoholic emulsions. Because of their greasy nature, anhydrous oil-based systems and water-in-oil emulsions are far less popular than the other two types and will not be considered further. Oil-in-water emulsions are perhaps most widely used since there are almost limitless opportunities in the design and formulation of these systems. Generally, they are sold in lotion form and are usually formulated to give good 'rub-in' and feel properties. The formulation possibilities are too extensive to be covered here and are discussed in Chapter 4.

The hydroalcoholic emulsions, sometimes referred to as bath satins, are worthy of mention in greater detail since they are quite a recent development. These products act in a similar way to any other emollient skin product in that they deposit a film of oil on the skin in order to reduce the rate of water loss. In addition, however, 'bath satins' give a cooling effect as the alcohol evaporates from the skin.

Basically these systems consist of an emollient oil emulsified in an aqueous alcoholic base. In order to produce a stable system, however, it is essential to include a sufficient quantity of an alcohol-compatible gum such as Carbopol or Klucel. A typical outline formulation is shown below.

	(16) *per cent*
Crodafos N-3 acid	0·5
Emollient oil	5·0
Carbopol 941	0·4
Denatured ethanol (95%)	40·0
Triethanolamine	*q.s.* to pH 6·5
Perfume	1–5
EDTA	*q.s.*
Water	to 100·0

Such a product could be made without heat by hydrating the Carbopol in the water with efficient mechanical agitation, followed by the addition of a mixture of emollient oil, perfume and Crodafos. Finally, the emulsion is completed by the addition of the alcohol in which the triethanolamine has been dissolved. EDTA or other suitable sequestrant is included to stabilize the Carbopol gel, since metal ions can depolymerize Carbopol and this can lead to loss in viscosity and hence emulsion instability.

Increased emolliency can be achieved by raising the level of emollient oil. This will produce a more opaque product and may destabilize the emulsion unless the levels of emulsifier and gelling agent are increased at the same time. Increasing the level of alcohol serves to provide greater perfume lift and quicker drying but produces a thinner, less opaque and sometimes a more unstable emulsion. Alcohol concentrations greater than about 50 per cent should be avoided since they tend to be unstable owing to coagulation of the Carbopol.

REFERENCES

1. FDC Reports, 'Pink Sheets', 19 October 1970; 7 December 1970.
2. Blakeway, J. M., Bourdon, P. and Seu, M., *Int. J. cosmet. Sci.*, 1979, **1,** 1.
3. British Patent 1 247 189, Unilever, 1971.
4. Beh, H. H. and James, K. C., *Cosmet. Toiletries*, 1977, **92,** 21.
5. Taylor, E. A., *J. invest. Dermatol.*, 1961, **37,** 69.
6. Taylor, E. A., *Arch. Dermatol.*, 1963, **87,** 369.
7. Knox, J. M. and Ogiva, R., *Br. med. J.*, 1964, **2,** 1048.
8. Knox, J. M. and Ogiva, R., *J. Soc. cosmet. Chem.*, 1969, **20,** 109.
9. Becher, P. and Courtney, D. L., *J. Soc. cosmet. Chem.*, 1966, **17,** 607.
10. Ross, S., Chen, E. S., Becher, P. and Ranauto, H. J., *J. Phys. Chem.*, 1959, **63,** 1681.

Chapter Eight

Skin Products for Babies

Introduction

During the first few years of life, the skin of the child undergoes extensive change and development—this is particularly so during the very earliest weeks after birth. It follows, therefore, that the skin of the young child differs from that of the adult and from that of the older child. It has been shown, for example, that the very young skin is very thin,[1] less cornified, less hairy and contains a relatively high proportion of water in comparison with the adult. Sebaceous glands are not only present in the newborn skin, but begin to function very early.[2] Apparently, however, transepidermal water loss at this time is lower than for adults (at least for some body areas).[3]

It is known that during the first few weeks of life the infant has very little capacity of its own to resist infection, its immunological protection being derived largely from antibodies passed on by the mother. It has often been argued that at this early stage the child is particularly susceptible to skin irritation and infection and that, being comparatively thin, the skin should be more permeable to topically applied agents. While it is certainly true that skin irritation and infection are not uncommon in very young children, the extent to which these are the result of the skin's special susceptibility due to its structure and how much the result, on the other hand, of the unique environment into which certain areas of the skin are placed, is debatable. Certainly, it has been shown that the skin surface of most babies at birth is far from sterile.[4]

Typically, the skin of babies and very young Caucasian children is pink, very soft and smooth to the touch.

Skin Problems in Babies

In spite of its histological differences from adult skin, the skin of young children is not exempt from the general rules governing skin care which apply universally. If exposed to excessive sunlight or very drying conditions, if subjected to abrasion and if allowed to accumulate grime and secretions, the very young skin reacts in the same way as adult skin and may become damaged. In addition, however, there is a hazard which applies exclusively to the very young and which becomes manifest most frequently as 'nappy' or 'diaper' rash. This condition arises because of the combination of close confining clothes and the uncontrolled urination and defaecation performed by the child at this age. As its name implies, nappy rash appears between and around the buttocks and groin, the area in which the excretions are contained by the close-fitting nappy, thus providing a damp and warm nutritive environment for the proliferation of bacteria. The metabolites of these (particularly of *Brevibacterium ammoniagenes*, which feeds on urea with the production of ammonia) combined with the abrasive nature of the nappy lead to the irritation and reddening of the skin which typifies nappy rash. If allowed to proceed to a severe form, nappy rash can result in an ulcerative condition with secondary eruptions being infected with pyrogenic organisms and causing extreme discomfort. *Candida albicans* is a frequent secondary invader and was isolated from 41 per cent of all napkins by one group of workers.[5] There is no doubt that badly laundered and inadequately rinsed napkins are a contributory cause to nappy rash if they are left with a rough, alkaline surface.

The infant skin is also susceptible to other forms of rash, notably infantile eczema (the origins of which remain somewhat controversial, but which may be related to diet) and impetigo neonatorum. These are clinical conditions, however, requiring medication which is not the province of cosmetic science.

It seems probable that most skin problems in babies stem from the tendency to wrap them up in tightly-fitting garments, thus providing a warm, stagnant environment for the growth of bacteria. Nappy rash is unknown in countries where infants are allowed to lie naked.

Functional Requirements of Baby Products

It follows from the foregoing considerations that, from the functional point of view, baby skin care products need to protect the skin from a hostile environment, to cleanse the skin thoroughly from sebum, grime and excreta and to keep the skin surface as dry as possible.

While there is no shortage of expert opinion on the best methods of cleansing baby skin, many of the views expressed are contradictory and confusing. The product types available are precisely the same as for cleansing products designed for older skins, namely soap and water, oils, emulsions and surfactant-containing gels. All these types are represented in the market although gels are relatively rare at the time of writing. In view of the association of nappy rash with bacterial growth, many formulators have been tempted to include a germicide in their products, the most frequently encountered examples being quaternary ammonium compounds such as cetyl trimethyl ammonium bromide, alkyldimethyl benzyl ammonium chloride, cetyl pyridinium chloride and benzethonium chloride. There appears to be no substantial evidence, however, that the incorporation of such active ingredients is of great benefit in the prevention of nappy rash. A more rational approach would seem to be to provide products which clean effectively and to treat skin infections of any kind with topical, pharmaceutical preparations (which are not themselves the province of the cosmetic chemist).

Most baby skins come into contact with soap and water within a few days of birth and subsequently at bath times thereafter. Although there is no evidence that ordinary soap has a bad effect on the baby, most baby soaps are white and free from perfume. An alternative not readily available in every country at the time of writing is the neutral or slightly acidic detergent bar.

Cleansing creams are not popular among baby products. This is perhaps due to the need to clean between the folds of baby fat for which a more liquid product is desirable. There is some evidence that oils and greasy materials can, by occluding the skin surface, predispose infants to prickly heat.[6,7] Thus, for cleansing purposes, lotions seem to be preferred. As with most baby products, these are only lightly perfumed or are not perfumed at all in deference to the potentially irritating effect of some perfume constituents.[8]

In view of the previously noted controversy over the use of occlusive oils on baby skin, it seems surprising that baby oils remain a popular product type. They represent a very convenient and relatively inexpensive method of cleansing the nappy area, and the residual layer which remains on the skin undoubtedly affords some protection to it from nappy contents. While baby oils composed of vegetable oils, lanolin derivatives, higher alcohols and esters have been cited, the most popular brands consist almost entirely of high purity mineral oil with, perhaps, a trace of perfume and solubilizer.

The protection which the baby skin requires has traditionally been given by zinc and castor oil creams or ointment. Zinc oxide is thought to have mildly antiseptic, astringent and anti-inflammatory properties; this accounts for its use in protective products, usually in the concentration range 2–10 per cent. Other raw materials frequently used to give a protective, occlusive barrier include petrolatum, castor oil, beeswax, lanolin, silicone oil and polyethylene wax. These may be used as anhydrous preparations or as the oil phase of protective baby creams. Inorganic salts of stearic acid and oleic acid are used to improve the water-repellent effect of creams while stabilizing the emulsion.

Baby powder is another traditional and valuable toiletry product. Its main function is to provide a dry and lubricated surface to skin that has been cleaned and protected with oils or lotions. The main constituent of baby powder is talc, but since this lacks the absorbency of other powders it is often blended with such materials as kaolin, hydrated aluminium silicate, magnesium and calcium carbonates, starches and pyrogenic silica. The grades of materials used are naturally important—particularly talc, which should be the very purest available and devoid of any fibrous materials. The adhesive power of baby powders as well as their water repellency can be improved by the incorporation of aluminium, zinc and magnesium stearates; cetyl and stearyl alcohols and zinc oxide perform a similar function.

One of the main problems associated with the use of talc is its susceptibility to contamination by micro-organisms. Various methods are available for the sterilization of talc, some being more suitable and successful than others.[9] Ethylene oxide treatment may leave irritating residues, while heat treatment is not always sufficient to sterilize the material completely.

The use of boric acid and borates in baby powders as a mild antiseptic and as a neutralizing buffer, although once popular, has now largely ceased because of the potentially toxic nature of these substances.[10,11]

Safety of Baby Products

The lack of resistance to bacterial attack in very young babies, together with the differences in histological appearance from that of the adult which baby skin exhibits, has already been discussed. Naturally, it is extremely important to ensure that all baby products are free from bacteria when sold and that they contain adequate preservative systems to prevent accidental contamination during use; such principles should apply equally to all cosmetic and toiletry products.

As the baby grows, however, there is an additional danger to which he is exposed and of which baby product formulators would do well to take note: namely, the possibility of poisoning from the ingestion of the contents of bottles and jars. The mouth is probably the most sensitive area of the body and the young child uses it to explore his environment. Moreover, liquids in bottles are associated with good things to drink. Cases of poisoning of very young children by ingestion of toiletry products are fortunately rare but, nevertheless, the major manufacturers of baby products report that they are frequently contacted by worried medical practitioners in search of reassurance about the contents of a product which has been swallowed in quantity by a lively youngster.

For this reason, if no other, it seems sensible to restrict the use of active constituents such as germicides to a minimum, since these can be potentially toxic to a baby. The other important safety aspect in protection against misuse of baby products concerns their packaging, in particular the ease with which the pack may be opened and the size of the contents available for the baby to eat or drink. While such matters are not normally in the hands of formulators to control, it is well that they should themselves be aware of the contribution to safety which such considerations may make.

The logical conclusions to be drawn from these considerations is that the raw materials used in baby products should, wherever possible, be chosen for their low toxicity as well as their non-irritating character when applied topically.

Example Formulations

The first two formulae for baby creams and lotions illustrate the use of a comparatively new type of mild, non-toxic emulsifier based on sucrose esters of palmitic and stearic acids.[12] These materials, which are known by their trade name 'Crodestas', are admixtures of mono-, di- and tri-esters giving a range of HLB values.[13] Example 1 is a lotion and example 2 a cream.

	(1) *per cent*	(2) *per cent*
Mineral oil	25·00	35·00
Cetearyl alcohol	—	0·50
Petrolatum	—	4·20
Lanolin alcohol	—	1·25
Crodesta F70	3·00	—
Crodesta F160	0·50	—
Crodesta F110	—	3·00
Hydroxyethylcellulose	0·20	—
Water	71·30	55·05
Glycerin	—	1·00
Perfume, preservative	*q.s.*	*q.s.*

More traditional baby creams and lotions are based upon the triethanolamine stearate (anionic) emulsifier system, of which there are many examples, examples 3 and 4 being fairly representative.

	(3) *per cent*	(4) *per cent*
Mineral oil	26·00	15·00
Lanolin	1·04	5·00
Stearic acid	0·94	2·00
Triethanolamine	0·52	1·00
Water	69·68	52·00
Stearyl alcohol	0·94	—
Cetyl alcohol	0·52	—
Sodium alginate	0·36	—
Isopropyl palmitate	—	2·00
Beeswax	—	8·00
Propylene glycol	—	5·00
PEG-400 stearate	—	10·00
Perfume, preservative	*q.s.*	*q.s.*

Examples 5–7 illustrate the use of some of the nonionic emulsifiers based upon sorbitol in baby creams and lotions.

	(5) *per cent*
Cetearyl alcohol	1·00
Mineral oil	4·00
Polysorbate 60	1·70
Sorbitan isostearate	1·00
Glyceryl stearate	1·00
Liquid lanolin	0·25
Water	83·35
Hydroxyethylcellulose	0·20
Glycerin	7·50
Perfume, preservative	*q.s.*

	(6) *per cent*
Mineral oil	35·50
Lanolin	1·00
Cetyl alcohol	1·00
Sorbitan oleate	2·10
Polysorbate 80	4·90
Dimethicone	5·00
Water	50·50
Perfume, preservative	*q.s.*

	(7) *per cent*
Petrolatum	20·00
Sorbitan isostearate	2·10
Microcrystalline wax	3·34
Mineral oil	10·55
Glycerin	3·14
Water	60·87
Perfume, preservative	*q.s.*

Examples 5 and 6 are of oil-in-water creams, whereas example 7 is water-in-oil. The last cream formula (example 8) illustrates the use of polyoxyethylene sorbitan lanolin derivatives, which are also thought to be fairly mild.

	(8)
	per cent
Mineral oil	15·00
Stearic acid	15·00
Beeswax	2·00
Lanolin	1·00
PEG-20 sorbitan lanolate	5·00
PEG-40 sorbitan lanolate	1·00
Sorbitol	10·00
Water	51·00
Perfume, preservative	*q.s.*

In view of its potentially irritating nature, the use of lanolin itself in baby products should be carefully considered. No doubt new information on the safety of this otherwise valuable raw material will be produced from time to time over the next few years and the cosmetic formulator will be well advised to study it.

In the transition from emulsions to baby oils, the use of anhydrous ointments (of which the zinc and castor oil cream in the *British Pharmacopoeia* is an example) should be considered. The following formulation is a little less sticky and more pleasant to use than zinc and castor oil, while still affording excellent barrier protection against excreta.[13]

	(9)
	per cent
Mineral oil	83·50
Acetylated lanolin alcohol	1·50
Silica	5·00
Zinc oxide	10·00

Procedure: Disperse the silica into the hot mineral oil/acetylated lanolin alcohol. Add the zinc oxide last and subject the whole to shear in a mill or by using a high speed rotor/stator device.

Baby oils are composed predominantly of a very pure grade of mineral oil. Small amounts of fatty acid esters, vegetable oils, lanolin derivatives and other compatible materials should be included only after careful consideration of safety and irritation potential.

Baby powders function, as has been noted, to lubricate, dry and perhaps to impart a slight perfume to the skin. The following formulae illustrate how the absorbency and adhesiveness of talc can be improved by the blending-in of other materials.

	(10) *per cent*	(11) *per cent*	(12) *per cent*	(13) *per cent*
Sterilized talc	80·00	74·00	95·00	90·50
Magnesium stearate	10·00	4·00	—	2·50
Calcium carbonate	10·00	—	—	—
Kaolin	—	20·00	—	5·00
Glyceryl stearate	—	1·00	—	—
Cetyl alcohol	—	1·00	—	—
Starch	—	—	5·00	—
Zinc oxide	—	—	—	2·00
Perfume	*q.s.*	*q.s.*	*q.s.*	*q.s.*

Cleansing of Nappies (Diapers)

Although nappies are not, strictly speaking, skin products for babies, or even cosmetics in the broad definition of the word, the cleansing of nappies plays such an important part in the care of baby skin that a short note on the subject here is appropriate.

The obvious requirement is for cleanliness, softness and sterility without the deposition of any potentially irritating residues. Most popular at present are the 'nappy soak' products which are dissolved or diluted in water to provide a cleansing/sterilizing medium in which the dirty nappies are merely allowed to soak for a few hours. This has obvious advantages for a busy mother. Formulae are based upon the bleaching and sterilizing properties of chlorine and hydrogen peroxide (the former being somewhat quicker-acting but potentially more irritating if traces are left behind). Liquid products may utilize hypochlorite but the more common powder formulae make use of chlorine or peroxide release agents. The following illustrative example employs sodium perborate in a powdered detergent base:

	(14) *per cent*
Sodium tripolyphosphate	30·00
Sodium carbonate	10·00
Sodium dodecylbenzene sulphonate (80%)	6·00
Sodium perborate tetrahydrate	20·00
Sodium sulphate	33·90
Optical brightener	0·10
Perfume	*q.s.*

After soaking, the nappies may be softened, if necessary, with a cationic clothes softener and—most important—thoroughly rinsed.

REFERENCES

1. Stuart, H. C. and Sobel, E. H., *J. Pediatr.*, 1946, **28,** 637.
2. Ramasastry, P., Downing, D. T., Pochi, P. E. and Strauss, J. S., *J. invest. Dermatol.*, 1970, **50,** 139.
3. Wildnauer, R. H. and Kennedy, R., *J. invest. Dermatol.*, 1970, **54,** 483.

4. Potter, R. T. and Abel, A. R., *Am. J. Obstet. Gynecol.*, 1936, **31,** 1003.
5. Dixon, P. N., Warin, R. P., English, M. P. and Grenfell, L., *Brit. med. J.* 1969, **2,** 23.
6. Perlstein, M. A., *Am. J. Dis. Child.*, 1948, **75,** 385.
7. Wrong, N. M., *Pediatrics*, 1952, **10,** 710.
8. Schimmel and Co., *Schimmel Briefs* No. 139.
9. Ferreira, J. M. and Freitas, Y. M., *Cosmet. Toiletries*, 1976, **91,** 48.
10. George, A. J., *Fd. Cosmet. Toxicol.*, 1965, **3,** 99.
11. Skipworth, G. B., Goldstein, N. and McBride, W. P., *Arch. Dermatol.* 1967, **95,** 83.
12. Chalmers, L., *Soap Perfum. Cosmet.*, 1977, **50,** 191.
13. *Croda Cosmetic/Pharmaceutical Formulary*, 1979.

Chapter Nine

Skin Products for Young People

Introduction

In the majority of the population, skin first comes into contact with cosmetic or toiletry products during the first few weeks of life. At this stage, there are few benefits which such products can provide other than those of cleaning and protecting the skin from its watery environment. During the early years, the skin of most children is healthy, soft, free from spots and blemishes (provided they enjoy a sensible diet) and requires little care other than regular cleansing. At the onset of puberty, however, the skin becomes susceptible to a range of problems, most of which can be attributed to over-activity of the oil-producing sebaceous glands. More than one survey among groups of adolescents in the age-range 12–18 years has indicated that the incidence of excessive spots, pimples, blackheads, acne and related conditions is greater than 50 per cent of the population.[1] Not surprisingly, therefore, the majority of skin products purchased by young people in this age group are concerned with the treatment, prevention or camouflage of blemishes or of oily skin. Additionally, there are other skin products bought and used by teenagers but which have the same composition as those sold on the general market, with perhaps a difference in presentation. These are dealt with in the appropriate chapters elsewhere in this book.

Adolescent Skin Problems

The sebaceous glands together with the muscular, nervous and vascular systems which are associated with them are collectively known as the pilosebaceous apparatus. It has long been realized that the pilosebaceous apparatus is largely under the control of endogenous hormones which are present in unusually high concentration in the blood during adolescence and puberty. The corresponding increase in activity of the sebaceous glands themselves gives rise to the production of excessive amounts of sebum. In itself, this causes an unpleasant oiliness of the skin, giving it a patchy, shiny appearance and (in girls) making it difficult or impossible to apply make-up to the affected areas. Unfortunately the condition is made worse by a simultaneous increase in the rate of keratinization of the skin's horny layer (the stratum corneum). In some young people, strips of dead keratinized cells can be removed from the face by simply rubbing with the fore-finger and this, naturally, adds to the problems of applying anything to the skin surface. Even more significantly, however, the horny cells lining the sebaceous follicles also proliferate; they become tightly-packed and can form an occlusive plug or comedone. This physical barrier, coupled with the increased production of sebum, leads to a rapid accumulation of back pressure and the stagnant sebum forms an ideal medium for the proliferation of bacteria (mainly *Staphylococcus aureus, Staphylococcus albus* and *Corynebacterium acnes*). When the plugged follicle eventually ruptures and allows the discharge of its contents, these will include the breakdown products of bacterial metabolism (including such irritants as fatty acids), causing local swelling and inflammation. Even then there remains the possibility that the exposed follicle cells will darken from the deposition of pigment from damaged cells in the deeper layers, giving rise to 'blackheads'. (Blackheads, it should be noted, are *not* due to the accumulation of dirt or debris.)

Products for Oily Skins

There is no topically active substance yet known capable of inhibiting local sebum production. The main treatment for oily skins therefore consists of careful and regular cleansing to prevent the accumulation of oil on the surface, the use of absorbent materials to soak up excessive oil and the application of products (particularly liquid foundations) capable of drying to a matt, non-shiny finish.

Frequent and adequate cleansing of the skin—particularly the face, neck, chest and back—is of paramount importance in the control of oily skin and the complications which often arise from it. Emulsion cleansers would not seem to be the best choice since they must, perforce, lay down additional oil on the skin surface. Oil-in-water emulsions with a low oil content are, nevertheless, marketed for the adolescent age group and, provided that the last traces of emulsion are removed by the subsequent use of astringent tonic or toner, such products may be satisfactorily used. A more traditional soap-and-water regime is very effective in the removal of surface oil and, although many experts believe that the prolonged use of anything as alkaline as soap can cause damage to the skin, there is no cheaper method of cleansing. As an alternative to soap, several varieties of detergent bar (of slightly acid pH) are now becoming available in soap-bar form.[2] Perhaps the most logical approach to the cleansing of oily skin is the simple aqueous solution of surfactant (which, for variety, may be gelled with a conventional organic gelling agent). Such cleansing products, containing no oils or harsh, alkaline materials, provide efficient cleansing power without exacerbating the oily condition of the skin. Additionally, they may be enhanced by the inclusion of ethanol as an astringent and a germicide to aid the control of acne-producing bacteria on the skin (this latter principle can also be applied to many of the other adolescent cleansing products already discussed). Two examples of the germicidal cleansing gel type of formulation may suffice to give the general principle:

	(1) *per cent*
Triclosan	3·00
Menthol	10·00
DEA-oleth-3 phosphate	2·50
Hydroxypropylcellulose	2·50
Amphoteric-1	5·00
Water	37·00
Ethanol (96%)	40·00

	(2)
	per cent
Phenoxyisopropanol	2·00
Sodium laureth sulphate	5·20
Propylene glycol	8·00
Quaternium-15	0·20
Hydroxyethylcellulose	1·00
Water	83·60
Perfume, colour	*q.s.*

Ranges of germicides, surfactants and gelling agents are available for substitution into the above formulae at the discretion of the formulator.

A second approach to the problem of oily skin consists in the development of products leaving a matt layer on the skin surface in order to combat shine or to absorb the excess oil. The most common approach is the incorporation into the product of pyrogenic silica which has the dual properties of oil absorbency and matt appearance.[3] The silica may be incorporated into the oil phase of an emulsion or, alternatively, applied as a gel or solution from aqueous or aqueous–alcoholic suspension. Other non-irritating powders may be used for a similar purpose, including polyethylene, talc and bentonite.[4]

Specific Treatments for Acne

In spite of much research, modern treatments for acne consist of containment until the condition clears up of its own accord. Successful treatment, limited though it may sometimes be, tends to be time-consuming and repetitive.[5] However, two lines of approach are simultaneously available. The first of these involves the use of 'peeling agents' for the rapid and effective removal of keratinized squamous cells of the horny layer which, if allowed to accumulate at the skin surface, make the formation of a comedone much more likely. Secondly, very thorough cleansing of the affected parts of the skin—particularly with some of the germicidal cleansing products already referred to—will help to keep the proliferation of the acne bacilli under control.

Among the 'peeling agents' commonly used are resorcinol, sulphur and benzoyl peroxide and these (particularly the latter) have been shown to be valuable in this limited role. Salicylic acid is sometimes incorporated, probably in an attempt to reduce the irritant effect of benzoyl peroxide. Guanidine and its compounds have also been used for the same purpose[6] although the extent of the irritation caused by the peroxide appears to depend on formulation variables, particle size and on the quality of the raw material itself.[7] A selection of published 'peeling' creams and lotions is given below. The formulator should avoid the use of organic amines and inorganic hydroxides since, like most organic peroxides, benzoyl peroxide decomposes in alkaline solution to give hydrogen peroxide.

Medicated vanishing cream	(3)
	per cent
Laneth-10	2·00
Lanolin alcohol	0·50

Medicated vanishing cream (cont.)

	per cent
Cetyl alcohol	5·50
Polawax	6·00
Myristyl myristate	2·00
Benzoyl peroxide	2·00
Resorcinol mono-acetate	0·20
Magnesium aluminium silicate	4·00
Methylparaben	0·20
Sulphur	1·40
Perfume	*q.s.*
Water	76·20

Procedure:
Dissolve the benzoyl peroxide in the propylene glycol and then add the rest of the oil phase ingredients. Add the magnesium aluminium silicate to the water at 75°C and disperse under shear, add the sulphur and methylparaben and shear again to disperse. Combine the phases and emulsify at 70°C, adding the perfume at 50°C.

Peeling lotion	(4)
	per cent
Resorcinol	3·50
Salicylic acid	2·00
Alcohol	17·00
Rose water	77·50

Acne cream		(5)
		per cent
A	Cetearyl alcohol	1·50
	Ceteareth-20	1·00
	Diisopropyl adipate	1·50
	Water	73·20
B	Cellulose	2·80
C	Benzoyl peroxide	5·00
	PEG-4000	5·00
	Water	10·00

Procedure:
Grind phase C together in a colloid mill. Add B to A and mix at 75°C. Add C at 40°C.

Some modern vehicles for benzoyl peroxide involve the use of gels.[8] The precise mode of action of benzoyl peroxide in aiding the removal of inflamed and superficial layers of skin does not seem to be known. Other oxygen-potentiating compositions have also been patented, however, notably one based upon N-acetyl-*dl*-methionine complexed with quaternary ammonium salts.[9] It is known that the primary acne-producing bacteria are anaerobic and cannot proliferate in the presence of oxygen.

Much interest is now being shown in vitamin A acid (retinoic acid or Tretinoin) in the control of acne. Vitamin A has long been known to influence the cells of the horny layer. Retinoic acid appears to stimulate epithelial growth so that a less adherent horny layer is formed. It may be applied directly to affected areas as a 0·025 per cent alcoholic solution or gel. Such formulations may be improved by the addition of antibiotics.[10] Antibiotics alone are not very effective for topical application, but there seems to be a complementary action between certain of them and retinoic acid, the latter helping to reduce the early aggregation of horny cells and the former inhibiting secondary infection. Tetracycline is an example of an antibiotic that can function in this way. It is also known that zinc is involved in the metabolism of vitamin A and some people with particularly severe acne improve as the result of being fed zinc sulphate. Other patents mention the use of retinoic acid in combination with lactic acid esters[11] and of desmethyl vinyl derivatives of retinoic acid.[12]

Further patents describing topical treatments for acne will no doubt continue to appear, although the nature of some of these will undoubtedly confer the status of pharmaceutical preparations on any products containing them. Much research is now needed into the problem of prevention rather than cure. There is evidence that some cosmetic ingredients can actually worsen or potentiate acne, thus giving rise to the term 'comedogenic'. Some clinical work seems to indicate that these raw materials tend to have a comedogenic effect on susceptible skins no matter what type of formulation they appear in.[13–16] Such research is still regarded as somewhat controversial but will undoubtedly continue and may prove a valuable aid to the cosmetic scientist interested in formulating products for the adolescent skin.

REFERENCES

1. Munro-Ashman, D., *Trans. Rep. St John's Hosp. Dermatol. Soc.*, 1963, **38,** 144.
2. Delmotte, A. *Arch. Belg. Dermatol. Syphiligr.*, 1960, **1,** 118.
3. US Patent 4 600 317, Colgate-Palmolive, 28 December 1976.
4. US Patent 4 164 563, Minnesota Mining and Manufacturing Co., 14 August 1979.
5. Parish, L. C. and Witkowski, J. A., *Acne, Update for the Practitioner*, ed. Frank, S. B., New York, Yorke Medical, 1979, pp. 7–12.
6. US Patent 4 163 800, Procter and Gamble, 7 August 1979.
7. Lorenzetti, O. J., Wernett, T. and McDonald, T., *J. Soc. cosmet. Chem.*, 1977, **28,** 533.
8. Anderson, A. S., Goldye, G. J., Green, R. C., Hohisel, D. W. and Brown, E. P., *Cutis*, 1975, **16,** 307.
9. US Patent 4 176 197, Dominion Pharmaceutical B Inc., 1979.
10. Kligman, A. M., Mills, O. H., McGinley, K. J. and Leyden, J. J., *Acta Derm. Venereol.*, 1975, **74** (Supplement), 111.
11. French Patent 1 551 637, Hoa, J. H. B., August 1979.
12. US Patent 3 882 244, University of California, May 1975.
13. Kligman, A. and Mills, O., *Arch. Dermatol.*, 1972, **106,** 843.
14. Fulton, J., *Cutis*, 1976, **17,** 344.
15. Frank, S. B., *Cutis*, 1974, **13,** 785.
16. Kligman, A., *J. Assoc. military Dermatol.*, 1976, **1,** 63.

Chapter Ten

Antiperspirants and Deodorants

Introduction

If a vote were taken to select the one cosmetic product that best illustrates the versatility of packaging, the deodorant/antiperspirant would likely be the unanimous winner,[1] for there is probably no other product that is sold in at least eight different kinds of package. Each was developed to meet a specific marketing and convenience need and each has inherent advantages and disadvantages.

Deodorants/antiperspirants are commonly packaged in:

Stick—solid	Pump sprays
Pads	Squeeze bottles
Dabber units	Creams
Aerosols	Stick—creams
Roll-ons	

With the confusion about the use and purpose of antiperspirant and deodorant products it is helpful to distinguish between the purposes they are intended to serve.

Antiperspirants are designed primarily to reduce (axillary) wetness. In the USA they are classified legally as drugs because their mode of action affects a body function, namely, eccrine sweating. Deodorants (except soaps) are designed to reduce axillary odour. Since this is considered a non-therapeutic purpose and a function of the body is not considered to be altered, they are classed as cosmetics.

Despite the avalanche of topical antiperspirants which has descended upon the consumer, and despite the implications of the advertising claims, there is not a single topical agent available today that eliminates axillary sweating in the hidrotic individual.[2]

Perspiration and its Control

The odour in the human skin is produced from the secretions of sebaceous and sweat glands.[3] Sebaceous glands are found with every hair, on the red surface of lips, in the nostrils, in the papillae, on the anus, and on the foreskin and labia minora.

The sebum secreted by these glands is made mainly of cholesterol and its esters, palmitic and stearic acid and their esters, and various other substances whose nature is not fully understood. Sebum is generally oily and may solidify on the skin surface. Pure sebum is not a critical factor in skin odour. The known constituents are odourless:[4] generally they are substances whose molecular weights are higher than those of odorous compounds.

To assess properly and to understand the action of deodorants and antiperspirants, a review of the physiology of sweating is essential.

Perspiration assists in the regulation of body temperature by dispelling heat through evaporation of moisture from the surface of the skin. It also functions in other capacities such as by eliminating lactic acid which is formed during muscular exercise and by protecting the skin from dryness.

It has been estimated that there are about 2 380 000 sweat glands distributed over the body surface. These sweat glands are of two types: the eccrine glands and the apocrine glands. The eccrine glands or small coil glands are the true sweat glands that occur over almost all of the body surface. They originate in the deeper layers of the dermis or in the subdermis and open via a thin duct directly on to the skin. The apocrine glands or large coil glands are those glands which are associated with sexual development, being post-pubertal in occurrence. They occur in relatively small numbers and are found in such areas as axillae, around the nipples, on the abdomen and in the pubic region.

Although the axilla is virtually an apocrine organ, the profuse flow of sweat we term hyperhidrosis is the result of intense activity of the eccrine rather than the apocrine sweat glands in this area. Numbering about 25 000 in each axillary vault, these eccrine glands can secrete large quantities of sweat. In hyperhidrotic individuals, each armpit may produce upwards of 12 grammes per hour. It is this heavy local outpouring which is so injurious to the affected individual's composure and clothes.

Laboratory studies indicate that both eccrine and apocrine sweat are sterile and odourless at the time of discharge.[5] The odour is produced later through the action of bacteria on primarily the apocrine sweat, which is rich in organic material and is an ideal substrate for bacterial growth. The far more abundant eccrine sweat is a highly dilute aqueous solution and has been shown to be much less important as a source of axillary odour.[6] However, the moisture from eccrine glands probably promotes odour production indirectly in two important ways: (i) the small amount of sticky, oily material from the axillary apocrine glands is dispersed over a wider surface; (ii) the moisture in the warm axillary vault completes an ideal environment for the rapid growth and proliferation of the resident bacteria feeding on this organic material. Axillary hair also has been found to promote the development of odour. It is thought that axillary hair acts as a collecting site for apocrine sweat and increases the surface area available for bacterial proliferation.

Decomposition of the sudoriferous and sebaceous gland secretions by the skin microflora and likewise decomposition of proteins on the surface of the skin give rise to numerous odorous substances often of strong smell. This is the mixture which produces the natural odour of human skin.[7] In it are the lower fatty acids (C_4–C_{10}) and macrocyclic systems, steroids, lactones, etc. Although these have no smell of their own, they serve to fix the odour potential. The basic skin odour of *Homo sapiens* is also dependent on the individual or group, from the combined action of food last eaten and physical and psychological conditions.

The actual odour of the human being is the sum of the natural and acquired odour: two women may smell differently although they are identically dressed, washed and perfumed. Human beings find it very difficult to recognize this difference, but a dog has no difficulty in such detection. Body odour is thus a completely individual property of a human being just like fingerprints or the characteristic sound of the voice. The intensity of body odour differs from person to person, depending upon personal circumstances, environment, social and psychological conditions.

From these findings these are several obvious ways to reduce or control axillary odour: (a) reduce apocrine sweating in the axillae; (b) remove the secretions from both types of sweat glands as quickly as practicable; (c) impede bacterial growth; (d) absorb body odours.

Many workers have taken it for granted that human emanations contain pheromones,[8] which have sexually pleasing, winsome effects and reflect the psychological state of the individuals. Good human odour can, therefore, be of great importance.

Antiperspirants act by limiting the magnitude of sweat gland secretion delivered to the skin surface. Consequently, the mechanism of action may involve a decrease in sweat production at the glandular level, formation of a blockage or plug in the sweat duct, alteration of the sweat duct permeability to fluids (as in a perforated water hose), or any one of several other theories involving concepts such as electrophysiological potential along the sweat duct. The many theories presented to explain the action of antiperspirants can be found in several papers and review articles.[9–11]

Despite the preponderance of mechanistic theories, the detailed mechanism of axillary anhidrosis is relatively unknown.

Papa and Kligman[12] have produced histological evidence with human subjects that aluminium chloride alters the physiological state of sweat ducts. Methylene blue iontophoretic sweat pore patterns suggested increased permeability of the sweat duct to water while adhesive tape stripping of the stratum corneum did not abolish the anhidrotic state produced. Both sets of data suggest that poral closure or obstruction (plug formation) does not occur when anhidrosis is produced by aluminium chloride. Furthermore Lansdown[13] has produced evidence that high concentrations of aluminium chloride result in epidermal damage in mammalian skin and, in addition, decompose phospholipids. On the other hand, Papa and Kligman have also reported that known protein precipitants, such as formaldehyde, produce superficial obstructions in the eccrine duct. Anhidrosis produced by several known protein precipitants was abolished by stripping of the stratum corneum.[14]

Partial or complete anhidrosis produced by anticholinergic drugs does not proceed via a mechanism involving such anatomical and histological factors. Several authors have reported that various drugs exhibiting anticholinergic activity suppress the secretion of sweat by direct action on the secretory process of the sweat gland.[9,15] Such drugs appear to inhibit the action of acetylcholine in stimulating the product of perspiration.

The vehicle from which anticholinergic drugs are delivered plays a significant role in influencing efficacy, as these compounds must penetrate the stratum corneum and epidermis in order to reach the active site.[15,16]

The modern story of topical antiperspirants for the axilla began with Stillians's observation in 1916 that a 25 per cent solution of aluminium chloride hexahydrate in distilled water, dabbed gently on the armpit every second or third day, will reduce excessive sweating.[17]

To date, the most detailed comprehensive review of the subject of antiperspirants is that of Fiedler[18] which contains 411 references. It is interesting to note today that Stillians's formulation of 1916 remains one of the most effective antiperspirants in use. It is not toxic and it is not allergenic. Nonetheless, it enjoys only a limited sale today because (a) it is irritating to the skin of some users and (b) its high acidity is damaging to clothing. One of the first major developmental changes occurred in the 1940s when it was found that a less acidic complex salt of aluminium—aluminium chlorhydroxide—could be substituted for aluminium chloride. This reduced irritation to the skin and markedly lessened the damage to clothing. Unfortunately it also reduced the antiperspirant effect.

Antiperspirant Ingredients

Several metal salts have astringent properties including those of aluminium, zirconium, zinc, iron, chromium, lead, mercury and several rarer metals.[19] Various attempts have been made to find the most effective antiperspirants from the salts of these metals. Obviously many had to be discarded straight away on grounds of toxicity, and the field has been narrowed to mainly aluminium and zirconium.

Zirconium

In 1955 sodium zirconyl lactate was used in deodorant sticks containing an alcoholic soap gel. In 1956 cases of granulomatous eruptions in the axillae of users of these products had been reported.[20]

Despite this, a whole series of patents was published over the period 1955 to 1961[21–24] covering the antiperspirant use of zirconium salts, sometimes in combination with aluminium compounds and/or buffers such as urea or glycine.

The next major effort started in 1968 with a Bristol-Myers patent[25] for a zirconium-based complex which was claimed to be both effective as an antiperspirant and also non-irritant. Several other patents followed and a number of aerosol antiperspirants containing zirconium complexes were introduced in USA from 1972 to 1975. These were claimed to be much more efficient antiperspirants than the aluminium compounds being used in commercial antiperspirants.

In 1973 Gillette withdrew their aerosol zirconium-based products because of 'mild inflammatory reactions in monkeys'.

In 1975 the US Food and Drug Administration was reported to be looking at a possible ban for zirconium-based aerosol antiperspirants, following suggestions of long-term hazards with such products. In 1977 the official FDA ban came, by which time no products were left on sale.[26] The FDA stated that there was no reason to ban zirconium-containing antiperspirants directly applied to the skin and aluminium–zirconium complexes are in Category I (safe and effective) for non-aerosol application at concentrations 20 per cent or less (on an anhydrous basis).

In 1978 three of the top four US roll-ons contained zirconium salts.[27]

Aluminium

Observed differences in the antiperspirant behaviour of aluminium chlorhydroxide and aluminium chloride have been attributed to differences in their interaction with skin.[28] The literature contains many references to methods for measuring the interactions of exogenous materials with skin. The electrical properties of skin have been used successfully as a means by which to describe this effect and it was thought appropriate to investigate this approach with respect to aluminium salts. Instrumentation and techniques for measuring the electrical impedance of excised epidermal membrane were developed. The effects of two aluminium salt antiperspirants on the impedance of guinea pig stratum corneum were measured. Aluminium chlorhydroxide reduced the impedance five times more than aluminium chloride. The results are in agreement with reported skin sorption behaviour for these salts and with their antiperspirant activities *in vivo*. The hypothesis that antiperspirancy is based, in part, on antiperspirant–skin interaction is supported by this study.

On 10 October 1978 the US Food and Drug Administration published the recommendations of the Advisory Review Panel on Over-the-Counter (OTC) Antiperspirant Drug Products as a proposed rule and expressed serious concern about the possible consequences of the inadvertent long-term inhalation of aerosolized antiperspirants. Aluminium-containing aerosol preparations have been placed in Category III by the OTC Panel (that is, insufficient available data to permit final classification at this time).[29]

While the nature of aluminium in many respects differs markedly from that of zirconium, particularly with its lack of potential antigenicity, its implication in possible granuloma formation under various conditions does not appear to be as clearly distinguishable.

Active Ingredients

The OTC Panel has developed a comprehensive and rigorous set of guidelines which is intended to serve as the standard protocol to be employed in chronic animal inhalation studies, designed to bring successfully-tested products into Category I classification (safe and effective).

Based upon the apparent awareness of the commercial availability of controlled-particle-size bulk aluminium antiperspirant powders intended for 'powder-in-oil' aerosol suspension use, the OTC Panel[30] has recommended that all marketed suspension-type aerosol systems in the USA should be formulated so that not less than 90 per cent of emitted particles are greater than 10 μm in diameter. Since the nose is considered to be the primary filter, there is virtually complete retention of particles in excess of 10 μm. Almost 50 per cent of 5 μm particles are retained, while almost all 1–2 μm particles penetrate beyond the nose. In general, particles below 5 μm are respirable and will penetrate into the lung.

Rubino *et al.*[31] describe the development of a controlled-particle aluminium chlorhydrate in which a minimum of 95 per cent of weight of the particles possess diameters of 10 μm or larger.

Table 10.1 sets out the OTC Panel's categories of active ingredients.

Table 10.1 Categories of Active Ingredients—US FDA OTC Antiperspirant Review Panel
I = Permitted. II = Prohibited. III = Temporarily permitted.

Ingredient	Non-spray	Spray
Aluminium bromohydrate*	II (S,E)†	II (S,E)
Aluminium chlorhydrates	I	III (S)
Aluminium chloride (15% or less aqueous solutions)	I	III (S)
Aluminium chloride (alcoholic solutions)	II (S)	II (S)
Aluminium sulphate	III (S,E)	III (S,E)
Aluminium zirconium chlorhydrates	I	II (S)
Buffered aluminium sulphate	I	III (S)
Potassium aluminium sulphate	III (S,E)	III (S,E)
Sodium aluminium chlorhydroxy lactate	III (E)	III (S,E)

* This ingredient has never been marketed in the USA for a material extent or material time and, therefore, cannot receive general recognition of safety and effectiveness.
† (S) refers to safety considerations; (E) refers to effectiveness considerations.

Category I Ingredients

1. Aluminium chlorhydrates at concentrations of 25 per cent or less, calculated on an anhydrous basis, in topical (non-aerosol) formulations for underarm use only.
2. Aluminium zirconium chlorhydrates at concentrations of 20 per cent or less, calculated on an anhydrous basis, in topical (non-aerosol) formulations for underarm use only.
3. Aluminium chloride in 15 per cent concentration or less, calculated on the basis of the hexahydrate form, in aqueous solution and for topical (non-aerosol) formulations for underarm use only. Alcoholic solutions are Category II because of excessive irritation noted in submitted data.
4. Buffered aluminium sulphate as an 8 per cent concentration of aluminium sulphate plus 8 per cent sodium aluminium lactate for topical (non-aerosol) formulations for underarm use only.

Category III Ingredients

1. All Category I ingredients described above except for zirconium salts, when used in aerosol formulations. The reason for this is the lack of sufficient evidence for long-term inhalation safety data.
2. Sodium aluminium chlorhydroxy lactate. The Panel concluded that this ingredient is safe, but lacked sufficient evidence of efficacy to permit final classification at this time. Evidence of efficacy is required as for all other antiperspirants.
3. Aluminium sulphate. The Panel found that this ingredient lacked sufficient evidence for safety unless its acidity is first reduced with sodium aluminium lactate. Also, its efficacy data was deemed insufficient due to lack of human test data.
4. Potassium aluminium sulphate (potassium alum). This ingredient was placed in Category III due to insufficient evidence for both safety and efficacy.

Antiperspirant compositions employing a starch-coated aluminium derivative as the active agent from the subject of two patents issued to L'Oreal.[32,33] The active agent described in one of the patents comprises microcrystals of a derivative of aluminium coated with degraded starch which gels in water at a temperature lower than 100°C so as to provide an atomizable gel with a starch concentration ranging between 5 to 30 per cent by weight. The starch coating is to prevent the aluminium compound reacting with perfumes and to reduce irritation to users with sensitive skins.

It is claimed that all these disadvantages can be avoided by using particles of a hygroscopic aluminium compound which are coated with a polymer that will dissolve in water at human body temperature sufficiently rapidly to permit quick release on contact with perspiration.

The second L'Oreal patent also employs a starch-coated aluminium compound but this time it is aimed at producing a delayed antiperspirant activity. It is claimed that conventional micronized antiperspirant derivatives of aluminium dissolve immediately on contact with perspiration, resulting in antiperspirant activity of only a short duration; by the use of a coated antiperspirant agent, the active material is progressively released during contact with perspiration and hence is active over a longer period of time.

Unilever published a patent in 1977[34] describing the use of moisture-absorbent organic polymers for absorbing superficial skin moisture. They can be applied in the form of an aerosol spray and include certain polysaccharides, polypeptides, vinyl carboxy polymers and copolymers. The preferred polymers are characterized by their ability to absorb an amount of moisture which is greater than their own weight and up to ten times their own weight after deposition of the composition on to the skin.

Evaluation of Antiperspirants

Efficacy of Antiperspirants

Since a product with a sweat reduction of 20 per cent promises only a barely perceptible antiperspirant effect, antiperspirants that achieve less than 20 per cent effectiveness in hotroom tests are probably worthless in terms of consumer benefit. The OTC Antiperspirant Review Panel[35] in the USA has proposed a statistical criterion that provides a reasonable assurance that only antiperspirant products that are likely to give 20 per cent sweat reduction in at least half of the subjects will be marketed.

The range of effectiveness (average percentage sweat reduction) in laboratory hotroom tests of OTC antiperspirants submitted to the Antiperspirant Review Panel is given in Table 10.2. One general conclusion that appears valid is that antiperspirants in aerosol form are generally not as effective as the other dosage forms.

Table 10.2 Range of Average Sweat Reduction—US FDA OTC Antiperspirant Review Panel

Dosage form	Average reduction (%)
Aerosols	20–33
Creams	35–47
Roll-ons	14–70
Lotions	28–62
Liquids	15–54
Sticks	35–40

Many factors influence antiperspirant activity.[36] A minor variation in formula composition is one of the most critical and is one which is occasionally not recognized. A formula additive may seriously inhibit antiperspirant activity or, in certain circumstances, may definitely enhance activity. Additives that reduce formula irritancy without adversely affecting antiperspirant activity are in this latter class.

There is marked variation of response between subjects. For example, unformulated aluminium chloride reduces sweating of some subjects by 40–50 per cent but increases that of others by a similar amount. Subjects showing marked increases in sweating (properspirant activity) will usually show visible axillary irritation. Majors and Wild[36] have also observed, however, many instances of samples that were effective on most subjects but exhibited no antiperspirant effect—or even showed properspirant activity—on some subjects, with no visual evidence of axillary irritation. This would indicate that there is some factor other than inactivation of antiperspirant activity by formula components which results in certain subjects' specificity of decreased individual efficacy.

The efficacy of an antiperspirant is best defined as the percentage reduction in the rate of sweating in the axilla that may be achieved after a realistic application or series of applications of the test product. The preferred methods for the determination of efficacy are gravimetry or the use of electronic hygrometers.

Gravimetric Method.[36] Panellists are required to abstain from the use of all antiperspirant materials for at least one week prior to initiation of the study. Sweat collections are carried out in controlled temperature rooms at 100 ± 2°F and about 35 per cent relative humidity. Sweat collections are made during two successive 20-minute periods using tared absorbent pads. These collections are preceded by a 40-minute conditioning period in the hot room during which the panellists hold unweighed pads in their axillae. A ratio of sweat produced by the left and right axillae is determined in the series of controlled collections. The effect of antiperspirant materials on the perspiration rate of each individual is determined by comparing the post-treatment ratio with the subject's average control ratio. For each individual the percentage change is calculated as

$$\text{reduction (\%) in sweat rate} = \frac{\text{post-treatment ratio}}{\text{average control ratio}} \times 100$$

Hygrometry. The most accurate methods available are those using electronic hygrometers. A cup is attached to the skin and the water from the enclosed area is evaporated by a constant stream of dry gas. The water content of this gas stream is monitored and the sweat rate is calculated.

Because the cell used to cover the skin is not very large and only encloses a small area in the axilla, the positioning of this probe is critical. Unless the cells are replaced in exactly the same position for each experimental session the difference in sweating of the different sites can be larger than any changes induced by the use of the antiperspirant products.

The forearm is probably a more suitable application site because (a) the even distribution of glands means that the positioning of the cell on the skin is not so critical and (b) the reflexes affecting sweating unilaterally are not so pronounced in the forearms or can be avoided completely by using two sites on the same forearm.

Mechanism of Deodorants and Deodorant Ingredients

Since axillary odour is largely produced by the action of bacteria on nutrients present in apocrine secretion, any compound which inhibits the growth of those micro-organisms found in the axillae will, in theory, exhibit deodorant properties. Antibacterial agents reported in the literature or employed by cosmetic and toiletry manufacturers include quaternary ammonium compounds such as benzethonium chloride (di-isobutyl phenoxyethoxy-ethyl dimethyl benzyl ammonium chloride monohydrate), cationic compounds such as chlorhexidine acetate (1,6-di-(N-*p*-chlorophenyldiguanido)hexane acetate and triclosan (2,4,4′-trichloro-2′-hydroxydiphenyl ether).

Before World War II cresols were the most popular bacteriostats. However, their objectionable odour limited their application. In 1941 researchers at Givaudan Corporation discovered that a halogenated bisphenol, hexachlorophene, exhibited bacteriostatic qualities when incorporated in soap. In mid-1971 the FDA issued a report that brain lesions in test animals could be produced by feeding high dosages of hexachlorophene (HCP). The FDA took its final position on hexachlorophene on 22 September 1972. This ruling completely banned the use of hexachlorophene in all non-prescription products. The major impact of this ruling was felt by Armour-Dial, the manufacturer of 'Dial' soap. Armour-Dial announced immediately that within one week it would change the 'Dial' soap formulation from 0·75 per cent HCP plus 0·75 per cent trichlorocarbanilide (TCC) to 1·5 per cent TCC alone.

A UK patent[37] relates to extended efficacy in use of hexamethylenetetramine which was previously utilized as a urinary antiseptic. An example of a deodorant composition is as follows:

	(1) *per cent*
Hexamethylenetetramine	14–20
Zinc oxide	16–23
Starch	16–23
Petroleum jelly	38–43
Perfume	0·5–1·2

The effectiveness of this deodorant is said not to decrease within 7–15 days.

The use of sodium bicarbonate (baking soda) as a deodorant has been known for many years.[38] Sodium bicarbonate is an acid salt which can act chemically as either a mild alkali or a mild acid. Underarm odours are largely caused by volatile acidic compounds which are absorbed by baking soda to form stable odourless salts. In 1975 aerosol products appeared in the USA containing baking soda and a patent specification was issued by Colgate-Palmolive[39] on aerosol compositions containing this material.

Metallic salts of ricinoleic acid,[40] particularly those of zinc and of the elements close to zinc in the periodic table, show a marked reactivity toward low molecular organic compounds with functional groups containing amino nitrogen and mercapto sulphur. This results in a deodorizing effect which can be intensified by adding small quantities of other derivatives of polyhydroxy fatty acids or resinic acids. A sensorial test does not reveal to any degree this deodorizing effect in the case of allied derivatives of other fatty acids.

L'Oreal published a patent[41] on deodorant compositions based on vegetable extracts, and after extensive investigations located the desired activity in the *Ungulina* species of fungus of the *Polyporaceae* family. The invention provides a composition suitable for application to the human body which comprises an extract from at least one fungus of the *Ungulina* species, a compatible vehicle and a perfume, the extract containing argariac acid. *Ungulina officinalis* is a parasitic wood-infesting fungus of larch trees, found mainly in alpine regions of Russia and Siberia.

Kabara[42] describes structure–function relationships of surfactants as antimicrobial agents. Nonionic surfactants, which in the past were considered not to have antimicrobial activity, were shown to be active when the monoesters were formed from lauric acid.

Ethyl alcohol, used as a vehicle in deodorant products, is also an active antibacterial agent.[43]

The antimicrobial properties of essential oils have been known for half a century. An extensive review and bibliography of the publications in this field up to 1955 was published by Cade.[44] These studies show that various essential oils exhibit significant antibacterial effect. The considerable variation in test results can be attributed to the fact that these are natural products in many cases; the variety of organisms and test methods employed also contribute, to a large extent, to aberration in results. An essential oil can vary considerably in its chemical composition, thus producing a corresponding lack of uniformity in antimicrobial activity. Some essential oils, such as thyme and clove, consistently show good antibacterial activity which is generally attributable to their high phenolic content, namely thymol and eugenol.

Kellner and Kober examined the antibacterial action of 175 oils against nine organisms and classified 21 of the most active oils according to chemical composition.[45] Terpenes were also found to have good antibacterial activity. Maruzella[46] conducted extensive investigations of the antibacterial and antifungal properties of essential oils, perfumes, and aromatic chemicals. He reported a high incidence of activity among these materials both as contact and vapour phase antimicrobial agents.

Assessment of Deodorants

In the case of deodorants, the techniques of assessment are relatively straightforward: conventional microbiological methods of analysing microbial content in properly designed experiments will supply data concerning the efficacy of deodorant compounds and products both *in vitro* and *in vivo*. The ultimate test, however, for any finished cosmetic deodorant product will involve well designed axillary sniff studies.[47] Although this may appear to be a primitive technique, it is extremely well adapted to a consumer product; for a useful account of these techniques see Rothwell.[48] Final product attributes are a function of the total formula in which the perfume plays a significant role. In the final analysis, it is the well trained nose capable of relating to consumer perception of odour that will aid the determination of the ultimate success or failure of a deodorant product.

Product Formulation—Antiperspirants

Aerosols

Powder-in-oil suspension aerosol antiperspirants using micronized powdered aluminium chlorhydrate and containing 3–4 per cent of active ingredients suspended in an oil base appeared in the USA in the mid-1960s. They have become by far the most preferred applicator type for antiperspirants.

Many combinations of raw materials are available for the formulation of aerosol antiperspirants and their selection must be carefully considered, since the surface chemistry of the system can affect sedimentation and dispersion characteristics of the formula.[49] In addition, formulations must provide maximum antiperspirant and deodorant effectiveness, maximum safety, cosmetic elegance and minimum staining.

A typical powder-in-oil formulation is:

	(2) *per cent*
Aluminium chlorhydrate (micronized)	4·50
Isopropyl myristate	3·70
Fumed silica	0·15
Perfume	*q.s.*
Propellants 11/12 (65:35)	to 100.00

The emollient or carrier for the aluminium chlorhydrate is used to produce a smooth feel on the skin and to help the powder to adhere. Commonly an ester such as isopropyl myristate is used although some products contain volatile silicones to reduce staining of clothes.

A suspending agent is added to prevent agglomeration of the aluminium chlorhydrate which could lead to valve blockage and leakage. Typical suspending agents are silica and 'Bentone' derivatives. Fumed silica has an extremely fine particle size and forms a coating over the powder particles to prevent the development of hard caking. Floyd[50] has reviewed the effects of montmorillonite clays and silicas on the rheological properties of antiperspirants in aerosol and other product forms.

Aerosols have been under continuing attack by environmentalists, particularly in the USA, because of the 'ozone' depletion theory. It is believed by some that fluorocarbon propellants react with and damage the ozone layer of the atmosphere. A major controversy now rages as to the truth behind the ozone depletion problem, but in the USA and FDA ruled that aerosols for non-essential uses containing fluorocarbons could not enter the market from April 1979 onward. In the USA aerosol antiperspirants and deodorants are now butane propelled.

Staining of Clothing by Aerosol Antiperspirants. Aerosol antiperspirant compositions in which the astringent material, for example aluminium chlorhydroxide, is dispensed as a solid suspended in an anhydrous hydrophobic liquid vehicle such as mineral soil, isopropyl myristate or isopropyl palmitate have been widely marketed. Although such compositions are effective in reducing perspiration they have a tendency to impart stains to clothing which remain after laundering.

In a patent by Gillette in 1974[51] it was stated that staining due to aerosol compositions could be substantially reduced by using as the liquid vehicle an ester which is miscible with the propellant and which is selected from the group represented by the formula

$$\begin{array}{c} CH_2COOR_1 \\ | \\ RO—C—COOR_1 \\ | \\ CH_2COOR_1 \end{array}$$

where R is a hydrogen atom or 2- to 3-carbon acyl group. Examples of compounds include triethyl citrate and acetyl triethyl citrate.

The effectiveness of these esters in reducing staining was demonstrated *in vivo* with about 100 males who were given coded cans of antiperspirants containing parallel triethyl citrate and isopropyl palmitate formulations and new cotton T-shirts. They were instructed to use one product under the right axilla only and the other under the left axilla only. The T-shirt was to be worn for at least four cycles of home laundry. After four weeks, the T-shirts were collected and evaluated. The T-shirt underarm areas in contact with the axilla under which triethyl citrate was used consistently showed substantially less staining than that under which isopropyl palmitate had been used. The confidence level of this observation was greater than 99·5 per cent.

A patent was published by Union Carbide in 1977[52] on the use of volatile cyclic silicone compounds which can be used in place of isopropyl myristate or other emollients commonly used in aerosol antiperspirants to reduce billowing (or clouding) effectively without staining clothing and which impart substantially reduced oiliness to the skin.

Several patents concerned with staining have been granted to Unilever Ltd.[53–55] In one of these it is claimed that a substantial reduction in the level of staining on clothing in repeated contact with antiperspirants can be achieved by incorporating into the antiperspirant composition certain polyalkylene glycols. The invention describes a non-staining aerosol antiperspirant composition of the powder-suspension type containing a colourless water-miscible polyalkylene glycol (such as polypropylene glycol and derivatives) in which a proportion of the hydroxyl groups of the glycols are butylated. The latter substances are supplied by Union Carbide Chemicals Co. under the proprietary names 'Ucon HB' and

'Ucon H'. The term 'polyalkylene glycol' also includes the block copolymers of ethylene oxide and propylene oxide which are supplied by the Wyandotte Chemicals Corporation under the proprietary name 'Pluronic'. A typical formulation described in the patent is as follows:

	(3) *per cent*
Aluminium chlorhydrate	3·50
Fumed silica	0·50
Ucon 50-HB-660	4·77
Pluronic L64D	1·50
Perfume	0·38
Propellants 11/12 (70:30)	to 100

Antiperspirant Sticks

The sodium aluminium chlorhydroxy lactate soap or cologne sticks have been available for several decades now and have often been called antiperspirants. In reality, however, they are deodorants since their efficacy is in the range of only 8–12 per cent sweat reduction. Within the last few years true antiperspirant sticks, utilizing as the active ingredient either the aluminium chlorhydrate propylene glycol complex or micronized aluminium chlorhydrate, have been introduced. These produce a sweat reduction of the order of 40 per cent.

The antiperspirant stick usually consists of a wax-like matrix which serves as a carrier for aluminium chlorhydrate powder and volatile silicone. A low-melting matrix and a high-boiling volatile silicone are necessary in order to conserve the latter during processing. For this reason stearyl alcohol is preferred over stearic acid because of its lower melting point (58·5°C vs 69·9°C).

Typical formulations are as follows:

	(4) *per cent*	(5) *per cent*
*Volatile silicone 7158	46	46
Aluminium chlorhydrate powder	20	20
Stearyl alcohol	24	24
Polyethylene glycol distearate 6000	6	6
*Carbowax PEG 1000	—	2
*Carbowax PEG 1540	4	2

*Union Carbide

Procedure: Heat the stearyl alcohol, Carbowax PEG 1000 and 1540 and polyethylene glycol distearate 6000 to 80°C. When melted, add the aluminium chlorhydrate and mix thoroughly. Cool to 70°C and rapidly mix in the volatile silicone 7158. When mixing is complete, pour the mixture into a stick container. Allow the mixture to cool undisturbed for 24 hours.

Dry compressed antiperspirant sticks using isostatic compaction of aluminium chlorhydrate and microcrystalline cellulose powders have been developed by the FMC Corporation.[56]

Dry antiperspirant stick

	(6) *per cent*
Powder phase	
Avicel PH-105 (FMC)	52·35
Italian talc	14·30
Aluminium chlorhydrate, Ultrafine (Reheis)	19·00
Dri-Flo Starch 4951 (National Starch)	7·30
Zinc stearate	1·90
Liquid phase	
Volatile silicone 7207 (Union Carbide)	4·80
Isopropylan 33 (Robinson-Wagner)	0·10
Perfume	0·25

Procedure: Blend the powder materials in a V-shell blender for 10 minutes, add the liquid phase and mix via intensifier bar for 5 minutes. Press the powder blend at 2000 psi (14 MPa) in a rigid die.

Avicel microcystalline cellulose is a pure spray-dried material which provides simple binder phase of active astringents.

The advantages of this type of stick are said to be:

(a) high levels of perceived antiperspirancy;
(b) smooth dry application;
(c) no staining or corrosive effect on fabrics.

Antiperspirant Creams

A US patent[57] describes anhydrous antiperspirant creams. While oil-in-water emulsions provide a convenient vehicle for storing and delivering antiperspirant actives, compositions of this type tend to produce an undesirable wet, cold and/or sticky sensation when they are applied to and rubbed into the skin. This can be minimized somewhat by utilizing compositions in anhydrous form. An example is as follows:

	(7) *per cent*
Isopropyl myristate	32·0
Bentone 38 (thickening/suspending agent)	7·0
Ethyl alcohol (gel-promoting agent)	3·0
Zirconium hydroxychloride/aluminium chlorhydroxide/glycine complex	47·0
Silicone (antisyneresis agent)	10·0
Perfume	1·0

This is a substantially anhydrous antiperspirant composition in the form of a cream and is resistant to syneresis.

A normal oil-in-water cream antiperspirant formulation[58] is:

		(8) *parts by weight*
A	* Neo-Fat 18–55	10·6
	Mineral oil	1·0
	Beeswax	1·0
	Glyceryl monostearate (pure)	6·4

		parts by weight
B	† Chlorhydrol (50% solution of aluminium chlorhydrate)	32·0
	Perfume	*q.s.*
C	Propylene glycol	5·0
	Sodium lauryl sulphate	1·3
	De-ionized water	to 100·0

* Armak Co., Chicago, USA.
† Reheis Chemical Co., Phoenix, USA.

Procedure: Heat A to 70°–80°C. Add C with agitation and cool to 35°–40°C. Add B and mix thoroughly.

Roll-on Antiperspirants

Roll-on antiperspirants have been on the market for many years. They are generally either emulsion products or aqueous alcoholic solutions thickened with cellulose gums. The aqueous alcoholic products generally dry quicker and are less sticky than the emulsion products. The viscosity of the final product is important to avoid leakage around the roll-ball.

In the following emulsion formulation (*ex* Reheis Chemical Co., USA) magnesium aluminium silicate is used as a thickening agent and emulsion stabilizer and glyceryl monostearate acts as an additional thickening agent and opacifier. The volatile silicone reduces the sticking of the roll-ball due to the drying out of the aluminium chlorohydrate.

		(9) *per cent*
A	Magnesium aluminium silicate	1·0
	De-ionized water	49·0
B	Glyceryl monostearate (acid stable)	8·0
C	Aluminium chlorhydrate (50% soln.)	40·0
D	Volatile silicone	2·0
E	Fragrance	*q.s.*

Procedure: Add magnesium aluminium silicate to water slowly, agitating continually until smooth, and heat to 70°C. Heat B to 75°C and add to 1; mix until temperature has fallen to 50°C. Heat C to 50°C and add to 2; mix until product has reached room temperature. Add D and E and stir for 15 minutes.

	Aqueous–alcoholic roll-on	(10) *per cent*
A	De-ionized water	29·10
B	Propylene glycol	4·00
	*Natrosol 250H (hydroxyethyl cellulose)	0·40
C	Aluminium chlorhydrate (50% soln.)	40·00
D	Alcohol 99% v/v	25·00
	Nonyl phenol ethoxylate (9 mol)	1·00
	Perfume	0·50

* Hercules Powder Co.

Procedure: Heat the water (A) to 70°C. Disperse the Natrosol in the propylene glycol and add this mixture (B) to the water with good agitation. Mix well until the Natrosol is fully hydrated. Add the aluminium chlorhydrate solution. Cool the batch to 30°C. Add D slowly with good agitation.

In this formulation the propylene glycol reduces crystallization of the aluminium salt on the roll-ball and the nonyl phenol ethoxylate is added to solubilize the perfume in the final preparation.

Product Formulation—Deodorants

Deodorant Soaps

The toilet soap market is one of major importance in all countries. The two largest deodorant soap brands worldwide are Armour-Dial's 'Dial', the leading US soap, and Unilever's 'Rexona', which is marketed in most of Europe.

In the USA deodorant soaps are classified as 'drugs'. When reviewing the use of antimicrobial agents used in soaps, the FDA's OTC advisory panel cited a series of studies[59] to support the position that these topical antimicrobials could lead to shifts in the microbial flora, placing the user at risk due to an overgrowth of Gram-negative micro-organisms. Not to be ignored, however, is the safety record of billions of antimicrobial soap bars sold in the USA during the last two decades. No significant medical problems have been reported, and there is no evidence to demonstrate an ecological shift in the skin flora due to the routine use of antimicrobial formulation in deodorant soap and similar products.

The most frequently used antimicrobial agents in soaps at the present time are trichlocarban (TCC), cloflucarban (CF_3) and triclosan (DP300). Prior to their ban, hexachlorophene (G11) and tribromsalan (TBS) occupied the key spots in this application. All but triclosan are active only against Gram-positive organisms when in the presence of soap; triclosan is active against both Gram-positive and many Gram-negative organisms.

Deodorant Sticks

Sodium stearate stick deodorants have been on the market for several years.[60] A typical formulation is given in example 11.

	(11) *per cent*
Sodium stearate	8·0
Ethyl alcohol	74·8
Propylene glycol	10·0
Isopropyl myristate	5·0
Triclosan	0·2
Perfume	2·0

Procedure: Slurry the soap in the cold with organic solvents and triclosan and then heat to 60°–75°C. Stir the mass while hot until clear. Add fragrance and colour as desired at 5°–8°C above the set point of the stick. When it is uniform, pour the soap solution into moulds and allow to cool. Sodium stearate can be prepared *in situ* but critical control is required to avoid excess alkali or fatty acid.

To avoid shrinkage which can occur with alcoholic sticks, particularly if the packaging is poor, non-alcoholic deodorant sticks can be prepared as follows (example 12).

	(12) *per cent*
Sodium stearate	8·0
Propylene glycol	10·0
Perfume	1·0
Coconut diethanolamide	5·0
PPG-3-myristyl ether	68·8
Triclosan	0·2
Water	7·0

The preparation is similar to that described for the alcoholic stick.

Aerosol Deodorants

Aerosol deodorants are based on alcoholic solutions of a bactericide. In some cases a product called a 'deo-cologne' and used as a body spray is based solely on an alcoholic solution of a perfume compound. A typical formula for an aerosol deodorant is given in example 13. Lacquered monobloc aluminium or tinplate containers can be used.

	(13) *per cent*
Triclosan	0·05
Propylene glycol	2·00
Alcohol (99% v/v)	57·45
Perfume	0·50
Propellant 12	40·00

REFERENCES

1. Glaxton, R., *Drug Cosmet. Ind.,* 1972, **110**(5), 64.
2. Shelley, W. B. and Hurley, H. J., *Acta Derm. Venereol.*, 1975, **55,** 241.
3. Sehgal, K., *Manuf. Chem. Aerosol News*, 1978, **49**(1), 43.
4. Fiedler, H. P., *Cosmet. Perfum.*, 1968, **84**(2), 25.
5. Shelley, W. B., Hurley, H. J. and Nicholls, A.C., *Arch. Dermatol. Syphilol.*, 1973, **68,** 430.
6. Hurley, H. J. and Shelley, W. B., *The Human Apocrine Sweat Gland in Health and Disease*, Springfield, Charles C. Thomas, 1960.
7. Geller, L., *Dragoco Rep.*, 1972, **19**(3), 54.
8. Comfort, A., *Dragoco Rep.*, 1973, **20**(3), 54.
9. Goodall, McC., *J. Clin. Pharmacol.*, 1970, **10,** 235.
10. Lansdown, A. B. G., *J. Soc. cosmet. Chem.*, 1973, **24,** 677.
11. Papa, C. M., *J. Soc. cosmet. Chem.*, 1966, **17,** 789.
12. Papa, C. M. and Kligman, A. M., *J. invest. Dermatol.*, 1967, **49,** 139.
13. Lansdown, A. B. G., *Br. J. Dermatol.*, 1973, **89,** 67.
14. Papa, C. M. and Kligman, A. M., *J. invest. Dermatol.*, 1966, **47,** 1.
15. McMillan, F. S. *et al.*, *J. invest. Dermatol.*, 1964, **43,** 362.

16. Grasso, P. and Lansdown, A. B. G., *J. Soc. cosmet. Chem.*, 1972, **23,** 481.
17. Stillians, A. W., *J. Am. Med. Assoc.*, 1916, **67,** 2015.
18. Fiedler, H. P., *Der Schweiss*, 2nd edn, Aulendorf, Cantor KG, 1968.
19. Bathe, P., *Manuf. Chem. Aerosol News*, 1978, **49**(7), 72.
20. Shelley, W. B. and Hurley, H. J., *Nature* (*London*), 1957, **180,** 1060.
21. British Patent 735 681, Carter Products, 1955.
22. US Patents 2 814 584, 2 814 585, Daley, E., 1957.
23. US Patent 2 906 668, Beekman, S., 1959.
24. US Patent 2 906 668, Beekman, S., 1959.
25. US Patent 3 407 254, Bristol-Myers, 1968.
26. Anon., *Manuf. Chem. Aerosol News*, 1977, **48**(12), 10.
27. Anon., *CTP Marketing*, 1978, (28), 5.
28. Floyd, D. T., *J. Soc. cosmet. Chem.*, 1978, **29,** 717.
29. *Federal Register*, 1978, **43**(196).
30. *Tentative Findings of the OTC Antiperspirant Panel, Draft Report*, US FDA, November 1977.
31. Rubino, A. M., Siciliano, A. A. and Magres, J. J., *Aerosol Age*, 1978, **23**(11), 22.
32. US Patent 4 080 438, L'Oreal, 1978.
33. US Patent 4 080 439, L'Oreal, 1978.
34. British Patent 1 485 373, Unilever, 1977.
35. *Federal Register*, **43**(196), 1978.
36. Majors, P. A. and Wild, J. E., *J. Soc. cosmet. Chem.*, 1974, **25,** 139.
37. British Patent 1 525 971, Hlavin, Z., 1978.
38. Anon., *Aerosol Age*, 1976, **21**(2), 32.
39. British Patent 1 476 117, Colgate-Palmolive, 1977.
40. Sartori, P., Lowicki, N. and Sidillo, M., *Cosmet. Toiletries*, 1977, **92,** 45.
41. British Patent 1 477 882, L'Oreal, 1977.
42. Kabara, J. J., *J. Soc. cosmet. Chem.*, 1978, **29,** 733.
43. Bandelin, F. J., *Cosmet. Toiletries*, 1977, **92**(5), 59.
44. Cade, A. R., *Antiseptics, Disinfectants, Fungicides and Chemical and Physical Sterilization*, Philadelphia, Lea and Febiger, 1957, Chapter 15.
45. Kellner, W. and Kober, W., *Arzneim. Forsch.*, 1955, **5,** 224.
46. Maruzella, J. C., *Am. Perfum.*, 1962, **77,** 67.
47. Dravnieks, A., *J. Soc. cosmet. Chem.*, 1975, **26,** 551.
48. Rothwell, P. J., paper presented to *Symposium on Sensory Evaluation*, Society of Cosmetic Scientists, 1980.
49. Jungermann, E., *J. Soc. cosmet. Chem.*, 1974, **25,** 621.
50. Floyd, D.T., *Cosmet. Toiletries*, 1981, **96**(1), 21.
51. US Patent 3 833 721, Gillette, 1974.
52. British Patent 1 467 676, Union Carbide, 1977.
53. British Patent 1 300 260, Unilever, 1972.
54. British Patent 1 369 872, Unilever, 1974.
55. British Patent 1 409 533, Unilever, 1975.
56. Raynor, G. E. and Steuernagel, C. R., *Manuf. Chem. Aerosol News*, 1978, **49**(4), 65.
57. US Patent 4 083 956, Procter and Gamble, 1978.
58. Anon., *Soap Cosmet. chem. Spec.*, 1975, **51**(9), 121.
59. *Federal Register*, **39**(179), 33103–33122, 1974.
60. Barker, G., *Cosmet. Toiletries*, 1977, **73**(7), 73.

Chapter Eleven

Depilatories

Introduction

Preparations for the removal of unwanted hair have been known for thousands of years. Among them was rhusma, a mixture of quicklime and arsenical pyrites in a ratio of 1:2 used in ancient times by the dancing girls of the East. Before use this product was reduced to a powder and mixed with an aqueous alkali, possibly obtained from wood ashes; another preparation which was used for the same purpose was orpiment which is, essentially, arsenic trisulphide.

There is no record of any development work having been carried out on this subject during subsequent centuries. The attitude taken was that, if one used anything, one used pumice stone and did not talk about such things.

In modern times, however, a rapidly increasing interest in depilatories has been noticed, brought about by changes in fashions, clothing and social customs.

While the term 'depilatory' has been applied to any preparation designed for the removal of superfluous hair (in particular hair occurring on the face and legs, as well as in the axilla) without causing injury to the skin, a distinction must be drawn between the mechanical removal of hair by either plucking it with tweezers or by embedding it in an adherent material which can then be pulled away from the skin bringing the hair with it (a process referred to as epilation), destruction of hair by electrolysis, and the removal of hair after it has been sufficiently degraded by chemical means.

Extensively bibliographed reviews of the historical development and technology of depilatories have been published.[1–4]

Epilation

Epilation has some following because the effect may be slightly longer lasting since, if the epilated hairs take with them the hair bulbs or the hair papillae, there may be a relatively long pause before the hair starts growing in the follicle and reaches the surface of the skin. It is, however, by no means painless and can often cause serious skin damage and subsequent infection, and is therefore frowned upon by doctors.

For many years epilatory preparations were based on mixtures consisting essentially of rosin and beeswax, modified in some instances by the addition of mineral oil and/or waxes. The following examples are illustrative:[5,6]

	(1)
	per cent
Rosin	75·0
Beeswax	25·0

	(2)
	per cent
Light coloured rosin	52·0
Yellow beeswax	25·0
Paraffin wax	17·0
Petrolatum	5·0
Perfume	1·0

Procedure: Melt the rosin and waxes, mix and add the petrolatum, then, when the temperature drops to about 60°C, add the perfume and pour the melted mass into suitable moulds. When this wax is used it is melted and painted over the surface to be dehaired.

In addition to rosin and waxes, mineral or vegetable oil may be included (for example at a level of about 15 per cent). Camphor is often included for its cooling effect which reduces the discomfort experienced when the hair is pulled off. A local anaesthetic, for example benzocaine, to enhance this effect and an antibacterial compound will reduce the chance of infecting the skin after damage or exposure.

There has been no dramatic development of a 'painless' epilatory in the industry. Such developments as have taken place have been concerned with modifications in the method of application, such as the provision of a flexible backing strip, and also the provision of a preparation that does not require melting prior to use but can be applied cold, the preparation being based upon a mixture of glucose and zinc oxide.[7] The use of a 'rubber solution', in which the solvent evaporates and the rubber film is stripped off, is covered in a US patent.[8]

Several more recent patents describe epilating compositions. A French patent[9] describes depilating waxes in strip form and quotes the following composition as an example:

	(3)
	parts
Rosin	1700
Vegetable oil	900
Triethanolamine	100
Benzoin	10
Balsam Tolu	10
Lemongrass bouquet	5
Butyl *p*-aminobenzoate	10
Alcohol	5

The wax is spread on the rough side of a strip of kraft paper, the smooth side of which is silicone-treated. Other patents describe the use of fresh lemon juice in a syrup-type epilatory paste;[10] a solution of cold water-soluble dextrin in glycerin;[11] an epilatory tape impregnated with a tacky substance;[12] a film-forming solution which is applied to skin, allowed to dry and peeled off;[13] a low-melting depilatory wax composition.[14]

Maxwell-Hudson[15] has described a 'home-made' depilating formula based on caramelized sugar, lemon juice and glycerin.

Gallant[16] has given a detailed account of professional techniques for depilatory waxing treatments.

Electrolysis

The mechanical methods mentioned above are temporary and often only relatively effective since the papillae and hair bulbs are not always removed and hair soon reappears. The most effective method of depilation is undoubtedly electrolysis which entails inserting a needle into the hair follicle and complete destruction of the hair root by means of a weak DC current. This method is practised in beauty salons and by some dermatologists but is expensive and time-consuming since every hair must be treated individually, and even a competent operator can only deal with 25–100 hairs per sitting.

Chemical Depilation

The term 'depilatory' as used nowadays refers to preparations intended for the chemical breakdown of superfluous hair without injury to the skin.

The advantage of such preparations is that they avoid any danger of cutting or abrading the skin in regions such as the underarms, where it is difficult to see the area clearly and even more difficult to guide a razor over the complicated contours. There is also a widespread belief that shaving increases the rate of hair growth or the coarseness of the hair. Although these beliefs are unfounded in fact, chemical depilatories have the apparent advantage that they discourage the regrowth of hair if they are applied regularly. There seems to be no scientific explanation for this, but possibly it arises from a gradual removal of keratinous debris from the mouth of the hair follicle, which allows removal of the hair at a deeper level.

Since the hair shaft is of similar composition to the skin (both are derived from keratin), a small degree of local damage may occur as the result of applying such preparations, particularly if the depilatory is kept in contact with the skin for any length of time and the pH is sufficiently high, when the horny layer of the skin will also be attacked.

Provided that the skin is reasonably healthy, that the time of application of the depilatory is not too long, and that this is correctly formulated, very little if any damage will result. In formulating depilatory preparations, therefore, care should be taken to ensure that they will react with the hair preferentially and that their effects will be sufficiently rapid to cause disintegration of the hair before causing any damage to the underlying and surrounding skin.

With the above aims in mind, the desirable requirements of a depilatory may be defined as follows:

(1) Non-toxic and non-irritant to the skin.
(2) Efficient in action, removing hair rapidly, preferably in 4–6 minutes.
(3) Preferably odourless.
(4) Stable on storage.
(5) Harmless to clothing.
(6) Preferably cosmetically elegant.

In line with the requirement for a rapid depilation, depilatory preparations usually contain as their active component an alkaline reducing agent. The latter will cause the hair fibres to swell and produce a cleavage of the cystine bridges between adjacent polypeptide chains as a preliminary to the complete degradation of the hair.

Sulphides

The use of sulphides has been known, as pointed out at the beginning of this chapter, for a very long time: however, it was not until 1885 that, in the United States, the first patent was taken out for the use of barium polysulphide for removing hair. In April 1891, the use of the monosulphide, polysulphide and sulphydrate of strontium for the same purpose was patented.[17] These preparations were, however, mixtures rather than creams, the first depilatory in cream form being developed in 1921.

Compositions based on alkali and alkaline earth sulphides are capable of producing rapid depilation, particularly if used together with a suspension of lime.

The alkali sulphides such as sodium sulphide were, however, found to be too drastic in action. Their depilatory action is linked to their hydrolysis and the formation of sulphydrates and sodium hydroxide. The latter acts as a primary irritant and will produce erythema. Even a 2 per cent aqueous solution of sodium sulphide will have a pH of 12. Although it will disintegrate hair within 6–7 minutes, it will simultaneously damage the stratum corneum. It is, therefore, no longer used in depilatory preparations on the market.

Strontium sulphide is a much milder depilatory, but must be used at a higher concentration than sodium sulphide to produce an equivalent dehairing action. Preparations containing strontium sulphide, although largely replaced by those based on thioglycollates, are still available. They are very effective and work within 3–5 minutes after application.

While some people are sensitive to such preparations (as indeed are some people to shaving soaps), the products appear to be innocuous if used according to the directions of the manufacturer. The main reason for their loss in popularity is that, in common with other sulphides, they generate the odour of hydrogen sulphide on application (and not infrequently on storage). This odour is most intense when the product is washed off, owing to hydrolysis of the sulphide. It is advisable, therefore, to remove the bulk of the product with a spatula before washing, and it is the usual practice to include such a spatula, made of wood or plastic, in the pack. It also serves to apply the product in the necessary thick layer (1–2 mm). Under no circumstances should the final washing be omitted.

In addition to the active agent, a depilatory preparation may contain a humectant such as glycerin or sorbitol. A thickening agent, for example methyl cellulose, may sometimes be incorporated, so as to thicken the solution sufficiently to allow it to remain in contact with the hair as long as necessary.

For a sulphide depilatory, the following formula will be found effective:

	(4) *per cent*
Strontium sulphide	20·0
Talc	20·0
Methylcellulose	3·0
Glycerin	15·0
Water	42·0

This may also be prepared using an emulsion base for smoothness and stability.

The formulation of depilatories depends upon very careful adherence to detail; slight departures from formulation in the process of manufacturing can produce remarkable differences between the efficacies of different batches of supposedly the same product. For this reason any formulation can be little more than a general guide.

Despite their disadvantages, sulphide-based depilatories are preferred by many black-skinned men for removing facial hair because of their comparatively rapid action. This subject is covered later in this chapter.

Stannites

In the 1930s considerable attention was devoted to 'soluble stannites'. For example, the use of sodium stannite as a preferred salt of tin with Rochelle salt as a stabilizer was covered in a US patent[18] published in 1933.

Another US patent[19] drew attention to the fact that, while the stannites have no appreciable odour, they suffer from instability, forming stannates in the presence of water. This patent proposed the use as stabilizers of water-soluble organic compounds having three or more carbinol hydroxy groups, or three hydroxyl groups other than those in a carboxyl group and also soluble silicates; specific examples are triethanolamine, dextrine, sugars and—in the case of the silicates—potassium or sodium metasilicate. A British patent[20] describes the addition of aqueous stannous chloride to aqueous sodium hydroxide containing sodium silicate to give a solution having pH less than 12·6 (12·3). A French patent[21] also claims a method for the preparation of a stable stannite. However, the stabilizers recommended in these patents were not found to be effective and did not produce stable preparations.

Substituted Mercaptans

The majority of depilatories available today are substituted mercaptans which are used in the presence of alkaline reacting materials, for example calcium thioglycollate, in conjunction with calcium hydroxide. These preparations possess less odour than the sulphide type, but take longer to act. They are safer on the skin than sulphides, and can therefore be used on the face—an area where superfluous hair can cause great distress and where women have a strong psychological aversion to using a razor. In general, thioglycollate preparations are more attractive than the sulphide type. However, their slowness in attacking the coarse resistant hair of the underarm has left a market open for sulphide depilatories used for this purpose only.

It is often said that depilatories can be used for smoothing the legs, but so much of the product is required to cover each leg that it becomes uneconomic for most users. In any case, the legs are easy to shave.

Thioglycollates

Thioglycollate-based preparations are non-toxic and stable at concentrations at which they are used, that is between 2·5 and 4 per cent. At concentrations of less than 2 per cent w/w they act too slowly to be of any use, while nothing will be gained in terms of effectiveness if their concentration is raised above 4 per cent. At the concentrations at which they are used, they may produce depilation in 5–15 minutes, this again depending on the pH of the preparation. This should not be less than pH 10·0 and should preferably be about pH 12·5 to produce depilation within a fairly short time and without irritating the skin.

The use of thioglycollate in depilatories stemmed from the research conducted by Turley and Windus[22] and was heralded by patents which were issued in France,[23] Britain[24] and the United States.[25]

The British patent[24] granted to Bohemen covered the use of calcium thioglycollate in a cream base. The base had the formula:

	(5) *per cent*
Stearyl alcohol	9·0
Sulphonated stearyl alcohol	1·0
Water	90·0

Sixty-six parts of the above base were mixed with 10 parts of hydrated lime and 4 parts of 90 per cent thioglycollic acid, whereupon the cream liquefied and was stiffened by the addition of at least 20 parts of precipitated chalk.

A US patent[26] issued in 1944 to Evans and McDonough covered the use of substituted mercaptans (thioglycollic acid) in conjunction with an alkaline-reacting material and a perfume. Substituted mercaptans having polar groupings are preferred. The patentees claim that in order to obtain a desirable depilating action, the depilating agent must conform to certain general rules:

(1) pH value should be between pH 9·0 and pH 12·5.
(2) Concentration of mercaptans should be between 0·1 and 1·5 mol per litre.
(3) The alkaline ingredient must have an ionization constant greater than 2×10^{-5}.
(4) In order to prevent skin damage, it is desirable that the concentration of the alkaline material in solution be not greater than twice the equivalent concentration of the mercaptans.
(5) Paste form is the most satisfactory.
(6) Natural gums are used to give stable formulations.

Basic formulae for preparing cream, semi-fluid and powder depilatories have been offered by Evans Chemetics:[27]

Depilatory cream	(6) *per cent*
Evanol	6·5
Calcium thioglycollate	5·4
Calcium hydroxide	7·0
Duponol WA paste	0·02
Sodium silicate '0'	3·43
Perfume	*q.s.*
Distilled water	to 100·0

Procedure: Heat the water to 70°C. With stirring add the Duponol and Evanol; continue stirring until melted and dispersed. Discontinue heating and cool/stir to room temperature. Add the calcium hydroxide and perfume. Add the calcium thioglycollate and stir until uniform.

Semi-fluid depilatories	(7) *per cent*	(8) *per cent*
Cream base		
Distilled water	60·0	60·0
Cetyl alcohol	6·0	6·0
Brij 35	1·0	1·0
Final product		
Distilled water	17·3	17·2
Calcium thioglycollate	5·4	5·4
Calcium hydroxide	6·6	10·4
Strontium hydroxide	3·7	—
Perfume	*q.s.*	*q.s.*
Cream base (as above)	67·0	67·0

Procedure: Prepare the cream base at 70°C and allow to cool to room temperature. Add the calcium thioglycollate to the bulk of the water and mix well; add the calcium hydroxide slowly with stirring, followed by the strontium hydroxide and any remaining water. Combine the two parts and stir well, adding the perfume at this point.

Powder depilatory	(9) *per cent*
Calcium thioglycollate	20·0
Calcium hydroxide	23·1
Strontium hydroxide $8H_2O$	8·9
Sodium lauryl sulphate powder	1·5
Cellosize QP.100M	1·0
Magnesium carbonate USP	45·2
Perfume	0·3

Procedure: Mix the calcium thioglycollate, calcium hydroxide, strontium hydroxide, sodium lauryl sulphate and Cellosize. Blend the perfume thoroughly with the magnesium carbonate. Add the latter to the former and blend thoroughly. (N.B. This formula contains a much higher level of thioglycollate than is permitted for sale with the EEC.)

The slower activity of thioglycollates has led to attempts to accelerate the depilatory action by incorporating substances which cause swelling of the hair fibres. Urea was considered for this purpose, but could not be used since it decomposes at the normal pH of depilatory preparations. Attempts were made, therefore, to find other compounds which could accelerate the rate of depilation in compositions containing mercaptans and which would remain stable at the pH at which these depilatory preparations are normally used.

In a L'Oreal patent,[28] melamine and dicyandiamide, or a mixture of these two compounds, were claimed to have an accelerating effect on depilation and it was proposed that they be used in depilatory compositions at levels ranging between 0·5 and 2 parts per part of (for example) thioglycollic or thiolactic acids or their calcium or strontium salts.

Other L'Oreal patents[29] suggest the use of lithium salts of thioglycollic and mercaptopropionic acids for improving the rate of depilation by virtue of their good solubility. The sodium and potassium salts are equally soluble but they are irritant to the skin. The quicker action of the lithium salts also permits them to be used at a lower concentration and at a lower pH than the calcium and strontium salts, which in turn will reduce the risk of irritating the skin.

A series of patents[30] proposes the use of sodium metasilicate with thiourea as synergistic accelerators for calcium thioglycollate. Morillère *et al.*[31] suggest the use of a blend of copolymers of N-vinyl lactam and a hemiester of an unsaturated dicarboxylic acid.

Another means of accelerating the rate of depilation is by increasing the temperature. Beecham workers[32] have proposed a two-part product contained in a tube-within-a-tube; one part is anhydrous and contains calcium oxide, the other contains water. Upon extrusion the two parts mix and generate heat.

An anhydrous depilatory composition in stick form has been patented[33] which is claimed to form, on contact with wetted skin, a cream capable of complete depilation in less than 10 minutes without producing any offensive odours during that time. The stick comprises a depilating agent, a solid basic substance, a solid base and a perfume. The depilating agent is a thiol derivative, preferably calcium thioglycollate; it constitutes 10–35 per cent by weight of the composition. The solid basic substance is an alkaline earth metal hydroxide or carbonate, or a mixture of these two; it is used in sufficient amount to provide a pH of 10·5–12·5 in a saturated aqueous solution.

The function of the solid base is to protect the sensitive facial skin from possible irritation by the active ingredient and to confer emolliency. It consists of the following: (1) a sterol, preferably an unsaponifiable lanolin fraction; (2) a solid inert organic filler to provide hydrophilic and emollient properties and body, and to prevent the penetration of externally applied water. The filler, which will usually constitute up to 75 per cent by weight of the solid base, may include petrolatum, paraffin wax, microcrystalline wax, fatty alcohols, spermaceti, beeswax and others; (3) a nonionic polyalkenoxy-type water-in-oil emulsifier, to provide homogeneity to the final composition and to supplement the emollient and hydrophilic properties of the organic filler. Commercially available compounds representative of this class are Polychols, Solulan, Atlas products G-1441, G-1471 and ethoxylated lanolin derivatives such as Lanogel and Ethoxylan.

Unlike depilatory pastes containing thiol derivatives, which are inherently unstable because of the decomposition of their active agents in the presence of water and give rise to an offensive odour, the depilatory compositions described in the patent, being anhydrous, are claimed to be indefinitely stable.

Another series of patents[34] describes an emulsified depilatory containing a Kritchevsky-type alkanolamide condensate which, it is said, may stay in contact with the skin for extended periods without causing irritation or other side-effects.

Sliwinsky[35] has described an aerosol depilatory which is sprayed onto the skin and then expands to a foam; this is based upon calcium/sodium thioglycollate and Polawax (Croda). A product which behaves similarly has been patented by Webster;[36] this utilizes lithium/sodium thioglycollate, thiolactate or thioglycerol together with sodium lauryl ether sulphate. International Chemical Co. have also patented[37] an aerosol depilatory foam product.

The EEC Cosmetics Directive[38] and the UK Cosmetic Products Regulations 1978[39] limit the use of thioglycollic acid in depilatories to a maximum of 5 per cent with a pH value not to exceed 12·65.

Other 'Thio' Compounds
Thioglycollic acid is the most economical and effective active agent of this type. However French legislation prohibits this material for products for home use. Several alternative 'thio' compounds have been used successfully and notable among these are 2-mercaptopropionic (thiolactic) acid, 3-mercaptopropionic acid, and 3-mercaptopropane-1,2-diol (thioglycerol).

An aerosol gel depilatory based on strontium thiolactate has been described.[40] A Swiss patent[41] claims a non-irritant, rapid-acting depilatory 'soap' bar:

	(10) *per cent*
Thiolactic acid	30·5
Urea-sorbitol complex	25·5
Strontium hydroxide	4·5
Calcium hydroxide	4·5
Sodium lauryl sulphate	35·0

A gelled depilatory based on thioglycerol with polyethyleneglycol ethers of fatty alcohols or alkylphenols has been described.[42]

Bristol-Myers took out a patent[43] in which the active depilating agent is a molecular complex (1 : 1, 0·5 : 1, or 1 : 0·5) of thioglycerol with a nitrogen base of the Deriphat or Miranol type. These are said to be highly effective depilators at pH 8·5–10·5.

A US patent[44] claims the use of 1,4-dimercaptobutanediol and related compounds in which the hydroxyl groups are substituted by alternative polar groups.

Enzymes
Depilatory preparations based on the enzyme keratinase have also been developed which do not have the unpleasant odour of sulphide or even thioglycollate depilatories, and are non-irritant. However, they are not quite as effective. Keratinase was isolated from *Streptomyces fradiae* by Noval and Nickerson[45] and found to be capable of digesting keratin.

Keratinase is used in depilatory preparations in a purified form, buffered to a pH within the range of 7–8, with an activity of 200 k units per mg. The k unit is defined as the amount of enzyme which will digest wool keratin so as to produce an increase in optical density of 0·04 at 280 nm.

A paste depilatory described in an embodiment example of a patent granted to Mearl Corporation[46] contained 3·3 per cent w/w of keratinase (200 k units/mg).

Facial Depilatories for Black Skin

The special problems of the black-skinned male in removing facial hair have been described by Shevlin[47] and de la Guardia.[48]

The facial hair of the black male is often curly and wiry. Shaving hairs leaves the exposed ends with sharp points and as the hairs regrow these sharp points can actually turn back onto and penetrate the skin, causing a clinical condition called *pseudofolliculitis barbae*. For this reason many black men will not shave but prefer to use a depilatory which not only gives a closer 'shave' but leaves the hair tip soft and blunt so that it does not puncture and re-enter the skin.

Conventional thioglycollate-based depilatories take 15–20 minutes to remove beard hair which is regarded as far too long. Effectiveness outweighs cosmetic elegance and the tendency is to use a powder depilatory which must be mixed with water before use but which effects adequate hair removal in 3–7 minutes.

In the USA the active ingredients commonly used in powder depilatories[48] are barium sulphide and calcium thioglycollate. The former is the most popular because of its effectiveness and despite its offensive odour. However, powders based on calcium thioglycollate are said to be gaining favour since they are less odorous and therefore more easily perfumed, though less effective.

It should be borne in mind, however, that in EEC countries barium salts are prohibited from use in cosmetics and that the upper limit for calcium thioglycollate (trihydrate) is 10 per cent (corresponding to 5 per cent thioglycollic acid).

The 'Ideal' Depilatory

The 'ideal' depilatory, one which would have no odour, would remove hair in about 1 minute, could be used regularly without any complaints of irritation; one that could replace the daily task of shaving has yet to be devised. When such a product arrives it will certainly have a market.

Some interesting avenues of exploration were opened up by studies on the depilating effect of certain unsaturated compounds.

Ritter and Carter[49] reported as follows:

> . . . When commercial polymerization of chlorobutadiene was undertaken in an aqueous emulsion, some of the employees began to lose their hair after a few weeks or months. The loss of hair could be much reduced by frequent and complete change of the air in which they were working. The hair loss was predominantly on the scalp, and hair always grew again when the employees were transferred to other occupations. Microscopical examination of the hair showed that it came out at the root, a bulb being present at one end of each hair shaft, but there was no evidence of change in the shaft itself. A composite product containing various cyclic and short-chain polymers of chlorobutadiene was prepared: when two drops of this were placed on the back of a mouse or guinea pig, the area covered by the solution became completely denuded of hair in 4–10 days. In about 3 weeks hair was again visible over the denuded area and in 6 weeks it had completely regrown . . .

The interesting part of this observation is that previous developments have been in connection with preparations which are intended to swell the hair near the mouth of the follicle and reduce its tensile strength to an extent that allows it to be 'cork-screwed' away very readily. Since the hair and skin are both composed of keratin there is not a very wide margin between what will attack the one and irritate the other. The above observation shows the possibility of a different method of attack: not damage to the hair (and skin) *per se*, but merely a detachment of the hair bulb from the hair follicle—such damage being, apparently, temporary since the hair grew again in from 3–6 weeks.

Another observation which could be of some relevance to the development of new depilatories was made by Flesch,[50] who found that squalene and a group of lipoid-soluble unsaturated compounds (with —C=C— groups) caused reversible complete baldness when applied to the skin of laboratory animals. These compounds include the synthetic dimers of chloroprene and certain esters of allyl alcohol, naturally occurring vitamin A, oleic and linoleic acids. Flesch postulated that the depilation is caused by the reaction between the —C=C— bonds and the sulphydryl group of epidermal protein. Although Flesch was not able to depilate human beings with any of these unsaturated depilatory agents, these findings suggested that human sebum may have an influence in hair growth.

Evaluation of Depilatory Efficacy

Yablonsky and Williams[51] described a procedure for determining the efficacy of depilatories. It involves the measurement of the cross-sectional diameter and the length of a hair immersed in a solution of a depilatory, and observing the time of maximum hair swelling. Sigmoid curves are obtained when both the length and width of swelling hair are plotted against time. The slope maxima of these sigmoid curves may be used to define an index of depilatory effectiveness *in vitro*. No test results were given. In a later paper,[52] Yablonsky describes a simplification of the method.

Elliot[53] designed a depilometer to simulate practical use conditions as closely as possible. In-use tests gave good correlation and rapid screening of formulation variables is possible using this technique.

Elliot found considerable variation in depilating times between individuals but in general leg hair is easier to remove than axillary hair, which is similar to head hair.

Table 11.1 Relative Efficacy of 'Thio' Compounds at 5 per cent Concentration, adjusted to pH 11·0–12·0

'Thio' compound	Average depilation time* (min)
2-Mercapto-ethanol	4·0
Thioglycerol	6·5
Thioglycollic acid	7·5
3-Mercaptopropionic acid	8·0
2-Mercaptopropionic acid	11·0
Thiodiglycol	15·0
Thiomalic acid	15·0

* Average of times for all alkalis used for neutralization. Some combinations were found to be more effective than others.

Various 'thio' compounds were studied together with the effect of different alkalis, pH values and concentration. Table 11.1 summarizes the relative efficacy of various 'thio' compounds at a concentration of 5 per cent with sufficient alkali to adjust the pH value to 11·0–12·0. Table 11.2 gives similar data found for different alkalis. Elliot showed convincingly that a concentration of 5 per cent 'thio' compound is sufficient; increased concentrations do not increase the speed of depilation appreciably.

An earlier L'Oreal patent[29] reported the results shown in Table 11.3.

Table 11.2 Relative Efficacy of Various Alkalis

Hydroxide	Average depilation time (min)
Sodium	5·7
Lithium	5·7
Potassium	6·5
Barium	6·5
Calcium	7·1
Strontium	8·3

Table 11.3 Depilatory Activity of Thioglycollic Acid and Thiolactic Acid Neutralized with Various Alkalis (mercapto–acid concentration 0·4M, pH 12·5)

Cation	Thioglycollic acid		Thiolactic acid	
	Depilation time (min)	Effect on skin	Depilation time (min)	Effect on skin
Calcium	7	0	10	0
Barium	12	0	7	0
Strontium	5	0	7	0
Sodium	4	+	$5\frac{1}{2}$	+
Potassium	$3\frac{1}{2}$	+	4	++
Lithium	$3\frac{1}{2}$	0	5	0

REFERENCES

1. Alexander, P., *Am. Perfum. Cosmet.*, 1968, **83**(10), 115.
2. Barry, R. H., Depilatories, in *Cosmetics Science and Technology*, 2nd edn, Vol. 2, ed. Balsam M. S. and Sagarin E., New York, Wiley-Interscience, 1972.
3. Rieger, M. M. and Brechner, S., Depilatories, in *The Chemistry and Manufacture of Cosmetics*, 2nd edn, Vol. 4, ed. De Navarre M. G., Orlando, Fla., Continental Press, 1975.

4. Rieger, M. M., *Cosmet. Toiletries*, 1979, **94** (3), 71.
5. De Navarre, M. G., *The Chemistry and Manufacture of Cosmetics*, New York, Van Nostrand, 1941, p. 532.
6. Anon., *Drug Cosmet. Ind.*, 1949, **65** (November), 575.
7. US Patent 2 417 882, Neary, L. J., 25 March 1947.
8. US Patent 2 067 909, Lucas, H. V., 19 January 1937.
9. French Patent 1 396 582 through *Recherches*, 1967, 16 (December), 241.
10. German Patent 2 115 423, Jung, W. A., 1972.
11. US Patent 3 563 694, Ganee, M., 1972.
12. US Patent 3 808 637, Lapidus, H., 1974.
13. German Patent 2 253 117, Henkel, 1974.
14. British Patent 1 348 760, Dupuy, J., 20 March, 1974; see also US Patent 3 711 371, 1973; German Patent 2 261 057, 1974; French Patent 2 137 112, 1972.
15. Maxwell-Hudson, C., *The Natural Beauty Book*, London, Macdonald and Jane's, 1976, p. 106.
16. Gallant. A., *Principles and Techniques for the Beauty Specialist*, London, Stanley Thornes, 1975, p. 188.
17. US Patent 450 032, Perl, J., 7 April 1891.
18. US Patent 1 899 707, McKee, R. H. and Morse, E. H., 28 February 1933.
19. US Patent 2 123 214, Stoddard, W. B. and Berlin, J. 12 July 1938.
20. British Patent 516 812, Stoddard, W. B. and Berlin, J., 9 July 1938.
21. French Patent 840 552, Stoddard, W. B. and Berlin, J., 11 July 1938.
22. Turley, H. G. and Windus, W., *Stiasny Festschrift*, Darmstadt, Edouard Roether, 1937, p. 396.
23. French Patent 824 804, Fletcher, W., 17 February 1938.
24. British Patent 484 467, Bohemen, K., 27 April 1938.
25. US Patent 1 973 130, Rohm & Haas, 11 September 1934.
26. US Patent 2 352 524, Sales Affiliates Inc. (Evans, R. L. and McDonough, E. G.), 1944.
27. Evans Chemetics, *Technical Bulletin: Depilatories*, April 1967; and Supplement 13 January 1969. (Now Evans Chemetics, Organic Chemical Division, W. R. Grace & Co., Darien, CT 06820, USA)
28. French Patent 1 345 572, L'Oreal, 4 November 1963; see also US Patent 3 271 258, 1966.
29. French Patent 1 359 832, L'Oreal, 23 March 1964; see also German Patent 1 236 728, 1967; British Patent 1 030 362, 1966; US Patent 3 384 548, 21 May 1968.
30. US Patent 3 981 681, Carson Chemical Co., 1976; British Patent 1 463 966, 9 February 1977.
31. French Patent 2 168 202, Morillère, G., Coupet, J. and Navarro, R., 1973.
32. German Patent 2 603 402, Hagarty, P. M. *et al.*, 1976.
33. British Patent 1 116 259, Chemway Corp., 1968; see also US Patent 3 194 736, 1965.
34. US Patent 3 154 470, Chemway Corp., 27 October 1964; see also French Patent 1 411 308, 1965; British Patent 1 064 388, 1967.
35. British Patent 1 142 090, Sliwinski, R. A., 1969.
36. British Patent 1 296 356, Webster, I. K. R., 15 November 1972.
37. British Patent 1 264 319, International Chemical Co. 23 February 1972.
38. EEC Directive 76/768/EEC, *Off. J. European Communities*, 1976, **19** (L262).
39. *The Cosmetics Products Regulations 1978*, SI 1978 No. 1354, London, HMSO.
40. French Patent 1 405 939, Roy, R., 1965.
41. Swiss Patent 525 674, Silvestri, R., 1972.
42. British Patent 1 488 448, Beecham (UK), 1977.
43. US Patent 3 686 296, Bristol-Myers Co., 22 August 1972.
44. US Patent 3 865 546, Collaborative Research Inc., 11 February 1975.

45. Noval, J. J. and Nickerson, W. J., *J. Bacteriol*, 1959, **77,** 251; see also US Patent 2 988 487, 1961; US Patent 2 988 488, 1961.
46. US Patent 2 988 485, Mearl Corp., 13 June 1961.
47. Shevlin, E. J., *Cosmet. Perfum.*, 1974, **89**(4), 41.
48. De La Guardia, M., *Cosmet. Toiletries*, 1976, **91**(7), 137.
49. Ritter, W. L. and Carter, A. S., *J. indust. Hyg.*, 1948, **30,** 192.
50. Flesch, P. *J. Soc. cosmet. Chem.*, 1952, **3**, 84.
51. Yablonsky, H. A. and Williams, R., *J. Soc. cosmet. Chem.*, 1968, **19,** 699.
52. Yablonsky, H. A., *Am. Perfum. Cosmet.*, 1970, **85**(2), 41.
53. Elliot, T. J., *J. Soc. cosmet. Chem.*, 1974, **25,** 367.

Chapter Twelve

Shaving Preparations

The typical man tends to regard the removal each day of 20 000–25 000 terminal hairs protruding 250–500 μm from the skin at angles of 30 to 60 degrees[1] and covering a facial area of 250 cm^2, as something of a chore. To minimize the trauma of shaving, a wide range of preparations is now available that prepare the beard and face for shaving, increase speed and comfort during the shave and confer a feeling of well-being after shaving. The choice of shaving preparation is highly individualistic; however, it is generally recognized that different forms of beard preparation are required for 'wet' (razor blade) and 'dry' (electric razor) shaving. This results from the contrasting mechanisms of hair cutting, which can be inferred from the appearance of the ends of hairs cut by the two implements. The description of shaving preparations has therefore been divided into three sections: wet shaving preparations, dry shaving preparations and after-shave preparations.

WET SHAVING PREPARATIONS

Introduction

The main functional requirements of a wet shaving preparation are to soften the beard, to lubricate the passage of the razor over the face and to support the beard hair. In addition, the preparation should be non-irritating to the skin, should assist in removing shaving debris from the face, should be stable over a range of temperatures, resistant to rapid drying out and collapse, non-corrosive to the razor blade and easily rinsed from the razor and face. There is good evidence for the hair softening and lubrication functions of the shaving preparation, but little has been reported on the hair-supportive role.

Beard Softening

Beard softening results from changes in the mechanical properties of hair by absorption of water. Hair absorbs 31 per cent of its dry weight of water at 100 per cent relative humidity; the relationship between water absorption and relative humidity is non-linear and the swelling of hair is highly anisotropic.[2] The force required to cut water-saturated beard hair is about 65 per cent less than that for dry hair.[3]

The hydration of hair is accelerated by increases in temperature; however, views differ on the time taken to hydrate hair completely. This ranges from 2 minutes at room temperature, as measured by the force required to cut beard hair,[3] through $2\frac{1}{2}$ to 3 minutes at 49°C (120°F), as measured by creep of scalp hair extrapolated to the thicker beard hair,[1] to 6 minutes at 43°C (110°F) as measured by changes in the elasticity of hair.[4]

The established view on beard softening is based on measurements of the creep and elasticity of hair, supported by practical shaving tests. This suggests that the rate of softening of the beard can be increased by the addition of a wetting agent to the water, increasing the pH of the water and the removal of sebum from the hair. More recent work[3] suggests that the force required to cut the hair is not reduced by the use of wetting agents, soap solutions or shaving creams below the value for water alone. Similarly, changing the pH over the range 4·0 to 9·1 and the presence of sebum on the hair do not influence the cutting force. The importance of the shaving preparation in beard softening clearly differs according to which set of results is accepted.

Skin Lubrication
There is little published work on the contribution of skin lubrication to the comfort, closeness and speed of wet shaving. Early work by Naylor[5] on the coefficient of friction of plastic materials on skin indicated that friction was lower if the skin was dry, greasy or very wet, but higher if the skin was merely moist. Other work has shown that skin friction is reduced by surfactant solutions, mineral oils[6] and silicone fluids.[7] The force of friction on skin is not a linear function of the normal load[7,8] as suggested by Amonton's law, the deviations being attributed to the elastic behaviour of skin.

The frictional force between a razor blade and facial skin has been measured using a razor with built-in strain gauges.[9] The frictional properties of dry skin were shown to be higher than for wet skin although the absolute values vary for different areas of the face. The type of shaving preparation used does influence the frictional properties to the extent that it is possible to distinguish between different aerosol shaving foam formulations. It is generally found that the second stroke of the razor over a given part of the face yields a higher frictional force than the first stroke, presumably because the first stroke effectively removes most of the shaving preparation. Shavers apparently adjust the applied load on the razor according to the shaving preparation used; the lower the frictional force between the razor and skin, the higher is the applied load and the closer the resultant shave.

One can only speculate on the mechanism of lubrication by the shaving preparation since it depends on the load applied to the razor, the area of contact with the face, the velocity of the razor across the face and the viscosity of the preparation. At high loads per unit area and low shaving speeds, boundary lubrication is likely to predominate so that for a low coefficient of friction the shaving preparation should have a high viscosity and form a condensed film which interacts strongly with skin to preserve the integrity of the lubricant film. At low loads per unit area and high shaving speeds, hydrodynamic lubrication is likely to predominate. The viscosity of the shaving preparation should be high enough to give a film thickness sufficient to prevent asperity contact, but thereafter the viscosity should be as low as possible.

Beard Softening Cream

For many, washing the face with soap and water is an adequate pre-shave preparation for the attainment of a satisfactory shave with a razor blade and shaving foam. Where this is felt to be inadequate, pre-shave preparations are available to wet and soften the beard and to lubricate the skin; these are often referred to as beard softeners. They are particularly helpful to those with easily abraded skin or large diameter beard hair (since the hydration time varies as the square of the radius of hair, assuming that the diffusion of water into hair obeys Fick's laws). Brushless, non-lathering shaving creams, although satisfactory in terms of their lubricating action, often do not soften the beard sufficiently quickly or adequately. The application of a beard softener containing soaps, synthetic surfactants or possibly urea prior to the application of a brushless cream will allow a more complete wetting and softening of the beard and ensure a close and smooth shave. Such a formulation may contain a lime soap dispersing agent to improve the wetting action in hard water, a soap-compatible anti-bacterial agent, menthol and a preservative.

A beard softening cream recommended by Keithler[10] has the following composition:

	(1)
	per cent
Stearic acid	13·8
Stearyl alcohol	2·0
Isopropyl palmitate	1·9
Paraffin oil	2·0
Lanolin	2·0
Tween 60	2·4
Span 60	1·0
Triethanolamine	1·0
Dupanol C	1·0
Water	72·4
Perfume	0·5

Bell[11] gives another example of a beard softening preparation with the following composition:

	(2)
	per cent
Coconut oil fatty acids, double distilled	4·20
Oleic acid (with low linoleic acid content)	5·60
Propylene glycol	5·00
Triethanolamine	2·85
Monoethanolamine	1·26
Tergitol NPX	2·00
Demineralized water	79·09

Procedure: Mix the fatty acids together and stir into propylene glycol. Add the amines and stir until a clear solution is obtained. Finally mix in the Tergitol and perfume if required, followed by the water.

Tergitol NPX (alkyl aryl polyethylene glycol ether) is used in example 2 to disperse insoluble lime soaps and to improve the wetting action in hard water. Pre-shave liquids, creams and gels based solely on synthetic surfactants have been developed from hair shampoo formulations and these are particularly effective beard softeners in hard-water areas.

Lather Shaving Cream

Criteria for a Good Lather Shaving Cream
The undoubted success of foamed shaving preparations is probably due to their economy in use and ability to supply water to the beard by drainage through the plateau borders formed at the junctions of bubbles in the foam, thereby maintaining the hair in a fully water-saturated condition. The requirements of a good lather shaving cream are as follows:

(1) It must produce a rich copious lather composed of small bubbles.
(2) It must be non-irritant.
(3) It must have good wetting properties.
(4) It should be smooth, soft and entirely free from lumps.
(5) It must adhere readily to both face and brush and yet be easily removed on rinsing.
(6) It must retain a satisfactory consistency and texture over all temperature conditions likely to be encountered in use.

When evaluating foamed shaving preparations, such as lather shaving cream or aerosol shaving foam, attention should be paid to be following points:

Ease of transfer to the face.
Ease of spreading on the face.
Wetting and drainage properties of the foam.
Comfort and closeness of shave.
Foam texture, rigidity and rheology.
Foam stability.
Ease of removal of the lather and shaving debris from the razor and basin.
Acceptability and compatibility of the perfume.
Compatibility with the container.
Effect on the life of the razor blade.

Formulation
Lather shaving creams are concentrated dispersions of alkali metal soaps in glycerol and water. To maintain the desired level of foamability, consistency and product stability, careful control of the manufacturing process is essential. Even the slightest change to the formulation or manufacturing procedure can result in a disastrous phase separation of the cream at slightly elevated temperatures. One must therefore be prepared for problems in the scale-up of laboratory formulations.

Lather shaving creams normally contain 30 to 50 per cent soaps. Formulations based on stearic acid alone do not produce a sufficiently voluminous lather and it is usual to add some coconut oil fatty acids. The ratio of stearic acid to coconut oil varies considerably in different products but the inclusion of about 25 per cent coconut oil with 75 per cent stearic acid will usually be found satisfactory. A mixture of sodium and potassium hydroxides is used to saponify the fatty acids. It has been suggested[12] that a 5:1 ratio of potassium hydroxide to sodium hydroxide with 3–5 per cent free fatty acids will give shaving creams of the correct degree of plasticity. A cream containing a high level of sodium soaps tends to be thick and stringy, from which it is often difficult to produce a good lather. Lather creams can be made with potassium soaps alone but these tend to be less stable.

In order to prevent premature drying out of the cream, up to 15 per cent of a humectant is usually added. This is normally glycerol, but sorbitol or propylene glycol can be used. Humectants also have the effect of making the creams softer; propylene glycol has the greatest influence on the texture of the cream. An improvement in the properties of the cream and the lather has been claimed[13] by replacing glycerol with 1,3-butylene glycol. Emollients such as lanolin, cetyl alcohol, mineral oil and fatty acid esters should be kept to a low level (1 per cent) if the lathering properties are not to be impaired. Other additives to lather shaving creams, such as synthetic surfactants as foam stabilizers, cooling agents, antibacterial agents, etc., are discussed more fully under aerosol shaving foams.

The pH value of a lather shaving cream is generally around 10. The apparent paradox of a high pH and free fatty acid can be explained in terms of the mixing process. Pockets of unneutralized alkali remain, even under well-controlled manufacturing conditions, because of the high viscosity of the product. As part of the quality control on lather shaving creams, the level of free fatty acid should be determined as a check on the efficiency of mixing during manufacture. The free fatty acid level is one of the factors which influences the maximum temperature at which the cream retains a stable consistency. Above this temperature, free fatty acid rises to the surface of the product, in a similar manner to the creaming of an emulsion, and the functioning of the cream is seriously impaired.

Free fatty acids and the less water-soluble stearate soaps are responsible for the characteristic pearlescence of lather shaving creams. The pearlescence is a result of the formation within the cream of liquid crystalline phases which, in addition to improving its appearance, can increase the stability of the foam generated. The rate at which the liquid crystalline phases form depends on the method of manufacture, in particular the rate of cooling and the amount of stirring. It is normal practice amongst manufacturers to store the cream for some time before packing to allow the structure to develop. The slight change in the consistency and foam stability during the immediate post-manufacturing period should be taken into account when testing production batches of lather shave cream.

Much of our knowledge on lather shaving creams is still empirical; however, further insight may be obtained by consulting the literature on soap manufacture. The following formulation will serve as a guide to experimentation, but the manufacture of a good shaving cream, which will stand up to various climatic conditions, is very much an art.

	(3)
	per cent
Stearic acid	30·0
Coconut oil	10·0
Palm kernel oil	5·0
Potassium hydroxide	7·0
Sodium hydroxide	1·5
Glycerin	10·0
Water	36·5
Perfume	*q.s.*

Procedure: Mix half of the stearic acid with the oils, melt by steam heat, and bring to a temperature of 75°–80°C. Run in the alkali, water and glycerin and stir well until saponification is complete. The remainder of the melted stearic acid is now added together with any water which may have been lost during manufacture. The perfume is added at 35°C.

An Atlas formula quoted by deNavarre[14] is as follows:

	(4)
	per cent
Stearic acid	36·0
Coconut oil	9·0
Potassium hydroxide	8·0
Sodium hydroxide	1·0
Sorbitol (70% solution)	3·0
Water	43·0
Preservative, perfume	*q.s.*

Procedure: Heat the coconut oil to 75°–80°C. Dissolve the alkalis in half of the water and add to the coconut oil. When saponification is complete, add melted stearic acid (70°C) in a thin stream followed by the sorbitol solution, preservative and the remainder of the water. The mixture is then cooled. The perfume may be added at 35°C, or after the emulsion has cooled to room temperature. A check is carried out for completeness of saponification and the free fatty acid content is adjusted to between 3 and 5 per cent. The product is eventually packed into tubes.

Lather Shaving Stick

A lather shaving stick can be prepared from a mixture containing 80 per cent fatty acid soaps, 5–10 per cent glycerol and 8–10 per cent water. The ratio of the fatty acids and the ratio of potassium to sodium soaps should be similar to those described under lather shaving creams. After mixing, the composition is chipped, dried and milled with any other components required, such as perfume, colour or an opacifier. The soap flakes are packed to the desired shape using a soap plodder.

Aerosol Shaving Foams

Aerosol shaving foams are oil-in-water emulsions in which propellant droplets, liquefied under pressure, form a substantial part of the oil phase. When the emulsion is discharged to the atmosphere, the dispersed propellant droplets vaporize, producing a foam consisting of propellant vapour bubbles surrounded by an aqueous surfactant phase.

Some of the early patents on aerosol shaving foams provide some useful pointers to the influence of the soap composition on the appearance and properties of the foam. The first aerosol shaving foam patent, granted to Spitzer,[15] protected the use of fluorocarbon propellants in aqueous soap solutions enclosed in a pressure-resistant container. Triethanolamine stearate at levels of 8–12 per cent was given as the preferred soap together with smaller amounts of triethanolamine soaps of coconut fatty acids to prevent gelling at low temperatures. Potassium soaps were also said to give satisfactory shaving foams but sodium stearate can only be used at very low concentrations because of its tendency to gel. A later patent granted to Colgate-Palmolive[16] suggested that triethanolamine soaps alone do not make a satisfactory aerosol shaving foam because of a tendency for the emulsion to foam inside the container. As a result, the dispensed foam contains large bubbles and a substantial proportion of the emulsion cannot be expelled from the container. The foams described in the patent contain from 4 to 15 per cent soaps, mainly triethanolamine stearate with minor proportions of potassium and sodium stearates, as in example 5.

	(5) *per cent*
Triethanolamine stearate	8·0
Sodium stearate	1·0
Potassium stearate	4·6
Water	72·5
Perfume	0·9
Borax	0·5
Propellant (fluorocarbon)	12·5

Another patent granted to Colgate-Palmolive[17] claimed that aerosol shaving foams containing less than 4 per cent of potassium soaps produced the best results in softening hair and reducing its resistance to cutting by the razor blade. Soaps of mono- and diethanolamines were also considered suitable for this purpose, but triethanolamine soaps were found to be ineffective. Since the foam produced by such dilute soap solutions was rather unstable, synthetic thickening agents were included in the compositions. Particularly preferred are water-soluble salts of polyacrylic acid and its derivatives, with a mean molecular weight between 100 000 and 200 000, used at concentrations ranging between 0·5 and 3 per cent. These polymers also provide additional lubrication for the razor blade on skin. A ratio of stearic to coconut fatty acid of 80:20 was claimed to give a better beard-softening effect than other fatty acid mixtures when used at low concentrations. A large proportion of free fatty acid is retained to improve lubricity and foam stability. Other lubricants such as cetyl alcohol or glycerol monostearate are also incorporated to improve the feel of the skin after shaving, while nonionic emulsifiers are present to enhance the emulsification of the free fatty acid.

An example of a shaving foam concentrate from the patent had the following composition:

	(6) *per cent*
Potassium soap from stearic acid/coconut oil fatty acids (80:20)	1·5
Potassium polyacrylate (polyacrylic acid mol. wt 100 000–200 000)	1·0
Polyvinylpyrrolidone	0·5
Stearic acid/coconut oil fatty acids (80:20)	3·0
Castor oil	3·0
Lauric acid diethanolamide	0·5
Polyoxyethylene sorbitan monolaurate	0·5
Perfume	0·5
Water	89·5

Guidance on Formulation

The following general guidance can be given in the formulation of aerosol shaving foams.

Fatty Acids. Saturated long-chain fatty acids containing 12 to 18 carbon atoms at a level of 7–9 per cent are the main components of aerosol shaving foams. Lower-molecular-weight fatty acids such as those found in unstripped coconut oil cause skin irritation. The ratio of the fatty acids can be varied widely to produce foams with different physical properties. The presence of stearic acid is not essential to an aerosol foam as might be inferred from the early patents. A high proportion of stearic acid in the fatty acid mixture tends to give stiffer foams and a reduction in the number of shaves per can. Replacing some of the stearic acid with lauric acid tends to produce softer foams and improves the expulsion characteristics.

Bases. Triethanolamine, potassium hydroxide or mixtures of the two are the preferred bases for the saponification of the fatty acids. Sodium hydroxide is rarely used and then only as a minor constituent. Mono- and diethanolamines are used occasionally but care is needed to avoid skin irritation. Triethanolamine soaps tend to give closer-knit foams than potassium soaps, particularly with fluorocarbon propellants.

It is common practice to adjust the quantity of base so that the formulation contains 1–3 per cent free fatty acid. The free fatty acid can improve the appearance and lubricity of the foam and, by complexing with the soap, increase foam stability. However, this may be at the expense of reducing the amount of available foam and increasing the rate at which the foam dries out on the face.

Surfactants. A wide variety of anionic and nonionic synthetic surfactants can be used in shaving foams to improve such properties as the emulsion stability (for example, self-emulsifying glycerol monostearate), the wetting properties of the foam (for example, sodium lauryl ether sulphate), the water dispersability of the foam and shaving debris (for example, polyethoxylated fatty alcohols, the foam stability (for example, lauric diethanolamide) and emolliency (for example, ethoxylated lanolins). Because of the complex nature of the interactions between surfactants, soaps and free fatty acids, their interfacial properties in the emulsion and foam are not easily predicted.

Humectants. Polyols such as glycerol, sorbitol or propylene glycol are usually added to shaving foam concentrates at a level of 3–10 per cent. By their ability to bind water, they reduce the tendency of the foam to dry out on the face.

Lubricants. To assist the passage of the razor over the face and to provide emolliency, additional lubricants such as mineral oils, silicone fluids, lanolin or isopropyl myristate can be included at a level of 1 to 2 per cent, to supplement the effects of the free fatty acid. Water-soluble polymers such as polyvinyl pyrrolidone, sodium carboxymethyl cellulose or polyacrylic acid and its derivatives can also improve lubrication and increase foam stability. Polyvinylpyrrolidone is said to act as an anti-irritant, that is, to reduce the irritancy caused by other compounds.

Propellants. Aerosol shaving foams contain either 7–10 per cent fluorocarbon propellant or 2·8–3·5 per cent hydrocarbon propellant. The fluorocarbon propellants are usually 40:60 to 60:40 weight ratio blends of dichlorodifluoromethane and dichlorotetrafluoroethane. The hydrocarbon propellants are mixtures of *n*-butane, isobutane and *n*-propane.

The higher the concentration of propellant, the lower the foam density, the stiffer the foam and the greater the number of shaves that can be obtained from a given weight of the emulsion. Foams having a density less than 65 $g\,l^{-1}$ are likely to be difficult to spread on the face and have little beard-softening capability.

In spite of the higher cost, fluorocarbon-propelled shaving foams became very popular, possibly because of the relative ease of forming close-textured, stable foams. Following the Rowland and Molina[18] hypothesis of stratospheric ozone depletion by fluorocarbon propellants and legislation in the USA, all US aerosol shaving foams are now based on hydrocarbon propellants. The best selling UK aerosol shaving foams are also based on hydrocarbon propellants.

Perfume. Soap-compatible perfumes are used at a level of 0·15–0·65 per cent.

Cooling Agents. Physiological cooling agents are often added to shaving foams to counteract the 'after-glow' associated with shaving. The most frequently used cooling agent is menthol at a concentration of 0·05–0·2 per cent. The volatility of menthol means that its cooling effect on skin is transient and its dominant odour is almost impossible to mask. A group of compounds ranging in chemical type from carboxamides to ureas to phosphine oxides have been shown to possess physiological cooling properties.[19] Many are as effective as menthol but without the disadvantages associated with the volatility of menthol. At a level of 0·1–0·2 per cent in shaving foams, the cooling effect can last for 5–15 minutes after application.

Colour. Foams may be coloured by the use of D&C or FD&C dyes. A very low concentration should be used to avoid staining the skin and towel.

Preservatives. Many shaving foams do not require a preservative; however, when necessary 0·2 per cent of a mixture of methyl and propyl *p*-hydroxybenzoate should suffice.

Antioxidants are sometimes required to avoid rancidity in formulations containing even low levels of unsaturated compounds.

Corrosion Inhibitors. Again these are not normally required with suitably lacquered containers. Borax (0·04 per cent, 10 mol) can be used with tinplate containers and 0·25 per cent of sodium silicate 35° Be solution with aluminium containers.

Bacteriostats, etc. 0·05 per cent trichlorohydroxydiphenyl ether (Irgasan DP300) and 0·05 per cent allantoin should reduce skin infections and promote healing of cuts.

Pilomotor Agents. It is claimed that a closer shave can be obtained by incorporating into the shaving preparation compounds having pilomotor activity—that is, ability to cause the contraction of the arrectores pilorum (hair follicle muscles). This contraction causes the beard hair to be pushed farther above the skin surface line by about 0·2–0·3 mm. A hair cut in the elevated position will retract below the skin surface as the follicle muscle returns to normal. Such patented compounds included: imidazolines,[20] for example, 2-(2′,5′-dimethoxy-4′,6′-dimethylbenzyl)-2-imidazoline; 2-amino-imidazolines;[21] morpholines,[23] for example 2-(3′-hydroxyphenyl)-morpholine; and 2-(phenylamino)-1,3-diazacyclopentenes-(2).[23]

A number of the above compounds can also be used to the same effect in lather shaving creams, brushless shaving creams and pre-electric shave lotions.

A statistical study of the formulation of aerosol shaving foams[24] examined the importance of a number of variables such as soap concentration, fatty acid type, free fatty acid concentration, polyol type and concentration, and propellant type and concentration. The concentrate was evaluated in terms of viscosity, pH, density and stability, while the discharge properties and foam were evaluated in terms of the number of shaves per can, residue in the can after discharge, foam density, foam strength, drying time and bubble size. A number of the findings of the study have been included in this section.

Example Formulations

	Fluorocarbon-propelled shaving foam	(7)
		per cent
A	Stearic acid	5·6
	Palmitic acid	2·2
	Isopropyl myristate	1·0
	Coco monoethanolamide	0·3
B	Sodium lauryl ether sulphate (40% solution)	3·5
	Triethanolamine	3·9
	Glycerol	5·0
	Water (deionized)	78·5
	Perfume	*q.s.*
	Concentrate	91·5
	Propellants 12/114 (40:60)	8·5

Aerosol shaving cream (Croda Chemicals Ltd[25]) (8)

		per cent
A	Stearic acid	4·0
	Lauric acid	2·0
	Liquid lanolin (Fluilan)	1·0
B	Cromeen*	3·0
	Triethanolamine	2·5
	Water (deionized)	87·5
	Perfume	*q.s.*
	Concentrate	92·0
	Propellants 12/114 (40:60)	8·0

* Cromeen (Croda Chemicals Ltd) is a substituted alkyl amine derivative of various lanolin acids.

Hydrocarbon-propelled aerosol shaving foam (9)

		per cent
A	Palmitic acid	5·0
	Lauric acid	1·0
B	Sodium lauryl sulphate	1·0
	Polyethylene glycol (400) monolaurate	0·5
	Polyacrylic acid (40% aq.) mol.wt 100 000	1·5
	Triethanolamine	2·0
	Potassium hydroxide	0·8
	Glycerol	5·0
	Water (deionized)	83·2
	Perfume	*q.s.*
	Concentrate	96·9
	Propellants, isobutane/propane	3·1

Procedure: The general procedure for making all aerosol shaving foams is to heat parts A and B separately to 75°C. Add A to B with vigorous stirring and allow to cool to 35°C, when the perfume is added. The aerosol container is charged when the concentrate has reached room temperature.

Consistent Aerosol Shaving Foam

The properties and expulsion characteristics of aerosol shaving foams change significantly as the container is emptied. Typically, the last 10–15 per cent of the product is wet and runny and may be poorly expelled. This is a familiar problem to the aerosol formulator and one which the consumer has reluctantly learned to accept. It is not possible to overcome this problem just by increasing the propellant content, since this causes the foam initially dispensed to be unacceptably stiff and dry. An interesting solution proposed by Mace and Carrion[26] involves the use of a concentrate formulation which absorbs just enough liquid propellant to give foam of the correct density. Excess propellant is added to form a discrete reserve layer which can supply all the vapour needed to fill the increasing head space. Hence the foam-forming composition remains unchanged as the container is emptied. This approach may impose constraints on the formulation and it would be necessary to educate the consumer not to shake the container before use. A versatile but relatively expensive means of obtaining a consistent product is provided by the use of a barrier pack, in which the propellant providing the driving force to expel the liquid concentrate is separated from the foam-generating composition by either a piston or a flexible bag. Foams with very uniform properties throughout the life of a barrier pack have been reported.[27] Such a container is used in the post-foaming gel example.

A novel means[28] of maintaining the consistency of an aerosol foam is to use propellant-swollen rubber as a source of additional propellant. Propellant vapour is released from the rubber only when the vapour pressure in the head space falls. Hence the properties of foam dispensed from a full container are unaffected by the presence of additional propellant. When the emulsion is expelled, the head space volume increases and there is a small but sharp drop in vapour pressure. The vapour pressure is restored to slightly below the original value by the release of vapour directly or indirectly from the reservoir and not solely from the emulsion. In this way the ratio of concentrate to propellant in the emulsion, which determines foam density, is not reduced to the same extent with the reservoir as would normally be the case.

The benefits of using a rubber reservoir have been demonstrated by the following formulation:[29]

	(10)
	per cent
Palmitic acid	5·0
Potassium hydroxide	1·0
Sodium lauryl sulphate	2·5
Lauric diethanolamide	1·5
Polyethylene glycol (6000) monostearate	2·0
Water (deionized)	88·0
Fill for a 6 oz aerosol container	*weight (g)*
Concentrate	177·0
Butane 40 propellant	7·6
Ethylene/propylene rubber	3·0

A conventional package without the reservoir would contain 5.1 g propellant (Butane 40) and 177 g concentrate.

After an appropriate period of storage to allow absorption of some of the propellant by the rubber, the total volume of usable foam was 25 per cent greater when the reservoir was used and the expulsion characteristics remained satisfactory until the container was empty. The proportion of usable contents was increased by the use of the reservoir from 79 per cent to over 95 per cent.

Heated Shaving Foam

The interest in heated shaving foams derives from the improvement in beard softening as a result of increasing the temperature. Heated shaving foam can be obtained either by an exothermic chemical reaction between components which are kept separate within the aerosol container or by bringing the foam into contact with a heat exchanger connected to a hot water supply. Products relying on an exothermic redox reaction require a dual dispensing aerosol value, the inner compartment containing hydrogen peroxide and the outer compartment containing the soap solution, propellant and a pyrimidine, a thiourea, a sulphite or a thiosulphite.

Whatever the means of heating the foam, conventional aerosol shaving compositions are generally unsuitable because at higher temperatures these form large unstable bubbles and the foam lacks the body necessary for satisfactory shaving. Compositions that are claimed to be suitable for heated aerosol shaving foams[30] are based on aqueous solutions of triethanolamine stearate, superfatted with free stearic acid. The presence of lower-molecular-weight fatty acids tends to reduce the stability of the foam at elevated temperatures. This can be compensated for to some extent by increasing the concentration of stearic acid which has the ability to thicken the heated soap solution. The inclusion of propylene glycol or of nonionic surfactants, which will produce much richer and more viscous foams at room temperature, tends to give foams of a watery consistency at elevated temperature.

Satisfactory formulations can be obtained with 8–15 per cent triethanolamine stearate, 1 per cent triethanolamine coconut oil soap and 2 or 3 per cent stearic acid. At lower levels of triethanolamine stearate (7–12 per cent) a satisfactory composition will be obtained with 2 per cent triethanolamine coconut oil soap and 3 or 4 per cent stearic acid.

Scum-free Aerosol Shaving Foam

Soap-based shaving preparations leave on the basin and razor an unsightly deposit that is not readily rinsed away. The deposit, known as lime soap curds or scum, is particularly noticeable in hard-water areas and consists of water-insoluble calcium and magnesium soaps and free fatty acids. The deposit can be reduced by the addition of lime soap dispersing surfactants to a conventional formulation. Even the best lime soap dispersing agent should be present at two to three times the concentration of soap to prevent scum formation. Formulations containing low concentrations of soap and high concentrations of surfactant produce unstable foams. It is possible to restore stability to the foam by the addition of long-chain fatty alcohols such as myristyl alcohol.[31] An example of a scum-free aerosol shaving foam containing a low concentration of soaps is as follows:

		(11) *per cent*
A	Palmitic acid	1·95
	Myristic acid	0·62
	Myristyl alcohol	2·10
B	Polyoxyethylene (20) cetyl ether	5·23
	Lauric diethanolamide	5·23
	Propylene glycol	0·82
	Glycerol	3·54
	Triethanolamine	1·54
	Water (deionized)	78·97
	Perfume	*q.s.*
	Concentrate	91·5
	Propellants 12/114 (40:60)	8·5
or	Concentrate	97·0
	Propellant (Butane 48)	3·0

Soap-free Aerosol Shaving Foams

The improvement in the foam stability in example 11 is brought about by molecular complex formation between the long-chain fatty alcohol and the polyoxyethylene fatty ether. The interaction of long-chain fatty acids and fatty alcohols with polyoxyethylene fatty ethers in fluorocarbon propelled shaving foams has been investigated by Sanders.[32] Many of these soap-free formulations showed increases in emulsion viscosity and stability and in foam stability and stiffness. However, it has been observed that many soap-free emulsions undergo an irreversible phase separation when maintained at 37°C for a few weeks, such that, even after vigorous shaking to re-emulsify the solid phase, the contents cannot be expelled as a foam.

Certain nitrogen-containing surfactants in combination with myristyl alcohol are claimed[33] to form stable soap-free aerosol shaving foams which do not undergo irreversible phase separation when stored at elevated temperatures. Such foams do not form a deposit in hard water and will often completely disperse the scum formed by the pre-shave wash with soap. With soap-free formulations it is possible to use hypoallergenic surfactants and the pH can be adjusted to the slightly acid value of skin.

An example of a soap-free aerosol shaving foam is as follows:

		(12) *per cent*
A	Myristyl alcohol	2·1
B	Dicarboxylated lauric imidazoline 40% solution (Cycloteric DL)	5·1
	Polyethylene glycol (1000) monolaurate	5·5
	Propoxylated polyol (Emcol CD-18)	0·7
	Glycerol	5·0
	Water (deionized)	81·6
	Perfume	*q.s.*
	Citric acid	*q.s.*
	Concentrate	97·0
	Propellant (Butane 48)	3·0

Procedure: Heat parts A and B separately to 75°C. Add A to B with vigorous stirring. The perfume is added after cooling to 35°C. The pH of the concentrate is adjusted to the desired value by the addition of citric acid.

A soap-free aerosol shaving foam containing anionic sarcosinate surfactants, given in Croda Chemicals Technical Literature,[34] is as follows:

		(13) *per cent*
A	Fluilanol	5·0
	Crodaterge LS	3·0
	Crodaterge OS	4·0
	Pentol mineral oil	1·0

		per cent
B	Propylene glycol	5·0
	Triethanolamine	1·0
	Water (deionized)	81·0
	Perfume	*q.s.*
	Concentrate	95·0
	Propellant isobutane/propane	5·0

Post-foaming Aerosol Gel

Products of this type are discharged from an aerosol container as a stable gel which, when spread on the face, is claimed to improve the wetting of the skin and beard. The foam is subsequently formed *in situ* on the surface of the skin by the vaporization of low-boiling-point aliphatic hydrocarbons. The product is packaged in an aerosol container with a barrier to separate the post-foaming gel from the propellant required for expulsion. The barrier pack ensures that a homogeneous gel is discharged, substantially free of bubbles, which can produce a self-generated lather of uniform consistency and density throughout the life of the product. The post-foaming gel consists essentially of an aqueous dispersion of soaps, water-soluble gelling agents and a post-foaming agent having a vapour pressure of 6–14 psi (41–96 kPa) at 37°C, for example saturated aliphatic hydrocarbons or halogenated hydrocarbons. Miscellaneous additives such as humectants, emollients, foaming aids, perfume, etc., can also be included in the formulation.

Two examples of post-foaming gels of different foam stiffnesses, given in a patent assigned to S.C. Johnson & Son Inc.,[35] had the following compositions:

	(14) per cent	(15) per cent
Stearic acid (95% purity)	2·000	2·250
Palmitic acid (97% purity)	5·800	6·500
Polyoxyethylene (2) cetyl ether	1·000	1·000
Hydroxyalkyl cellulose (Klucel HA)	0·067	0·075
Carbopol 934	0·180	0·225
Propylene glycol dipelargonate	2·750	2·750
Sorbitol (70% solution)	10·000	10·000
Propylene glycol	3·300	3·300
Triethanolamine	4·200	4·750
Water (deionized)	67·953	66·400
Fragrance, dye	*q.s.*	*q.s.*
n-butane	0·550	0·550
n-pentane	2·200	2·200
Foam stiffness	29·0 g cm^{-2}	69·8 g cm^{-2}

Procedure: Prepare the soap intermediate by adding an aqueous solution of sorbitol and triethanolamine to the fatty acids and polyoxyethylene (2) cetyl ether at 80°C. Add separate solutions of the Klucel HA in aqueous propylene glycol and Carbopol 934 in water to the soap intermediate at 27°C. Disperse the hydrocarbons in an equal volume of propylene glycol at 4°C and mix with the remainder of the formulation in such a way as to avoid trapping air in the gel.

The gel is immediately transferred to the inner compartment of a barrier aerosol dispenser and the valve crimped in place. The outer compartment is pressurized with about 10 ml of a mixture of propane and isobutane having a vapour pressure of approximately 46 psi (317 kPa) at 25°C.

Brushless or Non-lathering Cream

Brushless or non-lathering shaving creams are oil-in-water emulsions. They contain components similar to those in vanishing creams, the main difference being that the concentration of oils and emulsifying agents tends to be higher in the shaving preparations. Ideally, the cream should vanish on completion of the shave, leaving the face free from irritation and with a matt appearance. Since a too rapid disappearance of the cream would be deleterious to the comfort and closeness of the shave, it should be possible at the very least to rub any remaining cream into the skin after the shave.

The popularity of brushless creams is said to be due to their convenience, in that they eliminate the need for a brush and give a faster shave. The thick film of lubricant on the face can provide emolliency and protection to the skin by reducing razor drag during shaving. The lower pH value of brushless creams (7·5–8·5) lends support to the suggestion that they cause less irritation, particularly on broken skin, than lather creams of pH 10. The disadvantages of brushless creams are that more is required per shave than with a foam preparation, the cream is often difficult to rinse from the razor and it can leave the skin feeling greasy. Owing to the slower uptake of water from the emulsion by the hair, the beard-softening action of brushless cream is reported to be less effective than a foam preparation. This may result in a more rapid dulling of the blade edge. To promote beard softening, it is normally recommended that the face is washed with soap and water before applying the cream to the wet face.

Formulation

The oil phase of a typical brushless cream comprises: 4–10 per cent lubricant (for example mineral oil, long-chain fatty acid esters, petrolatum); 10–25 per cent stearic acid to provide a superfatting action and assist the characteristic pearlescent appearance of the cream; and 0–5 per cent emollient (for example lanolin, cholesterol, cetyl alcohol, stearyl alcohol, spermaceti). Spermaceti is widely recommended as a means of preventing the cream from vanishing too rapidly. The aqueous phase usually contains 1–5 per cent soaps (for example potassium or triethanolamine stearate), 0–5 per cent synthetic surfactant to improve emulsion stability, beard wetting and rinsibility (for example glycerol monostearate, sulphonated fatty alcohols, fatty acid amides), 0–1 per cent thickening agent which will also improve emulsion stability (for example gum tragacanth, sodium alginate, polyvinylpyrrolidone, polyacrylic acid and its derivatives), and 2–10 per cent humectant to prevent drying-out of the cream (for example glycerol, sorbitol, propylene glycol). It is normal to add a preservative, for example esters of *p*-hydroxybenzoic acid, to this type of formulation. Other additives, such as perfumes, cooling agents, bacteriostats, etc., as discussed under aerosol shaving foam preparations, can also be included.

The composition of a typical brushless shaving cream is as follows:

	(16) *per cent*
Mineral oil	9·0
Lanolin	0·5
Stearic acid	14·5
Carbopol 934	0·5
Triethanolamine	2·5
Triethanolamine lauryl sulphate	1·0
Glycerol	5·0
Water	67·0
Preservative, perfume	*q.s.*

Procedure: Heat the oil, lanolin and stearic acid to 75°C. Disperse the Carbopol 934 in cold water and add the glycerol, surfactant and triethanolamine. Add the oil phase to the aqueous phase at 75°C with vigorous stirring. Cool the mixture rapidly, the perfume being added at 45°C.

The composition of a brushless shaving cream quoted in *Seifen-Öle-Fette-Wachse*[36] is given in example 17.

	(17) *per cent*
Stearic acid	18·0
Lanolin	4·0
Propylene glycol monostearate	4·0
Isopropyl palmitate	4·0
Glycerol	2·0
Triethanolamine	1·0
Water	66·8
Perfume	0·2

The presence of silicone oil in a brushless shaving cream is said to impart a pleasant 'feel' on application and give enhanced razor glide. A formulation given in a *Union Carbide Bulletin*[37] had the following composition:

		(18) *per cent*
A	Stearic acid	18·0
	Mineral oil	5·0
	Silicone fluid L-45 (1000 cS)	1·0
	Polyoxyethylene (20) sorbitan monostearate	5·0
B	Sorbitol (70% solution)	5·0
	Borax	2·0
	Triethanolamine	1·0
	Water	63·0
	Preservative, perfume	*q.s.*

Procedure: Heat the oil phase to 90°C; heat the aqueous phase separately to 95°C and add to the oil phase with stirring.

Carbowax 1500 may be used in place of oils to formulate so-called non-greasy brushless shave creams:[38]

		(19)
		parts
A	Carbowax 1500	45·0
	Stearic acid	37·5
B	Triethanolamine	3·0
	Potassium hydroxide	1·6
	Water	2·0
	Sodium alginate	3·5
	Water	178·0
	Propylene glycol	12·0
	Carbitol	15·0
	Perfume	*q.s.*

Procedure: Heat A to 70°C and stir in B. Add the sodium alginate dispersed in water at 70°C followed by the remaining components. The perfume, dissolved in part of the Carbitol, is added when the cream has cooled to 50°C; pour the cream at 45°C.

As with lather shaving creams, the rate of cooling and the amount of stirring can affect the consistency of the product. Cooling the product under vacuum will help to reduce aeration during the manufacture of the cream.

It is possible to obtain a satisfactory brushless shaving cream without using high concentrations of fatty acids. Self-emulsifying glycerol monostearate at a level of 10–25 per cent is a suitable emulsifier for the oils and this produces more translucent creams with better emollient properties than standard creams. An example of this type of cream is as follows:[39]

	(20)
	per cent
Glycerol monostearate	10·0
Mineral oil	3·0
Lanolin	5·0
Glycerol	3·0
Stearic acid	2·0
Potassium hydroxide	0·1
Water	76·9

Brushless Shaving Stick

Thomas and Whitham[40] describe a brushless shaving stick which can be applied directly to the face. It is claimed that the continuous thin smear left on wetted skin provides adequate lubrication for the shaving operation. The stick is composed of fatty or waxy materials to which hydrophilic properties have been imparted by a suitable emulsifier, for example soap or a partial fatty acid ester of a polyhydric alcohol. This ensures that the product is readily wetted by, but is not more than very sparingly soluble in, water. Pigment, dyestuff or opacifier is incorporated to indicate the presence of the composition on the face.

The following example is quoted:

	(21) *per cent*
Sesame oil	35·35
Spermaceti	45·80
Stearin	7·60
Soap	5·00
Monoglycerides of coconut oil fatty acids	3·00
Titanium dioxide	2·00
Perfume	1·25

Novel Compositions for Wet Shaving

There are a number of products, often referred to as shaving assisting compositions, which function in a slightly different manner from conventional soap-based shaving preparations. These novel compositions place a greater emphasis on the protection of skin during shaving by the provision of an effective layer of lubricant. They are applied directly to the face or to the blade edge and can be used without any other type of wet shaving preparation. Alternatively, they can be considered as a pre-wet shaving preparation.

An example[41] of a composition based on mineral oil is as follows:

	(22) *per cent*
Mineral oil	95·97
Dioctyl sodium sulphosuccinate (Aerosol OT)	1·44
Lanolin	0·96
Silicone fluid (DC 400)	0·67
Octadecanol	0·48
Fragrance	0·48
Preservative	*q.s.*

The preparation is applied to moist (but not wet) skin that has been washed with soap and warm water. The oils and octadecanol provide lubrication and act as a barrier to the evaporation of water, thereby preventing dehydration of the skin and beard hair.

A synergistic action between nonionic and ionic surfactants and silicone fluids is claimed[42] to protect the skin during shaving, facilitate the cutting of the beard and to protect the blade from corrosion between shaves. The compositions can be applied as a lotion or cream and can be used with or without a conventional shaving preparation.

An example of an alcoholic lotion is as follows:

	(23) *per cent*
Methylphenylpolysiloxane (DC-555)	6·6
Dimethylpolysiloxane (DC-200, 350 cS)	6·6
Polyoxyethylene (20) sorbitan monooleate (Tween 80)	6·6
Polyoxyethylene (20) sorbitan monolaurate (Tween 20)	6·6
Dioctyl sodium sulphosuccinate	0·5
Water	25·8
Ethyl alcohol (95%)	47·3

Other preparations make use of the lubricating properties of water-soluble gums and polymers, to which synthetic surfactants, humectants, emollients and oils can be added. Assertions have been made that the anionic soaps and surfactants present in shaving creams tend to emulsify sebum, with the result that the skin is deprived of its natural protection and is left dry, exposed to adverse atmospheric conditions and to the abrading action of the razor blade.

Shaving preparations based on oil-in-water emulsions have been proposed by Clairol.[43] These compositions contains dimethylpolysiloxane (200–500 cS) emulsified with 0·3 to 0·7 per cent polyoxyethylene lauryl ethers in 87–95 per cent water which has been thickened with 0·2–1·0 per cent Carbopol neutralized by triethanolamine to pH 6·5–7·2. The silicone fluid is said to form a protective layer on the skin, thereby preventing the emulsification of sebum, reducing razor drag and minimizing skin irritations.

DRY SHAVING PREPARATIONS

Introduction

It is generally recognized that electric shavers do not cut the beard as close to the skin surface as a razor blade. This was confirmed in a study by Bhaktaviziam *et al.*[44] which also showed that the ends of hair observed 24 hours after shaving with an electric razor showed ragged edges and some vertical splitting of the hair shaft. Both electric and blade shaving result in the removal of skin, the amount removed for an individual being dependent on the pressure applied to the face. Generally, the closer the shave the greater the amount of skin damage. It has been suggested[45] that pre-electric shave preparations may not increase the quality of the shave but may assist in reducing skin damage.

In contrast to blade shaving where it is preferable to soften the beard, when using an electric razor the beard should be dry with individual hairs raised and stiffened so that they can be caught between the razor's combs and removed. The removal of the film of perspiration from the face reduces the friction between the razor and skin and prevents the beard from being slippery and elusive to the cutting edge of the electric razor. This is achieved in different ways by the two most popular forms of pre-electric shave preparation: the lotion, based on an alcoholic solution, and the talc stick. It should be noted that a completely contrary view of the function of a pre-electric shave preparation has been expressed in a patent granted to the Sunbeam Corporation,[46] where it is claimed that the removal of moisture from the skin and beard prior to electric shaving is not desirable: that in fact water softens the beard and by causing hairs to swell and to become elongated ensures a smoother, more efficient and closer shave. Furthermore, alcoholic lotions are claimed to cause shrinking of the hair into the follicle, making it more difficult to obtain a close, clean shave. The preparations claimed in the patent are oil-in-water emulsions containing 5–20 per cent by weight of fatty acid esters such as isopropyl myristate, and an emulsifying agent which is a mixed alkali metal/amine salt of polyacrylic acid.

Pre-electric Shave Lotion

In formulating a pre-electric shave lotion the following attributes are considered desirable:

(1) Adequate astringency to stiffen the beard and possibly to stimulate the hair follicle muscles.
(2) Quick drying to allow rapid evaporation of any moisture present on the face.
(3) A pH below the iso-electric point of keratin to prevent swelling of the hair (that is, pH 4·5–4·8).
(4) Provision of a coating on the skin on which the razor will glide, thereby preventing irritation of the skin and providing lubrication for the cutting edge of the electric razor.
(5) Freedom from any substances likely to corrode the cutting head.
(6) Absence of any lubricants likely to have an adverse effect on plastic components of the electric shaver.

The alcoholic pre-electric shave lotions may be either astringent or oily. The astringent lotions are intended primarily to dry and stiffen the hairs and, theoretically at least, to assist in raising them. The astringent effect of the alcohol can be further enhanced by the inclusion in the preparation of mildly astringent substances such as aluminium chlorhydroxide, zinc phenolsulphonate or lactic acid. Menthol or camphor may be included to give a cooling effect together with a suitable antiseptic and a low level of lignocaine as an analgesic. Compounds having pilomotor activity may also be added to pre-electric shave preparations.

Lotions of the oily type aim to deposit on the face a film of lubricant which reduces the drag of the cutting head against the skin. It has been shown[7] that a film of silicone oil substantially reduces the frictional force between skin and a smooth steel probe. The mechanism involved is hydrodynamic lubrication—that is, the frictional force is dependent on the viscosity of the lubricant. Perhaps the most frequently used lubricants for this type of product are the esters of higher fatty acids such as isopropyl myristate. By suitable choice of lubricant type and concentration, it should be possible to provide for a comfortable shave even in warm humid conditions without leaving the skin feeling oily. It is claimed by some that the oily type of preparation lengthens the life of the cutting edge of the electric razor because of its lubricating action.

Example Formulations

Astringent pre-electric shave lotions	(24)
	per cent
Ethanol	45·0
Sorbitol	5·0
Lactic acid	1·0
Water	49·0

	(25)
	per cent
Zinc phenolsulphonate	1·0
Distilled extract of witch hazel	40·0
Ethanol	40·0
Water	18·8
Menthol	0·1
Camphor	0·1

	(26)
	per cent
Aluminium chlorhydroxide (50%)	5·0
Isopropyl myristate	5·0
Ethanol	80·0
Perfume	*q.s.*
Colour	*q.s.*
Water	to 100·0

Lubricant pre-electric shave lotions	(27)
	per cent
Isopropyl myristate	20·0
Ethanol	80·0
Perfume	*q.s.*
Antiseptic	*q.s.*

	(28)
	per cent
Ethanol	77·0
Isopropyl myristate	13·0
Oleyl alcohol (cosmetic grade)	4·0
Perfume	1·0
Distilled water	5·0
Colour	*q.s.*

A pre-electric shave lotion containing a pilomotor agent was quoted in a patent assigned to E. Merck A–G.[20] It had the following composition:

	(29)
	per cent
2-(2′,5′-dimethoxybenzyl)-2-imidazoline	0·1
Citric acid	2·5
Polyvinyl pyrrolidone	0·5
Isopropyl myristate	3·5
Alcohol (96%)	80·0
Perfume	*q.s.*
Water	to 100·0

Pre-electric shave lotions may be applied directly to the face by a roll-on type of applicator. In such circumstances it may be necessary to adjust the viscosity and wetting properties of the lotion to prevent seepage round the ball when the applicator is inverted.

Collapsible Foam Pre-electric Shave Lotion

To aid the transference of the pre-electric shave lotion from the hand to the face, quick-breaking aerosol foams have been developed. The foam is quite stable and confined to a limited area as dispensed, but breaks to a thin liquid when sheared or warmed by body heat. The foam concentrate typically contains 55–70 per cent of an aqueous ethanol solution, 4–10 per cent of a lubricant, 0·5–5 per cent of a surfactant which should be soluble in only one of the miscible solvents but form a clear homogeneous solution on addition of 3–10 per cent of a liquefiable propellant to the concentrate. The persistence of the foam when left undisturbed on the hand can be varied from a few seconds to several minutes depending on the proportions of alcohol, water and propellant and the type and concentration of surfactant. The mechanism of foam stabilization is complex but appears to rely on the partial insolubilization and loose molecular structure formation by the surfactant in the aqueous ethanol solution once the propellant has evaporated. The foam collapses on shearing because the bubble walls are extremely thin compared with those of soap-based aerosol shaving foams.

The surfactants found to be suitable for most quick-breaking aerosol foams are nonionic emulsifiable waxes composed of polyethylene glycol ethers of cetyl and stearyl alcohol and auxiliary emulsifying agents, for example Polawax A-31 (Croda Chemicals Ltd), Promulgen Types D and G (Robinson Wagner Co.). The addition of lubricant oils to the concentrate can cause problems of instability; however, a limited number of compounds have been shown[47] to possess the right combination of dry lubricity, solubility in aqueous ethanol solutions and initial foam stability. Examples include di-isopropyladipate, dimethyl sebacate, diethyl succinate and propylene carbonate.

The composition of a collapsible foam pre-electric shave lotion disclosed in a patent assigned to Yardley[47] was as follows:

	(30) *per cent*
Di-(2-methoxy-2-ethoxy)ethyl adipate	2·4
Denatured ethyl alcohol (95%)	68·1
Polawax A-31	4·9
Water	21·9
Isobutane	2·7

Pre-electric Shave Gel Stick

Solid pre-electric shave sticks of the cologne type can be formed by gelling ethanol with sodium stearate in the presence of glycerol and a suitable lubricant.

Pre-electric Shave Talc Stick

Talc is used as the main component in some pre-electric shave preparations to absorb perspiration and sebaceous secretions from the skin and to confer its characteristic slip so that the head of the shaver will glide smoothly over the face. A reduction of 50 per cent in the frictional force between skin and polished steel was observed[7] after treating the skin with talc. Colloidal kaolin is usually present in the preparation to improve the moisture-absorbing capacity and adhesion to the skin. Zinc or magnesium stearate is included to enhance both adhesion and slip. Magnesium carbonate or precipitated chalk serves as the carrier for the perfume as well as increasing the absorbent properties. An important stipulation is that powders for pre-electric shave purposes should be free from grit to avoid abrading the cutting edge of the electric razor. This can be achieved by grinding the powders before use.

The most convenient way to apply the talc preparation to the face is to form it into sticks. The sticks can be moulded from an aqueous dispersion of the powders using colloidal magnesium aluminium silicate (Veegum) as the binder.

A formula for a pre-shave talc stick was given in a Technical Bulletin of the R.T. Vanderbilt Co. Inc.[48] It had the following composition:

		(31) *parts*
A	Veegum	1·9
	Water	30·0
B	Zinc stearate	4·7
	Light magnesium carbonate	1·9
	Perfume	*q.s.*
	Talc	91·5

Procedure: Add the Veegum slowly to water with continuous agitation to produce a smooth dispersion. Absorb the perfume with the magnesium carbonate, add the zinc stearate and disperse them in the talc. Add A to the powder blend B and mix to a smooth paste. Pour into moulds and allow to dry until hard. The sticks are finally dried in an oven.

It is claimed that the resulting sticks do not break easily and that they have excellent rub-off qualities. The degree of rub-off and the strength of the stick can be controlled by the level of the binder.

A method for the manufacture of talc sticks without the use of a binder in the talc was disclosed in a US patent[49] which quoted the following formula:

	(32) *per cent*
Talc	50
Zinc oxide	10
Chalk	10
Kaolin	10
Colloidal silica	20

The sticks are formed by compression moulding the powder mixture at pressures ranging between 450 and 600 psi (3–4 MPa), then coating, except on the end, with a suitable film-forming polymer to protect them against cracking or crumbling.

Pre-electric Shave Powder

A loose powder which can be used as a pre-electric shave preparation is illustrated by example 33.

	(33)
	per cent
Talc	50·0
Kaolin	14·0
Magnesium carbonate	12·0
ANM powder (etherified starch)	10·0
Cetyl alcohol	3·0
Glycerol monostearate	1·0
Zinc stearate	4·0
Zinc oxide	5·5
Perfume	0·5

To overcome the problems of handling the powder and to avoid spillage, pre-electric shave powders have been packed in aerosol containers. A low-pressure propellant mixture and careful selection of actuator are required to produce a soft, dry spray which does not result in excessive 'bounce' of fine particles from the target surface.

Zinc oxide, zinc stearate, kaolin and calcium carbonate all tend to agglomerate in the presence of propellant and cannot be used in powder aerosols. Minor portions of colloidal silica (Aerosil), magnesium carbonate, magnesium stearate and starch can be used to improve the dispersion of talc in the propellants. Isopropyl myristate or mineral oil (0·5–1 per cent) can be used as a lubricant and also to aid dispersion. The total powder content of these compositions rarely exceeds 15 per cent of the total weight in order to avoid blockage of the valve or actuator.

Example 34[50] illustrates the possible composition of a pre-shave powder aerosol:

	(34)
	per cent
Talc	80·0
Aerosil	5·0
Starch	4·5
Magnesium stearate	5·0
Light magnesium carbonate	5·0
Perfume	0·5

This powder base is passed through a 200 mesh sieve and packed into aerosol containers as follows:

	per cent
Powder base	15
Propellant 11	60
Propellant 12	25

AFTER-SHAVE PREPARATIONS

Wet or dry shaving causes the removal of skin as well as hair from the face. The total quantity of skin and hair removed can vary by a factor of 4 or more, depending on the individual. Similarly, the percentage of skin in the shaving debris can range between 25 and 75 per cent.[1] Much of the skin removed is the epidermal horny layer which would be shed naturally without shaving. The skin trauma associated with shaving occurs when the outer horny layer is penetrated. Damage is most likely to occur at follicular hairshaft openings.[1] A second source of irritancy in shaving is from the shaving preparation. The degreasing effect of soaps and synthetic surfactants can increase skin permeability and allow alkali and other irritants to reach the Malpighian cells.[51]

The purpose of an after-shave preparation is to relieve the slight irritation or 'after-glow' and confer a pleasant feeling of comfort and well-being after shaving. This is achieved by giving a slight coolness, anaesthesia, mild astringency or emolliency to the skin. At the same time, the preparation should be antiseptic and help to keep the skin free from bacterial infection during the short time it takes to recover from the slight degree of injury inflicted during the shaving operation. The extent to which these properties are emphasized depends on the type of formulation. The materials used in after-shave preparations will be discussed principally in the context of lotions.

After-shave Lotion

In its simplest form, an after-shave lotion is a clear aqueous ethyl alcohol solution containing a perfume. The desired balance of mild astringency and coolness is achieved by controlling the ratio of ethyl alcohol to water. Analysis of popular UK brands of after-shave lotion shows that they contain 50–70 per cent by weight ethyl alcohol. Other countries, notably Germany because of the tax structure, use much lower concentrations of alcohol. US sources recommend 40–60 per cent by volume of alcohol to obtain the balance of properties; however, manufacturers of popular brands tend to use alcohol levels similar to those in the UK.

The commercial success of an after-shave lotion is largely dependent on the perfume and the way in which the product is marketed. Many different perfume types, for example spicy, chypre, sandalwood, leather and tobacco, have been successful. The creation of a stable balanced perfume, which is free from components likely to cause skin irritation or sensitization, is the province of the perfumer.

The chemical composition of the perfume determines the maximum concentration at which it can be used in a particular water–alcohol mixture. It may be necessary with some perfumes to increase the alcohol content in order to achieve the required perfume level. In a situation where it is undesirable to increase the alcohol content or reduce the fragrance level, it is common practice to use a solubilizer to obtain a clear lotion. Nonionic surfactants with a hydrophile–lipophile balance (HLB) number in the range 15–18 are often found to be the most effective solubilizers, although anionic surfactants have also been used. For more detailed information, reference should be made to published work[52–54] on the effectiveness of surfactants in solubilizing specific perfume oils. The solubilization of a perfume does not appear to reduce its stability or to cause a deterioration in the odour.[52] Sugar-based surfactants, for example sucrose esters, sucroglycerides and ethoxylated sucroglycerides may make useful perfume solubilizers in after-shave lotions since they are said[55] to cause less defatting of the lipid layer than the more usual polyoxyethylene derivatives of fatty acids and alcohols and are therefore less likely to cause skin irritation.

A perfume oil containing resins, terpenes and certain crystalline materials is more difficult to solubilize and may require several times as much surfactant as perfume oil. A perfume based on terpeneless oils, alcohols, compounds of low molecular weight and polar compounds will require less surfactant. The type and level of solubilizer required is determined by dissolving the perfume and surfactant in alcohol and titrating with water to the required alcohol/water ratio. The optimum level of surfactant is that which just gives a clear micellar solution that remains stable over an appropriate range of temperatures (0°–40°C).

Humectants and emollients are frequently added to after-shave lotions at levels not exceeding 5 per cent. Polyols such as glycerol, sorbitol and propylene glycol help to maintain the water content of the skin. Glycerol has the best humectant properties of the group, but propylene glycol is often preferred because of its greater solvent power, lower viscosity and higher volatility. The feel of the skin can be improved by the addition of long-chain fatty esters, for example isopropyl myristate or lanolin. Quantities are often limited by their low solubility in aqueous alcohol solutions. Water-soluble lanolin derivatives can be used at higher levels to provide emolliency and to assist in the solubilization of the perfume oil.

Cooling of the skin results from the evaporation of the alcoholic solution. The effect can be augmented by the physiological cooling effect of menthol. The level of menthol should be kept below 0·1 per cent because of its lachrymatory properties and because its odour can upset the balance of the perfume. Odourless cooling agents[19] may be more appropriate for this type of product. Menthol is also said to cause slight surface anaesthesia of skin; however, it is preferable to achieve this effect with lignocaine at a level of 0·025–0·05 per cent.

The mild astringency of the alcohol can be supplemented by witch hazel extract or even zinc and aluminium compounds such as zinc phenolsulphonate, aluminium chlorhydroxide or alcohol-soluble aluminium chloride complexes.

In a study of the effects of after-shave lotions on skin flora, Theile and Pease[56] showed that an aqueous solution containing 55 per cent by weight ethyl alcohol reduced the facial flora count by over 90 per cent immediately after application, the count returning to the pre-application level after six hours. With the same alcoholic solution containing a perfume, the skin flora count was again reduced by over 90 per cent immediately after application and was still 35 per cent below the pre-application level after six hours. This accords with the well-established antimicrobial activity of many perfume oils. Although alcoholic solutions reduce facial flora, tests *in vitro* demonstrate that they are unable to inhibit the growth of bacteria and fungi on agar plates. Some inhibitory action is found with fragranced alcoholic solutions, but a much greater effect can be achieved with quaternary ammonium compounds. The addition of 0·1 per cent benzalkonium chloride to an after-shave lotion was found to double the zone of inhibition of facial bacteria growth.

Cationic surfactants are wide-spectrum germicides which can kill or inhibit growth of organisms over a wide pH range. Quaternary ammonium compounds, for example cetyl trimethyl ammonium bromide, can be used in after-shave preparations provided that anionic surfactants have been used to solubilize the perfume. Anionic surfactants have some activity against Gram-positive and yeast organisms but are rarely effective against Gram-negative bacteria. Nonionic surfactants are not considered to be germicidal. However, a significant development is the finding that monoesters formed from polyhydric alcohols and lauric acid (Lauricidin) have germicidal activity and they are GRAS materials.[57] This offers the possibility of perfume solubilization and germicidal activity in a single compound. More powerful bacteriocides and fungicides are available but these would require careful evaluation to ensure compatibility with the lotion and freedom from skin irritation or sensitization.

Soap-based shaving preparations tend to leave the skin slightly alkaline, whereas the pH value of normal skin is 5 to 6. Formerly it was suggested that boric acid was a useful component in after-shave lotions, both as an antiseptic material and as a neutralizer of any residual alkalinity. There is a slight possibility of boric acid intoxication by absorption through damaged skin and therefore it is preferable to use lactic or benzoic acid.

Allantoin is frequently added to after-shave lotions at a level of 0·1–0·2 per cent to promote wound healing.

Example Formulations

A basic after-shave lotion, not requiring solubilization of the fragrance, can be made up as in example 35.

	(35) *per cent*
Ethyl alcohol, specially denatured	60
Propylene glycol	3
Water, demineralized	36
Perfume	1

Procedure: Dissolve the perfume and propylene glycol in the alcohol and add the water slowly, stirring well to avoid locally high concentrations of water precipitating the less soluble components of the perfume. Allow the solution to stand for several hours at about 4°C, then filter.

Antiseptic after-shave lotion[58]	(36) *per cent*
Hyamine 10-X (25%)	0·250
Ethyl alcohol	40·000
Menthol	0·005
Benzocaine	0·025
Water	59·720
Perfume	*q.s.*

Astringent after-shave lotion	(37) *per cent*
Witch hazel extract	15·00
Ethyl alcohol	10·00
Alum	0·50
Menthol	0·05
Ethyl *p*-amino benzoate	0·05
Glycerol	5·00
Water	69·40

An aerosol after-shave lotion (example 38) is given in a Technical Bulletin[59] by Esso Chemicals.

	(38) *per cent*
Hexadecyl alcohol	0·8
Ethyl alcohol	53·7
Distilled water	33·0
Polawax A-31	2·0
Perfume	0·5
Propellant 12/114 (40:60)	10·0

A formulation containing colloidal alumina[60] gives a cooling astringent lotion with excellent lubricity:

		(39) *per cent*
A	Baymal alumina	2·10
	Water	57·90
B	Alcohol (74 OP)	36·950
	Polyethylene glycol (400) distearate	3·000
	Menthol	0·025
	Camphor	0·025
	Preservative, perfume	*q.s.*

Procedure: Parts A and B are prepared separately using heat as required to dissolve the PEG distearate in the alcohol. The two parts are mixed cold.

Some components of perfumes are notorious for their instability when exposed to ultraviolet light. It is not unusual to find that, after exposure to direct sunlight for a few months, an after-shave lotion in clear glass bottles develops a characteristic 'bottle-odour', quite unlike the original fragrance. Accelerated testing of the ultraviolet light stability of the packaged product should therefore be included in the evaluation of after-shave preparations. Many manufacturers circumvent the problem by using opaque or chromium-containing glass bottles to package the lotion.

Quick-break Foam After-shave

To aid the transference of the after-shave lotion from the hand to the face, quick-breaking aerosol foams have been developed. The principles of the action of this type of product have been discussed in the section on collapsible foam pre-electric shave lotion.

A quick-breaking after-shave lotion from the General Chemical Division of Allied Chemical[61] is given in example 40.

		(40) *per cent*
A	Polawax A-31	1·50
	Ethyl alcohol (SDA No. 40)	62·10

		per cent
B	Menthol	0·05
	Camphor	0·05
	Perfume	0·30
C	Emcol E-607	0·20
	Allantoin	0·10
	Water (distilled)	35·70
	Concentrate	92
	Propellants 12/114 (20:80)	8

Procedure: Warm part A to 45°C to dissolve the Polawax; cool to 37°C and add part B. Heat part C to 80°C to dissolve the components; cool to 37°C and add to the solution of A and B. Fill while still warm.

Crackling Foam Aerosol After-shave Lotion

After-shave preparations have been formulated that are dispensed from an aerosol container as a foam and which exhibit a crackling sound when subjected to shear during application to the face. Compositions of this type are said to be oil-in-water emulsions, the continuous aqueous phase containing suitable emulsifiers and the oil phase consisting of the liquefied propellant and propellant-soluble materials. The propellant, a fluorochlorocarbon, constitutes 75–95 per cent by weight of the emulsion.

An example[62] of a crackling aerosol after-shave lotion is as follows:

	(41) *per cent*
Di-isopropyl adipate	0·778
Perfume	1·090
Menthol	0·060
Tergitol-XD*	0·500
Water	9·572
Propellant 114	88·000

* Monobutoxy ether of polyethylene-polypropylene glycols mol. wt 2500 (Union Carbide).

Procedure: Blend the first three components and stir the resulting mixture into water containing Tergitol-XD. Cool this emulsion to 1°C and add 17·6 parts by weight of Propellant 114, pre-chilled to 1°C. Stir the resulting mixture until a uniform emulsion is formed, when the balance of the Propellant 114 (70·4 parts by weight) can be added. The resulting mixture is placed in pre-chilled (1°C) aerosol containers and the valve is crimped in place.

After-shave Gel

An aqueous alcoholic gel can be formed by neutralizing a carboxyvinyl polymer with a base. The amount of gelling agent (usually less than 1 per cent) and the degree of neutralization controls the stiffness of the gel. The gel may optionally contain a physiological cooling agent and an emollient which is soluble in the alcoholic solution.

Example 42 gives an after-shave gel from a B.F. Goodrich Chemical Co. Bulletin.[63]

	(42) *per cent*
Ethanol	45·1
Water	53·0
Carbopol 940	1·0
Menthol	0·1
Di-isopropylamine	0·8
Perfume oil	*q.s.*

Procedure: Dissolve the perfume and menthol in the alcohol and then slowly stir in most of the water (a solubilizer can be used to obtain a clear solution). Disperse the Carbopol 940 in the aqueous alcoholic solution. Reduce the speed of the mixer and slowly add the di-isopropylamine dissolved in the small quantity of retained water.

The resulting product should be a crystal-clear gel, which is subject to degradation by ultraviolet light. It is preferable, therefore, to package it in coloured or opaque containers. Alternatively, UV stabilizers can be added to the formulation.

A similar formulation,[64] containing Viscofas X 100 000 (1 per cent) neutralized by tri-isopropylamine (0·1 per cent) in place of Carbopol and di-isopropylamine, can be prepared by dispersing the Viscofas in water at 90°C, allowing it to cool to room temperature before adding the alcohol and then the tri-isopropylamine.

Salts of glycyrrhizic acid, derived from licorice roots, can be used to form transparent gels at pH values from 2 to 6. The strength of the gel can be increased by the addition of water-soluble metal salts, for example zinc sulphate, alum, zinc phenolsulphonate.

Astringent after-shave gel[65]	(43) *per cent*
Dipotassium glycyrrhizinate	1·0
Citric acid	0·5
Zinc sulphate	0·2
Zinc phenolsulphonate	0·2
Ethyl alcohol	10·0
Propylene glycol	5·0
Water	83·1
Perfume, preservative	*q.s.*

Procedure: A mixture of dipotassium glycyrrhizinate, alcohol and propylene glycol is prepared and the perfume added. Heat this mixture to about 50°C and add an aqueous solution of citric acid, the zinc salts and preservative. The resulting solution is cooled to around 10°C and allowed to stand overnight.

After-shave Cream and Balm

The astringency of after-shave preparations containing more than 50 per cent by weight alcohol can be irritating to skin which has been excessively damaged by shaving or over-exposure to sun and wind. Increasingly, manufacturers are introducing creams or balms into their range of after-shave preparations. The formulations used to obtain the soft-textured, oil-in-water emulsions are often similar to vanishing or moisturizing creams. There is some advantage to be gained from using soap-free compositions in that the pH of the emulsion can be adjusted to the slightly acid value of normal skin.

Soap-free after-shave cream[66]		(44)
		per cent
A	Glycerol monostearate S/E (Teginacid H)	10·0
	Mineral oil	10·0
	Petrolatum	6·0
	Tegiloxan 100	0·5
	Lanolin	3·0
	Cetyl alcohol	3·0
B	Glycerol	3·0
	Citric acid	0·2
	Potassium aluminium sulphate	0·1
	Water	64·2
	Perfume	*q.s.*

Opaque hydro-alcoholic balm[67]		(45)
		per cent
A	Amerchol L-101	5·0
	Isopropyl lanolate	1·0
	Polyethylene glycol (1540) monostearate	3·0
B	Carbopol 934	0·5
	Water	60·0
	Triethanolamine	0·5
	Ethyl alcohol	30·0
	Perfume	*q.s.*

Procedure: Add the Carbopol 934 slowly to the water at room temperature with rapid agitation. Mix thoroughly until a thin cloudy dispersion is obtained. Heat the Carbopol solution and the oil phase A separately to 75°C. Add the Carbopol solution to the oil phase and stir for 5 minutes to emulsify before adding the triethanolamine. Cool with stirring to 38°C. Add the alcohol and perfume and continue cooling.

After-shave Powder

The main purpose of after-shave powders, as of all other after-shave preparations, is to alleviate any discomfort produced by shaving and leave the face cool and refreshed. Additional possible functions of an after-shave powder are to cover minor skin defects and to mask any unacceptable shine produced by an excessively oily brushless shaving cream, leaving the face with a smooth matt appearance. Further functions can be added if desired, for example, the cooling effect produced by menthol, mild astringency by the incorporation of an aluminium salt, antimicrobial activity, agents to sooth irritation and to promote healing of minor wounds.

The important properties of an after-shave powder are slip, adherence and absorbency. Covering power is less important in after-shave powders than in the closely related face powder compositions. The main component of an after-shave powder is talc which provides slip and absorbency. Absorbency can be improved by the presence of kaolin or magnesium carbonate, while covering power is provided by precipitated chalk, zinc oxide or titanium dioxide. Colour is obtained from cosmetic-grade pigments. An ochre will produce a light tan effect, whereas an iron oxide will give a light pink colour. A mixture of these two pigments will produce a tone which will make the powder less conspicuous on the average Caucasian skin. Such a powder must adhere well to skin and so metallic soaps, for example zinc stearate, are usually added to the formulation.

The method of manufacture is very similar to that of face powders and talcum powders. Control of particle size is important and if necessary mixing should be preceded by a grinding process. Powder-grade perfumes should be absorbed by precipitated chalk before incorporation into the powder mix.

REFERENCES

1. Hollander, L. and Casselman, E. J., *J. Am. med. Assoc.*, 1939, **109,** 95.
2. Bendit, E. G. and Feughelman, M., *Encyclopaedia of Polymer Science and Technology*, New York, Wiley, 1968, p. 8.
3. Deem, D. E. and Rieger, M. M., *J. Soc. cosmet. Chem.*, 1976, **27,** 579.
4. Kish, A. B., *Drug Cosmet. Ind.*, 1938, **43**(6), 664.
5. Naylor, P. F., *Br. J. Dermatol.*, 1955, **67,** 239.
6. Highley, D. R., Coomey, M., Denbeste, M. and Woolfram, L. J., *J. invest. Dermatol.*, 1977, **69**(3), 303.
7. El-Shimi, A. F., *J. Soc. cosmet. Chem.*, 1977, **28,** 37.
8. Comaish, S. and Bottoms, E., *Br. J. Dermatol.*, 1971, **84,** 34.
9. Unpublished work, Wilkinson Match Research.
10. Keithler, W., *The Formulation of Cosmetics and Cosmetic Specialities*, New York, Drug and Cosmetic Industry, 1956.
11. Bell, S. A., *Cosmetics—Science and Technology*, ed. Sagarin, E., New York, Interscience, 1966.
12. Thomssen, E. G. and Kemp, C. R., *Modern Soap Making*, New York, MacNair-Dorland, 1937.
13. Harb, N. A., *Drug. Cosmet. Ind.*, 1977, **121**(4), 38.
14. Atlas *Bulletin* L D-69, quoted by deNavarre, *Chemistry and Manufacture of Cosmetics*, Van Nostrand, New York, 1941.
15. US Patent 2 655 480, Spitzer, J. G., 1953.
16. US Patent 2 908 650, Colgate-Palmolive Co., 1959.
17. British Patent 838 913, Colgate-Palmolive Co., 1960.
18. Rowland, F. S. and Molina, M. J., *Nature (London)*, 1974, **249,** 810.
19. Watson, H. R., Hems, R., Rowsell, D. G. and Spring, D. J., *J. Soc. cosmet. Chem.*, 1978, **29**(4), 185.
20. German Patent 1 150 180, E. Merck AG, 1962.
21. German Patent 1 147 712, Dr Karl Thomae GmbH, 1961.
22. US Patent 3 296 076, Boehringer Ingelheim GmbH, 1967.
23. US Patent 3 190 802, Boehringer Ingelheim GmbH, 1965.
24. Carter, P. and Truax, H. M., *Proc. sci. Sect. Toilet Goods Assoc.*, 1961, (35), 37.
25. Croda Chemicals Ltd, technical literature.

26. Mace, H. and Carrion, C., *J. soc. cosmet. Chem.*, 1969, **20**(9), 511.
27. Richman, M. D., Contractor, A. and Shangraw, R. F., *Aerosol Age*, 1968, **13**(3), 32.
28. US Patent 3 858 764, Wilkinson Sword Ltd, 1975.
29. Thompson, J. and Peacock, F., *Aerosol Age*, 1978, **23**(11), 14.
30. US Patent 3 484 378, Carter Wallace Inc., 1969.
31. British Patent 1 423 179, Wilkinson Sword Ltd, 1976.
32. Sanders, P. A., *Soap Chem. Spec.*, 1967, July, 68.
33. British Patents 1 479 706, 1 479 707, 1 479 708, Wilkinson Sword Ltd, 1977.
34. Croda Chemicals Ltd, *Cosmetic and Pharmaceutical Formulary, Supplement.*
35. British Patent 1 279 145, S.C. Johnson & Son Inc., 1972.
36. Schweisheimer, W., *Seifen Öle Fette Wachse*, 1960, **86,** 609.
37. Union Carbide, *Bulletin* CSB 45-185 4/64, Formula C-117.
38. Carbide and Chemical Corp., technical literature.
39. Kalish, J., *Drug Cosmet. Ind.*, 1939, **45,** 173.
40. British Patent 537 407, Thomas, R. and Whitham, H., 1939.
41. US Patent 3 178 352, Erickson, R., 1965.
42. US Patent 3 136 696, Corby Enterprises Inc., 1964.
43. US Patent 3 314 857, Clairol Inc., 1967.
44. Bhaktaviziam, C., Mescon, H. and Matoltsy, A. G., *Arch. Dermatol.*, 1963, **88,** 242.
45. Tronnier, H., *Am. Perfum.*, 1968, **83,** 45.
46. British Patent 1 011 557, Sunbeam Corp., 1962.
47. British Patent 1 096 753, Yardley & Co. Ltd, 1967.
48. R.T. Vanderbilt Co. Inc., *Technical Bulletin* No. 539.
49. US Patent 2 390 473, Teichner, R. W., 1941.
50. Du Pont, *Aerosol Technical Booklet*, Series No. 2, 60.
51. Bettley, F. R., *Br. med. J.*, 1960, **1,** 1675.
52. Moore, C. D. and Bell, M., *Soap Perfum. Cosmet.*, 1957, **30,** 69.
53. Alquier, R., *Soap Perfum. Cosmet.*, 1959, **30,** 80.
54. Angla, B., *Soap Perfum. Cosmet.*, 1966, **39,** 375.
55. Nobile, L., Rovesti, P. and Svampa, M. B., *Am. Perfum.*, 1964, **79**(7), 19.
56. Theile, F. C. and Pease, D. C., *J. Soc. cosmet. Chem.*, 1964, **15,** 745.
57. Kabara, J. J., *J. Soc. cosmet. Chem.*, 1978, **29,** 733.
58. *Schimmel Brief* No. 190, January 1951.
59. Esso Chemicals Ltd., *Technical Bulletin.*
60. Du Pont, *Baymal Colloidal Alumina—Use in Cosmetic and Toilet Preparations.*
61. Allied Chemical Corp., *Aerosol Age*, 1960, **5**(5), 50.
62. British Patent 1 170 152, Rexall Drug and Chemical Co., 1969.
63. B. F. Goodrich Chemical Co., *Technical Bulletin.*
64. Anon., *Specialities*, 1968, **4**(5), 2.
65. Cook, M. K., *Drug Cosmet. Ind.*, 1974, **114**(4), 40.
66. T. H. Goldschmidt AG, *Technical Literature*, Formula 5.
67. Amerchol *Laboratory Handbook for Cosmetics and Pharmaceuticals.*

Chapter Thirteen

Foot Preparations

Introduction

Although a large proportion of the adult population appears to suffer foot discomfort in one form or another, it is really amazing how little personal attention is lavished upon the much maligned human foot, although foot fatigue affects adversely both the physical and mental wellbeing of the sufferer, and leads to a sharp decline in the individual effort. Moreover, this effect is cumulative and ultimately often ends in various forms of stomach ailment plus the original foot discomfort (tired and aching feet lead to ill-humour, and meals eaten in a bad temper to indigestion).

This is not really surprising when we consider that the human foot has to maintain the whole weight of the body. In spite of the fact that it is equipped with a multitude of sweat glands, unequalled in any part of the skin surface except the palmar surfaces of the hands, it is, under the dictates and necessities of modern civilization, encased first in a stocking and then in a shoe or boot. The organic components of perspiration which accumulate in the shoes provide, particularly under warm and humid conditions, a good substrate for the growth of various types of micro-organism. Although socks and stockings are frequently changed and washed and also foot baths are taken, nevertheless bacteria remain in the shoes and clean socks may be reinfected. Such infection may further be promoted by poor ventilation as caused, for example, by impermeable rubber and synthetic resin soles and uppers, or by nylon stockings. The products formed by bacterial decomposition give rise to malodours, which are particularly pronounced when foot care is inadequate.

Influence of Footwear

Hole[1] reports on the interrelationships between modern footwear and the health of the feet. The disposal of sweat from the foot is inhibited by footwear and the moisture absorption and transmission properties of old and new shoe materials was studied. Leather is extraordinarily effective in absorbing and transmitting water vapour. Manmade materials, on the other hand, are relatively imperme-able to moisture vapour and thus to sweat and can have therefore a considerable influence on foot health. Wear trials have shown that it is the properties of the upper material which have by far the greatest effect on sweat accumulation in shoes.

Sweat accumulation in footwear has a direct effect:

(i) on mechanical properties of foot skin;
(ii) on physicochemical properties of the shoe material;
(iii) on encouragement of microbiological growth in the skin and the shoe materials.

It must be emphasized that there is no substitute for a well-fitting shoe, and a foot that is anatomically healthy. A sufferer from a foot ailment must select sensible and properly fitting footwear and, if suffering from what is commonly known as weak arch (a common complaint in those persons whose work entails long hours of standing, for example shop assistants, policemen, plant operatives and laboratory technicians), requires an adequate foot support in the form of an arch support or foot easer to be fitted by an operator experienced in this branch of work.

Foot Malodours

Sehgal[2] points out that hard skin, which is a layer of flat cells (stratified squamous epithelial cells) bonded together by desmosomes and penetrating tonofibrils, is largely keratin, and this together with perspiration and fungi is a feeding ground for the skin's resident micro-organisms; bacterial decomposition gives rise to the bad odours. The contribution of *Staphylococcus epidermidis* and *Trichophyton floccosum* in creating shoe and foot odours has been established.[3] The bacteria responsible for the breakdown of perspiration and fungi impart a musty or mouldy odour to the shoe and foot and this seems to be due to the skin rather than the shoe. However, other odours can arise by interactions with the material of the shoe.

Foot Ailments

Neglect of the feet may lead to one or more of the following unpleasant conditions (Schroder[4]): penetrating odour of sweaty feet, caused by bacterial decomposition of the sweat and skin debris; burning and itching sensation between the toes; painful, tired and swollen feet; softening of the toenail bed; moist skin irritation, creating ideal conditions for fungal infections.

Common foot disorders have been summarized by Chalmers:[5]

Corns—caused by friction and pressure, do not have roots and may be due to structural deformity.
Bunions—misaligned joints which become swollen and tender; basically due to weakness of the muscle structure, but heredity and ill-fitting shoes may contribute.
Calluses—protective hard skin growth over areas where repeated pressure or friction occurs; as most calluses are symptoms of some underlying disorder they cannot be eliminated permanently until the basic cause is corrected.
Verrucae—often mistaken for corns, are contagious and usually quite painful; they tend to spread if not treated.

In-growing toenails—usually due to improper trimming but heredity, injury and infection may contribute.

Athlete's foot—a common skin disease of fungal origin.

Foot Infections

Chalmers[5,6] has given excellent reviews of this subject. The feet spend long periods covered with socks or stockings and encased in footwear. This results in persistently warm and moist conditions which form an ideal environment for microbial proliferation and activity. In consequence, socks and stockings should be changed and laundered daily; what is not so well appreciated is that, whenever possible, shoes should not be worn for two consecutive days but allowed to dry out.

In a hospital study reported by Chalmers[6] the incidence of tinea pedis (including tinea unguium) was 40·8 per cent in 348 men examined. Of the infections, 50 per cent were due to *Trichophyton interdigitale*, 26 per cent to *Trichophyton rubrum*, 0·7 per cent to *Epidermophyton floccosum*, 4·1 per cent to non-dermatophytes (nails only), 11·7 per cent to mixed infections and 7·5 per cent were microscopically positive but not identified.

The use of communal washing and bathing facilities is a major cause of spread of such infections. Even laundering failed to eliminate *T. interdigitale* from bedsocks and in the hospital study mentioned above it was suggested that infected socks were an even more important source of cross-infection than bathroom floors. The supreme importance of hygienic cleansing of communal bathrooms, showers, etc., is clear; regular cleansing with germicidal products to remove skin debris and microbial contaminants is essential.

The most prevalent kinds of foot infection are caused by fungi—mycosis. The well known mycosis 'athlete's foot' has almost become an international disease. Athlete's foot, also referred to as tinea pedis, trichophytosis pedis and ringworm of the feet, is a complaint usually caused by the fungus *Trichophyton mentagrophytes* and less frequently by another fungus *Epidermophyton inguinale*. A chronic variety of tinea pedis, which often affects the nails and which is very resistant to treatment, is caused by *Trichophyton rubrum*. Athlete's foot is not as widespread as many people believe it to be. It is reckoned that in the temperate climatic conditions prevailing in the United Kingdom its incidence in the population as a whole is not greater than about 4 per cent. The complaint is more common among athletes, students, schoolboys and, in general, among those who bathe communally. In its commonest form the condition results in maceration and peeling between the toes, while in warmer weather a more active eruption of vesicles may occur in the interdigital areas and on the soles.[7] Tinea pedis is generally confined to one foot at a time.

There are other conditions resembling ringworm, and for treatment to be effective a correct diagnosis must be made, preceded by microscopical examination and if necessary by culturing. In this connection mention must be made of another affliction of the feet, namely dermatitis of the feet, which may be produced by dyes and chemicals from socks, shoe leather or rubber. Unlike tinea pedis, however, the eruptions are symmetrical and do not affect the interdigital areas but only the back and side of the feet.

Hyperhidrosis of the feet may also cause redness and maceration of the skin of the soles and between the toes, but again it can be differentiated from athlete's foot by its symmetry and the fact that it is not confined to the interdigital areas between the fourth and fifth toes as is the case in athlete's foot.

As far as athlete's foot is concerned, the sources of infection are usually flooring and bath mats in bathing establishments and shower rooms in factories and mines used by barefooted individuals suffering from fungal infections of the feet. Dermatophytes affect, in the first instance, the horny layers of the skin, but eventually produce inflammatory symptoms which vary in intensity, and which in the case of interdigital mycoses are not very easy to treat because causative organisms stay deeply in the crevices of the skin.

Foot Care and Hygiene

It is an undeniable truth that the feet, while needing more care and attention than most other parts of the body, in actual fact generally get far less. Good foot care extends to the choice and treatment of socks, stockings and, above all, shoes.

The feet should be washed at least once a day with soap and water, and after washing should be dried thoroughly, especially between the toes, and dusted with a talcum or foot powder. If infection is suspected or likely to have been encountered, both the soap and the powder should certainly contain an antiseptic agent.

The use of bactericidal compounds in toiletries is becoming widespread but in many products such use may be dubious. However, in foot products which must cope with conditions highly favourable to micro-organisms, the use of bactericides and fungicides is certainly justified.

There are many types of foot preparation in existence. Some are used to provide relief for tired and aching feet, others to soften cornified skin or to combat foot perspiration, and there are those used for alleviating skin irritations, eruptions and infections and for providing an antibacterial or antifungal effect.

Products of general toilet utility can also be used for these purposes and even marketed as foot products; the basic requirements are no different from those of similar products used for other parts of the body. For instance, foot lotions may be formulated on the lines of a skin tonic or mild astringent lotion.

Bathing the Feet

Being an extremity of the body, and not having adequate exercise, the human foot, particularly when constricted in ill-fitting shoes, or by tight elastic or other suspenders around the leg, is prone to an inefficient circulation. Bathing the foot in hot water stimulates the blood circulation, eliminates stale sweat secretions and temporarily reduces bacterial infection. If the bath is alkaline it softens the hard keratinous layer of the skin, corns, calluses, etc.

Proprietary bath salts for the feet consist essentially of alkaline salts, often in conjunction with an oxygen-releasing substance such as sodium perborate. In some of the oxygenated foot bath salts, catalysts such as manganese borate and enzymes have been used to accelerate the release of oxygen. Other types of preparation are formulated so that the final bath solution approximates in its composition to sea water. Chalmers[5] proposes the following formulae for bath salts for the care of the feet. Examples 1 and 2 are powder alkaline oxidizing foot fresheners; examples 3 and 4 are mineral foot baths for relieving aches and strains in tired feet:

	(1) *per cent*	(2) *per cent*	(3) *per cent*	(4) *per cent*
Sodium sesquicarbonate (needles or powder)	94·4	94·0	—	—
Sodium lauryl sulphate (powder)	0·2	0·5	—	—
Methyl salicylate	0·2	—	0·1	—
Menthol	0·2	0·1	—	0·1
Eucalyptol	—	0·2	—	—
Pineneedle oil	—	0·2	—	0·1
Sodium perborate	5·0	—	—	—
Sodium percarbonate	—	5·0	—	—
Sodium dichloro-isocyanurate (6% active chlorine)	—	—	0·1	0·1
Dowfax 2A	—	—	—	0·2
Sea salt crystals	—	—	99.7	—
Magnesium sulphate crystals (Epsom salts)	—	—	—	99·4
Perfume	—	—	0·1	0·1
Colour	*q.s.*	*q.s.*	*q.s.*	*q.s.*
Bath dose: grammes per 4 pints ($2\frac{1}{2}$ litres) water at 40°C	20	20	150	150

Artificial sea-salt powders are sometimes used in foot baths and relief is sometimes attained by their use. Usually the main sea-water constituents of cheap commercial quality are purchased separately and mixed in an approximation of sea-water, the powder being put into packets. This may be tinted and perfumed. Example 5 is such a product. One part of this power is sufficient for 20 parts of water for a full strength foot bath.

	(5) *parts*
Potassium iodide	1
Potassium bromide	2
Magnesium chloride	250
Calcium chloride	125
Magnesium sulphate	250
Sodium sulphate	500
Sodium chloride	1500
Colour, perfume	*q.s.*

A bubbling foot bath tablet which releases oxygen and carbon dioxide, may be formulated as follows:[8]

	(6)	
	per cent	
Lathanol LAL	26·0	
Tartaric acid powder	26·0	
Salicylic acid	1·0	
Sodium bicarbonate powder	25·0	
Sodium sesquicarbonate powder	9·0	
Sodium hexametaphosphate	4·0	
Sodium perborate	5·9	
Perfume	1·5	or *q.s.*
Calflo E	1·5	
Colour: FD & C Blue no. 1	0·1	or *q.s.*

'Luxury' liquid or cream foot baths with herbal and/or deodorant properties may be formulated along the lines of conventional foam baths:

Luxury foot bath[9]	(7)
	per cent
Tegobetaine L7	50
Aminoxid WS35	3
Tego 103S	5
Undecylenic diethanolamide	2
Lactic acid	5
Water	35
Perfume, colour	*q.s.*

Herbal foot bath[4]		(8)
		per cent
A	Zetesol 856T	10·0
	Water	52·8
	Extrapone Urtica Special	1·0
	Extrapone Witch Hazel Colourless Distillate Special	5·0
	Extrapone Chamomile Special	0·5
	Extrapone Alpine Herbs Special	2·0
	Salicylic acid 10% solution*	20·0
	Foromycen F10	0·3
B	Neo-PCL water-soluble	2·0
	Fungicide DA	5·0
C	Menthol recryst. puriss	0·4
	Perfume oil	1·0

* Salicylic acid 10% solution:

Borax	5·0
Salicylic acid	10·0
Water	85·0

Heat to near boiling point to dissolve the salicylic acid.

Deo-Foam bath for the feet[4]	(9) *per cent*
Texapon Extract N25	55
Comperlan KD	6
Salicylic acid 20% solution*	25
Foromycen F10	2
Water	7
Neo-PCL water-soluble	1
Perfume oil	4

* Salicylic acid 20% solution:

Borax	8
Salicylic acid	20
Water	72

Heat to near boiling point to obtain a clear solution.

Foot Powders

Lehman[10] states that foot powders are recommended in the US Army; they keep the feet dry and if they contain a fungicidal or fungistatic agent a better opportunity is presented for a more continuous application as perspiration dissolves the medicament. In shoes they help to prevent reinfection. Lehman quotes the following formula:

	(10) *per cent*
Thymol	1
Boric acid	10
Zinc oxide	20
Talc	69

Another simple fungicidal dusting powder for the feet is proposed by Goldschmiedt:[8]

	(11) *per cent*
Zinc undecenoate	10·00
Undecenoic acid	2·08
Pine oil	0·47
Starch	50·00
Light kaolin	37·45

The Quaker Oats Co.[11] suggests a foot powder containing an antiperspirant:

	(12) *per cent*
Talc	82·65
'Oat-Pro'	3·00
Microdry	10·00
Syloid 72	2·00
Ottasept extra	0·15
Zinc oxide	2·00
Perfume	0·2 or *q.s.*

Foot powders with various deodorant and antiperspirant additives have been offered in aerosol spray form, as in examples 13 and 14.

Aerosol foot powder[4]	(13)
	per cent
Talc	92·0
Zinc stearate	1·0
Irgasan DP.300	0·2
Santocel 54	2·8
Span 85	2·0
Isopropyl myristate	1·0
Perfume oil	1·0
Powder base	15
Propellants 11/12 (50:50 or 65:35)	85

Women's foot spray[12]	(14)
	per cent
Fragrance	0·35
Menthol crystals	0·10
Talc 'lo micron'	8·35
Acetol	1·00
Chloroxylenol	0·20
Propellants 12/11 (40:60)	90·00

Foot Sprays

Foot sprays are intended principally to cool and refresh the feet. They usually contain antimicrobial agents, sometimes an antiperspirant material, and may contain an absorbent powder. They may be sprayed through socks or stockings and therefore are suitable for 'emergency' treatment.

The following three formulae, two for aerosol sprays respectively from Croda and RITA, and one for a non-aerosol pump spray from RITA, illustrate modern products. Of example 16 it is claimed, 'This soothing, cooling spray provides relief for hot tired feet and leaves them feeling soft and smooth'; example 17 is described as: 'A clear liquid solution for packing in a manually operated spray dispenser which dries quickly and tack-free leaving the feet cool and refreshed'.

Antiperspirant foot spray[13]	(15)
	per cent
Talc USP	34·0
Cab-o-sil M5	2·0
Microdry	5·0
Procetyl AWS	4·0
PVP-VA Copolymer E-735	4·0
Menthol USP	0·5
Ethanol anhydrous	50·5
Concentrate	15·0
Propellant 114	35·0
Propellant 12	50·0

Aerosol foot deodorant spray[14]	(16)
	per cent
Ammonyx 4002	0·125
Menthol (racemic)	0·125
2-Ethyl-1,3-hexanediol	2·25
Laneto 100	0·25
Ethanol anhydrous	27·25
Perfume	*q.s.*
Propellants 11/12 (50:50)	70·00

Foot deodorant spray solution for pump spray[14]	(17)
	per cent
Ethanol 96%	28·00
Versene	0·03
Perfume	*q.s.*
Hyamine 10X	0·25
Laneto 50	2·00
Deionized water	68·62
Silicone fluid DC-556	0·10
Camphor	0·50
Menthol (racemic)	0·50

Procedure: Dissolve all of the ingredients in the ethanol and then add the water.

Foot Creams

Foot massage is extremely relaxing: in fact the Chinese believe that simply massaging and manipulating the feet can relieve all tension and even cure diseases.[15] Techniques for professional foot massage have been described by Gallant.[16]

Foot creams are a suitable adjunct to massage and may contain antimicrobials, antiperspirants, mildly keratolytic agents, vasodilators to stimulate circulation, cooling agents, as well as providing emolliency and skin-softening properties. A simple emollient and deodorant foot cream may be formulated as follows:

	(18)
	per cent
Glyceryl monostearate (self-emulsifying)	15·0
Lanolin	1·0
Sorbitol syrup 70%	2·5
Glycerin	2·5
Antimicrobial agent	0·25–0·50
Water	to 100·0

Many of the formulae given under foundation creams may be adapted for foot creams. In general the amount of fatty materials is restricted. Some spermaceti substitute or high-melting wax, or fatty acid such as stearic acid, may be included to give a waxy film rather than a greasy layer on the foot. Camphor and methyl salicylate are commonly included to give a cooling sensation; a trace of menthol may be used for a similar purpose. The formula in example 19 is, in general, along these lines. Three more sophisticated formulae are given in examples 20–22.

	(19) *per cent*
Glyceryl monostearate (self-emulsifying)	12·0
Mineral oil	2·0
Glycerin	5·0
Spermaceti substitute	5·0
Camphor	1·0
Methyl salicylate	1·0
Water	73·9
Preservative	0·1

Foot cream[17]

		(20) *per cent*
A	Ottasept extra	1·00
	Lexemul AR	16·00
	Cetyl alcohol	1·00
	Amerchol L-101	3·00
	Solulan 98	0·50
	Mineral oil	3·00
B	Alcloxa	0·25
	Propylene glycol	5·00
	Distilled water	69·95
C	Perfume	0·30

Procedure: Heat A and B to 75°C. With rapid stirring add B to A. Cool with steady stirring and add perfume at the creaming point.

Foot antiperspirant cream[17]

		(21) *per cent*
A	Stearic acid	2·00
	Genapol S200	8·00
	Cetyl alcohol	10·00
	PCL Liquid	2·00
	Cholesterol	0·30
B	Sorbitol syrup 70%	5·00
	Water, preservative	45·90
	Allantoin	0·20
	Aluminium hydroxychloride 23 (Hoechst)	20·00
C	Salicylic acid	5·00
	Fungicide DA	0·20
D	Perfume	0·20
	Irgasan DP.300	0·20
	Polyglycol 400	1·00

Procedure: Heat A to 70°C, B to 75°C. Add the salicylic acid and Fungicide DA to A just before emulsification. Add B to A with stirring. Cool with steady stirring adding mixture D at 40°C. Mill if desired.

Oil-in-water foot massage emulsion[4]		(22) *per cent*
A	Neo-PCL, self-emulsifying	25·0
	Propyleneglycol dipelargonate	3·0
	Hostaphat KL340N	0·5
B	Water	65·9
	Propylene glycol	5·0
	Preservative	0·3
	Borax	0·1
C	Perfume oil	0·2

Procedure: Heat A and B to 75°C and emulsify. Cool with stirring, adding the perfume at 40°C.

Corn and Callus Preparations

Corn cures or paints contain salicylic acid in collodion either with or without cannabis extract. Other preparations include lactic acid, trichloracetic acid, glacial acetic acid:

	(23) *per cent*
Salicylic acid	10
Lactic acid	10
Flexible collodion BP	80

	(24) *per cent*
Salicylic acid	10
Cannabis extract (BPC 1949)	10
Flexible collodion BP	80

Five per cent castor oil may be added to the above if desired, for its plasticizing effect.

For callus softening, alkaline products are used, such as:

	(25) *per cent*
Tribasic sodium phosphate	8
Triethanolamine	12
Water	80
Perfume (alkali stable)	*q.s.*

	(26) *per cent*
Potassium hydroxide	0·5
Glycerin	15·0
Water	84·5

Example 26 should be used with care and must not be left in contact with the skin as it is caustic.

A corn remover in gel form has been described:[37]

	(27)
	per cent
Salicylic acid	12
Benzoic acid	6
Pluronic F.127	47
Water	35
Perfume, colour	*q.s.*

Ten volume strength hydrogen peroxide applied on cotton wool for several minutes has mild but valuable skin-softening properties, and may be followed with advantage by a warm oil massage.

In the late 1950s products appeared which removed hard skin from feet, elbows, etc., by making ingenious use of the 'balling' property of wax-containing emulsions. The emulsion, presented as a milk or thin cream, is pressure-sensitive and breaks when rubbed on the skin. The solid disperse phase then aggregates to form large discrete particles which roll on the skin, collecting loose skin cells. The addition of latex and/or finely divided silica improves the efficiency. The product is effective on clean skin but the presence of oil or grease may cause it to 'smear' rather than 'roll'. The emulsion can be made pressure-sensitive by using minimal quantities of the correct type of emulsifier. The following formula will serve as a basis for experiment:

	(28)
	per cent
Stearic acid	1·0
Beeswax	4·0
Cetyl alcohol	0·5
Paraffin wax	12·0
Triethanolamine	0·6
Sorbitol syrup 70%	5·0
Veegum	0·5
Water	76·4

Chilblain Preparations

Chilblains are a condition resulting from poor circulation and an inadequate supply of blood to the hands and feet. As Gourlay[18] points out, the 'first sign of a chilblain developing is a local redness and irritation which comes on while in bed or on sitting in front of the fire'; subsequently the pain intensifies and 'broken' chilblains may occur. Gourlay has drawn attention to the fact that, while chilblains occur relatively commonly in Britain and Europe, they are rare in Canada and the USA, and he has recorded the successful treatment of some cases with orally administered nicotinic acid (believed to be effective because of its vasodilator action).

Winner and Cooper-Willis[19] tested an ointment having the composition given in example 29.

	(29) *per cent*
Phenol	1·0
Camphor	6·0
Balsam of Peru	2·0
Soft paraffin	25·0
Hard paraffin	7·5
Anhydrous lanolin	58·5

The instructions given were either (a) to immerse the affected part in hot water at bedtime, to dry it carefully and apply the ointment; or (b) to rub in the ointment night and morning. Medical officers reported that the ointment eased the pain and caused rapid healing, some describing it as the best remedy they had met. Winner and Cooper-Willis considered this ointment as an efficient palliative which probably acted by stimulating local circulation.

Although good results in the treatment of chilblains have been claimed by the administration of vitamin K,[20] it would appear that the best prophylactic measure to take, in our present state of knowledge, is to keep the feet and hands warm with woollen gloves, socks, etc.

Most proprietary products are based on the use of local anaesthetics in appropriate bases which will relieve irritation. They may also stimulate circulation.

Athlete's Foot Preparations

Treatment of tinea pedis is actually outside the sphere of interest of the cosmetic chemist, particularly if the condition is aggravated by eczema and by secondary bacterial infections. Certain preparations, however, such as those used to prevent the infection, or those used during the post-therapeutic stage to prevent re-infection, are within the scope of cosmetic chemistry.

There has been a marked increase in the number of preparations for the treatment of athlete's foot since the introduction of aerosol sprays and powders. The antiseptics that are mainly present in aerosol foot preparations are selected primarily for the control of fungi, but they have also an important function as deodorizing agents to inhibit the growth of odour-producing bacteria. A considerable amount of information has appeared on the treatment of athlete's foot, and good reviews of the subject have been published by Lesser[21] and Chalmers.[6]

Salts of certain fatty acids such as propionic and undecylenic acids, often used in conjunction with the free acids, have been claimed to give good results in the treatment of *Trichophyton* and other dermatophytes. Keeney[22] quoted the formulae given in examples 30 and 31. Little difference in the clinical efficacy of these two ointments was found.

Propionate ointment	(30) *per cent*
Sodium propionate	16·4
Propionic acid	3·6
Propylene glycol	5·0
n-Propyl alcohol	10·0
Carbowax 4000	35·0
Zinc stearate	5·0
Water	25·0

Undecylenate ointment	(31) *per cent*
Undecylenic acid	10·0
Triethanolamine	6·0
Propylene glycol	14·0
Carbowax 1500	10·0
Carbowax 4000	40·0
Water	20·0

These acids rarely cause any adverse reaction on the skin. Undecylenic acid is often used in conjunction with zinc undecylenate, which is similar to zinc stearate and, as claimed in a *Schimmel Brief*,[23] would be more suitable for use in aerosol powders than in pressurized lotions. A great drawback of undecylenic acid is its unpleasant odour, resembling that of perspiration malodour. This disadvantage can be overcome, however, by using derivatives of the acid, namely its monoethanolamide (Fungicide UMA, Loramine U185), diethanolamide (Fungicide DA, Loramine DU185), or monoethanolamidosulphosuccinate (Loramine SBU185). These substances combine the fungicidal effect of undecylenic acid with surfactant properties and are claimed to be non-irritant.

An example of an antifungal foot powder based on a mixture of salts of propionic and caprylic acids is illustrated by the following composition:

	(32) *per cent*
Calcium propionate	15·00
Zinc propionate	5·00
Zinc caprylate	5·00
Propionic acid	0·25
Talc	74·75

A foot product in gel form is illustrated by example 33.

	(33) *per cent*
Ethanol 96%	72·0
Water	21·0
Undecylenic acid	5·0
Carbopol 940	1·0
Di-isopropanolamine	1·0

Commercially available antifungal agents which have been shown to be effective against dermatophytes of the *Trichophyton* and *Epidermophyton* species include:

Bronopol (Boots)[24]
Dichlorophen (BP)
Fungicides DA and UMA (Dragoco)[25]
Hibitane (ICI)[26]
Irgasan DP.300 (Ciba-Geigy)[27]
Loramines DU.185, SBU.185 and U.185 (Rewo)[28]
Myacide SP (Boots)[29]
Tolnaftate (BP)[30]
Vancide 89RE (Vanderbilt)[31]

Other Developments

Among newer developments are 'odour-destroying insoles' with absorbent and deodorant properties,[32] a foot care kit with 'everything you need' for the care of the feet[33] and a synthetic fibre 'foot sponge' for use with soap to remove thick hard skin.[34]

A recent patent[35] describes a deodorizer for the foot, etc., based upon an ion-exchange material such as ion-exchange cotton.

Scholl have taken out a patent[36] for the use of vanillin for treating athlete's foot through various product forms, or in a controlled-release device.

REFERENCES

1. Hole, L. G., *J. Soc. cosmet. Chem.*, 1973, **24,** 43.
2. Sehgal, K., *Manuf. Chem.*, 1978, **49**(1), 43.
3. Russel, B. F., *Brit. med. J.*, 1962, **ii,** 815.
4. Schroder, B., *Dragoco Rep.*, 1976, (2), 35.
5. Chalmers, L., *Dragoco Rep.*, 1972, (11), 215; (12), 235.
6. Chalmers, L., *Manuf. Chem.*, 1972, **43**(1), 33.
7. Thompson, W. A. R., ed., *The Practitioner's Handbook*, London, Cassell, 1960, p. 398.
8. Goldschmiedt, H., *Soap chem. Spec.*, 1971, **47**(6), 31.
9. Ceccarelli, C, and Proserpio, G., *Goldschmidt Informiert*, English edition, 1973, (26/5), 18 (Th. Goldschmidt AG, Essen).
10. Lehman, A. J., *Soap Perfum. Cosmet.*, 1943, **16,** 435.
11. *Soap Cosmet. chem. Spec.*, 1977, **53**(5), 165.
12. Emery Industries Inc., Linden, NJ, *Malmstrom Chemicals Cosmetic and Proprietary Formulations*, 1976.
13. Croda Chemicals Ltd, *Cosmetic/Pharmaceutical Formulary*, 1978.
14. RITA Chemical Corp., Crystal Lake, Ill., *Lanolin and Lanolin Derivatives, Bioactive and Organic Substances for the Cosmetic Industry*, 1976.
15. Maxwell-Hudson, C., *The Natural Beauty Book*, 1976, London, Macdonald and Jane's, p. 107.
16. Gallant, A., *Principles and Techniques for the Beauty Specialist*, London, Stanley Thornes, 1975.

17. Nobel Hoechst Chimie, Puteaux, France, Brochure no. 2, *Allantoin, Typical Formulations in Cosmetics*.
18. Gourlay, R. J., *Brit. med. J.*, 1948, **i,** 336.
19. Winner, A. L. and Cooper-Willis, E. S., *Lancet*, 1946, **2,** 663.
20. Wheatley, D. P., *Brit. med. J.*, 1947, **ii,** 689.
21. Lesser, M. A., *Drug Cosmet. Ind.*, 1951, **69,** 468.
22. Keeney, E. L., *Med. Clin. North Am.*, 1945, March, 323.
23. *Schimmel Brief* No. 302, May 1960.
24. The Boots Company Ltd, *Bronopol—Boots*, Technical Bulletin issue 3, September 1979.
25. Nowak, G. A., *Dragoco Rep.* 1960, (4), 97.
26. Lawrence, C. A., *J. am. pharm. Assoc., sci. Ed.*, 1960, **49,** 731; ICI Ltd, Pharmaceutical Division Technical Data 1972, *Hibitane Chlorhexidine*.
27. Ciba-Geigy Ltd, Circular 2502, *Irgasan DP.300, General Information*, 1973.
28. Rewo Chemicals Ltd (formerly Dutton & Reinisch), Technical Bulletins: *Loramine Undecylenic Alkanolamides*; *Loramine SBU.185*.
29. The Boots Company Ltd, *Myacide SP*, Technical Bulletin, September 1979.
30. *Martindale, The Extra Pharmacopoeia*, 26th edn. ed. Blacow, N.W., London, Pharmaceutical Press, 1972, p. 782.
31. R. T. Vanderbilt Co., Technical Bulletin no. 3: *Vancide 89RE—Bactericide/Fungicide*.
32. Glaxton, R., *Drug Cosmet. Ind.*, 1979, **124**(3), 58.
33. Anon., *Drug Cosmet. Ind.*, 1979, **124**(5), 82.
34. *Cosmetic World News*, 1979, (66), 28.
35. US Patent 4 155 123, Klein, P. M., 1979.
36. US Patent 4 147 770, Scholl Inc., 1979.
37. Schmolka, I. R., *Cosmet. Toiletries*, 1977, **92**(7), 77.

Chapter Fourteen

Insect Repellents

Introduction

Insect repellents may legitimately be considered as toilet preparations, especially since they may be presented in cosmetic form or even in combination with other functional attributes, notably with sunscreens.

Much of the earlier work on insect repellents was concerned with the protection of military personnel in the field. More recent studies have been associated with insects as disease vectors in man and with the protection of animals, especially dairy cattle, to give substantially increased milk yields.

In a symposium on insects and disease,[1] Maibach *et al.*[2] reported that the relative attractiveness of man's skin to the mosquito, *Aedes aegypti*, depended upon a balance between the attractiveness of certain elements of the sweat and the repellency of skin lipids. Novak[3] reported on the various attractants for the mosquito: colours, intensity of light, humidity, temperature and odours. Wright and Burton[4] discussed the mode of action of insect repellents in a study on pyrethrum.

Current thoughts on the mode of action of mosquito repellents have been reviewed.[5] It is proposed that the mosquitoes 'home in' on convection currents from warm, living animals, and respond to increases in relative humidity. A defence therefore is to prevent the moisture sensors from functioning normally which, it is claimed, can be achieved by chemical repellents that physically block the pores of the cuticle. Molecular size and shape and absorption forces are the controlling factors.

The repellent factors in the skin lipids have been reported[5] to be straight-chain C_9—C_{20} unsaturated fatty acids among which 2-decenoic acid is one of the most potent.

The possibility of achieving insect repellency by oral means has been discussed;[1] one of the potential drugs is thiamine hydrochloride.[5]

Most efficacy studies on insect repellents have involved a 'time of protection' factor but Burton and his co-workers at the British Columbia Research Council[6–8] reported a new method whereby a chemical can be tested for 'intrinsic repellency' against the mosquito, *Aedes aegypti*, and other insects. The method depends upon the concentration of a repellent in air required to neutralize the effect of a 'standard attractive target'; results for some 47 compounds were reported.[7]

Repellent Materials

Prior to 1940 the commonly used insect repellent materials were such strongly odorous substances as citronella oil, clove oil and camphor, which by modern standards are relatively ineffective. Müller[9] reports the following essential oils as having insect repellent properties: bergamot, birch tar, cassia, cedrus atlantica, cedarleaf, citronella, eucalyptus, fennel, pine oil sylvestris, lavender, laurel leaf, melissa, clove, peppermint, pimento, pennyroyal, West Indian sandalwood, sassafras, tea-tree and wormwood.

The outbreak of World War II and the need to conduct operations in many tropical areas led to an exhaustive investigation[10] into the question of insect repellents; as a result of comprehensive screening tests for primary irritation and skin toxicity properties, some 4000 compounds were reduced, first to 42 compounds, and then, as a result of further intensive testing, to 18.

It must be pointed out that the requirements and conditions of use for insect repellents by military personnel differ from those for use by the civilian population, since the former are composed mainly of young adults under close medical supervision, whereas potential civilian users would include the very young, the aged, and the infirm. On the other hand, among the numerous requirements for military personnel was the use of the preparation under tropical conditions, that is, heavy perspiration and application to large areas of the body surface, and this is more rigorous than the requirement for use in a more temperate climate.

Among the materials considered safe for use were:

Dimethyl phthalate
2-Ethyl-1,3-hexanediol (Rutgers 612)
A mixture of the following composition:

Dimethyl phthalate	6 parts
2-Ethyl-1,3-hexanediol	2 parts
'Indalone' (butoxypyranoxyl)	2 parts

Di-isopropyl tartrate
Cyclohexyl acetoacetate
Hexahydrophthalic acid diethyl ester
Piperonyl ether butoxide

Among the better known substances considered safe for use was butyl-3,4-dihydro-2,2-dimethyl-4-oxo-2*H*-pyran-6-carboxylate (butoxypyranoxyl).

Martindale[11] points out that complete protection against biting insects (in the worst environments) requires application both to skin and to clothing. Repellents usable on the skin may also be used on clothing (except that caution must be exercised with rayon); the reverse is not necessarily so. Effective repellents for clothing include benzyl benzoate, butylethylpropanediol, dibutyl phthalate, and diethyltoluamide. When repellents are applied to the skin complete coverage must be achieved; mosquitoes will bite through any 'holes' in the protective film.

A WHO Report[12] recommends the most suitable repellents for many common biting insects.

Gilbert[13] reported that the best mosquito repellents were, for skin application: diethyltoluamide, chlorodiethylbenzamide, ethylhexanediol, dimethyl phthalate, dimethyl carbate and butoxypyranoxyl; for clothing: butylethylpropanediol, ethyl hexanediol and diethyltoluamide.

Gouck[14] reported the most effective tick repellents to include butoxypyranoxyl, diethyltoluamide, dimethyl phthalate and benzyl benzoate which gave 99 per cent protection on clothing. Diethyltoluamide and benzyl benzoate were most effective against fleas, giving better than 90 per cent protection on clothing. Diethyltoluamide, dimethyl phthalate and ethylhexanediol used on skin or clothing are effective against chiggers and mosquitoes.

Dimethyl Phthalate

Dimethyl phthalate is a colourless, almost odourless liquid, boiling point 282°–285°C, slightly soluble in water (1/250) and miscible with most organic liquids.

It is an effective repellent against blackflies, mosquitoes, midges, mites, ticks and fleas[11] and is usually applied as a cream or lotion containing upwards of 40 per cent; any less is ineffective. It is said to prevent the bites of mosquitoes for 3–5 hours on the skin unless washed off with profuse sweating, and for about a week on clothing—though dibutyl phthalate is preferred for this application as it is less volatile and less easily washed off.

Dimethyl phthalate is non-toxic but may cause local slight temporary smarting in sensitive persons. Smith[15] reported that the minimum effective dose for protection against *Aedes aegypti* is between about 1·15 mg and 3·5 mg per square centimetre of skin. Its main disadvantage is its solvent action for certain plastics materials and it should not be allowed to come into contact with rayon garments or with plastics spectacle frames.

Dimethyl Carbate

Chemical name: dimethyl-bicyclo-(2,2,1)-heptano-2,3-dicarboxylate. Dimethyl carbate is a white to straw-coloured crystalline solid, setting point 35°C, soluble in water (1·5 per cent), in mineral oil (5–6 per cent), and very soluble in esters and vegetable oils.[16]

It has been described as an effective insect repellent[17] and has been used by the US military forces, usually in combination with other materials, though it seems to find little use nowadays.

Ethylhexanediol

Chemical name: 2-ethylhexane-1,3-diol. *Other names:* Ethohexadiol (USP), octylene glycol, Rutgers 612, 6–12 (Union Carbide).

Ethylhexanediol is a clear, colourless, almost odourless (but reminiscent of witch hazel) oily liquid, boiling point 244°C, which is slightly soluble in water (1/50) and is miscible with alcohol, isopropyl alcohol, propylene glycol and many other materials.

Granett and Haynes[18] report on the historical development of ethylhexanediol at Rutgers University during World War II and describe its properties at length. It is stable under extreme storage conditions and its solvent action for plastics and synthetic fibres is weak. It is the subject of an early patent granted to Union Carbide.[19]

McClure[20] claimed ethylhexanediol to be among the best mosquito repellents. It will also repel biting flies, gnats, chiggers and fleas for about 4 to 8 hours when applied neat, but proportionately less when used in more dilute applications. Granett and Haynes,[18] reporting on comparative effectiveness studies against dimethyl phthalate, found ethylhexanediol to be almost always better against mosquitoes and to be remarkably effective against other insects. They showed a protection time of about eight hours on the skin and eight days on clothing, though the latter was affected by rain. It is stated to be particularly effective when used in conjunction with dimethyl phthalate and butoxypyranoxyl.[11]

In a study of its protective efficacy against *Aedes aegypti*, Smith[15] found that the minimum effective dose ranged from about 80 to 280 μg per square centimetre of skin. No synergistic effect has been found between ethylhexanediol and diethyltoluamide.[21] Extreme toxicity tests and large-scale use by military forces indicate its safety and lack of irritant properties.[18]

Butoxypyranoxyl

Chemical name: butyl-3,4-dihydro-2,2-dimethyl-4-oxo-2*H*-pyran-6-carboxylate. *Other names:* butyl mesityl oxide, 'Indalone' (US Industrial Chemicals Inc. proprietary name).

Butoxypyranoxyl is a yellow to pale reddish-brown liquid with a characteristic aromatic odour, insoluble in water and glycerin, but miscible with alcohol, propylene glycol, other glycols, light mineral oils and vegetable oils.

Butoxypyranoxyl is an effective repellent; lotions containing 20–45 per cent will give protection for 4–6 hours and even longer. It has been used mainly in conjunction with dimethyl phthalate and ethylhexanediol.

Butoxypyranoxyl also possesses modest sunscreening properties and it is claimed that a 0·1 mm film gives complete absorption of ultraviolet light up to 350 nm. However it is prudent to include a small amount of a more active ultraviolet absorber to achieve effective sunscreening.

Butoxypyranoxyl hydrolyses on storage in the presence of more than 10 per cent water and preparations containing it must therefore be formulated with a lower water content.

Diethyltoluamide

Chemical name: N,N-diethyl-*m*-toluamide. *Other names:* DET, DEET, Detamide, Delphene or Metadelphene (Hercules Powder Co.).

N,N-diethyl-*meta*-toluamide is a colourless liquid with a faint pleasant odour, almost insoluble in water and glycerin, but miscible with alcohol and isopropyl alcohol. The commercial material contains a minimum of 95 per cent of the meta-isomer.[22]

Diethyltoluamide was developed by the US Department of Agriculture for use by troops during the Korean War in 1951. Of the three isomers (*ortho*-, *meta*- and *para*-) the *meta*-isomer was found to be 10 per cent more effective. Gilbert, Gouck and Smith[23–25] found N,N-diethyl-*m*-toluamide and *o*-ethoxy-N,N′-diethylbenzamide superior or equal to the standard US Army repellent M2020. Diethyltoluamide containing about 70 per cent of the *meta*-isomer was found to be generally more effective than ethylhexanediol and other repellents both for skin and clothing.

Diethyltoluamide is effective against blackflies, chiggers, mosquitoes, ticks and fleas.[11] Clyde and Kingazi[26] found in laboratory studies that diethyltoluamide was an effective mosquito repellent for 18–20 hours compared with 4–4½ hours for a 40 per cent dimethyl phthalate cream. It has been reported to be the most effective repellent against *Aedes aegypti*.[21] Smith[15] showed that the minimum effective dose for protection against *Aedes aegypti* ranged from 50 μg to 77 μg per square centimetre of skin.

Among the properties of diethyltoluamide are claimed[27] its great persistence, resistance to wiping action, resistance to sweating and its non-oily nature.

Major and Hess[28] reported N-ethoxy-N-ethyl-*m*-toluamide to have considerable toxicity to mosquitoes *Aedes aegypti* and to repel stable flies *Stomoxys calculans* but not houseflies *Musca domestica*.

Two new toluamides, N-(*m*-toluyl)-2-methyl piperidine and N-(*m*-toluyl)-4-methyl piperidine, have been found particularly effective against stable flies.[29] They may be used in a variety of formulations ranging from ointments to sprays and are said to repel a large number of troublesome arthropods.

MGK Repellents

McLaughlin Gormley King have formulated a range of repellent compositions, using patented materials, under the designation 'MGK Repellents for Personal Use' which are said to utilize the latest developments in repellent materials and to provide maximum protection.[30] Advantages claimed over older formulae are:

(1) Best combinations available for stable flies, horse flies, deer flies and blackflies—frequently more of a problem than mosquitoes on beaches, golf courses, etc.
(2) Excellent mosquito repellency.
(3) Protection against ticks, fleas and chiggers.
(4) No adverse effects on synthetic fibres, except for a slight darkening of rayon.
(5) Plasticizing effect not as intense as compositions dependent upon diethyltoluamide alone or dimethyl phthalate.

'MGK Intermediates', a range of repellent compositions, are based upon blends of diethyltoluamide with MGK-264 (N-octyl bicycloheptene dicarboximide), an insecticide synergist; with MGK Repellent 11 (2,3:4,5-*bis*-(2-butylene)-tetrahydro-2-furaldehyde), primarily a fly repellent; and with MGK Repellent 326 (di-*n*-propyl isocinchomeronate), also primarily a fly repellent.

'MGK Intermediates' have the following compositions (per cent):

Intermediate No.	1995	2007	2020	5134	6339	5582
Diethyltoluamide	86	76·92	80	70	67	—
MGK-264	8	15·38	12	20	11	66·6
MGK Repellent 11	3	3·85	4	5	22	16·7
MGK Repellent 326	3	3·85	4	5	—	16·7
Typical usage level	25%	32·5%	25%	10–15%	9–15%	3–6%

1995 gives maximum protection against mosquitoes, less against other biting flies; 2007 has greater repellency to all flies and mosquitoes; 2020 is a general-purpose composition with excellent mosquito repellency; 5134 is a general-purpose repellent giving excellent repellency to mosquitoes and good repellency to all flies and insects; 6339 will provide excellent mosquito repellency as well as being particularly effective against such insects as blackflies and sand flies; 5582 is a fly repellent generally used in combination with diethyltoluamide or another repellent of choice.

MGK Repellent 326 should not be used in alcohol-based preparations when transesterification may occur; however, it is perfectly satisfactory in isopropyl alcohol. It sometimes shows instability in water-based systems, especially if alkaline.

Other Repellents

Benzyl Benzoate. This is the best repellent for chiggers and one of the best for fleas and ticks.[12] However, it should not be used on the skin as it may cause eruptions in sensitive persons. It is usually applied to clothing where it is very effective and its action persists after washing.

Dibutyl Phthalate. This is slightly less effective than dimethyl phthalate but preferred to the latter for impregnating clothing as it is less volatile and less easily removed by washing.

Butylethylpropanediol. This is used for the treatment of clothing and should not be used on the skin. It is an ingredient of M-1960 (butylethylpropanediol 30 per cent, butylacetamide 30 per cent, benzyl benzoate 30 per cent, emulsifier 10 per cent), used by the US Army for impregnating clothing; it is effective against blackflies, mosquitoes and other insects.[11]

Butoxypolypropyleneglycol. This is reported to be very effective against the house fly *Musca domestica*.[32] Under the name of Crag Repellent (Union Carbide Agricultural Chemicals) marketed for use on cattle, etc., but no longer approved for use on dairy and meat cattle in the USA.[31]

Di-n-butyl Succinate (Tabatrex). This is reported to be effective against the house fly *Musca domestica*.[32] It is used for agricultural purposes but no longer approved for use on dairy and meat cattle in the USA.[31] A patent[33] claims the use of the di-*n*-butyl succinate and, as a synergist, a fatty material such as oleic acid, ricinoleic acid, propyl oleate or benzyl oleate.

Undecenoic Acid. This is an effective insect repellent but its disagreeable odour is difficult to mask.[11]

Repellent 790 (E. Merck, Darmstadt). This is a proprietary repellent of undisclosed composition, possibly based upon caprylic acid diethanolamide.[34] It is said to have a broad spectrum of activity, to be highly potent, with a sustained action and well tolerated by the skin. It is claimed to be as effective as diethyltoluamide against mosquitoes and three times as effective as the latter against the house fly.[32]

Moskitox (Dragoco). This is a proprietary composition based on diethyltoluamide, dimethyl phthalate and hexylene glycol.

Other Developments

Ralston and Barrett[35] have reported that decyl, undecenyl and dodecyl alcohols, and aliphatic nitriles with 10–14 carbon atoms are highly repellent to flies. The inherent odour of these materials precludes their use in cosmetic preparations.

A Russian investigator, Nabokov,[36] reported that anabasine sulphate, an alkaloid from the plant *Anabasis aphylla* (*Chenopodiaceae*) which has been used in Russia as an agricultural insect repellent for many years, would confer protection for 10 hours when applied to the skin as a 5 per cent lotion. It was stated to be harmless to the skin and health of the person treated. The material was also said to be non-odorous.

Shambaugh and his co-workers[37] studied the repellency of some phenylphenols to houseflies; the most effective of these were biphenyl and 4-chloro-2-phenylphenol. A mixture of these with phenol, *o*-phenylphenol and 6-chloro-2-phenylphenol was more effective than any single compound.

Weaving and Sylvester[38,39] and Wright and Burton[4] have investigated pyrethrum as an insect repellent.

Quintana and his co-workers have carried out several studies on potential materials for use as long-lasting insect repellents: esters of undecanoic acid with phenols, especially resorcinol, hexachlorophene, 4-chloro-resorcinol and 4-chlorophenol;[40] esters of undecanoic acid with dihydroxyacetone, specifically 1,3-di-undecanoyl-oxyacetone and 1-undecanoyloxy-3-hydroxyacetone were shown to have long-lasting insect repellent activity on the skin;[41] the mono-hexanoate, mono-propanoate, mono-benzoate and mono-undecanoate esters of dihydroxyacetone were studied and the long-lasting insect repellency of dihydroxyacetone mono-hexanoate was particularly noteworthy.[42]

A patent[43] taken out by Stepanov *et al.* claims an insect repellent containing 20–90 per cent hexamethylenecarbamide to provide protection against mosquitoes, gnats, sand flies, blackflies, houseflies, fleas and ticks. In an example, an ointment containing a mixture of this compound (30 per cent) and dimethyl phthalate (52 per cent) was said to provide protection for 24–36 hours against mosquitoes and sand flies; an emulsion containing 75 per cent applied to cloth was said to protect for 5 months or more.

Two patents to Dow Corning[44,45] relate to derivatives of tetra-sila-adamantane as insect repellents.

Gualtieri *et al.*[46] investigated a number of acetals, amino-acetals, carboximide-acetals and aromatic esters on the skin for repellency to mosquitoes. The amino-acetals showed the highest degree of repellency but did not rival diethyltoluamide in duration of protection.

Klier and Kuhlow[47] evaluated a number of derivatives of N-disubstituted *beta*-alanine. Several N-alkyl esters of 3-(N-*n*-alkyl-N-acyl)-aminopropionic acid and of 3-(N-*n*-alkyl-N-carboxyalkyl)-aminopropionic acid were effective repellents for mosquitoes on the skin, equal to diethyltoluamide. One compound in particular, ethyl 3-(N-*n*-butyl-N-acetyl)-aminopropionate showed high mosquito repellency and extremely low toxicity and was well tolerated by the skin.

Formulation

In formulation, quantities of repellent of the order of 10 per cent or more are generally recommended and a trend towards the use of higher levels has been noted.[30]

Limitations of certain repellent materials must be taken into account when formulating, such as the tendency of dimethyl phthalate, especially, to attack some plastics and fibres, and the instability of butoxypyranoxyl in the presence of water levels above 10 per cent. MGK Repellent 326 and 'MGK Intermediates' containing it should not be used in alcohol-based compositions, though isopropyl alcohol is permissible, and water-based systems should be approached with caution. MGK Repellent 11 should not be used in the presence of 2-aminopropanediol (AMPD).[30]

Insect repellents may be formulated into the whole gamut of cosmetic product forms and formulae abound in the literature for lotions, oils, milks, creams, aerosol sprays, foams, 'quick-break foams', pump-sprays, towelettes, gels and sticks.

It has been pointed out[30] that under normal conditions of use—on the beach, golf-course and other holiday and pleasure pursuits—it is biting flies and other insects that are the problem and not mosquitoes, and products should be formulated accordingly. A significant trend is the logical combination of an insect repellent and a sunscreen into a single product to provide dual protection.

Formulae for a representative range of products follow. It is worth noting that more sophisticated products might be developed utilizing the more advanced formulation techniques available in the field of sunscreen preparations; product attributes and formulation problems are often similar.

Lotions

Lotions may be simply solutions of an insect repellent in alcohol with or without the addition of modifiers to moderate oiliness or to give a better skin feel. Aqueous–alcoholic products may be formulated using a solubilizer though clearly these will have considerably less wash-off resistance. Solutions may also be in other more or less volatile solvents (see later under 'Repellent Oils').

Alcoholic lotion	(1) *per cent*
Dimethyl phthalate	33
Alcohol 96%	67

Clear aqueous–alcoholic lotion[48]	(2) *per cent*
Tween 80	15
Repellent 790	10
Alcohol 96%	30
Water, purified	45

Aerosol Sprays

Examples 3,[32] 5[30] and 6[30] are insect repellents; example 4 is a combined sunscreen and repellent.

	(3) per cent	(4) per cent
Repellent 790	20	15·0
Eutanol G	15	15·0
Sunscreening substance 3573	—	2·5
Isopropyl alcohol	65	67·5
Concentrate	50	50
Propellant 11	25	5
Propellant 12	25	45

	(5) per cent	(6) per cent
MGK Intermediate 5734	15	10–25
Isopar E	30	70–55*
Isopropyl alcohol	51	
Nitrous oxide or carbon dioxide	4	—
A-46 Isobutane/propane	—	20

* All isopropyl alcohol, all Isopar E, or a blend of the two.

Pump Sprays

Owing on the one hand to current disquiet and the regulatory position regarding chlorofluorocarbon aerosol propellants, and on the other to the development of much improved spray pumps, an interest has developed in personal repellents which can be dispensed by this method. Tested formulae[30] include:

	(7) per cent	(8) per cent
MGK Intermediate 5734	15	—
MGK Intermediate 2007	—	32·5
Isopar E	15	15·0
Isopropyl alcohol	70	52·5

Repellent Oils

Since most repellent materials are oily by nature and there is a need to use fairly high levels to achieve efficacy, application as an oil is advantageous. Modern materials allow the formulation of products minimizing inherent oiliness on the skin to produce aesthetically acceptable products. Formulation experience in the sunscreen field, where this is a popular method of application, can be called upon. Pressurization to give aerosol-dispensed products is straightforward. An added dimension may be given by the inclusion of moderately volatile 'oils' such as the Isopars or volatile silicones.

Example 9 gives a simple general-purpose formula.

	(9) per cent
Mineral oil, light	40
Vegetable oil	30
Isopropyl palmitate	20
Repellent (of choice)	10

Hydrophilic repellent oil[48]	(10)
	per cent
Atlas-G–1086	7·5
Repellent oil	20·0
Mineral oil, light	42·5
Isopropyl myristate	29·0
Perfume	1·0

Aerosol sunscreen/insect repellent oils[49]	(11)	(12)
	per cent	*per cent*
Dimethyl phthalate	4	6·0
Moskitox	5	6·0
Prosolal S8	5	4·2
PCL Liquid	24	23·0
Arachis oil	10	—
Isopropyl myristate	40	60·0
Mineral oil, viscous	10	—
Antioxydol	1	—
Perfume oil	1	0·8
Concentrate	40	40
Propellants 11/12 (50:50)	60	60

Creams and Liquid Creams

Stable creams are less easy to formulate owing to the high levels of repellent material needed for effectiveness and their inherent emulsification problems. Once again, techniques learned from formulating sunscreens can be very useful.

The following formula for a midge repellent cream, avoiding emulsification, was proposed by the Scottish Scientific Advisory Committee:[50]

	(13)
	per cent
Dimethyl phthalate	67
Magnesium stearate	10
Zinc stearate	23

Procedure: The mixture is gelled by means of heat.

Another non-emulsified cream has the following formula:[51]

	(14)
	per cent
Ethylhexanediol	18·6
Butoxypyranoxyl	18·6
Dimethyl phthalate	55·8
Ethyl cellulose	2·7
Cellulose acetate-butyrate	2·3
Propyleneglycol monostearate	2·0

Examples 15 and 16 illustrate the use of ethylhexanediol in combination with zinc stearate and polyethyleneglycol.

	(15) *per cent*	(16) *per cent*
Ethylhexanediol	30·0	30
Zinc stearate	20·0	20
Stearic acid	4·8	—
Potassium hydroxide 41% aqueous solution	1·8	—
Water, purified	43·4	20
Carbowax 4000	—	30

A simple emulsion formula is proposed by Croda:[52]

		(17) *per cent*
A	Diethyltoluamide	20
	Polawax	7
	Ceto-stearyl alcohol	12
	Crillet 3 (Polysorbate 60)	1·5
B	Water, Purified	59·5
	Perfume, preservatives	*q.s.*

Procedure: Heat A and B separately to 70°C. Add B to A with high shear mixing. Cool with low shear mixing.

A pleasant, non-greasy oil-in-water lotion has been described[53] which uses Veegum as emulsion stabilizer:

		(18) *per cent*
A	Veegum	1
	Water	64
B	Diethyltoluamide	25
	Stearic acid	4
	Sorbitan monostearate	4
	Polysorbate 60	2
	Preservative	*q.s.*

Procedure: Add the Veegum to the water slowly, stirring continuously until smooth. Heat to 70°C. Heat B to 75°C, add B to A and mix until cool.

An elegant, glossy, soft sunscreen–insect repellent cream (example 19) suitable for squeeze-bottle or tube packaging has been described.[54] It is said to spread easily and rub in without whitening, to be without objectionable oiliness or tackiness and to provide good protection against the elements.

		(19) *per cent*
A	Amerchol L-101	3·50
	Modulan	1·00
	MGK Intermediate 5734	12·00
	Escalol 506	1·20
	Tinuvin P	0·05
	Arlacel 165	5·00
	Cetyl alcohol	1·50

B	Carbopol 940	0·21
	Water, purified	75·34
C	Triethanolamine	0·20
	Perfume, preservatives	*q.s.*

Procedure: Slowly add the Carbopol to the water with fast stirring. Mix until thoroughly dispersed. Add B at 75°C to A at 75°C with stirring. After emulsification, add the triethanolamine. Continue mixing and cool to 32°C.

Another elegant sunscreen–insect repellent lotion has been proposed by Malmstrom:[55]

		(20) *per cent*
A	Nimlesterol D	5·00
	Emerest 2400	5·00
	Mineral oil 70 visc.	18·00
	PEG-23 lauryl ether	5·00
	Methyl paraben	0·20
	Propyl paraben	0·10
	Cetyl alcohol	0·10
	Emersol 132	3·00
	Diethyltoluamide	10·00
	Escalol 506	1·20
B	Water	48·95
	Carbopol 934	0·25
	Emsorb 6915	2·00
C	Triethanolamine	1·20

Procedure: Heat A and B to 80°C. Add A to B slowly with stirring. Mix thoroughly and add C. Cool with stirring.

Gels

A good gel is an excellent way of presenting an insect repellent which may have an alcoholic, aqueous or oil basis. Certain of the formulae presented earlier in this chapter as creams are, in reality, opaque gelled oils; however, gels are usually thought of as being clear or almost so. Simple Carbopol-based gels may be prepared as in example 21.

Personal repellent gel[30]	(21) *per cent*
MGK Intermediate 5734	10–15
Isopropyl alcohol	50
Water	31–36
Carbopol 940	2
Ethomeen C-25	2

Procedure: Disperse the Carbopol smoothly in the isopropyl alcohol–water mixture using high shear mixing; low shear mix until the Carbopol is fully swollen. Add the MGK Intermediate, then add the Ethomeen, when a stiff gel will form.

The above formula will give a slightly hazy, pale green product. A clear, bright, soft gel, with additional emolliency, may be formulated as follows:[54]

	(22) per cent
MGK Intermediate 5734	10·00
Alcohol 96%	50·00
Carbopol 940	0·75
Water	28·25
Di-isopropanolamine, 10% aqueous	8·00
Solulan 98	3·00
Perfume, preservatives	*q.s.*

Procedure: Disperse the Carbopol in the water using high shear mixing. Mix the MGK Intermediate and the alcohol, reduce the speed of mixing and add. Add the di-isopropanolamine solution followed by the Solulan 98 and mix until uniform.

Sticks

Sticks have proved generally to be less effective owing to the difficulty of transferring enough repellency to the skin. However, two formulae are of interest as typical products; one is a soap–alcohol gel-stick and the other a wax-based product.

In example 23 an MGK Intermediate is used which does not contain Repellent 326 because the latter has been found to decompose rapidly in soap-based sticks. It is said to remain firm at elevated temperatures and to be semi-clear.

Personal repellent stick[30]

		(23) per cent
A	MGK Intermediate 6561	20·0
	Isopropyl alcohol	64·5
	Glycerin	5·0
	Sorbo sorbitol syrup	4·0
B	Sodium stearate	6·0
	Stearyl alcohol	0·5

Procedure: Heat A to 55°–65°C and stir until uniform. Add B and heat to 65°–70°C with stirring until clear. Immediately pour into moulds and cool.

Wax-type repellent stick[32]

		(24) per cent
A	Repellent 790	20
B	Lanolin	25
	Paraffin wax m.p. 68°–72°C	20
	Spermaceti (or substitute)	20
	Mineral oil, viscous	15
	Colour, perfume	*q.s.*

Procedure: Melt B to 75°C, mix and stir in A. Add perfume at 55°–60°C. Pour into moulds at 39°–40°C.

A US patent[56] disclosed a stick-type product (example 33) containing a high proportion of active repellent in a soap base. It was plasticized with glycerin and was claimed to be transparent, stable at the extremes of temperature normally encountered, non-brittle and easily applied to the skin.

	(33) *parts*
Dimethyl phthalate	18·2
Ethylhexanediol	18·0
Isopropyl alcohol (91%)	31·5
Sodium stearate (powdered)	20·3
Glycerin	12·1
Distilled water	3·5
Colour solution	1·2
Perfume	1·2

Procedure: Mix the first three ingredients together and add the sodium stearate, glycerin, water and colour solution to the mixture. Heat to 81°–82°C with occasional stirring. When a clear solution is obtained (usually 7–10 minutes), remove the heat source and allow to cool to 60°C. Add the perfume and pour into moulds; cool to room temperature.

In making such sticks, at least 30 per cent of repellent should be present; the gelling agent may consist of beeswax with the addition of other oils, fats and waxes to obtain the desired consistency.

Towelettes

Paper or fabric towelettes for applying various toiletry and cosmetic products have become increasingly popular, including their use for insect repellents. A suitable impregnating solution can be prepared as follows:[30]

	(34) *per cent*
MGK Intermediate 5734	10–25
Isopropyl alcohol	56–50
Water	34–25

REFERENCES

1. Symposium on Insects and Disease, *J. Am. med. Assoc.*, 1966, **196,** 236 *et seq.*
2. Maibach, H. I., *et al.*, *J. Am. med. Assoc.*, 1966, **196,** 263.
3. Novak, D., *Acta hyg.*, 1959, **7**(2), 84.
4. Wright, R. H. and Burton, D. J., *Pyrethrum Post*, 1969, **10**(2), 14.
5. *Norda Briefs* no. 469, 1975. Norda Inc., East Hanover, NJ 07936.
6. Kellogg, F. E., Burton, D. J. and Wright, R. H., *Can. Entomol.*, 1968, **100,** 736.
7. Burton, D. J., *Am. Perfum. Cosmet.*, 1969, **84**(4), 41.
8. British Columbia Research Council, University of British Columbia, Vancouver 8, Canada, reported in *Soap Perfum. Cosmet.*, 1968, **41,** 502; *Soap chem. Spec.*, 1968, **44**(9), 68.

9. Müller, A., *Internationaler Kodex der Ätherischen Öle*, Heidelberg, Alfred Hüthig, 1952.
10. Draize, J. H., Alvarez, E., Whitesell, M. F., Woodward, G., Hagan, E. C. and Nelson, A. A, *J. pharmacol, exp. Ther.*, 1948, **93,** 26.
11. *Martindale, The Extra Pharmacopoeia*, 26th edn. ed. Blacow, N. W., London, Pharmaceutical Press, 1972.
12. *Insect Resistance and Vector Control. Tenth Report of the Expert Committee on Insecticides,* Tech. Rep. Ser. World Health Organisation no. 191, 1960.
13. Gilbert, I. H., *J. Am. med. Assoc.*, 1966, **196,** 164.
14. Gouck, H. K., *Arch. Dermatol.*, 1966, **93,** 112.
15. Smith, C. N., *J. Am. med. Assoc.*, 1966, **196,** 236.
16. Harry, R. G., *The Principles and Practice of Modern Cosmetics*, vol. 2, *Cosmetic Materials*, rev. Myddleton, W. W., London, Leonard Hill, 1963.
17. Wright, R. H., *Nature (London)*, 1956, **178,** 638.
18. Granett, P. and Haynes, H. L., *J. econ. Entomol.*, 1945, **38,** 671.
19. US Patent 2 407 205, Union Carbide, 11 February 1943.
20. McClure, H. B., *Chem. Eng. News*, 1944, **22,** 416.
21. Gilbert, I. H., *J. Am. med. Assoc.*, 1966, **196,** 253.
22. McLaughlin Gormley King, Minneapolis, Minn. 55427, Technical Bulletin, *MGK Diethyltoluamide*, 1974.
23. Gilbert, I. H. and Gouck, H. K., *Florida Entomol.*, 1955, **38,** 153 (in *Chem. Abstr.*, 1956, **50,** 6732).
24. Gilbert, I. H., Gouck, H. K. and Smith, C. N., *J. econ. Entomol.*, 1955, **48,** 741.
25. Gilbert, I. H., Gouck, H. K. and Smith, C. N., *Soap chem. Spec.*, 1957, **33**(5), 115.
26. Clyde, D. F. and Kingazi, H., *East Afr. med. J.*, 1957, **34,** 185, reported in ref. 11 above.
27. Hercules Powder Company, Wilmington, Del., *Technical Bulletin no. 213*, 1957.
28. Major, R. T. and Hess, H. J., *J. Am. pharm. Assoc., sci. Ed.*, 1959, **48,** 485.
29. US Patent 3 463 855, S. C. Johnson, 26 August 1969.
30. McLaughlin Gormley King, Minneapolis, Minn. 55427, USA, Technical Bulletin: *MGK Repellents for Personal Use*, 1977.
31. McLaughlin Gormley King, *Dairy Stock Spray Bulletin*, 1972.
32. E. Merck, Darmstadt, Germany, *Product Information*: Repellent 790.
33. US Patent 2 991 219, Bruce, W. N., 1961.
34. German Patent 1 055 283, Merck, E., 7 September 1957.
35. Ralston, A. W. and Barrett, J. P., *Oil, Soap*, 1941, **18,** 89.
36. Nabokov, V. A., *Annu. Rev. Soviet Med.*, 1945, **2,** 449.
37. Shambaugh, G. F., Pratt, J. J., Kaplan, A. M. and Rogers, M. R., *J. econ. Entomol.*, 1961, **61,** 1485.
38. Weaving, A. J. S. and Sylvester, N. K., *Pyrethrum Post*, 1967, **9**(1), 31.
39. Sylvester, N. K. and Weaving, A. J. S., *Pyrethrum Post*, 1967, **9**(2), 18.
40. Quintana, R. P., Garson, L. R. and Lasslo, A., *Can. J. Chem.*, 1968, **46,** 2835.
41. Quintana, R. P., Garson, L. R. and Lasslo, A. *Can. J. Chem.*, 1969, **47,** 853.
42. Quintana, R. P., Lasslo, A., Garson, L. R., Smith, C. N. and Gilbert, I. H., *J. pharm. Sci.*, 1970, **59,** 1503.
43. US Patent 3 624 204, Stepanov, M. K. *et al.*, 30 November 1971.
44. British Patent 1 297 351, Dow Corning Corp., 22 October 1972.
45. British Patent 1 301 045, Dow Corning Corp., 29 December 1972.
46. Gualtieri, F., Johnson, H., Maibach, H. I., Skidmore, D. and Skinner, W., *J. pharm. Sci.*, 1972, **61,** 577.
47. Klier, M. and Kuhlow, F., *J. Soc. cosmet. Chem.*, 1976, **27,** 141.
48. Atlas Chemical Industries, *Atlas Manual for the Cosmetic and Pharmaceutical Industries*, 1978.

49. Dragoco (Great Britain) Ltd, data sheet, *Prosolal S.8 sun-screen agent*, 1966.
50. Todd, J. P., *Pharm. J.*, 1950, **165,** 2.
51. US Patent 2 404 698, Du Pont de Nemours, 4 January 1945.
52. Croda Chemicals Ltd, Cowick Hall, Snaith, Goole, N. Humberside DN 14 9AA, Technical Bulletin, *Crills and Crillets.*
53. R.T. Vanderbilt Co. Inc., New York, NY 10017, Bulletin no. 44, *Veegum.*
54. American Cholesterol Products Inc., Edison, NJ 08817, *Amerchol Laboratory Handbook for Cosmetics and Pharmaceuticals.*
55. Emery Industries Inc., Linden, NJ 07036, *Malmstrom Chemicals Cosmetic and Proprietary Formulations.*
56. US Patent 2 465 470, Onohundro, A. L, Neumeier, F. M. and Zeitlin, B. R., 1949.

Chapter Fifteen

Sunscreen, Suntan and Anti-sunburn Products

SUNLIGHT AND THE HUMAN BODY

Introduction

Exposure to sunlight can have both beneficial and harmful effects on the human body, depending on the length and the frequency of exposure, the intensity of the sunlight and the sensitivity of the individual concerned.

The most obvious effect of exposure to the rays of the sun is first of all erythema of the skin, followed by the formation of a tan which seems to have been adopted by the present civilized world as a symbol of physical health. In actual fact, development of a tan is a protective reaction of the human body to minimize any damaging effect of solar irradiation.

The intensity of erythema (reddening) produced on the skin following exposure to sunlight depends on the amount of UV energy absorbed by the skin. Erythema usually starts to develop after a latent period of 2–3 hours and reaches its maximum intensity within 10 to 24 hours after exposure.

Tanning

The tanning ability of an individual is genetically predetermined and depends on his capacity to produce melanin pigment within the melanocytes.

Tanning responses are stimulated by erythemogenic (as well as longer) wavelengths in the ultraviolet and visible radiation ranges. There are three types of tanning response:

(i) immediate tanning;
(ii) delayed tanning;
(iii) true tanning, also referred to as melanogenesis.

Immediate tanning is stimulated by energy between 300 nm and 660 nm and its maximum efficacy lies between 340 nm and 360 nm. It entails the immediate darkening of unoxidized melanin granules present in the epidermal layer of the skin, near its surface. It reaches a maximum about one hour after exposure to radiation and begins to fade within 2–3 hours after exposure.

Delayed tanning involves the oxidation of melanin granules present in the basal cell layer of the epidermis and their migration towards the surface of the skin. It may start as early as one hour after exposure, reaches a peak after some ten hours and then fades rapidly after 100 to 200 hours following exposure.

Delayed tanning and also true tanning are stimulated primarily by the so-called erythemogenic radiation, that is, between wavelengths 295 nm and 320 nm.

True tanning starts about two days after exposure and reaches a maximum about two to three weeks later.

Beneficial Effects of Sunlight

Moderate exposure of the human body to sunshine results, psychologically and physiologically, in a general sense of fitness, peace of mind and general well-being. Also it has certain definite beneficial effects on human health. It stimulates blood circulation, increases the formation of haemoglobin, and may also promote a reduction in blood pressure. Furthermore, it plays a vital part in the prevention and treatment of rickets by producing—through the activation of 7-dehydrocholesterol (provitamin D_3) present in the epidermis—vitamin D, which enhances the absorption of calcium from the intestine.

It has been used in the treatment of certain types of tuberculosis, such as the tuberculosis of glands and bones, and the treatment of certain skin diseases such as psoriasis. It is also believed to exert a beneficial influence on the autonomous nervous system and to reduce the susceptibility of individuals to various infections. Finally, by producing melanin and causing thickening of the skin, it plays an essential part in the formation of the body's natural protective mechanism against sunburn.

Adverse Effects of Sunlight

Solar irradiation can have both short-term and long-term adverse effects.

Sunburn

The short-term effect, as far as the skin is concerned, is a temporary damage of the epidermis, manifesting itself in the known symptoms of sunburn. These may range in severity from a slight erythema to painful burns and blistering accompanied in more severe cases, when large amounts of the skin have been affected, by shivering, fever and nausea, and sometimes pruritus.

According to Keller,[1] the symptoms of sunburn are the direct result of damage or destruction of cells in the prickle cell layer of the skin, possibly through denaturing of its protein constituents. Histamine-like substances released by the damaged cells are responsible for the dilation of blood vessels and erythema. They also cause swelling of the skin (oedema) and stimulate the basal cells of the skin to proliferation.

During the latent period preceding the appearance of sunburn symptoms, photochemical degradation products formed as a result of solar irradiation are believed to trigger a series of free-radical reactions leading to the formation of the biologically active substances referred to above, which diffuse into the dermal blood vessels and produce the symptoms described.

As a result of experiments conducted in the USA with exposure to mid-day sunshine in June, Luckiesh[2] arrived at the following definitions of four degrees of sunburn:

(1) *Minimal perceptive erythema*—a slight, but discernible red or pink coloration of the skin, produced in 20 minutes.
(2) *Vivid erythema*—a bright red coloration of the skin, not accompanied by any pain, produced in 50 minutes.
(3) *Painful burn*—characterized by both vivid erythema and pain ranging from mild to intense, produced in 100 minutes.
(4) *Blistering burn*—characterized by an extremely high level of pain accompanied by vivid erythema and possibly systematic symptoms with blistering and peeling, produced in 200 minutes.

Sunburn does not leave any scars. A slight burn protected from further exposure to sunlight will disappear within 24–36 hours. More severe burns will generally heal within 4–8 days. As the inflammation subsides, it will be followed by peeling of the skin.

Chronic Exposure

Chronic exposure to intense sunlight, to which sailors, farmers and construction workers are often subjected, entails more serious hazards such as, for example, the development of skin cancer. It may also produce degenerative changes in the connective tissue of the corium, and result in the so-called premature aging of the skin. This is evidenced by the thickening of the skin, the loss of natural elasticity and the appearance of wrinkles, all resulting from the loss of the skin's water-binding capacity. There is also an increased tendency to form skin blemishes.

Excessive exposure to solar radiation can also aggravate or be the direct cause of some skin diseases, ranging from a transient dermatitis to skin cancer. Certain types of dermatitis are produced by photo-sensitization following solar irradiation in the presence of certain dyestuffs and chemicals such as tetrachlorosalicylanilide. Another example is Berlock dermatitis, an irregular discoloration of the skin, resulting from the application of bergamot oil or cologne spirit to the skin followed by exposure to sunlight. Lerner, Denton and Fitzpatrick[3] have suggested that psoralens present in bergamot and other citrus oils are responsible for Berlock dermatitis.

There is some evidence to support the view that excessive sunlight is a major factor in the production of skin cancer and that the erythemogenic and carcinogenic wavelength limits of ultraviolet radiation coincide. Consequently, according to Piers,[4] a higher incidence of skin cancer would be expected to occur in regions with sunlight rich in the shorter ultraviolet rays.

Roffo[5] has shown that malignant growths occur mainly in those regions of the body that receive the greatest amount of light, such as the neck, head, arms and hands. Passey[6] stated that seamen who were exposed to intense sunlight over a number of years developed cancer of the skin more often than others. It is also known that fair-skinned people are far more susceptible to skin cancer than people with a deeper pigmentation, and that Negroes are noted for their resistance to skin cancer, an indication that a dark pigmentation protects non-whites against the harmful effects of sunlight.

Auerbach[7] found a constant rate of increase in the incidence of skin cancer when moving southwards towards the equator, which doubled for every 3°48′ reduction in latitude, the increase in incidence thus being related to an increase in the exposure of individuals to sunlight in southern latitudes over that in northern latitudes.

The realization of the existence of long-term hazards of solar irradiation led some dermatologists to advocate the adoption of precautionary measures on a wider scale than practised hitherto. Knox,[8] for example, suggested the inclusion of sunscreens in make-up bases, face powders and after-shave lotions, and claimed that 2,4-dihydroxybenzophenone in an alcohol and silicone oil vehicle affords excellent protection for photosensitive individuals, and will also prevent tanning.

Solar Radiation and its Effect on Skin

Solar radiation consists of a continuous spectrum of wavelengths ranging from the infrared through the visible light to the ultraviolet region. The infrared radiation comprises wavelengths longer than 770 nm, the visible light wavelengths are between 400 nm and 770 nm, while ultraviolet radiation comprises wavelengths between 290 nm and 400 nm.

The skin responds differently to radiations of different wavelengths. The reddening of the skin produced by visible and infrared radiation (390–1400 nm) will appear immediately and subside rapidly at the end of the exposure. Radiation between 320 nm and 390 nm induces pigmentation but is not erythemogenic. Erythema is essentially brought about by exposure to ultraviolet radiation between 320 nm and 290 nm but will also be induced by radiation at shorter wavelengths.

Many workers have concerned themselves with defining whether, and which, separate parts of the ultraviolet spectrum give rise to sunburn and suntan. In the interests of standardization many have used artificial sources of ultraviolet radiation such as arc lamps of various types. It must be remembered that the total radiation from these sources will include some radiation with a wavelength as low as 200 nm, and although the lower wavelengths are filtered from sunlight by ozone in the upper atmosphere, the cosmetician has at times to concern himself with tanning by artificial sunlamps, as well as by the sun's rays.

Luckiesh and Taylor[9] carried out some classical work in this field and established spectral curves of tanning and erythema production. Using a range of filters to isolate narrow bands of wavelengths from radiation generated by an electric arc lamp, behind a water screen to absorb heat, they found that the action on the exposed areas depended on the wavelength of the energy to which they had been exposed. Erythema and tanning were assessed immediately after exposure, on the next day and at weekly intervals thereafter. By examining the skin under ultraviolet light, they concluded that erythema and tan may be produced simultaneously but a strong erythema may obscure the tan.

With respect to wavelength they found that at wavelengths above about 330 nm the areas appeared somewhat brownish and tanned immediately after exposure. Later an erythema developed together with the tan. Wavelengths of

334·2 nm and 366·3 nm were especially effective in producing tan with minimum erythema. At wavelengths about 295–315 nm there was no immediate visible effect, but after a few hours a definite erythema set in. After a few days the erythema subsided and a tan formed. At 250–270 nm the erythema was quite superficial and disappeared in a few days with no resultant tan.

They found no difference in the rate of fading of erythema of tan produced by the different wavelengths.

The conclusions reached were that the spectrum erythemal curve cannot be separated from the tanning effectiveness curve at wavelengths above 295 nm, at least for those tanning effects which may be apparent within 12 hours, when the erythema has approximately reached its maximum.

When using intermittent sub-erythemal exposures, Luckiesh and Taylor found that a single exposure of 40 per cent less duration than one which produced neither tan nor erythema when applied intermittently, would produce erythema or tan. These results appeared to indicate that if tanning is desired, exposures should be long enough to produce some erythema.

Blum[10] classified radiation below 320 nm as erythemal and that of 300–420 nm as melanogenic.

To sum up the above and other works, the ultraviolet range can thus be subdivided into the following three bands:

(1) The UV–A range, also referred to as long-wave ultraviolet radiation, has wavelengths ranging from 320 nm to 400 nm with a broad peak at 340 nm. This range is believed to be responsible for the direct tanning of the skin without a preliminary inflammation, possibly due to the photo-oxidation of the leuco form of melanin already present in the upper layer of the skin; it is, however, weak in producing erythema.

(2) The UV–B range of the ultraviolet radiation lies within wavelengths of 290 nm and 320 nm. It is also referred to as the sunburn radiation or middle UV radiation, and it has a peak of effectiveness around 297·6 nm. This is the erythemogenic range of UV responsible for producing sunburn as well as for irritating reactions which lead to the formation of melanin and the development of tan.

(3) The UV–C range, also referred to as the germicidal radiation or short UV radiation, comprises wavelengths from 200 nm to 290 nm. Although it is damaging to tissue, it is largely filtered from sunlight by ozone in the atmosphere. It can, however, be emitted by artificial UV sources. While it is not effective in stimulating tanning, it can cause erythema.

The A, B and C bands of UV radiation emit different amounts of energy and produce an erythemal reaction at different time intervals after exposure. About 20–50 joules cm^{-2} of UV–A radiation are required to produce minimally perceptible erythema compared with only 20–50 mJ cm^{-2} of UV–B energy and 5–20 mJ cm^{-2} of UV–C energy.

In the case of UV–A energy, erythema of the skin produced as a result of exposure to this radiation attains its maximum intensity at about 72 hours after exposure, while in the case of UV–B radiation, the erythemal reaction reaches its maximum intensity within 6–24 hours after exposure.

The proportions of energy of different wavelengths, furthermore, will vary with many other factors, such as the time of day, season, altitude, latitude, humidity and presence of smoke or dirt particles in the atmosphere.

The E-viton Concept and the Minimum Erythema Dose

To quantify erythemal energy, Luckiesh and Taylor[9] adopted as a unit of erythemal flux, independent of the wavelength, the E-viton equivalent to 10 microwatts of radiant energy at 296·7 nm wavelength where the erythemal effect is greatest.

The unit of intensity of the erythemal flux is 1 E-viton cm^{-2} (also referred to as the Finsen).

The measurement of intensity is indirect and based on the premise that, to produce a minimum perceptible erythema on an average untanned skin, exposure of approximately 40 E-viton min cm^{-2} of skin is required. Thus a minimally perceptible erythema (MPE) is produced by one of the following exposures:

1 E-viton cm^{-2} acting for 40 min
10 E-viton cm^{-2} acting for 4 min
40 E-viton cm^{-2} acting for 1 min

The intensity of solar erythemal ultraviolet energy in E-viton cm^{-2} on a horizontal plane with hourly variations on three specific days is shown in Figures 15.1 and 15.2.

The energy required to produce a minimum perceptible erythema is nowadays referred to as the MED (minimum erythema dose). It has been determined for both monochromatic and polychromatic radiation. Rottier[11] used the time required for the ultraviolet erythema (as distinct from heat erythema) to

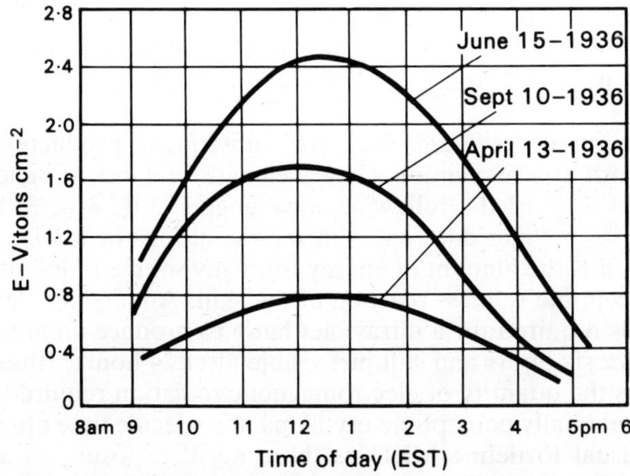

Figure 15.1 Erythemally weighted or antirachitic UV energy incident on a horizontal plane on three clear days in April, June and September. Exposure to approximately 40 E-viton min cm^{-2} produces a barely perceptible erythema on average untanned skin. (Courtesy *American Perfumer*)

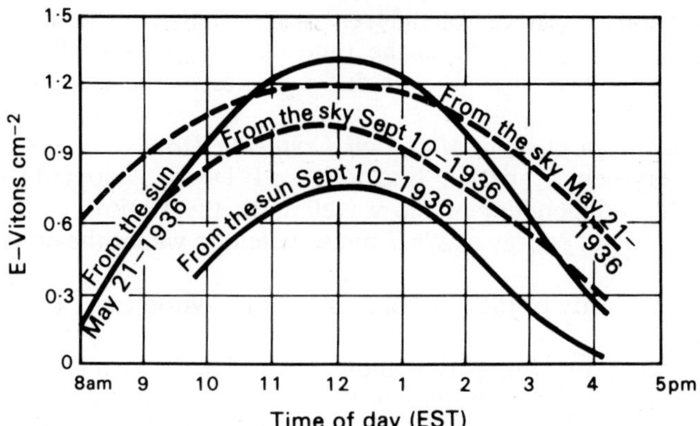

Figure 15.2 Erythemally weighted or antirachitic UV energy incident on a horizontal plane from the sun and from the entire sky on two clear days in May and September. (Courtesy *American Perfumer*)

develop and fade as a measure of the severity of the erythemal reaction. The UV erythema appears several hours after damage is done to the skin and the period of latency depends on the ultraviolet dose. Thus the latency associated with 1 MED is 8–10 hours, whereas with 8 MED it can decrease to 1–2 hours. The erythema produced by a dose greater than 3 MED may persist for several days, while 24 hours after such a dose there may be an oedema lasting several hours. Higher doses may cause severe oedema for days, while erythema may last for months, although it may be imperceptible under a heavy tan.

Each wavelength has a specific minimal erythemal dose, and a plot of log MED (s cm^{-2}) against wavelength (nm) for wavelengths 250 nm to about 550 nm gives an 'action spectrum' which shows skin to be most sensitive to radiation of wavelength 250–297 nm, and much less sensitive to longer wavelengths.

Kreps[12] studied the relative response of 'normal Caucasian skin' to monochromatic radiation and found that it varied markedly with the wavelength of the radiation.

The production of erythema and the subsequent production of melanic pigment are both at a maximum at a wavelength of 296·7 nm and decrease by factors of 10 at each of the following wavelengths: 307, 314, 330, 340 nm.

There have been slight differences in the definitions of MED. According to one definition, it is the amount of energy from any source required to produce a minimally perceptible redness reaction of the skin. Anderson[13] defined it as the time in seconds required for a ultraviolet lamp to produce an area of erythema developing after six hours and still just visible after 24 hours. Blum and Terus[14] define MED as the quantity of electromagnetic radiation required per unit area to produce a minimally perceptible erythema at a specific time after exposure. It is nowadays usual to define MED as the time of exposure to any particular source of UV (sun or UV lamp) required to produce an erythema which develops after six hours and is still just visible after 24 hours.

The energy quantification of sunburn-producing radiation proposed by Luckiesh is illustrated in Table 15.1, in which the degree of sunburn resulting from

Table 15.1 Energy Quantification of Sunburn-producing Radiation[9]

Degree of erythema	Exposure (min)	E-vitons ($s\ cm^{-2}$)	MED value
Minimally perceptible erythema	20	2 500	1·0
Vivid erythema	50	6 250	2·5
Painful sunburn	100	12 500	5·0
Blistering sunburn	200	25 000	10·0

different times of exposure is related to the intensity of erythemal flux and the minimum erythema dose (MED).

The length of time required to produce minimally perceptible erythema, and thus 1 MED, depends both on the quantity of energy emitted by the radiation source and on the response of a given individual's skin to sunlight, which in turn will depend on its pigmentation. Less time is required to produce a MED in light-skinned than in dark-skinned individuals. The MED for a dark-skinned Negro has been reported to be about 33 times higher than that for a Caucasian with a light complexion.

Reactions to exposure to sunlight also vary with season and the time of day. For example, at sea level the UV energy of sunlight is greatest between 10 a.m. and 2 p.m. in midsummer, with a maximum of UV–B energy falling on the skin at noon. In the early morning or late afternoon, when sunlight falls on the skin at a lower angle, the intensity of the solar energy is considerably lower and sunburn is unlikely to occur.

The different exposure times required to produce various degrees of sunburn in the average unprotected, untanned Caucasian can be seen in Table 15.2, which also illustrates the differences in exposure to produce the same effects at different latitudes. The time of exposure to produce sunburn may further be considerably shortened by reflection of additional UV light from snow and white sand.

Table 15.2 Exposure Times to Various Degrees of Sunburn[15]

Degree of sunburn	MED value	Length of exposure (min)	
		New Jersey (lat. 40°N)	Florida (lat. 25°N)
Minimal erythema	1	21	10
Vivid erythema	2	42	25
Painful sunburn	4	50	50
Blistering sunburn	8	165	120

Protective Mechanism of the Skin

The two factors which are mainly responsible for the skin's natural protection against sunburn are the thickness of the stratum corneum and the pigmentation of the skin.

A number of workers who investigated the nature of the protective mechanism of the skin have shown that solar radiation increases the mitotic rate of epidermal cells, causing a thickening of the stratum corneum in the course of 4–7 days, and making it thereby more impervious to the passage of erythemogenic radiation.

Some degree of protection against sunburn is conferred by an increased melanin content of the epidermis. Granules of melanin which are formed in the basal cell layer of the skin following the action of UV–B radiation migrate upwards towards the stratum corneum and the skin surface, where they are believed to be oxidized by radiation of the UV–A range. These granules are eventually shed during exfoliation, causing the skin to lose its immunity to sunburn.

The painful after-effects of solar irradiation of the unprotected skin which so often follow sunbathing can normally be prevented by a gradual exposure. The initial exposure (allowing for the sensitivity of the individual concerned) should not exceed 5–15 minutes, and should be progressively extended, the daily increase in exposure being of the order of 40 per cent over the preceding day. This should ensure development of full immunity to sunburn within 10–12 days. Maximum pigmentation can be achieved after above 100 hours of exposure.

It has been suggested by Hais and Zenisek[16] that urocanic acid present in the stratum corneum to the extent of 0·6 per cent may act as a natural physiological screening agent, in that it absorbs ultraviolet radiation within the 300–325 nm range. Its elution from the skin during bathing may explain the increased susceptibility of the skin to sunburn.

Rottier[11] proposed an arbitrary classification of people into three groups according to their reaction to sunlight.

Group 1. 'The insensitives' with good habituation and pigmentation.
Group 2. 'The sensitives' with bad habituation and no pigmentation.
Group 3. 'The diseased' with a pathological skin reaction to sunlight.

He pointed out that in Holland people belonging to Group 1 would acquire 1–3 MED 306 in 1–3 hours, at first exposure of the trunk to the summer sun. This would not harm their skin. Repetition of such doses on the following days would gradually produce a red-brown tan. The subjects after one week's exposure would easily tolerate 8–10 hours sunlight a day, even at Mediterranean latitudes.

People in Group 2 would acquire 4–10 MED 306 at the first exposure of one hour. This may cause a disagreeable sunburn in the evening. People in this group can never expose themselves to the sun for such long periods as those in the first group, and they remain as red as lobsters without much tanning.

Insensitive people do not require a high screening factor, and in order to acquire tan, they will need a fairly strong erythema. As sunscreen they will normally use an oil which does not screen out the short wave UV. Sensitive subjects, on the other hand, will require real protection against UV in order to withstand longer exposures without any disagreeable sunburn.

A more recent classification of skin types and the characteristics of appropriate sunscreen products for their protection are given below (page 240).

SUNSCREENS AND SUNTAN PREPARATIONS

Introduction

The purpose of sunscreens and suntan preparations is to prevent or minimize the harmful effects of solar radiation or to assist in tanning the skin without any painful effects.

Classification of Sunscreens According to Application
Depending on their intended application, sunscreens may be classified as follows.

(1) Sunburn preventive agents are defined as sunscreens which absorb 95 per cent or more of UV radiation within wavelengths 290–320 nm.
(2) Suntanning agents are defined as sunscreens which absorb at least 85 per cent of UV radiation within the wavelength range from 290 nm to 320 nm but which transmit UV light at wavelengths longer than 320 nm and produce a light transient tan. These agents will produce some erythema but without pain.

Sunscreens in both of these categories are chemical sunscreens absorbing a specific range of UV radiation. In some instances the same sunscreen may be employed in both product classes but at different concentrations (lower in a suntan product).

(3) Opaque sunblock agents aim to provide maximum protection in the form of a physical barrier. Titanium dioxide and zinc oxide are the most frequently used agents in this group. Titanium dioxide reflects and scatters practically all radiation in the UV and visible range (290–777 nm), thereby preventing or minimizing both sunburn and suntan.

It should be emphasized nevertheless that opaque sunblock agents based on inorganic materials are not the only compounds claiming to confer maximum protection against solar radiation. 'Supershade 15', a product of the Plough Corporation which contains a combination of 7 per cent octyl-dimethyl-*p*-aminobenzoic acid ester and 3 per cent oxybenzone is claimed to have a sun protection factor of 15 (see page 240) and to give complete protection against UV–B. It is also claimed that its regular use may prevent skin cancer.

Other products which will be mentioned in this chapter are palliatives and simulative preparations. Palliatives are designed to alleviate the pain and irritation resulting from excessive exposure to sunlight; many of them are purchased from the local chemist and druggist. Simulative preparations are designed for those who wish to feign a brown skin in the minimum of time and with the least possible pain or trouble. These are essentially preparations which stain the skin brown, or promote the synthesis of melanoid materials in the skin.

Sunscreen Agents

Sunscreens should either scatter the incident light effectively, or they should absorb the erythemal portion of the sun's radiant energy. Opaque powdered materials, when applied to the skin either in the dry state or when incorporated into suitable vehicles, will serve to scatter the ultraviolet light falling upon them. As discussed under Face Powders, in Chapter 18, zinc oxide is the most effective of such powders and is superior to titanium dioxide in this respect. Other powders which may be employed for this purpose with, however, very much less efficiency, are kaolin, calcium carbonate, magnesium oxide, talc, etc. Obviously the particle size of the powder employed is a factor of considerable importance in such preparations.

Although powders of this type rank fairly low in the sales of anti-sunburn preparations, when applied as a second line of defence over a suitable sun-screening foundation, their light-scattering possibilities should not be ignored.

The most important class of sunscreens are those which operate by absorbing the erythemal ultraviolet radiation. The properties necessary in a sunscreen are:

(1) It must be effective in absorbing erythmogenic radiation in the 290–320 nm range without breakdown which would reduce its efficiency or give rise to toxic or irritant compounds.
(2) It must allow full transmission in the 300–400 nm range to permit the maximum tanning effect.
(3) It must be non-volatile and resistant to water and perspiration.
(4) It must possess suitable solubility characteristics to allow the formulation of a suitable cosmetic vehicle to accommodate the requisite amount of sunscreening.
(5) It must be non-odorous or at least sufficiently mild to be acceptable to the user and be satisfactory in other relevant physical characteristics such as stickiness, etc.
(6) It must be non-toxic, non-irritant and non-sensitizing.
(7) It must be capable of retaining its protective capacity for several hours.
(8) It must be stable under conditions of use.
(9) It must not stain clothing.

Non-toxicity and dermatological acceptability are important because, as Draize[17] has pointed out, 'sunscreens are unique as cosmetics in that their manner of use may involve multiple and extensive daily applications to large areas of the body surface and, in addition, they may be applied to skin already damaged by sun or wind burn'. Moreover they may be used on persons of all age groups and in varying conditions of health.

Draize further pointed out that pharmacological and toxicologic tests should establish an 8-fold margin of safety and that this involves acute and sub-acute dermal toxicities and potential sensitization studies.

During World War II, systematic investigations were carried out of a number of materials likely to provide protection against sunburn to soldiers fighting in tropical countries and airmen shot down over tropical islands. This led to the introduction of many new organic materials which in turn had to be screened for efficacy and toxicity.

An extensive list of sunscreens was compiled by Klarmann.[18] It included the following substances:

Para-aminobenzoic acid and its derivatives (ethyl, isobutyl, glyceryl esters; *para*-dimethylaminobenzoic acid).

Anthranilates (i.e. *ortho*-aminobenzoates; methyl, menthyl, phenyl, benzyl, phenylethyl, linalyl, terpenyl and cyclohexenyl esters).
Salicylates (amyl, phenyl, benzyl, menthyl, glyceryl and dipropylene-glycol esters).
Cinnamic acid derivatives (menthyl and benzyl esters; alphaphenyl cinnamonitrile; butyl cinnamoyl pyruvate).
Dihydroxycinnamic acid derivatives (umbelliferone, methyl-umbelliferone, methylaceto-umbelliferone).
Trihydroxycinnamic acid derivatives (esculetin, methylesculetin, daphnetin, and the glucosides esculin and daphnin).
Hydrocarbons (diphenylbutadiene, stilbene).
Dibenzalacetone and benzalacetophenone.
Naphthosulphonates (sodium salts of 2-naphthol-3,6-disulphonic and of 2-naphthol-6,8-disulphonic acids).
Dihydroxy-naphthoic acid and its salts.
Ortho- and *para*-hydroxybiphenyldisulphonates.
Coumarin derivatives (7-hydroxy, 7-methyl, 3-phenyl).
Azoles (2-acetyl-3-bromoindazole, phenyl benzoxazole, methyl naphthoxazole, various aryl benzothiazoles).
Quinine salts (bisulphate, sulphate, chloride, oleate and tannate).
Quinoline derivatives (8-hydroxyquinoline salts, 2-phenylquinoline).
Uric and violuric acids.
Tannic acid and its derivatives (e.g. hexaethylether).
Hydroquinone.

Klarmann pointed out that isomerism plays an important role in determining absorptive capacity and illustrated this fact with obscuration (absorption) curves for *ortho*-, *meta*-, and *para*-aminobenzoic acids, which indicated the superiority of the *para*-isomer over the *ortho*- and *meta*-isomers. By way of contrast, *ortho*-hydroxybenzoic (salicylic acid) has a high absorption value for erythemal radiation, whereas *para*-hydroxybenzoic acid has practically none.

In experiments on protective skin preparations for the prevention of sunburn carried out on behalf of the US Army Air Forces, a large number of products were investigated.[19] It was felt that since men marooned on life rafts, or in the desert, might be subject to very severe solar exposure without shelter, and to either very high or very low temperature conditions as well as to spray and waves, it was essential that the substances chosen should give effective protection, should be stable at freezing and at tropical temperatures, should be free from rancidity, non-irritant, non-toxic, and should be waterproof. Although the requirements for such preparations were exceedingly stringent, and are not likely to be generally encountered in the case of toilet preparations, the results obtained in this investigation indicate a number of substances which are of very definite value. For use by the US Forces sunscreens were required to show an absorption of 99 per cent at 297 nm with a film thickness of 0·001 in.

The screening agents tested included titanium dioxide, zinc oxide, phenyl salicylate, yellow petroleum jelly, amber petroleum jelly, zinc oxide ointment, lotions containing menthyl salicylate, a proprietary sun-preparation, dark red veterinary petroleum jelly, and several other grades of petroleum jelly. Various types of bases included water-in-oil emulsions of lanolin and petroleum jelly, vanishing cream base, and petroleum jelly base.

A dark red petroleum veterinary jelly (Standard Oil Co., New Jersey) was quite opaque to erythemal energy and used on its own gave complete protection to the skin in an exposure equivalent to 20 hours of the strongest sunlight measured during a four-year period in Cleveland. This compound was found to be non-irritant and to adhere tenaciously to the skin. Phenyl salicylate (salol) was also found to be an excellent sunscreen when used at a level of 10 per cent in a suitable base, such as petroleum jelly, particularly in a jelly having erythemal screening properties of its own. Experiments showed phenyl salicylate to be non-toxic. Zinc oxide was also found of definite value in preventing sunburn, but not in conjunction with phenyl salicylate with which results were not so good when zinc oxide was added. Titanium dioxide was not found to be a very dependable protective judged on a specimen containing 20 per cent in yellow petroleum jelly. Yellow petroleum jelly was found to possess dependable screening properties against ultraviolet energy at wavelengths 296·7 nm and 302·2 nm, far more than those possessed by white petroleum jelly.

Of the very large range of compounds possessing satisfactory absorption characteristics which have been listed repeatedly in the literature, sunscreens have, in practice, been confined to *p*-aminobenzoates, *p*-dialkylaminobenzoates, salicylates and cinnamates, and frequently mixtures of these compounds have been employed. A combination of benzyl cinnamate and benzyl salicylate in an emulsion base was used in the first commercial sunscreen preparation marketed in the USA in 1928.

A number of proprietary products have been based upon menthyl salicylate. Unlike some salicylates, the menthyl ester of salicyclic acid is non-irritant, odourless and initially acts as a satisfactory sunscreen at concentrations of about 10 per cent. However, as pointed out by deNavarre,[20] menthyl salicylate undergoes chemical change on exposure to light with the result that its screening properties are considerably diminished. Menthyl anthranilate, marketed by Givaudan under a proprietary designation, was claimed to give maximum effectiveness at about 4 per cent concentration.

Other proprietary sunscreen preparations include 'Solprotex' of Firmenich, 'Prosolal 58' of Dragoco, the 'Antivirays' of A. Boake Roberts and 'Giv-tan F' of Givaudan. 'Giv-tan F', which is 2-ethoxyethyl-*p*-methoxycinnamate, was claimed by the manufacturers to have a sunscreen index of 14·4 (see Table 15.6); this is of the same order as ethyl-*p*-dimethylaminobenzoate which has an index of 14·8.

The following five sunscreen agents were recommended by the US Department of Health, Education and Welfare[21] for inclusion in 'a base such as mineral oil, cold cream or ethyl alcohol' at the levels stated:

	per cent
Cycloform (isobutyl *p*-aminobenzoate)	5·0
Propylene glycol *p*-aminobenzoate	4·0
Monoglyceryl *p*-aminobenzoate	2·5
Digalloyl trioleate	5·0
Benzyl salicylate and benzyl cinnamate (2% each)	4·0

Giese, Christensen and Jeppson[22] listed the extinction coefficients at 297 nm of a number of sunscreens, together with some of their physical properties (Table 15.3).

Pernich and Gallagher[23] examined transmission spectra of various sunscreens and found the following substances to be efficient sunscreens:

Ethyl *p*-dimethylaminobenzoate
Isobutyl *p*-aminobenzoate
Coumarin
8-Methoxycoumarin
5,7-Dihydroxy-4-methylcoumarin
6-Aminocoumarin
Umbelliferone
Benzyl-β-methyl-umbelliferone
Benzylacetophenone

Some other compounds, claimed to be effective sunscreens, have been disclosed in patent specifications.

One patent[24] claims that protection against erythema is obtained by the application to the skin of hydrazones of *ortho*- and *para*-aminobenzaldehydes, and of *ortho*- and *para*-aminoacetophenones. Acetylated aminocinnamates have been claimed as effective and stable sunscreens in another patent.[25]

Two other patents cover the use of organosilicon compounds with sunscreening properties. These are not easily washed off the skin. One example of such a compound is a reaction product of carbethoxyethyltriethoxysilane with *p*-aminobenzoic acid, which absorbs light in the range 260–310 nm.[26]

Another example[27] of an active UV-absorbing organosilicon compound is one produced by reacting ethyl-*p*-(N-butylamino) benzoate with gamma aminopropyltriethoxysilane. The resulting compound is said to absorb light in the 240–330 nm region.

The transmission spectrum obtained by Pernich and Gallagher for ethyl *p*-dimethylaminobenzoate is reproduced in Figure 15.3.

Stambovsky[28] expressed the view that esters of *para*-aminobenzoic acid in comparable concentrations possessed the highest absorption capacity of any chemicals then available. He stated that out of twenty-seven actively promoted suntan products on the US market in 1955, nineteen employed derivatives of aminobenzoic acid or salicylic acid.

Sunscreen compositions containing esters of *p*-dimethylaminobenzoic acid with C_5—C_{18} monohydric alcohols have also been patented.[29] The suitability of sunscreens for commercial application is determined by their screening efficacy, solubility and their stability in given formulations. The higher esters of *p*-dimethylaminobenzoic acid are claimed to have UV absorption screening properties superior to those of the lower alkyl esters of both *p*-aminobenzoic acid and dimethylaminobenzoic acid.

In general, esters of *p*-dimethylaminobenzoic acid are more effective sunscreens than esters of *p*-aminobenzoic acid and superior to them in respect of stability, storage and use. They are also less reactive when incorporated in the usual type of cosmetic formulations.

Table 15.3 Sunscreens—Extinction Coefficients at 297 nm[22]

Compound	State	Solvent	Molecular weight	Extinction coefficient	Concentration (%)
Ethyl *p*-dimethylaminobenzoate	Solid	Alcohol	194·13	27 000 27 300	1×10^{-4}
Ethyl *p*-aminobenzoate	Solid	Alcohol	165·19	21 750 21 450	2×10^{-4}
Isobutyl *p*-aminobenzoate	Solid	Alcohol	193·13	23 200 22 800	2×10^{-4}
Menthyl anthranilate	Liquid	Alcohol	275·374	941 956	5×10^{-4}
*Homo*menthyl salicylate	Liquid	Alcohol	390·38	6 720 6 720	$2{\cdot}5\times10^{-4}$
Phenyl salicylate	Solid	Alcohol	214·08	3 850 3 900	5×10^{-4}
Menthyl salicylate	Liquid	Alcohol	276·19	4 540 4 600	5×10^{-4}
Amyl salicylate	Liquid	Alcohol	208·12	4 150 3 970	5×10^{-4}
Isoamyl salicylate	Liquid	Alcohol	208·12	348 381	5×10^{-4}
Benzyl salicylate	Liquid	Alcohol	228·13	4 060 4 200	1×10^{-4}
Cinnamic acid	Solid	Alcohol	148·06	705 693	1×10^{-3}
Benzyl cinnamate	Solid	Alcohol	238·11	1 908 1 979	1×10^{-4}
*Homo*menthyl cinnamate	Liquid	Alcohol	300·32	505 530	5×10^{-4}
β-Methyl umbelliferone	Solid	Alcohol	176·066	8 510 8 560	5×10^{-4}

2-Naphthol-6-sulphonic acid	Solid	Water	304·28	3 310	5×10^{-3}
				3 240	
2-Naphthol-8-sulphonic acid	Solid	Water	304·28	2 010	1×10^{-3}
				1 830	
3-Hydroxy-2-naphthoic acid	Solid	Alcohol	188·17	3 470	5×10^{-4}
				3 330	
Acetanilide	Solid	Alcohol	135·08	162	1×10^{-1}
Violuric acid	Solid	Alcohol	157·05	4 800	1×10^{-3}
				4 700	
Benzylacetophenone	Solid	Alcohol	208·196	24 200	1×10^{-4}
				23 900	
Quinine sulphate	Solid	Alcohol	548·39	3 500	4×10^{-4}
				3 500	

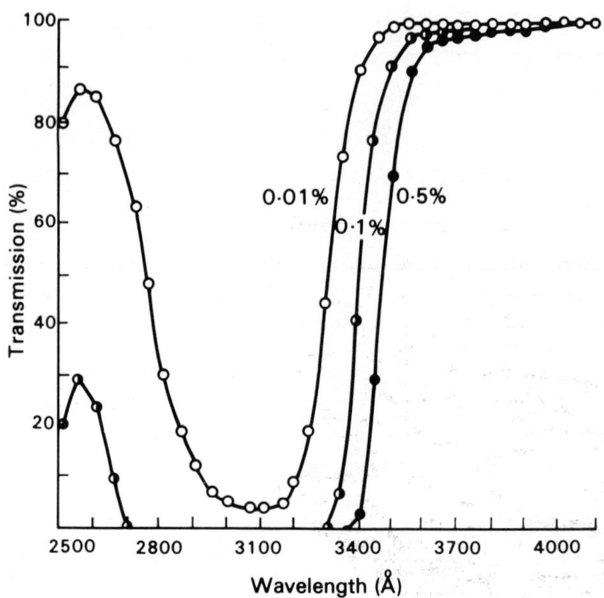

Figure 15.3 Transmission spectra of 0·5 cm layers of different concentrations of ethyl-*p*-dimethylaminobenzoate (in alcohol)

As far as esters of *p*-dimethylaminobenzoic acid are concerned, the lower esters are appreciably soluble in water, but insoluble in oils, and as a result of this are easily removed from the skin during bathing or by perspiration. With increasing molecular weight the esters become progressively less susceptible to removal by water, thus conferring longer lasting and more effective protection against erythema.

The higher alkyl esters (for example amyl, hexyl, heptyl and octyl esters) of *p*-dimethylaminobenzoic acid are oily liquids which are said to form continuous and adherent films which are not easily removed by water, exercise, abrasion or washing. Their solutions in mineral, vegetable and animal oils are claimed to remain completely homogeneous on storage for prolonged periods of time.

Sunscreening compositions fairly stable to actinic radiation, which are not readily removed from the skin, have been claimed by the GAF Corporation. As the active sunscreens, these compositions employ compounds prepared by condensing a benzaldehyde with a keto or thioketo hydrazine. Compounds specifically claimed are:

$$Cl-C_6H_4-CH=N-NHC(=O)NH_2 \quad \text{and} \quad Cl-C_6H_4-CH=N-NHC(=S)NH_2$$

These sunscreens may be used in combination with conventional additives.

Selection of a suitable base which will not be easily removed from the skin, and will therefore ensure a long period of protection against sunburn, has been the subject of a patent granted to Boots Pure Drug Co.[30] The patented compositions contain a sunscreen, preferably ethyl *p*-dimethylaminobenzoate, in a cosmetically acceptable diluent or carrier comprising not less than 5 per cent of castor oil.

Polymeric Sunscreen Materials

All the compounds conventionally used as sunscreens have a relatively low molecular weight and many of them are fairly quickly removed from the skin on contact with water, necessitating further applications of the product if protection against erythema is still required. An attempt to avoid the necessity of re-applications has led to the development of water-insoluble but alkali-soluble polymeric sunscreens.

Formulations described in a patent of the National Starch and Chemical Corporation[31] contain at least 1 per cent of an alkali-soluble, UV-absorbing polymeric sunscreen in a suitable vehicle. The polymer in question is produced from at least two essential comonomers:

(1) An ethylenically unsaturated compound capable of absorbing UV radiation of the erythemogenic range, but transmitting radiation which produces suntan (exemplified by certain substituted acrylates, methacrylates and benzoates as well as some ethers of 2,4-dihydroxybenzophenone, 2,2,4-trihydroxybenzophenone and ethers of benzotriazole derivatives).
(2) An acidic comonomer specified as an ethylenically unsaturated carboxylic acid containing at least one free carboxyl group (for example acrylic acid, methacrylic acid, itaconic acid, crotonic acid, etc.).

Compared with non-polymeric sunscreens, the polymeric sunscreens disclosed in the patent are very resistant to removal by fresh or sea water. By virtue of their free carboxylic groups, however, they may be readily removed from the skin, merely by the application of a dilute, mildly alkaline aqueous solution, such as soap and water, which converts the water-insoluble polymers into their water-soluble and readily removable alkaline salts.

Several more recent patents relate to sunscreens or sunscreen compositions which are substantive to the skin and resistant to water and perspiration, therefore providing a more lasting protection. An example of a compound with such properties mentioned in one patent[32] is a sunscreen comprising a 50 per cent by weight solution, in isopropanol, of a polymer produced from the 4-(3-acryloxy-2-hydroxypropyl) ether of 2-(2,4-dihydroxyphenyl) benzotriazole, vinyl acetate and ethyl hydrogen maleate.

Another active agent in a sunscreen composition, with a high absorption of UV radiation and capable of depositing a continuous film substantive to the skin, is a fluorescent concentrate recovered from dark green or red petrolatum.[33] 2-Hydroxy-4-methoxybenzophenone-5-sulphonic acid was the active ingredient of yet another patent.[34]

Sunscreen compositions have also been disclosed which contain as active agents highly substantive sulphonium salts[35] such as *para*-nitrobenzamide propyldodecylmethyl sulphonium bromide. These were quoted to have prominent absorption peaks within the wavelength range of 250 nm and 400 nm, which is wider than those of the majority of commercially available sunscreens; they also adhere well to the skin.

Another patent disclosed compositions containing sunscreens with absorption peaks between 280 nm and 320 nm used in a cosmetically acceptable vehicle in combination with an alcohol-soluble polyamide resin; these were capable of forming protective films, substantive to the skin and resistant to water and perspiration.[36]

Skin Types and Recommendations for the Choice of Sunscreens—
The Sun Protection Factor

The extent to which a sunscreen product protects from sunburn and other harmful effects of exposure to sunlight varies with the individual skin type. A classification system for sunscreening products[37] comprises five product category designations (PCD) to meet requirements of consumers with different types of skin.

Individuals can be classified into six groups according to skin type and tanning history:

I Always burns easily; never tans (sensitive).
II Always burns easily; tans minimally (sensitive).
III Burns moderately; tans gradually (light brown) (normal).
IV Burns minimally; always tans well (moderate brown) (normal).
V Barely burns; tans profusely (dark brown) (insensitive).
VI Never burns; deeply pigmented (insensitive).

The 'sun protection factor' (SPF) system has been developed by the Plough Corporation to define the relative effectiveness of sunscreen agents to protect the skin. It was subsequently recommended by the Over-the-Counter (OTC) Panel of the US Food and Drug Administration as a means of numerically identifying the efficacy of various sunscreen products and to provide for consumers a guide to the products suitable for particular types of skin. The SPF has been defined as the ratio:

$$\left(\begin{array}{l}\text{UV energy required to produce}\\ \text{a minimal erythemal dose (MED)}\\ \text{on protected skin}\end{array}\right) \Bigg/ \left(\begin{array}{l}\text{UV energy required to produce}\\ \text{one MED on unprotected skin}\end{array}\right)$$

or as the ratio between the UV exposure required to produce a minimally perceptible erythema on protected skin, and the exposure that will produce the same erythema on unprotected skin. The formal definition of SPF by the OTC Panel was

$$\text{SPF value} = \frac{\text{MED(PS)}}{\text{MED(US)}}$$

where MED(PS) is the minimum erythemal dose for protected skin after the application of 2 mg cm^{-2} or 2 μl cm^{-2} of the final formulation of the sunscreen product, and MED(US) is the minimum erythemal dose for unprotected skin, that is skin to which no sunscreen product has been applied. The larger the SPF, the greater the protection the sunscreen can confer.

The OTC Panel has proposed that all sunscreen products should be rated for the consumer according to the degree of protection they can provide, the rating numbers to range from 2–8. Products with a rating of 8 would thus provide the maximum protection to individuals who always burn easily and never tan, while products with a rating of 2 would be suitable for those who rarely burn and who tan profusely. Thus, for the skin-type groups listed above, sunscreen products with the following SPFs are recommended:

Skin type	*SPF*
I	8 or more
II	6–7
III	4–5
IV	2–3
V	2
VI	None indicated

The product category designations recommended to consumers in selecting the types of sunscreen product providing various SPFs are as follows.

PCD 1: Minimal Sun Protection Product	Provides an SPF value of 2 to under 4 and offers the least protection, but permits suntanning.
PCD 2: Moderate Sun Protection Product	Provides an SPF value of 4 to under 6 and offers moderate protection from sunburning, but permits some suntanning.
PCD 3: Extra Sun Protection Product	Provides an SPF value of 6 to under 8 and offers extra protection from sunburning, and permits limited suntanning.
PCD 4: Maximal Sun Protection Product	Provides an SPF value of 8 to under 15 and offers maximal protection from sunburning, permitting little or no suntanning.
PCD 5: Ultra Sun Protection Product	Provides an SPF value of 15 or above and offers greatest protection from sunburning, permitting no suntanning.

It has been pointed out that some people when first using this scale may misjudge the reactivity of the skin to sunlight. It has also been mentioned that elevated heat and humidity, sweating and swimming may lower the SPF value at any one time for an individual.

In practical terms, a person who usually gets red in the sun after 20 minutes should be able to stay in the sun for 120 minutes if he applies a sunscreen of extra protection (SPF 6), that is 20 minutes × 6, provided the product is not washed off or removed by sweat. A product in the maximal protection category (say,

SPF 8) would protect the average person who would get sunburn in 40 minutes or be exposed to sunlight in the dangerous sunburning hours between 10 a.m. and 2 p.m., for 40 × 8 = 320 minutes. However, once the skin has become accustomed to the sun (developed protection through pigmentation) the individual's self-protection period is longer, and since the risk of sunburn has become smaller he may gradually replace a product with a high PCD by a product with a lower PCD.

Highly sensitive individuals needing mainly protection against sunlight are recommended to use a product in the 'ultra protection' category (SPF 15 or more).

For the guidance of consumers the OTC Panel has recommended that the following labelling statements are prominently placed on the principal display panels of appropriate products:

1. For minimal sun protection products (SPF 2)—
   'Stay in the sun twice as long as before without sunburning'.
2. For moderate sun protection products (SPF 4)—
   'Stay in the sun 4 times as long as before without sunburning'.
3. For medium sun protection products (SPF 6)—
   'Stay in the sun 6 times as long as before without sunburning'.
4. For maximum sun protection products (SPF 8)—
   'Stay in the sun 8 times as long as before without sunburning'.
5. For ultra sun protection products (SPF 15)—
   'Stay in the sun 15 times as long as before without sunburning'.

Product Classification According to Safety or Efficacy

The OTC Panel has classified active sunscreen ingredients into three categories (Table 15.4). Products in Category I are generally recognized as safe and effective. Products in Category II are not generally recognized as safe or effective and are misbranded. The data relating to products in Category III are not yet sufficient to permit their final classification. A period of two years has been recommended to complete studies that may enable products in Category III to be switched to Category I.

In Europe, the proposed third amendment (1981) to the EEC Cosmetics Directive lists six sunscreen agents that may be contained in cosmetic products, and a further 36 sunscreen agents that may provisionally be contained in cosmetic products. These are given in Table 15.5.

Evaluation of Sunscreen Preparations

When formulating sunscreen preparations it is necessary to assess the efficiency both of the sunscreen and of the prototype products. This is done by examining their absorption characteristics spectrophotometrically in terms of concentration, thickness of liquid through which the light passes, and wavelength. The absorption characteristics may be expressed as a percentage of incident radiation absorbed or transmitted, or as the optical density. The last named is the logarithm of the ratio of the intensities of the radiation before and after passage through the solution. It has the advantage that it is directly proportional to the concentration and to the thickness of the material, hence calculation is simple.

Table 15.4 Classification of Sunscreen Ingredients—US FDA OTC Panel

I SAFE, EFFECTIVE

Compound	**Dosage limit (%)**
p-Aminobenzoic acid	5–15
2-Ethoxyethyl-*p*-methoxy cinnamate (Cinoxate)	1–3
Diethanolamine-*p*-methoxy cinnamate	8–10
Digalloyl trioleate	1–5
2,2-Dihydroxy-4-methoxybenzophenone (Dioxybenzone)	3
Ethyl-4-bis-(hydroxypropyl)aminobenzoate	1–5
2-Ethylhexyl-2-cyano-3,3-diphenyl acrylate	7–10
Ethylhexyl-*p*-methoxy cinnamate	2–7·5
2-Ethylhexyl salicylate	3–5
Glyceryl aminobenzoate	2–3
3,3,5-Trimethylcyclohexyl salicylate (Homosalate)	4–15
Lawsone with dihydroxyacetone { Lawsone	0·25
DHA	3·0
Menthyl anthranilate	3·5–5
2-Hydroxy-4-methoxy benzophenone (Oxybenzone)	2–6
Amyl-*p*-dimethylamino benzoate (Padimate A)	1–5
2-Ethylhexyl-*p*-dimethylamino benzoate (Padimate O)	1.4–8
2-Phenylbenzimidazole-5-sulphonic acid	1–4
Red petrolatum	30–100
2-Hydroxy-4-methoxybenzophenone-5-sulphonic acid (Sulisobenzone)	5–10
Titanium dioxide	2–25
Triethanolamine salicylate	5–12

II NOT SAFE, NOT EFFECTIVE

2-Ethylhexyl-4-phenylbenzophenone-2′-carboxylic acid
3-(4-Methylbenzylidene)-camphor
Sodium 3,4-dimethylphenyl glyoxylate

III UNCLASSIFIED

Allantoin combined with aminobenzoic acid
5-(3,3-Dimethyl-2-norbornylidene)-3-penten-2-one
Dipropylene glycol salicylate

The absorption characteristic of a chemical is frequently expressed as the molar extinction coefficient, i.e. the calculated optical density of a 1 cm layer of a molar solution, or the specific extinction coefficient, i.e. the calculated optical density of a 1 cm layer of a 1 per cent solution.

However, Stambovsky,[38] when discussing the technical causes of commercial failure of suntan preparations on the US market, warns against relying too much upon spectrometric data in selecting sunscreen agents and suggests that their value is limited to initial qualitative examination.

Kumler and Daniels[39] constructed a sunburn curve for which the ordinates were the products of the ordinates of the solar radiation curve and an erythematic curve derived from experimental data. This curve ranged from

Table 15.5 Classification of Sunscreen Ingredients—EEC Cosmetics Directive (Proposed Amendment)

Compound	Dosage limit (%)
SUNSCREEN AGENTS WHICH COSMETIC PRODUCTS MAY CONTAIN	
p-Aminobenzoic acid	5
3-(4-Trimethylammoniobenzylidene) camphor methosulphate	6
Homomenthyl salicylate (3,3,5-trimethylcyclohexyl salicylate)	10
Phenyl salicylate	4
2-Hydroxy-4-methoxybenzophenone	10
2-Amino-6-hydroxypurine (guanine)	2
SUNSCREEN AGENTS WHICH COSMETIC PRODUCTS MAY PROVISIONALLY CONTAIN	
N-propoxylated ethyl-*p*-aminobenzoate	5
Ethoxylated ethyl-*p*-aminobenzoate	10
Amyl *p*-dimethylaminobenzoate	5
Glyceryl *p*-aminobenzoate	5
2-Ethylhexyl *p*-dimethylaminobenzoate	8
2-Ethylhexyl salicylate	5
Benzyl salicylate	7
3,3,5-Trimethylcyclohexyl-N-acetylanthranilate (homomenthyl-N-acetyl anthranilate)	2
Potassium cinnamate	2
p-Methoxycinnamic acid salts (potassium and diethanolamine)	8 (expressed as acid)
Propyl *p*-methoxycinnamate	3
Salicylic acid salts (potassium and triethanolamine)*	5
Iso-amyl *p*-methoxycinnamate	10
2-Ethylexyl *p*-methoxycinnamate	10
2-Ethoxyethyl *p*-methoxycinnamate	5
Digalloyl trioleate	4
2,2′,4,4′-Tetrahydroxybenzophenone	10
2-Hydroxy-4-methoxy-4′-methylbenzophenone	4
2-Hydroxy-4-methoxybenzophenone-5-sulphonic acid and sodium salt	5 (expressed as acid)
2-Ethylhexyl-4′-phenylbenzophenone-2-carboxylate	10
2-Phenylbenzimidazole-5-sulphonic acid and its potassium and triethanolamine salts	8 (expressed as acid)
β-Imidazole-4(5)-acrylic acid and its ethyl ester	5 (expressed as acid)
2-Phenyl-5-methylbenzoxazole	4
Sodium 3,4-dimethoxyphenylglyoxylate	5
Dianisoylmethane	6
5-(3,3-Dimethyl-2-norbornylidene)-3-pentene-2-one	3
3-(3-Sulpho-4′-methylbenzylidene) camphor	6
3-(4′-Sulphobenzylidene) camphor	6
3-(4′-Methylbenzylidene)-d,l-camphor	6
3-Benzylidene-d,l-camphor	6
Methoxybenzylidene cyanoacetic acid and its n-hexyl ester	5
4-Isopropyldibenzoylmethane	5
p-Isopropylbenzyl salicylate	4
Cyclohexyl-*p*-methoxycinnamate	1
2-(*p*-Toluyl)-benzoxazole	10
ter-Butyl-4-methoxy-4-dibenzoylmethane	5

* pH of the finished product must be such that the acid is not liberated. Not to be used for children under 3 years of age.

296 nm to 326 nm with a maximum at 308 nm. In their view a compound had to fulfil two requirements to be regarded as an effective sunscreen. Firstly, it would have to superimpose on the entire sunburn curve. Secondly, it would have to possess high absorption properties at 308 nm. This latter qualification was later proposed by Kumler[40] as the basis for a simple and rapid method for the relative evaluation of sunscreens. He measured the optical density of 0·1 per cent solutions in a 0·1 mm silica cell at 308 nm and converted the results to a sunscreen index (SI) which corresponds to the OD of a 1 per cent solution in a 1 mm cell. Forty-five compounds are ranked in decreasing order of screening effectiveness in Table 15.6, headed by ethyl *p*-dimethylaminobenzoate.

If the absorption characteristics of the chosen material are known, the concentration required in a product to produce the desired effects can be calculated, taking into account the probable film thickness to be applied.

Film Thickness

Film thickness can, of course, be estimated readily by measuring the area covered by a known amount of the preparation applied in a practical manner.

Bergwein[41] considered that, when dealing with sunscreen preparations in the form of fatty ointments, the thickness would be within the range 0·007–0·01 mm, and in the case of oily and aqueous preparations it would be within the range 0·005–0·007 mm.

This gave rise to another concept, that of a critical layer thickness, which was proposed by Masch,[42] to provide an indication of the thickness of the protective film which would ensure a 90 per cent absorption by the suntan preparation. The critical layer thickness can be derived for any wavelength, but preferably at the wavelength corresponding to the maximum erythemogenic radiation, by the equation $S = 10/E$, where S is the critical layer thickness in μm and E is the extinction coefficient for a 0·1 per cent solution in a 1 cm cell.

The US Army specification of 99 per cent absorption at 297 nm by a film 0·001 in thick has already been stated; normal commercial sunscreens usually fall considerably short of this and transmit 15–20 per cent of the erythemogenic radiation.

Kreps,[43] who published a spectrophotometric method of evaluating the erythemal protection and tanning properties of suntan preparations, gave results for six commercial preparations which, he stated, transmitted between 1 and 17 per cent of the erythemal energy and from 63–85 per cent of the tanning energy.

Stambovsky,[44] who dealt very fully with the ultraviolet screening requirements of commercial suntan preparations, suggested that for persons of average ultraviolet tolerance a suntan film should not transmit more than 20 per cent of solar erythemal radiation, while for therapeutic-type products he suggested a transmission of not more than 5 per cent.

Having established the probability of the sunscreen's effectiveness *in vitro* the next step is to check its toxicity and dermatological acceptability before proceeding to examination *in vivo*.

Evaluation in vivo

It is general practice in examination *in vivo* to use an artificial sun-ray lamp with a filter and to test on a portion of the body which is not usually exposed to the sun and thus retains its sensitivity to erythemal radiation. Stambovsky's method[45] is typical. The filter used is practically impervious to radiation below 280 nm; at 295 nm it transmits 50 per cent and at 300 nm 72 per cent of the incident radiation.

The first stage in the evaluation is a calibration to determine the time required to produce the minimum perceptible erythema with the lamp at a given distance from untreated untanned skin. With a 500 watt lamp 12 in (305 mm) from the skin, one minute is typical. In a typical experiment a number of half-inch sectors are outlined on the inner surface of the forearm with strips of adhesive tape and numbered. The first sector, with the others covered, is exposed to the lamp for a unit period, say 30 seconds. Another sector is then exposed, with the first one, for a further period. This is repeated until all the sectors have been exposed for successively increasing periods. Readings are taken after about 10–12 hours at which time the test of the particular sunscreen is started.

In the second stage of the investigation the test product is applied to the other arm which is divided into sectors and exposed in the same way as before, except that, now the skin is protected, the units of exposure can be two or three times as long. After 10–12 hours the sector showing the minimum erythema can be seen and a comparison of the exposure times for the same effect with and without the test product will indicate the factor by which the sunscreen has lengthened the safe exposure. Several such tests with any given substance will increase the accuracy of determining its protective efficiency.

However, it must not be forgotten that there is no complete analogy between the erythemal radiation of the sun and that of the quartz mercury lamp. One important factor of dissimilarity is the occurrence in the solar radiation of infrared energy which induces hyperaemia in the skin; this, in turn, produces a stronger erythemic response than would be produced in a non-hyperaemic skin. Pre-heating the skin with an infrared lamp to induce hyperaemia will cause a 33 per cent increase in its sensitivity to ultraviolet radiation.

All the tests described have been rather protracted, involving strong exposure of the skin of human subjects to radiation and long periods over which the observations were made. This hampers development work to some extent.

A method developed by Master *et al.*,[46] using thin film techinques claims to allow the evaluation of sunscreens by means of transmission measurements without dilution, using the sunscreens either alone on quartz slides or applied to excised portions of human skin. To obtain more accurate measurement of the total radiation passing through the skin, a Cary spectrophotometer fitted with integrating spheres was used in place of the xenon arc monochromator, which only measures that portion of the radiation which is not scattered.

As far as the evaluation of sunscreens *in vivo* is concerned, it now seems likely that the evaluation procedure based on the SPF value system introduced by the Plough Corporation and mentioned earlier will become the standard procedure for testing the efficacy of sunscreens. The added advantage of this system, as has already been shown, is that it can be correlated with proposed product category designations to help consumers to select the sunscreen preparations most suitable for their particular types of skin.

Table 15.6 Sunscreen Index (SI) based on Optical Density, of a 1 per cent Solution at 308 nm

Compound	SI
Ethyl *p*-dimethylaminobenzoate	14·8
Ethyl *p*-aminobenzoate	9·6
Isobutyl *p*-aminobenzoate	9·2
Propyl *p*-aminobenzoate	9·0
n-Butyl *p*-aminobenzoate	8·0
β-Methyl umbelliferone	7·7
p-Aminobenzoic acid	7·4
Dehydroacetic acid	7·0
3-Carbethoxycoumarin	6·6
Benzilidine camphor	6·6
Heliotropine	6·5
Umbelliferone acetic acid	6·0
Salicylic acid	4·3
Sodium *p*-aminosalicylate	4·3
Menthyl salicylate	4·0
Salicylamide	3·9
Methylenedisalicylic acid	3·0
3-Carboxycoumarin	3·0
Thiosalicylic acid	2·7
Brightener W/450	2·7
p-Hydroxyanthranilic acid	2·6
Sodium salicylate	2·4
Digalloyltrioleate	2·3
α-Resorcylic acid	2·2
Salicylaldehyde	2·2
p-Aminosalicylic acid	1·9
Dipropyleneglycol salicylate	1·9
Pyribenzamine	1·8
Pyrianisamine maleate	1·7
Sodium gentisate	1·7
Fluorescent white	1·6
Ethanolamide of gentisic acid	1·5
Ethyl gallate	1·4
Sodium sulphadiazine	1·2
Ethyl vanillate	1·1
Sodium sulphapyridine	0·95
Lauryl gallate	0·85
Totaquine	0·80
Barbituric acid	0·19
Chlorophyll	0·15
Amberlite IR–4–B	0·15
Salicyl alcohol	0·05
Anisic acid	0·01
Uvitex RBS	0·01
Uvitex RS	0.005

Determination of SPF

The test procedure recommended by the OTC panel of the FDA[37] for determining the Sun Protection Factor of a sunscreen preparation after UV–B and UV–A irradiation of the skin includes the use of a specified standard homosalate solution, to ensure a uniform evaluation of sunscreen preparations by different laboratories. Test subjects may be exposed to sunlight or preferably to a xenon arc as the preferred source of artifical light.

Among advantages claimed for the use of xenon arc tests *in vivo* are:

(1) Its continuous emission spectrum simulates that of the sun in the UV region, with a comparable output in the 290–400 nm range.
(2) It provides a constant spectrum at a constant angle with a high output, and the spectrum is stable when used over a long period of time.

Among disadvantages arising from the use of a xenon arc are:

(1) The full solar spectrum 'is low in the visible and infrared wavelengths.'
(2) 'Its use is time-consuming if only one test site can be irradiated at a time.'

Advantages offered by the exposure of test subjects to sunlight include:

(1) Test conditions approximate more closely the actual conditions under which sunscreen preparations are used.
(2) Test subjects are exposed simultaneously to the full solar spectrum, heat and humidity.
(3) Several sunscreen preparations can be tested simultaneously.

However, these advantages can be offset by variable weather conditions, variable radiation intensity and variable heat-induced sweating and changes in the sun angle to the body surface. Although these variables can present difficulties in determining the total erythemal exposure they can be successfully overcome by the use of a recording radiometer such as the Robertson–Berger (R&B) meter which is capable of monitoring and reproducing solar erythemal exposures.

The R&B meter is said to record a measure of the cumulative amount of UV radiation passing through its filters and photosensors after each interval of 30 minutes. A count of approximately 400 has been estimated to produce 1 MED on 'typical' Caucasian skin.

Other recording radiometers used permit the continuous measurement of the sun's intensity in joules m^{-2}.

When using natural sunlight to test the efficacy of sunscreens, the exposure of individual subjects to the sun should be completed during one continuous exposure period even though the exposure of different subjects may not be on the same day. For any one test, solar exposure of all subjects must be completed within two weeks and must be conducted at the same geographical location. The sun intensity during each exposure should be measured continuously by an R&B meter or other recording radiometer and recorded in either joules m^{-2} or R–B counts.

According to the OTC Panel, 6×10^6 joules m^{-2} as measured by a recording radiometer will evoke 1 MED on subjects with skin types I and II when read 16–24 hours later.

If the recording R–B meter is used, 400 counts are equivalent to 1 MED in skin type III subjects and MEDs as low as 200 counts may be expected of skin type I subjects.

When using a solar simulator such as the xenon arc lamp, the measurement units used to obtain one MED for the calculation of the SPF value will be time units, usually seconds.

Regardless of the light source employed, differences may arise in interpreting the results of tests *in vivo* because the same dose of UV radiation will produce different intensities of erythema in different people. This will necessitate the determination of the MED for each subject irrespective of the radiation source.

Test Panel. Only fair-skinned individuals with skin types I, II and III as previously defined should be selected, and a group of at least 20 subjects should be used for each panel. The size of panel arises from the fact (among others) that testing of MED is done in 25 per cent increments of exposure which are reasonably close to standard deviations observed in test results; the standard error should not exceed ±5 per cent. Each test subject should be examined for the presence of suntan, sunburn or any dermal lesions. The test area is the back of subjects between the belt line and the shoulder blade and should be outlined with ink and subdivided into at least three test sub-areas of at least 1 cm^2. For each test usually 4–5 sub-sites are employed.

Both test sunscreen preparations and the standard sunscreen preparation are applied to the test sites in amounts of 2 mg cm^{-2} or 2 μl cm^{-2}, to ensure use of a standard film.

One series of exposures is administered to the untreated, unprotected skin to determine the test subject's inherent MED; the time intervals employed are a geometric series represented by $(1{\cdot}25)^n$ in which each exposure time interval is 25 per cent greater than the previous one.

The protected test sites (standard or test sunscreen preparations) are usually exposed to UV light the following day. The exact series of exposures to be given is determined by the MED of the unprotected skin. An example is quoted in the Federal Register to illustrate this point. The idea is to provide a series of exposures in which the shorter exposure times produce no effect on the skin, whereas the longer exposure times at 16–24 hours later produce light and moderately red exposure sites.

The SPF of the test product is then calculated from the exposure time interval required to produce the MED of the protected skin and from the exposure time interval required to produce the MED of the unprotected skin (control site).

Water Resistance and Perspiration Resistance of Sunscreens

The efficacy of sunscreen preparations also depends on their ability to form substantive films on the skin, which are resistant to removal by water or perspiration. It is therefore appropriate to include in the evaluation of sunscreen preparations tests for assessing their water and perspiration resistance.

The OTC Panel recommends the use of an artificial light source in such tests because of certain difficulties which are encountered when test subjects are exposed to sunlight, for example lack of protection of test subjects' untreated skin against sunburn or difficulty in determining the quantity of sunlight striking the skin when it is immersed.

A period of copious sweating of 30 minutes under controlled environmental conditions is considered to be an appropriate test for determining resistance to perspiration and for making substantivity claims for a sunscreen preparation. Such a claim will be allowed if the sunscreen preparation under test retains its original PCD after the test. The claim 'resists removal by sweating' is also appropriate if the product proves to be water resistant, since water immersion is a more severe test than sweating.

Tests for water resistance and waterproof claims are more easily conducted and more reproducible if carried out in an indoor pool. The test to determine the water resistance of a sunscreen preparation is carried out after 40 minutes of moderate activity in water, and that to test the claim that a given product is waterproof after 80 minutes of moderate activity.

Since sunscreens dissolve much more slowly in sea water than in fresh water, because of its salt content, the indoor pool should contain fresh water.

Efficacy of Sunscreens

The efficacy of a sunscreen preparation depends on the amount of harmful (erythemogenic) solar radiation which it is capable of absorbing, and hence it depends on the absorption range of the active sunscreen, its absorption peak, concentration employed, substantivity to the skin and its resistance to water and perspiration, and also on the nature of the solvent employed and other constituents present.

The molar absorptivity of a sunscreen in the erythemogenic and photosensitization ranges of the UV spectrum can also be used as a criterion of the relative effectiveness of a sunscreen agent. The larger its value at a specified wavelength, the greater the ability of the sunscreen agent to absorb UV radiation at this wavelength. The effectiveness of sunscreens with a low molar absorptivity can be increased by raising the concentration of the active ingredient in the preparation.

Other factors which can influence the effectiveness of a sunscreen preparation are:

(i) pH;
(ii) the solvent system employed;
(iii) the thickness of the residual film on the skin;
(iv) the stability of the product during the time it is employed.

A change in the pH, for example an increase in the H^+ ion concentration, will change the ratio of ionized and nonionized fractions of the sunscreen agent and result in a shift in the absorption range (and absorption peak) away from the erythemogenic range of 290–320 nm with a consequent reduction in efficacy.[47]

In practically all aromatic UV absorbers, addition of solvents can also produce a shift of their absorption range. Thus the absorption peak of *p*-aminobenzoic acid in 100 per cent isopropanol was moved from 287·5 nm to 267·5 nm in 100 per cent water.[48] Mineral oil is another example of a medium which can produce a shift of the absorption range of sunscreens to a lower wavelength. Change of solvent, can, therefore, adversely affect the protection given by the sunscreen agent.

On the other hand, 2-ethylhexyl palmitate is very stable to UV radiation and has practically no effect on the absorption peak.

Formulation of Sunscreens

Work on the lines already described in this chapter will have established the type and amount of active material necessary to yield a suitable product, and formulation becomes largely a matter of selecting a suitable vehicle. Sunscreens can be presented in almost any of the product forms which can be applied to skin so as to give a continuous film, ranging from aqueous or alcoholic lotions, through liquid and semi-solid emulsion products to non-aqueous lipid preparations. Included also are gels and aerosols.

Points which should be considered in building formulae are:

(1) Convenience in use, remembering that the product will often be used out of doors, will be carried to the beach and may be placed on irregular surfaces, which will influence the type of product, package and closure.
(2) The sunscreen should be present in sufficient quantity to be effective (see earlier in this chapter).
(3) The sunscreen and the vehicle should be compatible. From the point of view of the product the sunscreen may be dissolved in either the aqueous or non-aqueous phase, but it should be remembered that, in use, water and alcohol will evaporate from the product leaving the sunscreen dissolved or dispersed in the non-volatile portion of the cream, which may itself contribute to the sunscreen activity, or in the skin surface lipids.
(4) Consideration should be given to the properties desirable in the film of non-volatile material left on the skin. It is probably neither necessary nor desirable that the sunscreen should penetrate to any depth in the skin so that there is no special need for readily absorbed materials, except in so far as some absorbance may be required so that the product can contribute to keeping the skin supple. The choice between lipids of various types and hydrophilic materials such as glycerin or sorbitol is one for the manufacturer to make, based on such factors as the degree of greasiness, tack or wetness required, bearing in mind that the product may be used on sandy beaches, or may be required to adhere to the skin while bathing, or may come in contact with clothing.

With these points in view the formulator will be able to select from formulae given elsewhere in this book. Some formulae which have been recommended and which illustrate particular points or types of preparation are given below:

Lotion, primarily aqueous, with a small amount of thickener[49]	(1) *per cent*
Filtrosol B	7·0
Methyl cellulose	0·5
Glycerin	2·0
Ethyl alcohol	10·0
Water	80·5
Perfume	*q.s.*

Lotion, primarily alcohol, with oleyl alcohol as residual lipid[50]

	(2) per cent
Giv-tan F	1·5
Ethyl alcohol (denatured)	65·0
Oleyl alcohol	10·0
Water	23·5
Perfume	*q.s.*

Clear sunscreen lotion[51]

	(3) per cent
Isobutyl *p*-aminobenzoate	5·0
Tween 20	9·0
Alcohol SD-40	45·0
Water	41·0

Example 4 is a thickened aqueous alcoholic lotion with small amount of hexadecyl alcohol which is claimed to be a good vehicle for suntan oils, since it is relatively non-greasy and non-tacky after application to the skin.[52]

	(4) per cent
Isobutyl *p*-aminobenzoate	1·9
Hexadecyl alcohol	4·7
Isopropyl alcohol	52·6
Water	40·3
Carbopol 934	0·5
Di-isopropanolamine to pH 6·0	*q.s.*

Alcoholic lotion including silicone as the residual vehicle[53]

	(5) per cent
Sunscreen agent	1·0
L–43 Silicone (Union Carbide)	5·0
Ethyl alcohol	94·0
Perfume	*q.s.*

Gel with a hydrophilic residue[54]

	(6) per cent
Sunscreen (water soluble)	5·0
1,2,6-Hexanetriol	15·0
Distilled water	80·0
in addition:	
Carbopol 934	0·1
Sodium hydroxide 10%	0·42

Gel containing mineral oil and a high proportion of nonionic (lanolin) derivatice[55]

	(7) per cent
Filtrosol A. 1000 (Schimmel)	2·0
Crodafos N.3 neutral	6·80
Volpo N.3	4·06
Volpo N.5	2·72
2-Ethyl-1,3-hexanediol	3·42
Mineral oil 65/70 vis.	13·0
Distilled water	68·0
Perfume, preservative	*q.s.*

Oily product containing a mixture of vegetable and mineral oil[54]

	(8)
	per cent
Filtrosol A 1000	3·0
Lantrol	2·0
Isopropyl myristate	12·0
Olive oil	13·0
Mineral oil	70·0

Non-greasy oil based on isopropyl palmitate and silicone[53]

	(9)
	per cent
Sunscreen (oil-soluble)	1·0
Silicone fluid L–45 (100cS)	10·0
Isopropyl palmitate	89·0

Water-in-oil emulsion with an oil-soluble sunscreen[54]

	(10)
	per cent
Sunscreen (oil-soluble)	3·0
Mineral oil	34·0
Atlas G–1425	2·0
Avitex ML	5·0
Beeswax	2·0
Isopropyl myristate	0·5
Petroleum jelly	7·5
Water	46·0

Oil-in-water emulsion with a low oil content of non-greasy character[50]

	(11)
	per cent
Giv-tan F	1·5
Diethylene glycol monostearate	2·0
Stearic acid	3·5
Cetyl alcohol	0·4
Isopropyl myristate	4·0
Triethanolamine	1·0
Triethanolamine lauryl sulphate	0·75
Water	86·85
Perfume, preservative	*q.s.*

Cream with a moderate oil content, chiefly mineral[55]

		(12)
		per cent
A	Giv-tan F	2·0
	Glyceryl monostearate (self-emulsifying) (5% soap)	7·0
	Spermaceti wax	2·0
	Mineral oil	20·0
	Stearic acid (triple pressed)	2·0
	Polychol 5	1·0
B	Water	63·8
	Glycerin	2·0
	Preservative	0·2
C	Perfume	*q.s.*

Procedure: Add B to A at 70°C. Stir until the cream has cooled to 40°C.

Low oil content vanishing cream containing Veegum[49]

	(13) *per cent*
Filtrosol A	5·00
Stearic acid	6·00
Cetyl alcohol	0·50
Veegum	2·28
Water	85·42
Borax	0·50
Potassium hydroxide	0·30
Perfume, preservative	*q.s.*

Weakly basic products are said to promote tanning.

Oil-in-water cream with a moderate oil content of mixed character

	(14) *per cent*
Sunscreen, etc.	*q.s.*
Glyceryl monostearate (self-emulsifying)	16·0
Cetyl alcohol	1·0
Mineral oil	10·0
Seasame oil	10·0
Glycerin	7·0
Water	to 100·0
Antioxidant, preservative, perfume	*q.s.*

Water-in-oil cream with an oil: water ratio similar to example 14

	(15) *per cent*
Ethyl *p*-diethylaminobenzoate	1·0
Mineral oil	9·0
Ozokerite	2·0
Microcrystalline wax	1·0
Paraffin wax	5·0
Petroleum jelly	2·0
Lanolin	3·0
Isopropyl myristate	10·0
Arlacel 83	1·5
Glycerin	5·0
Preservative, perfume	*q.s.*
Water	to 100·0

Examples 16 and 17 are creams containing colloidal alumina, which has been claimed to enhance the water-resistance of suntan preparations. As with examples 14 and 15, one is water-in-oil and the other oil-in-water, but in this case the difference is due to the use of different emulsifiers and different amounts of oil.[56]

		(16) *per cent*	(17) *per cent*
A	2-Ethylhexyl salicylate	5·0	5·0
	Mineral oil	30·0	10·0
	Myverol 18–71	10·0	—
	Stearic acid	2·2	—
	Tween 80	—	3·5
	Span 80	—	1·5
	Preservative	0·2	0·2
B	Baymal	3·5	5·6
	Water	49·1	74·2

Examples 18 and 19 are aerosol sunscreen oils. Example 18[54] is based on mineral oil, while example 19[53] is alcohol-based, contains silicone and is non-greasy.

	(18) *per cent*
Filtrosol A 1000	3·0
Mineral oil	75·0
Eutanol G	8·0
Lantrol	2·0
Isopropyl myristate	12·0
Perfume	*q.s.*
Concentrate	40
Propellants 11/12 (50:50)	60

	(19) *per cent*
Sunscreen agent	5·0
Isopropyl myristate	25·0
L–43 Silicone	5·0
Perfume oil	1·25
Lanolin oil	2·5
Menthol Racemic USP	0·25
Absolute alcohol	61·0
Concentrate	20·0
Propellants 11/12 (75:25)	80·0

Aerosol quick-breaking foam, substantially non-greasy[53]	(20) *per cent*
Myristic acid	1·3
Stearic acid	5·0
Cetyl alcohol	0·5
Isopropyl myristate	1·3
Glycerin USP	5·0
Sunscreen agent	1·0
L–43 Silicone fluid	1·0
Triethanolamine	3·0
Distilled water	80·6
Benzyl alcohol	1·0
Perfume oil	0·3

Aerosol product that combines sunscreen activity with insect repellency[55]	(21)
	per cent
Dipropylene glycol monosalicylate	4·0
Polawax A.31	3·0
Ethyl alcohol 74 OP	47·0
Distilled water	31·0
Ucon lubricant 50 HB 660	5·0
Dimethyl phthalate	10·0
Perfume	*q.s.*
Concentrate	92·0
Propellants 12/114 (20:80)	8·0

Perfumes. Exposure to sun invariably results in an increased sensitivity of the skin, due to release of histamine. The irritant effect of many essential oils and other perfume components is well known, and is likely to be intensified under those conditions. It is essential, therefore, to exert very good care in the selection of ingredients, and also keep the perfume at a low concentration, preferably about 0·2 per cent.

Heavy Sunscreens

There exist preparations on the market, the object of which is to inhibit pigmentation completely. This class of products also includes preparations which are applied to the skin with the sole object of providing a mechanical barrier to sunlight, such as calamine lotion or heavy make-up. Many attempts have been made to develop cosmetic products which would inhibit the formation of freckles or any skin lesions resulting from exposure to sunlight or other UV light sources. Like other sunscreen preparations, these products are commercially available in cream, lotion or aerosol foam form. Among compounds which have been proposed for inclusion in such products by virtue of their sunscreening properties, the following have been mentioned:

In creams:

2-hydroxy-4-methoxy-4′-methylbenzophenone (Uvistat)
1 per cent of 3-benzoyl-4-hydroxy-6-methoxybenzophenone
3–15 per cent of *p*-aminobenzoic acid
5–10 per cent of aesculin
10 per cent of salol in yellow vaseline
10 per cent of methyl salicylate in an ointment base

In lotions:

10–15 per cent *p*-aminobenzoic acid in 70 per cent alcohol
5–10 per cent tannic acid in 25–30 per cent alcohol

PALLIATIVE PREPARATIONS

Preparations for the relief of sunburn may be formulated on the basis of calamine lotion, or other zinc preparations.

In connection with this, manufacturers should note that sun 'burns' are quite capable of producing exactly the same injury as steam burns, and the consequent risk of infection by bacteria and of absorption of the damaged proteins is still present. For this reason such preparations should be antiseptic.

Calamine lotion type	(22)
	per cent
Colloidal calamine	20·0
Glycerin	5·0
Water	75·0
Antiseptic	*q.s.*

Nadkarni and Zopf[57] have suggested the following improved calamine lotion:

	(23)
Zinc oxide	8·0 g
Prepared calamine	8·0 g
Polyethylene glycol 400	8·0 cm^3
Polyethylene glycol 400 monostearate	3·0 g
Lime water	60·0 cm^3
Water	to 100 cm^3

Triethanolamine stearate milks are also soothing:

	(24)
	per cent
Triethanolamine stearate	4·80
Liquid paraffin	10·00
Water	85·20
Antiseptic	*q.s*

If desired, 10 per cent colloidal calamine may be added to this mixture, which is prepared by heating the stearate and oil to 70°C and adding to the water at the same temperature, or preferably by preparing the triethanolamine stearate *in situ*. The choice of a suitable antiseptic is a matter of individual preference; many of the chlorinated bisphenols (see Chapter 35) are good germicides and are non-irritant and innocuous in ordinary concentrations. For badly burnt areas the inclusion of a local anaesthetic or analgesic is indicated, but in view of the dangers of absorption this should be chosen with care. Monash[58] studied the topical anaesthesia of unbroken skin using six different preparations in alcoholic, hydrophilic ointment base and petrolatum vehicles. A 2 per cent alcoholic solution acted within 45–60 minutes, while a 5 per cent concentration of active ingredients in the two ointments produced topical anaesthesia within 60–90 minutes. The hydrophilic ointment acted more rapidly than the petrolatum base. The effects lasted 2–4 hours. Removal of 10–15 cell layers using adhesive tape reduced the period for anaesthesia to between 1 and 4 minutes. The author suggested that the mechanism of penetration through untreated skin was entry via the follicles and lateral spreading through the follicle walls near or just below the lower level of the external barrier, and thence through the stratum mucosa to reach the papillary portions of the dermis. Aerosol, lotion and cream products containing benzocaine and the germicide triclosan are available.[59] The treatment of a serious burn should be carried out under medical supervision.

In general greasy preparations should not be used in the treatment of sunburn. They only retain the heat of the burn and prevent the use of an antiseptic capable of mixing with the secretions and preventing bacterial infection.

All such preparations should be either aqueous solutions or oil-in-water emulsions which are capable of exerting both a protective and cooling effect.

ARTIFICIAL SUNTAN PREPARATIONS

Stains

The enhanced colour of suntan may be regarded either as functional, to prevent skin damage by absorption of erythemal radiation, or as cosmetic, to indicate the health and well-being of the subject. In either case the effect may be obtained by the use of staining materials. From ancient times materials such as walnut juice have been used. Water-soluble dyes, however, will streak on exposure to rain or moisture, and oil-soluble materials have been recommended. A vegetable product may be obtained by extracting a mixture of cudbear and henna with ten parts of warm deodorized olive oil, the ratio of the vegetable dyestuffs being adjusted to give the desired shades.

A more modern practice is to use suitable oil-soluble dyestuffs in less greasy lipid material.

Heavily pigmented face powders or modifications of the 'cosmetic stocking' may also be used. In conjunction with suitable shades of make-up, excellent effects may be obtained.

Systemic Materials

None of the sunscreens mentioned above is able to speed the rate of skin tanning, although such a product, if safe in use, would obviously command a significant market.

In 1947, Fahmy and Abu-Shady[60] isolated the active principles from the plant *Ammi Majus* (Linn.), known in ancient folk-medicine practised in Egypt for restoring pigmentation in vitiliginous skin. Subsequently it was shown that taken orally or applied topically these extracts could bring about re-pigmentation after exposure of the skin to sunlight.

Among these compounds possessing photosensitizing properties are the alkoxypsoralens, and methoxsalen (8-methoxypsoralen) has been investigated extensively both as a treatment for vitiligo and also as a possible cosmetic tanning preparation.

Sulzberger and Lerner[61] in a report to the AMA Committee on Cosmetics, reviewed the introduction of orally administered drugs (psoralens) claimed to accentuate tanning. They reported that problems associated with the use of the oxypsoralens to cure vitiligo were far from solved, cosmetically satisfactory re-pigmentation being obtained in only one out of every seven patients treated. Local application, probably the most pigment-stimulating method of use, causes severe oedema, erythema, blistering and pain when the skin is subjected to natural or artificial ultraviolet radiation.

Increased pigmentation is said to occur after ingestion of 10–20 mg of methoxsalen when the skin is exposed to sunlight within 2–4 hours.

Daniels, Hopkins and Fitzpatrick[62] carried out a clinical trial using 106 subjects to assess the effectiveness of a daily dose of 10 mg of methoxsalen in producing tanning of the skin. It was demonstrated that these results did not discriminate between the methoxsalen tablets and a placebo tablet. Subsequently, Stegmaier[63] gave 20 mg daily doses of methoxsalen using a lactose placebo as a control. He concluded that this material ingested in sufficient dosage increased tanning (98 per cent confidence level) and decreased burn (90 per cent confidence level).

A suntan-promoting tablet was introduced on the USA market under the name TAN-IF-IC with 8-methoxypsoralen as the active principle. Other components of the tablet were carefully balanced vitamins to ensure the presence of tyrosine, copper and some of the vitamin B group. After ingestion, exposure to the sun produced, in the matter of a few days, a tan which was sometimes equivalent to the suntan previously achieved by a full summer's exposure. In addition, it was said to protect the skin by accelerating the body's production of melanin, the increased pigmentation acting in the normal way as a protection against sunburn. It was claimed as the only oral suntanning agent and sun protectant, but was later withdrawn.

Dihydroxyacetone (1,3-dihydroxy-2-propanone)

In 1959 there appeared on the USA market a colourless after-shave lotion which claimed to produce gradual tanning of the skin—the effect appearing within 6 hours of application. Examination of this product showed that the colouring of the skin was confined to the upper cell layers of the skin. The active principle used at a level of up to 2·5 per cent was dihydroxyacetone (DHA).

In 1960 a patent[64] was published covering the use of dihydroxyacetone for tanning the human epidermis. This patent included eight claims covering lotions, creams, ointments and dusting powder including dihydroxyacetone at a level of 0·2–4·0 per cent.

In the years which followed, numerous publications appeared and it seems that the reaction involved had first been described some years before in journals devoted to dental research.[65–67] Thus this product is an excellent example of cross-fertilization from one field of research to another.

More detailed investigation[68,69] showed that the DHA reacts with free amino groups in the skin proteins, and in particular[70] with the free amino group in arginine. It is suggested that the DHA exists in tautomeric isomerism with glyceraldehyde:

$$\begin{array}{c} CH_2{\cdot}OH \\ | \\ C{=}O \\ | \\ CH_2{\cdot}OH \end{array} \rightleftharpoons \begin{array}{c} CH_2{\cdot}OH \\ | \\ CH{\cdot}OH \\ | \\ CHO \end{array}$$

and that the aldehyde undergoes a Schiff-type reaction with amino or imine groups of keratin forming aldehyde-amino products which condense and polymerize to form dark-coloured melanoidins.

This scheme of reaction is in accord with later work by Laden and Zielinski,[71] who showed that the reaction is not specific to DHA but is general to many α-hydroxymethyl ketones. DHA is, however, one of the best compounds for giving colour (but see below—Erythrulose).

Although there were some reported cases of skin irritation on normal and abnormal skin by the use of products containing DHA,[68,72–76] considerable evidence soon accumulated to show that DHA is innocuous.[72]

Aqueous solutions of DHA do not spread well and produce relatively little tanning. Alcohols and surface-active agents increase the apparent rate of tanning.[77] The reaction decreases with increasing pH, solutions having a pH value in excess of 8·0 producing no colour reaction.

A DHA suntan lotion proposed by Fuhrer[78] consisted of:

	(25) *per cent*
DHA	4·0
Ethanol (95%)	28·0
Methyl *p*-hydroxybenzoate	1·0
Sorbitol syrup (70%)	3·0
Boric acid powder	1·0
Allantoin	0·3
Distilled water	60·7
Perfume	2·0

A US patent[79] assigned to Plough Inc. covered the use of dihydroxyacetone in conjunction with sunscreening agents which do not contain active amino groups, within a pH range of 2·5–6·0. These formulations were claimed to have a shelf-life of more than 6 months. They simulate skin tanning and also protect the skin from sunburn.

Dihydroxyacetone/Juglone or Lawsone Combinations. A mixture which stains the skin and then protects it against excessive sunburn was described by Fusaro and Runge.[80] It consists of a 3 per cent solution of dihydroxyacetone in 50 per cent isopropanol, also containing 0·035 per cent of juglone or lawsone. The former is chemically 5-hydroxy-1,4-naphthoquinone and is obtained from walnut shells; the latter is 2-hydroxy-1,4-naphthoquinone, derived from henna. On repeated applications to the skin, and after the skin has been washed with soap and water, this mixture produces a brownish tan colour, the shade of which can be governed by the number of applications. It has been claimed that the stain which is produced on the skin absorbs 95 per cent of the ultraviolet radiation and 20 per cent of the infrared radiation reaching the skin. Protection is conferred to the skin only after the colour has been developed, and is then maintained by re-application of the mixture at intervals from 2–7 days. Natural tan may still be acquired after the product has been applied to the skin, but only very slowly.

Erythrulose

Another compound which has been proposed in cosmetic preparations as an artificial tanning agent for the skin is butane-1,3,4-triol-2-one, $HO{\cdot}CH_2{\cdot}CO{\cdot}CH(OH){\cdot}CH_2OH$, also referred to as erythrulose. A L'Oreal patent covering its use[81] states that the preferable concentration of this compound is between 0·5 and 10 per cent by weight, depending on the degree of browning of the skin desired. According to the same patent, erythrulose may be used in combination with a UV absorber; alcohol and a surface-active agent may also be present to promote penetration into the skin. It can be presented in the form of a lotion, cream or gel as well as in compositions packed in pressurized containers. Formulation of an alcoholic spray of erythrulose for use in an aerosol pack is illustrated by example 26.

	(26)
Concentrate	
Erythrulose	3 g
Propylene glycol	0·2 g
Ethyl alcohol 99·8% q.s.	100 cm^3
Fill	*parts*
Concentrate	33
Trichloromonofluoromethane	33
Dichlorodifluoromethane	33

REFERENCES

1. Keller, P., *Strahlentherapie*, 1923/24, **16,** 537, 824.
2. Luckiesh, M., *Application of Germicidal, Erythemal and Infra Red Energy*, New York, Van Nostrand, 1946.
3. Lerner, A. B., Denton, C. R. and Fitzpatrick, T. B., *J. invest, Derm.*, 1953, **20,** 299.
4. Piers, F., *Br. J. Dermatol.*, 1948, **60,** 319.
5. Roffo, A. H., *Bol. Inst. med. Exp. Estud. Trat. Cancer Buenos Aires*, 1933–36.
6. Passey, R. D., *Report at the Annual Meeting of the Yorkshire Council of the British Empire Campaign*, 16 May 1938.
7. Auerbach, H., *Public Health Reports,* 1961, No.76, 345–348.
8. Knox, J. M., Guin, J. and Cockerell, E. G., *J. invest. Derm.*, 1957, **29,** 435.
9. Luckiesh, M. and Taylor, A. H., *Gen. Electr. Rev.*, 1939, **42,** 274.
10. Blum, H. F. and Kirby Smith, J. S., *Science*, 1942, **96,** 203.
11. Rottier, P. B., *J. Soc. cosmet. Chem.*, 1968, **19,** 85.
12. Kreps, S. L., *J. Soc. cosmet. Chem.*, 1963, **14,** 12.
13. Anderson, F. E., *Am. Perfum. Cosmet.*, 1968, **83**(1), 43.
14. Blum, H. F. and Terus, W. S., *Am. J. Physiol.*, 1946, **146,** 107.
15. Fitzpatrick, T. B., Pathak, M. A. and Parrish, J. A., in *Sunlight in Man*, ed. Pathak, M. A. *et al.*, Tokyo, Univ. of Tokyo Press, 1974, p. 751–765.
16. Hais, I. M. and Zenisek, A., *Am. Perfum. Aromat.*, 1959, **74**(3), 26.
17. Draize, J. H., *Arch. Dermatol. Syphilol.* 1951, **64,** 585.
18. Klarmann, E. G., *Am. Perfum. essent. Oil Rev.*, 1949, **54,** 116.
19. Luckiesh, M., Taylor, A. H., Cole, H. N. and Sollmann, T., *J. Am. med. Assoc.*, 1946, **130,** 1.
20. De Navarre, M. G., *Chemistry and Manufacture of Cosmetics*, New York, Van Nostrand, 1941.
21. US Dept of Health, Education and Welfare, *Health Information Series No. 1*, Public Health Service Publication No. 104.
22. Giese, A. C., Christensen, E. and Jepson, J., *J. Am. Pharm. Assoc. Sci. Ed.*, 1950, **39,** 30.

23. Pernich, P. and Gallagher, M., *J. Soc. cosmet. Chem.*, 1950, **2,** 92.
24. US Patent 3 058 886, Van Dyk & Co. Inc., 20 August 1957.
25. US Patent 3 065 144, Van Dyk & Co. Inc., 10 December 1959.
26. US Patent 3 068 152, Union Carbide, 13 November 1958.
27. US Patent 3 068 153, Union Carbide, 13 November 1958.
28. Stambovsky, L., *Drug Cosmet. Ind.*, 1955, **76,** 44.
29. US Patent 3 419 659, GAF Corporation, 31 December 1968.
30. US Patent 3 479 428, Boots Pure Drug Co., 18 November 1969.
31. US Patent 3 403 207, Kreps, S. I. and Ohlsson, E., 24 September 1968.
32. British Patent 1 177 797, Nat. Starch & Chem. Corporation, 14 January 1970
33. US Patent 3 532 788, Colgate-Palmolive Co., 6 October 1970.
34. US Patent 3 670 074, Miles Laboratories, 13 June 1972.
35. US Patent 3 864 474, Colgate-Palmolive Co., 4 February 1975.
36. US Patent 3 895 104, Avon Products Inc., 15 July 1975.
37. *Federal Register*, 1978, **43**(166), 38206.
38. Stambovsky, L., *Perfum. essent. Oil Rec.*, 1958, **49,** 181.
39. Kumler, W. D. and Daniels, F. C., *J. Am. pharm. Assoc. sci. Ed.*, 1948, **37,** 474.
40. Kumler, W. D., *J. Am. pharm. Assoc. sci. Ed.*, 1952, **41,** 492.
41. Bergwein, K., *Dragoco Rep.*, 1964, **6,** 123.
42. Masch, L. W., *J. Soc. cosmet. Chem.*, 1963, **14,** 585.
43. Kreps, S. I., *Proc. sci. Sect. Toilet Goods Assoc.*, 1955, (23), 13.
44. Stambovsky, L., *Perfum. essent. Oil Rec.,* 1958, **49,** 529.
45. Stambovsky, L., *New Jersey J. Pharm.*, 1947, **20,** 18.
46. Master, K. J., Sayre, R. M. and Everett, M. A., *J. Soc. cosmet. Chem.*, 1966, **17,** 581.
47. Torosian, G. and Lemberger, M. A., in *Handbook of Nonprescription Drugs*, 5th edn, Washington, Am. Pharm. Assoc., 1977.
48. Van Dyk & Co. Inc., Technical Bulletin, *Effect of Solvent on Escalol 507*, Belleville, 1978.
49. *Schimmel Brief* No.205, April, 1952.
50. Sindar Corporation literature.
51. Atlas Cosmetic Formulary, *Sunscreen and Fragrance Products*, Wilmington, 1970.
52. Enjay Corporation leaflet.
53. Union Carbide literature.
54. Janistyn, H., *Taschenbuch der Modernen Parfümerie und Kosmetik*, Stuttgart, Wissenschaftliche Verlagsgesellschaft, 1966.
55. Croda information sheet.
56. Du Pont de Nemours.
57. Nadkarni, M. V. and Zopf, L. C., *J. Am. pharm. Assoc. pract. Pharm. Ed.*, 1948, **9,** 212.
58. Monash, S., *Arch. Dermatol.*, 1957, **76**, 752.
59. *Chemist Druggist*, 1981, 3 January, 12.
60. Fahmy, I. R. and Abu-Shady, H., *Q. Pharm. Pharmacol.*, 1947, **20,** 281.
61. Sulzberger, M. B. and Lerner, A. B., *J. Am. med. Assoc.*, 1958, **167,** 2077.
62. Daniels, F., Hopkins, C. E. and Fitzpatrick, T. B., *Arch. Dermatol.*, 1958, **77,** 503.
63. Stegmaier, O. C., *Arch. Dermatol.*, 1959, **79,** 148.
64. US Patent 2 949 403, Andreadis, J. T., 16 August 1960.
65. Deakins, M. L., *J. dent. Res.*, 1941, **20,** 39.
66. Dreizen, S., Green, H. I., Carson, B. C. and Spies, T. D., *J. dent. Res.*, 1949, **28,** 26.
67. Dreizen, S., Gilley, E. J. and Mosny, I. J., *J. dent. Res.*, 1957, **35,** 235.
68. Maibach, H. I. and Kligman, A. M., *Arch. Dermatol.*, 1960, **82,** 505.
69. Flesch, P. and Esoda, E. C. J., *Proc. sci. Sect. Toilet Goods Assoc.*, 1960, (34), 53.

70. Wittgenstein, E. and Berry, H. K., *Science*, 1960, **132,** 894.
71. Laden, K. and Zielinski, R., *J. Soc. cosmet. Chem.*, 1965, **16,** 777.
72. Goldman, L., Barkoff, J., Blaney, D., Nakai, T. and Suskind, R., *J. invest. Derm.*, 1960, **35,** 161.
73. Blau, S., Kanof, N. B. and Simonsen, L., *Arch. Dermatol.*, 1960, **82,** 501.
74. Markson, L. S., *Arch. Dermatol.*, 1960, **81,** 989.
75. Anon., *Pharm. J.*, 1960, **184,** 420.
76. Anon., *Am. Perfum. Aromat.*, 1960, **75**(8), 49.
77. Buchter, J. N., Bandolin, N. and Kanas, F. J., *Am. Perfum. Aromat.*, 1960, **75**(12), 46.
78. Fuhrer, H., *Seifen Öle Fette Wachse*, 1960, **86,** 607.
79. US Patent 3 177 120, Plough Inc., 1 June 1960.
80. Fusaro, R. M., Runge, W. J., Lynch, E. W. and Watson, C. J., *Arch. Dermatol.*, 1966, **93,** 106.
81. British Patent 954 920, L'Oreal, 25 July 1961.

Chapter Sixteen

Skin Lighteners or Bleaches

Some knowledge of the physiology and biochemistry of skin colour and particularly of the pigmentation processes is essential for an appreciation of the mode of action of skin lightening agents. Readers are referred to Chapter 1 and to several authors providing excellent and detailed reviews of this subject[1–8] but a brief summary follows of our current understanding of the processes involved.

Colour of the Skin

Three main factors combine to give skin its colour. The dermis and epidermal cells provide a background natural yellowish-white colour, the dominance of which depends to some degree on skin thickness. The superficial blood vessels of the skin contribute a red to blue tone, the intensity depending upon the number and state of dilation of the blood vessels and their nearness to the surface and the colour upon the degree of oxygenation of the blood. By far the most important contribution, however, is that of the pigments carotene and, most important of all, the brown to black melanins which are principally responsible for racial colour differences.

Melanin is synthesized in dendritic cells known as melanocytes which are normally found in the epidermal basal layer. Within the melanocytes melanin is bound to a protein matrix to form melanosomes. The melanocytes transfer their melanosomes to surrounding keratinocytes and, having lost their melanin, these cells presumably migrate upwards through the epidermis.

The control of melanin production is due both to the direct stimulatory effect of ultraviolet light and to a hormone, the melanocyte stimulatory hormone (MSH), secreted by the anterior pituitary gland. Oestrogens also exert a localized effect which is especially evident during pregnancy.

It seems extraordinary that it was only in the late 1960s that we began to understand why Negroid skin was black and Caucasoid white. The number of melanocytes is similar in both. Black skin is produced by increased melanocyte activity associated with the production of melanosomes which are larger than those in white skin. Negroid melanosomes are usually disposed individually in keratinocytes whereas those in Caucasoids are usually complexed. Furthermore, Negro skin usually shows melanin granules as high as the horny layer whereas in Europeans melanin is seldom detectable above the epidermal basal layer, owing, it is suggested, to its chemical reduction to a leuco-base which may be reoxidized by exposure to sunlight (immediate pre-tanning). For an excellent review of this subject see Hunter.[8]

Roberts,[9] who studied the geographical and racial distribution of human skin pigmentation, found a close correlation with geographical factors, suggesting a very strong adaptive role. He concluded that protection against ultraviolet radiation in areas where this is intense, and increased synthesis of vitamin D where there is minimal ultraviolet radiation, provide the two most important selective roles. The protective function of melanin also seems to be two-fold. Firstly, in the short term it protects the deeper layers of the dermis against immediate damage by ultraviolet radiation and secondly, in the long term it affords protection against cancer.

Weiner,[10] Gibson,[11] Yuasa *et al.*[12] and Curry[13] describe instruments and procedures for measuring skin colour.

Chemistry of Melanin

The metabolic pathways leading to the production of melanin have been described by Riley.[14] The melanins are quinonoid polymers of somewhat uncertain structure, there being two main subdivisions: phaeomelanins, yellowish and reddish-brown pigments containing sulphur; and eumelanins, black or brown insoluble pigments derived from the polymerization of tyrosine oxidation products. Riley surveys the evidence that two types of oxidation are involved, namely oxygen addition to monophenols (cresolase activity) and dehydrogenation of diphenols (catecholase activity). Both processes form highly reactive intermediate quinones which may be important in cell metabolism.

A simplistic representation of the sequence of reactions believed to be involved in the conversion of tyrosine to melanin is given in Chapter 23 and the subject is discussed more fully by Mason,[15] Lerner and Fitzpatrick,[16] Lorincz,[2] Fitzpatrick, Brunet and Kukita,[17] Nicolaus and Piatelli[18] and Seiji, Bileck and Fitzpatrick[19]—not forgetting the early pioneer work of Raper and his co-workers.[20]

Mechanism of Depigmentation

The processes involved in the production and transfer of pigment granules can be interfered with and Bleehen[21] has summarized the possible modes of action of compounds which may do this. The compounds may:

- selectively destroy the melanocytes;
- inhibit the formation of melanosomes and alter their structure;
- inhibit the biosynthesis of tyrosinase;
- inhibit the formation of melanin;
- interfere with the transfer of melanosomes;
- have a chemical effect on melanin or enhance the degradation of melanosomes in keratinocytes.

A number of substituted phenols have been shown to have a specific melanocytotoxic effect: 4-isopropylcatechol etc. (Bleehen *et al.*[22,23]); hydroxyanisole (Riley[24]); hydroquinone monoethyl ether (Frenk and Ott[25]). Hydroquinone produces similar toxic effects on functional melanocytes affecting not only the formation, melanization and degradation of melanosomes but also producing membraneous cytoplasmic disruption. Studies on hydroquinone monomethyl ether and 4-isopropylcatechol[23,24] suggest that these compounds are converted by tyrosinase to highly toxic oxidation products, probably as free radicals, which then initiate a chain reaction of lipid peroxidation with consequent irreversible damage to the lipo-protein membranes of the melanocyte, producing death of the cell.

Products for Skin Lightening

The foregoing suggests two ways for lightening the skin colour by reducing pigmentation: decolorize the melanin already present and/or prevent new melanin from being formed. Present cosmetic practice usually achieves both of these objectives to varying degrees.

Negro skin differs from white skin in having substantial amounts of melanin pigment in the outer horny layer; this may be bleached either by oxidation with, for example, hydrogen peroxide, or more usually chemically reduced to its leuco-form which is colourless by the use of, for example, hydroquinone.

This epidermal layer will slowly be replaced by the natural process of keratinization. The leuco or reduced form of melanin is susceptible to reoxidation by ultraviolet light and therefore the presence of a sunscreening agent in a topical skin lightening preparation is highly desirable. The formation of new melanin in the basal layers of the skin may be inhibited by the application of suitable agents with the result that the newly generated epidermis has a lower pigment content and is therefore lighter in colour.

The scientific evaluation of the effectiveness of skin lightening preparations is complex and readers are referred to the work of Curry.[13]

Skin Lightening Agents and Formulations

Opaque Covering Agents

Suitable make-up products can achieve a remarkable temporary change in the colour of the skin and therefore should be mentioned here. The more effective products may be marketed especially as blemish covers to hide skin colour imperfections, to be used under normal make-up. In a wider sense, normal make-up with a higher than average content of white or pale pigments such as titanium dioxide, zinc oxide, talc, kaolin and bismuth pigments may be used. These products are more fully discussed elsewhere in this volume.

Oxidizing Agents

Creams containing hydrogen peroxide have had limited use in the past as skin bleaches.

A series of patents taken out by Fellows[26] claim a two-solution product, one containing hydroquinone monobenzyl ether, a non-irritant solvent such as octyl acetate and a lanolin absorption base, and the other sodium hypochlorite solution. The two are mixed before use and the mixture is claimed to be stable for about a week.

Mercury Compounds

The first active material, used over many years, was some form of mercury salt. Red mercuric oxide and mercurous chloride have been used, but mercuric chloride and ammoniated mercury give the most effective preparations.

Mercuric chloride is generally used in lotion form but, owing to the generation of hydrochloric acid when the salt reacts with the skin, causes sloughing of the epidermis. Ammoniated mercury—$NH_2 \cdot HgCl$—does not cause the same degree of exfoliation of the skin. Ammoniated mercury is thought to inhibit tyrosinase, possibly by replacing copper which is required for tyrosinase action, but there have always been doubts about its toxicity potential, despite claims by manufacturers that complaints were of the order of one per hundred thousand of sales.[27] In the United Kingdom, ammoniated mercury appears in the list of poisons and toxic substances which may be sold only by pharmacists and then with prescribed labels about toxicity. Furthermore, the UK Cosmetic Product Regulations 1978[28] prohibit the use of mercury compounds (other than specific exceptions) in cosmetics. This follows upon a similar prohibition within the EEC where mercury compounds are cited in Annex II, the list of prohibited substances, in the EEC Directive on cosmetics.[29] Following a study of the potential hazards of topically applied mercury compounds, the American Food and Drug Administration issued a statement of policy[30] which ruled that, because of the known hazards of mercury and its questionable effectiveness, there was no justification for the use of mercury in skin-bleaching preparations.

Marzulli and Brown[31] found significant skin penetration of mercury (^{203}Hg-labelled) from skin bleach creams containing 1 per cent and 3 per cent ammoniated mercury. All subjects showed symptoms consistent with mercurialism. They concluded that 'although (penetration is) extremely slow, over a long period of time without excretion, it can build up to significant levels'.

Barr, Woodger and Rees[32] presented evidence for renal damage in young healthy African women using mercury-containing skin lightening creams some of which contained 5–10 per cent mercury ammonium chloride.

In a modern context, ammoniated mercury can scarcely be considered to be a suitable material for use in a cosmetic product which will be applied regularly, day by day, despite its existence in pharmacopoeial formulae.

Hydroquinone

The most favoured material for use in skin bleach preparations currently is hydroquinone which was reported by Spencer[33] to be effective at $1\frac{1}{2}$–2 per cent in a vanishing cream in producing a temporary lightening of skin colour. Spencer found that a concentration of 5 per cent was liable to cause redness and burning. No sensitization was found. The bleaching action is slow and only becomes noticeable after application for some weeks or even months. It disappears when use of the product is discontinued. Despite Spencer's findings, a level of 5 per cent has commonly been used and some preparations contained 8–10 per cent. However, the EEC Directive of 1976[29] and the Cosmetic Productions Regulations 1978 (UK)[28] limit the permitted level to 2 per cent with specific labelling requirements.

Hydroquinone has been 'found to be safe and effective (for skin lightening) by a Food and Drug Administration (USA) expert medical panel'.[34] However, the panel recommended that all such products carry a level warning that exposure to sunlight can rapidly reverse the lightening effect.

Oettel[35] noted in 1936 that when hydroquinone was fed to black-haired cats their coats turned grey after 6–8 weeks. However the depigmenting effect of hydroquinone in man's skin was discovered by chance. It was found that a sunscreen cream containing it was being bought mainly for its skin bleaching effect.[36] Spencer[37] and Fitzpatrick *et al.*[38] have attested to the effectiveness of hydroquinone as a depigmenting agent for human skin.

The mechanism of action of hydroquinone was studied by Denton, Lerner and Fitzpatrick[39] who found that it could completely inhibit the enzymatic oxidation of tyrosine to 3,4-dihydroxyphenylalanine (dopa). Iijima and Watanabe,[40] on the other hand, found that hydroquinone inhibits the histochemical dopa reaction and postulated its direct action on tyrosinase.

Hydroquinone is a white crystalline material, melting point 170-1°C, soluble in water (1:14), freely soluble in alcohol and ether. Solutions become brown on exposure to air as a result of oxidation and must be stabilized. The oxidation is very rapid if the solution is alkaline.

In formulation, the usual stabilizers are sodium sulphite or metabisulphite, often with the addition of a little ascorbic acid. Compositions should preferably be slightly acid (pH 4–6). Manufacturing equipment must be stainless steel or glass-lined to avoid discoloration, and contact with air minimized.

Two formulae published many years ago by the then Goldschmidt Chemical Corporation still provide good basic products:

Hydroquinone bleach cream[41]	(1) *per cent*
Lexemul AS	15·0
Lexate TA	6·0
Stearyl alcohol	3·0
Silicone oil	1·0
Propyl paraben	0·1
Water, deionized	67·8
Propylene glycol	5·0
Hydroquinone	2·0
Sodium metabisulphite	0·05
Ascorbic acid	0·05

Hydroquinone lotion[41]	(2) *per cent*
Lexemul AS	5·0
Mineral oil light N.F.	2·5
Isopropyl myristate	2·5
Cetyl alcohol	1·0
Brij 35	0·9
Deionized water	79·85
Propylene glycol	5·0
Sodium lauryl sulphate 30%	0·9
Sodium metabisulphite	0·15
Ascorbic acid	0·1
Hydroquinone	2·0
Citric acid	0·1

Procedure: Add the oil phase to the water phase at 75°–80°C with high shear mixing. When a smooth emulsion has been formed, transfer to low shear mixing. Cool with stirring and at 55°–50°C add, in order, sodium metabisulphite, ascorbic acid, hydroquinone and citric acid. Perfume at 40°C; the perfume should not react with hydroquinone or bisulphite.

Shevlin[42] published the following formula for a medium viscosity lotion which will function as a sunscreen lotion as well as lightening and evening out skin tone.

Oil-in-water toning lotion for black skin[42]	(3) *per cent*
Veegum K	1·5
Titanium dioxide	0·2
Cetyl alcohol	1·8
Polychol 5	0·6
Glycerin	5·0
Uvinul D50	1·0
Sodium myristyl sulphate	1·0
Hydroquinone	3·0
Perfume	1·0
Sodium metabisulphite	0·2
Aluminium chlorhydroxyallantoinate	0·2
Water	to 100·0

A soft cream with a dry emollience may be formulated as follows:

Skin lightening cream[43]		(4) *per cent*
A	Arlacel 165	8·0
	Crodamol CSP	8·0
	Liquid base CB.3929	4·0
	Crodalan IPL	2·0
	Polychol 15	3·0
	Cetyl alcohol	1·0
	Nipasol M	0·05
B	Propylene glycol	5·0
	Kelzan	2·0
	Nipagin M	0·15
	Deionized water	64·4
C	Sodium metabisulphite	0·1
	Ascorbic acid	0·1
	Hydroquinone	2·0
	Perfume	0·2 or *q.s.*

Procedure: Heat A to 72°C; disperse the Kelzan in B and heat to 70°C. Add A to B with stirring until emulsified. Cool with stirring, adding the sodium metabisulphite, ascorbic acid and then the hydroquinone at 55°–50°C. Add the perfume at about 40°C. Adjust the pH value if necessary to 5·0–5·5 with citric acid.

The composition in example 5 gives added protection against exposure to sunlight.

Skin lightening cream[43]		(5) *per cent*
A	Arlacel 165	6·0
	Estol 1461	6·0
	Isopropyl palmitate	2·0
	Ceto-stearyl alcohol	3·0
	Silicone fluid DC.200/350 cS	1·0
	Crodamol CSP	3·0
	Tiona G	0·2
	Nipasol M	0·1
	Aduvex 2211	0·2
B	Propylene glycol	5·0
	Allantoin	0·2
	Empicol LZ powder	0·3
	Deionized water	70·55
C	Sodium metabisulphite	0·15
	Ascorbic acid	0·1
	Hydroquinone	2·0
	Perfume	0·2 or *q.s.*

Procedure: As described for example 4.

In 1974 Kligman took out a patent[44] claiming synergistic compositions for skin depigmentation comprising a mixture of hydroquinone, retinoic acid and a corticosteroid, for example dexamethasone. Subsequently, Kligman and Willis[45] showed that a composition containing vitamin A acid (tretinoin), hydroquinone and a corticosteroid could bring about complete loss of melanin from normal black skin and was highly beneficial in disorders of hyperpigmentation, notably melasma, freckles and excess pigmentation following inflammation. Each of the three components was essential for effectiveness and their relative parts in the depigmentation process were conjectured. Mills and Kligman[46] reported further clinical experiences with the cream shown in example 6. Black and hyperpigmented skin is much more readily lightened than white skin, which makes this composition especially useful for treating hyperpigmentation disorders.

	(6) *per cent*
Tretinoin	0·1
Hydroquinone	5·0
Dexamethasone	0·1
Hydrophilic ointment USP	to 100·0

Bleehen[21] reported the successful clinical use of a similar composition:

	(7) *per cent*
Hydroquinone	5·0
Hydrocortisone B.P	1·0
Retinoic acid	0·1
Butylated hydroxytoluene	0·05
Polyethylene glycol 300	47·0
Methylated Spirit 74 o.p.	to 100·0

Bristol-Myers Co. have taken out a patent[47] for a synergistic skin bleaching composition containing hydroquinone or an ether thereof, a skin irritant exfoliating agent, and an anti-inflammation corticosteroid, for example retinoic acid and fluorometholone, in a vanishing cream base.

In 1945, the Schering Corporation was granted two patents[48,49] for sunscreen preparations based on hydroquinone stabilized with laevo-ascorbic acid in a vanishing cream base. In 1973, Reckitt and Colman Products patented[50] a skin lightening preparation containing hydroquinone in, for example, glycerol; and Unilever Ltd took out a patent[51] for a product containing hydroquinone, ascorbyl palmitate and 2,2′,4,4′-tetrahydroxybenzophenone. In 1979, Helena Rubinstein was granted a patent[52] for a composition based on stabilized hydroquinone in a moisturizing base which also contains a sunscreen agent (amyl *p*-dimethylaminobenzoate).

Hydroquinone Monomethyl and Monoethyl Ether

Hydroquinone monomethyl ether (usually called hydroxyanisole in the clinical literature) has been extensively studied by Riley[24,53] whose work has been usefully reviewed by Hemsworth.[54] Brun[55] reported it to be a more rapid skin depigmenting agent than the monobenzyl ether while Sidi, Bourgeois-Spinasse and Planat[56] found the reverse.

Hydroquinone monoethyl ether is a potent depigmenting agent[54] and its mode of action has been studied by Frenk[57] and Frenk and Ott.[58] Prolonged treatment with the monoethyl ether causes irreversible depigmentation comparable to that observed in man after using the monobenzyl ether.[54] The unpredictability of its long-term effects has discouraged its use as a skin lightener.

Although no secondary effects have apparently been reported from the use of the monomethyl ether, the undesirable effects from both the monoethyl and the monobenzyl ether do not commend its general use.

Hydroquinone Monobenzyl Ether

The skin depigmenting activity of hydroquinone monobenzyl ether was first reported by Oliver, Schwartz and Warren[59] as an occupational disease in rubber workers. This activity has been confirmed, notably by Denton, Lerner and Fitzpatrick,[39] Dorsey,[60] Sidi, Bourgeois-Spinasse and Planat,[56] Becker and Spencer,[61] and Mosher, Parrish and Fitzpatrick.[62]

Calnan[63] reported findings in South Africa where skin lightening preparations are very widely used; Dogliotti *et al.*[64] found a sharp increase in the number of cases of leuco-melanoderma traceable to skin lightening creams following a formulation change from ammoniated mercury to the monobenzyl ether of hydroquinone (2 per cent), to which salicylic acid (2 per cent) was often added; Bentley-Philips and Bayles[65] stated that hydroquinone is much safer than the monobenzyl ether, which has now been prohibited in South Africa.

The unpredictable results obtained with this compound, its liability to cause dermatitis, sensitization and sometimes irreversible depigmentation, preclude its general use. The EEC Directive 1976[29] and the UK Cosmetic Products Regulations 1978[28] both prohibit the use of hydroquinone monobenzyl ether in cosmetics.

A series of patents taken out by Fellows[26] for a skin lightening preparation containing hydroquinone monobenzyl ether and an oxidizing agent were mentioned earlier.

Catechol and its Derivatives
Extensive screening tests by Chavin and Schlesinger[66,67] showed that catechol and some derivatives caused destruction of the pigment cells though with a lesser effect than hydroquinone.

Bleehen *et al.*[22] investigated thirty-three compounds, including catechol and derivatives, and found 4-isopropyl catechol to be a most potent depigmentating agent. However, levels of 3 per cent or more proved to be irritant to the skin and a sensitizer.[21]

4-Tert-butyl catechol may also cause skin depigmentation in man.[54,68]

A patent[69] covers the use of catechol, methyl- and carboxy-substituted catechol in a sunscreening composition. Another patent[70] claims a depigmenting composition containing a catechol substituted in the 4-position, for example 4-isopropyl catechol.

Ascorbic Acid and its Derivatives
Ascorbic acid itself has been listed[21] as an active constituent of skin bleaching preparations but its use is normally confined to that of an antioxidant stabilizer in hydroquinone-based products, where it assists in the inhibition of browning of creams.

Rovesti[71] reported on the successful use of creams containing 3 per cent and 5 per cent ascorbyl oleate for bleaching freckles in human skin, resulting also in a marked improvement in the condition of the skin, which became softer and more supple.

Takashima *et al.*[72] reviewed the chemistry of a number of esters of ascorbic acid and studied the stability of a few when incorporated into cosmetic creams. One in particular, the magnesium salt of ascorbic acid-3-phosphate, was shown to have a bleaching effect on pigmentation in human skin and to be useful in clinical practice.

A patent[51] described the use of a composition containing hydroquinone, ascorbyl palmitate and 2,2′,4,4′-tetrahydroxybenzophenone for skin bleaching.

Other Depigmenting Materials
Chavin and Schlesinger[66,67] found that, among other materials, several mercapto-amines were potent depigmenting agents in black goldfish. Two of these, 2-mercaptoethylamine hydrochloride and N-(2-mercaptoethyl)-dimethylamine hydrochloride, were potent depigmenting agents when applied to the skin of black guinea-pigs.[73] However, both are very malodorous and therefore unusable in man. See also Bleehen *et al.*[22] and Frenk.[57]

The use of mercapto-amines has been patented by Scherico[74] and by Marly.[75]

A method for the treatment of hyperpigmentation, described in a US Patent[76] granted to the Schering Corporation, entails the topical application to the affected area 1–4 times daily of a preparation comprising between 0·1 and 10 per cent of either *p*-amino-benzenesulphonic acid or its alkaline or alkaline-earth metal salts. It is claimed that such treatment will result in depigmentation of highly localized areas. The patent specification contains several example formulae.

A patent granted to Lever Brothers (USA)[77] claims a composition for simultaneously lightening skin and protecting it from the sun's rays, containing 0·2–10 per cent niacin and 0·1–10 per cent urocanic acid. A similar patent has been granted to Unilever Ltd (UK).[78]

The Natural Way

Limited skin lightening effects can be achieved with natural materials which have been used for centuries and which are useful for bleaching out fading suntan and freckles. Among the materials used are cucumber juice, lemon and lime juice, buttermilk, crushed strawberries and fresh horseradish. Some interesting preparations have been described by Buchman[79] and Maxwell-Hudson.[80]

REFERENCES

1. Masson, P., in *The Biology of Melanomas*, ed. Mineor, R. W. and Gordon, M., Vol. 4, New York, NY Academy of Sciences, 1948, p. 15.
2. Lorincz, A. L., in *The Physiology and Biochemistry of Skin*, ed. Rothman, S., Chicago, University of Chicago Press, 1954.
3. Montagna, W. and Hu, F., *Advances in Biology of Skin*, 8, *The Pigmentary System*, Oxford, Pergamon, 1958.
4. Montagna, W., *The Structure and Function of the Skin*, 2nd edn, New York/London, Academic Press, 1962.
5. Jarret, A., *Science and the Skin*, London, English Universities Press, 1964.
6. Fitzpatrick, T. B., Miyamoto, M. and Ishikawa, K., *Arch. Dermatol.*, 1967, **96,** 305.
7. Van Abbe, N. J., Spearman, R. I. C. and Jarrett, A., *Pharmaceutical and Cosmetic Products for Topical Administration*, London, Heinemann Medical, 1969.
8. Hunter, J. A. A., *J. Soc. cosmet. Chem.*, 1977, **28,** 62.
9. Roberts, D. F., *J. Soc. cosmet. Chem.*, 1977, **28,** 329.
10. Weiner, J. S., *Man*, 1951, **51,** 152.
11. Gibson, I. M., *J. Soc. cosmet. Chem.*, 1971, **22,** 725.
12. Yuasa, S., Morita, K. and Kaneko A., *J. Soc. cosmet. Chem. Japan*, 1976, **10**(1/2), 34, through *Cosmet. Toiletries*, 1977, **92**(4), 68.
13. Curry, K. V., *J. Soc. cosmet. Chem.*, 1974, **25,** 339.
14. Riley, P. A., *J. Soc. cosmet. Chem.*, 1977, **28,** 395.
15. Mason, H., *J. Biol. Chem.*, 1948, **172,** 83.
16. Lerner, A. B. and Fitzpatrick, T. B., *Physiol. Rev.*, 1950, **30,** 91.
17. Fitzpatrick, T. B., Brunet, P. and Kukita, A. in *The Biology of Hair Growth*, ed. Montagna, W. and Ellis, R. A., New York, Academic Press, 1958.
18. Nicolaus, R. A. and Piatelli, M., *J. Polymer Sci.*, 1962, **58,** 1133.
19. Seiji, M., Bileck, M. S. C. and Fitzpatrick, T. B. *Ann. N.Y. Acad. Sci.*, 1963, **100**(Part II), 15 February, 497.
20. Raper, H. S. and Wormall, A., *Biochem. J.*, 1925, **19,** 84. Raper H. S., *ibid.*, 1926, **20,** 735; 1927, **21,** 89. Duliere, W. L. and Raper, H. S., *ibid.*, 1930, **24,** 239. Heard, R. D. H. and Raper, H. S., *ibid.*, 1937, **31,** 2155.
21. Bleehen, S. S., *J. Soc. cosmet. Chem.*, 1977, **28,** 407.
22. Bleehen, S. S., Pathak, M. A., Hori, Y. and Fitzpatrick, T. B., *J. invest. Dermatol.*, 1968, **50,** 103.
23. Bleehen, S. S., in Riley, V., *Pigment Cell*, 1976, **2,** 108, Karger, Basel.

24. Riley, P. A., *J. Pathol.*, 1969, **97,** 185. Riley, P. A., *J. Pathol.*, 1969, **97,** 193.
25. Frenk. E. and Ott, F., *J. invest. Dermatol.*, 1971, **56,** 287.
26. British Patent 856 431, Fellows, W., 14 December 1960. British Patent 965 869, Fellows, W., 6 February 1963. Canadian Patent 610 726, Fellows, W., 21 February 1957. US Patent 3 060 097, Fellows, W., 15 March 1957.
27. Nealon, D. F., *Proc. Sci. Sect. Toilet Goods Assoc.*, 1944, (1), 7.
28. The Cosmetic Products Regulations 1978. Statutory Instrument 1978, No. 1354, London, HMSO.
29. EEC Directive 76/768/EEC, *Off. J. European Communities*, 1976, **19** (L262).
30. *Federal Register*, 30 June 1972, 37 F. R. 12967.
31. Marzulli, F. N. and Brown. D. W. C., *J. Soc. cosmet. Chem.*, 1972, **23,** 875.
32. Barr, R. R., Woodger, B. A. and Rees, P., *Am. J. clin. Pathol.*, 1972, **53,** 723.
33. Spencer, M. C., *Arch. Dermatol.*, 1961, **84,** 131.
34. Anon., *Soap, Cosmet. chem. Spec.*, 1978, **54**(12), 20.
35. Oettel, H., *Arch. exp. Pathol. Pharmacol.*, 1936, **183,** 319.
36. Arndt, K. A. and Fitzpatrick, T. B., *J. Am. med. Assoc.*, 1965, **194,** 962.
37. Spencer, M. C., *J. Am. med. Assoc.*, 1965, **194,** 114.
38. Fitzpatrick, T. B., Arndt, K. A., El-Mofty, A. M. and Pathak, M. A., *Arch. Dermatol.*, 1966, **93,** 589.
39. Denton, C. R., Lerner, A. B. and Fitzpatrick, T. B., *J. invest. Dermatol.*, 1952, **18,** 119.
40. Iijima, S. and Watanabe, K., *J. invest. Dermatol.*, 1957, **28,** 1.
41. Up-dated formula from Goldschmidt Chemical, Division of Wilson Pharmaceutical & Chemical Corporation, now Inolex Corporation, *Technical Bulletin No. 524*, 1 February 1968.
42. Shevlin, E. J., *Cosmet. Perfum.*, 1974, **89**(4), 41.
43. Private communication, Peter Reeves Creative Workshop, Wareside, Ware, Herts., England.
44. US Patent 3 856 934, Kligman, A. M., 1974.
45. Kligman, A. M. and Willis, I., *Arch. Dermatol.*, 1975, **111,** 40.
46. Mills, O. H. and Kligman, A. M., *J. Soc. cosmet. Chem.*, 1978, **29,** 147.
47. British Patent 1 349 955, Bristol-Myers Co. (USA), 10 April 1974.
48. US Patent 2 376 884, Schering Corporation, 1945.
49. US Patent 2 377 188, Schering Corporation, 1945.
50. British Patent 1 303 566, Reckitt and Colman (UK), 1973.
51. British Patent 1 319 455, Unilever Ltd (UK), 6 June 1973.
52. US Patent 4 136 166, Helena Rubinstein (USA), 1979.
53. Riley, P. A., *J. Pathol.*, 1970, **101,** 163.
54. Hemsworth, B. N., *J. Soc. cosmet. Chem.*, 1973, **24,** 727.
55. Brun, R., *Parfüm Kosmet.*, 1962, **43,** 44.
56. Sidi, E., Bourgeois-Spinasse, J. and Planat, P., *Presse Méd.*, 1961, **69,** 2369.
57. Frenk, E., *Bull. Soc. Fr. Dermatol. Syphiligr.*, 1971, **78,** 153.
58. Frenk, E. and Ott, F., *J. invest. Dermatol.*, 1971, **56,** 287.
59. Oliver, E. A., Schwartz, L. and Warren. L. H., *Arch. Dermatol.*, 1940 **42,** 993.
60. Dorsey, C. S., *Arch. Dermatol.*, 1960, **81,** 245.
61. Becker S. W. and Spencer M. C.: Evaluation of monobenzone, *J. Am. med. Assoc.*, 1962, **180,** 279.
62. Mosher, D. B., Parrish, J. A. and Fitzpatrick, T. B., *Br. J. Dermatol.*, 1977, **97,** 669.
63. Calnan, C. D., *J. Soc. cosmet. Chem.*, 1976, **27,** 491.
64. Dogliotti, M., Caro, I., Hartdegen, R. G. and Whiting, D. A., *S. Afr. med. J.*, 1974, **48,** 1555.
65. Bentley-Philips, B. and Bayles, M. A. H., *S. Afr. med. J.*, 1975, **49,** 1391.
66. Chavin, W. and Schlesinger, W. *Naturwissenschaften*, 1966, **53,** 413.

67. Chavin, W., Schlesinger, W. and Hu, F., *Advances in biology of skin* Vol. 8, Oxford, Pergamon, 1967, p. 421.
68. Gellin, G., Possick, P. A. and Perone, V. B., *J. invest. Dermatol.*, 1970, **55,** 190.
69. British Patent 1 107 072, Scherico Ltd, 24 April 1964.
70. British Patent 1 371 782, Maibach, H. (USA), 30 October 1974.
71. Rovesti, P., *Soap Perfum. Cosmet.*, 1968, **41,** 672.
72. Takashima, H., Nomura, H., Imai, Y. and Mima, H., *Am. Perfum. Cosmet.*, 1971, **86**(7), 29.
73. Frenk, E., Pathak, M. A., Szabo, G. and Fitzpatrick, T. B., *Arch. Dermatol.*, 1968, **97,** 465.
74. British Patent 1 107 071, Scherico Ltd, 24 April 1964.
75. Belgian Patent 513 023, Soc. Belge de l'Azote et des Produits Chimiques du Marly, SA, 16 November 1952.
76. US Patent 3 517 105, Schering Corporation, 23 June 1970.
77. US Patent 3 937 810, Lever Brothers (USA), 1976.
78. British Patent 1 370 236, Unilever Ltd (UK), 16 October 1974.
79. Buchman, D. D., *Feed Your Face*, London, Duckworth, 1973.
80. Maxwell-Hudson, C., *The Natural Beauty Book*, London, Macdonald and Jane's, 1976, p. 126.

Chapter Seventeen

Face Packs and Masks

Introduction

The use of face packs by women dates back to early antiquity when some of the earths used in them were credited with almost miraculous healing powers. These preparations are applied to the face in the form of liquids or pastes. They are then allowed to dry or to set with the object of improving the appearance of the skin, by producing a transient tightening effect as well as by cleansing the skin.

Their present vogue may be ascribed to their combined psychological and cleansing effect. The warmth and tightening effect resulting from their application produce the stimulating sensation of a rejuvenated face, while the colloidal and adsorptive clays and earth which are present in some packs will adsorb grease and dirt from the facial skin. When they are eventually removed from the face, skin debris and blackheads may be removed simultaneously.

Such a preparation should possess the following properties:

(1) It should be a smooth paste without gritty particles and without an 'earthy' or other objectionable odour.
(2) Applied to the face it should dry out rapidly to form an adherent coating on the skin but this coating should be capable of subsequently being removed either by peeling off the face or by gentle washing without producing any pain.
(3) It should produce a definite sensation of tightening of the skin after application.
(4) It should produce a significant cleansing of the skin.
(5) It must be dermatologically innocuous and non-toxic.

There are five basic systems which give products satisfying the above characteristics. These are based on wax, rubber, vinyl resins, hydrocolloids and earth respectively.

Wax-based Systems

Wax-based masks generally consist merely of paraffin wax of a suitable melting point, or may be mixtures of wax with the addition of a little petroleum jelly and polar materials such as cetyl and stearyl alcohols. The use of microcrystalline waxes may assist the continuity of a suitable 'wax mask'.

These products are solid at room temperature and for use have to be melted and brushed on hot. When the waxes harden, a sensation of tightness is felt. As the wax film forms a moisture-proof barrier, profuse perspiration is induced which helps to flush dirt and impurities from the follicular openings in the skin surface.

A little rubber latex may be included with the waxes to assist in the ease of removal. Even application can be facilitated by formulating the wax blend so that in the molten state—a few degrees higher than body temperature—the product is a thixotropic semi-solid. This can be achieved by the incorporation of a little organophilic bentonite. The following formula illustrates the use of this material:

	(1) *per cent*
Microcrystalline wax	13·0
Paraffin wax	60·0
Cetyl alcohol	5·0
Mineral oil	20·0
Bentone 38	1·4
Isopropyl alcohol	0·6

Rubber-based Systems

Mention must also be made of face packs mainly based on rubber latex. After drying out, these packs form a continuous, elastic and water-impermeable film on the face. By interfering with normal skin respiration, the film causes heat to be retained by the skin with a resulting rise in the temperature and increased blood circulation. Rubber masks are removed from the skin very easily, for example by mere pulling. After the removal, slight plumping of the skin becomes noticeable. This effect, however, is transient and disappears after skin respiration has been restored to normal.

Example 2 illustrates the composition of face packs based on rubber latex.

	(2) *per cent*
Latex emulsion	25
Sorbitol	5
Methylcellulose (low viscosity)	10
Kaolin	3
Borax	1
Water	56
Preservative	*q.s.*

Vinyl-based Systems

Vinyl-based systems are generally based on polyvinyl alcohol or vinyl acetate resin as film-former. The use of polyvinyl alcohol in face packs is illustrated by example 3.

	(3)
	per cent
Veegum	0·5
Kaolin	0·5
Titanium dioxide	0·3
PVA	12·0
Propylene glycol	8·0
Ethanol	20·0
Water	58·7

Procedure: Dissolve the Veegum in the water with rapid agitation and heat to 80°C. Add the kaolin and titanium dioxide. Mix the PVA and propylene glycol together, heat to 80°C, add to the Veegum mixture and homogenize. Cool the mixture to 40°C and very slowly add the alcohol.

The inorganic matter can be adapted to absorb both oleophilic and hydrophilic soil by use of nylon powders coated with titanium dioxide. Formulae quoted by Toida *et al.*[1] include:

	(4)
	per cent
Polyvinyl alcohol	14·00
Titanium dioxide/nylon 12 (1:7)	3·00
Glycerin	3·50
Preservatives	0·10
Polyoxyethylene sorbitan monolaurate (20 EO)	0·50
Water	to 100·00

Hydrocolloid-based Systems

Hydrocolloid-based systems can be presented either as high-viscosity sols which after application lose water and form a flexible gel film, or as solid gels which are melted before application to the face. The sensation of tightness is produced by the eventual shrinking of the gel on loss of further moisture. A variety of gums can be used; these include gum tragacanth, gelatin, casein, carragheen gum, sodium carboxymethyl cellulose, acacia and guar gum, as well as polyvinylpyrrolidone and many others. The film may be plasticized by the inclusion of humectants such as glycerol, propylene glycol or sorbitol.

The viscosity of liquid masks based on hydrocolloids can be varied considerably depending on the colloid used and its concentration in the face pack.

Masks based on hydrocolloids are sometimes preferred to masks based on earths and clays because they are easier to apply and because they dry more rapidly. Their cleansing action, however, is somewhat inferior because they do not contain a sufficient amount of solids to adsorb any dirt.

More rapid drying may be obtained by the incorporation of ethyl alcohol as part solvent. This restricts the choice of hydrocolloid somewhat. However, certain grades of methylcellulose, Carbopol 934 (Goodrich Chemical Co.) and polyvinylpyrrolidone are soluble in aqueous alcohol and can be used in alcoholic masks.

Face masks based on hydrocolloid systems may also contain small quantities of finely divided solids which act as opacifiers and sometimes facilitate application. Kaolin and bentonite may be used, for example, but in amounts preferably not exceeding 5 per cent since, if too much particulate matter is added, a discontinuous film which may lack mechanical strength will be formed.

The preparation of such face packs usually proceeds along the following lines. The preservative is dissolved in the appropriate amount of water, and humectant is added with stirring. The hydrocolloid is then sprinkled in slowly with continuous agitation to prevent the formation of aggregates which are difficult to disperse, if necessary with heating to assist solution. When dispersion has been completed and the hydrocolloid particles have started to swell, it is usual to reduce the rate of agitation so as to avoid undue aeration and the retention of air bubbles in the viscous solution or gel.

Janistyn[2] gave the following formula for a 'gelanthum' mask containing gelatin and tragacanth:

	(5)
	per cent
Gum tragacanth (best)	2·2
Glycerin	2·5
Gelatin (white)	2·3
Water	90·5
Zinc oxide	2·5

For a 'gelanthum mask with honey', Janistyn suggested the replacement of about 4.5 parts of the water in the above formula by honey. He also gave a formula for a casein-based mask:

	(6)
	parts
Casein (best quality)	20·0
Borax (powdered)	0·5
Glycerin	5·0
Water (distilled)	100·0
p-Hydroxybenzoic acid propyl ester	0·1

Procedure: Moisten the casein with the glycerin and dissolve in the aqueous borax solution (containing also the preservative) with the aid of heat.

Winter[3] suggested the following composition for a face pack:

	(7)
	parts
Gelatin	10·00
Water	50·0
Camphor (dissolved in a little alcohol)	0·05
Zinc oxide	3·00
Kaolin	5·00
Titanium dioxide	2·00

Procedure: First moisten the gum with glycerin, then add the gelatin and water, raising the temperature. (Solids should be mixed in the hot solution and their dispersion is facilitated by first moistening them with a drop of glycerin/or of surface-active agent.)

The preparations given in examples 6 and 7 are heated before use and applied in a warm state.

The use of alcohol as part solvent is shown in example 8.

	(8)
	per cent
Carbopol 940	2·0
Water	42·0
Alcohol	50·0
Glycerol	2·0
Di-isopropanolamine	4·0

The use of polyvinylpyrrolidone in face packs is illustrated by a model formula quoted from a technical bulletin[4]:

	(9)
	per cent
PVP K-15	3·0
Methylcellulose (low viscosity)	9·0
Glycerin	7·5
Water	80·5
Insoluble opacifiers, perfumes, preservative	*q.s.*

Earth-based Systems (Argillaceous Masks)

Earth-based systems are often referred to as paste masks. They include clay facial packs and the so-called mud packs and usually contain a high percentage of solids.

These products can either be presented in bulk, or packed in sachets, for mixing with water when required, or they can be presented ready mixed for use. In the latter case it is advisable to presterilize by the use of heat or ethylene oxide and to incorporate a suitable preservative as many of the natural earths are heavily contaminated with micro-organisms.

As the mask dries on the face it hardens and contracts, giving the sensation of mechanical astringency. The presence of absorbent clays such as bentonite produces a genuine cleaning effect, particularly on very greasy skins.

China clay, colloidal kaolin, fuller's earth, bentonite, etc., may be used as the 'argillaceous' material, the choice depending in part on the criteria which it is proposed to apply to the finished product.

If the colour of fuller's earth or bentonite is considered objectionable, this difficulty may be met by blending with kaolin, and/or the addition of zinc oxide or titanium dioxide.

Bentonite is a colloidal clay derived from volcanic ash found in certain parts of the United States, which is characterized by its strong affinity for water and its thixotropic properties. Certain bentonites will absorb up to fifteen times their volume of water, this property being greatly increased by the addition of a small quantity of magnesium oxide, or some other substance possessing a similar pH.

The wide variations in the analysis of bentonites reported in the literature[5] stem from the considerable variations in different beds in the Benton formation and even in different strata in the same beds.

The consistency of bentonite gels will vary with concentration and will be considerably influenced by the pH of the gels. According to Griffon,[6] a gel containing 6 per cent bentonite has the consistency of glycerin, while a 20 per cent gel has the consistency of lanolin.

Bentonite gels have been described as soothing to the skin,[7] and have been claimed to be of value in the treatment of eczema, abscesses, sores and wounds.[8] They have been used in a number of dermatological preparations.[9–11]

The nature of kaolin and its purification to a quality suitable for cosmetic purposes are described in Chapter 18. This type of electrolytically purified kaolin is equally suitable for use in face packs because of its qualities of fineness, softness, moisture adsorption and easy spreading.

Hydrocolloids such as the carragheen gums may be added to stabilize the suspension of solids while contributing to the mechanical strength of the dried film. Again, plasticizers such as glycerin may be added. Special attributes may be conferred by adding additional ingredients such as sulphur (see Chapter 9), astringents, bleaching agents, acids, etc.

The following formulae illustrate this type of product:

All-purpose masks	(10) *per cent*
Kaolin	35·0
Bentonite	5·0
Cetyl alcohol	2·0
Sodium lauryl sulphate	0·1
Glycerin	10·0
Nipagin M	0·1
Perfume	*q.s.*
Water	to 100·0

	(11) *per cent*
Glyceryl monostearate	3·0
Lanolin oil	2·0
Sodium lauryl sulphate	2·0
Veegum	8·0
Kaolin	10·0
Propylene glycol	7·0
Titanium dioxide	4·0
Ethanol	6·0
Isopropyl myristate	2·0
Water	56·0

	(12) *per cent*
Water	78·7
Nipagin M	0·2
Nipagin P	0·1
Titanium dioxide	1·0
Arlacel 83	0·2
Tween 60	0·3
Kaolin	9·5

Procedure: The cream emulsion is formed at 80°C, with half the water content. The kaolin is dispersed in the remaining water, heated to 80°C, and added to the emulsion.

Winter[3] gives the following two formulations for face packs for dry and greasy skins respectively:

Face pack for dry skin	(13) *parts*
Kaolin	80·0
Starch	10·0
Cold cream	20·0
Cetyl alcohol	2·0
Hydrophilic oil	5·0
Water, boric water or infusions	*q.s.*

Procedure: Melt the cold cream and cetyl alcohol in warm water. Next add the oil, the powders and then the water or other aqueous matter.

Face pack for greasy skin	(14) *parts*
Kaolin	80·0
Magnesium carbonate	15·0
Starch	5·0
Tragacanth gum (powdered)	1·0
Water	*q.s.*

A so-called 'oxygenated' face mask, based on kaolin and quoted by Bergwein[12] has the following composition:

	(15) *parts*
Colloidal kaolin	800
Salicylic acid	20
Aluminium lactate	5
Magnesium peroxide	200

'Oxygenated' masks have been recommended for use on oily skin, sallow skin or skin blemishes. However, because their use may cause irritation of sensitive skin, a preliminary test should be carried out on a small area of the skin.

Anti-wrinkle Preparations

Face packs are usually left on the face for about 10–25 minutes, to allow most of the water to evaporate and the resulting film to contract and harden, after which they are removed.

The so-called anti-wrinkle preparations which appeared on the market in the early 1960s are based on bovine serum albumin, and form an 'invisible' mask over the skin. They are allowed to remain in contact with the facial skin for about 6–8 hours, that is, for the period of time during which they remain effective, and are then removed by washing.

In the manufacture of bovine albumin, sodium citrate is added to fresh blood to prevent its coagulation, and the blood is centrifuged to remove the blood cells. The anticoagulant is then neutralized and the serum is defibrinated and spray dried. The resulting product is light in colour and completely water-soluble; its albumin content is about 80–95 per cent. Bovine albumin is available from several sources in three forms: as a 15 per cent sterile solution containing a suitable preservative and ready for immediate use without dilution; as a 30 per cent solution to be diluted with an equal volume of water prior to use; and also in the form of a freeze-dried powder which before use is reconstituted with water. The smoothing effect which follows the application of these preparations would seem to be a purely physical one entailing the 'filling-in' of facial wrinkles and the formation of a tight occlusive film which stretches the skin. How long this effect lasts will largely depend on how much the facial muscles are used. This has been clearly shown during investigations carried out by Kligman and Papa.[13]

Although bovine serum was the basis of most of the original preparations, there would appear to be no good reason why experimentation should not produce similar results from other forms of albumin such as egg white or even soluble casein.

In fact, a wrinkle-smoothing composition disclosed in an American patent[14] contains proteins obtained from cow's milk whey, namely α-lactalbumin and some β-lactoglobulin. The film produced with this composition is claimed to be effective for about four hours and can be re-activated by moistening it with a little water. Both the component proteins are undenatured and water-soluble, are present in the aqueous composition in a ratio of 4:1 and may constitute 10–30 per cent by weight of the total composition. Water may be present within a range of 70–90 per cent by weight, together with 2–4 per cent by weight (of the composition) of a non-toxic plasticizer, such as glycerin or propylene glycol, to increase the flexibility of the dried films. A preservative must also be included; it should be both bacteriostatic and fungistatic and must not precipitate the lactalbumin or the lactoglobulin. The pH of the final preparation is maintained within the 5–7 range.

Apart from the ability to re-activate films produced, other advantages claimed for this preparation are that when it is dry it is not shining in appearance, that it does not flake away after drying and that no make-up is required to conceal the film.

REFERENCES

1. Toida, H., Ishizaka, T. and Koishi, M., *Cosmet. Toiletries*, 1979, **94**(12), 33.
2. Janistyn, H., *Soap Perfum. Cosmet.*, 1937, **10,** 405.
3. Winter, F., *Alchimist (Boechut)*, 1947, **1,** 228.
4. General Aniline and Film Corporation, *PVP Formulary*, X-200/3, p. 16.
5. Ewing, C. O., Politi, F. W. and Shackelford, C. H., *J. Am. Pharm. Assoc. sci. Ed.*, 1945, **34,** 129.
6. Griffon, H., *J. Pharm. Chim. Paris*, 1938, **27,** 159.
7. Kulchar, G. V., *Arch. Dermatol. Syphilol.*, 1941, **44,** 43.
8. Davis, C. W., Vacher, H. C. and Conley, J. E., *Bentonite, Its Properties, Mining,*

Preparation and Utilisation, US Dept Interior, Bureau of Mines, Washington DC, 1940.

9. Fantus, B. and Dyniewicz, J. M., *J. Am. Pharm. Assoc. sci. Ed.*, 1938, **27,** 878.
10. Fantus, B. and Dyniewicz, J. M., *J. Am. Pharm. Assoc. sci. Ed.*, 1939, **28,** 548.
11. Hibbard, D. G. and Freeman, L. G., *J. Am Pharm. Assoc. pract. Ed.*, 1941, **2,** 78.
12. Bergwein, K., *Seifen Öle Fette Wachse*, 1967, **93,** 555.
13. Kligman, A. M. and Papa, C. M., *J. Soc. cosmet. Chem.*, 1965, **16,** 557.
14. US Patent 3 364 118, The Borden Co., 16 January 1968.

Chapter Eighteen

Face Powders and Make-up

FACE POWDER

Function and Properties

The function of face powder is to impart a smooth finish to the skin, masking minor visible imperfections and any shine due to moisture or grease either from perspiration or from preparations used on the skin. The object appears to be to make the skin look as though it would be pleasant to touch. The degree of opacity of the powder can vary from opaque and matt, as for example a clown's make-up, to almost transparent, which will have a type of shine due to the powder itself. Neither extreme is favoured, but between the limits, the pendulum of fashion will swing from time to time. Whatever the finish, it must possess reasonable lasting properties to avoid the need for frequent re-powdering, that is it must adhere to the skin, and be reasonably resistant to the mixed secretions of the skin. Finally, it should serve as a vehicle for a pleasing odour to be disseminated by intimate contact of perfume-laden particles over a warm and relatively large area.

No single substance possesses all the desired properties—covering power, slip absorbency, adhesiveness and bloom—hence a modern face powder is a blend of several constituents each one chosen for some specific quality. The various properties will now be considered in greater detail together with some of the principal ingredients, arranged according to their various functional contributions to the powder base. These properties are included in the section on face powders mainly for historical reasons. The popularity of face powders has declined considerably in recent years in favour of compact powders, foundation and liquid make-up. However, the materials and principles involved are equally applicable to these products so a careful study of the following sections will benefit the formulator in his or her work on more sophisticated products.

Covering Power

Good covering power is a very desirable attribute of face powders, its object being to conceal various defects of the facial skin including scars, blemishes, enlarged pores and excessive shine.

Titanium dioxide, zinc oxide, kaolin and magnesium oxide are the materials used to enhance the covering power of face powders.

Titanium Dioxide

Titanium dioxide has considerably greater covering power *per se* than zinc oxide, about 1·6 times that of the latter in air and about 2·9 times the latter in petroleum jelly. On a moist greasy skin, its covering power relative to zinc oxide is probably of the order of 2·5 times. Titanium dioxide is not astringent but is physiologically inert and may be found, in any rare cases of allergy to zinc compounds or in cases of dry skin, preferable to zinc oxide. Its sun-screening properties are, however, inferior to those of zinc oxide.

Difficulties sometimes encountered in blending titanium dioxide with other powder constituents may be overcome by using it in conjunction with zinc oxide.

Zinc Oxide

Zinc oxide is the other metallic oxide that is frequently employed in face powders to accentuate their covering power. It is also mildly astringent, mildly antiseptic and has soothing properties. Because of the last property it has been used in the therapy of minor skin irritations. It has been considered to give satisfactory covering power in powder formulations at a level of 15–25 per cent.

The measurement of the covering power of a pigment has been a controversial subject for a long time, since by varying the test conditions one can obtain widely different values.

Grady,[1] who investigated the characteristics of zinc oxide when used in face powders, also calculated the covering power of several face powder ingredients from the refractive indices of the pigments and of the various media in which they might be used for cosmetics. The values obtained are listed in Table 18.1. It will be noted that in the table the covering power of zinc oxide in each medium has been arbitrarily designated as 100; in fact, it decreases progressively from 100 to 37 to 21 as zinc oxide is placed in air, water and petrolatum respectively.

Table 18.1 Calculated Covering Power of Pigments

		Relative covering power		
Pigment	**Refractive index**	**In air** (n = 1·00)	**In water** (n = 1·33)	**In petrolatum** (n = 1·475)
TiO_2	2·52	166	232	292
Zinc oxide	2·008	100	100	100
Chalk	1·658	55	29	15
Talc	1·589	46	19	6

As well as the medium surrounding the pigment, it is necessary to consider the particle size. A reduction in particle size will obviously allow the material to be spread more thinly and thus give increased physical cover. At the same time, reduction in particle size is, in general, accompanied by increased light scatter, thus increasing the opacity of the powder and hence the optical covering power.

However, there is a limit, and a curve for light transmission for zinc oxide in water, published by Grady (Figure 18.1), shows a sharp increase of transmission (that is, decrease in opacity) below 0·25 μm when the particles have become small in comparison with the wavelength of light.

It should also be borne in mind that the covering power of a face powder will decrease as it absorbs moisture and sebum from the skin. However, under the

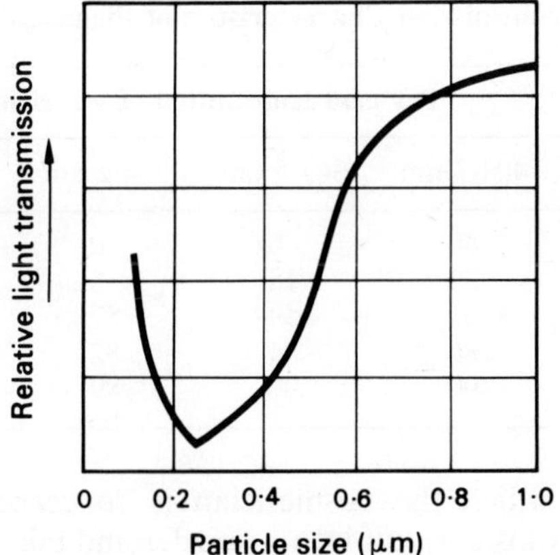

Figure 18.1 Calibration curve relating particle size to relative light transmission[1]

same conditions, pigments with a high refractive index will lose proportionally less opacity than materials of low refractive index, as shown in Table 18.2. This once again emphasizes the desirability of using materials of high refractive index in face powders.

Table 18.2 Relative Covering Power Retained after Wetting Dry Pigment

Pigment	With water (%)	With petrolatum (%)
TiO_2	51	37
Zinc oxide	37	21
Chalk	20	5
Talc	15	3

Grady also drew attention to the sunscreening action of zinc oxide which cuts off the ultraviolet more sharply than any other white pigment used in face powders. He considered, therefore, that zinc oxide and to a lesser degree titanium dioxide should be useful in preventing sunburn. The ultraviolet transmission characteristics of various pigments are listed in Table 18.3.

It is interesting to note that in the experiments carried out by Luckiesh *et al.*[2] on behalf of the US Army Air Force in December 1942 on protective skin coatings for the prevention of sunburn, zinc oxide was found to be of definite value in preventing sunburn whereas titanium dioxide was not found to be a very dependable protective judged on a specimen containing 20 per cent of this substance in petroleum jelly. In addition to titanium dioxide and zinc oxide, some grades of kaolin have a good covering power.

Various published articles have sought to correlate the weight of various face powder constituents with their covering properties or opacities. However, it is

Table 18.3 Ultraviolet Transmission Characteristics of Pigments[1]

	Per cent transmitted of wavelengths					
Pigment	435·8 nm	404·7 nm	365·5 nm	334·2 nm	313·1 nm	302·3 nm
Zinc oxide	46	40	0	0	0	0
TiO_2	35	32	18	6	0·5	0
China clay	63	61	59	57	55	54
Chalk	87	86	84	82	80	79
Talc	90	90	90	89	88	87

not the weight with which the cosmetician is concerned, but the volume, inasmuch as a woman dips her puff into a powder and takes out a volume which depends partially on the adherent properties of the powder, on the size and type of the puff employed and the depth and pressure with which the puff is applied to the face powder container.

To get a very rough idea of the average relative grade of opacity which prevails in cosmetic materials, the opacities of chalk, kaolin, magnesium stearate, rich starch, talc, titanium dioxide, zinc oxide, and zinc stearate were determined by applying these materials by means of a swansdown face powder puff to similar areas of black velvet paper. The whiteness or opacity produced on the smooth black adherent surface was (a) estimated visually and (b) recorded photoelectrically. For the materials employed, the only particle size specification was that they should all pass completely through a 200 mesh sieve. The order of opacity (starting with the least opaque) was found to be as follows:

Talc
Rice starch
Magnesium stearate
Chalk (light, precipitated)
Zinc stearate
Kaolin
Zinc oxide
Titanium dioxide

From what has already been said it will be apparent that to get a true comparison between the relative covering properties of the various face powder constituents it would be necessary to separate similar size particle fractions of each constituent and subject them to exact tests. However, the conditions of such a test, although scientifically more exact, would be so different from the conditions ruling in face powder formulation as to have little actual practical value. The opacities of various finished face powder formulae may be tested approximately by the above method. It should always be remembered that the criterion of effectiveness of a face powder in respect of its water absorption, grease absorption, and covering properties or opacity is judged by the interval elapsing between powdering and repowdering the face; and that a woman judges this by the evidence of shine, which is a complex phenomenon not related to any scientific test and often dependent upon a woman's skin, type of foundation cream used, and occupation. These facts should be adequately considered by consumer trials, properly carried out and statistically evaluated. Such consumer trials must be carried out scientifically, as ill-conceived trials can be made to prove anything and often prove nothing.

Absorbency

The second important function of face powders is to eliminate shiny skin in certain facial areas by absorbing sebaceous secretions and perspiration. The prime requirement of a material for this purpose is a high absorptive capacity. The components of face powders which confer this property are colloidal kaolin, starch, precipitated chalk and magnesium carbonate.

The water absorbent properties of face powders or face powder constituents may be determined by the method of Hewitt,[3] in which a known weight of the powder is shaken with excess water and filtered under a standard pressure through a Buchner funnel until no more water emerges. The wet powder is then transferred to a weighed, stoppered weighing bottle and the increase in weight determined. Methods involving the addition of water from a burette, until the powder becomes semi-fluid, are open to the objection that different observers do not obtain concordant results and the end point is not easily determined.

Water absorption is by no means the main characteristic required in a powder; it must also be absorbent for grease. If a person's face is inclined to dryness a more greasy foundation is usually employed. A powder which is not grease-absorbent will show a shiny nose or face which will necessitate re-powdering. Constituents of higher opacity such as zinc and titanium oxides tend to mask greasiness, while starch, chalks and kaolin absorb only a certain amount of grease.

Colloidal Kaolin

Kaolin, a hydrated aluminium silicate, is a naturally occurring compound. According to Halpern *et al.*,[4] kaolin is not a primary mineral but is a generic term applied to several hydrated aluminium silicates. Not all aluminium silicates, however, may be called kaolin. On the basis of X-ray and physical studies, Ross and Kerr[5] established that three different groups of clay are classified as kaolin. These clays (kaolinite, nacrite and dickite) have essentially the same formula ($Al_2O_3 \cdot 2SiO_2 \cdot 2H_2O$). Purified grades of kaolin that are light in colour and free from grit and water-soluble impurities should be used for face powders; the most suitable is electrolytically purified kaolin. Ordinary china clay is obtained by elutriation and on microscopic examination mica, quartz and felspar are readily discernible. Pharmaceutical grades of kaolin are obtained by a peptizing process in which the clay is suspended in water containing a suitable electrolyte (for example sodium pyrophosphate) which confers an electrical charge upon the clay and keeps the finer particles in suspension. Removal of the suspension of fine particles, followed by removal of the electric charge (by addition of another electrolyte or by means of an electric field), yields the finest forms of kaolin. One such grade is known as Osmokaolin.

Colloidal kaolin is used in face powders primarily because its high moisture absorption capacity enables it to absorb perspiration. It has also good covering power, excellent grease-resisting properties and it imparts greater skin adhesion properties to the finished product than does talc. Its relatively high density makes it a useful material for controlling the bulking properties of the powders in which it is used. It also helps to reduce the shine of talc which is present. However, it lacks slip, and is inclined to be somewhat harsh. Its proportion in face powders should therefore not exceed 30 per cent.

Starch and Modified Starches

At one time rice starch was used almost exclusively as the base of face powder formulae on account of its excellent absorptive properties, good covering power and the smoothness it imparted to the skin. The latter property was closely related to its small particle size, the average diameter of rice starch granules being 3–8 μm. However, objections were raised to the use of starch because of its tendency to cake when exposed to a humid atmosphere or in the presence of excessive skin secretions. McDonough[6] asserted that it readily forms a sticky paste when wet, clogging the pores, and that it is an ideal nutrient, when moist, for bacteria. In addition it coats the hair shaft and so accentuates the downy hair, otherwise unnoticeable, on a woman's face. It was also claimed that because of the tendency of starch to favour bacterial growth, it could give rise to skin irritation when in contact with the skin for any length of time. These assertions led eventually to the replacement of rice starch by talc as the powder base in face powders. However, it must be said that when any degree of bloom is required, there are few materials which can surpass starch.

Decomposition can be reduced in many cases by the addition of perfume; mention of clogged pores refers not to pores but to the openings of the hair follicles. (Pore openings are invisible by ordinary inspection.) There is no proof that starch can cause clogging of such openings.

Special grades of treated starch which will not swell up or agglutinate in the presence of moisture have been developed for the cosmetics industry. For example, ANM starch powders (Neckar-Chemie, GmbH) are starch ethers which are produced by reacting the hydroxyl groups of the starch molecule with tetramethylolacetylenediurea. These materials are claimed to have a good slip, good adhesive properties and covering power, and a high absorptive capacity for both water and oil. They are chemically inert and are also claimed to have some bactericidal properties by virtue of their small formaldehyde content. The ANM Rice 'K' grade was claimed, unlike untreated rice starch, not to swell in the presence of moisture or perspiration, and not to give rise to enlarged pores and bacterial decomposition. As in the case of untreated rich starch, it is said to confer a peach-like bloom to the skin, and to be superior to talc in terms of covering power.

Microcrystalline Cellulose (Avicel)

Avicel is a microcrystalline cellulose from the FMC Corporation.

It is interesting to note in this connection that in 1966 a powder was launched which, unlike conventional face powders consisting of talc or starch, was claimed to contain microporous cellulose derived from the centre of the corn cob, with an oil absorption rate many times higher than other powders.

Precipitated Calcium Carbonate

Precipitated chalk is yet another material that has been used in face powders because of its excellent absorption characteristics. Like kaolin it is also used to remove some of the inherent shine of talc. It has, however, a deleterious effect on the slip of the product and tends to impart an undesirable dry feel. Consequently (unless it is one of the special grades available) it should not be used in face powder formulations in amounts greater than 15 per cent.

The special grades of precipitated chalk are exceptionally fine and accurately balanced to prevent harshness. They possess good absorption and grease-resisting properties and are available in different densities, depending on the purpose intended. When such grades are used, it is possible to use considerably greater amounts of precipitated chalk than specified above. A specially treated grade of precipitated chalk is also available which is claimed to be unaffected in terms of its absorptive power for oils and grease, which does not dry the skin, and which is claimed to be particularly adhesive. These materials are guaranteed to conform to the USP specification for lead, arsenic, etc.

Magnesium Carbonate

Magnesium carbonate is a highly absorbent constituent of face powder formulations. Its absorbent power is about three times as great as that of precipitated chalk and its tendency to dry the skin correspondingly greater. Magnesium carbonate confers fluffiness to face powders and helps to prevent 'balling'. Light magnesium carbonate is the preferred substance for incorporating and maturing the selected perfume. It is subsequently blended with the bulk of the powder; 5 per cent magnesium carbonate is ample for this purpose.

Plastics

Powder bases made from plastics have been developed for use on the skin. These powders are available both in the form of solid spherical particles and in the form of a crushed foam. An example of the latter is 'Oracid'—a rigid urea-formaldehyde foam. Tables 18.4 and 18.5[7] show oil absorption and water

Table 18.4 Oil Absorption Capacities of Powder Bases[7]

Substance	**Oil take-up** (ml per g of substance)	**Saturation time** (min)
Oracid (urea-formaldehyde foam)	11·11	15
Aerosil	6·00	15
Magnesium carbonate	5·40	15
Magnesium oxide	3·30	15
Kieselguhr	2·80	15
Kaolin	2·70	15
Talc	2·50	15
Rice starch	2·10	15
Zinc stearate	0·40	15

Table 18.5 Water Absorption Capacities of Powder Bases[7]

Substance	Water take-up (ml per g of substance)	Saturation time (min)
Oracid (urea-formaldehyde foam)	16·60	30
Aerosil	8·70	45
Magnesium carbonate	4·03	28
Kieselguhr	3·20	12
Magnesium oxide	2·60	20
Titanium dioxide	2·30	30
Kaolin	1·50	5
Talc	1·40	10
Zinc oxide	1·10	18
Rice starch	0·75	15
Zinc stearate	0·05	120

absorption capacities of various powder bases, including Oracid, in terms of ml absorbed and the time taken for the saturation value to be reached.

Use of finely divided, highly crystalline, high density polyethylene as a substitute for talc in cosmetic powders has been described in a US patent.[8] Preparations containing polyethylene are claimed to be non-irritant and to have good adhesion, covering power and absorbency. The average particle size of the polyethylene used in cosmetic powders should preferably be not larger than 44 μm. For coloured compositions, for example powder rouge, the dye should preferably be incorporated in the molten polymer, rather than mixed with dry powder constituents.

Cosmetic powder preparations in solid, liquid or slurry form based on finely divided polymeric polyesters are claimed in a British patent.[9] These preparations are claimed to spread easily and to adhere tenaciously to the skin, giving a velvety matt finish. The polymers employed are high-molecular-weight polymeric linear polyesters such polyethylene terephthalate and isophthalates, or a copolymer of these two monomers, with a preferable average particle size of 1–10 μm. Cosmetic powder preparations based on mixtures of these polymeric particles preferably also contain one or more of the usual additives of cosmetic powders such as talc, kaolin, zinc oxide and metallic soaps to improve spreading characteristics, slip, adhesion and absorptive capacity for oily secretions and perspiration.

Polystyrene microspheres are a further example of polymeric materials developed for use in cosmetic powders.[10]

Slip

Slip is the quality of easy spreading and application of powder to produce a characteristic smooth feeling on the skin.

Slip is mainly imparted by talc and also by metallic soaps such as zinc stearate and, to a lesser extent, starch.

Talc

Talc is a hydrated magnesium silicate to which the formula of $3MgO{\cdot}4SiO_2{\cdot}H_2O$ has been assigned. In fact, the Mg/Si ratio appears to vary.

Talc may be obtained from Italy, France, Norway, India, Spain, USA, Australia, China, Egypt and Japan. Of these, the Italian, French, American, Australian and some Indian and Chinese grades can be used for face powders and compact powders. The grinding of raw talc is an important parameter in determining its suitability for make-up products. Talc must be very white and bright to allow products to cover the wide range demanded by the modern consumer and this can only be achieved by having the correct grinding process. Products can always be 'dulled off' with the use of colours but, to produce a 'bright' product, the talc must be bright in the first place. This phenomenon cannot be introduced by the use of other materials. One may even have to use two different types of talc to cover all products, for example one grade for make-up products and another cheaper grade for talcum powders, etc. Whatever the source of the talc it must be free from asbestos or 'amphibole material' and care must be taken with lower grade materials, since talc from certain countries may contain tetanus spores. In such cases it is essential that the talc is adequately sterilized (see also Chapter 8).

Talc is the major component of face powders, and in high class products it may be used in amounts of up to 70 per cent or more. Its main function in such powders is to impart to them slip and good adhesion. However, the covering power and the moisture absorbing capacity of talc are low, and it must therefore be combined with other powders to modify these deficiencies. Apart from its use in face powders, talc is, of course, used in talcum powders, baby powders and antiperspirant sticks, all of which are discussed under their appropriate headings (Chapters 7, 8 and 10).

The most suitable physical form of talc for use in cosmetics is the foliated variety, the flat platelets of which slide readily over each other, thus accounting for the high slip characteristics of the product.

In the USA, the standards of the Cosmetic, Toiletry and Fragrance Association stipulate that talc should be free from impurities such as carbonates and water-soluble iron and be neutral to litmus paper, so as to prevent any deterioration of colour and perfume in the finished product. In order to ensure the application of a smooth and even film, the talc used should also be free of any gritty particles and shiny specks of mica, and the bulk of it should pass through a standard 200 mesh sieve. It should also be free of asbestos, this requirement being emphasized in the USA by the OSHA regulations concerning asbestos-containing materials; a review of properties and specifications is given by Grexa and Parmentier.[11]

Adhesion

Adhesion is another important property of face powder constituents, determining how well the powder will cling to the face.

The property of adhesion is imparted to face powders by the inclusion of talc and some water-insoluble metallic soaps of stearic acid, such as zinc and magnesium stearates. The latter are used in face powders in amounts ranging between 3 and 10 per cent. In addition to increasing the adhesion of the powder to the skin, they also render the ultimate product soft and fluffy and furthermore impart to powders some water-repellent characteristics.

The adhesion of powders to the skin can also be improved by the incorporation of certain emollients such as cetyl or stearyl alcohols and glyceryl monostearate, usually in amounts varying between 0·5 and 1·5 per cent.

Various proprietary preparations have been marketed from time to time, having as their object the improvement of the adhesiveness of face powders. One British patent[12] describes the use of the zinc or magnesium salts of fatty acids containing an uneven number of carbon atoms, for example undecylic acid, such a base being employed in a proportion of 5–10 per cent in the finished powder. Many manufacturers prefer to correct any lack of adhesiveness in their product either by increasing the amount of zinc or magnesium stearates incorporated or by including in the powder mixture 2 per cent of petroleum jelly, mineral oil or cetyl alcohol. Further variations can be made using encapsulated mineral oil, especially when one is trying to produce a very light fluffy powder and yet achieve good adhesion.

Among other materials which have been suggested, mention may be made of powdered kapok proposed by Varma[13] as a potential face powder ingredient. Powder bases made from plastics, which were referred to earlier, have also been claimed to give improved adhesion.

Powdered Silica and Silicates

Very finely divided pure silica has been introduced for cosmetic purposes, for example Neosyl (Crosfield) and Aerosil (Degussa). It is claimed that the use of this substance obviates the need for zinc and magnesium stearates. The addition of 10 per cent to an ordinary powder mix exerts a marked effect in increasing its fluffiness. Up to 20 per cent may be used in a face powder or 30 per cent or more in talcum and baby powders. The incorporation of a small amount of such an ultrafine silica is very efficient as an anticaking agent in body powders.

In addition to silica, a number of ultrafine synthetic silicates which have extremely high oil and water absorption properties may also be incorporated into face powders.

Bloom

The materials chiefly used to impart bloom, the requirement for which may vary according to fashion, are chalk, rice starch and prepared starch. These have been described above.

Powdered Silk

A raw material for face powders (and other cosmetic preparations) which, apart from any unique or desirable properties it may have, will provide opportunities for the advertising agencies to produce rapturous copy writing, is powdered silk.

A British patent[14] describes a process for its preparation. Two further patents[15,16] have been taken out dealing respectively with a technique for pulverizing silk for producing silk powder and a method for breaking down silk by boiling successively in sulphuric and boric acids.

Powdered silk was also discussed in *Schimmel Briefs*,[17] where it was pointed out that raw silk consists of fibroin, a protein fibre, covered with a coating of a gummy material called sericin. The latter consists mainly of albuminoid substances with small amounts of fatty acids, resin and colouring matter. The silk is, therefore, first degummed by the conventional processes used in the textile industry, then treated with acid or alkali in order to bring about a partial hydrolysis of the protein molecules. At an appropriate stage it is washed and dried, and finally reduced to an impalpable powder by grinding.

It has been claimed[18] that the physical characteristics of silk powder make it well suited to serve as an ingredient of face powders, that it spreads easily and adheres tenaciously to the skin, producing a velvet matt finish, and that it possesses a very high absorbent power in that it will absorb as much as three times its volume of water and still retain the appearance of a powder. The addition of large amounts of such materials as kaolin or chalk is not advised, as it is claimed that they tend to make the final product too dense and compact.

Colour

Inorganic and organic pigments and organic lakes have all been used to confer colour to face powders. Water-soluble or oil-soluble dyestuffs should be avoided because of the danger of colour bleeding after application due to solubilization by sweat and lipid secretions. Inorganic pigments include natural and synthetic iron oxides which give yellows, reds, browns and black; ultramarines which give green and blue, and chrome hydrate and chrome oxide which give green.

Within the EEC all colours used in cosmetic products, together with their purity limits, are governed by the Cosmetics Directive of 1976.[19] In the USA the Food and Drug Administration controls the use of colours in cosmetics, but specifies purity limits only for the organic colours. However, it is recognized that inorganic pigments should be produced to a 'certifiable standard' which usually refers to heavy metal content and the Cosmetic, Toiletries and Fragrance Association have issued standards for these materials, for example iron oxides, that are in line with the Food and Drug Administration standards.

There is a considerable activity on the part of the Food and Drug Administration in the USA in regulating the materials to be used in, *inter alia*, toilet preparations. This involves not only the well known ranges of FD&C, D&C, and D&C (Ext.) colours, but also many inorganic materials both coloured and uncoloured which have been mentioned above either as additives (colours) or main materials. Present use is permitted pending the consideration by the FDA of applications for materials to be specifically on the permitted list.

Because of different proprietary names and numbers for various colouring matters, local regulations—as for example the use only of colouring matters certified by the FDA in the USA—and different opinions among firms themselves as to what particular combination of colours gives their 'particular' rachel, peach, tango, etc., no attempt has been made to list the multifarious colouring matters available. Useful information may be obtained by consulting the US Colour Regulations and the Colour Index[20] (see also Harry's *Cosmetic Materials*,[21] wherein ninety pages are devoted to listing the properties of these colouring matters). From a description of their solubilities, etc., it will be possible to rule out a number of these dyes; of the remainder belonging to the colour classification desired, a number will be found to be obtainable of similar composition in other countries[22] but without, in every case, the guarantee of various metallic and other impurity limits. The task of the formulator will often be simplified by consulting dyestuff manufacturers who will readily advise on suitable colouring matters, oxides and earths and give advice concerning proportions for various colour matches which may be required. Suitable blends of colours are readily available conforming to local regulations, and it is well worth while to evaluate them in one's own formulations before spending too much time on colour blending. The colouring of face powders has been discussed by Anstead.[23]

The choice of colour is usually a matter of taste. At one time it was considered, that 'naturelle' (clear pink) shades were suitable for blondes, and 'rachel' (more cream–yellow) shades for brunettes. Later, perhaps because of more outdoor exposure, it was realized that the natural skin colour tends more to a cream colour, and that the old 'naturelle' now suits only a few complexions which by reason of their transparent skin tend naturally toward blue. Conversely, florid complexions may be toned down by pale bluish green powders. The majority of the present-day range of colours are based on a cream–yellow–brown range.

As well as the actual complexion itself, varying colours of hair and dress all affect the apparent tint, which probably accounts for the multitude of colours in demand (see also Chapter 20).

It is advisable to keep colour formulations as simple as possible so that matching with fresh batches of raw material is made easier. The colour effect produced by a powder applied to the skin will depend *inter alia* upon the opacity of both tinted and white pigments, their particle size, the degree of dispersion, the thickness of the applied film and the colour of the skin.

The colour of the thin film of pigment (the undertone) may be different from the colour effect given by the powder viewed in bulk (the mass tone). It is thus important that the formulator assesses the performance of the product applied to the skin.[24]

Perfume

The importance of the perfume on the sales appeal of the product cannot be over-emphasized. Usually the powders are perfumed very lightly. The odour of the face powder must be fragrant and pleasant, and a preference is shown today for either a flowery fragrance or that of a synthetic bouquet. Unless the manufacturer has had wide experience of perfume manufacture he will be well advised to purchase the perfume from a reputable perfume manufacturer.

The compatibility of perfume with other constituents of the product must be carefully checked. Talc, for example, usually contains a little free lime, magnesia or iron which may adversely affect the perfume depending on the amount of these substances present. The perfume may also be affected by precipitated chalk, by kaolin, magnesium carbonate or a metal stearate, if these contain impurities, or indeed by some of the pigments used in colouring the powders. Finally, it should be remembered that the perfume note in a powder will be different from that, for example, in an alcoholic solution, particularly in the case of floral bouquets, and any tests on perfume acceptance must be carried out on the final product.

Formulation

The chemist familiar with the properties and functions of the various powder constituents and with the sources of supply will have no difficulties in formulating a satisfactory product. He should, however, be given details of the type of market for which the product is intended, the advertising story to be used and properties to be highlighted. He will then be able to judge what proportions of which constituents to use, and which materials to avoid in order to produce a suitable formula. Thus, for powders with a good covering power, he will use a higher proportion of either zinc oxide or titanium dioxide; for increased absorbency, the proportion of magnesium carbonate may well be raised at the expense of talc, and where a powder with good adhesion is required the amount of zinc or magnesium stearate may have to be increased.

Multifarious variations in formulae could be listed, many of which, under laboratory conditions, show slight differences in respect to opacity, slip, absorbency, water-resistance and grease-resistance, etc., but experience shows that many of these variations cannot be detected by the average woman under normal conditions of use. The following formulae exemplify various types of face powder, the variations in which are sufficient to be detectable.

The powder given in example 1 is very transparent, that is it is a 'light' powder, and is favoured by persons who wish to impart some colour and bloom to the face without appearing to be 'made up'. The starch may be replaced if desired with precipitated chalk.

	(1) *per cent*
Talc	80·0
Zinc oxide	5·0
Zinc stearate	5·0
Rice starch	10·0
Perfume, colour	*q.s.*

Example 2 has high opacity and gives a very definite opaque matt finish which tends to hide minor skin defects. It is more popular with certain people who like to have a definite powdered appearance without being over-powdered.

	(2) *per cent*
Talc	30·0
Zinc oxide	24·0
Zinc stearate	6·0
Precipitated chalk	40·0
Perfume, colour	*q.s.*

In between these two types of powder there are a number of popular variations such as the following:

	(3) *per cent*
Talc	65·0
Precipitated chalk	10·0
Zinc oxide	20·0
Zinc stearate	5·0
Perfume, colour	*q.s.*

	(4) *per cent*
Talc	60·0
Kaolin	20·0
Zinc oxide	15·0
Zinc stearate	5·0
Perfume, colour	*q.s.*

As previously stated, both starch and precipitated chalk tend to impart bloom; example 5 is a powder of medium weight, that is medium opacity or coverage and bloom.

	(5) *per cent*
Talc	50·0
Rice starch	15·0
Precipitated chalk	15·0
Zinc oxide	15·0
Zinc stearate	5·0

If it is desired to obtain maximum coverage and still maintain a high talc content, this may be achieved by replacing the zinc oxide in any of these formulae by about one-quarter of its weight of titanium dioxide.

The following formula is prepared from precipitated chalk and a high proportion of zinc stearate is incorporated. To provide some grease-resistance, kaolin and titanium dioxide are employed.

	(6) *per cent*
Waterproof chalk base	50·0
Zinc or magnesium stearate	10·0
Kaolin	20·0
Titanium dioxide	6·0
Talc	14·0
Perfume, colour	*q.s.*

Jannaway[25] states that the following formula gives a good medium powder, characterized by excellent slip, absorbency, adequate coverage, good velvety feel and adherence—and an indefinable improvement in mattness, etc., which is attributed to the rice starch:

	(7) *per cent*
Zinc oxide	16·0
Talc	37·0
Zinc stearate	5·0
Precipitated chalk (light)	18·0
Rice starch	8·0
Kaolin (best cosmetic grade)	16·0

Winter[26] has described a special kind of 'fatty powder' which he states is much favoured by persons afflicted with a rough or dry skin:

	(8) *parts*
Vaseline	50
White beeswax	40
Petroleum jelly	40
Stearin	20
Glyceryl monostearate	75

Procedure: Melt the above fatty materials together and add, while stirring constantly, 500 parts of hot water. Continue to stir until the emulsion has formed, then add 1000 parts of talc. Knead, allow to dry, rub to powder, pass through a sieve and perfume.

Other formulations are given by Keithler[27] and Hilfer.[28]

The modern trend with face powder is to apply it over foundation to achieve special effects, for example, matt or shimmer. The following complete formula illustrates a light shimmering effect:

	(9) *per cent*
Talc	77·00
Zinc stearate	5·00
Zinc oxide	2·00
Kaolin	5·00
Mica	10·00
Red iron oxide	0·36
Yellow iron oxide	0·36
Black iron oxide	0·03
Perfume	0·25

Manufacture

Mixing of the ingredients in face powders is usually carried out in a horizontal mixer with a screw agitator. If lakes are used they may be mixed with a small quantity of one of the constituents, chalk, zinc oxide or talc, and the colour concentrate so formed mixed with the main bulk. If water-soluble or alcohol-soluble dyes must be used they are best sprayed on to the mix or alternatively on to one of the constituents which possesses good absorbency such as chalk, kaolin or magnesium carbonate, this being then dried and mixed into the main bulk.

Machines are available which mix, sift and spray the perfume automatically. One method would be to add the perfume by a meter pump feeding a long tube with a multitude of small holes fitted along the top of the blender (for example, on to magnesium carbonate or chalk in the ribbon blender) before mixing with the remainder of the powder.

The use of water-soluble or oil-soluble colours should be avoided in a face powder since they lead to streaking, darkening and staining of the skin when applied over ordinary make-up. Chilson[28] has drawn attention to the pebble mill method of mixing which obviates the making of a colour base. All the ingredients, including the colour and perfume, are milled together in a pebble mill for six hours, discharged and then sifted. He states that such a mill delivers an excellent product and is widely employed in preference to mixers. This method, however, is too slow to be recommended for large-scale production. Micropulverizers are being increasingly employed since the material obtained is finely ground and uniformly mixed and only a rough pre-mixing is required. Various other pulverizers such as disintegrators, hammer mills, attrition mills, etc., may also be used.

Pin-disc mills give good results, especially in respect of colour dispersion, provided that the material is first given a rough pre-mix. Such mills are often applied in conjunction with a turbine sifter.

The Air Spun Process employed by Coty Inc. is described briefly by deNavarre[30] as follows:

> . . . Purified and cold air, under great pressure (100 psi; 700 kPa) is permitted to rush in a continuous stream into a closed drum-shaped chamber. This chamber or mill is called a micronizer because it reduces particles to micron (μm) size. The speed of the air stream, the manner in which it is directed and the shape of the chamber itself, cause air to revolve about this chamber at a rate somewhat in excess of 1000 mph. In this super cyclone all the ingredients are hurled against each other until they reduce themselves to the desired size and fluffiness. When this is attained such particles are emitted through a central exit while the heavier particles are forced to remain. It is claimed that the particle size was selected after numerous experiments had shown it to be the best for the appearance, effect and adhesiveness of a face powder.

Apart from this process, various physical methods exist for preparing powders of a desired particle size range. These depend upon elutriation by means of air or water. Control of such separated fractions may be carried out by microscopic examination of the fractions, but this method involves the visual or photomicrographic inspection of a very large number of fields in the microscope and is tedious. Various sedimentation tests, some based upon passage of a beam of light through the sedimentation column into a photoelectric cell, have also been suggested. One of the simplest methods for the cosmetician is that based upon the principles of Stokes's Law, described by Hinkley.[31] A useful variation is the method of Andreason in which volumes of the suspension, after definite times of settling (calculated from Stokes's Law) are pipetted into a tared dish, evaporated to dryness and weighed. Various air permeability methods are also widely used.

The most modern method of particle size analysis applicable to cosmetic powders entails the use of a Coulter Counter. A paper dealing with the subject was presented by Wood and Lines.[32]

Packaging is usually carried out by automatic vacuum filling devices, which minimize dusting.

In choosing powder boxes, care should be taken to see that these are prepared with an odourless glue, otherwise the fragrance of the powder will be ruined in storage.

COMPACT POWDER

Because they are convenient to use, compact powders enjoy wide popularity. They are nowadays prepared by either a damp or a dry compression process. The moulding process used originally, which entailed the use of plaster of Paris, has fallen into disuse.

In the damp process, the powder, intimately mixed with a suitable binding agent, is milled to the requisite plasticity, compressed into suitable containers, usually metal godets, and dried for the requisite period in a current of warm air. In the dry process the mass is subjected to compression without being wetted to any appreciable extent. This process, although difficult to achieve satisfactorily at first, is probably the best to use for manufacturing compacts on a large scale, because it can be rigidly adhered to once the mix and suitable conditions have been determined. Presses available for the manufacture of compact powders may be of the hydraulic or reciprocating mechanical type, varying in size, operating pressures and output. They range from foot operated presses producing one cake at a time, to fully automatic presses which may produce up to 60 units per minute.

The requirement of good covering power, adhesion and uniformity in compositions mentioned in respect of conventional face powders also applies to compact powders. The latter should, in addition, be easy to remove from the cake for application by means of a powder puff, without crumbling or breaking during handling; this requirement is met by conferring adequate binding properties to the powder mixture to be compressed. Furthermore, the powders used for compacts should be free-flowing so that they do not adhere to punches or dies during compression; otherwise, air pockets will be formed which will result in an uneven compression and cause the cakes to break. From this it can be seen that one of the main aims during manufacture of compact powders is to ensure that the compressed cakes are of uniform density.

The main difference between loose powders and compact powders lies in their binding properties. If these are inadequate, the compressed cakes are liable to crumble easily following compression. If they are excessive, the cake will form lumps and go greasy on application. Thus, satisfactory binding properties are essential for trouble-free compression and the production of good quality cakes over long manufacturing periods.

The actual pressing process can also affect the shade of the product so quality control can be a problem. In fact, if large volumes are anticipated it is always best to conduct an extended manufacturing trial rather than go straight from the laboratory bench into full-scale production.

The composition of compact powders is generally very similar to that of loose powders. The differences which exist arise from the need to meet the requirement of greater cohesion and are largely evident in terms of percentages of some of the components present. In compact powders, colloidal kaolin, zinc oxide and metallic stearates are usually present at a higher level than in loose face powders, and starch is sometimes incorporated to facilitate compression. If the powders are not sufficiently binding they will require the addition of a binding agent to improve their cohesion so that on compression a firm cake is produced. Water-soluble and water-insoluble binding agents may be used. The former are natural and synthetic gums which are used in amounts ranging from 0·1 per cent to about 3 per cent by weight of the product, and are usually mixed with component powders in the form of a 5–10 per cent aqueous solution. A very much favoured binding agent for this purpose is low-viscosity carboxymethylcellulose. A small amount of a humectant is usually added to the solutions. If a water-insoluble binding agent is employed, for example glyceryl monostearate, cetyl or stearyl alcohols, isopropyl esters of fatty acids, lanolin and its derivatives or ozokerite, paraffin wax and microcrystalline waxes, it is preferable to use it in the form of an oil-in-water emulsion so that it is uniformly distributed throughout the product.

In the dry compacting process, it is usual to employ zinc or magnesium stearates at a level of 5–15 per cent by weight, as well as a lubricant such as mineral oil in similar proportions.

Considerable changes have taken place in formulation during the last forty years or so; thus in 1932 Winter[33] recommended an ammonia–stearin–starch compound containing white petrolatum, ammonia, starch and stearin (example 10) for the manufacture of compact powder and rouge.

	(10)
Stearin	100 g
White petrolatum	20 g
Ammonium hydroxide solution, 0·97 s.g.	50 cm^3
Rice or maize starch	250 g

Procedure: Melt the stearin and the petrolatum together. Add the ammonium hydroxide solution and stir thoroughly while hot. Add the starch to the warm mixture with vigorous stirring (during the addition, thick lumps will form in the starch powder; with vigorous stirring and pressing, these break down and mix with the starch). Rub the resulting crumbly mass in a mortar and then pass it through a 70–80 mesh sieve. Cool well before sifting.

This compound is added, in a proportion of 13–15 per cent, to the powder base together with the colouring matter, and the mixture is subjected to pressure. It is stated that such pressing should not be carried out suddenly but by gradually increasing the pressure. Experiments have fully confirmed this opinion. Unless a little time is given for the air to escape gradually, it becomes entrapped in the compact with disastrous results.

As mentioned previously, rice starch has been used in compact powder manufacture to facilitate compression of the powders. There has, however, been some controversy regarding the maximum permissible proportion of starch in a compact. On one hand, views were expressed to the effect that the starch content should be low, otherwise there is a tendency to produce hard cakes and to make the removal of the powder, when the puff is applied, more difficult.[34] One formula quoted included only 2·5 per cent of rice starch.[35] Winter, on the other hand, considered that starch acts as a good binding agent and recommended a base containing about 13 per cent of starch.[33] In yet another article, up to 20 per cent of starch was regarded as helpful in binding a compact.[36]

Winter suggested the powder base mixture given in example 11, to which may be added 13–15 per cent of stearin-starch and the required colouring matters. Water is added to produce a dough-like paste which is dried, ground and passed through a 120 mesh sieve and the mass compressed into a suitable metal case or godet.

	(11)
	per cent
Talc	26·7
Kaolin	56·7
Zinc oxide	3·3
Rice starch	13·3

Modern manufacturing procedures and compositions vary appreciably from those just described. In the case of the damp process, any colours to be used are first ground with the powder constituents, and the resulting mixture is passed through a sieve. The powder is then moistened with the binder solution or emulsion and perfume is incorporated. After a thorough mixing, the blend is sieved once more, for example through a 60 mesh sieve. The produce is then dried at room temperature or in warm air, provided that the temperature used does not exceed that at which the perfume will volatilize. The product is then compressed and placed in suitable containers.

Example 12 gives a formula to which the damp compression process is applicable.

	(12)
	per cent
Kaolin	20·0
Zinc oxide	15·0
Precipitated chalk	25·0
Talc	32·0
Compact binder (e.g. soap)	8·0
Perfume, colour	*q.s.*

The second of the two modern compacting processes, that is the dry compression method, is particularly suitable for mass production, but necessitates the use of higher pressures than are employed in the damp compression process. The manufacture of a compressed face powder cake by the dry compression process has been described in some detail in a US patent[37]. In an embodiment example of this patent the following composition and procedure were given:

	(13)
	per cent
Talcum	61·25
Sodium lauryl sulphate	0·75
Titanium dioxide	7·50
Zinc stearate	11·25
Inorganic pigments	1·00
Mineral oil	4·50
Spermaceti	3·00
Cetyl alcohol	1·50
Lanolin	1·00
Glycerin	7·50
Hexachlorophene	0·25
Alkyldimethylbenzyl ammonium chloride—50%	0·20
Methyl-*p*-hydroxybenzoate	0·09
Propyl-*p*-hydroxybenzoate	0·09
Perfume	0·12

Procedure: Talc, sodium lauryl sulphate, titanium dioxide, zinc stearate and inorganic pigments are first mixed together in a ribbon mixer for about an hour. Next, hexachlorophene, quaternary ammonium base and preservative, thoroughly premixed, are added to the mixer and the batch is mixed for about two hours, before passing through a micro-pulverizer using a no. 0·013 screen. The temperature of the material should not be allowed to rise more than 10°C above room temperature during this operation. After cooling, the powder blend is repassed through the micro-pulverizer. The mineral oil, spermaceti, lanolin and glycerin, mixed with heating until liquid, are then sprayed into the dry batch which is mixed in the ribbon mixer for a further half hour. The batch is passed through a comminuter and once again through a pulverizer, using a 3/16 in screen and taking the same precautions about temperature as before. When cool, the powder is passed through the micro-pulverizer using a no. 0·027 screen, and finally, having been cooled once more, is passed through a 40 mesh screen, ensuring that no heat is generated during this final screening. The resulting material is filled into containers and pressure is applied, for example 40–50 psi (300 kPa), to convert the powder into cake. It is usually advisable to keep the powders for several days in a suitably humid atmosphere before pressing, to facilitate the escape of any trapped air, and to ensure that the powder blend will not be too dry when compressed. During the dry compression process, it is usual to apply a small pressure initially to squeeze out the air, and thus avoid the formation of air pockets in the powder cakes. Pressure is then gradually increased up to 150 psi or even more (1000+kPa) before the die is removed from the surface of the cake. It is possible, however, with careful formulation and preparation of the powder to press the cakes with immediate pressures of up to 600 psi (4 MPa) and outputs approaching two thousand units per hour.

CAKE MAKE-UP

The modern cake make-up originates from theatrical grease-paint and has acquired popularity because of its ease of application and stability and also because more product can be applied to the face and thus deeper shades and effects can be achieved. As in the case of compact powders, the cake make-up is also made from talc, kaolin, zinc oxide and precipitated chalk, but it contains, additionally, inorganic pigments such as titanium dioxide and iron oxides. Humectants such as sorbitol or propylene glycol may also be incorporated together with other additives such as sorbitan sesquioleate, lanolin or mineral oil and perfume. Humectants and other liquid constituents are usually combined and sprayed on to the powder constituents while these are mixed in the ribbon mixer. The resulting blend is granulated and finally compressed.

Patents[38,39] granted in the 1930s claimed a product in cake form which, when applied to the face using a damp sponge, dried to form a coherent water-repellent film of powder. The advantage of this product is that no foundation cream is required and make-up can be retouched very quickly and conveniently.

A patent by Max Factor[40] describes a dry cake-form make-up which can be applied with a moistened pad or sponge. The cake contains oily and waxy ingredients (0·8–24 per cent), a water-soluble dispersing agent (1–13 per cent) and fillers (35–80 per cent) and pigments (12–50 per cent) whose particles are coated with the oils and waxes to make them water-repellent. The make-up is prepared by adding the fillers (talc, chalk) and pigments (zinc, titanium and ferric oxides) to the oils and waxes dispersed in water, drying the mixture so formed, pulverizing the product, and compressing it into cake form.

Products made to this style of formula can be prepared without drying by using a much higher ratio of powder to emulsion, for example approximately 12 to 1, and mixing with a Beken Planetex or Duplex mixer. The mixing operation takes between 30 and 60 minutes and the mix is compacted after granulating through a fine screen.

The following formulae can be made in this fashion:

		(14) *parts*
Powder base (perfumed)		100
Emulsion:		
Stearic acid	34·5	8·5
Mineral oil	21·5	
Triethanolamine	10·5	
Water	33·5	

For deep shades with good coverage, zinc oxide should be replaced by titanium dioxide (this also facilitates better pressing characteristics):

	(15) *per cent*
Colour mix:	
Talc	89·75
Titanium dioxide	9·00
Yellow iron oxide	0·75
Red iron oxide	0·40
Black iron oxide	0·10

Base:	*per cent*
Stearic acid	15·0
Acetylated lanolin	3·0
Stearyl alcohol	3·0
Glyceryl monostearate (self-emulsifying)	2·2
Sulphonated castor oil	1·2
Triethanolamine	3·0
Mineral oil	35·0
Polyethylene glycol	10·0
Water	27·6
Completion ratio:	
Base	20·0
Colour mix	79·8
Perfume	0·2

Pressing of this product can be very rapid as for compact powders, although it is important to check that there is consistent dispersion of the base in the colour mix throughout long manufacturing runs.

A prototype cake make-up formula suggested in a cosmetic formulary of Atlas Industries[41] had the composition given in example 16.

	(16)
Pigment blend:	*per cent*
Talc	60·0
Kaolin	20·0
Zinc oxide	10·0
Titanium dioxide	5·0
Calcium carbonate, light	5·0
Iron oxide colours	*q.s.*
Completed formula:	
Pigment blend	81·4
Arlex (sorbitol)	4·1
Propylene glycol	2·4
Arlacel C (sorbitan sesquioleate)	9·7
Mineral oil or lanolin	2·4
Perfume, preservative	*q.s.*

Procedure: Mix the Arlex (sorbitol) with the propylene glycol and preservative and spray this mixture into the powder while it is being mixed in a dough mixer. Mix the Arlex C and mineral oil. If lanolin is used, melt the lanolin with the Arlacel C. Add the perfume and spray on to the powdered mass which is agitated in the dough mixer. Transfer to a Fitzpatrick tablet granulator or an equivalent machine. After granulation, the mass is ready for performing and compression. The same procedure is employed as in the case of compact powders, that is the initial application of a relatively light pressure (about 25–50 psi; 250 kPa) to remove air, followed by the application of a higher pressure of 100–150 psi or more (1000 + kPa).

Application of Cake Make-up

Maurice Seiderman, a pioneer of make-up in the motion picture industry, recommended the application of cake make-up in the following manner: 'To apply cake make-up correctly, wet and squeeze out a sponge and rub it lightly over the face. Smooth the cake make-up on the face and before it dries blend in the rouge. Carefully squeeze all water out of the sponge and rub it lightly over the face until the make-up is dry.' Macias-Sarria[42] adds that people who do not like a heavy make-up should blot off the excess of water with a dry towel or facial tissue.

It is claimed that cake make-up is satisfactory for the younger woman; in the case of persons in the later thirties, and even younger persons who suffer from dry skin, it is found to be rather drying. To cater for this market, to coincide with the general trend towards more greasy make-up as exemplified by many foundation creams and possibly also with a view to cashing in on the flat type of pack in which cake make-up is packed, a greasy type of compact containing pigments of high covering power in a waxy base has been marketed.

Such products appear to be poured instead of being compressed in the usual manner. Wetting agents may be incorporated so that these products may be removed with a damp sponge; alternatively, silicone oils may be used to improve the spreading characteristics and make the product suitable for finger-tip application.

MAKE-UP CREAM

Foundation make-up preparations in cream form are essentially suspensions of pigments in an emulsified lotion. The addition of the pigment is usually made at about 50°–55°C as the emulsion is slowly cooled with agitation. Janistyn quoted the following formula for a cream make-up.[43]

	(17) *parts*
Glycerol mono- and distearate (pure)	2·0
PEG 400 monostearate	1·0
Stearic acid	11·5
Cetyl alcohol	0·5
Isopropyl myristate or palmitate	2·0
Propylene glycol	12·0
Sorbitol syrup	2·5
Preservative	0·1
Titanium dioxide	2·2
Talc	8·0
Colour pigments	1·0
Water	57·4

The suspension is usually homogenized and milled.

LIQUID POWDER

So-called 'liquid powders' have sometimes been used as a base for ordinary powder or to replace such powder for evening wear, dances or similar occasions. They have also included a theatrical product, 'wet white', which was used for whitening the neck and arms. Its basic components were zinc oxide and bismuth oxychloride, subnitrate and carbonate incorporated into a liquid consisting of a mixture of glycerin and water in varying proportions. Glycerin was employed in amounts of up to 30 per cent and the water was often replaced by triple rose water or other fragrant waters; starch was also used sometimes to suspend heavier constituents. To prepare 'wet white' the constituent powders were mixed and glycerin was added. Surfactants could also be included to assist in dispersing the powders.

A 'wet white' formula on these lines, quoted by Poucher,[44] is given in example 18.

	(18) *per cent*
Bismuth subnitrate	5·0
Starch	5·0
Zinc oxide	10·0
Glycerin	15·0
Rose water	65·0

Other more modern formulae omit the bismuth salt and starch, and employ more pigments, fillers and colouring materials to simulate various shades of flesh colour. Examples 19 and 20 will serve as a basis for experiments:

	(19) *per cent*
Zinc oxide	3·0
Chalk (precipitated)	15·0
Kaolin	2·0
Colouring matter	*q.s.*
Glycerin	15·0
Water	65·0
Preservative, perfume	*q.s.*

	(20) *per cent*
Zinc oxide	10·0
Titanium oxide	10·0
Talc	10·0
Colouring matter	*q.s.*
Gum tragacanth (0·5% solution)	25·0
Glycerin	15·0
Water	30·0
Preservative, perfume	*q.s.*

Janistyn[43] quoted a liquid powder formulation of the following composition:

	(21) *per cent*
Sodium carragheenate (medium or high viscosity)	2·0
n-Propyl alcohol	2·0
Propylene glycol	5·0

	per cent
Water	68·5
Veegum HV	0·5
Talc	10·0
Magnesium carbonate	4·0
Titanium dioxide	1·0
Colour pigments	7·0
Perfume	*q.s.*

Procedure: Wet the carragheenate with propyl alcohol and then dissolve it in the propylene glycol–water mixture. Veegum is then dispersed in the solution, followed by the addition of mixed pigments.

Cosmetic Stockings

This type of aqueous make-up was used during World War II as a leg make-up or liquid cosmetic stocking. The requirement was that the appearance of a treated leg should be very similar to a leg encased in a normal stocking. It was essential that the preparation should not wash away in the rain, nor rub off on to clothing, yet should be easily removable by washing with soap and water.

A basic formula upon which to elaborate a liquid 'cosmetic stocking' is given in example 22.

	(22) *per cent*
Zinc oxide	6·0
Chalk (precipitated)	16·0
Methylcellulose	0·5
Glycerin	16·0
Water	61·5
Colouring matter, preservative, perfume	*q.s.*

More opacity in the finished make-up may be obtained by increasing the proportion of zinc oxide and/or including titanium dioxide, but an obviously artificial effect is undesirable. A little alcohol may be included in order to accelerate drying; the viscosity may be varied by the use of different viscosity grades of methyl cellulose or other cellulosic film-forming material; alternatively alginates or gum mucilages may be included, but care must be taken that the finished preparation is not sticky.

To increase the time during which the pigmented powder remains in suspension (after shaking) while the preparation is being applied to the leg, a few per cent of bentonite or other similar clays may be incorporated.

A certain amount of glycerin or similar non-evaporating and foundation-forming substance is desirable as this improves the adherence of the film of powder to the leg.

Colouring matters usually consist of mixtures of harmless yellow, red, and brown pigments according to the shade desired; in addition some soluble dyes may be included to produce a slight staining effect on the leg and also minimize the appearance of separation between the aqueous and powder phase in the container.

The following formula appeared in the *American Perfumer*:[45]

	(23) *per cent*
Zinc stearate	2·0
Titanium dioxide	3·5
Colloidal aluminium-magnesium silicate gel (Veegum)	20·0
Isopropyl alcohol	6·0
Umber	0·5
Yellow oxide	2·5
Red oxide	2·5
Propylene glycol	1·0
Methylcellulose (1500 viscosity grade)	0·5
Water, perfume, preservative	to 100·0

LIQUID MAKE-UP

Liquid make-up for cosmetic use is another development of the wetted powder which consists essentially of pigments dispersed in a viscous base.

The early liquid make-up preparations were suspensions of pigments in an aqueous alcoholic solution, which required vigorous shaking prior to use to ensure uniform distribution of the product during application.

The basic problem in the preparation of more elegant products of this type is to prevent the sedimentation of constituent pigments by dispersing them in a hydrocolloid base or in a liquid emulsion. The hydrocolloids used for thickening the preparations may be selected from cellulose derivatives, carragheenates, Carbopol 934 or 941, Veegum and others.

The pigments used in liquid make-up preparations are the usual components of powder bases such as talc, kaolin, zinc oxide, titanium dioxide, calcium or magnesium carbonates and others.

In emulsified products, raw materials used included propylene glycol monostearate, glyceryl monostearate, fatty alcohols such as cetyl or oleyl alcohols, isopropyl myristate, lanolin and its derivatives, polyethylene glycols, humectants and others; in general they resemble the make-up cream described in example 17 above.

A liquid make-up formula quoted by Shansky[46] is given in example 24. Viscosity may be adjusted by varying the amount of gum and bentonite.

	(24) *parts*
Propylene glycol	4·40
Polyethylene glycol 400 monostearate	1·92
Preservative	0·32
Gum tragacanth (0·175% solution)	76·68
Bentonite	0·96
White oil	1·20
Oleyl alcohol	6·72
Stearic acid	4·20
Triethanolamine	1·92
Perfume	*q.s.*
Titanium dioxide plus powdered pigments	*q.s.*

Pigments and titanium dioxide are usually in the formula at a level of 5–10 per cent, as in example 25 (the actual amount will depend on the shade desired).

	(25) per cent
Isopropyl lanolate	3·50
Isopropyl myristate	4·20
Squalane	1·40
Purcellin oil	2·10
Mineral oil	12·80
Sorbitan oleate	1·00
Veegum (5% solution)	30·00
Propylene glycol	8·00
CMC (1% solution)	20·60
Polyoxyethylene (20) sorbitan mono-oleate	4·00
Water	6·04
Titanium dioxide	4·60
Yellow iron oxide	0·56
Red iron oxide	0·55
Black iron oxide	0·10
Perfume	0·25
Preservative	0·30

STICK MAKE-UP

Liquid make-up products are by far the most acceptable products for face make-up, mainly due to their light application which suits the modern style and their convenience of use (bottle or tube). However, there is a small but significant proportion of the market who prefer a heavier make-up but with the convenience factor as well. This has led to the development of stick make-up, which in essence is a dispersion of pigments in a wax base. A typical formula is shown in example 26.

	(26) per cent
Mineral oil	47·65
Paraffin wax	3·50
Beeswax	1·50
Carnauba wax	4·00
Kaolin	9·00
Titanium dioxide	30·00
Yellow iron oxide	2·50
Red iron oxide	1·50
Black iron oxide	0·30
Perfume	0·05

Procedure: Mix the oils and waxes together and heat until a clear solution is obtained. Mix in the colours and pigments gradually with a high-speed Silverson-type mixer. Shade, dispersion and setting point should be checked prior to pouring the product into the appropriate containers.

This stick concept can be extended further to give the cover-up product, used to disguise birthmarks and blemishes. The product is heavily pigmented and is applied like a lipstick.

	(27)
	per cent
Lanolin alcohols	2·8
Ozokerite wax	8·0
Paraffin wax	6·0
Mineral oil	20·2
Isopropyl myristate	10·0
Lanolin	2·8
Titanium dioxide	36·8
Kaolin	8·0
Yellow iron oxide	2·5
Black iron oxide	0·6
Red iron oxide	2·0
Perfume	0·3

REFERENCES

1. Grady, L. D., *J. Soc. cosmet. Chem.*, 1947, **1,** 17.
2. Luckiesh, M., Taylor, A. H., Cole, H. N. and Sollman, T., *J. Am. med. Assoc.*, 1946, **130,** 1.
3. Hewitt, M. L., *Perfum. essent. Oil Rec.*, 1943, **34,** 35.
4. Halpern, A., Powers, J. V. and Bradney, C. H., *Proc. sci. Sect. Toilet Goods Assoc.*, 1950, (14), 4.
5. Ross, C. F. and Kerr, P. F., *US Geolog. Surv. Prof. Paper No. 165E*, 1934, p. 152.
6. McDonough, E. C., *Truth about Cosmetics*, New York, Drug Cosmetic Industry, 1937, p. 110.
7. Baumann, H., *Parfüm. Kosmet.*, 1959, **40,** 287.
8. US Patent 3 196 079, Phillips Petroleum, July 1965.
9. British Patent 1 093 108, ICI, November 1967.
10. Smith, R. L., *Manuf. Chem.*, 1967, **38,** (12), 35.
11. Grexa, R. W. and Parmentier, C. J., *Cosmet. Toiletries*, 1979, **94** (2), 29.
12. British Patent 433 142, IG Farbenindustrie, 1935.
13. Varma, K. C., *Soap Perfum. Cosmet.*, 1953, **27,** 505.
14. British Patent 482 269, Lawson, R. W., 1938.
15. British Patent 519 544, Brocklehurst Whiston Amalg. Ltd, 1940.
16. British Patent 555 044, Phelps, S. G., 1943.
17. *Schimmel Briefs*, 1948 (August), No. 161.
18. Morelle, J., *Parfum. Mod.*, 1947, (4), 29.
19. EEC Cosmetic Directive, 27 July 1976, 76/768/EEC.
20. *Colour Index*, 3rd edn, Bradford, Society of Dyers and Colourists, 1971.
21. Harry, R. G., *Cosmetic Materials*, London, Leonard Hill, 1962, Appendix III.
22. Carrière, G. and Luft, G., *Soap Perfum. Cosmet.*, 1966, **39,** 29.
23. Anstead, D. F., *J. Soc. cosmet. Chem.*, 1959, **10,** 1.
24. Russ, J., *Cosmet. Toiletries*, 1981, **96**(4), 25.
25. Jannaway, S. P., *Alchimist (Boechut*), 1948, **2,** 20.
26. Winter, F., *Alchimist* (*Boechut*), 1947, **1,** 188.

27. Keithler, W. R., *Drug Cosmet. Ind.*, 1955, **76,** 40.
28. Hilfer, H., *Drug Cosmet. Ind.*, 1953, **73,** 466.
29. Chilson, F., *Modern Cosmetics*, 1st edn, New York, Drug Cosmet. Ind., 1934, p. 68.
30. DeNavarre, M. G., *The Chemistry and Manufacture of Cosmetics*, New York, Van Nostrand, 1941, p. 361.
31. Hinkley, W. O., *Ind. Eng. Chem. anal. Ed.*, 1942, **14,** 10.
32. Wood, W. M. and Lines, R. W., *J. Soc. cosmet. Chem.*, 1966, **17,** 197.
33. Winter, F., *Handbuch der gesamten Parfümerie und Kosmetik*, 2nd edn, Vienna, Springer, 1932, pp. 615–620.
34. *Perfum. essent. Oil Rec.*, 1930, **21,** 9.
35. *Perfum. essent. Oil Rec.*, 1932, **23,** 362.
36. *Perfum. essent. Oil Rec.*, 1932, **23,** 203.
37. US Patent 3 296 078, Kay, M. and Amsterdam, M. J., August 1958.
38. US Patent 2 034 697, Max Factor & Co., 1936.
39. US Patent 2 101 843, Max Factor & Co., 1937.
40. British Patent 501 732, Max Factor & Co., 1939.
41. Atlas Powder Co., *Drug and Cosmetic Emulsions*, 1946, p. 29.
42. Macias-Sarria, J., *Am. Perfum.*, 1944, **46**(12), 48.
43. Janistyn, H., *Taschenbuch der modernen Parfümerie und Kosmetik*, Stuttgart, Wissenschaftliche Verlagsgesellschaft, 1966, p. 614.
44. Poucher, W. A., *Perfumes Cosmetics and Soaps*, Vol. 3, London, Chapman and Hall, 1950, p. 191.
45. *Am. Perfum. Essent. Oil Rev.*, 1945, **47**(6), 37.
46. Shansky, A., *Am. Perfum. Essent. Oil Rev.*, 1964, **79,** (10), 53.

Chapter Nineteen

Coloured Make-up Preparations

LIPSTICK

Introduction

Lipsticks, the lip cosmetics moulded into sticks, are essentially dispersions of colouring matter in a base consisting of a suitable blend of oils, fats and waxes.

Lipstick is used to impart an attractive colour and appearance to the lips, accentuating their good points and disguising any bad ones. Narrow bad-tempered lips may be widened, and broad sensual lips made to appear narrower by its use. In fact, if applied intelligently it is capable of entirely altering the apparent facial characteristics.

Since lips are considered to be more alluring when they possess a slightly moist appearance, this is always achieved by the use of a greasy base which also exerts an emollient action.

There is no doubt that the wide use of lipstick among women has led to a decrease in cracked and chapped lips, the crevices of which were always liable to bacterial infection. In addition, as in the case of many other cosmetics, it exerts a psychological effect difficult to assess, and induces a feeling of mental comfort.

Characteristics Required in a Lipstick

A good lipstick should have the following characteristics.

(1) It should have an attractive appearance, that is, a smooth surface of uniform colour, free from defects such as pinholes or grittiness due to colour or crystal aggregates. This should be retained during its shelf life and usage life—it should not exude oil, develop a bloom, flake, cake, harden, soften, crumble nor become brittle over the range of temperatures likely to be experienced.

(2) It should be innocuous, both dermatologically and if ingested.

(3) It should be easy to apply, giving a film on the lips that is neither excessively greasy nor too dry, that is reasonably permanent but capable of deliberate removal, and which has a stable colour.

It will be realized that a system of colouring matter dispersed in a plastic fatty medium is one most likely to satisfy the above requirements.

Ingredients of Lipsticks

Colouring Materials

The colour of a lipstick is one of the major selling points, but it is one which can only be dealt with in general terms, since the precise shades are dictated by ephemeral fashion. It is usual for the colour to contain some measure of red and this allows shades ranging between orange-yellow and purple-blue, although even greens are not unknown. Depth of colour and opacity are also variable and during periods when the fashion trend was to a 'no make-up' look, uncoloured lipstick base of high gloss has been seen under the name 'lip gloss'. 'Lip gleams' containing pearlescent materials are also known, and occasionally sticks with some degree of gold or silver lustre achieved by the use of finely divided metal (coloured aluminium) are presented. However, the main accent in this chapter will be on the predominantly red conventional shades, since these involve the basic principles of lipstick formulation.

The colour is imparted to the lips in two ways: (a) by staining the skin, which requires a dyestuff in solution, capable of penetrating the outer surface of the lips; (b) by covering the lips with a coloured layer which serves to hide any surface roughness and give a smooth appearance. This second requirement is met by insoluble dyes and pigments which make the film more or less opaque.

Typical proportions for the colours in a lipstick are as follows:

	per cent
Staining dyes (bromo acids)	$\frac{1}{2}$–3
Oil-soluble pigment	2
Insoluble pigment	8–10
Titanium dioxide	1–4

Staining Dyes. The most widely used staining dyes are water-soluble eosin and other halogenated derivatives of fluorescein which are generally referred to collectively as 'bromoacids', a term originally applied to acid eosin, tetrabromofluorescein.

Eosin, also known as D&C Red No. 21, is an insoluble orange compound which changes to an intense red salt when the pH value is above 4. When applied to the lips in the acid form, it produces a relatively indelible purple red stain on neutralization by the lip tissue.

Other halogenated fluoresceins can be used to give different staining colours and varying degrees of indelibility. Thus D&C Red No. 27 (tetrachlorotetrabromofluorescein) produces a brilliant bluish red stain, and D&C Orange No. 5 (dibromofluorescein) a yellow red, which often used in conjunction with D&C Red No. 21. D&C Orange No. 10 (di-iodofluorescein) is another derivative frequently used.

Unfortunately, cosin and some of its derivatives can give rise to sensitization or photosensitization, leading to cheilitis (inflammation of the red portion of the lips) or more general allergic reactions. Whether this is due to the bromoacid *per se*, or to impurities contained therein or even to the perfume in the lipstick is by no means clear, but the fact that it does occur with a small proportion of lipstick users (coupled with the facts that the skin of the lips is devoid of a horny layer, and that there is a possibility of ingestion of lipstick) has focussed attention on permissible colours. The USA and the EEC countries have defined lists of permitted colours, but these do not always coincide. Thus, for example, FD&C Red No. 2 is now banned in the USA whereas it is permitted in the EEC. The situation becomes even more confused when EEC directives clash with local laws. For example, D&C Orange No. 5 has been banned for some time in Germany, but is allowed for mucous membrane products by the EEC Cosmetics Directive. Incidentally, D&C Orange No. 5 is allowed in the USA with certain restrictions (see below). Formulators should check very carefully therefore to determine the use of the various colours in the individual countries in which the lip products are to be sold. A brief summary of the status of the most popular colours as used in the USA and the EEC is given in Table 19.1.

Table 19.1 Status of Colouring Materials for Lipsticks in the EEC and the USA

Colour index no.	Common name	Status under EEC Cosmetics Directive	Status under US FDA
12085	D&C Red 36	Allowed: Annex III Part 2	Provisionally listed 31 Jan 81; 3% allowed in lip products
15585	D&C Red 8 (Na)	Allowed: Annex III Part 2	Provisionally listed 31 Jan 81; 3% allowed in lip products
15630 15630 (Ca) 15630 (Ba) 15630 (Sr)	D&C Red 10 (Na) D&C Red 11 (Ca) D&C Red 12 (Ba) D&C Red 13 (Sr)	Allowed: Annex III Part 2	Delisted 13 Dec 77; contains β-naphthylamine
15850	D&C Red 7 (Ca)	Allowed: Annex III Part 2	Provisionally listed 31 Jan 81 for ingested and external use
16185	FD&C Red 2	Allowed: Annex III Part 2	Delisted 2 Dec 76
45170	D&C Red 19	Allowed: Annex III Part 2	Provisionally listed 31 Jan 81; 1·3% max. allowed in lip products
15880	D&C Red 34	Allowed: Annex III Part 2	Permanently listed for external use
45370	D&C Orange 5	Allowed: Annex III Part 2	Provisionally listed 31 Jan 81; 6% allowed in lipsticks
45380	D&C Red 21	Allowed: Annex III Part 2	Provisionally listed 31 Jan 81 for ingested and external use
45410	D&C Red 27	Allowed: Annex III Part 2	Provisionally listed 31 Jan 81 for ingested and external use
45425	D&C Orange 10	Allowed: Annex III Part 2	Provisionally listed 31 Jan 81 for external use only

Table 19.1 (*cont.*)

Colour index no.	Common name	Status under EEC Cosmetics Directive	Status under US FDA
45430	FD&C Red 3	Allowed: Annex III Part 2	Provisionally listed 31 Jan 81 for ingested and external use
77491	Iron oxide	Allowed: Annex III Part 2	Provisionally listed for ingested/external and eye area use
12075	D&C Orange 17	Allowed: Annex III Part 2	Provisionally listed 31 Jan 81; 5% max. allowed for lip products
15510	D&C Orange 4	Allowed: Annex III Part 2	Permanently listed for external use
15985	FD&C Yellow 6	Allowed: Annex III Part 2	Provisionally listed 31 Jan 81 for ingested and external use
19140	FD&C Yellow 5	Allowed: Annex III Part 2	Provisionally listed 31 Jan 81 for ingested and external use
77891	Titanium Dioxide	Allowed: Annex III Part 2	Permanently listed for ingested/external and eye area use
26100	D&C Red 17	Provisionally allowed: Annex IV Part 2	Permanently listed for external use
77163	Bismuth oxychloride	Provisionally allowed: Annex IV Part 2	Permanently listed for ingested/external and eye area use
15585 (Ba)	D&C Red 9 (Ba)	Provisionally allowed: Annex IV Part 2	Provisionally listed 31 Jan 81; 3% max. allowed in lip products
15800	D&C Red 31	Provisionally allowed: Annex IV Part 2	Permanently listed

Difficulties may be encountered when using bromoacid dyestuffs in obtaining entirely homogeneous dispersions of these dyes in the lipstick mass and this could result in shade variations in lipsticks. Novel compositions have been disclosed[1] which aim to overcome these difficulties and to ensure that the colours of the lipsticks sold are identical with those conferred by them on the skin. These compositions comprise amine salts of bromoacid dyestuffs and fatty substances which are at least partial solvents for such components. The amines employed in the preparation of these compositions are selected from non-aromatic primary, secondary or tertiary monoamines, in which the groups attached to the nitrogen atom contain at the most 6 carbon atoms. Some of them may contain hydroxy radicals, others may be heterocyclic compounds. Specifically mentioned in patent claims have been triethanolamine, diethanolamine, 2-amino-2-methyl-propanediol-1,3, monoisopropanolamine, trihydroxymethylaminomethane, diglycolamine and morpholine.

To prepare these compositions the bromoacids are dispersed in a fatty material, preferably soya lecithin, and the amine is added with mixing. The resulting mixture may be optionally heated and allowed to cool. The amine salts of the bromoacid dyestuffs are then dissolved in a mixture of waxes (for example carnauba wax, beeswax and ozokerite) and oils (for example various lanolin derivatives and vaseline).

Non-eosin staining dyes reported by Wilmsmann[2] are an interesting development in view of the restrictions applying to some of the fluorescein derivatives. It is claimed that water-soluble FD&C and D&C dyes, which are useless in lipsticks, when converted to the free sulphoacid form become water-insoluble, lipophilic, and suitable for use as staining dyes in lipsticks, covering a wide colour range.

Pigments. Both inorganic and organic pigments and metallic lakes are used to give intensity and variation of colour. When selecting lakes the possibility of reaction with the base, for example soap formation with free fatty acid, must be borne in mind.

Titanium dioxide, often used at levels up to 4 per cent, is the most effective white pigment for obtaining pink shades and giving opacity to the film on the lips. However, the use of titanium dioxide requires great care in the grade of material selected (anatase or rutile) and the surface treatment it has received to make it lipophilic, and also in the method of incorporation, if unexpected troubles such as oily exudation, streaking, dullness and coarse texture are to be avoided.

Two D&C colours, D&C Red No. 36 and D&C Orange No. 17, are so insoluble in both water and oil that they may be considered as pigments, although they are not in the form of metallic lakes. Similarly the amount of staining dye used often exceeds the solubility in the base and the insoluble portion will act as a pigment.

Lakes of many of the D&C colours with metals such as aluminium, barium, calcium and strontium are potential pigments for lipsticks. However, some strontium and zirconium lakes have to be avoided in most EEC countries, as they are banned. Aluminium lakes are not usually favoured because of their lack of opacity, but this very property would seem to suggest their use in transparent lipsticks.

The following lakes are considered to be the most useful lipstick colorants:

Calcium lakes of D&C Reds Nos 7, 31 and 34
Barium lakes of D&C Red No. 9 and D&C Orange No. 17
Aluminium lakes of D&C Reds Nos 2, 3, and 19 and FD&C Yellows Nos. 5 and 6

When the parent D&C colour is subject to restriction the lakes are also restricted in the same way.

As noted above, pigments and lakes are used at levels between 8 and 10 per cent or perhaps over a wider range, say 5–15 per cent.

Iridescent lipsticks utilize either mica coated with titanium dioxide or bismuth oxychloride at levels of up to 20 per cent, depending upon the effect desired.

Base

Apart from the colour, the quality of the lipstick during manufacture, storage and use will be determined for the most part by the composition of the fatty base. This quality is largely concerned with the rheology of the mixture at various temperatures. For instance, during manufacture (usually while warm) it must be possible to mill and grind the mass, and to pour and mould it while holding the insoluble colours evenly dispersed without settling. In the moulds it must set quickly with a good surface and good release properties. During shelf life and usage life the stick must remain rigid and stable, and generally in good condition. In use the stick must soften sufficiently in contact with the lips, and be sufficiently thixotropic to spread on the lips to form an adherent film which will not smear nor, ideally, transfer to cups or glasses.

Dyestuff Solvents. Although all the base ingredients must contribute to the physical and rheological properties, there is the additional requirement that some part of the base must act as the necessary solvent for the staining dyes. Many of the normal fatty materials which might be considered for use in the base are too non-polar to dissolve the dyestuffs, and it is convenient to consider first those ingredients which do have solvent properties for eosin, and which must form some part of the base.

Table 19.2 details the solubility of eosin in a number of fatty or lipophilic materials, not all of which could be included in lipsticks, and serves as a starting point for considering this type of material. In general terms, vegetable oils have the greater solvent power for eosin but suffer from degradation properties. Mineral oils are more stable but have poorer solvent properties.

Castor oil is a traditional material for dissolving bromoacid and it owes this property to its high content of ricinoleic (hydroxyoleic) acid, which is unique among natural oils. Its other properties include a high viscosity, even when warm, which delays pigment settling and the oiliness which helps with gloss and emollience, although too high a quantity causes drag during application and an unpleasant greasy film. As much as 50 per cent has been used, but a better quantity is probably about 25 per cent. The disadvantages of castor oil include an unpleasant taste and potential rancidity.

Fatty alcohols, four of which are included in Table 19.2, can be seen to have some solvent power for dyestuffs. Any of those quoted (lauryl—C_{12}, myristyl—C_{14}, stearyl—C_{18}, oleyl—C_{18} unsaturated), or cetyl—C_{16}, could be used according to the consistency contribution required. However, synthetic isomers of cetyl alcohol, such as the hexadecyl alcohol manufactured by the Enjay Chemical Co., which is actually β-hexyldecyl alcohol, are available. It is claimed that hexadecyl alcohol is a good solvent for bromoacid dyes, that lipsticks

Table 19.2 Solubility of Eosin in Various Solvents[3]

Solvent	Eosin in solution at 20°C (% approx.)
Polyethylene glycol 4000	12·0
Polyethylene glycol 1500	10·0
Polyethylene glycol 400	10·0
Hexa-ethylene glycol	9·0
Phenyl ethyl alcohol	8·0
Diacetone alcohol	6·5
Benzyl alcohol	6·0
Tetraethylene glycol	5·7
Hydroxycitronellal	4·5
Citral	4·5
Triethylene glycol	4·0
Acetone	3·5–4·0
Diphenyl ketone	3·5–4·0
Diethylene glycol	2·5
Terpineol	1·9
Ethylene ricinoleate	1·9
Cyclohexanol	1·6
Ethyl ricinoleate acetate	1·4
Oleyl alcohol	1·0
Ethylene glycol	1·0
Lauryl alcohol	0·75
Myristyl alcohol	0·57
Stearyl alcohol	0·5
Glycol oleate acetate	0·4
Cocoa butter	0·35
Lauric acid	0·3
Myristic acid	0·3
Cetyl acetate	0·3
Ethyl oleate	0·3
Castor oil	0·2–1·7
Ethyl stearate	0·2
Glycol oleate	0·1

Table 19.3 Solubility of Eosin in Various Compounds

Compound	Solubility (% at 20°C)
Isopropyl myristate	0·2
Oleyl alcohol	1·0
Diethyl sebacate	1·3
Di-isopropyl adipate	1·4
Propylene glycol	1·6

containing it can be applied with very little drag and do not bleed or smear and that its presence lessens any tendency to develop an unpleasant taste during storage.

Esters of various kinds have been proposed, including lower alkyl esters of fatty acids and dibasic acids such as adipic and sebacic, short chain-length acid esters of fatty alcohols, mono-, di- and mixed esters of glycol or glycerol. They have no specific virtues other than that they are lipophilic liquids of low oiliness giving lubricating and emollient effects with some dye solvent properties (see Tables 19.2 and 19.3). If too much is used the stick may sweat.

Glycols with two hydroxy groups are more polar than (for example) the fatty alcohols and might be expected to be better dye solvents; Table 19.2 shows this to be so. However, glycols are not particularly miscible with fatty materials and are of little importance.

Polyethylene glycols (Carbowaxes) also have good dyestuff solvent power (see Table 19.2). The solvent power correlates to some extent with water solubility and this is a detraction. However, correctly chosen, these materials would seem to have considerable possibilities in lipstick formulation.

Monoalkanolamides, for example Loramine Wax 101,[4] have been claimed to have good dye-dissolving properties and to have an advantage in that they have no action on the plastics material of the lipstick cases. It is suggested that dye can be pre-dissolved in Loramine Wax and kept as a concentrate for more convenient dispensing when required.

Other solvents proposed for bromoacids include Polychol 5 and Volpo N.3 (Croda Ltd); Polychol 5 is an ethylene oxide derivative of lanolin alcohols, while Volpo N.3 is a polyoxethylene oleyl ether. Both compounds are compatible with oleyl alcohol (for example Novol) and castor oil, and can be used in combination as a bromoacid solvent mixture consisting, for example, of

	parts by weight
Polychol 5 or Volpo N.3	10–20
Oleyl alcohol	20–10
Castor oil	40

to solubilize 1–3 parts by weight of bromoacid dyes.

Wetting agents, once used to some extent to 'solubilize' the dyestuff, nowadays find no application in lipsticks.

Other Base Ingredients. It will have been observed that few of the ingredients quoted so far have had the high melting points or hardness which are required to give satisfactory moulding properties, that is, quick setting and good release with a glossy surface and a rigid stick. This function is generally fulfilled by the inclusion of waxes or wax-like materials.[5] The importance of melting point as a criterion is emphasized by Gouvea,[6] who achieves the desired melting characteristics by working to a 'carnauba equivalent'.

Carnauba wax is a very hard vegetable wax used for raising the melting point, imparting rigidity and hardness and providing contraction properties in the moulding process.

Candelilla is another hard vegetable wax serving the same functions as carnauba wax but has a lower melting point and is less brittle.

Amorphous hydrocarbon waxes in mineral oil, for example ozokerite wax, give a short-fibred texture to the product.

Petroleum-based waxes, for example microcrystalline wax, are also used to modify the rheology of the product.

Beeswax is the traditional stiffening agent for castor oil, but it can give a grainy and dull effect if used in large quantities.

Cocoa butter might be thought an ideal material owing to its sharp melting point just below the human body temperature, which makes it so useful in other products. However, it cannot be used alone since it does not have all the required properties, and used in too large quantities it will cause the stick to 'bloom'.

Other more or less wax-like materials are hydrogenated vegetable oils which are more solid and less prone to rancidity than the unhardened oils. Of particular interest is hydrogenated castor oil, which in addition to wax-like properties still retains eosin solvent properties. Some other softer materials will be found to be of use in lipstick manufacture.

Lanolin and lanolin absorption bases are very useful ingredients up to about 10 per cent by virtue of their emollient properties. They are claimed to have eosin solvent properties and act as binding agents for the other ingredients, tending to minimize sweating and cracking of the stick and acting as plasticizers. Absorption bases in particular are recommended to enhance the gloss on the lips.

Petroleum jelly and the more viscous *paraffin oils* may be used to adjust consistency, act as lubricants and improve spreading properties. Large amounts tend to impair the adhesion properties and can be difficult to blend if much polar material such as castor oil is present.

Lecithin is another possible component which acts as a dispersing agent for pigments, in addition to facilitating the application of the lipstick and improving the adhesion to the lips.

Silicone waxes are included in improved lipstick compositions disclosed by Dow Corning.[7] These comprise essentially a wax (of which at least 15 per cent by weight is a silicone wax), a cosmetic solvent and a colouring agent. The preferred silicone wax compounds include organosilicon block copolymers, hydrocarbon silicone copolymers, and silphenylene copolymers.

The silicone waxes are largely insoluble in water, ethanol and organic fats and oils, and the compositions containing them have improved viscosity, stability and a sharper melting point. They are said to be superior to compositions with no silicone wax, retaining their form plus all the properties of the silicone wax over a wider temperature range. They are also said to have good shelf stability.

From the foregoing desciption of the properties of the various materials it will be seen that no one or two materials are able to provide all the properties and qualities required in a lipstick; this serves to explain the almost invariable complexity of lipstick formulae, which is evident in the examples given below.

Perfumes

Special attention must be paid to the choice of perfume, which is frequently used in relatively high amounts (2–4 per cent), from the point of view of consumer acceptance and freedom from irritation. The perfuming of lipsticks was discussed by Vasic[8] and was also the subject of an article in *Dragoco Report.*[9]

Perfumes selected should mask the fatty odour note of the base and should be non-irritant to the lips. Since the consumer is likely to apprehend the perfume in the mouth as well as the nose the flavour must be considered as well as the odour. Perfumes should be stable and compatible with the other constituents of the lipstick base.

The preferred perfumes are of the light floral or light sophisticated type with no single essential oil predominating. Rose alcohols and esters are often used as well as other essential oils, preferably terpeneless, such as aniseed, cinnamon, clove, lemon, orange and tangerine, although definite fruity flavours have not proved very popular when tried in lipsticks. Geranium, patchouli and petitgrain oils are not considered to be suitable components of lipstick perfumes.

Example Formulations

The following examples illustrate some of the many lipstick formulations in existence.

	(1) *parts*
Loramine O.M. 101	20·0
Lanolin	10·0
Cocoa butter	5·5
Refined beeswax	4·0
Ozokerite	18·0
Carnauba wax	4·2
Oleyl alcohol	7·0
Mineral oil (high viscosity)	29·3
Perfume	2·0 or *q.s.*
Insoluble lake	10·0 or *q.s.*
Bromo acid (or eosin salt)	2·0

(Cited from *Loramine Wax O.M. 101*, Dutton and Reinisch)

The use of silicones in lipstick base formulations is illustrated by example 2.

	(2) *per cent*
Castor oil or bromoacid solvents	30·0
Mineral oil	15·0
Beeswax	15·0
Paraffin	10·0
Carnauba wax	10·0
Ceresin wax	10·0
Union Carbide L-45 Silicone Fluid (1000 cS)	10·0
Perfume–flavour compound	*q.s.*

(*Union Carbide Bulletin* CSB 45–176 4/64, Formula C-106)

Two lipstick base formulae containing hexadecyl alcohol as bromoacid solvent, from technical bulletins of the Enjay Chemical Co., are given in examples 3 and 4.

	(3)
	per cent
Hexadecyl alcohol	26·0
Castor oil	20·0
Propylene glycol monolaurate	15·0
Anhydrous lanolin	2·0
Ceraphyl 28	5·0
Candelilla wax	32·0

Procedure: Heat the ingredients together until molten. Add 7 parts of pigment to 92 parts of base, and run the resultant mixture through a heated mill. Add 1 part of perfume and run the final blend into the moulds.

	(4)
	per cent
Hexadecyl alcohol	44·0
Butyl stearate	2·0
Isopropyl palmitate	3·5
Anhydrous lanolin	7·0
Petrolatum	12·0
Candelilla wax	11·0
Carnauba wax	11·0
Stearic acid, triple pressed	8·0
Cetyl alcohol	1·5
Nordihydroguaiaretic acid (antioxidant)	0·02
Citric acid	0·006

Lipstick formulae using Dehydag products (examples 5–8 below) have been suggested by Henkel.

	(5)
	per cent
Stearyl alcohol	7·00
Beeswax (bleached)	7·00
Stearic acid	1·75
Paraffin wax (MP 72°C)	12·25
Anhydrous lanolin	2·80
Carnauba wax	2·80
White mineral oil	1·40
Comperlan HS	20·00
Eutanol G	45·00
In addition:	
Bromoacid	1·50
Pigment colour	6·00

Procedure: Dissolve the bromoacid in Comperlan HS on the water bath at approximately 95°C. Blend or dissolve the pigment colour in Eutanol G. Melt the remaining fatty and waxy substances on the water bath and add to the two above-mentioned compounds. Pass the finished mixture through a roller mill while liquid, and finally pour into moulds. The finished sticks are then passed quickly through a small flame.

	(6)
	per cent
HD-Eutanol	20·0
Castor oil	24·8
Beeswax (bleached)	5·0
Carnauba wax	8·0
Ozokerite wax (70°–72°C)	11·0
Candelilla wax	3·0
Lanolin, anhydrous	8·0
Liquid paraffin	13·0
Bromoacid	0·2
Pigment colour	7·0
Antioxidant, perfume	*q.s.*

	(7)
	per cent
Eutanol LST	10·0
Castor oil	47·0
Cetyl alcohol	1·5
Beeswax (bleached)	5·0
Candelilla wax	12·0
Lanolin anhydrous	13·0
Bromoacid	0·2
Pigment colour	11·3
Antioxidant, perfume	*q.s.*

	(8)
	per cent
Eutanol LST	20·0
HD-Eutanol	14·0
Cetyl alcohol	5·0
Castor oil	20·0
Beeswax (bleached)	6·0
Lanolin, anhydrous	14·0
Ozokerite wax (70°–72°C)	15·0
Carnauba wax	5·0
Bromoacid	1·0
Antioxidant, perfume	*q.s.*

Procedure: Dissolve the bromoacid in the mixture of HD-Eutanol and Eutanol LST at 80°C. Melt the remaining fatty substances on a water bath, add the appropriate pigments and add the whole mixture to the bromoacid solution. Add perfume and antioxidant at 60°C. Pass the prepared mass two or three times through a roller mill, then remelt and pour into moulds at 80°C. For lipsticks without pigments, no rolling is necessary.

A deep-staining lipstick formula put forward by Croda, includes a polyoxyethylene oleyl ether, Volpo N.3 (Croda) as a bromoacid solvent, used in conjunction with oleyl alcohol and castor oil (example 9). The resultant lipstick is said to possess superior gloss (without flaming), plasticity and moulding properties. It is further claimed that the combination of Volpo N.3 and Novol in the stick provides pigment-dispersing action and helps to prevent titanium dioxide streaking.

	(9) *per cent*
Volpo N.3	10–20
Bromoacid dyes	1–3
Novol	20–10
Castor oil	40
Candelilla wax	10
Carnauba wax	10
Perfume	*q.s.*
Insoluble pigments and lakes	*q.s.*

Procedure: Dissolve the bromoacid in the Volpo N.3 with heating to facilitate solution; add the oleyl alcohol to the resulting solution, which remains clear. Now add the castor oil and the waxes at an elevated temperature; it is claimed that the bromoacid will not precipitate. Add the pigments and mill the mass.

Croda have also developed a series of synthetic waxes known as Syncrowaxes, which can be used instead of naturally occurring waxes in stick and salve-based cosmetics. Three lipstick formulae are given below:

	(10) *per cent*	(11) *per cent*	(12) *per cent*
Paraffin wax 145	—	3·0	—
Syncrowax HRC	5·0	—	5·0
Paraffin wax 125/130	—	—	13·0
Syncrowax PRLC (or ERLC)	6·0	6·0	7·5
Syncrowax HGLC	9·0	6·0	7·5
Castor oil	54·4	66·1	57·2
Crodamol IPP	15·0	6·3	—
Mineral oil	—	—	5·0
Timica brilliant gold	—	10·0	—
BHT	0·1	0·1	0·1
Perfume	0·5	0·5	0·5
Colour blend (in castor oil)	10·0	2·0	4·2

(cited from *Syncrowaxes in Cosmetics*, Croda Chemicals Ltd)

Procedure: Melt the waxes and oils and heat to 85°C. Remove from heat source; add colour blends, Timica and preservative. Re-warm with stirring to 70°C. De-aerate in 80°C water bath for 20–30 minutes. Add perfume. Pour into slightly warm mould. Leave to set.

It is also claimed that a softer stick with increased pay-off can be produced by replacing some of the PRLC or HGLC with HRC. The Crodamol IPP should be replaced by Crodamol OP or Crodamol ML when the final product is to be packed in polystyrene.

Example 13 is a lipstick based upon a non-self-emulsifying grade of glyceryl monostearate (Tegin 515).

	(13) *per cent*
Tegin 515	42·0
Castor Oil	36·0
Bromoacid	5·0
Mineral oil	6·0
White petroleum jelly	4·0
Carnauba wax	4·0
Lanolin	3·0

(Cited from *Emulsifiers*, Goldschmidt Chemical Corp.)

Another deep-staining lipstick formula suggested by Croda contains:

	(14) *per cent*
Polychol 5	20·0
Novol	20·0
Candelilla wax (68°/70°C)	10·0
Carnauba wax (85°C)	10·0
Castor oil	40·0
In addition:	
Indelible colours	1–3

Lipstick formulae proposed by Fishbach[10] contained:

	(15) *Deep stain lipstick* *per cent*	(16) *Creamy lipstick* *per cent*
Castor oil	60·0	65·0
Lanolin	5·0	10·0
Isopropyl myristate	—	5·0
Propylene glycol monoricinoleate	10·0	—
Polyethylene glycol 400	5·0	—
Beeswax	7·0	7·0
Candelilla wax	7·0	7·0
Carnauba wax	3·0	3·0
Ozokerite	3·0	3·0
In addition:		
p-Hydroxybenzoic acid propyl ester	0·2	0·2
Bromoacids	3·0	3·0
Colour lakes and pigments	12·0	12·0

Manufacture of Lipsticks

Lipstick manufacture is by no means a simple operation. The method employed will depend to some extent on the formulation and the plant available.

In general, the manufacture of lipsticks consists of three stages: (i) the preparation of component blends, that is the oil blend, colour dispersion and wax blend; (ii) the blending of these intermediates to form the lipstick mass; (iii) the moulding of the lipstick mass into sticks.

Preparation and Blending

The colours, that is the bromoacids, pigments and lakes, may be (a) blended separately with appropriate constituents of the base mixture, or (b) dispersed in the complete lipstick base—a procedure which is seldom employed, except when a proprietary complete base is used.

The object of the operation is to produce a completely homogeneous dispersion of colorants conducive to the preparation of a smooth stick. Although the colours used are usually received in a finely ground form, they nevertheless tend to be difficult to wet with the fatty mixture and it is necessary to employ some type of milling or grinding process. A variety of types of mill—ball mills, sand mills, roller mills, colloid mills, edge or end runner mills and others—has been used. It is desirable that the portion of the base used should be a plastic semi-solid during the milling and it may sometimes be necessary to mill under warm conditions.

The soluble dyes are first blended with the dye-solvent materials, using heat if necessary to secure solution. If solution is complete, this portion can be left on one side while the pigments are dispersed, but if not, the dye blend must at some point be ground with the pigment dispersion.

In preparing the pigment dispersion the colours are first milled with those constituents most likely to wet them, for example lanolin, polyglycol compounds or the like, and only later are the completely non-polar materials added.

The remainder of the base materials, for example the high melting waxes which may not have been used in the dye or pigment blends, are melted and blended together and then blended with the others. In order to secure an intimate mix a final milling operation is desirable.

Jakovics[11] suggested that vacuum processing of the lipstick melt facilitates dispersion of pigments by removing the film of adsorbed gas on the pigment particles, which otherwise acts as a barrier and causes incomplete wetting. It has been established elsewhere that petroleum jelly, for instance, will dissolve 10 per cent air by volume at 90°C but only 5 per cent at 20°C. If, therefore, the mix has been held at high temperature and allowed to absorb its full quota of air and then cools, the air is not released until the viscosity of the mass is too great to allow it to escape completely and it may collect round the pigment particles, displacing the oil and giving the effect of incomplete wetting.

The manufacture of lipsticks, therefore, should be conducted at the lowest temperature convenient for the process and it is very desirable that when mixing is complete the mixture should be transferred to a jacketed closed vessel in which it can be kept fluid and stirred gently while sufficient vacuum is applied to remove all occluded air. When the air has been removed, vacuum is shut off, the mass is stirred and the perfume is added and mixed through.

The mass is then ready for moulding immediately or can be set into blocks which are stored and moulded as required.

Lipstick manufacture in modern plant was also discussed by Daley.[12] In one of the two methods described, the wax blend is prepared and stored in a molten state while all constituent oils are metered into a storage tank fitted with a suitable mixer. A part of the oil blend is then mixed with the dry pigments in a pre-mix pan attached to a sand mill and the resultant slurry passed through the sand mill into the steam pan fitted with a combined mixer and pump. The remainder of the oil blend is then passed through the pre-mix pan and sand mill to flush any residual colour.

The wax blend is metered into the steam pan simultaneously and thoroughly mixed with the other constituents. The resulting base is tested for shade, perfume and flavour are added, and the lipstick mass is pumped into storage trays, using the mixer as a pump.

Moulding

When required, the lipstick mass is gently remelted in a small jacketed pan and agitated slowly for about 30 minutes, in order to allow any entrapped air to rise to the top and thus prevent pinholing of the finished product. The molten mass is then run into moulds for casting.

Moulds are usually filled to excess to prevent the formation of a depression in the centre of the stick. After they have been allowed to stand to allow this excess to congeal, the latter is scraped off and the mould is then carefully cooled to allow the mass to set, without overcooling which would delay the removal of the stick from the moulds.

The moulds are made of brass and aluminium. They may be of the vertical split type (with a split down the centre to allow for the easy removal of the sticks) or the automatic ejection type. A cooling cabinet is required with the split moulds. A preheating device is stated to be useful in that the moulds can then be raised to a temperature of about 40°C before filling with the lipstick mass, thus avoiding 'flow marks' which would otherwise be visible on the moulded stick. With the water-jacketed automatic ejection mould a cooling cabinet is unnecessary. The jacket is filled with warm water prior to pouring the lipstick mass, and then the cold water is introduced and kept running to chill the lipstick mass. When cooled, the moulds are opened and the sticks are pushed out automatically or with a rubber-covered finger.

It has been stated that no type of automatic ejection mould (whether air-cooled or water-cooled) will produce a good 'bullet'–shaped stick because a small ring or ridge is left near the tip of the stick caused by the thickness of the metal edge of the ejection plunger. It is considered that such moulds are best for the wedge-shaped stick. After moulding, the sticks should be stored for about a week before being filled into the lipstick holders, after which they are subjected to a procedure known as 'flaming', in which the lipstick is passed rapidly through a small gas flame to melt the surface layer, in order to remove any surface spots and to produce a bright, smooth and glossy surface. In large plants the procedure is fully mechanized.

A glazing apparatus for meltable products such as lipsticks was claimed in one British patent,[13] and a process for the manufacture of multi-coloured lipsticks was described in another.[14] A comparison between split moulding and automatic ejection moulding has been drawn by Dweck and Burnham.[15]

To overcome the problem of holding stocks and blending multiple materials, a lipstick base can be purchased ready blended from supply houses. This has adequate solvent powers for eosin dyes and it is only necessary to dissolve the dyes, mill in the pigments, melt, add perfume and mould.

Transparent Lipsticks

In a US patent published in 1964[4] and assigned to Yardley of London, a lipstick was described which does not contain any insoluble opaque pigments or lakes, but instead uses soluble or solubilized dyes. This allows light to shine through it, giving it a sparkle. The staining action of these dyes is enhanced by the use of suitable solvents such as a monoalkylolamide of mixed fatty acids (Loramine OM-101) or dipropylene glycol methyl ether.

Water-soluble dyes may be used for additional colour effect or instead of oil-soluble dyes. The compatibility of such dyes with the vehicle may be improved by using anhydrous lower alcohols such as ethanol or isopropanol as co-solvents. If used in amounts between 2 and 10 per cent, these alcohols are claimed to control syneresis, thereby improving the stability of the lipstick.

For moulding lipstick disclosed in the patent, polyamide resins with a molecular weight range of about 2000 to 10 000 are used, or preferably a blend of a solid polyamide resin of molecular weight of 2000–10 000 together with a small proportion of a liquid polyamide resin of a molecular weight of 600–800 to give a solid stick. Representatives of these resins are the Versamids of General Mills Inc. or the Omamids of Olin Mathieson Chemical Corporation.

To ensure that the resulting sticks are not brittle, these materials are used with softening agents such as the lower aliphatic alcohols in conjunction with other polyamide solvents such as fatty acid esters, for example the glycol ester of higher fatty acids, particularly those between C_{12} and C_{18}. Special mention is made in this connection of propylene glycol monolaurate, polyethylene glycol (400) monolaurate, castor oil, lauryl lactate and fatty alcohols. Four formulations are given in the patent to illustrate transparent lipsticks. In one of the examples, the lipstick composition is as follows:

	(17) *per cent*
Polyamide resin (average MW 8000)	20·0
Polyamide resin (average MW 600–800)	5·0
Propylene glycol monolaurate	28·0
Castor oil	12·6
D&C Red 21	0·3
Lanolin alcohols	8·0
Dipropylene glycol methyl ether	10·0
Ethoxylated lanolin alcohols (5 mol Et_2O)	10·1
Anhydrous isopropanol	5·0
Perfume	1·0

Lip Salves

The lip salve is used, not for decoration, but for protection against exposure to cold, in winter or sub-arctic conditions. The requirement is simply for a fairly substantial, flexible, adherent, moisture-resistant film on the lips and there is no need for staining dyes and hence none for dye-solvent materials. The base can be made largely from mineral oils, jelly and waxes, but it is necessary to include a proportion of a more hydrophilic material to promote adherence, perfume blending and general properties. In some cases a small amount of antiseptic can be added, and occasionally some users will prefer a coloured salve in which case the colour is provided by a small amount of oil-soluble dye or dispersed lake not necessarily of a staining type. Suitable base formulae are given in examples 18 and 19.

	(18)
	per cent
Paraffin wax	30·0
White petroleum jelly	35·0
Technical white oil	20·0
Beeswax	15·0

	(19)
	per cent
Cetyl alcohol	5·0
HD-Eutanol	30·0
Beeswax, white	25·0
Paraffin wax, 52°C	15·0
Liquid paraffin	25·0

A lip-protective pomade from a Union Carbide Bulletin (formula C-105) is made up as follows:

		(20)
		Parts
A	Mineral oil	25·0
	Petroleum	7·0
	Ceresin wax	24·0
	Beeswax	7·0
	Atlas G-2859	3·5
	Tween 60	0·5
	Union Carbide L-45 Silicone Fluid (1000 cS)	5·0
B	Water	28·0
	Preservative	*q.s.*
C	Perfume	*q.s.*

Procedure: Heat A to 65°C; heat B to 70°C and add to A with stirring. Continue stirring and add perfume at 50°C. Pour into heated moulds.

A lip salve which is used as a decorative item has become known in recent years as a lip gloss. This popular item can be applied to the lips without other make-up or over the normal lipstick. Lip gloss preparations take the form of a stick, or a salve for finger application, or a liquid for roll-on or brush application. These compositions may also contain a pearl pigment to confer a transparent sheen in addition to the normal gloss.

Finger-applied lip gloss	(21) *per cent*
Syncrowax HRC	5·0
Syncrowax HGLC	5·0
Timica silk white	8·0
Fluilan	8·0
Liquid paraffin	74·0
Perfume, antioxidant	*q.s.*

(Cited from *Syncrowaxes in Cosmetics*, Croda Chemicals Ltd.)

Procedure: Melt waxes. Add Fluilan and liquid paraffin. Disperse the Timica with low shear and de-aerate at 65°C in a water bath. Fill off at 50°C.

Liquid lip gloss	(22) *per cent*
Polybutene	30·0
Lanolin oil	20·0
Mineral oil	24·8
Oleyl alcohol	25·0
Saccharin	0·2
Perfume, antioxidant	*q.s.*

Liquid Lipsticks

In addition to conventional lipsticks, other methods for applying colour to the lips have also been proposed. Several patents have been issued, dealing for example with lip paint or paste applicator, lipstick brush applicator and others.[16–18] In 1959, a roll-ball lipstick was introduced in the USA.[19]

Liquid lipsticks have been developed with the object of providing more permanent films than can be obtained with conventional lipsticks. These liquid lipsticks consist of alcoholic solutions of alcohol-soluble dyes, suitable film-forming resins and plasticizers. The solvent employed is ethyl alcohol; the film-formers include ethyl cellulose, polyvinyl alcohols and polyvinyl acetate and, as plasticizers, triethyl citrate, dioctyl acetate, methyl abietate or polyethylene glycols have been used. The dyestuffs employed are alcohol-soluble halogenated fluoresceins and also other alcohol-soluble dyes.

The dyes and perfume are dissolved in the alcohol before combining with the film-former and other constituents of the preparation. The resulting solution is then filtered. The disadvantage associated with the use of these alcoholic preparations is 'smarting' produced on the lips on application. Excessive drying of the lips may also result. Persons affected in this way should discontinue the use of these preparations, apply occasionally a 40 per cent solution of aqueous glycerin and use an emollient lip salve.

The manufacture of liquid lipsticks based on ethyl cellulose is covered by a patent[20] from which the following illustrative examples are quoted.

	(23) *per cent*
Ethyl cellulose	3·1
Ethyl alcohol	68·4
Petroleum ether	20·0
Hydrogenated methyl abietate	7·5
Rhodamine	1·0

	(24)
	per cent
Ethyl cellulose	3·06
Bleached wax-free shellac	4·94
Ethyl alcohol	65·00
Petroleum ether	14·66
Hydrogenated methyl abietate	12·00
Fuchsine	0·30
Saccharin	0·04

According to the patent quoted, accessory agents such as perfumes, preservatives, antioxidants, etc., may be added. The preparations are preferably clear and transparent and are of a viscosity ranging between 3 and 500 cP, while best results are obtained in a range between 20 and 50 cP.

Another lipstick formulation put forward by Janistyn[21] had the following composition:

	(25)
	per cent
Ethyl cellulose	1·0
PVA (high viscosity)	3·0
Triethyl citrate	1·0
Alcohol-soluble lanolin	0·5
PEG 400	1·0
Isopropyl alcohol, pure	40·0
Bromoacid dyes	3·0
Flavour	0·5
Ethanol	50·0

ROUGE

Introduction

Rouge is one of the oldest types of make-up preparation used to apply colour to the cheeks. The Hittites used cinnabar for this purpose, the ancient Greeks coloured their cheeks with a root and the Romans were known to use a seaweed to impart a rosy tint to pale cheeks.

In Elizabethan days, red ochre, vermilion and cochineal were used as rouge, as well as extracts of sandalwood and brazilwood. Cochineal was still the standard material for rouge in the eighteenth century. In the late nineteenth and early twentieth centuries liquid rouge made from ammonia and carmine was popular, while in theatrical rouges the red pigment carthamine and the aluminium lake of the dye obtained from brazilwood were extensively used to enhance the facial colouring of actors under the glare of footlights on the stage.

In the early 1920s liquid rouges still consisted of carmine and ammonia, whereas grease rouges were composed of carmine dispersed in a tallow and ceresin base. Also available was dry rouge, prepared by mixing carmine solution and eosin with pumice, chalk and gum arabic. Another preparation employed alloxan in a wax base. The precursor of the modern compact rouge at the beginning of the twentieth century was a small book of thin paper leaves coated with various shades of red and white powder, which were detached and rubbed on the cheeks.

Such are the vagaries of the cosmetics industry that the term 'rouge' has become old-fashioned and somewhat passé. The more modern description currently in vogue is 'blusher', but in basic formulation terms the two are synonymous and, for the sake of continuity, we will continue with the older name. If a distinction had to be made between the two terms, it could be said that a rouge produces a bright red colour, whereas a blusher produces a duller, more brown effect. The preparations are available in liquid, cream and solid forms, of which the pressed powder or dry rouge is the most popular.

Dry Rouge (Compact Rouge)

Compact rouge differs from an ordinary compact powder in being more highly tinted. The desirable properties of compact rouges are therefore virtually identical with those of ordinary compact powders (Chapter 18). The finished product must be smooth and free from grittiness and should be easy to apply; it should have good adhesion to the skin and a good covering power. In addition however, because these products are highly tinted and may be conspicuous if not adequately blended, it is essential that component colours are distributed evenly throughout the product. Care must also be taken to ensure that any component liable to impart undesirable opacity is used in moderate amount so that it has no deleterious effect on the appearance of the product; consequently zinc oxide is used in compact rouges at lower concentrations than in ordinary compact powders. Similar stipulations apply to components such as kaolin, which by virtue of their hygroscopic nature are liable to produce colour streaking in damp weather.

Compact rouges may thus be manufactured broadly along the lines discussed for compact powders in Chapter 18. In addition, grinding and shade-matching stages are included in the manufacture of compact rouges. The component raw materials must be very finely divided to facilitate their intimate blending as a prerequisite to a uniform distribution of component colours. The constituents are, therefore, intimately mixed and simultaneously ground in hammer mills or attrition mills rather than just blended in ribbon mixers. This operation is carried out in the absence of large quantities of liquids, though the presence of a little water or oil will ensure better colour distribution. The binder is either sprayed into the powders while they are being mixed, or incorporated in dry form.

The raw materials used in the manufacture of compact rouges are talc, kaolin, precipitated chalk, magnesium carbonate, titanium dioxide, zinc stearate, inorganic oxides, certified colorants and perfumes. Zinc oxide, used to increase the adhesion of the rouge, confers opacity to the product and is usually employed in amounts of between 5 and 10 per cent.

Titanium dioxide has an appreciably greater covering power than zinc oxide, and it also gives brighter and more stable colour shades.

Metallic stearates are also essential components of compact rouges and improve the adhesion of the products to the skin. They are nowadays used extensively as dry binders for compact rouges. Amounts employed range between 4 and 10 per cent.

Compact Rouge Formulation
Typical compact rouge formulations are given in examples 26 and 27.

	(26) *per cent*
Talc	48
Kaolin	16
Zinc stearate	6
Zinc oxide	5
Magnesium carbonate	5
Rice starch	10
Titanium dioxide	4
Colours	6

	(27) *per cent*
Talc	67·5
Zinc stearate	5·0
Titanium dioxide	4·0
Red iron oxide	11·5
Black iron oxide	0·3
D&C Red 9—barium lake	0·2
Mica coated with titanium dioxide	6·8
Perfume	0·2
Base	4·5
Base:	
Beeswax	12·00
Lanolin	2·00
Mineral oil	86·00

The proportion of colouring matter may vary from $1\frac{1}{2}$ per cent in the case of the lighter shade to some 6 per cent with the deeper shades.

Colour lakes are the most extensively used pigments because of the wide range of shades that they provide; water-soluble and oil-soluble dyes, on the other hand, are not suitable as colorants in compact rouges.

For medium-red shades Keithler[22] suggested the following blend of pigments:

	parts
D&C Red No. 8 Na lake	2·2
D&C Red No. 19 Al lake	2·0
D&C Red No. 10 Na lake	2·0
D&C Red No. 11 Ca lake	1·1
D&C Red No. 34 Ca lake	0·7

D&C Red Nos 10 and 11 are now delisted but can be replaced by D&C Red No. 9—barium lake.

A variety of binding agents may be employed, as already mentioned in connection with compressed powders. For example Winter[23] has suggested 13–15 per cent of a starch–stearin mixture as binding agent, or alternatively a soap–oil mixture of about 3 per cent sodium stearate or triethanolamine stearate and 2 per cent oil may be used. However, compact rouges with incorporate water-soluble binders are subject to spotting by water. Consequently, water-repellant binders are preferred and particularly dry metallic stearates. If the latter are used, it will be necessary to employ higher pressures during the compression of powders.

Shade matching may be carried out in a ribbon mixer provided that colour blends are readily available in several concentrations.

In compressing rouges, the powder is pressed into godets by being subjected to a gradually increasing pressure.

Wax-based Rouge

Wax-based rouges are in many respects similar to lipsticks. Formulae for bases resemble the drier type of lipstick; shape is generally similar although frequently of larger diameter (about 2 cm) and colours are generally of a red or pink shade, although ochres and tans and even white are also used to disguise the contours of the face, when the products are known as 'gleamers'. The use of staining dyes of the bromoacid type is inadvisable, owing to the possibility of inducing eosin dermatitis with the greater exposure to light, and this renders it unnecessary to include any materials specifically for their eosin-solvent properties.

The following are examples of formulae for wax-based rouges:

	(28) *per cent*
Candelilla wax	8·6
Carnauba wax	5·4
Beeswax	4·0
Mineral oil	17·0
Lanolin	2·0
Isopropyl myristate	33·0
Inorganic pigments and colour lakes	30·0

	(29) *per cent*
Carnauba wax	10·0
Paraffin wax	2·5
Ozokerite wax	5·0
Petrolatum	4·0
Mineral oil	68·5
Titanium dioxide	1·9
Red iron oxide	1·0
Bismuth oxychloride	7·0
Perfume	0·1

Cream Rouge

Two types of cream rouge are available: anhydrous products and emulsified products.

Non-aqueous Cream Rouge

Although cream rouges are more difficult to apply than compact rouge, it is claimed that their use gives better results, in that there is no obvious line of demarcation when they are correctly applied. This gives a more natural effect.

An effective anhydrous cream rouge may be prepared to the following formula:

	(30) *per cent*
Petroleum jelly	70·0
Kaolin	30·0
Colour, perfume	*q.s.*

Procedure: Grind the colours with the powder constituent, and then mill this into the warmed fatty base; alternatively, mix all the ingredients and mill in a warmed ointment mill.

Hexadecyl alcohol is claimed to impart a smooth and soft feeling to the skin during application and use, and also facilitates the spreading of a cream rouge in which it is present. Such an anhydrous rouge may, for example, be prepared with the composition given in example 31.[24]

	(31) *per cent*
*Hexadecyl alcohol	27·0
Light mineral oil	23·5
Talc	10·0
Ozokerite	10·0
Carnauba wax	6·0
Titanium dioxide	20·0
D&C Red No. 9 lake	3·0
Perfume	0·5

*Enjay Chemical Co.

Procedure: Melt thc constituent waxes and blend them with the remaining constituents; pass through a roller mill and pour into containers.

Of the following formulae for cream rouge selected from the technical literature, example 32 is given by Thomssen.[25]

	(32) *per cent*
White beeswax	5·0
Stearic acid	7·0
Cetyl alcohol	3·5
Petroleum jelly (short fibre)	77·0
Mineral oil	7·5
Colours	*q.s.*

	(33) *per cent*
Eutanol G	25·0
Castor oil	45·0
Liquid paraffin	4·0
Beeswax, white	2·0
Ozokerite 70°–72°C	5·0
Carnauba wax	6·0
Candelilla wax	5·0
Pigment colours	8·0
Antioxidant, perfume	*q.s.*

(Cited from *Cosmetic Preparations based on Dehydag Products*, Henkel).

Emulsions

Emulsified products are either of the cold cream or of the vanishing cream type.

An emulsified form of cream rouge of the cold cream type may be prepared by the interaction of beeswax with an aqueous solution of borax:

	(34) *per cent*
Lanolin	5·0
Cocoa butter	5·0
Beeswax	14·0
Liquid paraffin	30·0
Cetyl alcohol	1·0
Water	44·2
Borax	0·8
Colour	*q.s.*

Procedure: Disperse the finely powdered colour base with the melted fats, add the hot aqueous solution of borax at about 75°C (both fats and water being at the same temperature to prevent crystallization), mix well and finally mill.

The cream rouge given in example 35 is of the water-in-oil type, the emulsifier being Arlacel C (sorbitan sesquioleate). It is claimed[26] that this formula provides a product which is soft and creamy, and which possesses excellent spreading properties. Because it is of the water-in-oil type it has less tendency to dry out than emulsified creams of the oil-in-water type.

Oil phase	(35) *per cent*
Arlacel C	2·0
Lanolin (anhydrous)	2·0
Mineral oil 65/75	16·0
Petrolatum	30·0
Preservative	*q.s.*
Aqueous phase	
*Brilliant red C-10-013	10·0
Arlex	5·0
Water	35·0
Perfume	*q.s.*

*Ansbacher

Cream rouge of the vanishing cream type may be prepared as in example 36.

	(36)
	per cent
Stearic acid	15·00
Water	76·32
Potassium hydroxide	0·50
Sodium hydroxide	0·18
Glycerin	8·00
Colour, preservative	*q.s.*

If water-soluble dyes are used in these emulsion products, they may be dissolved either in a portion of the water or in the glycerin, and added to the main bulk of cream after saponification in the normal manner (see Vanishing Creams, Chapter 4).

In example 37, diglycol stearate is employed as emulsifying agent.

	(37)
	per cent
Diglycol stearate	20·0
Water	70·0
Glycerin	10·0
Water-soluble colour	*q.s.* (from 0·6 to 1·0)
Preservative	*q.s.*

The difficulty in all cases where a water-soluble colour is introduced is to prevent evaporation, which leads to a darkening of the cream surface owing to concentration of the colour. When such colours are used it is necessary to incorporate sufficient hygroscopic constituents, such as glycerin, the glycols, etc; *d*-sorbitol would appear to offer interesting characteristics for this purpose, since it possesses a much narrower humectant range than glycerin, that is, it absorbs less moisture in a moist atmosphere, but holds it better in a dry one. A further improvement might be the packing of such products in a tightly capped tube.

In general, however, most of the colours employed in cream rouge are insoluble lakes, although sometimes oil colours are used. In the vanishing cream type, however, these lakes may be supplemented if desired by fluorescein dyestuffs when, as stated above, it is necessary to incorporate a non-volatile solvent.

A further example of a vanishing cream type of rouge, given by Janistyn,[27] has the following composition:

	(38)
	per cent
Cetyl alcohol	2·00
Stearin	18·00
Propylene glycol	4·00
Isopropyl myristate	8·00
Carbowax 4000	4·00
Potassium hydroxide	0·70
Water	54·65
Perfume	0·50
Preservative	0·15
Lakes	8·00

Liquid Rouge

Liquid rouges appear to maintain a degree of favour and are claimed to yield extraordinarily good results when skilfully applied. The difficulty of applying them has been overcome by fitting a small wick in the neck of the container, the rouge being readily transferred by capillary attraction to the lip or cheek.

The essential requirements of such a preparation are that it has sufficient viscosity to permit easy and even application and that it dries reasonably rapidly.

Aqueous preparations are prepared by dissolving the requisite amount of a suitable water-soluble dye, adding a gum or synthetic thickener to increase the viscosity of the solution and including a small proportion of a suitable wetting agent to promote easy spreading. A little glycerin may be included if desired. In example 39, the methylcellulose may be replaced by 0·4 per cent gum tragacanth.

		(39) *per cent*
Methylcellulose		2·0
Wetting agent		0·1–0·2
Water	to	100·0
Water-soluble red, preservative		*q.s.*

Another preparation has the following composition:

		(40) *per cent*
Sodium alginate		0·45
Calcium citrate		0·13
Wetting agent (dermatologically acceptable)		0·1–0·2
Water	to	100·0
Water-soluble red, preservative		*q.s.*

Procedure: Dissolve the sodium alginate, dye and wetting agent in about two-thirds of the water, form a slurry of the calcium citrate in the other third portion and add to the alginate solution as soon as thickening occurs on standing. By varying the proportion of the alginate and calcium citrate the viscosity may be altered at will.

In order to shorten the drying time, 10–20 per cent alcohol or more may be added to gum mucilages, but high concentrations must not be used with alginate products otherwise precipitation may occur.

The preparations that have already been mentioned in connection with liquid lipsticks, consisting of a suitable alcohol-soluble dye dissolved in an alcoholic solution of ethyl cellulose or alcohol-soluble resin, with a plasticizer to obtain the requisite flexible film, can also serve as liquid rouge. No wetting agent is necessary, of course, and by the use of suitable dyes more permanent preparations are obtainable than with aqueous mixtures. Since evaporation of alcohol will lead to an increase in the viscosity of the product, it is particularly important that containers should be provided with a well-fitted wick built into, or screwed into, the neck of the container, the whole being enclosed in a screw dust-cap. By this means, ready and even application is obtained with the minimum of care and with the least possible chance of covering the user's hands with sticky red varnish.

The resultant rouge film is readily removed with a little alcohol, to which may be added if desired a small proportion of a relatively high-boiling solvent to prevent too rapid evaporation of the rouge remover.

Preparations of the type just described are illustrated by the following formula:

	(41)
	per cent
Tincture of styrax, 5%	38·0
Alcohol, 95%	60·0
Castor oil	2·0
Spirit-soluble red	*q.s.*

The above solutions are prepared by simple mixture. Proportions of film-forming material, plasticizer and dye may be altered to suit particular requirements.

EYE MAKE-UP

Introduction

Eye make-up, too, has been used for thousands of years. The accepted eye make-up used by the women of many ancient civilizations was a black colouring, kohl, based on antimony trisulphide. In addition to kohl, Egyptians used malachite to confer a green tint, while Indian women tinted their eyelids with tsocco which has an antimony base. Chinese and Japanese women used Peruvian bark for eye make-up, and Phoenician women lengthened their eyebrows with a black paste composed of gum arabic, musk, ebony and powdered black insects.

Big eyes were also considered a mark of beauty, and it was the practice of women of these ancient civilizations to make their eyes appear larger and shinier.

Modern eye preparations include mascara, eyeshadow and eyebrow pencils, all of which must be applied sparingly and correctly if the effect is not to be ruined.

These preparations will now be discussed under their respective headings.

Mascara (Eyelash Cosmetic)

> . . . Egyptian women soon discovered that smearing the eyelids with unguent and stibium (that is, powdered antimony) not only eased the pain in the eyes and rested them but also added to the natural beauty of their faces. The whiteness of the whites of the eyes was emphasized by the darkened eyelids and the large dark pupils appeared like black pools in their midst, the whole effect being very striking . . .
>
> (*Dwellers of the Nile*, Sir Wallis Budge)

Mascara is a black pigmented preparation for application to the eyelashes or eyebrows to beautify the eyes. It has been claimed that by correctly applying mascara to darken the eyelashes and to increase their apparent length, the brightness and expressiveness of the eyes is enhanced.

Mascara is marketed in cake (block), cream and liquid forms, and must possess the following characteristics:

(1) It must be capable of easy and even application.
(2) It must have no tendency to run and thereby cause smudging.
(3) It must not cake, causing the eyelashes to stick together.
(4) It must not dry too rapidly to interfere with the evenness of the application, but should nevertheless dry fairly rapidly and be reasonably permanent once applied.
(5) It must be neither toxic nor irritant.

Under the US Food, Drug and Cosmetic Act, the use of coal tar colours in preparations to be applied in the area of the eyes is forbidden, even when those colours have been certified for use in cosmetics. The area of the eyes is defined as 'the area bounded by the supra-orbital ridge and the infra-orbital ridge, including the eyebrow, the skin below the eyebrow, the eyelids, the eyelashes, the conjunctival sac of the eye, the eyeball and the soft areolar tissue that lies within the perimeter of the infra-orbital ridge'. Thus defined, the 'prohibited area' extends from the top of the eye socket to the top of the cheekbone.

Thus in the USA only natural colours, that is vegetable colours and inorganic pigments and lakes (usually black and dark brown pigments), provided that these are in a highly purified form, may be employed in eye make-up preparations.

Pure vegetable colours which can be used include chlorophyll, except for the grade containing copper. The same applies to mineral and earth colours, but in the USA suppliers are required to guarantee the absence of certain impurities from their products; their arsenic content must not exceed 2 parts per million.

Black pigments used in eye make-up preparations are now restricted to black iron oxide (Fe_3O_4). This is sometimes used in conjunction with ultramarine blue to impart blue-black shades.

Umbers (brown ochres), burnt sienna (a mixture of hydrated ferric oxide $Fe_2O_3 \cdot H_2O$ with some manganic oxide) and synthetic brown oxides have been used for brown shades, and yellow ochres for yellow shades. For bluish shades soluble oil blue is employed, while for green shades chromium oxides and for red shades carmine, the aluminium lake of cochineal, have been used. Salts of cobalt may be used only if they are insoluble salts and will not react with other ingredients of the preparation to form soluble cobalt compounds. Finely powdered metals, such as silver and aluminium, have been widely employed in eyeshadow and may be used if they consist entirely of the pure metal. If they contain an appreciable content of copper they are considered to be harmful. White pigments such as titanium dioxide and zinc oxide may also be sometimes included to lighten shades.

Colours that may be used within the EEC for mucous membrane products (including eye products) are listed in Annex III Part 2 of the EEC Cosmetics

Directive. A list of provisionally approved colours is detailed in Annex IV Part 2. Both these lists contain coal tar colours, so the formulator must be careful in considering the colour combinations to be used for individual countries, especially for importation into the USA.

Cake (Block) Mascara

Cake mascara remains a very common form of product, although during the 1960s softer cream or liquid preparations gained in popularity with the development of special container–dispensers with a brush as part of the screw-on cap.

Earlier types of mascara were based on a mixture of soap and carbon black, with sufficient water added to produce a stiff paste which was dried, and then stamped into cakes.

These could be readily applied and were fairly permanent but, since they contained a water-soluble base, tears and even heavy rain were capable of producing black streaks and smudges. Moreover, a trace of soap base in the eye produced extreme irritation. The sodium soaps of coconut and palm oils used originally in these preparations were eventually replaced by the less alkaline and hence less irritant triethanolamine soaps, for example triethanolamine stearate. According to Jewel[28] these products consist essentially of a soap (generally triethanolamine stearate) modified in consistency with oils and waxes, and colour. Modern cakes may be produced by melting together the waxy materials, adding the colour and mixing them in thoroughly to obtain an even distribution. The resultant mixture is then cooled and worked in a heated roller-mill. It is subsequently re-melted in a pan and cast into cakes. A formula has been quoted (example 42) which is claimed to be very similar to a large majority of block mascara preparations sold in the USA.

	(42)
	parts
Stearic acid	27·0
Triethanolamine	12·0
Beeswax	30·0
Carnauba wax	50·0
Colour (bone black)	25·0

Other cake mascara formulae quoted from the technical literature are as follows:[29]

	(43)
	per cent
Triethanolamine stearate	40·0
Paraffin wax	30·0
Beeswax	12·0
Anhydrous lanolin	5·0
Lampblack	13·0

	(44) *per cent*
Glyceryl monostearate	60·0
Paraffin	15·0
Carnauba wax	7·0
Lanolin	8·0
Lampblack	10·0

Procedure: The various components are melted, mixed and then cast or extruded to the proper form, or after mixing they may be milled and passed through a warm plodder, after which the mascara strip is cut to the desired lengths.

Cake mascara formula including a silicone fluid[30]	(45) *per cent*
Triethanolamine stearate	50·0
Carnauba wax	24·0
Paraffin wax (MP45°C)	12·5
Anhydrous lanolin	4·5
*Silicone fluid L-43	5·0
Carbon black	3·8
Propyl-*p*-hydroxybenzoate	0·2

*Union Carbide

With the delisting of carbon black in the USA, even though it is allowed under EEC legislation, the material has become scarce. It can be replaced in most formulations by black iron oxide or with combinations of ultramarine blue and black iron oxide.

Cream Mascara

Cream mascara may be prepared by milling the pigment into a vanishing cream base, or by the use of a suitable oil-soluble dye, but in such cases it is often necessary to include a suitable wetting agent to lower the surface tension, otherwise the colour will not adhere to the brush. The coloured pigment may be incorporated into the base immediately after emulsification has been completed. Agitation is continued while the product is cooling, any entrapped air being allowed to escape. The product is then filled into the tubes.

Alternatively, a previously prepared base is melted in order to incorporate the pigment.

A reliable cream mascara which does not dry up in the tube and which adheres well to the eyelashes can, according to Richardson,[31] be made to the formula given in example 46. A modern equivalent of this formula would employ a synthetic hydrocolloid such as hydroxyethylcellulose to replace the quince seeds.

	(46) *per cent*
Mucilage of quince seeds	35·0
Sugar syrup	35·0
Gum arabic	7·5
Ivory black (or umber)	22·5

Procedure: Grind the gum and the ivory black (or umber, etc.) with the syrup (3 parts sugar to 2 parts water) and then add the mucilage containing a preservative, such as methyl-*p*-hydroxybenzoate.

More modern forms of cream mascara are shown in examples 47[31] and 48[32] and also in example 49[24] which is claimed to give an even application on eyelashes by being free from the tendency to run or cake, and also to have good performance properties.

	(47) *per cent*
Polyethylene glycol 400 stearate	10·0
Diglycol stearate	8·0
Lanolin	3·0
Stearyl alcohol	13·0
Isopropyl palmitate	2·0
Triethanolamine lauryl sulphate	1·5
Water	54·5
Colour	8·0

	(48) *per cent*
Triethanolamine stearate	45·0
Carnauba wax	15·0
Glyceryl monostearate	5·0
Anhydrous lanolin	10·0
Unbleached beeswax	5·0
Lampblack	20·0

	(49) *per cent*
Hexadecyl alcohol	7·3
Propylene glycol	9·1
Stearic acid	11·2
Glyceryl monostearate	4·5
Triethanolamine	3·6
Ultramarine blue	9·1
Methyl-*p*-hydroxybenzoate	0·2
Water	55·0

Another cream mascara formulation includes among its components a silicone fluid;[33] this is easily applied, gives smooth coverage and affords moisture protection:

		(50) *per cent*
A	Stearic acid triple pressed	9·1
	Petrolatum MP 43°C	5·5
	Mineral oil 65/75	4·1
	L-43 Silicone fluid	5·0
B	Triethanolamine	2·75
	Water	64·45
	Pigments	9·1

Procedure: Melt A and heat to 60°C. In a separate container heat B to the same temperature. Add B to A while stirring. Incorporate the pigments in a combined mix.

Greasy preparations, more resistant to removal, may be made by incorporating the pigment or dyestuff, about 10 per cent, into a greasy base, which is warmed slightly for application with a fine brush—examples 51 and 52.

	(51) *per cent*
Beeswax	4·0
Spermaceti	4·0
Cetyl alcohol	2·0
Cocoa butter	6·0
Petroleum jelly	64·0
Oil-soluble blue	20·0
Preservative	*q.s.*

	(52) *per cent*
Cocoa butter (odourless)	5·0
Petroleum jelly	50·0
Paraffin wax (uncrystallizable)	5·0
Lampblack	40·0

Procedure: Melt and mix.

The development of novel brushes and applicators has led to the introduction of speciality cream-type mascaras. For example, waterproof mascaras are very popular at present and there is a variety of applicators available which allow the lashes to be separated and coated evenly with the product without clogging.

Typical waterproof mascara	(53) *per cent*
Isoparaffins	54·7
Ozokerite wax	18·0
Carnauba wax	2·5
Aluminium stearate	2·0
Isopropyl myristate	2·5
Mineral oil	0·5
Kaolin	12·8
Black iron oxide	5·8
Yellow iron oxide	1·2

In some formulations the gelling properties of the aluminium stearate are supplemented with other materials such as bentone. The waterproof properties can be enhanced by the addition of polymer-type materials such as polyethylene or polybutene. Small rayon or nylon fibres can also be added to the formula to give lengthening and thickening effects to the eyelashes.

Liquid Mascara
Satisfactory liquid mascara formulae have long been available but they did not gain wide consumer acceptance until the advent of the special cylindrical container with a small-diameter brush integral in the screw-on lid. This dispenser enhances the convenience of the quick-drying liquid product and is also convenient to carry in the handbag.

The earlier preparations were prepared by suspending the pigment in a mucilage, as in example 54.

	(54)
	per cent
Gum tragacanth	0·2
Alcohol	8·0
Water	83·8
Lampblack	8·0
Preservative	*q.s.*

However, preparations of this type were not very popular, because they tended to be sticky and, being water-soluble, they tended to smudge.

In order to overcome these drawbacks, more modern preparations use an alcoholic solution of a resin in which carbon black is suspended. A little castor oil is often included in such formulae. These dry rapidly, producing a semi-permanent waterproof colour, but are irritant if dropped accidently into the eye. Such preparations may be illustrated by the following example:

	(55)
	per cent
Rosin (10% alcohol solution)	3·0
Castor oil	3·0
Ethyl alcohol	84·0
Lampblack	10·0

Other resins, or ethyl cellulose, may replace rosin if desired. Industrial methylated spirit is best avoided in such preparations as it tends to irritate.

Eyeshadow

Eyeshadows are marketed in a variety of shades, such as brown, green, blue and others. They are applied to the eyelids in order to produce an attractive 'moist'-looking background for the eyes. This is sometimes enhanced, particularly for evening wear under electric lights, by the inclusion of fine glittering metallic particles to produce a 'silver' or 'gold' effect. For this purpose gold leaf, bronze powder and powdered aluminium may be used. Various pearlescent pigments, for example those based on bismuth oxychloride or mica coated with titanium dioxide may also be used. Eyeshadow preparations are available, for example:

(i) as anhydrous creams of the liquefying or emulsifying type;
(ii) in the form of emulsions;

(iii) as stick eyeshadow;
(iv) in the form of liquid suspensions or dispersions;
(v) as pressed powders.

The anhydrous cream form and the stick eyeshadow have been very popular but at present pressed powder eyeshadows are the dominant force in the market place.

Cream Eyeshadow
Cream eyeshadow may be made by mixing first of all the selected colours (and, if pastel shades are required, including in the mixture also a white pigment such as zinc oxide or titanium dioxide) and blending these pigments with petrolatum in a roller mill. This mass is then stirred into the blend of fatty and waxy constituents to be used in the preparation, which have been previously blended by melting them together in a pan. After thorough agitation, the product is then poured into suitable containers while still in liquid form. Alternatively, the pigments in powder form may be stirred directly into the molten mass of fats and waxes, and passed through a roller mill to ensure thorough distribution of colours and remove any lumps from the formulation which may otherwise cause streaking of the product during application. A second mixing operation may follow before the product is poured into the containers. As in the case of mascara, the eyeshadow base should be capable of easy and smooth application. It should also be waterproof to avoid streaking. Sometimes, in order to facilitate liquid filling of the product into containers, the product requires to be kept warm in a jacketed steam-heated pan and, if necessary, gently stirred to prevent the sedimentation of any heavy pigments.

Eyeshadow creams of the anhydrous type may, for example, include cocoa butter, as in the following formulae.

	(56) *per cent*
Cocoa butter (odourless)	2·0
Beeswax	3·0
Spermaceti	5·0
Lanolin	5·0
Petroleum jelly	55·0
Zinc oxide	30·0
Cosmetic lake, preservative	*q.s.*

	(57) *per cent*
Petroleum jelly	75·0
Cocoa butter (odourless)	8·0
Lanolin	7·0
Cetyl alcohol	3·0
Paraffin wax (uncrystallizable)	7·0
Cosmetic lake, preservative	*q.s.*

A formula suggested by Miller[34] contained:

	(58) *per cent*
Bleached beeswax	4·5
Spermaceti	9·5
Lanolin absorption base	13·0
Paraffin of low melting point	73·0

Procedure: Mix 85 parts of this base with (say) 12 parts of titanium dioxide and 3 parts of a mineral pigment or earth colour.

To obtain a grey shade, Miller[34] suggested a mixture of 1 part of ultramarine blue, 1 part of lampblack and 2 parts of titanium dioxide: for a brown shade, a mixture of 3 parts of an umber and 1 part of titanium dioxide. He suggested heating the colours with the mixture of fats and waxes (example 58) and passing them through a roller mill. Another eyeshadow cream formula suggested by the same author contained:

	(59) *per cent*
White petroleum jelly	59·0
Glyceryl monostearate	17·0
Lanolin	4·0
Beeswax	8·0
Candelilla wax	4·0
Pigment	8·0

Emulsion-type eyeshadow creams are produced by mixing suitable pigments into an emulsion and distributing them evenly throughout the base. The products should be filled into containers when cold. Emulsions based on triethanolamine stearate can be used as bases for such preparations (example 60).

	(60) *per cent*
Stearic acid	1·5
Glyceryl monostearate	1·5
Lanolin	4·0
Isopropyl myristate	5·0
Veegum (5% solution)	30·0
Triethanolamine	0·75
Water	38·25
Propylene glycol	4·0
Ultramarine blue	4·5
Black iron oxide	1·2
Chrome hydrate	0·8
Mica coated with titanium dioxide	8·5

The anhydrous creams and the emulsion products were originally intended for finger-tip application. However, the modern trend is to use special applicators and the most popular of these is the sponge tip. It is attached to a plastic rod which is usually an integral part of the cap of the container, as in mascara packaging. The container itself can be either polyethylene or PVC and it is important that a good seal is obtained with the cap and the container to prevent the product, especially the emulsion products, from drying out.

Stick Eyeshadow

The sticks contain a fairly high proportion of waxes such as ceresin, ozokerite or carnauba. A stick eyeshadow suggested by Wetterhahn[35] contained:

	(61) *per cent*
Castor oil	43
Mineral oil 75/85	6
Hydrogenated cottonseed oil	5
White ceresin MP 76°C	26
Carnauba wax	4
Titanium dioxide	8
Iron oxide ochre shade	4
Iron oxide sienna shade	4

Croda (in *Syncrowaxes in Cosmetics*) have suggested the following formula:

	(62) *per cent*
Syncrowax HGLC	15·0
Syncrowax HRC	5·0
Mineral oil	35·2
Liquid base CB3896	15·0
PVP	0·8
Colour	9·0
Timica sparkle	20·0

Procedure: Melt waxes at 85°/90°C. Add liquid base and mineral oil. Maintain temperature at 80°C. Mill PVP and pigments into molten wax phase. Disperse the Timica with low shear agitation, allow to de-aerate and fill off.

Pressed Powder Eyeshadow

Pressed powder eyeshadows can be regarded as compact rouge with a different colour system. In general, they contain much higher colour levels than the rouge products and this has to be taken into account when manufacturing the products. For example, with products containing high levels of an iridescent material such as mica coated with titanium dioxide, care must be taken that, in dispersing the product, the platelets are not shattered by the hammer mills. The product is usually applied by means of a sponge-tipped plastic rod, although finger-tip application is popular—a point to remember when the hardness of the product is being determined during manufacture.

Typical formulae illustrating a matt shade (example 63) and a highly iridescent shade (example 64) are given below.

	(63) *per cent*	(64) *per cent*
Talc	82·5	41·7
Zinc stearate	6·0	7·0
Ultramarine blue	5·4	—
Black iron oxide	0·1	—
Chrome hydrate	2·0	—
Yellow iron oxide	—	2·0
Mica coated with titanium dioxide	—	40·0
Base (see example 27 above)	4·0	9·3

Liquid Eyeshadow
Liquid eyeshadows take the form of either a liquid suspension or a liquid dispersion. In the former case the pigments are suspended in a mixture of oils, but they usually settle and have to be shaken before use. The liquid dispersions are prepared, for example, from dilute solutions of alcohol in water thickened with a synthetic gum such as methylcellulose, suitably preserved and containing a wetting agent in which the pigments have been dispersed. However, liquid eyeshadows are not very popular.

Eyeliner

Eyeliners are preparations for use on eyelids, particularly the upper eyelids, close to the eyelashes, to help to accentuate the expressiveness of eyes. They are available in liquid, cake and pencil form. The compressed powder form is similar in composition to cake mascara. As an example of a dry eyeliner, Fishbach[10] quoted the following composition:

	(65) *per cent*
Mineral oil	5
Colour	30
Titanium dioxide	5
Talc	60

Liquid eyeliners may be used to apply colour to the tissue around the eyes. Usually a brown colour is considered a good colour for daytime wear.

A liquid eyeliner formula cited in *American Perfumer and Cosmetics*[36] contained:

	(66) *per cent*
Water	40·00
Methocel HG 60–50V	1·00
Veegum	1·00
Shellac	1·08
Oleic acid	0·50
Triethanolamine	0·40
Water	3·02
Pigment	18·20
Alcohol SD-40	5·00
Water, preservative	to 100·00

A formula which appeared in the *General Aniline Guide to Cosmetics* contained:

	(67) *per cent*
Veegum (Vanderbilt)	2·5
PVP K-30	2·0
Water	85·5
Pigment	10·0
Preservative	*q.s.*

Eyebrow Pencils

Eyebrow pencils may be either of a wax crayon type pigmented black or brown, in which the pigments are present at a higher level than in cream eyeshadow preparations, or they may be in the form of an extruded eyebrow pencil similar to an ordinary lead pencil, in which a crayon-type formulation has been enclosed in a wooden casing. In the latter form they are even more readily applied than the stick type. Eyebrow pencils are used either to accentuate the line of the eyebrows, or to modify their outlinc after plucking. The pigment is dispersed in a wax base of the lipstick type, in which the proportion of beeswax or other high-melting constituent may be optionally increased to produce a somewhat firmer pencil. However, such sticks must be capable of ready and uniform application and should not be brittle.

	(68) *per cent*
Hydrogenated castor oil	46·0
Hydrogenated cottonseed oil	12·0
Cocoa butter (odourless)	8·0
Castor oil	8·0
Lanolin absorption base	17·0
Black iron oxide	9·0

If a brown stick is required the black iron oxide may be partially replaced by the red and yellow versions.

An eyebrow pencil formula cited from the literature consisted of:

	(69) *per cent*
Ozokerite (MP 70°–75°C)	45·0
Unbleached beeswax	24·0
Cocoa butter	22·5
White petrolatum	6·0
Absorption base	1·5
Anhydrous lanolin	1·0
In addition:	
Pigment or lake colours	10·0
Oil-soluble colour	1·0

Procedure: Stir the colours into the molten fat phase and mill until uniform. Fill the warm batch into the moulds.

Dehydag Products (Henkel) have suggested the following formula:

	(70) *per cent*
Cetyl alcohol	5·0
Cutina GMS	5·0
HD-Eutanol	20·0
Beeswax, white	2·0
Candelilla wax	6·0
Carnauba wax	6·0
Ozokerite wax	24·0
Castor oil	20·0
Liquid paraffin	2·0
Pigment colour	10·0
Antioxidant	*q.s.*

Finally, a formula suggested by Keithler[22] may be prepared by combining the following constituents:

	(71) *per cent*
Beeswax	21·0
Carnauba wax	5·0
Paraffin	29·0
Cetyl alcohol	8·0
Vaseline	18·0
Lanolin	9·0
Pigments	10·0

REFERENCES

1. British Patent 1 206 542, L'Oreal, 23 September 1970.
2. Wilmsmann, H., *J. Soc. cosmet. Chem.*, 1965, **16,** 105.
3. Anon., *Soap Perfum. Cosmet.*, 1944, December, 924.
4. US Patent 3 148 125, Yardley of London, 8 September 1964.
5. Cadicamo, P. A. and Cadicamo, J. J., *Cosmet. Toiletries*, 1981, **96**(4), 55.
6. Gouvea, M. C. de B. L. F. de, *Cosmet. Toiletries*, 1978, **93**(1), 15.
7. British Patent 1 140 536, Dow Corning, 22 January 1969.
8. Vasic, V., *Manuf. Chem.*, 1958, **29,** 431.
9. *Dragoco Rep.*, 1965, (12), 25.
10. Fishbach, A. L., *J. Soc. cosmet. Chem.*, 1954, **5,** 242.
11. Jakovics, M., *Proc. sci. Sect. Toilet Goods Assoc.*, 1956, (26), 9.
12. Daley, P. D. W., *J. Soc. cosmet. Chem.*, 1968, **19,** 521.
13. British Patent 755 549, Metropolitan Vickers Elec. Co. Ltd, 19 February 1954.
14. British Patent 763 733, Chesa, J. M., 20 October 1953.
15. Dweck, A. and Burnham, C. A. M., *Int. J. cosmet. Sci.*, 1980, **2,** 143.
16. US Patent 2 763 881 , Lip Mate Corp., 23 December 1954.
17. US Patent 2 783 489, Bogoslowsky, B., 28 August 1952.
18. US Patent 2 774 984, Morelle, J. C., 7 June 1956.
19. *Drug Cosmet. Ind.*, 1959, **85,** 327.
20. US Patent 2 230 063, Gordon, M. M., 23 January 1939.

21. Janistyn, H., *Taschenbuch der modernen Parfümerie und Kosmetik*, Stuttgart, Wissenschaftliche Verlagsgesellschaft, 1966, p. 609.
22. Keithler, W. R., *The Formulation of Cosmetics and Cosmetics Specialties*, New York, Drug Cosmetic Industry, 1956.
23. Winter, F., *Handbuch der gesamten Parfümerie und Kosmetik*, 2nd edn, Vienna, Springer, 1932, p. 615.
24. Enjay Chemical Co., *Hexadecyl Alcohol for the Cosmetic Industry.*
25. Thomssen, E. G., *Modern Cosmetics*, 3rd edn, New York, 1947.
26. Atlas Powder Co., *Drug and Cosmetic Emulsions.*
27. Janistyn, H., *Taschenbuch der modernen Parfümerie and Kosmetik*, Stuttgart, Wissenschaftliche Verlagsgesellschaft, 1966, p. 616.
28. Jewel, P. W., *J. Assoc. off. agric. Chem.*, 1945, **28,** 741.
29. Westbrook Lanolin Co., *Lanolin—Formulary of Cosmetic and Toilet Preparations*, Section EP, July, 1965.
30. *Union Carbide Technical Bulletin* CSB 45–177 4/64.
31. Richardson. K. N., *Soap Perfum. Cosmet.*, 1945, **18,** 286.
32. Dragoco, Holzminden, *Cosmetic Products and their Perfuming*, p. 124.
33. *Union Carbide Technical Bulletin* CSB 45–178 4/64.
34. Miller, J., *Seifen Öle Fette Wachse*, 1960, **86,** 510.
35. Wetterhahn, J. and Slade, M., *Cosmetics: Science and Technology*, ed. Sagarin, E., New York, Interscience, 1957, p. 291.
36. Shansky, A., *Am. Perfum. Cosmet.*, 1964, **79**(10), 53.

Chapter Twenty

The Application of Cosmetics

Introduction

The use of cosmetic preparations has come a long way since the days when the detectable use of any cosmetic was the sign of an actress or of an 'abandoned woman'. Nowadays there are few women who totally abjure the use of any form of make-up, although the degree of use varies—from a little powder to reduce the shine on the nose, to the face that is such a work of art or craft that apparently every product described in this book must have contributed to it.

The advice offered by beauty columnists and in the various textbooks and cosmetics manufacturers' instructions is apt to prove confusing, but works by Kendall,[1] Anderson[2] and Young[3] are worthy of consultation. In this brief chapter, the technique for the use of beauty preparations has been stripped of all but essentials.

The correct use of cosmetics falls into two parts:

(1) *Skin care*, which has as its object the maintenance of a soft, supple and clean skin and the prevention of effects due to external causes such as excessive exposure to cold, heat, sun, wind, etc.
(2) *Decoration* to produce a pleasing appearance by minimizing facial defects of colour or shape and unobtrusively enhancing and directing attention towards better points.

The first depends to a large extent upon the type and condition of the skin. The possessor of a healthy normal skin is fortunate in that it will withstand many treatments and conditions which can have serious effect on skins which are definitely dry or greasy and which demand particular care and treatment. Even the healthy and normal skin may vary from time to time and need particular care to correct any departure from normality.

The second point—the amount of cosmetic decoration or make-up tolerable—is dependent chiefly on social conventions which vary from time to time and from society to society. The general trend for many years has been towards a greater tolerance, although there have been shorter swings in particular social sets to exaggerate effects in various directions; these are generally short-lived, but are assimilated, often in severely modified form, and contribute to the general trend.

Care and Cleansing of the Skin

It goes without saying that a healthy skin should be a clean skin, and in general a good toilet soap and water is the best way of achieving this cleanliness. A final rinse with clean cold water is an excellent measure to promote circulation and tone up the skin. At night such cleansing should be followed by the application of a cream of more or less greasy type according to the natural dryness or greasiness of the skin.

The cream should be applied by proper massaging movements, whether or not the massage is carried on for long, because friction increases the supply of blood to the skin and skilful massage assists in keeping the skin supple. Charts designed to show the paths which should be followed in facial massage and application of cosmetics have been prepared by cosmetic manufacturers (see Figure 20.1).

The regular use of a correctly formulated face cream in a suitable manner can prevent the premature (not ultimate) aging of the skin due to external causes and delay the appearance of wrinkles produced by loss of epidermal elasticity and subcutaneous moisture. It cannot, however, prevent the natural aging of the

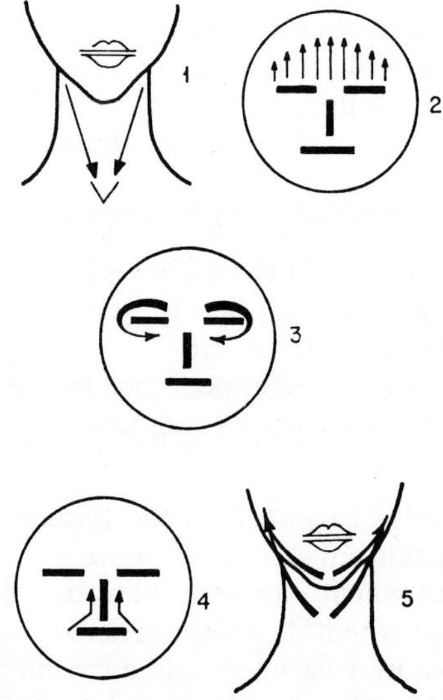

Figure 20.1 Facial massage (Courtesy Harriet Hubbard Ayer)

1. *For the neck*—Using alternately both hands cupped round the neck, firmly massage with downward movements.
2. *For the forehead*—Gently massage the forehead upwards from the eyebrows to the hairline with alternate movements of the hands.
3. *For the eyes*—Both hands working simultaneously, make circles round the eyes, ending by pats on the lower eyelid. Always work in the direction indicated on the diagram.
4. *To smooth the 'smile line'*—Massage with both hands with upward symmetrical movements over the 'smile line', finishing between the corner of the mouth and nostrils.
5. *Against sagging of the lower part of the face*—Both hands working alternately, taking the jaw line between the second and third finger, make upward movements from the point of the chin to the temporal muscle.

skin due to metabolic processes nor wrinkling caused by ill-health, psychogenic factors or certain diseases.

Wrinkles must be differentiated from the lines produced by emotions, for example constant scowling or sneering. The best cure for these is to cultivate a placid disposition.

In the case of excessively dry or greasy skin it may be necessary or advisable to cleanse it without the use of soap. The appropriate choice of material for dry skins would be complexion milk or a more greasy cleansing cream; for greasy skins, non-greasy cleansing lotions based on very mild alkalis such as sodium bicarbonate, or on certain buffered soapless detergents, may be used with advantage. Whatever method is adopted, the cleansing should be efficient and no trace of the day's grime or make-up should remain.

Cosmetic Application

Having secured a clean skin on which to work, the essentials of facial make-up include the application of:

(i) colour to the cheeks to give a pleasing healthy appearance and, if required, to modify the shape of the face;
(ii) powder to hide shine or greasiness and confer a matt bloom to the skin;
(iii) cosmetics in the region of the eyes to enhance the appearance and draw attention to the most attractive feature;
(iv) colour to the lips, to modify the shape of the mouth, again to attract attention and, by contrast, enhance the whiteness of the teeth and improve the smile.

It will be noted that all the objectives given are basically the natural attributes of youth; a lavish use of cosmetics, or a high degree of contrast, may produce a more formal or 'sophisticated' effect.

A little skin lotion with some astringency or greasiness, as appropriate to the skin condition, should be applied immediately after washing in the morning, using a pledget of cotton wool and smoothing in the same directions as indicated for massage. This should then be followed by the application of a light moisturizer to 'set the skin' ready for the make-up. This should not be applied immediately. The skin appears to require a period of acclimatization to the day, and if make-up is applied immediately on rising it is rarely satisfactory or long-lasting.

Foundation

When applying make-up, the foundation, with a degree of greasiness appropriate to the condition of the skin, is applied first of all, just sufficient to provide an adherent base for the rouge or blusher and powder. It is applied in the same directions as shown in Figure 20.1 for massage, but with more gentle smoothing movements, extending well on to the throat and further if desired, to avoid any lines of demarcation. It also hides skin imperfections and imparts a smooth, even appearance to the skin. In the case of normal or dry skin the foundation is usually a cream, but liquid foundation and cake make-up can equally be employed. In the case of oily skin, non-oily liquids, cake make-up and medicated lotions are more appropriate.

Rouge or Blusher
The application of rouge comes next. The rouge chosen should match the colour of the natural flush which appears through blushing or exertion. If it is applied correctly, it can accentuate the more attractive features and reduce the less attractive ones. Liquid or cream rouge is applied in sufficient quantity to the most prominent part of the cheek bones, just below the eyes and is spread with the finger tips, roughly into a triangle shape. The exact proportions and position of the triangles depend upon the shape of the face and the impression of length and breadth it is desired to convey. Naturally the edges of the coloured patch should blend easily into the remainder of the face with no hard outline. Powder rouge, however, is applied over the face powder.

Powder
The powder should be applied generously with cotton wool or a clean puff and the excess brushed off with a soft brush. It should be slightly darker than the foundation and have a covering powder appropriate to the type of skin—light for dry skin, and heavy for an oily skin. The correct shade is that which imparts an unobstrusive smooth, but not skiny, look to the skin. Thought should be given, as with all cosmetics, to the lighting under which the make-up will be seen; daylight demands shades nearer to natural, while for artificial light all colours should be a shade or two deeper, and slightly heavier application is advisable. To obtain a very transparent effect, which can look attractive especially for evening wear, two contrasting powders should be used, first a lighter one and then a deeper one. With two shades of powder used simultaneously, the same effect can be achieved as by the location of rouge, namely a change in the apparent shape of the face. Applying a powder of a darker shade across the lower part of the face will appear to shorten a long face. By application of a lighter-shade powder to either side of the jaw line, a heart-shaped face will appear wider.

Eye Make-up
For a long period eye make-up was practised much more discreetly by European or American women than by their Eastern sisters; even after the emancipation of make-up, lips were made up far more prominently than eyes.

However, the pendulum has now swung and all aspects of this most attractive feature of a woman's face are enhanced by the use of the appropriate cosmetics, outdoors by day as well as indoors by night. Thus, there are pencils to improve and enhance the colour and shape of the eyebrows, mascara to colour and lengthen the lashes, eyeshadow to draw attention to and re-shape the eyelids and to enhance the eye colour and eyeliners to outline and emphasize the eye itself. These are applied as follows.

Pencil. The eyebrows are first of all tidied, if necessary, by slightly plucking them with tweezers, mainly from below; a thin line should, however, be avoided. The eyebrows are then prolonged deftly towards the temples by means of an eyebrow pencil which should match the natural colour of the eyebrows. In order to obtain a neat result, a sharpened pencil should be used. The pencil can also be used to outline the upper eyelids just above the eyelashes, again extending the line slightly towards the temple.

Mascara. The function of mascara is to increase the natural charm of the features by darkening the eyelashes and increasing their apparent length. It is claimed that, by its judicious use, the brightness and expressiveness of the eyes is enhanced.

Mascara is available in either cake or cream form, and slightly different techniques are employed in their application. Cake mascara is applied by rubbing the wetted brush over the cake until sufficient colour has been imparted to it and until it is almost dry. It is then stroked over the upper lashes only, whereby the colour is concentrated on the tips and outer lashes in order to secure a natural effect. A fresh brush is used when the lashes are dry in order to separate them.

At one time cream mascara had to be applied by means of the fingers, and consequently was much less popular than blocks. However, the advent of a much more convenient applicator, in which the mascara as a thin cream is packed in a cylindrical container and applied by means of a small-diameter cylindrical brush, has increased the popularity of cream mascara to such an extent that it is now the dominant type. The brush, charged with mascara, is withdrawn from the tube and revolved against the outer side of the eyelashes, working away from the eyelid. The brush simultaneously deposits mascara and separates the eyelashes. Black mascara is suitable for black or dark brown eyelashes, while blue mascara when applied to the lashes of blue or grey eyes increases their apparent blueness.

Eyeshadow. Eyeshadow is used to impart more depth to the eyes and to intensify their colour. It is available in cream and stick forms, or as a pressed powder.

The shade is governed by the colour of the eyes, and may include colours ranging from black and blue to green and silver. The cream eyeshadow is applied just above the lashes on the centre of the eyelids and smoothed in with an outward movement of the finger tips. Pressed powder eyeshadows are usually sold with a sponge-tipped applicator and this is used to transfer the product from its compact onto the eyelids. Some powdered eyeshadows can also be applied with the finger. For daytime wear, eyeshadow should be applied lightly, and for evening wear the application should be only slightly more heavy. The use of two products at the same time is becoming very fashionable. A dark colour (blue, green or brown) is applied on the eyelid and lighter coloured product or 'high-lighter' (pearlized white or pink) is applied in the gap between the eyelid and eyebrow to merge and tone in with the original colour. Wearers of heavy-framed spectacles should refrain from heavy eye make-up.

Eye Liner. According to the instructions issued with one of the many liquid eye liner preparations, the brush provided with it is dipped into the liquid, held as one would a pencil, and then the desired line is drawn in one single brush-stroke. The same instructions also state that an eye opening effect can be achieved by broadening the line towards the centre of the eye.

Lipstick

The last touch to the make-up, but by no means the least, is the lipstick, which is used to impart colour and an attractive shape and appearance to the lips, and also to protect them. Intelligently used, it is capable of altering the apparent facial characteristics. Thus, by using a coat of darker lipstick with a lighter shade on top, narrow lips can be widened; by other methods wide lips can be narrowed, and the length of the lips can be brought into better proportion to the shape of the face.

Lipstick is best applied by means of a lipstick brush, which after a little practice will be found to give much better control than direct application. The desired mouth shape is first outlined and then filled in. The lips are blotted by biting on a tissue and the lipstick is applied once more. After a few seconds the lips are pressed together without tissue and the make-up is complete. If a more matt effect is desired, the lips should be lightly coated with powder before they are finally pressed together.

The colour of the lipstick can vary more than that of almost any other cosmetic and is chosen by the user not only on the basis of the tone of the skin, but also depending on the colours of the clothes worn. The fashionable colours for the seasons, determined each year by the fashion designers and the fabric manufacturers, lead to definite trends in the colours of lipsticks offered for sale. Thus, for a period, the great majority of lipsticks may have a bluish cast while the prevailing dress fabrics are greens, blues and mauves, whereas at another time an orange cast or pure reds will be needed to match yellow and brown outfits. The intelligent manufacturer will keep abreast of such fashion changes and change his lipstick range accordingly.

REFERENCES

1. Kendall, E., *Good Looks, Good Grooming*, London, Dent, 1963.
2. Anderson, E., *Be Beautiful*, London, Elek, 1971.
3. Young, D., *ABC of Stage Make-up for Men and Women*, London, French, 1979.

PART TWO

The Nails and Nail Products

Chapter Twenty-one

The Nails

Biology of the Nails

Structure

Nails [1,2] are translucent plates, composed of hard keratin, which protect the top surfaces of the end joint of each finger and toe. Each lies upon a *nail bed*, to which it appears to be fused, except at the far or *distal* end where there is a free margin. At the near or *proximal* end a paler *lunula* is visible, and the posterior margin is slightly overlapped by a narrow extension of the horny layer of the epidermis which is known as the *cuticle*.

Development and Formation

The nails start to develop in the nine-week human embyro from invaginations, or *nail folds*, of epidermis on the dorsal tip of each digit, and by the twentieth week they are complete. Exactly how the plate is formed and continues to grow has been the subject of some discussion. The traditional view is that the bed plays no part and that the nail is formed solely by proliferation of keratinizing cells from a *proximal matrix* which starts at the proximal limit of the fold, possibly involves a little of the adjacent roof, and extends below the nail to the distal margin of the lunula (Figure 21.1a). This interpretation was supported by autoradiographic studies in which tritiated glycine was injected into the squirrel monkey, a primate with flat nails very like human ones.[3] However, similar experiments on human volunteers suggested that there was some activity in the nail bed.[4]

A second view[5] is that the nail is formed in three layers. The dorsal and intermediate nail comes from the traditional matrix, but there is also a ventral layer which is formed from the nail bed distal to the lunula (Figure 21.1b). A third view[5] divides the nail bed into three zones, namely a proximal fertile matrix, an intermediate sterile bed, and a distal fertile matrix, the *solehorn*, which contributes a small quantity of material to the underside of the nail (Figure 21.1c).

The view of Samman[1,2] is that all three modes of nail formation can occur in normal subjects, but that the participation of the distal and intermediate parts of the bed is commoner in pathological conditions.

Histology, Ultrastructure and Composition

The epidermis of the nail bed is similar to that of the skin, but has no *stratum lucidum* or *stratum granulosum* and no hair follicles or sweat glands. It probably moves forward with the growth of the nail.[6] The epidermis of the matrix is thick and passes into the substance of the plate, which is formed by changes similar to keratinization in the superficial epidermis.[7] The appearance of the lunula is

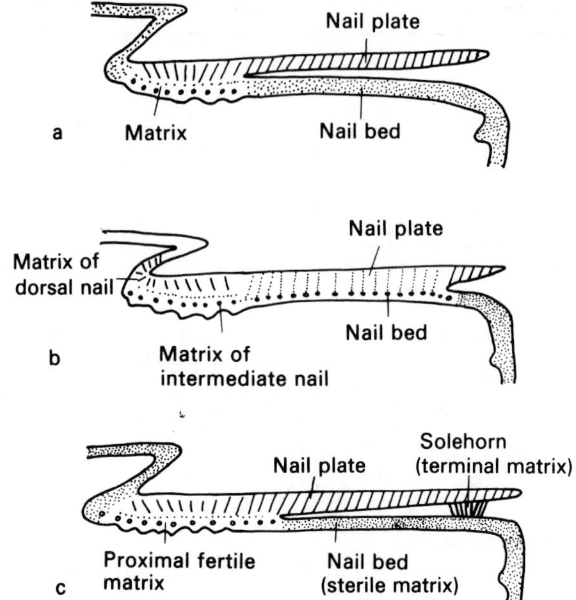

Figure 21.1 Possible methods of nail formation (after Samman[1])
a Traditional view: the nail plate is formed solely from a single proximal matrix
b Hypothesis that ventral layer of nail plate is formed from the nail bed
c Third view that distal terminal matrix, which is separated from proximal matrix by intermediate sterile matrix, contributes some material to underside of nail

probably due to a combination of incomplete keratinization in the nail plate and looseness of connective tissue in the nail bed.[8]

The nail plate is made up of impacted and adhering layers of flattened cells which have lost their nuclei. The cells contain hard keratin, similar to that of hair (see Chapter 23), with a high sulphur content, mainly in the form of cysteine, which comprises about 9–12 per cent of the weight of the nail.[1,9] The keratin fibrils are mainly orientated parallel to the nail surface from side to side.[10,11] There is a natural cleavage between the harder dorsal nail and the more pliable intermediate nail. Nail contains about 7–12 per cent of moisture[12] and 0·15–0·76 per cent of fat, a little more (1·38 per cent) in infants.[13] Calcium constitutes about 0·02–0·04 per cent of the weight, and could not contribute to the hardness.[10,14,15] Even this amount may be derived from the environment. Other trace elements have been detected and measured in nail clippings.[16]

The nail bed and matrix have a rich supply of blood from two arterial arches which lie below the nail plate.[1,2]

Rate of Nail Growth

Nails, unlike hairs, grow continuously throughout life. Le Gros Clark and Buxton[17] measured the rate in undergraduates and schoolchildren by notching the nail about 2 mm from the margin of the lunula and recording its progressive displacement in each succeeding month. They found no differences over the age range of 19 to 23 years, no sexual difference, and only a very slight difference between hands, growth on the right being faster. However, there were differences between fingers, growth being fastest on the third digit and least on the little finger. Dawber[18] has confirmed these observations. Bean,[19–22] by studying his own thumb for over 30 years, found some slowing-down followed by a levelling-out. The rate of fingernail growth varies between 0·5 and 1·2 mm per week.[2,17,18,23] On average, it is higher in psoriasis,[18] idiopathic oncholysis[24] and as a result of inflammatory change, and it may be temporarily depressed in infective diseases such as measles or mumps.[25] To replace a finger nail completely takes about 5½ months; toe nails need 12–18 months.[2]

Pathology of the Nails

Abnormalities of structure and appearance of the nails are of many kinds and can have many causes: genetic, infective or environmental. For a full account of the various conditions that affect nail growth and the ways in which they affect it, the reader is referred particularly to Samman.[2] Here abnormalities will be briefly surveyed, with the purpose of enabling the cosmetician to recognize the extent to which each may be ascribable to traumatic causes or local infection as distinct from underlying disease.

Absence of Nails

Complete absence of nails from birth, *anonychia*, is rare, and appears to be associated in several ways with other hereditary defects. In *nail patella syndrome*, complete absence of thumb nails, or a grossly defective nail less than half the normal length, is associated with reduction in size or absence of the knee caps and other skeletal anomalies.

Nail Shedding

Nails can be lost either by loosening at the base or by separation from the nail bed. Shedding can follow trauma or severe illness, or can be caused by drugs. More serious loss with scarring sometimes follows damage, defective circulation, lichen planus, epidermolysis bulbosa or drug eruptions.

Onycholysis

The separation of the nail from its bed, *onycholysis*, is fairly common. There are many causes but quite often none can be found. It may result from external damage, fungal infections, dermatitis or a drug eruption, or it may be associated with psoriasis, defective peripheral circulation, hypo- or hyper-thyroidism, or other disorders.

Of particular interest are reports of its causation by cosmetics. Nail hardeners containing formaldehyde,[26] nail varnish containing phenol[27] and dental acrylic polymerized *in situ* to strengthen the nail[28] have all been implicated.

Brittleness

Brittle nails are common. While sometimes they may be associated with impaired peripheral circulation as in Raynaud's disease, iron deficiency anaemia, congenital defects, diffuse alopecia or infections, the cause is often unknown.

Environmental factors are certainly important. Nails are kept pliable by their moisture content, and long thin nails are especially susceptible to very dry atmospheres. Frequent immersion of the hands is conducive, and Silver and Chiego[29] believed that soaps and detergents remove the protective lipids from the keratin.

The continuous use of nail varnishes and varnish removers has also been blamed.[30] Woolcott[31] investigated the reactions of twenty-five business women using the same brand of nail lacquer, and found that although nearly half of them showed brittleness and splitting of the nails, the rest were unaffected.

Tyson[32] reported in 1950 that oral ingestion of 7 g of gelatin per day for three months restored normal appearance and texture to fragile fingernails, and a number of other investigators have confirmed this.[33–37] In an extensive investigation, Michaelson and Huntsman[38] claimed that 2 g per day was enough to produce a significant increase in hardness within one month in five out of seven test subjects. Samman[2] simply recommends avoidance of frequent immersion and nightly use of a hand cream containing equal parts of salicylic acid ointment (2 per cent) and glycerin of starch.

Striations

Longitudinal striations are common in healthy nails, and become more prominent with aging, as well as in lichen planus, Darier's disease and defective peripheral circulation. An uncommon condition is median dystrophy, in which the nail plate has a longitudinal split, just off centre.

Regular transverse striations can occur as a developmental anomaly. Severe depressions, known as Beau's lines, indicate a period of severe systemic disability, such as measles, mumps, pneumonia or coronary thrombosis, which interferes with growth of the nail.

Koilonychia

In koilonychia or 'spoon nails' the nails are thin, soft, and depressed in the centre. The condition results from iron-deficiency, usually, but not invariably,[39] associated with anaemia. There are other causes; it is, for example, not uncommon in motor mechanics,[40] where perhaps it is due to softening with oils or soaps.

Splitting

The splitting of nails horizontally, so that pieces of the surface break away, is very common in women, and the main causes are probably repeated immersion in water and the use of varnish and varnish removers.

Pitting

Pitting of the nails is found in dermatitis and fungal infections, in association with alopecia areata and, most commonly, in psoriasis.

Leukonychia

Leukonychia is complete or partial whitening of the nail plate. Complete leukonychia occurs very rarely as an inherited abnormality; often epidermal cysts are associated with it. Partial leukonychia is very common, and may take the form of white spots or white transverse streaks. It may be associated with many types of illness and is found in chronic arsenical poisoning, but in the punctate cases there is usually no discernible cause.

A white appearance of the nails is not necessarily due to leukonychia; it may result from changes in the bed, not in the plate. Such a condition has been described in association with cirrhosis of the liver.[41]

Discoloration

Nails may be discoloured for a wide variety of reasons. External causes include hair or other dyes, nicotine and medicaments such as mercury, dithranol and picric acid. Tints may leak out of nail varnish and penetrate the nail.[42,43]

Abnormal formation or very slow growth of the nail will produce colour changes. Psoriasis, in particular, among other abnormalities, causes opaqueness and discoloration, and so can several other skin conditions.

In the *yellow nail syndrome* as described by Samman and White[44] the nails almost cease to grow and several months later become yellow or greenish. They may also be thickened and curved from side to side. The condition is accompanied by oedema and changes in the lymphatic system. Yellowish, green or grey nails can also occur as growth of nails slows down in old age.

Fungal or candidal infections which partly destroy the nail often cause brown discoloration. Infection by *Pseudomonas aeruginosa* under the nail will stain it black or blue.

Systemic drugs can alter the colour in many ways. Prolonged administration of tetracycline may occasionally turn nails yellow. Mepacrine makes them bluish, and chloroquine may produce blue-black pigmentation of the nail beds.[45]

REFERENCES

1. Samman, P. D., *Textbook of Dermatology*, 3rd edn, ed. Rook, A., Wilkinson, D. and Ebling, F. J. G., Oxford, Blackwell, 1979, p. 1642.
2. Samman, P. D., *The Nails in Disease*, 3rd edn, London, Heinemann, 1978.
3. Zaias, N. and Alvarez, J., *J. invest. Dermatol.*, 1968, **57,** 120.
4. Norton, L. A., *J. invest. Dermatol.*, 1971, **56,** 61.
5. Pinkus, F., *Handbuch der Haut und Geschlechtskrankheiten*, ed. Jadassohn, J., Berlin, Springer, 1927, p. 267.
6. Krantz, W., *Dermatol. Z.*, 1932, **64,** 239.
7. Lewin, K., De Wit, S. and Ferrington, R. A., *Br. J. Dermatol.*, 1972, **86,** 555.
8. Lewin, K., *Br J. Dermatol.*, 1965, **77,** 421.
9. Block, R. J., *J. biol. Chem.*, 1939, **128,** 181.
10. Forslind, B., *Acta Derm-Venereol.*, 1970, **50,** 161.
11. Forslind, B. and Thyresson, N., *Arch. Dermatol. Forsch.*, 1975, **251,** 199.
12. Silver, H. and Chiego, B., *J. invest. Dermatol.*, 1940, **3,** 133.
13. Langecher, H., *Hoppe Seylers Z. Physiol. Chem.*, 1921,**115,** 38.
14. Robson, J. R. K. and Brooks, G. J., *Clin. Chim. Acta*, 1974, **55,** 255.
15. Vellar, O. D., *Am. J. clin. Nutr.*, 1970, **23,** 1271.
16. Harrison, W. W. and Clemena, G. G., *Clin. Chim. Acta*, 1972, **36,** 485.
17. Le Gros Clark, W. E. and Buxton, L. H. D., *Br. J. Dermatol.*, 1938, **50,** 221.
18. Dawber, R., *Br. J. Dermatol.*, 1970, **82,** 454.

19. Bean, W. B., *J. invest. Dermtol.*, 1953, **20,** 27.
20. Bean, W. B., *Arch. intern. Med.*, 1963, **111,** 476.
21. Bean, W. B., *Arch. intern. Med.*, 1968, **122,** 359.
22. Bean, W. B., *Arch. intern. Med.*, 1974, **134,** 497.
23. Goodman, H., *Cosmetic Dermatology*, New York, McGraw-Hill, 1936, p. 412.
24. Dawber, R., Samman, P. D. and Bottoms, E., *Br J. Dermatol.*, 1971, **85,** 558.
25. Sibinga, M. S., *Paediatrics*, 1959, **24,** 225.
26. Lazer, P., *Arch. Dermatol.*, 1966, **94,** 446.
27. Dobes, W. L. and Nippert, P. H., *Arch. Dermatol.*, 1944, **49,** 183.
28. Fisher, A. A., Franks, A. and Glick, H., *J. Allergy*, 1957, **28,** 84.
29. Silver, H. and Chiego, B., *J. invest. Dermatol.*, 1942, **5,** 95.
30. Hollander, L., *J. Am med. Assoc.*, 1933, **101,** 259.
31. Woolcott, G. L., *Arch. Dermatol.*, 1940, **41,** 64.
32. Tyson, T. T., *J. invest. Dermatol.*, 1950, **14,** 323.
33. McGavack, T. H., *Antibiot. med. clin. Ther.*, 1957, **4,** IV.
34. Rosenberg, S., Oster, K. A., Kallos, A. and Burroughs, W., *Arch. Dermatol.*, 1957, **76,** 330.
35. Rosenberg, S. and Oster, K. A., *Conn. State med. J.*, 1957, **19,** 171.
36. Schwimmer, M. and Mulinos, M. G., *Antibiot. med. clin. Ther.*, 1957, **4,** 403.
37. Derzavis, J. L. and Mulinos, M. G., *Med. Ann D. C.*, 1961, **30,** 133.
38. Michaelson, J. B. and Huntsman, D. J., *J. Soc. cosmet. Chem.*, 1963, **14,** 443.
39. Comaish, J. S., *Newcastle med. J.*, 1965, **28,** 253.
40. Dawber, R., *Br. J. Dermatol.*, 1974, **91** (Supplement 10), 11.
41. Terry, R., *Lancet*, 1954, **i,** 757.
42. Calnan, C. D., *J. Soc. cosmet. Chem.*, 1967, **18,** 215.
43. Samman, P. D., *J. Soc. cosmet. Chem.*, 1977, **28,** 351.
44. Samman, P. D. and White, W. F., *Br. J. Dermatol.*, 1964, **76,** 53.
45. Tuffanelli, D., Abraham, R. K. and Dubois, E. I., *Arch. Dermatol.*, 1963, **88,** 419.

Chapter Twenty-two

Manicure Preparations

A complete manicure treatment can involve a number of different cosmetic preparations concerned with the cleansing and preparation of the nail and its decoration. These will be dealt with below, somewhat in order of their use.

CUTICLE REMOVER

When the free edge of the nail has been shaped, by mechanical means such as cutting or filing, the next problem is to shape the base of the nail. Where the skin adjoins the nail it becomes cornified and the dead cells, together with sebum, form an irregular appendage which grows thick and ragged and partially obscures the 'halfmoon' or lunule. Some improvement can be effected by mechanically loosening and pushing back the cuticle, but removal of excess cuticle by cutting is unsatisfactory and cuticle removers are used.

Cuticle removers are based principally on alkaline materials in liquid or cream form. One of the most effective and relatively cheap materials is potassium hydroxide and 2–5 per cent of this material in aqueous or aqueous–alcoholic solution formed the basis for many early preparations. The incorporation of humectants such as glycerin or propylene glycol, at a level of 10–20 per cent, served to counteract the irritation potential of alkali hydroxides and also to retard evaporation and increase viscosity. The latter effects can also be achieved by the use of suitable water-soluble gums or hydrocolloids.

Milder but correspondingly less effective preparations can be obtained by using alkaline polybasic salts such as trisodium phosphate or tetrasodium pyrophosphate, with the possible addition of 2–3 per cent sodium, or triethanolamine, lauryl sulphate.

An example of a basic formula is:

	(1)
	per cent
Trisodium phosphate	8·0
Glycerin	12·0
Water	80·0
Perfume	*q.s.*

Still milder products incorporate alkanolamines such as monoethanolamine or isopropanolamine at 8–10 per cent or triethanolamine at, say, 10–12 per cent.

To improve the convenience in use of cuticle removers attempts have been made to present them in cream form, which will also reduce the potential risk of damage to furniture, carpets, etc. Example 2 gives a basic formula suggested[1] for such a preparation:

	(2) *per cent*
Sorbitan monopalmitate	2·0
Polyoxyalkylene sorbitan monopalmitate	2·0
Mineral oil	15·0
Water	81·0
Cuticle remover	*q.s.*

Procedure: Heat the first three items to 65°C. Emulsify at this temperature with the water, except for a portion used to dissolve the cuticle remover, which should not, in this case, be alkali hydroxide. Mix the emulsion and the solution of cuticle remover when cold.

Special emulsions have to be prepared to allow the incorporation of highly alkaline materials. Example 3 is typical:

	(3) *per cent*
Cetyl alcohol	2·5
Myristyl alcohol	3·5
Polyoxyethylene (5) lauryl ether	1·0
Glycerin	4·0
Potassium hydroxide	1·6
Water	87·4

Procedure: The water-soluble ingredients are mixed together, heated to 75°C and then added to the mixture of alcohols that has also been heated to 75°C. The resultant cream is then cooled to 45°–50°C and should be filled at this temperature.

A combination of phosphate and caustic alkali can also be used with this type of cream emulsion, although the presence of phosphate can drastically affect the viscosity of the final product. This can be controlled, however, by careful homogenization of the cooled emulsion.

Care must be taken in packing all such alkaline preparations in glass containers which should be fitted with alkali-resistant stoppers such as rubber or plastic. Even so, the more alkaline preparations are liable to give rise to glass etching and silicate precipitates. The latter can be masked by the use of opaque containers or, it is claimed, prevented by the inclusion of potassium oleate or about 1 per cent of the tetrasodium salt of ethylenediaminetetra-acetic acid. Trends in legislation are now towards the requirement of warning labelling of these products, either by naming the caustic material (for example, 'Contains sodium hydroxide') or by specifying the pH of the product.

Cuticle Softeners

Another type of preparation which has been employed is the cuticle softener in cream form, which by its action and emolliency facilitates subsequent mechanical removal of the cuticle. Quaternary ammonium compounds are prominent in this class of preparation. Their softening action is the result of their affinity for protein, and increases with increasing molecular weight. Cetyl pyridinium chloride and stearyl dimethyl benzyl ammonium chloride are typical materials used at 3–5 per cent. They also exert a bactericidal action. Urea may be added to promote the swelling of keratin and enhance the cuticle softening and lanolin or isopropyl myristate will confer improved emolliency. Nonionic thickening agents such as methylcellulose or hydroxyethylcellulose will increase the viscosity.

NAIL BLEACH

Nail bleaches are solutions or creams used for the removal of ink or tobacco stains, vegetable stains, etc., from the nails. Such stains may yield either to oxidation or reduction, depending on the type of stain. Oxidation may be achieved by the use of hydrogen peroxide, either at 20 vols or diluted to about 1:4, by chlorine compounds such as the hypochlorites and the cyanurates and by sodium perborate and zinc peroxide (the latter two substances decompose in aqueous solution). Use of sulphites with dilute acid is probably the easiest method for achieving reduction in such preparations. These can be packed in two-solution form, one the stabilized oxidizing or reducing agent, the second containing enough acid or alkali to shift the pH sufficiently (when the two solutions are mixed) to render the agent to be employed unstable and hence active.

A one-solution nail bleaching lotion cited in literature[2] had the following composition:

	(4) *per cent*
Hydrogen peroxide (10 vols—3% w/w)	73·5
Ammonia (SG 0·93)	0·5
Rose water	25·0
Preservative	1·0

A number of stain-removing preparations are based on aqueous solutions or organic acids such as citric or tartaric acids and a 4 per cent solution of concentrated hydrochloric acid in water and glycerin has also been recommended[3] for such purposes.

The following formula for an abrasive nicotine stain remover has appeared in *Drug and Cosmetic Industry*:

	(5) *per cent*
Beeswax	10·0
Paraffin wax	5·0
Mineral oil	46·0
Pumice powder	8·0
Borax	0·5
Water	30·0
Perfume	0·5

NAIL CREAM

Nails are not porous in the normal sense of the word. They have no pores through which to 'breathe' or absorb any 'food', or which can become blocked by, for instance, nail varnish. Nevertheless, as explained in the previous chapter, nails can become brittle from various causes and it is natural to wish to apply some preparation to counteract this brittleness, which is probably due to dehydration.

One recommendation for the relief of brittle nails was to massage them with olive oil after having bathed them in warm water. It was stated, however, that nail plates will regain their normal consistency and elasticity only after several weeks.

A more practical preparation is an emollient cream containing a suitable humectant such as one of those already described in earlier chapters. Emollient creams for use as nail creams have been formulated on a lanolin absorption base or on beeswax–borax emulsions. These creams should be applied after soaking the hands in warm soapy water and thoroughly drying them, just before retiring for the night. This treatment should be carried out at least once and preferably 2–3 times per week (that is, on alternate nights) depending upon the condition of the nails. Wearing fingerstalls or gloves may assist the treatment.

Claims have been made that certain fatty substances, in particular cholesterol, appear to assist in maintaining the natural elasticity of the nail. In this connection the water-in-oil emulsifying properties of cholesterol should be borne in mind since it may assist in maintaining the requisite degree of moisture in the keratin. Although much remains to be discovered, it would appear that the use of a nail cream containing oils in an emulsified form with water (and possibly incorporating a mild antiseptic) would be of value.

Any of the cold, vanishing, foundation or all-purpose cream formulae given in the various chapters in this book may be modified slightly—if necessary—for use as a cream for brittle nails, by the inclusion of a little lanolin or other emollient and an adequate amount of humectant.

NAIL STRENGTHENER

The fact that dry fingernails tend to split and break off easily is bound to spoil their appearance and make manicuring difficult and often painful. This has led to the development of various products for strengthening brittle nails and for eliminating brittleness and dryness.[4]

A liquid composition consisting of an aqueous solution of 3 per cent stearyl trimethyl ammonium chloride, 1·5 per cent nonylphenolpolyoxethylene ($10Et_20$ per mol) ether and 0·5 per cent triethanolamine stearate was claimed in US[5] and British[6] patents, for eliminating brittleness and dryness of nails, counteracting the effect of solvents present in nail polish removers and improving the adhesion of nail lacquer applied subsequently.

Other nail-hardening preparations for increasing the resistance of nails to cracking, splitting and laminating were disclosed in another US patent.[7] They were all solutions of water-soluble metallic astringent salts such as, for example, aluminium sulphate, potassium, sodium and ammonium alums, zinc acetate and zirconium chloride. Treatment consists of wetting the nails by dipping in, or painting with, a 1–5 per cent aqueous solution of the metallic salt, and keeping the nails in contact with the solution for 5–10 minutes at an ambient temperature. If the recommended concentration is exceeded, drying and wrinkling of the skin of the finger tips may occur. These nail-hardening compositions preferably also contain 5–20 per cent of a humectant such as glycerin or propylene glycol to retard evaporation of the aqueous solvent, and to provide an even coating and a 'slightly enhanced penetration'.

A slight bactericidal effect has also been claimed for these salts, which may be further enhanced by the inclusion of other bactericidal substances such as formaldehyde. In one example the nail-hardening composition contained:

	(6) *per cent*
Potash alum	3·00
Glycerin	10·00
Formaldehyde	0·01
Menthol	0·001
Water	to 100·000

Partially polymerized formaldehyde resins have also been used as the cross-linking agent in nail-hardening compositions. However, attention must be drawn to a reaction to the formaldehyde component of a nail hardener reported by Lazer,[8] which resulted in nail damage including subungual haemorrhage, discoloration, subungual hyperkeratosis and dryness of the skin, and necessitated oral administration of a steroid to relieve the accompanying pain and oedema. The use of the preparation had to be discontinued to allow the nails and the skin to return to a normal condition.

In a French patent[9] the use of a solution of dimethylol- or diethylol-thiourea was suggested for a nail-strengthening composition, which also contained a weak acid polymerization catalyst; in an Australian patent application,[10] a nail-strengthening product was described containing a non-toxic cysteine derivative dispersed in a non-toxic carrier.

NAIL WHITE

Nail whites are preparations used to produce an even white edge to the nails. They are based on an inert white pigment such as zinc oxide, titanium dioxide, kaolin, talc or colloidal silica, of which the first two are the best. The pigment forms 20–30 per cent by weight of the preparation, which is generally presented as a stiff paste.

For general purposes a fatty base is preferable. Titanium dioxide is incorporated in the following example:

	(7) *per cent*
Titanium dioxide	38·0
Petroleum jelly	62·0

This simple preparation is readily produced by milling the titanium dioxide into the melted petroleum jelly. Various mixtures of suitable fats and waxes may replace the petroleum jelly as desired.

Nail white pencils have largely replaced nail white creams, since they are easier to apply, are less messy in use, and simply require moistening with water before application to the under-surface of the nail free edge, the pencil then being recapped.

A wax may be used as a base if desired:

	(8) *per cent*
Beeswax	55·0
Hydrogenated cottonseed oil	9·0
Cocoa butter	9·0
Castor oil	8·0
Lanolin base	9·0
Titanium dioxide	10·0

Such a base is ready for use and is not wetted before application. The usual type, which requires preliminary damping, is prepared by massing casein or gum arabic with the inert pigment, or by suspending it in soap, for example sodium stearate base.

The making of white pencils is a job for firms specializing in the manufacture of ordinary pencils and not for the ordinary cosmetic manufacturer. Consistency of the base will have to be adjusted according to plant operating conditions and hence any formulations given here probably bear little resemblance to those actually used by the pencil manufacturers.

NAIL POLISH

A distinction is drawn between those polishes which by abrasive action bestow a gloss on the nail surface, and a nail varnish. The former, by reaşon of the friction set up in the buffing process, draw the blood to the numerous capillaries of the nail bed and, by increasing the blood supply, may exert some slight stimulating effect upon the growth of the nail. The latter, which depend upon the deposition of a thin film of highly lustrous cellulose lacquer upon the nail plate, have increased enormously in popularity and have largely replaced the abrasive type, although abrasive polish may also be used between two successive coats of varnish to enhance the lustre, in the manner of the French polisher.

Abrasive polishes consist of a suitably finely powdered abrasive which is applied to the nail and then polished with a shaped chamois leather pad. The principal constituents of such powders are stannic oxide, talc, silica, kaolin, precipitated chalk, etc., the first being an excellent abrasive but rather more expensive than the others listed.

	(9) *per cent*
Stannic oxide	90·0
Powdered silica	8·0
Butyl stearate	2·0
Pigment, perfume	*q.s.*

Butyl stearate is included to render the product less gritty and may be replaced, if desired, by oleic acid. The method of preparation is by simple trituration in a mortar or revolving mill. By the addition of a suspending agent, methylcellulose or tragacanth together with glycerin or one of the glycol ethers, they may be prepared in liquid form or compressed into a paste or pencil.

Wax polishing pencils embodying a large proportion of abrasive powder replaced the older powder and block type. They were effective and could be carried and used without the danger of spilt contents:

	(10) *per cent*
Hardened palm kernel oil	22·0
Synthetic wax (high melting point)	7·0
Stannic oxide	71·0

A softer stick may be obtained by omitting the synthetic wax and making a simple mixture of abrasive powder with the hardened oil. Alternatively, a base of rosin, beeswax, ceresin and petroleum jelly, or other suitable combination of waxes, may be employed.

NAIL LACQUER (NAIL VARNISH)

Introduction

The largest and most important group of manicure preparations is that of nail varnish or lacquers. These names give some indication of the progress of this type of preparation from the transparent uncoloured or pale pink natural colours, which were the only ones tolerated at one time, to the vivid shades and even bizarre colours now accepted as a most essential part of modern beauty.

Requirements of a Nail Lacquer

A nail lacquer should fulfil the following requirements.

(1) It should be innocuous to the skin and nails.
(2) It should be easy and convenient to apply.
(3) It should be stable on storage as regards homogeneity, separation, sedimentation, colour and interaction of ingredients.
(4) It should give a film with satisfactory characteristics.

The characteristics of a satisfactory film are:

(1) Even thickness, which demands a satisfactory viscosity of the lacquer, neither too thin nor too thick, and good wetting and flow properties.
(2) Uniform colour, which demands a very finely divided pigment, intimately ground and wetted by the medium.
(3) Good gloss, which implies a very smooth surface and depends upon the properties of the medium.
(4) Good adhesion to the nail.
(5) Sufficient flexibility to avoid brittleness and cracking.

(6) Hard, non-tacky surface, resistant to impact and scratching, which will not adhere to other surfaces nor mark off colour on fabrics or paper.
(7) Satisfactory drying properties—drying time of about 1–2 minutes without development of bloom even in humid atmospheres.
(8) Maintenance of these properties for approximately 1 week.

Ingredients of Nail Lacquer

In practical terms, the main essentials for the manufacture of nail lacquer are a lacquer base with suspending properties and a colouring system. These two components can be broken down into necessary constituents, and these are described before manufacture of the product is considered.

Film-formers

The basic film-forming material in nail lacquers is nitrocellulose, a cellulose nitrate obtained by the reaction of mixtures of nitric acid and sulphuric acid with cotton. In this reaction all three of the alcohol groups in the cellulose ring can be esterified. The degree of esterification or substitution determines the intrinsic characteristics of the nitrocellulose and the degree of polymerization of the cellulose chain governs the viscosity of the product. The nitrocellulose used in nail lacquers has a degree of substitution of approximately two and is known as dinitrocellulose-pyroxylin.

Different grades of nitrocellulose are characterized by their viscosity in organic solvents, for example $\frac{1}{2}$-second or $\frac{1}{4}$-second nitrocellulose, using the US nomenclature based on the falling ball method of determining viscosity. In practice, the grades of nitrocellulose used to manufacture nail lacquers are those which give sufficiently fluid solutions to allow easy application on the nails. Peirano[11] considered a 15 per cent solution of $\frac{1}{2}$-second nitrocellulose to be suitable, giving a viscosity in the range 300–400 cP. Similarly, $\frac{1}{4}$-second nitrocellulose gives a viscosity of approximately 500 cP when dissolved in 20 per cent *n*-butyl acetate.

Films produced by nitrocellulose are waterproof, hard and tough and resist abrasion. However, when used alone nitrocellulose has some drawbacks, namely, poor gloss and a tendency to shrink and to become brittle; adhesion to most surfaces is only moderate. This has resulted in the use of modifying resins to impart adhesion and to improve gloss, and of plasticizers to impart flexibility and to reduce shrinkage. With the use of solvents and diluents one then has the complete range of ingredients used in nail lacquers.

Resins

Wing[12] has pointed out that plasticized nitrocellulose films do not possess a high lustre. Resins such as damar and alkyd resins which were added by manufacturers of industrial enamels were found to be unsuitable for nail lacquers, since they made the films more susceptible to water. In about 1938 resins of the aryl sulphonamide-formaldehyde type were introduced which gave good lustre to nail varnish films and improved their resistance in detergent solutions; such resins have been used in many nail lacquers since that time to impart gloss, improve adhesion and often to increase the hardness of resulting films.

Examples of two commercially available resins of the aryl sulphonamide-formaldehyde type are Santolite MHP and Santolite MS 80 per cent, of which the former is the harder while the latter produces films of greater flexibility. Both of these resins are said to impart high gloss and good flow properties and also to increase the hardness of nitrocellulose. They are also claimed to increase the moisture resistance of nitrocellulose films, and thus to reduce the incidence of water spotting and whiteness of such films. With a dry nitrocellulose: resin ratio of 2:1, the moisture permeability of a film has been claimed to have been reduced to about half of that of an unmodified nitrocellulose film of the same thickness.

These resins are moderately stable to light and soluble in the majority of lacquer solvents and diluents usually employed. It has been pointed out, however, that their solvent release of alcohols other than methyl and ethyl alcohols is slow and that in lacquers which contain these resins the amount of a higher-molecular-weight alcohol should be kept to a minimum. It should also be stressed that, because of the low viscosity of their solutions, relatively large amounts of aryl sulphonamide-formaldehyde resins can be used in many lacquer formulations without adversely affecting the ease of application of such formulations. These resins have the further advantage of making it possible to produce films of a given thickness with fewer applications and a correspondingly smaller solvent loss, thus allowing economies in the use of expensive solvents when manufacturing modified nail lacquer preparations.

Some users develop a sensitivity to aryl sulphonamide-formaldehyde resins; such persons should use a varnish which does not include this resin.

The use of other resins has also been suggested. Cosmetic Laboratories Inc. have claimed production of an improved film with quick-drying properties by an additional use of nylon in amounts of up to 4 per cent of solids. Maleic alkyd resins have also been used because of their compatibility with nitrocellulose and their enhancement of light stability, hardness and solvent release. Other resins that could be investigated include styrene alkyds, melamine-formaldehyde, urea formaldehyde and the acrylics.

Plasticizers

Nitrocellulose is too brittle to be used in lacquers on its own and even the inclusion of a resin will not impart the necessary flexibility to lacquer films. Plasticizers must therefore be included in nail lacquer formulations in order to ensure that the film which remains on the nails after the solvents have evaporated adheres well, is flexible and does not flake off. By virtue of their high boiling point, plasticizers will remain in the film after the solvents present in the formulation have evaporated and render the films pliable. Plasticizers, even at low concentration, will furthermore enhance the gloss of resultant films and will also improve the flow properties of lacquers.

There are two groups of plasticizers:

(i) solvent plasticizers which, as the name implies, are solvents for nitrocellulose;
(ii) non-solvent plasticizers, also referred to as softeners.

The first group, the members of which are true plasticizers, comprises mainly high-molecular-weight esters, with fairly high boiling points and low volatility.

The second group are not solvents for nitrocellulose and are not compatible with it. If they are used in the absence of solvent plasticizers they will form separate droplets on the film once the solvents have evaporated. They must therefore be used in conjunction with true plasticizers which will hold them in solution; under those conditions they will impart additional flexibility to the film.

The most common representative of this group is castor oil which, when used in combination with a true plasticizer in the proportion 1:1, at a level of about 5 per cent, produces a very flexible film.

A good plasticizer must:

(a) be miscible in all proportions with the solvent, the nitrocellulose (applies to true plasticizers) and the resins used;
(b) be dermatologically innocuous and free from any sensitizing properties;
(c) have a low volatility;
(d) improve the flexibility and adhesion of the lacquer;
(e) not cause any discoloration of the finished product, that is, it must have moderately good lightfastness;
(f) be stable and odourless, or it must have a pleasant odour, since it does not evaporate but remains in contact with the nail.

Provided that the plasticizer fulfils the above requirements, the criteria which govern the selection of a plasticizer for nail lacquers are its effect on viscosity, on the rate of drying, flexibility, adhesion and gloss. The mention of lightfastness primarily relates to the fact that certain plasticizers, such as tricresyl phosphate, tend to turn yellow when exposed to light, while others will darken.

The amount of plasticizer used in nail lacquers varies from about 25 to 50 per cent based on the dry weight of the nitrocellulose present, and depends on the degree of flexibility required. Some formulations contain a single plasticizer; in others two or more plasticizers may be present among which dibutyl phthalate is one of the most widely used. Phthalates, phosphates, phthalyl glycollates, sulphonamides and citrates constitute the main group of plasticizers used in nail varnishes.

For a list of plasticizers most commonly used in nail lacquer preparations and for a more detailed description of properties of some of them, the reader is referred to a review of nail lacquers in *Manufacturing Chemist*.[13]

Solvents

It will be seen from the list of desirable film properties given earlier that the properties of the solvent, particularly the evaporation characteristics, are of the utmost importance.

It is common practice to rank solvents by their boiling points, which also appear to correlate with the viscosities of the resultant nitrocellulose solutions and hence with the spreading characteristics.

The normal procedure in textbooks appears to be the listing of a number of solvents with their boiling points, from which the experimenter is expected, presumably, to select a judicious mixture of medium-, high-, and low-boiling solvents, which are generally differentiated as follows:

(1) Low-boiling solvents (with boiling points of up to 100°C) represented, for example, by acetone or ethyl acetate.
(2) Solvents of medium boiling point (100°–150°C) represented by *n*-butyl acetate, considered to be the best all-round solvent.
(3) High boiling point solvents (over 150°) exemplified by Cellosolve, Cellosolve acetate, butyl Cellosolve, and including all nitrocellulose plasticizers which are also solvents for nitrocellulose.

However, although the boiling point will play some part, it is the evaporation rate at temperatures not much above 30°C that is the real concern in this context. This will depend on a number of factors including specific heat, latent heat of evaporation, molecular weight, degree of association, to mention only a few.

In the case of cellulose lacquers, in which a mixed solvent is employed, the effect of the vapour pressure of the mixed solvents on each other, and that of any molecular attraction, has to be considered. Even this by no means exhausts the variable complex factors involved. It suffices, however, to show the difficulties of any purely theoretical selection.

In order to provide some simple approximate factors for evaporation rates of single solvents, a number of solvents were allowed to evaporate under uniform conditions and the times required for complete evaporation of standard volumes were compared with that for a sample of ethyl ether. The results are given in Table 22.1.

Table 22.1 Solvent Evaporation Coefficients based on the Evaporation Rate of a Standard Volume of Ethyl Ether

Solvent	Approximate evaporation coefficient	Boiling range (°C)
Ether, ethyl (0·72)	1·0	34–35
Ether, petroleum (40°–60°C)	1·3	40–60
Methyl acetate	2·0	56–59
Acetone	2·1	55–56
Ether, petroleum (60°–80°C)	2·5	60–80
Cyclohexane	4·2	81
Ethyl acetate	4·8	74–79
Methyl ethyl ketone	5·0	79
Carbon tetrachloride	6·0	77
Ethyl alcohol	9·0	78
n-Butyl acetate	12·8	124–128
Amyl acetate	13·0	137–142
Xylol	13·4	138
Isopropyl alcohol	22·0	80
n-Butyl alcohol	34·0	115–118
Diethylene glycol monomethyl ether	35·0	134
Diethylene glycol monoethyl ether	45·0	135
Ethyl lactate	90·0	150–160

The correct choice of the constituents of the solvents and their proportions is very important for the following reasons:

(1) The thin low-boiling solvents give the necessary mobility to enable the lacquer to spread easily and dry quickly, but if they are present in excess the lacquer may not wet the nail and consequently may spread unevenly.
(2) On the other hand, the higher-boiling, more viscous solvents give body to the lacquer and allow time for it to key on to the nail surface and to flow into an even film, but delay the drying and hardening times.
(3) Too rapid evaporation from the surface of a film may give rise to blushing or blooming, especially in a humid atmosphere.
(4) Evaporation from the surface of a thicker film may give a surface dryness with an underlying softness which can cause wrinkling of the film.
(5) Preferential evaporation of one part of a solvent blend may change the composition of the remaining liquid so much as to disturb the solvent properties, giving rise to premature or partial precipitation of the solid material, with an uneven final effect.

Diluents

Diluents, while not being actual solvents for nitrocellulose, are organic solvents miscible with nitrocellulose solvents. They are used as solvents for modifying resins used in lacquers, since true nitrocellulose solvents are expensive. They also help to stablilize the viscosity of lacquers, but their main value lies in reducing the overall cost of formulations. There are three classes of diluent: (i) the alcohols, (ii) aromatic hydrocarbons and (iii) aliphatic hydrocarbons.

There is a limit, however, to the amount of diluent that can be tolerated by the nitrocellulose solution, which may be defined by the so-called tolerance, or dilution ratio. This has been defined as the maximum diluent–solvent ratio that can be tolerated by a nitrocellulose solution without causing the latter to be precipitated. In addition to the possibility of nitrocellulose precipitating out of solution, the question of viscosity, which governs ease of application, must also be taken into account. Lacquers in which the solvent–diluent ratio approximates to the tolerance ratio have a considerably higher viscosity than those containing a smaller amount of diluent, and will cause difficulties in obtaining a smooth film. The actual tolerance ratio for any solvent–diluent system is determined by titrating a given volume of the nitrocellulose solution with a diluent until the solution becomes cloudy.

It has been pointed out elsewhere that, in the formulation of lacquers, use of a high boiling point diluent with a low boiling point solvent is to be avoided if precipitation of nitrocellulose and 'blushing' of the film are not to occur. In general, when dealing with compositions containing solvent and diluent mixtures, the rate of evaporation of the diluent should be greater than that of the solvent or solvent mixture employed, otherwise the diluent–solvent ratio in the film will steadily increase and eventually the tolerance ratio will be exceeded, causing nitrocellulose to precipitate on the film. As a result, instead of the smooth, clear and continuous film hoped for, a harsh, rough and cloudy film will be produced which will lack adhesion.

Alcohols, especially ethyl and butyl alcohols and isopropanol, are very efficient diluents, their tolerance ratio being 9:1. Butyl alcohol, with an evaporation rate which is neither too rapid nor too slow, is preferred to the more volatile ethyl alcohol, which may give rise to chilling. Butyl alcohol tends to lower the viscosity of nitrocellulose solutions. Isopropanol, with its lower cost and ready availability, is the most popular alcohol used.

In nail varnishes, aromatic hydrocarbons such as toluene and xylene are employed. Their tolerance ratio of 3:1 is much lower than that obtained with alcohols; consequently they are not such efficient diluents. They also tend to increase slightly the viscosity of nitrocellulose solutions and to impart poor flow properties to the film. To overcome these disadvantages, a mixture of toluene and isopropanol may be employed as the diluent.

The choice of solvent will also influence the gloss of the films produced. A high boiling point solvent will produce a brighter film than does a low boiling point solvent. For matt films, highly volatile solvents and diluents are still used, but care must be taken that 'blushing' of the film does not occur.

Colours

Insoluble lakes are incorporated in the majority of nail lacquers to produce a suitable shade, together with titanium dioxide in order to confer opacity and creaminess and to produce pastel shades.

Soluble dyestuffs such as erythrosine, carmoisine and rhodamine, which were used at one time in nail lacquers, were found to stain the surrounding skin and the use of such dyestuffs has been virtually abandoned.

As far as the required properties of these pigments are concerned, they are similar to the requirements specified for pigments used in other cosmetic products, namely that they should be non-toxic and non-sensitizing, that they should be non-staining, substantially insoluble in lacquer solvents, free from any tendency to bleed, compatible with other lacquer constituents and moderately stable to light. They should also be of very small particle size, free from gritty particles and should have satisfactory dispersing properties.

The colours used in nail lacquers must conform to appropriate national legislation. In the USA only colours certified by the Food and Drug Administration, or inorganic colours of suitable purity, may be used. In the EEC the colours used must be taken from the positive lists included in the EEC Cosmetics Directive of 1976 and its amendments. Surveys of permitted colorants are given in Chapter 19 and by Van Ham.[14] The most widely employed pigments include the following accepted colours:

D&C Red 6	Colour Index	15850 Na
D&C Red 30		73360
D&C Red 36		12085
D&C Red 9		15585:1
D&C Red 7		15850:1
FD&C Yellow 5		19140
FD&C Yellow 6		15985

As pointed out above, titanium dioxide has been employed to impart opacity and to produce pale shades, while iron oxides are used to produce brown and tan shades. A range of chemically inert brown pigments derived from dinitrobenzene has been suggested for use in nail lacquers instead of iron oxide.[15] These were said to give more brilliant colours, to have greater tinctorial strength than iron oxide and to be very fast to light; because of their lower specific gravity, they are less liable to settle out.

Pigments for use in coloured lacquers are often supplied in the form of dispersions. These dispersions are formed by dissolving 'colour chips' in the appropriate solvents. 'Colour chips' are mixtures of nitrocellulose, colour and solvents, processed on a roller mill by specialist operators to give solid colour particles that are easily handled. Provided that this processing is satisfactory, no further roller milling of the pigment will be necessary prior to its incorporation into the nail lacquer. The pigment content of coloured lacquers varies from product to product but does not normally exceed 5 per cent. For improved wear resistance, the pigment concentration in nail lacquers should not be lower than 3 per cent.

Pearlescent (Nacreous) Pigments

In addition to the conventional pigments, nail lacquers can incorporate iridescent materials Guanine (2-amino-6-hydroxy purine), a crystalline substance from the scales and body of various fish, has been one of the most frequently suggested. In its pure form it is non-toxic, but there have been references to this material being the cause of dermatitis arising from nail lacquers. This substance has a high refractive index (1.8) and its pearly lustre and brilliance are due (i) to the simultaneous reflection of light both from the upper layers of its transparent crystals and from crystal planes at different depths, and (ii) to the alignment of crystals in the same direction, made possible by their shape and dependent on the degree of dispersion of the crystals in the lacquer. Orientation of these crystals takes place when the nail surface is coated with the lacquer containing the pearl pigment and they remain in this alignment when the film hardens. Natural pearl essence is commercially available in the form of a suspension in a nitrocellulose–butyl acetate lacquer at a level of about 11 per cent and is used in clear or pigmented lacquers at a concentration of between 8 and 10 per cent.

Synthetic pearl pigments are also available; owing to the high cost and scarcity of guanine, these have become the most predominantly used materials. The most popular is bismuth oxychloride. It crystallizes in platelets, has a high refractive index and produces a high lustre and brilliance at low cost. However, the material is denser than guanine and this can lead to problems of settling on standing, within the finished product. Bismuth oxychloride is normally available as 12·5 per cent and 25 per cent suspensions in nitrocellulose–butyl acetate–toluene mixtures to which has been added a small amount of colloidal pretreated clay. This gives the system added suspension power to prevent the bismuth oxychloride settling out. Another pearlescent material gaining increasing popularity is mica coated with titanium dioxide. It is less dense than bismuth oxychloride and therefore can be suspended more easily. However, it is less brilliant and does not have the same covering power as bismuth oxychloride. It is commercially available in nitrocellulose–solvent suspensions and this method of addition to formulations is preferable to the use of the dry powder, which can be difficult to disperse without the use of specialist mixers.

Suspension Agents

The modern trend towards highly pigmented and pearlescent nail lacquers has led to a critical assessment of the ability of traditional nail lacquer formulae to suspend these materials at high concentration. Additionally, consumers have reacted against products showing 'settled out' materials, and thus systems developed to avoid sedimentation soon showed themselves both to be technically superior and to have enhanced consumer acceptance. The suspension properties are obtained by creating a thixotropic system with the use of pretreated colloidal clays such as benzyl dimethyl hydrogenated tallow ammonium montmorillonite (Bentone 27), dimethyl dioctadecyl ammonium bentonite (Bentone 34), or dimethyl dioctadecyl ammonium hectorite (Bentone 38). These clays increase the viscosity of the system to such an extent that the heavy oxide pigments remain in suspension. When a shear force is applied to the system by shaking, or by brushing the product across a nail, the viscosity drops sharply, allowing a smooth application. On standing, the system regains its initial high viscosity. A thixotropic system containing nacreous pigments has been described in a US patent.[16] In modern practice, bentone levels range from 0·5 to 2 per cent, since the viscosity of the system can be enhanced even further by adding small quantities of a polyvalent acid, for example, ortho phosphoric acid, which precludes the use of high levels of bentone and makes the system more controllable.

Formulation

In the preceding paragraphs the desirable properties of nail lacquer constituents have been outlined. These should be taken into consideration in the formulation of such lacquers.

Various tests must therefore be devised to ascertain the flow characteristics of formulations, the rate of drying of lacquer films, the compatibility of their constituents during drying, as well as their appearance, hardness, adhesion and resistance to soap and to detergents. If necessary, formula adjustments have to be made and the tests repeated until the required characteristics are obtained. A good indication of the behaviour of lacquers on fingernails can be obtained from the behaviour of a layer of the lacquer on a clean and clear glass plate. However, this does not obviate the need for studying the products by actual application to the nails. Tests should also be carried out on the flow and evenness of application of lacquer films, their hardness and gloss. Any drag observed on the applicator brush should also be recorded.

Abrasion tests may be used *in vitro* to assess the wear resistance of films. Ultimately, however, assessments of the products *in vivo* must be carried out, employing a large panel of subjects who have applied both the experimental lacquer and the control on different fingernails for a period of one week, and examining the treated nails daily. A summary of evaluation techniques has been given by Shansky.[17]

In connection with the drying of lacquer, the practice of blowing on freshly applied lacquer films to speed up their drying should be discouraged, so as to ensure that no moisture is present which could have an adverse effect on the adhesion and the gloss of the film.

It has already been pointed out that it would be preferable from the point of view of convenience and solvent economy to apply a film in a single application. Since the thickness of the film determines its gloss and resistance to wear, it is desirable to obtain as thick a film as is practical from the point of view of ease of application and the rate of drying. Practical limitations, however, sometimes make it necessary to apply two coats of the lacquer in order to obtain a film of the required thickness. If this is necessary, it is then good practice to allow the first film to dry completely before applying a second coat, otherwise the evaporation of solvent may produce small blisters in the upper layer and give rise to the 'orange peel' effect often observed with quick-drying lacquers.

Formulation of base nail lacquers can be illustrated by the following examples.

Typical nail lacquer base[12]	(11)
	per cent
Nitrocellulose (about ½-second viscosity)	10·0
Resin	10·0
Plasticizer	5.0
Alcohol	5.0
Ethyl acetate	20·0
Butyl acetate	15·0
Toluene	35·0

In more detail, a clear base lacquer can be described as:

		(12)
		per cent
Nitrocellulose (½-second)		18·28
Santolite MS 80	(resin)	6·43
Dibutyl phthalate	(plasticizer)	2·15
Camphor	(plasticizer)	1·88
Ethyl acetate	(solvent)	6·45
Butyl acetate	(solvent)	25·42
Toluene	(diluent)	29·56
Isopropanol	(diluent)	9·83

In a *Monsanto Bulletin*[18] the following nail lacquer base formulation was quoted:

	(13)
	per cent
RS ½-second nitrocellulose	10
Santolite MHP (resin)	10
Santicizer 8 or 160 (plasticizer)	5
Ethyl alcohol	5
Ethyl acetate	20
Butyl acetate	15
Toluene	35

Resins consisting of polymerization products of acrylic acid and its derivatives may be used in nitrocellulose compositions, and it has been claimed that this will overcome inherent disadvantages of nail varnishes such as shrinkage during drying, peeling at the edge of the nail, discoloration, thickening of the enamel by evaporation of volatile ingredients when stored in poorly stoppered containers, and failure to achieve proper gloss and durability. A typical formula cited is the following:

	(14) *parts*
Cellulose nitrate	3·2
Denatured alcohol	1·4
Butyl acetate	19·3
Ethyl alcohol	19·3
Toluene	19·3
Propyl methacrylate resin	15·0
Plasticizer	3·2

Both high-boiling and low-boiling solvents are said to be eliminated in the above formula and the enamel produced is said to afford excellent adhesion and to be remarkably resistant to soapy water, alkali, marking and scratching.

Keithler[19] gives an interesting formula containing no nitrocellulose and using a chlorinated biphenol plasticizer:

	(15) *parts*
Polyvinylchloride–acetate copolymer	10·7
Polymethylmethacrylate	7·0
Methyl ethyl ketone	21·5
Sorbitol	20·0
Arochlor 5460	15·0
Tricresyl phosphate	8·0
Ethyl acetate	22·3
Colours	0·5

Manufacture of Nail Lacquer

The preceding sections are to a certain extent theoretical in their approach to nail polish production. One would not, for example, make a nail lacquer by measuring into a mixing vessel the individual materials listed in example 11. In practice it is more usual to buy a base lacquer which incorporates the particular requirements for plasticizer, resin, etc., and mix with it pearlescent materials and pigments, etc. Because of the specialist approach needed to handle and process nitrocellulose and bentone, most cosmetic manufacturers handle nail lacquer production along these lines. Many just fill the finished product as supplied to them from a specialist contract manufacturer, although they themselves may well have developed the shades, etc.

An outline of all the manufacturing procedures involved in nail lacquer production is given in Figure 22.1.

Base Coats and Top Coats

The object of using a base coat is to prevent the chipping of the nitrocellulose film, by forming on the nails a film to which the nitrocellulose lacquer will closely

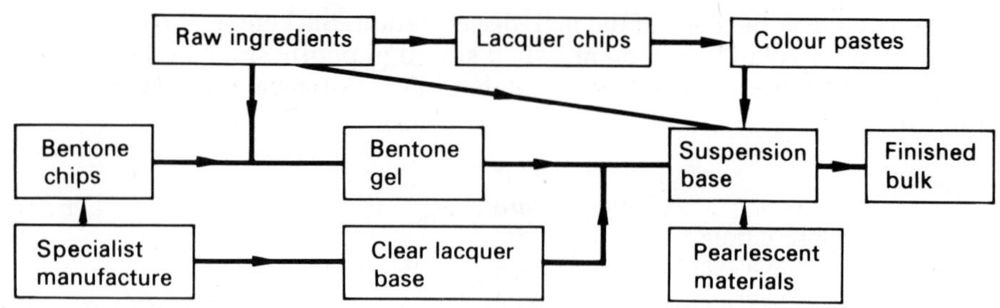

Figure 22.1 Manufacturing routes for nail lacquer production

adhere. Phenol-formaldehyde resins, sulphonamide-formaldehyde resins, alkyd resins and methacrylate are used for this purpose.

Base coats often contain a larger amount of resin for increased adhesion; they dry more rapidly and give a harder film that the normal lacquer film. Films built on a base coat are far less liable to crack than ordinary nail enamels. Clear and pigmented base coat formulations are available with a lower solid content and lower viscosity than normal lacquer formulations.

Top coats are usually clear lacquer films which are applied over the pigmented lacquer film in order to increase film thickness and thus resistance to abrasion and wear, and also to improve gloss. They have a higher nitrocellulose and plasticizer content and a lower resin content than normal nail enamel and they also have a relatively high diluent content and tend to dry rapidly. Three formulae for a base coat, a clear lacquer and a top coat, originally quoted by Peirano[11] and later cited in *Manufacturing Chemist*,[13] illustrate the differences in the composition of such products (examples 16–18).

	(16) Base coat *per cent*	(17) Clear lacquer *per cent*	(18) Top coat *per cent*
Nitrocellulose	10	15	16
Santolite resin	10	7·5	4
Dibutyl phthalate	2	3·75	5
Butyl acetate	—	29·35	10
Ethyl acetate	34	—	10
Ethyl alcohol	5	6·4	10
Butyl alcohol	—	1·1	—
Toluene	39	36·9	45

Enamel Remover

The products used as enamel removers usually consist of simple mixtures of solvents, such as acetone, amyl acetate or ethyl acetate, which may contain small amounts of fatty material to counteract any excessive drying action of the solvents on the nails. Originally castor oil was used, but nowadays esters such as butyl stearate or dibutyl phthalate, fatty alcohols or soaps may be used for the same purpose. An example of an enamel remover is provided by the following formula:

	(19)
	per cent
Butyl stearate	5·0
Diethylene glycol monoethyl ether	10·0
Acetone	85·0

Gamma-valeractone ($C_5H_8O_2$), a colourless and almost odourless solvent, has been suggested[20] by the manufacturers as a nail polish remover. Unlike the majority of organic liquids used for this purpose, GVL (as the compound is known) is completely miscible with water; a 50 per cent solution in water is suggested for use as a nail polish remover. It is stated to be non-irritating to the skin. Yet another solvent suggested for use in enamel removers is butyrolactone.

Another type of enamel remover is the so-called 'non-smeary' remover which is claimed to remove the nail lacquer without smearing the nails or the adjoining skin. These products consist of a mixture of water and water-miscible solvents such as ethyl acetate in which the water content is only of the order of 8–9 per cent.

Cream Nail Enamel Removers

A number of creamy nail enamel removers have been introduced, many of them patented. In general these consist of a fairly high-boiling solvent for the lacquer, introduced into a cream which may contain some lanolin or other emollient material; alternatively they may be composed of a lacquer solvent solidified by means of a suitable wax.

One example of a product of this type disclosed in a US patent[21] had the following composition:

	(20)
	per cent
Paraffin or beeswax	11·5
Lanolin	4·0
Sodium or potassium linoleate	2·6
Methyl ethyl ketone	to 100·0

Dreyling[22] described a creamy nail polish remover which contains about 70 per cent by weight of a substantially anhydrous organic solvent capable of dissolving nitrocellulose, the remainder of the preparation comprising a mixture in the ratios, by weight, of 10 parts of a solid fatty acid such as stearic or palmitic, 2 to 4 parts of ethyl cellulose, 2 to 10 parts of a cosmetic aid (castor oil, olive oil and lanolin) and 2 to 4·6 parts of concentrated ammonia water; example 21 is taken from the patent:

	(21)
	parts
Ethyl acetate	40·0
n-Butyl acetate	40·0
Castor oil	4·0
Perfume	0·1
Ethyl cellulose (60/80 seconds)	2·8
Stearic acid	11·0
Ammonium hydroxide (conc)	3·8

Procedure: Mix the above ingredients (with the exception of the ammonium hydroxide) until complete solution is obtained (heating at 50°–60°C will hasten rate of solution). When the composition forms a clear liquid, add the ammonium hydroxide slowly, stirring the mixture.

An important and critical part of the above invention is stated to lie in the balance existing between the ratios of ethyl cellulose, stearic acid, ammonia and water. Attention is drawn to the fact that, while ammonium stearate produced *in situ* is the emulsifying agent, it does not appear possible to produce a satisfactory preparation by the inclusion of a previously and separately prepared ammonium stearate.

Carter[23] described a cream to remove nail polish consisting of an intimate mixture of water-soluble soap, free stearic acid (in greater proportion than the soap, the two together forming at least 15 per cent of the cream) and a solvent for nitrocellulose. Ochs[24] described a type of preparation comprising 4 per cent lanolin, 11·5 per cent wax, 2–6 per cent alkali metal soap, 5 per cent water and 70–80 per cent of a solvent for cellulose nitrate (for example, methyl ethyl ketone).

In an earlier patent[25] Carter employed bentonite as an emulsifier. This patent points out that in mixtures of oil and water, bentonite collects at the interfacial boundary and acts as a protective colloid to stabilize the emulsion.

A suggested composition from this patent is given in example 22.

	(22) *per cent*
Sulphonated olive oil	10·0
Sodium hydroxide (10%)	1·0
Benzyl acetate	15·0
Acetone	15·0
Titanium dioxide	0·5
Bentonite	2·0
Water	56·5

Procedure: Mix the bentonite with 20 per cent of the water and allow to stand for about 12 hours. Then add the pigment, oil, organic solvents, alkali and the remainder of the water to the suspension in this order, with continuous agitation.

Example 23 gives a useful formula quoted in *Drug and Cosmetic Industry.*[26]

	(23) *per cent*
Stearic acid	9·5
Triethanolamine	3·5
Mineral oil	10·0
Butyl acetate	50·0
Butyl stearate	5·0
Carbitol	5·0
Water	17·0

Nail Drier

Nail driers are aerosol formulations which make use of the rapid evaporation of a propellant to speed up the drying of freshly enamelled nails, by drawing off the solvent present in the nail varnish. Sometimes drying is combined with the deposition of a transparent film of oil over the freshly applied enamel, to reduce its tackiness and to prevent it from smearing if touched.

PLASTIC FINGERNAILS AND ELONGATORS

Plastic fingernails and elongators are largely used to improve the cosmetic appearance of damaged or short stubby nails. They are produced by the polymerization or copolymerization of monomers in the presence of a polymer, a catalyst and a polymerization promoter.[27] A plasticizer, an opacifier, a pigment and a filler may also be included.

The most frequently employed monomer has been methyl methacrylate while poly(methyl methacrylate), polystyrene, polyisobutylene and cellulose acetate have been used as the polymer. If methyl methacrylate is copolymerized with a crosslinking agent such as ethylene dimethacrylate, more brittle polymers with improved solvent resistance are produced. The crosslinking agent polymer is usually employed in amounts ranging between 2 and 10 per cent by weight of the monomer. The actual level at which it is used largely determines the hardness and brittleness of the product. Compounds of improved flexibility may be produced by forming a copolymer containing 5–20 per cent by weight of a higher-molecular-weight ester of acrylic or methacrylic acid such as lauryl methacrylate.

Triphenyl or tricresyl phosphate, dimethyl, diethyl or dibutyl phthalates may be used as plasticizer in amounts ranging between 1 and 10 per cent by weight of the monomer.

Polymerization of methyl methacrylate is a free radical-catalysed reaction entailing the dissociation of the catalyst to free radicals which initiate polymerization. The catalysts which are preferred in the production of nail elongators are benzoyl peroxide or lauroyl peroxide, and they may be used in amounts of up to 3 per cent by weight of the solid constituent of the preparation; usually, however, 1 per cent of the catalyst on the weight of monomer will be adequate. A polymerization promoter is also usually present to induce free radical release and to allow polymerization of component monomers to take place at room temperatures, that is at temperatures appreciably lower than those at which polymerization usually takes place.

Among fillers, silica and aluminium silicate, mica and metallic oxides have been used. The filler may constitute up to about 25 per cent of the solid compositions. Pigments and opacifiers are generally used at a level of 0·1 per cent.

Irrespective of the procedure adopted, it is necessary to prepare the surface of the nail before applying the nail elongator composition by removing any excess cuticle and all traces of any old lacquer from the nail, as well as to smooth any ragged surfaces, so as to ensure good adhesion between the nail surface and the plastic material applied.

The skin surrounding the nail is protected from polymerization taking place on its surface by applying to the skin a material with little adhesion for the plastic, usually a water-soluble resin easily removable with water.

Viola[28] has described in detail the procedure used to form a nail elongator *in situ*. It entails successive applications, first of the liquid monomer and then of the powder containing the catalyst. This is repeated until the nail has been adequatedly lengthened. Once the tip of the nail has been strengthened, the nail is then built up from the centre of the nail to the outer edge.

The ease with which methyl methacrylate and also other vinyl monomers polymerize has led to the incorporation in such compositions of, for example, hydroquinone which gives rise to the actual polymerization inhibitor benzoquinone. The polymerization inhibitor reacts with the chain radicals as they are formed, thereby preventing the addition of monomer molecules. The inhibitor may be removed from the film by washing the monomer with a sodium hydroxide–salt mixture. It is also important to ensure that the catalyst and promoter are not present in excesive amounts, otherwise the inhibitor will not be able to react with all the excess free radicals produced.

To illustrate the composition of solid and liquid components, the following example is quoted from the literature:

	(24)
Solid component	*per cent*
Poly(methyl methacrylate) (granules)	75
Aluminium silicate	23
Benzoyl peroxide	2
Liquid component	
Methyl methacrylate (monomer)	83
Lauryl methacrylate (crosslinking agent)	15
Diethyl aniline (promoter)	2

Procedure: Mix 1·8 parts by weight of the powder component with 1 part by weight of the liquid component; polymerization to a hard plastic material takes place within 15–20 minutes at room temperature.

Compositions and the technique employed in producing natural-looking fingernails have also been described in another US patent;[29] the artificial nails are produced *in situ* by rapid polymerization or copolymerization of a constituent monomer at room temperature in the presence of a polymer and a redox catalyst system.

They are formed from two components, one of which is a primer composition containing a film-forming resin, for example a mixture of poly(methyl methacrylate) and a copolymer of vinyl chloride and vinyl acetate, and the oxidizing component of a redox catalyst system. This component is used to form coatings on the exposed nail bed and the surrounding skin and protects them from irritation by the monomer contained in the second composition. The film-forming solids content of the primer composition should preferably not exceed 25 per cent.

The second composition contains most of the resins for the artificial nails and consists basically of a solution of a polymeric film-forming resin, for example poly(methyl methacrylate), in a resin-forming monomer, for example methyl methacrylate. It also contains the reducing component of the redox catalyst. Both compositions are stable as long as they are stored separately, but when they are combined the monomer begins to polymerize.

The catalyst system employed in these compositions, as shown in embodiment examples of the patent, consists of benzoyl peroxide as the oxidizing component and N-dimethyl-*p*-toluidine as the reducing component.

NAIL MENDING COMPOSITIONS

Mending compositions described in a British patent[30] are basically mixtures of an adhesive, a fibre reinforcing material and a solvent. The first two components produce a film which forms a strong bond between the broken parts of the damaged fingernail, improving its appearance and preventing further damage, while the solvent, which permits easy application of the composition to the fingernail, will quickly evaporate and allow the reinforced adhesive film to set and to mend the damaged fingernail. These preparations are preferably applied in four coats, each coat being applied in a direction perpendicular to the previous one and allowed to become surface dry before applying the next coat. The nail polish is applied about one hour after the application of the final coat of the mending preparation, to allow all the coats to dry thoroughly.

The preferred organic film-forming material of the mending composition is nitrocellulose. A resin, for example an aryl sulphonamide formaldehyde, is also included to make the film produced by the composition more adherent, flexible and tough, and a plasticizer may be added to make the film flexible. Short rayon fibres (1·5 mm long) and of small diameter (1·5–5 denier) are used as reinforcing agent. Other fibres may also be used if they are not soluble in the mending composition. A gelling agent (preferably silica) is also present, to ensure that the fibres used remain suspended within the adhesive. Solvents specifically mentioned in the patent are ethyl acetate, butyl acetate and toluene used in a combination of 30–50 per cent, 5–20 per cent and 20–30 per cent respectively. Pigments or dyes may also be included to confer the desired colour. The compositions just described are illustrated by example 25. It has been claimed that nail polish can be applied to and removed from such mended nails without difficulty and without disturbing the 'repair'.

	(25) *per cent*
Nitrocellulose	10·3
Aryl sulphonamide formaldehyde resin	4·1
Silica	2·0
Dibutyl phthalate	0·5
Rayon fibres	0·5
Ethyl acetate	46·2
Butyl acetate	5·1
Toluene	31·3

Also patented have been keratin-containing compositions[31] which may be used either as nail varnishes or for the repair of chipped or cracked nails, as well as for producing artificial nails. They are intended to provide coatings that are not incompatible with natural nails. The keratin, which is resistant to various solvents including dilute acids and alkalis, is combined with a suitable water-insoluble bodying agent, for example cellulose nitrate, and dissolved in a suitable solvent, preferably methyl ethyl ketone, or a mixture of ethyl and butyl acetates. Synthetic resins may be included to increase adhesion, solids content and water resistance, and also to enhance gloss. Plasticizers may also be incorporated to avoid excessive brittleness of coatings.

The keratin is used in finely particulate form in the size range of 400 and 20 mesh, its amount according to embodiment examples varying between 5 and 9·5 per cent of the composition.

Coatings obtained with these compositions were claimed to give the necessary strength, body, hardness and adhesion for satisfactory performance of the coatings.

REFERENCES

1. Questions and Answers, *Am. Perfum.*, 1949, **54**(4), 282.
2. Dragoco, *Cosmetic Products and their Perfuming*, 1962, 105.
3. *Am. Perfum. essent. Oil Rev.*, 1938, **53**(5), 32.
4. Schlossman, M. L., *Cosmet. Toiletries*, 1981, **96**(4), 51.
5. US Patent 3 034 965, Drake, R. P. and Whitley, L. F., 12 June 1958.
6. British Patent 946 095, Drake, R. P. and Whitley, L. F., 12 June 1958.
7. US Patent 3 034 966, Williams E. W., 9 February 1959.
8. Lazer, P., *Arch. Dermatol.*, 1966, **99,** 446.
9. French Patent 1 388 164, Joos, M. B., 23 January 1964.
10. Australian Patent 284 933, Recherches Pharmaceutiques et Scientifiques, 20 July 1964.
11. Peirano, J., *Cosmetics: Science and Technology*, ed. Sagarin, E., New York, Interscience, 1966, p. 678.
12. Wing, H. J., *Proc. sci. Sec. Toilet Goods Assoc.*, 1947, (8), 9.
13. Alexander, P., *Manuf. Chem. Aerosol News*, 1966, **37**(6), 37.
14. Van Ham, G., in *International Cosmetic Regulations*, *9th Annual International Cosmetic Regulations Conference*, *Basle*, *1979*, Wheaton, Ill., Allured Publishing, 1979.
15. German Patent 1 143 483, Ciba, February 1963.
16. US Patent 3 422 185, Kuritzkes, C. M., 14 January 1969.
17. Shansky, A., *Drug Cosmet. Ind.*, 1978, **123**(5), 46.
18. *Monsanto Technical Bulletin* P. L. 320.
19. Keithler, W., *Drug Cosmet. Ind.*, 1957, **80,** 309.
20. *Schimmel Briefs* No. 142, 1947, January.
21. US Patent 2 286 687, Ochs, W. F., 16 June 1942.
22. US Patent 2 351 195, Dreyling, A., 6 June 1944.
23. US Patent 2 268 642, Carter, H. M., 6 January 1942.
24. Canadian Patent 412 396, Ochs, W. F., 16 June 1942.
25. US Patent 2 197 630, Carter, H. M., 16 April 1940.
26. Answers to Questions, *Drug Cosmet. Ind.*, 1956, **79,** 698.

27. US Patent 2 558 139, Knock, F. E. and Glenn, J. F., 26 June 1951.
28. Viola, L. J., *Cosmetics: Science and Technology*, ed. Balsam, M. S. and Sagarin, E., New York, Wiley-Interscience, 1972, p. 543.
29. US Patent 3 478 756, Inter-Taylor AG, 18 November 1969.
30. British Patent 1 177 420, Max Factor, 14 January 1970.
31. US Patent 3 483 289, Nicholson, J. B. and Criswell, A. F., 9 December 1969.

PART THREE

The Hair and Hair Products

Chapter Twenty-three

The Hair

Introduction

The profound psychological and social significance of hair in Man is in contrast to its complete lack of vital function.[1] For mammals, as a whole, fur provides an insulating coat for the conservation of body heat. Its properties can be adapted to seasonal changes in the environment by periodic moulting of old hairs and their replacement by new ones, and even human hair follicles remain endowed with such cyclic activity. However, Man evolved by the ability to keep cool. The ape, *Ramapithecus*, an ancestor of 10 million years ago, was probably hairy, but the early hominids, as they left the forest and moved to the savannah, started a march towards nudity; the hair on the body became sparser and shorter.

Not all was lost. The eyebrows and eyelashes remained, as did the hair on the scalp, perhaps as a protection from the midday sun for an animal beginning to stand—and walk—on its own two feet. The beard remained as a badge of maleness. And, in both sexes, the genital and axillary hair, which was probably associated with glandular scent-producing units, now became conspicuous.

The Hair Follicle

Development and Structure

The structure of the hair follicle, the method by which it manufactures hair and its cyclic activity are better made understandable by brief reference to its embryonic history.[2,3] Each follicle arises from an interaction between epidermis and dermis.

A thickened plaque of epidermis, lying over an aggregation of dermal cells, extends inwards to form a peg which eventually engulfs a small papilla of dermis to form the *hair bulb*. The epidermal cells surrounding the dermal papilla then proliferate to push outwards a column of keratinizing cells, which is the hair shaft invested in an inner root sheath. A hair canal is formed in the process.

The bulk of the hair shaft consists of elongated, keratinized cells cemented together and is known as the *cortex*. Some, though not all, hairs have a continuous or intermittent *medulla*. The cortex is surrounded by a *cuticle*, which arises from a single line of cells in the bulb, but becomes five to ten overlapping layers. From the outside the cuticular scales appear imbricated, like roof tiles, with the free edges directed outwards (Figure 23.1).

The inner root sheath interlocks with the overlapping cells of the cuticle of the growing hair and moves with it, but the keratinizing cells are desquamated as the hair emerges from the skin. Thus the outer surface of the inner root sheath glides against the stationary *outer root sheath*, which is the innermost part of the follicular wall.

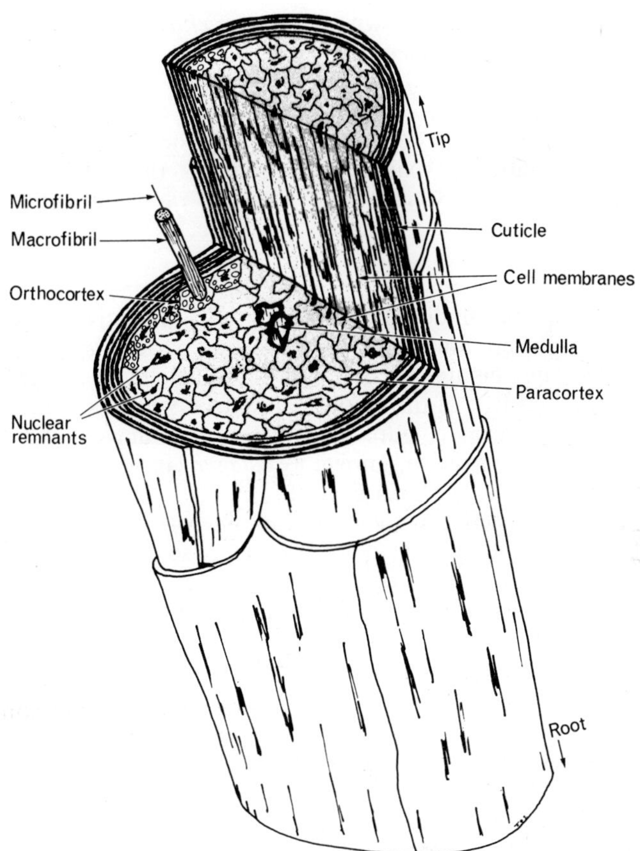

Figure 23.1 Section of a hair fibre (courtesy Unilever Ltd)

During the development of the peg, three bulges appear on its posterior wall. The outermost is the rudiment of the apocrine gland, which remains vestigial except in the genital, axillary and areolar regions, the next is the sebaceous gland, and the innermost is the attachment for the arrector muscle.

As the foetus grows, the first-formed follicles move apart and a new crop of secondary rudiments develop between them. But no new follicles are formed post-natally.[4] The total number of follicles in an adult man is about 5 million, of which about 1 million are in the head and perhaps 100 000 in the scalp.[5] A significant loss occurs with age;[6] young adults have an average of 615 per cm^2 on the scalp, but by the age of 80 the density has fallen to 435 per cm^2.[7] Some follicles are lost in baldness; a comparison over a wide range of ages gave averages of 306 per cm^2 for bald scalps as against 459 per cm^2 for hairy scalps.[7]

The first hairs to grow from hair follicles, which are fine, unmedullated and usually unpigmented, are known as *lanugo* and are normally shed *in utero* in the seventh or eighth month of gestation. Very rarely, lanugo-like hair is found in the adult, even on areas such as the nose and ears which are not normally hairy, either as an inherited trait known as *hypertrichosis lanuginosa*, or as the symptom of an underlying cancer.

Postnatal hair may be divided, at the extreme, into two kinds: fine, unmedullated, short *vellus* on the body, and longer, darker, *terminal hair* on the scalp. The infantile pattern is not definitive, for at puberty vellus is replaced by terminal hair in the pubic and axillary regions and, in the male only, on the face. This sexual hair continues to increase in area and rate of growth until the late twenties.[2]

Cyclic Activity

Virtually all hair follicles undergo cyclic activity.[8,9] An active phase, *anagen*, in which a hair is produced, alternates with a resting period, *telogen*, in which the fully formed *club hair* remains anchored in the follicle by its expanded base and the dermal papilla lies free of the epidermal matrix, which is reduced to a small, quiescent secondary germ (Figure 23.2). Between anagen and telogen is a relatively short transition phase, known as *catagen*, in which the newly formed club hair moves towards the skin surface.

The follicle becomes active again at the end of telogen by a downgrowth of the secondary germ to re-invest the dermal papilla, so that the matrix becomes reconstituted and a new hair starts to form (Figure 23.3). In effect, the follicle re-enacts its embryonic development. Ultimately the old club hair is shed, or *moulted*.

All hairs thus reach a terminal length which is determined mainly by the duration of anagen and partly by the rate of growth. These characteristics vary with site. In the scalp, anagen may occupy three years or more;[9,10] indeed the untrimmed locks of a young woman would require 6 or 7 years of continuous growth to reach her buttocks. The aging male scalp falls far short of such achievement. On the vertex of a man aged 60,[11] the growing period ranged from 17 to 94 weeks for coarse hairs and from 7 to 22 weeks for fine hairs. On the body, the period of the cycle is much less. In a young male[11] it ranged from 19 to 26 weeks on the leg, 6 to 12 weeks on the arm, 4 to 18 weeks on the finger, 4 to 14 weeks in the moustache, and 8 to 24 weeks in the region under the temple.

Animals show wide differences. For example, in the rat the dorsal hair is fully formed in three weeks and the shorter ventral hair in only 12 days,[12] and in the guinea-pig anagen lasts between 20 and 40 days.[13]

In many animals waves of follicular activity, followed by moulting, pass over the body in symmetrical patterns. In the rat, for example, replacement of hairs starts in the venter and bands of activity and shedding move over the flanks to the dorsum, subsequently spreading to the head and the tail regions.[12] In contrast, in the human scalp each follicle appears to be independent of its neighbours. At any one time an average of 13 per cent (4–24 per cent) of the follicles are in telogen[9,14,15] and about 1 per cent in catagen. Thus in a scalp containing around 100 000 follicles with an average cycle of 1000 days, about 100 club hairs should be shed each day. This is approximately the number recovered in practice.[9,10]

The guinea-pig has been said to resemble man in that moulting is continuous and at random. However, in the newborn animal all follicles are simultaneously active[13] and it seems that for at least 50 days, and probably for much longer,

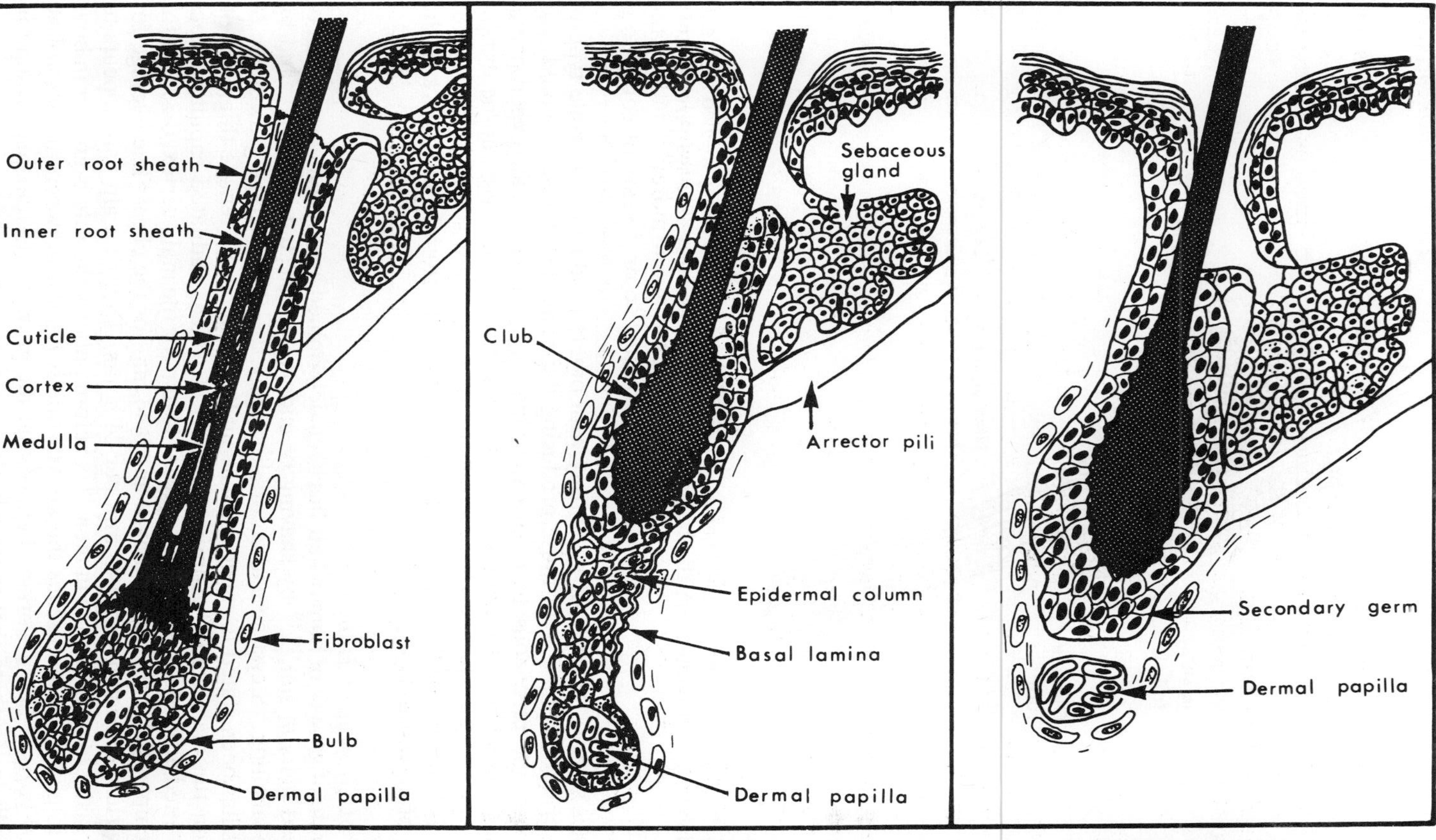

Figure 23.2 The hair follicle cycle

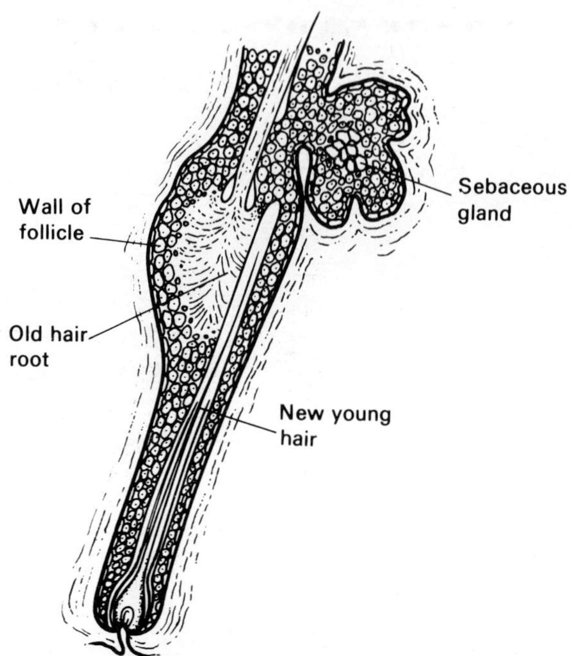

Figure 23.3 Hair follicle in early anagen, prior to shedding. Note newly formed hair, and old club hair still retained in the follicle (after Unna, courtesy Edward Arnold Ltd)

follicles show a measure of synchrony with others producing the same type of fibre, if out of phase with those producing different fibre types. It would thus appear to be erroneous to regard the guinea-pig follicle as an experimental model from which information relevant to the human condition can be drawn. However, at an early stage human follicles may also show some synchrony, for there is evidence of the passage of a wave of growth from front to back in the scalp of newborn infants.[10,16,17]

Rate of Growth of Hair

The rate of growth of human hair has been determined by direct measurement of marked hairs *in situ*,[18,19] by shaving or clipping at selected intervals[20] and by pulse labelling ^{35}S cystine and autoradiography.[21,22] The average growth per 24 hours has been stated to range from about 0·21 mm on the female thigh to 0·38 mm on the male chin.[20] Another study,[19] in which graduated capillary tubes were fitted around growing hairs, gavc 0·44 mm for the male vertex, 0·39 mm for the temple, 0·27 mm for the beard and 0·44 mm for the chest. The average for the vertex in women was 0·45 mm per 24 hours. Though scalp hair appears to grow faster in women than in men,[19,23] before puberty the rate is greater in boys than in girls.[15] In both sexes the growth rate is highest between the ages of 50 and 69 years.[24] Some workers believe that the growth rate remains constant in any follicle,[19] others that it usually decreases or increases.[21]

While daily variations in temperature have no effect on hair growth, it seems likely that there are longer-term seasonal changes; for example, the beard grows faster in summer than in winter.[25] It is generally agreed that shaving does not alter the rate of growth,[26] though prolonged irritation may affect the follicle.[27]

Rates may differ considerably between species. Hair growth can be more than 1·0 mm per 24 hours in the rat[12,28,29] and up to 0·6 mm in the guinea-pig, in which it has been clearly shown that the rate depends on the time for which activity of the follicle has been in progress.[30]

Hormonal Influences

There is ample experimental and clinical evidence that hormones influence hair growth. The analysis of their actions is, however, extremely complicated. In considering the extent to which animal models may illuminate the human condition, it is important to distinguish the control of the follicular cycle, concerned with the adaptive function of the moulting, from the control of sexual hair, concerned with adult social interactions. Moreover, hormones may act at more than one point; an action which modifies the structure of the hair, the rate of growth or the duration of the growing phase must be distinguished from one which simply reduces or prolongs the resting period. The failure to make such distinctions has led to many useless experiments and even more fallacious conclusions.

The control of the follicular cycle appears to be exercised at several different levels. First, experiments on rats have demonstrated that each follicle has an intrinsic rhythm which continues when its site is changed[31] or, at least for a period, when it is transplanted to another animal in a different phase of the moult.[32] When hairs are plucked from resting follicles, activity is initiated and at least for a time the follicles remain out of phase with their neighbours. Whether the intrinsic control of the follicle involves the build-up and dispersal of inhibitors[33] or the release of wound hormones when the follicle is epilated,[34] or both, continues to be debated.

The intrinsic cycle is, in turn, influenced by systemic factors, so that, for example, follicles on homografts gradually come into phase with their hosts.[32] Several hormones have been shown to affect the moult cycles in animals. Thus, the passage of the moult in the female rat is accelerated by removal of the ovaries or the adrenals, or by administration of thyroid hormone and, conversely, is delayed by administration of oestrogens or adrenocortical steroids, or by inhibition of the thyroid.[35–37] In a number of mammals moulting is influenced by seasonal changes, in particular of the photoperiod.[38] It seems likely that these act through the hypothalamus, the anterior pituitary and the endocrine system.

Hormones also influence the rate of hair growth and the duration of anagen in the rat. Oestrogen decreases both; thyroxine increases growth but reduces the growing period.[29] This finding suggests that the two hormonal systems act at different points.

Such experimental studies on animals have only a limited relevance in relation to human hair growth, but there are several points of interest. First, it appears that post-partum alopecia (see below) is caused by a disturbance of the hair follicle cycle brought about by circulating hormones, probably by the high level of oestrogens, in late pregnancy. Secondly, thyroid disturbances are frequently associated with diffuse alopecia. Thirdly, even the human scalp may not be free from environmental influences, for hair fall shows significant seasonal fluctuations.[11,39]

The growth of the sexual hair is brought about by androgens, that is, by male hormones. The first sexual hair to appear is that in the pubic region. In boys it reaches the classical 'female' pattern at an average age of 15·2 years,[40] in girls about $1\frac{1}{2}$ years earlier.[41] In about 80 per cent of men and 10 per cent of women the pubic hair spreads further over the abdomen in a sagittal, acuminate or disperse pattern; this growth may not be complete until the mid-twenties.[42] Axillary hair appears about two years after the start of pubic hair growth and continues to increase until the late twenties.[43] The rate of growth of the beard levels out after the age of 35, but the area continues to increase until the later decades of life.[43]

Castration virtually prevents the development of the beard if it is carried out before the age of 16 years.[43] Performed after the age of 21, it only reduces the extent, and the weight of hair growth per day.[43] There is little doubt that the rising levels of androgen at puberty account for the successive development of pubic and axillary hair in both sexes, and for the growth of the male beard. Equally, androgens are necessary for the development of female *hirsutism*, that is body and facial hair in part or whole of the male pattern. Either abnormally high levels of androgen or an increased response of the pilo-sebaceous apparatus, or both, may be involved. Anti-androgens, namely substances which block the attachment of male hormones to the receptors within the target cells, effectively reduce such hair growth.[44]

Paradoxically, androgens are also a prerequisite for the development of *male pattern alopecia* or *alopecia androgenetica* in subjects who are genetically disposed. Eunuchs retain their scalp hair, even when they have a family history of baldness, unless they are treated with testosterone.[45] However, castration of balding men, though it arrests the process, does not appear to restore luxurious hair growth to areas that are already bald.[46] The condition will be further considered below.

Nutritional Influences

The nutritional requirements of skin and the general consequences of dietary deficiency have been reviewed in Chapter 3. Attention was drawn to the necessity of certain vitamins, particularly some of the B complex, for normal hair growth and keratinization of the epidermis. On the other hand, evidence was presented that vitamin A inhibits the differentiation of the stratified squamous epithelium. Thus hyperkeratotic papular dermatitis is a symptom of vitamin A deficiency, and excess of the vitamin, for example when used in the treatment of psoriasis, appears to cause loss of hair.[47]

Protein deficiency is the cause of the condition *kwashiorkor* and has serious consequences for hair growth. The hair becomes sparse, thin and brittle and loses its pigment. The linear growth may be decreased by as much as one-half;

children with kwashiorkor produced only 59 μm^3 hair tissue per follicle per day compared to a normal figure of 514 μm^3, and the hair was weaker even when corrected for the difference in diameter.[48]

The changes in the hair reflect considerable alterations in the follicles themselves. Thus in a sample of Andean children with kwashiorkor the proportion of follicles in anagen was only 26±6 per cent compared with 66±6 per cent in healthy children of the same age, and even the anagen follicles were severely atrophied, with depletion of pigment and loss of internal and external root sheaths.[49] Even more severe changes were found in children suffering from marasmus; less than 1 per cent of the follicles were in anagen.[50] Marasmus appears to be a condition of chronic protein–calorie undernutrition in which the hair follicles have ceased activity in order to conserve nitrogen, whereas kwashiorkor is an acute condition following more or less normal growth.[50]

The hair bulb is rapidly affected by experimental protein malnutrition. In male volunteers, the mean root diameters became significantly reduced by the eleventh day of protein deprivation, pigmentation was visibly reduced by the fourteenth day, and there was progressive atrophy and loss of root sheaths.[51] The changes were reversible within days when protein was given. Thus while claims[52] that administration of amino acids will, in certain conditions, improve hair growth should be treated with great caution, the possibility should not be rejected out of hand.

Chemistry of Hair

Keratin

The greater part of hair and wool is made up of an insoluble protein material called keratin, which is formed as the ultimate product of the keratinization process which takes place in the follicle. There are also present the vestigial cell membranes, nuclei, etc., but these form a very small fraction of the substance of hair.[53] Small quantities of water-soluble substances are also present, such as pentoses, phenols, uric acid, glycogen, glutamic acid, valine and leucine.[54]

Keratin, like other proteins, is composed of amino acids, substances of the general formula

$$R\text{—}CH\begin{matrix} \diagup NH_2 \\ \diagdown COOH \end{matrix}$$

or in the zwitterionic form

$$R\text{—}CH\begin{matrix} \diagup \overset{+}{N}H_3 \\ \diagdown COO^- \end{matrix}$$

which give rise to the majority of the most characteristic properties of the proteins.

About 25 different amino acids are known to occur in proteins, and of these 18 are found in measurable amounts in keratin. They may be described and characterized by the side-chains R, which may be of three types:

(a) Inert chain as in the amino acid leucine:

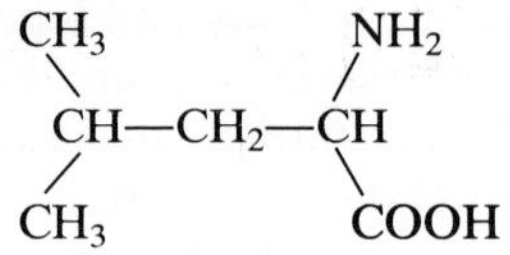

$$\begin{array}{ccc} CH_3 & & NH_2 \\ \diagdown & & \diagup \\ CH—CH_2— & & CH \\ \diagup & & \diagdown \\ CH_3 & & COOH \end{array}$$

(b) Basic side-chain as in the amino acid lysine:

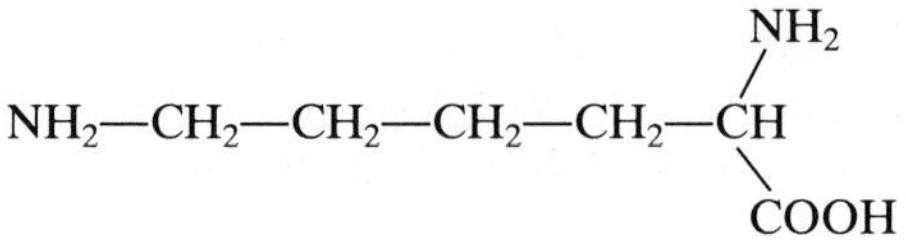

$$\begin{array}{ll} & NH_2 \\ & \diagup \\ NH_2—CH_2—CH_2—CH_2—CH_2— & CH \\ & \diagdown \\ & COOH \end{array}$$

(c) Acid side-chain as in the amino acid aspartic acid:

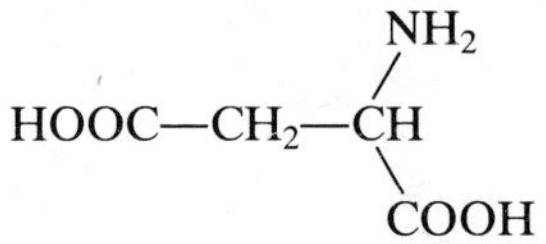

$$\begin{array}{ll} & NH_2 \\ & \diagup \\ HOOC—CH_2— & CH \\ & \diagdown \\ & COOH \end{array}$$

The incidence of amino acids in hair and wool is given in Table 23.1; the figures for wool are those of Spector[55] and those for hair are by Bell and Whewell.[56]

These amino acids can form large large condensed polymeric structures by the formation of amide links between the acid group of one amino acid and the amino group of another. The type of structure thus formed is a polypeptide of the following type:

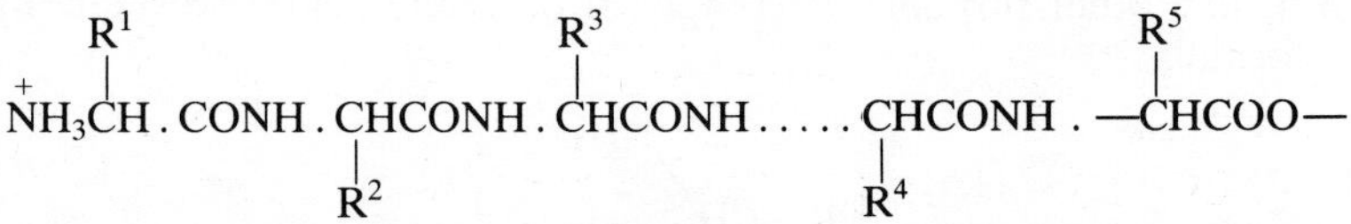

$$\overset{+}{N}H_3\underset{}{\overset{R^1}{\overset{|}{C}}}H\,.\,CONH\,.\,\underset{\underset{R^2}{|}}{C}HCONH\,.\,\overset{\overset{R^3}{|}}{C}HCONH\,.\,.\,.\,.\,.\,\underset{\underset{R^4}{|}}{C}HCONH\,.\,—\overset{\overset{R^5}{|}}{C}HCOO—$$

where R^1, R^2, etc., represent various types of side-chain.

Such a structure is common to all proteins, and in itself it is not sufficient to give the molecule the degree of stability and insolubility possessed by hair and wool.

For a protein molecule to have some organized and 'shaped' structure, the polypeptide chains must be very long, and there must also be other bonds to keep the chains in fixed relative positions to one another. These additional bonds can be set up in three ways.

1. *Formation of Hydrogen Bonds between Parallel Polypeptide Chains*. The hydrogen bonds are formed by the interaction of the NH group with a suitably placed CO group.

Table 23.1 Amino Acid Composition of Wool[55] and Hair[56]

Amino acid	Formula	Wool (%)	Wool (mole %)	Hair (%)	Hair (mole %)
Glycine	NH_2CH_2COOH	6·8	10·6	4·1–4·2	7·5
Alanine	$CH_3CH(NH_2)COOH$	4·0	5·3	2·8	4·25
Valine	$CH_3CH_2CH(NH_2)COOH$	5·4	5·4	—	—
Leucine	$(CH_3)_2CHCH_2CH(NH_2)COOH$	8·6	7·9	11·1–13·1	12·55
Isoleucine	$CH_3CH_2CH(CH_3)CH(NH_2)COOH$	4·3	3·8		
Phenylalanine	$C_6H_5CH_2CH(NH_2)COOH$	4·0	2·8	2·4–3·6	2·5
Proline	COOH / N / H (ring structure)	8·0	8·2	4·3–9·6	9·3
Serine	$HOCH_2CH(NH_2)COOH$	9·9	11·5	7·4–10·6	11·6
Threonine	$CH_3CH(OH)CH(NH_2)COOH$	6·5	6·4	7·0–8·5	8·8
Tyrosine	p-$HOC_6H_4CH_2CH(NH_2)COOH$	5·5	3·9	2·2–3·0	1·95
Aspartic acid	$HOOCCH_2CH(NH_2)COOH$	7·4	6·5	3·9–7·7	5·9
Glutamic acid	$HOOCCH_2CH_2CH(NH_2)COOH$	14·0	11·1	13·6–14·2	12·8
Arginine	$NH_2C(=NH)NH(CH_2)_3CH(NH_2)COOH$	10·6	7·0	8·9–10·8	7·7
Lysine	$NH_2(CH_2)_4CH(NH_2)COOH$	3·3	2·7	1·9–3·1	2·3
Histidine	$CH(NH_2)COOH$ / HN NH (ring structure)	1.1	0·8	0·6–1·2	0·85
Tryptophan	$CH_2CH(NH_2)COOH$ / N / H (ring structure)	1·5	0·7	0·4–1·3	0·55
Cystine	$HOOCCH(NH_2)CH_2SSCH_2CH(NH_2)COOH$	13·6	6·6	16·6–18·0	9·8
Methionine	$CH_3SCH_2CH_2CH(NH_2)COOH$	0·7	0·5	0·7–1·0	0·75
Cysteine	$SHCH_2CH(NH_2)COOH$	—	—	0·5–0·8	0·75

When a range of values is quoted, the mole per cent value is calculated from the mean of the range.

```
   R      H      O      R      H      O      R      H      O
   |      |      ‖      |      |      ‖      |      |      ‖
...CH—C—N—CH—C—N—CH—C—N—CH—C—N—CH—C—N—CH—C...
      ‖      |      |      ‖      |      |      ‖      |
      O      R      H      O      R      H      O      R
      :             :      :             :   hydrogen bonds
      :             :      :             :      :
      H      R      O      H      R      O      H      O
      |      |      ‖      |      |      ‖      |      |
...CH—N—C—CH—N—C—CH—N—C—CH—N—C—CH—N—C—CH —N ..
   |      ‖      |      |      ‖      |      |      ‖      |
   R      O      H      R      O      H      R      O      H
```

These bonds are very weak individually; but since they are extremely numerous they play a significant part in stabilizing the protein structure. However, the structural strength which they impart to a protein is limited by their property of lengthening to admit any other substance that can form hydrogen bonds, such as water, alcohols, phenols, amines, amides, etc. With water, for instance, the simple bond >N—H . . . O = C< becomes converted to the much more complex and weaker bond:

```
>N—H...O—H...O—H...O—H...O—H...O=C<
       |      |      |      |
       H      H      H      H
```

Such entry of water into the bonds is associated with solution of the globular proteins, and swelling characteristics of the more insoluble fibrous proteins.

2. *Formation of Salt Linkages between Acidic and Basic Side-chains*. As some of the side-chains of the polypeptide contain acidic groups and others contain basic groups, there is the possibility of salt formation between them if the groups are favourably placed; thus:

```
   (aspartic acid residue)
...—CO—CH—NH—...
        |
        CH2
        |
        COO⁻

        NH⁺
       (CH2)4
        |
...—NH—CH—CO—..
   (lysine residue)
```

3. *Formation of Disulphide Linkages*. The extreme strength and insolubility of hair keratin is attributed to its large cystine content. This amino acid contains two amino and two carboxyl groups; it can thus enter into two polypeptide

chains which are then linked together by a disulphide bond:

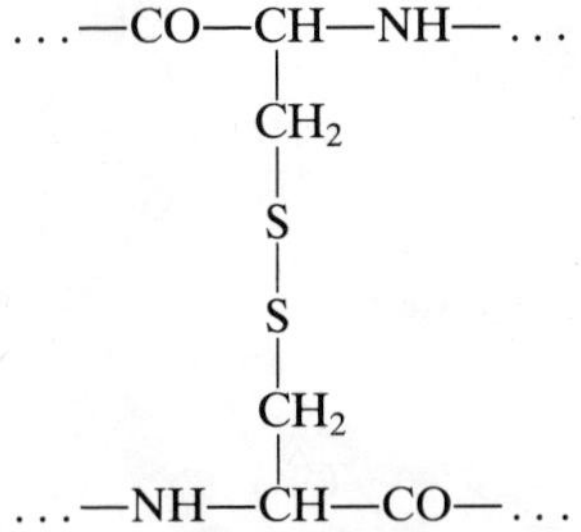

$$
\begin{array}{c}
\ldots\text{—CO—}\underset{|}{\text{CH}}\text{—NH—}\ldots \\
\text{CH}_2 \\
| \\
\text{S} \\
| \\
\text{S} \\
| \\
\text{CH}_2 \\
| \\
\ldots\text{—NH—CH—CO—}\ldots
\end{array}
$$

A few disulphide bonds are also believed to be along the main chains.[57,58] Other linkages such as ether cross-links between serine, threonine and tyrosine have been suggested, but there is little evidence for such linkages and the known chemical behaviour of hair can be explained in terms of hydrogen bonds, salt linkages and disulphide bonds.

Hair is thus an intensely cross-linked structure and can be considered as a series of submicroscopic fibrils with both parallel and linked polypeptide chains; X-ray studies show that a considerable proportion of hair has a crystalline structure (effectively a regular structure, though not necessarily crystalline in the sense associated with inorganic materials) known as the α-keratin structure. This has been described in X-ray diffraction terms by MacArthur[59] and Giroud and Leblond.[60]

Of the many structures proposed for α-keratin, the most widely accepted is that due to Pauling and his co-workers.[61–66] These authors propose a structure in which the polypeptide chains have assumed a helical form with 3·7 amino acids per turn of the helix. Each turn of the helix is located relative to the next by hydrogen bond formation between the carbonyl and imine groups of amino acids separated by two other residues (Figure 23.4). The diameter of the helix is roughly 10 Å (1 nm) and the translation per complete turn is 5·44 Å (0·544 nm). The side-chains of the amino acids all project outwards from the helical structure. A similar structure has been proposed for synthetic polypeptides such as poly-γ-methyl-L-glutamate.[67] The theoretical Fourier transform for such a helical structure has been calculated by Cochran and Crick[68] and found to agree closely with the observed patterns. Similar studies have been made by Yakel, Pauling, and Corey[69] and by Fraser *et al.*[70,71] However, Fraser notes from his work on porcupine quill tip that precise comparison is handicapped by interference effects due to the macromolecular organization of the keratin.

To explain some anomalous X-ray results, especially a calculated repeating unit at 5·2 Å (0·52 nm) spacing, Pauling and Corey suggested that the helical structure itself might have a slightly helical axis, and that under these conditions the more closely packed regions of fibres contain structures in which six of these screw-shaped helices have twisted about another, so as to form a compound helix rather like a seven-strand cable (Figure 23.5). The degree of twist in this way depends on the distribution of the side-chains, and the packing of these into the interstices of the 'cable'.

Electron microscopy has been used to study the structure of keratin,[72–75] and Orfanos and Mahrle[76] have used both the transmission and stereoscanning

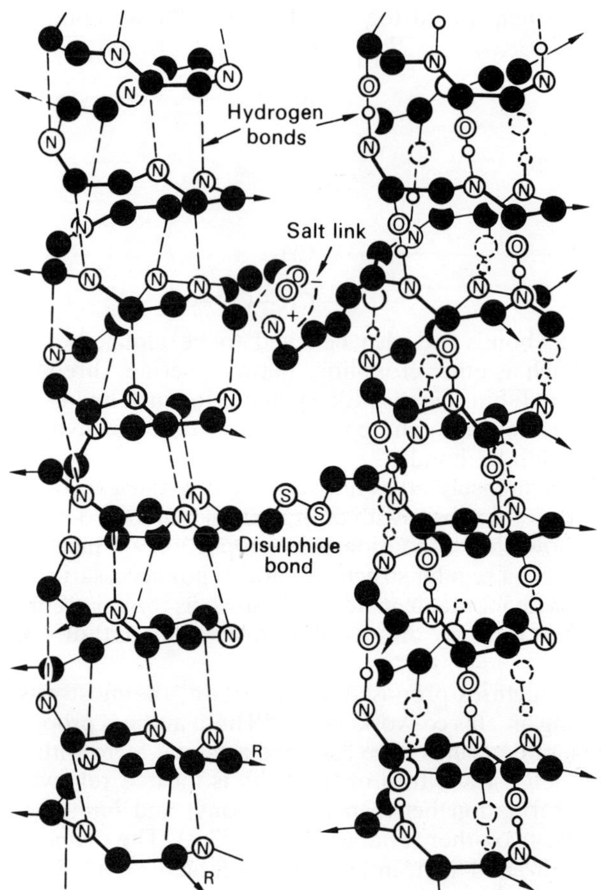

Figure 23.4 Pauling-Corey structure for α-keratin

modes to characterize the surface and structure of human hair following exposure to various hair products. Randebrock[77] has summarized the main morphological features observable in hair as:

	Diameter	
	(Å)	(nm)
Spindle cells	10 000–60 000	1000–6000
Macrofibrils	2000–9000	200–900
Microfibrils	60	6
Protofibrils	20	2
Peptide spirals	7	0·7

Not all the keratin structure is so well organized as the α-keratin structure, owing to irregularities in the material from which the keratin is formed. In

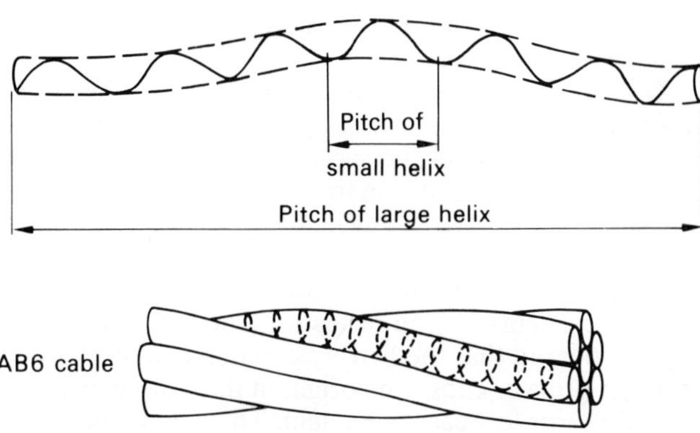

Figure 23.5 Structure of 'crystalline regions' of keratin

addition, the difficulty of packing the side-chains of 18 widely differing amino acids into any sort of regular structure over long sections of the polypeptide chain results in the existence of areas which lack any closely packed 'crystalline' character and which are referred to as the amorphous regions. These are very important in the chemistry of hair, as the amino acids in such regions are not protected from attack by the complexities of the compound helix. Some attempts have been made to estimate the relative amounts of 'crystalline' and 'amorphous' material in hair, but such measurements are difficult to interpret since there is no clearcut distinction between a slightly imperfect 'crystalline' region and a well organized 'amorphous' region. One can only assume that the helical polypeptide chains pack together where they meet in such a configuration as circumstances will permit, and one chain may pass through many crystalline and amorphous regions. The existence of several closely related keratins corresponding to various histological structural features of hair fibres has been postulated by Kuczera.[78]

A concise but detailed review of the amino acid composition, morphology and keratin-fibre histology of human hair and wool has been given by Leon.[79]

The cuticular surface of human hair has received scant attention in the literature, despite its obvious importance to those concerned with the formulation and performance of hair preparations. Keratin organization within the cortex contrasts with the non-helical cross-linked protein structure found in the surface cell layers. The results of amino acid analyses reported by Wolfram and Lindemann[80] show a high sulphur content in comparison with the cortex, accounting for the function of the cuticle as a chemical barrier and perhaps signifying a structural role in physical terms. The progressive loss of cuticle from root to tip has been described by Bottoms *et al.*[81]

Mineral Constituents of Hair

The mineral content of hair, in particular the trace metal content, has been a subject of interest for many years, especially since analytical techniques have been developed that permit the analysis to be carried out on the small samples of hair generally available. Apart from scientific curiosity there is always the possibility that metals present in trace amounts might influence the action of hair treatments such as bleaching or dyeing.

Bagchi and Ganguly[82] have determined the mineral constituents of human hair and point out that the amounts of carbon, hydrogen, nitrogen, sulphur and phosphorus are approximately of the same magnitude irrespective of age, race and sex. They also quote figures for the trace metal contents of the same samples of hair, but the data obtained are insufficient to allow generalizations to be made (Table 23.2).

Discoloration of hair during permanent waving is frequently attributed to the presence of metals, particularly iron or lead, in the hair. In cases known to one of the authors, it appeared that the hair was contaminated by iron from the use of old and worn hair grips or clips. The origin of the lead, which often amounted to 1000 ppm or more, was less easy to explain. The complaints, which concerned a slightly pinkish brown colour, generally came from subjects with perfectly white or naturally blonde hair, and it was not reasonable to suppose that they had used lead-containing hair colorants.

Tompsett[83] has shown that when the calcium metabolism of the body is disturbed, lead which is normally accumulated in the bones becomes mobile and

Table 23.2 Mineral Constituents and Trace Metal Contents of Various Samples of Hair[82]

	European girl (brown hair) (%)	**Hindu woman (black hair)** (%)	**Mixed hair of thirty male adults—from a hair cutting saloon** (%)
Carbon	44·03	44·20	44·60
Nitrogen	13·70	13·68	14·60
Hydrogen	5·58	5·60	5·40
Sulphur	3·80	1·50	3·80
Phosphorus	0·065	0·096	0·08
Chlorine	1·98	2·00	2·00
Water	3·96	4·20	4·10
	(mg/kg)	(mg/kg)	(mg/kg)
Lead	21·0	284·0	47·7
Copper	64·0	62·8	108·0
Arsenic	2·4	2·2	2·2
Zinc	116·0	182·0	212·0
Iron	133·0	126·0	141·0
Manganese	28·4	25·0	38·0
Cobalt	14·2	16·0	18·1
Nickel	5·4	5·5	8·2
Calcium	212·0	188·0	208·4
Aluminium	26·0	26·0	32·0
Silicon	188·0	178·6	150·4
Bismuth	nil	nil	nil
Silver	nil	nil	nil
Antimony	nil	nil	nil
Mercury	nil	nil	nil

can disperse into the blood stream, and hence into all the soft tissues and organs of the body. Unfortunately hair was not included in this investigation, but by analogy with arsenic, which has long been known to deposit in hair and nails, it seems highly probable that this could be the origin of lead in the hair. Further evidence for this theory comes from the work of Strain *et al.*[84] who showed that the zinc content of hair reflected the zinc content of body tissues. Similar results have been found with other metals and there seems little doubt that metals in the blood stream find their way into the hair.

The latest technique to be applied to the determination of the trace element content of hair is that of radioactivation analysis,[85,86] which enables a comprehensive analysis to be made even on a single hair. However, the technique has not, unfortunately, provided a definitive forensic technique for matching hairs.

Bate and various co-workers at Oak Ridge National Laboratory (USA) published a number of papers[87,88] on the subject and concluded that the method has serious handicaps owing to the ease with which hair adsorbs and desorbs traces of metals from externally applied solutions. Coleman *et al.*[89] at the Aldermaston Atomic Research Centre (UK) were also active in this field. They developed a standardized test procedure and a statistical method of evaluating the results. They claimed that by using a profile of 12 elements they could delineate the number of people in the population from whom a hair might have come; alternatively, they could associate the hair with one of a number of suspects whose hair could be sampled and compared. Coleman *et al.* determined the trace element content of human hair based on a random sample of some 800 subjects from England and Wales. They claimed that the difference between male and female hairs was significant and would enable unknown hairs to be 'sexed' with an accuracy of about 90 per cent.

Chemical Properties of Hair

While there is no space here to classify exhaustively all the chemical properties of hair, certain reactions are reviewed which have a direct bearing on the structural details already dealt with.

At the outset, it must be realized that about 50 per cent of the weight of hair keratin is made up of the side-chains of the amino acids, so these have a correspondingly great effect on the properties of the whole material. Because of the variety of these side-chains, the reactions are not clearcut, but the influence of certain groups may be detected in their contribution to the total chemical reactivity. For instance, if the disulphide bonds are broken the hair is weakened, but not destroyed as long as the salt links are intact. Similarly, the action of strong acids in breaking the salt links (by suppressing the ionization of the carboxylic acid groups) will not disrupt the hair unless the disulphide bonds are simultaneously broken. If the hydrogen bonds remain intact it is very difficult to carry out any other reactions with the hair, because it does not swell to admit any other reagents; it is certainly difficult to cause hair to react in non-polar solvents. Alexander[57] suggests, in fact, that most of the mechanical strength of dry hair resides in the hydrogen bonds.

Under normal conditions, however, the hydrogen bonds always contain some water adsorbed from the air, usually around 9 per cent—more or less, according to the humidity of the atmosphere, etc. In liquid water, hair can take up bound water to about 30 per cent of its own weight.[90]

Hair keratin is insoluble in aqueous salt solutions (except lithium bromide at concentrations over 50 per cent), in weak acids, weak alkalis and saturated neutral urea solution. In acid solutions between pH 1 and 2 moderate lateral swelling occurs because both hydrogen bonds and salt links are broken. The structure remains sound, however, because of the disulphide bonds. In alkaline solutions at pH 10 lateral swelling is intense for the same reasons, and at pH 12 the disulphide bonds begin to break down, lateral swelling becomes limitless and the hair passes into solution. (More details of the effects of alkali on the disulphide bonds will be found in Chapter 28 on permanent waving.)

All three types of bond can be affected by certain other materials such as sodium sulphide,[91,92] sodium thioglycollate,[93] mercaptoethanol,[94] urea-bisulphite,[95,96,97] potassium cyanide,[98] chlorine dioxide[99] and peracetic acid.[100] A mixture of phenol and thioglycollic acid will also dissolve hair keratin[101] as will formamide and urea at high temperatures (140°–160°C). Whewell[102] has reviewed the properties of keratin as a chemical and reports a comprehensive range of reagents that have been used for the modification of wool. Menkart[103] has compared the chemical properties of wool and human hair and has shown that wool is more reactive owing to its lower degree of cross-linking. This work enables the data given by Whewell to be placed in their proper perspective.

The table of amino acid analyses of hair and wool keratin shows that the total number of acid side-chains (from glutamic acid and aspartic acid) is approximately double the number of basic side-chains (from arginine, histidine, and lysine). This means that even if the positions of the amino acids in the keratin structure are favourable to salt linkage, there will still be a large excess of acidic side-chains. Normally these will be neutralized by ions such as ammonium, sodium, etc., and by the traces of other metals in the hair when these exist as cations, but these cations can be replaced by others, just as in an ion exchanger, if the circumstances are favourable. For instance, a high concentration of sodium ions (as in some shampoos) will tend to convert all the surplus acid side-chains into sodium salts; polyvalent ions will tend to replace monovalent ions (Ca^{2+} will replace Na^+; Al^{3+} will replace Ca^{2+}), and cations with some surface activity will replace the simpler inorganic ones. This last effect accounts for the readiness with which cationic detergents are adsorbed by the hair, and for the rapid adsorption of cationic dyestuffs such as methylene blue. An interesting example of the base-exchange characteristics of hair can be demonstrated with methylene blue. If hair is carefully washed with dilute sodium chloride solution for about 60 minutes, most of the acidic side-chains will become sodium salts. If such hair is dyed with methylene blue it will rapidly become a deep blue. If, however, the hair is soaked in aluminium sulphate solution, or preferably a more neutral aluminium salt, such as the chlorhydrate, before dyeing with methylene blue, it will only take on a pale blue tint, showing that many of the base-exchange sites are blocked by a fairly stable aluminium complex.

If hair is dispersed by sodium sulphide or 0·5 M sodium cyanide in 0·1 M sodium hydroxide, the solubilized material has a composition similar in amino acid content to the original hair. The isoelectric region of such a solubilized material is at pH 4·1–4·7, showing again the excess of acidic side chains.

Physical methods of evaluating hair damage such as extensometer tests and stress relaxation measurements[104] are well known. Later work has been aimed at obtaining suitable chemical methods for estimating hair damage and a number of different principles, including the swelling effect of lithium bromide, the effect of enzymes and the copper uptake of hair, have been evolved.[105–109]

Hair Colour

Melanin and Melanocytes

Hair colour is caused by pigment granules in the cells of the hair shaft. The effect is produced by the large amount of pigment in the cortex, but granules also occur in the medulla. The pigment is manufactured within melanocytes situated around the apex of the dermal papilla and transferred to the newly formed hair cells from the tips of their finger-like dendrites.[110–112]

The granules themselves are the end product of *melanosomes*. These originate as colourless *pre-melanosomes* in the region of the Golgi and become progressively darker as pigment is synthesized on them and they move peripherally. The granules themselves are ovoid or rod-shaped and vary in length from 0·4 μm to 1·0 μm and in breadth from 0·1 μm to 0·5 μm.[113,114] The darker the hair, the larger the average size of the granules,[115–117] and Negroids generally have larger and fewer granules than Caucasoids. Melanogenesis has been reviewed by Collins[118] and by Fitzpatrick.[119]

The shades of human hair result mainly from two kinds of pigment, *eumelanin*, which is brown or black, and *phaeomelanin*, which is yellow or reddish. Eumelanin is formed from the amino acid tyrosine by oxidation with the enzyme *tyrosinase*. Tyrosine is also the precursor of phaeomelanin, but it seems that the presence of an *o*-aminophenol derived from tryptophane is also necessary.[111,120,121]

Other pigments, which appear to be iron complexes, have been extracted from human red hair,[122–124] red rabbits and red mongrel dogs,[125,126] and given the name of *trichosiderins*. At least three distinct compounds have been isolated.[127] The importance of these pigments is questionable. Boldt[128] has shown that their colour is not due to the iron content, since it has been possible to isolate iron-free alcohols, named pyrrotrichols, which have the same colours as the iron complexes. Moreover, the red colour of human hair is not substantially altered by extraction of the trichosiderins. Thus, although it is admitted that red hair does contain greater amounts of iron than any other shade, its colour is now believed to be due to phaeomelanin.

Biochemistry of Melanogenesis

The first stage in the formation of either eumelanin or phaeomelanin is the oxidation of tyrosine to 3,4-dihydroxyphenylalanine (dopa) by the enzyme tyrosinase.[129–133] The second stage, also catalysed by tyrosinase,[134,135] is the dehydrogenation of dopa to form dopa quinone (Figure 23.6).[136]

Eumelanins involve the formation of indole-5,6-quinone. It now seems likely that eumelanin is not, as once believed, a simple polymer of indole-5,6-quinone units linked through a single bond, but a poikilopolymer of several types of monomer joined by multiple types of bonds, with the incorporation of an

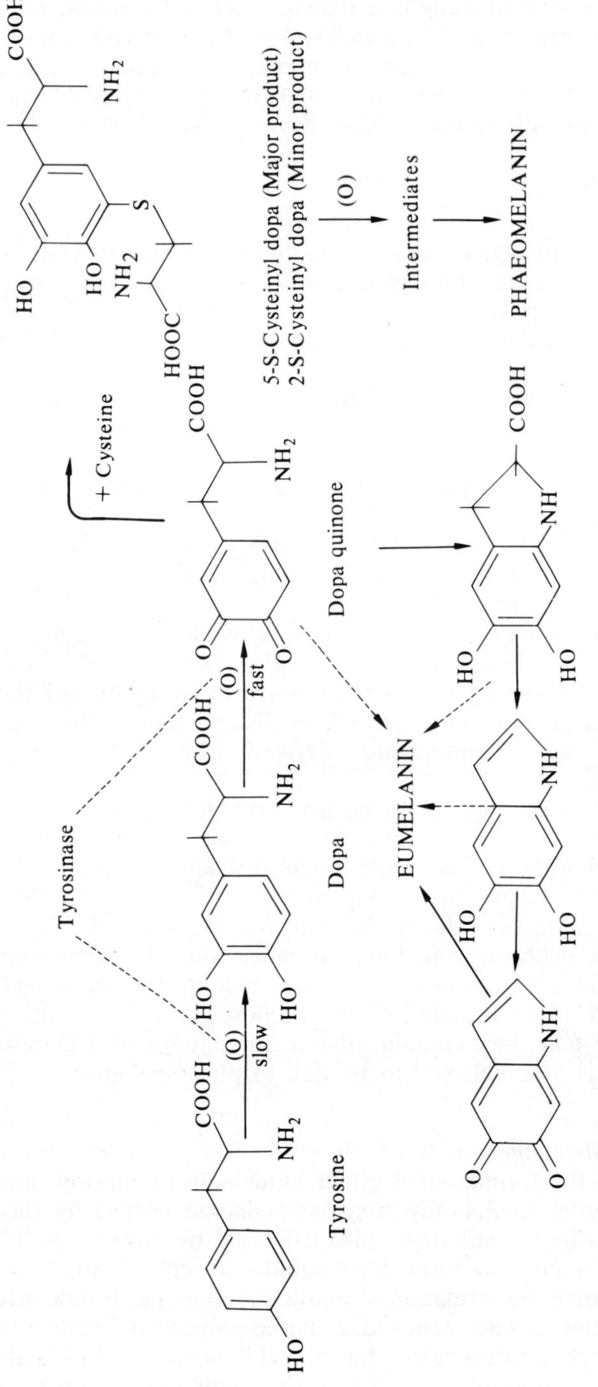

Figure 23.6 Simplified scheme for the formation of eumelanin and phaeomelanin (from Rook *et al.*,[136] courtesy Blackwell Scientific Publications)

assortment of oxidized intermediates. According to Dalgliesh[137] there are four sites at which the molecules could join together. The eumelanin becomes linked to protein.

Phaeomelanins are formed from dopa quinone by a different pathway. The dopa quinone reacts with cysteine to form 5-S- and 2-S-cysteinyldopa,[138] and further oxidation, which is probably non-enzymatic, produces intermediates which form the phaeomelanin polymers.

Tyrosinase contains trace amounts of copper, which can exist in either a cuprous or a cupric state. Tyrosinase activity is also shown by similar complexes containing other polyvalent ions such as cobalt, nickel and vanadium,[139] but their efficiency is lower than that of copper. The activiation of tyrosinase requires trace amounts of dopa to be present,[140] and this can be produced from dopaquinone by ascorbic acid.

A large number of materials can inhibit tyrosinase activity, and therefore pigmentation. Such materials include competitors for tyrosine (3-fluorotyrosine, N-acetyltyrosine, N-formyltyrosine, phenylalanine, 3-aminotyrosine), agents with a complexing activity with the copper prosthetic group of the enzyme (phenylthiourea, diethyldithiocarbamate, BAL, cysteine, glutathione, hydrogen sulphide, cyanides, etc.), compounds which alter the redox potential of the system (hydroquinone, indophenol, etc.), substances which combine with one or other of the intermediates in the melanin synthesis (aniline, *p*-aminobenzoic acid, *p*-phenylenediamine, etc.) and one or two materials (ethers of hydroquinone, pentachloronaphthalenes) where the mode of action is unknown. Useful reviews of such inhibitors are given by Lorincz[141] and Fitzpatrick.[111]

For obvious reasons, few of these materials have been tested *in vivo*. However, α-naphthylthiourea and phenylurea cause greying of hair in rats as long as they remain in the diet, and thiouracil has suppressed pigmentation in man.

Greying of Hair

Greying of hair involves a loss of pigment from the hair shafts and a progressive loss of tyrosinase activity from the hair bulbs. It must be considered as a normal part of aging; in Caucasoids white hairs first appear at the temples at the average age of 34, and by age 50 half the population has at least 50 per cent grey hairs.

Rapid greying of hair (as judged by its overall appearance) after severe emotional stresses has often been recorded, though the belief that the hair of Marie Antoinette and Sir Thomas More turned white overnight may stretch credulity. It is not conceivable that pigment could be destroyed in hair which has already grown. However, there are two possible explanations. One is that 'telogen effluvium' (see below, Hair Disorders) causes sudden shedding of dark hairs from a mixed population, leaving a predominance of white or grey hairs *in situ*; the other is that a traumatic event precipitates loss of hairs by an episode of 'alopecia areata'—the first hairs to regrow are invariably white. However, such a process would take weeks or months rather than hours. Green and Patterson[142] recorded such a case of an engine driver who fell from a stationary engine. His hair came out on the day of the accident, at first in a round patch, which was diagnosed as alopecia. Eventually it spread, until after four months he was completely bald. Two months later a few white downy hairs appeared.

Greying of hair can be experimentally produced in animals by copper or by pantothenic acid deficiency; greying due to copper can be reversed by pantothenic acid.[143] Though it has been claimed that huge doses of *p*-aminobenzoic acid will occasionally restore pigment to grey hair in man[144] there is no known medical treatment by which greying can be consistently reversed. Hair follicles can, however, sometimes produce pigmented or grey hair intermittently; one such condition is a rare anomaly known as *ringed hair* or *pili annulati*.

Hair Disorders

Introduction

Only a brief summary of the major conditions which affect the hair shaft and the hair follicle will be given here. For more comprehensive accounts see works by Orfanos[145] and Ebling and Rook.[136]

The cosmetician may need to be able to recognize several abnormalities of the hair shaft of genetic origin, each of which may be associated with sparse, brittle and often short hair. From these disorders must be differentiated various types of hair loss in which the shafts remain structurally normal.

Hair loss may be either rapid or gradual. Sudden shedding of hair is often, though not invariably, transient, whereas gradual loss, observed only by its long-term effect, is usually hopelessly chronic.

Rapid loss can be further subdivided into two types, according to whether the fallen hair is a club, or a growing hair shed from an active follicle.

Loss of club hairs is known as *telogen effluvium* and seems to have several possible causes. One, namely childbirth, is well established and the ensuing condition is known as *postpartum alopecia*. Loss of growing hairs is known as *anagen effluvium*. It occurs after cytotoxic drugs, and it seems likely that the shedding of hairs in patchy baldness, or *alopecia areata*, is a similar process.

Slowly developing hair loss, causing baldness in a symmetrical pattern, is well known in males, where it is named *male-pattern alopecia* or *alopecia androgenetica*. Diffuse alopecias have been attributed to several causes. It seems likely that much, probably most, diffuse alopecia in women is an inherited androgen-potentiated condition which is the female equivalent of male-pattern baldness.

Defects of the Hair Shaft

Monilethrix: a condition in which the hair shaft is beaded with elliptical nodes about 0·7–1·00 mm apart alternating with constricted internodes which lack any medulla. The hair is brittle and breaks off a centimetre or two above emergence.

Pili torti. The hair shaft is twisted on its axis at intervals, and usually breaks off short.

Trichorrhexis nodosa: a response to physical or chemical injury in which an apparently normal hair becomes swollen and split to form a node at which the hair subsequently fractures. In *trichorrhexis invaginata* the fractured hair telescopes to give a bamboo-like appearance.

Telogen Effluvium

Excessive shedding of normal club hairs after febrile illness (*post-febrile alopecia*) is well authenticated,[10] and other causes, for example antithyroid drugs, anticoagulants and psychogenic stress have been suggested. Kligman[10] has recounted the case of a prisoner who suffered a daily shed of 600–1500 hairs over a period of three months after conviction for murder, but subsequently recovered after a pardon had been granted.

Postpartum Alopecia

The shedding of hair within two or three months of parturition appears to be a telogen effluvium similar to post-febrile alopecia. The cause seems to be a prolongation of anagen by the hormonal conditions of late pregnancy. Lynfield[146] first described, and others have confirmed, that prior to childbirth as many as 95 per cent of the scalp follicles may be in anagen, but that the proportion falls to less than 70 per cent within 2–4 months after parturition, suggesting that the follicles have been precipitated into catagen.

Anagen Effluvium

In contrast to telogen effluvium, loss of anagen hairs is a rapid event. Cytotoxic drugs such as cyclophosphamide, adriamycin and vincristine can produce it in a few days.

Alopecia Areata

Alopecia areata, or patchy baldness, is usually easy to recognize. The lesions are sited asymmetrically, and each starts at a focal point and spreads outwards. The process may take two or more weeks or may, to quote Behrman,[147] be 'so rapid that a completely denuded plaque may appear overnight with the lost hairs found in a heap on the pillow in the morning'.

The margins of the lesions are characterized by the occurrence of short protruding club hairs with a frayed point, often named 'exclamation marks'. As the lesions progress, many follicles in a state of arrested anagen can be found. By plucking hairs successively from a series of concentric rings around the focus of the lesions, it has been shown that an area from which only club hairs can be removed develops at the centre and moves centripetally.[148]

The simplest hypothesis to explain these observations is that initially hairs are shed from all active follicles, leaving only the clubs unaffected. The shearing of hair from follicles moving into catagen produces the protruding stumps of the exclamation marks. However, it is quite possible that follicles are also precipitated into catagen. It should be added that the process does not always produce patchy lesions, but may sometimes occur diffusely, so-called 'diffuse alopecia areata'.

The cause of alopecia areata is a subject of debate. It seems likely that there are several types, probably four, of the disease, differing from each other in the age of onset, clinical features and prognosis.[149] Authors are divided about the role of heredity, some favouring a family history in 10–20 per cent of cases,[150] others finding it in none.[151] Some believe psychological factors are important,[152] others that they play no role.[153] It seems, however, difficult to deny that the condition is in some instances precipitated by mental shock.[154,155]

Various treatments such as ultraviolet light or miscellaneous irritants have been used for *alopecia areata*, but there is no objective evidence that they are of any value. One half of all patients recover within a year of the initial attack, though the incidence of relapse is high.[156] Corticosteroids, given systemically, have been shown to induce regrowth in many cases, though the growth is not always maintained after discontinuation of the treatment. More justifiable is the use of intralesional injections.[157]

Male-pattern Alopecia

No particular expertise is needed to recognize pattern alopecia[158] in the male, for the recession of the hair line, loss of hair from the crown, and the bald pate are only too familiar.[159] In the affected areas the hairs become steadily shorter and finer and, ultimately, cosmetically useless. About a third of the follicles may disappear. A reduction in the length of the growing period is reflected in the increased ratio of telogen to anagen in samples of plucked hair.

Pattern alopecia is inherited, apparently as an autosomal dominant trait, but it is only manifested in the presence of male hormone. Eunuchs retain their scalp hair, even when they have a family history of baldness, unless they are treated with testosterone.[45] Though a correlation with hairiness of the chest has been suggested,[158] baldness does not seem to be associated with other indices of masculinity, such as sebum secretion, muscle size and body hair in general.[160] While the finding that bald scalp has a greater capacity than non-bald scalp to convert testosterone to 5α-dihydrotestosterone *in vitro*[161] suggests that the key to understanding baldness lies in the field of steroid metabolism, no plausible hypothesis of how androgens promote baldness of the scalp but hair growth on the body has yet been proposed.

Diffuse Alopecia

Loss of hair is by no means uncommon in women. Occasionally such loss involves obvious recession of the hair line as in males, but usually it is diffuse.

A number of causes have been proposed. It has long been recognized that thin hair may be associated with hypothyroidism (myxoedema) as well as with hyperthyroidism.

There is some evidence pointing to other factors, such as, for example, iron deficiency[162] and ingestion of amphetamines for weight reduction.[163] However, in many cases (over half of the total in one series studied[163]) no possible cause could be found. It seems likely that these were androgenetic alopecias which had become manifested even within the normal range of female androgen levels.

Hirsutism

Hirsutism may be defined as growth in the female of coarse terminal hair in part or whole of the adult male sexual pattern. At one end of the scale the condition may be obviously associated with other signs of virilism, excessive androgen production and an endocrine pathology. But most cases show no signs of masculinization except hirsutism, and have only slightly or moderately raised androgen levels; some lie well within the range for normal, non-hirsute subjects. The condition appears to result either from abnormal levels of free androgens or from an abnormal response of the hair follicle, or both. Anti-androgens, such as the steroid cyproterone acetate given orally, have proved effective in management of the condition.[44]

Dandruff

Definition and Etiology

Dandruff is a condition of the scalp characterized by the massive desquamation of small flakes of stratum corneum. It has already been mentioned as a scaling disorder in Chapter 1, where its occurrence and debatable etiology were briefly reviewed.

Discussion about the cause of dandruff revolves around the relative roles of physiological, traumatic and infective factors. Many authors have attempted to correlate dandruff with body disorders or environmental factors. Thus Lubowe[164] has discussed the possible roles of hormones, metabolic faults, diet and nervous tension, as well as inflammatory reactions to topical medicaments and cosmetics. Sefton[165] noted that prisoners of war in Japanese camps in Singapore in 1942–45 had little dandruff and attributed this to shortage of fats in the diet, and it has been claimed that vitamin therapy is effective for some types of dandruff.[166]

The yeast-like organisms, *Pityrosporum ovale* and *Pityrosporum orbiculare* are common members of the scalp flora. *Pityrosporum* can be easily detected with stains such as methylene blue and Nile blue. It does not appear to exist in nature away from human beings and has proved difficult to culture, though this problem has been solved.[167,168] Its ultrastructure has been described by Swift.[169]

Pityrosporum ovale was first described in 1874 by Malassez who believed it to be the sole cause of dandruff.[170] This opinion was later supported by Reddish.[171] The role of *Pityrosporum* in dandruff has been much debated. What is the evidence? Koch laid down the original postulates for proving an organism is the specific cause of a disease as follows:

(i) the organism must be present in every case of the disease;
(ii) it should be collected and cultivated in pure culture;
(iii) inoculation from such culture must reproduce the disease in susceptible animals;
(iv) the organism must be re-obtained from such animals and again grown in pure culture.

There is little doubt that *Pityrosporum* is generally present in cases of dandruff; so are many other micro-organisms, including a number of bacteria and moulds.[172–174] The first postulate might thus appear to be satisfied, were it not for the fact that *Pityrosporum* is also present on scalps which do not have dandruff. The question therefore arises whether there is any relationship between dandruff and the level of infection. Some authors have failed to find any such correlation,[175] others have produced evidence that *Pityrosporum* is, in fact, more abundant in affected than in non-affected persons.[172,176]

The question of whether inoculation of cultured *Pityrosporum ovale* could produce dandruff was investigated by Moore and his associates.[177] Their results are quoted in Table 23.3. The evidence is not overwhelming, but it supports the view that *Pityrosporum* is a causative agent.

Table 23.3 Results of Experimental Inoculation with Pityrosporum ovale to produce Dandruff

	Reaction			
Type of innoculation	**Positive**	**Doubtful**	**Negative**	**Total**
Intradermal injection of whole cells in saline	22 (79%)	0	6 (21%)	28
Excoriations rubbed with whole culture	19 (40%)	8 (16%)	21 (44%)	48
Excoriations rubbed with whole culture + lipid mixture	12 (48%)	5 (20%)	8 (32%)	25
Exposure of intact skin to culture	9 (50%)	2 (11%)	7 (39%)	18
Application of culture to scalp	6 (100%)	0	0	6

Some further support comes from studies on the effect of antimicrobial preparations. When an aqueous solution of neomycin and nystatin was massaged into the scalp, both the scurf and the microbial flora were reduced.[178] Moreover, it appeared that a reduction in the yeast flora was more effective in controlling dandruff than a reduction in the number of bacteria.[179]

The circumstantial and experimental evidence thus implicates *Pityrosporum* as an agent in dandruff, but does not establish beyond doubt that it is the only cause. Indeed it may well be that the term 'dandruff' covers more than one condition and that dry dandruff needs to be differentiated from greasy dandruff. A further complication is the distinction between dandruff and seborrheic dermatitis. Spoor[180] believed that it was unrealistic to try to separate the conditions, whereas Kligman and his co-workers[181] strongly reject the view that they are in any way related.

Dandruff Therapy

The wide range of treatments that have been used for dandruff reflects the debate about its nature and etiology.

Inasmuch as dry dandruff might be related to external provoking agents such as unsuitable shampoos, alcoholic lotions or waving lotions used too close to the scalp, these should clearly be avoided.

Greasy dandruff has been assaulted by an armoury of materials. Before reviewing them, the difficulties of making an objective assessment of their efficacy should be mentioned. Properly controlled tests are essential, since regular washing, massage and anointment of the scalp will often alleviate dandruff, whatever the materials used. Alexander[182] examined the effects of three treatments:

1. Shampoo base
2. Shampoo base plus tar

3. Shampoo base plus tar plus the sodium salt of the sulphosuccinate of an undecylene alkanolamide

She found that washing with any of the shampoos reduced dandruff for a few days, the tar product reduced dandruff for a longer time than the base alone, while the product containing the additive was even better.

Procedures for clinical evaluation of antidandruff shampoos have been discussed by Van Abbé and Dean[183] and by Kligman and co-workers.[181]

Idson[175] and Lubowe[164] have reviewed the materials used in commercial preparations. Among the reported active ingredients in some forty-seven products were hexachlorophene, tar, salicylic acid, sulphur, resorcinol and cationic compounds.[164]

Germicidal treatments have included resorcinol, thymol and other phenols. Cationic germicides were introduced by Hodges[184] and Neville-Smith[185,186] in the form of therapy with 5 per cent Cetavlon solution, and a later account is given by Speirs and Brotherwood.[187] Sodium sulphacetamide has been used by Duemling[188] and Whelan[189] for the treatment of pityriasis circinata. Trichloromethylmercapto derivatives have been used by Ball[190] in the form of Vancide 89 (trichloromethyl-mercapto-4-cyclohexene-2,2-dicarboximide), again for the treatment of pityriasis circinata, and such materials have been used commercially in treatments for dandruff. Gross and Wright[191] used a 2·5 per cent suspension of tellurium dioxide, which they claimed was safe and effective in the control of pityriasis.

Elemental sulphur is one of the oldest treatments. The most effective form is colloidal sulphur, or milk of sulphur, which contains large amounts of polythionic acids, and these have been combined with cationic materials. Lubowe[192] reports success with 'Sarthionate' (*bis*-lauryltrimethyl ammonium polythionate). Cadmium sulphide suspensions have been used by Kirby.[193]

Selenium disulphide proved to be a very effective compound. Slinger and Hubbard[194] published tables showing complete control of seborrhoeic conditions in 73·3 per cent of severe cases, 84·6 per cent of moderate cases and 95·4 per cent of mild cases, and similarly good results were reported by Slepyan.[195] Further reports are given by Bereston,[196] Matson[197] and Caspers.[198] Brotherton has discussed the effect of selenium compounds on the sulphur metabolism of *P. ovale*,[168] suggesting one possible mode of action on the organism. However, Plewig and Kligman[199] have produced substantial evidence that the compound lowers the corneocyte count and postulated that it controls dandruff by its cytostatic capabilities.

One major difficulty with selenium compounds is that although they are effective they are not cosmetically attractive. Thus products containing selenium oxide or sulphide are invariably dark brown in colour and are not pleasant to use. In addition care must be taken to exclude contact with the eyes. Products containing selenium are considered more to be ethical products for use in severe cases of dandruff, rather than general toiletries, and should be used mainly under advice.

More recently, zinc pyridinethiol-N-oxide (zinc pyrithione, zinc omadine) has been introduced as a safe agent for incorporation into shampoos.[200] Kligman and his associates[181] compared shampoos containing 2 per cent zinc pyrithione or 2·5 per cent selenium sulphide with a detergent base. Both compounds significantly reduced the corneocyte count, as well as effecting clinical improvement of the dandruff, whereas the bland shampoos had little effect. The zinc pyrithione appeared, however, to act more slowly than the selenium sulphide.

An antidandruff effect slightly superior to that of zinc pyrithione has been claimed for a shampoo containing 1-hydroxy-4-methyl-6-(2,4,4-trimethylpentyl)-2-(1H)-pyridone monoethanolamine salt (Octopirox).[201]

REFERENCES

1. Ebling, F. J., *J. invest. Dermatol.*, 1976, **67,** 98.
2. Montagna, W. and Ellis, R. A., *The Biology of Hair Growth*, New York, Academic Press, 1958.
3. Montagna, W. and Dobson, R. L., *Advances in Biology of Skin*, Vol. 9, *Hair Growth,* Oxford, Pergamon, 1969.
4. Muller, S. A., *J. invest. Dermatol.*, 1971, **56,** 1.
5. Szabo, G., *Philos. Trans. R. Soc. London Ser. B*, 1967, **252,** 447.
6. Barman, J. M., Astore, I. P. L. and Pecoraro, V., *Advances in Biology of Skin*, Vol. 9, *Hair Growth*, ed. Montagna, W. and Dobson, R. L., Oxford, Pergamon, 1969, p. 211.
7. Giacometti, L., *Advances in Biology of Skin*, Vol. 6, *Ageing*, ed. Montagna, W., Oxford, Pergamon, 1965, p. 97.
8. Dry, F. W., *J. Genet.*, 1926, **16,** 287.
9. Kligman, A. M., *J. invest. Dermatol.*, 1959, **33,** 307.
10. Kligman, A. M., *Arch. Dermatol.*, 1961, **83,** 175.
11. Saitoh, M., Uzuka, M. and Sakamoto, M. *J. invest. Dermatol.*, 1970, **54,** 65.
12. Johnson, E., *J. Endocrinol.*, 1958, **16,** 337.
13. Jackson, D. and Ebling, F. J., *J. Anat.*, 1972, **111,** 303.
14. Barman, J. M., Astore, I. P. L. and Pecoraro, V., *J. invest. Dermatol.*, 1965, **44,** 233.
15. Pecoraro, V., Astore, I. P. L., Barman, J. M. and Araujo, C., *J. invest. Dermatol.*, 1964, **42,** 427.
16. Barman, J. M., Pecoraro, V., Astore, I. P. L. and Ferrer, J., *J. invest. Dermatol.*, 1967, **48,** 138.
17. Pecoraro, V., Astore, I. P. L. and Barman, J. M., *J. invest. Dermatol.*, 1964, **43,** 145.
18. Trotter, M., *Am. J. phys. Anthrop.*, 1924, **7,** 427.
19. Saitoh, M., Uzuka, M., Sakamoto, M. and Kobori, T., *Advances in Biology of Skin*, Vol. 9, *Hair Growth*, ed. Montagna, W. and Dobson, R. L., Oxford, Pergamon, 1969, p. 183.
20. Barman, J. M., Pecoraro, V. and Astore, I. P. L., *J. invest. Dermatol.*, 1964, **42,** 421.
21. Comaish, S., *Br. J. Dermatol.*, 1969, **81,** 283.
22. Munro, D., *Arch. Dermatol.*, 1966, **93,** 119.
23. Myers, R. J. and Hamilton, J. B. *Ann. N.Y. Acad. Sci.*, 1951, **53,** 562.
24. Pelfini, C., Cerimele, D. and Pisanu, G., *Advances in Biology of Skin*, Vol. 9, *Hair Growth*, ed. Montagna, W. and Dobson, R. L., Oxford, Pergamon, 1969, p. 153.
25. Eaton, P. and Eaton, M. W., *Science*, 1937, **86,** 354.
26. Lynfield, Y. L. and MacWilliams, P., *J. invest. Dermatol.*, 1970, **55,** 170.
27. Flesch, P., in *The Physiology and Biochemistry of the Skin*, ed. Rothman, S., Chicago, University of Chicago Press, 1954.

28. Priestley, G. C., *J. Anat.*, 1966, **100,** 147.
29. Hale, P. A. and Ebling, F. J., *J. exp. Zool.*, 1975, **191,** 49.
30. Jackson, D. and Ebling, F. J., *J. Soc. cosmet. Chem.*, 1971, **22,** 701.
31. Ebling, F. J. and Johnson, E., *J. Embryol. exp. Morphol.*, 1959, **7,** 417.
32. Ebling, F. J. and Johnson, E., *J. Embryol. exp. Morphol.*, 1961, **9,** 285.
33. Chase, H. B., *Physiol. Rev.*, 1954, **34,** 113.
34. Argyris, T. S., *Advances in Biology of Skin*, Vol. 9, *Hair Growth*, ed. Montagna, W. and Dobson, R. L., Oxford, Pergamon, 1969, p. 339.
35. Johnson, E., *J. Endocrinol.*, 1958, **16,** 351.
36. Johnson, E., *J. Endocrinol.*, 1958, **16,** 360.
37. Ebling, F. J. and Johnson. E., *J. Endocrinol.*, 1964, **29,** 193.
38. Ebling, F. J. and Hale, P. A., *Mem. Soc. Endocrinol.*, 1970, **18,** 215.
39. Orentreich, N., *Advances in Biology of Skin*, Vol. 9, *Hair Growth*, ed. Montagna, W. and Dobson, R. L., Oxford, Pergamon, 1969, p. 99.
40. Marshall, W. A. and Tanner, J. M., *Arch. Dis. Child.*, 1970, **45,** 13.
41. Marshall, W. A. and Tanner, J. M., *Arch. Dis. Child.*, 1969, **44,** 291.
42. Dupertuis, C. W., Atkinson, W. B. and Elftman, H., *Hum. Biol.*, 1945, **17,** 137.
43. Hamilton, J. B., *The Biology of Hair Growth*, ed. Montagna, W. and Ellis, R. A., New York, Academic Press, 1958, p. 399.
44. Ebling, F. J., Thomas, A. K., Cooke, I. D., Randall, V. A., Skinner, J. and Cawood, M., *Br. J. Dermatol.*, 1977, **97,** 371.
45. Hamilton, J. B., *Am. J. Anat.*, 1942, **71,** 451.
46. Hamilton, J. B., *J. clin. Endocrinol.*, 1960, **20,** 1309.
47. Mäusle, von R. and Zaun, H., *Fortschr. Med.*, 1972, **90,** 687.
48. Sims, R. T., *Arch. Dis. Child.*, 1967, **42,** 397.
49. Bradfield, R. B., Bailey, M. A. and Cordano, A., *Lancet*, 1968, **ii,** 1169.
50. Bradfield, R. B., Cordano, A. and Graham, G. G., *Lancet*, 1969, **ii,** 1395.
51. Bradfield, R. B., *Am. J. clin. Nutr.*, 1971, **24,** 405.
52. Jausion, H. and Benard, P., *Bull. Soc. fr. Dermatol. Syphiligr.*, 1949, **56,** 462.
53. Schuringa, G. J., Isings, J. and Ultée, A. J., *Biochim. Biophys. Acta*, 1952, **9,** 457.
54. Bolliger, A., *J. invest. Dermatol.*, 1951, **17,** 79.
55. Spector, W. S., *Handbook of Biological Data*, Philadelphia, Saunders, 1956, p. 90.
56. Bell, J. W. and Whewell, C. S., *A Handbook of Cosmetic Science*, ed. Hibbott, H. W., Oxford, Pergamon, 1963.
57. Alexander, P., *Ann. N.Y. Acad. Sci.*, 1951, **53,** 653.
58. Alexander, P., Hudson, R. F. and Fox, M., *Biochem. J.*, 1950, **46,** 27.
59. MacArthur, I., *Nature* (*London*), 1943, **152,** 38.
60. Giroud, A. and Leblond, C. P., *Ann. N.Y. Acad. Sci.*, 1951, **53,** 613.
61. Pauling, L. and Corey, R. B., *Proc. nat. Acad. Sci. Wash.*, 1951, **37,** 235, 241, 251, 261, 272, 282 and 729.
62. Pauling, L. and Corey, R. B., *Proc. nat. Acad. Sci. Wash.*, 1951, **38,** 86.
63. Pauling, L., Corey, R. B. and Branson, H. R., *Proc. nat. Acad. Sci. Wash.*, 1951. **37,** 205.
64. Pauling, L., Corey, R. B. and Branson, H. R., *Nature* (*London*), 1951, **165,** 550.
65. Pauling, L. and Corey, R. B., *Nature* (*London*), 1953, **171,** 59.
66. Pauling, L. and Corey, R. B., *J. Am. Chem. Soc.*, 1950, **72,** 5349.
67. Bramford, C. H., Brown, L., Elliott, A., Hanby, W. E. and Trotter, I. F., *Nature* (*London*), 1952, **169,** 357.
68. Cochran, W. and Crick, F. H. L., *Nature* (*London*), 1952, **169,** 234.
69. Yakel, H. L., Pauling, L. and Corey, R. B., *Nature* (*London*), 1952, **169,** 920.
70. Fraser, R. D. B., McRae, T. P. and Miller, A., *J. mol. Biol.*, 1964, **10,** 47.
71. Fraser, R. D. B., *Keratins: Their Composition, Structure and Biosynthesis*, Springfield, Charles C. Thomas, 1972.

72. Johnson, D. J. and Speakman, P. T. *Nature* (*London*), 1965, **205,** 268.
73. Dobb, M. G., *Nature* (*London*), 1965, **207,** 293.
74. Satir, B. and Satir, P., *J. theor. Biol.*, 1964, **7,** 123.
75. Johnson, D. J. and Sikorski, J., *Nature* (*London*), 1965, **205,** 266.
76. Orfanos, C. E. and Mahrle, G., *Parfüm. Kosmet.*, 1971, **52**(7), 203 and (8), 235.
77. Randebrock, R. J., *J. Soc. cosmet. Chem.*, 1964, **15,** 691.
78. Kuczera, K., *Seifen Öle Fette Wachse*, 1972, **98,** 115 and 152.
79. Leon, N. H., *J. Soc. cosmet. Chem.*, 1972, **23,** 427.
80. Wolfram, L. J. and Lindemann, M. K. O., *J. Soc. cosmet. Chem.*, 1971, **22,** 839.
81. Bottoms, E., Wyatt, E. and Comaish, S., *Br. J. Dermatol.*, 1972, **86,** 379.
82. Bagchi, K. N. and Ganguly, H. D., *Ann. Biochem.*, 1941, **1,** 84.
83. Tompsett, S. L., *Analyst*, 1956, **81,** 330.
84. Strain, W. H., Steadman, L. T., Lankau, C. A. Jr., Berliner, W. T. and Pories, W. J., *J. Lab. clin. Med.*, 1966, **68,** 244.
85. Cornelis, R. and Speecke, A., *Forensic Soc. J.*, 1971, **11,** 29.
86. Cornelis, R., *Medicine Sci. Law*, 1972, **12,** 188.
87. Bate, L. C. and Dyer, F. F., *Nucleonics*, 1965, **23,** 74.
88. Bate, L. C., *Int. J. appl. Radiat. Isotopes*, 1966, **17,** 417.
89. Coleman, R. F., Cripps, F. H., Stimson, A. and Scott, H. D., *Atom*, 1967, **123,** 12.
90. Valko, E. I. and Barnett, G., *J. Soc. cosmet. Chem.*, 1952, **3,** 108.
91. Jones, C. B. and Mecham, D. K., *Arch. Biochem.*, 1943, **2,** 209.
92. Olofson, B. and Gralen, N., *Proc. 11th int. Congr. pure appl. Chem.*, 1947, **5,** 151.
93. Gillespie, J. M. and Lennox, F. G., *Aust. J. biol. Sci.*, 1955, **8**(97), 378.
94. Jones, C. B. and Mecham. D. K., *Arch. Biochem.*, 1943, **3,** 193.
95. Friend, J. A. and O'Donnell, I. J., *Aust. J. biol. Sci.*, 1953, **6,** 630.
96. Ward, W. H., *Text. Res. J.*, 1952, **22,** 405.
97. Woods, E. F., *Aust. J. sci. Res.*, 1952, **A5,** 555.
98. Gillespie, J. M. and Lennox, F. G., *Biochim. Biophys. Acta*, 1953, **12,** 481.
99. Das, D. B. and Speakman, J. B., *J. Soc. Dyers Colour*, 1950, **66,** 583.
100. Alexander, P., Hudson, R. F. and Fox, M., *Biochem. J.*, 1950, **46,** 27.
101. Laxer, G., Sikorski, J., Whewell, C. S. and Woods, H. J., *Biochim. Biophys. Acta*, 1954, **15,** 174.
102. Whewell, C. S., *J. Soc. cosmet. Chem.*, 1964, **15,** 423.
103. Menkart, J., Wolfram, L. J. and Ma, O. I., *Am. Perfum. Cosmet.*, 1966, **81,** 92.
104. Wall, R. A., Morgan, D. A. and Dasher, G. F., *J. Polymer Sci.*, *Pt. C*, 1966, (14), 299.
105. Robbins, C., *Text. Res. J.*, 1967, **37,** 337.
106. Holmes, A. W., *Text. Res. J.*, 1964, **34,** 777.
107. Sagal, J. *Text. Res. J.*, 1965, **35,** 672.
108. Klemm, E. J., Haefele, J. W. and Thomas, A. R., *Drug Cosmet. Ind.*, 1965, **97,** 677.
109. Berth, P. and Reese, J., *J. Soc. cosmet. Chem.*, 1964, **15,** 659.
110. Barnicott, N. A. and Birbeck, M. S. C., *The Biology of Hair Growth*, ed. Montagna, W. and Ellis, R. A., New York, Academic Press, 1958.
111. Fitzpatrick, T. B., Brunet, P. and Kukita, A., *The Biology of Hair Growth*, ed. Montagna, W. and Ellis, R. A., New York, Academic Press, 1958.
112. Searle, A. G., *Comparative Genetics of Coat Colour in Mammals*, London, Logos Press, 1968.
113. Barnicott, N. A., Birbeck, M. S. C. and Cuckow, F. W., *Ann. hum. Genet.*, 1955, **19,** 231.
114. Mottaz, J. H. and Selickson, A. S., *Advances in Biology of Skin*, Vol. 9, *Hair Growth*, ed. Montagna, W. and Dobson, R. L., Oxford, Pergamon, 1967, p. 471.
115. Russell, E. S., *Genetics*, *Princeton*, 1949, **34,** 131.

116. Shackleford, R. M. *Genetics, Princeton*, 1948, **33,** 311.
117. Hutt, F. B., *Pigment Cell Growth*, ed. Gordon, M., New York, Academic Press, 1953.
118. Collins, G. M., *Arch. Biochem. Cosmetol.*, 1962, **4,** 12.
119. Fitzpatrick, T. B., *J. Soc. cosmet. Chem.*, 1964, **15,** 297.
120. Butenandt, A., Bickert, E. and Linzen, B., *Hoppe Seylers Z. physiol. Chem.*, 1956, **305,** 284.
121. Glassman, E., *Arch. Biochem. Biophys.*, 1957, **67,** 74.
122. Rothman, S. and Flesch, P., *Ann. Rev. Physiol.*, 1944, **6,** 195.
123. Flesch, P. and Rothman, S., *J. invest. Dermatol.*, 1945, **6,** 257.
124. Dutcher, T. F. and Rothmans, S., *J. invest. Dermatol.*, 1951, **17,** 65.
125. Flesch, P., *J. invest. Dermatol.*, 1968, **51,** 337.
126. Flesch, P., *J. Soc. cosmet. Chem.*, 1968, **19,** 675.
127. Boldt, P., *Naturwissenschaften*, 1964, **51,** 265.
128. Boldt, P. and Hermestedt, E., *Z. Naturforsch.*, 1967, **B22**(7), 718.
129. Raper, H. S., *Physiol. Rev.*, 1928, **8,** 245.
130. Mason, H. S., *J. biol. Chem.*, 1948, **172,** 83.
131. Nicolaus, R. A. and Piatelli, M., *J. Polymer Sci.*, 1962, **58,** 1133.
132. Nicolaus, R. A., Piatelli, M. and Fattorusso, E., *Tetrahedron*, 1964, **20,** 1163.
133. Mason, H. S., *The Pigmentary System*, ed. Montagna, W. and Hu, F., New York, Pergamon, 1967, p. 293.
134. Mason, H. S., *Nature (London)*, 1956, **177,** 79.
135. Mason, H. S., *Adv. Enzymol. relat. Areas mol. Biol.*, 1955, **16,** 105.
136. Ebling, F. J. and Rook, A., *Textbook of Dermatology*, 3rd edn, ed. Rook, A., Wilkinson, D. S. and Ebling, F. J. G., Oxford, Blackwell, 1979, p. 1733.
137. Dalgliesh, C. E., *Adv. Protein Chem.*, 1955, **10,** 31.
138. Prota, G. and Thomson, R. H., *Endeavour*, 1976, **35,** 32.
139. Kertesz, D., *Nature (London)*, 1951, **168,** 697.
140. Lerner, A. B. and Fitzpatrick, T. B., *Physiol. Rev.*, 1950, **30,** 91.
141. Lorincz, A. L., in *The Physiology and Biochemistry of the Skin*, ed. Rothman, S., Chicago, University of Chicago Press, 1954.
142. Greene, R. and Paterson, A. S., *Lancet*, 1943, **ii,** 158.
143. Singer, L. and Davis, G. K., *Science*, 1950, **111,** 472.
144. Zarafonetis, C., *J. invest. Dermatol.*, 1950, **15,** 399.
145. Orfanos, C. E., *Haar und Haarkrankheiten*, Stuttgart, Gustav Fischer, 1979.
146. Lynfield, Y. L., *J. invest. Dermatol.*, 1960, **35,** 323.
147. Behrman, H. T., *The Scalp in Health and Disease*, London, Henry Kimpton, 1952.
148. Eckert, J., Church, R. E. and Ebling, F. J., *Br. J. Dermatol.*, 1968, **80,** 203.
149. Ikeda T., *Dermatologica*, 1965, **131,** 421.
150. Muller, S. A. and Winkelmann, R. K. *Arch. Dermatol.*, 1963, **88,** 290.
151. Saenz, H., *Acta dermosifilogr.*, 1963, **54,** 357.
152. Feldman, M. and Rondón Lugo, A. J., *Med. cut.*, 1975, **7,** 95.
153. MacAlpine, I., *Br. J. Dermatol.*, 1958, **70,** 117.
154. Robinson, I. and Tasker, P., *Urol. Cutan. Rev.*, 1948, **52,** 467.
155. Anderson, I., *Br. med. J.*, 1950, **ii,** 1250.
156. Walker, S. and Rothman, S., *J. invest. Dermatol.*, 1950, **14,** 403.
157. Porter, D. and Burton, J. L., *Br. J. Dermatol.*, 1971, **85,** 272.
158. Baccaredda-Boy, A., Moretti, G. and Frey, J. R., *Biopathology of Pattern Alopecia*, Basel, S. Karger, 1967.
159. Hamilton, J. B., *Ann. N.Y. Acad. Sci.*, 1951, **53,** 708.
160. Burton, J. L., Ben Halim, M. M., Meyrick, G., Jeans, W. D. and Murphy, D., *Br. J. Dermatol.*, 1979, **100,** 567.
161. Bingham, K. D. and Shaw, D. A., *J. Endocrinol.*, 1973, **57,** 111.

162. Alexander, S., *Trans. St. John's Hosp. derm. Soc., London*, 1965, **51,** 99.
163. Eckert, J., Church, R. E., Ebling, F. J. and Munro, D. S. *Br. J. Dermatol.*, 1967, **79,** 543.
164. Lubowe, I. I., *J. Am. Pharm. Assoc.* 1967, **NS7,** 229.
165. Sefton, L., *Br. J. Dermatol.*, 1947, **59,** 159.
166. Varga Von Kibel, A., *Arch. Dermatol. Syphilol.*, 1942, **183,** 15.
167. Swift, J. A. and Dunbar, S. F., *Nature (London)*, 1967, **206,** 174.
168. Brotherton, J., *J. gen. Microbiol.*, 1967, **48,** 305.
169. Swift, J. A., in *Sixth International Congress for Electron Microscopy, Kyoto, Japan, 1966*, ed. Uyeda, R., Tokyo, Maruzen, 1966.
170. Malassez, L., *Arch. Physiol. norm. Path.* 1874, **1,** 203.
171. Reddish, G. F., *J. Soc. cosmet. Chem.*, 1952, **3,** 90.
172. Roia, F. C. and Vanderwyk, R. W., *J. Soc. cosmet. Chem.*, 1969, **20,** 113.
173. Ackerman, A. B. and Kligman, A. M., *J. Soc. cosmet. Chem.*, 1969, **20,** 81.
174. Van Abbe, N. J., *J. Soc. cosmet. Chem.*, 1964, **15,** 609.
175. Idson, B., *Drug Cosmet. Ind.*, 1965, **96,** 636.
176. Weary, P. E., *Arch. Dermatol.*, 1968, **98,** 40.
177. Moore, M., Kile, R. L., Engman, M. F. Jr. and Engman, M. F., *Arch. Dermatol. Syphilol.*, 1936, **33,** 457.
178. Vanderwyk, R. W. and Roia, F. C., *J. Soc. cosmet. Chem.*, 1964, **15,** 761.
179. Vanderwyk, R. W. and Hechemy, K. E., *J. Soc. cosmet. Chem.*, 1967, **18,** 629.
180. Spoor, H. J., *Am. Perfum.*, 1966, **81**(10), 81.
181. Kligman, A. M., Marples, R. R., Lantis, L. R. and McGinley, K. J., *J. Soc. cosmet. Chem.*, 1974, **25,** 73.
182. Alexander, S., *Br. J. Dermatol.*, 1967, **79,** 92.
183. Van Abbe, N. J. and Dean, P. M., *J. Soc. cosmet. Chem.*, 1967, **18,** 439.
184. Hodges, J. P. S., *Lancet*, 1951, **ii,** 225.
185. Neville-Smith, C. H., *Lancet*, 1951, **i,** 1016.
186. Neville-Smith, C. H., *Lancet*, 1951, **ii,** 314.
187. Speirs, R. J. and Brotherwood, R. W., *Practitioner*, 1957, **179,** 78.
188. Duemling, W. W., *Arch. Dermatol.*, 1954, **69,** 75.
189. Whelan, S. T., *Arch. Dermatol.*, 1955, **71,** 724.
190. Ball, F. I., *Arch. Dermatol.*, 1955, **71,** 696.
191. Gross, E. R. and Wright, C. S., *Arch. Dermatol.* 1958, **78,** 92.
192. Lubowe, I. I., *Am. Perfum.*, 1958, **71,** 43.
193. Kirby, W. L., *J. invest. Dermatol.*, 1957, **29,** 159.
194. Slinger, W. M. and Hubbard, D. N., *Arch. Dermatol.*, 1951, **64,** 41.
195. Slepyan, A. H., *Arch. Dermatol.*, 1952, **65,** 228.
196. Bereston, S., *J. Am. med. Assoc.*, 1954, **156,** 1246.
197. Matson, E. J., *J. Soc. cosmet. Chem.*, 1956, **7,** 459.
198. Caspars, A. P., *Can. med. Assoc. J.*, 1958, **79,** 113.
199. Plewig, G. and Kligman, A. M., *J. Soc. cosmet. Chem.*, 1969, **20,** 767.
200. Snyder, F. H., Buchler, F. V. and Winek, C. L., *Toxicol. appl. Pharmacol.*, 1965, **7,** 425.
201. Dietrich, G. and Böllert, V., *Arztl. Kosmetol.*, 1980, **10**(1), 3.

Chapter Twenty-four

Shampoos

Introduction

Nowadays shampoos constitute one of the main products for personal care used by all strata of the population (age, sex, . . .). In 1977 shampoos represented 43 per cent of the total US haircare market.[1]

Defined in earlier editions of this book as 'suitable detergents for the washing of hair, packaged in a form convenient for use', they have since undergone drastic changes in design and technology in order to respond to multiple requirements, from the most sophisticated to the simplest and which go much beyond the single purpose of cleaning.

As early as 1955, a panel reported that 'women want a shampoo to clean and also to rinse out easily, impart gloss to the hair and leave it manageable and non-drying'. At the same time, it was also stressed that 'the principal trend evident in shampoo formulation is toward surfactants which have a milder effect on the skin; complete elimination of any sting if the product gets into the eye is the objective'.[2]

These are the two basic attributes of the types of shampoo described here and it is these requirements, known as 'conditioning effects' and 'mildness', that shampoo formulators nowadays provide as the indispensable cosmetic counterpart to the original 'cleaning' function.

However, no matter how grand a shampoo may be, its primary function remains that of cleansing the hair of accumulated sebum, scalp debris and residues of hair-grooming preparations. Although any efficient detergent can do this job, cleansing should be selective and should preserve a quantity of the natural oil that coats the hair and, above all, the scalp. Undesirable side-effects have been shown to occur when using some of the best cleansers and indeed some authors include cleansing among the functions of shampoo only as an afterthought.[3,4] The view that shampoos should be 'inefficient' detergents arises mainly from the theory that the after-effects of shampooing—difficulty in combing the hair, roughness to the hand, lack of lustre and 'fly' when the dry hair is combed—are due to excessive removal of oil from the hair. This assumption is at first sight quite reasonable but further examination shows that it is very much oversimplified: if sebum somehow fulfils a natural function of protection, and enhances the lustre and lubricity of the hair, it also possesses the dangerous drawback of attracting and trapping dust and dirt and has a potentially deleterious effect on the maintenance of set and the 'feel' of hair.

Conventional detergents of the anionic type seem to cause unpleasant after-effects to the hair roughly in proportion to their grease-removing power, but many other materials which remove grease will cause no apparent deterioration in the hair condition. Thus, if the hair is extracted with ether or trichloroethylene it can rapidly be made essentially grease-free, but its condition will be found almost unchanged. The hair will still be smooth, lustrous, and easy to comb and set. Even among detergents, there will be found some which remove relatively little sebum, but leave the hair in bad condition, and others which wash quite thoroughly without harming the hair. So no cause–effect relationship is likely to exist between residual oil and hair condition[5] and it is up to the cosmetic chemist to find the right balance between adequate soil removal and desirable hair condition.

This balance between cleaning and conditioning must be chosen with care, and it must also take into account the type of market to be supplied. People with greasy hair are highly critical of a shampoo which produces effects lasting only for three or four days, while those with dry hair may be more easily satisfied. However, many dry-haired people also use dressings of an oily character which must be removed, and in any case there is a deep-seated feeling, among women in particular, that the process of shampooing is a cleansing, purifying activity, designed to free them from daily accumulation and maturation of grease, dirt, perspiration, cooking smells, dandruff, environmental pollution and so on. In fact, it seems sensible to retain the definition of a shampoo as a suitable detergent for washing the hair with the corollary that it should also leave the hair easy to manage and confer on it a healthy look.

DETERGENCY

The development of a detergent system adapted to hair is a complex problem in itself because of the variability of the substrate and the process, and it is further complicated by the ambiguity of the objective. The substrate to be cleaned is made of the relatively hard but porous keratin of the hair and of the soft keratin of the scalp, the latter being much more sensitive to drying and defatting. While there exists very large individual variation in fibre number and diameter, the average surface area of a female head of hair is calculated to be between 4 and 8 m^2—that is, 50 to 100 times the average scalp area.[6]

The kind of soil to be removed, either natural or captured, varies greatly according to the weather, the life style, the type of work, the physiological functions, the haircare practice and so on.

Although shampoo should 'remove more soil than oil',[7] the problem of cleaning hair is mainly one of grease removal. Hair presents a reasonably hard surface and, unlike cotton and some other textiles, it does not pick up particles of dirt without the intervention of a grease layer. As long as the grease can be removed, it is quite easy to remove the dirt. In order to remove grease from the hair, one must find some agent with a greater affinity for the grease. This function can be accomplished by absorbent solids such as fuller's earth or flour, but solutions of surface-active materials such as those discussed in Chapter 33 (Surface-active Agents) are much more convenient and more commonly employed.

The mechanism of detergent action involves a number of complex physical phenomena—wetting, foaming, emulsifying and peptization—several of which are imperfectly understood. It appears evident that detergency, that is the removal of dirt, involves the following processes:

1. The detergent solution must be able to wet both the dirt and the substrate which, in the case of a shampoo, is the keratinized hair fibre; hence it must lower the surface tension.
2. The interfacial tension must be reduced to such an extent that it will allow the dirt or oil particles to be displaced by the detergent solution.
3. The dirt particles must be kept dispersed in order that they may readily be washed away.

In a detergent the polar portion of the molecule must have some attraction to the surface to be wetted (in this case the hair) so that the detergent molecules in the interface between the water and the hair will 'drag' the water over the hair surface. In doing so, the detergent solution creeps under the oily layer and lifts it from the surface, eventually causing it to roll up into spherical particles which are then solubilized by the detergent.

The essential difference between a detergent and a simple emulsifier resides in the ability of the polar group of the detergent to displace oil from a surface, and in hair washing this is its most important property. The washing of cotton and similar textiles also requires the removal of strongly bonded metallic ions from the surface, and this complicates the process and the selection of suitable detergent materials.

The evaluation of shampoo detergency is a complicated and difficult process, and there are virtually as many methods as there are laboratories evaluating detergency. The method described by Barnett and Powers[8] is typical and depends on estimating gravimetrically the quantity of soil removed from wool yarn soiled in a standard manner.

Although the public commonly associates 'foam' with detergency, the two are by no means synonymous and many very effective detergents do not foam well. However, foam (or lather) is at least of psychological importance and a shampoo that will not foam adequately will be considered unsatisfactory.

EVALUATION OF DETERGENTS AS SHAMPOO BASES

While measurements such as surface tension and interfacial tension (see Chapter 33 on surface-active agents) may often be used as screening tests to eliminate unsuitable detergents, there seems to be no substitute for practical trials on heads in the selection of detergents for shampoos. There are several reasons why this should be so, the most important being the fact that the after-effects of the shampoo may often be the deciding factor. With the wide range of materials available it is a relatively easy task to find materials that will adequately clean hair and allow for adequate lather. The final criterion which then applies in the selection of the detergent is its effect on the hair.

These effects are best observed in comparative tests on the same head, as hair diameter, quantity, greasiness and previous treatments may all affect the results. Such comparative tests can be carried out by parting the hair in the middle and washing with one detergent preparation on one side, and the alternative on the other side.[9] Care should be taken to ensure that the amount of mechanical work applied by the operator's hands is the same on both sides, and obviously the temperature of the water, the amounts of rinsing, the water hardness and so on should be kept constant for each experiment. The following are the points to be looked for.

Ease of Spreading. Ease with which the shampoo can be distributed over the hair: some shampoos seem to 'sink into' the hair so that it is difficult to spread them all over the head and to raise a lather.

Lathering Power. An abundance of foam is usually required as a first sensory perception of efficacy, though this may be waived in consideration of achieving greater mildness. Foam has not only a psychological value: it allows assessment of the amount of shampoo necessary to ensure that all functions involved in cleansing are performed.[10] This means that several properties must be considered, such as the speed with which the lather may be generated, the volume, the consistency—whether loose or creamy—and the stability of the lather on the hair.

Efficient Soil Removal. Removal of grime, excess oil and scalp debris in soft and hard water: it has been proposed that cleaning agents should be prescreened *in vitro* by studying the detergent activity on hair switches soiled with synthetic sebum.[5,11]

Ease of Rinsing. Some shampoos rinse away very quickly, others continue to lather after what seems endless rinsing. This latter behaviour can be very annoying to a woman trapped with her head over a basin!

Ease of Combing Wet Hair. This evaluates the roughness and tangling tendency immediately after treatment with the detergent, under conditions where these defects are most apparent. The consumer associates this property with the cleansing action of the shampoo, although, as has been seen, the simple removal of oily material is not the whole story.

Lustre of the Hair. The importance of this to the average consumer is sufficient justification for taking notice of it, but hair that is left dull is also a sign of unsuitability or inefficiency of a detergent. Soaps in hard water, for instance, tend to leave dulling films of insoluble calcium and magnesium soaps on the hair; detergents with insufficient suspending power may redeposit dirt or grease with, again, a dulling effect. The shampooed hair must feel and smell clean and fresh.

Speed of Drying. Drying the hair is one of the most tedious operations in the normal shampooing process and, in the case of treatments carried out in hairdressers' salons, is the most costly in terms of time and equipment. Some detergents leave the hair very wet and slow to dry; others tend to leave behind a slightly hydrophobic surface which sheds water fairly quickly. There is a practical limit to the speeding-up of drying in that a great deal of the water taken up by the hair during shampooing is bound into the hydrogen bonds, and these are unlikely to be affected by such surface effects as the type of detergent. However, of the total weight of water removed by drying after a shampoo, as little as 20 per cent or as much as 50 per cent may be in the form of surface water, according to the efficiency of the mechanical drying or the type of detergent used.

Ease of Combing and Setting the Dry Hair. When the hair is dry, any roughness induced by the detergent during the shampooing operation appears as a resistance to combing and, more important, a tendency to produce static electricity as the comb is drawn through the hair. This static electricity, which is usually of positive charge (with ebonite or nylon combs, the hair is always positively charged and the comb negatively) can be a serious hindrance to setting, as it makes the hairs repel one another. The more the hair is combed or brushed to set it out in the desired style, the more it flies about and defeats this purpose.

Safety. The shampoo detergent must be safe for use on the scalp, and no irritation, reddening or other discomfort should be caused by its use.

Sorkin *et al.*[12] have described a method of evaluating shampoos in practice using most of the points listed above. The major difficulty with these evaluation procedures is that they depend on the subjective judgment of the operator applying the product and on the variability of the subjects. It is true that these effects can be minimized by using suitable statistical designs, but it would be far better to have applicable instrumental techniques. A review of instrumental methods available for the evaluation of shampoo performance has been given by Prall,[13] with special reference to the measurement of combing resistance of hair, and its correlation with perceived sensory evaluation.

More recently, factor analysis applied to panel evaluation of shampoos for normal, dry and greasy hair was shown by Baines[14] to be a powerful tool for drastically reducing the number of parameters required to describe product performance. At least 90 per cent of the variation of most of the 17 attributes for which shampoos were rated appeared to be accounted for by two orthogonal factors—rate of foam build-up and stability of the first lather, the hair condition being better described by the former—and an oblique factor corresponding to ease of rinsing.

RAW MATERIALS FOR SHAMPOOS

The types of ingredient that go to make a shampoo are the following:

- Surfactants (cleansing or foaming agents)
- Foam boosters and stabilizers
- Conditioning agents
- Special additives
- Preservatives
- Sequestering agents

Viscosity modifiers (thickening or thinning agents)
Opacifying or clarifying agents
Fragrance
Colour
Stabilizers (suspending agents, antioxidants, UV absorbers)

These ingredients may be classified more simply as:

Principal surfactants to provide detergency and foam
Auxiliary surfactants to improve detergency, foam, and hair condition
Additives to complete the formulation and to give special effects

A general coverage of this field is given in Chapter 33 and we deal here only with considerations particular to shampoo formulation.

Principal and Auxiliary Surfactants

Nonionic detergents have sufficient cleansing activity to be considered as shampoo detergents, but very few have sufficient foaming power. They are therefore more often used as auxiliaries; some are remarkable foam boosters and stabilizers, others are used in view of their emulsifying properties and their extreme mildness in non-irritating shampoos. However, new nonionics have been developed which possess good foaming properties *per se* and may be used as the main surfactant.

Cationic detergents would appear to be ideal for shampoos; they can be made to foam well and many of them have reasonable cleaning power. In addition, they also leave the hair in excellent condition—easy to comb and set, lustrous and free of electrostatic charge. Unfortunately they have suffered from two serious disadvantages: a tendency to weigh down hair and a somewhat injurious behaviour, particularly to the corneal eye tissue. However, non-irritating cationics are now obtainable and combination with suitable nonionics and ampholytics helps to reduce the risk of irritation even more.

The remaining two groups of detergents, the anionics and the ampholytics, are both suitable as bases for shampoos. Anionics are by far the most widely used surfactants because of their superior foaming properties and lower cost. However, ampholytics, which used to play only an auxiliary role by virtue of their good hair conditioning properties, now receive increasing favour because of their contribution to mildness.

Anionic Surfactants

Soaps are metallic or alkanolamine salts of fatty acids, mostly provided by saponification of vegetable oils and animal fats. They were the first bases for shampoos.

In soft water, soaps have most of the properties desirable in a shampoo detergent, but they suffer from the disadvantage that soap solutions, especially if they are to be clear, are always alkaline. The alkalinity tends to cause roughening of the scales in the hair cuticle, thus giving a dull appearance, and is potentially damaging to the scalp. These disadvantages may be overcome by a mild acid rinse or by using less alkaline alkanolamine salts.[15]

In hard water, soaps also cause dullness by the deposition of calcium and magnesium soaps on the hair shaft. This can be prevented by the inclusion of lime soap dispersant or of sequestering agents for calcium and magnesium ions, such as salts of ethylenediamine tetra-acetic (EDTA) or polyphosphates, but these agents have no effect on the essential alkalinity of the soap solution. Soap is cheap but its use in shampoos should be considered only for less sophisticated markets, unless as 'special additive'.

Paraffin Sulphonates. These materials were introduced as soap substitutes in Germany during World War II. They earned a bad name for harshness and drying action on the skin and hair, but superior materials of this class have become available.

The C_{12}–C_{15} compounds have good foaming power and solubility, in the form of the sodium salts, and when built up by the addition of monoalkanolamide materials they can form useful cheap detergents for liquid shampoos, but they have never really become accepted.

Alkyl Benzene Sulphonates. The sodium salts of alkyl benzene sulphonates in which the alkyl group is on average a linear C_{12} chain (and hence biodegradable) are produced in bulk for use in household washing powders. They are strong degreasing agents and can be used to prepare low-cost liquid shampoos by formulating with other active ingredients such as alkanolamides. However, they are reputed to leave the hair quite dry and rough, to cause problems of fly-away, and to give a high order of irritation.

TEA dodecyl benzene sulphonate has found use in some 'oily hair'-type formulae (3–5 per cent concentration).

Alpha Olefin Sulphonates. Newly available surfactants, the alpha-olefin derivatives are less irritating than those formerly available and are widely used in powdered bubble baths. The preferred 'cut' is C_{14}–C_{16}. They have the advantages of high foaming properties in the presence of sebum and even in hard water, low cloud point and excellent acid and basic pH stability, which should provide a broad range of use, particularly for building shampoos of low pH.

Alkyl Sulphates. The most widely used anionic detergents in current shampoos are the alkyl sulphates, especially those derived from lauryl and myristyl alcohols. These alcohols, $C_{12}H_{25}OH$ and $C_{14}H_{29}OH$ respectively, are obtained by the catalytic reduction of the fatty acids in coconut and palm kernel oils, and most commercial samples of the detergents contain a mixture of the alcohols in the proportions set by the source of the raw materials.

It is commonly accepted that lauryl sulphates give a greater volume of lather and myristyl sulphates greater richness, so a mixture of the two provides a satisfactory compromise. Cetyl sulphates are usually too insoluble to give a freely-foaming solution, and the octyl and decyl sulphates are definitely lather depressants, so it is preferable to distil the fatty alcohols before sulphation so as to use only the narrow cut containing the C_{12}, C_{14} and a little C_{16} alcohol.

The mixture of alkyl sulphates so obtained from natural fats is generally referred to as 'lauryl' sulphate. However, it should always be remembered that such materials are mixtures and may vary in their properties according to the source of supply and the narrowness of the cut of fatty alcohols.

Sodium lauryl sulphate tops all other salts for flash foam and foam volume. It is normally obtained as a white powder, consisting largely of the detergent with some sodium sulphate as diluent, or as pastes with varying detergent contents. This material is poorly soluble in cold water, but its solubility increases greatly with temperature so that at the normal shampoo temperature (35°–40°C) quite concentrated solutions can be made. Moreover the process of sulphation by sulphur trioxide now permits the preparation of a highly purified product, practically free of fatty alcohol and of inorganic salt, so that it is available as 28 per cent water solutions of suitable viscosity. Its high cloud point and high viscosity make it more specifically suitable for the paste type of cream shampoo or for powder shampoo; both types are passing out of favour.

Another criticism of the product pertains to its marked detergent and defatting effect and potentiality to be somewhat irritant. This is why alkanolamine salts, claimed to be milder and less defatting, are preferred; these compounds are much more soluble and show a better compatibility with other additives and a lower cloud point. They are usually sold as 30–40 per cent solutions and are yellow, rather viscous liquids with a tendency to darken on storage, particularly on exposure to light. They form the basis of many liquid and lotion shampoos, often with no more additions than colorant and perfume. Triethanolamine (TEA) salt is the more used; monoethanolamine (MEA) salt is usually lighter in colour—MEA is less liable to oxidative spoilage than TEA—and has a slightly lower cloud point at similar concentrations. Usual concentrations of lauryl sulphate in shampoos range between 7 and 15 per cent (anhydrous product).

With all the lauryl sulphates, the inorganic salt content has two main effects. Firstly, it raises the cloud point so that solutions with much inorganic sulphate tend to turn hazy in cold weather. This is due to salting-out of the detergent. Secondly, an increase in the inorganic salt content usually increases the viscosity of the detergent solution. This is often used deliberately to increase the viscosity of products. On balance, therefore, it is as well to specify a low level of inorganic salt so that salt can be added as a viscosity builder. Addition of fatty alcohol has a similar effect. In this respect, MEA lauryl sulphate is more sensitive to viscosity-modifying salts, especially chlorides, than TEA lauryl sulphate.

Ammonium lauryl sulphate has also achieved some popularity because of its lathering quality and good solubility and because it is more stable than the sodium salt with regard to hydrolysis at acid pH (up to 4·5), allowing formulation of shampoos with low pH.

Zinc, calcium and magnesium salts are also available, but they do not seem to have come into general use. Magnesium lauryl sulphate appears to be less hygroscopic than the sodium salt, which makes it attractive for powder shampoos which are liable to cake.

To complete the survey of alkyl sulphates used in shampoos, mention should also be made of those derived from fatty alcohols obtained by Fischer-Tropsch oxidation of paraffins, such as the so-called tridecyl sulphate which is very similar to lauryl sulphate in its properties.

Alkyl Polyethylene Glycol Sulphates (Alkyl Ether Sulphates). While sodium lauryl sulphate is less soluble than the amine salts, it is cheaper, and a considerable amount of work has been carried out on related sodium salts with greater solubility.

If, instead of lauryl alcohol, a polyethoxylated lauryl alcohol is sulphated, the resulting material is so much more hydrophilic that the sodium salts are soluble up to levels of 50 per cent or more.

Usually, 2–3 ethylene oxide moles are condensed with the alcohol, 3-ethylene oxide compounds showing better solubility and lower cost. Such materials, widely available, are almost water-white solutions containing about 28 per cent of detergent. They have good foaming properties but, compared with alkyl sulphates, the foam is lighter and more open, collapsing readily in the presence of grease, so that additions of foam booster and stabilizer are required.

They are good cleansers, good solvents for non-polar materials such as fatty additives and fragrances, and they can be adjusted within wide limits of viscosity by the addition of salts such as sodium chloride. They are rapidly gaining favour as principal surfactant in that they are stable in a broader range of pH than lauryl sulphates (though potentially liable to hydrolysis on storage); they display greater mildness, irritancy decreasing with increasing number of ethylene oxide (approaching zero in the Draize rabbit eye test for 7 EO content[16]); and they show reduced tendency to keratin degradation and 'dry hair'.

However, they leave the hair in a condition slightly poorer than that obtained from washing with triethanolamine lauryl sulphate at similar concentrations. Slightly better results on the hair may be obtained with the monoethanolamine, triethanolamine or ammonium salts; however, as with alkyl sulphates, these improvements are often so marginal that the extra cost may not be justified.

Magnesium salts find their use in some baby shampoos because of their lack of irritation.

Sulphosuccinates.

$$\mathrm{R{-}O{-}\overset{\overset{\displaystyle O}{\|}}{C}{-}\underset{\underset{\displaystyle SO_3^-}{|}}{CH}{-}CH_2{-}CO_2^-\quad 2M^+}$$

These surfactants, which contain both a carboxylate and a sulphonate group in the same molecule, are succinic hemiesters: the hydrophobic chain may come, for example, from a fatty alcohol—possibly polyethoxylated—or a polyethoxylated alkyl phenol or a fatty acid ethanolamide ($R{=}R'{-}CONH{-}CH_2{-}CH_2$). Very often they are average foamers and detergents, but they are mild to the skin and evince very low incidence of eye irritation and sting,[17] along with even some conditioning effect. So they find favour in the formulation of a number of mild shampoos, such as 'low pH', 'frequent' and baby shampoos.

Monoglyceride Sulphates.

$$\mathrm{R{-}\overset{\overset{\displaystyle O}{\|}}{C}{-}O{-}CH_2{-}\underset{\underset{\displaystyle OH}{|}}{CH}{-}CH_2{-}OSO_3^-\quad M^+}$$

These sulphation products of monoglycerides, such as monolaurin, have been described in a number of patents.[18] The ammonium salt of coconut acid monoglyceride sulphate was used as a basis for a widely popular US shampoo. In general, apart from the improved solubility of the sodium salt, these detergents seem to behave very much like the lauryl sulphates.

Fatty Glyceryl Ether Sulphonates. $R{-}O{-}CH_2{-}\underset{\displaystyle OH}{\underset{|}{CH}}{-}CH_2{-}SO_3^-\ M^+$

Protected by numerous patents,[19] these surfactants have been developed with an exclusive use. Their main advantage is a good hydrolytic stability at all pH values. They are also said to be mild to the skin and have excellent flash foam.

Isethionates. $R{-}COOCH_2{-}CH_2{-}SO_3^-\ M^+$

Isethionic acid, $HOCH_2CH_2SO_3H$, was one of the first materials successfully used by IG Farbenindustrie to convert fatty acids into synthetic detergents without passing through the fatty alcohol stage.

The esters, originally known as Igepon A (now Hostapon A), display qualities similar to those of the alkyl sulphates with similar chain-length, although the foaming power of 'coconut' isethionate (sodium salt) is not quite as good as that of sodium lauryl sulphate. Very mild for scalp and hair and practically unaffected by Ca salts, they become hydrolysed in solution, which limits their use to powder shampoos and to syndet bars.

Methyl Taurides. $R{-}CO{-}\underset{\displaystyle CH_3}{\underset{|}{N}}{-}CH_2{-}CH_2{-}SO_3^-\ M^+$

The fatty amides of methyl taurine, $CH_3{-}NH{-}CH_2{-}CH_2{-}SO_3H$, were also developed by IGF as the Igepon T (or Hostapon T) series. Less mild than the above esters, the detergents in this class are remarkable for the excellent condition in which they leave the hair after washing.

Acylsarcosinates. $R{-}CO{-}\underset{\displaystyle CH_3}{\underset{|}{N}}{-}CH_2{-}CO_2^-\ M^+$

These are condensation products of fatty acids with sarcosine (that is, N-methyl glycine, an amino acid). They offer very interesting properties. They foam well and impart a good feel to the skin and the hair, which makes them appreciated as auxiliaries for lauryl sulphates or amphoterics. Preferred compounds are sodium myristyl and lauryl sarcosinates. While quite stable at low pH, they somehow lose their foaming characteristics but effect a noticeable thickening.

Another advantage of these anionic compounds is their good compatibility with a wide range of cationics, of which they do not affect the conditioning or bactericidal properties.[20]

Acyl Peptides. $R—CO—NH\left[CH—CO—NH—CH\right]_n COO^- \quad M^+$

These are complex mixtures of peptide fatty amides obtained by reaction of a fatty acid chloride with protein hydrolysates. Earlier materials were soiled with large quantities of soaps derived from the unreacted acid chloride, which caused dulling of the hair and the formation of scum. The products available today—Maypons, Lamepons (the best known being the K salt of cocoyl hydrolysed animal protein)—no longer have these drawbacks but leave the hair lustrous, manageable and silky to the touch. They do not foam so well as alkyl sulphates but produce a soft, creamy, easily rinsed lather. They are good dispersing agents of Ca soaps and offer, like the above mentioned sarcosinates, a good compatibility with cationics. More expensive than alkyl sulphates, they are used in mixture with them so as to reduce irritation and impart fine foaming, mildness and conditioning properties.

Protein condensates with coco-fatty acid, oleic acid and abietic acid are more particularly recommended.[21]

Acyl Lactylates. $R—CO\left[O—\underset{}{\overset{CH_3}{|}}{CH}—CO\right]_n O^- \quad M^+ \quad n = 1 \text{ to } 3$

These anionic esters are condensation products of a fatty acid and lactic acid. Since they derive from compounds that occur naturally on the skin's surface, they are expected to be well tolerated. A further advantage lies in their substantivity, linked to a capacity to form complexes with proteins.[22] According to the length of the fatty chain, they may show cleansing, foaming, thickening, emulsifying, antistatic or conditioning properties; they are said to improve texture and manageability.[23] They could contribute as substantive humectants.

Polyalkoxylated Ether Glycollates. This class of substance covers ethers of an α-hydroxylated acid and polyalkoxylated fatty alcohols. They are mild and they offer some conditioning properties (increasing with decreasing pH); they are said to result in a creamier foam and to improve lubricity.[24]

The oldest known compounds are polyethoxylated glycollates[25] such as Sandopan DTC (from Sandoz AG—CTFA: Trideceth-7 carboxylic acid):

$$R—O—[CH_2—CH_2—O]_n—CH_2—COOH \quad R = C_{13}, \; n = 6{\cdot}5$$

More recently polyglycerolated glycollic ethers and thioethers have been patented; they may, in addition, contain oxyethylene or oxypropylene groups.[26]

Nonionic Surfactants

Fatty Acid Alkanolamides. These materials have no great use as shampoo detergents by themselves, but they are of great importance as additives to anionic detergents. The monoalkanolamides (usually monoethanolamides, $RCONHCH_2CH_2OH$, or isopropanolamides, $R—CONHCH_2CH(CH_3)OH$), of the C_{12} to C_{18} acids are waxy solids, insoluble in water but easily soluble in detergent solutions with gentle warming. They are commonly used with the lauryl sulphates, and the addition to shampoos of between 10 and 15 per cent of lauric monoethanolamide (based on the weight of the lauryl sulphate) has the following effects:

(a) The solubility of the lauryl sulphate is increased—for instance, 15 per cent of sodium lauryl sulphate, normally a paste, can be converted to a clear solution by the addition of 2 per cent of lauric monoethanolamide.
(b) The viscosity of the solution is increased: stearic ethanolamide is a pearlescent thickener.
(c) The after-effects on the hair are better since they exhibit softening properties. Oleic ethanolamides have been recommended as conditioning agents for shampoos.
(d) The volume and the richness of the lather are very much improved.

There is little point in exceeding the ratio of 15 parts of monoethanolamide to every 100 parts of detergent, as higher concentrations of the additive only serve to increase the viscosity without improving the lather or the condition of the hair.

The diethanolamides are sold as liquid products (up to myristic diethanolamide) or low-melting solids.

It is worth noting the amides of the Kritchevsky type, obtained by heating one mole of a fatty acid with at least two moles of diethanolamine (DEA), and the superamides prepared by a 1:1 molar ratio reaction between the methyl ester of a fatty acid (usually coconut or lauric acid) and DEA. The superamides, which largely outsell the other compounds, are mainly the amides $R—CO—N[CH_2—CH_2OH]_2$ but the first class of compounds—Kritchevsky type—are complex products also containing esters of DEA such as $R—COOCH_2—CH_2—NH—CH_2—CH_2OH$ and mixed ester amides.[27]

Diethanolamides are quite useful as shampoo additives although their thickening and lather synergist qualities may not be so great as those of the corresponding monoethanolamides; however, they offer better solubility—oleic diethanolamide is an emulsifier. Care should be taken with commercial products of the Kritchevsky condensation type as they usually have a high soap content and free DEA, and may thus lead to an alkaline reaction. There is also danger of the formation of toxic nitrosamines.

The lather-improving qualities of these compounds have not been adequately explained, but it is likely that they form complexes with lauryl sulphate ions at the air–water interface, perhaps through weak ionic attraction between the sulphate ions and the slightly cationic amide group. Fatty alcohols also form surface complexes of this type, but their influence on the lather is not nearly as pronounced as that of the amides. Goddard and Kung[28] have discussed the effect of adding long-chain polar materials to ionized surfactants.

Polyalkoxylated Derivatives. Though they may not be much used as main surfactant, mostly because of their low foaming power, these compounds nevertheless constitute an important class of auxiliaries for shampoos. They are obtained by polyaddition of an alkylene oxide to a hydrophobic compound containing a labile hydrogen atom, such as an alcohol, thiol or phenol.

Both ethylene oxide and propylene oxide may be used as polyadditive reagents. The oxyethylene unit imparts hydrophilic properties: for a given hydrophobic chain R, the number of oxyethylene units added will determine the water solubility, which gives wide freedom in adjusting the properties according to the hydrophobic chain and the rate of polyalkoxylation.

The compounds obtained may be wetting, emulsifying, dispersing or foaming agents, or detergents. The following have been used in shampoos:

Ethoxylated fatty alcohols, $R(OCH_2—CH_2)_nOH$: very stable whatever the pH, they bring stabilizing, emulsifying and opacifying characteristics.

Ethoxylated alkylphenols (mostly nonylphenol): low cost and very stable products; their use is limited by their high irritating tendency for eye mucosa.

Ethoxylated fatty amines and *fatty acid amides*.

Block polymers known as 'pluronics' (CTFA: poloxamers): they are polycondensates of ethylene oxide and propylene oxide, of general formula:

$$HOCH_2—CH_2(CH_2—CH_2O)_a—(\underset{\displaystyle CH_3}{\underset{|}{C}}H—CH_2O)_b—(CH_2—CH_2O)_c—H$$

Propylene oxide brings some hydrophobic properties. In these compounds the fatty chain R is replaced by a propylene oxide polyadduct, the hydrophilic properties being given by the adjunction to each end of a number of oxyethylene units (usually $a + c = 100$ to 200 ethylene oxide and $b = 15$ to 50 propylene oxide units).

The pluronics are very mild detergents imparting a good rinsability, and they may be used in high percentages in shampoos.

Sorbitol esters: the polyethoxylated sorbitol monoesters, known as Tweens, are excellent solubilizers and emulsifiers. Moreover, they are remarkably mild compounds. The laurate (Tween 20) is claimed to give neither irritation nor stinging. These compounds are therefore extensively used in non-irritating shampoos.

Highly ethoxylated (30–78 ethylene oxide moles per mole) *mono-* and *diglyceryl fatty acid esters* have been proposed within the same field of low irritation shampoo formulation.[29]

Polyglyceryl ethers: an important new class of nonionics has arisen in the last few years, characterized by the replacement of the usual 'oxyalkylene' units by hydroxylated units, thereby leading to considerable improvement in foam-generating capability in addition to mildness.

Two types of telomer have been patented; they derive from polyaddition of epichlorhydrin or glycidol on a compound containing labile hydrogen.

The first type is composed of hydroxymethylated oxyethylene units condensed on a fatty alcohol[30] (which may be polyethoxylated):

$$R—O{-}\left[CH_2—\underset{\displaystyle CH_2OH}{\underset{|}{C}}H—O\right]_n{-}H \qquad n = 1 \text{ to } 10$$

or on an α-diol:[31]

$$\mathrm{H{-}\!\left[O{-}CH_2{-}\underset{\displaystyle CH_2OH}{CH}{-}O\right]_p\!{-}\overset{\displaystyle R}{CH}{-}CH_2{-}O{-}\!\left[CH_2{-}\underset{\displaystyle CH_2OH}{CH}{-}O\right]_q\!{-}H} \qquad 1 < p + q \leqslant 10$$

The hydroxymethyl group may be replaced by a thioether or an α-hydroxylated sulphoxide[32] such as in the dihydroxypropylsulphinylmethyl group. Besides having foaming properties, these compounds are devoid of aggressiveness to the eye mucosa, even when mixed with cationic surfactants.

The second type of polyglycerylether is obtained through polyaddition of glycidol, either pure or prepared *in situ*[33] by acidic[34] or alkaline[35] catalysis, on thiols,[36] α-diols or alkanolamides with a hydrophobic chain:[37]

$$\mathrm{A{-}\!\left[CH_2{-}CHOH{-}CH_2O\right]_n\!{-}H} \qquad n \leqslant 10$$

$$\begin{aligned} A &= \mathrm{R{-}S(O)_m{-}} \\ \text{or}\quad & \mathrm{R{-}CHOH{-}CH_2O{-}} \\ \text{or}\quad & \mathrm{R{-}CO\underset{|}{N}{-}(CH_2{-}CH_2O)_p{-}} \\ R &= \mathrm{C_{12-14}} \end{aligned}$$

These nonionics offer such remarkable foaming properties that they may be used as main surfactants; they are also devoid of irritancy to skin and eye.

Amine Oxides. The third group of nonionics is often referred to as polar nonionic surfactants since they have a highly polarized link $N \rightarrow O$; they may even take on a true cationic character at low pH values. They have found use as foam boosters and antistatic agents.[38–39] Preferred are coconut and dodecyl dimethyl amine oxides, which are said to be more effective foam boosters than alkanolamides. Antistatic properties are only shown at relatively low pH values. They are often associated with ampholytics.

Amphoteric Surfactants

These surfactants are very fashionable in the development of mild shampoos. Their compatibility with other detergents, their ionic equilibrium together with anionic/cationic potentialities (depending on pH), allow for a large flexibility of use; in addition they may contribute to the conditioning effect on hair.

They can be classified into three groups:

Long-chain N-substituted amino acids
Long-chain betaines
Long-chain imidazoline derivatives

N-alkyl Amino Acids. These mainly consist of two types:

(i) β-amino acid derivatives, known as Deriphats (from General Mills—CTFA: sodium cocaminopropionate); these are products of the substitution of fatty amines with 1 or 2 carboxyalkyl residues, for example:

N-alkyl-β-amino propionates	$\mathrm{RNH{-}CH_2{-}CH_2{-}COOH}$
N-alkyl-β-imino propionates	$\mathrm{RN(CH_2{-}CH_2{-}COOH)_2}$

Their isoelectric point, which corresponds to the zwitterion state ($-\overset{|}{N}^{+}\sim\sim COO^{-}$), lies in the region of pH 4·3.

Best foaming properties are exhibited in slightly alkaline pH range where they display the anionic character of a carboxylic salt. On the other hand, cationic properties are assumed at lower pH values, provided by either a secondary or tertiary amino group; the optimum contribution to hair manageability is found at an acidic pH. Therefore a compromise must be found between foaming and conditioning properties, as required, by adjusting the pH value. Most often used is the coconut fatty acid derivative (Deriphat 151) at pH 5·5.

(ii) Asparagine derivatives.[40] These surfactants show good foaming, cleansing and conditioning properties:

$$\begin{array}{l} \quad\quad\quad\quad CH_2{-}CONH(CH_2)_n{-}N\begin{matrix} R_1 \\ R_2 \end{matrix} \\ \quad\quad\quad\quad\ | \\ R{-}NH{-}CH{-}COOH \end{array}$$

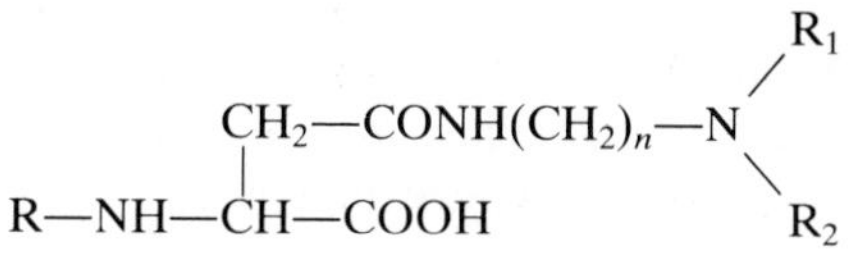

R_1 and $R_2 \leqslant 4$ carbon atoms
$n = 2$ or 3

Their added cationic valence imparts hair substantivity and easy untangling, while their ampholytic nature makes them compatible with both anionic and cationic surfactants.

Betaines

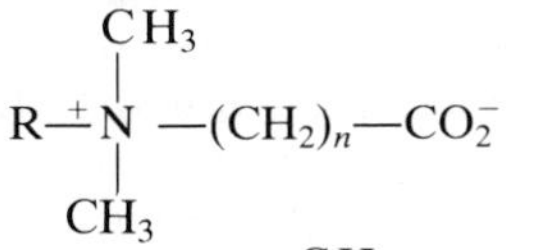

$$R{-}\overset{+}{\underset{\underset{CH_3}{|}}{\overset{\overset{CH_3}{|}}{N}}}{-}(CH_2)_n{-}CO_2^- \qquad \textbf{(A)}$$

$$R'{-}\overset{\overset{O}{\|}}{C}{-}NH{-}(CH_2)_3{-}\underset{\underset{CH_3}{|}}{\overset{\overset{CH_3}{|}}{N^+}}{-}(CH_2)_n{-}CO_2^- \qquad \textbf{(B)}$$

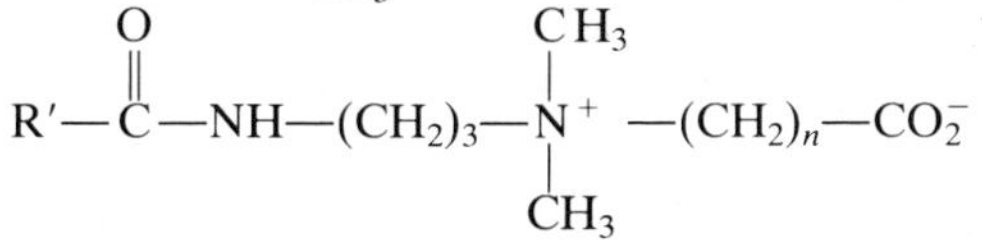

This generic term refers to zwitterionic compounds derived from trimethylglycine ($R = CH_3$, $n = 1$), known as betaine, one methyl group being replaced by a fatty C_{12-18} radical (**A**), or a fatty amido alkyl radical (**B**).

Cationic in acidic and anionic in alkaline solutions, and substantive to hair, these surfactants are mild and effective cleansers with high foaming properties, comparable to alkyl sulphates for flash foam characteristics. Moreover, foam performances are unaffected by any pH variation. Betaines are compatible with cationics, anionics and nonionics, and display additional interesting thickening properties, and therefore offer a wide range of use.

Amidobetaines are somewhat milder and therefore find a current surge of interest, more particularly the already widely developed cocoamidopropyl-betaine. Other available betaines are sulphobetaines (or sultaines) in which the carboxylic group is replaced by a sulphonic analogue (with $n = 3$), thereby resulting in further improved mildness for skin and eye with the same range of properties.

Alkyl imidazolines are also known as cycloimidates (CTFA: Amphoterics 1–20):

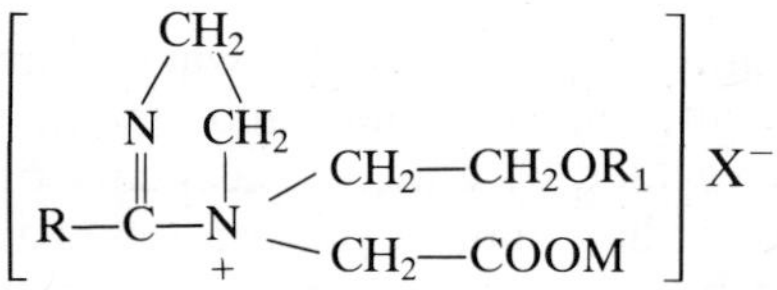

$$\left[\begin{array}{l} \quad\ \ CH_2 \\ \quad\ /\ \ \backslash \\ \ \ N\ \ CH_2 \\ \ \ \|\quad\ | \diagup CH_2{-}CH_2OR_1 \\ R{-}C{-}\overset{+}{N} \diagdown CH_2{-}COOM \end{array}\right] X^-$$

R_1 = H, Na or CH_2COOM
X^- = OH or a carboxylate (acid salt) or a sulphate (or sulphonate) anion from anionic surfactant.

They are obtained by condensing fatty acids with aminoethylcolamines and subsequent ring closure and quaternization;[41] these compounds have lately made a formidable gain related and directed to the formulation of fast-growing mild shampoos. First known under the trade name Miranol,[42] they are stable within a wide range of pH (2 to 12) and are present in most low-irritation and baby shampoos.

They exhibit fair foaming properties but they can be combined with practically all surfactants as well as with many electrolytes. They are reputed to improve foam stability while being quite innocuous to the eye mucosa.

Cationic Surfactants

Generally speaking, the cleaning and foaming properties of cationic surfactants are considerably inferior to those of anionics. Furthermore, their strong affinity for proteins such as keratin may induce the redeposition of dirt on the fibre during shampooing. As already mentioned, another drawback lies in a potential weighing-down effect on hair. Since in addition they are generally aggressive to the eye, above all with nonionics, and as they have gained a reputation of non-compatibility with anionics, their use has been limited to the beneficial development of their wet and dry combing, antistatic and lubricating properties as additives in small amounts (largely below 5 per cent). The most used compounds have been alkyl (C_{14-16}) trimethylammonium, stearyldimethylbenzyl ammonium, cetylpyridinium salts, and the less irritating double fatty-chained quaternary ammonium salts, which gained popularity with the development of rinse–conditioners, particularly distearyldimethyl ammonium (CTFA: Quaternium 5), dicetyldimethyl ammonium (Quaternium 31) and di-(hydrogenated tallow) dimethyl ammonium chlorides (Quaternium 18).

By and large, this class of surfactants has been kept for a long time in the background on account of their ill-famed properties. More recently, some very interesting breakthroughs have taken place in this field: compatibility with anionic surfactants has been shown to be possible and foam depression to be avoidable by proper formulation and suitable choice of cationics.[43]

Under the pressure of a need for extra effects in anionic shampoos, and despite (or thanks to) the rise of anionic-compatible cationic polymers, new cationics for anionic bases have flourished on the market. Examples include quaternized fatty acid amides:

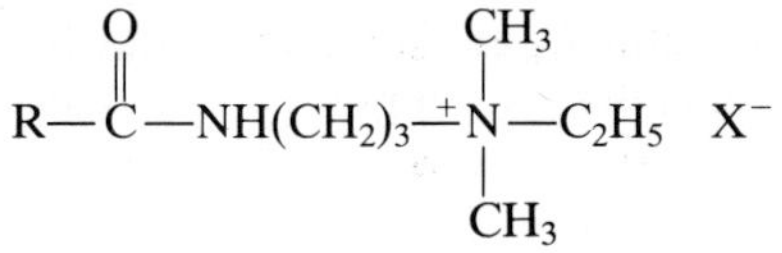

$$R{-}\overset{\overset{\displaystyle O}{\|}}{C}{-}NH(CH_2)_3{-}\overset{+}{\underset{\underset{\displaystyle CH_3}{|}}{\overset{\overset{\displaystyle CH_3}{|}}{N}}}{-}C_2H_5\quad X^-$$

derived from isostearic acid (Schercoquat—Scher Chemicals Inc.) and lanolin acid (Lanoquat—Malmstrom)[44] and proposed to impart body and softness; polyoxypropylene methyldiethyl ammonium chlorides (Witco—CTFA: Quaternium 6, 20, 21) with antistatic and foam stabilizing properties; N-stearoylcolaminoformylmethyl pyridinium chloride (Emcol 607 S, Witco—CTFA: Quaternium 7); and benzalkonium saccharinates, cyclamates and phthalimidates.[45]

On the other hand, adequate combinations of cationics with ampholytics and nonionics have been shown to alleviate aggressiveness while providing effective conditioning effects. The insertion of polar connecting groups between the fatty chain and the cationic end has also helped to diminish considerably the potential for causing irritation; typical examples are fatty acid monoesters of quaternized aminopropane diol, which are claimed to be particularly well tolerated by eye mucosa while providing greater beneficial effects in manageability, feel and appearance of the hair.[46]

More recently, new polyglycerolated cationics have been patented which seem to offer the ideal characteristics of a surfactant for shampoo, that is good foaming, cleansing and conditioning properties with a fairly low risk of irritation.[47]

Additives

Numerous materials other than detergents are incorporated in shampoos to complete the formula or to impart specific properties. Some play a prominent part in the composition, others are optional adjuncts.

Conditioning Agents

Conditioning agents have been a main focus of study in recent years. Besides specifically designed surfactants, these agents include a wide variety of materials such as fatty materials (lanolin, mineral oil), natural products (polypeptides, herbal additives, egg derivatives) and synthetic resins. They are intended to influence favourably manageability, feel and lustre of the hair, covering the whole range of magnitudes according to the nature of the hair, its condition, various compatibilities and the particular expectations of the consumer (including the quality–cost ratio). As seen above in reviewing surfactants, everything points to cationics and more precisely to quaternary ammonium compounds as conditioning agents. The great improvement in recent years has been the introduction into shampoos of cationic compounds, the hydrophilic functional groups of which are no longer at the end of a fatty chain but inserted within a polymeric structure, whereby potential for irritation is far reduced as compared with conventional products. Such resins, formerly used in setting lotions for their hair-holding properties, can be combined with amphoteric and nonionic surfactants and some show remarkable compatibility with anionics, which is the most required property both for good foaming and cleaning power and for economic reasons.

The first resin to be introduced into an anionic shampoo in 1972 was a strongly cationic cellulose derivative known as 'Polymer JR' (CTFA: Quaternium 19) from Union Carbide.[48] It is obtained by reaction of hydroxyethylcellulose with epichlorhydrin followed by quaternization by trimethylamine.[49] The average molecular weight of JR resins is between 250 000 and 600 000; they display strong affinity for keratin. Hairs treated with shampoos containing such a resin and then rinsed have been shown to be uniformly coated by it.[50]

Subsequently a number of other highly substantive resins have been successfully introduced into shampoos to impart wave-retention properties, as well as manageability and combability. Thus dimethylsulphate quaternized poly(diethylaminoethylmethacrylate)[51] and water-soluble phosphate salts of either polyacrylic acid aminoethylester or a copolymer of the latter with hydroxypropylacrylate, or the terpolymer aminoethylacrylate–hydroxypropylacrylate–acrylamide[52] have been reported in amphoteric shampoos. Similarly, condensation products of polyamines, bifunctional polyalkyleneglycol derivatives and epichlorhydrin[53] have been used in anionic-based shampoos. Cationic polymers derived from piperazine,[54] added to anionic, nonionic or ampholytic bases, have been claimed to increase body and set-retention qualities.

Other conditioning resins specifically claimed to provide easy wet combing are a dimethylsulphate quaternized copolymer of vinylpyrrolidone and dimethylaminoethyl methacrylate, known as Gafquat 755 resin (from General Aniline—CTFA: Quaternium 23),[55] quaternized[56] or cross-linked[57] polyaminoamides, and diallyl dimethylammonium chloride cyclopolymers[58] known as Merquat resins (from Merck Co.). To avoid excessive build-up on the hair with subsequent shampoos, less substantive resins have been proposed; examples are a cationic polymer grafted on the end of a cellulosic chain (Product 78-4329, National Starch & Chemical Corp.), which has been found highly efficient in dissipating static charges generated by combing and brushing, and a highly conductive interpolymer aminoethylacrylate phosphate–acrylic acid (Catrex, National Starch & Chemical Corp.).[1]

Another class of interesting compounds for these various purposes comprises polyquaternaryammonium (that is 'polyazonia') chains; the simplest elements correspond to completely quaternized substituted polyalkyleneimines such as poly(dimethylbutenylammonium chloride)-*α*,*ω*-*bis*(triethanolamine chloride) (Millmaster Onyx Corp: Onamer).[59]

A great variety of patterns and combinations within the following general structure has been patented, which may provide a wide scope of extra effects and modulation possibilities:[60]

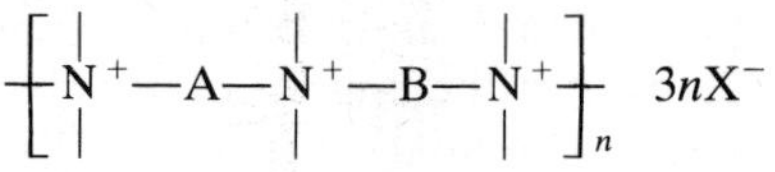

$$\left[-\overset{|}{\underset{|}{N^+}}-A-\overset{|}{\underset{|}{N^+}}-B-\overset{|}{\underset{|}{N^+}}- \right]_n \quad 3nX^-$$

A typical newly available product is the proprietary Mirapol A 15 (from Miranol Co.):

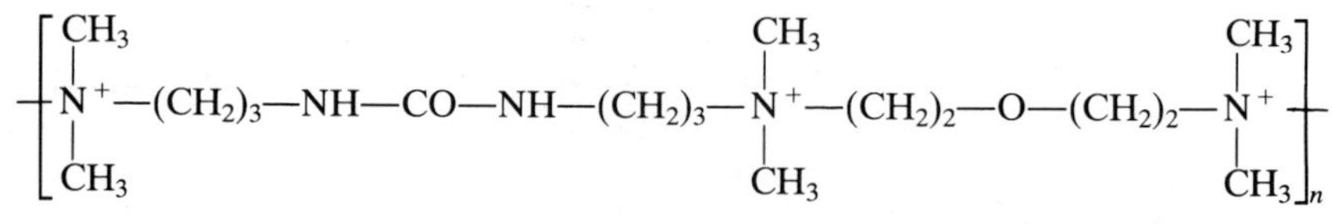

$$\left[-\overset{CH_3}{\underset{CH_3}{N^+}}-(CH_2)_3-NH-CO-NH-(CH_2)_3-\overset{CH_3}{\underset{CH_3}{N^+}}-(CH_2)_2-O-(CH_2)_2-\overset{CH_3}{\underset{CH_3}{N^+}}- \right]_n$$

Its average molecular weight is 2260 (about six repeated units); it is compatible with anionic surfactants and is thought to absorb on the hair and link with the surfactants, whereby the hair becomes tangle-free, smooth, shiny and antistatic.

Another compound on the borderline between surfactants and polyquaternary compounds, stearylpentamethylpropylenediammonium dichloride, has been claimed to enhance the mechanical properties of hair when used at a concentration of 1 per cent in an anionic shampoo.[61]

However, the issue of conditioning is not limited to polyammonium (or azonia) compounds or resins which, furthermore, are not easy to manage and to adjust so as to monitor the deposition on the hair. Thus, alkylated[62] and partly formylated[63] polyethyleneimines have been claimed to impart softness and combability. Water-soluble proteins, preferably hydrolysed collagen (average molecular weight 500–10 000), and quaternized derivatives have been proposed for improving curl retention and wet combing, as well as for protective and synergistic mildness effects. The more damaged the hair the higher the protein absorption onto hair and the lower the pH at which it occurs. Virgin hair absorbs little, with a maximum at pH 9–11; waved hair absorbs the most, followed by bleached hair with a maximum uptake at pH 9 → 6, decreasing with increased level of bleaching. The polypeptidic fraction with average molecular weight of 1000 shows highest substantivity.[21,64]

More recently, a serum containing antibodies ('hair anti-serum') has been claimed to display similar properties with improved substantivity.[65] A mixture of proteins and polysaccharides from waste liquid beer sludges has been patented as conferring body,[66] as have natural wood rosins.[67]

All these 'texturizers' more or less help to improve the feel and appearance of hair after shampooing. Another compound, the triethanolamine salt of alginic acid, added to anionic base, has been claimed to give more specifically a 'feel of a velvety nature'.[68] Smoothness and gloss can be obtained by including so-called superfatting agents, that is, oily materials such as ethoxylated lanolin derivatives, silicones, Ucon fluids (PEG and PPG alkylethers), PEG-modified polysiloxanes,[69] mineral oil, mink oil, sesame oil, jojoba oil, or other vegetable or animal oils, which deposit onto the keratin fibre during shampooing and lubricate it. Benzyl and phenethyl alcohols have also been claimed to impart extra effects when added to anionic–amphoteric systems.[70]

Addition of honey has been patented as a lubricating agent.[71] Lustre performance may also be enhanced by depositing a specific resin or resin–oil film on the hair surface, as suggested by different patents; thus, some data have been given on the beneficial incorporation in anionic shampoos of water-soluble anionic linear polymers—such as polymethacrylic acid (MW 10 000–100 000) or hydrolysed copolymers of ethylene (or methyl vinyl ether)–maleic anhydride[72]—or of various resins either combined with mineral oil[73] or coacervated with dimethylpolysiloxane.[74]

Miscellaneous Additives

Among the parameters susceptible to influence by specific additives, the drying time should also be mentioned; it usually increases with the state of damage of the hair. Apart from aluminium soaps,[75] the introduction of particular water-repellent film-forming fluorinated polymers in low amounts (0·003 per cent) has been claimed to reduce significantly the time needed for drying;[76] a typical resin is the copolymer of N,N-diethyl aminoethylmethacrylate and hexafluoroisopropyl (or perfluoro-octyl) methacrylate. Similar fluorinated compounds[77] have been shown to be efficient in delaying migration of sebum on the hair and therefore checking the subsequent greasy relaxation of set hair.

Another approach to preventing the hair from regreasing too rapidly has been patented, using thiocompounds such as cysteine, glutathione and amino-alkanethiol derivatives.[78–83]

Lastly, it is worth noting the use of compounds—notably antioxidants—to prevent the production of unpleasant odours arising from sebum maturation on the hair and scalp.[84–86]

Viscosity Modifiers

Thickening of a shampoo may be achieved by inclusion of various types of compound, such as:

(a) electrolytes: 1–4 per cent (w/w) ammonium or sodium chloride in alkylether sulphates increases the viscosity steadily;
(b) natural gums (karaya, tragacanth), alginates;
(c) cellulose derivatives (hydroxyethyl, hydroxypropyl, carboxymethyl) which protect the hair against redeposition and soften foam;
(d) carboxyvinyl polymers (Carbopol 934 and 941 from Goodrich—CTFA: Carbomer) which in addition promote stability of the shampoo.

However, attention must be paid to the temperature and concentration dependence of most of these effects.

Ethoxylated fatty acid diesters (for cream shampoos), phosphate esters, amidoamine oxides, TEA soaps, polyvinyl pyrrolidones and polyvinylalcohols and the previously mentioned alkanolamides are other thickeners.

Reduction in viscosity may be obtained through the addition of small amounts of solvents (alcohols), or polyoxyalkylene compounds or of sodium xylene sulphonate, which also provides clarity.

Opacifying and Clarifying Agents

Opacity or pearlescence is provided by:

(a) alkanolamides of higher fatty acids (stearic, behenic);
(b) glycol mono- and distearates, propyleneglycol and glycerol monostearates and palmitates;
(c) fatty alcohols (cetyl, stearyl), which also contribute smoothness;
(d) milky emulsions of vinyl polymers and latexes;
(e) insoluble salts—for example, magnesium, calcium or zinc—of stearic acid;
(f) finely dispersed zinc oxide or titanium dioxide;
(g) magnesium aluminium silicate (for example, Veegum—Vanderbilt Co.), which also prevents sedimentation in the product.

Pearl-like effects depend on the size, shape, distribution and reflectance of the opacifier crystals included.

Transparency may be improved and stabilized by addition of solubilizing alcohols (for instance, ethanol, isopropanol, propyleneglycol, hexylene glycol and dimethyloctyne diol—Surfynol 82: Air Products), phosphates, or nonionic solubilizers (for example, polyethoxylated alcohols and esters).

Sequestering Agents

The function of these compounds is to prevent the formation and deposition onto the hair of Ca and Mg soaps when rinsing with hard water. Ethylene diaminetetra-acetic (EDTA) salts or polyphosphates are mostly used. Nonionic surfactants may also display efficacy by peptizing Ca salts.

Preservatives

Preservation is an extremely important aspect of shampoo formulation. The early soap shampoos were not hospitable media for moulds and bacteria, but as products became milder in their action on the skin, they became milder in their action on bacteria. The introduction of nonionic surfactants and more recently of some 'natural additives' has appreciably increased the risk of microbial contamination.[87] Most modern shampoo materials are liable to mould attack unless preserved with agents such as the hydroxybenzoate esters (see Chapter 36 on preservatives).

More serious in many ways is the growth of bacteria in shampoos, since they can lead to the breakdown of the detergent and discoloration of the product. Many shampoo surfactants will support the growth of Gram-negative organisms of the *Pseudomonas* type, and in fact it seems likely that these widespread bacteria have adapted themselves to prolific growth in unpreserved shampoo solutions, where they can produce off-odours (especially in diluted solutions) and cloudiness as a result of the production of mycelium.

The question of preservatives for toilet preparation in general is discussed in Chapter 36, but it may be said here that the surface-active agents in shampoos tend to interfere with the activity of germicides—thus it was shown that the bactericidal action of quaternary ammonium surfactants was markedly reduced in the presence of nonionics[88]—so that higher concentrations of a preservative are often necessary in a shampoo compared with that in simple solutions.

The preservation of shampoos is therefore a specialized subject and its peculiar aspects have been reviewed by various authors.[87,89] The most simple broad-spectrum germicide, that is formaldehyde, is about the most effective, together with phenyl mercuric salts. Although formaldehyde is unaffected by surfactants and is efficient in amounts of 0·1–0·15 per cent, it is however not compatible with certain ingredients, particularly protein hydrolysates, which furthermore present problems of preservation.[90] Esters of *p*-hydroxy benzoic acid are quite effective against moulds, but inactive against *Pseudomonas* and inactived by nonionics. Other recommended preservatives are 5-bromo-5-nitrodioxan, known as Bronidox,[91] and a formaldehyde-releasing quaternary ammonium compound known as Dowicil 200,[87] neither of which is adversely affected by anionic or nonionic surfactants.[92,93] Dimethyloldimethylhydantoin also releases formaldehyde and is of interest because of its efficiency and compatibility.

Perfumes

Long considered as an incidental additive, perfume is becoming an important feature for sales appeal with the increasing impact of the 'back to nature' concept. Simple descriptive odours—herbal, fruity, floral—are increasingly sought to provide the evocative whiff of natural freshness.[94] Some believe that fragrance will play an increasing functional role of quality-communication,[95] with higher levels of concentration and a thrust towards 'substantive odours' (residual notes for long-lasting fixatives).

Shampoo fragrance must first comply with basic technical requirements such as solubility, compatibility—that is, no effect on viscosity and stability—non-discoloration of formula (or hair), and non-irritation. Moreover, the introduction of conditioning agents into shampoos has increased possible interactions and often added to the need for a blend of covering fragrances.[96]

FORMULATION OF SHAMPOOS

Many of the detergents available for shampoos have already been described. The examples of shampoos given below are not exhaustive, but represent the various types of formulae. Where a formula is based on some particular detergent, it may usually be assumed that other detergents or mixtures can be used in its place, allowing for such matters as solubility, foaming capacity, etc. By this means the simple formulae quoted may be used as bases for further formulation.

The principles involved in formulating shampoos (see also Cook[97] and Mannheim[98]) have been reviewed by Zviak[10] and Markland;[99] an excellent survey of materials and formulations has been published by Alexander[38] (see also Kass[100a] and Donaldson and Messenger[100b]).

It is important to note that consumers in different countries have different ideas of the ideal concentration for a shampoo. In the UK, for instance, most people prefer to use a shampoo of medium concentration, whereas in some other countries the consumer expects a shampoo to have high active level. This difference in social habits and requirements makes it difficult to recommend detergent levels that will be universally suitable. Formulators should make themselves aware of the requirements of the market they are working for. Most of the formulae that follow comply with the UK pattern of 12–20 ml per head.

Many shampoos are available in three types: for normal, dry and oily hair. Those for oily hair often have a higher percentage of surfactant or a blend that is more active in emulsifying sebum out of hair; those for dry hair usually contain a higher level of conditioner.

The main types of shampoo on the market are the following:

Clear liquid shampoos
Liquid cream shampoos
Solid cream shampoos
Oil shampoos
Powder shampoos
Aerosol foam shampoos
Dry shampoos

Alternatively, shampoos may be classified according to function, more particularly:

Conditioning shampoos
Anti-dandruff shampoos
Baby shampoos
Acid-balanced shampoos

Clear Liquid Shampoos

This is at present the most popular type, and is subject to the greatest variety of formulation and presentation. While there is no very clear picture of what members of the public expect from a clear liquid shampoo, it seems that the formulae can be divided roughly into those bought largely on the grounds of their cleansing power for greasy hair (which can be categorized as 'cleansing shampoos') and those bought because, along with their promise of cleansing, they carry the suggestion that the hair will still be left in good condition after shampooing. Predominantly transparent or translucent, the latter are popular among consumers, particularly women with dry or normal hair, and are often called 'cosmetic shampoos'.

The 'cleansing' type are easily formulated, as they required only a suitably presented solution of a detergent, such as triethanolamine (TEA) lauryl sulphate or a lauryl ether sulphate, providing a low cloud point and therefore ensuring clarity even at low temperature; TEA lauryl sulphate is usually sold as 30–33 per cent solution, and 50 parts of this, perfume, colour, and water to 100 parts will make a mobile, clear solution with good foaming power. For a more viscous product, the ether sulphate may be used as in example 1.

	(1) *per cent*
Sodium lauryl ether sulphate 30%	45·0
Sodium chloride (according to the viscosity required)	2·0–4·0
Perfume, colour, water, preservative	to 100·0

The 'cosmetic' type of liquid shampoo can be formulated by selecting those detergents recommended for their good after-effects, such as methyl taurides, amphoterics, etc., in admixture with lauryl sulphate and alkanolamide additives:

	(2) *per cent*
Triethanolamine lauryl sulphate (33%)	45·0
Coconut monoethanolamide	2·0
Perfume, colour, water	to 100·0

	(3) *per cent*
Lauryl amino propionic acid (Deriphat 170 C)	10·0
Triethanolamine lauryl sulphate (33%)	25·0
Coconut diethanolamide	2·5
Lactic acid to give pH 4·5–5·0	*q.s.*
Preservative	*q.s.*
Fragrance, colour, deionized water	to 100·0

Dry hair shampoo (Alcolac Inc.[87])	(4)
	per cent
TEA lauryl sulphate	49·0
TEA oleate (50%)	9·8
Propylene glycol	2·0
Oleyl alcohol	1·0
Water	38·2

High quality low-cost shampoo base (Witco Chemical Corp.[87])	(5)
	per cent
Sodium C_{14-16} olefin sulphonate	25·0
Coconut monoethanolamide	3·0
Conditioners, fragrance, preservative	*q.s.*
Water	to 100·0

Liquid Cream or Lotion Shampoos

Liquid cream shampoos really form an extension to the class of 'cosmetic' shampoos, since users expect them to be very mild in their action on the hair. The appearance of the liquid creams is calculated to suggest emollience, though it is unwise to include very much fatty material in such a product, or the hair will become greasy again very soon after use. Example 6, from the American Alcolac Corporation, is typical.

	(6)
	per cent
Sodium lauryl sulphate 30%	25·0
Polyethylene glycol 400 distearate	5·0
Magnesium stearate	2·0
Distilled water	68·0
Fatty acid alkanolamide (for thickening)	*q.s.*
Oleyl alcohol (for conditioning)	*q.s.*
Perfume	*q.s.*

The opacifier used in this example to convert clear liquid shampoo into liquid cream shampoo is a glycol distearate. The addition of insoluble magnesium stearate is desirable because the glycol esters tend to redissolve in the shampoo in hot weather, thus leaving the shampoo only hazy instead of creamy.

DeNavarre[101] has discussed the formulation of liquid cream shampoos and notes that the fatty alcohol sulphates are the best foaming agents, and are usually used at concentrations of 25–45 per cent of the finished shampoo. He recommends the use of an opacifier such as magnesium stearate in admixture with a mucilaginous thickener such as a solution of polyvinyl alcohol, methyl cellulose, solubilized methacrylate, alginate, Irish moss or carboxymethyl cellulose at a concentration of about 0·5 per cent. He notes that higher alcohols or alcohol esters can also be used for opacifying purposes, and lanolin and glycol or glyceryl laurate at a level of 1 or 2 per cent will further opacify and thicken the emulsion. Pantaleoni *et al.*[102] quote a formula based on the above principles:

	(7) *per cent*
Fatty alcohol sulphate paste	30·0
Magnesium stearate	1·0
Polyvinyl alcohol, 10% solution	20·5
Methyl cellulose, 3% solution	9·0
Water	38·0
Lanolin	0·5
Glyceryl monolaurate	1·0

Procedure: Mix the magnesium stearate with the fatty alcohol sulphate, then add the polyvinyl alcohol, methyl cellulose and water, each separately, to make the water phase. Agitate the mixture while heating to 71°C. Bring the oil phase to the same temperature and add the water phase with agitation.

In this class fall most of the shampoos described as containing egg, milk, cream, coconut, and other 'magic' ingredients. The oldest and still the most popular additive is egg. Eggs can, of course, be used to clean the hair; after rinsing (with cold water!) they will have removed a surprising amount of dirt and leave the hair with a smooth, albeit rather greasy, appearance. The action is mainly suspension by the albumen acting as a protective colloid followed by the absorption of the cationic phosphatide lecithin from the egg-yolk. All this is a far cry from the extravagant claims made for minimal amounts of egg-yolk churned up in an excessive amount of detergent, as in most modern 'egg' shampoos. However, if these are made, it is better to avoid the egg-white, which is liable to solidify, and use dried or frozen yolk. The dried material can often be dispersed easily if it is first slurried with a little nonionic detergent or with perfume.

Egg shampoo (Clintwood Chemical Co.)	(8) *per cent*
Sodium lauryl sulphate 30%	20·0
Coconut diethanolamide	5·0
Glycol stearate	1·0
Preservative:	
Methyl *p*-hydroxybenzoate	0·1
Formaldehyde	0·1
Sodium benzoate	0·1
Phosphoric acid to pH 7·5–8·0	*q.s.*
Sodium chloride	0·25
Yellow dye	*q.s.*
Perfume	*q.s.*
Egg, powdered (or whole)	2·0
Water	to 100·0

Solid Cream and Gel Shampoos

Solid cream and gel shampoos are usually intended for use from jars or collapsible tubes, and should therefore be of a suitable consistency to stay in an open tube without running out of the nozzle. They are traditionally made with sodium lauryl sulphate pastes, or other detergents with a low solubility at room temperature but a higher solubility slightly above room temperature, gelled up with a little sodium stearate or other soap.

	(9) *per cent*
Sodium lauryl sulphate (100%)	20·0
Coconut monoethanolamide	1·0
Propyleneglycol monostearate	2·0
Stearic acid	5·0
Sodium hydroxide	0·75
Water, perfume, colour if desired	to 100·0

		(10) *per cent*
A	Sodium lauryl sulphate (90%)	20·0
	Sodium lauryl ether sulphate (27–30%)	20·0
	Coconut diethanolamide	0·6
	Anhydrous lanolin	0·6
	Sodium chloride	1·2
B	Stearic acid	4·0
C	Sodium hydroxide	2·12
	Colour, perfume, preservative	*q.s.*
	Water	to 100·0

Procedure: Dissolve A and C in the water and heat to 70°C. Melt B and add at the same temperature with agitation. Cool the mixture and add the perfume.

The major trouble with these pastes is the effect of temperature. The consistency depends on a mass of crystals, and in warm weather a large proportion of these may dissolve, making the product runny and translucent. When the product recrystallizes, the new crystals may be large and cause either lumps or fibrous characteristics in the shampoo. Similar considerations apply to all cream shampoos in which the opacity or the consistency depends on the detergents forming a paste of incompletely dissolved material. Since the 1950s cream paste shampoos have receded to a minor position except for some medicated forms in the anti-dandruff field.

Clear gel shampoos gained some popularity in concentrated form, notably in beauty shops, but the interest has largely declined. These may be formulated using detergents alone or in admixture with soaps. By varying the amounts of triethanolamine–coconut soap, triethanolamine lauryl sulphate and sodium lauryl sulphate, preparations of different consistencies and textures may be obtained. By variation in the type and amount of sequestrant the properties of this gel may be modified. Gelation can also be achieved by thickening a conventional clear liquid shampoo with hydroxyalkyl or methyl cellulose ethers or alkanolamides, or by combining anionic and amphoteric surfactants.

Clear gel shampoo	(11) *per cent*
Miranol C2M conc.	15·0
Triethanolamine lauryl sulphate (40%)	25·0
Coconut diethanolamide	10·0
Methocel (hydroxypropylmethyl cellulose)	1·0
Water	to 100·0
Colour, perfume, preservative, etc.	*q.s.*

Oil Shampoos

The development of sulphonated oils (that is, oils which have been treated with sulphuric acid or other sulphonating agents under the influence of heat and then neutralized with alkali) led to their introduction many years ago as hair shampoos. They are effective detergents in that they remove soil and oil from the hair, but they are nearly devoid of foaming properties. For this reason, they never really successfully competed with the more conventional surfactants which were becoming available at the time. However, there has been a tendency to use them in the formulation of 'oil shampoos' in which the head is treated with a mixture of sulphonated oils (mainly castor and olive oil) which is allowed to remain on the head for a short time and then rinsed off. Example 12 is representative (see also Alexander[38]).

	(12) *per cent*
Sulphonated olive oil	16·0
Sulphonated castor oil	16·0
Water	68·0
Colour, perfume	*q.s.*

The modern version of an oil shampoo is a two-phase separable shampoo in which the upper layer consists of an oily material, while the lower layer is a conventional shampoo. The product is shaken before use to give an unstable oil-in-water emulsion.[103,104]

Powder Shampoos

Powder shampoos have fallen out of favour, partly because of the trouble involved in using them, but mainly because they tend to leave the hair in poor condition.

They consist of a detergent powder extended with easily soluble non-hygroscopic substances. Sodium and magnesium lauryl sulphates are suitable materials for such cheap shampoos, and about 3–4 g is sufficient per head.

As diluent, sodium pyrophosphate or bicarbonate or, more generally, sulphate is used (example 13).

	(13) *per cent*
Sodium lauryl sulphate	25
Sarcoside	5
Sodium bicarbonate	10
Sodium sulphate	60

Aerosol Shampoos

Aerosol shampoos are not a special type of formulation but represent an alternative way of applying the product. The only extra condition that such products must satisfy is that they must be stable in the presence of the propellant gas; it is usual to use a clear liquid shampoo of medium viscosity so as to allow easy mixing with the propellant, while maintaining it in an emulsified form long enough for use after shaking. Expensive, and having more drawbacks than advantages, they never achieved much success.

Dry Shampoos

Dry shampoos are powder compositions which allow the hair to be cleaned simply by sprinkling absorbent powder onto the greasy hair, leaving it for about 10 minutes and then brushing it off. Their main attraction is that they do not involve the use of water. There is no fear that the process of shampooing will remove the set, and there is no loss of curl.

However, powder shampoos are not very efficient, because it is mechanically difficult to get the particles of the powder into contact with the grease. It is also difficult to brush the powder out of the hair, with the result that a hasty dry shampoo treatment can leave the hair dirtier than it was before. Consequently these shampoos are useful only for hair that tends to become rapidly weighed down with sebum, for those who have no time to set and dry wet hair and for use in times of acute water shortage.

They consist of a mixture of absorbent materials, such as starch, borax and silicas. A powder composition modified to retain or impart 'bounce' and gloss might be based on the following mixture:[105]

	(14) *per cent*
Insolubilized rice starch (tetramethyl acetylendiurea reaction product)	30
Boric acid	7
Finely divided silica	25
Starch	23
Talc	15
Perfume oil	*q.s.*

Dry shampoos can also be based on solvents such as white spirit, isopropyl alcohol, ethylene dichloride, etc. They are efficient at dissolving the grease, but the main difficulty is to find some effective and aesthetic way of removing the solution from the hair. Another problem is the hazardous nature of most of the good organic solvents—the alcohols, etc., are flammable, and the chlorinated solvents are usually narcotic and sometimes extremely toxic.

An alternative approach has been to combine the two methods and to use a powder as a carrier for the organic solvents, preferably fluorinated hydrocarbons applied from an aerosol dispenser. This mixture gives adequate cleaning and products of this type have achieved limited popularity.

Another formulation, based on powdered chitin as a suspension in a volatile liquid carrier such as ethyl alcohol, has been shown to be efficient in removing sebum.[106]

Conditioning Shampoos

As pointed out by Gerstein,[107] the distinction is not always obvious between what are generally called 'conditioning shampoos' and 'shampoo conditioners', particularly since they are similar in application. However, the former are intended primarily to cleanse and secondarily to improve manageability and promote desired feel and appearance; the latter are mainly designed to enhance untangling of wet hair and improve manageability of both wet and dry hair.

The development of shampoo–conditioners is mainly the result of exploiting the benefits and optimizing the properties of synthetic cationic polymers, of which the efficacy is linked to the characteristics and amount of film deposited on the hair.

Their main drawback, the counterpart of their substantivity, is the excessive build-up on the hair which may occur following successive applications and which may lead to an unexpected behaviour of the hair after other treatments. However, the adsorption of some cationic polymers on hair has been demonstrated to be notably reducible by addition of small amounts of electrolytes (such as sodium chloride) which may also partially desorb the absorbed polymer.[108]

The ways in which conditioners act, as well as the effects obtained, are various; it should always be borne in mind that they involve a number of extremely complex interactions, both among the compounds themselves—which in most cases are statistical mixtures—and with the hair. As a consequence, the adjustment and the control of these effects—stiffening, softening, untangling, combability, body, springback, antistatic, etc.—require a careful and skilled development according to the type of hair concerned and the surfactant system used.

Examples are given below—see also Kass.[100] In example 16, mildness and foaming properties are given by the mixture of ethoxylated nonionics with amphoteric and anionic surfactants. Conditioning properties and thickening are given by cationic cellulose.

Shampoo leaving hair soft and combable, both wet and dry[63]	(15) *per cent*
Sodium lauryl sulphate or lauryl sarcosinate	10·0
Lauric diethanolamide	5·0
Hexylene glycol	3·0
Methyl *p*-hydroxy benzoate	0·1
Boric acid	1·0
Sodium chloride	2·0
Polyethyleneimine-ethylformate polyamide	2·5
Ethyl alcohol	15·0
Water	to 100·0

Mild, thickened shampoo with conditioning properties[109]	(16)
	per cent
Coconut amidopropyl-3-dimethylamino betaine	5·4
Sodium lauryl sarcosinate	5·2
Ethoxylated tridecyl alcohol (20 EO)	14·0
Cationic cellulose (Polymer JR: Union Carbide)	0·5
Water	to 100·0

Conditioning clear shampoo giving easy wet combing (Sandoz)	(17)
	per cent
Ammonium lauryl ether sulphate	25·00
Cocoamino betaine 31–32%	25·00
*Sandopan TFL conc. 48%	4·20
Adipic acid/dimethylamino hydroxypropyl diethylenetriamine copolymer (Cartaretine F4)	3·33
Citric acid anhydrous	0·98
Water	*q.s.* to 100·0

*Sandoz—CTFA: Amphoteric 7 (sulphoamidobetaine)

One-step shampoo-creme rinse[45]	(18)
	per cent
Potassium cocoylhydrolysed protein (Maypon 4C)	25·0
Lauraminopropionic acid (Deriphat 170C)	15·0
Laurylmonoethanolamide	3·0
Stearyldimethylbenzylammonium saccharinate	0·3
Ethanol	3·0
Quaternary polymer of vinyl pyrrolidone/di (lower alkyl) aminoacrylate (Gafquat 755, MW = 1 000 000)	1·0
Colour, perfume	*q.s.*
Water	to 100·0

Shampoo conditioner[110]	(19)
	per cent
Polymer JR 30 M (CTFA: Quaternium 19) (MW = 30 000)	1·5
Miranol C2MSF (CTFA: Amphoteric 2) 70%	11·0
Sandopan DTC acid (CTFA: Trideceth-7-carboxylic acid) 90%	15·0
Ethyleneglycol distearate	2·0
Methyl *p*-hydroxybenzoate	0·2
Propyl *p*-hydroxybenzoate	0·05
Protein hydrolysate	0·5
Perfume oil	0·3
Water	to 100·0

Some conditioning shampoos incorporate oil to improve the quality of the deposit on hair: thus a single-phase formula uses 1·5 per cent olive oil, 0·5 per cent extra light mineral oil, together with an amide, an amine oxide, an amphoteric surfactant, and a quaternary ammonium such as 'tallow' amidopropyl dimethyl hydroxyethylammonium chloride;[105] a two-phase shampoo combines hydrocarbon oil, giving a creamy emulsion top layer, and a mixture of cationic and amphoteric surfactants as a clear aqueous alcoholic lower layer.[111]

Among a number of patented compositions, it may be worth mentioning high-lathering conditioning shampoos that combine tertiary amine oxides, alkyl sulphobetaines, and triethanolamine soaps,[112] and also the use of triethanolamine salts of 'lipoproteins' with anionic and amphoteric surfactants to provide shampoos effective in repairing split ends of hair,[113] which has been identified as a significant problem of young women with long hair.

Baby Shampoos

In baby shampoos the overriding requirement of blandness towards eye, skin and hair (and even towards the digestive tract if they are accidentally ingested) fully warrants some sacrifice in foaming and cleaning performance. Their implicit safety and innocuity explain the increasing favour they encounter among adults, and also contribute to their liberal and frequent—if not daily—use.

Mildness is provided by choosing non-irritating surfactants that produce limited detergency—most commonly amphoteric imidazoline derivatives and the fatty sulphosuccinate esters and amides, which are known to be almost irritation-free and to exhibit anti-irritant properties when associated with laurylsulphate.[16] Imidazolines are usually combined with ethoxylated sorbitan or mannitan esters to give sting-free compositions. This association was the basis of the success achieved by Johnson and Johnson who, in the 1950s, initiated baby shampoos: for 'no more tears', Tween 20 was combined with a complex obtained from tridecyltriethoxysulphate and N-(2-cocoamidocthyl) diethanolamine.[114] It has recently been shown that foam stability may be improved with no loss of mildness by changing the last compound for betaines or sulphobetaines and by increasing the number of ethylene oxide moieties in sodium tridecylether sulphate (to 4·4) and in sorbitan ester (over 20).

A typical composition is given in example 20.[115]

	(20) *per cent*
3-Cocoamidopropyldimethyl betaine 30%	17·1
Tridecylether sulphate salt 4·4 EtO 65%	8·3
Polyoxyethylene (100) sorbitan monolaurate	7·5
Preservatives, perfume, dye	*q.s.*
Water	to 100·0

Generally speaking, a high ethoxyl value is recommended to provide mildness; thus the association of 20 per cent lauryl ether sulphate with 12 moles of oxyethylene and 8 per cent imidazoline—both as 100 per cent active bases—has been claimed to produce an exceptionally low level of ocular irritation while providing good foaming characteristics.[116]

Anti-dandruff and Medicated Shampoos

Any of the types of product described above is suitable as the basis for the preparation of medicated or anti-dandruff shampoos; the most popular are clear liquids and opaque lotions.

Whatever the etiology of the dandruff disease (defined as a chronic, non-inflammatory scaling of the scalp), the problem remains of removing the scurf produced by a cutaneous medium in poor condition; an efficient but mild and non-drying shampoo is required. As dandruff is commonly associated with a marked microbial proliferation it is usually recommended to add a germicide (with particular effect against *Pityrosporum ovale*) as active controlling agent. As the shampoo remains on the scalp and hair for only a very short time, the germicide must be of the substantive type so that it is left behind on the scalp to exercise its action.

Various agents have been used for this purpose, such as quaternary ammonium surfactants, thymol, chlorinated phenols, trichlorocarbanilide, halogenated salicylanilides, 5,7-dichloro 8-hydroxyquinoline,[117] zinc undecylenate, undecylenic acid ethanolamides, polyvinyl pyrrolidone–iodine complex and Captan.[118]

A typical formula is given in example 21.

	(21) per cent
Triethanolamine lauryl sulphate	15·0
Lauric diethanolamide or mixture of lauric and myristic diethanolamide	3
Germicide	0·5–10·0
Colour, perfume	*q.s.*
Water	to 100·0

In recent years the greatest developments have been in the derivatives of 2-pyridinethiol-N-oxide; the first to be introduced into a shampoo was the highly insoluble zinc salt, used in the early 1960s at a 2 per cent concentration in the very successful 'Head and Shoulders' shampoo. Its high acceptance and efficiency[119] led to a number of studies and patents concerning the best conditions of formulation and covering the range of compounds of the same family. Attention was particularly focused on substantivity and tolerance. One of the great advantages of this type of substance lies in its great affinity for the hair. The absorption of zinc salt has been shown to reach a maximum at a concentration of about 1 per cent in shampoo,[120] but the nature of the vehicle may notably affect the absorption rate as well as the bactericidal activity. Thus, polyethylene glycol 400 laurate was shown markedly to reinforce activity, but not polyethoxylated lanolin alcohols.[121]

Another interesting observation was the enhancing effect of cationic resins and particularly polyethyleneimine resins (antimicrobial *per se*[122]) on particulate deposition onto the hair, whereby the concentration of zinc pyridinethiol-oxide (ZPTO) may be reduced and efficiency maintained longer.[123] ZPTO has been found to cause severe irritation of the scalp in some cases, and it has been claimed that the conditioning agent stearyl dimethylamine oxide exerts a strong counter-irritant effect when added to sodium lauryl sulphate–ZPTO systems.[124]

With regard to other 2-pyridinethiol-1-oxide salts, the use should be mentioned of the corresponding disulphide which, when combined with metal salts such as $CaCl_2$ and $MgSO_4$, is soluble in water, allowing the formulation of clear products of comparable efficiency.[125]

Among other patented compounds for anti-dandruff shampoos are the following: the chemically related 2-mercapto-quinoxaline-1-oxide,[126] 2-mercapto-quinoline-1-oxide[127] and polyene derivatives related to vitamin A.[128]

Details of other germicidal additives, including soluble and insoluble materials, are given in Chapter 23 (The Hair) and Chapter 35 (Antiseptic Materials).

Acid-balanced Shampoos

Among proposals to minimize damage to the hair and the skin, there should be mentioned a trend to promote shampoos of lower pH, the so-called acid-balanced shampoos. Mild acidity prevents swelling and promotes the tight flattening of scales against the hair shaft, thereby reducing penetration and inducing sheen. It offers favourable conditions for the display and use of conditioning properties of cationics, amine oxides and amphoterics, but is detrimental to foaming and cleansing performance.

Moreover, acidic pH often entails instability for surfactants, which may result in changes of viscosity of shampoo with time. These disadvantages may be overcome either by the use of surfactants unaffected by pH values between 5 and 7—such as ammonium lauryl sulphate and lauryl ether sulphate—or by the development of synergistic associations including amphoterics and nonionics, or the addition of fair amounts of foam boosters. Examples 22 and 23 illustrate acid-balanced shampoos.

	(22)
	per cent
Miranol C2M (Amphoteric 2)	20
Cocamidopropylamine oxide	5
Lauric diethanolamide	2·5
Polyethoxylated lauryl alcohol	8
Lactic acid	*q.s.* to pH 6
Water	to 100

	(23)
	per cent
Ammonium lauryl sulphate	40
Cocamidopropyl betaine	15
Lauric diethanolamide	4
*PPG 5 Ceteth 10 Phosphate	3
Hydrolysed animal protein	1·5
Perfume, preservatives, dyes	*q.s.*
Citric acid	*q.s.* to pH 5·4
Deionized water	to 100

*CTFA name.

SAFETY OF SHAMPOOS

The question of the dermatological safety of shampoos is extremely important. Shampoos need to be safe to the skin and eyes as well as being generally non-toxic. The evaluation of products for dermatological safety is discussed in Chapter 2, while the evaluation of general detergent toxicity is dealt with in Chapter 33. The subject of eye safety will be discussed in detail here.

In the process of shampooing the hair it is not unusual for some of the shampoo to trickle down the face and gain access to the eyes. When the shampoo is diluted adequately this is not usually harmful, but in some cases it has, in fact, been found that contact of any of the undiluted shampoo with the eyeball may (depending upon the type of detergent employed, its concentration and the formulation) involve a very serious hazard and even lead to scarring and opacity of the cornea of the eye.

The first shampoo to be recognized as hazardous to the eyes consisted of a nonionic detergent, described as a polyethylene oxide alkyl-phenol ether in combination with a quaternary ammonium salt, which caused clouding of the cornea and impaired vision. Eckermann and Gessner[129] reported cases of eye damage caused by cosmetics including some egg shampoos; the damage was reversible.

Smyth *et al.*[130] tested the effects of some synthetic surfactants on the eyes of rabbits, any damage to the cornea being observed by staining with fluorescein.

The basic technique of evaluating eye safety is that described by Draize *et al.*[131] In this method a 0·1 ml portion of the preparation under test is instilled into the conjunctival sac of an albino rabbit with the left eye being allowed to serve as a normal control. After a 4 second period of contact, both the treated and control eyes are flushed with 25–30 cm^3 of physiological saline to remove the test solution. Observations for pathological symptoms are made 1 hour after instillation, again after 2 hours, and thence daily for 35 days.

Draize and Kelley[132] examined a number of cationic, anionic and nonionic surface-active agents used in cosmetic preparations, for eye irritation. The 'maximal tolerated concentrations' (Table 24.1) are those that produced no corneal or iris lesions which lasted 7 days.

Martin, Draize and Kelley[133] pointed out in a later publication that a number of detergents are capable of causing anaesthesia of the eye and this permits increased pain levels to be tolerated. If these detergents also damage the eye then this is extremely dangerous. Such materials included polyoxyethylene condensates and fatty acid amine condensates, and formulations containing these compounds must be assessed with extreme care. Numerous variations and modifications of the Draize test have been reported and it is best to examine these in a historical order. The method described in a BIBRA bulletin[134] recommends the original test[135] to be changed from 9 rabbits' eyes with 3 rinsed after 2 seconds, 4 rinsed after 4 seconds and the remainder not rinsed, to 6 rabbits' eyes with none being rinsed.

One of the major difficulties with the test is the variability of the results. This can be overcome to some extent by modifying the scoring system, but this is not completely satisfactory. Weltman *et al.*[136] attempted to overcome this by using large numbers of rabbits, but this is a somewhat lengthy and expensive

Table 24.1 Eye Irritation Studies with Surface-active Agents

Trade name	Chemical designation	Maximal tolerated concentrations of active ingredients (%)
A. CATIONICS		
BTC®	Lauryldimethylbenzylammonium chloride	0·5
Hyamine 1622®	*p*-Diisobutylphenoxyethoxyethyldimethyl-benzylammonium chloride	0·5
Roccal®	Alkyldimethylbenzylammonium chloride	0·5
Isothan Q–15®	Laurylisoquinolinium bromide	0·8
Ethyl Cetab®	Cetylethyldimethylammonium bromide	1·0
Triton X–400®	Stearyldimethylbenzylammonium chloride	1·0
B. ANIONICS		
Nacconol NRSF®	Alkylaryl sulphonate	5·0
Miranol SM®	Capryl imidazoline derivative	10·0
Aerosol OT®	Dioctylester of sodium sulphosuccinic acid	15·0
Duponol WA®	Sodium lauryl sulphate	20·0
Orvus WA®	Sodium lauryl sulphate	20·0
Triton X–200®	Sodium salt of alkylated aryl polyether sulphate	28·0
C. NONIONICS		
Triton X–100®	Alkylated aryl polyether alcohol	5·0
Nonic 218®	Polyethylene glycol *tert*dodecyl thioether	10·0
Alrosol®	Fatty acid amide condensate	10·0
Neutronyx 600®	Aromatic polyglycol ether condensate	15·0
Detergent 1011®	Secondary amide of lauric acid	15·0
Ninol 2012®	Fatty acid alkanolamine condensate	20·0
Span 20®	Sorbitan monolaurate	100·0
Tween 20®	Sorbitan monolaurate polyoxyethylene derivative	100·0
Span 80®	Sorbitan mono-oleate	100·0
Tween 80®	Sorbitan mono-oleate polyoxyethylene derivative	100·0
Arlacel A®	Mannide mono-oleate	100·0

procedure. A special application device was described by Battista and McSweeney[137] and its use results in a reduction of the variability of the results.

Gaunt and Harper[138] pointed out that there is poor correlation between rabbits and man, and they have found that 10 shampoos available to the public with no deleterious effects caused irritation to rabbits. Bonfield and Scala[139] reported similar results on 13 brands of shampoo. Regarding animal species, an interesting study was carried out by Gershbein and McDonald[140] who tested commercial shampoos and surfactants, according to the standard Draize method, on rabbits, guinea-pigs, rats, mice, hamsters, dogs, cats, monkeys, and chickens. In all cases, albino rabbits and mice yielded much higher corneal irritation scores than the other species, while cats, monkeys and chickens exhibited the best resistance to corneal involvement.

The whole subject of the evaluation of eye safety has been reviewed by Beckley,[141] Alexander[142] and more recently in a comprehensive analysis by McDonald and Shadduck[143] who concluded that assessment methods in eye irritation testing are still to be improved. Later, reviewing objective methods for the assessment of eye irritation—that is, slit lamp examination, corneal thickness, intra-ocular pressure, corneal curvature, conjunctival and corneal weight, histological studies—Heywood and James[144] were led to conclude that the eye test system is subject to such wide variations that precise measurements are most unlikely to be achieved. Consequently they consider that the most satisfactory evaluation is provided by the subjective Draize clinical appraisal, supplemented by measurement of corneal thickness and of intra-ocular pressure, for which applanation tonometry[145] seems best suited.[146]

Over a period of time a wide variety of materials has been tested using the rabbit's eye test and for some time it has been realized that, in general, the eye irritancy potential of surface-active agents follows the order of cationics > anionics > nonionics. There are, however, exceptions to this order which suggest that factors other than this classification by chemical nature may be involved, or that there may be a correlation between eye irritation and some more specific properties of these agents. With this in mind, certain physical and chemical properties were studied by Hazleton[147] for correlation with capacity to produce eye injury, as well as potency in such activity.

To confine the survey to general trends, Hazleton made no attempt to review the minimum exposures required to produce injury. In turn, injury was interpreted broadly rather than specifically and included generalized erythema, oedema, necrosis, vascularization of the sclera and/or cornea, and opacity. Also, injury could vary in onset, severity and duration. For quantitative comparison the maximum concentration (percentage of active ingredient in aqueous solution) that would produce only moderate irritation was used, if available. The capacity to induce injury was the irritant potential for a specific material. Potentials of 1–10 per cent were considered as severe irritants, 10–20 per cent as moderate and above 20 per cent as mild.

A good supporting example is that of the imidazoline amphoteric surfactants of 'Miranol' type, which are rather severe irritants *per se* in high concentrations and display super-mildness and anti-irritant properties when neutralized and diluted below 20 per cent w/w.[16,148]

The compounds used in Hazleton's study are given in Table 24.2 and the results of the physical measurements and irritation studies in Table 24.3, and summarized below.

Surface Activity at 10 per cent

Of the materials included in the report, the maximum variation in surface tension of the 1 per cent solution was from 27 to 46 dyne/cm. Determinations were made on the Cenco-duNouy Tensiometer at 25°C. In view of the wide variations in irritant potential there appears to be very little possibility of a correlation between this and surface activity.

pH at 10 per cent

The pH of aqueous solution of the surface-active agent is not necessarily characteristic of the compound, but is influenced by manufacturing process.

Table 24.2 Chemical Nature of Cationics, Anionics and Nonionics

Commercial name or Laboratory No.	per cent active	Chemical nature
CATIONICS		
G–271®	35	*N*-soya-*N*-ethyl morpholinium ethosulphate
Hyamine 1622®	100	Di-isobutyl phenoxy ethoxy ethyl dimethyl benzyl ammonium chloride
Ceepryn®	100	Cetylpyridinium chloride
Emcol E607®	100	*N*(acyl colamino formyl-methyl) pyridinium chloride
Roccal®	50	Mixture of high-molecular-weight alkyl dimethyl benzyl ammonium chlorides
Hyamine 2389®	50–52	Alkyl (C_9H_{19} to $C_{15}H_{31}$) tolyl methyl trimethyl ammonium chlorides
Isothan Q–15®	20	Lauryl isoquinolinium bromide
ANIONICS		
Duponol WAT®	50	Triethanolamine salt of lauryl sulphate
Ultrawet 60L®	60	Triethanolamine salt of alkyl aryl sulphonate
Armour's 600		
KOP Soap®	100	Potassium coconut oil soap
Ivory Soap®	100	
NONIONICS		
Span 20®	100	Sorbitan monolaurate
Span 80®	100	Sorbitan mono-oleate
Tween 20®	100	Polyoxyethylene sorbitan monolaurate
Tween 40®	100	Polyoxyethylene sorbitan monopalmitate
Tween 60®	100	Polyoxyethylene sorbitan monostearate
Tween 65®	100	Polyoxyethylene sorbitan tristearate
Tween 80®	100	Polyoxyethylene sorbitan mono-oleate
G–7569J®	100	Polyoxyethylene sorbitan monolaurate
Myrj 45®	100	Polyoxyethylene monostearate
Myrj 52®	100	Polyoxyethylene monostearate
Brij 30®	100	Polyoxyethylene lauryl ether
Brij 35®	100	Polyoxyethylene lauryl ether
G–2132®	100	Polyoxyethylene lauryl ether
G–3721®	100	Polyoxyethylene 2-butyl octanol
Renex®	100	Polyoxyethylene esters of mixed resin and fatty acids
G–1690®	100	Polyoxyethylene ether of alkyl phenol
G–1790®	100	Polyoxyethylene lanolin condensate
G–1441®	100	Polyoxyethylene sorbitol lanolin derivative

The pH of the various solutions tested was determined on a Beckman Model G pH meter. The range of the pH was from 3·93 to 9·52, with most of the values falling in the range pH 6·0 to 7·0; there appeared to be little correlation between pH and irritancy.

Wetting Power (Draves Method) at 0·1 per cent

Within the group of materials tested, six were moderately good wetting agents with values below 51 seconds, and the balance were 85 or above; there appeared to be very poor correlation between wetting power and irritation.

Table 24.3 Surface-active Agents and their Physical Properties
Wetting power was determined by the Draves method and foam height by the Ross-Miles method

Trade name	pH at 10%	Surface tension at 10% (dyne/cm)	Wetting power at 0·1% (s)	Foam height at 1·0% (mm)	Irritation potential
CATIONICS					
G–271®	7·11	33	>300	217	1
Hyamine 1622®	7·01	36	142	268	1
Ceepryn®	4·79	41	258	240	1
Emcol E607®	3·48	37	300	213	1
Roccal®	7·28	33	234	110	1
Hyamine 2389®	7·30	31	85	292	1
Isothan Q–15®	3·93	34	300	150	1
ANIONICS					
Duponol WAT®	7·11	32	31	273	20
Ultrawet 60L®	7·10	30	10	302	10
Armour's 600 KOP Soap®	9·52	27	>300	375	10
Ivory Soap®					>10
NONIONICS					
Span 20®		28	>300	14	100
Span 80®		30	>300	13	100
Tween 20®	6·81	36	203	173	100
Tween 40®		40	>300	87	100
Tween 60®		43	>300	81	100
Tween 65®		31	>300	27	100
Tween 80®	6·46	40	>300	59	100
G–7596J®		33	203	162	100
Myrj 45®		33	>300	15	100
Myrj 52®		44	>300	26	100
Brij 30®	4·70	28	51	18	20
Brij 35®	6·02	42	174	220	
G–2132®	5·98	28	13	247	20
G–3721®	4·39	27	7	287	<20
Renex®	6·95	39	>300	77	100
G–1690®		31	8	235	<10
G–1790®	5·83	46	>300	23	100
G–1441®		43	>300	31	100

However, among the nonionics there did appear to be some correlation between the wetting properties and irritant potential. None of the group having a value of 174 seconds or longer exhibited irritation either undiluted or in solution. Four of the nonionics had values of 51 seconds or less, and each was rated for irritation at 20 per cent or less. No similar correlation could be established within either the limited anionic or cationic groups studied.

Foaming Power (Ross-Miles Test) at 1·0 per cent
Foam heights for non-injurious undiluted materials ranged from 13 to 220 mm, and even higher for materials with an irritant potential rating of 20 per cent or above. There is thus some evidence that most of the irritant chemicals are good foamers but, fortunately for the formulator, the converse that good foamers are irritants is not true.

Wetting and Foaming
Since it is indicated that there may be some slight correlation between wetting power or foaming power and eye irritation, the combination of these two properties is obviously suggested. The data here suggest that good wetting plus good foaming may be an irritant combination; for example:

	Wetting (s)	*Foaming* (mm)	*Irritant potential* (%)
Alkyl aryl sulphonate	10	302	10·0
G–3721®	7	287	<20·0
G–1690®	8	235	<10·0

That these two properties are not essential to highly irritant material is shown by the following:

	Wetting (s)	*Foaming* (mm)	*Irritant potential* %
Hyamine 1662®	142	268	1·0
G–271®	>300	217	1·0

Hopper *et al.*[149] tested a number of commercial detergents for corneal irritation in rabbits. However, these investigators used solutions of 1 per cent concentration only, that is, very much weaker than those which might enter the eye during shampooing. Nevertheless, according to these investigators, butyl diphenyl sodium sulphonate, dodecyl benzene sodium sulphonate and decyl benzene sodium sulphonate produced severe inflammation of the cornea.

Bellows and Gutman[150] found solutions containing 1 per cent or less of Aerosol OS (sodium isopropyl naphthalene sulphonate) harmless to the conjunctiva and cornea but 2 per cent solutions produced inflammation. In concluding the above brief discussion of the potential hazards of detergent solutions to the eyes, attention should be drawn to the warning of the Division of Pharmacology of the US Food and Drug Administration that mixtures of surface-active agents may be more irritating to the eyes than could be predicted from the ocular toxicity of the individual components. However, the reverse is also true[16] and a number of compounds have been shown to reduce considerably the damaging effects of irritant surfactants when adequately combined, for example N-hydroxyethylacetamide with sodium lauryl sulphate.

The only way to ensure the safety of a given shampoo formulation is to have the finished formulation, including any antiseptics, perfumes, etc., tested for corneal irritation by instillation into the eye of animals in the way recommended for the Draize tests. Most shampoos on the market today are irritant to the eyes, in the sense that they produce stinging or even acute pain if the undiluted shampoo enters the eye; most of them produce some inflammation of the eyelid. While this is undesirable, it can at least be considered as normal, and a hazard not confined to shampoos—alcohol, for instance, would have the same effect.

What must be guarded against is the use of materials which, even if misused grossly, could produce corneal opacity of any sustained duration.

REFERENCES

1. Riso, R. R., *Soap Cosmet. chem. Spec.*, 1978, **54**(6), 56.
2. Elder, T. H., Jr., and Pacifico, C., *Drug Cosmet. Ind.*, 1955, **77,** 622.
3. Powers, D. H. and Fox, C., *Soap Perfum. Cosmet.*, 1959, **32,** 393.
4. Powers, D. H., *Cosmetics, Science and Technology*, ed. Sagarin, E., New York, Interscience, 1957.
5. Rader, C. A. and Tolgyesi, W. S., *Cosmet. Perfum.*, 1975, **90**(3), 29.
6. Cottington, E. M., Kissinger, R. H. and Tolgyesi, W. S., *J. Soc. cosmet. Chem.*, 1977, **28,** 219.
7. Markland, W. R., *Kirk-Othmer Encyclopedia of Chemical Technology*, 2nd edn, New York, Interscience, 1966, Vol. 10, p. 769; *Norda Briefs*, No. 487, July–August 1978.
8. Barnett, G. and Powers, D. H., *J. Soc. cosmet. Chem.*, 1951, **2,** 219.
9. Myddleton, W. W., *J. Soc. cosmet. Chem.*, 1953, **4,** 150.
10. Zviak, C. and Lachampt, F., *Problèmes capillaires*, ed. Sidi, E. and Zviak, C., Paris, Gauthier-Villars, 1966, p. 221.
11. Spangler, W. G. *et al.*, *J. Am. Oil Chem. Soc.*, 1965, **42,** (8), 723.
12. Sorkin, M., Shapiro, B. and Kass, G. S., *J. Soc. cosmet. Chem.*, 1966, **17,** 539.
13. Prall, J. K., *Proceedings of the VIth Congress of the International Federation of Societies of Cosmetic Chemists*, September 1970.
14. Baines, E., *J. Soc. cosmet. Chem.*, 1978, **29,** 369.
15. Wassell, H., *J. Soc. cosmet. Chem.*, 1953, **4,** 282.
16. Goldemberg, R. L., *J. Soc. cosmet. Chem.*, 1977, **28,** 667.
17. Sass, C., Society of Cosmetic Chemists Meeting, 9–10 May, 1974.
18. US Patent 2 928 772, Colgate-Palmolive Co., 15 March 1960; US Patent 3 001 949, Colgate-Palmolive Co., 28 September 1961.
19. US Patent 2 989 547, Procter and Gamble Co., 20 June 1961; US Patent 3 024 273, Procter and Gamble Co., 6 March 1962; German Patent 1 205 221, Procter and Gamble Co., 28 August 1961.
20. Hart, J. R. and Levy, E. F., *Soap Cosmet. Chem. Spec.*, 1977, **53**(8), 31.
21. Johnsen, V. L., *Cosmet. Toiletries*, 1977, **92**(12), 29.
22. US Patent 3 728 447, Patterson Co., 17 April 1973.
23. Baiocchi, F., *et al.*, *Cosmet. Perfum.*, 1975, **90**(9), 31.
24. Barker, G., *et al.*, *Soap Cosmet. chem. Spec.*, 1978, **54**(3), 38.
25. US Patent 2 623 900, Sandoz, A. G., 30 December 1950; Stache *et al.*, *Tenside*, 1977, **14,** 237.

26. US Patent 3 959 460, L'Oreal, 25 May 1976; US Patent 3 983 171, L'Oreal, 28 September 1976.
27. Kroll, H. and Lennon, W. J., *Proc. Sci. Sect. Toilet Goods Assoc.*, 1956, (25), 37.
28. Goddard, E. D. and Kung, H. C., *Soap chem. Spec.*, 1966, **42**(2), 60.
29. Anon., *Soap Perfum. Cosmet.*, 1978, **51**(5), 206.
30. US Patent 3 578 719, L'Oreal, 11 May 1971; US Patent 3 666 671, L'Oreal, 30 May 1972; US Patent 3 865 542, L'Oreal, 11 February 1975; US Patent 3 877 955, L'Oreal, 15 April 1975.
31. US Patent 3 708 364, L'Oreal, 2 January 1973; US Patent 3 880 766, L'Oreal, 29 April 1975.
32. US Patent 3 906 048, L'Oreal, 16 September 1975; US Patent 3 998 948, L'Oreal, 21 December 1976.
33. US Patent 4 105 580, L'Oreal, 8 June 1978.
34. British Patent 1 385 060, L'Oreal, 17 November 1972.
35. US Patent 3 821 372, L'Oreal, 28 June 1974; US Patent 3 928 224, L'Oreal, 23 December 1975; US Patent 3 966 398, L'Oreal, 29 June 1976; US Patent 4 087 466, L'Oreal, 2 May 1978.
36. US Patent 3 984 480, L'Oreal, 5 October 1976; US Patent 4 058 629, L'Oreal, 15 November 1977.
37. British Patent 1 519 378, L'Oreal, 22 October 1976.
38. Alexander, P., *Manuf. Chem. Aerosol News*, 1971, **42**(12), 27.
39. Jungerman, E. and Ginn, M. E., *Soap Perfum. Cosmet.*, 1964, **37**(9), 59.
40. US Patent 3 272 712, L'Oreal, 22 April 1966; US Patent 3 290 304, L'Oreal, 6 December 1966.
41. Koeber, A., *et al.*, *Soap Cosmet. chem. Spec.*, 1972, **48**(5), 86.
42. US Patent 2 773 068, H. Mannheimer, 1956; US Patent 2 781 354, H. Mannheimer, 1957.
43. Schoenberg, T. G., *Cosmet. Perfum*, 1975, **90**(3) 89.
44. US Patent 4 069 347, Emery Industries, 17 January 1978.
45. US Patent 4 001 394, American Cyanamid Co., 4 January 1977.
46. US Patent 3 331 781, L'Oreal, 18 July 1967; US Patent 3 337 548, L'Oreal, 22 August 1967; US Patent 3 534 032, L'Oreal, 13 October 1970.
47. US Patent 3 879 464, L'Oreal, 22 April 1975; US Patent 4 009 255, L'Oreal, 22 February 1977; US Patent 4 096 332, L'Oreal, 20 June 1978.
48. US Patent 3 816 616, Warner-Lambert Co., 11 June 1974.
49. US Patent 3 472 840, Union Carbide Corp., 14 October 1969.
50. Gerstein, T., *Cosmet. Perfum.*, 1975, **90**(3), 35.
51. US Patent 3 313 734, Procter and Gamble Co., 11 April 1967.
52. US Patent 4 009 256, National Starch and Chemical Corp., 22 February 1977.
53. US Patent 3 987 162, Henkel et Cie., 19 October 1976.
54. US Patent 3 917 817, L'Oreal, 4 November 1975; US Patent 4 013 787, L'Oreal, 22 March 1977.
55. French Patent 2 077 417, General Aniline and Film Corp., 29 January 1977.
56. French Patent 1 583 363, Sandoz AG, 16 September 1978.
57. British Patent 1 494 915, L'Oreal, 29 November 1974; British Patent 1 494 916, L'Oreal, 27 May 1977; French Patent 2 368 508, L'Oreal, 2 March 1977.
58. British Patent 1 347 051, Gillette Co., 13 February 1974.
59. US Patent 3 874 870, Millmaster Onyx Corp., 1 April 1975; US Patent 3 931 319, Millmaster Onyx Corp., 6 January 1976.
60. British Patents 1 513 671 and 1 513 672, L'Oreal, 15 May 1975; French Patent 2 389 374, Ciba Geigy AG, 2 May 1978; British Patent 1 546 162, L'Oreal, 20 March 1979.
61. Canadian Patent 908 056, Unilever Ltd, 22 August 1972.

62. Canadian Patent 920 512, Colgate Palmolive Co., 6 February 1973.
63. US Patent 3 862 310, Gillette Co., 21 January 1975.
64. Bonadeo, I. and Variati, G. L., *Cosmet. Toiletries*, 1977, **92**(8), 45.
65. US Patent 3 987 161, Procter and Gamble Co., 19 October 1976.
66. US Patent 3 998 761, Bristol-Myers Co., 21 December 1976.
67. US Patent 3 950 510, Lever Bros Co., 13 April 1976.
68. US Patent 3 988 438, American Cyanamid Co., 26 October 1976.
69. US Patent 3 957 970, American Cyanamid Co., 18 May 1976.
70. Canadian Patent 905 852, Beecham Group Ltd, 25 July 1972.
71. US Patent 4 070 452, Borchorst, B., 24 January 1978.
72. US Patent 3 969 500, Lever Bros Co., 13 July 1976.
73. US Patent 3 932 610, Lever Bros Co., 13 January 1976.
74. US Patent 3 964 500, Lever Bros Co., 22 June 1976.
75. US Patent 3 976 588, Center for New Product Development, 24 August 1976.
76. US Patent 3 972 998, Lever Bros Co., 3 August 1976.
77. US Patent 3 959 462, Procter and Gamble Co., 25 May 1976.
78. US Patent 3 976 781, L'Oreal, 24 August 1976.
79. US Patent 3 984 569, L'Oreal, 5 October 1976.
80. British Patent 1 391 801, L'Oreal, 23 April 1975.
81. US Patent 3 968 218, L'Oreal, 6 July 1976.
82. US Patent 4 002 634, L'Oreal, 11 January 1977.
83. US Patent 3 862 305, L'Oreal, 21 January 1975.
84. US Patent 3 984 535, L'Oreal, 5 October 1975.
85. US Patent 3 903 257, Kao Soap Co., 2 September 1975.
86. US Patent 4 009 253, Monsanto Co., 22 February 1977.
87. Jablonski, J. I. and Goldman, C. L., *Cosmet. Perfum.*, 1975, **90**(3), 45.
88. Thoma, K. and Will, K., *Pharm. Z.*, 1975, **120,** 1013.
89. Bryce, D. M. and Smart, R., *J. Soc. cosmet. Chem.*, 1965, **16,** 187; Walker, G. T., Seifen Öle Fette Wachse, 1967, **93,** 134.
90. Tuttle, E., *et al.*, *Am. Perfum. Cosmet.*, 1970, **85**(3), 87; Schuster, G., *et al.*, *Am. Perfum. Cosmet.*, 1966, **81**(6), 39.
91. British Patent 1 250 725, Henkel et Cie., 21 April 1970; Lorenz, P., *J. Soc. cosmet. Chem.*, 1977, **28,** 427.
92. See references by Croshaw, B., *J. Soc. cosmet. Chem.*, 1977, **28,** 3.
93. Mihaljev, B. and Jujnovic, N., *Farm. Vestu* (*Ljubljana*), 1977, **28**(2), 141. C.A. **88,** 11727k.
94. Carsch, G., *Soap Cosmet. Chem.*, 1977, **53**(2), 52.
95. Unger, L., *Soap Perfum. Cosmet.*, 1976, **49**(2), 45.
96. Pajaujis, D., *Drug Cosmet. Ind.*, 1973, **112**(1), 30; 1973, **112**(4), 42.
97. Cook, M. K., *Drug Cosmet. Ind.*, 1966, **99**(2), 52.
98. Mannheim, P., *Soap Perfum. Cosmet.*, 1965, **38,** 348.
99. Markland, W. R., *The Chemistry and Manufacture of Cosmetics*, 2nd edn, ed. de Navarre, M. G., Orlando, Continental Press, 1975, Vol. 4, p. 1283.
100. (a) Kass, G. S., *Cosmet. Perfum.*, 1975, **90**(3), 105; (b) Donaldson, B. R. and Messenger, E. T., *Int. J. cosmet. Sci.*, 1979, **1,** 71.
101. de Navarre, M. G., *Am. Perfum. essent. Oil Rev.*, 1950, **55**(8), 109.
102. Pantealoni *et al.*, *Proc. sci. Sect. Toilet Goods Assoc.*, 1949, (12), 9.
103. British Patent 1 133 870, Unilever Ltd, 15 June 1965.
104. US Patent 3 808 311, Colgate Palmolive Co., 30 April 1974.
105. Wells, F. V., *Soap Cosmet. chem. Spec.*, 1976, **52**(10), 54.
106. US Patent 4 035 267, American Cyanamid Co., 12 July 1977.
107. Gerstein, T., *Cosmet. Toiletries*, 1978, **93**(2), 15.
108. Faucher, J. A., *et al.*, *Text. Res. J.*, 1977, **47,** 616.

109. US Patent 3 962 418, Procter and Gamble Co., 8 June 1976.
110. US Patent 3 990 991, Revlon Inc., 9 November 1976.
111. British Patent 1 414 243, Colgate Palmolive Co., 19 November 1975; US Patent 3 810 478, Colgate Palmolive Co, 14 May 1974.
112. US Patent 3 755 559, Colgate Palmolive Co., 28 August 1973.
113. Canadian Patent 998 613 (19.10.76), Unilever Ltd, 19 October 1976; Anon., *Chemist and Druggist Supplement*, 1972, April 15, p. 4.
114. US Patent 3 055 836, Johnson and Johnson Co., 25 September 1962.
115. US Patent 3 950 417, Johnson and Johnson Co., 13 April 1976.
116. *Norda Briefs*, 1977, March, No. 419; Finkstein, A. and Ardita, M., Society of Cosmetic Chemists, Scientific Meeting, December 1976.
117. US Patent 3 886 277, Schwarzkopf GmbH, 27 May 1975.
118. US Patent 3 671 634, Vanderbilt, R. T., Co., Inc.
119. US Patent 3 236 733, Procter and Gamble Co., 1966; Orentreich, N., *J. Soc. cosmet. Chem.*, 1972, **23,** 189.
120. Okumura, T., *et al.*, International Federation of Societies of Cosmetic Chemists, VIII International Congress, UK, August 1974.
121. Elkhouly, A. E., *Pharmazie*, 1974, **29,** 1.
122. US Patent 3 769 398, Colgate Palmolive Co., 30 October 1973.
123. US Patent 3 761 417, Procter and Gamble Co., 25 September 1973.
124. US Patent 4 033 895, Revlon Inc., 5 July 1977.
125. US Patent 3 818 018, Olin Corp., 18 June 1974; US Patent 3 890 434, Olin Corp., 17 June 1975.
126. US Patent 3 733 323, Colgate Palmolive Co., 15 May 1973.
127. US Patent 4 041 033, Colgate Palmolive Co., 9 August 1977.
128. US Patent 4 021 574, Hoffmann-La Roche, 3 May 1977.
129. Eckermann, M. and Gessner, L., *Z. Arztl. Fortsbild.*, 1966, **60**(10), 612.
130. Smyth, H. F., Jr., Seaton, J. and Fischer, L., *J. Indust. Hyg.*, 1941. **23,** 478.
131. Draize, J. H., Woodward, G. and Calvery, H. O., *J. Pharmacol.*, 1944, **82,** 377.
132. Draize, J. H. and Kelley, E. A., *Proc. sci. Sect. Toilet Goods Assoc.*, 1952, (17), 1.
133. Martin, G., Draize, J. H. and Kelley, E. A., *Drug Cosmet. Ind.*, 1962, **91,** 30.
134. Anon., *Bibra Bull.*, 1963, **2,** 284 and 422.
135. Anon., *Bibra Bull.*, 1962, **1,** 16.
136. Weltman, A. S., Sparber, S. B. and Jurtshuk, T., *Toxic. Appl. Pharmacol.*, 1965, **7,** 308.
137. Battista, S. P. and McSweeney, E. S. J., *J. Soc. cosmet. Chem.*, 1965, **16,** 119.
138. Gaunt, I. F. and Harper, K. H., *Bibra Bull.*, 1963, **2,** 602.
139. Bonfield, C. T. and Scala, R. A., *Proc. sci. Sect. Toilet Goods Assoc.*, 1965, (43), 34.
140. Gershbein, L. L. and McDonald, J. E., *Food Cosmet. Toxicol.*, 1977, **15,** 131.
141. Beckley, J. H., *Am. Perfum. Cosmet.*, 1965, **80**(10), 51.
142. Alexander, P., *Specialities*, 1965, **1**(9), 35.
143. McDonald, T. O. and Shadduck, J. A., *Dermato-toxicology and Pharmacology*, ed. Marzulli, F. N. and Maibach, H. I., Washington, Hemisphere, 1977, pp. 169–191.
144. Heywood, R. and James, R. W., *J. Soc. cosmet. Chem.*, 1978, **29,** 25.
145. Ballantyne, B., Gazzard, M. F. and Swanston, D. W., *J. Physiol.*, 1972, **226,** 128.
146. Walton, R. M. and Heywood, R., *J. Soc. cosmet. Chem.*, 1978, **29,** 365.
147. Hazleton, L. W., *Proc. sci. Sect. Toilet Goods Assoc.*, 1952, (17), 5.
148. Ciuchta, H. P. and Dodd, K. T., *Drug Chem. Toxicol.*, 1978, **1,** 305.
149. Hopper, S. H., Hulpieu, H. R. and Cole, V. V., *J. Am. Pharm. Assoc., sci. Edn.*, 1949, **38,** 428.
150. Belows, J. G. and Gutman, M., *Arch. Ophthal.*, 1943, **30,** 352.

Chapter Twenty-five

Hair Setting Lotions, Sprays and Dressings

Use and Purpose of Hair Dressings

The main aims of hair dressings, whether they are for men or women, are to improve the control and manageability of hair, to impart some lustre, and to maintain a hairstyle in spite of the movement involved in daily activities and despite the various environmental conditions to which hair is submitted (wind, humidity, dryness, cold, heat, sun, etc.).

The relative importance of these factors varies from one country to another, from age to age, and according to the state of the hair. It also differs notably between male and female users: men generally consider adequate control as the prime requisite for a hair dressing, with gloss as secondary, and as such men's products have been classically based on the use of oleaginous materials. Women first look for products that give a pleasing appearance to the hair, but at the same time they require good hold; they do not want products which render hair heavy and which tend to make it lank or greasy.

Because of the different demands of men and women, and in spite of the fact that the evolution of hair styles and a common desire for a 'natural look' have brought them closer during the last few years, hair dressing products will be considered separately in this chapter according to the sex for which they are designed.

The trend in women's fashions for softer and freer styles and the increasing use of the brushing technique have led to an increase in the demand for products that enhance disentangling and a greater need for products that protect, strengthen and/or improve the condition of the hair; this has occurred at the same time as the extraordinary development of conditioners. Though there may exist some overlap for some new compounds intended to impart set, manageability and sheen, products which more specifically treat the condition of hair will be dealt with in the next chapter, 'Hair Tonics and Conditioners'. The present chapter discusses products that are intended to set and maintain the hair style.

WOMEN'S HAIR DRESSINGS

Setting Lotions

Setting lotions are intended to strengthen and maintain for an extended period a temporary deformation imparted by waving. An important characteristic of these lotions is that they are applied to wet hair. They differ from permanent waving compositions in that they do not affect in general the internal structure of the hair; although a number of hydrogen bonds and ionic sites in the hair may be involved, the lotions mainly provide a mechanical means of holding the set and a means of restraining water uptake. The principle is to deposit on the shampooed hair a solution of a polymeric material which after setting and drying leaves a flexible film that ensures the cohesiveness and hold of the hair style and protects it from the effects of humidity.

The oldest products of this type were simple aqueous or hydroalcoholic solutions of natural polymers, including tragacanth, karaya gum, arabic gum, shellac, etc. The lotions based on these mucilages are now obsolete. They work by sticking the hairs together and present major disadvantages; they yield dull, brittle films, which crumble into dust and become sticky in humid air because they are very hygroscopic.

They have been replaced by lotions based on synthetic polymers which are soluble in hydroalcoholic solutions. The first resin to be used was polyvinylpyrrolidone (PVP) which, owing to its solubility in many solvents including water, and to its non-ionic character, offered a wide range of use.

Two product types have appeared: the clear lotion and the aerated gel. Both product types represent a considerable advance over the older gum-based products; they form clear, non-greasy and quite glossy films while considerably improving the set retention of the hair. Their mode of action is somewhat different from that of the gum-based products in that, instead of just sticking the hair together, they tend to form a plastic sheath round each individual hair so that the setting effect is not completely lost on combing out the hair. Formulating a setting lotion, though simple in principle, is complex in reality. The film formed on the hair has to withstand, without breaking or flaking, repetitive combing and brushing while maintaining good adhesion in order to preserve the set.

The setting product should be compatible with the wetness of the hair at the time of application and be waterproof after drying, while remaining flexible. It should be easily shampooed out, and should not be sticky or 'cardboard-like'. The flexibility of the film and its resilience to extensive handling are imparted by adding amounts of plasticizers.

Generally speaking, a setting lotion contains, in a hydroalcoholic solution, film-forming polymers, plasticizers for these polymers, perfumes, colorants, and additives to enhance sheen, softness and disentangling. Example 1 gives a clear lotion, and example 2 an aerated gel.

	(1) *per cent*	(2) *per cent*
PVP K.30	4·0	0·8
Polyethyleneglycol	2·0	1·5
Carboxyvinyl polymer	—	1·5
Diisopropanolamine or triethanolamine	—	1·5
Alcohol	40·0	—
Water	54·0	94·7
Colour, perfume	*q.s.*	*q.s.*

PVP is too hygroscopic, however, and has been replaced by vinylpyrrolidone–vinyl acetate copolymers; the higher the proportion of vinyl acetate, the higher the resistance to water, but the hardness of the film increases concurrently and the addition of plasticizer is necessary. The ratios of vinylpyrrolidone to vinyl acetate and of water to alcohol, and the amount of plasticizer, are chosen in relation to the desired physical properties of the film. Plasticizers are polyethyleneglycols, lanolin derivatives, silicones, esters, and polysiloxanes, which modify the hardness and flexibility of films as well as the water resistance.

	(3) *per cent*
PVP–VA (60:40)	2·5
Alcohol	50·0
PEG 40 castor oil	0·3
Water	to 100·0
Perfume	*q.s.*

A number of patents have appeared describing the use of a wide range of water-soluble or alcohol-soluble polymers for use in hair setting. The majority of these materials have been claimed for use in aerosol sprays; they are used in lower concentrations in setting lotions. Examples are resins with carboxylic groups such as copolymers of vinyl acetate and crotonic acid[1,2] or copolymers of alkylvinylethers[3,4] or vinylesters[5,6] and maleic acid hemiesters. Representative commercial resins are Resyn 28-1310, Resyn 28-2930 and Luviset CE 5055 for the former, Gantrez ES 225 and ES 425 for the latter. Typical formulae are given in examples 4–6.

	(4) *per cent*
Resyn 28-2930	3·00
2-Amino-2-methylpropanol (AMP)	0·25
Carbowax 1000 (PEG 20)	0·80
Ethanol	47·95
Distilled water	48·00
Perfume	*q.s.*

	(5) *per cent*
Luviset CE 5055	2·50
AMP (80% neutralization)	0·21
Ethanol or Isopropanol	50·00
Distilled water	47·29
Perfume	*q.s.*

	(6) *per cent*
Gantrez ES.255 or ES.425	4·2
Alcohol	21·2
Water	74·0
PEG.40 castor oil	0·3
2-Amino-2-methyl-1,3-propanediol (AMPD)	0·3
Perfume	*q.s.*

More recently, sulphonated polystyrenes, which are strongly anionic polymers, have been proposed.[7]

The addition to the setting lotions of various ingredients (quaternary ammonium compounds, proteins, panthenol, etc.) that are not rinsed off, which more or less blend with the film and interact during drying and heating, yields 'body' and lubrication. The advent of cationic polymers was revolutionary for hair dressings; these compounds are substantive to keratin and when added to setting lotions they improve the appearance of the hair and promote disentangling. Such products are copolymers of quaternized dialkylaminoalkylacrylate and polyvinylpyrrolidone[8,9] such as Gafquat 755, as well as Polymer JR resins which are esters of hydroxyethylcellulose and 2-hydroxypropyltrimethylammonium chloride; they permit the formulation of setting strengtheners–conditioners which provide at the same time hold, manageability and sheen. The following formula has been cited:

Lotion with holding and grooming properties[10]	(7) *per cent*
Polymer JR 400	0·30
Carbowax 400	4·00
Diisopropyl adipate	1·00
Dipropyleneglycol	0·70
Benzethonium chloride	0·10
Perfume	0·40
Colour	0·37
Ethanol	50·00
Water	43·13

Polymers with dipolar aminimide[11] and betaine-type groups[12] have also been claimed as effective setting–conditioning compounds. Polyamide epichlorhydrin resin added to PVP has been said to impart stronger resistance to moisture attack.[13,14] Other additives specified in patents are finely divided silica or aluminium oxide,[15] silicates[16] and perfluorinated polymers, the hydrophobic and oleophobic properties of which are used to render styling longer lasting.[17–21]

Heated Curlers and Blow Drying

The major difficulty with the use of setting lotions is that, since the products contain water, the hair has to be dried after use. This problem can be avoided by the use of aerosol sprays, but this introduces its own problems.

One of the earliest methods of introducing waves into hair was the use of a heated comb or tongs. Such treatments gave the hair a very good set but, since it was easy to overheat the hair, were often damaging. The heat applied was such that the hair was modified chemically and this accounted both for the good set and for much of the damage. For further details on the theory of setting see Chapter 28.

The possibility of using a controlled heating system, which would not change the hair chemically but would only affect the water set of the hair, has been considered for some time and a number of different types of heated curlers have appeared. The majority of these are heated electrically, and are usually fitted with some form of control device to prevent overheating. However, many of the models currently available are somewhat cumbersome in use and are rather expensive.

The trend of fashion towards softer, freer, loosely moving hair styles found expression in the brushing technique which combines mechanical action for shaping the hair with thermal blowing for setting. The desire and skill of women to style their hair themselves induced the development of an array of appropriate devices (blow combs and brushes). These new techniques have also required the availability of specific products—in the field of conditioners—to promote disentangling and manageability of hair when blow drying.

The following formula has been given:[22]

	(8) *per cent*
Quaternized VP–dimethyl amino ethyl methacrylate (CTFA—Quaternium 23)	3·50
Castor oil amidopropyldimethylamine	0·40
Citric acid monohydrate	0·10
Perfume	0·20
Hydrolysed animal protein	1·00
2-Bromo-2-nitro-1,3-propanediol	0·10
Water, dye	to 100·0

Procedure: Combine all ingredients except perfume, heat to 40°C with stirring until dissolved. Add perfume to batch at 40°C. Stir and cool batch to 30°C, add water to make up evaporation loss, and fill at 30°C.

A novel proposal in this line is a colloidal aqueous suspension of fibrillatable polytetrafluorethylene resin particles, said to generate on brushing a random network of microscopic fibres which holds the hair in a natural set.[23]

Hair Sprays

Hair sprays meet the requirement for a quick-dry preparation which imparts just sufficient rigidity to the set to keep it in place and control loose ends during the day, while not detracting from the natural sheen of the hair.

The early sprays were not entirely satisfactory because the spray units gave little control over the spread of the product. The advent of the pressurized container changed the situation radically. The fine spray and easy control obtained from pressurized packs made uniform coverage of the hair possible. As a consequence, the popularity of hair setting sprays of all kinds experienced a phenomenal increase and they were developed until they became the most important hair setting product, if not indeed the most important hair product. However, the success reached its peak around 1970 when the sales started to level off and decline with the growing fashion among young people for soft and free hair.

A further marked decline was seen in 1975 when Rowland and Molina proposed that the amount of stratospheric ozone (a shield against UV radiation) was progressively reduced by the action of the chlorinated propellants released from aerosols. A widely-publicized campaign against the use of these propellants culminated in 1979 with the banning by the US Food and Drug Administration of the chlorofluorocarbons currently used in aerosols. Most European countries have not followed the FDA recommendation, particularly because the theory about the depletion of ozone has not yet been confirmed and is the subject of controversy amongst scientists. However, the trend in European countries is setting towards a limitation of the concentration of fluorocarbons. These battles and threats against propellants have compelled investigators to consider new formulations and technology for aerosols and to find acceptable alternatives.

The objective of a hair spray (when combing is finished and a hair style established) is to deposit onto the dry hair an invisible film to protect it against all external agents that are likely to change its desirable features. Requirements for such a product are stringent and numerous:

(a) The spray must be very fine.
(b) The force of the spray should be gentle; it should not be wet, which means that the characteristics of the spray should allow it to be of a fineness not to wet the hair but also not to dry before reaching the hair.
(c) It should spread over a wide area in a short period and dry quickly.
(d) The film should be relatively flexible to follow movement of the hair without breaking.
(e) The film should be substantive to the hair but should be easy to eliminate without becoming powdery under a mere brushing. Any accumulation would render hair heavy and dirty.
(f) It should be soluble enough to be shampooed off easily.
(g) The film should not be sticky nor tacky to the touch and should hold the hair, but at the same time leave it free to move.

A good hair spray should be a compromise between adhesion and elimination, tightness and lightness. Building a hair spray is a matter of careful adjustment depending on technological qualities both of the spraying system and of the formulation. The new restraints imposed for propellants have further increased the difficulties of such an adjustment.

Hair Spray Resins

The earliest material used in hair lacquers was shellac, a natural resin composed of polyhydroxy acids and esters. Alcoholic solutions (8–15 per cent) were sprayed onto the hair out of soft plastic containers. Shellac produced hard and insoluble films. To make the shellac more water-soluble and therefore able to be shampooed away, an alkali such as borax or ammonia solution was used to form a salt. The following formula and procedure have been quoted:[24]

	(9) *parts by wt*
Shellac (bleached, dewaxed)	3·5
Ammonium hydroxide (28%)	0·5
Water	32·0

Procedure: Add the ammonia and about half the water to the shellac and warm with constant stirring until solution is obtained. Add the rest of the water (the solution should be clear; if it is not, more ammonia should be added); filter and cool. This solubilized shellac may then be diluted for use.

As early as 1948 a hairdresser thought of applying the aerosol technique to an alcoholic solution of shellac by adding to the solution chlorofluorocarbon propellants 11 and 12, and by packaging it in an aerosol can to obtain a spray that was much improved compared with the current sprays.

The use of anhydrous alcohol to prevent hydrolysis of propellants and the use of purified shellac helped in the formulation of aerosol sprays with good holding properties,[25] but the films formed after drying were hard, brittle and sticky, and rendered the hair rather harsh; furthermore they were so insoluble in water that they could not be shampooed off. The addition of lanolin, isopropyl myristate and other plasticizers to soften the film generally had the effect of making the film greasy. The first significant improvement in formulating hair sprays was achieved in 1953 with the use of a polyvinylpyrrolidone, a synthetic resin that is soluble in water and easily shampooed off. It shows some substantivity for keratin and yields clear and flexible films which, however, are deficient in hardness and therefore in holding power. Its main drawback is its hygroscopicity, which tends to make the film rather tacky in a highly humid atmosphere. A remedy is to add some hydrophobic materials or a resin less sensitive to moisture. PVP K.30 (MW 40 000) is now the most commonly used grade.

At about the same time the dimethylhydantoin formaldehyde resin (DMHF) was proposed; it yields a film of low hygroscopicity but which has a low holding power and is easily powdered.

In order to overcome the hygroscopicity of the PVP films, copolymer resins of vinylpyrrolidone and vinyl acetate (PVP–VA) were developed. They provide good clear films with good hair-holding properties. As mentioned above, the higher the ratio of vinyl acetate, the less susceptible to atmospheric humidity and the harder the film, but consequently it is less easily shampooed off. The copolymers used contain between 30 and 70 per cent PVP. These resins became very popular and nearly superseded PVP, but their use has somewhat declined with the development of resins with free carboxylic groups which give harder films and better adhesion, with a higher moisture resistance.

The advance brought about by carboxylated resins lies, above all, in the flexibility they provide in their neutralization, allowing the building of a variety of formulations adapted to multiple-hold requirements. By controlling the degree of neutralization of the free acid groups, the hardness, solubility and hygroscopicity of the films may be closely monitored. The earliest polymer of this type was Dicrylan 325, which is a polymer of acrylic acid and esters. But the most favoured resin was the copolymer of vinyl acetate and crotonic acid with the brand name Resyn 28-1310. To solubilize it in alcohol, a hydroxylated amine should first be added, such as 2-amino-2-methyl-1,3-propanediol (AMPD) to partially neutralize the acidic groups. Another resin, derived from 28-1310, was proposed later; it offered the advantage of being soluble in a hydroalcoholic medium without previous neutralization. This is Resyn 28-2930, which is a terpolymer of vinylacetate (75 per cent), crotonic acid (10 per cent) and vinyl pivalate (15 per cent). This resin, in addition to improved solubility, brings a better curl-retention under varying humidity.

More recently, another copolymer of vinyl acetate and crotonic acid called Luviset CE 5055 has been proposed; its characteristic is a lack of odour which may be advantageous compared with some earlier compounds which could involve problems of perfuming. Another category of carboxylic polymers developed for hair sprays is the Gantrez series, which are copolymers of methylvinylether and alcoholized maleic anhydride, and particularly the ethyl and butyl hemiesters called ES 225 and ES 425; they are poorly soluble in water even when neutralized, and they yield films with firm hold.

Other marketed resins include a terpolymer of polyvinylpyrrolidone–methylmethacrylate–methacrylic acid, sold under the name 'Stepanhold R-1', the VEM resin which is a carboxylic acrylic resin, and a complex resin called Quadramer which results from the copolymerization of four different monomers (acrylamide–N-t-butyl acrylamid–acrylic acid–vinylpyrrolidone) and which contains fewer free acid groups than the others (about half as many as Resyn 28-2930 and seven times less than Gantrez ES 425).

The preferred neutralizers for these carboxylic resins are 2-amino-2-methyl propanol (AMP), the above cited AMPD, and triisopropanolamine (TIPA). In addition to controlling solubility and hardness, the neutralizers act as internal plasticizers, thus preventing the film from scaling and imparting to it homogeneity and softness; they also help to compensate for a possible sensitivity to non-volatile additives in the preparation.[26]

Another resin, launched later, is Amphomer, an amphoteric acrylic resin which possesses both carboxylic and cationic groups; it is described as offering, through a three-dimensional matrix, a higher hold than conventional carboxylic resins, and thus it can be used in lower concentration to provide a soft and natural hold.[26]

Many more than 100 patents have been published since 1966 which claim an array of polymers to be used in hair sprays. Only the most notable commercial proposals have been cited above. In addition to resins, formulation of hair sprays requires the use of plasticizers, softening agents, glossing agents, perfumes, solvents, and propellants. Plasticizers are mainly aimed at making the film more flexible, at modifying the adhesion and preventing brittleness. Introduced at an average of 5 per cent compared with the weight of the resin, they may be derivatives of lanolin, silicones (which also exert a lubricating effect and bring sheen), esters (isopropyl myristate, diacid esters), polyols (glycerol, polyglycols) and glycol ethers. Other much used additives are protein hydrolysates.[27–29] Polysiloxanes[30] and more recently fluorinated polymers[31] have also been suggested.

Solvents have an important role in the spreading of the film, the rate of drying and the compatibility of the ingredients. The best of all is ethanol, which is highly taxed in Europe, so that for reasons of cost either the alcohol phase is reduced (and thus the resin concentration increased) or isopropanol is substituted where this is permitted (as in Germany and Switzerland).

The common ratio between Freon propellant and solution for a spray is 40–70 propellant to 60–30 solution.

Perfumes should be unobtrusive and just covering because they are enhanced by spraying.

Typical formulae are given in examples 10–13.

	(10)
	per cent
PVP K.30	1·25
Lanolin	0·20
Silicone	0·10
Perfume	0·20
Alcohol	28·25
Propellants 12/11 (35:65)	70·00

	(11)
	per cent
PVP–VA 70/30	1·50
Lanolin derivative	0·05
Silicone	0·10
Isopropyl myristate	0·05
Perfume	0·10
Alcohol	28·20
Propellants 12/11 (35:65)	70·00

Source: Baudelin[31]	(12)
	per cent
Resyn 28-1310	3·45
AMP	0·38
Lanolin derivative	0·10
Isopropyl myristate	0·40
Dipropyleneglycol	0·10
Perfume	0.10
Ethanol	95·57
Concentrate	35·00
Propellants 11/12 (65:35)	65·00

	(13)
	per cent
Amphomer	0·75
AMP	0·123
Perfume	0·075
Ethanol	39·052
Propellants 12/11 (40:60)	60·00

New Spraying Systems

Fluorinated propellants may be considered the ideal means of dispensing minute droplets onto and into the hair, thereby inducing a good hold of the hair by welding individual hairs together through quick drying.

The FDA ban on chlorofluorocarbons created critical issues from a technological point of view. Three alternatives were developed: hydrocarbon aerosols, pump sprays, and carbon dioxide aerosols.

Pumps. In addition to the price, which is about two or three times higher than that of an aerosol, pumps are encumbered with a number of disadvantages such as a spray that is too wet, clogging, leakage, manual dependence, which render them unsuitable for dispensing a spray of quality.

Carbon Dioxide. The main problem with the use of this propellant is the wetting effect of the spray dispensed. Essentially it is a technological issue since it requires extremely small holes at the valve together with a device to prevent clogging.

Hydrocarbons. As liquefiable and compressible gases, propellants which came forward to replace fluorocarbons were mixtures of propane, butane and isobutane. However, comparatively, they exhibit a number of drawbacks:

(i) They are not such good solvents for components of sprays, which means that the ratio of propellant has to be reduced and other solvents have to be introduced, such as methylene chloride and water.
(ii) They are inflammable, which demands the use of special valves[32] to limit the rate of delivery of the product and means increasing the relative amount of methylene chloride and water to cut down flammability.
(iii) Sprays are more wetting because they carry more solvent, which necessitates the incorporation of break-up spray actuators and an additional gas inlet to obtain a better spray pattern and finer droplets for fast drying.

Moreover, these changes in the spray device and the ratios between propellants and solvents entail their own drawbacks and incompatibilities. It is extremely difficult with these new systems to regain the quality of sprays propelled by fluorocarbons.

Among the proposed alternatives to fluorocarbons it is also worth noting dimethyloxide (dimethyl ether) which is, apparently, comparable with regard to toxicity and flame propagation when it is adequately formulated.[33] Aerosol products containing dimethyl ether are sold in several countries.

The use of alternative spraying systems is discussed further in Chapter 40.

A number of resins cannot be adjusted to these new systems of dispensing; however, at least PVP–VA, PVP K.30, Gantrez ES, Amphomer, Stepanhold, Luviset, and Resyns 28-2930 and 28-1310 can be formulated according to the three systems. Owing to its higher solubility in a hydroalcoholic medium, Resyn 28-2930 is more adaptable than Resyn 28-1310 in the new types of hair spray. Hard resins such as Gantrez and Amphomer allow the concentration of film-forming agent to be reduced and give less sticky and less tacky films, which can be more easily eliminated and which leave the hair softer.

Typical formulations are given in examples 14 and 15,[32] 16 and 17[34] and 18.[35] The solvent in examples 14 and 16 is anhydrous ethanol, or ethanol/water (90:10), or ethanol/methylene chloride (80:20). It should be noted that each propellant system requires specific adjustment of perfume.

	(14) *per cent*
Gantrez ES 225 or ES 425	4·00
AMP	0·08
Polyethoxylated (75 EO) lanolin	0·10
Perfume oil	0·10
Solvents	75·72
Isobutane/propane (90:10)	20·00

Concentrate for carbon dioxide	(15) *per cent*
Gantrez ES 225 or ES 425	8·00
AMP	0·40
Dioctyl sebacate	0·50
Perfume	0·10
Ethanol	91·00

	(16) *per cent*
PVP–VA	4·00
Benzyl alcohol	0·10
*Dimethicone copolyol	0·05
Perfume	0·10
Solvents	75·75
Isobutane/propane (90:10)	20·00

*CTFA; polymer of dimethylsiloxane with polyoxyethylene and/or polyoxypropylene side-chains.

	(17) *per cent*
Resyn 28-2930	2·25
AMP	0·18
*Dimethicone copolyol	0·12
Perfume	0·10
Methylene chloride	20·00
Ethanol	72·85
Carbon dioxide	4·50

*CTFA; polymer of dimethylsiloxane with polyoxyethylene and/or polyoxypropylene side-chains.

	(18) *per cent*
Resyn 28-2930	1·50
AMPD	0·38
Alcohol-soluble lanolin	0·90
Isopropyl myristate	0·40
Dipropyleneglycol	0·10
Perfume oil	0·35
Ethanol	96·37
Concentrate	75
Isobutane/propane (90:10)	25

Evaluation of Hair Sprays

The great number of resins and formulations used entails the development of techniques for their evaluation and selection. The following properties need to be measured.

(a) for the film, *in vitro:* hardness, clarity, stickiness, tackiness, flaking, gloss, solubility, ability to be shampooed off, moisture uptake;
(b) on hair: hold, ease of combing (softness, disentangling), powdering, sheen, stickiness, tackiness, harshness.

Procedures have been described for measuring the more important properties[36,37] as follows:

Hardness	Walker Steel Swinging Beam Apparatus or Sward Hardness Rocker
Clarity	Visual assessment
Curl retention	Curl height determined under controlled humidity conditions
Moisture uptake	Weight gain under different controlled humidity conditions

A schedule[38] for the evaluation of hair sprays at a number of important development stages is given in Table 25.1

Table 25.1 Tests to Evaluate Hair Sprays[38]

Purpose	Test
Screening of new resin	Hardness Tackiness and stickiness Switch test to evaluate affinity for hair
Evaluation of modified propellant and solvent systems	Compatibility in transparent glass aerosol Blow-about Headspace pressure Spray pattern
Evaluation of new components e.g. valves and actuators	Spray pattern Cone angle
New formulae	Spray on glass plates and hair pieces Curl-retention Salon and home user tests
Routine assessment	Spray pattern Blow-about Curl-retention Plate testing to determine tack and moisture resistance

Another important factor to be considered concerning non-fluorinated propellants is the necessity to comply with official flammability standards. Two techniques are commonly used: the flame extension test and the closed drum test.

A review of the recent methods for assessment of aerosols (flammability, particle size, corrosion, conservation and spray pattern) is given by Whyte.[39]

Toxicity of Hair Sprays

The question of potential toxicity of sprays has given rise to numerous studies and kept many controversies alive, mainly linked to particular conditions of exposure and new toxicological risks due to dispensing as an aerosol. Three important issues were successively raised:

The toxicity of the resins *per se*
The size of the sprayed particles in relation to the potential penetration by inhalation
The toxicity of propellants

The first doubt against resins was raised in 1943 when Schwartz[40] drew attention to the fact that, when certain resins were substituted for shellac, cases of dermatitis were noted. These resins contained maleic anhydride. Patch tests confirmed that sprays containing the resins were the cause of dermatitis which was apparently due to sensitization (rather than to primary irritation) caused by a residual monomer. Since then, reactions to sprays based on synthetic resins have been extremely rare. However, before any new resin is incorporated in a hair spray, it should be tested for both primary irritation and sensitization.

A major dispute started in 1960 about the potential hazard of inhaling particles of resin emitted by a spray. It was initiated by the report of two cases of thesaurosis ascribed to hair sprays, and by the statement by Bergmann *et al.* that the inhalation of solid PVP formulation could induce the formation of foreign-body granulomatous inflammation in the lungs.[41] These results, supported by others,[42–44] led to a number of works—of which 70 were published mainly between 1962 and 1966—trying to demonstrate whether there is a correlation between thesaurosis and hairsprays.[45] Cambridge has given a very good review on these studies.[46]

The analysis of clinical findings suggests that the quoted cases could be due to an abuse, over a long period, of sprays.[47–49] The cause of the observed affections could not be convincingly established. Epidemiological studies dealt with two small samples of population of high exposure level (hairdressers, cosmetologists) to provide useful data.[47,50–56] Experimental inhalation studies carried out on various animal species by numerous authors and under multiple conditions have failed to demonstrate that prolonged exposure to hair spray aerosol could cause granuloma.[57–60] Even under extreme exposure conditions involving deposition of resin in tissues, it has not been possible to provoke pulmonary lesions.[61]

Brunner *et al.*[62] submitted animals to large exposure of aerosols based on copolymers PVP–VA, VA–crotonic acid and vinyl alkyl ether–maleic ester; they concluded that it was impossible to repeat the 'syndrome of granulomatosis' as alleged in human cases. Kinkel and Eder[63] failed to find evidence of thesaurosis in the monkey, the rabbit or the hamster when submitted to excessive dosage for 70 to 100 days; they noted that inhaled PVP–VA resins were eliminated from the lungs within five days by phagocytosis. In another study carried out on guinea pigs which were exposed for a year to repeated sprays of an aerosol dispensing particles of about 1 μm size, Cambridge[46] could not evince any histological alteration of the pulmonary tract.

From these investigations, it appears that the likely risk of thesaurosis by aerosol sprays is low if not nil. However, it does not preclude the risk, by inhalation, of exciting or even enhancing pre-existing disturbances, particularly in cases of allergic background;[64,65] therefore, taking into account the large surface area of the lungs (25 times that of the skin), it appears highly necessary to include tests of toxicity by inhalation in the evaluation of safety of aerosol hair sprays. A good review on inhalation testing, including preliminary screening methods, the precautions to be taken as well as the criteria used to assess toxicity, is given by Wells.[66] An important factor to consider in trying to reduce a potential ill-effect is the particle size distribution of the spray;[67] it is commonly admitted that particles above 10 μm are unlikely to reach the pulmonary tract.[46,69] If the particles have a low probability of penetrating, there is little risk that a particular toxicity will appear after inhalation.

A number of devices allow assessment of the portion of the spray said to be 'respirable', that is, having a high risk of penetrating into the lungs.

In the early 1970s, another cause for concern arose when propellants were believed to be responsible for cardiac attacks following the use of aerosol bronchodilators.[70,71] Since then propellants, on their own or in formulated products, have been subjected to thorough toxicological studies.[72–75] Although no signs of alteration were observed in the lungs,[76–78] it appeared that chlorofluorinated hydrocarbons, at very high dosage, could have untoward effects, notably cardiac arrhythmia. However, in view of the quantities concerned, it may be considered that their use in aerosol sprays does not offer any risk, even under extreme conditions as experienced in a salon.[79,80] Potential adverse effects of alternative propellants were also studied closely, both individually and in mixtures.[81–83] Potentially beneficial (detoxifying) interactions of ethanol or isopropanol with methylene chloride and hydrocarbon propellants have been noted.[84]

MEN'S HAIR DRESSINGS

Hair dressings are used by men firstly to keep their hair in order and secondly to enhance its lustre. Adequate control of the hair, so important for men's grooming, requires two properties from a product—initial setting and long-term control.

The initial set requires a temporary softening of the hair, and for this there is no better ingredient than water. Water exerts a reversible plasticizing effect on hair which allows the early morning tousle to be brushed into shape, but this effect is only transient and disappears when water evaporates. It can be helped to remain longer by adding so-called moisturizing agents which fix water, thereby protecting the hair against drying. However, these do not provide durable control of the hair.

Alternatively, the hair may be loaded with oil or grease to secure the initial set. This method has the disadvantage that much more oil is needed to get the hair in order than is needed to maintain the set during the day. Once distributed over the hair, oil helps the comb to slide, but initially the oil spreads badly and does not plasticize the hair, so that combing is rather difficult. Because of this, many users of solid brilliantines and similar products wet their hair before applying the dressing.

Another approach is to apply a mixture of oil and water—such as an emulsion—so that first water is partially absorbed by the hair, facilitating combing, control and the setting of hair, and the style is then maintained by the film of oil left on the hair shaft after the water has evaporated. But this requires that the emulsion deposits a continuous film when applied, in order to reduce efficiently the loss of bound water. Oil is enough to ensure the control throughout the day. Adequately chosen, it yields desirable sheen, provides reasonable fixative action, lubricates and protects the hair.

Gum-based products have become less popular because their effect is destroyed when the hair is combed, owing to flaking off of the gum film.

For these reasons, most of the popular hairdressings consist of combinations of oil and water in various forms or, for the less sophisticated market, simple presentations of oil or grease.

Formulation

Initially, vegetable and animal oils were used but, because of their susceptibility to oxidation, they have been replaced by mineral oils which are colourless, odourless, stable and available in a wide range of viscosities. Since antioxidants are now available to overcome rancidity, vegetable and animal oils, the constitution of which is nearer to human sebum, have regained some favour; they are generally thought to have a less greasy feel. Those most commonly used are olive, sesame, peanut, almond, apricot, avocado, peach kernel, and sunflower seed oils. The greater the viscosity of the oil, the better the grooming properties but the more difficult it is to spread the oil thinly and uniformly. The mixture of oils should be carefully chosen and dosed according to the qualities of adhesion, spreading, hold, sheen and combability desired.

Deodorized kerosene is sometimes added to improve spreading. Other main ingredients used to obtain good grooming properties are film-formers such as resins (PVP and copolymers), rosins, and various waxes because of their adhesive and fixative effects, together with substances intended to enhance compatibility and to produce emollience, lubricity, pleasant feel and sheen; such compounds are lanolin and its derivatives (esters, acids, alcohols, acetylated derivatives),[85,86] glycerides, fatty alcohols (cetyl, oleyl, isostearyl), ethoxylated and propoxylated fatty alcohols or esters, fatty alcohol lactates, and dialkyl sebacates and adipates. More conditioning attributes are imparted by fatty acidamides, protein derivatives and cationic compounds.

Some hairdressing compositions described as 'hair foods' also contain vitamins, lecithins and other 'natural' ingredients, so that they should be considered more as hair tonics (see Chapter 26). The idea of a non-greasy product is attractive and many attempts at formulation have been made by partially or totally replacing mineral oil by polyethylene or polypropylene glycols or by blends of branched chain esters—such as 2-ethylhexyl hydroxystearate and palmitate[87]—thereby reducing greasiness and replacing an excessively lustrous appearance by a more natural sheen.

Brilliantines

Non-aqueous Semi-solid Fats

Brilliantines are probably the oldest type of hair dressing and have been in use for hundreds if not thousands of years. They have virtually passed into history, leaving only such memories as 'bear's grease', and the name 'antimacassar' for the cloth used to protect upholstered chairs from the macassar oil used in hairdressings in Victorian days. Coconut oil is still used in some of the less sophisticated parts of the world.

Solid Brilliantines (Pomades)

Pomades usually consist of suitably thickened, tinted and perfumed mineral or vegetable oil. Stiffening to the desired consistency may be obtained through waxes such as paraffin wax, ozokerite, spermaceti, beeswax, ceresin, carnauba wax and microcrystalline waxes. According to what is sought, consistency may vary from relatively hard wax to butter-type cream; the pomade may be greasy, rapidly absorbed or stringy to the touch.[88]

A simple mixture which is satisfactory is given in example 19, and slightly more complex mixtures in examples 20, 21,[88] 22[87] and 23.[31]

	(19) *per cent*
Petroleum jelly	90·0
Paraffin wax (uncrystallizable)	10·0
Colour, perfume	*q.s.*

	(20) *per cent*
Carnauba wax	5·0
Paraffin wax	5·0
Petroleum jelly	70·0
Mineral oil	20·0

	(21) *per cent*
Petroleum jelly	87·0
Paraffin wax	3·0
Lanolin, anhydrous	2·0
Lanolin alcohol	3·0
Mineral oil 65/75	5·0
Perfume	*q.s.*

	(22) *per cent*
Petrolatum	65·0
2-Ethylhexyl palmitate or isostearate	25·0
Arachidyl propionate	5·0
Myristyl myristate	5·0

	(23) *per cent*
Castor oil	20·0
Almond oil	2·0
Isopropyl myristate	8·0
Petrolatum	46·0
Lanolin, anhydrous	12·0
Alcohol	12·0

Solid brilliantines can also be made with mixtures of mineral oil and 12-hydroxy stearic acid (from hardened castor oil fatty acids). This acid sets the oil to a crystalline mass which is liable to syneresis; this property can be corrected by the addition of a small proportion of wax, such as paraffin or microcrystalline wax. A suitable formula is given in example 24.

	(24) *per cent*
Mineral oil	90·0
12-Hydroxy stearic acid	3·0
Microcrystalline wax	7·0
Colour, perfume	*q.s.*

Procedure: Melt the materials together and stir until the mixture is homogeneous.

Transparent solid brilliantines can be made by gelling mineral oil with a suitable agent. The earliest materials were certain metallic soaps[89] particularly those of lithium and aluminium. Such a composition is as follows:

	(25) *per cent*
Aluminium tristearate or isostearate	5
Paraffin oil	95
Castor oil, perfume, etc.	*q.s.*

Procedure: Heat the mixture to 130°C and stir until homogeneous. Then the optimum method of cooling is to allow the temperature to fall from 130° to 50°C over 1 h.

A number of other materials have been described which are suitable for gelling mineral oil: these include derivatives of glyceryl tris-12-hydroxy stearate,[90] dialkylamino propionic acids[91] and polyethylene.[92] Thau and Fox[93] state that, whereas high molecular weight (about 20 000) polyethylene has previously been recommended for gelling mineral oil, involving heating to about 130°C and a complicated cooling process, more satisfactory gels can be prepared by the use of lower molecular weight (1500–2000) polyethylene wax which is dissolved in about seven times its weight of mineral oil at 90°–95°C. After solution is achieved the solution is mixed with a high shear stirrer while it cools to 65°C (about 10° below the cloud point) after which it is stirred gently without air entrapment down to 45°C, at which temperature it is filled into jars where it finally sets. Products prepared in this way have an improved temperature stability.

Liquid Brilliantines

The day of the heavy brilliantine has passed. The aim today is to secure a glossy sheen and light fixative action without the suggestion that the hair has been plastered down. Formerly it was not possible to incorporate a suitable oil in a volatile mineral base, but by the use of deodorized kerosene, or esters such as isopropyl myristate, a fine film of the oil may now be deposited on the hair shaft.

By the addition of deodorized kerosene the viscosity of light petroleum oil may be further reduced, and such preparations lend themselves to spraying on the hair:

	(26) *per cent*
Light mineral oil	60·0
Deodorized kerosene	40·0
Colour, perfume	*q.s.*

If desired, vegetable oils may replace part of the mineral oil. Because of the tendency of most vegetable oils to rancidity, the addition of a suitable antioxidant is necessary. Alternatively, isopropyl fatty esters may be employed together with, if desired, a small proportion of lanolin, as in examples 27 and 28.[31]

	(27) *per cent*
Isopropyl myristate	24·0
Lanolin	1·0
Light mineral oil	75·0
Colour, perfume	*q.s.*

	(28) *per cent*
Isopropyl palmitate	5
Petrolatum	5
Liquid petrolatum	15
Acetylated lanolin alcohol	15
Benzyl alcohol	24
Propyleneglycol butylether	26
Alcohol, anhydrous	10

Alcoholic Brilliantines

Solutions of oil in alcohol offer the advantage of allowing a good and uniform distribution of oil on the hair. They also give a feeling of freshness and a stimulating effect for the scalp.[31]

Traditionally these were solutions of castor oil in alcohol, suitably coloured and perfumed. These do not possess very marked fixative properties although, if the hair is combed with a damp comb immediately after application, the precipitation of the castor oil in a finely dispersed form gives sufficient body to act as a fixative for hair that is not very dense—examples 29 and 30.[31]

	(29) *per cent*
Castor oil	20·0
Industrial alcohol	80·0
Perfume, colour	*q.s.*

	(30) *parts by wt*
Castor oil	12
Almond oil	4
Polypropyl 40 butyl ether	16
Dihydroabietyl alcohol	5
Isopropyl myristate	15
Butyl stearate	2
Alcohol	44
Water	12

Such a formulation, although not smelling of castor oil when suitably perfumed, has a tendency to develop an odour when castor oil is deposited in the wad of the bottle or along the threads of the bottle neck. In addition, unless the hair is washed frequently, a slight castor oil smell may also develop on the user's head. This may be overcome to some extent by substituting oleyl alcohol for castor oil. Such a preparation will also develop a somewhat fatty odour in time, particularly around the bottle neck and on combs and brushes, but it is not as objectionable as castor oil. An even more satisfying product, as far as odour is concerned, may be obtained by the substitution of isopropyl myristate for castor oil. Unfortunately both oleyl alcohol and isopropyl myristate lack fixative properties though they are useful with suitable additions of perfume and antiseptic in a friction lotion or as a dressing for men with thinning hair.

	(31) *per cent*
Isopropyl myristate	20·0
Cetyl alcohol	2·0
Alcohol	78·0
Colour, perfume	*q.s.*

The next products of this type were those which used hydrophilic derivatives of beeswax and lanolin, for example ethoxylated lanolin. These resemble the alcoholic brilliantines except that they are essentially aqueous. The increased solubility of the fatty material permits more formulation flexibility.

Then came new clear liquid preparations, based on the use of polyethylene or polypropyleneglycols, and esters thereof, which largely superseded previous products. They consist basically of a solution of polyethyleneglycol (10–30 per cent) in water. They may also contain other ingredients including resins (for example PVP) to improve their hair-holding characteristics, and germicides. These products are much less greasy than mineral-oil-based products, but do not give as much gloss. A typical formula is given in example 32.

	(32) *per cent*
Polyethyleneglycol	20·0
Water	20·0
Alcohol	59·0
Resin (PVP)	1·0
Colour, perfume	*q.s.*

Yet another variant on the same theme is the use of the Ucon oils. These are polypropylene–ethylene glycol copolymers, some of which are completely soluble in water. They are much less greasy than mineral oil, but give less gloss. One of the problems with this type of product is the mobility of the liquid. This necessitates the use of a bottle with a narrow orifice. A basic formula is:

	(33) *per cent*
Ucon LB-1715	20·0
Water	20·0
Ethanol	60·0
Colour, perfume	*q.s.*

Separable Brilliantines

Separable preparations are two-layer systems prepared by adding a solution of the desired perfume in the requisite amount of alcohol to non-miscible mineral or vegetable oil. By being vigorously shaken before application the mixture forms a temporary dispersion which may then be spread onto the hair. A drawback is the rapid reversion to the initial state which leads to spreading on the hair of disproportionate amounts of one or the other phase. Adding an emulsifying agent to either phase helps to remedy the problem to some extent. It is usual to tint both the oily and alcoholic layers a similar colour. Formulae are given in examples 34 and 35 (see also Baudelin[31]).

	(34) *per cent*
Industrial alcohol	60
Light mineral oil	40
Spirit-soluble yellow dye	*q.s.*
Oil-soluble yellow dye	*q.s.*
Perfume	*q.s.*

	(35) *per cent*
Sesame oil	32
Olive oil	10
Alcohol, anhydrous	58
Antioxidant, colour, perfume	*q.s.*

The product may be varied by altering the ratio of alcoholic to oily phase, and may be cheapened by using aqueous alcohol provided that the strength is sufficient to dissolve the colour and perfume or alternatively that the perfume is formulated in the oily phase.

Non-oily Fixatives

The oldest type of non-oily fixative was based on the use of mucilages, particularly gum tragacanth. These possess very good fixative properties and do not produce greasy stains. They are particularly effective in controlling unruly hair and they are often preferred to oils to comb into light or grey hair.

However, on brushing the hair the setting characteristics are lost and the particles of gum that flake off tend to accumulate on the scalp and clothes, often giving the impression of dandruff. Moreover, they leave the hair with a harsh feel and a dull appearance, although this disadvantage may sometimes be overcome by adding the proper oil. A typical formula is given in example 36. If desired, about 10 per cent oil may be added to this cream. The addition of a little tincture of benzoin, tolu or styrax assists in increasing the whiteness of the product.

	(36)
	per cent
Gum tragacanth	1·0
Alcohol	6·0
Castor oil	2·0
Water	90·0
Glycerin	1·0
Preservative	*q.s.*

Procedure: Disperse the gum in the alcohol and glycerin and add the perfume and any oils; run in the water with continuous agitation. Allow the cream to stand to thicken, restir, filter through muslin and bottle.

Sodium carboxymethylcellulose,[94] hydroxypropylcellulose[95] and carboxyvinyl polymers and various resins may also be used for such gum-like dressings. Representative formulae are given in examples 37[94] and 38.

	(37)
	per cent
Sodium carboxymethylcellulose 2%	47·85
Alcohol	39·70
Polyethyleneglycol 600 laurate	10·00
Propyleneglycol laurate	2·00
Tincture of capsicum	0·05
Cetyldimethylethyl ammonium bromide	0·10
Perfume	0·30

	(38)
	per cent
Carbopol 940	0·6
Water	55·3
Triethanolamine	0·6
PVP K.30	2·5
Alcohol	28·0
Ethoxylated oleyl alcohol	3·0
Polyethylene glycol	10·0

Aerosols

Aerosol sprays for men tend to fall into two classes: those which contain resins very similar to those used in women's hair sprays, and those which contain only mineral oil. Such products are extremely clean in use, but they do not give such good initial control as some of the more conventional products and the gloss obtained is often not very good. A further drawback is that men tend to regard aerosol hair sprays as essentially feminine products, and thus before such products become generally accepted there is a need to overcome this inherent prejudice.

Emulsions

Water-in-oil Creams

Water-in-oil creams have achieved considerable popularity, particularly in Western Europe and the USA. It is claimed that they spread easily and offer greater fixatives properties, owing to their oil content and more uniform coating, than do similar creams of the oil-in-water type; moreover they impart a better gloss to hair.

Such creams are usually prepared using mixtures of calcium hydroxide with fatty acids to produce the emulsifying agent. Alternatively, nonionic emulsifiers such as sorbitan esters may be used.

Example 39 is a typical formula of the calcium oleate type. It is convenient to incorporate the calcium hydroxide in the form of limewater which contains approximately 0·14 per cent of calcium hydroxide. By the use of sugar solutions, higher amounts of calcium salts may be rendered soluble in the form of calcium saccharate. If the calcium hydroxide is added in the form of limewater, it should be subjected to a simple titration before use to establish its content of calcium hydroxide.

	(39) *per cent*
Beeswax	1·50
Calcium hydroxide	0·07
Olcic acid	1·00
Mineral oil	40·60
Perfume	1·00
Water	53·83
Magnesium sulphate crystals, 25% aqueous solution	2·00

Procedure: Warm the beeswax, oleic acid, and oil to 70°–75°C and add the limewater with constant stirring. Finally add the magnesium sulphate solution, which helps to stabilize the emulsion, followed by the perfume. The mixture should be kept in continuous agitation until it has cooled down to about 20°C.

It may be added that the manufacture of a stable water-in-oil cream, which is also pourable at ordinary temperatures and suitable for use as a hair dressing, is not easy. Water-in-oil emulsions cannot lose their water, either by evaporation or by absorption into the hair, as rapidly as oil-in-water emulsions, and if emulsion breakage is not rapid the product remains unpleasantly white on the hair after application.

To make a good hair dressing, a water-in-oil emulsion must be formulated as a compromise product. It must have sufficient stability towards temperature changes and transport shocks to possess adequate shelf life, yet it must break easily when submitted to the mechanical action of rubbing it on the hands or hair. This combination of thermal stability and mechanical instability is best achieved by using an emulsifier which produces a solid interface between the oil and the water—hence the usefulness of calcium soaps. Once such an interface is disturbed mechanically, it is entirely destroyed as a barrier to emulsion breakdown.

The formulator must decide on the balance to be achieved between thermal stability and mechanical instability. If greater stability is required, however, it may be obtained by the addition of wool wax alcohols or cholesterol at a level of 0·1–0·5 per cent, or with such emulsifiers as sorbitan sesquioleate at a similar level. Alternatively, small amounts of soluble soaps may be incorporated. Such a formula is given in example 40.

	(40) *per cent*
Beeswax	2·0
Limewater	59·0
Mineral oil	30·0
Petroleum jelly (short-fibre)	8·0
Stearic acid	0·2
Wool wax alcohols (refined)	0·3
Perfume	0·5

For short (crew cut) hair styles:[86]

	(41) *per cent*
Oil phase	
Petrolatum	40·0
Amerchol CAB (petrolatum and lanolin alcohol)	29·0
Microcrystalline wax	10·0
Sorbitan sesquioleate	6·0
Water phase	
Water	14·0
Tween 81 (Polysorbate)	1·0
Perfume, preservative, colour	*q.s.*

The following formula is based on the use of beeswax–borax as emulsifier, assisted by a nonionic ester in small proportions:

		(42) *per cent*
A	Petroleum jelly	7·5
	Mineral oil 65/75	37·5
	Lanolin, anhydrous	3·0
	Sorbitan sesquioleate	3·0
	Beeswax	2·0
B	Borax	0·5
	Water	46·5

Procedure: Bring A to 75°C. Bring B to 75°C. Add B to A slowly with moderate but thorough agitation. Add perfume at 45°C and agitate until cold.

Oil-in-water Creams

Oil-in-water creams are reputed to give a less greasy feel than water-in-oil creams. This probably arises from the fact that they do not give as much gloss when they are first applied to the hair and that they can be diluted with water.

Stable oil-in-water creams of varying viscosities may be obtained by the use of triethanolamine stearate, together with, if higher viscosity is desired, some fatty alcohol, such as cetyl alcohol. A simple formula is given in example 43; by adding 1–3 per cent cetyl alcohol the viscosity may be increased.

	(43) *per cent*
White mineral oil	45·0
Stearic acid	3·5
Triethanolamine	1·5
Water	50·0
Preservative, perfume	*q.s.*

Examples 44[96] and 45[87] are representative of more complicated formulae; in example 45 mineral oil is replaced by a branched-chain ester.

	(44) *parts*
Mineral oil	33·00
Beeswax	3·00
Stearic acid	0·50
Cetyl alcohol	1·30
Borax	0·25
Triethanolamine	1·85
Water	59·80
Chlorocresol	0·10
Perfume	*q.s.*

Procedure: Melts the oils, fats and waxes at 60°–65°C and add them to the aqueous phase containing the triethanolamine and borax at the same temperature. Stir the cream until cool.

		(45) *per cent*
A	2-Ethylhexyl palmitate	20·00
	Beeswax	10·00
	Cetyl palmitate powder	5·00
	Cetyl alcohol	5·00
	Arlacel 165	8·00
	Propyl paraben	0·15
B	Demineralized water	51·65
	Methyl paraben	0·20

Procedure: Heat A to 70°C. Heat B to 70°–72°C. Slowly add B to A using high-speed mechanical agitation such as homomixer. Mix for approximately 20 min and begin to cool while maintaining agitation. Package at 35°–40°C.

High molecular weight polyacrylamido-sulphonate has been claimed to impart lubricity.[97]

A tendency amongst products of this type has been the inclusion of additives which are intended to give special effects. Thus oil-in-water creams are used as suitable bases for anti-dandruff ingredients. Other additives such as alcohol, which is intended to give a refreshing effect to the scalp, can also be added.

Gels

The semi-solid brilliantines described above have been succeeded by gel products of two types: the micro-gels, which are really transparent oil-in-water emulsions in which the oil droplets are so small that the emulsion appears clear, and the true gels which are based on the use of aqueous polyethylene glycol solutions in conjunction with a cellulosic thickener.

The advantage of the micro-gels is that they are miscible with water and feel much less greasy than conventional creams. Examples 46,[98] 47[99] and 48[100] have been proposed.

		(46) *per cent*
A	Sorbitan monolaurate	12·0
	Deodorized light petroleum distillate	45·0
	Light mineral oil	5·0
B	Mannitol monolaurate (modified)	19·0
	Water	19·0

Procedure: Add B to A. As the aqueous solution is mixed with the oils the mixture becomes first thin and cloudy, then thick and clear.

	(47) *per cent*
Mineral oil 65/75	20·0
Lauric diethanolamide	6·0
Lanolin alcohols	2·7
Water	62·5
Polyoxyethylene (3 EO) lauryl alcohol	4·0
Polyoxyethylene (23 EO) lauryl alcohol	4·8

		(48) *per cent*
A	Polyoxyethylene (10) oleyl ether	15·5
	Polyoxyethylene fatty glyceride	15·5
	Light liquid petrolatum	13·7
	Propylene glycol	8·6
	*Atlas sorbitol solution	6·9
B	Perfume	*q.s.*
C	Water	to 100·0

* Atlas Chemical Company.

Procedure: Heat A to 90°C and C to 95°C. Add C to A with stirring and add B at 70°C. Cool and pour at 55°–60°C. On further cooling a ringing gel of a sparkling clarity is obtained.

The true gels have the same advantages as the micro-gels. A basic formula would consist of polyethylene glycol (30 per cent), alcohol (30 per cent), water

(40 per cent) and a small amount of cellulosic thickener such as isopropyl cellulose. A British patent[101] has described the preparation of transparent hair gels. Such products give good fixative properties and control hair satisfactorily, but they do not give a very good gloss.

REFERENCES

1. US Patent 3 984 536, L'Oreal, 5 October 1976.
2. US Patent 3 716 633, L'Oreal, 12 February 1973.
3. British Patent 1 321 836, ICI, 4 July 1973.
4. US Patent 3 862 306, Gillette, 21 January 1975.
5. German Patent 2 404 793, BASF, 14 August 1975.
6. German Patent 2 453 629, Henkel, 13 May 1976.
7. US Patent 3 972 336, National Starch and Chemical Co., 3 August 1976.
8. US Patent 3 910 862, GAF, 7 October 1975.
9. US Patent 3 954 960, GAF, 4 May 1976.
10. US Patent 3 959 463, Bristol-Myers, 25 May 1976.
11. US Patent 3 904 749, Ashland Oil, 9 September 1975.
12. French Patent 2 110 268, Minnesota Mining and Manufacturing, 7 July 1972.
13. US Patent 4 007 005, Redken Laboratories, 8 February 1977.
14. Japanese Patent 72 49 702, Nippon Osker, 13 December 1972.
15. German Patent 2 053 505, Wella, 10 May 1972.
16. German Patent 2 542 338, Wella, 24 March 1977.
17. US Patent 3 993 745, Alberto Culver, 13 November 1976.
18. US Patent 4 015 612, Minnesota Mining and Manufacturing, 5 April 1977; US Patent 4 044 121, Minnesota Mining and Manufacturing, 23 August 1977.
19. US Patent 4 059 688, Clairol, 22 November 1977.
20. French Patent 2 054 478, Biechier, F. J. and Martineau, J. E., 28 May 1971.
21. German Patent 2 314 659, Ugine Kuhlmann, 4 October 1973.
22. Schoenberg, T. G. and Scafidi, A. A., *Cosmet, Toiletries*, 1979, **94**(3), 57.
23. US Patent 4 047 537, Harshaw Chemical, 13 September 1977.
24. De Navarre, M. G., *Am. Perfum. Aromat.*, 1950, **56**(1), 23.
25. Martin, J. W., *Aerosol Age*, 1966, **11**(8), 14.
26. Koehler, F. T., *Cosmet. Toiletries*, 1979, **94**(4), 75.
27. David, L. S., *Drug Cosmet. Ind.*, 1965, **97,** 502.
28. Lange, F. W. and Muller, J., *Seifen Öle Fette Wachse*, 1965, **91,** 165.
29. Riso, R. R., *Proc. sci. Sect. Toilet Goods Assoc.*, 1965, (42), 36.
30. German Patent 2 325 645, Unilever, 6 December 1973.
31. Baudelin, F. J., *Cosmet. Toiletries*, 1979, **94**(3), 51.
32. Root, M. J., *Cosmet. Toiletries*, 1979, **94**(3), 37.
33. Bohnen, L. J. M., *Aerosol Rep.*, 1979, **18**(3), 70.
34. Murphy, E. J. and Bronnsack, A. H., *Aerosol Rep.*, 1978, **17**(6), 171.
35. *Manf. Chem. Aerosol News*, 1977, **48**(10), 66.
36. McFarland, J. H. and Scott, R. J., *Drug Cosmet. Ind.*, 1966, **98**(2), 41.
37. Root, M. J. and Bohac, S., *J. Soc. cosmet. Chem.*, 1966, **17,** 595.
38. Blackmore, N. F. E., *Specialities*, 1964, **1**(4), 11.
39. Whyte, D. E., *Aerosol Age*, 1979, **24**(5), 29 and **24**(6), 31.
40. Schwarz, L., *Public Health Rep. Wash.*, 1943, **58,** 1623.
41. Bergmann, M., Flance, I. J. and Blumenthal, A. T., *N. Engl. J. Med.*, 1958, **258,** 471.
42. Edelston, B. G., *Lancet*, 1959, **2,** 112.

43. Nevins, N. A., *et al.*, *J. Am. med. Assoc.*, 1965, **193,** 266.
44. Ripe, E., *et al.*, *Scand. J. respir. Dis.*, 1969, **50,** 156.
45. Ludwig, E., *Aerosol Rep.*, 1964, **3**(2), 27.
46. Cambridge, G. W., *Aerosol Rep.*, 1973, **12**(7), 273.
47. Gowdy, J. M. and Wagstaff, M. J., *Arch. environ. Health*, 1972, **25,** 101.
48. Bergmann, M., *et al.*, *N. Engl. J. Med.*, 1962, **266,** 750.
49. Törnell, E., *Sven. Laekartidn.*, 1963, **25,** 1819.
50. Larson, R. K., *Am. Rev. respir. Dis.*, 1964, **90,** 786.
51. Sharma, O. P. and Williams, M. H., *Arch. environ. Health*, 1966, **13,** 616.
52. Haug, H. P., *Dtsch. Med. Wochenschr.*, 1964, **89,** 87.
53. Favez *et al.*, *Int. Arch. f. Gewerbepath. Gewerbehyg.*, 1965, **21,** 268.
54. Garibaldi, R. and Caprotti, M., *Medna. Lav.*, 1964, **55,** 424.
55. McLaughlin, A. I. G., Bidstrup, P. L. and Konstam, M., *Food Cosmet. Toxicol.*, 1963, **1.** 171.
56. Epsom, J. E., *Med. News*, 1965, **89,** 10; *Food Cosmet. Toxicol.*, 1965, **3**(1), 136.
57. Draize, J. H., *et al.*, *Proc. sci. Sect. Toilet Goods Assoc.*, 1959, (31), 28.
58. Shelanski, M. V., *Parfum. Kosmet.*, 1958, **39**(9), 614.
59. Calandra. J. and Kay. J. A., *Proc. sci. Sect. Toilet Goods Assoc.*, 1958, (30), 41.
60. Giovacchini, R. P., *et al.*, *J. Am. med. Soc.*, 1965, **193,** 298.
61. Lowsma, H. B., Jones, R. A. and Prendergast, J. A., *Toxicol. appl. Pharmacol.*, 1960, **9,** 571.
62. Brunner, M. J., *et al.*, *J. Am. med. Assoc.*, 1963, **184,** 851.
63. Kinkel, H. and Eder, H., *Int. Arch. f. Gewerbepath. Gewerbehyg.*, 1966, **22,** 10.
64. Gelfand, H. H., *J. Allergy*, 1963, **34,** 374.
65. Cares, R. M., *Arch. environ Health.*, 1965, **11,** 80.
66. Wells, A. B., *Int. J. cosmet. Sci.*, 1979, **1,** 135.
67. Sciarra, J. J., *Aerosol Age*, 1971, **16**(7), 40.
68. Mitchell, R. I., *Am. Rev. respir. Dis.*, 1960, **82,** 627.
69. Sciarra, J. J., *J. pharm Sci.*, 1974, **63**(12), 1892.
70. Taylor, G. J. and Harris, W. S., *J. clin. Invest.*, 1971, **50,** 1546; *J. Am. med. Assoc.*, 1970, **212,** 2075.
71. Flowers, N. C. and Horan, L. G., *J. Am. med. Assoc.*, 1972, **219,** 33.
72. Aviado, D. M., *J. clin. Pharm.*, 1975, **15**(1), 86.
73. Belaj, M. A. and Aviado, D. M., *J. clin. Pharm.*, 1975, **15**(1), 105.
74. Paulet, G., *Aerosol Rep.*, 1977, **16**(1), 22.
75. Paterson, J. N. and Sudlow, M. F., *Lancet*, 1971, **2,** 565.
76. Quevauviller, A., *et al.*, *Ann. Pharm.*, 1963, **21,** 727.
77. Vashkov, V. I., *et al.*, *Gig Sanit.* 1964, **29,** 61.
78. Paulet, G., *et al.*, *Arch. Mal. Prof. med. Trav.*, 1969, **30,** 194.
79. Gulden, W., *Aerosol Rep.*, 1973, **12**(6), 248.
80. Bower, F. A., *Résumés 9e Congrès International Aerosol Montreaux*, 24–28 Septembre 1973, p. 17.
81. Aviado, D. M., Zakhari, S. and Watanabe, T., in *Non-fluorinated Propellants and Solvents for Aerosols*, ed. Goldberg, L., Cleveland. CRC Press, 1977.
82. Rampy, L., *Aerosol Age*, 1977, **22**(6), 20.
83. Johnsen, M. A., *Aerosol Age*, 1979, **24**(6), 20.
84. Aviado, D. M., in *Non-fluorinated Propellants and solvents for Aerosols*, ed. Golberg, L., Cleveland, CRC Press, 1977.
85. Fleischner, A. M. and Seldner, A., *Cosmet. Toiletries*, 1979, **94**(3), 69.
86. Conrad, L., *Am. Perfum. Cosmet.*, 1968, **83**(10), 63.
87. Calogero, A. V., *Am. Perfum. Cosmet.*, 1979, **94**(4), 77.
88. Goode, S. T., *Am. Perfum. Cosmet.*, 1979, **94**(4), 71.
89. British Patent 745 688, Laboratories Scientifiques de Neuilly, 29 October 1952.

90. Kline, C. H., *Drug Cosmet. Ind.*, 1964, **95,** 895.
91. Berneis, K. I., *Am. Perfum. Cosmet.*, 1967, **82**(5), 25.
92. US Patent 2 628 187, Research Products, 10 February 1953.
93. Thau, P. and Fox, C., *J. Soc. cosmet. Chem.*, 1965, **16,** 359.
94. US Patent 2 771 395, Colgate-Palmolive, 29 June 1953.
95. US Patent 3 210 251, Hercules Powder, 8 February 1963.
96. Druce, S., *Manuf. Chem.*, 1949, **20,** 534.
97. US Patent 4 065 422, General Mills Chemical, 27 December 1977.
98. US Patent 2 402 373, United Rexall Drug, 1944.
99. US Patent 3 101 300, Siegel B., Petgrave, R. and Thau, P., 1963.
100. *Manuf. Chem.*, 1967, **38**(7), 49.
101. British Patent 1 002 466, Chesebrough Ponds, 14 February 1963.

Chapter Twenty-six

Hair Tonics and Conditioners

Introduction

No toilet preparation has aroused such derision as the so-called hair 'tonic'; the use of the word 'tonic' is unfortunate in this respect since even in therapeutics it is viewed with grave suspicion. Much of this scepticism is the result of the claims made in the past by the manufacturers of such preparations—widely based on old recipes inspired by 'Nature'—coupled with the lack of success experienced by the user.

However, there are many purposes for which a lotion provides an ideal vehicle and the increasing success of modern rinses and conditioners is outstanding evidence of this.

Two categories of hair 'tonic' should, in fact, be distinguished: medicated products which deal with specific problems of the hair and scalp—greasy hair, dandruff, loss of hair—and the so-called conditioners, used primarily by women, which aim at improving, restoring and maintaining the condition of the hair. To this class should be added rinses which, as their name does not make clear, are usually followed by a water rinse, and which aim to combat tangles and to render combing easier. Rinses and conditioners look similar to the older type of women's hair dressings and mostly take the form of a clear lotion or a liquid cream. The border between rinses and conditioners tends to become ever vaguer owing to the development of clearer rinses with increased conditioning properties, warranting the name of clear rinse–conditioner. However, the main requirement is to provide the conditioning effect.

MEDICATED PRODUCTS

The objective of medicated products is to cure, to reduce, to restrain or to offset the unaesthetic phenomena or disabilities resulting from some abnormality in the functioning of the scalp. Although in the past use was often made of irritant, keratolytic, rubefacient compounds, the present trend is that the treatment should rather bring about a return to a normal state, promote balance and alleviate and restore the disturbed substrate. The main conditions with which medicated products should deal are dandruff, seborrhea and hair loss.

Dandruff

Dandruff results from an excessive scaling of the scalp, without clinical signs of inflammation.[1] Rare among children and the elderly, it usually reaches a peak in winter and eases off in summer. Although a relationship may exist with the general state of health (nervous, digestive or metabolic factors, state of tiredness), the causes of this scaling are still not clear. Dandruff has been noticed to be very often accompanied by a proliferation of some micro-organisms, and more particularly a sporulating yeast, *Pityrosporum ovale*, once considered to be responsible for the condition. The scalp is a lush environment for the growth of micro-organisms and high scaling conditions increase their niches and the nutrients beneficial to their development. Indeed, the scalp flora appears to be ubiquitous, but only one micro-organism experiences a dramatic increase in the dandruff condition: the prevalent resident, *P. ovale*—75 per cent of the population, versus 45 per cent under normal conditions.[2] However, as stressed by Kligman following investigations with specific germicides,[1] it is generally believed today that the increase in *P. ovale* is not causative but is a secondary effect resulting from increased scaling.

In the absence of knowledge of the cause of dandruff, the trend is to treat the phenomenon; thus the compounds added to antidandruff lotions generally meet two requirements:

(i) antimicrobial, to prevent proliferation, an aggravating factor of the local disorder;
(ii) keratolytic or exfoliative, to cleanse the scalp by promoting the elimination of dead skin.

As germicides, quaternary ammonium compounds (such as benzalkonium, cetyltrimethylammonium and cetylpyridinium salts), chlorophenols, PVP–iodine complexes, oxyquinoline and 5,7-dichloro-8-hydroxyquinoline[3,4] in particular have been used. But the compounds which have gained the dominant position because of their efficacy are the derivatives of 2-pyridine-thiol-N-oxide; according to Kligman these compounds, although strongly antimicrobial, may exert rather a cytostatic activity, depressing the turnover rate of epidermal cells and inducing more complete keratinization.[5,6] Soluble derivatives[7–10] are preferred for lotions. Related compounds from pyridazines,[11] quinoxalines[12] and quinolines[13] have been patented.

As keratolytic ingredients, colloidal sulphur, resorcinol, salicylic acid[14,15] and selenium disulphide have been used most often. The latter, which owing to its toxicity is largely confined to prescription uses, has been shown to act rather as a powerful cytostatic agent.[16] Coal tar is also known to provide good results; it may act similarly by restraining the multiplication of epidermal germinative cells, as will corticosteroids, or affect cohesiveness of desquamating cells.[1] Other patented materials are polyene derivatives[17] and zinc hydroxymethylsulphinate.[18]

Antidandruff products can be simple solutions of an antimicrobial agent in a 1:1 aqueous–alcoholic medium or a setting or rinse composition (examples 1–4).

Anti-dandruff lotion	(1)
	per cent
*Bis(2-pyridyl-1-oxide) disulphide soluble complex	0·15
Camphor	0·10
Menthol	0·05
Ethanol or isopropanol	50·00
Water	*q.s.* to 100

*Magnesium sulphate or calcium chloride complexes (Olin Mathieson).

Anti-dandruff lotion[19]	(2)
Cationic polymer from epichlorhydrin piperazine	0·5 g
Lauryl isoquinolinium bromide	1·3 g
Lactic acid	*q.s.* to pH 5·0–5·3
Ethyl alcohol	55 ml
Menthol pantothenate	0·1 g
Perfume	0·3 g
Water	*q.s.* to 100 g

Anti-dandruff hair-dressing[20]	(3)
	per cent
Polypropylene glycol (40) monobutylether	1.75
Polypropylene glycol (33) monobutylether	0·50
Polyoxypropylene (12) polyoxyethylene (16) monobutylether	15·75
Alkyl benzyldimethylammonium saccharinate	0·52
Ethanol	63·05
Perfume	*q.s.*
Water	*q.s.* to 100

Clear anti-dandruff rinse	(4)
	per cent
Stearyl dimethylbenzylammonium chloride	5·0
Cetyltrimethylammonium chloride	0·25
Tallow amido propylamine oxide	3·00
Propylene glycol	3·00
Hydroxypropylmethyl cellulose	0·6
Distilled water	to 100

Methods have been developed for quantitative evaluation of anti-dandruff products by counting corneocytes, bacteria and yeasts on the hairy scalp.[6,21]

Greasy Hair and Scalp

Seborrhea results from an excessive secretion from the sebaceous glands. Discharged into the follicular duct, sebum is then excreted onto the surface of the skin where it may undergo a number of transformations by aerial oxidation and under the action of the resident micro-organisms. The mechanisms which lead to the setting up of a seborrheic state remain largely unknown. In addition to the fact that it may induce a loss of hair, its consequences on the hair are particularly unaesthetic: hair becomes greasy very quickly after shampooing, forms bunches and catches dust, it becomes heavy and the benefits of a setting operation are lost soon afterwards.

The phenomenon of regreasing is itself a source of much controversy.[22–25] A method of reducing sebum uptake by hair is to apply an oil-repellent treatment, that is, to deposit a lipophobic ingredient onto the hair; this approach is illustrated by the use of some hydrophobic and lipophobic perfluorinated compounds[26–29] such as $CF_3(CF_2)_x(CH_2)_yZ$, where Z is a water- or oil-solubilizing group which may be anionic, cationic, nonionic or amphoteric

(x = 7–11, y = 0–4). These substances[26,27] are used in 70:20 aqueous–alcoholic lotions at concentrations between 0·05 and 0·2 per cent; commercial compounds are Fluorad (3 M Co.) and Zonad (Du Pont).

The formula given in example 5 has been claimed to retard oil uptake by 85 per cent.

	(5) *per cent*
Ethanol	30.0
Water	69·7
Polyoxyethylene polymethylsiloxane	0·1
Perfluorononanoic acid	0·1
Perfume	0·1

The use should also be mentioned of oil-adsorbing materials such as finely divided starches[30] and silicas to keep hair looking clean for longer periods.

Another approach consists in a topical treatment in order to reduce secretion or excretion of sebum; it has given rise to a multitude of patents describing a large number of possible molecules. Most of the proposed substances are rather indiscriminately recommended for seborrhea and hair growth. Classes of compounds with a seemingly more selective activity are thio-ether or sulphoxide derivatives of cysteine, cysteamine, glutathione, pyridoxine and hydroxylated aminothiols;[31] some, such as salts of 2-benzylthioethylamine, have been reported to act specifically on lipid synthesis processes,[32,33] thereby reducing the production of sebum. Other claimed inhibitors of sebum synthesis are thiolanediol derivatives,[34] polyunsaturated acids such as eicosa-5:8:11:14-tetraynoic acid[35,36] and 5-phenylpentadienoic acid,[37] as well as N,N′-sebacoyl dimethionine.[38] Another more specifically medicated proposal, which tends to counteract androgen-induced sebum secretion, is to apply steroid[39,40] or non-steroid anti-androgens such as flutamide.[41] However, a local action, that is a non-systemic effect, of such compounds is questionable.

It is also worth mentioning antioxidant-based lotions—for example alkyl gallates, *t*-butyl hydroxyanisole—to prevent peroxidative transformation of sebum on hair and scalp.[42,43] Antimicrobial agents can be added to reduce or control enzyme (for example, lipase)—releasing bacteria or yeasts. The after-shampoo lotion represents the mode of application offering the best hope of producing activity and somewhat modifying the refatting process. Alcohol content should not be too high; application of 70 per cent isopropyl alcohol as hair tonic has been shown to displace scalp lipids from the scalp on to the hair.[44]

Hair Loss

It remains essentially true that there is no substance known to science which, when applied externally to the human scalp, will cause—in a statistically significant number of cases—the regrowth of normal hair on a bald head.

Causes of hair loss vary widely and are usually complex (see also Chapter 23). It may be more or less profuse, more or less localized; it may be congenital or acquired. It may present an acute state or be a transient event resulting from an affective disturbance, a trauma, an infectious disease, the effect of some medicines, endocrinological troubles, anaemia or an ionizing radiation treatment. It may be shown to be progressive, chronic, ineluctable, linked or unlinked to seborrhea or to a hormonal imbalance. The tensions of modern life and the evolution of societies with high living standards are the reasons that hair loss appears increasingly among women and with psychological, even social significance, so that the resort to tonics may have little promise of success.

Many points raised by earlier reviewers[45,46] remain to be explained; knowledge is lacking for the understanding of the processes and the biological factors that elicit or aggravate hair loss. In this field the problems of investigation lie in the time needed to collect statistically significant data; the development of trichogram and macrophotographic studies should help to collate observations on the mystery of evolving hair. Chase[47] suggested five methods by which hair growth might be increased, but concluded that only the last offered any real hope of practical realization:

1. Increase of the anagen growth of hair.
2. Production of new follicles or more multiple follicles.
3. Lengthening of the anagen stage or shortening of the telogen stage.
4. Prevention or delay of telogen.
5. Initiation of anagen in follicles in the telogen state.

In normal male-pattern baldness, it is known that there is a shortening of the hair cycle and a progressive thinning until what is called 'club hair' remains. In alopecia areata, in contrast, most of the hair roots are in the anagen state; the hair falls out for a variety of reasons during anagen phase and the scalp is left bald during the period corresponding to catagen and telogen, and then more hair grows during the next anagen period. If the disease has passed off during the bald period these new hairs will be completely healthy. If not, they will be stunted and thin, and will again fall out during anagen. Alopecia areata is considered today to be an autoimmune disease which may be due to a deficiency of T-cells; in support of this is the success of treatment with a solution of the potent sensitizer chlorodinitrobenzene in acetone.[48,49]

Hair loss has long been held to be linked to poor 'irrigation' of the scalp and a number of preparations recommended as 'anti-loss' were based on the use of irritating, rubefacient, 'exciting' substances to provoke an abundant blood supply of growth-promoting nutrients. It now appears that blood circulation does not play a determinant role and it may be preferable to reduce the flow rate in an already well vascularized region. The development of hypoxic conditions has been favourably noted for the reduction of a seborrheic alopecia.[39]

Evidence suggests that the process of loss more probably involves a lack of balance in the growth and division of germinative cells of the hair; this imbalance may be due to a deficiency of some nutrients and to the influence of certain hormonal factors. The role played by a metabolite of testosterone—the 5-α-dihydrotestosterone—is highly suspect, and treatments based on anti-androgen[40,50,51] or on some quinones and phenols[52] have been recommended, which should inhibit the production of the 'guilty' metabolite by the enzyme α-reductase.

Aristotle pointed out that eunuchs do not go bald, and in general this is true. Though there is no clear-cut relationship between the incidence of baldness and the quantity of circulating male hormones in normal subjects,[53] it is significant that the removal of any one of several glands (hypophysis, adrenals, pituitary) from rats will have a profound effect on the quantity or quality of their hair, and it seems most likely that a similarly complex system governs hair growth in humans.

It still seems to be true that the only sure preventative for baldness is the proper choice of parents.

Formulation of Medicated Hair Tonics

A number of completely unrelated compounds have been recommended for use in hair tonics and some of the formulae, both old and new, are given below.

To build up a good tonic, it seems sensible to incorporate the following ingredients:

Nutrients necessary to the biosynthesis of keratin (vitamins, sulphur-containing amino acids).
Regulating factors for the secretion from sebaceous glands.
Antiseptics to control scalp conditions which may interfere with hair growth.
Soothing ingredients (to reduce itching).

The main adversary for a tonic is the time required before an effect other than that attributable to placebo may be detected. Considering the relative slowness of the process of rejecting mature hair, it is most likely that the application of a product is perceived as an increase of loss of hair at first; possible effects may not be identifiable before two months. To encourage persistence in use it may be desirable to add to the active ingredients elements offering preliminary benefits such as compatible additives which impart a healthy look, sheen and volume.

'Stimulating Agents'

Most of the compounds used as stimulants have been drawn from nature, being irritant agents which supposedly have stimulating effects on the growth of hair. The compounds mostly used have been cantharides tinctures (which contain a very irritating lactone, cantharidin); capsicum tinctures (containing a pungent principle, capsaicin); jaborandi tinctures with pilocarpine as the active agent; and miscellaneously, arnica and red quinquina tinctures, Urginea maritima,[54] rhubarb extract,[55] camphor, β-naphthol, pilocarpine, quinine salts, turpentine oil. Esters and salts of nicotinic acid, notably ethyl and tetrahydrofurfuryl esters and pyridoxine nicotinate, have also been claimed as rubefacient and penetrating agents, and brassidic acid and 1-citronellol have been claimed as acanthogens.[56]

Sulphur Derivatives

The use of topically applied amino acids in hair growth has been investigated by Edwards.[57] He applied radio-labelled methionine (S^{35}) to the skin of guinea pigs and found a 1 per cent incorporation of sulphur in the hair, but when the methionine was fed to the guinea pigs, 2·5 times as much sulphur was incorporated. Graul *et al.*[58] applied radio-labelled sulphur, methionine and methionine esters (S^{35}) to the skin of guinea pigs and found that the uptake of sulphur was greatest for elemental sulphur. These results indicated the importance of sulphur and sulphur-containing amino acids in hair growth, and a number of patents have appeared claiming the use of such materials in topically applied preparations. Cysteine itself, although very important in normal growth, is not a very stable compound and attempts have been made to overcome this drawback. Magnesium cysteinate in conjunction with magnesium dehydrocholate and magnesium salts of unsaturated fatty acids have been claimed[59] to stimulate growth of hair, nails and skin. Various derivatives suitable for topical delivery, with protecting groups thought to release free cysteine *in situ*, have been patented; they include 4-thiazolidine carboxylic acid,[60] and S-aralkyl and S-carboxymethyl cysteines,[61] the latter having in addition been shown to be efficient in reducing the greasy condition of the scalp.

Tars

Vegetable tars resulting from the carbonization of specific woods (pine tar oil, cade, cedar and birch oils) have proved to be of interest in cases of seborrhea, dandruff and dry scalp. Antiseptic and astringent, they are complex mixtures of polyphenols, high molecular weight acids or alcohols, waxes, ketones, etc.

Vitamins

The topical action of vitamins is controversial but numerous compositions are based on their use. Group B vitamins (B_1, B_2, B_6, and B_{12}), vitamins A and E, extracts of wheat germ which contain a cocktail of vitamins, the so-called vitamin F composed of essential (unsaturated) fatty acids, biotin, and vitaminic factors such as *p*-amino-benzoic acid, pantothenic acid and the related alcohol panthenol,[62] are widely used. They aim at stimulating cellular synthesis.

The use of amniotic liquid[63] and placenta extracts[64,65] also tends to provide nutrients and stimulating elements that are necessary to keratinization.

Miscellaneous Materials

Sericine (silk glue) is a complex serine-rich protein that has been noted to provide an interesting tonic and anti-seborrheic effect.

Among other miscellaneous substances recommended as hair growth promoters are egg yolk, ginseng,[66] snake serum,[67] ferulic acid,[68] cholic acids,[69] 4-iodo-3,5-dimethyl-2-cyclohexylphenol,[70] steroids,[71] hydroperoxides[72] and an organic silicon compound—chloromethyl silatrane.[73]

Examples of Formulae

Cinchona lotion	(6)
	per cent
Tincture of cinchona	2·0
Birch tar oil	0·3
Glycerol	6·0
Resorcinol	0·3
Jamaica rum	15·0
Ethanol	50·0
Perfume, dye	*q.s.*
Water	*q.s.* to 100

Quinine-based lotion	(7)
	per cent
Quinine sulphate	0·01–0·2
Alcohol	30·0
Water	69·0
Perfume, colour	*q.s.*

Cantharidin lotion	(8)
	per cent
Cantharidin	0·002
Alcohol	30·000
Malic acid	0·500
Water	*q.s.* to 100

Jaborandi hair tonic	(9)
	per cent
Tincture of jaborandi	0·5
Tincture of capsicum	0·5
Isopropyl alcohol	60·0
Salicylic acid	0·1
Panthenol	0·2
Distilled water	38·7

Vitamin lotion[74]	(10)
	per cent
Vitamin E nicotinic acid ester	0·1
Isopropyl myristate	3·0
1-Menthol	1·0
Calcium pantothenate	0·05
Pyridoxine chlorhydrate	0·05
Irgasan DP 300	0·2
Perfume	0·5
Ethanol 95%	80·0
Water	15·1

Vitamin lotion		(11)
		per cent
A	Ethanol	50·0
	Birch cambium sap	7·5
	Boric acid	0·5
	Glycerin diacetate	2·0
	Water	40·0
B	Linoleic acid ethyl ester	10 mg
	Arachidonic acid ethylester	10 mg
	Ascorbyl palmitate	180 mg
	Vitamin D_3	10 000 I.U.
	Calcium pantothenate	10 mg
	p-Amino benzoic acid	5 mg
	Inositol	10 mg

Procedure: Add B to each litre of product A.

Oily hair lotion[75]	(12)
	per cent
Benzylthio-2-ethylamine malate	1·0
Benzalkonium chloride	0·8
Ethanol	45·0
Water	*q.s.* to 100

Anti-loss lotion[30]	(13)
	per cent
Amniotic fluid	10
Chlorohydroxy-aluminium allantoinate	10
Glycerol	1
Pine tar	2
Lemon essence	0·1
Ethanol	*q.s.* to 100

CONDITIONERS

It may be said that conditioners have their origin in the fundamental need of women for attractive and healthy-looking hair. Conditioners should give to hair life, spring, softness, volume, body, sheen, a silky touch, fly-away control and ease of styling.

Today's conditioners are the modern sophisticated and scientific expression of what was looked for earlier in egg yolk and white, marrow and vegetable oils. They may be formulated to be applied either as a pre- or a post-treatment. They are mainly intended for disturbed or weakened hair resulting from chemical treatments such as bleaching or permanent waving, too frequent shampooing, handling abuse (blow-drying, brushing), weathering (severe sun exposure) or internal causes. Such hair types, usually gathered under the description 'dry', tend to look dull and are harsh, porous and brittle, in addition to having increased sensitivity. To make up for a natural deficiency of sebum, a conditioner must smooth, soften, 'texturize', restore the protective sheath, fill in gaps, flatten or tighten cuticle scales, alleviate sensitivity and provide resilience, bounce, control and ease of combing.

Another reason why conditioners have been one of the fastest growing segments of the hair care market[76] is the trend towards a more natural hair style. From this arose the need for a clear, water-white agent to be applied after a shampoo, without the need for rinsing as after conventional cream rinses, while imparting a better mix of the above mentioned characteristics of a healthy hair, and obviously accompanied by a natural appearance.

To a great extent, conditioning is based on the concept of substantivity, that is the sorption of adequate ingredients to modify surface properties and perhaps hair texture. Generally speaking, substantivity is higher when hair is more damaged or more porous. Keratin is an anionic resin, and therefore it shows preferential affinity for cationic substances. Some treatments, such as bleaching, markedly enhance this feature by creating strongly anionic sulphonic acid sites; long-term exposure to sun and the environmental atmosphere have similar effects, though to a lesser extent.

Thus, not surprisingly, the conditioners first marketed in 1945 were based on the use of a cationic surfactant, cetyltrimethylammonium chloride; it is still used. Then followed alkyldimethylbenzylammonium, alkylisoquinolinium and alkylpyridinium halides. The best conditioning properties were shown to be contributed by the longer-chain alkyl compounds.[77] In small amounts these quaternary ammonium derivatives[78] improve manageability, prevent fly-away by neutralizing negative charges in the scalp and, to some extent, promote 'body'.

A wide variety of other cationic surfactants (see Chapter 24) have since been proposed and introduced into conditioners. The emphasis is usually on a lower irritancy (or 'non-irritancy') for hair and skin compared with the earlier surfactants; they also offer greater flexibility in formulation and impart particular properties—enhanced body, softness, volume and spring—to hair. Examples are ethoxylated quaternary ammonium phosphates[79] (for example, Dehyquart SP—Henkel), quaternized fatty acid amino-amides derived from lanolic acids[80–82] (Lanoquats—Emery) and mink oil,[83] and N-acyl colaminoformylmethylpyridinium derivatives (Emcol E607S—Witco):

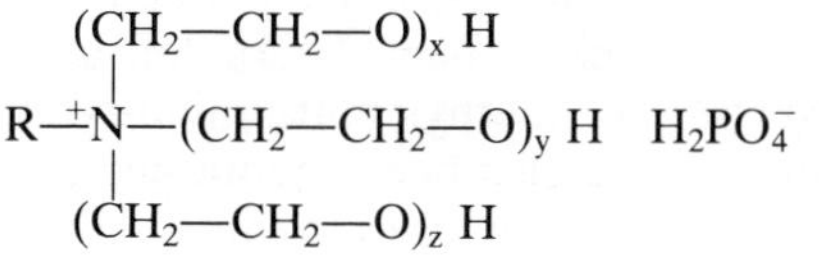

$$R\overset{+}{-}N(CH_2-CH_2-O)_xH\,(CH_2-CH_2-O)_yH\,(CH_2-CH_2-O)_zH \quad H_2PO_4^-$$

Ethoxylated quaternary ammonium phosphate

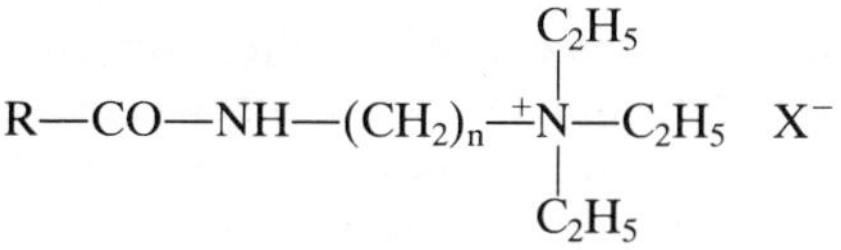

$$R-CO-NH-(CH_2)_n\overset{+}{-}N(C_2H_5)_3 \quad X^-$$

Quaternized fatty acid amino-amide

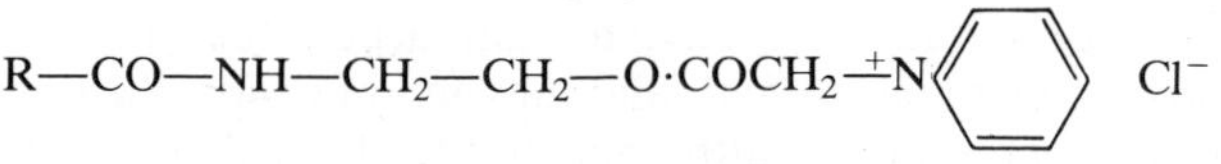

$$R-CO-NH-CH_2-CH_2-O\cdot COCH_2\overset{+}{-}NC_5H_5 \quad Cl^-$$

N-acyl colaminoformylmethylpyridinium derivative

Non-quaternized fatty acid amino amides salts, for example stearoylamidopropyldimethylamine lactate, have been said not to differ substantially from related quaternary ammoniums in respect of conditioning properties and to have the additional advantage of creating no build-up.[84] More recently, similar salts whose anionic part is also a surfactant have been patented.[85]

Some cationic polymers have been proposed (see Chapter 24—conditioning agents in shampoos) to provide clear formulations yielding high conditioning effects; among them are cationic cellulose derivatives (Polymer JR), vinylpyrrolidone–quaternized dimethylaminoethyl methacrylate copolymer (Gafquats),[86] 'polyazonia' compounds,[87–90] epichlorhydrin cross-linked adipic acid–diethylenetriamine polymers,[91] polyethyleneimines, piperazine copolymers[92] and cationic silicones.[93] The high substantivity of these polymeric materials, however, leads to some problems—such as excessive build-up—and they need careful control. Sodium chloride may help to reduce the substantivity.[94]

Another class of compounds long used in conditioners to protect, enrich or repair hair fibres is that of partially or totally hydrolysed proteins, particularly hydrolysates of collagen, ichthyocolla, keratin (horn, horsehair, hoof, hair), and milk casein. The sorption of enzyme-hydrolysed collagen has been widely studied;[95,96] virgin hair takes up only low levels of protein, and maximally at pH

9–11; waved hair absorbs very high levels, which can penetrate deeply into the cortex. Generally speaking, sorption is higher and is obtained at lower pH when the hair is more damaged (pH 6 for strongly bleached hair). Sorption is controlled by ionic equilibrium phenomena and therefore is highly sensitive to pH modifications;[97] it seems to increase rapidly with concentration up to a value of about 5 per cent.[98] Highest substantivity is obtained with polypeptidic fractions of average molecular weight 1000.[99]

Evidence of the restorative and healing effects on the hair shaft may be found in the 95 per cent improvement in split ends obtained with a leave-on protein conditioner;[100] similarly a cream rinse formulation containing 5 per cent hydrolysed animal protein (Lexein X-250—Inolex Group) has been shown to repair 50 per cent of split ends versus 25 per cent for the same rinse without protein.[100]

New hydrolysates are now available in which free amino groups have been quaternized[102,103] to increase protein substantivity. Combined elastin and collagen hydrolysates have been claimed to impart volume and softness to hair.[104] Oleic acid–protein condensate and abietic acid homologue have been said respectively to improve wet combing and to thicken hair.[99] Condensation products of proteins with amines—such as diethylenetriamine—and epoxide,[105,106] and protamines from fish milt[107] have also been mentioned.

Oily materials are the third type of ingredient classically used to improve hair condition. Since antiquity they have been used to supply lubrication and lustre. They help to reduce fibre friction and the abrasive effects of handling, and they improve the condition of hair either by levelling out the scaly surface or by coating it with a material of high refractive index. In this respect, silicone oils offer interesting properties. Lanolin derivatives are still highly praised and widely used for their emollient qualities; hydroxylated and acetylated lanolins, and lanolin esters, add lustre and a pleasant feel, and they tend to reproduce on dry hair the beneficial effect of sebum on healthy hair. Other fatty compounds used with this aim include fatty alcohols, natural waxes (beeswax, spermaceti), ethoxylated alcohols, fatty acids and waxes, fatty acid esters, partially sulphated fatty alcohols such as lanette wax, polyethyleneglycols, lecithins, and some oils (almond, avocado pear, mink, castor, wheat germ). Fatty acid lactylates have been suggested as efficient restoring agents for damaged hair.[108]

A new approach to the lubrication of hair has recently been proposed: it is based on the use of perfluorocarbon resins such as tetrafluoroethylene and hexafluoropropylene polymers.[109]

Products for mending damaged hair have been more specifically dealt with in Chapter 28, but it is relevant to mention here the association of quaternized vinylpyrrolidone, cationic surfactant, hydrolysed collagen and a calcium salt to restore strength and elasticity to bleached hair.[110]

Examples of Formulae

Conditioner to be applied without rinsing		(14)
		per cent
Gafquat 734		0·5
Cetyl trimethyl ammonium chloride		0·3
Ethanol		10·0
Water	*q.s.* to	100·0

Conditioner to be applied without rinsing		(15)
		per cent
PVP–VA S630		1·0
Cetyldimethyl(hydroxyethyl)ammonium chloride		0·5
Silicone oil		0·3
Ethanol		25·0
Water	*q.s.* to	100·0

Old-fashioned cream for care and restoration of dry or damaged hair[111]	(16)
	per cent
Beef marrow	35
Almond oil	40
Mink oil	15
Deodorized castor oil	10

Thick creamy hair conditioner[112]		(17)
		per cent
A	Glycerol stearate	5·0
	Cetyl alcohol	3·0
	Lanolin quaternary (50% active)	5·0
	Synthetic beeswax	2·0
	*PEG 75—Lanolin and hydrolysed animal protein	3·5
	Propylene glycol	2·5
	Sesame oil	1·5
	Stearic acid	1·5
	Lecithin	1·0
	Hydroxypropyl cellulose (2% aqueous solution)	7·5
	Methyl paraben	0·1
	Propyl paraben	0·2
B	Demineralized water	67·2

*Protolate WS (Malmstrom).

Procedure: Heat A to 75°C and B to 60°C. Add A to B with constant agitation until homogeneous. Cool to room temperature with gentle mixing.

Hair conditioning lotion[112]		(18)
		per cent
A	Propylene glycol	5·0
	Glyceryl stearate	3·0
	Sorbitan stearate	1·5
	Polysorbate 20	2·5
	Mineral oil, 70 visc.	3·5
	Cetyl alcohol	3·0
	Stearic acid	1·5
B	Lanolin quaternary	5·0
	PEG 75 lanolin oil hydrolysed animal protein	5·0
	Propyl paraben	0·1
	Methyl paraben	0·2
C	Demineralized water	69·7

Procedure: Heat A and B separately to 75°C. Add B to A with stirring then add C (preheated to 60°C). Continue mixing while cooling to room temperature.

Conditioner for fine, thin, limp hair[113]	(19) *per cent*
Water phase	
*PPG 20 methylglucose ether	2·0
Water	60·0
Hydroxyethylacetamide	15·0
Hydrolysed animal protein	5·0
Oil phase	
Glyceryl stearate	3·0
PEG 100 stearate	5·0
Triton X 400	5·0
Cetyl alcohol	2·0
Stearyl alcohol	1·0
Acetylated lanolin alcohol	2·0
Perfume, preservative, colour	*q.s.*

*Glucam-P20 (Amerchol).

Rapid-acting conditioner to impart body, elasticity and lustre[113]	(20) *per cent*
Water phase	
Water	89·8
Cetyltrimethylammonium chloride	1·5
Alumina	0·5
Oil phase	
Petrolatum	1·5
Glyceryl stearate	0·2
Acetylated lanolin alcohol	2·0
Lanolin alcohol	2·0
Stearyl alcohol	2·5

Flowing emulsion conditioner (manageability, appearance)[113]	(21) *per cent*
Water	54·7
Propylene glycol	7·5
Alcohol	6·0
Triton X 400	4·8
Stearic diethanolamide	5·0
Lauric diethanolamide	4·0
Hydrolysed animal protein	2·0
PPG 20 methylglucose ether	1·5
Polyethoxylated (16) lanolin alcohol	2·0
Sorbitan sesquioleate	5·0
Polyethoxylated (4) dodecyl alcohol	5·0
Sorbitan stearate	1·5
Polysorbate 40	1·0
Perfume, dye, preservatives	*q.s.*

Protein conditioner[84]

	(22) per cent
Stearylamidopropyldimethylamine	1·5
Cetyl alcohol	2·0
Stearyl stearate	1·0
Lactic acid (88%)	0·7
Sodium chloride	0·5
Hydrolysed animal protein	1·5
Methyl paraben	0·15
Propyl paraben	0·05
2-Bromo-2-nitro-1,3-propanediol	0·05
Perfume, dye	*q.s.*
Water	*q.s.* to 100

Procedure: Dissolve sodium chloride in water, add remaining components except dye and perfume and heat to 70°–75°C. Stir until homogeneous and cool. Add dye and perfume at 45°C and cool. Fill batch at 30°C.

Balsam conditioner[84]

	(23) per cent
Stearylamidopropyldimethylamine	1·60
Cetyl alcohol	1·80
Phosphoric acid (85%)	0·90
Sodium chloride	0·30
Preservatives	*q.s.*
Perfume CS 18479 (Albert Verley)	*q.s.*
Water	*q.s.* to 100

Lotion hair conditioner, balsam type[83]

	(24) per cent
Cetyl lactate	2·0
Isopropyl linoleate	2·0
Glyceryl stearate	4·0
PEG 40 stearate	1·0
*Cetyl stearyl alcohol and its polyethoxylated (20) derivative	2·0
Cetyl alcohol	1·0
Deionized water	83·8
Ethyl cellulose	0·3
Quaternium 22†	1·0
Quaternium 26†	2·5
Lactic acid	0·3
BTC 2125 M†	0·1
Colour, perfume oil (V-2374/2)	*q.s.*

*Promulgen D (Robinson-Wagner).
†See also under 'Rinses' for the use of quaternary compounds.

Pretreatment of hair before anionic shampoo[114]	(25) *per cent*
Stearyl dimethylbenzylammonium or N-lauroylcolamino formylmethyl pyridinium chloride	2·5
N-ethanol acetamide	15·0
Polyethyleneimine (60 000 HW) (40% active)	1·5
Nonionic emulsifier	2·0
Formic acid (90% active)	1·4
Hydroxyethyl cellulose	0·37
Methylparaben	0·1
Perfume	*q.s.*
Water	*q.s.* to 100

An original mode of application has been suggested according to which conditioning polymer is delivered from a moulded comb, composed of the watcr-soluble polymer, for example, cationic Polymer JR, and a water-insoluble carrier polymer, for example, high-density polyethylene.[115]

Evaluation of Conditioning

The objective evaluation of hair conditioning properties is difficult. The relation of sensory perception of hair condition factors to quantifiable physical properties remains a problem that requires a great deal of inventiveness to develop appropriate experimental techniques. A good review of this issue has been published by Breuer *et al.*[116] A powerful tool now available to provide precise data on hair condition in relation to treatment is the scanning electron microscope (SEM).

Hair Thickeners

Hair thickeners are a variant of conditioners designed to give a temporary appearance of thicker and more full-bodied hair.[117] These preparations are usually oil-in-water emulsions which combine synthetic or natural polymers used in hair sprays or setting lotions and various ingredients described above, together with bulking agents (Veegum, Bentonite) and thickening agents (Carbopol resins). Cationic polymers are more particularly used (such as Merquats, Gafquats and Polymer JR) to deposit a substantive and stiffening film coating which thickens the hair and imparts body.

The formula in example 26 has been proposed.

		(26) *per cent*
A	Deionized water	44·00
	Methyl paraben	0·20
	Carbopol 940	1·00
	Triethanolamine	2·00

B	Deionized water	25·20
	Merquat 550	5·00
	Ucon HB 660	8·00
	Propyl paraben	0·10
	PVP K90	1·50
	Titanium dioxide	0·50
	Carbowax 6000	7·00
	Polyethoxylated (5) lanolin alcohol	3·00
C	Dowicil 200 (10%)	2·00
D	Perfume	0·50

Procedure: Sprinkle Carbopol into rapidly agitated water. Heat to 70°C and add triethanolamine. In a separate vessel, combine all ingredients of B and heat to 70°C. Add B to A and vacuum mix until 25°C. Add C and D.

RINSES

If the development of rinses is a comparatively recent phenomenon, the idea and the need were evident long ago. The forebears of rinses are nothing more than the vinegar and lemon juice used by women to remove the 'lime soap scum' left on the hair by the rough soaps used in the early days. In addition to the dissociation of calcium salt, acidic rinsing brings the pH of hair nearer to the isoelectric point and helps to maintain its integrity; hydroxyacids have been widely used in this respect. However, modern rinses came to life with the use of a cationic surfactant, stearyldimethylbenzylammonium chloride, combined with fatty alcohols; the best known compound is Triton X-400 (Rohm and Haas), an aqueous solution containing 20 per cent of stearyldimethylbenzylammonium chloride and 5 per cent stearyl alcohol. This combination improves hair manageability and wet and dry combability, and gives a smooth feel to hair. The rinse was formulated as a cream not for technical reasons but with a view to marketing, that is, because of the need to cope with emollience and give the impression that the hair was 'creamed'. The first formulations generally consisted of 3 per cent stearyldimethylbenzylammonium chloride diluted about ten times when applied to the hair.

	(27) *per cent*
Triton X-400	12·5
OPE-1 (octylphenoxyethanol)	1·0
Perfume, colour	*q.s.*
Water	*q.s.* to 100

Procedure: Mix thoroughly the OPE-1 and warmed X-400. Heat the water to 70°C and add to the paste with thorough stirring. Cool to 50°C, add perfume and colour. Viscosity can be adjusted by 0·1–0·2 per cent sodium chloride; pH can be adjusted as desired by citric acid.

Good creamy emulsions are obtained with various nonionic emulsifiers such as glycerol, glycol or diethyleneglycol stearates, polyethoxylated fatty alcohols, polyethyoxylated sorbitan esters, methyl glucoside esters and ethoxylates. In addition to fatty alcohols, the most popular additives to promote feel and lustre are lanolin derivatives and silicone oils, some of which are interesting because they are more soluble in cold water than in hot water and so resist rinsing; hydroxyalkylcelluloses and perfluorinated compounds[26,27] have also been suggested.

Other basic cationics for rinses are N-stearyl colaminoformylmethylpyridinium (Emcol E607S—Witco), ethoxylated quaternary ammonium (Dehyquart—Henkel) and quaternized fatty amidoalkyldialkylamides. The last-named, as non-quaternized tertiary amines, offer the advantage of possible neutralization into salts by the choice of the most suitable mineral or organic acid.

Rinse balsam (*to disentangle, and give smoothness and sheen*)	(28) *per cent*
Stearyldimethylbenzylammonium chloride	1·5
Stearyl alcohol	0·75
Cetyl alcohol	0·75
Polawax GP200 (Croda)	3·0
Oleyl alcohol	1·0
Preservatives, perfume	*q.s.*
Citric acid	*q.s.* to pH 4–5
Deionized water	*q.s.* to 100

Procedure: Heat the first five ingredients to 80°–85°C, then add water with stirring. After cooling to 45°C, add perfume and preservatives under thorough stirring. When the temperature reaches 30°C, add citric acid to adjust to pH. A fluid cream of 500 cS viscosity is obtained, to which protein hydrolysate may be added at a concentration of 0·5–1·0 per cent.

Pearly cream rinse[84]	(29) *per cent*
Stearamido propyldimethylamine	1·5
Cetyl alcohol	1·0
Sodium chloride	0·5
Lactic acid 88%	0·5
Water	to 100·0
Dye, perfume, preservative	*q.s.*

Procedure: Add components, except dye and perfume, to water and heat to 65°–70°C. Blend until homogeneous. Add dye and perfume at 45°C and fill at 35°C or less.

Pearlescent rinse[118]	(30) *per cent*
Stearyldimethylamine oxide (25% active)	7·5
HCl 36%, 1:1 dilution in water	0·23
Colour, perfume, preservative	*q.s.*
Water	92·27

Procedure: Heat together amine oxide and water and mix thoroughly. Add HCl in small increments until pH 5·5. Mix for 15 min, then cool gradually with moderate agitation.

Pearlescent repairing rinse (with cationic polymer)[119]	(31)
	per cent
Stearyldimethylbenzylammonium chloride (25% active)	5·0
Cetyl alcohol	0·3
Glyceryl monostearate	0·5
Cationic cellulose derivative (Quaternium 19)	1·0
Preservative	0·1
Water	93·1

Hair rinse–conditioner[120]		(32)
		per cent
A	Arlacel 165	4·0
	Emcol E 607S	2·5
	Stearyl alcohol	2·0
	PPG 30 lanolin ether	2·5
	2-Ethylhexyloxystearate	0·5
	Propyl paraben	0·1
B	Demineralized water	87·75
	Methyl paraben	0·15
	Glucose glutamate	0·5
C	Perfume, colour	*q.s.*

Procedure: Heat separately phases A and B to 75°–80°C. Slowly add A to B at 75°–80° using high-speed mechanical agitation for approximately 15 minutes. Begin to cool and continue agitation. Add perfume and colour at 40°–45°C. Continue mixing until temperature reaches 35°–40°C and then fill containers.

Foaming cream rinse[121]	(33)
	per cent
Ammonium lauryl sulphate	15·00
Cocoamido betaine	10·00
Lauric diethanolamide	2·00
*PPG 5 Ceteth-10-phosphate	2·00
Stearyl alcohol, polyethoxylated (10) and (20) stearyl alcohol	2·50
Sucrose cocoate	12·00
Stearyldimethylammonium hydrolysed animal protein	3·00
Magnesium silicate	0·45
Fragrance, preservatives	*q.s.*
Water	53·05

*CTFA name; Complex mixture of phosphoric esters of polyoxypropylene (5) polyoxyethylene (10) ether of cetyl alcohol.

Rinse–conditioner[83]		(34)
		per cent
A	Hydroxyethylcellulose (Cellosize QP 30.000)	0·3
	Water, deionized	92·5
B	*Quaternium 26	2·5
	Ethanol (anhydrous)	2·5
	Standapol OLP	1·0
	Perfume	*q.s.*

		(34)
		per cent
C	†Quaternium 22	1·0
	Citric acid (30% aq)	0·1
	BTC 2125 M	0·1

*CTFA name (Ceraphyl 65—Van Dyk).
†CTFA name (Ceraphyl 60—Van Dyk).

A great improvement in the field of rinses was the incorporation of quaternary ammonium compounds with two fatty chains, as exemplified by distearyl dimethylammonium chloride. These compounds, such as commercial products known under the name Arquad-2C, 2T, 2HT and 2S, are very efficient against tangles and rather less irritant.

Cream rinse with di fatty-chained 'quat'	(35)
	per cent
Distearyl dimethylammonium chloride	3·0
Cetyl alcohol	1·0
Partly acetylated ethoxylated lanolin alcohols	0·1
PEG 600 distearate	1·0
Water	*q.s.* to 100·0
Perfume, colour, preservative	*q.s.*

Cream rinse (to be left on the hair)[27]		(36)
		per cent
	Water	83·20
A	Anti-foam AF	0·05
B	Arquad 2HT (10% active)	15·50
C	Arquad S50	0·10
D	Perfluorinated hydrophobic–lipophobic compound	1·00
	Perfume	0·15

Procedure: Disperse A in the water with stirring and then add B. Mix the perfume with C and add to the previous mixture. Add D slowly with stirring over a period of about 20 minutes.

The development of conditioners led to the evolution of rinses in a water-white form and to the promotion of a clear hair rinse–conditioner; it was argued that an oil-free composition should lead to a non-oily feeling on the hair. For this purpose clear soluble 'quats' are used, such as cetyltrimethylammonium and benzalkonium chlorides; an oleyldimethylbenzylammonium salt was designed with this specific aim.[122]

Interesting complements to these formulations are amine oxides, giving the benefit of cationic properties at low pH. They are said to contribute to manageability, to reduce fly-away and to impart a soft feel to hair. The search for conditioning properties explains the introduction into rinses of a wide variety of other ingredients such as proteins (whose concentration should be comparatively high to comply with 'instant' action as required in the rinse concept), cationic polymers such as Gafquats, Polymer JR (bearing in mind that body and lubrication are often antagonistic characteristics), sterols, ethoxylated cholesterols, lipoaminoacids,[123,124] polyethyleneimines, silicones, Ucon fluids and anti-dandruff compounds.

Clear rinse[122]	(37) *per cent*
Oleyl dimethylbenzylammonium chloride	4·0
Hydroxyethyl cellulose (3% aqueous soln)	40·0
Ethoxylated (10) lanolin acetate	0·5
Distilled water	*q.s.* to 100

Clear protein rinse	(38) *per cent*
Stearyldimethylbenzylammonium chloride	5·0
Hydrolysed animal protein	5·0
Ethoxylated cetyl alcohol	0·5
Ethanol 90%	5·0
Water	*q.s.* to 100

Clear rinse[84]		(39) *per cent*
A	Castor oil amidopropyldimethylamine	1·0
B	Cocodiethanolamide	0·3
C	Hydroxyethyl cellulose	0·8
	Citric acid	0·3
	Perfume	0·1
	Deionized water	*q.s.* to 100
	Preservative, dye	*q.s.*

Procedure: Disperse C in water, heat to 50°C and stir for 20 min. Add A, B, citric acid, and cool. Adjust to pH 5·0–5·5. Preservative, dye and perfume can be added as desired.

Clear rinse[112]	(40) *per cent*
*Quaternium 33	1·5
Hydroxypropylethyl cellulose	1·0
Polyoxyethylene 20 sorbitan monolaurate (Polysorbate 20)	1·0
Methyl paraben	0·2
Demineralized water	96·3

*CTFA name (Lanoquat DES—Malmstrom)

Procedure: Dissolve methyl paraben in water with heat. Disperse the cellulose gum and mix until clear. Then add the quaternary and Polysorbate 20.

Clear conditioner, to be rinsed (gives volume, sheen, spring, untangling, smoothness)	(41) *per cent*
Hydroxyethyl cellulose (Cellosize QP 4400 H)	1·0
Oleyldimethylbenzylammonium chloride	2·5
Gafquat 755	1·0
Crotein Q	0·6
Parabens, water-soluble perfume	*q.s.*
Water	*q.s.* to 100·0

Procedure: Dissolve Cellosize in water and warm. Then add the other ingredients and adjust the pH to 5·0.

REFERENCES

1. Leyden, J. J. and Kligman, A. M., *Cosmet. Toiletries*, 1979, **94**(3), 23.
2. McGinley, K. J., Leyden, J. J., Marples, R. R., Path, M. R. C. and Kligman, A. M., *J. invest. Dermatol.*, 1975, **64,** 401.
3. German Patent 1 617 836, Schwarzkopf, 25 January 1973.
4. *Parfum. Kosmet.*, 1976, **57**(7), 190.
5. Kligman, A. M., *et al.*, *J. Soc. cosmet. Chem.*, 1974, **25,** 73.
6. Leyden, J. J., McGinley, K. J., Shorter, A. M. and Kligman, A. M., *J. Soc. cosmet. Chem.*, 1975, **26,** 573.
7. French Patent 2 156 768, Olin, 17 October 1972.
8. French Patent 2 183 647, L'Oreal, 2 October 1972.
9. Gerstein, T., *J. Soc. cosmet. Chem.*, 1972, **23,** 99.
10. Nowak, G. A., *Parfum. Kosmet.*, 1975, **56**(2), 30.
11. Japanese Patent 72′ 22598, Yoshitomi Pharmaceutical, 24 June 1972.
12. US Patent 3 733 323, Colgate-Palmolive, 15 May 1973.
13. US Patent 3 862 151, Ciba-Geigy, 21 January 1975; US Patent 3 961 054, Ciba-Geigy, 1 June 1976.
14. Huber, C. and Christophers, E., *Arch. dermatol. Res.*, 1977, **257,** 293.
15. Windhager, K. and Plewig, G., *Arch. dermatol. Res.*, 1977, **259,** 187.
16. Plewig, G. and Kligman, A. M., *J. Soc. cosmet. Chem.*, 1969, **20,** 767.
17. British Patent 1 504 350, Hoffmann-La Roche, 23 September 1975.
18. German Patent 2 634 677, Henkel, 2 August 1976.
19. US Patent 4 013 787. L'Oreal, 22 March 1977.
20. German Patent 2 228 355, Colgate-Palmolive, 21 December 1972.
21. Heilgemeier, G. P., Dorn, M. and Neubert, U., *Arztl. Kosmetologie*, 1978, **8,** 127.
22. Eberhardt, H., *J. Soc. cosmet. Chem.*, 1976, **27,** 235.
23. Gloor, M., Rickotter, J. and Friederich, H. *Fette Seifen Anstrichtmittel.*, 1973, **75**(3), 200.
24. Ludwig, E., *J. Soc. cosmet. Chem.*, 1969, **20,** 293.
25. Maes, D., *et al.*, *Internat. J. cosmet. Sci.*, 1979, **1,** 169.
26. US Patents 3 993 744 and 3 993 745, Alberto-Culver, 23 November 1976.
27. US Patent 4 013 786, Alberto-Culver, 22 March 1977.
28. German Patent 2 537 374, Procter and Gamble, 11 March 1976.
29. US Patent 4 044 121, Minnesota Mining and Manufacturing, 23 August 1977.
30. Dahl. T., *Household pers. Prod. Ind.*, 1978, **15**(10), 43.
31. US Patents by L'Oreal: 3 671 643, 20 June 1972; 3 879 560, 22 April 1975; 3 968 218, 6 July 1976; 3 950 532, 13 April 1976; 3 984 569, 5 October 1976; 4 073 898, 14 February 1978. British Patents by L'Oreal: 1 050 870, 27 January 1965; 1 310 759, 23 July 1970; 1 397 623, 6 August 1974; 1 391 801, 21 April 1972;

1 161 349, 22 December 1966; 1 290 602, 4 September 1970; 1 397 624, 6 August 1974.
32. Laporte, G., *Am. Perfum. Cosmet.*, 1970, **85**(2), 47.
33. Aubin, G., Brod, J. and Manoussos, G., *Am. Perfum. Cosmet.*, 1971, **86**(12), 29.
34. British Patent 1 367 841, L'Oreal, 22 January 1975.
35. Summerly, R., Woodbury, S. and Yardley, H. J., *Br. J. Dermatol.*, 1972, **87,** 608.
36. Burton, J. L. and Shuster, S., *Br. J. Dermatol.*, 1972, **86,** 66.
37. US Patent 3 755 604, Mead Johnson, 28 August 1973; US Patent 3 886 278, Mead Johnson, 27 May 1975.
38. French Patent 2 159 183, Fabre Pierre, 27 July 1973.
39. Marechal, R. E., *Cosmet. Toiletries*, 1979, **94**(4), 85.
40. European Patent 3 863, Scherico, 5 September 1979.
41. Lutsky, B. N., *et al.*, *Cosmet. Toiletries*, 1977, **92**(2), 57.
42. German Patent 2 137 036, L'Oreal, 27 January 1972.
43. US Patent 3 984 535, L'Oreal, 5 October 1976.
44. Gloor, M., Fichtler, C. and Friederich, H. C., *Arch. dermatol. Res.*, 1973, **248,** 79.
45. Chase, H. B., *The Biology of Hair Growth*, ed. Montagna, W. and Ellis. R. A., New York, Academic Press, 1958.
46. Heald, R. C., *Am. Perfum. Cosmet.*, 1964, **79**(9), 23.
47. Chase, H. B. and Montagna, W., *Proc. Soc. exp. Biol. Med.*, 1951, **76,** 35.
48. Daman. L. A., *et al.*, *Arch. Dermatol.*, 1978, **114,** 1036.
49. Happle, R., *et al.*, *Arch. Dermatol.*, 1978, **114,** 1629.
50. US Patent 4 088 760, Richardson-Merrell, 9 May 1978.
51. US Patent 4 161 540, Schering, 17 July 1979.
52. Belgian Patent 865 031, Marechal, R. E., 1977; Japanese Patent 76′ 57838, Yamashita, Noriko, 20 May 1976.
53. Bingham, K. D., *Int. J. cosmet. Sci.*, 1981, **3,** 1.
54. Japanese Patent 76′ 09731, Kaminomoto, 26 January 1976.
55. German Patent 2 715 214, Giannopoulos, Sotirios, 13 October 1977.
56. German Patent 2 312 091, Cosmital, 12 September 1974.
57. Edwards, L. J., *Biochem. J.*, 1954, **57,** 542.
58. Graul, E. H., Hundeshagen, H. and Steiner, B., *Atompraxis*, 1959, **5,** 265.
59. German Patent 1 131 847, Marbert Pharm. and Kosmet. Spec. Prap. Roeber and Sendler, 22 September 1958.
60. German Patent 1 149 858, Schwarzkopf, 30 November 1960.
61. British Patent 1 051 870, L'Oreal, 27 January 1965.
62. Rubin, S. H., Magid, L. and Scheiner, J., *Proc. sci. Sect. Toilet Goods Assoc.*, 1959, (32), 6.
63. US Patent 3 733 402, L'Oreal, 15 May 1973.
64. German Patent 2 540 971, Hoechst, 17 March 1977.
65. German Patent 2 602 882, Vibert, F., 5 August 1976.
66. German Patent 2 604 201, Mesh Casa Cosmetica, 4 February 1976.
67. French Patent 2 226 155, Aizawa, H., 18 April 1973.
68. French Patent 1 547 573, Ota, M., 21 October 1968.
69. German Patent 2 758 484, Also Laboratori Br. P. Sorbini et Co., 28 December 1977.
70. German Patent 2 297 611, Aries, R., 13 August 1976.
71. German Patent 2 523 820, Wagner, O., 11 January 1976.
72. French Patent 2 248 825, Menyailo, A. T. *et al.*, 23 October 1976.
73. German Patent 2 615 654, Irkutskij Inst. Org. Chim. Sibir. Otd. Akad., 9 April 1976.
74. Japanese Patent 72′ 47 663, Pola Chemical Ind., 2 September 1971.
75. French Patent 2 160 286, Metzinger, A. M., 3 August 1973.
76. Nowak, F., *Cosmet. Perfum.*, 1973, **88**(2), 31.

77. Finkelstein, P. and Laden. K., *Appl. Polymer Symp.*, 1971, **18,** 673.
78. Garcia, M. L. and Diaz. J., *J. Soc. cosmet. Chem.*, 1976, **27,** 379.
79. Moxey, P., *Soap Perfum. Cosmet.*, 1976, **49**(1), 18.
80. US Patent 4 069 347, Emery Industries, 17 January 1978.
81. Schlossman, M. L., *Seifen Öle Fette Wachse*, 1977, **103**(7), 187.
82. McCarthy, J. P. *et al.*, *J. Soc. cosmet. Chem.*, 1976, **27,** 559.
83. US Patent 4 012 398, Van Dyk, 15 March 1977.
84. Schoenberg, T. G. and Scafidi, A. A., *Cosmet. Toiletries*, 1979, **94**(3) 57.
85. US Patent 4 168 302, Richardson, 18 September 1979.
86. US Patent 4 035 478, American Cyanamid, 12 July 1977.
87. US Patent 4 157 388, Miranol Chemical, 5 June 1979.
88. French Patent 2 389 374, Ciba-Geigy, 2 May 1978.
89. British Patents 1 513 671 and 1 513 672, L'Oreal, 15 May 1975; British Patent 1 546 162, L'Oreal, 20 March 1979.
90. US Patent 4 155 994, Kerwanee Industries, 22 May 1979.
91. German Patent 2 546 638, L'Oreal, 5 June 1975.
92. French Patent 2 162 025, L'Oreal, 15 August 1973.
93. Goldemberg, R. L., *Drug Cosmet. Ind.*, 1981, **129**(1), 20.
94. Chalmers, L., *Soap Perfum. Cosmet.*, 1979, **52**(3), 116.
95. Karjala, S., *et al.*, *J. Soc. cosmet. Chem.*, 1966, **17,** 513.
96. Karjala, S., *et al.*, *Proc. sci. Sect. Toilet Goods Assoc.*, 1966, (45), 6.
97. Bonadeo, I. and Variati, G. L., *Cosmet. Toiletries*, 1977, **92**(8), 45.
98. Karjala, S. A., Johnsen, V. L. and Chiostri, R. F., *Am. Perfum. Cosmet.*, 1967, **82**(10), 53.
99. Johnsen, V. L., *Cosmet. Toiletries*, 1977, **92**(2), 29.
100. Di Bianca, S. P., *J. Soc. cosmet. Chem.*, 1973, **24,** 609.
101. Cannell, D. W., *Cosmet. Toiletries*, 1979, **94**(3), 29.
102. Benzyltrimethyl or Stearyltrimethyl Ammonium Hydrolysed Animal Protein, Croda.
103. Japanese Patent 79′ 08688, Lion Oil and Fat, 21 June 1977.
104. German Patent 2 804 024, Freundenberg, C., 31 January 1978.
105. German Patent 2 151 739, Henkel, 26 April 1973.
106. German Patent 2 151 740, Therachemie, 26 April 1973.
107. US Patent 3 917 816, General Mills, 4 November 1975.
108. Murphy, L. J., *Cosmet. Toiletries*, 1979, **94**(3), 43; Murphy, L. J., *Drug cosmet. Ind.*, 1978, **122**(5), 35; US Patent 3 728 447, Patterson, C. J., 17 April 1973.
109. US Patent 4 062 939, Widner College, 13 December 1977.
110. German Patent 2 324 797, L'Oreal, 29 November 1973.
111. French Patent 2 167 407, Waltham, R. C. M., 28 September 1973.
112. McCarthy, J. P. and Laryea, J. M., *Cosmet. Toiletries*, 1979, **94**(4), 90.
113. Fleischner, A. M. and Seldner, A., *Cosmet. Toiletries*, 1979, **94**(3), 69.
114. US Patent 4 061 150, Alberto-Culver, 6 December 1976; US Patent 3 980 091, Alberto-Culver, 14 September 1976.
115. Faucher, J. A. and Rosen, M. R., *Cosmet. Toiletries*, 1977, **92**(8), 35.
116. Breuer, M. M., Gikas, G. X. and Smith, I. T., *Cosmet. Toiletries*, 1979, **94**(4), 29.
117. Garlen, D., *Cosmet. Toiletries*, 1979, **94**(4), 66.
118. US Patent 4 007 261, Millmaster Onyx, 8 February 1977.
119. *Polymer JR for Hair Care*, Publication F-46004, Union Carbide, 1976.
120. Calogero, A. V., *Cosmet. Toiletries*, 1979, **94**(4), 77.
121. *The Croda Cosmetic and Pharmaceutical Formulary*, New York, Croda Inc., 1976, p. 51.
122. Smith, L. R. and Weinstein, M., *Soap Cosmet. chem. Spec.*, 1977, **53**(4), 50.
123. German Patent 2 731 059, S. C. Johnson, 19 January 1978.
124. Netherlands Patent 76 04794, American Cyanamid, 23 November 1976.

Chapter Twenty-seven

Hair Colorants

Introduction

The colouring of hair is one of the most important acts of adornment among those made by men and women since the origins of man. The quest for a change of external appearance has found expression in various ways in all civilizations. This desire, far from being a secondary and futile activity, as it is sometimes considered today, can be linked to a whole series of fundamental attitudes in the life of the individual and can even be linked to the sexual instinct of the species and thus reflect its inevitable presence.

Leaving on one side the most ancient attempts and the empirical formulae used, the matter has now entered the phase of a major industrial development, in that since World War II there has been a great advance in the discovery and utilization of a series of new synthetic colorants for hair. Indeed, it is estimated that today 30–40 per cent of women in the industrialized countries are users of colouring products, either in the home or at the hairdresser.

The usual reasons for colouring the hair are the following: to change the natural colour, to colour the white hairs which begin to appear with age, or to change the colour of the hair temporarily on a particular occasion. To satisfy all these purposes it is necessary to use a range of colorants with varying composition and behaviour which will be described later. First, in order to provide a more complete picture, the various hair colouring systems, the characteristics of an ideal hair colorant and the various practical hair colouring processes are described.

HAIR COLOURING SYSTEMS

Modern systems of hair colouring may be divided into three categories in terms of the duration of the presence of the colour on the hair after the operation has been carried out.

Temporary Colouring

These are fugitive colours which can be removed at the first shampooing. One finds in this category the commercial products commonly designated 'colour rinses'. These products utilize colours of a high molecular weight which are in effect deposited on the surface of the hair without being able to penetrate into the cortex.

Semi-permanent Colouring
These are colours which resist several shampooings (three to six), but whose fastness is poorer than that of permanent colours. The colours used in this case are direct dyes of low molecular weight, having a good affinity for hair keratin. Because of this, they are capable of penetrating the cortex.

Permanent Colouring
As the name indicates, this category provides effectively permanent coloration, resistant to shampooing and other external factors such as brushing, friction, light, etc. This is the process most widely used and represents at least 80 per cent of the total colourings effected.

In this system, uncoloured intermediates are used, which then by a series of chemical reactions produce *in situ* in the hair the desired colour. The process is one of oxidation (almost always effected by hydrogen peroxide), followed by coupling and further oxidations, as will be examined in detail later.

CHARACTERISTICS OF AN IDEAL HAIR COLORANT

The ideal hair colorant should possess the following properties:

Harmlessness

(a) It should be non-injurious to the hair shaft, but should colour the hair without impairing the natural texture and gloss.
(b) It should possess no primary irritant action and be free from sensitizing properties, i.e. it should not be a dermatitic agent.
(c) It should produce no toxic effect when it comes in contact with the skin.

As regards points (a) and (b), thanks to the steady improvement in the purity of raw materials used (dyestuffs, surfactants, polymers, etc.) and thanks to systems of formulation which permit rapid reaction and elimination of the intermediate products possibly responsible for various dermatological consequences, colorants in fact no longer pose a major problem. On the other hand, the question of systemic toxicity has recently been raised, taking account of the small quantities of some ingredients which can penetrate across the skin to the interior of the body.[1–6]

Thus the problem of evaluating the potential degree of risk to health which hair colorants could present can be seen as a part of the much more general problem of the possible danger to health presented by introducing chemical substances into man's environment. The specific problem may be represented in terms of the three following possible effects of chemicals:

Mutagenicity—the risk pertaining to future generations
Carcinogenicity—the lifetime risk to the individual
Teratogenicity—the risk linked to conception

The relation of these risks to that of hair colorants is currently a matter of considerable discussion and a conclusion has not yet been reached. Major work is in hand in various research institutes and groups, as well as in the hair products industry.

Very briefly summarized, one can say that inconsistent results have been obtained across the experimental variables:

(a) of methods used and their significance;
(b) of operating variables (concentrations, methods of application, duration etc.);
(c) of biological system used (bacteria, yeasts, animals);
(d) of the different research workers themselves.

A similar situation is found with the epidemiological studies carried out.

In our opinion, these inconsistencies, imprecisions and contradictions demonstrate that hair colorants do not present a real danger to health. Of course, no chemical product can be shown to be completely non-toxic to man. But whenever a product does possess real toxicity, this has been recognized by science rapidly and without any doubt. This is not the case for hair colorants.

Adequate Physical and Chemical Stability on the Hair
The colour of the dyed hair should be stable to air, sunlight, friction (rubbing, brushing) and sweat.

Compatibility with Other Hair Treatments
It should not change colour, nor bleach out, on the application of toilet preparations such as brilliantines, setting lotions, hair lacquers, hair waving preparations, soaps or shampoos.

Stability in Solution
Colorants should be stable over time in the aqueous solutions and formulated products in the forms in which they are sold and used.

Absence of Selectivity
Because it is always necessary to use a mixture of dyestuffs, the phenomenon of selectivity assumes a very great importance. More precisely we are concerned with hairs to be tinted that are very heterogeneous both individually and in their 'history' (the ends damaged by air and sun, previous perms or other treatments, roots compared with shaft, etc.) and thus trying to avoid:

(a) different coloration on different parts of the same hair;
(b) different fastness over time of different dyestuffs on the hair *vis-à-vis* external agents.

The problem of selectivity plays a most important part in the technology of hair colorants. One can try to avoid it by utilizing dyestuffs belonging to more or less the same chemical class from the point of view of their physicochemical behaviour. It is necessary to check the solutions by preliminary tests on switches in the laboratory, and then, and above all, by tests on hair on the head.

Affinity for Hair Keratin
The physicochemical characteristics of affinity, in conjunction with the penetration of the dyestuff into the hair shaft, can be seen to be very important when

account is taken of the technical limitations controlling the dyeing of hair, such as: temperature not greater than 40°C; short time of contact of hair with the dye, up to say 40 minutes; very weak dye solutions, etc.

The problem is tackled by the use of dyes with small molecular dimensions or by making use of formulation techniques such as use of solvents, swelling agents, alkalis, etc., with the object of improving penetration or of modifying the partition coefficient between water and hair.

Much scientific research has been devoted to studying the mechanism of hair dyeing, in conjunction with the much greater volume of work on wool and on the structure of hair keratin. Thus Evans[7] pointed out that X-ray measurements of the keratin lattice structure indicate that molecules much larger than ethylene glycol penetrate keratin slowly if at all. The interatomic and interchain distances in hair can be altered by causing hair to swell in aqueous solutions at high pH values. Such swelling will promote absorption and diffusion of solute molecules into the hair shaft. But the degree of reversible swelling that may be resorted to without damage to the hair is limited and therefore the molecular size of the dyestuff that can diffuse within is also limited. Alexander[8] has reviewed the information available on the effect of molecular size on the suitability of a molecule for dyeing hair. He concludes that the smaller the dye molecule the less critically does the penetration depend on the molecular size.

Holmes[9] has discussed the theoretical aspects of the diffusion of dye molecules into hair and fibres and concludes that the mechanism can be explained in terms of passage of a molecule across a barrier which contains holes. Thus molecules below a certain critical size can pass through rapidly, whereas larger molecules pass through slowly or not at all. He also notes that size is by no means the only critical factor and that the basicity of the dye is also of vital importance.

It is interesting to note that in the most successful dyeing systems the majority of the ingredients, including ammonia, water, hydroxyl ion, hydrogen peroxide, *p*-phenylenediamine and resorcinol, are all small molecules and can penetrate the hair shaft readily.

Zviak[10] has discussed the effect of solvents on the penetration of molecules into hair. He concludes that the presence of solvents which swell hair results in the penetration of molecules of much larger size. Thus for dry hair the largest molecule that can enter the hair has an apparent diameter of 5 Å (0·5 nm). In the presence of water the diameter is increased to 40 Å (4·0 nm) and in the presence of polar solvents molecules of still larger diameter can enter the hair. This accounts in part for the fact that many colorant systems contain solvents which help the dyes to penetrate the hair shaft.

Considerable information is available on the dyeing of textile fibres and Walker[11] has discussed the theory of hair dyeing in relation to the theories of wool dyeing. However, such information is only of limited value, since the dyeing of human hair differs in a number of important aspects from wool dyeing.[12]

THE PROCESS OF HAIR COLOURING

Hair colorants provide a range of commercial products, capable of colouring the hair in various shades and tints, ranging from very light blonde to black, passing through a range of tones: golden, ash, reddish, mahogany, violets, etc. The number of shades constituting such a range can exceed sixty.

All these products use and are based on strictly limited technical factors, of which a summary follows.

Commercial Products are Mixtures
The dyeing solutions contain mixtures of several single dyestuffs, say 3 to 10. In fact each particular colour is the overall result of the superposition of individual colorations (red, yellow, violet, blue, etc.) supplied by each of the dyestuffs in the commercial mixture.

Concentrations of Dyestuffs
The total quantity of all the dyestuffs used to obtain a shade is small and limited. It can range between 0·01 and 5 per cent by weight of the tinting medium applied to the head. The concentration is a function both of the dyestuffs used and of the procedure involved.

Duration of Colouring Process
The time of contact of the dyeing solution with the scalp and the hair is of the order of 5 to 40 minutes.

Quantity of Solution Applied
The amount of dyeing solution applied to a female head varies between 15 and 100 ml.

Frequency of Application
For temporary colorants, this is of the order of once a week. On the other hand, for more permanent colorants, it is about once a month. In fact this is controlled by the regrowth rate of hair, which is about 1 cm per month.

Treatment after Coloration
Colorants must be conceived and formulated so as to avoid to the maximum extent the staining of the scalp. This is, moreover, assisted by abundant rinsing with water which is obligatory after every application of permanent dye and above all by carrying out one or more shampooings, which will lift the greater part of the dye which has not been absorbed by the hair.

These details of the colouring process have been cited over and above the description of the procedure itself, because they are elements that must be considered in designing toxicological protocols so as to extrapolate the findings to a true appreciation of their significance to the human being. Indeed, colorants have for some time been investigated more and more under conditions in which these procedures are 'forgotten' by the experimenters, and could thus yield results which are erroneous or unrealistic.

Uptake of Dyestuff by the Scalp
Semi-permanent and Temporary Dyes. In order to achieve the desired colour, it is sufficient to place on the head a maximum of about 60 g of dye solution. This quantity is spread over a total surface of 700 to 1000 cm^2 of scalp and about

50 000 cm^2 representing the surface of the hair. (Normally the number of hairs is given as 150 000, with a mean diameter of 60 μm and a typical length of 20 cm.) One can thus easily calculate that 1·2 mg of solution is applied per square centimetre of scalp or hair. Referring back to the dyestuffs themselves and for a concentration of 2 per cent dyestuffs in the solution, that is 24 $\mu g\ cm^{-2}$ with which the scalp or hair comes in contact. In other words, the scalp in its totality comes in contact with 1·2 g of solution or with 24 mg of dyestuff.

Taking as the mean percentage for passage of a dyestuff across the scalp the figure of 0·2 per cent normally quoted,[1,5] one finds an actual transfer of 0·05 $\mu g\ cm^{-2}$ (50 $ng\ cm^{-2}$) of dyestuff or otherwise a total transport of 50 μg for the whole scalp and hence the body.

Permanent Colours (Oxidation). The numerical data remain unaltered, but there is one difference to note: it concerns the first stage of about 10 to 15 minutes, during which half the quantity of product is applied to the roots of the hair, the first 2 cm, close to the scalp, before applying the remainder to the rest of the hair and leaving in contact for a further 15 minutes.

TEMPORARY HAIR DYES

Dyestuffs

The dyestuffs used are generally basic dyes, acid dyes, disperse dyes, pigments or metallized dyes, belonging largely to the chemical classes: azo, anthraquinone, triphenylmethane, phenazinic, xanthenic or benzoquinoneimine. A list of some 145 dyes capable of being used for this purpose has been submitted by the European manufacturers of hair colorants to the European Economic Community authorities with the object of their acceptance in the appropriate directives.

Table 27.1 lists dyestuffs according to category together with the Colour Index Number where known. This list is obviously not exhaustive. Thus one could add a whole series of certified dyestuffs such as may be found in the examples of formulation given by Daniels.[13] Furthermore, many basic dyes may also be used

Table 27.1 Dyestuffs used in Temporary Hair Dyes

Chemical class	Colour	Colour Index Number
ACID DYES		
Azo	Yellow	13 065
	Yellow	19 140
	Red	14 720
	Red	15 620
	Red	16 185
	Red	16 250
	Red	17 200
	Orange	15 575
	Orange	16 230
	Brown	14 805

Table 27.1 (*cont.*)

Chemical class	**Colour**	**Colour Index Number**
Triphenylmethane	Green 22	42 170
	Violet acid 49	42 640
	Acid blue	42 735
Xanthene	Acid violet	45 190
Azine	Nigrosines (violet)	50 420
Anthraquinone	Acid violet 43	60 730
	Blue 62	62 045
BASIC DYES		
Azo	Yellow 57	12 719
	Orange	11 270
	Orange	11 320
	Red 76	12 245
	Brown 16	12 250
Triphenylmethane	Basic blue 5	42 140
	Violet 14	42 510
	Violet 3	42 555
Azine	Red	50 240
Indoaniline	L'Oreal[14–22]	
Indophenol	L'Oreal[23–25]	
Indamine	L'Oreal[26–32]	
DISPERSE DYES		
Azo	Yellow	11 885
	Orange solvent 45	11 700
	Orange solvent 9	11 005
	Disperse red 17	11 210
Anthraquinone	Orange 11	60 700
	Red 15	60 710
	Violet	60 725
	Blue 3	61 505
	Blue	62 500
	Black acetoquinone	—
	Black celliton	—
METALLIZED DYES		
Azo	Cibalane blue F.B.N.	—
	Solvent yellow 90	—
	Solvent brown 43	—
	etc.	

in the form of their leuco-derivatives, in which form they penetrate the hair better and develop their colour later by aerial oxidation.[33–41]

Types of Commercial Temporary Product and their Formulation

Temporary hair colouring can be achieved by two principal types of product: rinses and coloured setting lotions.

In rinses the dyestuffs are used in the form of simple aqueous or aqueous–alcoholic solutions. In order to increase the substantivity to hair, various assistants (organic acids, special solvents) are added[10] or the hair may be pretreated with cationic compounds.[42–43] It should be noted that such tinting solutions may be sold as ready for use or alternatively be prepared by the user from concentrated products by simple dilution in water.

The second approach, that of coloured setting lotions, consists in applying the dyestuffs to the hair through the medium of solutions containing transparent polymers, and following the technical requirements of hair sprays and setting lotions. Such a medium can be produced by dissolving 3 per cent polyvinylpyrrolidone (K30 grade) in water (or aqueous alcohol for quicker drying). Any innocuous dyestuff may be added to this base.

Alternatively such a product may be packed in an aerosol can. In this case the use of any water in the product is not advisable in view of possible corrosion of the can (see Chapter 40 on aerosols), and care should be taken that no solid matter precipitates which could block the valve.

The main problem in using such polymer sprays is the tendency of the polymer to flake off the hair, carrying the dyestuffs with it into pillows, towels and clothes. Interested readers should experiment with other hair spray materials, such as ethylcellulose, and PVP–VA copolymers. It has been suggested that such problems might be overcome by grafting the dye on to a polymer skeleton.[44–54]

Crayons have been used from time to time for temporary colouring, and are employed like mascara—either rubbed direct on to the wet hair or transferred to the hair with a brush. They can be formulated with soaps and waxes, so as to give a product rather like a lipstick (example 1).

	(1) *per cent*
Stearic acid	15·0
Triethanolamine	7·5
Glyceryl monostearate	4·0
Beeswax	46·0
Paraffin wax	10·0
Microcrystalline wax	10·0
Coconut diethanolamide	7·5
Colour	*q.s.*

SEMI-PERMANENT COLORANTS

Dyestuffs

These dyes occupy an important place in formulation practice, not only because they permit the creation of this important class of hair colorants, but also because their presence is often indispensable in the formulation of the permanent colours, the oxidation dyes. Certain of these dyes are capable of providing hair shades ranging from yellow to orange, shades which, in terms of efficacy, are practically impossible to obtain with oxidation dyes and whose contribution is necessary to provide the complete range of shades such as the copper tones.

The great majority of these dyes belong to the following chemical classes:

Nitrophenylenediamines
Nitroaminophenols
Aminoanthraquinones

A list of some 50 dyes which could be used has been submitted by the European manufacturers to the EEC authorities for approval. In making the selection, consideration has been given to a great number of physiological criteria. It should be noted that it is also possible to utilize a certain number of the dyestuffs already quoted for temporary dyes such as some nitroazo dyes and some basic dyes.

Nitrophenylenediamines
Nitrophenylenediamines can be described by the general formula:

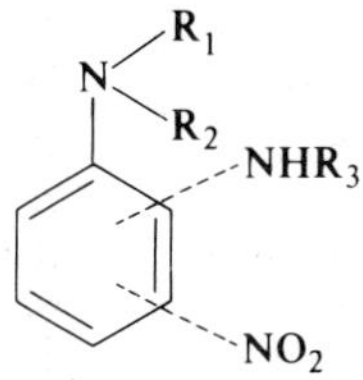

$N(R_1)(R_2)$, NHR_3, NO_2

where R_1, R_2 and R_3 may be the same or different, and represent H or substituted or non-substituted alkyl groups such as —CH_3, —CH_2CH_2OH, —$CH_2CH_2NH_2$, —CH_2COOH, —CH_2CONH_2, etc. According to the position occupied by the —NO_2 and —NHR_3 groups these dyes can be regarded as derivatives of 4-nitro-*o*-phenylenediamine, 2-nitro-*p*-phenylenediamine or 4-nitro-*m*-phenylenediamine.

By successive alkylations (R_1, R_2, R_3), starting from these nitrophenylenediamines which are themselves excellent dyes, one arrives at interesting increases in depth of colour and thus enriches the range.[55–66] Examples are given in Table 27.2. Note also that this class of substitute nitranilines can be enriched:

(a) by using other substituted derivatives in the aromatic ring, by weak electron donors such as —CH_3 or —OCH_3;[67–68]
(b) by using some derivatives of nitrodiphenylamine[69–70] such as:
4[*bis*-(2-hydroxyethyl)]amino-3-nitro-4′-methylamino-diphenylamine (blue)
disperse yellow 1 (Colour Index No. 10385)
2-nitro-4-[*bis*-(2-hydroxyethyl)]amino-diphenylamine
2-nitro-4-methoxy-diphenylamine
2-nitro-4-amino-diphenylamine

Table 27.2 Shades Produced by some Nitrophenylenediamine Compounds

Compound	Shade
nitro-*p*-phenylene diamine	Orange red
4-amino-3-nitro-N-methylaniline	Purple
4-amino-3-nitro-N-(2-hydroxyethyl)aniline	Violet red
4-(2-hydroxyethyl)amino-3-nitro-aniline	Violet red
4-(2-hydroxyethyl)amino-3-nitro-N-(2-hydroxyethyl)aniline	Violet
4-(2-hydroxyethyl)amino-3-nitro-N,N-[*bis*-(2-hydroxyethyl)]aniline	Blue violet
4-methylamino-3-nitro-N,N-[*bis*(2-hydroxyethyl)]aniline	Violet blue
4-methylamino-3-nitro-N-methyl-N-(2-hydroxyethyl) aniline	Blue violet
4-nitro-*o*-phenylene diamine	Orange yellow
2-amino-4-nitro-N-(2-hydroxyethyl)aniline	Orange
2-(2-hydroxyethyl)amino-4-nitro-N-(2-hydroxyethyl)-aniline	Orange
2-amino-4-nitro-N-[tris-(hydroxymethyl)] methyl aniline	Orange
4-nitro-*m*-phenylene diamine	Yellow

Nitroaminophenols

Dyes of this class can be represented by the general formula:

OR_1, NHR_2, $(NO_2)_n$

in which R_1 and R_2 may be the same or different and represent —H or a lower alkyl group, substituted or not, such as $—CH_3$, $—CH_2CH_2OH$, and where n is 1 or 2.

According to the positions of the nitro and amino groups, a series of dyes can be produced, of which the most important are given in Table 27.3. Other derivatives with different substituents have also been synthesized.[71–75]

Aminoanthraquinones

Aminoanthraquinones form a whole range of dyes based on amino- and hydroxy-anthraquinones with various substituents.[76–83] Examples of interesting dyes in this range include the following:

1-amino-4-methylamino anthraquinone (disperse violet 4/solvent violet 12/Colour Index No.61105)

Table 27.3 Shades Produced by some Nitroaminophenol Compounds

Compound	Shade
2-amino-4-nitro-phenol	Orange
2-amino-4,6-dinitro-phenol (picramic acid)	Deep orange
2-amino-5-nitro-phenol	Orange yellow
2-(2-hydroxyethyl)amino-5-nitro-phenol methyl ether	Yellow
2-(2-hydroxyethyl)amino-5-nitro-phenol-2-hydroxyethyl ether	Yellow
4-amino-2-nitro-phenol	Salmon pink
4-methylamino-2-nitro-phenol	Rose
4-methylamino-2,6-dinitrophenol (isopicramic acid)	Rose
4-amino-3-nitro-phenol	Deep orange
4-(2-hydroxyethyl)amino-3-nitro-phenol	Red
4-(2-hydroxyethyl)amino-3-nitro-phenol methyl ether	Orange
4-amino-3-nitro-phenol-2-hydroxyethyl ether	Orange

1,4-diamino-5-nitro anthraquinone (disperse violet 8/Colour Index No.62030)
1,4,5,8-tetra amino anthraquinone (disperse blue 1/solvent blue 18/Colour Index No.64500)
1-methylamino-4-(2-hydroxyethyl)amino anthraquinone
1-hydroxy-2,4-diamino anthraquinone

Various Dyes
Numerous other dyes can be used in formulating semi-permanent colorants, in general as auxiliaries serving to modify the shade: for example, *heterocyclic azo-derivatives*,[84–85] also derivatives of *diazamerocyanines*[86–87] and *quaternary derivatives of aminophenoxazinium*.[88–89]

Reactive Dyes. The use of reactive dyes has been a relatively new approach in the field of textile dyeing. These include materials such as dichlorotriazines (for example Procion dyes *ex* ICI), monochlorotriazines (for example Procion dyes *ex* ICI and Cibacron dyes *ex* Ciba) and trichloropyrimidines (Reactone dyes *ex* Geigy, Drimaren dyes *ex* Sandoz). These dyes work by actually reacting with the fibre and thus the dye part of the molecule is firmly held by the fibre.

Broadbent[90] has discussed the possibility of using reactive dyes for the colouring of human hair. Shansky[91] has reviewed the mode of action of such dyes and also considers their application to human hair. However, he concludes that considerable modification is required before they are as acceptable for human hair as the more conventional oxidation dyes. Patents[92–94] have appeared describing the use of reactive dyes, but little practical experience of their use is available.

Other types of reactive dye have been developed recently:

(a) after a pretreatment of the hair with a reducing agent, a reactive dye of the class —C(=S)—S—R is applied;[95]

(b) after pretreatment with a dialdehyde, various amino–aromatic compounds can be applied, such as aminonitrophenols, *p*-amino-diphenylamine, etc.[96–97]
(c) a mixture of dihydroxyacetone and aliphatic or aromatic amines.[98]

The use of reactive dyestuffs for the dyeing of hair is not as revolutionary an idea as it may at first seem. In fact it is now believed that conventional oxidation dyes such as *p*-phenylenediamine develop their final colour by interacting with reactive sites in the hair. Oxidation dyestuffs may thus be regarded as being a particular type of reactive dyestuff.

Metallized Dyes. The preparation of such dyes may be based on the formation of complexes *in situ* in the hair by means of nickel or cobalt ions and various complexing agents.[99–101] In addition, one may use the metallized dyes already offered for wool dyeing, such as the Neolans, Irgalans, Cibalans, etc.[10]

Azo dyes Obtained by Coupling on the Hair. Diazonium salts, together with various coupling agents, may be applied to the hair, as described in a series of patents.[102–105]

Commercial Semi-permanent Products and their Formulation

Semi-permanent colorants can be presented either in the form of foaming lotions (anionic, cationic or nonionic) or as anionic or cationic shampoos. According to the type of medium chosen and the product performance sought, various approaches to formulation have been developed, all having the goal of promoting the penetration of the dyes into the hair. The great number of patents describing various procedures and compositions may be classified as five types:

1. Procedures based on the simultaneous or successive use of thiols and particularly thioglycollic acid.[106–109]

2. Procedures based on the use of various solvents: work at Leeds University and elsewhere has shown the possibility of cold dyeing of textiles by using metallized acid dyes (such as the Cibalan and Neolan ranges) in the presence of a solvent such as butanol, which is small enough to penetrate the fibre and which will set up a partition of dyestuff between water and itself. A patent[110] describes a similar process and its application to hair, where the dyestuffs are selected from almost every class and the recommended solvents cover a number of different alcohols. A number of other patents have appeared on the use of solvent-assisted dyeing systems and these have been described in a *Schimmel Brief*.[111] In addition Alexander[8] and Heald[112] have reviewed the patent literature. The work of Peters and Stevens[113–114] is of particular importance in the development of solvent-assisted dyeing systems and its application to the colouring of hair at low temperatures, and is worthy of special note.

In a US patent[115] the use of boosters for anionic direct dyestuffs has been suggested.

Many other solvents or mixtures have been put forward. Suggestions include

Aryl ethers of the formula $Ar(OCH_2CH_2)_{1\text{-}4}OH$[116]
N-substituted formamides[117]

Phenoxyethanol[118]
Phenoxyethanol and ethylene glycol acetate[119]
N,N-dimethylamides of monocarboxylic acids C_{5-9} and N,N-N′,N′-tetramethylamides of dicarboxylic acids C_{9-19} or linoleic acid dimer[120]
Alkyl glycol ethers[121]
Benzyl alcohol and its lower carboxylic acid esters or cyclohexanol[122]
A mixture of urea and benzyl alcohol[123] or benzyl alcohol and N-alkyl pyrrolidone[124]

3. The selection of a range of shades for a semi-permanent dye product depends largely on the market in question. It is usually sufficient to provide brown shades for light and dark brown hair with adequate intensity to cover up 10 per cent grey hair, and a small range of decorative shades such as auburn, chestnut, copper, burgundy, etc.[125] Greys for levelling the colour of hair which is more than 90 per cent white can also be included. It is essential to note, however, that such products should have a bluish-violet tone, as natural grey hair has a yellowish tint and application of a natural or blue-grey will result in a pale-green colour.

4. A popular medium for such dyes is a colour shampoo, formulated as a cream or liquid cream product containing the solubilized dyestuffs[126–132] (see Chapter 24 on shampoos for suitable formulations). However, it must be remembered that the inclusion of detergents increases the affinity of the dyestuffs for the medium and hence decreases the amount of 'active' dyestuff available to colour the hair. The concentration of dyestuffs should be higher in a colour shampoo than in a plain aqueous solution. Testing of such products can only be carried out on actual hair, preferably on the head. If wigmaker's hair is used, care should be taken not to use white hair that has been bleached with sulphur dioxide, as this hardly behaves like hair at all.

5. Finally, so-called anionic–cationic complexes may be used, that is to say the formation of a complex from acid dyes, for example azo or azinic, and a quaternary surfactant, followed by solution in a, generally, nonionic surfactant.[133–135]

PERMANENT HAIR DYES

Permanent hair dyes are based almost exclusively on the use of oxidation dyes, the so-called *para*-dyes, which are substances that are colourless at the time of their application to the head (the precursors) and are transformed into a coloured material *in situ* on the hair as a consequence of chemical reactions set in motion by the execution of the coloration.

The precursors can be classified into two categories: the compounds called oxidation bases or primary intermediates and those called couplers or modifiers.

The chemical reactions in the formation of dyes are oxidation reactions and couplings or condensations, effected at alkaline pH (essentially due to the presence of ammonia) by the action of an oxidizing agent, almost exclusively hydrogen peroxide or one of its solid derivatives—urea peroxide[136] or melamine peroxide.[137] The choice of hydrogen peroxide is justified not only by its action on the precursors, but also by its ability to promote the simultaneous decoloration of the hair to be tinted.

In fact, the hydrogen peroxide, which is used for this purpose in quantities much greater than are necessary to effect the oxidation of the precursors, is capable of acting on a part of the melanin pigments of the hair, which are the origin of the natural colour, by oxidizing and so solubilizing them—that is to say, decolorizing the hair.

This bleaching, which occurs simultaneously with but independent of the dyeing, results in the hair being rendered lighter and so permits it, in accordance with the aesthetic goal, to be given new shades with the help of the new pigments which this system of dyeing is capable of creating in an almost infinite variety. It can thus be seen why this system of coloration is unique and irreplaceable; this explains its widespread use and its name of colour lightening ('teinture eclaircissante').

To sum up, in order to create colour by this process it is necessary to have three types of reactive chemical:

Base or primary intermediates
Couplers or modifiers
Oxidizing agent, almost always hydrogen peroxide

The chemical nature of the bases and the couplers and the possible mechanism of the formation of colours and pigments are now considered.

Bases

The bases are aromatic compounds, almost exclusively benzene derivatives, substituted by at least two electron-donor groups, such as NH_2 and/or OH, these being *para* or *ortho* to each other; this confers the property of easy oxidation.

The most important compounds of this class are thus *p*-phenylenediamine and *p*-aminophenol, and *o*-phenylenediamine and *o*-amino phenol, to which one could add the *p*- and *o*-dihydroxybenzenes.

Starting from these basic compounds and proceeding by various substitutions or by drawing on other aromatic systems, chemists have been able to enlarge considerably the number of 'bases' usable in oxidation dyes. The field is so vast that only recent references can be given, such as the following:

(a) proceeding from alkylation of the $—NH_2$ and its transformation into $—NR_1R_2$ (where R_1, R_2 are the same or different, and are H or lower alkyls) one such enlargement of the number of bases is obtained.[138–139]
(b) Another increase arises from substitutions in the benzene ring by weak electron donors such as $—OCH_3$, $—CH_3$, $—NHCOCH_3$, etc., which can produce bases having special or different properties.[140–145]
(c) Also other aromatic rings can be used, such as pyridine, pyrimidine, quinoline, indole, pyrazolone, benzimidazole, etc., giving rise to a new series of oxidation bases.[146–160]

In fact the most important oxidation 'bases' are:

p-phenylenediamine
p-toluenediamine(2,5-toluenediamine, sometimes called *p*-toluylenediamine or *p*-tolylenediamine)
p-aminodiphenylamine
p-aminophenol
p-diaminoanisole
o-phenylenediamine
o-aminophenol

Couplers or Modifiers

Couplers or modifiers are aromatic compounds, almost exclusively benzene derivatives, substituted by the same groups (—NH_2 and —OH) as the 'bases', but this time in *meta* position to each other. In this position it should be noted that the couplers do not have the property of easy oxidation by H_2O_2.

The range of couplers can also be enlarged:

(a) by introducing weak electron donors such as —OCH_3, —$NHCOCH_3$, etc., with or without various alkylations of the OH and NH_2 groups by alkyls and hydroxyalkyls.[161–178]
(b) by using heterocyclic rings, such as pyridine, quinoline, indazole, benzimidazole, benzoxazine, pyrazolone.[179–188]

The most usual couplers are among the following:

m-phenylenediamine
2,4-diaminoanisole
resorcinol
m-chlororesorcinol
m-aminophenol
1,5-dihydroxynaphthalene
6-methyl-3-aminophenol
2-methylresorcinol

Formation of Colours in the Hair

A very large amount of work has been devoted to elucidating the mechanism of oxidation dyeing and to the structure of the dyes. But all is not yet clear.

In fact, the number of parameters affecting the overall process is very great. For example, consider the influence of pH on the speed of reaction, the presence of the hair keratin itself which can affect the orientation of reactions, the complexity of the reaction mixtures (it is not unusual to use up to 10 couplers and bases), the possible hydrolyses of intermediate products, etc. All these variables make it practically impossible to specify exactly all the compounds which could potentially be formed. Thus empiricism in formulation still plays a very important part in the technology of oxidation dyeing, and no mean effort is required to arrive in practice at a formula that gives commercially reproducible performance.

Be that as it may, the general picture of formation of colours lies in a series of oxidation and coupling reactions in which one can schematically distinguish the three following stages:

(i) *Formation of Quinoneimines*. This phase consists of the oxidation of the bases under the action of alkaline H_2O_2 with the formation of quinone monoimines from *p*- and *o*-aminophenols, and quinone diimines from *p*- and *o*-phenylenediamines.

These reactions can be typified by the scheme for *p*-phenylenediamine and *p*-aminophenol shown in Figure 27.1a. In the same manner the structures of other quinone immonium cations, derived from other bases, can be represented.

(ii) *Formation of Diphenylamines*. The quinone immonium cations formed in the first stage readily undergo a Michael-type conjugate addition with the pseudo-carbanions of the couplers, to give an N-substituted *p*-phenylenediamine: in other words, a differently substituted diphenylamine.

Nucleophilic compound structures capable of addition at the —NH of the quinone imines by attacking the nitrogen atom not only include the *meta*-structure couplers but also the original *para*-bases not yet oxidized and which then function as couplers for their own imines. As an example, consider the formation of diphenylamines from the reaction of *p*-phenylenediamine with *m*-phenylenediamine in its carbanion form (Figure 27.1b).

In the same way, a whole series of variously substituted diphenylamines can be obtained starting from the quinoneimines of other *para*-bases and reacting with other non-oxidized bases.

(iii) *Formation of Colour*. The previously formed transitory diphenylamines can be seen in their turn as new oxidation bases, in which one of the benzene rings is at least trisubstituted (in positions 1, 2, 4 or 1, 2, 5) by electron donor groups. By virtue of this, they possess the same two reaction potentialities as the original *para*-base from which they derive, namely oxidizability and ability to couple—and this to an enhanced degree.

Thus, either they go to be oxidized and transformed into the appropriate indoamines, indoanilines or indophenols—in fact into a first group of dyes—or they act themselves as couplers and are involved in an attack on the quinoneimines from the original *para*-bases, which continue to be formed in the reaction medium, thus leading to a 'double' phenylamine. These new compounds, easily oxidized in their turn, give rise in their oxidized form to a new group of dyes with three benzene rings.

This process of addition of the initial quinoneimines on to the transitory aromatic forms of compounds which are more and more condensed, followed by further oxidations, can lead to still new dyes with more than three benzene rings. Because these dyes have other reaction capabilities, such as intramolecular cyclization or partial hydrolysis, they can partially change into azines or oxazines, that is to yet more dyestuffs.

All these dyes and pigments, of which all the structures are not yet completely elucidated, make up the third group of dyes formed in the hair.

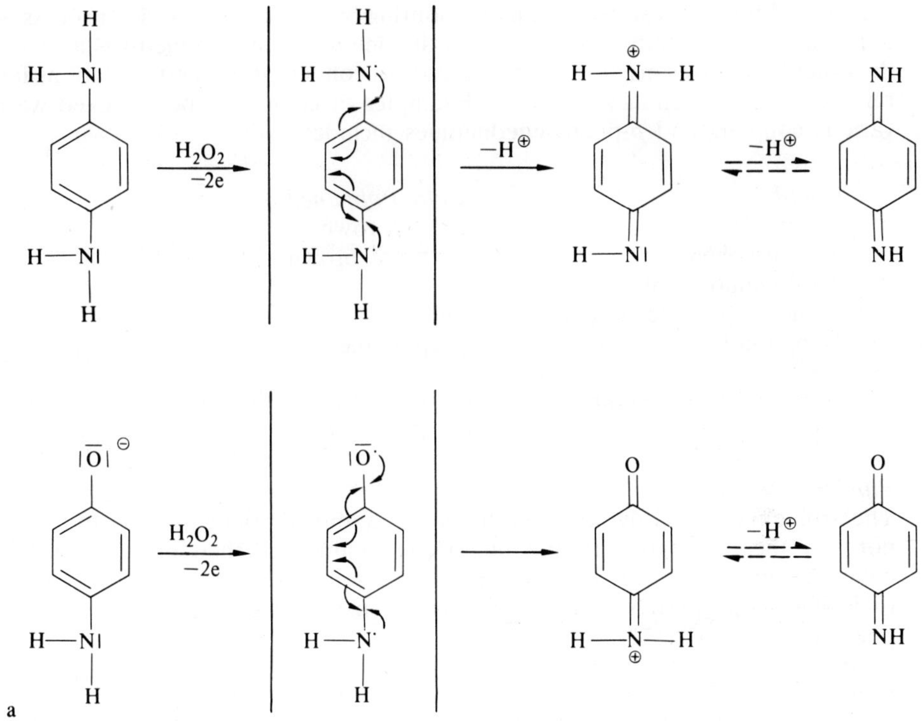

a

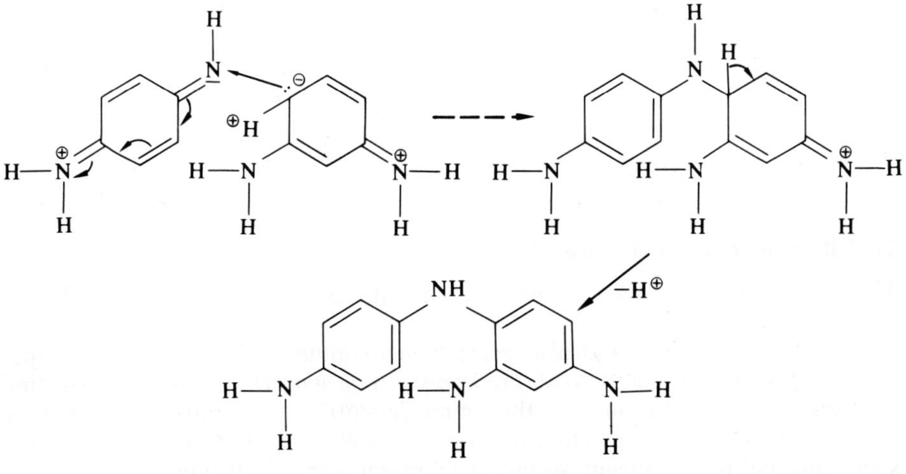

b 4,2′,4′-triaminodiphenylamine

Figure 27.1 *a* Formation of quinoneimines from *p*-phenylene diamine and *p*-aminophenol
b Formation of 4,2′,4′-triaminodiphenylamine

It should thus be restated that hair colouring by the permanent dye process is the result of a competition between, on the one hand, indoamine dyes and, on the other hand, dyes originating in a cascade of condensations and oxidations far remote from the primary reactions. Examples of colours to be achieved with various couplers and *p*-phenylenediamines include:

Coupler	*Colour obtained*
resorcinol	green/brown
m-aminophenol	magenta/brown
2,4-diaminoanisole and *m*-phenylenediamine	blue
1-naphthol	purple-blue

The structures of a number of these dyes have been studied,[189–191] but within the scope of this text it is not possible to discuss this subject in depth.

Bandrowski Bases

The problem of Bandrowski bases should be examined briefly in order to clarify not only their formation from *p*-phenylenediamine and their presence in the colorant mixture, which have given rise to some confusion, but also their toxicological properties which have, by extension, been attributed to *para*-phenylenediamine.[192–193]

At one time these trimers of *p*-phenylenediamine were thought to be a principal intermediate in the formation of the permanent dyes, but it should be made clear that in hair-dyeing conditions, i.e. *in the presence of hair* and in a process of oxidation of the order of 30–45 minutes:

(a) Bandrowski's base is not formed to any appreciable extent during the oxidation of *para* by hydrogen peroxide—the formation of the Bandrowski base is a slow process;
(b) in the presence of the various couplers which accompany *para* in a hair dye mix, and because these couplers are considerably more reactive towards quinoneimine than the oxidized *para* itself, no Bandrowski base is ever formed in quantities detectable by analysis.[194–197]

Toxicity and Dangers of Para Dyes

There are two problem areas, that of skin irritation or sensitization and that of systemic toxicity.

p-Phenylenediamine, and also *p*-toluylenediamine, which are the principal components of an oxidation dye, are known to be sensitizers capable of causing contact dermatitis. Because of this, many countries have introduced laws or regulations requiring users to carry out prophetic patch tests before dyeing, specifying limits to concentrations of ingredients in the formulae, etc.

For example, in the UK the Pharmacy and Poisons Act (1933) stipulated that all hair dyes containing phenylenediamines or tolylenediamines or other alkylated benzene diamines or their salts must be labelled with the words: 'Caution. This preparation may cause serious inflammation of the skin in certain persons and should be used only in accordance with expert advice.' The required labelling was repeated in the Poisons Rules, 1970.

In the USA, Sections 601 (a) and 902 (a) of the Food, Drug and Cosmetic Act require, among other things, that a coal tar hair dye which may be injurious to users shall bear the statement: 'Caution. This product contains ingredients which may cause skin irritation on certain individuals, and a preliminary test according to the accompanying directions should first be made. This product must not be used for dyeing the eyelashes or eyebrows; to do so may cause blindness.'

Section 602 (c) requires that the caution statement appear on the label in a prominent and conspicuous place. This requirement is met if the caution statement appears conspicuously in a colour that contrasts with the background and the remainder of the printed matter. The caution statement should appear on the main panel of the label with the name of the product.

In a notice to 'Manufacturers of Hair Dye Preparations', dated 17 October 1937, the Food and Drug Administration of the US Department of Agriculture, Washington, DC, states: 'It is necessary, as provided in the caution statement, that all such hair dyes bear labelling prescribing adequate preliminary tests. For the benefit and guidance of those interested, there are published herewith directions which, in the light of present knowledge and information, are regarded as acceptable. This information is given merely as a guide and does not mean that other tests may not be acceptable' (see Table 27.4).

Table 27.4 Directions for Making Skin Test (US FDA Recommendations)

1. The hair dye contained in this package must never be used for dyeing the hair unless a preliminary skin test has been made. The skin test must be made each and every time before the hair is to be dyed, regardless of whether or not a skin test has been made at some time previously.
2. The dye used for the preliminary test must be a portion of the article intended to be used for dyeing the hair.
3. The sample of dye to be used for the preliminary skin test should be mixed and prepared in exactly the same manner and according to the directions applicable to the actual use of the hair dye itself.
4. By means of a suitable applicator (clean camel's hair brush, cotton-tipped applicator or other applicator) a streak of dye not less than a quarter of an inch wide and at least one-half inch long is made on the skin and scalp behind one ear. It is important that the streak of dye extend into the hair portions of the scalp as well as that portion of the skin that is hairless.
5. The streak of dye should be permitted to remain for at least 24 hours. The test should be read between 24 and 48 hours after application. Preferably the test area should not be covered with any type of dressing and contact with comb, hats, spectacles or any other object should be avoided.
6. Warning: If redness or burning, or itching, or small blisters, or any type of eruption appears in the general area used for the skin test during the first 24 hours the individual is sensitive to the dye, and under no circumstances should it be used for dyeing the hair. Hair dyes should not be used when there is any disease or eruption present anywhere on the skin or on the scalp.

Even if all these precautions are necessary and desirable, it should just the same be noted that the technology of hair dyes has, with time, made considerable progress, with the result that cases of allergy or dermatitis have become very rare. Thus in 1969, the Committee of Cutaneous Health and Cosmetics of the American Medical Association[198] estimated that 'allergic dermatitis due to PPD in hair dyes would be one reaction in every 50 000 applications'. Dermatologists themselves consider that the level of these incidents has been decreasing over the years.

On the basis of their statistics, the producers estimate that the level of allergic incidents lies at one case of allergy per 1 million units of dye applied.

The improvements in dermatological behaviour of dyes since the beginning of their use can be found in the purity of raw materials used in the formulations, in the improvement in the nature of products formed and, finally, in the associated use of shampoos which remove all remaining substances at the end of the dyeing process.

As to systemic toxicity, a large series of toxicological and epidemiological studies is now in progress. Final conclusions cannot yet be drawn, although a number of indications are beginning to appear.[1–6]

Formulation of Permanent Hair Dyes

Cook,[201] while noting that in practice they are never used alone, has listed the colours obtainable from the main oxidation bases and coupling agents (Table 27.5).

Brown,[199] Cook[200–201] and Zviak[10] have reviewed the main factors which have to be considered in developing suitable formulations, and note that the following are critical:

1. The formulation base: solution, emulsion, gel, shampoo or powder.
2. The selection of the dye components: the oxidation base, the coupling agent or the addition of a direct colorant.
3. The selection of an alkali: ammonia is usually used.
4. Antioxidants: usually a sulphite or ammonium thioglycollate is used to prevent oxidation of the dye before the product is to be used.
5. The pack: required to be attractive and convenient in use.

Considerable work has been carried out on selection of components but the other major approach is to improve the formulation base so as to achieve more efficient and more convenient colouring.

The most convenient medium for colouring the hair is a shampoo. The system consists of a solution of the dye precursors together with an ammonium oleate soap or other surfactant, to which is added at the time of use another solution of stabilized hydrogen peroxide contained in a separate vessel. This mixture is applied direct to the hair and left in contact for 20 to 40 minutes. Afterwards the hair is rinsed with water and then washed again.

By use of various other additives, such as fatty alcohol sulphates, fatty acid dialkanolamides, nonionic or amphoteric surfactants, fatty alcohols, amine oxides, fatty amines and cationic surfactants, a whole range of emulsions can be

Table 27.5 Hair Dye Colours Obtained Using the Main Oxidation Bases and Coupling Agents[201]

	Pale blonde	Honey blonde	Ash blonde	Red blonde	Red	Copper	Auburn	Light brown	Red brown	Medium brown	Dark brown	Black	Light drab	Dark drab
p-Phenylenediamine	+	+	+	+	+	+	+	+	+	+	+	+	+	+
p-Aminophenol hydrochloride	+	+	+											
2-Amino-4-nitrophenol	+		+											
4-Nitro-*o*-phenylenediamine	+	+	+	+				+						
o-Aminophenol	+	+	+	+		+	+	+	+	+	+	+		
Resorcinol	+	+	+			+	+	+	+	+	+	+	+	+
Pyrogallol	+	+	+	+		+	+		+	+		+	+	+
Hydroquinone	+													
2,4-Diaminophenol	+	+												
p-Aminodiphenylene base	+	+											+	
2-Nitro-*p*-phenylenediamine	+	+	+	+	+	+	+	+			+			
4,4-Diaminoanisole sulphate					+							+	+	+
4-Nitro-*o*-phenylenediamine				+			+		+		+			
p-Aminophenol						+		+		+				+
p-Toluylenediamine hydrochloride														+
m-Aminophenol													+	
2,6-Diaminopyridine														+
6-Chloro-4-nitro-2-aminophenol	+													

formulated in order to produce the dyes in the form of creams, gels, shampoos, etc.

The use of antioxidants is necessary to protect the system from aerial oxidation during manufacture and packing and also to retard this oxidation during the preparative procedures and the application of the colouring mixture with hydrogen peroxide. Sodium sulphite or thioglycollic acid with or without hydroquinone can be used as antioxidant. The use of ascorbic acid has also been recommended[202–203] as has that of pyrazolones.[204]

The hydrogen peroxide, essential to the reaction as already detailed, can be used in various procedures. At one time it was common practice among hairdressers to ensure evenness of oxidation dyeing by bleaching the hair first with a mixture of hydrogen peroxide and ammonia and then applying the dyeing solution in mixture with a new minimum quantity of peroxide. This technique is now rarely used.

The modern version of this process is the combined bleach and dye, in which oxidation dyes are used in the presence of enough hydrogen peroxide to bleach the hair while the dyes are penetrating. In this way, dark mousy hair can be dyed to a warm blonde colour several shades lighter than the original colour, yet not apparently bleached. The difficulties found with early formulae have been overcome and suitable products are now available for both home and professional use.

The formulae which may be used are similar to those used for conventional oxidation dyeing, except that more alkali and hydrogen peroxide are used. The degree of lightening depends on the concentration of hydrogen peroxide and ammonia. Thus, for example, if 15 per cent ammonia and 20 per cent hydrogen peroxide are used a lightening of one shade may be achieved (say from black to brown), while if 15 per cent ammonia and 30 per cent hydrogen peroxide (20 vol) are used, then much greater effects can be produced (for example, dark auburn to blonde).

Research has been conducted over many years into the details of final formulation and additives, the objects being improvement in technical performance of the dye mixes themselves and also a widening of the role of the product, which is more and more devoted to functions additional to dyeing. In fact, those skilled in the art now concern themselves not only with the synthesis of higher-performance colorants, obtaining a wider range of colours, greater stability and less risk of skin or systemic toxicity, but also with the preparation, evaluation and manufacture of formulations which provide an almost ideal medium for the dyestuffs, increasing the advantages and reducing the inconveniences to a minimum. This preoccupation, already noticeable in the years 1950–60 in the use of better surfactants, emollients, alkalis and solvents and hair-conditioning agents, has become marked since 1970. Thus the number of hair colorants containing various additives has grown considerably.

One can identify two endpoints of this work on formulation:

1. *The improvement of essential dyeing properties*:
   Storage stability
   Ease of application
   Shortening of time of application

Increased covering power and penetration, fixation in the hair
Reduction in potential for skin damage
Protection against eventual hair damage

2. *The conferring of new properties on the hair colorant*:
Protection of hair structure
Improvement of aesthetic quality of the hair, brightness, bulk, combability, general appearance
Addition of antiseptics, antidandruff agents, antiseborrhea agents for the scalp, deodorants, etc.
Addition of substances specific to operations other than dyeing, for example, film-formers, etc.

Some details of various types of these improvements are given below.

The classic permanent hair dye formulations demand a strong basicity; this has generally been obtained by addition of ammonia, which has the disadvantage of strong odour and is aggressive. It is proposed therefore to replace it partially or totally by alkalis that are less injurious and more pleasant in use. Eugene-Gallia replaces the ammonia by an alkali metal or ammonium carbonate, or an amino acid associated with an organic base (morpholine, mono-, di-, or tri-ethanolamine).[205] Bristol-Myers propose the addition of an aminohydroxyl compound such as the lower alkanolamines, tris-hydroxymethyl-aminoethane etc.[206] Procter and Gamble propose guanidine or arginine derivatives or peptides rich in arginine together with a NaH_2PO_4-Na_2HPO_4 buffer.[207]

Another type of improvement concerns the increase of the light stability of a toluenediamine–resorcinol preparation by addition of diamino-1,2- or 1,3-benzene substituted in positions 4,5 or 4,6 by identical groups that are not proton-donors (F or CH_3),[208] or further addition to the dye mix of a UV-filter such as benzylidine-camphor[209] or mixtures of the isomeric branched-chain dodecyl benzotriazoles.[210]

Additives designed to inhibit premature oxidation of *para* dye bases include indazolone sulphonic acid and its salts,[211] sodium sulphite or dithionite, or ascorbic acid.[212] The use of enzymes as additives has been known for a long time; recently Procter and Gamble have extended the range in covering the use of a whole series of oxidases.[213] In the same area, L'Oreal have proposed the incorporation of peroxide dismutase in oxidation hair colorants.[214]

In the surfactant field, an extension has been proposed to the number of agents used by the employment of surface-active oligomers of the polyhydroxy-lated polyether type as colouring vehicle.[215]

It is well known that some of the necessary conditions for achieving a good permanent wave, a good dyeing or bleaching can cause damage in the long run to the structural integrity of the hair. Thus strong alkalinity and use of oxidizing agents can attack the hair shaft. To inhibit eventual damage or to restore damaged hair, the following substances have been proposed for incorporation in hair dyes: keratin hydrolysates,[216] keratose obtained by hot treatment of keratinous materials with aqueous bases,[217] methylol derivatives,[218] an alkyl imidazolone associated with an amidobetaine or with Miranol,[219] polyesters of polycarboxylic cycloaliphatic or aromatic acids and of polyalkylene oxides.[220] The application has also been proposed of an after-dye lotion containing alkoxymethyl esters of carboxylic acids[221] or an oxazolidine[222–223] in order to improve the structure of damaged hair.

Another line of formulation has as its goal not only the protection and/or restoration of damaged hair, but at the same time improvement of the aesthetic qualities (combability, lustre, bulk, general appearance). For this purpose, such formulae contain conditioning agents. Accordingly there are proposals for the incorporation of: quaternary ammonium compounds on their own,[224] or together with polyethoxylated fatty amides[225] or a film-former such as polyvinyl pyrrolidone (PVP); formaldehyde-dimethyl hydantoin resin,[226] or a whole series of cationic polymers, among which may be mentioned the quaternized copolymers of N-vinyl pyrrolidone–dimethylaminoethylmethacrylate–polyethylene glycol;[227] quaternary polymers of the ionene type;[228–231] cationic cellulose ethers; piperazine polycondensates;[232–233] cyclopolymers of diallyldimethylammonium and its copolymers with acrylamide and diacetone acrylamide.[234–236]

A special type of formulation has been developed to permit cosmetic operations other than dyeing to be carried out by means of specific additions. As example can be quoted compositions in the form of aqueous creams or gels containing direct and/or oxidation dyes together with film-formers such as PVP, acrylic resins, PVP–vinyl acetate, basic polymers of acrylates and methacrylates.[237] Also there are those intended for colouring and defrizzing which contain lithium, sodium or potassium hydroxide for the softening of the keratin.[238]

It has been further envisaged that additives could be incorporated with the object of reducing the concentration of active matter in cosmetic preparations (including colorants). A little before use an addition is made of 1–10 per cent by weight of organic compounds which dissociate in alkaline media and possess ester groups and/or halogens, and which can form an acid on dissociation (ethyl acetate, ethyl lactate, chloracetamide).[239]

OTHER DYES FOR HAIR

Considerable work has been reported on the development of oxidation dyes which can be used under ambient conditions and which do not require chemical oxidation. A large number of different materials have been examined and these can be classified according to their basic chemical structures.

Aromatic Polyhydroxy Compounds

Aromatic polyhydroxy compounds include trihydric phenols such as 1,2,4-trihydroxybenzene, 2,4,5-trihydroxytoluene and 1,2,4-trihydroxy-5-chlorobenzene. A British patent[240] describes the use of such materials in a formula which includes sulphite or mercaptan reducing agent. The use of compositions free from sulphites or other reducing agents is described in two later patents;[241,242] the polyhydric phenols are used in conjunction with short-chain aliphatic amines. A range of shades from blonde through reddish-brown and auburn to chestnut and bluish-black is available.[243–245]

There has been, for some time, an interest in the use of 3,4-dihydroxyphenylalanine as an oxidation dye, mainly because it is the precursor of melanin, the natural pigment of hair. The use of the next material in the chain, 5,6-dihydroxyindole has been described in a British patent,[246] but it appears that the melanocyte does a better job in producing a satisfactory colour. Light shades ranging from light ash to blonde have been reported[247,248] and are based on the use of methyl derivatives of 5,6-dihydroxyindole. The use of 3,4-dihydroxyphenylalanine in admixture with other phenols such as hydroquinone is described in a US patent,[249] but the colours produced are again rather pale. Other dihydroxybenzene derivatives have also been reported as being suitable. Thus *ortho*-dihydroxybenzenes (i.e. catechol and some of its derivatives),[250] dihydroxyaminobenzenes[251,252] and N-substituted 2,4-dihydroxyaminobenzenes[253] have been reported as giving colours ranging from brown and auburn to grey and black.

Vegetable Hair Dyes

Henna

Of the vegetable hair dyes, only henna is of any real importance today. It consists of the dried powdered leaves of *Lawsonia alba*, *Lawsonia spinosa* and *Lawsonia inemis*, which are removed from the plants prior to flowering.

Henna owes its hair-dyeing properties to the presence of 2-hydroxy-1,4-naphthaquinone, often termed *lawsone*, which is soluble in hot water and is, in acid solution, a substantive dye for keratin. In dyeing hair with henna, a paste of the powdered henna and hot water, slightly acidified with citric, adipic or other suitable acids to an optimum pH of about 5·5, is applied to the washed hair.

This 'henna pack', which is kept in place by means of a towel, is allowed to remain on the head for the required time, which may vary from five to sixty minutes, after which the hair is thoroughly shampooed, rinsed and dried. The treatment time depends upon the shade desired, the texture and condition of the hair, the activity of the henna, the acidity of the paste and the temperature at which the pack is applied.

Henna has the advantage that it is neither a primary irritant nor a sensitizer and possesses no local or systemic toxicity. The colour obtained is relatively stable and is deposited in the hair shaft, whereas metallic dyes coat the hair shaft. Unfortunately it suffers from a number of disadvantages—in particular, that it is messy to use, and the range of colours produced is limited to reddish auburn shades. Contact with the finger-nails must be avoided or the keratin of the nails will also be stained. Repeated dyeing with henna tends to spoil some of the effect and produces a somewhat hard auburn colour.

In addition to its use as a hair dye, henna extract is incorporated in certain acidic rinses. The difficulty here lies in the fact that any rinse strong enough to tint the hair is equally likely to tint the nails during application unless they have been previously protected.

Henna Reng

By the addition of other substances to henna, shades other than auburn may be obtained; for example, a mixture of powdered indigo leaves and henna produces blue-black shades and such mixtures are known as *henna rengs*.

Chamomile

Of the various species of chamomile, only *Anthemis nobilis* (Roman chamomile) and *Matricaria chamomillae* (German chamomile) have a cosmetic use; they appear to be equally useful in tinting hair. The active ingredient in these flowers is 1,3,4-trihydroxyflavone, known also as apigenin. Either an aqueous extract or a paste of the ground flower heads may be used. To lighten the hair, a paste consisting of 2 parts chamomile and 1–2 parts kaolin mixed to a thin cream with hot water is applied to the head for a period varying from 15 minutes to 60 minutes depending on the shade required.

Chamomile is also used as a constituent of hair-brightening rinses and shampoos. Some doubt has arisen as to whether chamomile is really effective. The point is that at least 5 per cent chamomile, or its equivalent in extract, must be present to produce any effect at all, but the azulene also present in chamomile probably contributes to the brightening effect.

Metallic Hair Dyes

In metallic dyes, compounds of lead are the most frequently used; compounds of silver, copper, iron, nickel and cobalt are sometimes employed, and less frequently salts of bismuth.

It is not quite certain whether the colours produced by such compounds are due to sulphides formed by a reaction between the sulphur in the keratin and the metallic salts, or to metallic oxides formed by the keratin reducing the metal salts. It is possible that both reactions occur to some extent. Whatever the mode of action, the result obtained is the deposition of a coloured film along the hair shaft which, eventually, gives the hair a characteristic dull metallic appearance, renders the hair brittle and diminishes the efficiency of a subsequent permanent wave.

Lead Dyes

The active ingredient in these dyes is usually lead acetate together with some precipitated sulphur, glycerin and water. Example 2 is a typical formula. Sodium thiosulphate may be incorporated in place of precipitated sulphur. Such preparations have poor stability.

	(2) *per cent*
Precipitated sulphur	1·3
Lead acetate	1·6
Glycerin	9·6
Rose-water	87·5

The action of lead dyes is slow and progressive, and the shades produced in grey hair usually pass from yellow through brown to black. The shades achieved depend upon the concentration of lead salts in the preparation, the number of applications, the original colour of the hair and the time during which the colour has been allowed to develop.

Probably because of interaction with skin proteins, such lead solutions are relatively non-toxic under normal conditions of use. However, it must be remembered that these preparations are usually sold for home use and systemic effects may follow if lead remains on the hands and contaminates food. The ingestion by children might be fatal and preparations containing lead should carry adequate precautionary notices.

Other Metallic Dyes
Other metallic hair dyes, such as bismuth, silver, copper, nickel and cobalt salts, have been proposed, but pose important problems of toxicity and of necessary precautions.[254–256]

HAIR DYE REMOVERS

The removal of hair dyes is sometimes necessary either because of a mistake or because the user wishes to have a lighter shade. In the case of metallic dyes it is usually dangerous to remove the dye by chemical means since the metals catalyse many reactions and may cause a violent production of heat which will probably damage both hair and scalp. The only remedy for an unwanted metallic dye is to let the hair grow.

Oxidation dyes can be removed, more or less successfully, by treating the hair with reducing agents such as sodium hydrosulphite or sodium formaldehyde sulphoxylate, usually at a concentration of 5 per cent. The use of formamidine sulphinic acid has also been described[257] and a formula[258] has been quoted as follows:

	(3) *per cent*
Formamidine sulphinic acid	1·5
Polyvinylpyrrolidone	5·0
Ethylene glycol monobutyl ether	5·0
Ammonium carbonate	1·0
Ammonia (25%)	0·5
Carboxymethylcellulose	2·5
Water	to 100·0

Semi-permanent dyes can often be removed by vigorous washing with shampoos, particularly if a little ammonia is added. Some dyestuffs, however, prove very resistant and a mixture of shampoo, reducing agent and bleach in the proportions 1:1:2 has been recommended for such cases.[259]

BLEACHING AND LIGHTENING

Obviously no description of hair dyes would be complete without reference to the method of producing blonde shades. This is accomplished by bleaching the hair in the usual manner to the palest possible shade of blonde with ammonia and hydrogen peroxide. The hair is afterwards given a rinse with a blue rinse containing about 1:100 000 of methylene blue (Ext. D&C Blue No. 1), or other suitable blue certified colour. The addition of the blue colour is necessary since the human eye considers a substrate which is very slightly blue in colour as whiter than white.

Bleaching of the hair must be thorough or the combination of a deep yellow hair shaft and a blue rinse may give the hair a distinctly greenish appearance.

Cook[260] has reviewed the bleaching treatments available for human hair and discusses the effects of factors such as controlled oxygen release and the use of additives to help overcome the damage caused to the hair by such treatments. He particularly recommends the use of substantive proteins either added to the oxidant or used before treatment.

To obtain better control in the application of peroxide to the hair, and extend the bleaching time, powder products have been introduced. These powders vary from inert materials such as kaolin or magnesium carbonate, used with peroxide and ammonia, to powders which themselves provide ammonia and some form of active oxygen, when wetted with water or hydrogen peroxide. These powders were at one time called 'white henna', which was misleading because they contained no henna, but was convenient because it indicated their method of use. Typical early formulae contained up to one-third of sodium perborate or percarbonate, the balance being kaolin and/or magnesium carbonate. More complex formulae are given in examples 4 and 5.

	(4) *per cent*	(5) *per cent*
Ammonium persulphate	3·0	—
Potassium persulphate	—	8·0
Potassium hydrogen tartrate	3·0	—
Potassium hydrogen oxalate	—	8·0
Sodium carbonate	3·0	13·0
Surfactant	1·0	1·0
Thickener	5·0	—
Magnesium hydroxide and/or aluminium hydroxide	to 100·0	to 100·0

Procedure: Mix the powder to a paste with hydrogen peroxide (10–40 vol) before use, and spread evenly over those parts of the hair which it is wished to bleach.

All the preparation types described above are essentially for the professional hairdresser because of the degree of skill and care required to get satisfactory colour results with a minimum amount of damage.

For home use, products are formulated on much more simple lines. A suitable two-bottle pack would be:

		(6) *per cent*
A	Hydrogen peroxide 20 vol	98·6
	Tartaric acid	0·8
	Sodium stannate	0·6
B	Ammonia	4·5
	Surfactant, e.g. ammonium soap	3·0
	Water	92·5

Procedure: Mix equal volumes of A and B before use.

Even more simple products, popular in the 1950s, were known as hair lighteners or brighteners. These were often simple solutions of hydrogen peroxide at concentrations between 1 and 3 per cent (3–10 vol) stabilized as solution A in example 6. The solution is combed through the hair and left to react slowly. Since the reaction does not take place in alkaline conditions, and there is no ammonia present, the solutions cause very little damage to the hair, and no red or brassy off-shades.

A bleaching composition for human hair described in US patent[261] is particularly intended for localized bleaching, e.g. bleaching new hair growth close to the scalp without damaging any previously bleached hair and without skin irritation. The bleaching composition described is prepared by combining a dry mixture of an anhydrous silicate, an alkali or ammonium salt of a peracid, for example a persulphate, with hydrogen peroxide and a liquid alkaline ammonium soap within a pH range of 9·3 to 10·0, when this mixture will gel. By mixing hydrogen peroxide with a constant amount of the liquid ammonium soap but different amounts of the dry powder mix, it is possible to produce a range of bleaching compositions of different strengths in which the pH value remains within a desired narrow range. The function of the soap is to contribute to the alkalinity of the compound, to help in attaining the desired consistency of the gel and to function as a shampoo in the removal of the bleach. The peracid salts, preferably a mixture of potassium and ammonium persulphate, assist by virtue of their oxidizing action in the bleaching process and thus permit less hydrogen peroxide to be used. Together with the sodium meta-silicate and the ammonium soap, they are claimed to buffer the bleaching composition within the desired pH range throughout the bleaching operation, and to produce a gel of the desired consistency.

REFERENCES

1. Kiese, M. and Rauscher, E., *Toxicol. appl. Pharmacol.*, 1968, **13,** 325.
2. Kiese, M., Rachor, M. and Rauscher, E., *Toxicol. appl. Pharmacol.*, 1968, **12,** 495.
3. Frenkel, E. P. and Brody, F., *Arch. environ. Health*, 1973, **27,** 401.
4. Maibach, H. I., Leaffer, M. A. and Skinner, W. A., *Arch. Dermatol.*, 1975, **111,** 1444.
5. Hruby, R., *Food Cosmet. Toxicol.*, 1977, **15,** 595.
6. Yare, R. and Garcia, M., *Arch. Dermatol.*, 1977, **133,** 1610.
7. Evans, R. L., *Proc. sci. Sect. Toilet Goods Assoc.*, 1948, (10), 9.
8. Alexander, P., *Manuf. Chem.*, 1964, **35**(9), 70.
9. Holmes, A. W., *J. Soc. cosmet. Chem.*, 1964, **15,** 595.
10. Zviak, C., *Problèmes Capillaires*, ed. Sidi, E. and Zviak, C., Paris, Gauthier Villars, 1966.
11. Walker, G. T., *Seifen, Öle, Fette, Wachse*, 1967, **13,** 319.
12. Menkart, J., Wolfram, L. J. and Mao, I., *J. Soc. cosmet. Chem.*, 1966, **17,** 769.
13. Daniels, M. H., *Drug Cosmet. Ind.*, 1958, **82,** 158.
14. French Patent 2 117 662, L'Oreal, 10 December 1971.
15. French Patent 2 050 990, L'Oreal, 9 June 1970.
16. French Patent 2 101 603, L'Oreal, 12 July 1971.
17. French Patent 2 254 557, L'Oreal, 11 December 1974.
18. French Patent 2 139 385, L'Oreal, 9 May 1972.

19. French Patent 2 121 101, L'Oreal, 29 December 1971.
20. French Patent 2 234 276, L'Oreal, 21 June 1974.
21. French Patent 2 234 277, L'Oreal, 21 June 1974.
22. French Patent 2 338 036, L'Oreal, 19 January 1976.
23. French Patent 2 148 103, L'Oreal, 28 July 1972.
24. French Patent 2 047 932, L'Oreal, 24 June 1970.
25. French Patent 2 119 990, L'Oreal, 20 December 1971.
26. French Patent 2 106 661, L'Oreal, 18 September 1970.
27. French Patent 2 089 423, L'Oreal, 8 April 1971.
28. French Patent 2 106 660, L'Oreal, 18 September 1970.
29. French Patent 2 106 662, L'Oreal, 18 September 1970.
30. French Patent 2 097 712, L'Oreal, 18 January 1971.
31. French Patent 2 051 802, L'Oreal, 15 July 1970.
32. French Patent 2 122 442, L'Oreal, 14 January 1972.
33. French Patent 2 056 799, L'Oreal, 10 August 1970.
34. French Patent 2 145 724, L'Oreal, 13 July 1972.
35. French Patent 2 165 965, L'Oreal, 19 December 1972.
36. French Patent 2 262 024, L'Oreal, 21 February 1975.
37. French Patent 2 262 023, L'Oreal, 21 February 1975.
38. French Patent 2 262 022, L'Oreal, 21 February 1975.
39. French Patent 2 261 750, L'Oreal, 21 February 1975.
40. French Patent 2 359 182, L'Oreal, 21 July 1976.
41. Japanese Patent 72 47 666, Arimino Kagaku, 1 December 1972.
42. US Patent 2 359 783, Orelup, J. W., 10 October 1944.
43. US Patent 3 194 735, Warner Lambert, 13 July 1965.
44. French Patent 1 309 399, L'Oreal, 5 October 1961.
45. French Patent 1 484 836, L'Oreal, 10 May 1966.
46. French Patent 1 482 993, L'Oreal, 20 April 1966.
47. French Patent 1 498 464, L'Oreal, 22 July 1966.
48. French Patent 1 517 862, L'Oreal, 9 January 1967.
49. French Patent 1 527 405, L'Oreal, 14 June 1967.
50. French Patent 1 604 203, L'Oreal, 14 June 1967.
51. French Patent 2 361 447, L'Oreal, 12 August 1976.
52. Japanese Patent 71 14 360, Katsuraya, K.K., 17 April 1971.
53. US Patent 3 743 622, Wagner, E.R., 3 July 1973.
54. US Patent 3 790 512, Wagner, E.R., 5 February 1974.
55. US Patent 3 743 678, Clairol, 3 July 1973.
56. Canadian Patent 899 888, Bristol-Myers, 9 May 1972.
57. Canadian Patent 900 490, Bristol-Myers, 13 January 1967.
58. British Patent 1 228 604, Gillette, 15 April 1971.
59. German Patent Application 2 149 467, Therachemie, 12 April 1973.
60. German Patent Application 2 207 683, Therachemie, 30 August 1973.
61. US Patent 3 646 216, Shulton Inc., 29 February 1972.
62. US Patent 3 742 048, L'Oreal, 26 June 1973.
63. British Patent 1 363 937, Gillette, 21 August 1974.
64. US Patent 3 726 635, L'Oreal, 10 April 1973.
65. French Patent 2 290 186, L'Oreal, 5 November 1974.
66. French Patent 2 349 325, L'Oreal, 30 April 1976.
67. Canadian Patent 935 094, Bristol-Myers, 26 February 1968.
68. French Patent 2 348 911, L'Oreal, 19 April 1977.
69. US Patent 4 021 486, Clairol, 3 May 1977.
70. Canadian Patent 989 862, Bristol-Myers, 25 May 1976.
71. German Patent Application 2 204 026, Therachemie, 28 January 1972.

72. German Patent Application 2 219 225, Therachemie, 20 April 1972.
73. French Patent 2 315 256, L'Oreal, 26 June 1975.
74. French Patent 2 290 186, L'Oreal, 5 November 1974.
75. French Patent 2 349 325 L'Oreal, 30 April 1976.
76. US Patent 3 817 698, L'Oreal, 18 June 1974.
77. French Patent 2 106 264, L'Oreal, 6 September 1971.
78. British Patent 1 205 365, Gillette, 16 September 1970.
79. US Patent 3 661 500, Shulton Inc., 9 May 1972.
80. French Patent 2 290 185, L'Oreal, 8 November 1974.
81. French Patent 2 349 325, L'Oreal, 30 April 1976.
82. US Patent 3 168 441, Clairol, 2 February 1965.
83. Kalopissis, G., Bugaut, A. and Bertrand, J., *J. Soc. cosmet. Chem.*, 1964, **15,** 411.
84. French Patent 2 189 006, L'Oreal, 18 June 1973.
85. French Patent 2 285 851, L'Oreal, 26 September 1975.
86. French Patent 2 282 860, L'Oreal, 29 August 1975.
87. French Patent 2 140 205, L'Oreal, 2 June 1972.
88. British Patent 1 249 438, Gillette, 13 October 1971.
89. French Patent 2 099 399, L'Oreal, 30 July 1971.
90. Broadbent, A. D., *Am. Perfum. Cosmet.*, 1963, **78**(3), 23.
91. Shansky, A., *Am. Perfum. Cosmet.*, 1966, **81**(11), 23.
92. British Patent 1 009 796, Partipharm AG, 24 April 1963.
93. Canadian Patent 731 512, Turner Hall Corp., 16 March 1961.
94. British Patent 1 113 661, Unilever Ltd, 25 January 1965.
95. British Patent 1 309 743, Unilever Ltd, 14 March 1973.
96. US Patent 3 904 357, Avon Products, 9 September 1975.
97. US Patent 3 871 818, Avon Products, 18 March 1975.
98. Japanese Patent 70 21 399, Pias K.K., 20 July 1970.
99. German Patent Application 2 327 987, Henkel, 2 January 1975.
100. German Patent Application 2 327 986, Henkel, 2 January 1975.
101. German Patent Application 2 327 985, Henkel, 2 January 1975.
102. German Patent Application 1 927 959, Therachemie, 3 December 1970.
103. French Patent 2 363 323, Ciba-Geigy, 1 September 1977.
104. German Patent Application 2 757 866, Ciba-Geigy, 29 June 1978.
105. German Patent Application 2 807 780, Ciba-Geigy, 31 August 1978.
106. British Patent 721 831, Ashe Laboratories, 19 May 1952.
107. US Patent 2 776 668, H. Rubinstein Inc., 28 June 1951.
108. British Patent 906 526, County Laboratories, 15 June 1960.
109. Japanese Patent 75 09 852, Hohyu Shokai, 16 April 1975.
110. British Patent 1 159 331, L'Oreal, 23 July 1969.
111. Anon., *Schimmel Brief*, 1965, 365.
112. Heald, R. C., *Am. Perfum. Cosmet.*, 1963, **78**(4), 40.
113. British Patent 826 479, Rapidol Ltd, 6 January 1960.
114. Peters, L. and Steven, C. B., *J. Soc. Dyers Colour.*, 1956, **72,** 100.
115. US Patent 3 482 923, Therachemie. 9 December 1969.
116. French Patent 1 603 028, L'Oreal, 24 June 1968.
117. British Patent 1 236 560, Revlon, 23 June 1971.
118. German Patent Application 2 022 676, Gillette, 19 November 1970.
119. US Patent 3 822 112, L'Oreal, 6 May 1975.
120. US Patent 3 586 475, Colgate-Palmolive, 22 June 1971.
121. US Patent 3 632 290, Lovenstein Dyes and Cosmetics, 4 January 1972.
122. US Patent 3 933 422, Avon Products, 20 January 1976.
123. Belgian Patent 840 879, Kindai K.K., 16 August 1976.
124. Japanese Patent 73 23 911, Shiseido, 17 July 1973.

125. US Patent 3 733 175, Clairol, 15 May 1973.
126. Japanese Patent 71 13 278, Sanwyo, 6 April 1971.
127. British Patent 1 241 832, Beecham Group, 4 August 1971.
128. British Patent 2 096 377, Helene Curtis, 18 February 1972.
129. Japanese Patent 76 151 341, Shiseido, 25 December 1976.
130. French Patent 1 113 505, L'Oreal, 4 November 1954.
131. British Patent 741 307, L'Oreal, 30 November 1955.
132. US Patent 2 848 369, L'Oreal, 19 August 1958.
133. French Patent 1 113 505, L'Oreal, 4 November 1954.
134. British Patent 741 307, L'Oreal, 30 November 1955.
135. US Patent 2 848 369, L'Oreal, 19 August 1958.
136. German Patent Application 2 028 818, Gillette, 17 December 1970.
137. German Patent Application 2 119 231, –232, Therachemie, 26 October 1972.
138. French Patent 2 362 112, L'Oreal, 20 August 1976.
139. British Patent 1 482 170, Bristol-Myers, 31 December 1974.
140. US Patent 3 884 627, Clairol, 20 May 1975.
141. US Patent 3 970 423, Clairol, 20 July 1976.
142. French Patent 2 017 995, L'Oreal, 4 September 1969.
143. German Patent Application 1 907 322, Therachemie, 3 September 1970.
144. French Patent 2 364 888, L'Oreal, 17 November 1976.
145. US Patent 3 658 454, American Cyanamid, 25 April 1972.
146. German Patent Application 2 518 393, Henkel, 4 November 1976.
147. German Patent Application 2 357 215, Henkel, 22 May 1975.
148. German Patent Application 2 424 139, Henkel, 27 November 1975.
149. German Patent Application 2 359 399, Henkel, 12 June 1975.
150. German Patent Application 2 524 329, Henkel, 16 December 1976.
151. German Patent Application 2 523 045, Henkel, 2 December 1976.
152. German Patent Application 2 523 629, Henkel, 9 December 1976.
153. German Patent Application 2 516 118, Henkel, 21 October 1976.
154. German Patent Application 2 516 117, Henkel, 21 October 1976.
155. German Patent Application 2 613 707, Henkel, 13 October 1977.
156. German Patent Application 2 441 598, Henkel, 11 March 1976.
157. German Patent Application 2 526 313, Henkel, 23 December 1976.
158. German Patent Application 2 527 791, Henkel, 30 December 1976.
159. German Patent Application 2160 317, Therachemie, 7 June 1973.
160. German Patent Application 2 554 456, Henkel, 16 June 1977.
161. German Patent Application 2 622 451, Henkel, 8 December 1977.
162. British Patent 1 484 638, Bristol-Myers, 1 September 1977.
163. British Patent 1 484 639, Bristol-Myers, 1 September 1977.
164. German Patent Application 2 617 739, Henkel, 10 November 1977.
165. French Patent 2 233 984, L'Oreal, 21 June 1974.
166. French Patent 2 233 982, L'Oreal, 21 June 1974.
167. French Patent 2 233 983, L'Oreal, 21 June 1974.
168. German Patent Application 2 509 152, Henkel, 9 September 1976.
169. German Patent Application 2 509 096, Henkel, 23 September 1976.
170. French Patent 2 315 255, L'Oreal, 26 June 1975.
171. French Patent 2 017 164, L'Oreal, 13 August 1969.
172. French Patent 2 012 986, L'Oreal, 11 July 1969.
173. US Patent 3 834 866, Alberto-Culver, 10 September 1974.
174. French Patent 2 364 204, L'Oreal, 9 September 1976.
175. German Patent Application 1 949 749, Therachemie, 8 April 1971.
176. German Patent Application 2 628 999, Henkel, 5 January 1978.
177. French Patent 2 362 116, L'Oreal, 20 August 1976.

178. French Patent 2 362 118, L'Oreal, 20 August 1976.
179. German Patent Application 2 629 805, Henkel, 12 January 1978.
180. German Patent Application 1 949 748, Therachemie, 15 April 1971.
181. German Patent Application 2 160 318, Therachemie, 7 June 1973.
182. German Patent Application 2 334 738, Henkel, 30 January 1975.
183. German Patent Application 2 603 848, Henkel, 11 August 1977.
184. German Patent Application 2 625 410, Henkel, 15 December 1977.
185. German Patent Application 2 632 390, Henkel, 26 January 1978.
186. German Patent Application 2 623 564, Henkel, 15 December 1977.
187. French Patent 2 013 346, L'Oreal, 13 June 1969.
188. US Patent 3 817 995, L'Oreal, 18 June 1974.
189. Corbett, J. F., *Rev. Progr. Color. relat. Top.*, 1973, **4,** 3.
190. Shah, M. J., Tolgyesi, W. S. and Britt, A. D., *J. Soc. cosmet. Chem.*, 1972, **23,** 853.
191. Corbett, J. F., *J. Soc. cosmet. Chem.*, 1973, **24,** 103.
192. Ames, B. N., Kammen, H. O. and Yamasaki, E., *Proc. nat. Acad. Sci.*, 1975, **72,** 2423.
193. Venitt, S., *IARC Scientific Publication No. 13, Inserm Symposia Series*, 1976, **52,** 263.
194. Dolinsky, M., Eilson, C. H., Wisneski, H. H. and Demers, F. X., *J. Soc. cosmet. Chem.*, 1968, **19,** 411.
195. Altman, M. and Rieger, M. M., *J. Soc. cosmet. Chem.*, 1968, **19,** 141.
196. Tucker, H. H., *J. Soc. cosmet. Chem.*, **18,** 609.
197. Cox, H. E., *Analyst*, 1933, **58,** 738.
198. Rostenberg, I. and Kass, G. S., *Hair Coloring: AMA Committee of Cutaneous Health and Cosmetics*, American Medical Association, 1969.
199. Brown, J. C., *J. Soc. cosmet. Chem.*, 1967, **18,** 225.
200. Cook, M. K., *Drug Cosmet. Ind.*, 1966, **99**(11), 52.
201. Cook, M. K., *Drug Cosmet. Ind.*, 1966, **99**(10), 50.
202. US Patent 3 488 138, Iscowitz, 6 January 1970.
203. British Patent 979 266, Wella, 1 January 1965.
204. British Patent 1 065 223, L'Oreal, 12 April 1967.
205. Belgian Patent 777 516, Eugene-Gallia, 15 December 1970.
206. British Patent 806 252, Bristol-Myers, 23 December 1958.
207. US Patent 3 861 868, Procter and Gamble, 21 January 1975.
208. German Patent Application 2 158 670, Unilever, 29 June 1972.
209. US Patent 3 840 338, L'Oreal, 8 October 1974.
210. US Patent 3 983 132, GAF Corp., 28 September 1976.
211. German Patent Application 1 934 766, Therachemie, 21 January 1971.
212. US Patent 3 893 803. Procter and Gamble, 8 July 1975.
213. US Patent 3 957 424, Procter and Gamble, 18 May 1976.
214. French Patent 2 287 899, L'Oreal, 14 October 1975.
215. French Patent 2 359 165, L'Oreal, 19 July 1976.
216. US Patent 3 842 848, Wilson Sinclair, 22 October 1974.
217. German Patent Application 2 338 518, Henkel, 13 February 1975.
218. French Patent 2 027 178, L'Oreal, 5 December 1969.
219. German Patent Application 1 941 100, Schwarzkopf, 25 February 1971.
220. French Patent 2 161 782, Cincinnati Milacron, 29 November 1971.
221. German Patent Application 2 657 613, Henkel, 22 June 1978.
222. German Patent Application 2 657 689, Henkel, 22 June 1978.
223. German Patent Application 2 657 715, Henkel, 29 June 1978.
224. US Patent 4 096 243, Bristol-Myers. 9 February 1976.
225. US Patent 3 930 792, Bristol-Myers, 6 January 1976.
226. US Patent 3 653 797, Reiss, 4 April 1972.

227. French Patent 2 312 233, L'Oreal, 26 May 1976.
228. French Patent 2 279 851, L'Oreal, 11 July 1975.
229. French Patent 2 316 271, L'Oreal, 2 July 1976.
230. French Patent 2 331 323,–324, L'Oreal, 12 November 1976.
231. French Patent 2 333 012, L'Oreal, 15 May 1975.
232. French Patent 2 331 325, L'Oreal, 12 November 1976.
233. French Patent 2 280 361, L'Oreal, 2 August 1974.
234. US Patent 3 986 825, Gillette, 19 October 1976.
235. US Patent 4 027 008, Gillette, 31 May 1977.
236. French Patent 2 331 325, L'Oreal, 12 November 1976.
237. German Patent Application 2 157 844, Wella, 28 September 1971.
238. German Patent Application 2 014 628, Wella, 14 October 1971.
239. German Patent Application 2 349 050, Wella, 24 April 1975.
240. British Patent 710 134, Union Francaise Commerciale et Industrielle, 9 June 1954.
241. British Patent 745 531, Gillette Industries Ltd, 29 February 1956.
242. British Patent 745 532, Gillette Industries Ltd, 29 February 1956.
243. British Patent 824 519, Monsavon-L'Oreal, 2 December 1959.
244. British Patent 827 439, Monsavon-L'Oreal, 3 February 1960.
245. British Patent 889 812, Monsavon-L'Oreal, 21 February 1962.
246. British Patent 797 174, Monsavon-L'Oreal, 25 June 1958.
247. British Patent 823 503, Monsavon-L'Oreal, 11 November 1959.
248. US Patent 3 194 734, L'Oreal, 13 July 1965.
249. US Patent 2 875 769, Apod. Corps, 3 March 1959.
250. British Patent 831 851, Monsavon-L'Oreal, 6 April 1960.
251. British Patent 857 070, Monsavon-L'Oreal, 29 December 1960.
252. British Patent 951 509, L'Oreal, 4 March 1964.
253. British Patent 899 051, Monsavon-L'Oreal, 20 June 1962.
254. German Patent Application 2 435 578, Combe Inc., 18 November 1974.
255. British Patent 1 518 874, Beecham Group, 26 July 1978.
256. German Patent Application 2 617 162, Zikuda, G., 9 March 1978.
257. German Patent 1 151 242, Vogt, G., 31 January 1961.
258. Anon, *Schimmel Brief*, 1964, 365.
259. US Patent 3 206 364, Thompson, H.L., 14 November 1965.
260. Cook, M. K., *Drug Cosmet. Ind.*, 1966, **99**(1), 46.
261. US Patent 3 378 444, Rayette Faberge, 16 April 1968.

Chapter Twenty-eight

Permanent Waving and Hair Strengtheners

Introduction

From the time the earliest Egyptian woman curled her hair by means of wet mud, and through the Roman and Grecian era up to the present day, the desire of all women has been to possess an attractive and beautiful coiffure. In general, even allowing for the vicissitudes of fashion, curly, or at least wavy, hair is more attractive than straight hair, and at the same time provides more opportunities for rearrangement in suitable and fashionable styles. Hence any process by which curls or waves could be introduced into hair was bound to affect to a very marked degree the trend of modern hairdressing.

Suter[1] has pointed out that '. . . until around 1910, hair was always curled by means of a curling iron or by boiling it in water. This, however, never resulted in a lasting wave . . . Three pioneers, Charles Nessler, E. Frederics and Eugene Suter, each working independently, found that the addition of chemicals such as borax to the boiling water would curl hair so that it would survive several washings. Around 1924 ammonium hydroxide first came into use in connection with borax . . .'.

Subsequently chemical methods of heating and the cold wave have been developed; these are discussed later in this chapter.

The statement that naturally curly hair is observed to be oval or flattened when examined microscopically while straight hair is usually circular is incorrect. Many naturally curly hairs are oval, but many others cannot be differentiated from ordinary straight hair. Kneberg[2] has shown that hairs of widely varying cross-sectional form may or may not curl, while Danforth[3] states that there would seem to be no reason why a shaft that is uniform in structure should curl merely because its cross-section is elliptical. If, however, the upper and lower sides of the shaft were of different density or if the transverse axis did not bisect the vertical, a differential would be provided which might be expected to form curls readily.

The natural shape of the hair is determined during the keratinization stage, as the almost fluid cells produced by the papilla are shaped by the follicle wall and converted into keratin. The shape of the hair is thus a very deep-seated structural characteristic, not easily altered by subsequent treatments of the fully keratinized shaft. Hence all the processes described in this chapter, although they are called 'permanent' are subject to gradual relaxation, as the hair returns to its normal straightness or kinkiness. The time taken for this relaxation varies with the process, the hair and the environment, and may be anything from a few weeks to many months.

The process of straightening hair involves deformation of wavy hair to an uncurled condition and depends on similar chemistry to the waving process. It is largely confined to one section of the consumer market and is discussed in Chapter 29.

Chemistry of Hair Waving

Hair is both strong and elastic, and any process for changing its shape must depend on softening or plasticizing the keratin, reshaping it in the required way while it is soft, and hardening it again while it is in the new shape.

The simplest softening process is the application of water, which enters the hydrogen bonds of the hair and gives it greater flexibility. Reshaping is possible by the deformation of the wet hair, and after drying the new shape will be held. However, the hair will not deform more than a limited amount when wet, and it is still very elastic; more importantly, water in a shampoo or even moisture in the atmosphere will soften the hair again, so that it returns to the shape that its follicle and gravity decide. Setting lotions may give it a little mechanical help, but their effects do not last beyond the next shampoo.

If the water is applied as steam, the softening process goes further and the resultant set is not easily disturbed by water at ordinary temperatures. If the water is made alkaline, softening occurs below 100°C, and adequate conditions have been proposed to obtain the setting of hair thereby.[4] These treatments are effective because they affect not only salt electrostatic or hydrogen bonds of the hair but also a number of covalent disulphide cross-links, which are known to exert a prevailing influence in the cohesion of the keratinic structure.

Alkaline Agents

Disruption and transformation of keratin disulphide linkages have been shown to occur through the action of hydroxyl ions. The mechanism of this action, as well as the nature of the newly created bonds and their contribution to the mechanochemical properties of keratin, have been much discussed and investigated for over 30 years.[5–15] The formation of a thioether derivative of cystine was suggested as early as 1933[16] and confirmed in the 1940s, when lanthionine was isolated in hydrolysates of wool treated by sodium carbonate at boiling point.[17]

Among the reaction sequences proposed to explain lanthionine formation, two appear to prevail: one entails a direct attack on the disulphide bonding, implying a bimolecular nucleophilic substitution mechanism:

$$>CH-CH_2-S-S-CH_2-CH< + B^- \rightleftharpoons >CH-CH_2-S^- + B-S-CH_2-CH<$$

$$\rightleftharpoons BS^- + >CH-CH_2-S-CH_2-CH<$$

The other involves a reaction of β-elimination induced by the attack of a hydrogen located on the carbon atom in β-position to the disulphide bond and leading to the intermediate formation of a dehydroalanyl group—see (**B**) below; this highly reactive group undergoes further addition with thiol and amine functions.[18,19] Thus the cysteine group (**C**) released in the preceding reaction and the lysine group existing in the keratin side-chains lead to the formation of lanthionine and lysinoalanine residues, respectively:

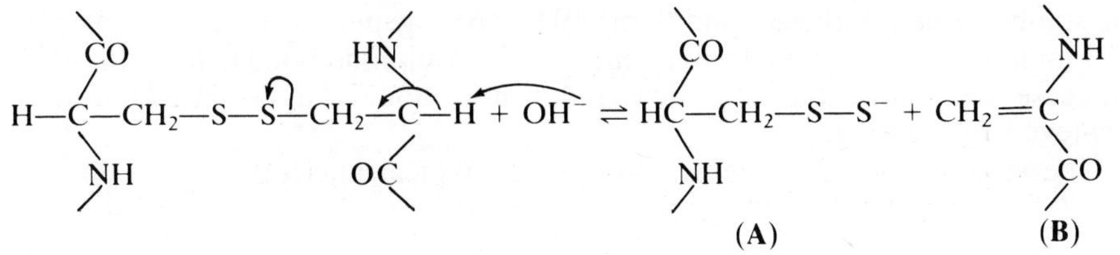

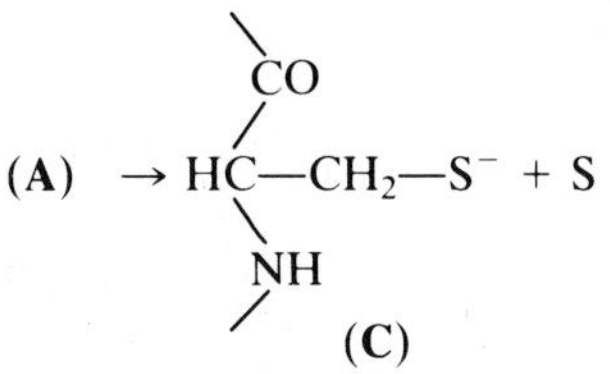

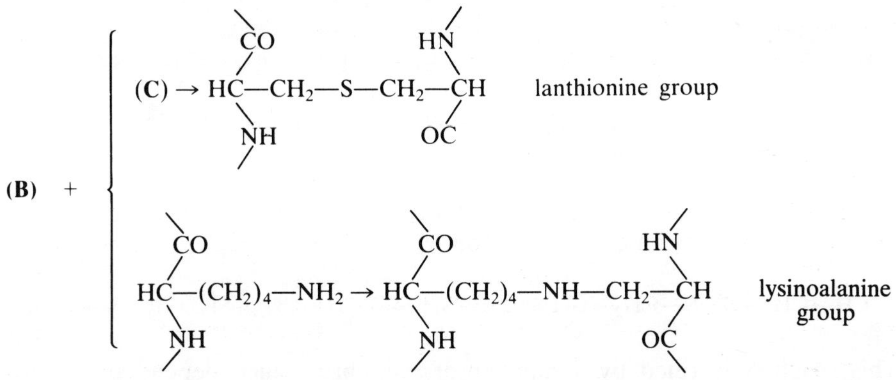

$$\mathrm{H{-}C(CO)(NH){-}CH_2{-}S{-}S{-}CH_2{-}C(HN)(OC){-}H + OH^- \rightleftharpoons \underset{(\mathbf{A})}{HC(CO)(NH){-}CH_2{-}S{-}S^-} + \underset{(\mathbf{B})}{CH_2{=}C(NH)(CO)}}$$

$$(\mathbf{A}) \rightarrow \mathrm{\underset{(\mathbf{C})}{HC(CO)(NH){-}CH_2{-}S^-} + S}$$

(**B**) +

$$(\mathbf{C}) \rightarrow \mathrm{HC(CO)(NH){-}CH_2{-}S{-}CH_2{-}CH(HN)(OC)} \quad \text{lanthionine group}$$

$$\mathrm{HC(CO)(NH){-}(CH_2)_4{-}NH_2 \rightarrow HC(CO)(NH){-}(CH_2)_4{-}NH{-}CH_2{-}CH(HN)(OC)} \quad \text{lysinoalanine group}$$

This mechanism seems the most likely to occur when hair is treated by alkali, as shown by the fact that lanthionine formation in alkaline medium is greatly enhanced by increasing the ionic strength.[4,20] An interesting evidence of the transformation of disulphide bonds into a bond stable against reducing agents, without hydrolysis of peptidic bonds, may be found in the low solubility or insolubility of 'lanthionized' fibres in urea-bisulphite or phenol-thioglycollic acid media.

This leads to another means for disrupting disulphide bonds and thereby considerably lowering the temperature of softening: the use of reducing agents.

Reducing Agents

In normal hot waving practice, the reducing agent is invariably a sulphite; in cold waving it is usually a thiol compound, more particularly thioglycollic acid, which since it was introduced in 1940 has remained in favour with the user.

The reaction of sulphites with disulphide bonds is not simple because of the reversibility of the basic processes involved and its high sensitivity to pH. The maximum reductive action of sulphite ranges between pH 3 and 6[21] but there may in fact be two maxima, one at pH 3·25–3·50 and the other about pH 5.[22] In practice, an acidic medium is not often used because of the instability of sulphites under these conditions; pH $\geqslant$ 6 is generally used, and at pH 7 (30 minutes, 30°C) about 15 per cent of hair disulphide bonds are reduced. To increase the reduction rate the equilibrium should be displaced, which is achieved by heating.

The sulphite reaction seems to involve a typical nucleophilic reaction:[23,24]

$$\rangle CH{-}CH_2{-}S{-}S{-}CH_2{-}CH\langle + SO_3H^- \rightleftharpoons \rangle CH{-}CH_2{-}S{-}SO_3H + {}^-S{-}CH_2{-}CH\langle$$

$$\downarrow$$

$$\rangle CH{-}CH_2{-}S{-}SO_3^- + HS{-}CH_2{-}CH$$

Bunte salt

In alkaline medium the keratocysteine formed is ionized and the formation of lanthionine is promoted:

$$\rangle CH{-}CH_2{-}S{-}SO_3^- + {}^-S{-}CH_2{-}CH\langle \rightarrow \rangle CH{-}CH_2{-}S{-}CH_2{-}CH\langle + S_2O_3^=$$

The displacement of the previous equilibrium towards the right should be improved by this reaction but the excess of negative charges prevails against sulphitolysis.

Thiol is, beyond any doubt, the best reactive agent for cold reduction of disulphide bonds through the reaction:

$$\rangle CH{-}CH_2{-}S{-}S{-}CH_2{-}CH\langle + 2\,R{-}SH \rightarrow 2\,\rangle CH{-}CH_2{-}SH + R{-}S{-}S{-}R$$

This reaction is ruled by a number of equilibria which depend on several parameters (pH, stress during treatment, swelling, protein charge, concentration, time, etc.) but the overriding factor controlling the equilibrium cleavage level of keratin disulphide is the pK of the thiol relative to the keratin pK, 9·8:[21,25] in other words, the relative concentration of thiolate ions RS^- in relation to pH. If the relative pK of the thiol is above 9·8, the cleavage level will be optimal at alkaline pH; if it is below 9·8, the optimal cleavage rate will be reached at neutral or even acidic pH.

The main body of the studies refers to the action of thioglycollic acid, with its high pK of 10·4, or more precisely to its ammonium salt.[26] The first step of the reaction leads to the formation of a mixed disulphide (**A**) which was first proved by Schöberl:[27]

$$\mathrm{K{-}CH_2{-}S{-}S{-}CH_2{-}K + {}^{-}S{-}CH_2{-}CO_2^- \rightleftharpoons}$$

'keratocystine' thioglycollate

$$\mathrm{K{-}CH_2{-}S^- + \underbrace{K{-}CH_2{-}S}_{(b)}{-}\overbrace{S{-}CH_2{-}CO_2^-}^{(a)}}$$

'keratocysteine' (b) (**A**)

(K = keratinic chain)

However, it may further evolve in two different ways, as shown by the experiments of Boré and Arnaud;[28] the assessment of the modifications occurring in the hair confirm the hypothesis of a nucleophilic substitution:

scheme (a):

$$\mathrm{(\mathbf{A}) + {}^{-}S{-}CH_2{-}CO_2^- \rightleftharpoons K{-}CH_2{-}S^- + {}^{-}O_2C{-}CH_2{-}S{-}S{-}CH_2{-}CO_2^-}$$

(a) dithiodiglycollate

scheme (b):

$$\mathrm{(\mathbf{A}) + K{-}CH_2{-}S^- \rightarrow K{-}CH_2{-}S{-}CH_2{-}K + {}^{-}S{-}S{-}CH_2{-}CO_2^-}$$

(b) $\mathrm{(S + {}^{-}S{-}CH_2{-}CO_2^-)}$

Any reduction of keratin by thioglycollic acid in weakly alkaline medium (pH 9·5) leads to the formation of cysteine and lanthionine with a lowered level of sulphur; a linear relationship exists between the amounts formed of these two compounds ($\frac{2}{3}$, $\frac{1}{3}$). In 1950 Schöberl[29] reported an increase of the sulphur level, but this finding may be due to the use of unclean or old thioglycollic acid containing polythioglycollides which could fix a thiol group on the lysine group of hair. This factor may be entirely ignored in practice.

The cleavage level of disulphide bonds depends on the concentration of reducing agent and on the stress applied to the hair. But, whatever the conditions, no more than 65–70 per cent of keratocystinic bonds may be disrupted—it is noteworthy that this value amounts to the estimated quantity of amorphous keratin in the hair. In practice, the reduction level ranges between 19 and 43 per cent.[30]

Optimal conditions seem to be reached with 5 per cent thioglycollic acid and the ratio cysteine:lanthionine quickly falls below 1 when the concentration is less than 1 per cent; it is also noteworthy that thioglycollic acid has some keratolytic effect which increases with stress.[28] A decrease in pH improves the formation of mixed disulphide (**A**), as was suggested in the previous equilibrium—scheme (a)—by the occurrence of a low level of thiolate $\mathrm{{}^{-}O_2C{-}CH_2{-}5}$.[27,30,31] Similarly, the addition of disulphide (for example, dithioglycollic acid) would displace the equilibrium of step (a) towards the formation of mixed disulphide (**A**).

The nature of the thiol compound, mainly expressed by the pK, is therefore critical to the rate at which the above equilibria are established and to the competitive reactions which could occur according to the pH.

Formulae of thiol compounds mentioned below are given in Table 28.1. Those containing carboxylic acid functions have necessarily a high pK because the carboxylate anion prevents the ionization of the thiol group, as for β-mercapto propionic acid, thiolactic acid, cysteine and its N-acylated derivatives,[32,33] dimercapto adipic acid and thiomalic acid, which is a diacid monothiol.

All these thiol compounds are rather less efficient than thioglycollic acid, but their use is claimed to be more suitable for hair that has already been damaged by previous treatments, and with regard to odour.[25,34,35]

The blocking of carboxylic acidity by esterification or amidification significantly lowers the pK (10·4 → 7·8) and allows very effective reductions at neutral pH,[25,34,36] for example with glycol thioglycollate, glycerol thioglycollate and thioglycollamides.

Table 28.1 Formulae and pK Values of Some Thiol Compounds

		pK value
Thioglycollic acid	$HS{-}CH_2{-}CO_2H$	10·4
β-Mercapto propionic acid	$HS{-}CH_2{-}CH_2{-}CO_2H$	10·4
Thiolactic acid	$HS{-}CH(CH_3){-}CO_2H$	10·4
Cysteine	$HS{-}CH_2{-}CH(CO_2H)(NH_2)$	8·3; 10·8
Dimercapto adipic acid	$HS{-}CH(CO_2H){-}(CH_2)_2{-}CH(CO_2H){-}HS$	
Thiomalic acid	$HS{-}CH(CO_2H){-}CH_2{-}CO_2H$	
Thioglycollamides	$HS{-}CH_2{-}CONH_2$	8·4
	$HS{-}CH_2{-}CONH{-}NH_2$	8·0
Glycol thioglycollate	$HS{-}CH_2{-}COOCH_2{-}CH_2OH$	7·8
Glycerol thioglycollate	$HS{-}CH_2{-}COOCH_2{-}CHOH{-}CH_2OH$	7·8

Unfortunately, in addition to the problem of hydrolysis with time, which may alter the efficiency of the compounds, there seems to be some inverse relationship between the potential irritation to the scalp and the pK of the thiol compound.[34] Conditions and very specific associations had therefore to be investigated, but their possible development was limited.[37] Despite its disadvantages (odour, only average efficacy, alkali pH for activity), thioglycollic acid remains a reliable 'partner' since it offers an excellent compromise between activity and tolerance. However, it is not necessarily irreplaceable: among a number of new proposals are some dithiol compounds[38] which may be used at low concentration and at neutral or even acidic pH; they have only faint odour and limited aggressiveness. Their efficacy is linked to their ability to yield a cyclic disulphide by oxidation, which results in a complete displacement of the equilibrium thiol–disulphide towards formation of keratocysteine through an internal thiol–disulphide exchange reaction on thc mixed disulphide:

$$\text{K—CH}_2\text{—S—S—CH}_2\text{—K} + {}^-\text{S}\sim\text{SH}$$
$$\rightleftharpoons$$
$$\text{K—CH}_2\text{—SH} + \text{K—CH}_2\text{—S—S}\sim{}^-\text{S} \rightarrow \text{K—CH}_2\text{—S}^- + \text{(cyclic S—S)}$$

Other interesting thiol compounds have been patented for permanent waving:[39] the polythiolated polymers obtained by addition of an aminothiol compound to a maleic anhydric–vinyl ether copolymer:

$$\left[\text{—CH(R')—CH}_2\text{—CH(CO}_2\text{H)—CH(CO—NH—R—SH)—}\right]_n$$

They may reduce keratinic disulphides, thereby being oxidized to an insoluble disulphide polymer with various degrees of cross-linkage. By setting within the keratinic structure, the polymer is believed to strengthen hair and setting and to protect them from environmental effects, particularly those of water. Furthermore, these compounds are expected to give a much better skin tolerance than monomeric thiol compounds.

Use of enzymes has also been suggested for disrupting the disulphide bonds of keratin, particularly transhydrogenases and reductases,[40] but thiol compounds are usually required as hydrogen donors.

Many other reducing agents have been proposed, such as thioglycerol, hydrosulphide,[41] formamidine sulphinic acid or thiourea dioxide,[42] dithiocarbamates,[43] esters of trithiocarbonic acid,[44] borohydrides,[45] phosphinic derivatives,[46] chlorothioformates,[47] 2-aminoethanol and the so-called keratin, that is, reduced keratin.[48]

Re-oxidation Step

The oxidation of hair after any of these reducing treatments is a necessary stage in hardening the structure into the new imposed shape (straightened, curled or waved). The basic process is a simple re-oxidation of keratocysteine into keratocystine but at the same time it should induce the formation of cross-linked fibres and restore the former mechanophysical properties of hair:

$$\rangle\text{CH—CH}_2\text{—SH} + \text{HS—CH}_2\text{—CH}\langle$$
$$\downarrow [\text{O}]$$
$$\rangle\text{CH—CH}_2\text{—S—S—CH}_2\text{—CH}\langle$$

It means that pairs of keratocysteine groups happen to be in a favourable position to re-establish a disulphide bridge under the action of oxygen—otherwise keratocysteine may undergo peroxidation to the stage of sulphinic or sulphonic acid. Similarly, side compounds formed during the reducing step may evolve into different oxidation levels. So, in fact, the oxidation process is no simpler than the reduction process considered above, the more so because it is operated with an excess of oxidant (H_2O_2, bromate, iodate, etc.).

A number of studies have been carried out to explain the various possible reactions involved,[26,29,49] but they were often biased because of analytical artefacts. For example, a certain amount of cysteic acid was said to appear; it is now known that, although this degree of oxidation may be reached to some extent with continued treatments,[30] it is usually produced during the analytical process[28] as a result of acidic hydrolysis of lanthionine oxidation species. The studies of Boré[28] have cleared up most of the mystery and settled some of the earlier proposals, by giving a consistent assessment of the two phases in cold waving process: under mild conditions of setting (20 minutes, 25°C) transformation of keratocysteine into disulphide is optimal (96–99 per cent) without prejudicial keratolytic effect either with '6 volume' hydrogen peroxide at pH 3 or with 18 per cent sodium bromate solution at pH 5. The lanthionine formed during the reduction phase—which may represent 30 per cent of the hair cysteine—is oxidized, at least 70 per cent, to sulphoxide and sulphone (the same applies to the hair methionine):

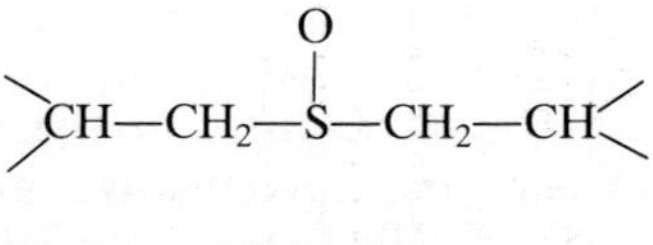

$$>CH-CH_2-\overset{\overset{\displaystyle O}{|}}{S}-CH_2-CH<$$

sulphoxide

$$>CH-CH_2-\underset{\underset{\displaystyle O}{|}}{\overset{\overset{\displaystyle O}{|}}{S}}-CH_2-CH<$$

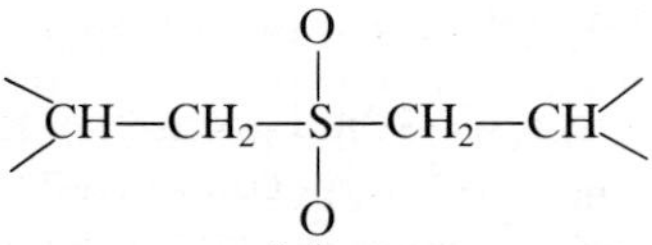

sulphone

On the whole, permanent wave processing with thioglycollic acid shows a loss of 10–30 per cent of cysteine transformed into lanthionine and its oxidation derivatives.

As to whether lanthionine may form bridges between the chains or contribute to the quality of permanent wave, discussion is still open, although the considerable decrease of solubility in urea–bisulphite medium when lanthionine increases favours the cross-linkage theory.

Sulphite-reduced hair is usually hardened through the same procedure: the thiosulphate and the thiol groups are transformed into disulphide by hydrogen peroxide, but by a much slower reaction process and therefore not completely.

Cross-linking of reduced hair may be effected by other means than through peroxides, which may involve a risk of lightening the hair; thus polythionates[50] allow the rebuilding of keratocystine without the side-effects mentioned with the former compounds. A formaldehyde-producing agent such as hexamethylene tetramine may lead to another kind of cross-linkage: 'methylene dithio' linkage. Azadioxabicyclooctane derived from tris(hydromethyl) aminomethane and formaldehyde is recommended for this purpose.[51]

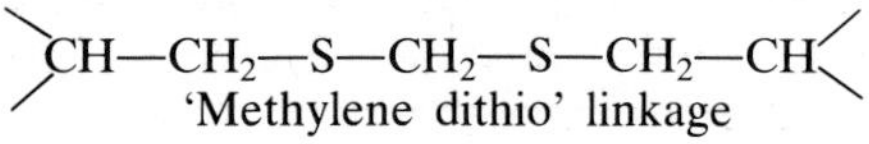

$$>CH-CH_2-S-CH_2-S-CH_2-CH<$$

'Methylene dithio' linkage

Alkylene dihalides[52] also yield dithioether compounds but the conditions of the reaction and the potential skin irritancy of these alkylating substances seem to be hardly compatible with human use. The use of particular salts has been claimed for cross-linking: triethanolamine titanate,[53] the zirconium salt of an organic hydroxy acid,[54] barium salts[55] which produce S—Ba—S linkages, divalent metal mercaptide of N-methylol thioglycolamide with double cross-linking potentialities;[56] sulphinamides were also described as suitable neutralizers.[57] More recently, acylthiosulphates[58] were proposed as cross-linking or alkylating agents, as well as unsaturated compounds such as maleates.[59]

Another patent[60] reports the insertion of polymers into the keratinic structure by polymerization *in situ* of a vinylic monomer by a catalyser in reduced hair.

More recently a rather different chemical reaction was patented to wave hair[61] whereby hair undergoes oxidation, for example by alkaline monopersulphate, without prior reduction. This reaction, better performed with chelating agents (EDTA or its salts), is believed to modify hair without yielding visible discoloration.

In addition, in spite of the theoretical and practical reasons for using sulphites in hot waving and thio-compounds in cold waving, successful cold products for the achievement of 'soft' permanent waves are now available which employ sulphite at high pH (about pH 10) and hydrogen peroxide neutralization. Such products have been discussed by Markland.[62]

Evaluation of Permanent Waving

It may be understood from the foregoing that the chemistry of the waving process is extremely complicated and therefore the investigative method, however sophisticated, cannot encompass all the physicochemical modifications involved, particularly the state of hair after treatment. As with the other hair processes, permanent waving may only be assessed *in vivo*.

A number of techniques *in vitro* may, however, give some valuable information on the evolution of the mechanophysical properties during and after the process. They thus help in the screening of potential agents.

The instrument most commonly used to investigate the mechanical properties of hair is the extensometer, which allows a graph to be drawn relating the extension of the hair to the load applied. Typical curves obtained in this way are shown in Figure 28.1. It may be seen that there is a sharp change in the slope of the extension curve at A. The yield point corresponds to an extension of slightly less than 2 per cent and section OA represents an elastic-like response. Beyond A, hair is seen to stretch easily with a plastic-like response up to an extension of almost 25 per cent (B). Beyond B there is a new change in the slope of the curve which shows increasing resistance to stretching. C corresponds to the disruption point.

The precise physical interpretation of the various sections of the curve is still controversial, particularly in relation to the contribution of amorphous and crystalline regions of the hair. Speakman,[63] studying wool, demonstrated that salt links had no effect on these properties. Hamburger and Morgan[64] attributed the OA portion of the curve to the breaking of hydrogen bonds and unfolding of keratin spirals and stated that the main effect of permanent waving agents was to

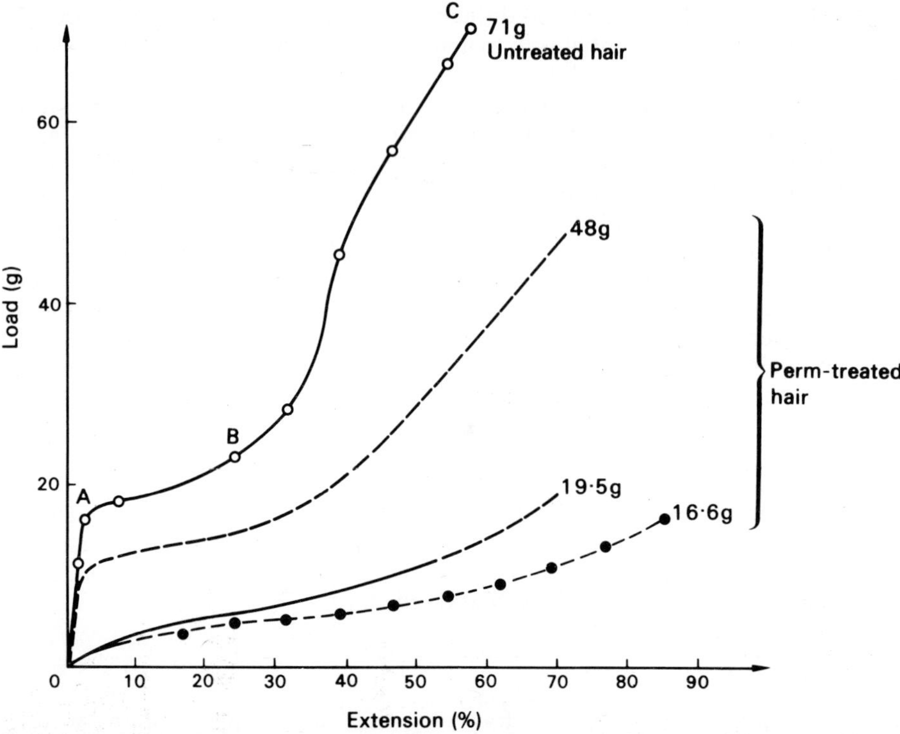

Figure 28.1 Percentage extension of hair related to load applied

lower the slope of the BC section, which is said to represent the resistance of disulphide bonds to extension.

Patterson *et al.*[65] showed that the total work of extension to 30 per cent was lowered to 65 per cent of its original value after reduction with thioglycollic acid and could be brought back to its former value either by oxidation or by substituting bulky groups on to the cysteine residues, but this appears to be a somewhat idealized picture.

These studies, more or less adapted from investigations into wool, may not adequately explain the effect of a process performed on hair which, although of similar structure, is of different composition and sensitivity. The most suitable extension conditions for studying the setting of hair cover the range of the mechanical strain usually set up in hair, that is 0–10 per cent; in setting, the strain ranges from 0 to 1·5 per cent and in permanent waving it lies around 2 per cent. The plastic region (Figure 28.2) is thus suitable; the rates of reduction and of setting and a mechanical overall assessment well correlated with analysis[28] are obtained by comparing the stresses necessary to effect a given extension. It is noticed that in the end the permanent waving process always leads to a notable loss of mechanical properties;[30] the oxidative step restores only part of the initial strength of the hair fibre. However, owing to a lack of systematic studies a direct relationship with loss of keratocystine has not been established.

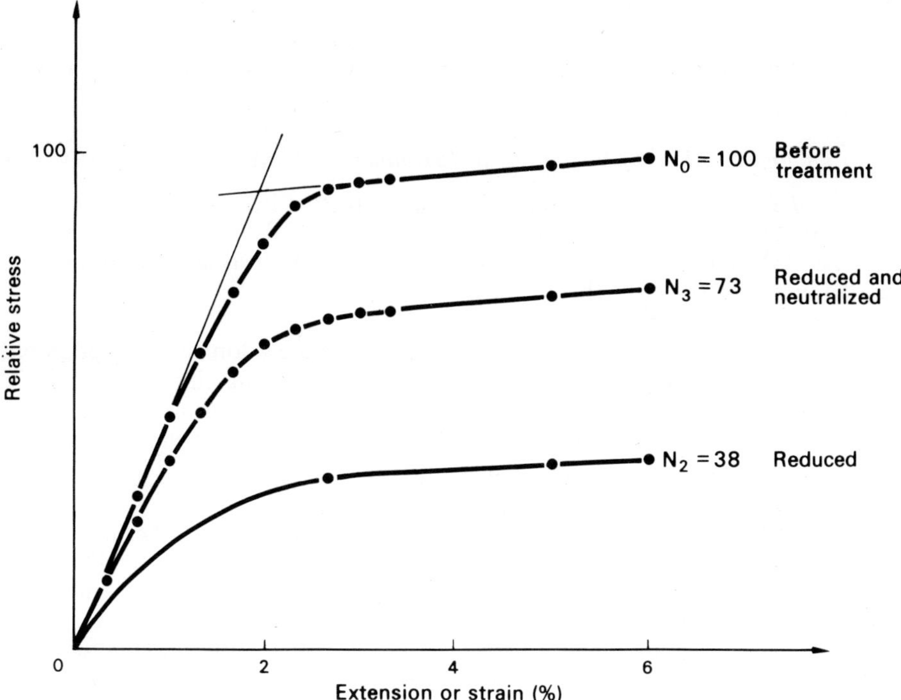

Figure 28.2 Relative stress, N, for a 6 per cent strain at stages of the permanent waving process; $N_0 = 100$ for untreated hair

Formation of lanthionine may contribute to weakening of the mechanical properties of the fibre but it is most likely that intra-keratinic chain cystine links may be formed in the neutralizing step, which reduces the level of cross-linking, that is, the rebuilding.

Other techniques for evaluation *in vitro* are the following:

Measurement of disruption loads, extension at disruption point and disruption energy (point C of curve 1, Figure 28.1).

Examination of curve of extension at constant load, hair being immersed in the reducing solutions to be compared.

Evaluation of perming efficacy, that is the ratio of diameters between curlers and final curl.[37]

Measurement of fibre swelling during reducing step.

The effect of permanent waving on the surface of hair can be studied by scanning electron microscopy.

However, the final evaluation lies in the hands of the 'man of the art', whose sensitivity supersedes any other means of assessment. Permanent waving is a meticulous operation in which a number of parameters must be taken into account in practical trials. Any variation in one of these can result in significant changes in the tightness and permanence of the resulting wave, and the usual precautions must be taken in carrying out comparison work to make sure that observed variations are really due to the factor under investigation (for example, the strength of the lotion) and not to an artefact.

Main Factors Leading to Variations in Permanent Waving

Choice of Lotion. This is the most significant factor; it is independent of the degree of curling desired but depends on the quality and structure of the hair to be treated. It may even be chosen according to whether it is applied to the root or to the tip of the hair (bleached hair).

Temperature Variation. The ambient temperature in the salon has a considerable effect on the cold waving process. This factor will be nullified if tests are carried out on either side of the same head, but other sources of variation such as the heat of the hair and possible heat loss by evaporation can be quite significant if the lotions being examined are vastly different in type.

Processing Time. This should be short, since if it is extended the result may be excess softening of the hair without further efficacy of the treatment. In hot or tepid waving the processing time may be reckoned to begin either when the apparatus is switched on or when the heating pads are activated.

In the usual method of cold waving, the hair is wetted with lotion before and after winding the curls. A salon operator cannot wind a head of hair in much less than 30 minutes, and the home user may take as long as 90 minutes. In view of this time interval between the application of lotion to the first and last curls, the so-called processing time after rewetting becomes far less significant.

The equilibrium between cystine in the hair and thiols in the lotion is reached after about four minutes when the hair is immersed in a large volume of lotion, but this process is probably slower on an actual head. Most of the lotions are studied at setting times, after winding on natural hair, varying between 10 and 20 minutes.

Some manufacturers (see example 5 below) suggest that the lotions applied before and after winding should be of different strengths. Others recommend a modified process in which the hair is wetted with water only before winding, then treated with lotion from the outside of each curl. This, of course, raises in an acute form the problem of penetration, but it is often of use for the treatment of bleached or otherwise damaged hair, as it allows the use of a really short processing time entirely within the control of the user.

Penetration Rate. This depends not only on the presence or absence of wetting agents, etc., but also on the nature of the hair to be waved. Fine hair will obviously be penetrated more rapidly than coarse hair, but there may also be differences in the porosity of the hair. After bleaching, for instance, the hair has usually lost a good many cross-linkages and will swell very much more rapidly than normal hair, thus taking up the waving lotion more completely. Washing the hair with a shampoo beforehand removes sebum which might hinder penetration but some detergents are strongly absorbed on the fibre and may be capable of changing the permeability. This point should not be overlooked when adding detergents to the lotion itself—cationic detergents, for instance, show a definite inhibiting effect when present at concentrations as low as 1 per cent.

Choice of Curlers. The curl diameter obviously depends on the diameter of the curler. Some methods make use of very large diameter curlers or rollers, and some depend on making a pincurl with the aid of a waving lotion. Another factor is the amount of hair on each curler, which depends on the number of curlers per head and the amount of hair on the head, as the outside layer of hair has a curl diameter equal to the curler diameter plus the thickness of hair.

Neutralizing Step. This is a rather important operation since it determines the set of the permanent waving and the rebuilding of keratin. Hair must be very carefully rinsed beforehand to eliminate reducing lotions. Then the setting lotion must be applied in two separate stages: first, two-thirds of the lotion are applied to the wound hair, impregnating each curl for 5 minutes and left for 5 minutes; then, after the hair has been unwound without stretching, the remaining lotion is applied to the tips and left for 5 minutes before being carefully rinsed out.

Hot Waving Processes

The procedure adopted by the professional hairdresser in hair waving is as follows:

1. Any grease is removed from the hair by shampooing.
2. The hair is then divided and wound around a suitable roller under slight tension.
3. A sachet or absorbent strip dipped in a suitable solution is wound over the hair and the whole encased in an electric heater, and the hair is steamed for the required period.

It must be understood that the above method is modified somewhat according to whether *sachet* waving or *oil* waving is being employed and according to the particular type of wave desired. Usually heat is supplied to the rollers by means of an electric current; in the wireless system the heaters are preheated and allowed to cool for the desired period. Chemical heating methods are also used, in which moisture from the wrapper induces an exothermic chemical reaction when in contact with a suitable mixture. The advantages claimed for this latter system are that there is no risk of electric shock and the individual has freedom of head movement.

Permanent waving demands, in addition to suitable reagents, considerable professional skill and experience. The selection of suitable methods of winding and of strengths and types of reagents, times of steaming, etc. (according to the type and condition of hair treated) is of paramount importance.

It has been claimed by hairdressers that the success of permanent waves can be affected by the state of health of the subject. Thus after pregnancy, after an operation involving an anaesthetic or during periods of the menstrual cycle the effectiveness is reported to be reduced. In view of our lack of knowledge of the changes occurring in the skin and hair during menstruation and pregnancy and following parturition and surgical operations, there is no definite proof either way.

Goodman[66] considers that hair which has been keratinized and has erupted beyond the limits of the hair follicle is outside the sphere of influence of anything taking place inside the body. Only the freshly growing part of the hair within the follicle—the part not yet keratinized—is under the influence of the body. He points out that hairdressers have successful salons in hospitals and that the actual causes of failure of permanent waving are:

Taking too large a section of hair
Taking too small a section of the hair
Too much or too little alkali
Too tight or too loose winding
Too short or too long steaming time

Goodman suggests that reasons other than these provide excuses for the careless hairdresser, the hurried technician and the cut-rate operators who do not make a test curl.

Permanent Waving Solutions
Permanent waving solutions are almost invariably strongly alkaline in reaction, since the presence of alkalis considerably shortens the time necessary to produce a satisfactory wave. Those recommended for such purposes include lithium hydroxide; sodium, potassium, ammonium carbonates, borax, ethanolamine or neutral or alkaline solutions of sulphites (sodium, ammonium, mono- or triethanolamine, morpholine sulphites).

The following formulae give good results on average hair in about 10 minutes steaming times (this being adjusted to suit the type of hair):

	(1) *per cent*
Monoethanolamine	6·0
Potassium sulphite	1·5
Potassium carbonate	1·5
Ammonium carbonate	2·5
Borax	0·5
Sulphonated castor oil	1·0
Distilled water	87·0

	(2) *per cent*
Ammonium hydroxide (s.g. 0·88)	20·0
Sodium carbonate	4·0
Potassium sulphite	2·0
Water	74·0

	(3) *per cent*
Monoethanolamine or ammonium hydroxide (s.g. 0·88)	14·0
Borax	4·0
Potassium sulphite	2·0
Water	80·0

The use of agents that activate the lanthionization (NaCl, sulphates, cationic surfactants) and of boosters (urea, amides, lithium bromide) helps to reduce the concentration and lower the pH.[4]

A gel to be applied at 40° for 30 minutes is given in example 4. However, hot waving, because it is aggressive to the hair and uncertain in its results, is not much used nowadays.

	(4)
	per cent
Hydroxyethyl cellulose WP 4400	4·0
Lithium hydroxide, $LiOH.H_2O$	2·0
Sodium chloride	17·5
Water	*q.s.* to 100·0

Chemical Heating Methods (Heating Packages)

Chemical heating methods have been introduced, in which the heat required for the normal waving process is obtained without the aid of electricity. Such methods depend upon heat evolved by the reaction of an exothermic material with a moistening medium as a result of one of the following:

Oxidation and reduction
Hydration
Neutralization

This method originated in England in 1923[67] with the use of quicklime to generate the heat. Combining ammonia sulphate, agar or ammonia with the moistening agent was shown to delay further the development of heat.

Since that time numerous chemical mixtures have been recommended and patented, including as active agents aluminium and its chloride and sulphate, ammonium salts of various organic acids, barium salts, copper carbonate or nitrate, iron filings and other salts, etc.[68]

Cold Waving Processes

Cold waving has to a large extent replaced the older hot waving process, particularly in the more sophisticated parts of the world where the replacement is total. The operation is performed at room temperature without input of heat energy. The hair is shampooed thoroughly, and divided into sections for ease of handling. Locks of hair of such size as to give 35–50 curlers (6–14 mm diameter) per head are moistened with the waving lotion, and then wound on to croquignole curlers. Because the nape hair needs a good curl to stop it 'wisping' and because it also seems rather resistant to perming, it is usual to start at the nape and work forward.

When winding is complete, the hair is left to 'process' for 10–40 minutes. Some manufacturers recommend set times for various types of hair, while others advise the examination of a test curl by the user, who then decides how much longer to leave on the lotion. Hairdressers usually use the test-curl method. The complications of the home perm have brought about an interesting social situation, in that about 60 per cent of UK users have someone to help them with the process. After the processing with reducing agents, the hair is rinsed and neutralized by application of an oxidizing solution to the wound curls. After 5–10 minutes the curls are unwound and usually a further application of neutralizer is made. The hair is then rinsed and set into the desired style.

The Reducer in Cold Waving

As explained above, most permanent waving formulations are based on thioglycollic acid. The simplest lotions are ammonium thioglycollate at pH 9·2–9·8. The curve shown in Figure 28.3 represents solutions with the same curling potential (average lotion for normal hair).[37]

In practice, pH 9·3–9·5 is used for concentrations of thioglycollic acid between 7·5 and 11 per cent, but the strength of the lotion is essentially dictated by the quality of the hair to be waved; the average concentrations for use in a salon are as follows:[37]

	Thioglycollic acid per cent
Difficult natural hair	8–9
Medium and easy natural hair	7
Slightly bleached hair	5
Medium bleached hair	3
Heavily bleached hair	1

For use at home, concentrations should be decreased by about one-third.

The use of bases other than ammonia has been much discussed. Sodium and potassium hydroxides have been said to render the hair too soft to take a good wave;[69] ammonia and organic amines have been proved to hydrolyse peptide less than alkali. In the event, only monoethanolamine seems to be as good as ammonia, and this is used together with ammonia to reduce the odour.

The main problem is to maintain the reducing activity of the thiol during the whole softening phase—that is, to maintain pH despite the volatility of

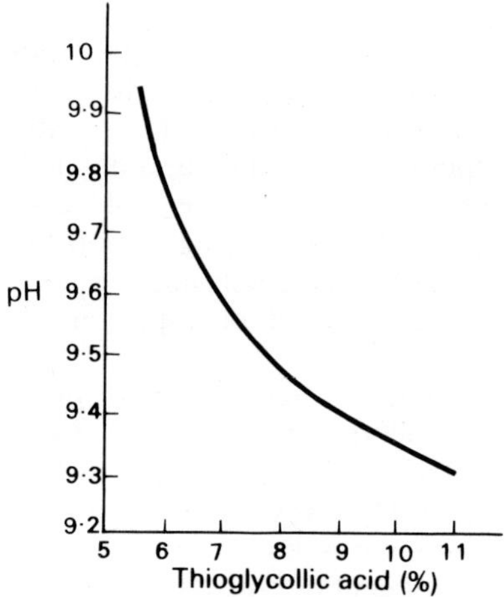

Figure 28.3 Solutions of thioglycollic acid with the same curling potential

ammonia; to achieve this, buffering with ammonium carbonate, sesquicarbonate or bicarbonate is often employed, and this also allows the use of a lower pH (<9).

Another ingenious method of ensuring the base supply and obtaining more regular curling and reduced aggressiveness is to generate ammonia *in situ* under controlled conditions by introducing urea and the enzyme urease which, in the presence of water, transforms urea into ammonia.[70]

Other bases have been suggested: basic amino acids such as arginine,[71] carbonates of alkali amino acid salts such as sodium glycinate carbonate,[72] alkanolamines 'specific for bleached hair',[73] and guanidine or its carbonate[74,75] which, in addition, contribute to the swelling of the hair by disrupting hydrogen bonds, although the problem of safety has to be dealt with.

In order to maintain the reducing activity during the stretching and shaping process it has also been proposed to introduce reducing agents in two phases, the second lotion containing a more potent reducer (ester of thioglycollic acid) which can give stronger bonds,[76] at the same time lowering the pH to reduce damage to keratin (example 5).

	(5) *Parts by weight*
Ammonium mercaptoacetate (50% aqueous)	9·4
Ammonium bicarbonate	4·2
Ammonium carbonate	0·5
Urea	3
Perfume	0·4
Oleic acid pentaerythritol polyglycol ether	0·4
Water	62·1

Procedure: 40 g of this lotion is applied to the hair. Another 40 g is mixed with 4 g mono mercaptoacetic glycerol ester and the hair is treated for 10–15 minutes, then washed with water and fixed with hydrogen peroxide.

Many other ingredients may be used in permanent waving lotions: for example, hardening agents to reduce the necessary thioglycollic acid concentration and pH, and boosters such as urea, lithium bromide or 2-imidazolidinone (ethyleneurea) which penetrate by breaking hydrogen bonds. Example 6 using such a booster,[77] has a pH value of 8.

	(6) *Parts by weight*
Ammonium thioglycollate (50% aqueous)	18
2-Imidazolidinone	4
Ammonium bicarbonate	4
Perfume	0·3
Polyoxyethylene octylphenol	0·5
Water	73·2

Other ingredients include:

Alcohols (ethanol and isopropanol) and sodium sulphite to strengthen the lotions.

Wetting or foaming agents.

Softening agents: animal and vegetable oils, protein hydrolysates, lanolin derivatives.

Complexing agents, mostly for iron which colours and also catalyses the formation of hydrogen sulphide.

Opacifiers, to give an impression of softness and gentleness through their milkiness—acrylic, vinyl and styrene polymers have been claimed as opacifiers;[78] stable emulsions of ammonium thioglycollate have been reported,[79] but a cream formulation makes rinsing difficult.

The application of permanent waving products as a quick-breaking foam from an aerosol dispenser has been described[80] and such preparations have been marketed.

More recently, cationic compounds and, chiefly, resins have been introduced so as to impart improved strength, body or elasticity to the waved hair. They are cationic PVP–VA polymers,[76] polyamide–epichlorhydrin resins,[81] quaternary ammonium salts[82] and protein hydrolysates.

For hair which has been weakened by previous treatments, particularly by bleaching and tinting, less potent reducers may be used such as thiomalic or thiolactic acids[83] or the so-called 'acid permanent waving' formulations (pH between 5·5 and 7) said to 'protect' hair,[84] and which today enjoy some popularity. Following this trend, glycerol monothioglycollate has undergone some revival.

The Neutralizer in Cold Waving

Most hairdressers use hydrogen peroxide as neutralizer because it is not expensive and is easy to handle. It can be obtained as a concentrated solution to be diluted or as a ready-to-use solution. In either case it is acidified by adding citric, tartaric or lactic acid for stability. Usually agents such as polyoxyethylene fatty alcohols or cationic compounds are added to improve wetting, together with softening agents (self-emulsifying waxes, lanolin derivatives).

To speed up the neutralizing step, activators of the decomposition of inorganic percompounds have been suggested.[85] Self-warming systems have also been proposed—for instance, the addition to hydrogen peroxide of either sulphite or thiourea,[86] or anhydrous calcium chloride or magnesium sulphate,[87] elicits an exothermic reaction, producing an available oxidizing bath of about 40°–45°C.

Catalase[88] was suggested to degrade the potential hair-decoloring excess of hydrogen peroxide. As in the reducing stage, cationic polymers and surfactants are added to enhance the setting.[89,90]

For home use, dry powdered sodium perborate or percarbonate is often used in the UK and sodium or potassium bromate in the USA. The bromates have the advantage that stable solutions in water can be made and stored, but they are extremely toxic and should be labelled 'Poison'. These persalts should be used at a neutral pH; bicarbonate, monosodium phosphate or carbonate is added to settle the pH after solution is made, together with foaming agents.

It used to be claimed that the neutralizing step could be achieved without the intervention of a chemical agent, by merely leaving the hair in curlers for more than 6 hours and trusting the air oxygen to do the job; this process is obviously only suitable for home use. Evidence now exists to show that this does not work and a still more convincing proof is that a hair switch, previously reduced under standard conditions by thioglycollate, was not significantly reoxidized even after more than a year of atmospheric exposure![91]

Formulae of both setting lotion and neutralizer for home and professional use, quoted by Shansky,[75] are given in examples 7 and 8.

(7)

Setting lotion (pH 9·4–9·5)				
Ammonium thioglycollate (52%)	17	US gal	64·3	litres
Caustic soda (76%)	20	lb	9·0	kg
Ammonia (28%)	6·4	lb	2·9	kg
Nonionic detergent	3·0	lb	1·4	kg
Resin cloud	3·0	US gal	11·3	litres
Water	150	US gal	568	litres
Neutralizer				
Hydrogen peroxide	158·5	lb	72	kg
Citric acid	6·75	lb	3·0	kg
Polyoxyethylene lauryl ether	16·75	lb	7·6	kg
Resin opacifier	2·5	lb	1·1	kg
Water	400	US gal	1514	litres

(8)

Setting lotion (pH 8.7)				
Ammonium thioglycollate (52%)	15	US gal	56·8	litres
Ammonium hydroxide (28%)	8·0	lb	3·6	kg
Ammonium carbonate	8·0	lb	3·6	kg
Resin opacifier	2·0	US gal	7·6	litres
Nonionic detergent	1·8	lb	0·8	kg
Water	100	US gal	378·5	litres
Neutralizer				
Sodium bromate	200	lb	90·7	kg
Polyglycol 400 laurate	50	lb	22·7	kg
Diglycol stearate	10	lb	4·5	kg
Glyceryl monostearate	10	lb	4·5	kg
Sodium cetyl sulphate	25	lb	11·3	kg
Polyglycol 400	5	lb	2·25	kg
Water	700	US gal	2650	litres

Single-operation Waving

A method has also been patented for hair waving or straightening in a single operation.[39,92] It had not previously been possible to effect permanent deformation in a single phase, because one process involves reduction of the hair and the other oxidation: if both active agents were to be combined, they would interact and their separate effects would be inhibited. The process for the permanent deformation of hair by a single stage entails subjecting the hair to the action of a mixture of a thiol and a disulphide for 20–30 minutes at ambient temperature, after which the hair is rinsed and the curlers are removed. The optimum molar ratio of disulphide to thiol is a function of the pK value of the thiol used. The pH value of the applied composition should preferably lie within a range of 8·5–9·5 in order to permit the disulphide present to act as an oxidizing agent.

It is believed that when hair keratin, itself a disulphide of a complex nature, is brought into contact with the mixture of thiol and disulphide (the latter at a higher concentration), part of the thiol reduces part of the keratin, resulting in the formation of the corresponding disulphide, thereby increasing the concentration of disulphide in relation to the thiol and bringing the reduced keratin into contact with the disulphide. The disulphide then regenerates the original keratin, the corresponding thiol being formed at the same time. In other words, it is a cascade reaction—disulphide breakage and rebuilding—based on an alternated displacement of chemical equilibria and thereby allowing a smooth styling of the hair.

New conditioning systems for permanent waving agents have been proposed, including strips, squares, or continuous rolls from which the agents may be released by wetting,[93] and microcapsules soluble in appropriate solvents,[94] which might in the future constitute a means of applying hydrogen peroxide, reducing agents and ammonium salts all together.

Tepid 'Warm Air' Wave

Thioglycollates have been used at slightly elevated temperatures to obtain a better result with less active material, either employing a special version of the standard electric apparatus or using hood dryers. Cold waving lotions are usually diluted by 30 per cent and sodium sulphite is often added so as to reduce thiol concentration even more. However, this technique may be regarded as obsolete nowadays.

Roller and Pin Permanent Waves

Ordinary permanent waves are normally intended to remain in curl for 4–6 months, and may therefore start with the hair much more tightly curled than is desirable. After the process the hairdresser has to set the hair in a looser style by water waving. This resetting takes a good deal of salon time, which accounts for most of the cost of a professional permanent wave.

It is possible, of course, to set the hair in looser styles while still using a permanent waving process, as long as the client does not expect the curls to last more than 6–8 weeks, and this is, in fact, done with roller and pin permanent waves.

The first method uses large-diameter curlers ($1\frac{1}{2}$ in or 64 mm), while the second employs ordinary hairpins to make flat curls about 1 in (25 mm) in diameter. A reducing lotion is applied as usual, the hair is processed and neutralized and the client then goes straight under the dryer, and needs only brushing to complete the set when her hair is dry.

Some manufacturers recommend a further saving by drying the hair immediately after the waving lotion is rinsed out, but this tends to produce waves that fall out as soon as moisture is present. It is doubtful whether this process, carried out every 6 to 8 weeks, is good for the hair and furthermore these soft waves require constant attention at the expert hands of the hairdresser.

Instant Permanent Waves

Instant permanent waves are not really different from those that are chemically neutralized in the ordinary way, except that the 'processing time' after winding is nil. The hair is first wetted with the reducing lotion, then wound onto the curlers; as soon as winding is finished, setting is carried out without processing time. It saves time but the loss of efficacy must be compensated by the use of a stronger lotion; as a result, the waving is often uneven and the quality of the hair suffers.

Perfuming of Thioglycollate Lotions

An extremely good review of this subject is given by Sagarin and Balsam,[95] who list the results of their own tests with over 200 materials. Unfortunately it is impossible to mask the smell of thiols completely, particularly during the application stages when they are spread over the whole head, and the inherent smell of the thiol is enhanced by the odour of the reduced hair. Another difficulty is to choose perfumes that can be stable in ammoniacal reducing medium. The cosmetic industry and the public are still awaiting new methods of permanent waving using completely odourless materials;[96] a significant reduction in odour is claimed for the cold 'soft perms' which use sulphite as the reducing agent.[62]

Toxicity

The toxicity of permanent waving lotions has been reviewed in considerable detail by Norris.[97] It must be accepted that, owing to the pH of use and in view of the mode of action on hair, compounds such as thioglycollate will be toxic if they get into eyes, or irritant if they are left on the skin for a considerable period of time. The inclusion in the reducing lotion of anti-irritation agents such as antihistaminic imidazoles[98] or thiamine[99] has been claimed to be effective. In all cases care must be taken to prevent contact of such lotions with eyes and skin.

Hair Strengthening Preparations

Weakened hair, which may result from sensitization due to hair treatment or overprocessing or from environmental or internal causes, usually lacks bulk, tensile strength, lubricity, sheen, body and a barrier to penetration. It is often stringy, oversensitive to moisture, or even flimsy and, as a result, difficult to handle. The prevailing factor in the weakening of hair is the breaking of too many disulphide links, as may be the case with hair that is too frequently and excessively waved, bleached or dyed.

A process was proposed in 1967[100] for improving the strength and elasticity of weakened hair, which entailed treatment of the hair with solutions or dispersions of dimethylol urea or dimethylol thiourea for 15 minutes at a temperature of 30°–40°C in the presence of an acid catalyst (for example, glycerophosphoric acid) in an organic or aqueous medium at a pH between 1·5 and 6·0. These compounds react with the free amino groups of hair keratin, producing cross-links between the hair fibres, and polymerize in the presence of the acid catalyst to form a resin within the hair fibres. The resin strengthens the hair and is said to make the hair virtually insoluble in water and various organic solvents.

It has been found, however, that these compounds when used alone are markedly unstable and can release a certain amount of formaldehyde while in solution or in contact with the hair and, to some extent, even during storage. In many countries legislation restricts the amount of free formaldehyde that may be applied to the scalp, and thus the commercial use of such compounds. The release of free formaldehyde has subsequently been shown to be efficiently reduced by the inclusion of stabilizing compounds such as urea, dicyandiamide, melamine or ethylene urea in the methylol compositions; improved compositions for strengthening damaged or weakened hair[101] contain, more specifically, monomethylol dicyandiamide and its methyl ether, methylolethylenethiourea, and methylolated melamines as well as the 'methylols' previously used.

Further compositions for strengthening degraded hair, which comprise at least one alkylated compound, were claimed to be so stable that virtually no formaldehyde was released during application. The methylol ether compounds proposed include mono- and dialkoxymethylureas or ethyleneureas and corresponding thioureas, tris(alkoxymethyl) melamines, N-alkoxymethylcarbamates or adipamides.

Other stable, easily obtainable yet more water-soluble compounds are condensation products of the former linear or cyclic methylolated substances with secondary amines such as N,N′-*bis*-(morpholinomethyl) urea or N,N′-*bis*-(ureidomethyl) piperazine[102] and methylol derivatives of glyoxalurea or thiourea condensation products.[103]

Lastly, addition of fairly small quantities of sulphites to any of the previous compounds has been claimed to prevent almost any occurrence of free formaldehyde.[104]

The amount of methylolated compound included in strengthening preparations—preferably between 1 and 4 per cent by weight—will depend on the product used and on its water solubility, as well as on the type of hair to be treated. Hydroxylated acids, acetic, phosphoric and hydrochloric acids (or salts thereof, such as acid phosphates) may be used as polymerization catalyst. The remarkable strengthening effect given by these compositions is confirmed by the notable reduction they produce on the solubility of hair in alkali, which has been suggested as a criterion for evaluating physical and chemical changes in the human hair.[105]

Many new compounds have been proposed to enrich hair, enhance its protection against subsequent hair treatments, impart more body and bulk, eliminate a slimy feel on wetting and add lubricity, at the same time restoring it. Particularly interesting are, on the one hand, cationic resins with methylol groups[106] and, on the other hand, methylolated compounds or derivatives with functional groups that may be transferred into the hair or onto its surface so as to modify its sensitivity to chemicals. The above-mentioned functional groups are disulphides[102,107]—such as in N,N′-*bis*(hydroxymethyl)dithiodiglycolamide and 2,2′-dithiodiethyl-*bis*-(morpholinomethylurea); tertiary amines—such as in diethylaminoethylurea and methylimino-*bis*-(3-propylurea)methylols; and quaternary ammonium compounds—such as in alkyldimethylammonium acetamide chloride and diethyl(ureido ethyl) ammonium acetamide chloride methylols.[102–105,107,108] Dicarbonyl compounds such as glyoxal,[109] glutaraldehyde, quinone, etc.[110] and amino-dialdehydes[111] have also been claimed to have strengthening properties.

Repetitive bleaching or retinting will also cause non-uniform porosity of hair with an adverse effect during subsequent treatments. A process for uniform waving of damaged hair[112] entails pretreatment of the damaged hair with an aqueous acidic oxidizing agent, before it is subjected to the action of an aqueous alkaline thioglycollate.

In permanent waving the degree to which the hair is waved will depend on the amount of waving lotion that gets into the hair, and this in turn is governed by the porosity of the hair. The more porous the hair, the more waving lotion has to be applied. This will cause end sections to be well waved while root sections will only be weakly waved. If waving solutions are then applied that are strong enough to wave root sections of, for example, bleached hair, they may cause damage and break the more porous end sections of the hair. However, by using a pre-neutralizing agent, it is possible to apply a waving solution of average strength to obtain a uniform wave irrespective of the variations in the porosity of the treated hair, that is the degree of damage. This will be mainly absorbed at the most porous end section of hair, as with cold waving lotion, and will inactivate it by oxidation, simultaneously reducing its pH and thereby blocking its action. The concentration of the pre-neutralizing agent, which is generally an acidic persulphate, for example, potassium or ammonium persulphate, may vary over a wide range. In the majority of cases proposed formulations contain between 0·5 and 5·0 per cent of active oxygen and have a pH value of between 3·5 and 7·0. Treatment time is preferably about 10 minutes. The cold waving solution is subsequently applied for a further 10 minutes, whereupon the hair is rinsed and neutralized.

More recently, a number of pre-treatments with cationic compounds or resins have been recommended for protection of the weakened hair shafts by preventing penetration and excessive softening by the reducing lotions.

REFERENCES

1. Suter, M. J., *J. Soc. cosmet. Chem.*, 1948, **1,** 103.
2. Kneberg M., *Am. J. phys. Anthrop.*, 1935, **20,** 51.
3. Danforth, C. H., *Physiol. Rev.*, 1939, **19,** 99.
4. French Patents 2 220 243 and 2 220 244, L'Oreal, March 1974.
5. Schöberl, A., *Biochem. Z.*, 1942, **313,** 214.
6. Schöberl, A., *et al.*, *Biochem. Z.*, 1940, **306,** 269; 1964, **317,** 171.
7. Mitzell, L. R. and Harris, M., *J. Res. nat. Bureau Stand.*, 1943, **30,** 47.
8. Cuthbertson, W. R. and Phillips, H., *Biol J.*, 1945, **39,** 7.
9. Elliot, R. L., Asquith, R. S. and Hobson, M. A., *J. Text. Inst.*, 1962, **51,** T 692.
10. Nicolet, B. H. and Shinn, L. A., *Abstracts of 103rd Meeting of the American Chemical Society*, April 1942.
11. Tarbell, D. S. and Harnish, D. P. *Chem. Rev.*, 1951, **49,** 1.
12. Swan, J. M., *J. Text. Inst.*, 1960, **51,** T 573.
13. Parker, A. J. and Kharasch, N., *Chem. Rev.*, 1959, **59,** 583.
14. Zahn, H., *et al.*, *J. Text. Inst.*, 1960, **51,** T 740.

15. Vassiliadis, A., *Text. Res. J.*, 1963, **33,** 376.
16. Speakman, J. B., *Nature*, 1933, **132,** 930.
17. Horn, M. J., Jones, D. B. and Ringell, S. J., *Biol. Chem.*, 1941, **138,** 141; 1942, **144,** 87.
18. Ziegler, K. J., *Biol. Chem.*, 1964, **239,** 2713.
19. Asquith, R. S., *et al.*, *J. Soc. Dyers Colour.*, 1974, **90,** 357.
20. Crewther, W. G. and Dowling, L. M., *Proc. internat. Wool Textile Res. Conf. Paris* (CIRTEL), 1965, **2,** 393.
21. Wolfram, L. J. and Underwood, D. L., *Text. Res. J.*, 1966, (36) 947.
22. Parra, J. L., *et al.*, *Proc. 5th internat. Wool Textile Res. Conf. Aachen*, 1975, **III,** 113.
23. Miro, P. and Garcia-Dominguez, J., *J. Soc. Dyers Colour.*, 1968, **84,** 310.
24. Zahn, H., Chimia, 1961, **15,** 378.
25. Haefele, J. W. and Broge, R. W., *Proc. sci. Sect. Toilet. Goods Assoc.*, 1961, (36), 31.
26. Henk, H. J., *Fette Seifen Anstrich.*, 1963, **65,** 94; Zahn, H., *et al.*, *J. Soc. cosmet. Chem.*, 1963, **14,** 539; Randebrock, R. and Eckert, L., *Fette Seifen Anstrich.*, 1965, **67,** 775.
27. Schöberl, A. and Grafje, H., *Fette Seifen Anstrich.*, 1958, **60,** 1057.
28. Boré, P. and Arnaud, J. C., *Actual. Dermopharm.*, 1974, **6,** 75.
29. Schöberl, A., *Naturwissenschaften*, 1950, **40,** 390.
30. Gumprecht, J. G., *et al.*, *J. Soc. cosmet. Chem.*, 1977, **28,** 717.
31. Asquith, R. A. and Puri, A. K., *Text. Res. J.*, 1970, **40,** 273.
32. US Patent 3 242 052, Mead Johnson and Co., 21 September 1962.
33. German Patent 2 717 002, Kyowa Hakko Kogyo Co., 19 April 1976; Japanese Patent Kohai 77' 125 637, Tanabe Seiyaku, 10 April 1976; Japanese Patent Kohai 73' 14 934, Tanabe Seiyaku, 27 September 1968.
34. Voss, J. G., *J. invest. Dermatol.*, 1958, **13,** 273.
35. Finkelstein, P., *et al.*, *J. Soc. cosmet. Chem.*, 1962, **13.** 253.
36. French Patent 1 197 194, L'Oreal, 29 May 1958; German Patent 2 255 800, Saphir J. and H., 15 November 1972.
37. Zviak, C., *Problèmes Capillaires*, ed. Side. E. and Zviak, C., Paris. Gauthier-Villars, 1966,pp.191–213.
38. US Patent 3 459 198, Collaborative Research Inc., 10 March 1966; French Patent 2 005 648, L'Oreal, 5 April 1969.
39. US Patent 3 693 633, L'Oreal, 26 September 1972.
40. German Patents 2 141 763 and 2 141 764, Henkel, 20 August 1971.
41. Japanese Patent 72' 46 333, Yamazaki, 5 June 1970.
42. US Patent 2 403 937, E. I. du Pont de Nemours, 1946.
43. British Patent 771 627, Van Ameringen Haebler Inc., 1 February 1954.
44. US Patent 2 600 624, Parker, A., 15 March 1950; British Patent 672 730, Henkel, 1949.
45. British Patent 766 385, Gillette Co., 15 September 1953.
46. US Patent 3 256 154, Gillette Co., 18 October 1963.
47. Walker, G. T., *Seifen Öle Fette Wachse*, 1963, **89,** 349.
48. German Patent 2 345 621, Henkel, 10 September 1973.
49. Robbins, C. R. and Kelly, C., *J. Soc. cosmet. Chem.*, 1969, **14,** 555.
50. French Patent 1 309 816, L'Oreal 15 April 1960.
51. US Patent 4 013 409, IMC Chemical Group Inc., 22 March 1977.
52. French Patent 1 011 152, Amica, 15 December 1948.
53. British Patent 745 179, National Lead Co., 21 June 1960.
54. US Patent 2 707 697, Horizons Inc., 3 May 1955.
55. British Patent 453 701, Speakman J. B., 1934; German Patent 2 421 248, Shiseido Co., 4 May 1973.

56. US Patent 3 674 038, L'Oreal, 4 July 1972; US Patent 3 803 138, L'Oreal, 9 April 1974.
57. US Patent 3 253 993, Shulton, 29 July 1964.
58. US Patent 3 906 021, Gillette Co., 29 March 1974.
59. Japanese Patent 72' 37015, Tanabe Seiyaku, 9 January 1970; Japanese Patent 76 128907, Tanabe Seiyaku, 30 April 1975.
60. German Patent 2 025 452, Colgate Palmolive Co., 25 May 1970.
61. German Patent 2 349 048, Procter and Gamble Co., 2 October 1972.
62. Markland, W. R., *Norda Briefs*, 1979, (492), 1.
63. Speakman, J. B., *J. Text. Inst.*, 1947, **38,** T 102.
64. Hamburger, W. J., and Morgan, H. M., *Proc. sci. Sect. Toilet Goods Assoc.,* 1952, (18), 44.
65. Patterson, N. L., *et al.*, *J. Res. nat. Bureau Stand.*, 1941, **27,** 89.
66. Goodman, H., *J. Am. med. Assoc.*, 1943, **123,** 743.
67. British Patent 225 256, Sartory, P., 1925.
68. McDonough, E. G., *J. Soc. cosmet. Chem.*, 1948, **1,** 183.
69. Heilengotter, R. and Komarony, R., *Am. Perfum. Aromat.*, 1958, **71,** 31.
70. German Patent 1 124 640, Wella AG, 17 December 1959; German Patent 1 229 980, Schwarzkopf, 30 August 1960.
71. Japan Patent 76' 15 639 Ajinomoto Co., 24 July 1974.
72. German Patent 2 111 959, Eugène Gallia, 16 December 1971.
73. British Patent 1 020 919, Rayette Inc., 28 December 1961.
74. Bogaty, H. and Giovacchini, P., *Am. Perfum. Cosmet.*, 1963, **78**(11), 45.
75. Shansky, A., *Am. Perfum. Cosmet.*, 1965, **80**(3), 31.
76. German Patent 2 263 203, Wella AG, 23 December 1972.
77. German Patent 2 614 724, Wella AG, 6 April 1976.
78. US Patent 2 464 281, Raymond Laboratories Inc., 6 August 1946.
79. US Patent 2 479 382, Ronk, S. O. and Hunter, L.R., 16 August 1949.
80. British Patent 959 772, Procter and Gamble Co., 11 September 1959; German Patent 1 492 164, Schwarzkopf, 21 March 1974; US Patent 3 433 868, Warner-Lambert, 18 March 1969.
81. US Patent 3 981 312, Redken Laboratories Inc., 21 September 1976.
82. US Patent 4 038 995, Helene Curtis, 2 August 1977.
83. Shansky, A., *Soap Cosmet. chem. Spec.*, 1976, September, 32.
84. US Patent 3 847 165, Redken Laboratories Inc., 12 November 1974; Australian Patent 408 443, Summit Laboratories Inc., 27 January 1970.
85. German Patent 1 800 069, Henkel, 1 October 1968.
86. German Patent 2 316 600, L'Oreal, 18 October 1973.
87. German Patent 2 317 140, Wella AG, 5 April 1973.
88. Japan Kokai 75' 95 436, Tanabe Seiyaku, 28 December 1973.
89. German Patent 1 492 163, Schwarzkopf, 7 March 1964.
90. US Patent 3 964 499, Wella AG, 22 June 1976.
91. Arnaud, J. C., unpublished results.
92. British Patent 1 119 845, L'Oreal, 17 July 1968; US Patent 3 768 490, L'Oreal, 30 October 1973.
93. US Patent 3 837 349, Avon Products Inc., 24 September 1974.
94. French Patent 2 033 289, L'Oreal, 30 January 1970; German Patent 2 120 531, Schwarzkopf, 24 April 1971.
95. Sagarin, E. and Balsam, M., *J. Soc. cosmet. Chem.*, 1956, **7,** 480.
96. Shansky, A., *Drug Cosmet. Ind.*, 1975, **116**(4), 48.
97. Norris, J. A., *Food Cosmet. Toxicol.*, 1965, **3,** 93.
98. Japan Patent 70' 32079, Shiseido, 8 March 1965.
99. Japan Patent 70' 26875, Tanabe Seiyaku, 13 August 1965.

100. British Patent 1 196 021, L'Oreal, 19 April 1967.
101. British Patents 1 197 031 – 1 197 038, L'Oreal, 1967–1969.
102. British Patent 1 249 477, L'Oreal, 25 October 1968.
103. British Patent 1 267 846, L'Oreal, 3 December 1969.
104. British Patent 1 236 463, L'Oreal, 26 July 1968.
105. Erlemann, G. A. and Beyer H., *J. Soc. cosmet. Chem.*, 1972, **23**(12), 791.
106. Shansky, A., *Drug Cosmet. Ind.*, 1977, **121**(7), 27.
107. US Patent 3 694 141, L'Oreal, 1 August 1968.
108. British Patent 1 234 408, L'Oreal, 1 August 1968; US Patent 3 642 429, L'Oreal, 23 October 1968.
109. German Patent 2 052 780, Roberts, 28 October 1970.
110. Canadian Patent 900 358, Bristol-Myers, 30 July 1968.
111. US Patent 3 812 246, L'Oreal, 21 May 1974.
112. US Patent 3 395 216, Clairol, 30 July 1968.

Chapter Twenty-nine

Hair Straighteners

Introduction

Hair straighteners are a need for several types of consumer, some of whom want tightly curled hair made straight or very gently waved, while others want essentially kinky hair opened up and made more controllable.

The idea held at one time that curly or kinky hair is elliptical in cross-section whereas straight hair is circular is now completely discarded. According to recent work there are two forms of keratin which make up the orthocortex and paracortex of the hair shaft (see Figure 23.1); Negro hair appears to have a much higher proportion of the reactive and highly differentiated orthocortex.

There are several types of hair straightening preparation on the market:

Hot comb–pressing oil methods
Caustic emulsion
Methods entailing the use of keratin reducing agents

Hot Comb Method

In the hot comb method, hair is straightened by the use of petroleum jelly and a hot metal comb. This procedure is referred to as 'hot pressing'. Alternatively a mixture of petrolatum and paraffin may be used, in which petrolatum is the major component, acting as a heat transfer agent between the comb and hair, lubricating the latter to allow the comb to slide through it without drag. The hair is washed and dried before the pressing oil is applied and the heated metallic comb is used to straighten the hair.

Quite apart from the fact that considerable stress is applied to the hair, causing a high incidence of breakage, the set is not very permanent and tends to be destroyed by rain or even by perspiration so that the hair reverts to its original state. Silicone-containing sprays have been marketed as barriers against moisture, but they did not appear to bring any really significant improvement.

Caustic Preparations

Caustic alkali preparations, usually in cream form, are the second type of hair straightening preparation still fairly widely employed. The use of caustic lye involves risks such as irritation of the scalp and even accidental eye damage.

The viscosity of these products varies according to the softening and melting points of the cream base. The amount of active constituent employed is between 2 and 9 per cent; it is frequently about 4–5 per cent. The more alkali present, the more rapid the action but at the same time the greater the risk of hair solubilization and the care required in using the product.

The addition of activators of lanthionization helps to reduce the temperature, the time and the pH value of treatment[1] (cf. permanent waving).

The selection of ingredients for the cream base to act as vehicle for the caustic obviously requires considerable care if incompatibility is to be avoided. The emulsifier should be neither an acid nor an ester, nor should it be an anionic of a type that would be salted out by the caustic alkali, such as soap. The fatty phase, too, should be neither an acid nor an ester and is best formulated from waxes, more particularly from fatty alcohols, with mineral oils and petroleum jelly as protecting agents and polyoxyalkylated lanolin as emollient.[2] Generally, the cream base also contains sodium lauryl sulphate or lauryl ether sulphate as wetting agent and emulsifier.

A typical composition is given in example 1.[3] This paste is applied by combing and allowed to remain on the hair for 30 minutes. The hair is then washed well with water to remove all the paste.

		(1) *parts by weight*
A	Gum tragacanth	2
	Boric acid	1
	Water	40
B	Sodium carbonate	1
	Potassium hydroxide	1
	Glycerin	2
	Water	8

Procedure: Unify A as a paste and then stir it into the previously dissolved B.

The straightener with activator given in example 2[1] is applied to the hair for 30 minutes at a temperature of 40°C.

		(2) *grammes*
Hydroxyethylcellulose WP 4400		4·0
Lithium hydroxide		2·0
Sodium chloride		17·5
Water	to	100·0

The simple lye straighteners described above are mainly designed for people who want their hair straightened, not styled. There are, however, more complicated multi-component straighteners on the market which provide a more sophisticated process that is more beneficial to the scalp and hair and the final styling.

Multi-component lye straighteners conform to a general pattern of five components:

(i) A soft pomade based on mineral oil–jelly–wax applied to the scalp as a protective pretreatment.

(ii) The 'relaxer', which consists of an oil-in-water emulsion containing about 3 per cent caustic soda and about 40 per cent fatty material on the same general lines as the simple straighteners described above.

(iii) A cream shampoo which is used to follow the relaxer. The constitution is usually 12 per cent alkyl sulphate, 2 per cent fatty acid diethanolamide, together with some unspecified emulsified fatty material.
(iv) A dilute oil-in-water emulsion containing about 2 per cent fatty esters and a cationic wetting agent referred to as the 'neutralizer', preferably slightly acid.
(v) A stiffer pomade containing petroleum jelly, fatty ester and lanolin to provide the final set and dressing.

It has also been recommended to follow the alkaline treatment by an application of an alkaline solution or suspension containing an alkaline earth metal, particularly calcium as in quicklime, so as to form chelates of this metal with keratin and thus firmly set in the new hair shape; keratin may then be released from the chelates by means of a complexing agent (EDTA) and a surfactant.[4]

Compositions that are both straightening and colouring have been patented; they consist of alkali at pH 12–13·8, dyestuffs together with shading agents in a vehicle containing cetyl stearyl alcohol, sodium lauryl sulphate and a carboxyvinylic polymer.[5]

Chemical Hair Reducing Agents

Hair straightening preparations of the third type contain a chemical keratin-reducing agent as the 'relaxer' which effects the softening and straightening of the hair. The active agents are frequently thioglycollates, that is, the same compounds which are used in the opposite process of permanent waving but usually at a slightly lower concentration.

Unlike permanent waving, where the hair is kept in curlers throughout the treatment, in hair straightening the hair is free and kept in shape while combing by virtue of the high viscosity of the product. As a result, a cream-like formulation is no longer a disadvantage but becomes highly desirable, so that the majority of the preparations are oil-in-water emulsions, gels or thickened liquids.

This type of straightener is sold in two containers, one holding the 'relaxer', the other the 'neutralizer', although, just as with 'home perms' there are preparations that dispense with the neutralizer and rely on supposed 'aerial oxidation'.

Relaxer

In most existing examples, the relaxer is an ammoniacal thioglycollate and the pH value of the product is adjusted to 9·0–9·5. Organic bases such as monoethanolamine and an alkali carbonate of an amino acid[6] may partly replace ammonia. Cream products are based on glycerol, glycol stearates, or on cetyl and cetyl stearylic alcohols emulsified in water by polyoxyethylated (usually 20 to 25 ethylene oxide) cetyl or oleyl alcohols. Gels or viscous liquids are obtained by means of carboxyvinyl polymer or copolymers. In example 3,[6] imidazoline is included with the aim of speeding up straightening.

	(3) *per cent*
Emulsion base:	
Demineralized water	to 100·0
Cetyl alcohol emulsified by oxyethylated cetyl alcohol	22·0
Demineralized water	30·0
Sodium carbonate glycinate	5·0
Ammonium thioglycollate or thiolactate (50% aqueous soln)	12·0
EDTA (disodium salt)	0·3
Sodium *p*-hydroxybenzoate methyl ester	0·05
Monoethanolamine	2·0
Imidazoline	0·2
Perfume	0·2

Neutralizer

Neutralizers are the usual materials, either sodium perborate or hydrogen peroxide. It is important to remember that hair which has already been damaged during hot combing treatment or by the use of lye straighteners should not be subjected to a thioglycollate straightener until several weeks have elapsed. The thioglycollate cream is applied liberally to the hair, which is then combed rapidly until the hair no longer has a tendency to curl. When the hair is sufficiently straight the cream is rinsed off and the neutralizer is applied to give a permanent set. Care should be taken to ensure complete neutralization so as to avoid hair damage. Again a cream-like composition is preferentially used to weight down the hair and improve the maintenance of a straightened shape.

To avoid irreversible damage, due in particular to the solubilizing action of thioglycollic acid on the keratin in highly alkaline medium and on already treated hair, different systems have been proposed. The first—rather complicated—consists in introducing the basicity usually required at a late stage in the action of the reducer and entails a series of four applications:

(i) A cream containing 3–8 per cent of thioglycollic acid in an oil-in-water ointment base adjusted to pH 7·0–8·6 with ammonium hydroxide (28 per cent): this can be left on the hair for 45–90 minutes.

(ii) A cream containing ammonium hydroxide or ethanolamine or ethanolamine carbonate in an oil-in-water base adjusted to pH 10–12·5, designed to help straighten the hair: the amount to be applied depends on the thickness and the degree of frizziness of the hair.

(iii) The hair is then washed and treated with an aqueous composition comprising an oxidizing agent, for example potassium bromate or sodium perborate and a buffering agent, for example gelatin (included to ensure that the hair is kept firmly stretched for about 15–20 minutes during the reoxidation process, while the buffering agent maintains the pH of the solution during the treatment).

(iv) The hair is then washed and rinsed with an acid rinse of maleic acid to neutralize any residual alkali. Unlike citric, tartaric and boric acids which have been found to cause curling-up of straightened hair, maleic acid helps to maintain the hair in a straightened condition.

Another technique, based on the use of less potent reducers such as sulphites and bisulphite of ammonium or alkaline metals, offers the advantage of being effective between pH 6·5 and 8·5. In addition, such preparations contain alkali or ammonium carbonate, wetting or jellifying agents (sodium lauryl sulphate, Carbopol) and hair swelling agents. They may also contain a chelating agent to complex metals which could prevent the action of sulphites or, as in example 4, a soap which is said to prevent discoloration and embrittlement of the hair.[7]

	(4) *per cent*
Ammonium carbonate	4·5
Sodium bisulphite	2·2
Sodium lauryl sulphate	4·2
Tallow soap	5·6
Oleic acid	1·0
Water	82·5

These compositions are said to be mild towards scalp and skin and less damaging for hair.[8,9]

Another suggested process derives profit from the delayed use of swelling agents to complete and to enhance the straightening action, thereby reducing the contact time to the minimum with a highly alkaline medium. The so-called swelling agents are compounds such as urea and its derivatives, thiourea, formamide, lithium bromide, lower aliphatic alcohols, benzyl alcohol, sulphoxide and sulphone, which are thought to break hydrogen bonds and/or weaken hydrophobe interactions, penetrating keratin easily and thereby dramatically altering the network of non-electrostatic interactions.

The proposed process consists in applying the keratin-softening agent (alkali, sulphite or thioglycollate) to the hair, and, when the action is substantially complete, in combing out surplus agent; then, in a second step, exposing the hair to a swelling agent and smoothing the hair with repeated combing, and lastly fixing or neutralizing the hair.[11] A formulation for this two-step process is given in example 5.[11]

		(5) *grammes*
A	Thioglycollic acid	8·0
	Cetyl alcohol	5·4
	Paraffin oil	1·8
	Adduct of oleyl alcohol and 20 EO units	3·9
	Colloidal silicic acid	1·5
	Ammonia 25%	12·3
	Water	67·1
B	Cetyl stearyl alcohol	17·0
	Sodium lauryl sulphate	2·0
	Water	59·0
	Thiourea	7·0
	Sulpholan	15·0

Procedure: Apply Cream A (pH 9·6), strand by strand, on the previously shampooed hair and distribute uniformly. After 5 minutes reaction time, carefully comb out the agent; then apply Cream B for 5–7 minutes with repeated combing. Thoroughly rinse the hair with water and fix in the usual way with 100 ml 2 per cent hydrogen peroxide solution.

Example 6 gives another two-stage composition, in gel form, of which cream A is left for a 15 minute action period on the hair.

		(6) *grammes*
A	Ammonium sulphite 35% aqueous soln ($22{\cdot}7$ g SO_2 per 100 ml)	36·5
	Carbopol 960 (ammonium salt of a carboxyvinyl polymer)	3·5
	Water	60·0
B	Carbopol 960 (Goodrich)	0·5
	Water	69·5
	Urea	3·5
	2,2-dimethyl-1,3-propanediol	10·5
	Isopropyl alcohol	15·0

Straightening is still a delicate process and should be carried out with great care, taking into account the nature of the hair so as to avoid dryness, degradation and breaking of hair.

The addition of natural polymers has been patented to promote the setting of the straightened shape; they include soluble proteins, corn or rice starch,[12] fruit galactomannan.[13] More recently, the application of vinyl polymers used for wave sets[14] has been claimed.

Lastly, it is to be noted that another chemical straightening process, entailing an oxidation by alkaline monopersulphates ('oxone') is said to alter the shape of natural hair without appreciably modifying its colour.[15]

REFERENCES

1. French Patents 2 220 243 and 2 220 244, L'Oreal, 7 March 1974.
2. US Patent 3 017 328, Hair Strate, 30 January 1957.
3. Shansky, A., *Am. Perfum. Cosmet.*, 1968, **83**(10), 71.
4. US Patent 3 973 574, Umezawa, F., 10 August 1976.
5. German Patent 2 014 628, Wella AG, 26 March 1970.
6. Belgian Patent 763 084, Eugène Gallia SA, 8 January 1969; British Patent 1 314 625, Eugène Gallia SA, 26 April 1973.
7. US Patent 2 865 811, Irval Cosmetics, 28 December 1958; US Patent 3 171 785, Irval Cosmetics, 21 December 1958.
8. US Patent 3 864 476, Altieri, F. J., 4 February 1975.
9. Spoor, H. J., *Cutis*, 1975, **16,** 808.
10. British Patent 996 279, Scherico, 15 August 1962.
11. German Patent 1 955 823, Wella A G, 6 November 1969.
12. German Patent 2 319 240, Schweifer, J., 16 April 1973.
13. French Patent 2 067 649, Gonzales, J. I., 12 November 1969.
14. US Patent 3 568 685, Scott, H. L., 9 March 1971.
15. German Patent 2 349 058, Procter and Gamble, 18 April 1974.

PART FOUR

The Teeth and Dental Products

Chapter Thirty

The Tooth and Oral Health

Introduction

The function of the cosmetic chemist is that of maintaining or enhancing the appearance of the external surface of the body. The emphasis changes when the human mouth is considered. In civilized communities the prevalence of tooth decay and other oral diseases has come to be accepted as normal. Oral preparations are therefore much more orientated to the prevention and control of oral diseases such as caries and gum disorders. This situation has arisen in the last 30 years as developments in oral biochemistry have increased the understanding of the problems.

THE TOOTH AND ITS SURROUNDINGS

The field of oral hygiene is concerned not only with the tooth, but with the whole of its environment. This involves an understanding of the biochemistry of the whole mouth.

Structure of the Tooth

The tooth is distinguished macroscopically by the *crown* (that portion of the tooth above the gum) and the *root* (the portion embedded in the gum); the constricted portion separating these is termed the *neck*.

Enamel

The outer surface of the crown of the tooth is composed of enamel, a hard tissue, thickest at the apex of the tooth and thinner at the neck. The root is protected by a thin layer of cementum.

Enamel is the hardest tissue in the body. It is composed chiefly of hydroxyapatite ($3Ca_3(PO_4)_2.Ca(OH)_2$) which accounts for about 98 per cent of the composition, the rest being keratin and water. Hydroxyapatite is capable of ion exchange, and anions such as F^- and CO_3^{2-} may replace the OH^- group, while cations such as Zn^{2+} and Mg^{2+} may replace Ca^{2+}. Caries susceptibility may be influenced by this ion exchange and, for example, the extent to which OH^- has been replaced by F^- has an effect on the vulnerability of the enamel.

Enamel in fact contains a wide variety of other elements which are present systemically or by ion exchange. In addition to those mentioned sulphur (270 ppm), chlorine (4400 ppm), potassium (370 ppm) and strontium (56 ppm) are all present at levels above 10 ppm together with a host of other elements at levels below this. Their biological significance is not known.[1]

The hydroxyapatite of which enamel is largely composed is present in the form of crystallitcs which make up rods orientated at right angles to the surface. The theoretical Ca:P molar ratio for pure hydroxyapatite is 1·67, but biological apatites such as enamel are frequently below this level. This departure from stoichiometry has stimulated speculation, but the subject is still not completely resolved. Adsorption of ions which are too large to enter the crystal lattice or which have an inappropriate charge, and ionic exchange, are two mechanisms which may account for non-stoichiometry. A summary of the current position is given by Jenkins.[2]

Dentine

The layer of material beneath tooth enamel is the dentine. It, too is composed of hydroxyapatite to the extent of about 70 per cent, the remainder being collagen and water. The dentine matrix is perforated by a number of tiny canals which radiate from the pulp cavity to the surface. These are the *dentine tubules*.

The apatite of dentine may also contain a variety of other elements and these must of course be of systemic origin since dentine is not exposed to oral fluids. The most remarkable element present is fluorine—remarkable because its content varies with the fluoride content of the water supply and because it is higher than the level in enamel:[3]

	Fluoride concentration (ppm)		
Water supply	0·0–0·3	1·1–1·2	2·5–5·0
Enamel	100	130	340
Dentine	240	360	760

(The enamel figures are average for bulk enamel—the fluoride concentration at the surface of the enamel may be as much as 10 times the average figure.) Thus the softer and more vulnerable tissue, dentine, has a much higher fluoride content than enamel and this fluoride must be of systemic origin. The significance of these facts will be discussed later in this chapter.

The major differences between enamel and dentine are set out in Table 30.1.

Table 30.1 Major Differences between Dentine and Enamel[4]

	Dentine	**Enamel**
Hardness		
Knoop's scale	55·60	250–300
Moh's scale	2	4
Specific gravity	2·14	2·9–3·0
Inorganic matter (%)	68	96
Phosphorus (%)	11·5	16·5
Calcium (%)	24	35
Fluorine (ppm)	240	100
Protein (%)	20	1
Protein type	Collagen	Keratin

Saliva

Saliva is part of the environment of the tooth and is a major factor in the maintenance of a healthy mouth. Since saliva is continuously being produced, the tooth environment is dynamic not static and this complicates the study of the chemistry of the mouth.

Saliva is produced by three pairs of large glands and the smaller glands of the oral mucosa (labial, lingual, buccal and palatal). The secretions differ from one another in composition and may differ themselves according to the rate of flow, time of day, etc. There are also differences between individuals. It is impossible, therefore, to give any meaningful figures for the composition of saliva; tables of published data are probably the best guide.[5]

Saliva contains bacteria, mucopolysaccharides, proteins, enzymes and inorganic materials such as calcium, sodium, potassium, chloride and phosphate ions.

The organic constituents of saliva are thought to be responsible for the development of acquired pellicle and plaque. The presence of calcium phosphate is believed to be important both in the control of dental caries and in calculus formation. An inadequate flow of saliva has been associated with increased caries susceptibility, probably because of the poor removal of food debris, a loss of buffering capacity and a reduction in Ca^{2+} and PO_4^{3-} concentration. The complexities of saliva composition are fully discussed by Jenkins.[6]

Acquired Integuments of the Tooth

Leach[7] has drawn attention to the confusion in dental literature over the nomenclature of the various acquired integuments of human dentition. If a tooth surface, either *in vivo* or *in vitro*, is thoroughly cleaned with a toothbrush and toothpaste it is possible to remove the various integuments until the individual crystallites of the surface enamel are exposed. If this surface, again either *in vivo* or *in vitro*, is exposed to saliva for an order of time measured in minutes, an acellular, bacteria-free layer is deposited on the tooth surface.[8] This is defined as the 'acquired pellicle' or 'pellicle' and must be derived from the saliva. In the course of time, now measured in hours, bacteria begin to become deposited on the surface of the pellicle, at the same time surrounding themselves with a matrix that is distinctly different from the acquired pellicle. This aggregation of bacteria and its surrounding matrix, once it is present in sufficient quantity to be recognized, is defined as 'dental plaque'. The electron microscope is able to detect this material at an earlier stage of formation than can histological staining which, in turn, is more sensitive than ordinary visible detection as observed clinically. In certain areas of the mouth, crystals of various calcium phosphates begin to appear in localized areas both in the acquired pellicle and in the extracellular matrix of the dental plaque. The order of time for this material to form is measured in days, and it is clinically recognizable as 'calculus' when sufficient of it forms for the individual crystals to be sufficiently closely packed together for the aggregates to become resistant to deformation. The electron microscope reveals the initial stages of calculus formation considerably in advance of this condition[9] at a stage before the crystals become closely packed together, and is able to differentiate between the initial calcification of the matrix and the subsequent intracellular calcification of the plaque bacteria.[10]

The nomenclature of the various integuments and their nature have been clearly tabulated by Jenkins.[11]

Acquired Pellicle

There is ample evidence that there is a physical film on the surface of the teeth and this is the acquired pellicle. This layer is usually 1–3 μm thick and is free from bacteria. It also has staining reactions different from those shown by plaque.

The pellicle film can be removed from the surface of an extracted tooth by dissolving the surface layer of the underlying enamel in 2–5 per cent HCl Material which has been removed in this way has been found to consist of protein high in glutamic acid and alanine and low in S-containing amino acids. It also contains carbohydrates and muramic acid which is a constituent of bacterial cell walls.[12]

It would appear that pellicle is formed by the selective adsorption of some of the salivary proteins by the apatite of the enamel.

It has been suggested[13,14] that pellicle has some protective action against the attack of acids on the enamel surface but this is by no means certain.

Plaque

On all parts of the teeth other than those cleaned by erosion, i.e. the biting surfaces, there is a mucous coating of varying thickness and this is known as dental plaque. Plaque is now considered to be the prime etiological factor in the development of caries;[15] it is also implicated in periodontal disease.

The amount of plaque formed is very variable, but 10–20 mg per day is typical. It is now believed to consist of salivary proteins, perhaps modified by bacterial enzymes, with varying quantities of polysaccharide of bacterial origin. The composition may vary not only among individuals, but also in different parts of the mouth, but typically it contains about 82 per cent water. Other components are:

	% dry weight
Protein	40–50
Carbohydrate	13–17
Lipid	10–14
Ash	10
	μg/mg dry weight
Calcium	8
Phosphorus	16
	ppm dry weight
Fluorine[16]	20–100

Perhaps the most interesting and significant of these components is the fluorine which, at a typical level of 50 ppm on dry weight, is at a higher concentration than in any food or drink passing through the mouth.

The carbohydrate present is most likely to arise from the dietary intake of sugars. These may be partly metabolized by plaque bacteria to produce extracellular polysaccharides such as dextran and levan.[17,18] This sugar is also metabolized to form acids which it is believed are the initial cause of caries. This effect—the production of acid in plaque—can be easily demonstrated by measuring the fall in plaque pH after a sugar mouth rinse.[19,20] It is of great significance in the study of caries.

Calculus

The term *tartar* was commonly used to describe the mineralized deposits formed on neglected teeth. The origin of the term was of course the crystalline deposits formed in wine.

The deposit is now more correctly called *calculus*. It may occur both above (supragingival) and below (subgingival) the gum. The two forms differ in properties and are possibly of different origin. Supragingival calculus is certainly of salivary origin and occurs near the orifices of the main salivary ducts.

Calculus varies in composition but always contains about 80 per cent of inorganic material containing calcium, magnesium, phosphorus and other elements. The calcium and phosphorus are present in early plaque as Brushite ($CaHPO_4 \cdot 2H_2O$) and octacalcium phosphate ($Ca_8(HPO_4)_2(PO_4)_4$). Whitlockite ($Ca_3(PO_4)_2$) may also be found, but the ultimate stage is probably apatite ($3Ca_3(PO_4)_2 \cdot Ca(OH)_2$). This transition is helped by the presence of fluorine which is present to the extent of about 400 ppm.[21,22]

The remaining 20 per cent of calculus is an organic matrix containing carbohydrate, protein, lipid and bacteria. Typical total compositions[23,24] of supragingival calculus are:

Density	1·09–1·33
Organic matter (%)	11–20
CO_2 (%)	2·0–3·7
Calcium (%)	26·9–28·0
Phosphorus (%)	14·9–16·0
Sodium (%)	2·09–2·58

Calculus is first observed when the plaque on the enamel surface begins to mineralize. There are a number of theories of calculus formation. One of the earliest was that loss of CO_2 from the saliva caused a rise in pH and a consequent precipitation of calcium phosphates. A change in pH may also be caused by ammonia formation. Phosphatase concentration in the plaque has also been proposed as a calculus controlling mechanism. Today, the most generally accepted explanation is that calculus is formed by a seeding mechanism, but this does not explain differences between individuals which are often considerable. The role of bacteria cannot be disregarded, but calculus can form in germ-free animals. The subject is thoroughly discussed by Jenkins.[25]

It has been recognized for some time that there is an association between the presence of calculus and the incidence of periodontal disease. The commonly accepted theory is that calculus irritates the gingivae and encourages the formation of a pocket between tooth and gingivae, in which food debris and bacteria may lodge.

An inverse relationship between calculus and caries has also been noticed[26] and this would seem logical since calculus should only be deposited under non-acidic conditions.

Mechanical treatment by a dental hygienist is the only means of removing calculus, and until recently there has been no means of preventing its formation. Within the last few years, however, a number of patents have claimed to prevent calculus formation.[27]

The secret of preventing calculus formation must surely be in preventing plaque deposition, since it is from plaque that calculus is precipitated. Regular brushing will, of course, help to reduce plaque and there are now claims that certain toothpaste or mouthwash additives will reduce plaque formation. Among compounds suggested probably most work has been done on chlorhexidine.[28]

Food Debris and Materia Alba

Food debris, unless impacted between the teeth, is usually easily removed by a stream of water. Materia alba is a white, diffuse, loosely attached layer which is also easy to remove. It is usually a mixture of bacteria with a high proportion of extracellular polysaccharide.

Since normal oral hygiene removes these deposits easily, they are not thought to have any significant effect on oral disease. They may, of course, contribute to mouth odours.

MAJOR PROBLEMS OF ORAL HEALTH

Magnitude of the Problem

Dental diseases are primarily diseases of civilization and the incidence in Western communities is appalling. Surveys conducted by the UK Department of Health and Social Security[29,30] give a complete and depressing picture of the dental health of the population. Among adults the proportion edentulous (that is, without natural teeth) was found to be as given in Table 30.2. Although an improvement is evident between 1968 and 1978, the proportion of the UK population with no natural teeth is still high. Among children the picture is equally depressing (Table 30.3). It has been estimated that 4 tons of teeth are extracted from British children each year, which corresponds to about 4 million lost teeth. In the USA it is estimated that the adult population (111 million) has the staggering total of 2·25 billion ($2{\cdot}25 \times 10^9$) decayed, missing and filled teeth, and that at any one time there is a backlog of 700 million teeth awaiting treatment.[31]

Teeth which are not extracted because of caries may later be attacked by periodontal disease, which is the major cause of tooth loss after the age of 35 (see Figure 30.1).[32]

Despite these appalling statistics, serious research into the causes of dental disease has only been undertaken in the last 30 years. This is all the more

Table 30.2 Edentulous Adult Population of England and Wales[29]

Age group	Percent edentulous 1968	1978
16–24	1	—
25–34	7	3
35–44	22	12
45–54	41	29
55–64	64	48
65–74	79	74
75+	88	87

Table 30.3 Active Decay in Children's Teeth, England and Wales[30]

Age	Per cent with decayed teeth
5	63
6	69
7	73
8	78
9	76
10	69
11	66
12	61
13	61
14	62
15	57

surprising since the probable cause of caries was described as long ago as 1890 by an American dentist who first showed the relationship between dietary sugar and caries.[33] A recognition that the primary cause of caries was biochemical stimulated research by chemists and the last 30 years has seen a massive research input both in university dental schools and in the laboratories of commercial companies. This research has now reached the stage where the cause of caries and the mechanism of its development are fairly well understood and accepted. The clinical treatment of disease is no part of the cosmetic chemist's function—the prevention of disease is, and in order to be able to formulate prophylactic products it is necessary for him to understand the disease that he is trying to prevent.

Dental Caries

A conference held at the University of Michigan in 1947 to evaluate and correlate all available information on the aetiology and control of dental caries,

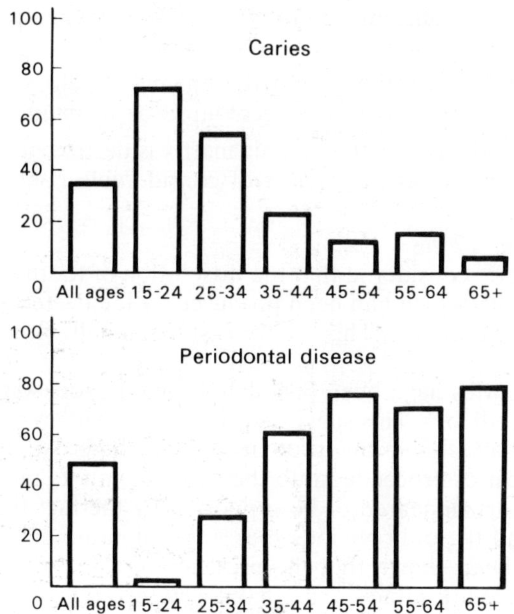

Figure 30.1 Teeth extracted because of caries and periodontal disease

unanimously adopted the following definition:[34]

> . . . Dental caries is a disease of the calcified tissues of the teeth. It is caused by acids resulting from the action of micro-organisms on carbohydrates, is characterized by a decalcification of the inorganic portion and is accompanied or followed by a disintegration of the organic substance of the tooth. The lesions of the disease predominantly occur in particular regions of the tooth and their type is determined by the morphological nature of the tissue in which they appear . . .

The following general observations may be used to describe the process:

(a) Carious lesions are localized, generally beginning in a small circumscribed area. They are found most frequently in pits and fissures of occlusal surfaces.

(b) Demineralization begins with the enamel, usually beneath an intact enamel layer. This demineralization is often accompanied either by a discoloration or a change in the opacity of the enamel, that is, development of a so-called 'white spot'.

(c) The demineralization penetrates deeper into the tissue and spreads within it, roughly assuming the shape of a cone with its apex orientated towards the dentino–enamel border and the base still below the enamel surface. At this stage of the process there still may be no visible break or cavitation evident on the enamel surface. It is generally accepted that the dentine begins to demineralize or soften before an actual cavity becomes visible. Gradually, the tissue softens sufficiently so that it 'breaks up' when pressure is applied with a dental probe.

(d) The decay process continues and the enamel develops a lesion which is visible. The cavitation is often discoloured or pigmented.

(e) During the decay process the organic matrix is destroyed. The exact step at which this occurs is still a subject of considerable controversy.

Theories of Dental Caries

In the 1890s Miller[33] considered that the dissolution of tooth substance was due to the action of acids which had been produced by the bacterial fermentation of carbohydrates in the mouth. This theory is still generally accepted as the basic cause of caries.[34]

Alternative theories have been postulated such as, for example, the proteolytic theory of Gottlieb.[35] This suggested that the initial effect on enamel might be a proteolytic attack on the protein content. This would be even more important if the caries proceeded into the dentine with its much higher protein content. This theory appeared to be supported by the fact that carious lesions are pigmented and the only obvious source of colouring matter is the protein. However, it has been shown that the presence of Ca^{2+} salts actually protects dentine from proteolytic attack[36] and that it is necessary to demineralize dentine with acid before proteolysis can occur.

Schatz and his colleagues[37] have postulated a 'proteolysis–chelation' theory which suggests that some of the products of the action of oral bacteria on saliva, food debris, enamel and dentine may be capable of forming complexes with calcium.

It is now generally accepted that the acid theory is most probably correct and that the initial step in the carious process is the presence of fermentable carbohydrate in the diet. It is important to recognize that caries is a step-wise process:

Bacteria in the mouth release
enzymes which attack
fermentable carbohydrates releasing
acids which attack
enamel thus producing a
carious lesion

This is a simplified statement of the series of events which lead to the development of caries.

Control of Caries

In the light of the above discussion it is possible to suggest methods for the control of caries:

Reduction of Fermentable Carbohydrate Intake. There is a mass of clinical evidence to support the association of dietary sugar with caries. In separate papers Marthaler[38] and König[39] have reviewed the evidence for this association.

The improved dental health of Norwegian school children during World War II[40] was attributed to the reduction of dietary sugar. The dental health of the natives of Tristan da Cunha fell rapidly when they were exposed to Western diet.[41] The World Health Organization[42] has now accepted the fact that sucrose plays a dominant role in the development of caries.

It is impracticable to prevent fermentable carbohydrate intake, as sugar has become an important part of the Western dietary habit. Nevertheless, there is an excellent case for reducing sugar intake, particularly when it is taken between meals, for example in the form of sweets (candy).

Removal of Fermentable Debris from the Mouth. There is ample evidence that the removal of food debris from the mouth reduces the incidence of caries. Fosdick[43] found a 50 per cent reduction by after-meal tooth brushing. Certain foods such as apples, carrots, nuts, etc., help in the removal of debris by mechanical means and by stimulation of salivary flow, which additionally buffers plaque acids resulting from previously eaten sugary foods (Figure 30.2).[44]

Brushing the teeth after eating is not always practicable and a twice-a-day routine is more realistic. Clark *et al.*[45] developed a tooth cleaning tablet which stimulated salivary flow and was shown to be as effective as tooth brushing with water. Slack *et al.*[46] showed a small but significant benefit in children using the tablets as an adjunct to tooth brushing.

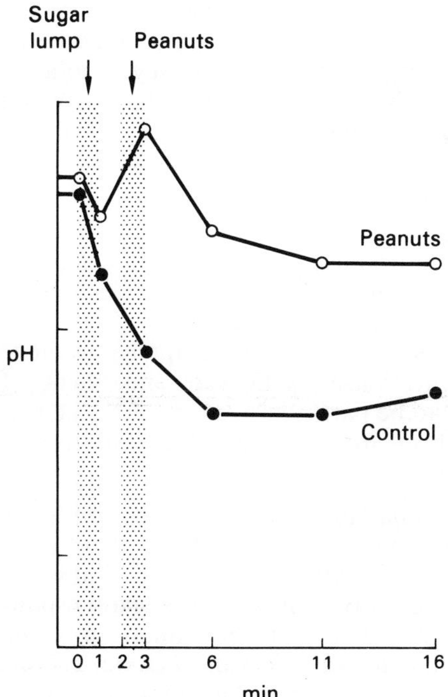

Figure 30.2 Effect on plaque pH of eating peanuts after consumption of sugar

Reduction in Bacterial Activity. Bacteria present in the mouths of most human subjects will promote fermentation in dilute sugar solutions causing a pH drop to a level of about 5·3. This fall in pH is the classical 'Stephan curve'[47] which can be reproduced *in vitro* and *in vivo*. Certain organisms have been identified as specific causative agents and these include *Lactobacillus acidophilus* and *Streptococcus mutans*. This immediately suggests that it should be possible to produce a vaccine which would confer caries immunity on subjects. A considerable volume of work has been done on this assumption and though vaccination of monkeys has produced positive effects, the results are not as spectacular as was hoped. Bowen *et al.*[48] and Lehner[49] have used this technique on monkeys with some success.

An obvious technique for reducing the bacterial population of the mouth would be to use an antibiotic in a mouthwash or in a toothpaste. This method was tried by Zander[50] using penicillin. Although this technique appeared to cause a reduction in the incidence of caries it is potentially harmful. It could lead, for example, to the development of penicillin-resistant organisms and it might sensitize significant numbers of the population to penicillin. For these reasons it is unlikely that antibiotics will ever be components in routine prophylactic oral products.

It is, of course, possible to reduce the bacterial population of the mouth by the regular use of chemical bacteriostats. There is some evidence that this may be effective in preventing caries.[51]

Interference with Enzyme System. A further obvious method of interrupting the step-wise development of the carious process would be to inhibit the enzymes responsible for the glycolytic breakdown of sugars to acids. This approach was used by Fosdick who showed that sodium N-lauroyl sarcosinate specifically inhibited the enzyme *hexokinase*. This subject is dealt with in a later section.

Decrease of Susceptibility of Tooth to Attack. It should be possible to strengthen enamel by topical treatment with fluorides which should ion-exchange with the hydroxyl group in hydroxyapatite, producing the more resistant fluorapatite. This may be the mechanism by which fluoride-containing toothpastes and mouthwashes exert their effect.

Alternatively, a physical barrier may be applied to the tooth surface; this has been explored by Irwin, Walsh and Leaver[52] and others.[53] Films polymerized by UV radiation appear to be particularly effective.[54] The use of these polymers as fissure sealants has been found useful in reducing occlusal caries.[55]

Fluoridation of Water Supplies. In 1952 it was established in the USA that there was an inverse relationship between the incidence of caries in children and the fluoride content of the local water supply (Dean[56] and see Figure 30.3). Weaver[57] confirmed these findings in Britain. An investigation sponsored by the Government showed beyond doubt that the American findings were valid. Experimental fluoridation in Great Britain was started and after eleven years' experience it was concluded that the addition of 1 ppm of fluorine to water supplies was a safe and effective way of reducing dental decay.[58]

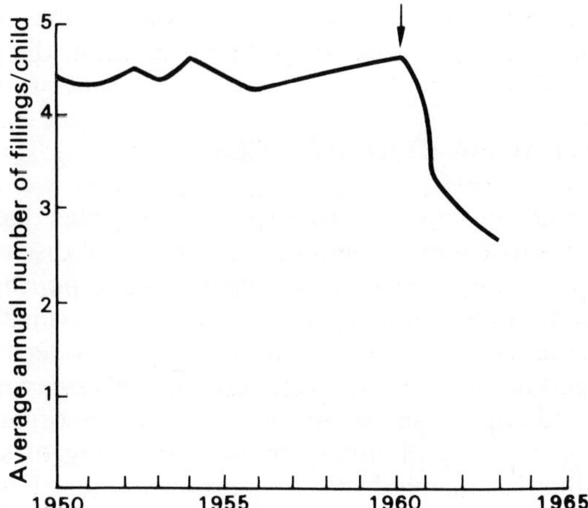

Figure 30.3 Average number of fillings per child per year, School Dental Service, Gothenburg, Sweden. Fluoride rinsing was started in 1960

The two aspects of effectiveness and safety have been further emphasized by the World Health Organization[59] and by the Royal College of Physicians.[60] No scientist can deny the validity of the conclusions reached by these bodies. It could be argued indeed that the artificial fluoridation of water is merely reinforcing the natural situation where both the enamel and dentine of human teeth collect and retain fluorine and thereby increase their resistance to attack.[3]

The effect of the fluoridation of water in Sweden is shown dramatically by Figure 30.3 which plots the incidence of fillings in children's teeth before and after fluoridation. It was originally thought that this benefit only accrued to children born in the area of fluoridated water—in which case the fluoride content of teeth would be of systemic origin. Later studies indicate that adults may benefit also—presumably by a topical effect.[61]

The whole subject of the caries mechanism has been summarized by Jenkins[62] and by Grenby.[63] The vast accumulation of knowledge on this subject has reached the stage at which it can be stated that caries is a preventable disease, or at the worst, its incidence can be enormously reduced.

Periodontal Disease

Teeth are attached to the basal bones of the jaw through the periodontal tissues. There are many abnormalities of these tissues which can be regarded as periodontal diseases. The World Health Organization (1961) has divided periodontal diseases into three broad classes: inflammatory processes, degenerative processes and neoplastic processes. Periodontal diseases involving inflammation are by far the most common and they are the easiest to prevent and to treat.

Gingivitis is an inflammation of the gums and is rapidly caused by poor oral hygiene. It appears to be caused by substances produced in plaque which irritate the gum membranes. Periodontitis in addition to gum inflammation is manifested by a pocket formation and destruction of the collagen fibres of the periodontal ligaments. It is encouraged by calculus at the gum margin.

USE OF PROPHYLACTIC TOOTHPASTES

The review of the mechanism of the caries process indicates that toothpastes and mouthwashes may play a part in the prevention of oral diseases. The primary function of a toothpaste as a cosmetic for cleaning and polishing the teeth has now been extended. A toothpaste in fact is an ideal vehicle for conveying prophylactic ingredients to the tooth surface. Most modern dentifrices have accordingly changed their emphasis and are formulated with the object of controlling caries and improving oral hygiene. A properly organized clinical trial thus becomes an essential preliminary to the marketing of such a product.

Advertising claims for prophylactic oral products must be substantiated by clinical evidence that is acceptable to dental authorities. It is worth noting that if such claims are made to dental practitioners the product must come within the scope of the Medicines Act in the UK or corresponding regulations such as those of the US Food and Drugs Administration.

Active Ingredients

Chlorophyll

Earlier editions of this work devoted considerable space to chlorophyll which was once popular as a toothpaste ingredient. It is perhaps indicative of the increased knowledge of the chemistry of the mouth and the volume of clinical research performed, that chlorophyll has almost disappeared as a toothpaste ingredient.

Bacteriostats

Although it would seem obvious to prevent caries by destroying the bacteria in the mouth, the ideal bacteriostat has yet to be found. Such a compound should be adsorbed strongly onto mucous membranes and should continue to exert its influence for some time; it should be non-toxic and non-irritant. Among compounds which have been used in toothpastes and mouthwashes are hexachlorophene (G11), benzethonium chloride and other cationics and, most recently, chlorhexidine (Hibitane, ICI). Probably most research has been done on this last compound as it has been shown to be effective in reducing plaque and hence in improving gingival health. It may also be expected to have an effect on the incidence of caries. Unfortunately, chlorhexidine (in common with many other cationics of the same type) produces a slight brown staining of the teeth. A symposium devoted to the subject[64] presents a very clear picture of the whole subject.

Enzyme Inhibitors

Another obvious route to preventing or reducing the production of acid from sugar breakdown is to inhibit the enzymes concerned with glycolysis. Fosdick[65] proposed the use of sodium N-lauroyl sarcosinate ($C_{11}H_{23}CO\,.\,N(CH_3)$

CH_2 . COONa) which is an excellent inhibitor of the enzyme hexokinase. This was found to be clinically effective in reducing caries but this has not been confirmed. More recently Mühlemann[66] has shown that this substance prevents the reduction of plaque pH to the danger level and Tomlinson[51] has shown it to be effective in a mouthwash.

Sodium N-lauroyl sarcosinate has mild detergent properties and its use in dentifrices is described in a number of patents.[67]

Enzymes
In contrast to the use of enzyme inhibitors, some workers have proposed that enzymes are added to toothpastes to assist in the breakdown of proteins, starch and lipids. Patents exist for oral products containing such enzymes.[68]

Dextrans and levans are gummy metabolic products of bacterial attack on carbohydrates and the presence of these substances in plaque is thought to play a part in maintaining the plaque film on teeth. It is not surprising therefore to find that it has been suggested that dextranases should be added to toothpastes.[69]

Leach[70] has reviewed the relationship between dextrans in the plaque and caries. It is unlikely that this route would lead to a dramatic solution to the problem.

Ammonium (Salt)/Urea
Dentifrices containing an ammonium ion and urea have been proposed[71] on the grounds that these compounds might neutralize acids produced in dental plaque or inhibit their formation. Clinical trials have not established this approach as valid[72] and Peterson[73] does not believe that these compounds have a future in prophylactic dentifrices.

Fluorides
It is now well established that the fluoridation of water supplies is an excellent public health measure since it is effective, safe and economical. It is an obvious step therefore to add fluorine-containing compounds to toothpastes and thus convert the social habit of toothbrushing into a prophylactic treatment. Among compounds which have been used for this purpose are sodium fluoride, stannous fluoride, sodium monofluorophosphate and amine fluorides.

The addition of compounds containing the F^- ion to dentifrices is difficult. Most toothpaste abrasives are calcium salts and calcium fluoride is one of the most insoluble substances known. Loss of fluoride ion on aging is therefore one of the great problems of fluoride-containing dentifrices. It has been overcome by the use of highly insoluble abrasives (calcium pyrophosphate), by the use of non-calcium-containing abrasives (silica, plastics, insoluble sodium metaphosphate) and by the use of sodium monofluorophosphate (Na_2FPO_3) which does not release free F^- ions except after hydrolysis.

Since the introduction and widespread use of fluoride toothpastes in the 1950s the Council on Dental Therapeutics of the American Dental Association has included certain fluoride-containing toothpastes in 'Accepted Dental Therapeutics'. In order to qualify for acceptance products have to show evidence of clinical effectiveness. A similar position has been taken by the British Dental

Association which will endorse fluoride-containing toothpastes of which the clinical effectiveness can be demonstrated.

Thus, in a period of less than 25 years, toothpastes have moved from the position of being cosmetic products to the current position in which many of them have been professionally endorsed because of their clinically-proven prophylactic effect.

Simple Inorganic Fluorides

The two simple inorganic fluorides commonly used are sodium fluoride and stannous fluoride. Provided the problems of inactivation can be overcome, both these compounds have shown clinical effectiveness in reducing the incidence of caries. The mechanism is thought to be that the free F^- ion reacts with the hydroxyapatite present in tooth enamel, thus converting it to the less soluble fluorapatite:

$$\underset{\text{hydroxyapatite}}{3Ca_3(PO_4)_2 \cdot Ca(OH)_2} + \underset{\text{fluoride ion}}{2F^-} \rightleftharpoons \underset{\text{fluorapatite}}{3Ca_3(PO_4)_2 \cdot CaF_2} + \underset{\text{hydroxyl ion}}{2OH^-}$$

Such a reaction would be favoured by acid conditions and the first stannous fluoride toothpaste produced in 1955 had a pH value well below neutrality.

This and other fluoride dentifrices have been extensively tested clinically. Reductions of 20–30 per cent in the incidence of caries in children are commonly reached and, though this is by no means a solution to the problem, it is a step in the right direction.

The volume of clinical testing done is so enormous that no single reference is adequate. Duckworth[74] has reviewed the mechanism of clinical trials and Von de Fehr and Møller[75] have recently reviewed caries-preventive fluoride dentifrices.

Dentifrices containing inorganic fluorides normally contain about 0·4 per cent of SnF_2 or 0·2 per cent of NaF. Both these levels correspond to about 1000 ppm of F in the toothpaste. At this concentration one of the inevitable consequences of use is that CaF_2 will be precipitated in the plaque or on the surface of the tooth. This should itself resist acid attack and may produce a very low F^- ion concentration near the surface of the tooth. This condition is known to promote the remineralization of early carious lesions[76] and it may be that some of the clinical effect is due to this.

Low concentrations of F^- ion are also known to have effects on enzymes and it may be that acid-producing enzymes in the plaque are inhibited by F^- ions present.

Thus the reasons for the clinical effectiveness of F-containing toothpastes are not fully understood. It may be that improvements in clinical efficiency still await the fuller understanding of the caries preventive mechanism.

Sodium Monofluorophosphate

Sodium monofluorophosphate (Na_2FPO_3) has the advantage over sodium and stannous fluorides in that it ionizes to give not the F^- ion, which will immediately precipitate in the presence of calcium ions, but the FPO_3^{2-} ion of which the calcium salt is soluble. It was this thought which prompted Ericsson to propose the use of sodium monofluorophosphate in a calcium carbonate dentifrice.[77] Subsequent clinical trials showed this substance to have an anti-caries effect.[78] Patents on various formulation combinations are well known.[79]

The mechanism of the protective action of sodium monofluorophosphate is not fully understood. The commercial material is made by the fusion of NaF and $NaPO_3$:

NaF	$+ NaPO_3$	$= Na_2FPO_3$
sodium fluoride	sodium metaphosphate	sodium monofluorophosphate

This reaction yields a product about 95 per cent pure and consequently the commercial material is always contaminated by up to 2 per cent NaF. It is not possible therefore to say with certainty which component is responsible for the clinical effect, particularly since it is known that low F^- concentrations may have a significant effect.

Ingram[80] suggested that the FPO_3^{2-} ion exchanged with the HPO_4^{2-} ion in the apatite lattice of calcium-deficient enamel, thus rendering it more resistant to acid. Duff[81] suggested that the FPO_3^{2-} ion exchanged with HPO_4^{2-} on the enamel surface and that the subsequent dissolution of this under acid conditions favoured the precipitation of fluorohydroxyapatite.

Pearce and More[82] showed that only free F^- is taken up by the enamel and that this arises only from the free F^- impurity and by hydrolysis of Na_2FPO_3. Tomlinson[51] showed that purified Na_2FPO_3 (that is, free of F^-) did have a clinical effect and that the mechanisms involving the whole FPO_3^{2-} ion were therefore possible.

Whatever the mechanism, there is no doubt of the clinical effectiveness of sodium monofluorophosphate when incorporated in a dentifrice.

It is not possible to be dogmatic about the relative effects of stannous fluoride and sodium monofluorophosphate. Both have about the same level of effectiveness and both appear to show their greatest effect on teeth which erupt during the course of a clinical trial, that is on teeth which have been exposed to a fluorine-containing compound from the moment of eruption. A review of a number of trials is given in a special issue of the British Dental Journal.[83] Duckworth[74] states that the clinical evidence obtained from UK trials supports the view that 'certain fluoride-containing dentifrices can be of value when used in a conscientiously applied programme of oral hygiene and regular professional care'.

Organic Fluorides

Work by Mühlemann and Marthaler in Zurich suggested that certain amine fluorides had more protective action against caries in rats than had sodium fluoride. Subsequent clinical trials on children[84] showed a substantial reduction in the incidence of caries. Compounds of this type would be expected to adsorb onto tooth and mucous membrane surfaces and thus retain an effect in the mouth for some time. They will, of course, present many problems in toothpaste formulation. A number of patents exist for compounds of this type.[85]

Other Fluorides

A number of other metallic fluorides have been proposed as toothpaste components and their use is protected by patents. These include ferric and zirconium fluorides, stannous and indium fluorozirconates, manganese fluoride and aluminium fluoride.[86] The fluoride and monofluorophosphate salts of chlorhexidine (Hibitane, ICI) have also been patented. So far as is known, none of these compounds is in commercial use at the present time.

Other Metal Compounds

A number of metals have been proposed as toothpaste additives for various purposes.

Thus water-soluble indium salts are said to increase the activity of fluorides in toothpastes.[87] Organo–tin compounds have been patented as bactericides.[88] Aluminium salts are astringent and are claimed to have some effect on gum health.[89] Zinc salts have also been used for this purpose and have some clinical evidence to support the claim.[90]

The presence of molybdenum in water supplies in concentrations as low as 0·1 ppm appears to have an effect in reducing caries and this has been confirmed in animal experiments. The mechanism is not understood; the whole subject has been well reviewed by Jenkins.[91]

Strontium in the form of strontium chloride has been used to reduce the sensitivity of exposed dentine, for example because of gum recession in older subjects. Reports vary on the clinical effectiveness of this procedure;[92] nevertheless, patents exist for this additive and it has been used commercially.[93]

Anticalculus Ingredients

Although there has been no large-scale promotion of an anticalculus toothpaste there has been a considerable volume of work on this subject.

Dissolution of existing calculus is a problem because of its chemical similarity to tooth substance. For this reason sequestering agents of the EDTA type cannot be used. Research has concentrated instead on means to inhibit crystal growth and means of preventing adhesion to the teeth. Patent literature is a particularly useful source of information on developments in this field.[94]

Sturzenberger[95] has shown that disodium etidronate (a diphosphonate) inhibits crystal growth of apatites, and zinc salts are also claimed to have a similar effect.

Organic titanates[96] are also claimed to prevent the adherence of calculus to tooth surfaces.

Remineralizing Ingredients

Within recent years it has become clear that early carious lesions may become repaired by a process which has been termed remineralization. The effect has been described by Levine[76] and the phase conditions necessary for remineralization to occur have been worked out by Duff.[97] It is not surprising, therefore, that the concept of remineralization has become part of toothpaste formulation; indeed it is possible that the clinical effect of some existing fluoride toothpastes is due, in fact, to their remineralizing effect.

Remineralization requires the presence of very low concentrations of F^-, PO_4^{3-} and Ca^{2+} ions; because these components would precipitate, it is customary to have a two-component system.

The patent literature is a valuable source of information[98] but so far no commercial product has been promoted with a strong remineralization claim.

REFERENCES

1. Jenkins, G. N., *Physiology and Biochemistry of the Mouth*, Oxford, Blackwell Scientific Publications, 1977, p. 79.
2. Jenkins, G. N., *Physiology and Biochemistry of the Mouth*, Oxford, Blackwell Scientific Publications, 1977, pp. 72–74.
3. McClure, F. J. and Likins, R. C., *J. dent. Res.*, 1951, **30,** 172.
4. Sutton, M. M., MSc Thesis, Salford University, 1979.
5. Mason, D. K. and Chisholm, D. M., *Salivary Glands in Health and Disease*, London, Saunders, 1975.
6. Jenkins, G. N., *Physiology and Biochemistry of the Mouth*, Oxford, Blackwell Scientific Publications, 1977, p. 284.
7. Leach, S. A., *Br. dent. J.*, 1967, **122,** 537.
8. Lenz, H. and Mühlemann, H. R., *Helv. Odont. Acta*, 1963, **8,** 117.
9. Leach, S. A. and Saxton, C. A., *Arch. oral Biol.*, 1966, **11,** 1081.
10. Schroeder, H. E., *Helv. Odont. Acta*, 1964, **8,** 117.
11. Jenkins, G. N., *Physiology and Biochemistry of the Mouth*, Oxford, Blackwell Scientific Publications, 1977, p. 361.
12. Jenkins, G. N., *Physiology and Biochemistry of the Mouth*, Oxford, Blackwell Scientific Publications, 1977, pp. 361–368.
13. Darling, A. I., *Proc. R. Soc. Med.*, 1943. **36,** 499.
14. Meckel, A. M., *Arch. oral Biol.*, 1965, **10,** 585.
15. McHugh, W. D., ed., *Dental Plaque*, Symposium University of Dundee, September 1969, Edinburgh, Livingstone, 1970.
16. Jenkins, G. N., *Physiology and Biochemistry of the Mouth*, Oxford, Blackwell Scientific Publications, 1977, p. 372.
17. Critchley, P., Wood, J. M., Saxton, C. A. and Leach, S. A., *Caries Res.*, 1967, **1,** 112.
18. Wood, J. M., *Arch. oral Biol.*, 1967, **12,** 849.
19. Jenkins, G. N. and Kleinberg, I., *J. dent. Res.*, 1956, **35,** 964.
20. Ludwig, T. G. and Bibby, B. G., *J. dent. Res.*, 1957, **36,** 56.
21. Jenkins, G. N. and Speirs, R. L., *J. dent. Res.*, 1954, **33,** 734.
22. Grøn, P. *et al.*, *Arch. oral Biol.*, 1967, **12,** 829.
23. Little, M. F. *et al.*, *J. dent. Res.*, 1961, **40,** 753.
24. Little, M. F. *et al.*, *J. dent. Res.*, 1963, **42,** 78.
25. Jenkins, G. N., *Physiology and Biochemistry of the Mouth*, Oxford, Blackwell Scientific Publications, 1977, pp. 402–409.
26. Baines, E., MSc Thesis, Salford University, 1979.
27. British Patent 1 419 692, Procter and Gamble, 9 March 1973; British Patent 1 421 064, Procter and Gamble, 9 March 1973; British Patent 1 432 487, Procter and Gamble, 29 June 1973; British Patent 1 536 261, Monsanto, 28 October 1977.
28. Symposium: Hibitane in the Mouth, *J. clin. Periodont.*, 1977, **4**(5).
29. Todd, J. E. and Walker, A. M., *Adult Dental Health*, Vol. 1, *England and Wales 1968–1978*, London, HMSO, 1980.
30. *Children's Dental Health in England and Wales 1973*, London, HMSO, 1975.
31. Steele, P. F., *Dimensions of Dental Hygiene*, London, Henry Kimpton, 1966, p. 22.
32. Pelton, W. J., Pennell, E. H. and Druziva, A., *J. Am. dent. Assoc.*, 1954, **49,** 439.
33. Miller, W. D., *Micro-organisms of the Human Mouth*, Philadelphia, S. S. White, 1890; *Dental Cosmos*, 1904, **46,** 981; *Dental Cosmos*, 1905, **47,** 18.

34. Michigan Workshop Conference, *J. Am. dent. Assoc.*, 1948, **36,** 3.
35. Gottlieb, B., *Dental Caries*, Philadelphia, Lea and Febiger, 1947.
36. Evans, D. G. and Prophet, A. S., *Lancet*, 1950, **1,** 290.
37. Schatz, A. *et al., Proc. Pennsyl. Acad. Sci.*, 1958, **32,** 20; Schatz, A. and Martin, J. J., *J. Am. dent. Assoc.*, 1962, **65,** 368.
38. Marthaler, Th., *Forum Medici* (Zyma Nyon SA, Switzerland), 1971, **13,** 22.
39. Ǩonig, K., *Forum Medici* (Zyma Nyon SA, Switzerland), 1971, **13,** 28.
40. Toverud, G., *Br. dent. J.*, 1964, **23,** 149.
41. Barnes, H. N. V., *Br. dent. J.*, 1937, **63,** 86; Sognnaes, R. F., *J. dent. Res.*, 1941, **20,** 303; Gamblen, F. B., *J. R. Soc. med. Serv.*, 1953, 252; Holloway, P. J., *Br. dent. J.*, 1963, **115,** 19.
42. WHO Technical Report Services 1972, No. 494.
43. Fosdick, L. S., *J. Am. dent. Assoc.*, 1950, **40,** 596.
44. Slack, G. L. and Martin, W. J., *Br. dent. J.*, 1958, **105,** 366; Geddes, D. A. M., Edger, W. M., Jenkins, G. N. and Rugg-Gunn, A. J., *Br.dent. J.*, 1977, **142,** 317; Rugg-Gunn, A. J., Edgar, W. M. and Jenkins G. N., *Br. dent. J.*, 1978, **145,** 95.
45. Clark, R., Hay, D. I., Schram, C. J. and Wagg, B. J., *Br. dent. J.*, 1961, **111,** 244.
46. Slack, G. L., Millward, E. and Martin, W. J., *Br. dent. J.*, 1964, **116,** 105.
47. Stephan, R. M., *J. Am. dent. Assoc.*, 1940, **27,** 718.
48. Bowen, W. H. *et al., Br. dent. J.*, 1975. **139,** 45.
49. Lehner, T. *et al., Nature*, 1975 **254,** 517.
50. Zander, H. A., *J. Am. dent. Assoc.*, 1950, **40,** 569.
51. Tomlinson, K., *J. Soc. cosmet. Chem.*, 1978, **29,** 385.
52. Irwin, M., Walsh, J. P. and Leaver, A. G., *N.Z. dent. J.*, 1957, **76,** 166.
53. Ripa, L. W. and Cole, W. W., *J. dent. Res.*, 1970, **49,** 171.
54. Buonocore, M., *J. Am. dent. Assoc.*, 1970, **80,** 324.
55. Williams, B., Price, R. and Winter, G. B., *Br. dent. J.*, 1978, **145,** 359.
56. Dean, H. *et al., Pub. Health Rep.*, 1942, **57,** 1155.
57. Weaver, R., *Br. dent. J.*, 1944, **76,** 29; *Br. dent. J.*, 1944, **77,** 185; *Br. dent. J.*, 1950, **88,** 231.
58. Dept. of Health and Social Security. *Report on Public Health and Medical Subjects*, No. 122, London, HMSO, 1969.
59. World Health Organisation, *Fluorides and Human Health*, Geneva, WHO, 1970.
60. Royal College of Physicians, *Fluoride, Teeth and Health*, Tonbridge Wells, Pitman Medical, 1976.
61. Jackson, D. and Weidmann, S. M., *Br. dent. J.*, 1959, **107,** 303.
62. Jenkins, G. N., *Physiology and Biochemistry of the Mouth*, Oxford, Blackwell Scientific Publications, 1977, pp. 414–461.
63. Grenby, T. H., *Chemistry and Industry*, 1968, 1266; *Chemistry in Britain*, 1971, **7,** 276.
64. Symposium: Hibitane in the Mouth, *J. Clin. Periodontol.*, 1977, **4**(5).
65. Fosdick, L. S. *et al., J. dent. Res.*, 1953, **32,** 486.
66. Mühlemann, H. R., *Helv. Odont. Acta*, 1971, **15,** 52.
67. British Patent 753 979, Colgate-Palmolive, 2 November 1954.
68. British Patent 1 031 838, Warner-Lambert, 3 September 1963; British Patent 1 033 229, Iptor Pharmazeutische Preparate AG, 4 September 1963.
69. Bowen, W. H., *Br. dent. J.*, 1968, **124,** 347; British Patent 1 265 468, Blendax Werke, 1 September 1969; British Patent 1 270 200, Blendax Werke, 23 October 1970; British Patent 1 272 454, Colgate-Palmolive, 9 August 1968.
70. Leach, S. A., *Br. dent. J.*, 1969, **127,** 325.
71. US Patent 2 588 324, Ammident Inc., 1952; US Patent 2 588 922, Schlaeger, J.R., 1952; Japanese Patent 65/14440, Lion Dent. Co., 12 August 1961.
72. Davies, G. N. and King, R. M., *J. dent. Res.*, 1954, **30,** 645.

73. Peterson, J. K., *Ann. NY Acad. Sci.*, 1968, **153,** 334.
74. Duckworth, R., *Br. dent. J.*, 1968, **124,** 505.
75. Van Der Fehr, F. R. and Møller, I. J., *Caries Res.*, 1978, **12**(Supplement 1), 31.
76. Levine, R. S., *Arch. oral Biol.*, 1972, **17,** 1005; *Arch. oral Biol.*, 1973, **18,** 1351; *Br. dent. J.*, 1974, **137,** 132; *Br. dent. J.*, 1975, **138,** 249.
77. Ericsson, Y., *Acta Odont. Scand.*, 1961, **41,** 19.
78. Naylor, M. N. and Emslie, R., *Br. dent. J.*, 1967, **123,** 17.
79. British Patent 907 417, Ericsson, Y., 3 August 1960; British Patent 1 004 039, Colgate-Palmolive, 4 March 1963; British Patent 1 005 089, Colgate-Palmolive, 3 July 1964.
80. Ingram, G. S., *Caries Res.*, 1972, **6,** 1.
81. Duff, E. J., *Caries Res.*, 1973, **7,** 79.
82. Pearce, E. I. F. and More, R. D., *Caries Res.*, 1975, **9,** 459.
83. Slack, G. L., Berman, D. G., Martin, W. J. and Young, J., *Br. dent. J.*, 1967, **123,** 9.
84. Marthaler, T. M., *Br. dent. J.*, 1966, **119,** 153.
85. British Patent 896 257, Gaba, 13 July 1957; British Patent 1 003 595, Procter and Gamble, 15 February 1962; US Patent 3 201 316, Procter and Gamble, 28 September 1962.
86. Berggren, A. and Welander, E., *Acta Odont. Scand.*, 1964, **22,** 401; US Patent 3 266 996, Indiana University, 27 June 1963; US Patent 3 495 002, Indiana University, 10 February 1970.
87. US Patent 3 175 951, Procter and Gamble, 8 November 1963.
88. US Patent 3 495 002, Indiana University, 10 February 1970.
89. British Patent 935 703, Unilever, 4 September 1963.
90. British Patents 1 290 627 and 1 296 952, Unilever, 28 January 1971; British Patent 1 319 247, Beecham, 6 May 1973; British Patent 1 373 001, Unilever, 28 January 1971.
91. Jenkins, G. N., *Br. dent. J.*, 1967, **122,** 435, 500 and 545.
92. Anderson, D. J. and Matthews, B., *Arch. oral Biol.*, 1966, **11,** 1129; Cohen, A., *Oral Surg.*, 1961, **14,** 1046; Maffert, R. M. and Hoskins, F. W., *J. Periodont.*, 1964, **35,** 232; Smith, B. A. and Ash, M. M., *J. Am. dent. Assoc.*, 1964, **68,** 639; *J. Periodont.* 1965, **35,** 223.
93. Canadian Patent 714 032, Block Drug Co., 4 February 1964; British Patent 990 957, Stafford Miller, 24 February 1964.
94. British Patents 1 394 034 and 1 394 035, Henkel, 18 May 1972; British Patents 1 419 692 and 1 421 064, Procter and Gamble, 9 March 1973; British Patent 1 432 487, Procter and Gamble, 29 June 1973; British Patent 1 536 261, Monsanto, 28 October 1977.
95. Sturzenberger, O. P. *et al.*, *J. Periodont.*, 1971, **42,** 416.
96. US Patent 3 317 396, Tamas, I., 25 June 1964.
97. Duff, E. J., *J. inorg. nuclear Chem.*, 1970, **32,** 3707; *Caries Res.*, 1976, **10,** 234; *Caries Res.*, 1973, **7,** 70.
98. British Patent 1 408 922, Blendax Werke, 2 February 1972; British Patent 1 452 125, Procter and Gamble, 2 October 1973; British Patent 1 468 149, NRDC (Levine) 26 February 1974; British Patent 1 477 823, Colgate-Palmolive, 10 January 1974.

Chapter Thirty-one

Dentifrices

BASIC REQUIREMENTS OF A DENTIFRICE

The original author of this work laid down the minimum requirements of a dentifrice. Over the years these requirements have changed in emphasis and in content. They are listed below—not necessarily in order of priority:

1. When used properly with an efficient toothbrush it should clean the teeth adequately, that is, remove food debris, plaque and stains.
2. It should leave the mouth with a fresh, clean sensation.
3. Its cost should be such as to encourage regular and frequent use by all.
4. It should be harmless, pleasant and convenient to use.
5. It should be capable of being packed economically and should be stable in storage during its commercial shelf life.
6. It should conform to accepted standards such as British Standards[1] in terms of its abrasivity to enamel and dentine.
7. If prophylactic claims are made, these should be substantiated by properly conducted clinical trials.

Over the past two decades toothpastes have been in a transition stage between cosmetic and prophylactic products. At present the bulk of toothpaste sales is in the prophylactic range, in complete contrast to the situation 25 years ago.

This change has made life more difficult for the formulation chemist who now has the problem, not only of formulating a good cosmetic product, but also of incorporating into it an active ingredient frequently incompatible with normal ingredients.

Over the same period television advertising has grown enormously and the television authorities require that prophylactic and therapeutic claims shall be substantiated. This in turn has meant that dentifrice manufacturers have moved into the area of clinical research in order to produce evidence for their claims. Nevertheless there is still a place for the purely cosmetic product. Indeed, such a product, properly used, probably has some anti-caries effect and should certainly help to prevent gum disorders by improving oral hygiene.

Oral products have appeared in many physical guises, but in convenience terms the most important is the semi-solid paste packed in a collapsible tube. Powders, solid blocks and liquid products can, of course, also be made.

TOOTHPASTES

Basic Structure

The primary function of a dentifrice is to remove adherent soiling matter from a hard surface with minimal damage to that surface. This is a common domestic cleaning situation which is normally solved by using a mildly abrasive powder to which a surface-active agent should be added. The function of the surface-active agent is to aid in the penetration and removal of the adherent film and to suspend removed soiling matter. The foam produced also has a psychological effect in making tooth cleaning more pleasurable.

This cleaning function must be achieved in a short time—say, under two minutes—and at body temperature. The basic formulation would in fact be a simple tooth powder.

The requirement of convenience in packing and in use determines that this basic product should be made into a paste. It thus becomes necessary to add liquids which should have humectant properties to prevent the toothpaste drying out at the tube nozzle. In order to maintain a high-solids suspension in a stable viscous form, it also becomes necessary to increase the viscosity of the liquid phase by the addition of a gelling agent.

Finally it is necessary to add flavours and possibly preservatives, colours and active ingredients, and all these components must be non-toxic and non-irritant under the conditions of use.

The total product should maintain its consistency over a temperature range from 0°C to 37°C (that is, it should have a relatively flat viscosity–temperature curve). It should also be capable of being stored without physical or chemical change over the same temperature range. Most large manufacturers have international sales and may have to take into account local conditions in many countries.

From the manufacturer's point of view the product should be made from the least expensive freely available raw materials compatible with good product quality.

A simple cosmetic toothpaste formulated in this way may have to be modified if an active ingredient is incorporated. In such a case the product virtually becomes a vehicle for the active ingredient and this may affect the basic formulation.

Ingredients

A balanced formula can only be achieved by considering all the ingredients together since many of them may have a dual function or may interact with one another. Cost and availability as well as local laws, regulations and even local habits may cause formulations to vary from country to country.

Abrasives

The abrasive used in a toothpaste must always be a compromise between the ability to clean the surface and the necessity to avoid damage to the tooth surface. In the words of the Council on Dental Therapeutics of the American

Dental Association: 'a dentifrice should be no more abrasive than is necessary to keep the teeth clean—that is, free of accessible plaque, debris and superficial stain. The degree of abrasivity needed to accomplish this purpose may vary from one individual to another'.[2] The abrasivity and cleaning action of abrasives are governed by size, shape, brittleness and hardness.[3,4] The work of Wright[5,6] has led to a clearer understanding of the mechanics of the effect of abrasives on the wear rate of teeth.

The most commonly used abrasives are precipitated calcium carbonate and dicalcium phosphate dihydrate. Other materials include tricalcium phosphate, calcium pyrophosphate, insoluble sodium metaphosphate, various types of alumina, silica and silicates. Particles of plastics may also be used.

Calcium Carbonate. Chalk, or, as it is normally purchased, precipitated calcium carbonate, is available in a number of grades varying in crystalline form, particle size and surface area. Detailed information is available from suppliers and specifications have been established by the Cosmetic, Toiletry and Fragrance Association of America and the Cosmetic, Toiletry and Perfumery Association (formerly the Toilet Preparations Federation) of the United Kingdom.

By varying the conditions of precipitation, precipitated chalks of different densities and crystal habit may be obtained. The two common crystal types are aragonite (orthorhombic) and calcite (rhombohedral) and particle sizes in the range 2–20 μm are normally used.

Chalk is an efficient cleaner but does not produce good lustre on teeth. Grades containing a proportion of particles over 20 μm can also produce scratching on enamel surfaces. Probably the best compromise is to use a small proportion of chalk with a larger proportion of one of the less abrasive phosphates.

Waterworks chalk has been employed in toothpastes but quality is not always uniform and the higher level of water-soluble calcium may cause problems in some formulations.

All chalks give an alkaline reaction to toothpastes and it may be necessary to protect aluminium tubes from corrosion by adding sodium silicate.

Calcium phosphates. The varieties of calcium phosphates used in dentifrices are:

Dicalcium phosphate $CaHPO_4.2H_2O$
Dicalcium phosphate anhydrous $CaHPO_4$
Tricalcium phosphate $Ca_3(PO_4)_2$
Calcium pyrophosphate $Ca_2P_4O_7$

Synthetic apatites have also been proposed as toothpaste abrasives.[7]

Dicalcium phosphate dihydrate (DCP) is the phosphate most commonly used in dentifrices. The pH value of a toothpaste made with DCP is normally in the range 6–8. The taste of DCP-based toothpastes is normally better than that of chalk-based products and the flavour stability is improved.

DCP is in the metastable state and reverts to the anhydrous form with consequent hardening of the paste. This change is accelerated by the presence of fluoride ions. The DCP normally supplied is stabilized to delay or prevent this change. Trimagnesium phosphate, tetrasodium pyrophosphate and calcium sodium pyrophosphate are common stabilizers.[8,9]

Anhydrous DCP is more abrasive than the dihydrate and should only be used in smaller quantities. It is less soluble than the dihydrate and this can be an advantage in fluoride-containing pastes.

Tricalcium phosphate (*TCP*) is not used to a large extent. It too is less soluble than DCP.

Calcium pyrophosphate (*CPP*) was originally developed as the abrasive of choice for products containing sodium or stannous fluoride. The particular form used is patented in many countries. It is claimed that the low availability of soluble calcium ions contributes to the stability of the fluoride.[10,11]

Insoluble sodium metaphosphate (*IMP*) is a particularly useful abrasive for fluoride-containing dentifrices since it contains no calcium ions. It has the minor disadvantage that it contains a small proportion of soluble phosphates.

Other Abrasives. There have been a number of developments in abrasive systems. These have arisen because of particular demands:

(i) the desire to avoid calcium salts in fluoride pastes so as to increase fluoride stability;
(ii) the recent development of transparent dentifrices, that is, dentifrices in which the refractive index of the abrasive is the same as that of the liquid medium in which it is suspended

The first requirement has led to the wider use of hydrated alumina[4] and of synthetic plastics.[12–15] The second requirement led to the use of silica in the form of a hydrated xerogel[16] and sodium aluminium silicates.[17]

Zirconium silicate has also been used in small quantities to impart lustre to the teeth.

Detergents

Tooth cleaning is essentially a detergent process and all toothpastes incorporate a surface-active agent. Soap was the earliest detergent used, but the obvious disadvantages (high pH, taste and incompatibility with other components) has led to its replacement by synthetic detergents.

The detergent must of course be tasteless, non-toxic and non-irritant to the oral mucosa. The foaming qualities are important since they have a significant influence on the subjective assessment of toothpaste performance.

Some surface-active agents may have intrinsic prophylactic or therapeutic properties but in this section they will be considered solely by their detergent function.

Sodium Lauryl Sulphate (*SLS*). Sodium lauryl sulphate is probably the most widely used detergent for oral products and satisfies almost all requirements. In this context 'lauryl' denotes that the alkyl radical R in $ROSO_3Na$ is derived from a narrow cut alcohol predominantly C12 but with some C14. The original sources are palm kernel or coconut oil fatty acids. Various grades are available from manufacturers and particular attention should be paid to taste. This is influenced by the free alcohol content. A low content of inorganic salts is also desirable. Recrystallized grades are excellent in quality but are expensive.

Sodium N-lauroyl Sarcosinate: $R.CO.N(Me)CH_2COONa$. This compound has been widely used in accordance with a Colgate patent [18] which claims prophylactic effects as a consequence of its anti-enzyme properties. It is particularly useful in oral products because of its high solubility.

Sodium Ricinoleate and Sodium Sulphoricinoleate. Sodium ricinoleate (castor oil soap) has been used in dentifrices where it has the advantage of high solubility, but is vulnerable (as are all soaps) to the presence of calcium ions. Sodium sulphoricinoleate (Turkey Red Oil) has also been used.

Other Detergents. Sodium lauryl ether sulphate ($R(OC_2H_4)_nOSO_3Na$), coco monoglyceride sulphate, alkane sulphonates and alkyl polyether carboxylates have all been proposed as surface-active agents for dentifrices.

Humectants

As mentioned above, it is necessary to incorporate a component with humectant properties to prevent a dentifrice from drying out. This is most likely to happen if the cap is left off the tube. Twenty years ago the only humectant used was a 50 per cent solution of glycerin in water. This is the perfect humectant in the sense that it is stable, non-toxic, has some solubilizing properties and contributes an element of sweetness.

More recently glycerin has been partly or wholly replaced by 70 per cent sorbitol syrup which has similar properties and is usually less expensive. It is available in crystallizing and non-crystallizing grades.

Propylene glycol has also been used as a third component of the humectant system.

Gelling Agents

As mentioned above, it is necessary to incorporate a gelling or binding agent in order to maintain a high-solids suspension in a stable form. The gelling agent also modifies the dispersibility, foam character and 'feel' in the mouth. Gelling agents used in toothpastes are hydrophilic colloids which disperse in aqueous media. These include natural gums such as Irish moss and gum tragacanth, synthetic cellulosic products and silica.

Gum Tragacanth. This gum was extensively used at one time and satisfactory pastes can be made with it. The final product may be variable because of the natural origin of the gum.

Carragheen. This is the generic name given to gums derived from the seaweed *Chondrus crispus* or Irish moss. The purified colloid consists of a mixture of two sulphated polysaccharides and the gelling properties may be controlled by the extent to which the metal ions present—sodium, potassium, calcium and magnesium—have been interchanged by ion exchange.

Commercial carragheens are standardized products of uniform and reproducible quality. Though very commonly used 20 years ago, they have been largely replaced by the cellulose derivatives.

Cellulose Derivatives. These are now the most commonly used gelling agents for toothpastes. Since they are largely man-made they can be tailored to suit any requirement in terms of solubility, gel strength, etc. They are non-coloured, non-toxic and relatively tasteless. Their behaviour in toothpastes has been reviewed by Watson.[19]

Carboxymethyl cellulose (CMC) or, more strictly, sodium carboxymethyl cellulose (SCMC) is prepared by the action of sodium chloracetate on alkali cellulose. The physical properties may be controlled by adjusting the degree of breakdown of the cellulose before substitution and by the degree of substitution.

SCMC gels are anionic and sensitive to pH values outside the range 5·5–9·5. They are reasonably stable in the presence of electrolytes and calcium ions and in general are suitable for most toothpaste formulations. SCMC is indeed the most commonly used gelling agent for toothpastes.

Because it is anionic SCMC is not suitable for toothpastes containing cationic agents such as certain antibacterials. For these a non-ionic cellulose derivative must be used.

One minor disadvantage of SCMC is its possible breakdown if toothpaste is infected with the organism *Penicillium citrinum* but this is a rare occurrence.

Commercial grades of SCMC include Celacol and Courlose (British Celanese), Cellofas and Edifas (ICI), FMP and FHP (Hercules), Tylose (Hoeschst). Official specifications are TPF47 and the CTFA specification (formerly TGA34).

Cellulose ethers are generally the methyl or hydroxyethyl ethers of cellulose. As with SCMC these ethers can be tailor-made to give prescribed properties by varying the degree of substitution. They are of course nonionic and are stable over wide pH ranges and unaffected by metal cations. They are most valuable in formulations that contain antibacterials which are cationic.

Methylcellulose is more soluble in cold than hot water, but since toothpastes can normally be made by cold processes this is no particular disadvantage. Methylcellulose is somewhat incompatible with glycerin and this can be a drawback.

Hydroxyethylcellulose (HEC) has the general characteristics of the cellulose ethers but does not have the inverse solubility/temperature characteristic of methylcellulose. Toothpastes made with HEC are slower to disperse than those made with SCMC so that foam and flavour are slower to develop. Nevertheless HEC is probably nearest to the ideal binding agent for toothpastes, particularly products that contain cations.

Methylcellulose is sold under names such as Celacol (British Celanese), Methocel (Dow), Methofas (ICI), Tylose (Hoechst). Specifications are TPF60 and the CTFA specification (formerly TGA30).

Hydroxyethylcellulose is sold under names such as Cellosize (Union Carbide) and Natrosol (Hercules).

Miscellaneous Gelling Agents. Starch ethers have been used in toothpastes and are satisfactory.

Two synthetic resins, Polyox (an ethylene oxide polymer) and Carbopol (a carboxy-vinyl polymer)—both made by Union Carbide—have also been suggested for use in toothpastes.

A recent newcomer to the field of gelling agents is Laponite (Laporte) which is a synthetic clay of the Hectorite type. This has received some attention in the literature.[20]

Flavours

The flavour of a toothpaste is one of the most important characteristics influencing consumer acceptance. Apart from the matter of consumer reaction the flavour may account for as much as 25 per cent of the unpacked cost of the product. For these reasons it is essential to choose a flavour with great care.

Contrary to popular belief, pleasant-tasting flavours such as fruit, chocolate (even whisky has been suggested!) are not popular. The consumer demands a flavour that is conventionally acceptable (and this varies in different countries) and which leaves a fresh sensation in the mouth and a lasting awareness that the mouth has been cleansed.

Conventionally, flavours have usually been based on the oils of spearmint and peppermint. These are often fortified with a trace of menthol to give a cooling effect. They are also modified with clove (or eugenol) wintergreen (or methyl salicylate), eucalyptus, aniseed, etc.

Wintergreen-type flavours are common in the United States but are less acceptable in Europe, possibly because of the association with embrocations.

All flavours require sweetening and saccharin is the sweetener of choice. Cyclamates are now banned and though a number of other synthetic sweeteners have been proposed none has yet found common acceptance. An important flavour additive has been chloroform which not only has a sweet taste, but promotes a 'flavour burst' sensation in use.

The other components of the toothpaste contribute to the flavour pattern, for example DCP-based pastes have usually a superior flavour to those based on chalk. The flavour may also be modified by the presence of an active ingredient (such as chlorhexidine) and even by the pH of the product.

The nature of the foam and the dispersibility of the paste also affect the flavour impact in the mouth.

These problems can best be solved by creating a flavour panel of expert tasters who can describe and quantify flavour sensations and thus construct a flavour profile.

Other Ingredients

Preservatives. It used to be common practice to add preservatives to the formulation of a toothpaste to protect it from the effect of micro-organisms. The gelling agent for example may be particularly vulnerable. Formalin and sodium benzoate and *p*-hydroxy benzoates were commonly used for the purpose.

The use of preservatives is now less common for a variety of reasons. Formalin is now banned by EEC regulations and sodium benzoate is not effective at neutral and higher pH values. Flavour components themselves have some anti-bacterial action, as have some of the active ingredients now used.

Overriding these considerations, the product should be manufactured in conditions such that the final product is sterile and it should then not be necessary to add preservatives.

Corrosion Inhibitors. Sodium silicate is often added to high pH chalk-based toothpastes to prevent attack on aluminium tubes. Some phosphates also reduce the corrosion risk with alumina-based toothpastes.[21]

Chloroform and high levels of electrolytes can also promote corrosion. Increasing the glycerin level in the water phase will often reduce the risk of this type of corrosion.

Colours. Colours are sometimes added to toothpastes. These must be chosen with care as colour fading, particularly at the nozzle, is not uncommon. The range of colours available is now restricted by EEC regulations.

A novel development was the production of a striped toothpaste by Lever Bros in the USA;[22] this was done by mixing white and red pastes in an ingenious nozzle fitment.

Bleaches. To enhance the whitening effect of toothpastes and powders and to assist in the removal of stains, oxidizing agents are often added to the product. These included sodium perborate, magnesium peroxide, hydrogen peroxide–urea compounds, stabilized hydrogen peroxide compounds, etc. It is doubtful if such compounds remain active after storage and their use has diminished.

Formulation of Toothpastes

The raw materials described above differ not only in constitution but also in their effect on the physical properties of the final product. It is pointless, therefore, to present general formulations except in fairly wide terms. Example 1 gives a general picture of the components of a standard toothpaste.

	(1) *per cent*
Gelling agent	1·0
CMC	
HEC	
Irish moss	
Gum tragacanth	
Humectant	10–30
Glycerin	
Sorbitol 70%	
Propylene glycol	
Abrasive	15–50
$CaCO_3$	
$CaHPO_4 \cdot 2H_2O$	
$CaHPO_4$	
$Al_2O_3 \cdot 3H_2O$	
$Ca_2P_4O_7$	
$MgHPO_4 \cdot 3H_2O$	
SiO_2	
$(NaPO_3)_x$	

(1—*continued*)

	per cent
Sweetener (saccharin)	0·1–0·2
Flavour	1·0–1·5
Spearmint	
Peppermint	
Menthol	
Vanillin	
Eugenol	
Wintergreen	
Anethole	
Anise	
Eucalyptus	
Cinnamon	
Surface-active agent	1·0–2·0
Sodium lauryl sulphate	
Sodium N-lauroyl sarcosinate	
Monoglyceride sulphate	
Preservative (*p*-hydroxy benzoates)	0·1–0·5
Prophylactic agent	0·1–1·0
NaF	
SnF_2	
Na_2FPO_3	
Amine fluorides	
etc.	
Colour (see, for instance, EEC list)	*q.s.*
Water	to 100·0

The addition of fluorides (sodium or stannous) or sodium monofluorophosphate to a toothpaste presents problems so that the above general formula may have to be modified. For example, if free fluoride ions are present in a calcium carbonate formula they will quickly be precipitated as calcium fluoride and the cariostatic activity will be lost. In these cases the abrasive must be chosen with care to prevent or reduce this effect.

Insoluble sodium metaphosphate (IMP) and special grades of calcium pyrophosphate (CPP) are usually used with compounds releasing fluoride ions (for example SnF_2, NaF). Sodium monofluorophosphate is less of a problem since the ion is FPO_3^{2-} and not F^-. In this case DCP and even precipitated calcium carbonate may be used. Great care must be used in all fluoride formulations to ensure that the fluoride activity remains at a high level throughout the life of the product.

The level of fluoride ingredient used has conventionally been such that there is 1000 ppm of fluorine in the final product. This corresponds to 0·2 per cent of NaF, 0·4 per cent of SnF_2 and 0·76 per cent of Na_2FPO_3. In the EEC a proposed amendment to the Cosmetics Directive gives an authorized total concentration of fluoride of 1500 ppm.

Manufacture of Toothpastes

Two basic processes are involved in toothpaste manufacture—the hydration of the gelling agent and the dispersion of the abrasive in the gel. The hydration of the gel is normally done by adding the solid gelling agent to the glycerin and part of the water under conditions of vigorous agitation. It is not necessary to heat the mixture if CMC is used, but heating to 60°C is usual with Viscarin-type gelling agents. Over-stirring of CMC gels results in an irreversible diminution of viscosity and should be avoided.

Gel hydration can be continuous by means of an eductor (supplied by Hercules) in which the gel powder is introduced gradually into a stream of cold water which is then forced through a nozzle. The vigorous agitation produced gives a smooth uniform gel.

The powder addition may be done in a variety of types of vessel capable of heavy-duty mixing, such as the Petzholdt, Fryma and Unimix vessels. The final mixing is always done under vacuum so as to de-aerate the product. It is usual practice to add the active ingredient (if present) late in the mixing cycle and to add the surface-active agent and the flavour last of all. This is done to avoid excessive foaming and to reduce loss of flavour during evacuation.

The degree to which de-aeration is complete can be checked by density measurement. For a general formula such as that described above, a density of 1·55–1·60 would be expected.

TOOTHPOWDERS

Toothpowders are the original, the simplest and the cheapest compounded forms of dentifrice. Powders have been replaced very largely by the more convenient pastes, but they still hold a small share of the market. Formulation problems are not as severe since interaction between components is unlikely in the absence of water. Fluorides and oxidizing agents, for example, are likely to retain their effective concentration longer than they would do in a paste formulation.

Other formulation problems are likely to be concerned with physical characteristics such as the preparation of ingredients of fairly uniform size so as to prevent separation on shaking, and in ensuring that the product does not cake on storage.

Typical formulations are given in examples 2–5.

	(2) *per cent*
Precipitated calcium carbonate	95·0
Sodium palmitate	5·0
Flavour, sweetener	*q.s.*

	(3) *per cent*
Dicalcium phosphate dihydrate	79·0
Precipitated calcium carbonate	20·0
Sodium lauryl sulphate	1·0
Flavouring, sweetener	*q.s.*

Oxygenated powder	(4) *per cent*
Precipitated calcium carbonate	96·0
Sodium lauryl sulphate	2·0
Magnesium peroxide	2·0
Flavouring, sweetener	*q.s.*

Fluoride powder	(5) *per cent*
Dicalcium phosphate dihydrate	75·0
Precipitated calcium carbonate	23·0
Sodium lauryl sulphate	1·0
Sodium monofluorophosphate	0·8
Flavouring, sweetener	*q.s.*

Manufacture of Toothpowders

The manufacture of powders is very simple. The sweeteners and flavour, together with a little alcohol if desired, are made into a pre-mix concentrate with part of the abrasive powder. This is then mixed with the rest of the powders in a conventional powder mixer.

SOLID DENTIFRICE

Solid dentifrice is essentially a soap in which the abrasive powder is mixed. The proportion of soap may vary within fairly wide limits from about 10 to 30 per cent depending on the glycerin content of the finished product and the hardness desired. It is usual to add a higher proportion of flavour than in paste dentifrices and the product is usually coloured. Solid dentifrices, like tooth powders, have largely been replaced by toothpastes. A typical formula would be:

	(6) *per cent*
Dental soap	18·0
Precipitated calcium carbonate	79·0
Glycerin	3·0
Colour, flavour, sweetener	*q.s.*

The soap and abrasive materials are milled with the glycerin and sufficient water to give a plastic mass. Colour and flavour are added and the product is then plodded and extruded in a conventional soap plodder, cut into billets and stamped.

The abrasive nature of the product demands specially fabricated plodders and cutters, and in general the manufacture of solid dentifrices presents a number of problems.

PERFORMANCE TESTS

The clinical claims made for dentifrices and the restraints on advertising have helped to increase the volume of work done in recent years on the performance of dentifrices. Biological material in the form of extracted human and animal teeth is readily available and some experimental work can even be done on teeth *in situ* in the mouth. Chemists, physicists and dentists have all contributed to the vastly increased knowledge of the performance of oral products.

Abrasive Action

The cleaning properties of a dentifrice depend primarily on the nature and quantity of abrasive present; the design of the toothbrush may play a part and even the detergent, but their effects are insignificant compared with that of the abrasive.

During cleaning, food debris, plaque, acquired pellicle, stains and calculus should be removed from the tooth surface, if possible without damage to the underlying enamel. Dentifrice abrasives are a compromise between the desire for perfect cleaning and the desire to avoid enamel wear; evaluation methods and the standards adopted reflect this compromise.

There is now no doubt in the minds of experts in this field that abrasion studies must be done on human dentine and enamel and not on other substrates. The use of a metal substrate can often produce misleading results. The most obvious method of measuring abrasion should be by weight loss, but this requires excessive abrasion and would lead to wear far in excess of that met with in a real life situation.

A variety of techniques has been proposed for the measurement of the abrasive quality of toothpastes, for example the shadowgraph method,[23] the surface profile method,[24] interference microscopy[25] and replication techniques.[26] There is now almost universal agreement that the technique most nearly approximating to natural conditions is the radio-tracer method first described by Grabenstetter *et al.*,[27] further developed by Wright[5,6] and finally incorporated in a British Standard.[1]

Specimens of tooth crown (enamel) and root (dentine) are bombarded with neutrons which change a minute fraction of the phosphorus atoms present from ^{31}P to ^{32}P. After brushing under standard conditions the slurry of toothpaste used is dried and counted for β emission. By comparison with a standard toothpaste of fixed composition it is possible to give a relative abrasivity rating to any toothpaste. The British Standard sets the figure of 100 as the abrasivity rating of the standard (a conventional chalk dentifrice) and sets the maximum allowable abrasivity of a toothpaste as 200 against dentine and 400 against enamel. This reflects the greater vulnerability of the much softer dentine.

In fact almost all conventional toothpastes fall comfortably within this standard; abrasivity ratings of 50–100 against dentine and 50–120 against enamel would probably cover most commercial toothpastes.

There is no evidence to show that any conventional dentifrice, *when properly used*, has caused excessive wear of enamel or dentine. Cervical erosion, that is, erosion at the neck of tooth following gum recession, which occurs in older subjects, is due to bad brushing technique.

The amount of material lost by abrasion with a relatively abrasive paste has been calculated as $1{\cdot}2 \times 10^{-8}$ g per brush stroke from enamel and 98×10^{-8} g per brush stroke from dentine.

Using the radio-tracer technique it is possible to assess the relative wear rates of different particle sizes of different abrasives. Within the fairly narrow limits of conventional toothpaste formulations there is an almost linear relationship between (a) particle size and wear rate and (b) percentage concentration of abrasive and wear rate.

Hardness, crystallinity of particles and particle shape all play some part in determining wear rate, but with the knowledge and experience available to the expert it is now possible to pre-set the abrasivity of a toothpaste within fairly narrow limits. Probably the best compromise would be to have a relatively high proportion of large soft particles (e.g. DCP 5–10 μm in size) and a small proportion of small hard particles (e.g. zirconium silicate or silica 1 μm in size). The large soft particles should remove most of the adherent soiling matter from teeth and the small particles should give some degree of polish without visible scratching.

Lustre (Gloss or Polish)

The measurement of lustre is complicated by a number of factors. Hunter[28] has pointed out that it is impossible to measure specular reflectance and diffuse reflectance as separate entities in any but an approximate way, and describes six different kinds of gloss:

1. *Specular gloss*—shininess.
2. *Sheen*—surface shininess at grazing angles.
3. *Contrast gloss*—contrast between specular reflectance of different areas.
4. *Absence-of-bloom gloss*—the absence of reflection haze or smear adjacent to highlights.
5. *Distinctness-of-reflected-image gloss*—distinctness of images reflected in surfaces.
6. *Absence-of-surface-texture gloss*—lack of surface texture and surface blemishes.

This is a highly sophisticated analysis of the problem of lustre determination and is probably too complex for measurement on human teeth. The simplest assessment of lustre on teeth is a subjective one and what is required is a simple objective procedure which will duplicate subjective assessment.

In the paint and lacquer industry a test method has been described[29] the basic principle of which is the illumination of the test specimen with a parallel beam of light and the measurement of the reflection of this beam at a predetermined angle. It appeared to Tainter[30] and co-workers that this method could be modified for the measurement of directional reflectance. Considerable refinement of apparatus and technique was necessary to adapt this method to measurements of gloss on teeth because of their small size and the curvature of the enamel surface. As a result of their experiments Tainter *et al.* published their quantitative method for measuring the polish produced by dentifrices.[31]

Phillips and Van Huysen[32] reported the results of an investigation into the action of dentifrice polishing agents on the tooth surface. Two methods were employed: (i) visual observation of lustre changes by comparison with a series of standards and (ii) microscopic study of the tooth before and after brushing. As a result of their work it was concluded that calcium carbonate tended to dull enamel surfaces. Calcium phosphates (DCP and TCP) had little effect on lustre, while a mixture of metaphosphate and calcium phosphate appeared to be superior to other abrasives. It would be unwise to be too dogmatic on this subject since the degree of lustre on the one hand and dulling on the other are strongly influenced by mean particle size and the range of particle size of the abrasive concerned.

The depth of scratches measured by profilometer techniques has been found to correlate with polishing power and this again emphasizes the importance of particle size. Manly *et al.*[33] used a similar technique in measuring the scatter of a laser beam and Schiff and Shaver[34] have adapted Tainter's methods to measurement of polish *in vivo*. Probably the simplest assessment of polish and cleaning power is the technique described by Wilkinson and Pugh[4] who showed that there is a direct relationship between abrasion and cleaning power. It should thus be possible to devise an abrasive system to give both adequate cleaning and minimum abrasion. A mixture of large (10 μm) soft crystals with a minor proportion of small (1 μm) hard crystals should achieve this result.

THE TOOTHBRUSH AND TOOTHBRUSHING

The toothbrush and the mechanism of toothbrushing play an important part in oral hygiene.

Surveys of toothbrushing habits[35,36] show that 60 per cent of adults claim to brush their teeth twice a day, usually on waking and before going to bed, and that on the whole women are more conscientious than men. This is a very superficial picture, however; the statistics of toothpaste sales in different countries show clearly that a large number of people either do not brush their teeth at all, or do it rarely.

It has never been shown unequivocally that toothbrushing alone is instrumental in reducing dental decay. Fosdick[37] has shown that regular brushing with a cosmetic dentifrice reduces the incidence of decay among susceptible subjects, while Smith and Striffler[38] in a review of the literature found some evidence to the contrary. What is surely beyond question is that the frequent use of a fluoride dentrifrice does reduce the incidence of decay. This has been demonstrated in a large number of clinical trials.

It is also clear that regular toothbrushing is effective in reducing or preventing periodontal disease. The removal of food debris and the massaging of the gums are a part of good oral hygiene.

A case can therefore be made for regular toothbrushing with a good prophylactic toothpaste:

(i) It is aesthetically satisfying to produce a clean fresh sensation in the mouth.
(ii) A good prophylactic toothpaste will certainly reduce the incidence of caries.
(iii) The removal of food debris should improve mouth odour.
(iv) Regular brushing will help to prevent periodontal disease.

The introduction of the electric toothbrush has stimulated tests designed to show whether mechanical brushing is superior to hand-brushing. Unfortunately, there are no accepted criteria for evaluating the effectiveness of a toothbrush. Prevention or removal of plaque and calculus, the gingival index and the absence of staining are all common criteria. The evaluation is further complicated in that toothbrushes vary not only in head size and basic design, but also in the bristle pattern and the nature and stiffness of the bristle. In a review of this subject, Ash[39] concluded that electric toothbrushes were no more effective than manual toothbrushes for the average subject, but that individuals might find one or other to be more effective.

McKendrick *et al.*[40] performed a two-year study on dental students comparing manual and electric toothbrushes and found a lower periodontal index in all subjects but no significant difference between groups.

Muhler[41] found that the frequency of use of electric toothbrushes rose from 1·04 to 2·90 times per day after two months and declined thereafter. After one year half the subjects gave up this method of cleaning teeth. During the same period, the subjects using manual brushes did not alter their frequency of brushing.

Other devices have been suggested for cleaning teeth. An elongated brush with a flexible abrasive string attached has been proposed,[42] as have water jet rinsing devices[43] and water jets containing abrasives.[44] These methods have been reviewed[45,46] and it has been found that though they may be effective in reducing the population of micro-organisms on the tooth, they are not as effective as toothbrushes for the removal of oral debris.

It can be assumed that most brushes, properly used, have some effect in improving oral hygiene and hence periodontal disease, but no case can be made for their having any anti-caries effect. In general, professional dental care, proper diet, the use of a prophylactic dentifrice and education in brushing technique are essential features in the prevention of oral diseases.

DENTURE CLEANSERS

Denture cleansers are marketed either in powder or tablet form or as liquids. Though the solid products may differ widely in composition, they comprise essentially an oxidizing agent, an electrolyte and an alkali.

The oxidizing agent used is normally sodium perborate or sodium percarbonate, though hypochlorites, trichlorisocyanuric acid and its salts and persulphates have also been used or proposed.

Sodium percarbonate is more soluble in water than sodium perborate, but is not quite so stable, though in the solid form its stability is adequate.

Sodium perborate is available in two forms:

(i) *Sodium perborate tetrahydrate*, usually written as $NaBO_3 \cdot 4H_2O$ but more properly $NaBO_2 \cdot H_2O_2 \cdot 3H_2O$, has an active oxygen content of 10·38 per cent and is sold on the basis of 10 per cent active oxygen.

(ii) *Sodium perborate monohydrate*, usually written as $NaBO_3 \cdot H_2O$ but more properly $NaBO_2 \cdot H_2O_2$, has an active oxygen content of 16 per cent and is sold on a basis of 15 per cent active oxygen.[47]

Sodium percarbonate, like sodium perborate, is not a true persalt and should be written as $2Na_2CO_3 \cdot 3H_2O_2$.

The purpose of a denture cleanser is to loosen debris, which consists of saliva and food particles, to remove stains and to sterilize the denture.

Solid products are normally dissolved in water to form a solution in which the denture is immersed. Bubbles of oxygen form and help mechanically to loosen food debris which is itself partly solubilized by the alkali present. Electrolytes such as sodium chloride have some solubilizing action on mucous deposits. The combined effect is to loosen the debris so that, after a suitable period of soaking, it is easily brushed away.

In addition the denture is sterilized and stains are removed. Regular use of denture cleansers will also prevent the build-up of calculus on the denture surface.

Whatever the form of oxidizing agent used, care must be taken to ensure that the product does not alter the colour of the dental plate, though this danger is much less now than formerly. Modern plastic dental plates normally retain their colour well.

The amount of sodium perborate or percarbonate used is usually in the range 20–50 per cent; corresponding amounts of other oxidizing agents may be employed with the proviso that active chlorine-producing compounds should be used at such a level as to leave the denture, after rinsing, without an unpleasant chlorine after-taste.

The electrolyte used is invariably sodium chloride and the alkali is most commonly anhydrous trisodium phosphate, though sodium carbonate or bicarbonate and other alkalis may be used.

Typical formulae are given in examples 7 and 8.

	(7)
	per cent
Sodium perborate	40·0
Sodium chloride	30·0
Trisodium phosphate	30·0
Flavour, colour	*q.s.*

	(8)
	per cent
Sodium percarbonate	40·0
Sodium chloride	40·0
Sodium carbonate	20·0

Tablets can also be prepared, either from conventional mixtures as in examples 7 and 8, or as, for instance, in example 9.

	(9)
	per cent
Sodium percarbonate	88·0
Sodium chloride	10·0
Sodium silicate and/or other binders	2·0
Flavour, colour	*q.s.*

Liquid products are normally dilute solutions of sodium hypochlorite to which additional sodium chloride may be added. Such solutions of course lose some of their available chlorine over time.

It is common to promote effervescence in some solid products. This helps in the break-up and dissolution of tablets and promotes the concept of activity. Effervescence may be provided by the addition of conventional carbonate–acid mixtures (tartaric or citric acids) or by the incorporation of a peroxide-decomposing catalyst such as a trace of a copper salt. In the latter case great care should be exercised in checking the stability of the finished product.

The inclusion of proteolytic and amylolytic enzymes in denture cleansers[48] has been proposed. Other new developments are most easily found in patent literature.[49]

The manufacture of these products is not as simple as it appears. It is vital to keep moisture out of the product and to protect it from atmospheric humidity. The powder products must not cake on storage and solid products should dissolve quickly to give a clear solution.

Stringent age testing should be performed on the packed product and the continued performance of the product under usage conditions should be monitored.

REFERENCES

1. *British Standard Specification* BS 5136: 1981.
2. American Dental Association, *J. Am. dent. Assoc.*,1970, **81,** 1177.
3. Swartz, M. L. and Phillips, R. W., *Ann. NY Acad. Sci.*, 1968, **153,** Art. 1, 120.
4. Wilkinson, J. B. and Pugh, B. R., *J. Soc. cosmet. Chem.*, 1970, **21,** 595.
5. Wright, K. H. R., *Wear*, 1969, **14,** 263.
6. Wright, K. H. R. and Stevenson, J. I., *J. Soc.cosmet. Chem.*, 1967, **18,** 387.
7. British Patent 1 460 581, Colgate-Palmolive, 6 January 1977; US Patent 4 048 300, Colgate-Palmolive, 13 September 1977.
8. British Patent 914 707, Victor, 2 January 1963.
9. British Patent 1 143 123, Albright and Wilson, 19 February 1969.
10. British Patent 810 345, Hedley, 20 August 1956.
11. British Patent 746 550, Indiana University, 8 March 1953.
12. British Patent 939 230, Procter and Gamble, 3 November 1959.
13. British Patent 995 351, Procter and Gamble, 28 October 1960.
14. US Patent 3 151 027, Procter and Gamble, 7 June 1961.
15. British Patent 1 055 784, Unilever, 18 January 1967.
16. British Patent 1 186 706, Unilever, 2 April 1970.
17. British Patent 1 305 353, Colgate-Palmolive, 31 January 1973.
18. British Patent 728 243, Colgate-Palmolive, 8 October 1952.
19. Watson, C. A., *J. Soc. cosmet. Chem.*, 1970, **21,** 459.
20. Neumann, B. S. and Sansom, K. G., *J. Soc. cosmet. Chem.*, 1970, **21,** 237.
21. British Patent 1 277 586, Unilever, 14 June 1972.
22. British Patent 956 377, Unilever, 29 April 1964.
23. Manly, R. S., *J. dent. Res.*, 1944, **23,** 59.
24. Ashmore, H., Van Abbé, N. J. and Wilson, S. J., *Br. dent. J.*, 1970, **133,** 60.
25. Ashmore, H., *Br. dent. J.*, 1966, **120,** 309.
26. Brasch, S. V., Lazarou, J., Van Abbé, N. J. and Forrest, J. O., *Br. dent. J.*, 1969, **127,** 119.

27. Grabenstetter, R. J., Broge, R. W., Jackson, F. L. and Radike, A. W., *J. dent. Res.*, 1958, **37,** 1060.
28. Hunter, R. S., *Methods of Determining Gloss,* US National Bureau of Standards, Research Paper RP 958, January 1937.
29. American Society for Testing Materials, D 523–44 T, 1946.
30. Tainter, M. L., *Proc. sci. Sect. Toilet Goods Assoc.*, 1944, (1), 24.
31. Tainter, M. L., Alford, C. E., Henkel, E. J. Jr., Nachod, F. C. and Priznor, M., *Proc. sci. Sect. Toilet Goods Assoc.*, 1947, (7), 38.
32. Phillips, R. W. and Van Huysen, G., *Am. Perfum. Essent. Oil Rev.*, 1948, **63,** 33.
33. Manly, R. S., Brown, P. W., Harrington, D. P., Crane, G. L. and Schichting, D. A., *Arch. oral Biol.*, 1975, **20,** 479.
34. Schiff, T. and Shaver, K. J., *J. oral Med.*, 1971, **26,** 127.
35. Cohen, L. K., O'Shea, R. M. and Putnam, W. J., *J. oral Ther. Pharmacol.*, 1967, **4,** 229.
36. Survey of Family Toothbrushing Practices, *J. Am. dent. Assoc.*, 1966, **72,** 1489.
37. Fosdick, L. S., *J. Am. dent. Assoc.*, 1950, **40,** 133.
38. Smith, A. J. and Striffler, D. F., *Public Health Dentistry*, 1963, **23,** 159.
39. Ash, M., *J. Periodont.*, 1969, **10,** 35.
40. McKendrick, A. J. W., Barbenel, L. M. H. and McHugh, W. D., Paper presented to IADR (British Division), April 1968.
41. Muhler, J. C., *J. Periodont.*, 1969, **40,** 268.
42. British Patent 1 471 435, Thornton, 27 April 1974.
43. British Patents 1 469 399 and 1 469 400, Halstead, 6 April 1977.
44. British Patent 1 480 594, Laing, 20 July 1977.
45. Toto, P. D., Evan, C. L. and Sawinski, V. J., *J. Periodont.*, 1969, **40,** 296.
46. Jann, R., *Periodont. Abst.*, 1970, **18,** 6.
47. Laporte Chemicals Ltd, *Sodium Perborate and Sodium Percarbonate.*
48. British Patent 1 391 318, Miles Laboratories, 23 April 1975.
49. British Patent 1 470 581, Rossbrook, 14 April 1977; British Patents 1 483 501 and 1 483 502, Colgate-Palmolive, 24 August 1977.

Chapter Thirty-two

Mouthwashes

Introduction

In principle a mouthwash appears to be the ideal means for the application of any form of medication to the mouth, the gums or the teeth—provided, of course, that the tooth surface has been cleaned.

McCormick[1] has reviewed the literature on mouthwashes. In general they may be any of three types: antibacterial, which deal with the bacterial population of the mouth; fluoride, which help to reinforce the fluoride layer of the enamel of teeth; and remineralizing, which help to repair early carious lesions. This chapter deals only with mouthwashes of the antibacterial type. Such products are expected by the consumer to give a healthier and fresher mouth and to provide some assurance of good breath odour. Specific studies on the caries-preventive properties of mouthwashes containing fluoride have been described by Birkeland and Torell.[2]

Antiseptic mouthwashes have not attained a high level of sales in the UK, but they are popular in other countries, particularly in the USA where sales value in 1978 amounted to $269 million.[3] In general, the American mouthwashes have been formulated to be ready for use, while European mouthwashes have tended to be used after dilution.

Mouthwashes have normally been marketed on the basis of a social necessity for clean breath, though recent work with chlorhexidine has shown that they may have anti-plaque properties[4] and indeed even anti-caries properties.[5]

It is of course neither possible nor desirable to aim at complete sterility in the mouth. The use of antibiotics, for example, might destroy normal bacteria and thus permit the growth of undesirable organisms such as *Candida albicans*. It is possible, however, to reduce the bacterial population and maintain it at a lower level by the use of antibacterials which are adsorbed onto the mucous membrane. Cationic antibacterials normally have this property to some degree. The Buccal Ephithelial test of Vinson and Bennet can serve as a standard technique.[6]

The total effect exerted is a combination of three factors: (a) the mechanical effect of rinsing food debris from the mouth; (b) the effect of the antibacterial agent on the oral flora; and (c) the effect of the flavour present. Since many flavouring substances have an antibacterial effect, there may be synergism between (b) and (c).

Mouthwashes which merely claim to promote general oral hygiene should nevertheless be rigorously tested for absence of toxic and irritant properties. Prophylactic and therapeutic claims would require even more stringent clinical tests and should be regarded in the light of the regulations of the Federal Drug Administration of the USA and the Medicines Act (1968) in the United Kingdom. There are also EEC regulations covering individual ingredients.

Stock formulations are available from recognized sources such as the Pharmaceutical Formulary (Chemist and Druggist), Extra Pharmacopoeia (Martindale), etc. Products discussed below are examples that are known to be acceptable to consumers.

Choice of Antibacterial Agent

The antibacterial agents usually employed in mouthwashes include phenols, thymol, salol, tannic acid, chlorinated thymols, hexachlorophene and quaternary ammonium compounds.

Chlorinated Phenols

*Para*chlormetacresol and *para*chlormetaxylenol are both suitable for use in mouthwashes, both for their antibacterial properties and their flavour. They are not very soluble in water but may be solubilized with terpineol (or other suitable solubilizer) and soap to give a 1 per cent solution of active material. Such solutions, for example Liq. Chloroxylenol BPC, are used at 10–20 per cent dilution.

Soap-based Mouthwashes

Soap-based mouthwashes normally have no antibacterial properties and are used to clean and freshen the mouth.

	(1) *per cent*
Powdered soap	2·0
Glycerin	15·0
Alcohol	20·0
Water	63·0
Flavour	*q.s.*

Synthetic detergents which are non-irritant may be added to give extra foaming and of course standard antibacterials may also be added to this basic formula.

Thymol (Isopropyl Metacresol)

Thymol is not very soluble, but may be solubilized in the normal way, for example with suitable alcohols, or used in aqueous solution with borax as in example 2. This product is used diluted to between 5 and 20 per cent concentration.

	(2) *per cent*
Thymol	0·03
Alcohol	3·00
Borax	2·00
Sodium bicarbonate	1·00
Glycerin	10·00
Flavour	*q.s.*
Water	to 100·00

Hydrogen Peroxide
Hydrogen peroxide is an excellent non-toxic antibacterial agent for use in mouthwashes. It can be used for cleansing ulcers and abscesses in the mouth, etc. A solution of one part hydrogen peroxide (10 vol.) diluted with 8 parts of water is useful as a mouthwash, or twice the strength can be used for septic cavities. Because of its instability it is not normally used in proprietary mouthwashes.

Sodium perborate, however, is a stable powder which on dissolving in water gives an alkaline solution of hydrogen peroxide. In practice 17 g of such a powder with 6 g of citric acid will, on the addition of 80 ml water, give a solution of 10 vol. strength which should be further diluted 1:8 before use.

Such products should be used only sparingly and to combat specific conditions, since the citrate present could lead to decalcification of the teeth.

Hexachlorophene
Hexachlorophene is substantive to the mucous membrane and is an effective antibacterial agent. The suggested concentration is 0·02 per cent in a 25 per cent alcohol–water mix. Some reservations have been expressed with regard to possible toxicity.

Quaternaries
The use of quaternaries is now well established in mouthwashes. These compounds combine antibacterial and substantive properties and many of them are non-toxic and non-irritant at the concentrations normally used. Because of their antibacterial properties many of them are effective against plaque. Benzethonium chloride is used for this purpose, but probably the most effective antibacterials are of the chlorhexidine type. Unfortunately chlorhexidine, in common with most cationics, can produce a brown stain on teeth with continued use. However, this stain is easily removed with good toothbrushing.

A product of this type has been presented in the form of a gel, which, though intended for brushing onto the teeth, is really a mouthwash rather than a toothpaste since it contains no abrasive.[7]

Other Mouthwash Components
Tannic acid, alum and zinc salts have all been used in mouthwashes because of their astringent properties. It is generally assumed that because of this property they have anti-bleeding effects on the gums.

A dilute solution of sodium hypochlorite is also commonly used because of its antibacterial effect. Formalin was formerly used, but is now prohibited under EEC regulations.

Flavouring of Mouthwashes

An essential feature of a good mouthwash is its flavour, since the consumer must be aware of the freshness of the mouth after use. Money spent on market research is well invested since there are national preferences for particular flavours; for example, methyl salicylate (oil of wintergreen) is more popular in the USA than in the United Kingdom.

Peppermint, menthol, eugenol, etc. are commonly used flavours and all leave the mouth with a feeling of freshness. Chlorinated phenols always have a characteristic flavour which, though not unpleasant, is difficult to cover. Cationics normally are slightly bitter and must be covered. The small manufacturer would be well advised to use the services of a flavour house to design a compounded flavour unique to his product.

Aerosol Mouth Fresheners

Aerosol mouth fresheners, a natural development of aerosol products, are marketed in the USA and in Europe. They are recommended for freshening the breath after eating, drinking or smoking and usually contain only flavouring agents, though antibacterials could be added. Aerosol mouth fresheners are thus presented rather as an alternative to chewing gum than as mouthwashes.

A typical $\frac{1}{2}$ oz (14 g) pack is fitted with a metered valve and contains sufficient product for 200–300 applications. A metered valve is not essential, but does protect the consumer from excessive amounts of product entering the mouth.

One significant advantage of this form of presentation is that the container is small enough to be carried in the handbag or pocket, so that its use is not confined to the bathroom.

Some manufacturers market breath freshener dispensed as a liquid. One drop placed on the tip of the tongue is claimed to produce instant freshness and the removal of breath odour.[8]

REFERENCES

1. McCormick, S., *Ann. NY Acad. Sci.*, 1968, **153,** Art. 1, 374.
2. Birkeland, J. M. and Torell, P., *Caries Res.*, 1978, **12,** Suppl. 1, 38.
3. *Drug Topics*, 1979, 1 June, 70.
4. Symposium: Hibitane in the Mouth, *J. Clin. Periodont.*, 1977, **4**(5), 1.
5. Tomlinson, K., *J. Soc. cosmet. Chem.*, 1978, **29,** 385.
6. Vinson, L. J. and Bennet, A. G., *J. Am. pharm. Assoc. sci. Edn.*, 1958, **47,** 635.
7. Corsodyl, ICI Ltd. COR-MA-L/9173/254/April 1977.
8. *Chemist Drug.*, 1967, **187,** 250.

PART FIVE

Product Ingredients and Manufacture

Chapter Thirty-three

Surface-active Agents

Introduction

The fundamental phenomenon of surface activity is *adsorption* which can lead to two quite distinct effects: (a) lowering of one or more of the boundary tensions at interfaces in the system, and/or (b) stabilization of one or more of the interfaces by the formation of adsorbed layers.[1]

A surface-active agent (surfactant) is a material which, by use of this phenomenon, has the property of altering the surface energy of a surface with which it comes into contact.[2] This lowering of surface energy can easily be observed in, for example, foaming, the enhanced spreading of a liquid on a solid, the enhanced suspension of solid particles in a liquid medium and the formation of emulsions.

The use of surfactants is well established in cosmetics and toilet products and falls into five main areas depending on the surface-active properties required:

1. *Detergent*. Where the main problem involves the removal of soiling matter, surface-active agents with detergent properties are needed, for example in shampoos and toilet soaps.
2. *Wetting*. In products where good contact is required between a solution and a substrate, good wetting properties are required, for example in the application of hair colorants and permanent waving lotions.
3. *Foaming*. Some products need to have a high level of foam in use, and for these products special surface-active agents are used, for example in shampoos and foam baths.
4. *Emulsification*. In products where the formation and stability of an emulsion is a vital feature, surface-active agents with good emulsifying properties are required, for example in skin and hair creams.
5. *Solubilization*. Products in which it is necessary to solubilize an insoluble component need a surface-active agent with the appropriate properties, for example the solubilization of perfumes and flavours.

These qualities are not mutually exclusive; they are shared to some degree by all surface-active agents. Experience has shown the value of particular products for various end uses, but there is a large degree of overlap.

Classification of Surfactants

All surface-active agents have one structural feature in common: they are all amphipathic molecules; that is, the molecule has two distinct parts—a hydrophobic unit and a hydrophilic unit.

Hydrophobic units are usually hydrocarbon chains or rings or a mixture of the two. Hydrophilic units are usually polar groups such as carboxylic, sulphate or sulphonate groups, or, in nonionic surfactants, a number of hydroxyl or ether groups. The dual nature of these molecules allows them to adsorb at interfaces and this accounts for their characteristic behaviour.

Surfactants may be classified on the basic of the uses to which they may be put, on the basis of their physical properties or on the basis of chemical structure. None of these is entirely satisfactory, but probably the most logical is to classify them according to their ionic behaviour in aqueous solution. Using this procedure there are four types of surfactant—anionic, cationic, nonionic and ampholytic surfactants. In addition, the different structures of the hydrophobic and hydrophilic groups have to be considered. Schwartz and Perry,[2] McCutcheon,[3,4] and Moillet, Collie and Black[1] have all used this classification system.

Anionic Surfactants

Anionic surfactants are those molecules in which the surface-active ion is negatively charged in solution. The classic example is soap: $C_{17}H_{33}COO^-Na^+$ (sodium oleate). The anionic surfactants are further subdivided according to the manner in which the anionic group is attached to the hydrophobic part of the molecule (Table 33.1).

Cationic Surfactants

Cationic surfactants are characterized by the fact that the surface-active ion is positively charged in aqueous solution (Table 33.2).

Nonionic Surfactants

Nonionic surfactants are characterized by the fact that the hydrophilic part of the molecule is usually made up from a multiplicity of small uncharged polar groups, for example hydroxyl groups or the ether linkages in ethylene oxide chains. The same linkages are used to reinforce the hydrophilic character in certain anionic surfactants, for example alkyl ether sulphates, $R(OCH_2CH_2)_nOSO_3^-\ M^+$ (Table 33.3).

Ampholytic Surfactants

Ampholytic surfactants are characterized by their ability to form a surface-active ion with both positive and negative charges (Table 33.4).

Properties of Surface-active Agents

The change in surface properties as the concentration of an aqueous solution of a surfactant rises is characteristic of most surface-active molecules. For example, as concentration rises the surface tension of an aqueous solution of, say, sodium dodecyl sulphate ($C_{12}H_{25}OSO_3Na$) falls rapidly (Figure 33.1), with corresponding changes in the physical properties such as interfacial tension, electrical conductivity, etc. At a certain concentration level a discontinuity occurs and surface tension and other properties no longer fall. The concentration at which this discontinuity occurs is called the *critical micelle concentration* (CMC).

Table 33.1 Anionic Surfactants

R denotes a hydrophobic chain usually of 12 to 18 carbon atoms, or a ring or system of rings.

M represents a suitable cation, usually sodium, potassium, ammonium or an organic base.

Anionic groups connected directly to the hydrophobic unit	
Fatty acid soaps	$RCOO^- M^+$
Alkyl sulphates	$ROSO_3^- M^+$
Alkyl sulphonates	$RSO_3^- M^+$
Alkyl aryl sulphonates	$RC_6H_4SO_3^- M^+$
α-Sulphonyl fatty acids	$RCHCOO^- M^+$ with $SO_3^- M^+$ on the CH
Secondary alkyl sulphates	$RCH(OSO_3^-)R' M^+$
Alkyl phosphates	$ROPO_3^{2-} 2M^+$
Anionic groups connected through ester links	
Monoglyceride sulphates	$RCOOCH_2CHOHCH_2OSO_3^- M^+$
Dialkyl sulphosuccinates (R usually C_8—C_{10})	$ROCOCH_2$ / $ROCOCHSO_3^- M^+$
Polyethyleneglycol ester sulphates	$RCO(OCH_2CH_2)_nOSO_3^- M^+$
Isethionates	$RCOOCH_2CH_2SO_3^- M^+$
Anionic groups connected through ether links	
Alkyl ether sulphates	$R(OCH_2CH_2)_nOSO_3^- M^+$
Phenol ether sulphates	$RC_6H_4(OCH_2CH_2)_nOSO_3^- M^+$
Alkyl ether carboxylates	$R(OCH_2CH_2)_nOCH_2COO^- M^+$
Anionic groups connected through amide links	
Alkanolamide sulphates	$RCONHCH_2CH_2OSO_3^- M^+$
Taurines	$RCONHCH_2CH_2SO_3^- M^+$
Sarcosinates	$RCON(CH_3)CH_2COO^- M^+$
Anionic groups connected through amidine links	
Imidazole sulphates	ring: N—CH_2, RC=N, CH_2, N bearing $CH_2CH_2OSO_3^- M^+$

The discovery of this discontinuity and the reason for it were first described by McBain[5] in the 1920s and there has been a considerable volume of work on the subject since then (see Moillet *et al.*[1] and Schwartz *et al.*[2] on micelles and also Hartley[6]).

McBain postulated that surface tension fell as the concentration of single anions increased (for instance $C_{12}H_{25}OSO_3^-$ in the example given) until at the CMC the single ions began to associate into groups which he called micelles. These micelles may be in the form of spheres of molecular size, in which the hydrophobic tails of the anions are oriented to the centre of the sphere, while the

Table 33.2 Cationic Surfactants
R denotes a hydrophobic chain usually of 12 to 18 carbon atoms or an aromatic ring. X represents a suitable anion, usually chlorine or bromine.

Simple quaternary ammonium salts in which the nitrogen is attached directly to the hydrophobic unit	
Alkyltrimethyl ammonium salts	$R\overset{+}{N}(CH_3)_3\ X^-$
Dialkyldimethyl ammonium salts	$RR^1{-}\overset{+}{N}(CH_3)_2\ X^-$
Alkyldimethylbenzyl ammonium salts	$R\overset{+}{N}(CH_3)_2\ X^-$ with $CH_2C_6H_5$ on N
Ethoxylated alkyldimethyl ammonium salts	$R{-}\overset{+}{N}(CH_3)_2\ X^-$ with $(OCH_2CH_2)_n$ on N
Cationic group separated from the hydrophobic group	
Quaternized amides of ethylenediamine	$RCONHCH_2CH_2\overset{+}{N}(CH_3)_3\ X^-$
Quaternized amides of polyethyleneimine	$RCONH(CH_2CH_2N)_nCH_2CH_2\overset{+}{N}(CH_3)_3$ with $CH_3\ X^-$ on N
Cationic group located in a heterocyclic ring	
Alkyl pyridinium salts	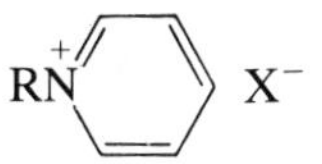 $R\overset{+}{N}$ (pyridine ring) X^-
Alkyl morpholinium salts	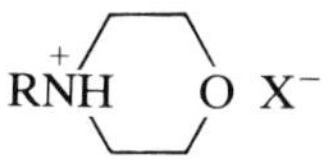 $R\overset{+}{N}H$ (morpholine ring) O X^-
Alkyl imidazolinium salts	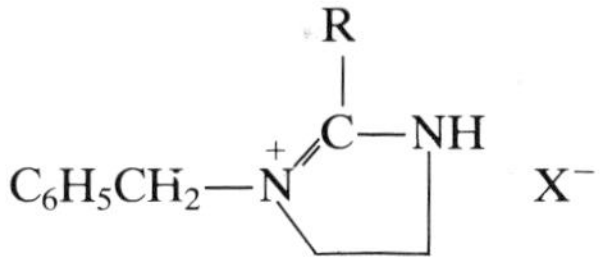 $C_6H_5CH_2{-}\overset{+}{N}$ (imidazoline ring, C—NH, with R on C) X^-
Non-nitrogenous cationic surfactants	
Sulphonium salts	$R{-}\overset{+}{S}(CH_3)_2\ X^-$
Phosphonium salts	$R{-}\overset{+}{P}(CH_3)_3\ X^-$
Dicationic surfactants	
Quaternized diamine salts	$R{-}\overset{+}{N}(CH_3)_2CH_2CH_2{-}\overset{+}{N}(CH_3)_3\ 2X^-$

hydrophilic heads are at the outer surface. Thus a spherical micelle of sodium dodecyl sulphate would consist of a group of $C_{12}H_{25}$ tails pointing towards the centre of the sphere, with OSO_3^- heads at the surface. This micelle would correspond very roughly to a droplet of dodecane of molecular size. In fact, micelles do have the property of dissolving water-insoluble organic matter. This phenomenon is called solubilization and is one of the characteristics of surface-active agents important to the cosmetic chemist.

Table 33.3 Nonionic Surfactants
R denotes a hydrophobic chain usually of 12 to 18 carbon atoms.
n is a whole number.

Alkanolamides	
Fatty acid alkanolamides	$RCONHCH_2CH_2OH$ (ethanolamides)
Fatty acid dialkanolamides	$RCON(CH_2CH_2OH)_2$
Polyethyleneglycol derivatives	
Alkyl polyglycol ethers	$R(OCH_2CH_2)_nOH$
Alkyl aryl polyglycol ethers	$RC_6H_4(OCH_2CH_2)_nOH$
Polyglycol esters	$RCO(OCH_2CH_2)_nOH$
Thioethers	$RS(CH_2CH_2O)_nH$
Polyethyleneimine derivatives	
Alkylpolyethyleneimine	$R(NHCH_2CH_2)_nNH_2$
Polyethyleneimine amides	$RCONH(CH_2CH_2NH)_nH$

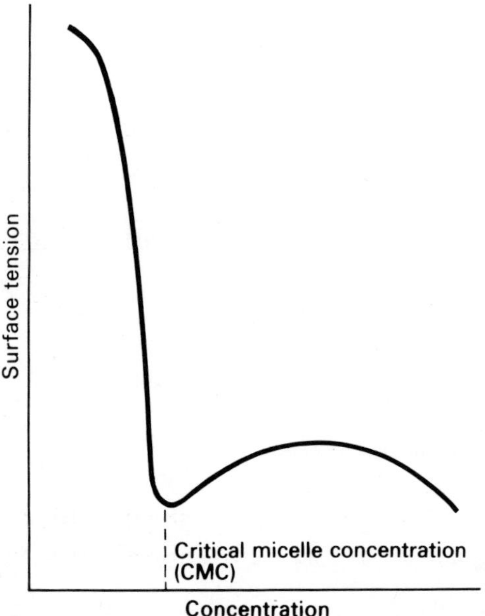

Figure 33.1 Typical curve of surface activity versus concentration for a surface-active agent

The properties of surface-active agents can be described very broadly in terms of Figure 33.1. As surface tension falls, foaming and wetting properties are usually increased. A fall in surface tension is usually accompanied by a fall in interfacial tension which gives better emulsifying and detergent properties. Finally, at concentrations above the CMC, all surface-active agents have some solubilizing properties. These properties all overlap to some degree.

Table 33.4 Ampholytic Surfactants
R denotes a hydrocarbon chain of 12 to 18 atoms.

Alkylamino acids	
Alkyl β-aminopropionates	$R\overset{+}{N}H_2CH_2CH_2COO^-$
Betaines	$R\overset{+}{N}(CH_3)_2CH_2COO^-$ and $R\overset{+}{N}(CH_3)_2CH_2CH_2COO^-$
Acylamino acids	
Acyl β-aminopropionates	$RCO\overset{+}{N}H_2CH_2CH_2COO^-$
Acyl peptides	$RCO\overset{+}{N}H_2\,C(R')—CHCONH\,CH(R'')·COO^-$ R′ and R″ = low molecular weight alkyl group
Alkyl imidazolines	$R·C$ (=N–CH_2–CH_2–) ring with ^{+}N bearing CH_2CH_2OH and CH_2COO^-

Selection and Use of Surface-active Agents

Detergency

Detergency is a complex process which involves the wetting of a substrate (hair or skin), the removal of greasy soiling matter, the emulsification of the removed grease and the stabilization of the emulsion.

For skin cleansing, soap is still an excellent detergent. Custom dictates that a high level of foam is necessary, though it performs no function. Increased foaming can easily be achieved by superfatting with long-chain fatty acids (as in shaving soaps).

Hair washing is more complex and here foam volume does appear to play some part. Sodium lauryl ether sulphate (SLES) is a common component of shampoos, and foaming is frequently enhanced by the addition of alkanolamides. Ampholytic surface-active agents are used for specialized shampoos.

Wetting

All surface-active agents have some wetting properties. Short chain (C_{12}) alkyl sulphates, alkyl ether sulphates and alkyl aryl sulphonates are all commonly used.

Foaming

See detergency. High foam volume and stable foams are generally achieved by the use of SLES reinforced by an alkanolamide.

Emulsification

Usually a good emulsifying agent requires a slightly longer hydrophobic unit than does a wetting agent. Soap is still used as an emulsifying agent in cosmetic products, very often because of ease of preparation. If a fatty acid is incorporated in the oil phase and the alkali in the aqueous phase, then stable oil-in-water emulsions are easily formed *in situ* by simple mixing. Water-in-oil emulsions (such as in certain hair creams) are frequently stabilized by calcium soaps.

Nonionic surface-active agents are also of value in emulsions.

The theoretical basis of emulsification, choice of emulsifiers and methods of forming stable emulsions are considered in detail in Chapter 38.

Solubilization

All surface-active agents above the CMC have solubilizing properties. This is important when it is required to incorporate a perfume or an insoluble organic component into a clear product, for example a shampoo. Soaps, alkyl ether sulphates and indeed most surface-active agents have been used for this purpose. It is of course necessary to use high concentrations to give good solubilization.

All the above properties may be modified by the presence of electrolytes. In general electrolytes tend to lower the CMC and this should improve solubilization. They may also tend to break emulsions and in general electrolytes should not be added to cosmetic products containing surface-active agents until their effects on surface-active properties have been fully checked.

Other Properties

In addition to the surface-active properties listed, some surface-active molecules have special features.

All cationic products adsorb strongly at protein and other negatively charged substrates. They are thus used to modify the surface of a substrate, for example to improve the feel and appearance of hair. Cationics have some antimicrobial properties also and may be used as components of special shampoos and of mouthwashes (for example chlorhexidine) (Schwartz *et al.*,[2] p. 204).

Sodium N-lauroyl sarcosinate is known to inhibit the enzyme *hexokinase* (which is involved in the glycolytic breakdown of sugars in the mouth) and has been used in toothpastes.

Different surface-active agents should not be mixed in a product without prior testing, as one may modify the behaviour of another. Cationics and anionics should not, of course, be mixed as they give rise to the formation of a large cationic–anionic salt which is usually insoluble. (This is in fact the basis of the analytical method for estimating surface-active agents.) Even anionics can have an effect on each other; for example, the foam produced by SLES can easily be destroyed by soap (both anionic). This property is made use of in the formulation of low foam detergents.

General

There is no short cut to the selection of a particular surface-active agent for a particular end use. The measurement of surface tension, interfacial tension, foam volume, detergency, wetting, emulsifying power, etc., are all useful indicators of surface activity, but none will predict precise requirements. In particular, behaviour under laboratory conditions may not be paralleled by behaviour of a product in use. For example, foam volume in a shampoo is considerably modified when tested in the presence of grease. Products containing surface-active agents should, therefore, always be tested under the conditions in which they will be used.

Biological Properties of Surface-active Agents

By definition surface-active agents are adsorbed at surfaces and may, therefore, modify surfaces. It is not surprising to find that as a consequence they may have biological effects. All cosmetic products containing surface-active agents should be checked rigorously to ensure that they do not have harmful effects on users.

Dermatological Effects

Surface-active agents wet the skin and may remove grease from the surface of the skin. When wrongly used they may create chapping, cracking and dryness of the skin. The C_{12} moiety seems particularly active in this respect and the C_{12} sulphate, for example, is used to create chapping artificially. Fortunately the effects are easily reduced by mixture with sulphates of other chain length, by addition of ethylene oxide (as in SLES) and by other means.

Cationic surface-active agents are strongly adsorbed on protein surfaces and care should be taken before cationics are incorporated in products which may come into contact with the eyes or the mouth.

In general, all cosmetic products containing surface-active agents should be patch tested (and if appropriate eye-tested) to ensure that they produce no adverse reaction.[7–10]

Biodegradation

The increasing use of synthetic detergents in place of soap in domestic washing has led to problems in sewage works because certain synthetic detergents are not broken down by sewage bacteria. The use of surface-active agents in the cosmetics industry is very small compared with the use in domestic washing. Nevertheless it is good practice to use only biodegradable surface-active agents. In some countries this is compulsory.

Branched-chain alkyl aryl sulphonates are not biodegradable, but the corresponding straight-chain compounds are, as are all soaps and alkyl sulphates. Most manufacturers of surface-active agents have details of the biodegradability of their products.

A Standing Committee has been reviewing the position in Britain for a number of years.[11]

Toxicological Effects

Surface-active agents are not, as a class, compounds of high toxicity. Nevertheless, since they may be ingested either accidentally or from a toothpaste or mouthwash, it is wise to check oral toxicity of cosmetic products containing them.

Of the products, cationics are the most toxic and have LD_{50} values of the magnitude of 50–500 mg per kg body weight; anionics are roughly in the range 2–8 g per kg and nonionics range upwards from about 5 g per kg[12] (and see Schwartz *et al.*,[2] p. 368). Cosmetic products containing surface-active agents should therefore be reasonably safe from toxic hazards.

REFERENCES

1. Moillet, J. L., Collie, B. and Black, W., *Surface Activity*, London, Spon, 1961.
2. Schwartz, A. M., Perry, J. W. and Berch, J., *Surface Active Agents and Detergents*, New York, Robert E. Krieger, 1977.
3. McCutcheon, J. W., *Synthetic Detergents*, New York, McNair-Dorland, 1950.
4. McCutcheon, J. W., *Detergents and Emulsifiers*, New Jersey, MC Publishing Co., 1979.
5. McBain, J. W., *Third Colloid Report of the British Association*, 1920.
6. Hartley, G. S., *Aqueous Solutions of Paraffin Chain Salts*, Paris, Herman, 1936.
7. Geotte, E. K., *Kolloid-Z.*, 1950, **117,** 42–7.
8. Fiedler, H. P., *Fette Seifen*, 1950, **52,** 721–4; Jacobi, O., *Fette Seifen*, 1954, **56,** 928–32.
9. Stüpel, H., *Soap Perfum. Cosmet.*, 1955, **28,** 58 and 300.
10. Faucher, J. A., Goddard, E. D. and Kulkarni, R. J., *J. Am. Oil Chem. Soc.*, 1979, **56,** 776.
11. Standing Technical Committee on Synthetic Detergents, *Progress Reports*, London, HMSO (19 reports have so far been issued).
12. Valico, E. I., *Ann. NY Acad. Sci.*, 1946, **XLVI,** 451; Alexander, A. E., *Colloid Chem.*, 1950, **7,** 211.

Chapter Thirty-four

Humectants

Introduction

Humectants are hygroscopic materials that have the property of absorbing water vapour from moist air until a certain degree of dilution is attained. This dilution depends on the character of the humectant used and the relative humidity of the surrounding air. Equally, aqueous solutions of humectants can reduce the rate of loss of moisture to the surrounding air until equilibrium is attained.

Humectants are added to cosmetic creams, particularly of the oil-in-water type, to reduce drying out of such creams on exposure to air. In addition, the hygroscopic properties of the film of humectant which remains on the skin on application of the product may be an important factor in influencing the texture and condition of the skin.

A humectant helps to provide control in use by reducing the rate at which water disappears and viscosity decreases. It is believed to minimize 'balling' and 'rolling' of a product in use.[1]

Drying Out

Drying out of a cosmetic product may occur at any time between manufacture and final use by the consumer. It is determined by the temperature of the product, its degree of exposure to air and the relative humidity of the air to which it is exposed. It is essentially a rate process which proceeds towards the equilibrium state in which the water vapour pressure of the product is equal to that of the surrounding air.

The nature of the container in which the product is packed, and particularly the means of closure, are clearly vital in preventing drying out on storage. With an efficient closure, the humectant is of less importance since there is only a small space above the product to be saturated with water vapour.

In the case of emulsion products the type of emulsion is critical. Water-in-oil emulsions lose water at a much lower rate than oil-in-water emulsions, because of the lower water content and the fact that the external phase is oil. Oil-in-water creams are very difficult to maintain in a factory-fresh state even with a screw cap and compressible wad.

Toothpaste packed in a metal tube and with a screw cap presents a slightly different problem. The product will not normally dry out if the cap is kept in position, but if the cap is left off after use the drying out at the nozzle can cause blocking of the orifice. This can be serious in pressure-packed products where there is no easy means of clearing the obstruction. Fortunately, toothpastes can tolerate high concentrations of humectant and a level of 30 per cent of glycerol is not unknown.

Humectants certainly do reduce drying out, as is shown by published evidence,[2,3] but their effect should not be exaggerated. The concentration of humectant in the water phase of a typical cosmetic product is normally much too low for it to be in equilibrium with average atmospheric humidity. All that the humectant can do is to reduce the rate of water loss to the atmosphere and this effect can and should be reinforced with an effective pack closure.

Griffin, Behrens and Cross[4] relate the factors of hygroscopicity and environment, and summarize the properties of the ideal humectant (Table 34.1). No humectant totally satisfies all these criteria and the final choice of humectant is usually a compromise dictated chiefly by the requirements of the product of which the humectant is a part.

Table 34.1 Properties of the Ideal Humectant (after Griffin *et al.*[4])

Hygroscopicity	The product must absorb moisture from the atmosphere and retain it under normal conditions of atmospheric humidity.
Humectant range	Within the normal r.h. range, change of water content should be small in relation to r.h. changes.
Viscosity	A low-viscosity humectant is easily mixed into a product, but conversely a high viscosity helps to prevent creaming or separation of emulsions, or settling of suspensions.
Viscosity index	The viscosity–temperature curve should be relatively flat.
Compatibility	The humectant should be compatible with a wide range of raw materials; solvent or solubilizing properties are desirable.
Colour, odour, taste	Good colour, odour and taste are essential.
Toxicity	The humectant should be non-toxic and non-irritant.
Corrosion	The humectant should be non-corrosive to normal packing materials.
Stability	The humectant should be non-volatile and should not solidify nor deposit crystals under normal temperature conditions.
Reaction	The humectant should preferably be neutral in reaction.
Availability	Humectants should be freely available and should be as inexpensive as possible.

Types of Humectant

There are three general classes of humectant: inorganic, metal–organic and organic.

Inorganic Humectants

Calcium chloride is typical of inorganic humectants, which are quite efficient but which fail badly on corrosion and compatibility. They find only limited use in cosmetic products.

Metal–Organic Humectants

The principal metal–organic humectant is sodium lactate which has, in fact, greater hygroscopic powers than glycerin. However, it is incompatible with some raw materials, can be corrosive, has a pronounced taste and may discolour. It has not been widely used in cosmetics but has been recommended for use in skin creams,[5] particularly because lactates occur naturally in the body and there is no risk of toxicity or dermatitis. The problem of pH can be overcome by admixture with lactic acid which is also fairly hygroscopic. Buffered solutions can be obtained between pH 7·1 and pH 2·2 at 5 per cent sodium lactate/lactic acid.

Organic Humectants

Organic humectants are the most widely used type; they are usually polyhydric alcohols, their esters and ethers. The simple unit is ethylene glycol and by progression up the series the most common products are:

Glycerol (trihydroxypropane)
Sorbitol (hexahydrohexane)

A series can be built up by the addition of ethylene oxide to a basic unit or just to itself. This produces, for example, polyethylene glycols of varying molecular weight which often have useful cosmetic properties of their own. The multiple ether linkages reduce hygroscopic properties which depend primarily on the ratio of —OH groups to C atoms.

In general the type of organic humectant used is determined primarily by availability. The soap industry inevitably produces glycerol as a by-product and this can also be synthesized from petroleum building blocks. This makes for availability and price stability and glycerol is probably the most popular humectant used in cosmetics though this accounts for only a small percentage of total use.[6,7]

Sorbitol (in the form of 70 per cent syrup) has recently replaced or partially replaced glycerol in many cosmetic products. The replacement increases the water content of the final product. This is not important in most cosmetic products, but it may be vital in toothpastes which normally have a low water content.

Thus the compounds most generally used in cosmetic products for hygroscopic purposes are:

Ethylene glycol
Propylene glycol
Glycerol
Sorbitol
Polyethylene glycol

Hygroscopicity

The method most frequently used to determine hygroscopic qualities is to construct a curve of relative humidity of atmosphere against humectant concentration in equilibrium. This is done by exposing small weighed amounts of solutions of known composition in atmospheres of controlled humidity and weighing periodically. The controlled humidities can be achieved in small desiccators charged with crystals wetted with their own saturated solutions (Table 34.2). Other humidities in the lower range are best achieved over sulphuric acid solutions of known concentration; the humidities are given in standard tables.

It is both unnecessary and inadvisable to continue weighing until equilibrium is attained because this can be a lengthy process, and moreover the more concentrated solutions do not mix well without stirring. After one, or at most two weighings at intervals of a few hours, at each humidity there will be found a division between the more concentrated solutions which gain weight by attracting water and the more dilute solutions which lose water. It is within this gap that the required concentration in equilibrium with the particular humidity lies and a second experiment with solutions covering the smaller range of concentrations will pinpoint the exact value.

Figures taken from curves for ethylene glycol, glycerin, sorbitol, propylene glycol, 2,3-butylene glycol and sodium lactate determined by a similar method to that described are given in Table 34.3.

The extensive survey of Griffin, Behrens and Cross[4] of many hygroscopic materials included a comprehensive table which is reproduced in part as Table 34.4. They also revised a graphical method proposed by Livengood[8] for choosing humectants for a desired equilibrium hygroscopicity and for calculating the effect of combinations of humectants. Their revised form of the Livengood graph, with a corrected family of curves over a limited range where the relationship holds with an accuracy of about ±5 per cent water for most organic humectants, is shown in Figure 34.1.

To use the graph the water content of the chosen humectant at 50 per cent r.h. is determined by reference to Table 34.4. This value is then transferred to the 50 per cent relative humidity ordinate on Figure 34.1. A curve is drawn through this point parallel to the nearest curves on the graph. The appropriate

Table 34.2 Suitable Crystals for Establishing Atmospheres of Controlled Humidity

	Relative humidity (%)
$K_2Cr_2O_7$	98
$Na_2SO_4.10H_2O$	94
$BaCl_2.2H_2O$	88
NaCl	75
KI	71
$NaNO_2$	66
$NaBr.2H_2O$	58
$NaHSO_4.H_2O$	52
$Na_2Cr_2O_7.2H_2O$	52
KCNS	47
$CaCl_2.6H_2O$	33
CH_3COOK	20

Table 34.3 Concentration Related to Relative Humidity for Various Humectants

	Ethylene glycol	Glycerin	Sorbitol	Propylene glycol	2,3-Butylene glycol	Sodium lactate
Relative humidity (%)	**Concentration of humectant (%)**					
90	40	35	49	35·5	44	24
80	57	50	65	58	59	36
70	69	62	73	66	70	44
60	77	71	78	75	79	48
50	84	78	83	83	85	52
40	89	84	—	91	91	57
30	93	89	—	96	94	64
20	96	94	—	99	96	—
10	98	97	—	99	97	—
Humectant (%)	**Relative humidity (%)**					
10	98	97	98·5	97·5	97	96
20	96	95	97	94	96	92·5
30	93	92	95	93	94	86
40	90	87	93	91	91	76
50	85	80	89·5	86·5	85	55
60	78	72	84	77·5	79	35
70	69	61	73	66	70	25
80	57	48	57·5	54	59	22
90	38	28	57·5	42	44	—

equilibrium moisture content can then be read off the curve for humidities between 25 and 75 per cent.

Blends of humectants do not always show hygroscopicities exactly in accord with the arithmetic average of their individual hygroscopicities. But, in general, it is sufficiently accurate for practical purposes to consider the individual quantities of humectant present as acting independently of each other.

However, the moisture content at equilibrium is far from being the whole story. It will be seen from Table 34.3 that to be in equilibrium with normal humidities of 70–75 per cent the amount of humectant in the aqueous phase needs to be of the order of 60–70 per cent, which is clearly impractical for various reasons.

Bryce and Sugden[3] confirmed the findings of Griffin *et al.*[4] that a humectant could show different efficiencies at different concentrations and, in addition to minima, reported a maximum efficiency in the region of 1 per cent humectant in a typical vanishing cream formula. This can only be attributed to surface-active effects which predominate at very low concentrations but become insignificant at higher concentrations where hygroscopic effects predominate.

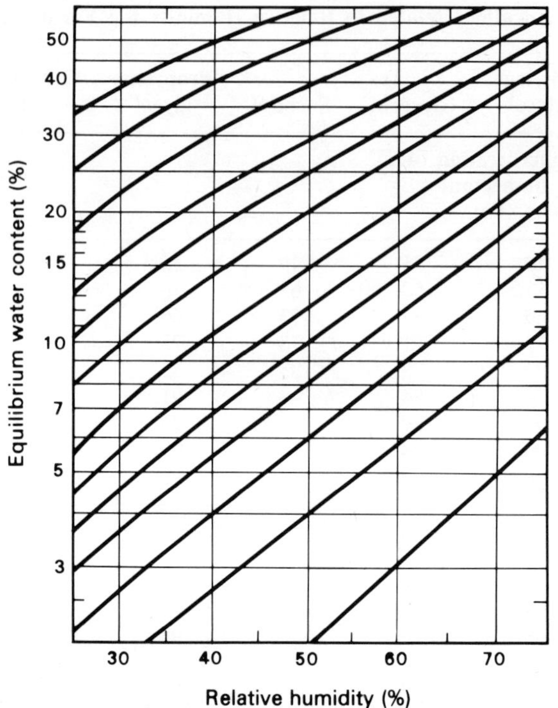

Figure 34.1 Humectants—estimation of equilibrium moisture content from 50 per cent r.h. value (Courtesy Atlas Powder Co. and *Journal of the Society of Cosmetic Chemists*[4])

Thus it may well be considered that the levels, particularly of glycerin, traditionally used are a reasonable compromise between optimum performance in terms of other properties.

Water Loss from Oil-in-water Emulsions

Griffin, Behrens and Cross[4] prepared two similar oil-in-water creams, one emulsified with soap, the other with nonionic materials, with different levels of propylene glycol, glycerin or sorbitol (0, 2, 5, 10, 20 per cent). The different creams were exposed at 30, 50 and 70 per cent humidities and the weight loss monitored for 48 hours; the weight loss was about 10 per cent of the cream, which is from 10–15 per cent of the net water content and adequate for the estimation of rates of loss. The fact that the nonionic cream lost weight slightly faster was attributed to the surface crust which formed on the soap-based cream and retarded the water loss but, at the same time, made the cream unfit for use as a cosmetic. The results, given in Table 34.5, show discrepancies which are attributed to changes in texture and consistency, and crust formation, which are not wholly dependent on water loss but are, in addition, a function of the type of emulsifier used.

Table 34.4 Physical Properties of Humectants

* Per cent w/w solids. † 25°C. —Data not available. X Crystalline.

Humectant	Freezing point (°C) vs concentration*				Viscosity (cP) vs concentration* at 20°C				
	100%	75%	50%	25%	100%	85%	70%	50%	25%
Ethylene glycol	−11·5[10]	−65[5]	−37[5]	−11[5]	19[5]	12[5]	7[5]	4[6]	1·8[5]
Diethylene glycol	−10·45[6]	—	−28[5]	−8[5]	38[5]	17[5]	9·5[5]	4·6[6]	2·0[5]
Triethylene glycol	−5[9]	—	−24[5]	−6[5]	45[5]	—	20[6]	6·8[6]	2·3[5]
Polyglycol 400	4 to 10[12]	—	—	—	—	—	(75%) 43[3]	14[4]	1·5[3]
Polyglycol 600	20–25[12]	—	—	—	—	—	—	14[4]	1·5[3]
'Carbowax' 1000	35–40[12]	—	—	—	solid[3]	—	—	21[4]	1·5[5]
'Carbowax' 4000	50–55[12]	—	—	—	solid[3]	—	solid[3]	—	17[3]
Propylene glycol	—	—	−33[5]	−10[5]	55[5]	21[5]	11·2[5]	5·2[6]	2·0[5]
Dipropylene glycol	—	—	—	—	—	—	—	—	—
Glycerin	17·0[8]	−29·8[8]	−23·0[8]	−7·0[8]	1499[11]	112·9[11]	22·94[4]	6·050[11]	2·095[11]
Polyoxyethylene glycerin	—	—	—	—	—	—	—	—	—
Alpha methyl glycerin	—	—	—	—	viscous liquid[7]		—	—	—
Xylitol	—	—	—	—	—	—	—	—	—
Sorbitol (ARLEX) (85% soln.)	—	—	—	—	—	3500†	110†	9†	2†
Sorbitol (SORBO) (70% soln.)	—	—	—	—	—	—	110†	9†	2†
Mannitol	166	—	—	—	—	—	—	—	—
Sorbitan (A-810)	—	—	—	—	—	1000†	—	—	—
Sorbide (A-815)	—	—	—	—	—	200†	—	—	—
Polyoxyethylene sorbitol (G-2240)	—	—	—	—	(95%) 975	250†	53†	12†	—
Polyoxyethylene sorbitol (G-2320)	—	—	—	—	—	—	—	—	—
Glucose	—	—	—	—	—	—	—	—	—
Propylene glycol glucoside (A-850)	—	—	—	—	—	—	—	—	—
Triethanolamine	21·2[1]	−40	−14[1]	−2[1]	1013[1]	135[1]	45[1]	10[1]	4[1]
Sodium lactate	—	—	—	—	—	—	—	—	—
Triethanolamine lactate	—	—	—	—	—	—	—	—	—
Urea	132·7	—	—	—	—	—	—	—	—

Table 34.4 (*cont.*)

	Viscosity (cP) vs **temperature** (°C:100%)					**Equilibrium hygroscopicity** (% solids vs % r.h.)		
	0	25	50	75	100 (120)	30	50	70
Ethylene glycol	50[5]	17[5]	6[5]	2·0[5]	0·35[2]	88	75+	56−
Diethylene glycol	100[5]	28[5]	8[5]	5[5]	—	90	82	60[4]
Triethylene glycol	—	35[5]	14[5]	7[5]	—	91+	84	63[4]
Polyglycol 400	solid[3]	—	—	—	—	95−	89	79+
Polyglycol 600	solid[3]	—	—	—	—	96−	90	80−
'Carbowax' 1000	solid[3]	solid[3]	—	—	—	98−	92	—
'Carbowax' 4000	solid[3]	solid[3]	solid[3]	—	—	99+	99−	96
Propylene glycol	—	40[5]	12·5[5]	5[6]	—	91	82	68
Dipropylene glycol	—	—	—	—	—	96[4]	89[4]	77[4]
Glycerin	solid[8]	945[8]	—	—	—	89+	80	65+
Polyoxyethylene glycerin	—	—	—	—	—	94	87+	76+
Alpha methyl glycerin	—	—	—	—	—	—	—	—
Xylitol	—	—	—	—	—	92	84−	71−
Sorbitol (Arlex) (85% soln.)	—	—	—	—	—	96+	87+	75+
Sorbitol (Sorbo) (70% soln.)	—	—	—	—	—	X	X	75−
Mannitol	—	—	—	—	—	X	X	X
Sorbitan (A-810)	—	—	—	—	—	95	87−	75+
Sorbide (A-815)	—	—	—	—	—	97	91+	80
Polyoxyethylene sorbitol (G-2240)	—	—	—	—	—	—	88−	—
Polyoxyethylene sorbitol (G-2320)	—	—	—	—	—	91−	90+	—
Glucose	—	—	—	—	—	94	88+	79
Propylene glycol glucoside (A-850)	—	—	—	—	—	89−	84+	77−

Triethanolamine	—	(35°C) 280[1]	115[1]	32[1]	17[1]	92[9]	80[9]	52
Sodium lactate	—	—	—	—	—	84–	68–	50–
Triethanolamine lactate	—	—	—	—	—	—	81–	—
Urea	—	—	—	—	—	X	X	(80%) 51

[1] Carbon and Carbide Co., *Amines*, New York, 1944.
[2] American Maize Products Co., *Average Temperature and Humidity* (twelve maps—months of year), New York, 1939.
[3] Carbon and Carbide Co., *Carbowaxes*, New York, 1946.
[4] Dow Chemical Co., *Dow Glycols*, 1947.
[5] Carbon and Carbide Co., *Glycols*, New York, 1941.
[6] Hodgman, C. (Ed.), *Handbook of Chemistry and Physics*, Cleveland, Chemical Rubber Publishing Co., 1939, p. 351.
[7] US Patent 2 483 418, Kamlet, J., 4 October 1949.
[8] Lawrie, J. W. *Glycerol and the Glycols*, ACS Monograph No. 44, New York, 1928, pp. 155, 369.
[9] Livengood, S. M., *Chem. Ind.*, 1948, **63,** 948.
[10] Perry, J. H. (Ed.), *Chemical Engineering Handbook*, New York, McGraw-Hill, 1941, p. 271.
[11] Sheely, M. L., *Ind. Engng Chem.*, 1932, **24,** 1060.
[12] Carbon and Carbide Co., *Synthetic Organic Chemicals*, New York, 1945.

Table 34.5 Water Loss from Soap-based and Nonionic-based Oil-in-water Emulsions[4]

	r.h.	Sorbitol	Glycerin	Propylene glycol
Soap-based o/w cream	30%	Concs. as low as 2% provide protection against drying out.	At least 5% required to protect. Below this level the cream loses more than in absence of humectant.	At least 10% required to protect. Below this level, greater loss than in absence of humectant.
	50%	Inhibited water loss at all concentrations.	None of the concentrations used afforded protection against drying out and at lower concentrations the creams lost more than in absence of humectant.	As for glycerin
	70%	Inhibited water loss at all concentrations.	Ineffective at 10% or less	Ineffective at 5% or less
Nonionic-based o/w cream	30%	All concentrations of all humectants inhibited weight loss. Order of effectiveness at 2%, best first: sorbitol, propylene glycol, glycerin. Same order persisted at 5% and 10% but to lesser degree and differences had almost disappeared at 20%.		
	50%	All concentrations of all humectants inhibited weight loss with no appreciable difference between humectants at equivalent concentrations.		
	70%	All concentrations of all humectants inhibited weight loss with no appreciable difference between humectants at equivalent concentration.		

Water Loss from Water-in-oil Emulsions
In water-in-oil creams the aqueous phase is completely enveloped in oil, and water loss will not be as great as with oil-in-water creams. Nevertheless it is generally believed that the addition of about 5 per cent humectant on the cream or 8 per cent on the water phase may play a part in reducing water loss. Griffin, Behrens and Cross, however, made the eminently sensible suggestion that a humectant for water-in-oil cream should be selected more for its desirable properties when left on the skin than for the reduction of moisture loss which is adequately controlled if the product is an efficient water-in-oil emulsion.

Stability of Emulsions

Experiments carried out by deNavarre[9] indicate that the polyols glycerin, sorbitol and propylene glycol were not interchangeable in that in a water-in-oil cream the sample containing propylene glycol and the control sample had the greatest stability while that containing glycerin showed the most oil separation.

Conversely, in an oil-in-water emulsion the sample containing glycerin and the control remained fluid while those containing propylene glycol and sorbitol would not flow after storage. On the basis of such observations it would appear that glycerin promoted the formation of an oil-in-water emulsion whilc propylene glycol favoured a water-in-oil emulsion, their use in the wrong types leading to instability. Sorbitol was apparently midway, in such properties, between these other two humectants. Cessna, Ohlmann and Roehm[10] also reported that the three polyols glycerin, propylene glycol and sorbitol undoubtedly affected the fundamental emulsion properties of the preparations tested in different manners, since glycerin and propylene glycol appeared to have opposite effects in emulsions of the same type.

The humectant will affect stability through its viscosity and in addition by its chemical nature. It is possible that in some complex systems, especially those containing monoglycerides, glycerin can positively promote stability.

Safety

The three humectants widely employed in the cosmetic and toilet industry at the present time—and which have been discussed in this chapter—namely, glycerin, sorbitol and propylene glycol, are non-toxic and dermatologically innocuous.

Ethylene glycol is not considered safe, since it is oxidized in the body to oxalic acid and any absorption through the skin might lead to renal calculus; for the same reason diethylene glycol is considered toxic. The mono-ethyl ether of diethylene glycol (Carbitol) has been widely used in cosmetic and toilet preparations and because of its ether grouping does not constitute, so far as is known, a hazard when used externally in cosmetic and toilet preparations.

Glycerin, in particular, has been questioned because of the hygroscopicity of pure glycerol which had been considered to be capable of drawing water from the skin.[11,12] This may be irrelevant since the concentrations necessary to produce this effect are never used in practice and the equilibrium is always approached from the other side.

Skin Moisturizing

This has been discussed in general in Chapter 4. In the present state of knowledge, it is not possible to define satisfactorily the role of humectants in skin care.

It would appear that the presence of a humectant may be expected to stabilize the water content of the residual film from a cream on the skin and prevent excessive drying out. However, the manner in which adjustment is made with changes in the ambient humidity, and whether the skin is dried or moistened, will depend on the relative transfer rates of water between the atmosphere and the film and the skin. The measurement *in vivo* of transepidermal water loss is discussed by Idson.[13]

On balance, therefore, the traditional belief in the use of humectants in skin products both for the probable benefit to the product while it remains in its container, and the possible benefit to the skin during use, appears to be justified.

REFERENCES

1. Kalish, J., *Drug Cosmet. Ind.*, 1959, **85,** 310.
2. Henney, G. C., Evanson, R. V. and Sperandio, G. J., *J. Soc. cosmet. Chem.*, 1958, **9,** 329.
3. Bryce, D. M. and Sugden, J. K., *Pharm. J.*, 1959, **183,** 311.
4. Griffin, W. C., Behrens, R. W. and Cross, S. T., *J. Soc. cosmet. Chem.*, 1952, **3,** 5.
5. Osipow, L. I., *Drug Cosmet. Ind.*, 1961, **88,** 438.
6. *Chem. Mark. Rep.* 1978, 22 May.
7. Miners, C. S. and Dalton, N. N., *Glycerol*, ACS Monograph No. 117, 1953.
8. Livengood, S. M., *Chem. Ind.*, 1948, **63,** 948.
9. deNavarre, M. G., *Proc. sci. Sect. Toilet Goods Assoc.*, 1945, (4), 22.
10. Cessna, O. C., Ohlmann, E. O. and Roehm, L. S., *Proc. Sci. Sect. Toilet Goods Assoc.*, 1946, (6), 20.
11. Powers, D. H. and Fox, C. J., *J. Soc. cosmet. Chem.*, 1959, **10,** 109.
12. Rovesti, P. and Ricciardi, D., *Perfum. essent. Oil Rev.*, 1959, **50,** 771.
13. Idson, B., *J. Soc. cosmet. Chem.*, 1978, **29,** 573.

Chapter Thirty-five

Antiseptics

Introduction

An examination of products in current use indicates that a substantial proportion of toilet preparations applied to the body for normal hygienic and cosmetic purposes are formulated as medicated products which range from soaps and shampoos to mouthwashes and toothpastes.

Antibacterial agents are included in toilet preparations mainly to alleviate commonly occurring conditions such as halitosis, body odours and minor skin infections including secondary infections associated with acne. Although there is inevitably some overlap, these products should be distinguished from pharmaceutical products used for treatment of pathological conditions which may contain antibiotics and other agents not normally considered suitable for general purposes of hygiene.

The major types of personal hygiene product, with examples of antibacterial agents they may contain, are given in Table 35.1 which also includes various pharmaceutical and cosmetic products sometimes used for specific purposes such as 'first aid' and treatment of minor skin ailments and for 'hand degerming' by medical personnel.

The terms 'antiseptic' and 'germicidal' are predominantly used to describe preparations applied to living tissue to prevent infection although 'germicidal' is often applied to antibacterial soap bars. Although the term 'disinfectant' is more correctly used to describe preparations for treatment of inanimate objects such as floors, toilets, drains and so on, the term 'skin disinfectant' is often applied to products used by medical and other personnel to prevent transmission of infection in hospitals, the food industry and other high risk areas.

Use of antiseptics in toilet preparations should be distinguished from use of preservatives, in that the former are expected to render the product active against micro-organisms present on the skin or scalp or in the mouth, whereas the function of preservatives (often the same antibacterial agent) is to maintain the product in a satisfactory condition during its shelf life and use.

The benefits of including antimicrobial agents in toilet preparations have in recent years been seriously questioned with the result that health and welfare authorities in many parts of the world have made recommendations or introduced laws governing the use of certain germicides. These usually limit the choice available to manufacturers, and in some cases specify the concentration considered safe. These laws and recommendations differ in different countries and moreover they are subject to change as evidence accumulates about the safety of individual antiseptics in relation to the benefits of their use—hexachlorophene, until recently the most widely used of all germicides, is an example.

Table 35.1 Antibacterial Agents Commonly Used in Toilet Preparations

Type of product	Antibacterial agents
Antibacterial soap bars	Coal tar TCC TBS Irgasan DP300 (Hexachlorophene)
General disinfectants and antiseptics	Chlorhexidine/cetrimide Chloroxylenol
Formulated emulsions, etc.	Chlorhexidine Irgasan DP300 Iodophors (Hexachlorophene)
Antiseptic creams and ointments	Chlorhexidine Cetrimide Resorcinol Phenol Sulphur etc.
Antidandruff and medicated shampoos	Zinc pyridinethione Selenium sulphide Coal tar Irgasan DP300 (Hexachlorophene)
Deodorants and antiperspirants (including feminine hygiene products)	Zinc phenolsulphonate Chlorhexidine Irgasan DP300 Quaternary ammonium compounds (Hexachlorophene)
Toothpastes and mouthwashes	Quaternary ammonium compounds Chlorhexidine (Hexachlorophene)

In 1974 the US Food and Drug Administration published a report based on an evaluation of existing data on the use of antimicrobials in soap and cosmetics. The report classified antimicrobial agents for topical use into three categories:

(1) safe and effective,
(2) not safe and/or effective and
(3) insufficient evidence.[8]

Microbial Flora of the Body

The normal flora of the body surface comprises two distinct groups of organisms—the resident flora and the transient flora. Resident organisms that proliferate on the skin are mainly non-pathogenic, that is, Gram-positive staphylococci, micrococci and corynebacteria, although in moist areas such as the axilla and groin Gram-negative organisms such as acinetobacter may be present. It has been estimated that, for 35–50 per cent of the population, the resident skin flora also includes *Staphylococcus aureus* although a report by Armstrong-Esther *et al.*[9] indicates that, for many of these subjects, presence of *Staph. aureus* is only intermittent and that a much higher proportion of the population harbours this organism occasionally.

It should be noted that the population of bacteria varies considerably on different parts of the body; the hair, face, axilla and groin harbour the greatest numbers of organisms, whereas colonization of more exposed and dry areas such as legs, arms and hands is relatively less extensive. Most resident organisms are found on the superficial skin surface but 10–20 per cent of the total flora is concentrated in hair follicles, sebaceous glands and so on where lipid and superficial cornified epithelium make their removal difficult. It is generally accepted that skin washing is relatively ineffective in removal of resident organisms and that significant reductions in skin flora can only be achieved by application of antibacterial agents.

Fortunately the resident skin flora is predominantly of low virulence although minor skin infections, particularly pyogenic infections, are relatively common. More serious septic infections usually result only where organisms are introduced into the body by injury or surgical procedures.

Various areas of the body (although mainly the hands) also contain, in addition to the resident flora, transient flora consisting of contaminants picked up continuously from the environment and other body areas such as the nasal mucosa and gastrointestinal tract. This flora may contain any number of different organisms including pathogenic strains of *Pseudomonas*, *Enterobacter*, *Salmonella*, *Shigella* and *Escherichia coli*. In general, however, these transient contaminants survive for only relatively short periods owing to insufficient moisture and the presence of bactericidal substances such as fatty acids on the skin surface. In contrast to the resident flora, these organisms are only loosely attached to the skin and may be removed in substantial numbers by washing and bathing.

A more detailed treatment of the microbial flora of the human body surface is given by Skinner *et al.*[10]

Effects of Antibacterial Agents on Body Flora

From the fairly extensive investigations carried out over the past 10–20 years there is little doubt that, whereas a substantial reduction in skin flora can be achieved by washing with soap and water alone, this effect may be significantly increased by the use of antiseptics. In contrast, however, there is considerable disagreement as to whether the strength of evidence indicating that use of antiseptics for routine hygiene is associated with reduction of infection outweighs the possibility of harmful effects resulting from continual application of these agents to the skin.

It has become increasingly apparent that the effectiveness of antiseptic preparations depends not only on the properties of the antimicrobial agent but also on the nature of the formulation, which may be a soap bar, emulsion, liquid soap or detergent formulation. A typical set of results obtained by Lilly, Lowbury *et al.*[1,2] (Figure 35.1) shows that application of chlorhexidine formulations produces rapid and immediate reduction in skin flora, whereas phenolic compounds such as hexachlorophene and Irgasan DP300 produce limited effects after a single application, maximum activity from these latter compounds being obtained only after prolonged use.

Detailed investigations illustrating the way in which the activity of a number of antibacterials *in vivo* may be affected by the nature of the formulation have been reported by Lilly, Lowbury *et al.*,[1–7] Gibson[11] and Ojajarvi.[12]

In the development of effective antiseptic toilet preparations, therefore, it is necessary that products are formulated according to the desired effect (that is, reduction in resident or transient flora) or the need for immediate or progressive and prolonged reduction in bacterial flora. Medicated toilet preparations used for routine washing and bathing are intended largely to protect the individual against minor skin infections by both resident and transient bacteria and to assist in the control of conditions such as body odour and halitosis, whereas routine handwashing associated with toilet visits, food hygiene and the handling of

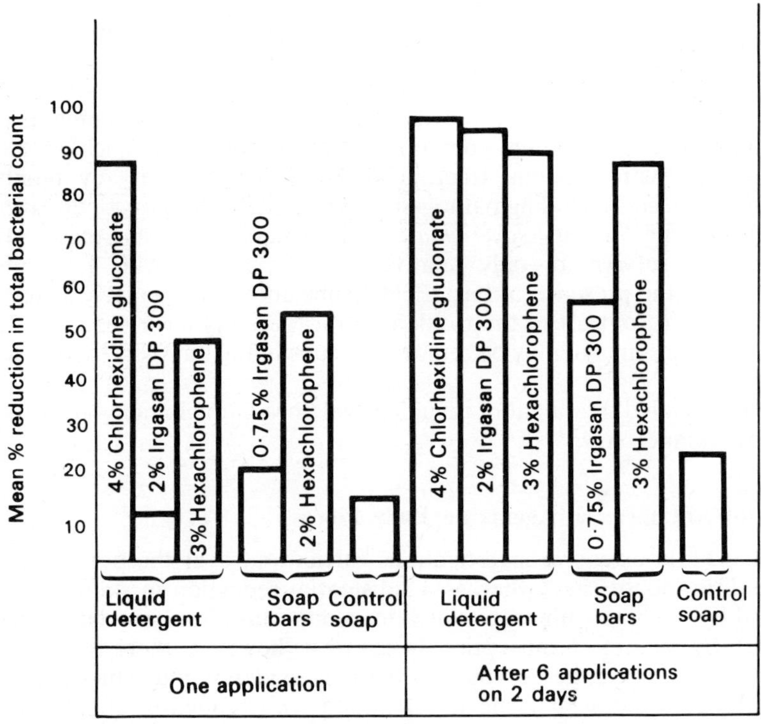

Figure 35.1 Mean percentage reduction of bacteria on skin as determined by handwashing tests associated with use of skin disinfectants (data reproduced by kind permission of Dr H. A. Lilly, Dr E. J. Lowbury, *British Medical Journal* and *Journal of Applied Bacteriology*[1–7])

newborn infants and sick persons is intended for removal of transient organisms from the skin to prevent transmission of infection.

The use of antiseptic preparations for routine hygiene is considered below. More detailed consideration of deodorants, medicated shampoos, toothpastes, mouthwashes and baby preparations is given in Chapters 10, 24, 31, 32 and 8 respectively.

Antibacterial Soap Bars and Other Skin Degerming Preparations

Routine Hygiene Usage

Investigations by a number of workers indicate clearly that persistent reduction in skin flora may be obtained by routine use of soap bars containing salicylanilides, carbanilides, various halogenated phenols and related compounds. Percentage reduction of skin flora may be determined by standard handwashing techniques involving bacterial counts taken before and after washing with antibacterial soap over a specified period.[3,13,14] Results of a number of investigations using different antibacterial soap bars are summarized in Table 35.2.

Table 35.2 Reduction in Skin Flora with Use of Antibacterial Soaps

Antibacterial agent	No. of days of test	Reduction in microbial flora (%)	Source
Hexachlorophene 2%	2 (6 applications)	87·7	Lilly *et al.*[7]
Irgasan DP300 0·75%	2 (6 applications)	56·2	
Irgasan CF3 0·5%	3	90·0	Guklhorn[14]
Irgasan DP300 1%	3	92·0	Furla *et al.*[15]
Irgasan DP300 2%	3	95·0	
Irgasan DP300 0·6%	4	60	Ojajarvi[12]
Hexachlorophene 2%	7	81	
Bithionol 2%	7	77	
TMTD 1%	7	76	Hurst *et al.*[16]
TCC 2%	7	72	
TBS 0·5%	7	65	
TCS 0·5%	7	84	
TCC 1%	12	91·7	Roman[17]
2%	12	97·8	
Hexachlorophene 2%	28 (minimum)	63–90	Wilson[18]
TCC 2%	28	89–97	

It is suggested that the sustained effects of medicated soaps are due to the substantive properties of antibacterial agents which remain on the skin after handwashing. Methods used to test skin substantivity of antibacterial agents are described by Gibbs *et al.*[19] while studies using radioactive labelled hexachlorophene and trichlorocarbanilide are reported by Taber *et al.*[20]

Under normal conditions, the healthy person is adequately resistant to organisms present on the skin surface but minor skin infections are relatively common even in healthy families, and particularly where there are young children, and it is possible that routine use of antibacterial soap bars, medicated shampoos, etc., may assist in their control. In relation to this, a number of studies involving military personnel and prisoners[21–23] have shown reductions in incidence of pyogenic skin infections associated with routine use of antibacterial soap bars for periods of up to nine months.

Although these workers report no evidence of adverse skin reactions, various investigators have from time to time expressed concern regarding possible dermatitic effects, skin drying and other effects which might result from prolonged use of these products. Other workers have demonstrated that suppression of the normal Gram-positive skin flora by use of the various agents with more selective action against these organisms may promote colonization by Gram-negative bacteria;[21,25,26] however, this is probably of little concern compared with the significant overgrowth which may accompany antibiotic usage.[27]

Antiseptic preparations are generally used to prevent or alleviate minor skin infections in the user but there is little doubt that the body surface and particularly the hands play an important part in transmission of infection in the community.

Although removal of transient skin flora can be substantially increased by application of suitable antiseptic preparations such as liquid detergent formulations containing iodophors or chlorhexidine, which produce rapid and immediate bactericidal action,[1,4,7,12] investigations show little evidence that disinfectant usage for regular washing or for rapid 'skin degerming' before patient contact by hospital personnel is accompanied by any significant reduction in incidence of cross-infection.[28,29] It therefore appears that, although skin washing is vital in controlling dispersal of infection, disinfectant usage is probably not justified except in certain 'high risk' hospital areas and that the substantial removal of transient contamination achieved by soap and water washing[4] is adequate for most purposes of food and toilet hygiene.

Antiseptics in First Aid

Antiseptic creams or other skin disinfectants are used in first aid treatment of cuts, burns and other wounds to prevent infection during healing. They are also used for treatment of minor skin infections, particularly those associated with the face. For first aid purposes, rapid acting compounds such as iodophors, chlorhexidine and chloroxylenol products are recommended[4] although certain workers have suggested that antiseptics are unnecessary in this situation and may damage healthy skin cells, thereby delaying normal healing.

Antimicrobial Agents Commonly Used in Antiseptic Products

Space permits only an outline of the more important properties of antimicrobial agents used in antiseptic products and of investigations of their activity *in vivo*. Detailed information on studies of bacteriostatic and bactericidal activity *in vitro* may be obtained from extensive reviews by Guklhorn[14] and Block.[30]

Phenols and Cresols

A very considerable number of phenol and cresol derivatives are known to have antibacterial activity although in general these compounds are more active against Gram-positive than Gram-negative bacteria. For antiseptic purposes, concentrations of 0·1–5 per cent are used but, since many of the compounds are only sparingly water-soluble, it is necessary to use soaps or other surface-active agents to achieve concentrations sufficient for optimal activity. Although toxicity may be fairly low, many phenolics are irritant in high concentrations.

For these reasons these compounds are not much used in toilet preparations, having been largely superseded by the bisphenols, salicyanilides and carbanilides.

Although DCMX (2, 4-dichlor-*sym*-metaxylenol) has been investigated for use in antibacterial soap bars,[16,31–33] chloroxylenols are mainly used in antiseptics for treatment of minor injuries or in disinfectants, including pine oil disinfectants for hospital or domestic disinfection.[4]

Chlorocresol BP (PCMC, parachlorometacresol) (Figure 35.2) has similar antimicrobial power to chloroxylenol BP (PCMX, parachlorometaxylenol) although its Rideal Walker coefficient is lower. It has irritant action at high concentration and a persistent and characteristic odour demanding careful selection of perfumes in use. Like the chloroxylenols it is seldom used except as a preservative.

Bisphenols

Of the very large number of phenolic antiseptics that have been synthesized, the halogenated phenols are amongst the most potent. Several halogenated diphenol derivatives have gained wide usage in toilet preparations particularly because of their compatibility with soap. Of these hexachlorophene (2,2′-methylene-*bis*-(3,4,6-trichlorophenol)), dichlorophene (2,2′-methylene-*bis*-(4-chlorophenol)), bithionol (2,2′-thio*bis*-(4,6-dichlorophenol)) and Irgasan DP300 (2,4,4′-trichloro-2″-hydroxydiphenylether) (Figure 35.2) have been more widely used than others. Like other phenolics these compounds are generally more active against Gram-positive than Gram-negative bacteria and fungi. As with all the substituted phenols, since these compounds are only sparingly soluble in water, aqueous preparations are formulated to contain surface-active molecules in order to achieve satisfactory concentrations for activity. For antiseptic purposes, concentrations of 0·5–2·0 per cent are generally used. All these compounds are incompatible with cationic compounds and incompatibility of hexachlorophene with a number of anionics and nonionics has also been reported.[16,32–44]

Hexachlorophene (G11) (Givaudan). The activity of soap bars, liquid soap and detergent emulsion formulations containing from 2–3 per cent hexachlorophene

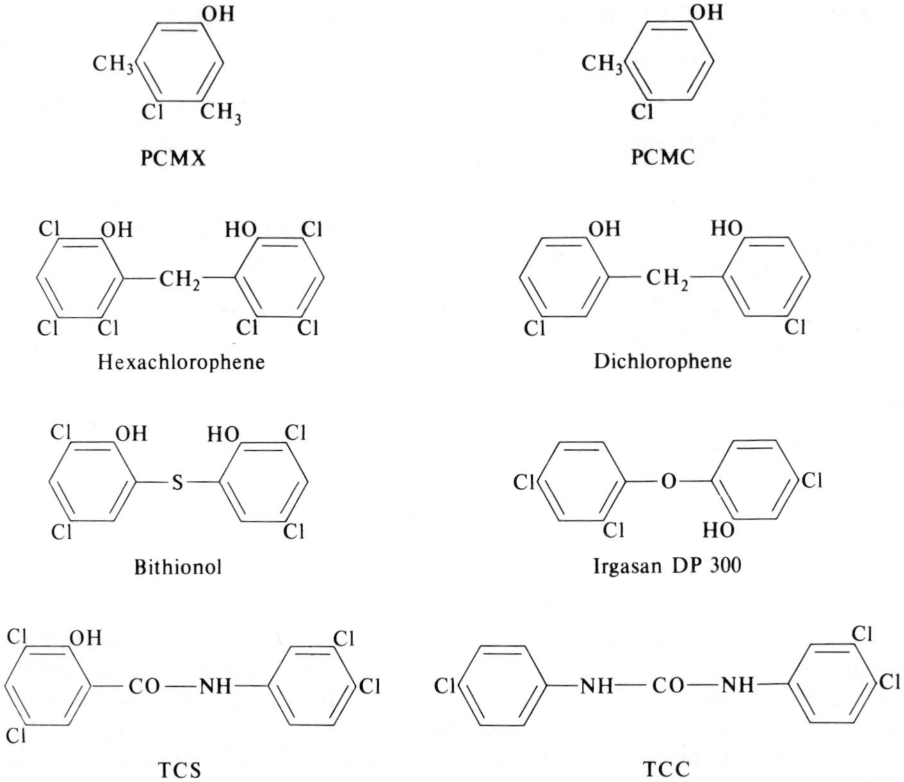

Figure 35.2 Formulae of some anionic-compatible antiseptics

has been investigated by Lilly and co-workers,[3–5,7] Gibson,[11] Ojajarvi,[12] Hurst *et al.*,[16] Wilson[18] and Sprunt.[29] In general these workers show that hexachlorophene preparations have only limited activity after a single application although extensive reduction in skin flora may be demonstrated after prolonged application over several days. Lilly *et al.*[5] and Ojajarvi[12] showed that the immediate effects associated with a single application of 3 per cent hexachlorophene liquid soap preparation could be increased by addition of 0·3 per cent chlorocresol.

One important disadvantage associated with use of hexachlorophene is its selective activity against Gram-positive organisms with the result that Gram-negative organisms may be found growing in these products.[35] A number of investigations also indicate that prolonged use of hexachlorophene may be associated with increased skin colonization by Gram-negative bacteria.[26,36–38]

Until recently, hexachlorophene commanded wide usage in toilet preparations and was considered to have a remarkably good safety record. However, its widespread use in soaps, toothpastes, mouthwashes, deodorants, feminine hygiene products, medicated shampoos, hair cream and baby products gave rise to fears that undesirable accumulation of the germicide in the body could occur and, following a series of animal toxicity studies, the use of hexachlorophene was restricted in several countries. Studies on the toxicity of hexachlorophene have been summarized by Kimbrough.[39]

In September 1972, however, following reported gross contamination by hexachlorophene of a baby powder associated with fatalities in a number of infants in France, the US Food and Drug Administration announced that it would limit the use of hexachlorophene to products available on prescription.[40] This followed an earlier statement of policy which permitted the use of up to 0·1 per cent of hexachlorophene as a preservative where there was no suitable alternative.

In the United Kingdom, the Committee on Safety of Medicines has recommended that medical personnel should be informed of a possible hazard in using hexachlorophene, especially with respect to infants, that products containing it should bear a cautionary label and that certain of these should be recommended for use only on medical advice.[41]

No evidence of hazard to health in man has been seen in the many years of use of hexachlorophene at recommended levels, but because of the observations in animal tests and the tragic events which occurred in France, the future of this once well respected germicide must now be regarded as uncertain.

Dichlorophene (G4) (2,2′-methylene-*bis*-(4-chlorophenol)) (Givaudan). Dichlorophene has also been used in soaps and toilet preparations but to a much lesser extent than hexachlorophene. Investigations by Lowbury *et al.*[3] indicate that use of a liquid soap preparation containing 2 per cent dichlorophene over a period of four days was less effective than 2 per cent hexachlorophene in liquid soap in reduction of resident skin flora.

Bithionol (Actamer, Vancide BL) (Hilton Davis Chemicals Co., USA). The properties of bithionol as an alternative to hexachlorophene have been discussed by Powell *et al.*[42] Although bisphenols are rarely irritant to the skin and only occasionally produce allergic reactions, evidence of photosensitization caused by bithionol has been accumulating in the last few years, and in 1968 the US Food and Drug Administration issued an order preventing the further introduction in the USA of products containing bithionol.

Irgasan DP 300 (*Geigy*). Irgasan is a relatively new antiseptic which is now used widely in antibacterial soap bars, skin degerming preparations, medicated shampoos and deodorant products.

Although it is very effective even at low concentrations against several Gram-negative bacteria and the usual range of Gram-positive bacteria, it has no activity against *Pseudomonas* and is relatively ineffective against fungi. For this reason it cannot be recommended as the sole preservative in products prone to microbial spoilage but, because of its action against the resident and transient bacteria on the skin, it is suitable for use in antiseptic products.

Irgasan DP 300 has been subjected to a very thorough toxicological examination and has been found to be practically non-toxic.[14,43] Animal tests have shown very low local irritant or systemic toxic effects and no sensitizing activity was observed in tests carried out on guinea-pigs. Extensive trials on humans have also been carried out and no cases of sensitization or photosensitization have been reported.

Concentrations of 0·5–2 per cent are recommended for use in products which are rinsed from the skin after application, such as toilet soaps, bath additives, liquid shower soaps, etc. For products designed for application to the skin without removal, such as deodorant sticks, creams, aerosol sprays, intimate hygiene products, etc., concentrations of 0·05–0·2 per cent are suggested.

Irgasan DP 300 is unusual in being compatible with soaps and other anionic systems while at the same time being very effective against the Gram-negative bacteria likely to be present on the skin. Savage[44] has pointed out its value in combating the organisms responsible for certain characteristic malodours on the body, and has reported residual bacteriostatic activity on the hair against both Gram-positive and Gram-negative organisms after use of shampoos containing 0·15–0·3 per cent of Irgasan DP 300.

The activity of bar soap and a bactericidal washing cream containing respectively 0·6–0·75 per cent and 2 per cent Irgasan DP 300 has been investigated by Lilly and co-workers[2,7] and Ojajarvi.[12] Like hexachlorophene, these products produced only limited effect after a single application, optimum activity being achieved only after a period of use (Figure 35.1). Although the reduction in skin flora following prolonged usage of 2 per cent Irgasan DP 300 washing cream compared favourably with other preparations, in general activity of this antiseptic was somewhat less than that of hexachlorophene. Investigations by Furla *et al.*[15] indicate that activity of soap bar formulations can be substantially improved by increasing the concentration of Irgasan DP 300 to 1–2 per cent (Table 35.2).

Fentichlor (*Bis*-(2-hydroxy-5-chlorophenyl) sulphide) (Cocker Chemical Co.). Fentichlor is an antifungal agent as well as an antibacterial agent which has been in use by dermatologists for many years for the treatment of skin diseases. Recently it has been used in a number of medicated cosmetics and possible applications include foot powders, industrial hygiene creams and medicated soaps.

Fentichlor is claimed to be active in particulate form and is more effective in the acid pH range than under alkaline conditions. Fentichlor suspensions of 2 per cent have been used for the successful treatment of ringworm, favus of the scalp, athlete's foot and barber's rash; although there has been a reasonably good history of dermal tolerance to this antiseptic, Burry[45] has reported cases of photosensitization. Some sources have ascribed this to an impurity, *p*-chlorophenol, which is often present. Further investigation will therefore be necessary before Fentichlor can be recommended for use in mass market products.

Salicylanilides and Carbanilides

Halogenated Salicylanilides. Homologues of both salicylamide and salicylanilide have been used, particularly for their antifungal and antibacterial activity. These materials have been tested and found to be effective as additives in soaps. The compounds mainly concerned are the following:

4′,5-dibromosalicylanilide (DBS)
3′,4′,5-trichlorosalicylanilide (Anobial)
3,4′,5-tribromosalicylanilide (Temasept IV, Tuasal 100, TBS)
2,3,3′-5-tetrachlorosalicylanilide (TCS)
3,3′,4,5′-tetrachlorosalicylanilide (Irgasan BS200, TCS)

Although TBS (Theodore St Just and Co. Ltd) and particularly TCS (Geigy and Co.) (Figure 35.2) have been shown to be highly active germicidal agents both *in vitro* and *in vivo*,[16] one of the factors that have hindered the use of halogenated salicylanilides is the confused literature about their safety. There can now be little doubt that tetrachlorosalicyanilide is potentially dangerous because of its ability to induce photosensitization. Wilkinson,[46] Calnan,[47] Vinson and Flatt,[48] Anderson[49] and Baer[50] have all provided evidence that TCS causes photodermatitis and it is now no longer used in soaps, as the incidence of this kind of reaction is much too high for widespread use.

Although reactions to the polybrominated salicylanilides have been reported (for example dibromosalicylanilide by Behrbohm and Zschunke[51] and tribromosalicylanilide by Epstein,[52] Harber[53] and Osmundsen[54]), these reports require interpretation before manufacturers can decide whether to use these materials or not. Many acceptable antiseptics provoke occasional reactions in hypersensitive individuals, but it is not a question of whether or not positive reations will be obtained if a certain substance is used in a new product, but whether a large number of people will be injured by its use. Complete exclusion of all materials which provoke any form of sensitization is impossible, and testing either on animals or humans with the aim of being absolutely certain is unrealistic since at least 30 000 persons would have to be patch tested before it could be predicted that less than 1 in 10 000 people would be sensitized by the product.

Peck and Vinson[55] have used the Schwartz-Peck and Draize-Shelansky patch tests, modified to include UV irradiation, to test tribromosalicylanilide and dibromosalicylanilide on 150 subjects. No photosensitization was revealed and the authors claim that the result confirms that these materials have a very low photosensitization potential. Moreover, they point out that these germicides have been used in toilet soaps for many years in the USA and that hundreds of millions of bars have been used with excellent consumer acceptance.

A fluorinated derivative, 3,5,dibromo-3′-trifluoromethylsalicylanilide (Fluorophene) (Stecker Chemicals) is claimed to be safer in use than many other germicides of similar potency. It exhibits good light stability, and at 2 per cent in soap does not discolour even after long periods of use. As with the other halogenated salicylanilides, there is likely to be a risk of photosensitization, but no reliable data are available.

Carbanilides. Trichlorocarbanilide (3,4,4′-trichlorocarbanilide) (Monsanto), also known as TCC or Trichlorocarban, is a highly active antibacterial agent which, like Irgasan DP 300, is widely used at concentrations of 1–2 per cent in antibacterial soaps and other toilet preparations. TCC is almost insoluble in water but can be solubilized by certain nonionics. Unlike some of the bisphenols, it does not discolour on exposure to light but, like most phenolic antibacterials, it is more active against Gram-negative than Gram-positive bacteria. The activity *in vivo* of antibacterial soap bars containing TCC has been investigated by a number of workers (see Table 35.2).

TCC has a reasonably clear record of safety and although isolated cases of photosensitization have been reported there is a fairly general consensus of opinion about its safety for use in shampoos, deodorants, skin products and soaps. Care must be taken, however, to avoid high temperatures in manufacture, since it breaks down to form chloroaniline which is highly toxic.

Another carbanilide, 3-trifluoromethyl-4-4′-dichlorocarbanilide (Irgasan CF3, Anobial TFC) has also been found to have good activity *in vivo* and *in vitro*[12,14,45] (Table 35.2). This compound has not been found to show evidence of photosensitization. Voss[56] showed that *ad lib* use of soaps containing 1·0 per cent TCC and 0·5 per cent Irgasan CF3 over a period of two to seven months reduced the prevalence of *Staph. aureus* on the skin. This study also indicated that partial inhibition of the Gram-positive flora was not accompanied by any increase in Gram-negative species. Ojajarvi[12] showed that an emulsion containing 2 per cent Irgasan CF3 and 0·1 per cent β-phenoxyethanol exerted a sustained antibacterial effect similar to that observed with hexachlorophene preparations, and suggested that this formulation could offer an alternative to hexachlorophene where long-term antisepsis is needed.

Cationic Surface-active Antibacterials

Cationic surface-active antibacterials are widely used in mouthwashes, deodorants, feminine hygiene products, baby products, antidandruff conditioners or rinses, hairdressings and tonics, and astringent lotions. Their antibacterial activity has been widely studied and there has been much confusion in the literature about their comparative effectiveness, mainly because of the different techniques used to evaluate them. Some of the problems which have arisen in testing quaternaries are associated with the bacteriostatic 'carry over' of effective germicides which cannot be eliminated by simple dilution, as in the case of other antiseptics, because quaternaries tend to make bacteria form clumps, and they also adhere strongly to the surface of cells without necessarily killing them. For this reason an effective chemical quenching agent must be added and many workers have failed to do this. In general the quaternaries are more effective against Gram-positive than against Gram-negative bacteria, although the difference is small in the case of some compounds.

It should be noted, however, that Gram-negative organisms, particularly the pseudomonads, are quite frequently found to be resistant to these agents. On a number of occasions *Pseudomonas*, *Enterobacter* and other Gram-negative organisms have been found growing in solutions of benzalkonium chloride, cetrimide and chlorhexidine.[57,58]

Cationics in general are incompatible with a considerable range of materials, especially anionic compounds (including soaps). These compounds are therefore seldom used in creams and lotions unless nonionic emulsifiers are used, although it should be noted that they may also be inactivated by high concentrations of nonionics (see Chapter 36). Incompatibility with other materials such as silicates, alginates, methylcellulose, lanolin and so on has also been reported.[59–61] The optimum pH range for antimicrobial activity is 7–8 although there is some evidence for greater activity in the alkaline range.[62,63]

Quaternary Ammonium Compounds (QACs). Examples of some of the most widely used of these compounds are given in Table 35.3 and Figure 35.3. From toxicity data summarized by Guklhorn[14] and Block[30], it is concluded that concentrations used for antiseptic purposes are relatively non-toxic and non-irritant when used externally, except to the mucous membranes of the eye where concentrations of 1–2 per cent or above can cause permanent opacity of the cornea.

Concentrations of the order of 0·5 per cent QAC are used in hair-rinse products which may come into contact with the eye, and have been found safe. For use on the skin, concentrations of between 0·5 per cent and 1·5 per cent are commonly used, while in mouthwashes, owing to the somewhat bitter taste of most quaternaries, lower concentrations are normally advisable. In deodorants

Table 35.3 Quaternary Ammonium Antibacterial Agents

Trade name	Chemical group
Benzalkonium chloride Marinol (Berk) Vantoc CL (ICI) Roccal (Bayer) Zephiran (Bayer) Zephirol (Bayer)	Alkyl-dimethyl-benzyl ammonium chloride
Arquad 16 (Armour Hess)	Alkyl-trimethyl ammonium chloride
Vantoc AL (ICI)	Alkyl-trimethyl ammonium bromide
Cetrimide CTAB Cetavlon (ICI) Morpan CHSA (Glovers)	Cetyl-trimethyl ammonium bromide
Domiphen bromide Bradosol (CIBA)	*β*-Phenoxyethyl-dimethyl-dodecyl ammonium bromide
Benzethonium chloride Phemerol (Parke Davis) Octaphen (Ward, Blenkinsop) Hyamine 1622 (Rohm and Hass)	*p*-tert-Octylphenoxyethoxyethyl-dimethyl-benzyl ammonium chloride
Fixanol VR (ICI) Vantoc B (ICI)	Tetradecyl-pyridinium bromide
Fixanol C (ICI) Ceepryn (Merrell)	Cetyl-pyridinium bromide or chloride
Diometam (British Hydrological Ltd)	Di-(*n*-octyl)-dimethyl ammonium bromide
Isothan Q (Onyx Chem.)	Alkyl-isoquinolinium bromide

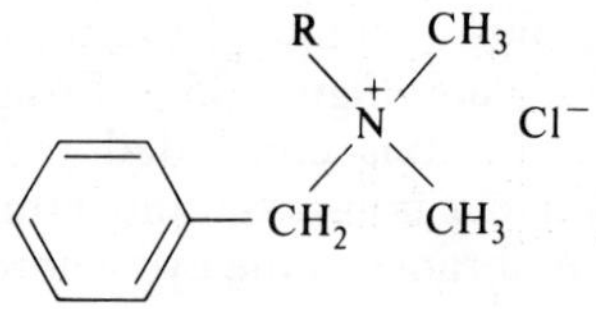

Benzalkonium chloride

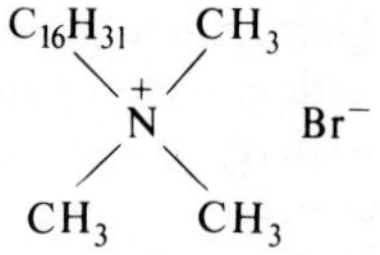

Cetavlon

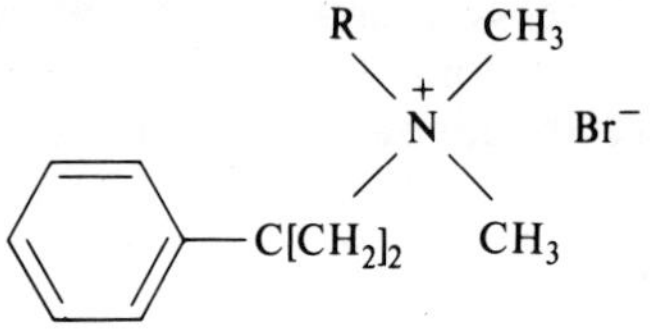

Domiphen

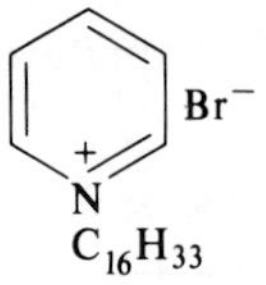

Ceepryn

(R is long-chain alkyl or other group)

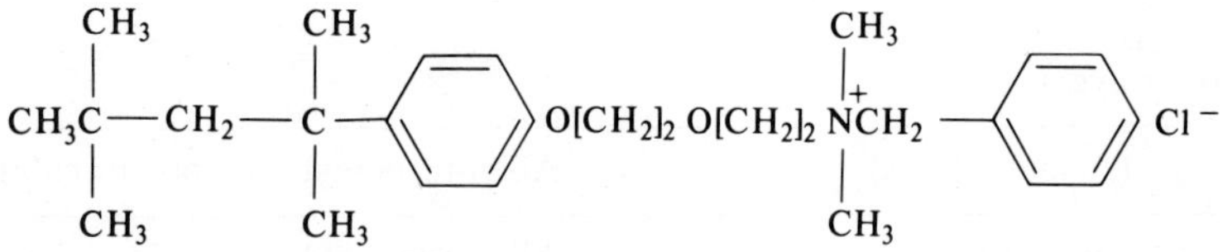

Benzethonium chloride

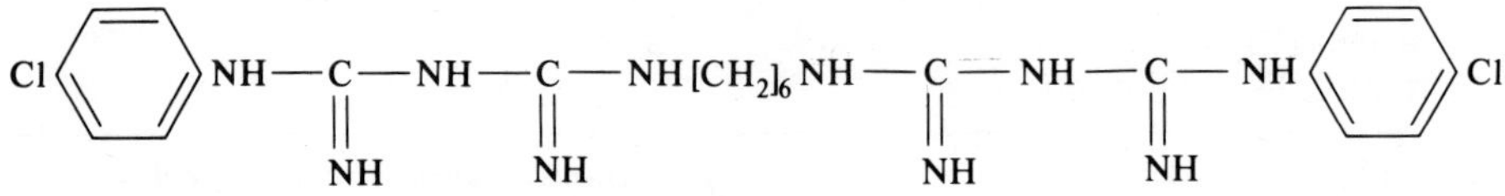

Chlorhexidine

Figure 35.3 Formulae of some cationic surface-active antiseptics

and antiseptic baby powders, levels of 0·1–0·2 per cent are frequent, and rinse products for treating babies' nappies (diapers) usually contain 0·2 per cent at use dilution.

Many of these compounds, notably cetrimide, Domiphen bromide, benzalkonium chloride, Ceepryn, Phemerol and Zephiran, are mentioned in British Pharmacopoeias. Studies of the effectiveness of Zephiran, Ceepryn, cetrimide and other compounds which have been widely used in medical practice for surgical hand disinfection and preparation of skin prior to surgery are summarized by Block.[30]

Since about 1970 there has been an increasing number of hospital reports in which aqueous QACs have been implicated as the source of infection resulting either from intrinsic contamination of the antiseptic solution or its lack of effectiveness against certain pathogens. As a result, the use of these products in medical practice has been seriously questioned by a number of workers.[58,64]

Dowicil 200. 1-(3-Chloroallyl)-3-5-7-triaza-1-azoniaadamantane chloride (Dow Chemicals) is a quaternary ammonium compound which acts by the slow release of formaldehyde in aqueous solution. Dowicil 200 is a broad-spectrum antimicrobial agent but is considered to be rather more effective against bacteria than against fungi and yeasts. Of particular interest is its activity against *Pseudomonas aeruginosa*.

It is highly soluble in water, although solutions discolour on aging and have a characteristic odour. Activity is retained in the presence of both anionic and nonionic detergents and is not markedly affected by pH.

Although the manufacturers report no evidence of primary irritation on human subjects with solutions up to 2 per cent, the possibility of sensitization to breakdown products of this germicide should not be overlooked.

Chlorhexidine. An antiseptic which, although not truly a quaternary compound, nevertheless resembles one in being strongly inhibited by anionic material, has gained prominence in recent years and is probably more effective than most of the true quaternaries. This compound, Hibitane or chlorhexidine (ICI) (1,6,di(N-*p*-chlorophenylguanidino)hexane) (Figure 35.3), which is available as a diacetate and the more soluble digluconate, has good all-round antibacterial properties, very low toxicity and shows no evidence of irritation or sensitization.[65] A number of materials reduce the effectiveness of chlorhexidine, and the presence of free chloride, sulphate, phosphate or carbonate ions causes precipitation. It is also reported to be incompatible with sodium carboxymethyl cellulose, gum tragacanth, alginates, beeswax and formaldehyde. Chlorhexidine salts show optimal activity at pH 6–8.[66]

The activity of aqueous and alcoholic solutions, creams and detergent formulations containing between 0·5 and 4 per cent chlorhexidine used for disinfection of hands and for pre-operative skin disinfection has been extensively investigated.[2–7,12,25,65,67] In general these investigations indicate, as illustrated in Figure 35.1, that chlorhexidine is a rapid-acting antibacterial agent which can be used to produce both immediate reductions in skin flora following a single application and also further reduction after repeated usage.

It should be noted, however, that investigations by Ojajarvi *et al.*[67] indicate that use of 4 per cent chlorhexidine detergent scrub by nursing staff in a neonatal unit for a period of more than a week was associated with increases in skin flora and these workers suggest that more attention should be paid to long-term handwashing tests to assess the efficacy of antiseptics under in-use conditions.

In medical practice, chlorhexidine formulations are widely used as general disinfectants in surgical procedures, in treatment of burns and wounds, and in the prevention of cross-infection. Use of 1 per cent chlorhexidine cream in treatment of wounds and burns is described by Soendergard,[68] Fowler[69,70] and Grant.[71] A combination of 0·05 per cent chlorhexidine and 0·5 per cent cetrimide is an antiseptic and detergent system that is also used for wound cleansing. In baby powders, 0·1 per cent chlorhexidine hydrochloride confers mild antiseptic properties, while concentrations of less than 0·5 per cent are now used in feminine intimate hygiene products.[72]

The effect of chlorhexidine on the oral flora has also been studied and Löe *et al.*[73,74] have found that a mouthwash containing 0·2 per cent of chlorhexidine digluconate reduces the amount of bacterial plaque which normally forms on the teeth. This had the effect of reducing the formation of calculus and also prevented the onset of gingival inflammation in a panel of students who did not clean their teeth for 3–4 weeks, during which time they rinsed their mouths twice a day with the test mouthwash.

A more extensive review of the properties and clinical usage of chlorhexidine antiseptics is given by Senior[75] and Madsen.[76,77]

Amphoteric Surface-active Compounds

The amphoteric surface-active compounds, or ampholytes, are a group of compounds that combine detergency in their anionic group with bactericidal power in their cationic moiety. The most widely used agents in this group are the Tego compounds which are made up of the amino acid glycine substituted with a long-chain alkyl amine group: $RNH(CH_2CH_2NH)_2CH_2COOH$ where R is an alkyl group, usually C_{10}–C_{16}. Schmitz and Harris[78] showed that the greater the number of nitrogen groups in the surface-active ion, the greater the activity; dodecylglycine is only about one-tenth as active as dodecyl-(aminoethyl)-glycine.

These compounds are claimed to be virucidal and fungicidal as well as bactericidal. One particularly useful property is their surface activity which is associated with good wetting and soil penetrating power.

Because these materials are not as susceptible to inactivation by proteins as the quaternary ammonium compounds, they are widely used in industrial products. Their effectiveness is, however, reduced by soaps and other anionic detergents and in some cases by nonionic detergents. The amphoteric bactericides, particularly Tego 103S ($C_{12}H_{25}NH(CH_2)_2NH(CH_2)_2NHCH_2CO_2H.HCl$ as a 15 per cent solution), have been used for skin disinfection and other medical applications.[79]

The toxicity of Tego 103S and other Tego compounds, according to the manufacturer's literature, is very low and there is little or no evidence of skin irritancy or sensitization.

Miscellaneous Antimicrobial Agents

Bronopol (2-Bromo-2-nitropropan-1,3-diol) (The Boots Co.) (Figure 35.4) is a highly water-soluble compound which shows approximately equal activity against Gram-positive and Gram-negative bacteria including *P. aeruginosa.*[80,81] It is also active against fungi at low concentrations and its effectiveness does not vary very much over the pH range 5–8. It is not adversely affected by anionic and nonionic surfactants.[82,83]

Bronopol does not appear to be a primary irritant at the concentrations normally employed (0·2–0·5 per cent) and tests on guinea-pigs have shown no signs of sensitization. Aqueous solutions of Bronopol gradually decompose under alkaline conditions.

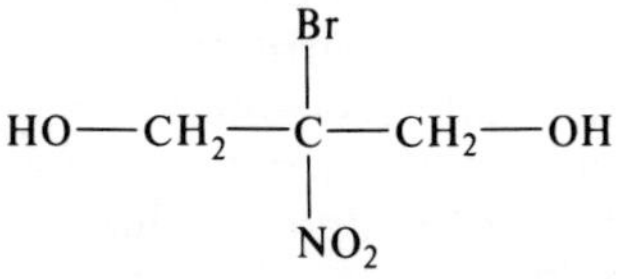

$$HO-CH_2-\underset{NO_2}{\overset{Br}{C}}-CH_2-OH$$

Bronopol

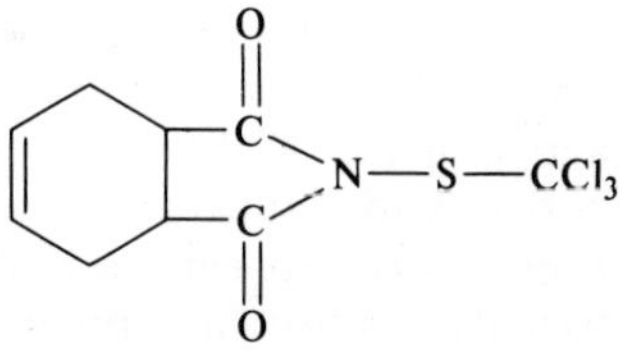

Captan

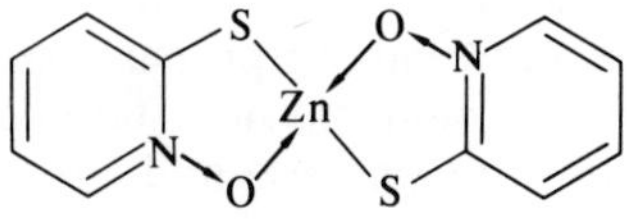

Zinc pyridine-2-thiol-1-oxide

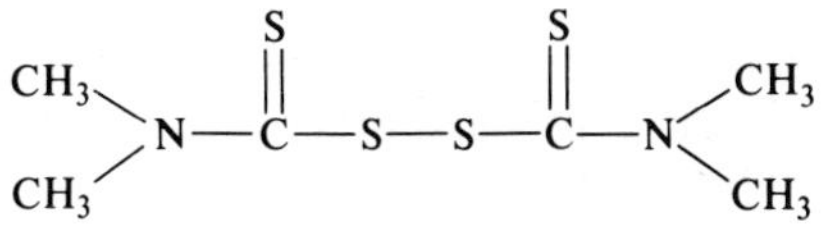

$$(CH_3)_2N-\overset{S}{\overset{\|}{C}}-S-S-\overset{S}{\overset{\|}{C}}-N(CH_3)_2$$

TMTD

Figure 35.4 Formulae of some miscellaneous antiseptics

Captan. n-Trichloromethylthio-4-cyclohexene-1,2-dicarboximide (Figure 35.4) is also known as Vancide 89RE (R.T. Vanderbilt Co.) and is a water-insoluble solid which has been used at 0·1–0·25 per cent in medicated powders. It has also been recommended as an antidandruff agent but there is only scanty evidence of its effectiveness for this purpose. Vancide 89RE is unstable in alkaline conditions but has a reasonably wide spectrum of activity against bacteria and fungi in the acid state. The manufacturers report satisfactory results of tests for corneal damage carried out on aqueous suspensions, and human patch tests have shown no primary irritation after 24 hours application as a 50 per cent aqueous paste.

Dioxin. 6-Acetoxy-2,4-dimethyl-*m*-dioxane (Givaudan) is a clear amber liquid which, when added to water, hydrolyses to produce acetic acid, resulting in a lowering of pH. Dioxin is claimed to be active over a wide pH range. It is particularly effective against Gram-negative bacteria and has been used as a preservative at concentrations between 0·1 per cent and 0·2 per cent. Its main disadvantage is its characteristic odour which is difficult to mask in cosmetic products. Dioxin lotions of 1 per cent cause no irritation or damage to rabbits' eyes and patch tests at this concentration have been carried out on human subjects with satisfactory results.

Germall 115 (Imidazonidyl Urea) (Sutton Laboratories, Inc., Roselle, NJ) is a water-soluble antibacterial which is generally non-toxic and non-irritant and is active against Gram-negative and Gram-positive bacteria and also some yeasts and moulds. It is effective over a wide pH range and retains activity in the presence of proteins and surfactants. The properties of this compound, including toxicity studies, are further discussed by Berke *et al.*[84]

Halogens. Although iodine solutions, as such, are now little used because of their irritant and skin staining properties, they have largely been replaced by the iodophors. Iodophors are mixtures of iodine with surface-active agents which are used as antiseptics at concentrations between about 0·5 and 1 per cent available iodine. They have a low vapour pressure and almost complete lack of odour, low irritant properties,[85] and are non-staining. They are active against Gram-negative and Gram-positive bacteria and also have fungicidal, sporicidal and virucidal activity. Optimum activity is observed in acid solution (pH 3–4).

Iodophors may be formulated with anionic, cationic and nonionic surface-active agents and the resulting solubilized products have the added advantage that the system acts as a skin cleanser as well as an antiseptic. Some commercial products are made with compounds such as polyvinylpyrrolidone and polyethoxyethanol derivatives. Trade names include Betadine (Berk), Wescodyne (Bebgue), Virac (Ruson Labs) and Povidone-Iodine (Berk).

Iodophors are widely used in medical practice for skin-degerming by hospital personnel and for pre-operative and post-operative skin disinfection and antisepsis. The activity of iodophors compared with chlorhexidine and other skin disinfectants has been investigated by a number of workers.[3–6,12,86] These investigations indicate that iodophors, like chlorhexidine formulations, have the advantage of producing good immediate effects after a single application with further reductions in skin flora after repeated or prolonged usage.

Other pharmaceutical/cosmetic iodophor formulations that are available include shampoos, mouthwashes, and skin and scalp cleansers.

Sodium hypochlorite and the various organic chlorine-releasing compounds such as Chloramine T are also highly active bactericidal, fungicidal and virucidal agents and are used extensively as general disinfectants in public health and the domestic environment. Although used widely for disinfection of wounds in hospitals they are not generally employed as skin disinfectants, probably because of their unpleasant smell and their tendency to produce skin irritation at concentrations of more than about 0·5 per cent available chlorine.

Mercury Compounds. Inorganic mercury compounds such as mercuric chloride and mercuric nitrate are still used in medical practice for skin disinfection but are scarcely used at all in toilet preparations. Organic compounds such as phenylmercuric nitrate, borate and acetate are used in shaving creams and antidandruff lotions in the United Kingdom although their use is restricted by the Poisons Rules 1978 in which a third schedule entry (exempted from provisions of Poisons Rules provided that a prescribed limit is not exceeded) applies to toilet, cosmetic and therapeutic preparations containing not more than 0·01 per cent of phenylmercuric salts. The toxic effects of mercury compounds are further considered in Chapter 36. Compounds of mercury are highly active against both

Gram-negative and Gram-positive organisms but, despite their excellent antibacterial properties, their use in products for mass-market sale is difficult to justify.

Pyridine N-oxides. The pyridine N-oxides, also known as cyclic thiohydroxamic acids or pyridinethiones, are highly active antibacterial and antifungal agents. Cox[87] has described their preparation and properties and reviewed their application as antibacterials and fungicides, while Snyder *et al.*[88] have reviewed the safety aspects. Snyder states that of some 1350 compounds screened in his laboratory, one of the most active antifungal and antibacterial materials examined was zinc pyridine-2-thiol-1-oxide (ZnPTO) (Figure 35.4), also known by the trade name Omadine (Olin Mathieson Chemicals).

Brauer *et al.*[89] claim that the compound is several hundred times more effective against *Staphylococcus aureus*, *S. albus* and *Pityrosporum ovale* than many of the traditional materials used in antidandruff treatments. Shampoos containing 2 per cent of ZnPTO appear to be extremely effective against dandruff and are on sale in the USA, the UK and Europe. A men's hair dressing containing 0·5 per cent of ZnPTO has also been shown by Brauer *et al.*[89] to be effective against dandruff.

Cadmium, titanium and zirconium pyridinethiones have also been prepared and found to be effective antimicrobial agents.

Tenenbaum *et al.*[90] have compared the antimicrobial properties of many agents used in antidandruff treatments and have concluded that, although ZnPTO is an exceptionally potent antimicrobial agent, it must owe its dramatic effects against dandruff to a property other than its antimicrobial activity, since other antiseptics which were more potent *in vitro* were significantly less effective against dandruff in clinical trials.

Snyder *et al.*[88] found rodents to be particularly susceptible to oral ingestion of low concentrations, which produced paralytic symptoms. In tests on dogs, they found no effects attributable to a shampoo formulation containing 2 per cent, but dogs were the most susceptible species to the effects of the compound itself, and after oral administration as a water-in-oil emulsion at a level of 2·5 mg per kg, ocular manifestations were observed. However, the same dose had no ocular or other toxic effects when given to monkeys.

Opdyke *et al.*[91] claim that, in their experience over a period of eight years, ZnPTO does not penetrate the skin and does not produce toxic reactions when used with normal precautions. Collom *et al.*[92] and Coulston *et al.*,[93] however, demonstrated partial absorption of zinc and sodium omadine through the skin of rats, rabbits and monkeys.

In using these compounds in toilet preparations it is therefore important that possible toxic hazards to process workers and consumers are carefully considered.

Tetramethylthiuram Disulphide. TMTD (3,4,5-tetramethylthiuram disulphide) (Figure 35.4) is an antibacterial agent which has also been used in antibacterial soaps. Handwashing tests indicate that 1 per cent TMTD in soap produces 76 per cent reduction in count of skin bacteria over seven days (Table 35.2).

Although insoluble in water, TMTD has high activity against Gram-positive organisms, while its effect against Gram-negatives is higher than many of the other compounds previously mentioned. Drawbacks to its use in cosmetic products are its tendency to discolour and its lack of stability, which leads to liberation of thio odours. The compound is also known to produce irritant effects although it is recognized that this may be due to an oxidation product.

Synergism

The use of synergistic combinations of antimicrobial agents both for preservation and antiseptic purposes is widely reported in the literature. The term 'synergism' used here refers to an antibacterial effect greater than the sum of the antibacterial effects of the separate components.

Noel *et al.*[94] describes the enhancement of antibacterial activity by combining halogenated bisphenols with halogenated aromatic anilides or with halogenated carbanilides. Synergistic effects were specifically demonstrated for 3,4,4′-trichlorocarbanilide and 3,3′,4-trichlorocarbanilide with hexachlorophene, the sulphur analogue of hexachlorophene, tetrachlorophene, bithionol and 2,2′-thio*bis*-(4-chloro,6-methylphenol). Other synergistic combinations comprised 2-hydroxy-5-chlorobenzoic acid, 3′,4′-dichloroanilide and 2-hydroxy-5-chlorobenzoic acid, 3′4″-dichloroanilide and 2-hydroxy-5-chlorobenzoic acid 4′-chloroanilide in mixtures with hexachlorophene and the sulphur anologue of hexachlorophene (bithionol) and tetrachlorophene. These synergistic pairs of antiseptics are covered by US and other patents.[95]

Several other examples of synergistic combinations are given by Casely *et al.*[96] who have shown the ability of various substituted ureas, other than the well-known trichlorocarbanilides, to form combinations with synergistic activity with hexachlorophene. Moore and Hardwick[97] demonstrated synergistic effects of certain ampholytes with cetrimide, while Barr *et al.*[98] have shown that organic mercury compounds can be used to potentiate the activity of bisphenols. Further examples of synergism given in Chapter 36 refer mainly to compounds used as preservatives.

REFERENCES

1. Lilly, H. A. and Lowbury, E. J. L., *Br. med. J.*, 1974, **2,** 1792.
2. Ayliffe, G. A. J., Babb, J. R., Bridges, K., Lilly, H. A., Lowbury, E. J. L., Varney, J. and Wilkins, M. D., *J. Hyg. (Camb.)*, 1975, **75,** 259.
3. Lowbury, E. J. L., Lilly, H. A. and Bull, J. P., *Br. med. J.*, 1963, **1,** 1251.
4. Lowbury, E. J. L., Lilly, H. A. and Bull, J. P., *Br, med. J.*, 1964, **2,** 230.
5. Lilly, H. A. and Lowbury, E. J. L., *Br. med. J.*, 1971, **2,** 674.
6. Lowbury, E. J. L. and Lilly, H. A., *Br. med. J.*, 1973, **1,** 510.
7. Lilly, H. A. and Lowbury, E. J. L., *Br. med. J.*, 1974, **2,** 372.
8. US Food and Drug Administation, *OTC Topical Antimicrobial Products and Drug and Cosmetic Products*, Federal Register, 39(179) Part II, 33102, 1974.
9. Armstrong–Esther, C. A. and Smith, J. E., *Ann. hum. Biol.*, 1976, **3,** 221.
10. Skinner, F. A. and Carr, J. G., *Society for Applied Bacteriology Symposium Series No. 3,* London, Academic Press, 1974.
11. Gibson, J. W., *J. clin. Path.*, 1969, **22,** 90.

12. Ojajarvi, J., *J. Hyg.* (*Camb.*), 1976, **76,** 75.
13. Price, P. B., *Antiseptics, Disinfectants, Fungicides and Sterilization*, ed. Reddish, London, Henry Kimpton, 1954.
14. Guklhorn, I. R., *Mfg. Chem.*, 1969, **40**; 1970, **41**; 1972, **42** (*passium*).
15. Furla, T. E. and Schenkel, A. G., *Soap chem. Spec.*, 1968, **44,** 47.
16. Hurst, A., Stuttard, L. W. and Woodroffe, R. C. S., *J. Hyg.* (*Camb.*), 1960, **58,** 159.
17. Roman, D. P., *Proc. sci. Sect. Toilet Goods Assoc.*, 1957, **28,** 12.
18. Wilson P. E., *J. appl. Bact.*, 1970, **33,** 574.
19. Gibbs, B. M. and Stuttard, L. W., *J. appl. Bact.*, 1967, **30,** 66.
20. Taber, D., Lazanas, J. C., Fancher, O. E. and Calandra, J. C., *J. Soc. cosmet. Chem.*, 1971, **22,** 369.
21. Leonard, R. R., *Arch. Dermatol.*, 1967, **95,** 520.
22. Mackenzie, A. R., *J. Am. med Assoc.*, 1970, **211,** 973.
23. Duncan, W. C., Dodge, B. G. and Knox, J. H. *Arch. Derm.*, 1969, **99,** 465.
24. Forfar, J. O. J., Gould, J. C. and MacCabe, A. F., *Lancet*, 1968, **2,** 177.
25. Aly, R. and Maibach, H. I., *Appl. environ. Microbiol.*, 1976, **31,** 931.
26. Bruun, J. N. and Solberg, C. O., *Br. med. J.*, 1973, **2,** 580.
27. Ehrenkranz, N. J. D., Taplin, D. and Butt, P., *Antimicrob. Agents Chemother.*, 1966, **255,** 1967.
28. Steere, A. C. and Mallison, G. F., *Ann. intern. Med.*, 1975, **83,** 683.
29. Sprunt, K., Redman, W. and Leidy, G., *Paediatrics*, 1973, **52,** 264.
30. Block, S. S. *Disinfection, Sterilization and Preservation*, Philadelphia, Lea and Febiger, 1977.
31. Lord, J. W. and Parker, E., *Soap Perfum. Cosmet.*, 1953, **26,** 463.
32. Gump, W. S. and Cade, A. R., *Soap sanit. Chem.*, 1952, **28,** 52.
33. Gump. W. S. and Cade, A. R., *Manuf. Chem.*, 1953, **24,** 143.
34. Ehrlandson, A. L. and Lawrence, C. A., *Science*, 1953, **118,** 274.
35. Sandford, J. P., *Ann. intern. Med.*, 1970, **72,** 283.
36. Light, I. J., Sutherland, J. M., Cochran, L. and Sutorius, J., *N. Eng. J. Med.*, 1968, **278,** 1243.
37. Johnson, J. D., Malachowski, B. A. and Sunshine, P., *Paediatrics*, 1976, **58,** 354.
38. Knittle, M. A., Eitzman, D. V. and Baer, H., *J. Paediatr.*, 1975, **86,** 433.
39. Kimbrough, R. D., *Paediatrics*, 1973, **51**(suppl.), 391.
40. *Food, Drug and Cosmetic Reports* (The Pink Sheet), **34**(39), Hexachlorophene Special Supplement.
41. *Hansard*, 18 February, 1972.
42. Powell, H. C. and Lampert, P. W., *Paediatrics*, 1973, **52,** 859.
43. Lyman, F. and Furia, T. E., *Ind. Med. Surg.*, 1969, **38,** 45.
44. Savage, C. A., *Proceedings of the Sixth Congress of the International Federation of Societies of Cosmetic Chemists, Barcelona*, IFSCC, 1970.
45. Burry, J. N., *Arch. Dermatol.*, 1967, **95,** 287.
46. Wilkinson, D. S., *Br. J. Dermatol.*, 1961, **73,** 213.
47. Calnan, C. D., *Br. med. J.*, 1961, **2,** 1266.
48. Vinson, L. J. and Flatt, R. S., *J. invest. Dermatol.*, 1962, **38,** 327.
49. Anderson, I., *Trans. St. John's Hosp. Dermatol. Soc.*, 1963, **49,** 54.
50. Baer, R. L., *Arch. Dermatol.*, 1966, **94,** 522.
51. Behrbohm, P. and Zschunke, E., *Berufsdermatosen*, 1966, **14,** 169.
52. Epstein, S., *J. med. Assoc.*, 1965, **194,** 1016.
53. Harber, L. C., Harris, H. and Baer, R. L., *Arch. Dermatol.*, 1966, **94,** 255.
54. Osmundsen, P. E., *Br. J. Dermatol.*, 1968, **80,** 228.
55. Peck, S. M. and Vinson, C. J., *J. Soc. cosmet. Chem.*, 1967, **18,** 361.
56. Voss, J. G., *Appl. Microbiol.*, 1975, **30,** 551.
57. Bassett, D. J. C., *Proc. R. Soc. Med.*, 1971, **64,** 980.

58. Dixon, R. E., Kaslow, R. A., Mackel, D. C., Fulkerson, C. C. and Mallison, G. F., *J. Am. med. Assoc.*, 1976, **22,** 2415.
59. Lawrence, C. A., *Soap Perfum. Cosmet.*, 1954, **27,** 369.
60. Richardson, G. and Woodford, R., *Pharm. J.*, 1964, **192,** 527.
61. Deluca, P. P. and Kostenbauder, H. B., *J. Am. Pharm. Assoc.* (*Sci. Ed.*), 1960, **40,** 430.
62. Gershenfeld, L. and Perlstein, D., *Am. J. Pharm.*, 1941, **113,** 306.
63. Lawrence, C. A., *Surface Active Quaternary Germicides*, New York, Academic Press, 1950.
64. Hussey, H. H., *J. Am. med. Assoc.*, 1976, **236,** 2433.
65. Rosenburg, A., Alatary, S. D. and Peterson, A. F., *Surgery Gynec. Obstet.*, 1976, **143,** 789.
66. Richards, R. M. E., *Lancet*, 1944, **1,** 42.
67. Ojajarvi, J., Makela, P. and Ratsalo, I., *J. Hyg.* (*Camb.*), 1977, **79,** 107.
68. Soendergard, M. W., *J. Hosp. Pharm.*, 1969, **26,** 53.
69. Fowler, A. W., *Lancet*, 1963, **1,** 387.
70. Fowler, A. W., *Lancet*, 1963, **1,** 769.
71. Grant, J. C., *Br. med. J.*, 1968, **4,** 646.
72. Morris, G. M. and Maclaren, D. M., *Br. J. clin. Pract.*, 1969, **23,** 349.
73. Loe, H. and Schiott, C. R., *J. periodont. Res.*, 1969, Suppl. No. 4, 38.
74. Loe, H. and Schiott, C. R., *Dental Plaque*, ed McHugh, W. D., Edinburgh, Livingstone, 1970, p. 247.
75. Senior, N., *J. Soc. cosmet. Chem.*, 1973, **24,** 259.
76. Madsen, W. S., *J. Hosp. Pharm.*, 1969, **26,** 53.
77. Madsen, W. S., *J. Hosp. Pharm.*, 1969, **26,** 79.
78. Schmitz, A. and Harris, W. S., *Manuf. Chem.*, 1958, **29,** 51.
79. Frisby, B. R., *Lancet*, 1961, **2,** 829.
80. Croshaw, B., Groves, M. J. and Lessel, B., *J. Pharm. Pharmacol.*, 1964, **16,** 127T.
81. Saito, H. and Onoda, T., *Chemotherapy* (*Tokyo*), 1974, **22,** 1461.
82. Brown, M. R. W., *J. Soc. cosmet. Chem.*, 1966, **17,** 185.
83. Allwood, M. C., *Microbios*, 1973, **7,** 209.
84. Berke, P. A. and Rosen, W. E., *Am. Perfum. Cosmet.*, 1970, **85,** 55.
85. Shelanski, H. A. and Shelanski, M. V., *J. int. College Surg.*, 1956, **25,** 727.
86. Crowder, U. H., Welsh, J. S., Bornside, G. H. and Cohn, I., *Am. Surgeon*, 1967, **33,** 906.
87. Cox, A. J., *Manuf. Chem.*, 1957, **28,** 463.
88. Snyder, F. H., Buehler, E. V. and Winek, C. L. *Toxicol. appl. Pharmacol.*, 1965, **7,** 425.
89. Brauer, E. W., Opdyke, D. L. and Burnett, C. M., *J. invest. Dernatol.*, 1966, **47,** 174.
90. Tenenbaum, S. and Opdyke, D. L., *Proc. sci. Sect. Toilet Goods Assoc.*, 1967, **47,** 20.
91. Opdyke, D. L., Feinberg, H. and Burnett, C. M., *Drug. Cosmet. Ind.*, 1967, **101**(10), 48.
92. Collom, D. and Winek, C. W. L., *J. pharm. Sci.*, 1967, **56,** 1673.
93. Coulston, F. and Golberg, L., *Toxicol. appl. Pharmacol.*, 1969, **14,** 97.
94. Noel, D. R., Casely, R. E., Linfield, W. M. and Harriman, L. A., *Appl. Microbiol.*, 1960, **8,** 1.
95. US Patent 3 177 115 Armour and Co., 12 June 1958.
96. Casely, R. E., Brown, J. and Taber, D., *J. Soc. cosmet. Chem.*, 1968, **19,** 159.
97. Moore, C. D. and Hardwick, R. B., *Manuf. Chem.*, 1958, **29,** 194.
98. Barr, F. S., Moore, G. W. and Gragg, B. J., *J. pharm. Sci.*, 1970, **59,** 262.

Chapter Thirty-six

Preservatives

Introduction

Many of the materials used in the manufacture of toilet preparations are susceptible to biological degradation by micro-organisms. In this chapter methods of combating biodeterioration of products will be discussed.

In the past the contribution of the microbiologist to the cosmetics industry was considered of minimal importance.[1] More recently there has been an increasing awareness of the problems of microbial spoilage of cosmetic and toiletry products. Formerly the emphasis was placed on loss of aesthetic appeal of the product with a consequential loss of profitability, while the health hazard to the consumer was largely ignored. Since micro-organisms are ubiquitous, the body is continually exposed to them and it therefore seemed unnecessary for toilet preparations to be sterile. Although it was accepted that products should not constitute any greater microbial hazard than that presented by the normal environment, little was done to show that this was achieved.

Normally healthy individuals have considerable resistance to infection by bacteria and fungi commonly found on their skin and in their usual environment, but in susceptible individuals, for example the newborn, the very old, those in ill health or under drug therapy, there is an increased probability of infection developing. It should be remembered that a product may contain a growing bacterial population even though there is no visible evidence of this. Such a product placed in intimate contact with the skin, particularly if it is broken or damaged, may give rise to infection.[2] Examples in the literature which demonstrate the importance of considering the possible health hazard of a contaminated product will be discussed later.

Preservatives are, therefore, added to products for two reasons: firstly to prevent spoilage, that is, to prolong the shelf-life of the product, and secondly to protect the consumer from the possibility of infection.[3] It is recognized that products may require protection from contamination during manufacture, although preservation must never be used to hide bad manufacturing procedures.[4] Duke[3] quotes examples illustrating how products containing preservatives can be microbiologically unacceptable because of bad manufacturing practice. It is also recognized that cosmetic products, possibly more so than pharmaceutical products, are liable to consumer abuse. Although products cannot be protected against extremes of abuse, such as the use of saliva in the application of eye make-up, the manufacturer should anticipate misuse when the product is formulated.[3]

It is therefore imperative that the microbiologist be involved in the formulation of a product from its development stage onwards.[5] It can no longer be assumed that a manufacturer has fulfilled his obligation to the consumer if the product in question leaves his premises in a microbiologically satisfactory condition. If, however, products are to be expected to remain resistant to the introduction of extraneous micro-organisms during their period of use, this will impose a greater strain on the preservative system chosen and will make it even more important for manufacturers to carry out thorough and realistic tests both during the development of new products and in the production and quality control of existing lines.

The Council of the British Society of Cosmetic Chemists has published a monograph on the hygienic manufacture and preservation of toiletries and cosmetics.[6] Further indications of the importance currently attached to ensuring the adequate preservation of all toilet preparations and cosmetics are evident in the number of investigations into contaminated preparations and the possible clinical significance of contamination. Some of these reports are discussed later in the chapter.

Microbial Metabolism

Micro-organisms grow and multiply by utilizing the materials in their immediate environment. In considering spoilage problems created by micro-organisms, the variety of chemical reactions they can carry out and the rate at which these can occur must be taken into account.

Bacteria and fungi are widely distributed in nature and there are few places on or near the surface of the earth that are free from them. They occur, for example, in such unlikely places as hot mineral springs, the effluents from gas works, stagnant salt lakes and even in the essentially anhydrous environment of diesel fuel. The only places where micro-organisms are not found are those in which a sterilizing influence prevails or in the interior tissues of healthy animals and plants. In growing, bacteria and fungi can cause rapid and profound changes in their immediate environment and in the synthesis of new protoplasm many complex chemical reactions are accomplished within a remarkably short period of time. The organisms carry out these reactions by means of enzymes and some of the basic reactions that can occur are as follows:

Hydrolysis — The addition of H_2O to the molecule. The next step in molecular breakdown is often thus facilitated so that molecules are split at the hydrolysed link.

Dehydration — The removal of H_2O from one or more molecules.

Oxidation — The removal of hydrogen or the addition of oxygen to the molecule; also a process which involves an increase in the number of positive charges on an atom, or a decrease in a number of negative charges.

Reduction — The removal of oxygen or the addition of hydrogen; the reverse reactions to those of oxidation.

Decarboxylation — The removal of CO_2.

Deamination — The removal of $—NH_2$.

Phosphorylation	The esterification of the molecule with phosphoric acid. This is usually accomplished by the transfer of the phosphate radical from some substance other than phosphoric acid itself.
Dephosphorylation	The removal or hydrolysis of phosphoric acid from phosphorylated compounds.

In a product supporting the growth of different kinds of micro-organisms a large variety of end products will be produced, and in mixed populations there is, of course, a certain amount of competition for essential nutrients between the different organisms. Those that can convert the environment satisfactorily survive, while those which only live with difficulty die out, providing a further source of convertible substrate for those that remain. In utilizing the substrate, the metabolic processes of some organisms result in the formation of acid end products which have a limiting effect on growth, and in some cases the changes brought about are sufficient to inhibit further growth. Most organisms are, however, capable of carrying out neutralizing reactions and in this way some degree of stabilization of the environment can be achieved.

The speed with which micro-organisms can propagate, and the variety of reactions which they can accomplish, indicate how necessary it is to inhibit their growth in products whose physical characteristics must remain unchanged during long periods of storage and use in the hands of the consumer.

The manifestations and mechanisms of microbial spoilage have been discussed by Smart and Spooner.[7] Microbial contamination may be made apparent by visible growth of the contaminant. For example, moulds and fungi frequently can be detected as they grow macroscopically, usually on the surface of the product or the lining of the pack. Micro-organisms may also be visible in liquid preparations as turbidity or sedimentation. Colour changes may occur as a result of alterations in pH or redox potential or because of pigment production by the contaminating organism, for example the blue-green to brown pigments produced by *Pseudomonas* genus.

The metabolic processes of some organisms cause the formation of gas which may be seen as bubbles or frothing in liquid preparations. Frequently spoilage is indicated by odour productions. In addition, micro-organisms may cause cracking of emulsions, alterations in rheological properties or loss of texture in topical preparations. Although direct evidence is lacking, spoilage may be detected as an allergic reaction to the application of foreign protein to the skin from a heavily contaminated product.

All these effects can be brought about quite quickly if large numbers of organisms are present or if the product possesses properties which favour rapid multiplication. Inadequate preservation may, however, only reveal itself after many months if the conditions are such that growth can only take place after the organisms have adapted themselves to the environment. The process of adaptation may involve gradual alteration of pH by the organism to a level at which growth can take place at a greater speed, or may necessitate the use of metabolic processes not normally employed by the organism under optimum conditions.[8] It may also involve development of increased resistance to, or metabolism of, the preservative. *Pseudomonas* species have been found to develop total resistance to parabens and benzalkonium chloride, used as preservatives in detergent-containing products. In both products the *Pseudomonas* was found to metabolize the detergent.[9] The contamination of a product with *Cladosporium resinae* was discovered to occur as a result of the ability of this fungus to hydrolyse the preservative methyl parabens to *p*-hydroxybenzoic acid.[10]

Clinical Significance of Contamination

A number of independent surveys have been conducted on the type and extent of contamination in used and unused cosmetics (see Table 36.1, Baker,[11] Ahearn and Wilson[20]). Table 36.2 lists some of the genera of micro-organisms which have been isolated from cosmetic and toiletry preparations. Gram-negative organisms, in particular the pseudomonads, seem to be the most frequently isolated organisms in unused cosmetics. Used cosmetics are most frequently contaminated with staphylococci, diphtheroids, micrococci, fungi and yeasts. While it is appreciated that contaminated cosmetics may lose their aesthetic appeal, less is known about the potential danger of these contaminants to the consumer. Certain cosmetic preparations, such as hand creams and lotions, are extensively used in hospitals; because patients are more likely to be susceptible to infections than healthy individuals, the microbial state of these cosmetics may have important implications. Perhaps the most striking example of this was the investigation by Morse *et al.* [21] of a septicaemia outbreak caused by *Klebsiella pneumoniae* in a hospital intensive care unit, the source of which was found to be a dispensing bottle of contaminated lanolin hand cream. These workers in a later survey[22] within a particular hospital examined 26 brands of hand cream, used and unused. Four brands were found to contain a variety of Gram-negative bacteria and it was suggested that the increase in numbers of Gram-negative infections in this hospital was related to the use of contaminated products.

There is increasing evidence implicating contaminated cosmetics in the production or persistence of eye infections. Ocular cosmetics become contaminated with resident bacterial flora from the skin and eyes, plus animal-associated yeasts and saprophytic moulds. These organisms may grow in the cosmetics and be inoculated in high numbers into the outer eye.[14,17] A correlation between organisms found in the outer eye and those found in the cosmetics of the user has been shown.[14] In one case an association between contaminated mascara and keratomycosis was demonstrated, the causative fungus, *Fusarium solani*, being isolated from the mascara. In another case *Staphylococcus aureus* was isolated from the eyelid margins of a woman with blepharitis and from the mascara which she was using daily, thus perpetuating the condition. Other observations indicated that potential eye pathogens can establish themselves in a cosmetic within a week with only moderate use.[17] Ahearn *et al.*[20] were able to correlate use of contaminated cosmetics with four cases of staphylococcal blepharitis or conjunctivitis when the conditions were cured by withdrawal of the cosmetic.

Since new eye cosmetics are less frequently contaminated with fungi or bacteria, it appears that contamination of the product by the user is the principal problem.[14]

Table 36.1 Summary of Some Investigations into Product Contamination

Products	Used/ Unused	Contaminated (%)	Organisms Most Frequency Isolated	Source
250 products covering wide range	Unused	24·4	*Pseudomonas* species and other Gram-negative rods. Most frequently contaminated products were hand and body lotions, liquid eye liners and cake eye shadows.	Wolven and Levenstein (1969)[12]
169 hand and body lotions and creams	Unused	19·5	*Pseudomonas* species and other Gram-negative rods.	Dunnigan and Evans (1970)[13]
428 eye cosmetics	Used	12	Fungi—mainly *Penicillium* and *Cladosporium*, and yeasts—mainly *Rhodotorula rubra* and *Candida parapsilosis*.	Wilson *et al.* (1971)[14]
		43	Bacteria—mainly Gram-positive micrococci. 17 products contained Gram-negative rods, 6 of which were *Pseudomonas aeruginosa*.	
58 eye cosmetics	Unused	3·4	Gram-negative rods, one of which was *Escherichia coli*.	
223 products covering wide range	Unused	3·5	*Pseudomonas* species and other Gram-negative rods.	Wolven and Levenstein (1972)[15]
165 products covering wide range, of which 23 were eye cosmetics	Unused	12	Diphtheroids, staphylococci and aerobic spore-forming bacilli. None of the eye cosmetics were contaminated.	

Table 36.1 (*cont.*)

Products	Used/ Unused	Contaminated (%)	Organisms Most Frequency Isolated	Source
222 products covering wide range, of which 79 were eye cosmetics	Used	49	Staphylococci. 35 per cent of the eye cosmetics were contaminated. For both used and unused cosmetics, Gram-negative bacilli were isolated from an insignificant number of products.	Myers and Pasutto (1973)[16]
19 medicated cosmetics	Used	52	Staphylococci. Indicates that contamination is possible even in products with high concentrations of antimicrobial agents.	
29 applicators	Unused	27·5	Staphylococci and diphtheroids.	
37 applicators	Used	100	Staphylococci.	
200 mascaras	Unused	1·5	—	Ahearn *et al.* (1974)[17]
	Used	60	Most common bacteria were *Staphylococcus epidermidis* and *Micrococcus* species. Most common fungi and yeasts were *C. parapsilosis* and *Cladosporium* species.	
172 products covering wide range	Unused	<50	Gram-negative rods. Most heavily contaminated products were specific brands of eye make-up (especially liquid eyeliner), bath detergent and complete make-up.	Jarvis *et al.* (1974)[18]
147 products covering wide range	Unused	32·7	Principally aerobic spore-bearers and Gram-positive cocci, but Gram-negative rods, in particular the pseudomonads, also isolated.	Baird (1977)[19]

Table 36.2 Some Micro-organisms Isolated From Toilet Preparations

Fungi	Bacteria	Yeasts
Absidia	Acinetobacter	Candida*
Alternaria*	Alcaligenes	Monilia
Aspergillus*	Bacillus*	Torula
Citromyces	Diphtheroids	Zygosaccharomyces
Cladosporium*	Enterobacter	
Dematium	Enterococcus	
Fusarium	Escherichia	
Geotrichum	Klebsiella*	
Helminthosporium	Micrococcus	
Hormodendrum	Proteus	
Mucor*	Pseudomonas*	
Paecilomyces	Sarcinia	
Penicillium*	Serratia	
Phoma	Staphylococcus*	
Pullularia	Streptococcus*	
Rhizophus*		
Stemphylium		
Thamnidium		
Trichothecium		
Verticillium		

* Several different species of these genera have been reported.

Origins of Contamination

Raw Materials

If the raw materials used in manufacturing of cosmetics are heavily contaminated then it is almost inevitable that the finished product will be similarly contaminated, and any preservative system present will be unnecessarily strained. This can be avoided by careful monitoring of raw materials.

The water used in product manufacture is possibly the most frequent source of contamination, often supporting large numbers of micro-organisms. Mains water contains low numbers of organisms, usually less than 300 ml^{-1}. Softened, distilled or, in particular, deionized water is capable of supporting the growth of some bacteria, and numbers can rise to 10^6–10^7 ml^{-1} when water is stored.[23–25] Storage tanks are thus frequently responsible; Baker[11] describes several cases of serious product deterioration which were traced to heavy contamination in the beds of ion exchangers or water tanks. More recently Duke[3] describes the contamination of a hair rinse with waterborne Gram-negative organisms which were using cetrimide in the product as a nutrient. Correlations between shampoo spoilage and waterborne organisms have been discussed.[26]

Fats, waxes and refined oils contain relatively few organisms whereas natural materials such as gums and herbs are frequently very heavily contaminated by a variety of fungi, yeasts and bacteria. The natural gums, tragacanth, karaya and acacia, are always very heavily contaminated; the synthetic gums are often almost sterile. Other naturally occurring materials such as talc, kaolin, chalk and rice starch frequently carry large numbers of bacteria, particularly those that are capable of forming spores. Duke[3] reports contamination of a face preparation by a pigment contaminated with bacteria. The contaminants were killed by addition of a preservative.

The containers of raw materials—drums, sacks, cartons, etc.—may also be a source of contamination prior to manufacture.

Environment

A further possible source of contamination is the air, which contains mainly mould and bacterial spores and skin cocci. Environmental control is facilitated by covering of containers and reduction of air-currents over a product.[23] Routine monitoring of the air and selected surface sites within the production area is advised, so that deviations from normal standards of cleanliness are immediately detected.[1]

A survey of the types of organisms found in manufacturing environments and how these varied over a nine-month period has been conducted by Bruch.[27]

Equipment

During manufacture the product can easily become contaminated by organisms which accumulate in the plant as a result of faulty or inadequate cleansing. Pieces of equipment with inaccessible joints, pipes and pumps are often difficult to clean properly and washing with detergent solutions may only result in dilution of the product to form stagnant foci where bacteria and fungi thrive. Plant should be designed to facilitate easy cleaning and disinfection, and inaccessible grooves and dead-ends in all items which come into contact with the product should be avoided wherever possible (see Chapter 43).

An example of product deterioration resulting from plant contamination is given by Duke.[3] A broken sealing joint in a bulk tank used for the manufacture of a shampoo allowed shampoo and contaminated cleansing water to enter a false bottom of the tank. As the preservative had been diluted, the contaminants survived and were washed into each new batch of shampoo. Using the shampoo base as nutrient, the bacteria then contaminated the whole product.

The need for good housekeeping procedures cannot be stressed too strongly. Of paramount importance is the education of plant operatives to appreciate the necessity of properly performed plant cleaning operations.

Advice on methods of plant cleaning and disinfection are given by Davis[28] who states that 150–200 ppm of hypochlorite will sanitize clean metal and glass in two minutes and sterilize in 10 minutes. The corrosive nature of this treatment is a disadvantage and formaldehyde is frequently used in preference. Hot water, or preferably live steam, has been recommended as the best disinfecting agent.[23] Detergent sanitizers containing quaternary ammonium compounds or detergent–iodine mixtures are also useful, but it is essential to remove product residue from the plant and to rinse thoroughly with hot water before utilizing any of the above disinfection procedures since many are inactivated by residues of organic material. Air locks in the plant may also prevent certain parts coming into contact with the cleaning and sterilizing fluids and should be avoided.

Microbiological tests should be made on equipment and filling lines before and after sanitization to ascertain the effectiveness of cleaning procedures. RODAC plates and swabbing techniques may be used for this purpose together with sampling of the final rinse fluid after sanitization to determine the cleanliness of internal surfaces.[1]

Packaging Materials

Since most products are exposed to further contamination during filling into containers, cleaning and disinfection of filling equipment is also important (see Chapter 43). Containers and closures should be dust-free and microbiologically clean. This can be achieved by a filtered air blast which is probably more effective than the use of detergent and water.[23] Caps and liners are notorious for an abundant fungal flora and much of the growth on the surface of cosmetic creams is directly attributable to cap linings. Quite frequently creams containing preservatives, which are otherwise adequately preserved, break down following the introduction of large numbers of fungal spores from a cap lining. These organisms germinate initially in the microfilm of water on the lining and spread gradually to areas of slight separation around the edge of the cream. Many toilet preparations, in addition to having a long shelf-life, may be in use by the consumer for many months, so that spoilage does not become apparent until the contents are more than half used. Products packed in wide mouthed jars, and flexible bottles which draw air back into them, are more liable to contamination than those packed in collapsible tubes and bottles with small orifices.

As plastic materials are not subject to biodegradation, use of such materials instead of cellulosic materials should be associated with a reduction in microbial spoilage. Unlike paper, card and cork, plastics are microbiologically pure but as they are porous to oxygen and carbon dioxide and encourage water condensation, spoilage by mould spores will be facilitated.[7]

Personnel

Probably the greatest microbial hazard to the product during manufacture or packaging is from the operators.[23,29] Operators should be properly instructed to appreciate that they are a potential source of contamination and trained to maintain high standards of personal hygiene and cleanliness. The wearing of protective overgarments is advisable.[23]

Microbial Growth in Products

A number of factors determine whether micro-organisms will survive and propagate in a product and hence these influence the need for preservation. Some of the more important are examined below.

Water Content

Because micro-organisms depend on water for the synthesis of cellular components, the physical and chemical characteristics of the water phase in an emulsion, for example, are dominant factors in determining whether growth occurs. In single-phase non-emulsion products, however, the amount of growth that will occur is determined by the pH, osmotic pressure, surface tension and oxygen tension of the system.

In general, emulsions with a water-continuous phase are more susceptible to bacterial attack than those with an oil-continuous phase although bacteria have frequently been isolated from inadequately preserved water-in-oil emulsions. Until fairly recently it was presumed that micro-organisms would only survive in aqueous environments but De Gray and Killian[30] stated that certain bacteria and fungi could survive for extended periods of time in hydrocarbons free of any separate aqueous phase. In addition, anhydrous cosmetic products have been shown to support the growth of contaminating micro-organisms.[17] In this instance it was suggested that moisture had entered the products either via the user or through condensation. In emulsions there is likely to be migration of organisms from the oil phase to the water phase and since the oil phase may not be entirely anhydrous, migration in the opposite direction cannot be ruled out. Some micro-organisms are able to degrade triglycerides in emulsions, a process which is facilitated by the marked adsorption of organisms at the oil–water interface. Fatty acids and glycerol are liberated, which can then be metabolized for growth purposes.[31] Bennett,[32] in a study of the preservation of emulsions, found that the oil-to-water ratio had a significant effect upon the magnitude of growth. Equal numbers of *Pseudomonas aeruginosa* cells were inoculated into a series of emulsions with different oil-to-water ratios and it was found that the amount of growth increased with the increase in water content.

The nutritive value of the aqueous phase of any product will contribute to the amount of growth that will occur and the presence of such materials as carbohydrates, proteins and, for example, phospholipids will increase the need for adequate preservation. Sorbitol, glycerol and even surface-active agents (particularly the nonionics, but also to a lesser extent anionics) when present at low concentrations can all be metabolized by micro-organisms. Barr and Tice[33] reported the splitting of the ester linkages of certain nonionic surfactants by *Pseudomonas aeruginosa*, *Aspergillus niger*, *Penicillium notatum* and *Monilia albicans*. Their results indicated that the organisms were capable of growing in the surfactant esters and splitting them and that the rate of growth and ester splitting was dependent on the number of organisms in the inoculum. Anionic surfactants are also capable of acting as sources of energy for micro-organisms. Their chemical structure controls their susceptibility to attack and certain bacteria are capable of oxidizing terminal methyl groups to carboxyl groups. According to Sawyer and Ryckman[34] the alkyl sulphates, sulphonated fatty acids, amides and esters, and the low-molecular-weight polyethylene glycol derivatives are rapidly broken down, while the alkyl aryl sulphonates, alkyl phenoxypolyoxyethanols and high-molecular-weight polyethylene glycol derivatives are attacked more slowly. Yu-Chih Hsu[35] discovered six strains of *Pseudomonas* which were capable of splitting sodium alkyl sulphates.

The nutritive value of many vegetable gums utilized as thickeners is also well known. Polysaccharides may be attacked by extracellular enzymes and thus depolymerized. Starch may be degraded by amylases and carboxymethylcelluloses by cellulases. It has been reported that, of the range of celluloses available, the most resistant to attack are methyl- and ethyl-celluloses.[31]

pH Value
The pH value of a product will affect the degree of ionization of utilizable materials, influence the electrical charge at the bacterial and fungal cell walls, determine enzyme production and activity and hence regulate the availability of nutrients and the ease with which they are assimilated by the microbial cell. However, since growth tolerance limits for pH differ widely for various micro-organisms, pH itself should never be regarded as a factor likely to contribute to self-sterilization of a product. *Pseudomonas* species, which are extremely common contaminants of toilet preparations, can exist over as wide a pH range as 3·0–11·0 and although many fungi grow most prolifically at acid pH they have been known to survive on vanishing creams at pH 9.

Osmotic Pressure
The living semi-permeable membranes which surround the bodies of all micro-organisms can be ruptured by changes in osmotic pressure and this can lead to membrane shrinkage and dehydration of the organism. For this reason osmotic pressure can have a limiting effect on growth. Concentrations of between 40 and 50 per cent of glycerin and sorbitol are inhibitory to micro-organisms by reason of the osmotic pressure, and high concentrations of electrolytes can exert a similar limiting action. Thus very concentrated products are likely to be self-preserving but those which may be diluted and left standing before use can deteriorate if left for periods of several days. Shampoos, for example, which are frequently sold to professional hairdressers as concentrates for dilution before use, are often susceptible to bacterial degradation as the concentrates may be diluted with contaminated water and left uncovered for long periods during use.

Surface Tension and Oxygen Tension
Quite apart from the nutritive value of low concentrations of some surfactant molecules, surface tension is itself a factor influencing growth. Many Gram-negative bacteria, and the coliforms in particular, grow well in environments abounding in surfactants, while most Gram-positive organisms do not grow well at surface tension levels much below 50 dyn/cm (0·05 N/m). Gram-negative organisms, particularly the pseudomonads, flourish in shampoos and are also common contaminants of the aqueous phases of emulsions. Cationic surfactants are toxic to many organisms, anionics to a few, and nonionics to hardly any; thus surface tension *per se* will not be a gross limiting factor but will have an effect in association with the presence or absence of toxic groups in the surfactant molecules. Most organisms, bacteria and fungi which contribute to product spoilage are aerobic and depend on the availability of oxygen for their metabolism. The microclimate in most products, with perhaps the exception of those in pressurized packs, will almost invariably provide sufficient oxygen for the growth of micro-organisms provided that all other factors are favourable.

Antimicrobial Spectrum of the Preservative System
In a cosmetic, as in the natural environment, there is competition between organisms for nutrients so that one organism will survive and grow more successfully than others. Thus where a preservative is active against a limited spectrum of organisms, the product may be left open to spoilage by less susceptible organisms.[36]

Temperature
The susceptibility to microbial attack will vary with the temperature of storage, so that a cosmetic kept at room temperature will be liable to spoilage by different organisms from those that flourish in a product kept in a hot environment (for example one left in the sun or in a hot car). Bacteria generally prefer temperatures of 30°–37°C and fungi and yeasts 20°–25°C.[36]

Preservative Requirements

The 'ideal' preservative which is both safe and effective in all kinds of toilet preparations has not yet been discovered and this means that the composition of each new product must be studied in detail before selecting a suitable preservative. In order to avoid preservative failure a careful analysis must be made of the factors in the product that are likely to favour the growth of micro-organisms, the ingredients that are likely to be contaminated before use, and also those that are likely to influence adversely the efficiency of whatever preservative is ultimately selected. The essential requirements of a preservative are:

(i) freedom from toxic, irritant or sensitizing effects at the concentrations used on the skin, mucous membranes and, in the case of orally administered products, on the gastro-intestinal system;
(ii) stability to heat and prolonged storage;
(iii) freedom from gross incompatibility with other ingredients in the formula and with the packaging material, which could result in loss of antimicrobial action.

Other requirements are that the preservative should be active at low concentration; should retain its effectiveness over a wide range of pH; should be effective against a wide range of micro-organisms; should be readily soluble at its effective concentration; should have no odour or colour; should be non-volatile; should retain its activity in the presence of metallic salts of aluminium, zinc and iron; should be non-corrosive to collapsible metal tubes and non-injurious to rubber.

Table 36.3 lists some of the preservatives used in cosmetics and toilet preparations. Compounds closely related to those listed in Table 36.3 have also been used, and inspection of the variety of preservatives available shows that a thorough understanding of the factors that can influence the efficiency in a particular system is necessary before a selection can be made.

The properties of some individual preservatives have been discussed by Gucklhorn,[37] Rosen and Berke,[36] Croshaw[38] and Cowen and Steiger.[39] Table 36.4 indicates the advantages and disadvantages of some of the established groups of preservatives.

Factors Influencing the Effectiveness of Preservatives

Dissociation and pH
Formulations of cosmetics and toiletries encompass a wide pH range and since micro-organisms of one sort or another are capable of growing between pH 2 and pH 11, ideally a preservative should be effective over this range.[38] In

Table 36.3 Some Preservatives Used in Cosmetics and Toilet Preparations

p-Hydroxybenzoic acid	Phenol
Benzoic acid	Cresol
Sorbic acid	Chlorothymol
Dehydroacetic acid	Methylchlorothymol
Formic acid	Chlorbutanol
Salicylic acid	*o*-Phenylphenol
Boric acid	Dichlorophene
Vanillic acid	Hexachlorophene
p-Chlorobenzoic acid	Parachlormetaxylenol
o-Chlorobenzoic acid	Parachlormetacresol
Propionic acid	Dichlormetaxylenol
Sulphurous acid	*p*-Chlorphenylpropanediol
Trichlorphenylacetic acid	*β*-Phenoxyethylalcohol
	β-*p*-Chlorphenoxyethylalcohol
Methyl *p*-hydroxybenzoate	*β*-Phenoxypropylalcohol
Ethyl *p*-hydroxybenzoate	Potassium hydroxyquinoline sulphate
Propyl *p*-hydroxybenzoate	8-Hydroxyquinoline
Butyl *p*-hydroxybenzoate	*p*-Chlorphenylglyceryl ether
Benzyl *p*-hydroxybenzoate	Formaldehyde
	Hexamine
Benzethonium chloride	Monomethylol dimethyl hydantoin
Benzalkonium chloride	2-Bromo-2-nitro-1,3-propanediol
Cetyltrimethyl ammonium bromide	1,6-*Bis*-*p*-chlorophenyl diguanidohexane
Cetylpyridinium chloride	Phenyl mercury acetate
Dimethyldidodecenyl ammonium chloride	Phenyl mercury borate
β-Phenoxy-ethyl-dimethyl-dodecyl ammonium bromide	Phenyl mercury nitrate
Tetramethylthiuramdisulphide	Sodium ethyl mercurithiosalicylate
1-(3-Chloroallyl)-3,5,7-triazonia-adamantane chloride	Tetrachlorsalicylanilide
5-Bromo-5-nitro-1,3-dioxan	Trichlorsalicylanilide
6-Acetoxy-2,4-dimethyl-*m*-dioxan	Trichlorcarbanilide
Imidazolidinyl urea	
Vanillin	
Ethyl vanillin	

practice many preservatives are pH-dependent, the majority of them being more active in the acidic than alkaline range. Some preservatives with a wide pH profile have the disadvantage of being chemically highly reactive compounds (for example formaldehyde and formaldehyde donors) which react with other components of the formulation.[39] pH may also have an effect on the microbial cell surface[40] and may affect the partitioning of an antimicrobial agent between the cell and the product.[41]

For many preservatives the most pronounced effect of pH on activity is on the antimicrobial agent itself. Many weak acids are used as preservatives, their activity depending upon the amount of undissociated acid, which in turn depends on the dissociation constant and pH of the system (see Tables 36.5 and 36.6). It has been suggested that anions of acids may be inactive as a result of

Table 36.4 Advantages and Disadvantages of Some Established Preservatives

Preservative	Advantages	Disadvantages
Alcohols, e.g. ethyl alcohol, isopropyl alcohol	Broad spectrum	Volatile High concentrations (15–20 per cent) required
Quaternary ammonium compounds	Better as active agents, e.g. in deodorants	Ineffective against some pseudomonads except at high concentrations which may be irritant Incompatible with proteins, anionics and nonionics
Acids, e.g. benzoic, sorbic, dehydroacetic	Active against fungi	pH-dependent because of dissociation
Formaldehyde	Broad spectrum Cheap Water-soluble Retains activity in presence of surfactants	Irritant (banned in some countries) Volatile Unpleasant odour Highly reactive chemically Incompatible with proteins
Parabens (*p*-hydroxybenzoates)	Low toxicity Relatively non-irritant at use concentrations Relatively effective over wide pH range	More active against fungi and Gram-positive bacteria than against Gram-negative bacteria Low water-solubility Partition in favour of the oily phase Inactivated by nonionics, proteins
Organic mercurials, e.g. phenyl mercuric salts	Broad spectrum Stable	High toxicity and irritancy Inactivated by proteins, anionics but to much less extent by nonionics Low water-solubility.
Phenolics	Useful as packaging preservatives and as active agents.	Low water-solubility Partition into the oil phase Volatile Incompatible with anionics above critical micelle concentration and with nonionics May be irritant

Data reproduced by kind permission of Miss B. Croshaw and *International Journal of Cosmetic Science*.[38]

Table 36.5 Dissociation Constants of Acids used as Preservatives

Preservative		Dissociation constant
Sulphurous acid	H_2SO_3	$1{\cdot}70 \times 10^{-2}$
o-Chlorobenzoic acid	*o*-Cl.C_6H_4COOH	$1{\cdot}20 \times 10^{-3}$
Salicylic acid	*o*-HO.C_6H_4COOH	$1{\cdot}06 \times 10^{-3}$
Formic acid	H.COOH	$1{\cdot}80 \times 10^{-4}$
p-Chlorbenzoic acid	*p*-Cl.C_6H_4COOH	$1{\cdot}05 \times 10^{-4}$
Benzoic acid	C_6H_5COOH	$6{\cdot}30 \times 10^{-5}$
p-Hydroxybenzoic acid	*p*-HO.C_6H_4COOH	$3{\cdot}00 \times 10^{-5}$
Sorbic acid	$CH_3CH{:}CHCH{:}CHCOOH$	$1{\cdot}73 \times 10^{-5}$
Propionic acid	C_2H_5COOH	$1{\cdot}40 \times 10^{-5}$
Dehydroacetic acid	$OCOCH(COCH_3)COCH{:}C(CH_3)$ (ring closure bracket)	$5{\cdot}30 \times 10^{-6}$

Table 36.6 Percentage of Preservative Undissociated Related to pH Value

Preservative	pH 2	pH 3	pH 4	pH 5	pH 6	pH 7
Benzoic acid	99	94	60	13	1·5	0·15
Boric acid	100	100	100	100	100	100
Dehydroacetic acid	100	100	95	65	15·8	1·9
p-Chlorobenzoic acid	99	91	52	9·7	1·06	0·107
Propionic acid	100	99	88	42·0	6·7	0·71
Salicylic acid	90	49	8·6	0·94	0·094	0·0094
Sorbic acid	—	98	86	37	6·0	0·6

repulsion from the negatively charged microbial cell wall. Benzoic acid is an excellent preservative in its undissociated form but is strongly pH-dependent, so that at pH 6 approximately 60 times as much benzoic acid is required as at pH 3. Similarly, sorbic acid is present mainly in the undissociated (active) form at pH 4, but only 6 per cent of this form is present at pH 6. A chlorine addition product of sorbic acid[42] is claimed to be more effective than sorbic acid itself and less affected by high pH.

Dehydroacetic acid enolizes to give a weak acid and has been studied extensively as a preservative, particularly for foodstuffs.[43–45] It has a very low dissociation constant which indicates that it retains activity at higher pH values than most other organic acids and for this reason it commands wider usage. At pH 6, 16 per cent of the dehydroacetic acid is in its undissociated form, which is more than benzoic acid at pH 5. The anion of dehydroacetic acid is said to be weakly antimicrobial so that at pH 7 this acid will retain some activity.[36]

Phenols, which include the Parabens, behave as weak acids and consequently are less dramatically affected by pH than stronger acids. For example, methyl Paraben at pH 8.5 is approximately 50 per cent undissociated.

The relationships between pH and effectiveness of a range of preservatives have been studied by a number of workers.[46–48] Simon[48] showed that the biological activity of weak acids is influenced in a regular manner by changes in pH; in a study of 90 pH experiments involving a wide range of acids and test organisms he showed that the relationship between pH and activity is quite general for this type of preservative. When the pH of the environment is below the pK_a, changes of pH are of little consequence, but as the pH is increased above the pK_a higher concentrations are required to produce a standard response.

Other preservatives, for example cationics, are active only in the ionized form. Activity of cetrimide increases with pH as a result of increased cellular uptake.[49] Quaternary ammonium compounds are active at alkaline pH, but activity is progressively lost at lower pH values.[39]

Some preservatives are pH-dependent by virtue of their chemical instability. For example, 2-nitro-2-bromo-propanediol (Bronopol) loses activity as a result of degradation above pH 7 more rapidly than at pH 4. On the other hand hexamethylene tetramine is stable and inactive above pH 7 since it relies on chemical breakdown with the production of formaldehyde for antimicrobial action.[39]

Cowen and Steiger[39] give the optimum pH range for a number of preservatives.

In order to utilize a preservative economically and effectively it is necessary to know whether there is a correlation between pH and activity. If this aspect of formulation had been studied more closely in the past, a great deal of money could have been saved by the many manufacturers who used preservatives which could not be effective under the pH conditions prevailing in their products.

Concentration of Preservative

There cannot be any hard and fast rules about the optimum concentrations at which various preservatives should be used. This will be obvious from the foregoing paragraphs which mention the multitude of factors that contribute to the growth of organisms in products and the effectiveness of preservatives in those environments. Some products, by virtue of the concentration of materials in their aqueous phase, are virtually self-preserving without any addition of preservatives being necessary, whereas others may provide a nutritious environment for the growth of micro-organisms and thus require a fairly high concentration of a powerful preservative.

Effective concentrations of preservatives range from as little as 0·001 per cent in the case of organic mercurial compounds to 0·5 per cent or even 1 per cent of such materials as the weak acids, depending on the pH of the product.

The availability of the preservative to the micro-organisms it is required to inhibit is probably more important than the overall concentration itself. 'Availability' in this context can be defined in accordance with the mechanism of action of the particular preservative, and may depend on the permeability across the cell wall (if this is the mechanism), the flux (if diffusion rate is important) or degree of adsorption (if the preservative acts by coating the surface of the organism). The availability of the preservative is also influenced by the distribution, or partition, of the preservative between phases of the product. Partitioning properties of preservatives are discussed later in the chapter.

There are certain advantages in using preservatives in combination rather than singly, which are as follows: (a) a broadening of the antimicrobial spectrum of activity; (b) the use of a lower concentration of each of the preservatives, thus avoiding problems of toxicity or insolubility; (c) a reduced probability of survival of an organism resistant to one of the preservatives, provided that the other preservative(s) in the system act by a different mechanism; (d) thc antimicrobial activity of the combination may be greater than the additive effects of the individual preservatives.

Frequently the esters of *p*-hydroxybenzoate are used in combination, the methyl ester in the aqueous phase of an emulsion and the propyl ester in the oil phase. Addition of the preservative to the oil phase is not so much to prevent the multiplication of organisms in this phase, as this seldom occurs, but to prevent diffusion or partition of the methyl ester from the aqueous phase into the oil phase; the presence of propyl ester in the oil phase will tend to stabilize the distribution between the phases. Bean *et al.*[50] have examined the activity of phenol against *E.coli* in oil–water dispersions and determined the distribution of phenol between oil and water in liquid paraffin and arachis oil dispersions. Determination of the extinction times of the organism in the systems showed that the bactericidal activity was governed by the concentration of phenol in the aqueous phase and the proportion of oil to water.

The Parabens have been used in combination with other preservatives, for example phenoxyethanol. This combination (marketed as Phenonip) is claimed to have a wide antimicrobial spectrum of activity, the activity being retained in the presence of surfactants and proteins.[38] A synergistic effect has been reported between Phenonip and hexachlorophene, cetylpyridium chloride, thiomersal or dichlorophene.[51] Imidazolidinyl urea has been found to act synergistically with other preservatives including methyl and propyl Parabens, increasing the antimicrobial capacity of the preservative system and the spectrum of antimicrobial activity.[36]

Synergism has also been found to occur with combinations of benzalkonium chloride or chlorhexidine with some aromatic alcohols.[52] Hugbo[53] has reported synergism between *p*-chloro-*m*-cresol and benzalkonium chloride, *m*-cresol and phenylmercuric acetate, and benzalkonium chloride and phenylmercuric acetate.

Partition Coefficient

The preservation of formulations containing oil and water is complicated by the ability of preservatives to distribute themselves between these two phases. Since micro-organisms only grow in the aqueous phase it is important that the preservative does not distribute itself in such a way as to leave an ineffective concentration in this phase. Ideally a preservative should have high water solubility and low oil solubility, that is, have a low oil–water partition coefficient. For simple systems where no emulgent is present the concentration of the preservative in the aqueous phase (C_W) can be calculated from the following equation:

$$C_W = \frac{C(\phi + 1)}{(K_w^o\phi + 1)}$$

where C is the total preservative concentration, ϕ is the oil–water ratio and K_w^o is the oil–water partition coefficient.[54] The concentration of preservative in the aqueous phase is influenced by the oil–water ratio. As a general rule, when $K_w^o < 1$ the aqueous concentration is increased by increasing the proportion of oil, and when $K_w^o > 1$ an increase in the proportion of oil decreases the aqueous concentration.

The partition coefficient itself varies with pH and the nature of the oil. Some oils are predominantly hydrocarbon, whereas others, for example vegetable oils, contain oxygen atoms. Preservatives such as chlorinated phenols form hydrogen bonds with the latter type of oil giving them a high partition coefficient and thus rendering the preservative unsuitable for systems containing this type of oil. However, chlorinated phenols are suitable preservatives for formulations based on oils that are predominantly hydrocarbon.[39] The solubilities of some preservatives in oil and water are given by Cowen and Steiger.[39]

Various workers have shown that the addition of propylene glycol to the water phase of an emulsion reduces the partition coefficient and thus makes more preservative available in the water phase. De Navarre[55] found propylene glycol to be a reliable preservative at 16 per cent in many cosmetic products and stated that its antimicrobial properties were three or four times greater than those of an equivalent amount of glycerin. Propylene glycol does not appear to act solely by virtue of its osmotic effect, showing toxic effects on some micro-organisms at high concentrations.

For systems containing oil and water phases and an emulsifying agent, the preservative concentration in the aqueous phase may be further reduced by binding or solubilization of the preservative by the surfactant. A voluminous literature exists on the inactivation of preservatives by surfactants, particularly the nonionics, and some aspects of inactivation are dealt with in a subsequent section of this chapter.

Susceptibility of Organisms to Preservative

Several nonionic surfactants, notably Tween 80, polyethylene glycol 1000 monocetyl ether and polyethylene glycol 400 laurate, have been found capable of exerting a 'protective' effect on micro-organisms. Judis[56] showed that Tween 80 protected *E. coli* from the lethal effects of *p*-chloro-*m*-xylenol by preventing in part the leakage of the cell contents, as indicated by the release of radio-labelled glutamate which had previously been added to culture media in which the organisms were grown. Similar phenomena have been observed by Wedderburn[57] using cells of different bacteria suspended in nonionic solutions and then washed thoroughly in saline before exposure to media containing 0·1 per cent of esters of *p*-hydroxybenzoate. Tween 80 and various polyethylene glycol esters protected Gram-negative organisms from the inhibitory effects of the preservative.

Interaction between Ingredients and Preservatives

Apart from chemical incompatibility between the ingredients used in products and the preservatives, physical factors—such as solubilization, adsorption or bonding with active sites—can render preservatives inactive in otherwise chemically compatible systems.

Surface-active agents. Certain cationic surfactants have strong antimicrobial properties and when they are used in combination with other antiseptics or preservatives their effect is additive. The antimicrobial effectiveness of cationics varies according to the length of the hydrophobic chain, the most effective compounds having an alkyl chain length of about twelve to fourteen carbon atoms.

Soaps and anionic surfactants exert mild antimicrobial influences at high concentrations but tend to support the growth of Gram-negative bacteria and fungi at low concentrations.

The preservation of emulsions stabilized with either soap or anionic surface-active agents has not, in general, presented many problems, because, when these agents are used as emulsifiers in creams, the concentration of surfactant in the aqueous phase is tolerably high and usually presents an environment that is hostile to the growth of micro-organisms. Nevertheless, these materials decrease the activity of many preservatives to some extent and this is the result of solubilization of the preservatives in the surfactant micelles. Below the critical micelle concentration (CMC) of a soap or anionic detergent solution, preservatives and antiseptics tend to be potentiated in their action, whereas above the CMC activity is diminished. Bean and Berry[58,59] explain the physicochemical phenomena associated with the efficiency of preservatives in aqueous solutions of soaps and anionic detergents.

Nonionic surfactants are now widely used as emulsifiers for creams and also as solubilizers for perfumes in non-emulsified products. The relationship between these materials and preservatives is thus of great importance. Nonionic surface-active agents inactivate preservatives to a far greater extent than soaps and anionic or cationic detergents and, unlike the other surfactants, most nonionics have no growth-inhibiting properties, thus increasing the necessity for adequate preservation of systems that contain them. They have no denaturing effect on bacterial proteins and many can be utilized by bacteria and fungi as sources of energy. For this reason alone the absence of an effective preservative in nonionic-containing products often becomes unpleasantly apparent in a remarkably short time. However, some nonionic surfactants, notably the more hydrophobic octyl and nonyl phenols, have been shown to possess growth-inhibitory properties.[60] Allwood[60] has shown that combinations of polyoxyethylene octyl and nonyl phenols and some antibacterials, for example 2-nitro-2-bromo-propanediol, produced synergism.

There is a large literature on the subject of nonionic–preservative incompatibility. The fundamental theory has been discussed by Kostenbauder[61,62] and has been recently developed by Kazmi and Mitchell[63–65] and others;[66,67] Manowitz,[68] Wedderburn,[69] Russell[40] and Schmolka[70] have also reviewed this subject.

The hydrophile–lipophile balance (HLB) of nonionic surfactants influences their effect on preservative efficiency. The more oil-soluble nonionics, having HLB values of about 3–6, which are often used in water-in-oil emulsions, have a greater inactivating effect on commonly used preservatives than those with higher HLB values.[71]

The mechanism of the interaction between nonionic surfactants and preservatives has attracted a great deal of attention and there is evidence to support the view that the interaction is attributable in part to micellar solubilization of preservatives by nonionic surfactants and also to complex formation. Complexes appear to be formed by hydrogen bonding between the phenolic hydroxyl group in certain preservatives and the basic oxygen in the ether group of the ethylene oxide adducts. This mechanism cannot, however, account entirely for nonionic–preservative interactions since a high degree of hydrogen bonding, and consequently inactivation, would be expected when carboxymethyl cellulose and gum tragacanth are present, but these materials do not inactivate preservatives to anything like the same extent as the high-molecular-weight polyethylene glycol ester, for example.

Coates and Richardson[72] have examined the activity of cetylpyridinium chloride in aqueous solutions of polyethylene glycol and found that, except for very high concentrations, activity is reduced by the presence of the glycol, although not as much as predicted by binding data. Since polyethylene glycol does not form micelles, these workers suggest that the interaction may be due to an attraction between the electron-deficient pyridinium ring of the antibacterial agent and the electron-rich polyether linkages of the glycol.

Preservatives that appear to be much less affected by the presence of nonionic surfactants are formaldehyde, sorbic acid, benzoic acid and dehydroacetic acid.[57]

Nonionic surfactants form micelles in aqueous solutions at very low concentrations and for this reason, when used either as emulsifiers or solubilizers, will always be present at concentrations well above their CMC. In order to be effective, preservatives must be in solution and 'available' in the aqueous phase of a product and the hydrophile–lipophile characteristics of the preservative will influence its relationship with the nonionic. The more lipophilic preservatives appear to be bound to a greater extent than the more water-soluble compounds, and Patel and Kostenbauder[73] have studied the effect of Tween 80 on methyl *p*-hydroxybenzoate and propyl *p*-hydroxybenzoate. The propyl ester was found to have a far greater affinity for Tween 80 than the methyl ester. At 5 per cent Tween 80, 22 per cent of the methyl *p*-hydroxybenzoate existed as free preservative; under equivalent conditions, only 4·5 per cent of the propyl *p*-hydroxybenzoate existed in a free state. The interaction between the nonionic surfactant cetomacrogol and benzoic acid, *p*-hydroxybenzoic acid, methylparabens, propylparabens and chloroxylenol has been investigated and partition and binding phenomena discussed by Kazmi and Mitchell.[65] Konning[66] describes the interaction between phenol or chlorocresol and polysorbate 80 in arachis oil–water systems, and discusses the effect of altering the surfactant concentration and the proportion of oil on the concentration of preservative in the aqueous phase.

Baley *et al.*[67] have investigated the bactericidal properties of some quaternary ammonium compounds in dispersed systems. The concentration of quaternary in the aqueous phase was varied by using different hydrocarbons, different concentrations of hydrocarbons and different surface-active alcohols, and it was demonstrated that the bactericidal activity corresponded to the concentration of free preservative in the aqueous phase.

Attempts have been made to describe the systems mathematically in order that the quantity of preservative required to produce effective preservation in a surfactant solution or emulsified system may be calculated. Kazmi and Mitchell[63] have derived separate equations for surfactant solutions and for emulsified systems which were found to correlate with experimentally determined values. In another paper[64] they developed the theory of capacity. In a solubilized system micelles act as a reservoir of preservative. Any loss of preservative from the aqueous phase—for example due to interaction with micro-organisms, product ingredients or packaging—will lead to an adjustment of preservative concentration in the other phases until the equilibrium is re-established. A method for calculating the capacity of surfactant solutions and emulsions is given. However, since these systems are complex and many variable factors are involved, the concentration of preservative given by mathematical considerations can be regarded only as a starting concentration which must be subjected to microbiological evaluation within the product.

Schmolka[70] has investigated the effects of nonionic surfactants on cationic antimicrobial agents and gives methods by which the interaction can be influenced to produce a satisfactorily preserved product.

Parker[74] has reviewed the methods that have been used to measure the interaction between preservatives and macromolecules used in creams.

Hydrophilic polymers, including high-molecular-weight polyethylene glycols, gum tragacanth, methyl cellulose, carboxymethyl cellulose and polyvinyl pyrrolidone, have only a marginal effect in reducing the efficiency of the majority of preservatives. Quaternary ammonium compounds lose activity in the presence of lanolin and methylcellulose.[36] Figure 36.1 summarizes some of the work showing the degree of binding of methyl *p*-hydroxybenzoate by various macromolecules.

Several workers have reported the successful addition of certain materials to the aqueous phase of emulsions to minimize the inactivating effect of nonionic surfactants on preservatives. Materials such as propylene glycol, glycerin and hexylene glycol change the partition coefficient of the preservative between the phase of the emulsion, thus making more preservative available in the aqueous phase. Ethyl alcohol, propanediol, butanediol and methyl-pentanediol have also been reported as being useful for this purpose.[75]

Influence of Solid Particles on Preservatives. A large number of different insoluble solids are used in cosmetic and toilet preparations. These include talc, kaolin, titanium dioxide, tartaric acid, zinc oxide and chalk, as well as the insoluble solids used to colour creams and the natural and synthetic pigments, all of which present surfaces on to which adsorption of preservative will occur. The extent of this adsorption depends upon the nature of the solid, the type of preservative and the pH of the system. For any particular solid, knowledge of the surface electrical charge under particular conditions in the product, of the total surface area presented to the aqueous phase and of any ion exchange mechanisms which might operate, should enable reasonable predictions to be made about the quantity of preservative lost to the surface.

McCarthy[76] has studied the effects of particulate solids on various commonly used preservatives and has quantified the loss of active material due to adsorption and to factors associated with changes in pH. Clarke and Armstrong[77] showed that kaolin will adsorb benzoic acid and that the extent of the adsorption can be adjusted by regulation of the pH.

Surfactants adsorb on to solid surfaces so that the order of addition of ingredients during manufacture can influence preservative adsorption. If the preservative is dissolved in a slurry carrying the suspended solid particles,

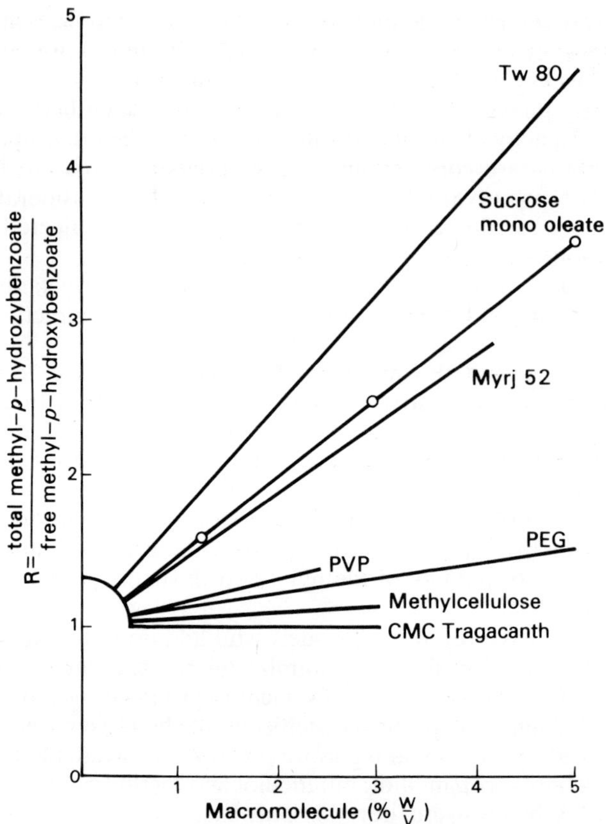

Figure 36.1 Ratio of total to free preservative concentration in the presence of different concentrations of macromolecules (*R* value of 1·0 corresponds to no measurable binding[61])

greater adsorption will occur than if it is added after the particle surfaces have become coated with surfactant.

The activity of a preservative may be reduced by interaction with, or loss through, the container or closure. Interaction of preservatives with rubber has been well documented.[4,40] There is increasing interest in the interactions between preservatives and plastics. Phenolic and quaternary ammonium compounds have been reputed to react with polyurethane.[39] Parabens, benzoic, sorbic and salicylic acids are taken up by nylon, polyvinylchloride and polyethylene.[78]

Selection of a Preservative

Although it is not possible to lay down a set of 'rules' to which a formulator should adhere, weeks of work can often be saved by considering on a theoretical basis the factors that are likely to influence the preservation of a new product.

This approach, coupled with simple laboratory tests on various combinations of the formula components, while not a substitute for thorough microbiological testing of the finished formula, can save time and frustration.

The complexity of modern formulae often means that there is a variety of materials present, some of which will act in favour of good keeping qualities while others will act against whatever preservative is chosen; the relative hostility of micro-organisms or the nutritive value of the formula itself is also of importance.

Obvious steps to follow before selecting a preservative are listed below.

1. Check ingredients for the likelihood of contamination (for example water, materials of natural origin, packaging, etc.).
2. Consider which materials might provide sources of energy for microbial growth (for example glycerin, sorbitol, etc., at concentrations below 5 per cent; nonionic surfactants at almost any useful concentration; soaps and anionic surfactants at concentrations below about 15 per cent, proteins, carbohydrates, cellulose derivations and natural gums).
3. Determine the pH of the aqueous phase of the product before attempting to use any of the preservatives that depend strongly on being in an undissociated form for their activity (see Table 36.6 for percentages of preservatives undissociated related to pH). Consider changing the pH to provide enhanced antimicrobial activity.
4. Depending on the ratios of water and oil present in the formula, estimate whether certain preservatives will be partitioned between the two phases, possibly leaving insufficient in solution in the aqueous phase to be effective. Decide whether any of the materials in solution in the aqueous phase are likely to reduce the partition coefficient (for example propylene glycol and hexylene glycol) and thus tend to help the effectiveness of the preservative or alternatively, increase the partition coefficient (for example surface-active agents), thus reducing its effectiveness. Consider the possibility of adding agents which will alter the partition coefficient or the CMC; for example, urea increases the CMC of nonionic surfactants, thus reducing the number of miscelles and the degree of preservative inactivation.[70]
5. As a guide, estimate the approximate ratio of total to free preservative in the presence of macromolecules in the formulation, and multiply the normally effective concentration by the appropriate factor (see Table 36.7). Formulae derived by Kazmi and Mitchell,[63] as described earlier, may be used to estimate the total concentration of preservative to be added to solubilized systems or emulsions stabilized by nonionic surfactants to give the required concentration of free preservative in the aqueous phase. Equations relating preservative capacity to surfactant concentration and the interaction between surfactant and preservative may also give useful data.[64]
6. Choose the least toxic of the possible preservatives for the system, in order that sufficient quantity can be included to provide a long shelf-life.

Although it is tempting to consider first in any new formula those preservatives that have stood the test of time and are well known, several less obvious choices should be tried in systems which appear difficult to preserve. Some such

Table 36.7 Approximate Ratios of Total to Free Preservative in the Presence of Macromolecules

Preservative	2% Tween 80	5% Tween 80	2% Myrj 52	5% Myrj 52	2% PEG 4000	5% PEG 4000	2% Methyl cellulose	5% Methyl cellulose
	Factor by which normal concentration of preservatives should be multiplied							
Methyl *p*-hydroxybenzoate	2·5	4·5	2·0	3·0	1·2	1·5	1·05	1·25
Ethyl *p*-hydroxybenzoate	5·0	11·0	3·0	5·0	1·3	1·6	—	—
Propyl *p*-hydroxybenzoate	12·5	27·0	6·0	13·5	1·4	1·7	—	—
Butyl *p*-hydroxybenzoate	30·0	63·0	18·0	40·0	—	—	—	—
Phenol	1·6	2·5	—	—	1·2	1·25	—	—
Sorbic acid	1·8	2·9	1·7	2·7	1·1	1·2	—	—
Cetylpyridinium chloride	38·0	60·0	—	—	—	—	—	—
Benzalkonium chloride	3·0	5·5	—	—	—	—	—	—

materials are:

Bronopol (2-bromo-2-nitro-1,3-propanediol) which, according to Croshaw *et al.*,[79] is active at low concentrations against *Pseudomonas* species, is only slightly reduced in activity by nonionics and has low toxicity.

Chlorhexidine (*bis*(*p*-chlorophenyl-diguanido)hexane), is a wide-spectrum antimicrobial agent with a good record of safety.

Dehydroacetic acid, which is suitable for formulae of low pH, is relatively unaffected by the presence of high levels of nonionic emulsifiers and appears to be safe for use on the skin.

Imidazolidinyl urea, which is not pH-dependent, has high water solubility, is non-toxic, non-irritating and non-sensitizing. It is active against Gram-positive and Gram-negative bacteria, but selectively active against yeasts and moulds. It retains its activity in the presence of many cosmetic ingredients, including surfactants, and frequently acts synergistically with other preservatives, for example Parabens.[36]

Mixtures of preservatives that are effective against different micro-organisms are also often useful. For example *β-p*-phenoxy-ethyl alcohol, which is highly active against Gram-negative bacteria and fungi, can be used with a quaternary ammonium compound such as benzalkonium chloride which acts against Gram-positive bacteria at very high dilutions. In addition there are advantages in using combinations which are not only active against a wide range of organisms but also act synergistically, for example imidazolidinyl urea and Parabens.

Safety Aspects

Preservatives are commonly expensive ingredients and it is always advisable to use the lowest effective concentration. The cost, however, is secondary to the more important question of safety to the consumer.

If tests to measure the effectiveness of certain preservatives show that several times the usual concentration is required to achieve the desired antimicrobial action (because of increased partitioning into the non-aqueous phase, physico-chemical binding or factors influencing dissociation), it is wise to consider the toxicity of the preservative at the higher concentration before proceeding. Although a preservative may be partially bound in a product and the remaining fraction in the aqueous phase may represent no more than is safely used in other formulae, the ratio of bound to free preservative is unlikely to remain unchanged when the product is actually in use. Thus from a toxicity point of view the *total* amount is important rather than only that fraction which is acting as a preservative in the particular vehicle. Application of the product to the skin, for example, will disturb the original preservative equilibrium between the various phases of the product and will almost certainly result in liberation of the preservative previously bound. Evaporation of water will increase the concentration of the preservative available to the skin and may result in primary irritation or, in some cases, sensitization.

However, rather than a sharp dividing line between a toxic and a non-toxic concentration of preservative there is, instead, a reasonably continuous spectrum of toxicity, ranging from the very low concentrations to which a few people may show an adverse reaction, to the high levels where both primary irritance and allergic responses will be more numerous. The toxicology of the *p*-hydroxybenzoate esters has been throughly studied and no primary irritation following their use at concentrations up to about 0·3 per cent has been reported. Levels of between 5 and 10 per cent have been used in powders, ointments and solutions to treat athlete's foot and, even at these levels, adverse reactions have not been numerous. Cases of sensitization to the *para*-hydroxybenzoate esters have been reported by Hjorth and Trolle-Lessen[80,81] and also by Sarkany.[82] They showed that it was necessary to use higher concentrations of the preservative than are normally employed in products in order to identify true skin allergies in their standard 48-hour patch test. These results were subsequently confirmed by Schorr[83] in the USA, who found that the allergy usually occurred when chronic contact dermatitis caused by other chemicals was being treated with creams and lotions containing *para*-hydroxybenzoate esters. After an extensive study he concluded that these preservatives are relatively safe in comparison with others commonly used on the basis of their sensitization potential. However, saturated solutions may cause irritation of the eye.[36]

Sorbic and benzoic acid have also been used in products at concentrations far in excess of those required for normal preservation. Benzoic acid appears to have a reasonably clear bill of health but sorbic acid has caused primary irritation characterized by erythema and itching at concentrations below 0·5 per cent.[84]

Sensitization to sorbic acid has been reported by Hjorth and Trolle-Lassen,[80] a well as by Schorr[83] who carried out a large comparative study. In this he found that the incidence of sensitization to sorbic acid was somewhat lower than to the esters of *para*-hydroxybenzoic acid and concluded that if used at concentrations of about 0·2 per cent it was unlikely to constitute a safety hazard.

There is little in the literature to indicate adverse effects brought about by dehydroacetic acid; its widespread use as a food preservative, and the fact that it is relatively little affected by the presence of nonionic emulsifiers, indicate that it merits more consideration as a toilet preparation preservative than it has commanded in the past.

The organic mercury compounds are, of course, recognized poisons. Although they have for a number of years apparently been safely used at concentrations below 0·01 per cent, they present a toxicity hazard to those who have to handle them in concentrated form in factories. Several leading scientists have advised against the use of phenyl-mercuric acetate, borate, and nitrate and also methiolate, on the grounds of their ability to penetrate the skin and endanger the liver and kidneys.

In the USA the Food and Drug Administration has ruled[85] that mercurial preservatives should not be used in cosmetics. An exception is made in the case of eye-area cosmetics because the mercury compounds (up to a limit of 50 ppm) are very effective in preventing *Pseudomonas* contamination; *Pseudomonas* infection of the eye can lead to serious injury or blindness.

The quaternary ammonium compounds have been extensively tested for skin irritation and sensitizing properties. At concentrations below 0·1 per cent most of those commonly used as preservatives appear to cause little or no irritation;

higher concentrations can cause erythema and drying of the skin. Cases of sensitization to cetrimide at concentrations of about 1 per cent have been reported by Cruickshank and Squire.[86] Their substantivity to the human skin has caused concern about plant safety.[87]

Formaldehyde is well known to be a skin irritant, and for this reason, and for reasons of volatility and odour, it has not been used extensively as a preservative. In some countries such as Japan and Sweden its use is banned.[36] Slow formaldehyde-release agents appear to possess less sensitizing potential.[39]

The toxic thresholds of preservatives will depend not only upon the concentrations at which they are used but also upon the vehicle. A certain concentration of a particular preservative may be quite harmless in one system while the same level might evoke adverse skin responses in another because of the presence of substances which increase its penetration through the skin.

Tests for Preservative Effectiveness

Initial Screening Tests

Agar plate tests can be used to obtain a rough indication of whether a particular preservative is likely to be effective but the results should be interpreted with caution.

The test normally involves the use of agar plates seeded with a variety of micro-organisms; wells are cut in the seeded agar and small amounts of the product under test are placed in the wells before incubation of the plates. Gram-negative and Gram-positive bacteria, together with fungi typical of those which frequently contaminate toilet preparations, are listed in Table 36.8 and confirmatory tests for preservation should be carried out using organisms of these or similar kinds.

In this type of test some of the preservative will inevitably diffuse into the agar leaving a lower concentration than was originally present in the product in the well. Although this kind of test can give a rapid indication of whether the preservative shows any likelihood of being effective in the product, it is not by any means definitive and should not be used as a substitute for longer-term and more rigorous evaluation.

Table 36.8 Typical Contaminants of Toilet Preparations

Gram-positive bacteria	Gram-negative bacteria
Staphylococcus aureus	*Pseudomonas aeruginosa*
Streptococcus mitis	*Escherichia coli*
Bacillus subtilis	*Enterobacter aerogenes*
Fungi	**Yeasts**
Aspergillus niger	*Monilia albicans*
Penicillium chrysogenum	*Candida*
Alternaria solani	

Because all the factors governing preservative effectiveness are not at present completely understood, it is almost impossible to predict with certainty whether a particular preservative in a particular system will be efficient. Consequently during the fairly early stages of formulation some microbiological testing must be undertaken to check the compatibility of the preservative in the new system.

Inoculation Tests

Methods by which known numbers of bacteria or fungi are introduced into the product and samples taken at intervals to estimate survival are by far the most reliable. Various test procedures have been proposed[6,88,89] and have been reviewed by Cowen and Steiger.[90] In general, ascertaining the resistance of a product to bacterial contamination involves inoculating a suitable size sample (for example 10 g) of the product with the test organism to give a final concentration of 10^5-10^7 organisms g^{-1}. The number of survivors in the sample is determined at intervals after storage at room temperature. The standard which must be met in order that a product may be considered effectively preserved varies depending on its intended use. Differences in opinion regarding interpretation of results are reflected by the different standards set by the USP test, the Society of Cosmetic Chemists test and the Toilet Goods Association test.

Longer-terms tests, in which smaller numbers of micro-organisms are used and the samples are observed for changes in their physical characteristics over several months, may be more meaningful. Products inoculated with spore-bearing organisms can only be observed for physical changes, as microbiological sampling is unrealistic since spores which might remain dormant in the product will germinate when transferred to nutrient culture medium. In samples inoculated with vegetative organisms, a gradual diminution in numbers can be traced over a period of time if the preservative is effective, but with fungi and spore-bearing organisms one can only wait for the appearance of visible spoilage and this sometimes takes several months to occur.

Inoculation tests have been known to produce misleading results when organisms, artificially introduced into a product, have died off in a short time and the product has been considered to be adequately preserved. Then, later, perhaps during manufacture at the factory or on storage, the product has shown the effects of contamination. One reason for this is that the wrong types of organism were used for the test, and time and the right conditions were not provided for adaptation of the organisms to their environment. Test organisms should be chosen to represent the types of organism that are known to be frequent product contaminants, for example *Pseudomonas* species, and those with which the product is likely to come into contact, for example *Staphylococcus* species. In addition, organisms isolated from the manufacturing environment, raw materials and, where possible, from contaminated products (preferably of the same or similar formula) should be used. In the testing of shampoos, for example, tap water provides a suitable source of test organisms.[8]

The maintenance of test organisms is crucial since their resistance is influenced markedly by the medium on which they are grown. To ensure suitable resistance, test organisms may be grown in a medium containing the preservative or product in low concentration.[91] Product contaminants may be kept in unpreserved or inadequately preserved preparations.[88,92]

Mixed cultures may be used initially to reduce the amount of testing required to assess the adequacy of the preservative system, whilc pure culture challenge may be employed to give more detailed information about preservative adequacy against specific organisms.[93]

Most tests run for a minimum of 28 days, the product being sampled for viable organisms at various intervals, depending on the probable frequency of usage during this period. Slow adaptation of micro-organisms to their environment makes it essential to test for long enough to determine whether inoculated bacteria and fungi will grow after a dormant period, and in some instances tests lasting as long as six months may be too short.

Tests should also be performed on products that have been stored for specified time intervals at temperatures and humidities which the product is likely to meet during use, in order to ensure that adequate preservative activity is retained throughout the shelf-life.[92] A concurrent chemical assay of the preservative gives additional valuable information.

Products may be tested using a single inoculation or using an inoculation–sampling cycle.[8] The latter method, in which the sample is subjected to more than one challenge, has been advocated by several workers[90] since it is considered to be more representative of in-use conditions and has the advantage of indicating at what point the preservative system will fail. A criticism of the capacity test is that it may lead to excessive preservation, with consequential use of dermatologically unsafe preservative concentrations. The balance between over-preserving and under-preserving a product will depend on the number of challenges, which can only be chosen after a period of experimentation.[90]

In all cases the challenge should take place with the product in the container in which it will be used by the consumer.[92]

Current UK Regulations Relating to Microbial Quality Control of Cosmetics and Use of Preservatives

Microbial limits and guidelines for cosmetic products have been issued by several national trade associations including the Cosmetic, Toiletry and Perfumery Association (CTPA) in the UK. In 1967 the Society of Cosmetic Chemists of Great Britain appointed a working party to draw up a Code of Practice for the cosmetics industry.[6] This Code of Practice, which was published in 1970, covers many aspects of quality assurance including design and operation of manufacturing processes to minimize contamination, microbiological testing procedures and factors influencing choice of preservatives. The working party also initiated several surveys of microbiological quality of UK products; based on this information, the CTPA published in 1975 a set of recommended microbiological limits and guidelines.[94] These indicate that cosmetics and toiletries should comply with the following limits:

Aerobic bacteria	Less than 1000 colony-forming units per g or ml.

Yeasts and moulds	Less than 100 colony-forming units per g or ml.
Products intended for use in the eye area and for use on babies	Less than 100 colony-forming units per g or ml.
Harmful organisms	Wherever significant numbers of colonies are observed, the presence of harmful organisms should be excluded.

Preservatives

A proposed amendment to the EEC cosmetics directive[95] contains a list of preservatives for use in cosmetics and toiletries. This list is divided into two parts. The first part includes eleven preservatives that are considered acceptable for use in cosmetics, while the second part is a 'provisional list' of 56 preservatives which may be considered acceptable subject to provision of further information regarding their safety over the period up to 1982. At the time of writing the amendment has not been incorporated into the directive.

REFERENCES

1. Rodgers, J. A., Berka, S. Y. and Artest, E. G., *Developments in Industrial Microbiology*, Vol. 15, New York, Plenum Press, 1974, p. 217.
2. Dunnigan, A.P., *Drug. Cosmet. Ind.*, 1968, **102**(6), 43.
3. Duke, A. M., *J. appl. Bact.*, 1978, **44,** Sxxxv.
4. Allwood, M. C., *J. appl. Bact.*, 1978, **44,** Svii.
5. Halleck, F. E., *Developments in Industrial Microbiology*, Vol. 12, New York, Plenum Press, 1971, p. 155.
6. Van Abbé, N. J., Dixon, H., Hughes, O. and Woodroffe, R. C. S., *J. Soc. cosmet. Chem.*, 1970, **21,** 719.
7. Smart, R. and Spooner, D. F., *J. Soc. cosmet. Chem.*, 1972, **23,** 721.
8. Flawn, P. C., Malcolm, S. A. and Woodroffe, R. C. S., *J. Soc. cosmet. Chem.*, 1973, **24,** 229.
9. Breach, G. D., *J. Soc. cosmet. Chem.*, 1975, **26,** 315.
10. Sokolski, W. T., Chidester, C. G. and Honeywell, G. E., *Developments in Industrial Microbiology*, Vol. 3, New York, Plenum Press, 1962, p. 179. .
11. Baker, J. H., *J. Soc. cosmet. Chem.*, 1959, **10,** 133.
12. Wolven, A. and Levenstein, I., *TGA Cosmet. J.*, 1969, **1,** 34.
13. Dunnigan, A. P. and Evans, J. R., *TGA Cosmet. J.*, 1970, **2,** 39.
14. Wilson, L. A., Kuehne, J. W., Hall, S. W. and Ahearn, D. G., *Am. J. Ophthal.*, 1971, **71,** 1298.
15. Wolven, A. and Levenstein, I., *Am. Cosmet. Perfum.*, 1972, **87,** 63.
16. Myers, G. E. and Pasutto, F. M., *Can. J. pharm. Sci.*, 1973, **8,** 19.
17. Ahearn, D. G., Wilson, L. A., Julian, A. J., Reinhardt, D. J. and Ajello, C., *Developments in Industrial Microbiology*, Vol. 15, New York, Plenum Press, 1974, p. 211.
18. Jarvis, B., Reynolds, A. J., Rhodes, A. C. and Armstrong, M., *J. Soc. cosmet. Chem.*, 1974, **25,** 563.
19. Baird, R. M., *J. Soc. cosmet. Chem.*, 1977, **28,** 17.
20. Ahearn, D. G. and Wilson, L. A., *Developments in Industrial Microbiology*, Vol. 17, New York, Plenum Press, 1976, p. 23.

21. Morse, L. J., Williams, H. L., Green, F. P., Eldridge, E. E. and Rotta, J. R., *New Engl. J. Med.*, 1967, **277,** 472.
22. Morse, L. J. and Schonbeck, L. E., *New Engl. J. Med.*, 1968, **278,** 376.
23. Sykes, G., Microbiological control during manufacture, Society of Cosmetic Chemists of Great Britain Symposium, *Microbiological Safety of Toiletries and Cosmetics,* February 1976.
24. Goldman, C. L., *Drug Cosmet. Ind.*, 1975, **117**(7), 40.
25. Favero, M. S., Carson, L. A., Bond, W. W. and Peterson, N. J., *Science*, 1971, **173,** 826.
26. Malcolm, S. A. and Woodroffe, R. C. S., *J. Soc. cosmet. Chem.*, 1975, **26,** 277.
27. Bruch, C. W., *Drug Cosmet. Ind.*, 1972, **110**(6), 32.
28. Davis, J. G., *J. Pharm. Pharmacol.*, 1960, **12,** Suppl.29T.
29. Anon, *Drug Cosmet. Ind.*, 1968, **103,** (12), 53.
30. De Gray, R. J. and Killian, L. N., *Developments in Industrial Microbiology*, Vol. 3, New York, Plenum Press, 1962, p. 296.
31. Beveridge, E. G., *Microbial Aspects of the Deterioration of Materials*, London, Academic Press, 1975, p. 213.
32. Bennett, E. C., *Developments in Industrial Microbiology*, Vol. 3, New York, Plenum Press, 1962, p. 273.
33. Barr, M. and Tice, L. F., *J. Am. pharm. Assoc. sci. Ed.*, 1957, **46,** 480.
34. Sawyer, C. N. and Ryckman, D. W., *J. Am. Water Wks Assoc.*, 1957, **49,** 480.
35. Yu-chih Hsu, *Nature*, 1965, **207,** 385.
36. Rosen, W. E. and Berke, P. A., *J. Soc. cosmet. Chem.*, 1973, **24,** 663.
37. Gucklhorn, I. R., *Manuf. Chem. Aerosol News*, 1969–1971, **40–42** (*passim*).
38. Croshaw, B., *J. Soc. cosmet. Chem.*, 1977, **28,** 1.
39. Cowen, R. A. and Steiger, B., *Cosmet. Toiletries*, 1977, **92,** 15.
40. Russell, A. D., *Microbios*, 1974, **10,** 151.
41. Bean, H. S., *J. Soc. cosmet. Chem.*, 1972, **23,** 703.
42. British Patent 998 189, Farbwerke Hoechst AG, 5 October 1960.
43. Wolf, P. A. and Westveer, W. M., *Arch. Biochem.*, 1950, **28,** 201.
44. Von Schelhorn, M., *Dtsch. Lebensm. Rundsch.*, 1952, **48,** 15.
45. Von Schelhorn, M., *Fol. Techol., Champaign*, 1953, **7,** 97.
46. Bandelin, F. J., *J. Am. pharm. Assoc. sci. Ed.*, 1958, **47,** 691.
47. Entrekin, D. N., *J. pharm. Sci.*, 1961, **50,** 743.
48. Simon, J., *New Phytol.*, 1952, **51–52,** 163.
49. Sykes, G., *Disinfection and Sterilization*, London, E. & F. N. Spon, 1965.
50. Bean, H. S., Richards, J. P. and Thomas, J., *Boll.chim.-farm.*, 1962, **101,** 339.
51. Boehm, E. E. and Maddox, D. N., *Am. Perfum. Cosmet.*, 1970, **85,** 31.
52. Richards, R. M. E. and McBride, R. J., *J. pharm. Sci.*, 1973, **62,** 2035.
53. Hugbo, P. G., *Can. J. pharm. Sci.*, 1976, **11,** 17.
54. Bean, H. S., Konning, G. H. and Malcolm, S. A., *J. Pharm. Pharmac.*, 1969, **21** 173S.
55. De Navarre, M. G., *Chemistry and Manufacture of Cosmetics*, Vol. 2, 2nd edn, Princeton, NJ, Van Nostrand, 1962, p. 257.
56. Judis, J., *J. pharm. Sci.*, 1962, **57,** 261.
57. Wedderburn, D. L., *J. Soc. cosmet. Chem.*, 1958, **9,** 210.
58. Bean, H. S., and Berry, H., *J. Pharmacol.*, 1951, **3,** 639.
59. Bean, H. S. and Berry, H., *J. Pharm. Pharmacol.*, 1953, **5,** 632.
60. Allwood, M. C., *Microbios*, 1973, **7,** 209.
61. Kostenbauder, H. B., *Am. Perfumer. Arom.*, 1960, **75**(1), 28.
62. Kostenbauder, H. B., *Developments in Industrial Microbiology*, Vol. 3, New York, Plenum Press, 1962, p. 286.
63. Kazmi, S. J. A. and Mitchell, A. G., *J. pharm. Sci.*, 1978, **67,** 1260.

64. Kazmi, S. J. A. and Mitchell, A. G., *J. pharm. Sci.*, 1978, **67,** 1266.
65. Kazmi, S. J. A. and Mitchell, A. G., *J. Pharm. Pharmacol.*, 1971, **23,** 482.
66. Konning, G. H., *Can. J. pharm. Sci.*, 1974, **9,** 103.
67. Baley, G. J., Peck, G. E. and Banker, G. S., *J. pharm. Sci.*, 1977, **66,** 696.
68. Manowitz, M., *Developments in Industrial Microbiology*, Vol. 2, New York, Plenum Press, 1962, p. 65.
69. Wedderburn, D. L., *Advances in Pharmaceutical Science*, Vol. 1, London, Academic Press, 1964, p. 196.
70. Schmolka, I. R., *J. Soc. cosmet. Chem.*, 1973, **24,** 577.
71. Tilbury, R. H., *Specialities*, 1965, **1**(11), 3.
72. Coates, D. and Richardson, G., *J. appl. Bact.*, 1973, **36,** 240.
73. Patel, N. K. and Kostenbauder, H. B., *J. Am. pharm. Assoc. sci. Ed.*, 1958, **47,** 289.
74. Parker, M. S., *J. appl. Bact.*, 1978, **44,** Sxxix.
75. Poprzan, J. and De Navarre, M. G., *J. Soc. cosmet. Chem.*, 1959, **10,** 81.
76. McCarthy, T. L., *J. Mond. Pharm.*, 1969, **4**(12), 321.
77. Clarke, C. D. and Armstrong, N. A., *Pharm. J.*, 1972, **211,** 44.
78. Armstrong, N. A., *Am. Cosmet. Perfum.*, 1972, **87,** 45.
79. Croshaw, B., Groves, M. J. and Lessel, B., *J. Pharm. Pharmacol.*, 1964, **16,** suppl. 127T.
80. Hjorth, N. and Trolle-Lassen, C., *Trans. Rep. St. John's Hosp. Derm. Soc., Lond.*, 1963, **49**(10), 127.
81. Hjorth, N., *Acta derm.-vener., Stockh.*, 1961, **41,** (Suppl. 46), 97.
82. Sarkany, I., *Br. J. Dermatol.*, 1960, **72**(10), 345.
83. Schorr, W. F., *Am. Perfum. Cosmet.*, 1970, **85,** 39.
84. Fryklof, L. E., *J. Pharm. Pharmacol.*, 1958, **10,** 719.
85. Federal Register, 37 F.R. 12967, 30 June 1970.
86. Cruickshank, C. N. D. and Squire, J. R., *Br. J. ind. Med.*, 1949, **6,** 164.
87. Smith, J. L., *Cosmet. Toiletries*, 1977, **92,** 30.
88. Halleck, F. E., *TGA Cosmet. J.*, 1970, **2**(1), 20.
89. *United States Pharmacopeia* XIX, Easton, Pa., Mack Publishing Company, 1975.
90. Cowen, R. A. and Steiger, B., *J. Soc. cosmet. Chem.*, 1976, **27,** 467.
91. Cowen, R. A., *J. Soc. cosmet. Chem.*, 1974, **25,** 307.
92. Moore, K. E., *J. appl. Bacteriol.*, 1978, **44,** Sxliii.
93. Cosmetic Toiletry and Fragrance Association Preservation Sub-Committee, *CTFA Cosmet. J.*, 1973, **5**(1), 2.
94. Cosmetic, Toiletry and Perfumery Association Ltd, *Code of Good Practice for the Toiletry and Cosmetic Industry; Recommended Microbiological Limits and Guidelines to Microbiological Quality Control*, 1975.
95. *Off. J. European Communities*, 1979, No. C 165/52, 2 July.

Chapter Thirty-seven

Antioxidants

Introduction

The ability of atmospheric oxygen to act as an oxidizing agent for fats, fatty acids and many other organic materials is of commercial importance. In some cases the phenomenon may be advantageously employed, but in cosmetics the effects of oxidation are normally deleterious and can lead to complete spoilage. Although the literature on the chemical and physical changes involved in oxidation is extensive and can be traced back to the eighteenth century[1] it is only in the past thirty-five years that the mechanisms involved have been understood with any degree of certainty.

Two of the problems associated with an understanding of the general oxidation reactions have been the very wide spectrum of organic materials which are subject to this type of decomposition and, secondly, the large number of factors which can effect both the rate and course of the reactions. Amongst these latter may be numbered the effects of humidity, oxygen concentration, temperature, UV irradiation and the presence or absence of anti- and pro-oxidants. In the early studies of oxidative reactions many of the above environmental conditions were not recognized as being important and were therefore not controlled, so that the results obtained are in many cases invalidated. Another problem with the generalized oxidation theories which were developed from this early work was that it was found difficult to apply the results from simple reference substances such as methyl oleate or methyl linoleate to more complex naturally occurring materials such as sunflower seed oil or soya bean oil, particularly at the high temperatures sometimes reached during the purification or processing of these products. Other difficulties were encountered when the general oxidation scheme was applied to organic materials other than fats and their derivatives.

General Autoxidative Theory

The bulk of development work on a generalized theory of autoxidation was concerned with the reactions of olefinic materials and consequently the study of the reactivity of the carbon–carbon double bond is of paramount importance. This reaction was first studied by Schönbein[2] using almond oil; it was not until almost half a century later, when the effect of organic peroxides on oxidation was studied by Engler and his co-workers,[3,4] that the modern oxidative theory began to be formulated. Engler believed that the reaction was due to molecular rather than atomic oxygen producing a substance $R_1—O—O—R_2$ which in turn could oxidize other oxidizable substances when the loosely held 'molecular activated' oxygen was released. This led Fahrion[5] to assume the formation of a cyclic peroxide which could at later stages of the oxidative reaction rearrange to form dihydroxyethylenic or hydroxyketo compounds along with more stable ethers. Until the development of the modern chain reaction theory of oxidation this ring peroxide theory was generally accepted by most workers in the field. However, the mechanism was never fully developed and increasingly it became apparent that the autoxidation mechanism was more complicated.

At a fairly early date Staudinger[6] had suggested, from studies using *asym*-diphenylethylene, $(C_6H_5)_2C{=}CH_2$, that ring peroxide formation was the second stage of the reaction and that primary oxidation involved the addition of a molecule of oxygen across the ethylene bond to form a moloxide which subsequently rearranged to the cyclic peroxide:

$$(C_6H_5)_2\,C{=}CH_2 \rightarrow (C_6H_5)_2\,\underset{\underset{\underset{\textstyle O}{\|}}{O}}{C{-}CH_2} \rightarrow (C_6H_5)_2\,\underset{O-O}{\overset{\displaystyle C-CH_2}{|\quad\;\;|}}$$

The problem with the cyclic peroxide oxidation theory was that only indirect evidence for the existence of such compounds could be obtained. Most of the evidence for their existence came from determinations of carbonyl, diene, hydroxyl, iodine, peroxide and saponification values. These were considered alongside determinations of molecular weight and the total oxygen uptake. The problem was that the analytical results obtained were unreliable because the oxidizing material was usually a non-fractionated natural product which would, of necessity, show batch-to-batch variations in the proportion of individual components and, as a consequence, exhibit variable oxidative behaviour. It was also not realized that many of the analytical techniques used were not strictly quantitative.

Alongside the ring peroxide theory existed the suggestion due to Fokin[7] that the initial stage in the oxidation was the formation at the ethylene bond of an epoxide of the type which can be isolated from many oxidizing peracids and other oxidizing mixtures, but it is doubtful if these represent primary products in the normal oxidation reaction.

The development of the modern theory of autoxidation may be said to start from the isolation of a cyclohexene peroxide by Stephens in 1928.[8] On the basis of theories then current he assumed that this was a saturated product with a possible formula:

$$\begin{array}{ccc} & CH_2 & \\ CH_2 & & CH \\ | & & |\!\!>O \\ CH_2 & & CH \\ & CH_2 & \end{array}$$

However, Farmer and Sundralingam[9] showed that this interpretation was incorrect and that Stephens' compound was an unsaturated hydroperoxide:

$$\begin{array}{ccc} & CH\text{—}OOH & \\ CH_2 & & CH \\ | & & \| \\ CH_2 & & CH \\ & CH_2 & \end{array}$$

Farmer and his co-workers[10–15] developed these observations into the now generally accepted theory of autoxidation by means of a free radical mechanism involving the formation of hydroperoxides. According to this theory the reaction involved the addition of one molecule of oxygen at a carbon atom α- to the double bond; while more recent work has suggested that this is probably not the primary reaction, it would appear that it is the reaction responsible for the propagation of the chain mechanism. Hargrave and Morris[16] have suggested that an alternative path for reaction between a peroxy radical and an olefine molecule may exist and that oxygen would be partitioned between hydroperoxide and diperoxide groupings. Breakdown of the olefin during oxidation is thought to lead also to the formation of carbonyl and carbinol compounds.

From these studies of individual oxidizable compounds, a general kinetic scheme for autoxidation was developed[17,18] and may be represented as:

$$\begin{array}{lll} \text{Initiation} & RH \xrightarrow{r_i} \text{free radicals } (R^{\cdot}, RO_2^{\cdot}, \text{etc.}) & \\ \text{Propagation} & R^{\cdot} + O_2 \xrightarrow{k_2} RO_2^{\cdot} & \\ & RO_2^{\cdot} + RH \xrightarrow{k_3} ROOH + R^{\cdot} & \\ & RO_2^{\cdot} + RO_2^{\cdot} \xrightarrow{k_4} & \\ \text{Termination} & RO_2^{\cdot} + R^{\cdot} \xrightarrow{k_5} & \text{non-propagating products} \\ & R^{\cdot} + R^{\cdot} \xrightarrow{k_6} & \end{array}$$

where RH is an olefine and $RO_2^{\cdot}$ and $R^{\cdot}$ are hydroperoxide and olefinic free radicals.

This scheme shows a typical free radical reaction mechanism where free radicals of outside origin may catalyse or inhibit the reaction and where the rate may be shown to be dependent upon the square root of light intensity.

It can be shown that the reaction $R^{\cdot} + O_2 \rightarrow RO_2^{\cdot}$ is extremely rapid, which means that at most normal oxygen pressures, the termination reactions $R^{\cdot} + R^{\cdot} \rightarrow$ and $R^{\cdot} + RO_2^{\cdot} \rightarrow$ may be ignored as the concentration of $RO_2^{\cdot}$ will greatly exceed $R^{\cdot}$.

The determined experimental rate equations for olefinic oxidation under a number of conditions showed that a common equation was applicable:

$$\text{Rate} = R_1^{1/2}k[\text{RH}] \cdot \frac{[O_2]}{k'[\text{RH}] + [O_2]}$$

where R_1 is the rate of formation of chain carriers and k, k' are constants. It has been shown that because $(k_4k_6R_1)^{1/2}$ is negligible in long-chain-length reactions there would be complete correlation between the experimental and theoretical equations:

$$\frac{-\mathrm{d}[O_2]}{\mathrm{d}t} = R_1^{1/2} \cdot \frac{k_3}{(k_4)^{1/2}} \cdot [RH] \frac{k_2(k_4)^{1/2}[O_2]}{k_3(k_6)^{1/2}[RH] + k_2(k_4)^{1/2}[O_2] + (k_4k_6R_1)^{1/2}}$$

Uri[19] has applied stationary state conditions and produced the following kinetic equations:

$$\mathrm{d}[R^{\cdot}]/\mathrm{d}t = r_1 - k_2[R^{\cdot}][O_2] + k_3[RO_2^{\cdot}][RH]$$
$$\mathrm{d}[RO_2^{\cdot}]/\mathrm{d}t = k_2[R^{\cdot}][O_2] - k_3[RO_2^{\cdot}][RH] - k_4[RO_2^{\cdot}]^2$$
$$\mathrm{d}[ROOH]/\mathrm{d}t = (r_1/k_4)^{1/2} \times (k_3[RH])$$

At low oxygen level termination reactions involving $R^{\cdot}$ become significant and these kinetics would have to be modified to take account of this fact. Uri therefore proposes a more general rate equation equivalent to that of Bolland[17] and Bateman[18] for the formation of peroxide:

$$\mathrm{d}[ROOH]/\mathrm{d}t = (r_1/k_4)^{1/2} \,.\, k_3[RH] \frac{k_2(k_4)^{1/2}[O_2]}{k_3(k_6)^{1/2}[RH] + k_2(k_4)^{1/2}[O_2]}$$

He further suggests that provided $k_6 = k_4$ this may be simplified to:

$$\mathrm{d}[ROOH]/\mathrm{d}t = (r_1/k_4)^{1/2} \,.\, k_3[RH] \frac{k_2[O_2]}{k_3[RH] + k_2[O_2]}$$

Oxidation of Mono-unsaturated Systems

Because of the slow speed of autoxidation in mono-unsaturated systems, it is normal to study the catalysed reaction. If a comparison is made between the oxidation rate of compounds with one, two and three double bonds respectively then the ratios are of the approximate order 1:12:20.[19] One of the earliest studies was by Farmer and Sutton[12] who managed to extract a pure hydroperoxide from oxidizing methyl oleate using a technique of molecular distillation and chromatography. This work was extended by other workers[20,21] to obtain hydroperoxide concentrates from oxidizing methyl oleate. Originally Farmer suggested that a variable mixture of mono- and dihydroperoxides was formed at the eighth or eleventh carbon atom of the oleate chain. Privett and Nickell[22] showed that the α-hydroperoxides predicted by the general theory are formed in approximately equal proportions. It is not certain that only hydroperoxides are formed from mono-unsaturated compound oxidation, as polarographic analysis[23] indicates a significant wave in a region not normally associated with hydroperoxides. It has been suggested that this wave corresponds to cyclic hydroperoxides.

The problem with the hydroperoxide chain mechanism for mono olefine oxidation is that a large amount of energy is required to break an α-methylenic C–H bond and Farmer[14] and other workers[24,25] suggested that there must be a small primary attack at the double bond and that the products of this attack initiated the normal α-methylenic chain reaction.

Because of resonance around the double bond the hydroperoxide can be shown to produce a double bond shift[15] in the oxidizing mono olefine.

Oxidation of Nonconjugated Polyunsaturated Systems

The rate of oxidation of nonconjugated polyunsaturated systems is very much more rapid than that of those containing only a single double bond. This is due to the activation brought about by the presence of a methylene group adjacent to two double bonds. The presence of this type of compound in the fats and oils used in cosmetics is the main source of oxidative rancidity.

While Farmer[26] had observed that in the early stages of oxidation the oxygen was used to form peroxide and that the double bonds remained unchanged, it was also soon found that the originally nonconjugated double bonds were now showing a conjugated formation[12] which was not due to the peroxide structure. The degree of conjugation found (70 per cent) was taken as evidence of random attack on the resonating free radicals from the linoleate nucleus; however, Bergström[27] found it impossible to isolate the 11-hydroxystearate from linoleate oxidation, although the hydrogenated oxidation product did allow the isolation of the 9- and 13-hydroxystearates. One of the major problems was that the UV evidence was based on comparison with *trans*, *trans*-10,12-octadecadienoic acid with an extinction coefficient of 32 000. However, it can be shown that *cis*, *trans* isomerism can occur and also that the peroxide group and its position can effect the extinction coefficient. These problems suggested that the value of 70 per cent conjugation might be incorrect. Privett[28] and co-workers suggest that up to 90 per cent of the hydroperoxide is conjugated and consists of the *cis*, *trans* isomers. However, the effect of temperature must also be considered, as at 24°C the *cis*, *trans* isomer appears to rearrange and result in the production of a more stable *trans*, *trans* form. This led Holman[29] to suggest an oxidation mechanism which accounted for these observations (see Figure 37.1).

The oxidation of these compounds is therefore by the normal general hydroperoxide chain reaction developed by Farmer, but it can be shown that an initiating free radical is required as highly purified linoleates and similar compounds exhibit a very long induction period.[30]

With linoleate radicals which contain three nonconjugated double bonds there is a more complicated resonance and in the later stages of oxidation diperoxides could be formed. However, at low temperatures approximately 60 per cent of monomeric *cis*, *trans* conjugated diene monohydroperoxides appear to be formed.[31]

Oxidation of Conjugated Polyunsaturated Systems

Although not studied to the same extent as in the previous systems, the oxidation of conjugated polyunsaturated systems is important and the available evidence suggests that this oxidative mechanism may differ in a number of respects. The oxidative products appear to be noncyclic polymeric peroxides produced by the addition of oxygen to the diene system.[32,33] The attack of oxygen on these compounds produces both 1,2- and 1,4- addition apparently alongside a proportion of α-hydroperoxides. However, fairly simple isomeric differences in the material studied appear to produce profound changes in the products of oxidation.

Thus Allen and his co-workers,[34] studying the oxidation of 9,12- and 10,12-methyl linoleate, found that whereas in the early stages of the reaction of the former compound all the oxygen was in the form of peroxide, in the oxidation of

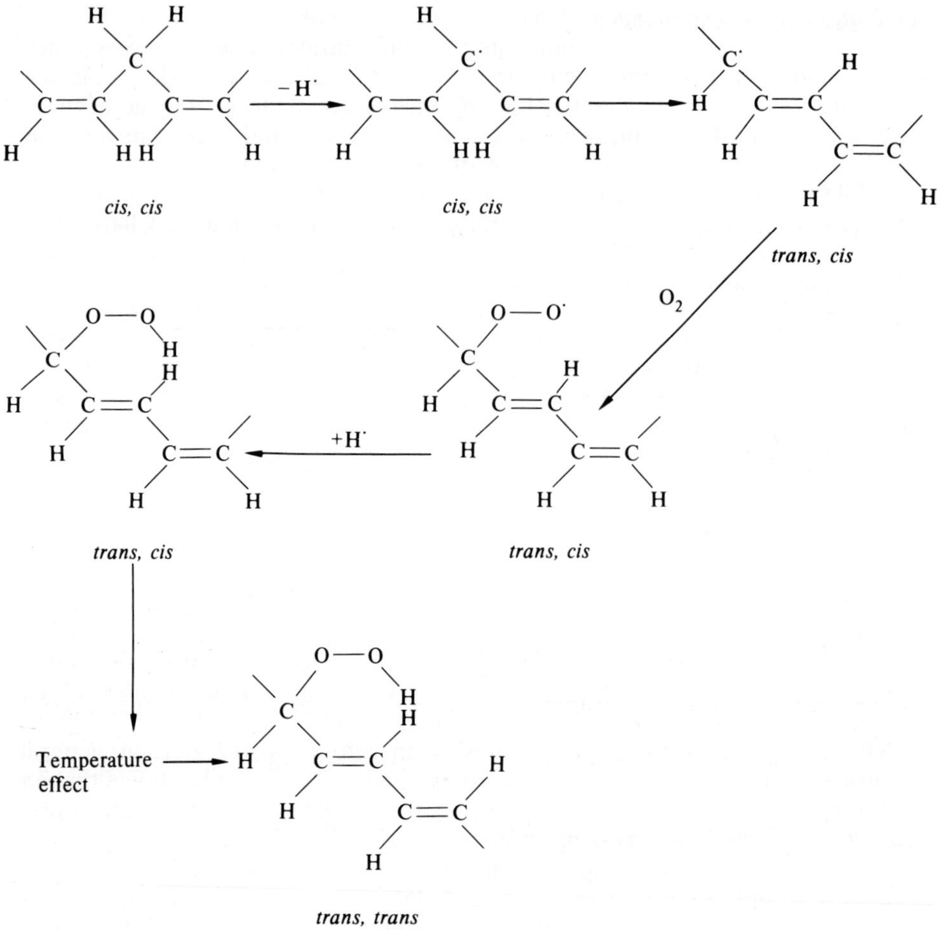

$-H^{\cdot}$

cis, cis *cis, cis* *trans, cis*

O_2

$+H^{\cdot}$

trans, cis *trans, cis*

Temperature effect

trans, trans

Figure 37.1 Oxidation of nonconjugated polyunsaturated systems (after Holman[29])

the latter compound no peroxide occurred in the early stages. The reaction was accompanied by the disappearance of double bond conjugation which was directly related to the amount of oxygen absorbed, suggesting some form of carbon-to-oxygen polymerization. Metal catalysts have also been shown to have much smaller effects on the oxidation of conjugated systems, suggesting that the decomposition of peroxides did not play a significant role in their oxidation.[35] The relative rate of oxidation of the corresponding conjugated and nonconjugated systems is open to doubt and reports have appeared suggesting both different and similar rates of oxidation, depending upon experimental conditions.[34,36,37] The formation of polymer appears to cause steric hindrance, thus reducing further oxidation, although the corresponding increase in viscosity does not, as such, prevent oxidation even though it slows down the rate of diffusion of oxygen into the system. The products of oxidation from these systems are much more variable. Eleostearate oxidation appears to produce a cyclic peroxide with formula:

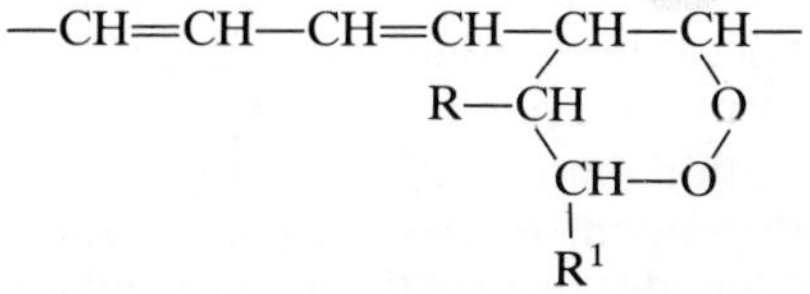

$$\begin{array}{l} —CH{=}CH—CH{=}CH—CH—CH— \\ \qquad\qquad\qquad\quad R—CH \quad\; O \\ \qquad\qquad\qquad\qquad\;\; CH—O \\ \qquad\qquad\qquad\qquad\;\; | \\ \qquad\qquad\qquad\qquad\;\; R^1 \end{array}$$

although further oxidation is possible.[38] The overall kinetics of oxidation appear to simplify to:

$$dO_2/dt = k[\text{Product}]^{1/2} \,.\, [\text{Ester}]$$

These kinetics suggest that propagation is by polymerization, but other studies have suggested an autocatalytic mechanism[36,39] while O'Neill,[40] studying the UV light oxidation of methyl eleostearate, found a complex mixture of products suggesting that both polymeric and autocatalytic processes were taking place simultaneously.

Oxidation of Saturated Systems

Under the normal conditions of oxidation 'experiments' most saturated compounds appear to be inert, but many unsaturated materials undergo a slow oxidative process at elevated temperatures. If the oxidation rates of the compounds methyl stearate: methyl oleate: methyl linoleate: methyl linolenate are compared at 100°C then the reaction ratio is 1:11:114:179.[19]

The rate of oxidation of saturated hydrocarbon compounds increases with chain length and, while the primary product is a hydroperoxide, the position of attack by the oxygen appears to be random.[41,42] As secondary products, alcohols and ketones, which can themselves undergo further oxidation, are formed at an early stage.

The oxidative behaviour of a number of saturated fatty acids and their derivatives has been studied, at elevated temperatures, by Paquot and de Goursac[43] and it was shown that under their experimental conditions approximately 50 per cent of the material oxidized. β- attack occurred to produce oxalic acid and shorter-chain-length fatty acids. Minor amounts of lactones and methyl ketones were also detected.

Autoxidation of Aldehydes

Aromatic aldehydes, and in particular benzaldehyde, appear to have been the most popular material for study. Almquist and Branch[44] showed that initially a peroxide was formed, the quantity of which subsequently decreased owing to the catalytic effect of the build-up of benzoic acid. This catalytic effect speeded up the reaction of benzaldehyde and the initially formed perbenzoic acid. These two compounds form an adduct whose rate of decomposition into benzoic acid determines the order of reaction:

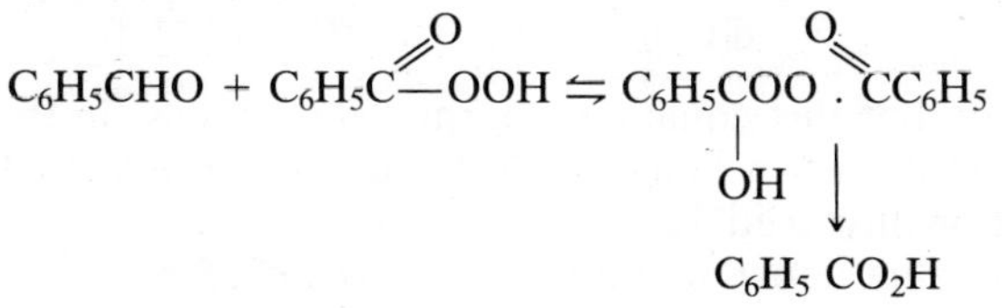

$$C_6H_5CHO + C_6H_5C({=}O){—}OOH \rightleftharpoons C_6H_5C(OH)OO\,.\,C({=}O)C_6H_5 \rightarrow C_6H_5\,CO_2H$$

The possibility that the perbenzoic acid was formed by a chain reaction which could be influenced by light was discussed by Bäckström[45] while later it was shown that heavy metal catalysis could also occur.[46] Waters and Whickham-Jones[47] found that the rate of oxidation was proportional to the square of the benzaldehyde concentration and to the square root of dibenzoyl peroxide concentration, and independent of oxygen concentration.

That other organic aldehydes oxidize through a similar free radical mechanism has been shown by Cooper and Melville[48] using decanal. They showed that the oxygen molecule reacted directly with the aldehyde and that with photoelectric initiation the light was directly responsible for the formation of free radicals.

Oxidation of Ketones

The oxidation of ketones has been less studied than that of aldehydes and appears to require a high temperature to produce decomposition. A temperature above 100°C was found by Sharp[49] to be required and the initially produced hydroperoxide rapidly decomposed to give a mixture of acids and aldehydes.

Antioxidants

General Mechanism

If the general chain propagation reactions for oxidation are valid then it is conceivable that the suppression of oxidation could occur either by suppressing the formation of free radicals or by the introduction into the system of material that would react with free radicals as they were formed, and so prevent a build-up of reaction chains. The formation of free radicals cannot be wholly prevented and therefore substances which behave as free radical acceptors—antioxidants—are important.

Bolland and ten Have[50] studied the effect of hydroquinone (AH) on the oxidation of ethyl linoleate (RH). They suggested that this compound reacted with the free radicals to give:

$$\left.\begin{array}{l} R^{\cdot} + A\dot{H} \xrightarrow{k_7} \\ R\dot{O}_2 + AH \xrightarrow{k_8} \end{array}\right\} \text{inactive products}$$

This led to the rate equation:

$$r_a = -d[O_2]/dt = r_i k_3[RH]/k_2[AH]$$

where r_a is the rate of oxidation in the presence of the antioxidant and r_i is the rate of initiation of chains, and where the reaction $R^{\cdot} + AH$ is ignored as being of little importance in the antioxidant reactions. However, Davies[51] has assumed that the free radicals initially formed are $R^{\cdot}$ and thus the rate equation in the presence of antioxidants would become:

$$r_a = -d[O_2]/dt = r_i(1 + k_3[RH]/k_2[AH])$$

The possibility that this equation is significant, at least in the early stages of oxidation, has been cast in doubt by Bolland and ten Have who showed that, when the reaction involved $R\dot{O}_2$ radicals, the value $(r_a/r_u{}^2)$[AH][RH] was a constant with a value depending on the reaction coefficients $k_4/k_3.k_2$; r_u is the rate of oxygen uptake in the absence of antioxidant. A similar treatment for reactions involving $R^{\cdot}$ radicals did not produce a constant. These authors[50,52] were able to show that with hydroquinone (now designated AH_2) the reaction occurred in two stages involving the intermediate formation of a semiquinone radical. Thus:

$$RO_2^{\cdot} + AH_2 \xrightarrow{k_9} ROOH + AH^{\cdot}$$
$$AH^{\cdot} + AH^{\cdot} \xrightarrow{k_{10}} A + AH_2$$

The rate constant (k_9) has been shown to be approximately the same for different $RO_2^{\cdot}$ radicals and is therefore of possible importance in the determination of antioxidant efficiency.[48] Except in the initial induction period and the final stages of oxidation it was found that a plot of $1/r_a$ versus time was a straight line and that the rate of initiation was composed of elements due to temperature and light intensity, so that:

$$r_i = k_i[RCHO][O_2] + I$$

where I was the light intensity.

A mechanism involving the intermediate formation of a complex between the free radical $RO_2^{\cdot}$ and the antioxidant has been proposed by Boozer and his co-workers.[53] This would be followed by a subsequent rate-controlling reaction with further $RO_2^{\cdot}$ radicals:

$$RO_2^{\cdot} + AH \xrightarrow{\text{rapid}} [RO_2 + AH]\cdot \xrightarrow[\text{rate controlling } (k_{in})]{RO_2^{\cdot}} \text{products}$$

The rate equation from this reaction would become:

$$r_a = -d[O_2]/dt = \{k_3[RH]/(r_i/k_{in}[AH])\}^{1/2}$$

These kinetics were shown to account for the kinetics of oxidation of tetralin in chlorobenzene with phenol and N-methylaniline as inhibitors and azoisobutyronitrile as initiator. However, Boozer's mechanism appears to occur only with weak antioxidants where the removal of hydrogen by the $RO_2^{\cdot}$ radical from the antioxidant AH is too slow in comparison with its removal of hydrogen from the oxidizing species RH.

Comparison of Antioxidant Efficiency

The comparison of many practical antioxidants was reviewed by Olcott,[54] Banks,[55] Lovern[56] and Lea.[57] Most of the early data were based on the extension of the induction period rather than on the kinetics of inhibition.

Rosenwald[58] has shown that the antioxidant effect per unit of concentration appears to depend on the concentration of antioxidant. An equation:

$$\log \text{induction period} = r + s \log[AH]$$

(where r and s are constants which may have either a positive or negative value) was found. Uri[59] has pointed out the danger of extrapolating comparative results obtained at elevated temperature to lower temperatures or of extrapolating results in an homogeneous phase to two-component emulsified systems.

Many chemical factors have been investigated in the comparison of antioxidant efficiency. Although it can be shown that there appears to be a correlation between oxidation-reduction potential and the antioxidant efficiency of phenols it is possible that this is fortuitous. The activation energy is a better criterion for chemical reactions, and before oxidation-reduction potential can be used it is important to show that it runs parallel with the activation energy changes. Another problem is that while some phenolic systems show well defined but reversible end products, for example quinone from hydroquinone, others have irreversible or partially irreversible end products.[60]

Bolland and ten Have[52] related antioxidant efficiency, as measured by the rate constant for the reaction $RO_2^{\cdot} + AH \rightarrow ROOH + A^{\cdot}$, to the oxidation-reduction potential and by plotting log(relative efficiency) against oxidation-reduction potential obtained an approximate straight-line relationship. They suggest that efficiency increases in proportion to a decrease in the bond dissociation energy A–H and that a limit will eventually be reached when $AH + O_2 \rightarrow HO_2^{\cdot} + A^{\cdot}$ becomes a significant reaction. This view has been challenged by Uri[59] who considers that the reaction $A^{\cdot} + O_2 \rightarrow AO_2^{\cdot}$ is far more critical.

The chemical structure, in particular the form and position of substituents, is also important when considering relative antioxidant efficiency.[51] The basic observation is that electron-repelling groups will increase antioxidant efficiency while the inclusion of electron-attracting groups in an antioxidant will decrease its efficiency. The effect of poly-substitution appears to be additive, but steric factors may interfere with direct comparison, particularly when substitution is in the *ortho* position. It must also be remembered that the addition of oxygen to a simple phenolic —O—H bond would lead to an impossible —O—O—O$^{\cdot}$ structure.

Special problems also exist in the choice and relative efficiency of antioxidants in emulsified and solubilized preparations. With emulsions it is an advantage for the antioxidant to be present at the interface between the oil drop and the continuous aqueous phase. Therefore the antioxidant should exhibit a suitable balance between lyophilic and lyophobic groupings. If it exists in both phases, then in the aqueous phase it will decompose to give free radicals which may initiate oxidation of oil.

Synergism

Synergism is said to occur when two or more antioxidants present in a system show a greater overall effect than can be accounted for by a simple addition of their individual effectiveness. Although this phenomenon is well known, most of the systems have been studied on an empirical basis. The phenomenon is associated with two separate systems: (a) mixed free radical acceptors, and (b) the metal chelating agents.

Mixed Free Radical Acceptors. It would appear that the effect of mixed free radical acceptors is due to both steric factors and activation energy changes.

In a synergistic system involving a material such as ascorbic acid (BH) which has a low steric factor and hydroquinone (AH) in which steric factors would not be important, Uri[59] has suggested that the following reactions will take place:

$$RO_2^{\cdot} + AH \longrightarrow RO_2H + A^{\cdot}$$
$$A^{\cdot} + BH \longrightarrow B^{\cdot} + AH$$

The possible disappearance of $A^{\cdot}$ by reaction with oxygen is thus eliminated and the effective antioxidant AH is regenerated. On its own BH would not produce any significant antioxidant effect as the reaction:

$$RO_2^{\cdot} + BH \longrightarrow RO_2H + B^{\cdot}$$

would be prevented by steric factors.

Metal Chelating Agents. The normal effect of metal chelating agents is to bond with pro-oxidant metallic ions and thus prevent their catalytic effect on the normal oxidation chain reaction. This reaction does not therefore prevent normal oxidation taking place, but only slows down the formation of peroxide while at the same time extending the induction period. Metallic pro-oxidants that are already present as part of complex organic structures are not usually affected by chelating agents. Stabilization has been achieved by reaction of the metal with organic acids of the tartaric or citric acid type or with materials such as ethylenediaminetetra-acetic acid (EDTA).

Recent reviews of antioxidant–synergist systems have been published by Prosperio[61] and Lozonczi.[62]

Typical antioxidants and synergistic systems used in cosmetics are given in Table 37.1.

Measurement of Oxidation and the Assessment of Antioxidant Efficiency

Tests to measure oxidation and to assess antioxidant efficiency may be similar and in general are designed to measure either the rate of oxidation (by direct measurement of oxygen uptake or the formation of decomposition products) or the extension of the induction period. Many of the tests used are artificially accelerated by the use of UV irradiation or elevated temperature and the extrapolation of such results to normal shelf storage conditions is suspect owing to possible changes in the oxidation reactions under the accelerating conditions. A secondary problem is that many antioxidant combinations are tested on the pure oil or fat and no account is taken of other materials present in the formulation which may materially alter the overall efficiency of the system. Whatever indications accelerated tests give, it is imperative that long-term storage tests are used to confirm the choice of preservative.

One difficulty is that common techniques such as the determination of hydroxyl or iodine values are frequently misleading owing to the interference of other products of the oxidation system, particularly peroxides, and methods of assessing antioxidant efficiency based on such values may give completely false impressions of efficiency. Even the normal estimations of peroxide values must be viewed with a degree of suspicion as they are not very specific and the reactions involved may be non-stoichiometric. However, this is the most common method of measuring oxidation and antioxidant efficiency, although the peroxide measured is the undecomposed peroxide and really indicates that

Table 37.1 Antioxidants for Use in Cosmetic Systems

Aqueous systems		
Sodium sulphite		Ascorbic acid
Sodium metabisulphite		Isoascorbic acid
Sodium bisulphite		Thioglycerol
Sodium thiosulphate		Thiosorbitol
Sodium formaldehyde sulphoxylate		Thioglycollic acid
Acetone sodium metabisulphite		Cysteine hydrochloride
Non-aqueous systems		
Ascorbyl palmitate	Butylated hydroxyanisole	
Hydroquinone	α-Tocopherol	
Propyl gallate	Phenyl α-naphthylamine	
Nordihydroguaiaretic acid	Lecithin	
Butylated hydroxytoluene		
Synergistic systems		
Antioxidant	*per cent*	*Synergists*
Propyl gallate	0·005–0·15	Citric and phosphoric acid
α-Tocopherols	0·01–0·1	Citric and phosphoric acid
Nordihydroguaiaretic acid (NDGA)	0·001–0·01	Ascorbic, phosphoric, citric acids (25·50% NDGA content) and BHA
Hydroquinone	0·05–0·1	Lecithin, citric acid and phosphoric acid, BHA, BHT
Butylated hydroxyanisole (BHA)	0·005–0·01	Citric and phosphoric acids, lecithin, BHT, NDGA
Butylated hydroxytoluene (BHT)	0·01	Citric and phosphoric acids up to double the weight of BHT and BHA

peroxides are being formed faster than they are broken down. This condition does not necessarily apply in the later stages of oxidation which may only show small peroxide values.

Determination of Peroxides
A large number of methods is available for the determination of peroxides, but the results, while reproducible within a given set of experimental conditions, are difficult to compare from worker to worker as differing experimental techniques will give discordant values even on the same substrate. The technique normally involves the liberation of iodine from sodium or potassium iodide in the presence of peroxide. Lea[63] showed that the system should be acidified during this release, while Knight[64] showed that the presence of other functional groups did not interfere and Swift[20] showed that one mole of iodine was liberated by one mole of methyl oleate hydroperoxide. Lea's method[65] of heating with glacial acetic acid and chloroform in the presence of solid potassium iodide and a nitrogen atmosphere is claimed to detect as little as 10^{-6} equivalents of peroxide per g of fat. The reacted mixture when cooled was added

to 5 per cent potassium iodide solution and titrated with 0·002 N sodium thiosulphate. A modification which is in common use is due to Wheeler[66] who, while still using a chloroform–glacial acetic acid solvent (50 ml), used 1 ml of saturated potassium iodide solution and 3–10 g of the oil under investigation.

Both of these methods, and many other variations, gave conflicting peroxide values. It has been found that the solvent used and the acid condition will cause variation,[67] and a modified Wheeler technique using sulphuric acid and identical sample weights is claimed to have given the most reproducible values.[68] The importance of constant sample size has been demonstrated by numerous workers as also has the presence of an inert atmosphere to prevent further oxidation during the determination.[69–71]

Other Methods of Analysis

Chemical. Other chemical methods have been employed for the detection of rancidity but all suffer to some degree from the problem of non-reproducibility and difficulty of interpreting the results. However, they are still occasionally used and therefore worthy of mention.

The Kreis test, first described in 1902, is the one most frequently employed to determine oxidative rancidity. In this test 1 ml of the oil (or melted fat) is shaken with 1 ml concentrated hydrochloric acid for one minute; then 1 ml of a 0·1 per cent solution of phloroglucinol in ether is added and shaking is continued for a further minute. A pink or red coloration of the lower acid layer is considered to be indicative of rancidity, the amount of which is approximately (although not exactly) proportional to the intensity of the colour. A preferable modification of this method was suggested by the Committee of the American Oil Chemists' Society, in which the colour is measured by means of glass colour standards in a Lovibond tintometer.

This red coloration is produced by epihydrin aldehyde,

$$\underbrace{CH_2 - O - CH}_{} - CHO$$

but it should be remembered that a fat is not necessarily rancid if it responds to this test. Fresh, crude vegetable oils may give similar colours, although these usually disappear on refining. Essential oils and aldehydes which may be present in certain toilet articles and cosmetic products can give positive results; hence such tests should be carried out on the unperfumed product.

A modified Kreis test applicable to cosmetic preparations, in which aeration is employed together with a modified absorption reagent, has been described by Jones.[72] This test eliminates the interference of many other substances. Directions are given in the original article for testing various types of cosmetics.

The Schiff test for aldehydes appears to be slightly more sensitive, but the colorimetric evaluation is not so easy with this test and it offers no particular advantage. Since it may be readily found in any textbook, the exact procedure is not described.

Schibsted[73] has developed this test to be specific for aldehydes of a high molecular weight. However, the stipulated conditions must be carefully followed, and for these the original paper should be consulted.

Lea[74] has devised a method by which the relatively minute amounts of aldehydes in a rancid fat may be measured by a simple titration with sodium bisulphite. Some preliminary tests with cotton-seed oil appeared to indicate that this method correlated more closely with the organoleptic test than did the Kreis test.

Another test which has been suggested for the determination of oxidative rancidity is that employing a 0·025 per cent solution of methylene blue in alcohol. About 2 ml of this reagent are added to 20 ml of the oil or fat, which is shaken and the amount of colour reduction taken as a measure of its rancidity.

Lea[75] states that the aerobic ferric thiocyanate method, although it gave high results in the presence of atmospheric oxygen and values too low in its absence, showed excellent reproducibility, required much less material than the iodometric method and under any one set of conditions gave results which appeared to be directly proportional to the iodometric values.

The most commonly used accelerated test is that known as the aeration test, active oxygen method or Swift stability test.[76] The sample is kept at 97·8°C and is aerated with a standard flow of air. From time to time samples are taken and the degree of rancidity is assessed, either organoleptically or chemically, until a predetermined value is reached. The peroxide value is generally chosen as the chemical criterion. Modifications of this involve other temperatures, for example 100°C, at which temperature the time required to attain a given degree of rancidity is about 40 per cent of that in the Swift test. Becker, Gander and Hermann[77] describe a type of Swift test, and Lea[78] describes an accelerated autoxidation test in which fat and water are kept in the closest possible contact.

Chromatography. The presence and concentration of butylated hydroxyanisole (BHA), butylated hydroxytoluene (BHT) and alkyl gallates may be determined by thin layer chromatography which permits the identification of 2 μg of BHA, 4 μg of BHT and 1 μg of alkyl gallate with diazotized *p*-nitroaniline.[79] A chromatographic method has also been used by Dooms-Goosens[80] for the identification of antioxidants. He separated his material by the addition of anhydrous sodium sulphate followed by solubilization with petroleum ether and finally extraction with acetonitrile. A silica gel plate was used in conjunction with a petroleum ether–benzene–acetic acid developing solution.

Determination of Oxygen Uptake. The direct determination of oxygen uptake may be made using the Warburg constant volume respirometer or the Barcroft differential manometer. Detailed methods for the use of these instruments are given by Umbreit, Burris and Stauffer.[81] In essence the method involves the measurement of oxygen taken up in the presence and in the absence of antioxidant and allows both the rate of oxidation and the length of the induction period to be measured. The end of this latter function may be difficult to determine and the time for an arbitrary uptake of oxygen is often used to signify the end of the induction period. Spetsig[82] and Lew and Tappel[83] have both described the use of the Warburg apparatus for the measurement of oxidation rate in a stable emulsion, while Carless and Nixon[84,85] have used the method to study oxidation in both emulsified and solubilized oils of cosmetic interest. The effect of antioxidant activity and the concentration of peroxide present in a linoleic acid system have been studied by this method.[86] A detailed study of the accelerated oxidation of hemin in an emulsion at 45°C was made by Berner *et al.*[87] The antioxidant was added to the fat prior to emulsification and the oxygen uptake then measured. The effect of antioxidants was to increase the induction period. The age and purity of the hemin, pH and peroxide value of the fat and temperature all affect the induction period. The activity of BHA, propyl gallate, *tert*-butyl hydroquinone, tocopherol and synergists (EDTA, ascorbic acid and citric acid) were studied.

Spectrophotometry. Ultraviolet analysis was first used by Mitchell[88] and the method was improved and extended. The fatty acids with conjugated unsaturation absorb at 230–375 nm with diene unsaturation being at 234 nm and triene at 268 nm. Chipault and Lundberg[89] found a direct relationship between peroxide value and ε at 232·5 nm.

Of possibly more value is infrared spectroscopy in that it may be used to identify hydroxyl, hydroperoxide, carboxy and many other groupings. However, the technique appears to have been little used for quantitative studies of peroxidation although Morris[90] has reviewed its application. The 3·0, 6·0 and 10·0 μm bands appear to be the principal regions involved and the principal use appears at the moment to be the determination of stereo isomeric changes during the course of the oxidation. The infrared spectra of some alkyl hydroperoxides are characterized by weak absorption in the 11·4–11·8 μm region.[91]

Choice of Antioxidant

The ideal antioxidant should be stable and effective over a wide pH range and be soluble in its oxidized form, and its reaction compounds should be colourless and odourless. Other obvious and essential requirements are that it should be non-toxic, stable and compatible with the ingredients in the products and their packages.

The list of effective antioxidants permitted in the USA for use in foodstuffs includes the materials shown in Table 37.2.

Table 37.2 Effective Antioxidants for Foodstuffs Permitted in the USA

Guaiacum resin
Tocopherols
Lecithin
Propyl gallate
Butylated hydroxyanisole (BHA)
Butylated hydroxytoluene (BHT)
Trihydroxybutyrophenone
Ascorbic acid
Ascorbyl palmitate
Monoisopropyl citrate
Thiodipropionic acid
Dilauryl thiodipropionate

Phenolic Antioxidants

Guaiacum Resin
Guaiacum resin is largely phenolic in character, but is a less effective antioxidant than most of the other phenolics mentioned above. It is more effective in animal than in vegetable oils and possesses an advantage over some other antioxidants in that it is equally effective in the presence and absence of water and it is not seriously affected by heating.

Nordihydroguaiaretic Acid
NDGA shares many of the properties of guaiacum resin but is more effective weight for weight. Higgins and Black[92] summarize their studies on the protection of lard with NDGA as follows: pure lard of low initial peroxide value was protected against development of oxidative rancidity by 0·003 per cent of NDGA, compared with the requirement of 0·006 per cent of propyl gallate. The stabilizing effect was proportional to the antioxidant concentration over the range of 0·003–0·03 per cent. A synergistic effect occurred with 0·003 per cent NDGA and 0·75 per cent citric acid. This, of course, arises from the sequestering effect of citric acid on heavy metals. NDGA is soluble in fats at levels up to about 0·05 per cent at 45°C and does not crystallize out very much on cooling. It was removed from the US permitted list in 1968.

Tocopherols
These natural materials are not widely used in practice because of their high price. They have some antioxidant effect with animal fats such as tallow and with distilled fatty acids, particularly in the presence of a synergist such as citric acid, lecithin, or phosphoric acid, but are of little value for the preservation of vegetable oils. Issidorides[93] has shown that the action of citric acid with tocopherols is not only a sequestering effect, but derives also from regeneration of the tocopherol in the reduced state. Sisley[94] gives a method for obtaining tocopherols mixed with lecithin as a synergist by extracting wheat germ oil with dichloroethylene.

Gallates
Gallates constitute one of the most important classes of antioxidants. The propyl ester is the only one permitted in foodstuffs in most countries but methyl, ethyl, propyl, octyl and dodecyl gallates are commonly used in cosmetics. Gallic acid itself is a powerful antioxidant, but tends to turn blue in the presence of traces of iron.

The valuable antioxidant properties of the esters of gallic acid have been described by many investigators. Boehm and Williams[95] found that for practical and commercial considerations of ease of solubility at low temperatures, acidity and colour, and from the point of general effectiveness, normal propyl gallate (that is, normal propyl-3,4,5-trihydroxybenzoate) was the outstanding antioxidant amongst the gallic acid esters investigated. These workers found that the protection afforded by 0·1 per cent of the normal propyl gallate in lard was equal to that obtained by ten times as much Siam benzoin and probably greater than that of thirty times as much Sumatra benzoin. Peredi[96] showed a tenfold increase in storage time of lard in the presence of 0·01 per cent of propyl or ethyl gallate. However, Tollenaar[97] suggests that gallates are not suitable for the antioxidation of vegetable oils.

Further recommendation of the esters of gallic acid has been given by Stirton, Turer and Riemenschneider.[98] They compared the antioxidant activities of *nor*dihydroguaiaretic acid (NDGA), propyl gallate, benzyl hydroquinone, alpha-tocopherol and their synergistic combinations with citric acid, *d*-iso ascorbyl palmitate and lecithin in a number of fat substrates: methyl oleate, methyl linoleate, methyl linolenate and the distilled methyl esters of lard. *Nor*dihydroguaiaretic acid and propyl gallate surpassed the other substances in antioxidant activity. Citric acid showed marked synergism with each antioxidant; the most effective combinations were those of citric acid with *nor*dihydroguaiaretic acid and with propyl gallate.

The possible toxicity of ethyl gallate has been thoroughly investigated, and no symptoms of toxicity have been observed in mice which received, orally or subcutaneously, massive doses of ethyl gallate, in concentrations far greater than could ever be approached by human beings when receiving foods stabilized against oxidation by the ester.[99]

Boehm and Williams[95] report that 0·5 g of propyl gallate was administered orally to one of them on six consecutive days. An examination of the urine during this period, and for a further six days, showed that no albumen was present, abnormal sedimental contents were not observed, there being a complete absence of red blood corpuscles and casts of any kind. They also cite pharmacological tests received from the Pharmacological Laboratories of the College of the Pharmaceutical Society (University of London) which concludes as follows:

1. Normal propyl trihydroxybenzoate is less toxic than pyrogallol, when administered orally to mice (acute tests).
2. There are no acute skin effects observable when a 10 per cent solution of normal propyl gallate in propylene glycol is left in contact with shaven guinea-pigs for 48 hours, or the human skin for 24 hours. This contrasts favourably with the erythematous effect produced under similar conditions by a 10 per cent solution of pyrogallol in propylene glycol.

Williams[100] gives solubilities of gallates in various oils (Table 37.3).

Table 37.3 Solubilities of Gallates in Various Oils at 20°C

	Almond oil (%)	**Castor oil** (%)	**Mineral oil** (%)	**Groundnut oil** (%)
Gallic acid	—	—	—	0·01
Methyl gallate	0·30	—	—	—
Ethyl gallate	0·40	—	—	0·01
Propyl gallate	2·25	22·0	0·5	0·05
Octyl gallate	3·00	18·0	0·005	0·30
Dodecyl gallate	3·50	21·0	0·01	0·40

Mixtures of octyl and dodecyl gallates with BHT and BHA (see below) have been recommended for fat stabilization by Peereboom[101] and in general it may be said that propyl gallate at a level of 0·01–0·1 per cent is superior to equal amounts of NDGA, tocopherol, guaiacum resin, sesamol, lecithin, or hydroquinone for the preservation of vegetable fats. Gearhart[102] showed that propyl gallate gave greater protection in the presence of BHT.

Butylated Hydroxyanisole (BHA)

BHA consists mainly of two isomers, 2- and 3-*tert*-butyl-hydroxyanisole. It is seldom used alone, as its activity in most systems is less than that of propyl gallate, but it forms a number of very useful synergistic mixtures with the gallate esters. Thus a mixture of 20 per cent BHA, 6 per cent propyl gallate, 4 per cent citric acid, and 70 per cent propylene glycol is commonly used in both the food and cosmetic industries. If such a mixture is used at levels of about 0·025 per cent total antioxidant, most animal and vegetable oils can be protected, as can the fatty esters such as methyl oleate.

Olcott and Kuta[103] have observed an interesting synergistic effect with BHA and amines such as octadecylamine, tri-iso-octylamine, and proline. No synergistic effect was found with BHT. Proline was particularly useful in the treatment of vegetable oils.

Butylated Hydroxytoluene (BHT)

BHT is 2,6-di-*tert*-butyl-4-methylphenol, sold as BHT by Kodak Chemical Co., as Topanol O and OC (purified grade) by ICI, and as Ionol and Ionol CP by Shell Chemical Corporation. It may also be called di-*tert*-butyl-*p*-cresol, DBPC.

BHT is widely used as an antioxidant for fatty acids and vegetable oils, and possesses several advantages over the other phenolic antioxidants in its freedom from any phenolic smell, its stability towards heating, and its low toxicity. It was approved for use in foods in the USA in 1954, at levels not exceeding 0·01 per cent. Normally, in cosmetics containing unsaturated materials, a level of 0·01–0·1 per cent should be used, with addition of a suitable sequestering agent such as citric acid or EDTA. BHT is not synergized by gallate esters.

Auto-oxidation of fatty materials takes place with a logarithmic velocity coefficient, so that it is important to stop such oxidation as early as possible in the life of a material. Some manufacturers of fatty acids, etc., will now supply their products with BHT or other suitable antioxidants already added, so that oxidation is checked immediately after manufacture.

Trihydroxybutyrophenone

Knowles *et al.*,[104] report that some of the 2,4,5-trihydroxyphenones, especially the butyrophenones, have outstanding effects with lard, groundnut oils, and tallow. Although the product is a recognized food additive in the USA, it does not appear to be widely used in the cosmetics industry.

Non-phenolic Antioxidants

Many non-phenolic antioxidants are chelating agents. Ascorbic acid and ascorbyl palmitate appear to act by stopping the free-radical oxidation process.

Ascorbyl esters are particularly effective in vegetable oils, and make an excellent synergistic mixture with phospholipids such as lecithin and tocopherol.[105]

Among the sequestering agents, the thiodipropionates are widely used, usually in conjunction with phenolic antioxidants. The esters of fatty alcohols have greater solubility in oils.

Monoisopropyl citrate has similar sequestering action to citric acid itself, but a greater fat solubility.

Lecithin is an effective synergist for many phenolic antioxidants, mainly because it is an oil-soluble phosphate with excellent sequestering properties. Members of another class of oil-soluble sequestering agents are MECSA (mono-octadecyl ester of carboxymethylmercapto-succinic acid) and METSA (mono-octadecyl ester of thiodisuccinic acid). Under some conditions these materials can function as effective antioxidants in concentrations as low as 0·005 per cent—see the review by Evans *et al.*[106] They possess the disadvantage that they decompose on heating, and must therefore be added, like perfume, during the cooling phase of manufacture.

In general, the effect of any true (that is, chain-stopping) antioxidant can be enhanced by the proper choice of a suitable sequestering agent to slow down the initiation of chain reactions at the outset. Citric, phosphoric, tartaric and ethylenediamine tetra-acetic acids should always be considered as possible additives to a system that is insufficiently protected against oxidation, before including more phenolic material. The use of such sequestering agents is cheaper, and it is less likely to lead to discoloration or odour development, than the use of high concentrations of phenols. Of the phenolic materials, BHT is probably the most universally useful, but each system has its peculiarities which must be studied at first hand.

Photo-deterioration

Another form of deterioration sometimes encountered is that caused by light in the visible or the UV spectrum. Such photo-deterioration generally manifests itself as fading of the colour of the product or development of off-colours.

Packaging in opaque containers or wrappers so as to exclude all light is, of course, an obvious way to avoid this type of deterioration, but this is not always desirable, nor even necessary. It is frequently possible to wrap or pack in transparent material suitably coloured or containing a UV absorber to screen out the offending portions of the spectrum. In some cases where UV energy is causing the deterioration, the UV absorber can be incorporated in the product. Mecca was reported[107] as having found that uric acid at 0·02–0·5 per cent would protect solutions coloured with FD&C Blue No.1, D&C yellow No.10, FD&C Green No.8 and cochineal, exposed to direct sunlight, while control solutions without uric acid were completely bleached.

UV-induced deterioration frequently involves the presence of traces of metals, particularly iron, and when this is the case the screening agent may be reinforced, or in some cases replaced, by a chelating agent such as ethylenediamine tetra-acetic acid (EDTA). The permeability of the cell walls of some bacteria, notably the Gram-negative *Pseudomonas aeruginosa*, is altered by EDTA; Smith[108] has found that the addition of concentrations in the region of

0·05 per cent greatly enhances the antibacterial potency of phenolic antiseptics. Small amounts of this material might thus serve two useful purposes in protecting systems prone to deterioration by UV and susceptible to the omnipresent *Pseudomonas* species.

REFERENCES

1. Chastaing, P., *Ann. chim. Phys.*, 1799, **11,** 190.
2. Schönbein, C. F., *J. makromol. Chem.*, 1858, **74,** 328.
3. Engler, C. and Wild, W., *Ber. Dtsch. Chem. Ges.*, 1897, **30,** 1669.
4. Engler, C. and Weissberg, J., *Ber. Dtsch. Chem. Ges.*, 1898, **31,** 3046, 3055.
5. Fahrion, W., *Chem.-Ztg.*, 1904, **28,** 1196.
6. Staudinger, H., *Ber. Dtsch. Chem. Ges.*, 1925, **58,** 1075.
7. Fokin, S., *Z. angew. Chem.*, 1909, **22,** 1451.
8. Stephens, H. N., *J. Am. chem. Soc.*, 1928, **50,** 568.
9. Farmer, E. H. and Sundralingam, A., *J. chem. Soc.*, 1942, 121.
10. Farmer, E. H., *Trans. Faraday Soc.*, 1942, **38,** 340, 356.
11. Farmer, E. H., Bloomfield, G. F., Sundralingam, A. and Sutton, D. A., *Trans. Faraday Soc.*, 1942, **38,** 348.
12. Farmer, E. H. and Sutton, D. A., *J. chem. Soc.*, 1943, 119, 122.
13. Farmer, E. H., Kock, H. P. and Sutton, D. A., *J. chem. Soc.*, 1943, 541.
14. Farmer, E. H., *Trans. Inst. Rubber Ind.*, 1945, **21,** 122.
15. Farmer, E. H. and Sutton, D. A., *J. chem. Soc.*, 1946, 10.
16. Hargrave, K. R. and Morris, A. L., *Trans. Faraday Soc.*, 1956, **52,** 89.
17. Bolland, J. L., *Q. Rev. chem. Soc.*, 1949, **3,** 1.
18. Bateman, L., *Q. Rev. chem. Soc.*, 1954, **8,** 147.
19. Uri, N., *Autoxidation and Antioxidants*, ed. Lundberg, W. O., New York & London, Interscience, 1961, Vol.1,p. 66.
20. Swift, C. E., Dollear, F. G. and O'Connor, R. T., *Oil Soap*, 1946, **23,** 355.
21. Privett, O. S., Lundberg, W. O. and Nickell, C., *J. Am. Oil Chem. Soc.*, 1953, **30,** 17.
22. Privett, O. S. and Nickell, C., *Fette Seifen*, 1959, **61,** 842.
23. Willits, C. O., Ricciuti, C., Knight, H. B. and Swern, D., *Analyt. Chem.*, 1952, **24,** 785.
24. Bolland, J. L. and Gee, G., *Trans. Faraday Soc.*, 1946, **42,** 236.
25. Gunstone, F. D. and Hilditch, T. P., *J. chem. Soc.*, 1946, 1022.
26. Farmer, E. H. and Sutton, D. A., *J. chem. Soc.*, 1942, 139.
27. Bergström, S., *Nature*, 1945, **156,** 717.
28. Privett, O. S., Lundberg, W. O., Khan, N. A., Tolberg, W. E. and Wheeler, D. H., *J. Am. Oil Chem. Soc.*, 1953, **30,** 61.
29. Holman, R. T., *Progress in the Chemistry of Fats and Other Lipids*, ed. Holman, R. T., Lundberg, W. O. and Malkin, T., London, Pergamon Press, 1954, Vol. 2, p. 51.
30. Nixon, J. R. and Carless, J. E., *J. Pharm. Pharmacol.*, 1960, **12,** 348.
31. Privett, O. S., Nickell, C., Tolberg, W. E., Paschke, R. F., Wheeler, D. H. and Lundberg, W. O., *J. Am. Oil Chem. Soc.*, 1954, **31,** 23.
32. Kern, W., Heinz., A. R. and Höhr, D., *Makromol. Chem.*, 1956, **18/19,** 406.
33. Privett, O. S., *J. Am. Oil Chem. Soc.*, 1959, **36,** 507.
34. Allen, R. R., Jackson, A. H. and Kummerow, F. A., *J. Am. Oil Chem. Soc.*, 1949, **26,** 395.
35. Jackson, A. H. and Kummerow, F. A., *J. Am. Oil Chem. Soc.*, 1949, **26,** 460.

36. Myers, J. E., Kass, J. P. and Barr, G. O., *Oil Soap*, 1941, **18,** 107.
37. Holman, R. T. and Elmer, O. C., *J. Am. Oil Chem. Soc.*, 1947, **24,** 127.
38. Allen, R. R. and Kummerow, F. A., *J. Am. Oil Chem. Soc.*, 1951, **28,** 101.
39. Brauer, R. W. and Steadman, L. T., *J. Am. chem. Soc.*, 1944, **66,** 563.
40. O'Neill, L. A., *Chem. Ind.* (*London*), 1954, 384.
41. Benton, J. L. and Wirth, M. M., *Nature*, 1953, **171,** 269.
42. Wibaut, J. P. and Strang, A., *Proc. Koninkl. Ned. Akad. Wetenschap.*, 1952, **55B,** 207.
43. Paquot, C. and de Goursac, F., *Oléagineux*, 1950, **5,** 349.
44. Almquist, H. J. and Branch, G. E. K., *J. Am. chem. Soc.*, 1932, **54,** 2293.
45. Bäckström, H. L. J., *Z. physik. Chem.*, 1934, **25B,** 99.
46. Cook, A. H., *J. chem. Soc.*, 1938, 1768.
47. Waters, W. A. and Wickham-Jones, C., *J. chem. Soc.*, 1951. 812.
48. Cooper, H. R. and Melville, H. W., *J. chem. Soc.*, 1951, 1984, 1994.
49. Sharp, D. B., Whitcomb, S. E., Patton, L. W., and Moorhead, A. D., *J. Am. chem. Soc.*, 1952, **74,** 1802.
50. Bolland, J. L. and ten Have, P., *Trans. Faraday Soc.*, 1947, **43,** 201.
51. Davies, D. S., Goldsmith H. L., Gupta, A. K. and Lester, G. R. *J. chem. Soc.*, 1956, 4926, 4931, 4932.
52. Bolland, J. L. and ten Have, P., *Discussions Faraday Soc.*, 1947, **2,** 252.
53. Boozer, C. E., Hammond, G. S., Hamilton, C. E. and Sen, J. N., *J. Am. chem. Soc.*, 1955, **77,** 3233, 3238.
54. Olcott, H. S. and Mattill, H. A., *J. Am. chem. Soc.*, 1936, **58,** 2204.
55. Banks, A., *J. Soc. chem. Ind.*, 1944, **63,** 8.
56. Lovern, J. A., *J. Soc. chem. Ind.*, 1944, **63,** 13.
57. Lea, C. H., *Research*, 1956, **9,** 472.
58. Rosenwald, R. H. and Chenicek, J. A., *J. Am. Oil Chem. Soc.*, 1951, **28,** 185.
59. Uri, N., *Autoxidation and Antioxidants*, ed. Lundberg, W. O., New York & London, Interscience, 1961, Vol. 1, p. 133.
60. Fieser, L. F., *J. Am. chem. soc.*, 1930, **52,** 5204.
61. Prosperio, G., *Riv. Ital. Essenze, Profumi, Piante Off., Aromi., Sopeni, Cosmet., Aerosol.*, 1977, **59,** 424.
62. Lozonczi, B. and Lozonczi, Mrs. B., *Olaj. Szappan Kosmit.*, 1977, **26,** 115.
63. Lea, C. H., *Food Investigation*, Special Report No. 46, Dept. Scientific and Industrial Research, 1938.
64. Knight, H. B. and Swern, D., *J. Am. Oil Chem. Soc.*, 1949, **26,** 366.
65. Lea, C. H., *Proc. R. Soc.* (*London*), 1931, **108B,** 175.
66. Wheeler, D. H., *Oil Soap*, 1932, **9,** 89.
67. Nakamura, M., *J. Soc. chem. Ind.* (*Japan*), 1937, **40,** 206, 209, 210.
68. Stansby, M. E., *Ind. Eng. Chem., Analyt. Edn.*, 1941, **13,** 627.
69. Lea, C. H., *J. Soc. chem. Ind.* (*London*), 1946, **65,** 286.
70. Stuffins, C. B. and Weatherhall, H., *Analyst*, 1945, **70,** 403.
71. Volz, F. E. and Gortner, W. A., *J. Am. Oil Chem. Soc.*, 1947, **24,** 417.
72. Jones, J. H., *J. Am. Oil Chem. Soc.*, 1944, **21,** 128.
73. Schibsted, H., *Ind. Engng. Chem., Analyt. Edn.*, 1932, **4,** 204.
74. Lea, C. H., *Ind. Engng. Chem., Analyt. Edn.*, 1934, **6,** 241.
75. Lea, C. H., *J. Sci. Fd. Agric.*, 1952, **3,** 286.
76. Wheeler, D. H., *Oil Soap*, 1933, **10,** 89.
77. Becker, E., Gander, K. F. and Herman, W., *Fette Seifen Anstr.-Mittel*, 1957, **59,** 599.
78. Lea, C. H., *J. Soc. chem. Ind.* (*London*), 1936, **55,** 293T.
79. Guven, K. C. and Guven, N., *Eczacilik Bul.*, 1974, **16,** 93.
80. Dooms-Goosens, A., *J. Pharm. Belg.*, 1977, **32,** 213.

81. Umbreit, W. W., Burris, R. H. and Stauffer, J. F., *Manometric Techniques* (Rev. Edn), Minneapolis, Burgess Publishing Co., 1957.
82. Spetsig, L. O., *Acta Chem. Scand.*, 1954, **8,** 1643.
83. Lew, Y. T. and Tappel, A. L., *Fd. Technol., Champaign*, 1956, **10,** 285.
84. Carless, J. E. and Nixon, J. R., *J. Pharm. Pharmacol.*, 1957, **9,** 963 and 1960, **12,** 348.
85. Nixon, J. R., Ph.D Thesis, London, 1958.
86. Ozawa, T., Nakamura, Y. and Hiraga, K., *Shokuhin Eiseigaku Zasshi*, 1972, **13,** 205.
87. Berner, D. L., Conte, J. A. and Jacobson, G. A., *J. Am. Oil Chem. Soc.*, 1974, **51,** 292.
88. Mitchell, J. H., Kraybill, H. R. and Zscheile, F. P., *Ind. Engng. Chem., Analyt. Edn.*, 1943, **15,** 1.
89. Chipault, J. R. and Lundberg, W. O., *Hormel Inst. Univ. Minn., Ann. Rpt.*, 1946, 9.
90. Morris, S. G., *J. Agr. Food Chem.*, 1954, **2,** 126.
91. Williams, H. R. and Moser, H. S., *Analyt. Chem.*, 1955, **27,** 217.
92. Higgins, J. W. and Black, H. C., *Oil Soap*, 1944, **19,** 277.
93. Issidorides, A., *J. Am. chem. Soc.*, 1951, **73,** 5146.
94. Sisley, J. P., *Perfum. Essent. Oil Rev.*, 1955, **46,** 117.
95. Boehm, E. and Williams, R., *Q. J. Pharm, Pharmacol.* 1943, **16,** 232.
96. Peredi, J., *Hung. Tech. Abstr.*, 1955, **7,** 24.
97. Tollenaar, F. D. and Vos, H. J., *J. Am. Oil Chem. Soc.*, 1958, **35,** 448.
98. Stirton, A. J., Turer, J. and Riemenschneider, R. W., *Oil Soap*, 1945, **22,** 81.
99. Hilditch, T. P. and Lea, C. H., *Chem. Ind.*, 1944, 70.
100. Williams, R., *Am. Perfum. Aromat.*, 1959, **73**(2), 39.
101. Peereboom, J. W. C., *Am. Perfum. Aromat.*, 1959, **73**(2), 27.
102. Gearhart, W. M. and Stuckey, B. N., *J. Am. Oil Chem. Soc.*, 1955, **32,** 386.
103. Olcott, H. S. and Kuta, E. J., *Nature*, 1959, **183,** 1812.
104. Knowles, M. E., Bell, A., Tholstrup, C. E. and Pridgen, H. S., *J. Am. Oil Chem. Soc.*, 1955, **32,** 158.
105. Lundberg, W. O., *Manuf. Confec.*, 1953, **33**(4), 19, 67.
106. Evans, C. D., Schwab, A. W. and Cooney, R. M., *J. Am. Oil Chem. Soc.*, 1954, **31,** 9.
107. Anon, *Drug Cosmet. Ind.*, 1965, **97,** 97.
108. Smith, G., *J. med. Lab. Technol.*, 1970, **27,** 203.

Chapter Thirty-eight

Emulsions

Introduction

Every cosmetics laboratory worker knows that emulsions are relatively stable mixtures of oils, fats and water and are made by mixing oil-soluble and water-soluble substances together in the presence of an emulsifying agent. Emulsions—creams and lotions—form a very important part of the cosmetics market; much time is spent in the development of new raw materials by both suppliers and cosmetics companies. Formulae for good, stable cosmetic emulsions are available from books or suppliers' literature and it is not difficult for even a novice laboratory technician to make a satisfactory emulsion by following simple written instructions. Clearly, however, no cosmetics chemist can consider himself competent until he understands how to formulate emulsions of his own and how to incorporate into them certain desired characteristics. In order to do this, he must learn something of the fundamentals of emulsion technology. How far this is pursued will depend upon the inclinations of the individual, but the aim of this chapter is simply to provide a sufficiently detailed account of these fundamentals to allow the reader to experiment for himself with understanding and to form ideas of his own. More detailed information can be obtained from the references cited at the end of the chapter.

Basic Principles

The starting point in this study is the recognition that certain substances show an 'affinity' for each other and others do not. A simple illustration of this point is that water and ethanol are completely miscible. Their molecules are quite happy to exist side-by-side together and show no tendency to separate into discrete areas populated largely or exclusively by their own kind. These two materials show an 'affinity' for each other which is obviously not shared by (say) mineral oil and water. This idea of 'affinity' plays an important part in emulsion technology and has to do with the manner in which individual molecules appear to attract their neighbours in a given environment. Molecules of the same substance—for example, water—exert an attractive influence on each other and were it not for the fact that under normal circumstances each molecule is in turn attracted by many others around it in all directions, any two molecules might be pulled together. This phenomenon is called 'cohesion' and the force of cohesion between molecules is attributed to their 'cohesive energy'. The magnitude of these cohesive forces depends not only upon the size of the molecules taking part but also upon their chemical make-up. The basic principle is that 'like attracts like'. Related molecules such as water and ethanol show no tendency to separate because the cohesive forces between water and ethanol molecules are similar in magnitude to those between water and water or ethanol and ethanol. When mineral oil is introduced into water, however, the cohesive forces between water and mineral oil are negligible compared with those between the molecules of the two materials themselves, and separation rapidly occurs.

'Affinity' manifests itself not only as solubility but also in the concept of the 'phase'. When two or more materials in contact with each other co-exist as overtly different and separate entities, each is referred to as a 'phase' (Table 38.1). In two-phase systems one phase may be distributed as a large number of distinct and separate entities in the other. Under these circumstances, the former is known variously as the 'internal', 'disperse' or 'discontinuous' phase and the latter as the 'external' or 'continuous' phase. When one material is dispersed in a finely divided state within another in this way, the area of contact between the two phases is exceedingly large. It is not surprising, therefore, that many of the characteristics exhibited by such a system depend primarily on the chemical or physical natures of the two surfaces and the interaction between them. This is certainly true of cosmetic emulsions.

Table 38.1 Some Common Two-phase Systems

Continuous phase	Disperse phase	System
Gas	Solid	Smoke
Gas	Liquid	Aerosol
Liquid	Gas	Foam
Liquid	Solid	Dispersion
Liquid	Liquid	Emulsion
Solid	Gas	Foam

Properties of Surfaces

In general, the outermost layers of materials exhibit very different properties from the bulk, this being entirely due to the environment in which the surface molecules find themselves. In Figure 38.1, A represents a molecule some way in the interior of a liquid. A is surrounded by other molecules which, if close enough, exert an appreciable attraction on it. These will be contained in a sphere, centre A, with a very small but finite radius (represented in the figure by an exaggerated dotted line around A)—the 'sphere of molecular activity'. Since there are as many molecules attracting A in any one direction as attract it in the opposite direction, there is no resultant cohesive force on A. This is far from true of a second molecule, B, situated in the liquid surface. Here, forces of attraction below B are not precisely cancelled out by the attraction of other molecules above it and a resultant force F is exerted on B, tending to pull it into the interior of the liquid. (It should be emphasized that B is very close to the surface and that Figure 38.1 is greatly exaggerated for clarity.)

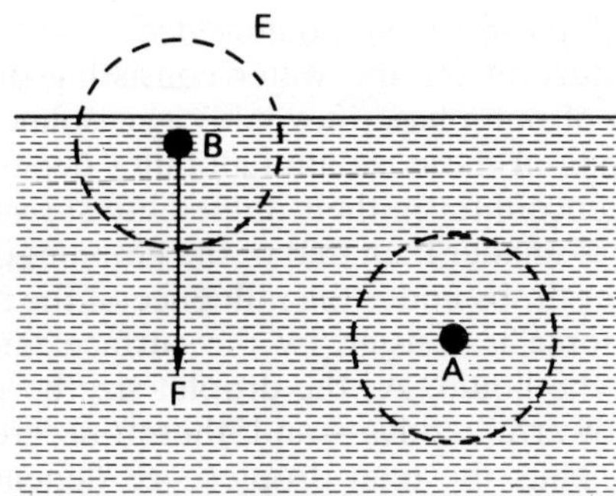

Figure 38.1 Molecular explanation of surface tension

This same consideration applies to all surfaces, whether gas, liquid or solid, although it is in gases and liquids, in which the molecules have considerable mobility, that its effect is clearly seen in determining the shape of the material. Since the inward forces on the molecules in a liquid surface tend to make the molecules move towards the interior, the surface tends to contract and become as small as possible, so that the surface area will be the minimum for a given volume of liquid. Since a sphere is the shape which has the minimum surface area for a given volume, if no other forces were acting on the liquid, it should be expected to assume a spherical shape. Relatively large masses of liquid are subject to proportionately large gravitational forces; for this reason they will assume the shape of any retaining vessel (for example, a beaker or measuring cylinder). For very small masses, however, such as may be found in droplets of internal phase in cosmetic emulsions, gravitational forces play a relatively minor role and the droplets may be spherical to a high level of approximation.

Because of the inward resultant force on surface molecules, all surfaces are said to possess 'surface tension' and the magnitude of the surface tension depends on F. Of course, if the molecules of material above the surface of the liquid have little or no measurable attraction for B, then the magnitude of F depends primarily on the properties of the liquid itself. Should the air, gas or solid above the liquid surface show some affinity for B, then F and the surface tension will be lower. Each surface is a boundary or interface whose characteristics, including the surface tension, depend upon the influence of the materials on both sides.

Principles of Emulsion Stability

It is a very commonplace observation that an oil and water mixture can be emulsified by shaking. The more vigorous the shaking, the finer the size of the droplets of dispersed phase. Sooner or later, however, the droplets of dispersed phase become noticeably larger as they coalesce until the initial two layers are re-formed ('phase separation'). It is important to understand the reasons for emulsification and for coalescence. There are two complementary ways of describing these phenomena, both of which will be discussed briefly. One viewpoint is mechanical and the other thermodynamic; they are equally important.

Mechanical Model of Emulsification Coalescence

When a quiescent mixture of oil and water consisting of two simple layers is shaken, large volumes of one phase inevitably get isolated and trapped within the other phase. The fate of these isolated globules depends partially upon the turbulence which they encounter in their immediate surroundings. If the size of the local eddy currents is smaller than that of the globule, the latter will break up into a number of smaller drops under the influence of the shear force exerted by the eddy. This shear force is resisted by the surface tension at the interface between the drop and the liquid. As the droplet size becomes reduced it must find smaller, more powerful eddys—and therefore greater turbulence—to become even smaller. Thus the final droplet size depends almost exclusively upon the surface tension at the interface and the degree of turbulence set up in the continuous phase.

Even while the droplets of dispersed phase are being broken up, however, they are simultaneously coalescing. The coalescence process can be considered in a number of stages. Firstly, the droplets must have sufficient mobility to move through the continuous phase to find each other. When they do bump into each other, very few collisions result in immediate coalescence. The thin film between two colliding droplets may cause them to rebound. If they do not, they may adhere to each other; this is the next essential stage in the process and is sometimes referred to as 'aggregation' or 'flocculation'. Finally, the intervening film of continuous phase must drain away to the point where it can rupture, allowing the contents of the two drops to combine to form a larger droplet with a smaller total surface area.

The rate of coalescence is determined by the slowest of these processes. If the viscosity of the external phase is high and the total volume of internal phase low, then the low mobility of the scattered drops of dispersed phase may wholly determine the rate of coalescence. If the droplets of internal phase are uniformly small, adhesion may determine the coagulation rate. The forces governing collision and adhesion for liquid droplets are the same as for those governing solids suspended in liquids or in other solids. The ease of adhesion increases with the particle size of the largest of the two adhering entities—thus a few large globules in an emulsion of otherwise small droplet size can markedly increase the rate of coalescence. When clumps of aggregated internal phase either rise to the top or fall to the bottom of an emulsion, the effect is often called 'creaming'. At this stage, further agitation can re-disperse these aggregates and rescue the emulsion. Once the aggregates have coalesced, however, and phase separation has occurred, re-formation of the emulsion is made more difficult. It is often pointed out that the rate at which particles sink or float in liquids—whether they be single particles or agglomerates—is predicted by Stokes's Law, one form of which is as follows:

$$K = \frac{2}{9} \cdot g \cdot \frac{r^2(d_1 - d_0)}{v}$$

where K is the terminal velocity of a sphere, radius r and density d_1, falling through a liquid of density d_0 and viscosity v. Although Stokes's Law can apply only very approximately to most emulsions, it does serve as a very simple model for the movement of internal phase droplets through the external phase.

Thermodynamic Description of Emulsification and Coalescence
When the surface area of a liquid is increased (for example, by agitation), molecules from the interior rise to the surface. They do so against the force of attraction of neighbouring molecules and hence some mechanical work or energy is always required to increase the surface area. The surface also tends to become cooled and thus heat flows into it from the surroundings. Hence there is an increase in surface energy equivalent to the sum of mechanical energy expended and the heat energy absorbed. The relationship between the increase in surface energy, ΔS, associated with an increase in surface area, ΔA, is as follows:

$$\Delta S = T \, . \, \Delta A$$

where T is the interfacial surface tension between the liquid and its surroundings. Thus it can be seen that surface tension is no more than the increase in surface energy associated with a unit increase of surface area.

It is a well-known principle in mechanics that an object is in stable equilibrium when its potential energy is at a minimum. Given the opportunity, therefore, the emulsion will lose its considerable energy excess to its surroundings in the form of heat by coalescence of the droplets of internal phase and phase separation.

Stabilization of Cosmetic Emulsions

The problem facing the cosmetic formulator, having decided upon an emulsion, is how to prevent this thermodynamically unstable system from separating into layers. Based on the considerations of the last few paragraphs, the following recommendations can be made.

(a) By increasing the viscosity of the external phase, he will decrease the mobility of internal phase droplets making it more difficult for them to collide with each other.
(b) By ensuring that the internal phase is of the smallest and most uniform drop size possible, he will decrease the likelihood of adhesion between two drops.
(c) By increasing the mechanical strength of the interface, he will make this less susceptible to rupture with the resulting coalescence of adhering drops.
(d) By decreasing the surface tension, he will decrease the thermodynamic 'driving force' for coalescence.

It should be noted that the increase in stability which results from the formation of internal phase droplets of very small size represents an apparent anomaly. It has already been shown that decreasing the droplet size causes a rapid increase in surface area and also that a large surface area can only be achieved, in a given system, by a larger energy input. Such a system should therefore possess a high excess energy content—which seems to be in conflict with the rule about high energy systems being less stable than those of low energy content. Apparently, therefore, the stabilizing effect of a low probability of adhesion between internal phase droplets far outweighs the influence of excess free surface energy in bringing about coalescence.

Surfactants and Emulsifiers
Returning to the recommendations listed above, it can be seen that suggestions (c) and (d) relate directly to the interface between the droplet of internal phase and its environment. Although the effect was discovered by accident (well before the theories of chemistry and physics had been developed), it has proved possible to stabilize emulsions by providing a physical barrier at the interface which not only reduces the likelihood of its rupture but which may actually prevent droplets from touching each other while at the same time making emulsification easier by reducing the interfacial surface tension. The possibility of finding materials which will migrate and live in an oil–water interface stems from the idea of chemical affinity; all that is necessary is that at least part of the material should show an affinity for oil and part for water (although neither affinity must be strong enough to overwhelm the other). Any such materials would be bound to migrate to the interface in order to satisfy these predispositions—and so, apparently, they do. Inevitably, materials which possess these characteristics have been called '*surface-active agents*' and this has been shortened to 'surfactant', a word which may be used either as a noun or as an adjective. Surfactants have a large number of uses in industry other than in the formation and stabilization of cosmetic emulsions; they may be used, for example, as solubilizing, wetting or spreading agents. These functions are all related to their role in emulsification but when designed and used for the latter purpose they should be referred to as emulsifying agents or 'emulsifiers'.

Cosmetic emulsions are almost invariably stabilized with emulsifiers and may be thought of as 'oil-in-water' (with water as the continuous phase) or 'water-in-oil' (where water is the internal phase). With very few exceptions, such a simple picture is only very approximate since the two phases will have some mutual affinity for each other, causing the probable formation of other phases of intermediate composition. However, for the sake of clarity the oil-in-water and water-in-oil models will be adopted in subsequent discussion. It is important to appreciate, however, that even these crude emulsion systems can no longer be considered to consist of two phases once the emulsifier has been added (although, regrettably, this error is often committed). The interface, containing the emulsifier, must now be regarded as a third phase.

Figure 38.2 illustrates the same liquid surface described in Figure 38.1 but now a thin layer of surfactant, S, has occupied the interface. The affinity of S for the surface molecule B is much greater than that of the original environment, E, and

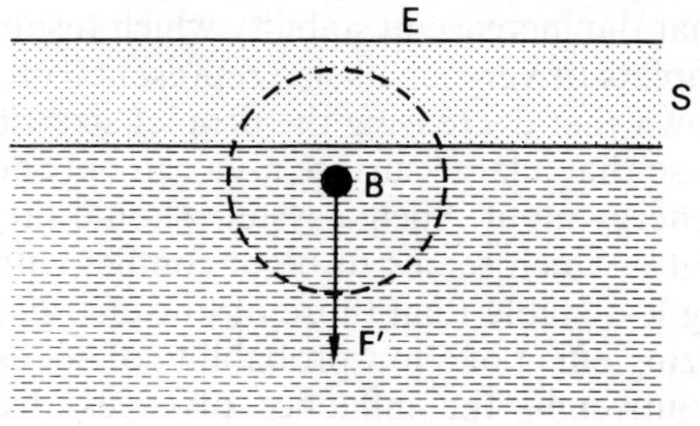

Figure 38.2 Reduction of surface tension by a surfactant

the resultant inward force, F', is less than before. The surface tension between the liquid and S is therefore lower than that between the liquid and E. This is referred to as 'a lowering of surface tension between the liquid and E', although we can see that this is not strictly accurate. At the same time, the surface tension between S and E is lower than that between the liquid and E for similar reasons.

Types of Emulsifier
Such is the number and variety of emulsifying agents now commercially available that their classification must be considered a daunting task—indeed, there are manuals devoted exclusively to it. Fortunately, the problem of choosing from among the bewildering variety of products is made easier by the classification of emulsifiers according to their chemical type and their mode of action.

Before proceeding to describe the emulsifier types of major commercial importance, however, a distinction should be made between these (which function at a molecular level) and certain finely divided solids which have also been shown to exhibit emulsion-stabilizing properties. Such solids undoubtedly function by migrating to the emulsion interface, forming a barrier against coalescence: it follows that the surface of such solids must not be predominantly water-wettable (hydrophilic) or oil-wettable (lipophilic). Such powders have little value as cosmetic emulsifiers but since many cosmetic emulsions also contain suspended powders (liquid foundations, for example) it is as well to bear in mind that these could possibly play some part in deciding the stability of the product.

Figure 38.3 is a diagrammatic representation of a molecule of the more conventional type of surfactant. The molecule can be considered as being composed of two parts—a water-loving or hydrophilic group at one end (H) and an oil-loving or lipophilic group at the other end (L). Since the lipophilic group is usually a hydrocarbon chain, it is often represented diagrammatically as a 'tail' as in Figure 38.3. It is easy to see how such a molecule would behave if dispersed in a single liquid: in Figure 38.4a, the surfactant has been dispersed in oil. Since the cohesive forces between the hydrophilic portion of the molecule and the oil molecules is negligible compared to those between hydrophilic ends of the molecules among themselves, the molecules orientate as indicated in clusters or 'micelles'. The lipophilic parts of the surfactant molecules, experiencing comparatively large cohesive forces from the oil molecules, are happy to extend outwards into the oily environment.

Figure 38.4b indicates the opposite orientation, encountered when the same molecules are dispersed in water or hydrophilic media. The same rules apply here as before except, as would be expected, it is the lipophilic ends of the

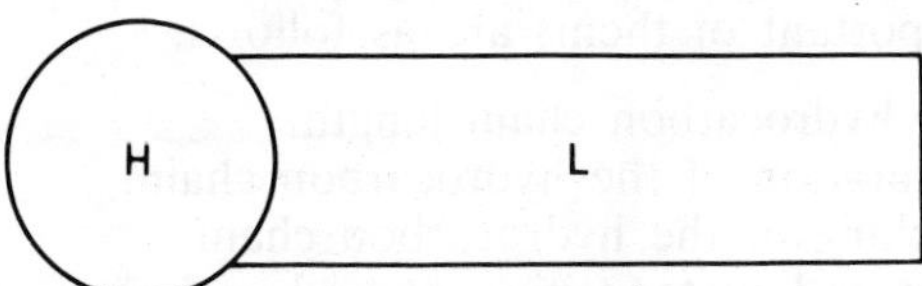

Figure 38.3 Surfactant molecule

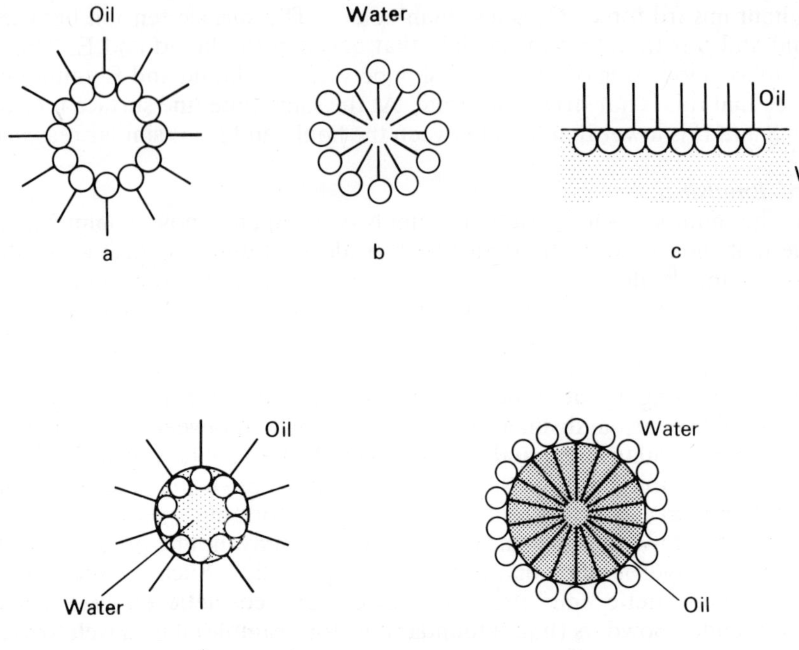

Figure 38.4 Behaviour of surfactants in various environments

molecule which cluster closely together in order to escape from the watery environment.

In Figure 38.4c oil has been added to the water and, as expected, the surfactant has migrated to the interface. If now an emulsion is formed, each spherical internal phase droplet will be covered with orientated surfactant molecules. (Figure 38.4d illustrates a water-in-oil and Figure 38.4e an oil-in-water emulsion.)

It is easy to verify, by experiment, that such interfacial surfactant layers can stabilize emulsions and this is primarily because the interface is made less liable to rupture. To understand this more clearly, however, it is first necessary to examine the chemistry of surfactants in greater detail.

Classification of Emulsifiers
There is only a limited number of chemical variations which can be played on the surfactant theme. Those relating to the lipophilic end of the molecule (or, at least, the more important of them) are as follows:

1. Variation in the hydrocarbon chain length.
2. Degree of unsaturation of the hydrocarbon chain.
3. Degree of branching of the hydrocarbon chain.
4. The introduction and juxtaposition of aryl groupings in the hydrocarbon chain.

For the hydrophilic end of the molecule the following variations are possible:

1. Introduction of ionizable anionic end groups.
2. Introduction of ionizable cationic end groups.
3. Introduction of amphoteric groups.
4. Introduction of other water-soluble but 'nonionizing' groups such as hydroxyl or ethoxyl.

All these variations have been used in practice and the classification of surfactant type depends, customarily, only upon the nature of the hydrophilic end of the molecule. Emulsifiers may thus be 'anionic', 'cationic', 'amphoteric' or 'nonionic'.

Hydrophilic–Lipophilic Balance of Surfactants

It is rarely the case that the affinity which the lipophilic end of the surfactant molecule has for the oil phase is equal to the affinity which the hydrophilic end of the molecule has for the aqueous phase. Clearly, the ratio of these affinities should play an important part in deciding the performance of the emulsifier in an emulsion system, and it is therefore fortunate that a relatively simple means of assessing this balance is available—at least, for certain emulsifier types. The fundamental point is that water-wetting or oil-wetting power seems to be a colligative property of certain atoms or chemical groups in the surfactant molecule. In other words, these entities contribute to wettability in a predictable way so that the value of each one may be added to the whole in order to obtain a composite value. For example, in nonionic emulsifiers consisting of alkyl chains coupled to polyoxyethylene chains, each oxygen atom is equivalent in water-wetting power to the oil-wetting power of three CH_2 groups. One ethylene oxide group ($—CH_2 . CH_2—O—$) is thereby balanced by each $—CH_2—$ group in the alkyl chain. This important point is illustrated by reference to the following generalized formula for the condensation products of ethylene oxide and stearyl alcohol, namely polyethylene glycol ethers of stearyl alcohol:

```
   O
  / \      ethylene oxide
CH2—CH2
```

$CH_3(CH_2)_{16}CH_2OH$ stearyl alcohol

$CH_3(CH_2)_{16}CH_2 \vdots (OCH_2CH_2)_nOH$ polyethylene glycol ethers of stearyl alcohol

(n varies between 2 and 30 in most commercial forms). The dotted line indicates the 'balance point' of the molecule, the groups to the left being oil-soluble and those to the right being water-soluble. Counting each $—CH_2—$ group or $—CH_3—$ group as unity, the lipophilic end of the molecules adds up to 18 and the hydrophilic end to $3 + n$ ($O = 3$; $—CH_2CH_2O— = 1$). If n is less than 15, therefore, the lipophilic tendencies of the molecule outweigh its hydrophilic properties, while at values of n greater than 15 the reverse is true. Moreover, the greater the difference in numerical value between the two sides, the greater the imbalance in the relative affinity for the two phases exhibited by the molecule.

The hydrophilic–lipophilic balance is an important property of the emulsifier since it determines the type of emulsion it tends to produce. The simple colligative nature of this phenomenon as applied to nonionic emulsifiers led the Atlas Chemical Company (as it then was) to devise a linear scale which enables the overall balance of each emulsifier to be expressed as a single number—the HLB (hydrophilic–lipophilic balance) number.[1] This is merely the weight per cent of the hydrophilic content of the molecule divided by an arbitrary factor of 5. Thus if a nonionic emulsifier were 100 per cent hydrophilic (which, of course, is impossible) it would have an HLB value of 100/5 or 20. The HLB scale therefore stretches (in theory) from 20 for a completely hydrophilic molecule to 0 for a completely lipophilic one. In the case of the polyethylene glycol ethers of stearyl alcohol cited above, HLB would be calculated as follows:

Molecular weight of the lipophilic stearyl chain, $C_{18}H_{37}$ = 253.

Molecular weight of the hydrophilic end of the molecule = $44n$ (ethoxyl groups) + 17 (remaining oxygen and hydrogen).

If $n = 3$,

$$\text{total molecular weight} = 253 + [(44 \times 3) + 17] = 253 + 149 = 403$$

$$\text{HLB} = \frac{149}{403} \times \frac{100}{5} = 7{\cdot}4$$

If $n = 20$,

$$\text{total molecular weight} = 253 + 897 = 1150$$

$$\text{HLB} = \frac{897}{1150} \times 20 = 15{\cdot}6$$

The HLB value can only be determined in this simple way for nonionic emulsifiers of known and definite composition. The concept is also applicable to anionic or cationic emulsifiers, although it is possible to exceed the theoretical upper limit of 20 with these materials. This does not detract from the practical merits of the HLB system, but it does mean that alternative methods of determining HLB values have to be used. This will be described later.

Figure 38.5 summarizes the applications of surfactants of various HLB values.

Stabilizing Influence of the Surfactant Phase at the Interface

In Figure 38.2 it was seen that the interfacial film of surfactant—the third phase of a stabilized emulsion—produced two new interfacial surface tensions, one between the surfactant, S, and the original liquid (which we now identify as the water phase) and one between S and E, the environment outside the water phase, which now corresponds to the oil phase. For clarity, these new surface tensions will be designated T_{ws} and T_{so} respectively. Clearly, since both of these tensions depend upon the affinity of the appropriate ends of the orientated surfactant molecules for their watery or oily environment, the ratio T_{so}/T_{ws} is directly related to the HLB value of the surfactant. If the HLB is high (greater

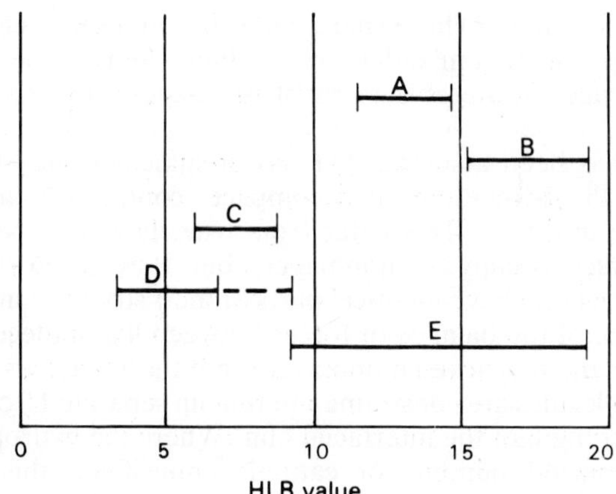

Figure 38.5 Practical applications of surfactants of various HLB values
A Detergents
B Solubilizers
C Surface wetters
D Water-in-oil emulsions
E Oil-in-water emulsions

than 10), then T_{so} is larger than T_{ws}. Remembering that the net inward force, F, on molecules in a surface is directly proportional to the interfacial surface tension, the surfactant surface has a propensity to curve towards the side having the greater surface tension—in this case, towards the oil phase.[2,3] With mechanical agitation, therefore, such a system would produce overwhelmingly an internal phase of oil dispersed in water. For surfactants having a low HLB value, exactly the reverse is true and a water-in-oil emulsion is the most probable end result.

Thus it can be seen that the relative lowering of the surface tension on each side of the surfactant interface helps to determine the nature of the emulsion and the ease of emulsification. It does not, however, determine the stability of the emulsion—this is a point which deserves great emphasis. The essential factors governing the integrity of the interfacial film and its resistance to rupture are its extension, its compactness and its electrical charge.[4] A lowering of interfacial tension is not vital for the stability of an emulsion.

Factors Contributing to the Strength of the Interfacial Surfactant Film

Perhaps the most obvious requirement is that there must be sufficient emulsifier present to form at least a monolayer over the surface of the internal phase droplets—and this, in turn, will depend upon the droplet size.[5,6] It is easy to verify in the laboratory that even the most stable cosmetic emulsion can be rendered unstable by progressive reduction of the quantity of emulsifier used in its production. Indeed, stability can usually be improved by a certain excess of emulsifier over this essential minimum. Evidence exists[7] that when there is a danger of a break in the interfacial film, the interfacial tension at the threatened point rises and signals to the reserve emulsifier molecules in the continuous phase that there is a fault. These spare emulsifier molecules close in and repair the damage. It appears that mixed interfacial films (films formed from more than one emulsifying agent), are able to resist breakage in this manner even more easily.[5]

Until now it has been assumed that, given sufficient emulsifier, there is no impediment to the formation of a compact, continuous film of emulsifier molecules in the interface. This is far from true, however. Not only do these identical molecules occupy a monolayer, but they are orientated so that neighbouring molecules have identical parts of their structure in close proximity. As a consequence of the balance of forces between like molecules at very close proximity, such a tightly packed monolayer is not thermodynamically stable, and the emulsifier molecules are constrained to remain separated from each other, so weakening the strength of the interfacial film. Where the hydrophilic ends of the molecules are ionized (anionic or cationic emulsifiers) this separation and weakening is further exaggerated by the repulsion of juxtapositioned like electrical charges.

The second effect tending to interrupt the continuity of the interfacial monolayer is simple steric hindrance. Consider, for example, the problem experienced by neighbouring molecules such as the polyoxyethylene sorbitan unsaturated esters indicated in Figure 38.6. Obviously it would not be easy for such molecules to pack closely together in an orientated monolayer. The importance of these steric effects can also be judged by the fact that whereas soaps of monovalent metal ions tend to form oil-in-water emulsions, the similar soaps with polyvalent metal ions produce largely water-in-oil emulsions. In this case, steric hindrance actually dictates the direction of curvature of the interface.

One way of overcoming these problems is to incorporate one or more additional species of emulsifier molecule into the interfacial film—in other words, to use a mixed emulsifier system. It is relatively simple to show, experimentally, that mixed emulsifiers produce more stable emulsions from a given oil–water mixture than single emulsifiers—provided only that the emul-

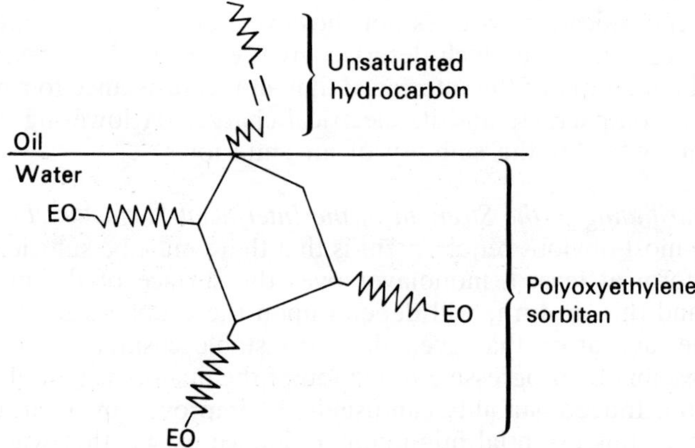

Figure 38.6 Oil–water interface showing orientation of a molecule of a polyoxyethylene sorbitan unsaturated ester

sifiers chosen are chemically and physically compatible with each other. The choice of a combination of anionic and cationic molecules, for example, would be unwise because of chemical combination and electrochemical neutralization of charges. This does not mean, however, that they have to be of similar HLB values. In fact, better results are often achieved by a combination of surfactant molecules having widely differing HLB values combined in such quantities as to produce a resultant HLB close to the optimum for the system to be emulsified (this latter point will be discussed more fully later).[8] The reason for this phenomenon is clearly that the two dissimilar molecular types can, by alternating in the interface, form a much more closely packed, condensed interfacial film. In Figure 38.7a the internal phase has been surrounded by an interface of surfactant molecules of low HLB value which, because of steric hindrance and the mutual repulsion of chemically identical entities, has formed only a discontinuous film. Figure 38.7b shows this surfactant partially replaced by two others with widely different HLB values and chemical composition allowing a close-packed, continuous film to be produced which acts as a much better mechanical barrier to coalescence. Sometimes the second emulsifier can have very little emulsifying potential at all—that is, it can have an HLB value close to zero—just so long as it can migrate to the surface and 'insulate' the other molecules from each other. Cetyl and oleyl alcohols will function in this manner and so will glyceryl monostearate.

Influence of Electrical Charge on Emulsion Stability

Consider an oil droplet stabilized with sodium stearate in an oil-in-water emulsion. The negatively charged carboxylate groups project outwards from the interfacial film into the water phase while the non-polar hydrocarbon chains enclose the oil droplet. Thus the presence of electrical charges at the surface of each droplet is explained and the mutual repulsion of like charges on all the droplets helps to prevent coalescence. The optimum stability conditions exist, not surprisingly, when the interfacial film is completely covered with charges.[5,6] It is also obvious that, since opposite charges will neutralize one another, the combination of anionic with cationic emulsifiers can only result in a decrease in the stability of the emulsion.

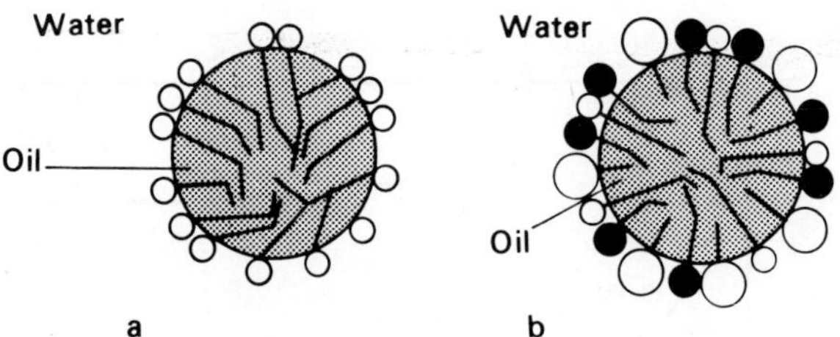

Figure 38.7 Effect of mixed emulsifiers on emulsion stability
a Poor stability *b* Good stability
○ Low HLB—unsaturated
● Medium HLB—unsaturated
◯ High HLB—saturated

More surprisingly, it has been shown that charges at the interface also occur when a totally nonionic emulsification system is used. This has been accounted for by frictional forces originating in the movement of oil droplets in the continuous phase. It has been shown that electrical charge can result when two liquids with different dielectric constants are mixed and that the one with the higher dielectric constant is always positively charged while the one with the lower dielectric constant is always negatively charged.[9] (Thus in oil-in-water nonionic emulsions, the oil droplets are always negatively charged.)

Although no method exists for measuring surface potentials in an emulsion directly, the 'zeta potential' can readily be evaluated by measuring the velocity of charged droplets in an applied DC field. The measured values of emulsion zeta potentials for nonionic emulsifier combinations have proved to be surprisingly high (about 40 mV); perhaps even more interesting is the fact that, when zeta potential is plotted graphically against the HLB of the emulsifier combination, the maximum zeta potential is always found to coincide with the optimum HLB value for the particular system studied.

If mobile ions are present in the external phase of an emulsion, they are attracted by the charged droplets of the internal phase (if these have an opposite charge) giving rise to the formation of an electrical double layer. The nature and effect of this double layer are markedly different in oil-in-water emulsions and water-in-oil emulsions (Figure 38.8).[10] The thickness of the double layer around oil droplets in oil-in-water emulsions amounts to only 10^{-3} to 10^{-2} μm. Electrical repulsion therefore occurs at very short inter-globular distances and this results in a very considerable electrical barrier which must be overcome before two droplets can coalesce. On the other hand, the electrical double layers around water droplets in oil are very diffuse (several μm in size) and the electrical potentials of adjacent droplets overlap, lowering the potential barrier. The stability of water-in-oil emulsions cannot, therefore, be attributed to electrical repulsion of charged droplets.[11]

Other Factors Affecting the Stability of Emulsions

It has been shown how the physical barrier afforded by the condensed molecular layer of emulsifiers at the interface of an emulsion can help to prevent coalescence and how this same layer can, by electric repulsion and steric

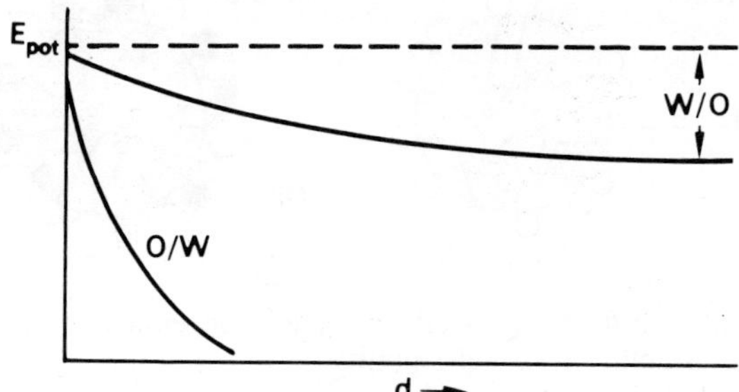

Figure 38.8 Potential energy (E_{pot}) of repulsion due to electrical double layers plotted against distance (d) between internal phase droplets

hindrance, prevent droplets coming together. Both these phenomena have a stabilizing influence on the emulsion but there are other factors which may also affect the stability for better or worse.

The simple picture given by Stokes's Law shows that mobility of internal phase droplets is affected by the viscosity of the continuous phase, the difference in density between the oil and water phases and the size of the disperse phase droplets. Of these, the differential density between the two phases is least amenable to experimental control, although it is obvious that the smaller this difference is, the less is the likelihood that the internal phase will float to the surface or sink to the bottom of the emulsion.

Viscosity of the Continuous Phase

Viscosity is an important parameter, because it can be readily varied—usually by the addition of a thickening or gelling agent (provided that these are compatible with the emulsifier system). Although a large number of such agents is available, the formulator is limited in his choice because the rheological behaviour of most emulsions is almost entirely determined by that of the external or continuous phase. Thus the viscosity, thixotropy and the 'feel' on application of the total emulsion may be affected by the thickener chosen for the continuous phase. The mode of action of many thickening agents is two-fold: firstly, by forming gels consisting of long, interlinked molecules, they physically hamper the flow of continuous phase and of particles of internal phase within it; secondly, they compete with the internal phase chemically for the available external phase. For example, sodium carboxymethylcellulose forms a gel-like dispersion in water, trapping free water and oil droplets within its interstices in oil-in-water emulsions. At the same time, the polymer chains absorb water and become swollen, so decreasing the amount of continuous phase available to the oil droplets. In the same way, free stearic acid (or simple complexes of it) will gradually crystallize in emulsions stabilized with sodium or triethanolamine soaps, resulting in a gel structure which gradually builds up over several hours after manufacture. If, moreover, excess cetyl alcohol over the quantity required to help form a condensed, monolayer interfacial film is present, this will increase the viscosity of the continuous phase by micelle formation, trapping water within each micelle and crowding the available continuous phase. In the same way, pigment suspensions can increase the continuous phase viscosity, although this is a rather special case since the degree of viscosity increase is also very dependent upon the surface characteristics of the pigments involved.

Ratio of Oil Phase to Water Phase

Although the proportion of oil phase to water phase has a marked effect on such parameters as the 'feel' and overall viscosity and appearance of the emulsion, it can also influence the stability. The higher the proportion of internal phase, the greater the number of droplets. The chances of collision are thereby increased and the average distance which one droplet must travel to collide with another (the 'mean free path') is reduced. All this increases the likelihood of coalescence.

Temperature
In has already been stressed that optimum stability is achieved by the correct choice of emulsifier combination. The chosen emulsifiers must be compatible, of correct HLB value and of correct chemical type. These last two characteristics are heavily dependent upon the relative solubility of the hydrophilic and lipophilic ends of the surfactant in the water and oil phases respectively. Solubility, however, is very temperature-dependent. It is unlikely that as the temperature of an emulsion changes, the relative solubilities of both ends of all its emulsifier system will change in strict proportion. In other words, HLB is to some extent a temperature-dependent property itself. Variation of temperature can therefore decrease the stability of an emulsion. This is obviously something which must be borne in mind when formulating products for differing climates. However, the stability of emulsions is often tested in the laboratory by storage at elevated and refrigerator temperatures (and sometimes by cycling between these two extremes) in the hope that such methods will give a rapid indication of the stability on prolonged storage at average temperature. Such a practice is highly questionable. The stability testing of emulsions is discussed more fully later in the chapter.

Concentration of Ions in the Water Phase
The dielectric constant of the oil phase of an emulsion is not great enough to allow any ionizable chemical species to dissociate to any great extent and it has already been seen that this reduces the stabilizing influence of any electrical double layer when oil is the continuous phase. Dissociation is an important factor in oil-in-water emulsions, however, not only because of the influence on ionic emulsifiers but also because of the effects of other soluble ionizable species in solution including hydrogen and hydroxide ions.

The pH value of emulsions is a parameter which is often discussed for a variety of reasons, not least because of its effect on stability. It is necessary to remind ourselves, however, that pH is a measure of hydrogen ion activity in an aqueous environment. Whether the term should strictly be applied to an oil-in-water emulsion is a moot point since the influence of the oil phase and emulsifiers on hydrogen ion activity is probably unknown. However, common usage dictates that the measurement of the pH of emulsions, when the continuous phase is aqueous, will continue. Under no circumstances can the concept of pH be applied when oil is the continuous phase.

Nevertheless, the influence of hydrogen ion concentration in oil-in-water emulsions is dramatic whenever an ionizable emulsifier system is used because of the change of species which can be brought about. Anionic emulsifiers are converted to non-ionizable salts in acidic media and the reverse is true of cationic emulsifiers. In both instances, water solubility and therefore all emulsifier activity can be lost. The effect of pH on amphoteric emulsifiers is less dramatic but obviously dictates whether the anionic or cationic form predominates.

The presence of other mobile ions in solution in the external phase also has an important influence. They are attracted by the charged droplets of the disperse phase, as we have seen. The zeta potential falls as more electrolyte is added and this can decrease emulsion stability. Even at relatively high electrolyte concentrations and zeta potentials close to zero, many emulsions remain stable. This must be due to steric stabilization and the mechanical barrier afforded by a good interfacial film.

Practical Aspects of Emulsifier Choice

Having now accounted for the stabilization of emulsions, it will be apparent that the cosmetic formulator has a large number of options to choose between when considering the composition of a new emulsion. Many of these questions will be settled by his brief—the purpose for which the emulsion is to be designed and the desirable properties it is to have. A detailed account of the influence of the oil-soluble or water-soluble components other than the emulsifier system is outside the scope of this chapter, so that our starting point in choosing a suitable emulsifier system must be that the compositions of both phases have already been chosen.

Probably the first question that needs to be settled is the chemical classification of emulsifier system to be used. This, in turn, can depend upon the content of the other two phases. If the product is to be alkaline, cationics should not be considered. It would be equally unwise to use an anionic emulsifier in an emulsion of low pH and, if electrolyte concentration in the aqueous phase is to be high, nonionics are the best choice. On the whole, members of this latter group are probably least affected by incompatibilities with the remainder of the formulation—apart from the well known ability of polyoxyethylene chains to deactivate certain preservatives. Unfortunately, complete guidance on the choice of emulsifier type is not possible because of the many and varied factors involved; the formulator must experiment for himself to gain good practical experience in order to be able to come to a quick decision.

Determination of Required HLB

The optimum or 'required' HLB value of the emulsifier system for a given composition of oil and water phases provides a useful starting point in the selection of emulsifiers which will give an emulsion of good stability. The determination of the optimum HLB value is based upon a series of practical experiments in which a set of emulsions is produced, identical in every way except for variation in the ratio of a pair of emulsifiers. These emulsifiers are a matched pair, one lipophilic and one hydrophilic, of known HLB values—for example, sorbitan monostearate (HLB 4·7) and polyoxeythylene sorbitan monostearate (HLB 14·9). These are mixed in ratio to give combined HLB values according to the formula.

$$\text{HLB} = xA + (1 - x)B$$

where x is the proportion of a surfactant having an HLB value of A and the other surfactant has a value of B. This is a straight-line relationship and can therefore be computed graphically. The HLB values of the series of mixtures are chosen to differ by an increment of 2 throughout the range bounded by values of the two emulsifiers chosen. In the example above, a range of 4·7, 6, 8, 10, 12, 14·9 might be considered. For each of the test emulsions, an excess of emulsifier (approximately 10 per cent of the weight of the oil phase) is used and all the emulsions are made in precisely the same way. Usually, one or more of the emulsions will give better stability than the others. Should they appear to be uniformly good, however, the experiments must be repeated using less emulsifier; more emulsifier should be used if the set is uniformly bad. Occasionally, two combinations having widely different HLB values show outstanding ability. In this case the low value probably relates to a water-in-oil and the high value to an oil-in-water emulsion. This trial-and-error process has now enabled the experimenter to arrive at an idea of the optimum HLB value for his system. At this stage, a more accurate determination can be achieved by producing a second set of emulsions using the same emulsifier pair but combined to give HLB values in smaller increments close to the value obtained in the first series of experiments. For example, if the initial value was found to be 8, then a series 7·4, 7·6, 7·8, 8·0, 8·2, 8·4 might be used.

Determination of the Best Chemical Type

Having found the optimum for the HLB value, it is now necessary to discover the best chemical type of emulsifier to use. Since this has to do with the greatest cohesive energy of each end of the surfactant molecule for its appropriate phase, some preliminary theoretical selections can be made on the basis of the simple rule 'like attracts like'. For example, if the oil phase is to contain a high proportion of unsaturated or highly branched molecules, then a choice of emulsifier based on oleates or 'iso' esters might be appropriate. Eventually, however, the final choice must depend upon trial-and-error. It should be remembered that best stability will be obtained with a mixed emulsifier system of the optimum HLB value.

Limitations of the HLB System

The above account of the use of the HLB system and the earlier sections on its theoretical basis give no more than an outline of the concept. Many studies on the limitations of the system and on the possibility of improving it have been made, none of which detract from its general usefulness. The reader is urged to study this concept in more depth.[12–15]

Orientation of Phases

Three factors combine to determine which phase will be continuous and which phase will be disperse in a cosmetic emulsion: the type of emulsifier system used, the volume ratio of light to heavy phase and the method of manufacture. All three factors are interrelated, but any one of them can exert a controlling influence—at least, during the initial formation of the emulsion. It is well known, however, that spontaneous changes in the orientation of phases can occur—a phenomenon known as 'phase inversion'.

The effect of the HLB value of a chosen emulsifier system has already been discussed. Clearly, the phase having the greatest interfacial surface tension tends to produce a concave surface so that, other factors allowing, it becomes the internal phase of the emulsion. If the surface tensions on both sides of the interface are equal—or nearly so—then inversion might be expected to take place readily. It has been shown that there is an HLB value at which inversion takes place most easily.[16]

Volume Ratio
Theoretical calculation shows that the maximum volume which can be occupied by uniform spherical particles is 74 per cent of the total liquid volume. Emulsions may, however, be prepared with internal phases amounting to 99 per cent of the total liquid volume.[17] This is possible because the spherical droplets can become distorted in shape. In such emulsions, the internal phase particles become increasingly angular as they are crowded together. Not surprisingly, the effect of high internal phase concentrations is to produce emulsions of greatly increased viscosity and their ability to remain stable depends primarily on the mechanical strength afforded by a highly condensed and structured interfacial surfactant film.

Method of Manufacture
Although there are exceptional cases, it is generally difficult for a phase to be dispersed in a stirred tank if it occupies more than 74 per cent of the total liquid volume. Either liquid may, however, be dispersed over a wide range of relative volumes (the ambivalent region) and for systems containing no emulsifier the choice of dispersed phase is often dependent on the manner in which dispersion is initiated. If a simple, two-layer mixture of water and oil phases is agitated it will tend to form an oil-in-water system if the agitator is sited in the water phase or a water-in-oil system if the agitator is immersed entirely in the oil phase, because the disperse phase is most likely to be the one that is drawn into the other. For the same reason, if the vessel is initially filled with one phase (prior to the addition of the second phase), this initial phase will be the continuous one.

It appears that the orientation of the emulsion is also affected by the type of agitator used and its speed. For a given stirrer speed, there exists a volume ratio (of light to heavy phase) above which the heavy phase is dispersed, and a region at lower volume ratio below which the lighter phase is dispersed. Between these limits lies the ambivalent or metastable region where either phase may be dispersed—but this shows a strong hysteresis effect. Thus if, at a constant stirrer speed, water is added to a stable water-in-oil emulsion the system will eventually invert at the lower volume ratio limit. If oil is now added to this emulsion, reinversion will not occur until the ambivalent region has been traversed and the upper limit reached. This ambivalent region can be increased by adding solutes which are partially soluble in both phases, such as surfactants.

As the stirrer speed is increased, the inversion points of all volume ratios tend to increase asymptotically to a constant value which is dependent upon the stirrer design. It has also been noted that for equal phase volumes at high stirrer speeds, the heavy phase tends to be the continuous one.

Another interesting point is that droplet size of dispersed phase varies with volume ratio and is greater when the lighter phase is dispersed. Thus, on inversion, a step change in droplet size occurs.[18]

These observations provide the background to various methods of practical emulsion manufacture.

Mechanism of Phase Inversion
Phase inversion is a spontaneous process and must therefore be accompanied by a decrease in the total energy content of the system. The power input at

inversion does not appear to change, therefore the emulsion must be examined for a change of energy from within the system itself. It has been seen that there may be increase or decrease in the drop size and interfacial area, hence interfacial energy may increase or decrease. It does not seem possible, therefore, that inversion is concerned with minimizing interfacial energy.

Figure 38.9 is a schematic representation of an oil-in-water to water-in oil inversion. As increasing amounts of oil phase are added, agglomeration of oil droplets proceeds, enclosing small quantities of water between them. Eventually, at the inversion point, the interfacial film at the points of contact of agglomerating droplets reorientate so as to form water droplets of unusual shape which can then float away.[11]

The main observable change at the inversion point is a sudden marked decrease in the viscosity of the system as the close-packed disperse phase droplets suddenly become the continuous phase. It seems likely, therefore, that

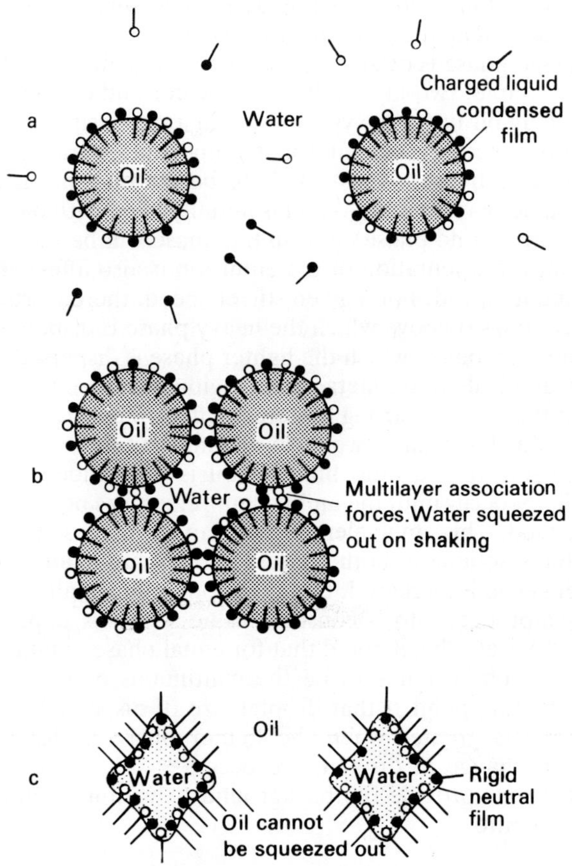

Figure 38.9 Schematic representation of phase inversion
○ Na cetyl sulphate
● Cholesterol

the energy change which brings inversion about has to do with flow and turbulent behaviour.

Except when used purposely, phase inversion is not often encountered in cosmetic production provided that the production chemist is aware of the conditions under which it can occur.[16,19] It is probably true to say that at the present time inversion cannot always be predicted but that the conditions which increase the risk can be recognized and it can be explained when it does happen.

Assessment of Emulsion Stability

Although all emulsions will eventually lose their excess energy by breaking down, it is obviously important that any commercial product should maintain its integrity throughout its useful life. In the case of cosmetic creams and lotions, the formulator has to bear in mind that they may well have to wait six to nine months on a shelf in a variety of conditions of temperature and humidity before they are purchased. The purchaser should then be able to expect these same products to withstand unfavourable storage conditions and microbiological insult for a further period of three to six months (longer in some cases) while they are in use.

The importance of the packaging in protecting the product cannot be overestimated and no emulsion should ever be placed on sale until the complete compatibility of product and pack can be assured. Even before this, however, the chemist needs information about the relative stability of his emulsion to guide him during the initial stages of formulation and pilot manufacture. Clearly, it is impossible to wait twelve months before he can arrive at a conclusion; so it is that so-called 'accelerated' storage test procedures have been devised and used in formulation laboratories. These tests take two complementary forms: those which are designed to speed up the aging process of emulsions and those designed to detect aging and measure it in an objective way.[20–25]

Because no emulsion can be separated from its environment, the influence of such factors as temperature variation, light, mechanical vibration, atmospheric oxygen and microbiological contamination cannot be ignored in any assessment of stability. For this reason, almost every cosmetic emulsion is certain to be subjected to one or more of the following accelerated aging processes at some time during its development.

(a) Storage at ambient temperature for up to nine months in glass or plastic containers.
(b) Storage at 35°–40°C for up to three months in glass or plastic containers.
(c) Storage in partially filled containers at ambient or elevated temperature.
(d) Storage at low temperatures (−5°C to +5°C) for up to three months.
(e) Storage in freeze–thaw cycle cabinets (−5°C to +30°C, two cycles per 24 hours).
(f) Centrifugation tests.
(g) Microbiological challenge tests.

It has already been pointed out that since the HLB value of a surfactant molecule is temperature-dependent, the fact that an emulsion breaks down rapidly at elevated temperature is not much of a guide to its performance at more normal temperatures over a longer period of time. Such accelerated test procedures should therefore be looked upon only as an assessment of resistance to temperature changes, and at best can only be regarded as an indication of normal ambient stability.

Centrifugation speeds up the rate of sedimentation by increasing the value of *g* according to Stokes's equation. Whether or not it has a measurable effect on the likelihood of cohesion once it has pushed particles close to each other is uncertain. Nevertheless, centrifugation provides a simple and rapid method of assessing the potential stability of various emulsion formulae.[26,27] Each laboratory has its own detailed methodology; a good emulsion should be able to stand up to 5000–10 000 rpm in a standard laboratory centrifuge for 30 minutes without showing signs of separation.

Increasingly, attention is being paid to those methods of emulsion assessment which give a measurement of the rate of the aging process. The oldest and probably the most widely used of these is examination by light microscopy. The size, distribution and shape of the disperse phase droplets can, to the experienced eye, tell a great deal about the emulsion and its likely stability. In particular, uneven size distribution and the aggregation of droplets are danger signs which should be looked for when comparing emulsifier performance in a given system; those giving droplets of the smaller particle size under identical conditions are obviously to be preferred.

Another approach to the monitoring of emulsion breakdown is to monitor its composite dielectric constant[25] or electrical conductivity.[28–31] In particular, the conductivity of oil-in-water emulsions can be expected to decrease with the increasing drop-size of the dispersed phase. The conductivity of a water-in-oil emulsion should be zero, but once the disperse phase particles have reached a critical size (depending on the composition of the emulsion and the applied voltage) continuous paths of conducting water phase allow a measurable current to flow. It may be said that, although not widely used, conductivity measurements have some value in the study of water-in-oil and oil-in-water emulsions.

The apparent viscosity of an emulsion depends partly upon the size distribution of the internal phase droplets. Change of viscosity is therefore another parameter by which changes likely to affect the stability of emulsions can be monitored.[32]

Generally, these storage and monitoring techniques can be valuable help in formulation, in the development of manufacturing procedures and in production control.

Characteristics of Emulsions

Having discussed the factors affecting the stability of cosmetic emulsions in some detail, it is appropriate to turn to the other characteristics by which they are judged by the user and the means by which these can be controlled.

Of prime importance when considering cosmetic emulsions is their appearance, since this can help to determine their customer appeal. Emulsions may vary tremendously in appearance from glossy opaque whiteness through a grey translucence to sparkling clarity. Opacity is due to two interrelated factors: the size of the internal phase droplets and the difference between the refractive indices of the internal and external phases. Light is reflected and refracted at each interface between droplet and continuous phase. Such changes in direction are so numerous (because of the large number of droplets) that much of the light escapes from the emulsion surface in the same direction that it entered—that is, back towards the viewer. If, however, the refractive indices of both phases are identical, or nearly so, no such reflections and refractions take place; light travels unhindered through the emulsion which has a sparkling clear appearance. This applies no matter what the size of the internal phase droplets. If the droplets are large, however, each ray of light encounters only a small number of interfaces during its passage through the emulsion. Sufficient light is reflected back towards the viewer to make the presence of the droplets obvious, but the bulk of the light, which is refracted, can find its way through. This accounts for the globular appearance of emulsions in and advanced stage of aggregation and separation. As the particle size of the internal phase diminishes, the familiar milky-whiteness appears: as the size reduction continues, the colour takes on a blueish hue, becoming grey, semitransparent and finally transparent. These changes in appearance occur as the particle size of the droplets approaches that of the wavelength of light itself. The probability that a light ray will collide with (and be reflected by) a tiny particle is enormously reduced once the particles become so small that they are comparable in size to the wavelength of light. Under these circumstances the majority of rays pass through the emulsion without being reflected or refracted and the emulsion appears to be transparent.

As the droplet size approaches that of the wavelengths at the red end of the spectrum, the reflected or refracted light is made up increasingly of smaller wavelengths at the blue end of the spectrum until, eventually, the droplets become too small for interaction at all.

In practice, it is difficult to formulate emulsions in which both phases have similar refractive indices—micro-emulsions are far more frequently found, although even these are not common (Table 38.2).

The gloss of the emulsion is a function of the microscopic smoothness of its surface. For ultimate smoothness and gloss, the internal phase particles must be relatively small and even in distribution and there must be no inclusions in the external phase such as large crystallites of stearic acid or inorganic matter of large particle size.

Table 38.2 Effect of Particle Size of Internal Phase on Emulsion Appearance

Internal phase droplet size	Emulsion appearance
≥0·5 mm	Globules clearly visible
0·5 mm to 1 μm	Milky-white
1 μm to 0·1 μm	Blue-white
0·1 μm to 0·05 μm	Grey, semi-transparent
<0·05 μm	Translucent or transparent

Rheological Properties

The rheological behaviour of emulsions is an important subject, not only because of its influence on the 'feel' and acceptability to the consumer but also because of its impact on the manufacturing process. The science of rheology relates to matter which is being deformed or made to flow by applied forces.

Figure 38.10 represents an emulsion flowing steadily under a constant force through a pipe. The layer A in contact with the pipe is practically stationary but the central part C of the emulsion is moving relatively fast; in this sense, the emulsion is not only flowing, but is being deformed. At other layers between A and C (such as B) the emulsion has a velocity less than at C, the magnitude of the velocities being represented by the length of the arrowed lines in the figure. Since the velocities of neighbouring layers are different, a frictional force is generated between them just as in the case of two solid surfaces moving over each other.

It was Newton who first suggested that this frictional force, F, was proportional to the area of surface considered and the velocity gradient in the part of the liquid at the point of interest. Thus:

$$F = \eta A \times \text{velocity gradient}$$

where A is now an area and the proportionality constant, η, is known as the coefficient of viscosity.[33]

When materials are subjected to deformation of the type illustrated in Figure 38.10, they are often said to be under the influence of 'shearing' forces. Thus the quantity F/A is known as the 'shear stress' per unit area. To be consistent, the velocity gradient is similarly referred to as the 'rate of shear'. In this way, the viscosity of emulsions and other liquids can be defined as the shear stress divided by the rate of shear:

$$\eta = \frac{F'}{S}$$

where $F' = F/A$ and S is the velocity gradient, or rate of shear.

Evidently, for all fluids conforming to the above equation, the viscosity is independent of the rate of shear; if the emulsion is forced twice as hard through the pipe, it will flow at twice the rate. Such materials are said to exhibit

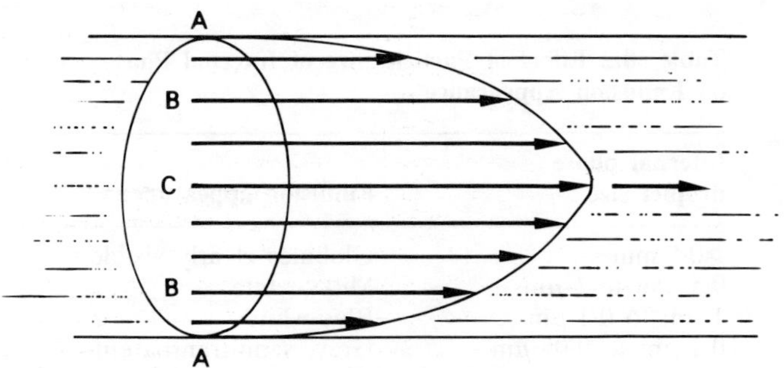

Figure 38.10 Different flow velocities of layers of emulsion in a pipe

'Newtonian' behaviour and include water, hydrocarbon oils and certain other liquids such as low-viscosity silicone oils. Many liquids, however, including the vast majority of emulsions, show deviations from this simple model and are therefore to be regarded as 'non-Newtonian'.

In Figure 38.11 curve A illustrates the relationship between viscosity and rate of shear for a Newtonian fluid; curve B illustrates the case where the viscosity apparently falls as the rate of shear increases. In some instances, a certain force must be applied before any shear (or flow) takes place at all—such materials are said to exhibit 'plastic' behaviour. Many fluids, on the other hand, merely exhibit a decrease of viscosity as rate of shear increases from zero, and these are referred to as 'pseudoplastic'. Dilatant materials 'firm up' as rate of shear increases; emulsions of this kind are infrequently encountered. Much more common are emulsions which show a degree of 'thixotropic' behaviour. Thixotropic materials exhibit reversible behaviour—in other words, after a lowered viscosity caused by increased rate of shear, a subsequent reduction of shear results in a corresponding increase in apparent viscosity. This increase may not be immediate, and recovery can be slow. The majority of emulsions are non-Newtonian and show some degree of thixotropic behaviour, although the complete recovery of the initial viscosity is not always achieved.[34]

Two factors contribute to the viscosity of emulsions: the viscosity of the external phase—which has been dealt with already—and the ratio of internal to external phase. Apparent viscosity increases with the proportion of internal phase. In extreme cases, where this exceeds 74 per cent of the total volume, the emulsion can be transformed so that it has a paste-like consistency and examination under the microscope shows that the usual spherical shape of the internal phase droplets has become angular and distorted.[17] Such viscosity is referred to as 'structural' viscosity and may be achieved with emulsions of either type.[34]

Finally, it should be noted that air trapped in the emulsion can cause a considerable increase in apparent viscosity, particularly if it is very finely divided.

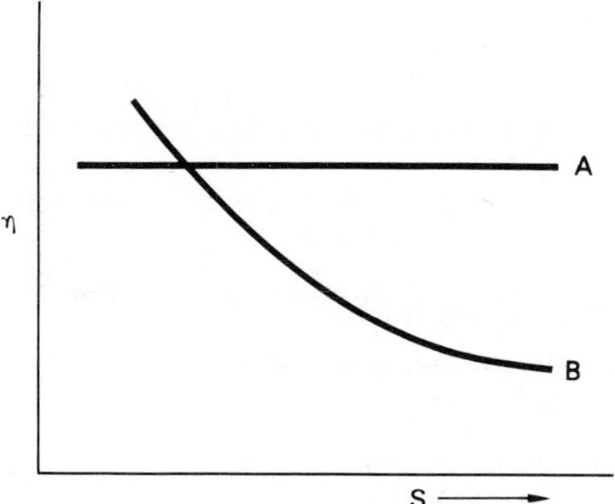

Figure 38.11 Viscosity (η) versus rate of shear (S) for a Newtonian fluid (A) and for a pseudoplastic material (B)

Application Properties
The in-use properties of cosmetic emulsions can be thought of as those which are apparent during its application to the skin or hair (the 'feel') and the after-effects once the product has been applied. Both types of property are important since even the most effective products will not appeal to the consumer if the 'feel' on initial application is unpleasant.

The initial feel of an emulsion is largely dependent on that of the external phase; thus an oil-in-water emulsion will feel like water, whatever is dispersed in the aqueous phase. Water-dispersible thickeners and additives such as glycerin, sorbitol and glycols will all exert some effect. Water-in-oil emulsions will feel oily—but whether or not they are sticky, for example, depends upon the choice of oil-phase ingredients. Viscosity also plays an important part in the initial impact of an emulsion: high viscosities tend to give a cream 'richness'.

During the application, some emulsifiers tend to promote the appearance of a foam-like whitening, often referred to as 'soaping'. Anionic emulsifiers are particularly prone to this and the effect is not always considered advantageous since it prolongs the application time.

As the water and other volatile ingredients evaporate, the 'feel' changes. Oil-in-water emulsions invert; this may happen abruptly or gradually but the difference in 'feel' as this happens, while being easy to detect, is somewhat difficult to describe in words.

Finally, the after-effects are determined by the choice of oil-phase ingredients (which may be greasy or non-greasy) and any non-volatile water-phase ingredients.

Determination of Emulsion Type

Many methods have been proposed for the determination of the identity of the external phases of emulsions. Although each has its drawbacks, a combination of all three of the following methods may be expected to give a reliable indication of the orientation of an emulsion.

(a) The emulsion is subjected to an electric voltage.[28] If no current flows, the external phase is non-conducting (that is, oily). If an appreciable current flows, the external phase is conducting (that is, water). If a small current flows, this may indicate a dual emulsion or a gradual inversion.
(b) Oil-in-water emulsions will disperse easily in water, water-in-oil emulsions will disperse easily in oil.
(c) Water-soluble dyes will spread through oil-in-water emulsions, oil-soluble dyes through water-in-oil emulsions.

Quality Control and Emulsion Analysis

The following properties of emulsions are most commonly examined for the purposes of analysis and quality control.

Colour, Odour and General Appearance.

Weight per Millilitre. Several varieties of pyknometer are commercially available for use with emulsions.

Apparent Viscosity. Several methods of determining this property are used in the cosmetics industry. Better still is the measurement of apparent viscosity at two rates of shear (preferably with a ten-fold different in shear rate) since the ratio of the two readings gives an indication of the degree of non-Newtonian behaviour (Figure 38.11).

Hydrogen Ion Concentration. The measurement of pH has been dealt with earlier in the chapter.

Water Content. Probably the best method is by Karl Fischer titration.

Volatile Content. This is usually measured by weight loss over 24 hours in an oven at 110°C.

Stability. This has already been covered.

Chemical Identity of Separated Phases. This may be required for the purposes of evaluating emulsions of unknown composition or to check that certain key ingredients (for example, preservatives) have been added to the emulsion under examination. This is a rather specialized topic and is normally reserved for the expert analytical chemist, although analysis of this type is occasionally reported in the general literature.

REFERENCES

1. Griffin, W. G., *J. Soc. cosmet. Chem.*, 1949, **1,** 311.
2. Bancroft, W. D., *J. Phys. Chem.*, 1913, **17,** 501.
3. Bancroft, W. D., *J. Phys. Chem.*, 1915, **19,** 215.
4. King, A., *Trans, Faraday Soc.*, 1941, **37,** 168.
5. Schulman, J. H., and Cockbain, E. G., *Trans, Faraday Soc.*, 1940, **36,** 651.
6. Alexander, E. A., and Schulman, J. H., *Trans. Faraday Soc.*, 1940, **36,** 960.
7. Hildebrand, J. H., *J. Phys. Chem.*, 1941, **45,** 1303.
8. Griffin, W. G., Emulsions, in *Encyclopaedia of Chemical Technology*, ed. Kirk R. E. and Othmer, D. F., Vol. 8, 2nd edn, New York, Interscience, 1965, p. 137.
9. Coehn, A., *Ann. Phys. Chem.*, 1898, **66,** 217.
10. Lange, H., *J. Soc. cosmet. Chem.*, 1965, **16,** 697.
11. Schulman, J. H. and Cockbain, E. G., *Trans. Faraday Soc.*, 1940, **36,** 661.
12. Neuwald, F., *Sci. Pharm.*, 1964, **32**(2), 142.
13. Riegelman, S. and Pichon, G., *Am. Perfum.* 1962, **77**(2), 31.
14. Sherman, P., *Rheology of Emulsion*, Proc. Symp. Brit. Soc. Rheology, Harrogate, 1962.
15. Riegelman, S., *Am. Perfum.* 1962, **77**(10), 59.
16. Becher, P., *J. Soc. cosmet. Chem.*, 1958, **9,** 141.
17. Griffin, W. G., Emulsions, in *Encyclopaedia of Chemical Technology*, ed. Kirk, R. E. and Othmer, D. F., Vol. 8, 2nd edn, New York, Interscience, 1965, p. 121.
18. Marsland, J. G., *Heterogeneous Liquid Systems*, Notes for Postgraduate Course in Mixing Technology, Bradford University, 1976.
19. Haynie, F. H. Jr., Moses, R. A. and Yeh, G. C., *A.I.Ch.E.J.*, 1964, **10,** 260.
20. Lachman, L., *Am. Perfum.*, 1962, **77**(10), 59.
21. Wilkinson, J. B., *Am. Perfum.*, 1962, **77**(10), 105.

22. Kennon, L., *J. Soc. Cosmet. Chem.*, 1966, **17,** 313.
23. Sherman, P., *Soap. Perfum. Cosmet.* 1971, **45**(11), 693.
24. Jass, H. E., *J. Soc. cosmet. Chem.*, 1967, **18,** 591.
25. Kaye, R. C. and Seager, H., *J. Pharm. Pharmacol.*, 1965, **17,** Suppl. No. 12, 92S.
26. Rehfeld, S. J., *J. Phys. Chem.*, 1962, **66,** 1966.
27. Vold, R. D. and Groot, R. C., *J. Phys. Chem.*, 1962, **66,** 1969.
28. Holzner, G. W., *Seifen Öle Fette Wachse*, 1966, **92**(12), 299.
29. Mrukot, M. and Schmidt, M., *Parfum. Kosmetik*, 1976, **57,** 337.
30. Brandau, R. and Bold, K. W., *Fette Seifen Anstrichmitt.*, 1977, **79**(9), 381.
31. Ludwig, K. G. and Hameyer, P., *Parfum. Kosmetic*, 1974, **55**(9), 253.
32. Sherman, P., *J. Soc. cosmet. Chem.*, 1965, **16,** 591.
33. Minard, R. A., *Instrum. Control Syst.*, 1959, **32**(6), 876.
34. Shaw, A. M. *Chem Engineering*, 1950, **116,** Jan.

Further Reading—General

35. Manegold, E., *Emulsionen*, Heidelberg, Verlag Strassenban Chemie und Tecknik, 1952.
36. Sumner, C. G., in *Theory of Emulsions and Their Technical Treatment*, ed. Clayton, London, Churchill, 1954.
37. Becher, P., *Emulsions: Theory and Practice*, New York, Reinhold, 1965.
38. Adam, N. K., *Physics and Chemistry of Surfaces*, London, Oxford University Press, 1941.
39. Davies, J. T., *Interfacial Phenomena*, New York, Academic Press, 1963.

Further Reading—Emulsion Stability

40. Lange, H. and Kurzendörfer, C–P. *Fette Seifen Anstrichmitt.*, 1974, **3,** 116.
41. Woods, D. R. and Burrill, K. A., *J. Electroanal. Chem.*, 1972, **37,** 191.
42. Garetti, E. R., *J. Pharm. Sci.*, 1965, **54**(11), 1557.
43. Lin, T. J., Kurihara, H. and Ohta, H., *J. Soc. cosmet. Chem.*, 1973, **24**(13), 797.
44. Sonntag, H., Netzel, J., and Klare, H., *Kolloid*, 1966, **211**(1–2), 111.
45. Rimlinger, G., *Am. Per.*, 1967, **82**(12), 31.
46. Miller, A., *Drug Cosmet. Ind.*, 1965, **97**(11), 679.
47. Vold, R. D. and Groot R. C., *J. Soc. cosmet. Chem.*, 1963, **14,** 233.
48. Boyd, J., Parkinson, C. and Sherman, P., *J. Colloid Interface Sci.*, 1972, **41**(2), 359.

Chapter Thirty-nine

The Manufacture of Cosmetics

Introduction

It is probably true to say that processing in the cosmetics industry has evolved largely through practical experience and principles gleaned by analogy with other industries rather than by many fundamental studies. While the literature abounds with reports from workers in the areas of product development and efficacy, little seems to have been reported on new production technology over the last few decades. This does not necessarily indicate that cosmetics manufacturers see no need for improving their production facilities; on the contrary, production problems—especially those associated with relatively large-scale processes—continue to cause difficulties and delays even in the largest and best equipped factories.

The greatest single obstacle to process improvement in most plants is the enormous variety of product types, each with its own set of physical and chemical characteristics, which must be dealt with during the course of a year. The need for flexibility is usually of great importance and this leads towards compromise and, except in the largest manufacturing units, away from equipment especially designed to perform specific tasks.

It is particularly important, therefore, that chemical engineers and production chemists in the cosmetics industry should understand the basic principles and characteristics of the plant at their disposal and that they should be vigilant in the search for new equipment which will perform with even greater efficiency the tasks for which they are responsible.

While cosmetics manufacture is concerned with a very broad range of processes, there are enough common elements to allow a relatively simple overall view of the subject; this helps considerably in a study of the basic principles of cosmetic production technology.

The first step in this simplification procedure is the division of the subject into two parts: bulk manufacture and unit manufacture.

Manufacture of Bulk Product

The whole subject of bulk cosmetics manufacture can be satisfactorily described with reference to three types of process: mixing, pumping and filtering (the process of heat transfer, as will be seen, can be legitimately regarded as a mixing process). Of these processes, by far the most important is mixing.

Table 39.1 represents a convenient way of classifying the mixing processes most commonly found within the cosmetics industry. Every single cosmetics manufacturing process contains at least one mixing operation and often more than one type is involved. For example, the manufacture of a pigmented

Table 39.1 Scope of Mixing Operations within the Cosmetics Industry

Type of mixing	Examples
1. Solid/Solid	
(a) Segregating	None
(b) Cohesive	Face powders, eye shadows and all dry mixing
2. Solid/Liquid	(i) Dissolution (of water-soluble dyes, preservatives, powder surfactants, etc.) (ii) Suspensions and dispersions (pigments in castor oil and in other liquids)
3. Liquid/Liquid	
(a) Miscible	(i) Chemical reactions (formation of soaps from acid and base) (ii) pH control (iii) Blending (spirituous preparations, clear lip gloss products)
(b) Immiscible	(i) Extraction (none) (ii) Dispersion (emulsions)
4. Gas/Liquid	(i) Absorption (none) (ii) Dispersion (aeration and de-aeration)
5. Distributive	
(a) Fluid motion	Heat transfer (during emulsion and other manufacture)
(b) Limited flow	Pumping (pastes and other highly viscous products)

emulsion-based foundation cream may include:

(i) Preliminary dry blending of pigments and excipient (type lb).
(ii) Dissolution of oil-soluble and water-soluble materials separately in their appropriate phase (type 2 example i and type 3a).
(iii) Dispersion or suspension of pigments in the oil or water phase (type 2, example ii).
(iv) Mixing of the two phases to form an emulsion, possibly with the formation *in situ* of a soap as part of the emulsifier (types 3a and 3b).
(v) Adjustment of pH (type 3a).
(vi) De-aeration of the bulk (type 4).
(vii) Cooling to ambient temperature and pumping into a storage vessel (type 5a).

Not only are all these operations different from each other, but at each stage the characteristics of the bulk are quite different and require a different set of processing characteristics to achieve the optimum economic process. Not surprisingly, therefore, the optimum is rarely achieved.

The subject of pumping is not clearly separated from that of mixing since pumping implies the forced flow of product. Any flow will naturally introduce an element of mixing if the product is not already homogeneous. Further, since flow is a common element of both processes, the same product characteristics (for example, rheological behaviour) must be taken into account.

Filtering is not usually a unit operation of major importance in cosmetics manufacture except in the production of spirituous preparations (colognes, aftershave and perfumes). It is possible to regard filtering as un-mixing and certainly the flow characteristics of the filtered product are again of prime importance. The use of sub-micrometre filters for the sterilization of water is discussed elewhere in the book.

Unit Manufacture

Most cosmetic products are filled from bulk in machines specifically designed to handle the units of a particular product type. While it is true that great care must be taken in the choice and the setting up of such machines, the main problems encountered are often concerned with the characteristics of the machines themselves rather than with the manufacture or processing of the product. There are at least two areas, however, where special understanding of the product units and their characteristics are essential for the achievement of efficient production: these are the moulding processes (lipsticks, wax-based sticks, alcohol–stearate gels) and compression processes (compressed eyeshadow, blushers and face powders).

A description of unit manufacture could include all filling and packaging operations; for the purposes of this chapter, however, discussion will be confined to bulk manufacture.

MIXING—AND THE MANUFACTURE OF BULK COSMETIC PRODUCTS

Definition of Terms

The object of a mixing operation is to reduce the inhomogeneities in the material being mixed. As Table 39.1 shows, inhomogeneity may be of physical or chemical identity or of heat. Further, in the processes demanded by cosmetic manufacture, the mixing is designed to be permanent—or as permanent as it is possible to make it—as distinct from those operations (such as extraction and stripping) which rely on eventual un-mixing in order to achieve the desired objective.

Clearly, the degree to which inhomogeneity can be reduced depends on the efficiency of the mixing apparatus used and also on the physical characteristics of the materials constituting the mixture. For miscible liquids, homogeneity can be produced at a molecular level whereas for mixture of powders homogeneity is limited to the sizes of the powder particles themselves. When examing a mixture for quality, therefore, the *scale of scrutiny*—the magnification at which the mixture is examined—must vary from product to product. At an acceptable scale of scrutiny, *perfect mixing* implies that all samples removed from the mixture will have exactly the same composition. This is rarely achievable. *Random mixing* is achieved if the probability of finding a particle of a given component in a sample is the same as the proportion of that component in the whole mixture. Random mixing is the aim of all industrial mixing operations and whereas samples removed from such a mixture will not be identical, the variations should be very small. If the scale of scrutiny is reduced sufficiently, however, this may no longer be true.

Mixing can only occur by relative movement between the particles of the constituent components of the mixture. Three basic mechanisms for achieving this relative movement have been identified: bulk flow, convective mixing and diffusive mixing. *Bulk flow* (which includes shear mixing, cutting, folding and tumbling) occurs in pastes and solids, when relatively large volumes of mixture are first separated and then redistributed to another part of the mixing vessel. *Convective mixing* involves the establishment of circulation patterns within the mixture. Finally, *diffusive mixing* occurs by particle collisions and deviation from a straight line. In miscible liquids of sufficiently low viscosity, the thermal energy which is possessed by the constituent molecules may be enough to achieve a good mixture quality by thermal diffusion without additional energy being applied, although this process is usually too slow for industrial purposes.

It is incorrect to assume, however, that the relative movement between mixture particles brought about by these mechanisms always results in an improved mixture quality; on the contrary, many mixing problems arise from the tendency of mixture particles to segregate during attempts to mix them. *Segregation* is defined as the preference of the particles of one component to be in one or more places in a mixer rather than in other places. The size of the non-uniformities in an imperfect mixture is sometimes referred to as the 'scale of segregation' and the difference in composition between neighbouring lumps or volumes is the 'intensity of segregation'. Segregation is not, fortunately, a major problem in cosmetics manufacture although it does manifest itself occasionally (as, for example, in the flotation of pigments during lipstick processing).

SOLID–SOLID MIXING

Table 39.1 distinguishes between two types of solid–solid mixing operation: those concerned with segregating powders and those with non-segregating or cohesive powders. The essential difference between these two categories relates to the properties of the powders themselves and, in particular, to the freedom which individual particles have to move independently of their neighbours. Free-flowing powders exhibit many process advantages (such as easy storage, easy flow from hoppers, smooth flow of product), but have the disadvantage that they tend to segregate unless all the constituent particles are of very similar shape and size. Cohesive powder, on the other hand, lacks mobility, and individual particles are bonded together and move as clumps or aggregates. Although segregation does not appear to be a problem (except, as will be seen, at very small scales of scrutiny), cohesive powders are difficult to store and do not easily flow from hoppers.

In a powder mass, there are forces at work which tend to make the particles bond to each other and these are balanced by the gravitational masses of the particles which cause them to fall apart again. Although the bonding forces, for a given powder, are largely independent of particle size, their gravitational mass is obviously not. For this reason, particles will stick together only when they are small enough for the gravitational forces acting on them to be much smaller than the bonding forces. Powders composed primarily of such particles exhibit cohesive characteristics and those consisting of larger particles tend to be free-flowing. To a first approximation, therefore, the division between the two types of powder is one of size and the critical size is approximately 50 μm: below this particle size, powders are cohesive.

Figure 39.1 shows the particle size range of commercial grades of some powders commonly used in cosmetics production; by inference it will be noted that they are all predominantly cohesive in nature.

Nature of Inter-particle Bonds

The nature of the bonding forces between powder particles is of fundamental importance to many industries and these are now well-understood.[1–4] The characteristics which it is essential for the cosmetics production chemist to understand, however, are as follows:

1. These forces operate over very short distances. The particles must be brought into very close contact to obtain maximum agglomerate strength (as in pressing).
2. These forces are greatly enhanced by the presence of any liquid—particularly if it is easily capable of wetting and spreading over the particle surfaces.

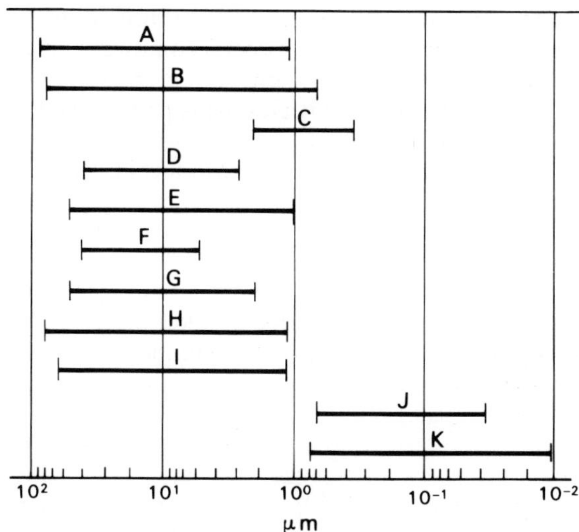

Figure 39.1 Range of particle size of some powders commonly used in cosmetics
A Titanium dioxide
B,C Magnesium carbonates
D Mica
E Zinc stearate
F,G,H Micas coated with titanium oxide
I Talc
J Organic pigments
K Inorganic pigments

3. The forces are very much weaker than those holding the particles themselves together; that is, it is much easier to break up agglomerates than it is to break up a primary particle.
4. The probability of a small particle bonding to a larger one is much greater than that of a particle bonding to another of the same size.

Manufacture of Pigmented Powder Products

Powder eyeshadows, face powders and powder blushers are commonly composed of the following types of material:

Talc
Pigments
Pearl agents
Liquid binder
Preservative

The order in which these ingredients are mixed and the process by which the mixing is carried out depend largely upon the type of equipment that is available. A satisfactory powder product, when examined under high magnification, is seen to consist of small agglomerates or single particles of the pigments adhering to and covering the surface of the larger talc particles. Improperly processed powders contain larger agglomerates of pigments existing as discrete entities and separate from any talc particles. When rubbed, for example between finger and skin surface, such improperly processed powders change hue as these agglomerates are broken and the smaller pigment groups so released follow their natural tendency to coat the larger particles. This process is often referred to as the 'extension' of pigments onto talc.

The processing of bulk pigmented powder products is dominated by the need to achieve adequate 'extension' on an industrial scale. Of all the devices which have from time to time been tried, none has proved more popular than the hammer mill (Figure 39.2). The hammer mill was designed as a comminution machine. It consists of a fast rotating shaft fitted with freely swinging hammers mounted in a cage which is equipped with a breaker plate against which the feed is disintegrated, chiefly by impact from the hammer. The very high speed at

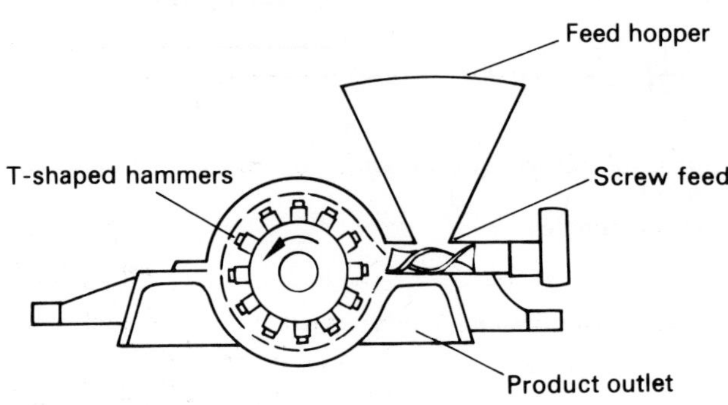

Figure 39.2 Hammer mill

which the hammers move (60–100 m s^{-1}) increases the chance of a hammer making contact with each particle and the dwell-time of particles within the chamber is increased by the placement of a variable size screen over the exit.

Hammer mills are very efficient in the comminution of brittle particles in the range of 1500–50 μm but below this size their efficiency (the probability of direct impact) falls off rapidly. This is fortunate, since it means that cosmetic talcs and micas can be passed through without being substantially altered. At the same time, however, the very high rotational speed of the hammers and the air flow within the chamber ensure that there are enough weak secondary impacts (particle–wall and particle–particle) to break the much weaker pigment agglomerates—which may be up to 50 μm in diameter. The disintegrated agglomerate fractions then stabilize by becoming coated on to larger talc particles and should not be further changed by subsequent passes through the mill.

Nevertheless the hammer mill, in its role as an extender of pigments onto talc, has certain disadvantages. For example, most of the extremely high energy which it makes available is wasted and is largely dissipated in heating the powder. From the viewpoint of energy consumption, therefore, a hammer mill used in this way is very inefficient. The feed-rate and therefore the processing time for all but the smallest batch sizes of powder is very slow. Attempts to speed up the process by the substitution of exit screens of large diameter often result in inadequate extension, necessitating a second or third pass through the mill. On the other hand, increasing the residence time of powder within the grinding chamber by decreasing this mesh size can cause the screen to become blocked with compacted powder, resulting in overheating and damage to machine and product. Perhaps the biggest disadvantage of all, however, is that the hammer mill is a continuous processing device being used for batch processing. For this reason, it must be fed with a powder mixture which has already been effectively mixed, otherwise the colour of the milled product changes as each section of unmixed bulk passes through. This preliminary mixing must be efficient although it is not necessary for any extension to be achieved at this stage.

Since the pre-mix is an additional operation and adds to the processing time, the mixer which is selected must be as efficient as possible. Probably the most widely used is the 'ribbon blender' which comprises a horizontal drum containing a rotating axial shaft which carries ribbon-like paddles. In such a device, the pre-mix can take anything between 20 and 60 minutes. Other mixers are now available which utilize higher energy input but are quicker. Table 39.2 summarizes the properties of some of the more conventional powder mixers. Since it is relatively easy to achieve good mixture quality (at a large scale of scrutiny) in cohesive powders, any mixing device will eventually produce a satisfactory even distribution of components provided that it contains no dead spots where mixing does not take place.

It is usual to add the liquid binder during this preliminary mixing stage. The binder may be poured into a suitable orifice in the mixer although many production chemists prefer to spray it into the mixer cavity as an aerosol through a venturi or similar device. This procedure helps to distribute the liquid more evenly and avoids the formation of wet, lumpy areas in the powder body. The

Table 39.2 Conventional Powder Mixers

Type of mixer	Batch/ continuous	Main mixing mechanism	Speed of mixing	Ease of cleaning	Energy consumption	Quality of extension
Horizontal drum	B	Diffusive	Poor	Good	Low	Poor
Löedige-type	B	Convective	Good	Fair	Medium	Fair
Ribbon blender	B or C	Convective	Poor	Fair	Low	Poor
Nauta mixer	B	Convective	Good	Poor	Low	Poor
V-mixer (with cutters)	B	Diffusive	Poor	Good	Medium	Unknown
Airmix	B	Convective	Good	Fair	Low	Unknown

separation of large agglomerates which takes place subsequently in the mill normally assures the completion of the wetting process provided only that the binder is correctly chosen. Should the binder still appear to be unevenly distributed after the passage of the powder through the mill, the product can often be rescued by passing it through as fine a mesh sieve as possible.

Pearl agents, especially the titanium-coated micas, present a special problem. Many of these brittle materials, which depend on their size to achieve the desired effect, are prone to disintegration in the hammer mill. For this reason, they usually have to be mixed into the bulk after its passage through the mill, necessitating an additional mixing operation. The pearl may be added in to the bulk in the same device used to perform the preliminary coarse mixing and it may sometimes be necessary to pass the bulk finally through a sieve to break up agglomerates of pearl and to ensure its even distribution.

Batch Colour Correction

It is not unusual for the bulk powder product, even though it has been correctly processed, to require colour correction in order to obtain a satisfactory match to the standard. Since any addition of pigment or talc needs to be extended, a passage through the mill is necessary. A common procedure is as follows. After the preliminary coarse mix has been completed, a small amount of the bulk (usually about 5 kg), which is assumed to be representative of the whole, is passed through the mill. This is examined in the laboratory, and if necessary, a pigment addition specified. This correction is added to the 5 kg of milled product, mixed in roughly by hand and the 5 kg is re-milled. The twice-milled sample is returned to the remainder of the bulk and re-mixed in the original mixer. A further 5 kg is then removed and the process is repeated until a match is obtained.

There is a number of minor variations to this procedure which are adopted to suit individual companies; the most important of these is the use of pigments previously extended on talc and stored as such. This has the merit of speeding up the correction process.

When pearl agents are part of the formulation, unless an un-pearlized standard is provided, the pearl must be added in the correct proportion to the laboratory sample before colour can be assessed. Pearl is only added to the bulk in the last stage of the manufacturing procedure.

Alternatives to the Hammer Mill

The drawbacks to the hammer mill as used for powder extension have led to a search for other machines which can fulfil this function more satisfactorily. The ideal equipment would probably have the following properties:

(i) It would be capable of breaking up weak particles in the size range 50–0·5 μm without damaging talc or mica particles of similar diameter.
(ii) It would be a low energy device, consuming little power itself without heating the powder mixture excessively.
(iii) It would be a batch processing device capable of mixing and extending in one operation.
(iv) It would be rapid: processing times of less than 10 minutes would be acceptable.

(v) It would not cause excessive aeration of the powder (since this causes further processing problems in later processing).
(vi) It would be easy to clean.
(vii) Its efficiency would not vary with the cohesiveness of the powder; it would not be affected by poor flow characteristics.
(viii) It would be quiet and clean in operation.

Other comminution devices have been shown to produce extension, particularly pin mills and fluid energy mills, yet none seems to work as efficiently as the hammer mill. In recent years, however, the development of high-speed powder mixers which are also capable of producing some degree of extension has brought the industry closer to the ideal. Two types in particular are worthy of mention. The first of these is best described as the vertical vortex mixer. The powder mixture is placed in a vertical, cylindrical chamber and is then accelerated outwards and upwards into a fluidized vortex motion. The motion may be produced by compressed air blasted sequentially from a series of nozzles contained in a lower cone-shaped section; alternatively, a propeller-shaped tool of 'poor aerodynamic' design may be used which rotates rapidly in the dished base of the mixing bowl. Mixing and dispersion occur at the point of conversion of the powder particles (in the upper point of the mixing bowl) by particle–particle collisions. The second type of high-speed mixer is often referred to as a 'plough–shear' device, because of the unusual shape of the mixing paddles which rotate on an axial shaft in a cylindrical horizontal mixing chamber. These paddles cause the powder from all parts of the chamber to be thrown about in such a way that it all passes rapidly through a zone occupied by a series of rapidly revolving blades on a separate shaft, referred to as a 'chopper'. The chopper is largely responsible for the powder extension and may be switched on or off independently of the main axial drive.

Both types of mixer have been used as partial or complete replacement for the traditional blender–hammer mill combination. The plough–shear type may also be used for wet-processing.

Storage of Cosmetic Powders

Two factors have an important effect on stored cosmetic powders: moisture and pressure. It is not always appreciated that a small increase in relative humidity can give rise to sufficient moisture in the stored powder to change the main mechanism of particle–particle bonding, increasing the bond strength of agglomerates by a factor of 2 or more. Such an increase in cohesiveness can make the handling and flow problems already inherent in cosmetic powders perceptibly worse and can change the processing characteristics of (say) an eyeshadow to the point where all the pressing machine settings may have to be altered to compensate.

In the same way, powder bulk that has been stored in large vertical containers exhibits increasingly difficult flow characteristics as the container gradually empties. The lower layers, having been compressed by the weight of powder above them, become increasingly cohesive as the bottom is approached. For this reason it is far better to store powder in a large number of small well-sealed containers than in loosely covered large bins.

MIXING PROCESSES INVOLVING FLUIDS

Apart from the 'dry' powder processing already discussed, the remaining processes listed in Table 39.1 involve liquids present in sufficiently large quantities as to impose fluid characteristics on the mixture. Although there are similarities between the flow of powders and the flow of liquids it is obviously easier to set up and sustain flow patterns in the latter. On the whole this makes the mixing processes easier to perform and a much larger variety of equipment is consequently available to choose from.

Even for liquids, however, the science of mixing has not yet been sufficiently developed to enable the optimum mixer to be designed for a given task from purely theoretical calculations. Much of the knowledge we have is empirical, and has been accumulated from trial-and-error practical experience; we have little detailed knowledge of many of the mixing processes at work.

General Principles of Fluid Mixing

Not only is there a great variation in the physical form and properties of substances which the cosmetics industry needs to mix, but there is also a divergence of purpose. Some mixing operations can be thought of as simple blending—for example the blending of colour solutions into miscible bulk liquids and the blending of oils, alcohol and water in perfumes and colognes. On the other hand, the formation of an emulsion, the suspending of a gelling agent and the distribution of pigment agglomerates in a viscous liquid all involve the breaking up of one of the constituents of the mixture into finer particles during the mixing process. For this reason, it is referred to as 'dispersive' mixing to distinguish it from simple blending.

On the industrial scale, mixing occurs as the result of forced bulk flow within the mixing vessel. Two types of flow can be distinguished, laminar and turbulent. Laminar flow occurs when the fluid particles move along streamlines parallel to the direction of flow. The only mode of mass transfer is by molecular diffusion between adjacent layers of fluid (Brownian motion). In turbulent flow, the fluid elements move not only in the parallel paths but also on erratic and random paths, thus producing eddies which transfer matter from one layer to another. For this reason, turbulent mixing is rapid compared with other mixing mechanisms.

When a quiescent liquid is slowly stirred the flow is laminar but as the velocity increases it may become turbulent; thus the velocity is a significant factor in determining the type of flow set up in the mixing vessel. A valuable aid in describing the critical point at which laminar flow becomes turbulent is due to Reynolds who, in 1883, first demonstrated turbulence. The dimensionless number which bears his name, Re, can be calculated for agitated vessels as follows:

$$\mathrm{Re} = \frac{D^2 N \rho}{\eta} \tag{1}$$

where D is the diameter of the impeller, N the impeller speed (rpm), ρ the density of the mixture and η its viscosity.

Although little is known about the mechanism of turbulence, experience has shown that in agitated tanks the onset of turbulence occurs at Reynolds numbers of about 2×10^3. For fully developed turbulence, Reynolds numbers greater than 10^4 are required and are found in many cosmetic mixing processes. It can easily be seen that it becomes more difficult to achieve turbulence as the viscosity increases. Below 10–100 poise, turbulent flow can be achieved without need for an excessive amount of power—and this viscosity range covers many cosmetic products. For highly viscous pastes, however, mixing raises certain problems since the flow pattern in the mixer is invariably laminar. Under these circumstances, distributive mixing (cutting and folding) is more applicable than turbulent mixing. Turbulence not only provides rapid mixing, but also influences dispersion: it is important to understand why.

It has been pointed out in Chapter 38 that when liquids flow there is a simple relationship between the force causing the movement, F, and the velocity gradient between the layers of liquid moving at the point of measurement and the stationary or slower moving layers adjacent to them:

$$F = \eta A \times \text{velocity gradient} \qquad (2)$$

where A is the cross-sectional area of the mass of liquid under investigation and η is the proportionality constant, called the coefficient of viscosity. F/A is commonly referred to as the 'shear stress' and the velocity gradient as the 'rate of shear'. If the equation is to be believed, as the velocity of flow increases, so does the shear stress and this is the force which breaks up the weak bonds holding together pigment aggregates, or another immiscible phase into droplets. Of course, shear forces are also produced when liquids flow under laminar conditions, but under these circumstances the energy used to generate flow is dissipated largely as heat.

During turbulent flow, the energy is dissipated in disorder; eddies are produced in size and intensity depending upon the viscosity of liquid and upon F. Size reduction can only occur effectively if the eddy is smaller than the drop or aggregate. It follows that, for a liquid of given viscosity, the drop size of an emulsion or the fragmented size of dispersed pigment agglomerates depends primarily on the energy input from the agitator, the velocity gradient and the nature of the forces holding together the disintegrating entities.

Unfortunately, the simple model shown in equation (2) has limited application in cosmetics manufacture. The majority of products exhibit non-ideal (non-Newtonian) behaviour which can often be more appropriately described by the expression

$$F = (\eta_{\text{app}})^n A \times \text{velocity gradient} \qquad (3)$$

In this case η_{app} is termed the 'apparent viscosity' and n usually has a value between 0 and 1. The name given to this type of behaviour is 'pseudoplastic' and the basic difference between materials exhibiting this property and ideal or 'Newtonian' fluids is illustrated in Figure 39.3. As can be seen, pseudoplasticity is manifested by a fall in viscosity with increasing shear rate at constant temperature. Many cosmetic liquids exhibit this behaviour—especially emulsions and suspensions of particles of the order of 1 μm or less in size. Pseudoplasticity is usually reversible to some extent—in other words, when left unstirred for long enough, the fluid will recover some or most of its original

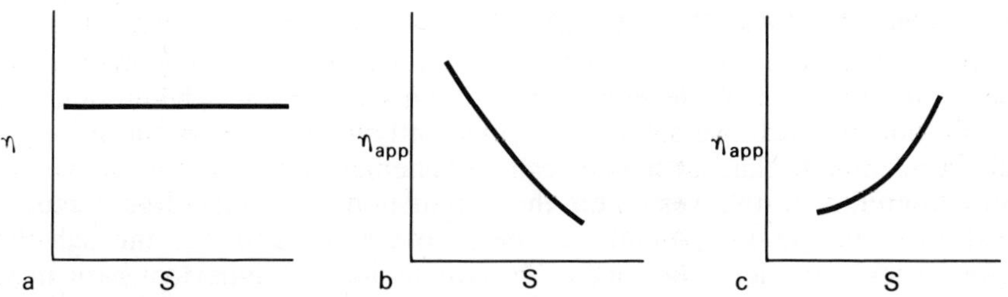

Figure 39.3 Rate of shear (S) plotted against viscosity (η) or apparent viscosity (η_{app})
a Newtonian material
b Pseudoplastic material
c Dilatant material

viscosity. The magnitude of the pseudoplastic effect is variable with the identity of the fluid, although a fall of 25 per cent in viscosity when the rate of shear is doubled is not unusual.

A related type of non-Newtonian behaviour is the property of 'elasticity'. As the name implies, elastic fluids have the ability, on being deformed by the action of shear, to recover their structure rapidly, with a consequent regain of the energy absorbed during deformation. It is probably accurate to regard elastic behaviour as an extreme of pseudoplasticity in which the recovery time is very short. Certainly it is true that all pseudoplastic fluids exhibit some degree of elasticity.

Three other types of rheological behaviour are also worth noting, although they are less frequently encountered in cosmetics processing. Truly 'plastic' fluids exhibit viscosity versus shear rate curves similar to those of pseudoplastic materials, but in this case a certain force must be applied before any shear (or flow) takes place. 'Dilatant' materials show the opposite effect, viscosity increasing with shear rate (Figure 39.3).

The term 'thixotropic' is often used erroneously to describe pseudoplastic behaviour. Thixotropic liquids exhibit a fall of viscosity with time at *constant* shear rate.

Mixing Equipment For Fluids

In the mixing of fluids, all three mixing mechanisms—bulk flow, turbulent diffusion and molecular diffusion—are usually present. As viscosity increases, however, and turbulence becomes correspondingly more difficult to establish, the parts played by turbulent and molecular diffusion become less important. Mixing equipment can therefore be divided into two categories, depending on whether or not turbulent conditions prevail, as follows:

Laminar shear/distributive mixers	*Turbulent mixers*
Helical screw/ribbon blenders	Turbine-agitated vessels
Two-blade mixers	Pipes
Kneaders	Jet mixers
Extrusion devices	Sparged systems
Calenders	High-speed shear mixers
Static mixers: low Re	Static mixers: high Re

Fluids with Low or Medium Viscosity
By far the most common form of mixing in liquids of low or medium viscosity is achieved by forced convection in stirred or agitated vessels. The motion of the liquid produced in the vessel must be sufficiently intense to sustain turbulence. Since it is unlikely that turbulence can be generated uniformly throughout the whole contents of the vessel on the production scale, the liquid must be circulated continuously around the vessel so that it all passes through those regions where turbulence develops. Thus the number of important basic mixing parameters to be considered is two: the extent of turbulence and the circulation rate of the contents.

The liquid motion is very often produced by a mechanical mixer—usually a rotating impeller of some kind: two distinct types may be distinguished. In the first of these, exemplified by the rotating disc, momentum is transferred from the impeller to the liquid primarily by shear stress. As the agitator rotates, the layer of liquid immediately adjacent to it rotates with it. Viscous drag then causes the next layers to move and so on until the entire vessel contents are in motion. In order to produce efficient mixing, such systems need to be operated at high Reynolds number and develop comparatively high shear, and hence one of the main applications is in emulsification.

Far more common, however, are mixers of the second group which transmit their momentum through the pressure exerted by the impeller on the liquid (that is, in the direction of flow). Included in this group are paddles, turbines and propellers which exert pressure on the liquid in front of them as they rotate. This results in the displacement of part of the liquid in their path into the surroundings and the formation of a rotational flow pattern in the liquid. Additionally the decrease in pressure behind the blades entrains liquid from the surroundings, the effect of which is to produce turbulent eddies around the blades, especially at the blade tips. Such turbulence is, of course, extremely localized and therefore of limited value. As the impeller speed rises, however, the centrifugal force acting on the liquid increases, leading to a flow of liquid away from the periphery of the impeller, displacing and entraining other liquid and inducing further turbulence. The exchange of momentum of the flowing liquid with its surroundings results in a loss of velocity as the distance from the impeller increases, and thus the flow pattern and mixing efficiency vary according to the viscosity and flow behaviour of the fluid, the design of the impeller and vessel and the rotational speed of the impeller itself.

Flow Patterns. Flow patterns in agitated vessels can be resolved into three principal types: tangential, radial and axial.

In tangential flow, the liquid moves parallel to the direction of the impeller. Movement of liquid into the surroundings is small and there is little movement perpendicular to the blades except in eddies near the tips. Tangential flow may be observed in paddle mixers operating at low speeds or in liquids of sufficient viscosity to prevent centrifugal flow from being developed (Figure 39.4).

During radial flow, the liquid is discharged outwards from the impeller by centrifugal force. If the moving liquid strikes the wall of the vessel, it splits into two sections, circulating back towards the impeller where it is re-entrained. Further turbulence and mixing are induced by the splitting of the flow at the

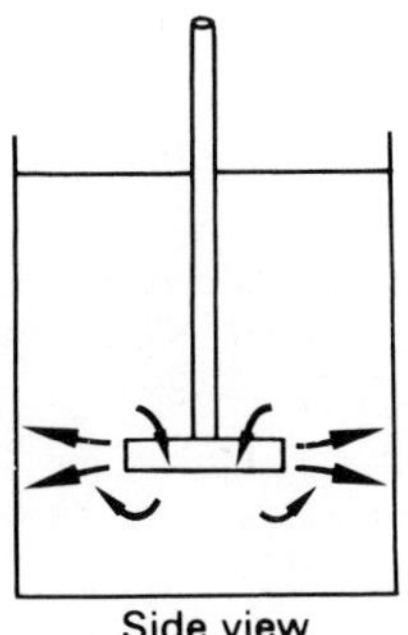

Figure 39.4 Tangential flow

wall. There is usually some element of radial flow in agitated vessels, but flat-blade turbines produce flow patterns that are primarily radial (Figure 39.5a).

Axial flow, as the name implies, takes place parallel to the axis of rotation. Usually the impeller or propeller blades are pitched so that liquid is discharged axially—the direction of flow may be from top to bottom of the vessel or vice versa (Figure 39.5b).

Generally speaking, the most difficult flow pattern to maintain is one of axial flow.

Impellers for Liquids of Low and Medium Viscosity

Paddle mixers are simple and cheap but very inefficient for all but very low viscosity liquids. They produce mainly tangential flow and are usually mounted centrally because of their large diameter compared with that of the tank.

Turbines are probably the most common impeller type used in cosmetics processing since they can cope with a wide range of viscosities and densities. For liquids of low viscosity, the flat-blade impeller is sometimes used (Figure 39.6a and b). For very viscous materials the blades may be curved backwards in the direction opposite to the rotation, since these require a lower starting torque and seem to give better energy transfer from impeller to liquid (Figure 39.6c).

Figure 39.6d illustrates a fixed-pitch *axial flow impeller*. Used without baffles, however, the axial component generated by such turbines remains secondary to the radial flow component. Typically, impellers of this kind are used at rotational speeds of 100–2000 rpm as distinct from the low speed (15–50 rpm) of paddles.

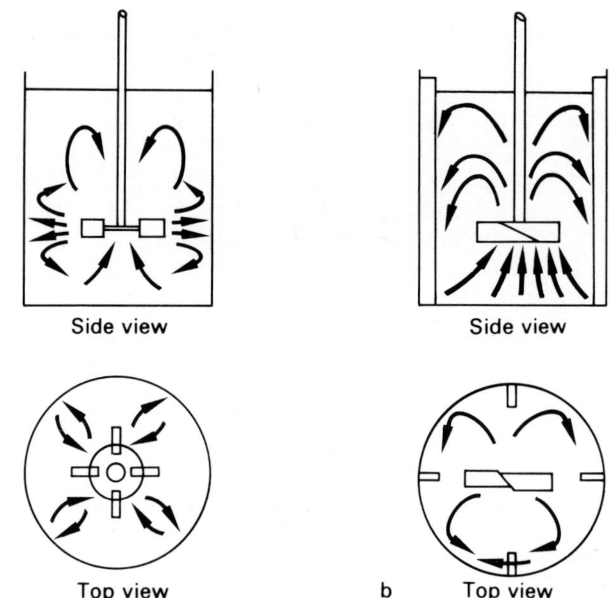

Figure 39.5 *a* Radial flow *b* Axial flow

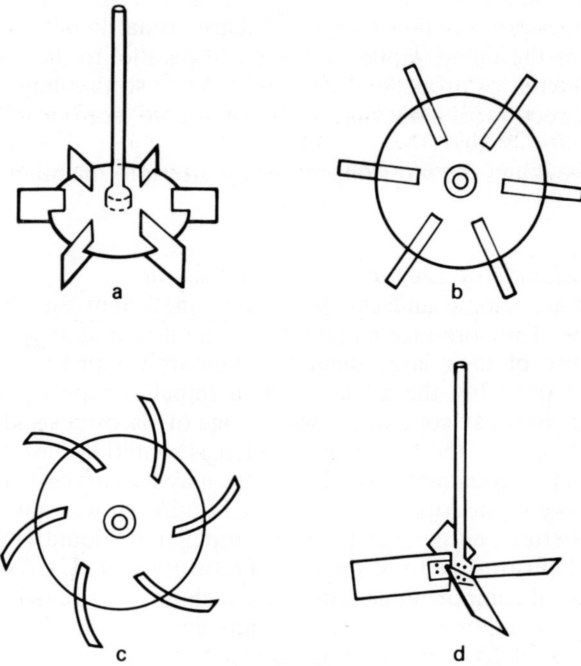

Figure 39.6 Various designs of turbine impeller

Propeller mixers are restricted to use with low viscosity fluids. They have pitched blades of which the blade angle varies along the length from centre to tip. Flow patterns developed by propeller mixers have a high axial component and the rate of circulation is high. They are usually of relatively small diameter, typically three-bladed, and are used at speeds between 450–2500 rpm. Such stirrers are used extensively in the cosmetics industry for simple blending operations but are not suitable for the suspension of particles which settle rapidly or for the dissolution of sparingly soluble heavier materials.

Many portable mixers are of the propeller type. If the mixer is mounted centrally in the mixing tank (Figure 39.7a), because of the entrainment of liquid above the impeller, the surface becomes depressed and a vortex is formed. Generally, vortices are to be avoided because of the low order of turbulence and the air-entrapment which they cause. When they are mounted eccentrically, however (Figure 39.7b), turbulence is increased and vortices avoided.

Influence of Vessel Shape

In general, the absence of smoothness and rounded corners in mixing vessels contributes to turbulence and would be expected to improve mixing times. An extreme case is the introduction of baffles around the circumference of vessels of cylindrical cross-section (Figure 39.7c). It is easy to visualize how the introduction of baffles could interfere with the flow patterns generated by tangential and radial flow impellers and cause the suppression of vortex formation in centrally mounted mixers. Experience shows that baffling improves axial flow and increases turbulence.

It will also be evident that the ratio of tank dimensions can be an important factor in determining the efficiency of any mixing process. For example, it is sensible to perform simple blending, such as alcohol–water, in tall cylindrical vessels of small cross-section using a propeller (with high axial flow). In the production of an emulsion of medium viscosity using a radial flow turbine, it would evidently be desirable to keep the height of the vessel as small as possible for a given capacity.

Mixing in Non-Newtonian Liquids of Low or Medium Viscosity

Many liquids—perhaps the majority—encountered in cosmetics processing are of the shear-thinning and/or elastic rheological type. Naturally, if the liquid is

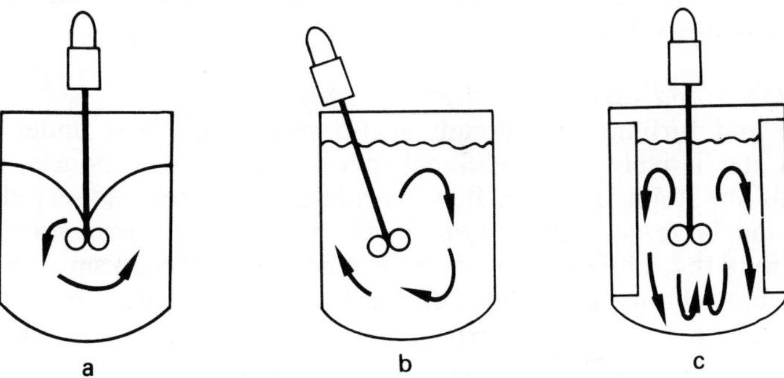

Figure 39.7 Portable mixers

already of low viscosity, the effect of shear-thinning may not be noticeable. On the other hand, more viscous liquids showing these characteristics present considerable problems to the cosmetics processor. The fluid close to the rotating impeller of a mixer is sheared at a high rate and so becomes relatively mobile, but as this is pumped away from the impeller it encounters regions of less intense flow and hence of much higher viscosity. Turbulence is therefore rapidly damped out, decreasing the turnover in the vessel and slowing down the mixing process. Moreover any elasticity shown by the liquid results in the absorption of energy by deformation of a recoverable variety, thus damping out turbulence even further.

Fluids of High Viscosity
As the viscosity of the mixture increases, it becomes increasingly difficult—and finally impossible—to produce turbulent flow within the mixer. At viscosities of 1000 poise or above, flow is inevitably laminar, power consumption high, and the rate of mixing exceedingly low. In such systems, the input power is not used to create disorder but to create heat. The rate of temperature rise is dependent upon the energy input, the thermal conductivity of the mixer and the efficiency of the cooling surfaces, but the range 1°C to 30°C per minute would include many cosmetic mixing processes of this kind. Generally it is difficult or impossible during industrial mixing of highly viscous materials to dissipate heat faster than it is generated. This is particularly true if the mixer is of large capacity (in which the ratio of volume to heat-exchange surface is high) and if appreciable films of chilled liquid are allowed to build up on the walls of the vessel, so insulating the contents from further cooling. The increase in temperature associated with such processes has both obvious advantages and disadvantages. On the one hand a rise in temperature might cause a decrease in viscosity, making mixing more efficient, and might also help in the melting or dissolution of some of the components of the mixture. Taken too far, on the other hand, the decrease in shear stress caused by the fall in viscosity can decrease the efficiency of stress-dependent processes (such as the breaking up and dispersion of pigment agglomerates), and also the rise in temperature may damage the product by causing the thermal degradation of heat-sensitive components such as preservatives and perfumes.

The relatively high energy input required to mix viscous materials also influences the mechanical construction of the mixing machinery and the method by which mixing is achieved.

Impeller Types and Mixers for High Viscosity Fluids
Propellers and turbines, as already mentioned, work best under turbulent conditions at relatively high rotational speeds. In viscous products, given that such speeds are attainable at all, flow is confined to the regions very close to the impeller, and large stagnant regions in the mixer exist where no mixing can occur without the employment of some secondary mechanism. To eliminate these stagnant regions, large impellers such as paddles, gates, anchors and leaf impellers may be used; these sweep a much greater proportion of the vessel and produce more extensive flow. Usually such impellers are designed to have close clearances with walls, giving a degree of wall-scraping. This helps to eliminate

build-up of unmixed materials at walls, provides a region of high shear for dispersing aggregates and lumps, and may improve the wall heat transfer to and from the bulk.

Such impellers provide extensive flow but only of the tangential and radial variety. Axial flow—and therefore top-to-bottom mixing—is almost totally absent. For this reason, more complex designs such as the helical screw and helical ribbon have been introduced (Figure 39.8). These are more efficient for viscous mixing but their performance is poor compared with that of more conventional impellers for medium and low viscosity mixtures. Consequently, they are rarely employed in cosmetics manufacture.

Axial flow cannot be achieved by the introduction of baffles as when mixing lower viscosity fluids, but some success has been achieved by the use of impeller–draught tube combinations. As the name implies, a draught tube is a tubular, axially orientated enclosed space within the main mixing chamber containing an impeller or some other means of forcing the flow of mixture along it. Small impellers or helical screws designed to fill most of the cross-section of such a tube have been successfully used to promote axial flow in most liquids, even those of very high viscosity.

An alternative approach to the problem created by lack of flow in viscous media is the use of impellers which progressively sweep the whole contents of the vessel while the mixture remains stationary. Examples of this include the 'Nauta'-type mixer in which a helical screw sweeps the wall of a conical mixing chamber.

For even more viscous products such as mascara and very thick pastes, equipment which exhibits a greater degree of distributive mixing may be utilized. Such mixers are designed to produce bulk flow and laminar shear by spatial redistribution of elements of the mixture. Perhaps the most commonly encountered mixers of this type are of the single or double action planetary type or the two-blade 'dough' mixer. Their essential feature involves the cutting and

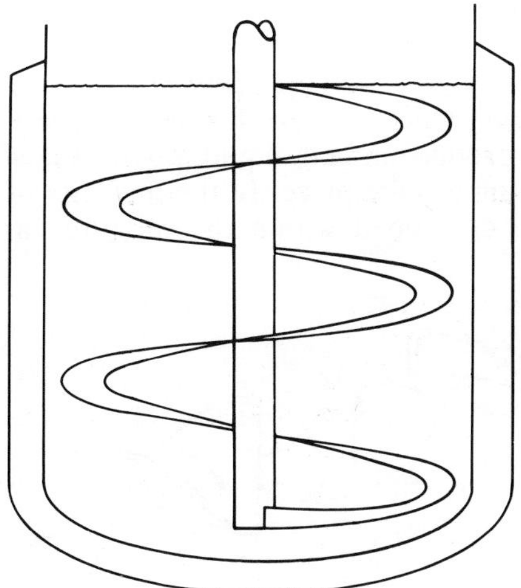

Figure 39.8 Helical impeller

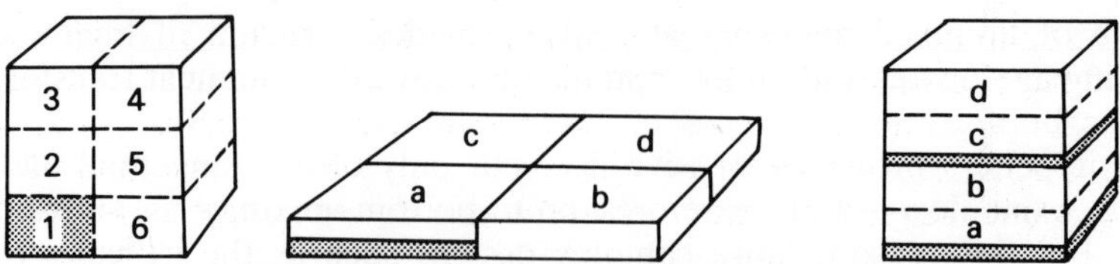

Figure 39.9 Distributive mixing mechanism

folding of a volume of the mixture and the physical replacement of it into another part of the mixer where it is cut and folded again. An example of this distributive mechanism is illustrated in Figure 39.9, in which for clarity a volume of a mixture has been isolated and divided into six equal segments, one of which consists of a black minor component.[5] The cube is compressed to one quarter of its initial height, cut and reassembled as shown. Redistribution of the minor component has been achieved which, were the process to be repeated often enough, would eventually achieve the desired level of homogeneity.

A more recent innovation is the so-called static mixer, of which several designs are now commercially available. Static mixers are essentially in-line mixing devices in which mixtures flowing through a pipe are cut and folded by a series of helical elements in a circular tube (Figure 39.10). These elements (which do not move—hence the name 'static') turn the flowing mixture through an angle of 180°. Since alternative elements have opposite pitch and are displaced 90° to each other, this causes the bulk flow to reverse direction at each junction, and thus the leading edge of each element becomes a cutting device, splitting and re-folding the mixture in on itself.

Finally, mention must be made of extruders, in which a helical screw forces the bulk mixture to flow down a tube. Here, the pressure generated can be enormous, as in soap-plodding, and such energy can cause materials with the viscosity of toilet soap to undergo laminar flow. The actual flow pattern produced is complex, being a combination of pressure and drag flow within the tube.[6]

High Shear Mixers and Dispersion Equipment

The mixing equipment which has so far been covered in this chapter is designed primarily to produce in bulk liquid mixtures flow patterns of sufficient intensity to allow mixing to take place. In the majority of cases, the pattern of shear and turbulence developed within the mixture varies according to the

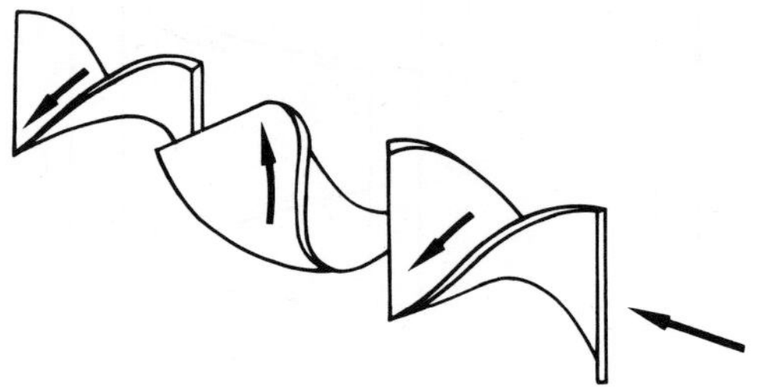
Figure 39.10 Static mixer

viscosity of the bulk, the method of producing flow and the volume within the mixture under consideration. For certain applications, however, it is desirable to generate a very intense degree of shear stress in the mixture, and for this purpose specialized equipment is available. The uses to which such machines are put in cosmetics processing include the breaking up of pigment agglomerates and their dispersion in liquids, the rapid fracture and dispersion of gelling agents (for example, bentones, cellulose derivatives and alginates) and the size reduction of internal phase droplets in emulsion products.

The basic principle on which high shear mixers work is the forcing of the mixture through a very narrow gap at the highest possible velocity. This may be illustrated with reference to a widely used device in which the mixture is entrained by a high-speed rotor, moving in very close proximity to an enveloping stator, which may or may not contain perforations through which the mixture is forced.

Figure 39.11 is a diagrammatic representation of one of the blades of the turbine separated by a very small gap, h (a few thousandths of an inch), from the stator. If the velocity of the rotor is v, then the shear rate, γ, is given by

$$\gamma = \frac{v}{h} \tag{4}$$

(γ has a typical value of 100–500 s^{-1}). It follows from the basic definition of viscosity given in equation (2) that for a Newtonian liquid having a viscosity η,

$$\eta = \frac{h\tau}{v} \tag{5}$$

or

$$\tau = \frac{\eta v}{h} \tag{6}$$

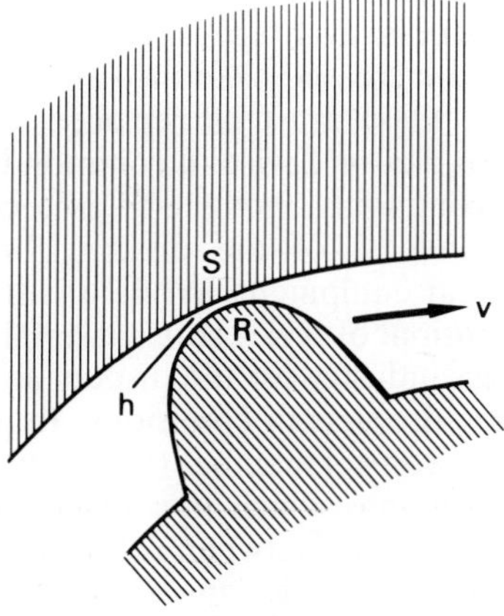

Figure 39.11 Principle of a rotor/stator high shear mixer
S = stator R = rotor

where τ is the shear stress. Thus it may be seen that shear stress required to disperse agglomerates or to reduce droplet size may be increased with increasing viscosity of the mixture, increasing velocity of the rotor and decreasing clearance between rotor and stator. For non-Newtonian liquids for which the apparent viscosity, η_{app}, is given by

$$\eta_{app} = \frac{\tau}{\gamma^n} \tag{7}$$

the corresponding shear stress is

$$\tau = \eta_{app}\left(\frac{v}{h}\right)^n \tag{8}$$

This simple treatment does not give the complete picture. Shear stress can be increased still further by the perforation of the stator, and the flow pattern over and through the shearing head may be drastically altered by the design of the assembly enclosing it. Usually, an interchangeable choice of stator designs is available with the mixer, allowing the most appropriate type to be used for a particular task. With some designs, it is possible to make use of the considerable pumping capacity afforded by the high-speed rotor.

Generally, a high shear rotor–stator mixer may be used either for batch processing in a mixing tank, or as an in-line device when it is encased in a suitable all-enveloping chamber. Used as a batch mixer, it is capable of generating considerable turbulence because of the great velocity with which fluid is pumped out of the mixing head. As with other devices, however, this high energy is increasingly converted into heat with increasing viscosity of the mixture. A serious disadvantage for certain processes is the tendency of the mixer to cause aeration when used in the top-entry mode. For this reason, such devices are often incorporated into the bottom of processing vessels.

The rotor–stator is generally more efficient when it is used as an in-line mixer, particularly as the mixing time required to ensure that every part of the mixture has passed through the mixing head is generally reduced (especially for higher viscosity mixtures). Another disadvantage is that the outer tips of the rotor may wear out rapidly, leading to increased clearance and decreased efficiency. Since no adjustment is possible, these must be replaced at considerable cost.

Another high shear rotor–stator device in common use is the colloid or stone mill. Such equipment is commonly thought of as a comminution device; this serves to illustrate the fineness of the dividing line between mixing and comminution with high shear equipment. While it is true that colloid mills may be used for the comminution of very soft materials in a slurry, they find application in the cosmetics industry for the dispersion of pigments and the size reduction of internal phase droplets in emulsions. In principle, the colloid mill consists of a rapidly rotating conical member (which may be toothed or grooved) and a similarly coned stator into which the former fits. The fluid mixture is forced through the small clearance between rotor and stator (0·5–0·05 mm) as before. Figure 39.12 illustrates the design of colloid mills in greater detail.

Colloid mills are used exclusively as an 'in-line' or continuous device. They may be water-cooled and can be adjusted as the moving parts wear down.

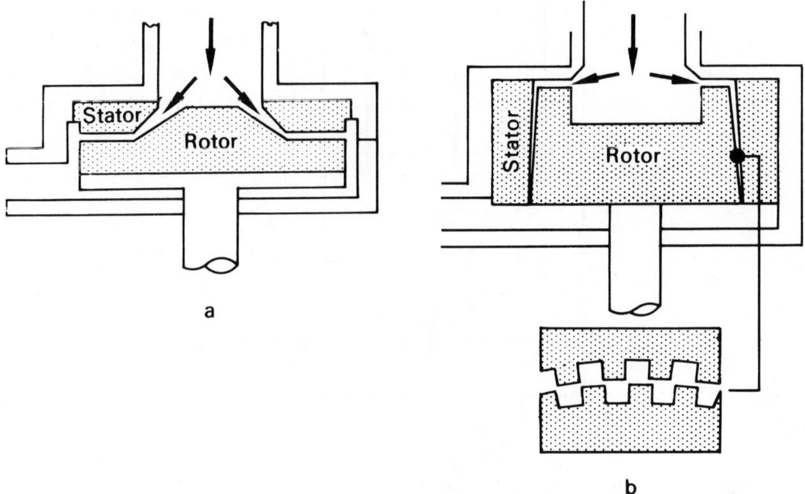

Figure 39.12 Colloid mills
a Stone *b* Toothed

For more viscous products, an alternative device is the triple roll mill (Figure 39.13). The device consists of three steel rollers which rotate in the directions indicated in the diagram. Each roller is water-cooled and machined to great accuracy so that the gaps between each pair of rollers can be set very fine. The product is applied at the top of roller A, passes between A and B and round the underside of B between rollers B and C. As it traverses each gap, the product is subjected to enormous compression forces which are particularly effective in breaking down pigment agglomerates. Strictly speaking, therefore, the triple roll mill is not a high shear device but a compression device. It is included here, however, as a real alternative to the rotor–stator mixers in bringing about effective pigment dispersion in liquids—particularly in viscous liquids.

Perhaps the highest shear stress of all is generated by a valve homogenizer, which is still extensively used in the production of emulsions with very fine internal phase droplets. A valve homogenizer (Figure 39.14) is simply a high-pressure pump which forces the product through a small orifice at pressures of up to 350 atmospheres (354 bar).

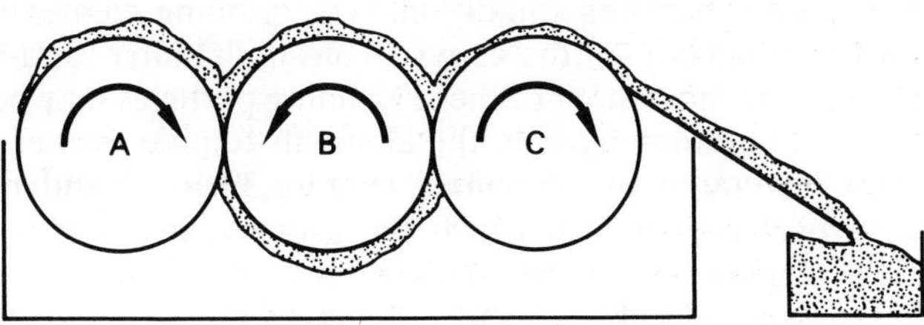

Figure 39.13 Triple roller mill

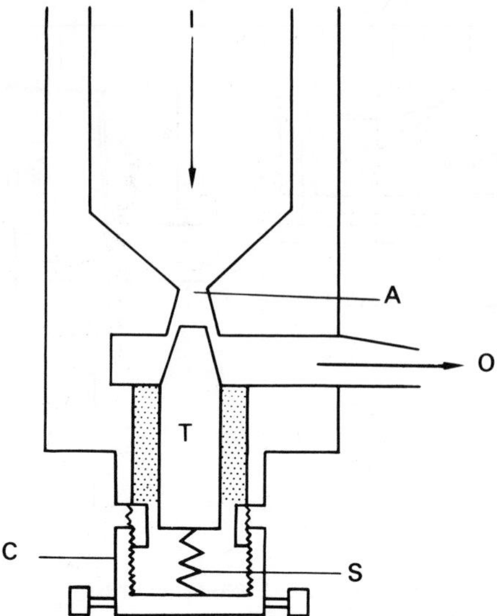

Figure 39.14 Valve homogenizer
The liquid at I is forced through the valve seating at A and leaves via O. T is a tapered shaft of which the position can be adjusted with screw head C. S is a powerful spring against which the product is pushed through the narrow valve orifice

An interesting alternative to the valve homogenizer, and one which is finding increasing use in cosmetics manufacture, is the ultrasonic homogenizer. When high intensity ultrasonic energy is applied to liquids, a phenomenon known as cavitation occurs. Cavitation is complex and not fully understood. As ultrasonic waves are propagated through the fluid, areas of compression and rarefaction are formed and cavities are produced in these rarefied areas. When the wave passes on, these cavities collapse and change to an area of compression and it has been demonstrated that the pressure in these cavities just before their collapse can be as much as several thousand atmospheres. Most of the effects of ultrasonic radiation in liquids are attributed to the powerful shock waves produced immediately following the collapse of such cavities. One design of ultrasonic homogenizer is illustrated in Figure 39.15.

No description of dispersing equipment would be complete without mention of ball mills and sand mills. In both these devices the breakdown of agglomerates is achieved by attrition between rapidly moving grinding elements which take the form of pebbles, balls or (in the case of sand-mills) finer sand-like particles < 1 mm diameter. The movement of these grinding particles may be achieved in a number of ways. In a tumbling mill, the elements tumble over each other as a horizontal drum is rotated on trunnions (Figure 39.16). Sand mills may be horizontal or vertical cylinders in which the grinding medium is stirred by a rotating agitator (Figure 39.17), whereas in vibration mills movement of the whole chamber may be caused by eccentric cams, by out-of-balance weights on a drive shaft, or electrically.

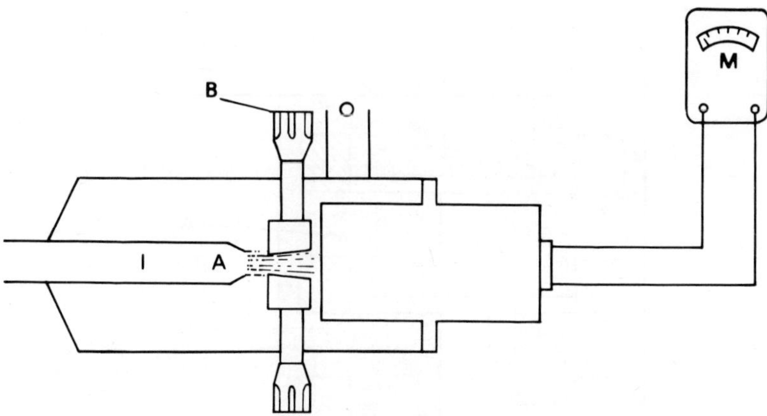

Figure 39.15 Ultrasonic homogenizer
The roughly premixed product enters at I and, on passing through orifice A, is subjected to intense ultrasonic energy by vibrating blade B. The treated product leaves via O. The meter, M, and 'tuning' devices at B combine to allow the operator to achieve maximum effect with each different product type

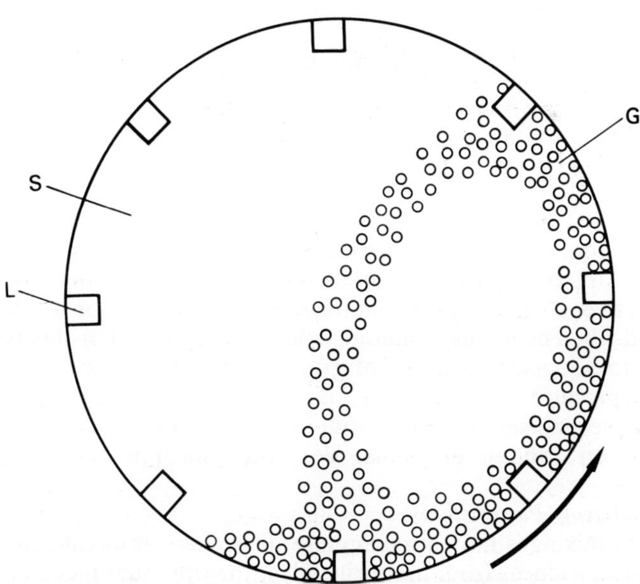

Figure 39.16 Ball mill showing ball pattern in rotating drum
G Grinding medium
L Lifter elements
S Slurry

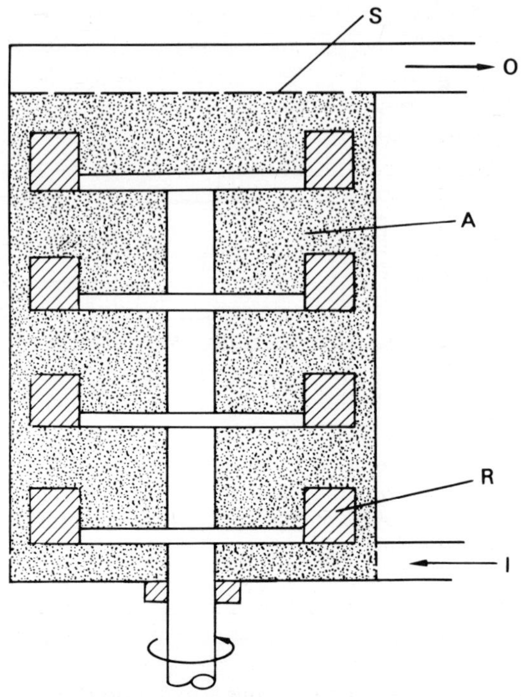

Figure 39.17 Sand mill
I Slurry in
O Slurry out
S Screen
R Rotating agitator
A Glass beads, sand or shot

Small ball mills and sand mills are used extensively in the dispersion of pigments into liquids (as, for example, in the production of castor oil lipstick pastes) and for the dispersion of bentone into nail varnish media. Although they are very effective, their chief disadvantage is the extremely protracted cleaning time required when changing from one colour to another. For this reason many users prefer to keep separate sets of grinding media for each different colour they wish to produce.

The basic mechanism by which mills of this kind produce their effect is attrition between grinding elements. Very little shear is developed.

Measurable Mixing Parameters

The science of mixing is far from complete. Designers of mixing equipment are not yet able to produce, from first principles, the optimum piece of equipment for a specific job even if they are in possession of all the parameters and fundamental characteristics of the process and mixture which they desire. One of the reasons for this is that the complete mathematical description of the flow pattern of fluid within each mixing vessel is extremely difficult and complex to achieve. Progress, however, is being made using the mathematical tool of dimensional analysis.[7–10] The serious cosmetics processing chemist is to be encouraged in pursuit of such knowledge leading, as it does, to a more thorough understanding of the equipment which may be available to him and the processes with which he must work. As an illustration, however, of the practical usefulness of the data that emerge from this analytical approach, a brief description of the relationship between some of the relevant parameters may prove useful.

Power Consumption. Power consumption is of great relevance to the economics of the mixing process. The choice of the wrong equipment can lead to the consumption of vastly greater quantities of power than are necessary to achieve the desired end result. On the other hand, sufficient power must be available and applied to the fluid to ensure that the end-point of the mixing process can be achieved in a sensible time.[8–10] Dimensional analysis requires that power input is described in the form of a dimensionless number, P_o, which is analogous to a friction factor or drag coefficient:

$$P_o = \frac{P}{D^5 N^3 \rho} \tag{9}$$

where P is the power imparted to the fluid of density ρ, and D is the diameter of the impeller which has a rotational speed of N. For many pieces of mixing equipment, P is measurable from electrical consumption data (provided that friction losses in gearboxes and bearings are ignored) and hence P_o can be calculated.

It has already been pointed out in equation (1) that another dimensionless term, the Reynolds number (Re), is useful in describing the onset of turbulence. The relationship between P_o and Re has the general form indicated in Figure 39.18. The value of graphs such as this lies in the insight they give into the interdependence of flow and mixing behaviour and the design and operating characteristics of the mixer. This should help the process engineer to choose not only the best equipment for a particular task but also the best conditions under which to operate it.

Figure 39.18, for example, shows that when flow in the mixing vessel is non-turbulent, the power which the agitator applies to the mixture falls rapidly as the agitator speed increases until, at a Reynolds number of about 200, it flattens out. If the vessel is unbaffled (curve 5), the power input then decreases again as vortex formation takes place. Here, the maximum power input that can be achieved occurs when the vortex just reaches the turbine.

The other curves show the effect of four different baffle widths. By the time the increased agitator speed has taken Re to 10^4, turbulence has fully developed. Notice that at this point the power input becomes independent of Reynolds number and dependent upon the extent of baffle. The task of the production chemist is to choose the Reynolds number and degree of baffle which will achieve the desired end result with the minimum power input.

Mixing Time. Another important measurable parameter is the mixing time, t_m. This is the time taken to achieve the desired degree of homogeneity in the mixer.

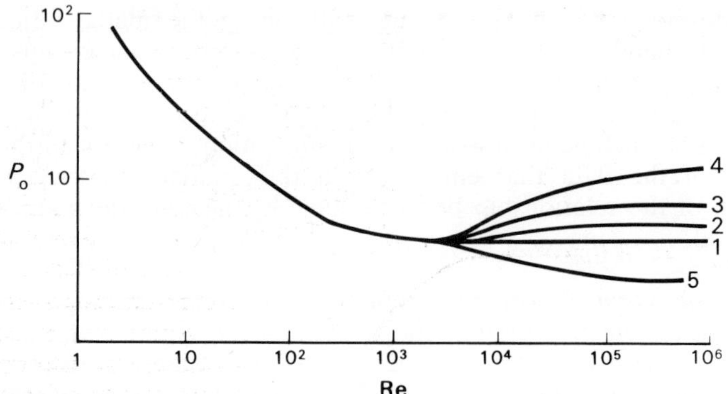

Figure 39.18 Relationship between power number, P_o, and Reynolds number, Re
1,2,3,4 represent increasing degrees of baffle
5 represents an unbaffled tank

There are many methods by which this characteristic may be measured but perhaps the most obvious is the time taken for a soluble dye to become uniformly dispersed throughout the mixing vessel (as, for example, in the manufacture of a coloured shampoo). The relationship between mixing time, t_m, and the degree of uniformity can clearly be shown if some index of mixing can be established.

A simple example of this would be the ratio of colour intensity between the top and bottom of the mixer contents at intervals after dye is added to the top (so that uniformity is achieved as the mixing index, M, approaches unity). This comparatively simple experiment should give rise to a curve similar to that shown in Figure 39.19. Since the approach of M to unity is asymptotic, t_m is difficult to measure accurately unless a colorimeter or other optical colour measuring device is available.

Once t_m has been established, however, more useful insight into the parameters controlling mixing rates may be gleaned from relationships such as that illustrated in Figure 39.20.[8,11] In Figure 39.20a (which relates to a viscous liquid in which turbulence is not established), t_m, the mixing time, has been replaced by

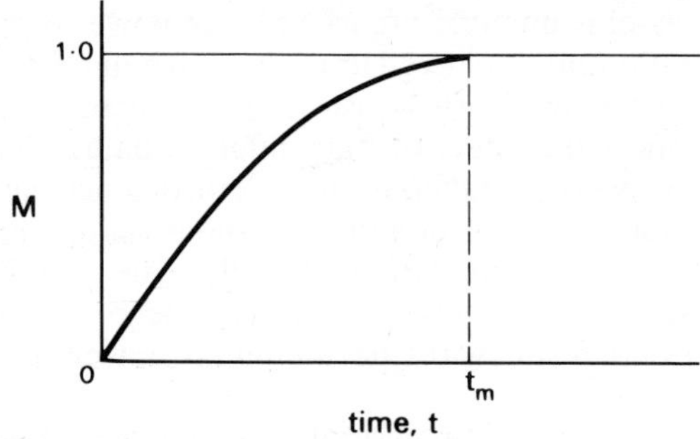

Figure 39.19 Mixing index, M, plotted against time to give mixing time, t_m

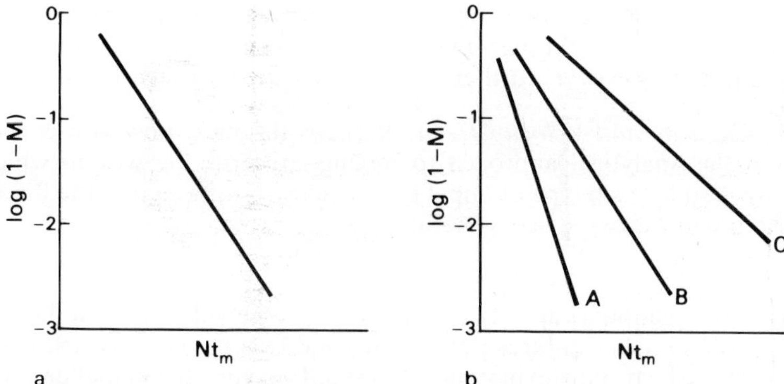

Figure 39.20 Mixing index, M, plotted against number of revolutions, Nt_m
a Newtonian fluids *b* Non-Newtonian fluids

the product of rotational speed N and t_m, that is, by the number of revolutions of the impeller.

It is interesting to note that for Newtonian fluids precisely the same plot is produced whatever the viscosity of the medium and speed of impeller. In other words, only the number of impeller revolutions determines the change in mixing index. This is not true of non-Newtonian fluids—plots A,B and C in Figure 39.20b represent liquids showing increasing divergence from Newtonian behaviour. This illustrates the difficulties, already mentioned, of mixing non-Newtonian media, in which flow is damped out rapidly by regions of high viscosity away from the vicinity of the impeller blade.

The mixing time may be converted into another dimensionless group, for example $t_m ND^3/v$, where v = volume. The relationship between mixing time and the development of turbulence can then be elucidated. Figure 39.21 shows the general shape of curves most commonly obtained by plotting dimensionless

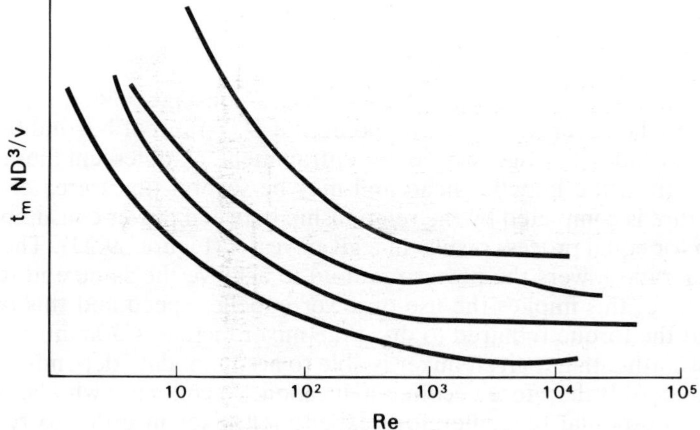

Figure 39.21 Dimensionless mixing time versus Reynolds number

mixing time against Reynolds number.[9,10] Mixing time is seen to decrease rapidly as Re increases through the laminar region. Once full turbulence has been established, however, further increase of impeller speed has little effect.

Pumping Capacity and Velocity Head. Perhaps the most powerful concept to arise from the analytical approach to mixing concerns the way in which the power provided by each type of impeller is actually transmitted to the fluid. This relationship can be expressed generally as

$$P \propto QH \tag{10}$$

where Q is the pumping capacity of the impeller (the volume of fluid displaced directly by the impeller in litres per minute) and H is the velocity head—this is related to the shear rate experienced by fluid leaving the impeller. A large slow-moving impeller might produce, for example, a large pumping capacity and a low-velocity head, while a small impeller operating at high speed might produce a lower volume of fluid pumping but at a much higher velocity head. Some cosmetics production processes—perhaps the majority—require a high pumping capacity; others require a high shear rate or velocity head. It is therefore useful to know the parameters that affect both these functions and how they interrelate.

For simple blending operations (the manufacture of shampoos or colognes, for example) pumping capacity is often of the greatest significance. Under conditions of laminar flow, the number of complete circulations of the bulk ('turnover') required to bring about homogeneity is approximately three. For a turbine operating in turbulent conditions, this is reduced to about 1·5. However, given that only a fixed amount of power is available from the motor, it is not likely that a relatively small turbine will have sufficient pumping capacity to push a fairly viscous product around by even this amount.

Not surprisingly, the factor which determines whether power is used as pumping capacity or velocity head is the ratio of impeller to tank diameter (D/T). Experiments have established the following relationship over a wide range of conditions

$$\left(\frac{Q}{H}\right)_P \propto \left(\frac{D}{T}\right)^{2\cdot 66} \tag{11}$$

where the first term is the ratio of Q to H at a constant power. As shown in Figure 39.22, however, there is little point in a D/T ratio of beyond 0·6.[12] The total flow includes flow generated by entrainment of quiescent liquid in the direct flow from the impeller head and may be several times greater than Q.

The picture is completed by the relationship between power consumption and D/T ratio for equal process results in a given vessel (Figure 39.23). The use of a larger D/T ratio lowers the power required to achieve the same end result. At the same time, this implies the use of lower impeller speed and this inevitably means that the torque required to drive the mixer increases dramatically.[12] The amount of torque that a given mixer is able to accommodate depends largely on its construction. It therefore becomes a question of economics whether to invest in a more substantial (and therefore expensive) mixer in order to reduce the power consumption needed to achieve a given mixture quality.

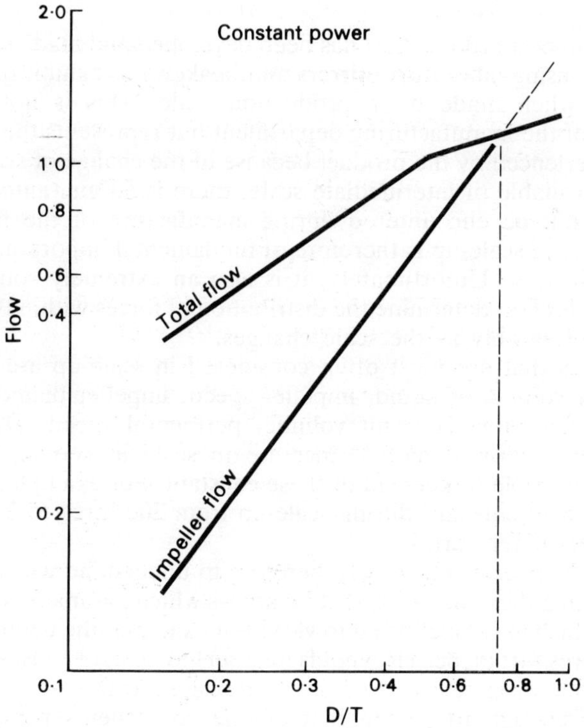

Figure 39.22 Impeller flow and total flow as functions of D/T

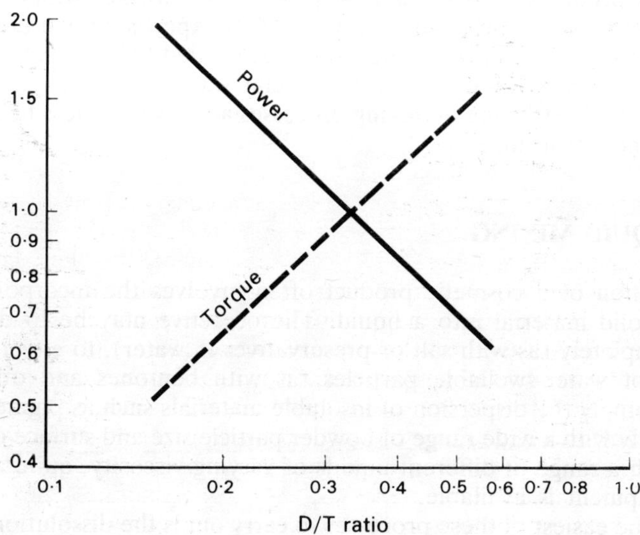

Figure 39.23 Effect of D/T on power consumption and torque

Scale-up

It is not unusual for a product that has been developed and made successfully in the laboratory (using laboratory stirrers and beakers) to exhibit quite different characteristics when made on a production scale. This is not due to the incompetence of the manufacturing department but represents the difference in conditions experienced by the product because of the change of scale. Even if a pilot plant is available of intermediate scale, there is no guarantee that severe problems will not be encountered during manufacture of the first full-scale batch. The study of scale-up is therefore of fundamental importance to efficient cosmetics production. Unfortunately, it is also an extremely complex subject since the variables that determine the distribution of forces within the processing vessel vary considerably as the scale changes.[12]

The quantities that are most often considered in scale-up are total power, power per unit volume of liquid, impeller speed, impeller diameter, the flow from the impeller, flow per unit volume, peripheral speed, D/T ratio and Reynolds number. For a ten-fold increase in scale in geometrically similar vessels, it is impossible to keep all of these constant. For example, if power per unit volume is held constant during scale-up from 200 litres to 2 tonnes, then every other parameter varies.

The task of the process chemist is therefore to understand which characteristics are controlling the process, so that he knows which parameters must be kept constant and which may be allowed to vary. For example, the manufacture of an emulsion requires turbulence (Reynolds number) to be held constant in order to achieve consistent droplet size of the internal phase. At the same time, the flow must be great enough to ensure that all the constituents pass through the turbulent regions.

In general, it may be said that there is not always sufficient awareness on the part of the development chemist of the problems arising from scale-up. There is no point in producing a superb new product in the laboratory which it is impossible to make in the factory. A pilot plant experiment can be a valuable tool in choosing the equipment and processes needed for full scale production but not if the study is made in a superficial way. This is an area where a fundamental understanding of mixing processes and a willingness to experiment will prove most fruitful.

SOLID–LIQUID MIXING

The production of a cosmetic product often involves the incorporation of a powdered solid material into a liquid. The objective may be to dissolve the powder completely (as with salt or preservatives in water), to effect a colloidal dispersion of water-swellable particles (as with bentones and other gelling agents) or simply the dispersion of insoluble materials such as pigments. To do this efficiently with a wide range of powder particle size and surface characteristics and with a range of different liquids of varying viscosity, quite a variety of mixing equipment is available.

Perhaps the easiest of these processes to carry out is the dissolution of a fairly large size, smooth-faced solid, such as salt. The initial incorporation of each crystal into the liquid involves the complete replacement of the air–solid surface with a liquid–solid surface. This may be considered to be a three-stage process of adhesion, immersion and spreading (Figure 39.24). Immersion is complete when all the air has been displaced and the surface of the crystal has been completely wetted by the liquid. This process is aided by a low surface tension and low contact angle between the liquid and solid.

Not all powders used in cosmetics, however, have such favourable size and surface characteristics. The majority are of extremely small particle size and, as has already been noted, highly agglomerated. Each agglomerate will have a complex structure with an uneven surface and will be perforated by cavities of irregular shape. The complete wetting of such structures, involving the penetration of liquid into all the crevices and cavities together with the expulsion of air, is very much more difficult. It should be noted, for example, that penetration into cavities requires a low contact angle but a high surface tension—in conflict with conditions for easy wetting.

Even more complex is the immersion of powders which swell in the liquid to form dispersions of colloidal size, since the particles on the outside of each agglomerated mass tend to swell and adhere to each other, slowing down penetration of the liquid to the still-dry core.

Powders that are small enough in particle size to form agglomerates are by far the most commonly used in cosmetics processing. In the dry state they entrap an enormous quantity of air (a bag of cosmetic-grade titanium dioxide, for example, contains only 25 per cent of powder together with 75 per cent of air). The majority of this air must be expelled if a smooth uniform mixture is to be obtained.

Immersion is only the first stage in the production of cosmetic quality dispersions. Even if the agglomerates were evenly distributed, the larger of them would give rise to 'grittiness'. Further, for pigments maximum colour can only be developed when these agglomerates are broken up and the maximum possible surface area of pigment is exposed. Disagglomeration is therefore the next step in the production process.

The forces holding these agglomerates together are precisely the same as those described earlier in the chapter in the section on powder mixing. The obvious difference, of course, is that these agglomerates are situated in a fluid medium, the physicochemical characteristics of which may enter into the reckoning of bond strength, ease of separation and likelihood of reagglomeration. Consequently, the theoretical treatment of particle–particle interaction in

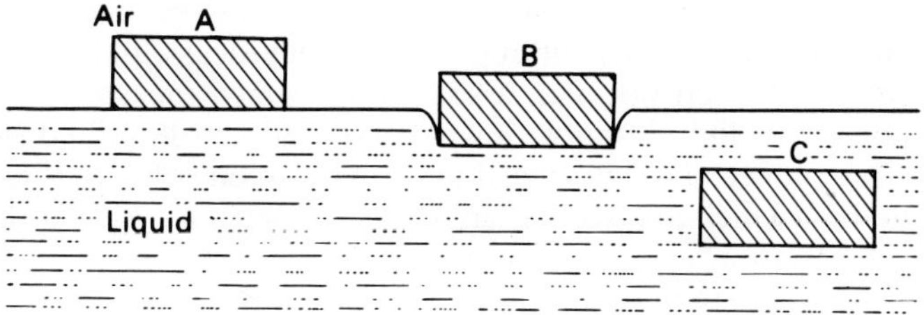

Figure 39.24 Immersion of a solid in a liquid
A Adhesion B Immersion C Spreading

liquid media is even more complex than for dry solids—although this has been extensively studied and written about.

In cosmetic processes, the disagglomeration of solid particles in liquid media can be brought about by a variety of machines. In lipstick processing, for example, pigments are 'ground into' castor oil by preparing a coarse mixture which is then passed over a triple roll mill, or further ground in a colloid mill, ball mill or sand mill. These machines are used specifically because they will deal effectively with media of lipstick-paste viscosity.

For less viscous media (for example, the dispersion of pigments into the aqueous phase of an emulsion), a high shear device of the rotor–stator type is frequently used. In this case, processing time can be shortened by ensuring that the whole contents of the vessel are brought into the catchment area of the shearing head by secondary stirring. As with all disagglomeration, shear stress is largely responsible for the partial disintegration of the agglomerate.

For soluble powders, the enormous increase in the solid–liquid interface brought about by immersion and disagglomeration will ensure that the actual process of dissolution can proceed at the maximum possible rate. For insoluble powders, however, there remains the problem of maintaining a good stable dispersion.

Disagglomeration is usually a reversible phenomenon and it can usually be assumed that the opposite process—'flocculation'—will be simultaneously taking place.

It was noted when considering powder–powder dispersion that stabilization could be achieved by the introduction of particles of larger size to which the disintegrated agglomerates could adhere. In some cases this can be applied to solid–liquid systems—for example, by pre-extending pigments onto talc before adding them to a liquid foundation base—but in many instances all the solid particles are of too small a size for this to be done. Under these circumstances rules similar to those used in emulsion technology can be applied. Thus the rate of flocculation can be slowed by some or all of the following means:

(i) The use of surface-active agents (sometimes as polymer coating of the powdered solid) to inhibit flocculation by steric hindrance.
(ii) The manipulation of electrostatic charges on the surfaces of the powder particles.
(iii) The manipulation of the viscosity of the dispersion.

Surface-active agents have a part to play at two stages in the process of manufacturing a stable dispersion. It has already been seen that the lowering of the solid–liquid contact angle speeds up the wetting process. In practice, the best results are often achieved not with a surfactant, which measurably lowers the surface tension of the liquid, but with what is sometimes described as a 'surface activator' which reduces the interfacial tension between solid and liquid. These surface activators (which are also described as 'dispersants' or 'wetting agents') can, if correctly chosen, cause an immediate improvement in the quality of the dispersion which is made manifest by a sudden increase in the colour intensity.

The rules for choosing a wetting agent or dispersant are similar to those used for surfactants in emulsions: part of the molecule must have affinity for the liquid medium and part for the solid. Clearly, if those parts of the molecule which have affinity for the liquid are large and present in sufficient numbers, by spreading out around each solid particle they can form a physical barrier that prevents particles from coming close enough together to reagglomerate. In this way an iron oxide dispersion in castor oil, for example, could be stabilized by the addition of a wetting agent having a hydrophilic end which would adhere to the oxide surface, and a long unsaturated fatty acid end which would be able to extend out into the castor oil medium surrounding each particle.

Where the liquid medium has a sufficiently high dielectric constant, other types of dispersion can be stabilized by surface activators having a residual electrostatic charge associated with them, thus preventing the flocculation of particles by the mutual repulsion of like charges.

A significant difference exists, however, between emulsions (where the two phases must perforce have very different chemical affinities) and dispersions of solids in which, for example, a hydrophilic surface can be dispersed in water. When this happens, this matching of affinities is an advantage. This has led to the processing of powders in order to coat their surfaces with a chemical (usually polymeric) of suitable characteristics to allow easy wetting and dispersion. Nowhere is this more clearly illustrated than in the case of water-wettable and oil-wettable grades of titanium dioxide. In this instance, the same grade of titanium dioxide can be coated with different resins to modify the surface in such a way as to make it wettable by either water or oil.

In aqueous media, the possibility exists for the particle surfaces to become electrostatically charged. This is an important consideration in the discussion of flocculation. For example, the effect of pH on the quality of pigment dispersions in emulsion-based products is often overlooked until it causes a problem by giving rise to an unexpected colour change. For those dispersions, particularly oxide surfaces, in which electrostatic repulsion is part of the stabilization mechanism, flocculation can occur rapidly as the isoelectric point is approached, if the pH is allowed to vary during manufacture.

Naturally, since particles have to move towards each other in order to flocculate, the viscosity of the medium through which they must move has a part to play in the rate of flocculation. It is necessary, however, to distinguish between the viscosity of the total dispersion which, as will be shown, is influenced by the solids content, and the intrinsic viscosity of the liquid medium itself; it is the latter which predominantly influences the rate of flocculation in low solids-content dispersions. The addition of thixotropic gums to nail lacquers and of colloidal thickeners to the aqueous phase of emulsions serves to slow down flocculation without materially influencing the basic flocculation process itself. For this reason the reheating of liquid pigmented foundation products sometimes results in unexpected changes of hue which are often erroneously ascribed to phase-inversion. The truth is that the rate of flocculation of an intrinsically unstable dispersion has been speeded up because of a drop in viscosity caused by the heating process.

Notwithstanding the viscosity of the liquid phase, it is generally true that the viscosity of the dispersion increases with the solids content, as does the difficulty of maintaining a good dispersion. For a given solids content, viscosity decreases with the particle size of the solid phase, and thus disagglomeration is usually accompanied by a fall in viscosity as well as a deepening of colour. In the same way, the substitution of one pigment by another with a different particle size can lead to a change of viscosity, and so it is, for example, that in order to achieve the optimum viscosity for the dispersion of pigments into castor oil, the ratio of pigment to oil may vary from one colour to another.

Suspension of Solids in Agitated Tanks

If a particulate solid is dispersed in a liquid in which it does not dissolve and the suspension so formed is allowed to stand undisturbed in a vessel, provided that the densities of the two components are dissimilar some degree of settlement or flotation will eventually take place. Where the particles are present in sufficiently low concentration to have a negligible effect on the viscosity of the suspension, re-suspension can be achieved by the establishment in the liquid of flow patterns of sufficient turbulence. The suspension of solids in agitated tanks is frequently encountered in cosmetics processing, as an aid to dissolution or as a means of obtaining a good dispersion of particles prior to a change of viscosity of the liquid medium by gelling or cooling. Although the theory concerning flow patterns in agitated tanks has already been discussed, it is necessary to reiterate that it is axial flow which is of prime importance in the movement of solid particles away from the top or bottom of a tank and that this can best be achieved with the aid of baffles. These are essential for effective particle suspension.

Three conditions can be recognized during the production of a suspension, namely complete suspension, homogeneous suspension and the formation of bottom or corner fillets.

Complete suspension[13–17] exists when all particles are in motion and no particle remains stationary on the bottom or surface for more than a short period. Under these conditions, the whole surface of the particles is presented to the fluid, thereby ensuring the maximum area for dissolution or chemical reaction. Complete suspension is achieved when the agitator has reached 'complete suspension speed' N_s. At this point there is normally a decrease in concentration of solids with tank height, with a clear liquid zone at the top. The concentration decays and clear liquid depth increases rapidly with increasing particle size and density difference. A summary of the factors contributing to the establishment of N_s and complete suspension can be obtained from the following correlation:[14]

$$N_s = \frac{S\nu^{0\cdot1}d^{0\cdot2}[(g\Delta\rho)/\rho]^{0\cdot45}X^{0\cdot13}}{L^{0\cdot85}}$$

where S = a constant

ν = viscosity

d = particle size (cm)

g = force due to gravity

$\Delta\rho$ = density difference between particle and fluid

ρ = fluid density

X = percentage w/w of solids in suspension

L = agitator diameter (m).

Homogeneous suspension exists when the particle concentration and (for a range of sizes) the size distribution are the same throughout the tank. The homogeneous suspension speed is always considerably higher than N_S and more difficult to achieve and to measure. Nevertheless, homogeneous suspension is very desirable for certain types of cosmetics applications and particularly so for continuous processing. In practice, for such processes the requirement is only that the particle size distribution and concentration in the discharge and the vessel are the same.

Sometimes heavier particles are allowed to collect in corners or on the bottom of the vessel in relatively stagnant regions to form *fillets*.[18] This may have the practical advantage of very large saving in power consumption compared with the energy that may be needed to achieve complete suspension (provided, of course, that this saving offsets the loss of active solids in the fillets).

In general, it may be said that propeller or 45°-angle turbines offer the best advantage for rapid suspension for low power consumption—particularly if draught tubes are introduced. If, on the other hand, radial flow agitators need to be used, these should be of relatively large length-to-diameter ratio, be placed close to the bottom of the tank and have turbine blades extending to the shaft to prevent problems with central stagnant regions.

LIQUID–LIQUID MIXING

As indicated by Table 39.1, it is convenient to consider separately the case in which the liquid components are all mutually soluble and the case in which some or all of them can coexist as separate phases (that is to say, they are sparingly or partly soluble in each other).

Miscible Liquids

The mixing of miscible liquids (blending) represents perhaps the simplest mixing operation in cosmetics manufacture. Several examples have already been cited and no further elaboration is needed except to reiterate that it is important to choose the mixing apparatus best suited to the viscosities of various components in order to carry out the operation efficiently.

Immiscible Liquids

Practically the only representatives of this category of mixing operation are emulsions. The theory of emulsions is covered fully in Chapter 38; briefly, all cosmetic emulsions consist of two major immiscible liquids, one dispersed as fine droplets in the other and separated by a layer of surface-active agent at each liquid–liquid boundary. (This is a simple view. In practice the two main phases are not always liquid at room temperature, and the total number of composite phases may be greater than three—this in no way invalidates a general discussion of the process available for the manufacture of emulsions.)

The Emulsification Process

The two major phases (these will be referred to as 'oil' and 'water') together with the emulsifier are brought together under turbulent conditions. Depending on the prevailing conditions, one major phase is broken up into droplets (predominantly by the action of shear stress imparted by turbulent eddies) and distributed throughout the other major ('continuous') phase.

While the droplets remain larger than the eddies, they will continue to break up into ever smaller droplets. Eventually a point is reached in this process when the available power giving rise to the turbulence cannot provide the shear stress necessary to reduce the droplet size any further. All this stage there exists an emulsion containing droplets of a certain mean diameter but over a range d_{min} to d_{max}. Provided that it is correctly chosen, the emulsifier prevents the rapid coalescence of these droplets and a stable emulsion is formed.

In order to obtain products of maximum stability that can be made consistently from batch to batch, it is desirable to keep the droplet size range as small as possible. In an agitated tank, droplet size is smallest near the impeller in the region of greatest turbulence whereas the maximum droplet size is to be found in any quiescent region of the tank. Thus it can be seen that d_{min} is fixed by the power available to generate turbulence and d_{max} depends on the efficiency of the mixing machinery fitted in the tank in producing a good circulation rate (so as to bring all the contents through the region of maximum turbulence).

Superimposed on the effect of circulation patterns in the vessel is a further factor affecting the range of particle sizes of the droplets. Theoretical considerations show that the mean droplet size is proportional to $N^{-6/5}$ (where N is the agitator speed) whereas d_{min} is proportional to $N^{-3/4}$. For a given vessel and agitator, therefore, the effect of increasing the stirrer speed should be to reduce the particle size range to a minimum after which a further increase in speed should give rise to instability and rapid coalescence (Figure 39.25). In practice, coalescence does not take place if a suitable emulsifier is present; nevertheless the graph shows the importance of attaining the correct agitator speed in reducing the particle size range to a minimum.

Orientation of Phases. In any emulsion the orientation of the phases (that is, whether the oil or the water phase is continuous) is determined principally by the choice of emulsifier and the volume ratio of oil to water. Usually, however, there is a range of volume ratio over which either phase may be dispersed, depending upon the method of manufacture. If a quiescent mixture of two phases coexisting as two simple layers (one upon the other) is agitated, the phase into which the agitator is placed is most likely to form the continuous phase in the resulting emulsion. In other words, drops are drawn into the phase in which the impeller is placed. If initially only one phase is present in the mixing vessel containing the impeller, an added second phase will inevitably form the disperse or discontinuous phase. If, however, the continued addition of the second phase combined with the choice of emulsifier eventually leads to a volume ratio at which the system is more stable with the added phase being continuous, then the emulsion will spontaneously invert to achieve this end result. When inversion takes place, it is very often accompanied by a change in droplet size. Where this

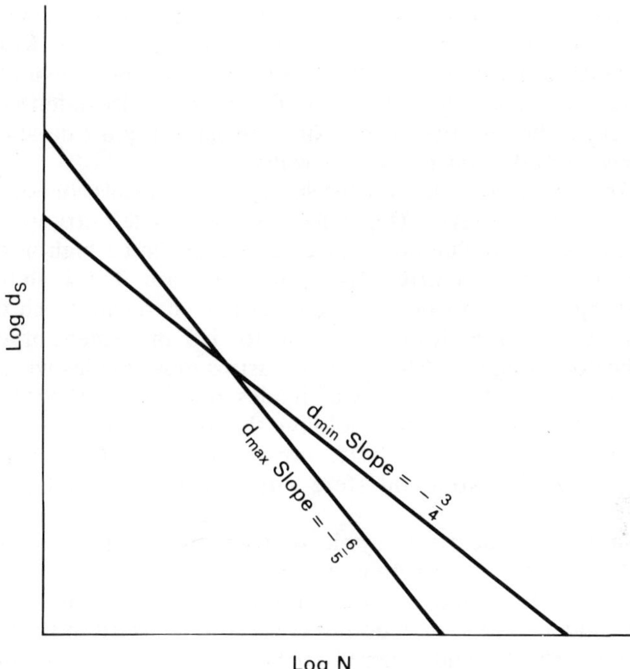

Figure 39.25 Relationship between drop size, d_s, and stirrer speed, N

change is a decrease, then inversion leads to a more stable emulsion and gives rise to a valuable method of manufacture.

Addition of Surfactant. In a batch manufacturing process for emulsions, there are four possible methods of adding the emulsifier. The first of these involves dissolving (or dispersing) the emulsifying agent in water, to which the oil is added. An oil-in-water emulsion is initially produced but inversion to water-in-oil may take place if more oil is needed.

Alternatively, the emulsifier may be added to the oil phase; the mixture may then be added directly to water to form an oil-in-water emulsion or water may be added to the mixture to form a water-in-oil emulsion. Many emulsions, on the other hand, are stabilized by soaps which are formed at the interface between the two phases. In this case, the fatty acid end of the soap is dissolved in the oil and the alkaline component is dissolved in the water. The two phases can be brought together in any order.

Finally, a less used method is one in which water and oil are added alternately to the emulsifying agent. Usually, the improvement in product quality obtained by the use of this method does not warrant the complication it causes in the manufacturing procedure.

Batch Processing Equipment. It will be evident from the discussion so far that there are at least two important elements of emulsion processing, namely shear (for the emulsification and particle size reduction process) and flow (in order to bring the whole contents of the vessel through the region of high shear). Flow is also important in the heating and cooling of the emulsion. Most emulsion processing vessels are equipped with a jacket through which steam or hot water can be circulated to heat the contents and cold water circulated to cool them. Evidently then, to be effective, the mixing mechanism must be able to provide adequate flow to and from the vessel walls.

For these reasons, most emulsion batch processing vessels contain a high shear turbine or rotor–stator device (typically bottom or side entry rather than top entry, to decrease the likelihood of air entrapment) and a high flow, low shear mixing device which may be driven by a separate motor. This high flow device is of variable design, the most popular being a gate stirrer in which the arms are inclined at about 45° to the horizontal so as to give an element of axial flow. In more complicated designs, a central shaft carries more blades which sweep the area between the first set, the sets of blades rotating in opposite directions. Whatever the design, the frame holding the outer blades normally carries spring-loaded plastic scraper blades to prevent the build-up of product on the inner vessel wall, which would interfere with efficient heat exchange across the surface (Figure 39.26).

The main motor may be driven by electricity or by air (up to about 100 psi or 9 bar). The advantages of air-driven motors are that they are infinitely variable in speed, torque-sensitive (and therefore less likely to become damaged when subjected to sudden loads), they do not constitute a hazard in the processing of low flash-point materials and generally they require less maintenance. Electric motors can be built to match some of these advantages (with slipping clutches and flameproofing), but only at considerable expense.

This traditional gate-type impeller system suffers from the grave disadvantage of limited axial flow. This is not noticeable in smaller vessels (below 600 litres capacity) but becomes a major problem in large tanks. One approach to the problem is to provide top-to-bottom transfer of the contents by means of a pump and an external pipe. A more satisfactory arrangement for the manufacture of emulsions of medium and low viscosity is to replace the gate stirrer by one or more axial flow impellers mounted centrally on a single central shaft. Although it becomes more difficult to provide wall-scrapers, the excellent flow around the vessel walls makes scraping less necessary.

The problem of ensuring that all the product passes through the region of high shear has led to the idea of passing the batch through an external circuit containing an in-line homogenizer; this may be a rotor–stator device, colloid mill or valve homogenizer.

Continuous Processing. In view of the difficulties encountered in manufacturing large batches of emulsion, a logical extension of an external circuit with in-line homogenizer is a continuous processing plant. A simple form of such a plant is illustrated diagrammatically in Figure 39.27. Such a plant is more correctly referred to as 'batch-continuous' since in essence a single batch is manufactured at a time. For long-run products, a truly continuous plant would be suitable, such as that illustrated in Figure 39.28. In this case, the addition of second vessels A′ and B′ (which are exact duplicates of A and B respectively) together with the three-way valves V_1 and V_2 means that a second batch of each phase can

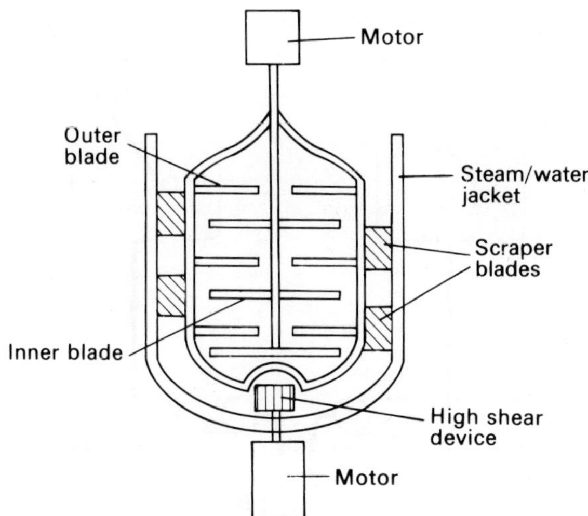

Figure 39.26 Batch emulsion processing plant

be prepared while the first is being used. In this way, a continuous supply of each phase is assured by the turning of a valve. In reality, continuous plants tend to be slightly more complex than is illustrated in the diagrams with the inclusion of take-off points for sampling and other sophisticated features. Nevertheless, continuous manufacture is a very practical and, for some applications, extremely economical method of processing emulsions.

Emulsion Temperature. The primary reason for raising the temperature of the phases during emulsion manufacture is to ensure that both are in the liquid state. In particular, the oil phase may contain fats and waxes that are solid at room temperature; there is very little point in raising the temperature of the oil phase much above that at which these liquefy. Excessive heating of the phases during manufacture prolongs the manufacturing time and wastes energy.

If the water phase is liquid at room temperature, it is customarily heated to approximately 5°C above the temperature chosen for the oil phase (so as not to cause the sudden solidification of the latter on blending). There is, however, an interesting alternative—namely, emulsification between hot oil phase and cold water phase. The plant for this procedure is illustrated in Figure 39.29, which shows that mixing of the phases and homogenization take place simultaneously. The obvious advantage of such a method is the saving of time and energy in not having to heat the aqueous phase.

REFERENCES

1. London, F., *Z. Phys. Chem*, 1930, **11,** 222.
2. Lippschitz, E. M., *Soviet Phys. JEPT*, 1956, **2,** 73.
3. Manegold, E., in *Kapillarsysteme*, Heidelberg, Verlag Strassenbau, Chemie und Technik, 1955, p. 524.

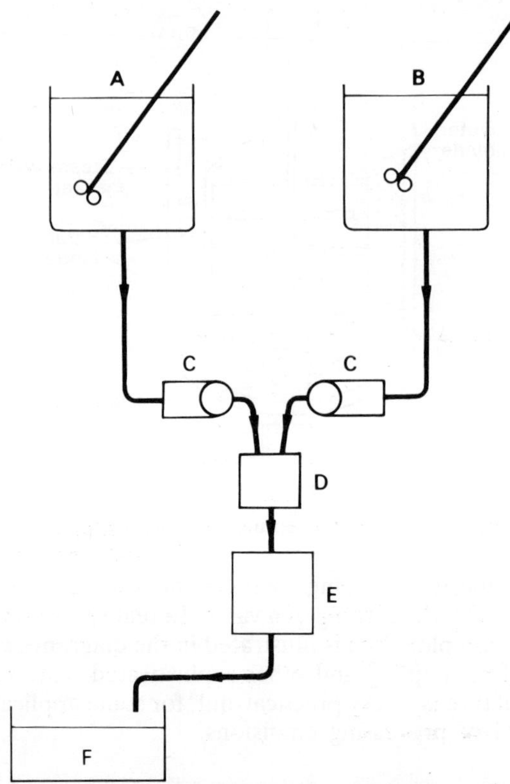

Figure 39.27 Simple continuous-processing emulsion plant
The two phases are prepared separately in tanks A and B, then pumped in correct proportions via metering pumps, C, into an in-line premixer, D (such as a static mixer), and then through a homogenizer, E. Finally, the formed emulsion is pumped into F, which may be a storage tank or the hopper of a filling vessel. A heat exchanger can be incorporated between E and F for rapid cooling

4. Rumpf, H., Strength of Granules and Agglomerates, in *Agglomeration*, ed. Knapper, New York, Wiley, 1974.
5. Spencer, R. S. and Wiley, R. M., *J. Colloid Sci.*, 1951, **6,** 133.
6. Mohr, W. D., Saton, R. L. and Jepson C. H., *Ind. Eng. Chem.* 1957, **49,** 1857.
7. Bridgeman, P. W, *Dimensional Analysis*, New Haven, Yale University Press, 1931.
8. Hoogendoorn, C. J. and den Hartog, A. P., *Chem. Eng. Sci.*, 1967, **22,** 1689.
9. Rushton, J. H., Costich, E. W. and Evrett, H. J., *Chem. Eng. Progress*, 1950, **46,** 395.
10. Rushton, J. H., Costich, E. W. and Evrett, H. J., *Chem. Eng. Progress*, 1950, **46,** 467.
11. Nagata, S., Yanagimoto, M. and Tokiyama, T., *Mem. Fac. Eng., Kyoto Univ.*, 1956, **18,** 444.
12. Oldshue, J. Y. and Sprague, J., *Paint Varn. Prod.*, 1974, **64,** 19.
13. Zweitering, T. N., *Chem. Eng. Sci.*, 1958, **8,** 244.

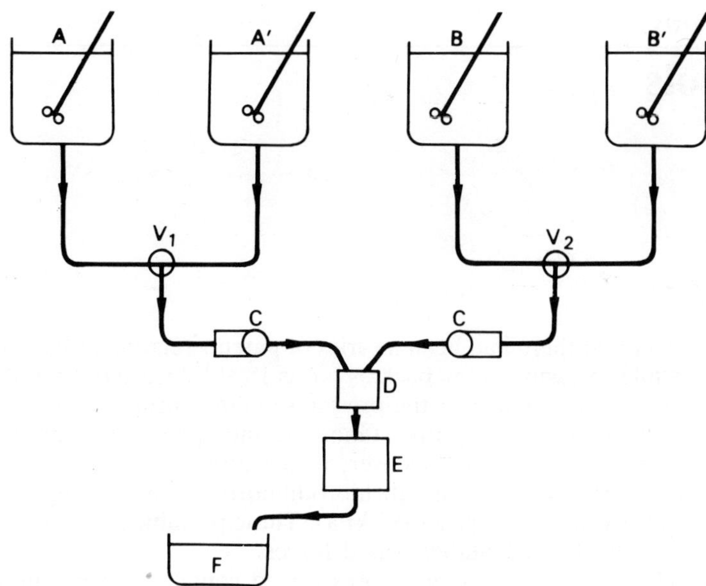

Figure 39.28 Continuous-processing emulsion plant for long-run products

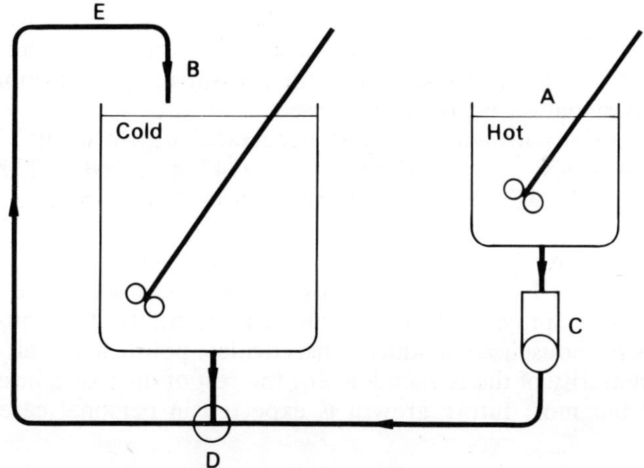

Figure 39.29 Hot/cold processing emulsion plant
The hot oil phase from tank A and the cold water phase from tank B are pumped into an in-line homogenizer at D and thence into the main tank B at E

14. Nienow, A. W., *Chem. Eng. Sci.*, 1968, **23,** 1453.
15. Pavlushenko, I. S., *Zh. Prikl. Khim.*, 1957, **30,** 1160.
16. Oyama, Y. and Endoh, K., *Chem. Eng.(Tokyo)*, 1956, **20,** 66.
17. Weisman, J. and Efferding, L. E., *A. I. Ch. E. J.*, 1960, **6,** 419.
18. Oldshue, J. Y., *Ind. Eng. Chem.*, 1969, **61,** 79.

Chapter Forty

Aerosols

Introduction

Despite the fact that there had been a series of patents relating to the packaging of products under pressure going back as far as 1899,[1] it was not until the 1940s that the principle was applied for the presentation of consumer goods. In 1943, as a result of research on pest control, Goodhue and Sullivan were granted a US patent[2] for a portable aerosol dispenser, in the form of a heavy gauge metal container filled with insecticide and dichlorodifluoromethane having an internal pressure of approximately 70 psig (483 kPa). These portable aerosols were used at the time by the United States armed forces.

In 1947, USA statutes relating to pressurized containers were modified to permit the use of thin-walled containers more suitable for the consumer market. A rapid development of containers and valves then followed, giving rise to a new packaging industry. Because the parent pack (the wartime insecticide) dispensed its product as a true aerosol, all the descendant pressure packs, irrespective of the physical form of the product dispensed, have been subsequently called 'aerosol'. While for many products the term 'pressurized pack' is more accurate, the original designation is still widely used.

The new form of packaging enjoyed a spectacular growth rate. Thus in the USA a market of just over 5 million units in 1947 increased to 2000 million in 1967. World fillings[3] were close to 4000 million in 1968, and increased to 6000 million in 1978.

Over 6500 million were filled in 1980,[4] just over one-third in the USA and another one-third in Europe. In these two geographical areas, personal products—mainly hair sprays and deodorant/antiperspirant—form just over half the market, with household products (insecticides, polishes and air fresheners) forming the majority of the remainder. For the rest of the world insecticides are predominant but most future growth is expected in personal care products.

THE AEROSOL

Every aerosol consists of a gastight container, a valve closure, an actuator button and a protective cap; in the majority of packs dip tubes are present.

The operation of an aerosol is based on the discharge of its contents by the pressure of a compressed gas or the vapour phase generated by a propellant present in the container as a liquid. The container may be constructed of metal, glass or plastic, the most extensively used materials being tinplate and aluminium.

Valves differ in design and are mostly operated by downward finger pressure; some valves operating by a tilt action (sideways pressure) are also available. A dip tube which reaches almost to the bottom of the container is usually attached to the inlet side of valves to enable the liquid to flow to the valve when the container is in an upright position. Packs not fitted with dip tubes are intended to operate in an inverted position.

The design of the actuator button attached to the valve, particularly the shape and size of the orifice, largely determines how the product is dispensed. When the valve is actuated, the pressure inside the container forces the liquid concentrate through the dip tube, past the valve, into the actuator button and then out into the atmosphere in the form of a stream, mist, spray or foam.

CONTAINERS

Containers are available in tinplate and aluminium, plastic-coated glass and plastics. In America seamless drawn steel cans, usually internally lacquered and fitted with a tinplate base, can be found.

Tinplate

More than half of the world's aerosols are packed in three-piece tinplate containers with welded or soldered side seams. They are available in many sizes from 4 oz (150 ml) to 30 oz (1000 ml). The amount of tincoating applied varies with the component—body, cone or dome—and the country of origin, E2·8 (2·8 g m^{-2}) being common for bodies in America and Europe.

There are three types of solder generally used for making side seams:

1. The 2/98 tin/lead system (mixed solder or standard solder).
2. The Duocom (antimony/tin) system.
3. The 100 per cent tin system (plain solder).

The advantage of using plain (100 per cent tin) solder is that it does not give rise to the internal or external corrosion which may occur with mixed solder; however, because of its inferior mechanical strength it is liable to give rise to troubles at high internal pressures.

Only containers with welded seams are made now in the United Kingdom[5] and welding is gradually replacing soldering in other countries. The process is based on technology rather than craftsmanship and has therefore potential for further development. The weld is stronger than the body plate, allowing higher pressure formulations to be packed. The seam width is only a quarter that on soldered cans and thus it is possible to get nearly all-round decoration.

Aluminium

Aluminium cans are of two main types:

1. One-piece containers, the monoblocs, available in sizes of up to 36 oz (1200 ml).
2. Two-piece containers in sizes of up to 20 oz (750 ml).

These also differ in the neck profile, that of the monobloc being in the form of a rolled bead, while two-piece containers have solid beads which are not as consistent as rolled beads and are more prone to ridges and dents.

The two-piece cans have an aluminium body and a seamed-on base of either aluminium or tinplate. The latter type has the advantage of allowing the use of a magnetic water bath, but this is to some extent offset by a greater likelihood of corrosion. Both tinplate and aluminium frequently require protection from corrosion by the contents of the container. This is generally achieved by lacquering or, in the case of aluminium, by anodizing (see later in this chapter—Corrosion).

Uncoated Glass

Uncoated glass aerosol containers are attractive, with freedom of shape, but as a safety measure[6] only low pressures (between 15 and 20 psig; 103–138 kPa) should be used.

Plastic-coated Glass

The use of plastic-coated glass containers is restricted by pressure considerations. Vinson[7] recommended a PVC coating with a thickness of not less than 0·015 in (0·38 mm).

Valves for these containers are more expensive, since bottle valves are used instead of standard one inch valves and they are not produced in large quantities. The crimping method is different, and involves external crimping or swaging. The rejection rate of bottle valves is large. Advantages of glass packs are that they are non-reactive and cause no corrosion problems, since valves are not in contact with metal.

Selections of the various types of metal and glass containers available at present are illustrated in Figures 40.1, 40.2 and 40.3.

Plastic Packs

Plastic containers for aerosols are a more recent addition to this field of technology. They combine the advantages of safety and freedom from corrosion. However, they are rather expensive and in some cases interactions may occur between the perfume and the plastic or the plasticizer present. The plastic materials used include polyacetal and polypropylene.

Safety Considerations

Modern metal containers, especially those without a side seam, can be used with safety for pressures of up to 100 psig (690 kPa) at 21°C but glass packs containing liquefied gas propellants are only suitable for lower pressures. Even then they may require to have a metal, cardboard or plastic protective cover, which may be integral with the glass pack.

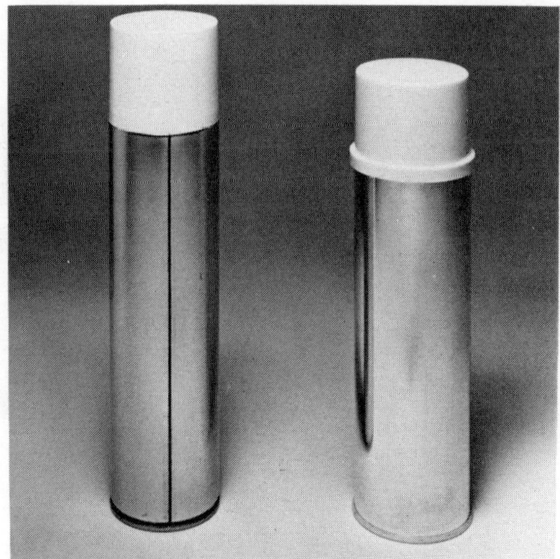

Figure 40.1 Tinplate aerosol containers
Trimline can with cover and Regular can with Top Hat cover
(Courtesy Metal Box Ltd)

Figure 40.2 Aluminium aerosol containers

Figure 40.3 Glass aerosol containers
(Courtesy Max Factor Ltd)

VALVES

The most important part of the dispenser is the valve. There are many types of valve available, the major manufacturers being the Precision Valve Corp., Aerosol Research and Development, the Risdon Manufacturing Co., the Newman-Green Valve Co., Seaquist, Ethyl Corporation and Coster Tecnologie Speciali S.p.A.

The details of construction and operation of some of the main types of valve are outlined below (see Figure 40.4, p. 819).

The valve is the mechanism for discharging the product from the container, and in aerosol technology the term refers to the entire assembly which is sealed on to the aerosol pack, including the valve stem, spring and valve housing or body, the metal cup in which it is mounted, the dip tube (when present) connecting the valve with the contents of the container, and finally the actuator button or spray tip through which the product is discharged into the atmosphere. Figure 40.4 illustrates the intricacies of a typical valve assembly, and shows the propellant injection paths during the filling stage. The gasket, which may be of one of many materials (for example, nitrile, neoprene, butyl or Viton), plays a vital role in the operation of the aerosol and maintenance of pack integrity throughout its life. It is important, therefore, to apply deliberate care and attention to the correct choice of gasket, in order to achieve complete compatibility with the product and to ensure optimum valve performance. The valve determines the rate of discharge of the product, the form in which it is discharged—spray, foam, cream or powder—and, in the case of sprays, the fineness, the spray profile and spray pattern. By using metering valves, it is also possible to discharge a predetermined amount of product each time the valve is operated. The choice of valve and its various components is just as important in determining the success of the product as the selection of the propellant composition and the choice of product constituents.

Components

The basic components of the valve mechanism are:

1. The dip tube through which the product is delivered from the bottom of the container into the housing. It is usually made of polyethylene or polypropylene.
2. The valve housing or body which holds all the valve components together, and which itself is held in the mounting cup.
3. The valve, which transmits the product from the housing to the actuator button or spray tip and which acts as an expansion chamber.
4. The stem, a component of the valve, upon which the button is placed.
5. The gasket, which seals the stem orifice when the valve is closed.
6. The spring, which returns the stem to close the valve when the pressure upon the actuator button is released.
7. The actuator button (spray tip), which operates the valve and which controls product discharge. It can be made from various plastics, including nylon, polyethylene, polypropylene and acetal. The orifice size of the button will vary within a fairly wide range depending on the product to be dispensed.
8. The mounting cup, which is crimped around the housing and mounted on the container.
9. The cup lining or inner gasket, which is either flowed into, or placed in, the rim of the valve cup to form a seal when the valve cup is crimped over the neck of the container.

Operation

The presence of a compressed or liquefied gas in the sealed container gives rise to a pressure differential between the head space inside the container and the surrounding atmosphere. When the valve is opened, the product is pushed up from the base of the container through the dip tube into the valve housing, then through the open seal, up the valve stem, through the valve stem orifice into the expansion chamber formed by the valve stem and the actuator passage, and finally emerges into the atmosphere as a jet, or a fine or coarse spray or foam, depending on the design of the valve and actuator button, as well as on the propellant system.

Types of Valve

There are many types of valve designed to meet various specialized needs and purposes. Some of the variations are described below.

Standard Valves

These valves are designed to fulfil the function of controlling the discharge of the product from the container in a simple on–off manner by breaking or making a seal within the valve. The rate of discharge is governed by either the orifice at the seal or that at the final exit, whichever is the smaller, and this can be selected according to requirements. There are several variations of mechanism which come within the description of standard valves; these include vertical or tilt action buttons, integral or separate movable valve stems, and various designs of springs and seals.

Foam Valves

The valves used for dispensing foams are essentially standard valves fitted with wide unobstructed passages on the exit side of the seal orifice. This passage, which ends in a foam spout, serves as the expansion chamber in which the foam forms. The type of foam is governed by the composition and proportions of product and propellant.

Powder Valves

These also are little different from standard valves and the principal requirement is a smooth flow passage on both sides of the valve seal for the slurry of powder and propellant, so that there is a minimum possibility of powder deposition to interfere with the valve operation. Correct choice of powder ingredients and propellant is essential.

The Risdon Company[8] claimed to have overcome the problem of valve blockage by powders by means of a 'slide-and-clean' valve in which the internal metering orifice is in a movable stem, and which cleans itself by passing through the seal each time the actuator button is released. The stem wall carrying the orifice is thin so that the powder film in the orifice is disrupted by the pressure when the valve is opened. A number of other valve companies, notably the Precision Valve Corporation, successfully introduced long-stroke valves to prevent powder build-up and blockage of the orifice.

Mechanical Break-up Spray Valves

The standard valve serves primarily to open and close the container to the atmosphere and has no special provision for breaking up the product other than the vaporization of the propellant. Additional break-up can be provided by means of an expansion or swirling chamber in the actuator button. The product, with its flow accelerated by means of a constriction, passes into the chamber tangentially so that it moves in a spiral fashion, and the droplets break up by collision with each other and with the chamber walls before leaving the chamber through a small orifice, at high velocity.

Harris and Platt[9] pointed out that a product leaving the actuator in a neat condition will form a wide hollow cone, while if it still contains a proportion of miscible propellant it will form a narrower solid cone.

Break-up valves are used for dispensing aerosol perfumes and for hair lacquer, particularly those based on propellant 12 only.

Valves for Water-based Sprays
Water-based sprays with hydrocarbon propellants, for example insecticides and air fresheners, tend to require more intense break-up. This can be provided by special valves which have two entrances to the chamber in the valve housing. One entrance, the vapour phase tap, admits propellant vapour and the other, which is constricted, admits the product via the dip tube, which may be capillary. The restricted orifice is necessary to prevent propellant vaporizing in the dip tube, which would cause cavitation in the liquid. The stem has a large internal orifice to lead away the expanded volume of droplets and vapour.

Metering Valves
When it is desired to dispense a fixed amount of product for each actuation of the button, this is achieved by providing a chamber with sealed entrance and exit orifices. With the button in the closed position the metering chamber exit is closed and the entrance open, thus allowing the chamber to fill from the container. Depression of the button closes the entrance, trapping a metered volume of product (usually of the order of 50 μl) and then opens the exit to allow the escape of the product for use.

Controlled Discharge Rate Valves—Vapour Phase Taps
A valve is basically an on–off mechanism but it is sometimes possible to control the amount discharged, either by means of the extent to which the actuator is depressed or by using a variable discharge valve. The restricting orifice in a standard valve is usually the internal metering orifice, which normally has a diameter of not less than 0·010 in (0·254 mm).

Lower discharge rates can be secured by the following methods:

1. Replacing the normal dip tube (bore about 0·15 in or 3·8 mm) by a capillary tube with bore 0·04–0·10 in (1·01–2·54 mm) or microcapillary of 0·014–0·017 in (0·35–0·43 mm).
2. Using a low-delivery housing in which the orifice at the bottom of the housing is restricted to 0·008–0·025 in (0·02–0·63 mm).
3. Using a vapour phase tap. This is a small orifice in the side of the housing, connecting the head space of the container with the inside of the housing, for introducing propellant vapour into the valve at this point. It may be used to improve break-up (*cf.* the valve described under 'water-based sprays') or to allow a rapid return of the liquid phase left in the dip tube and minimize possible blockage by products containing dispersed solids. A vapour phase tap has advantages in that:

   (i) it allows the aerosol to be operated in both upright and inverted positions, when the propellant and product each enter the housing through the orifice normally serving the other;
   (ii) it lessens the cooling effect by reducing the discharge rate and by reducing particle size, which accelerates evaporation before the spray reaches the target area;
   (iii) it allows faster filling in the presence of a low-delivery housing which normally demands under-cup fillling, except with the new generation of valves.

The vapour phase tap may, however, result in the loss of too much propellant from the head space with the risk of incomplete discharge of the product, or a changed spray pattern as the container is emptied. This can be prevented by ensuring that the tap is of a suitable size and that there is an adequate amount of propellant present.

Fast Filling Valves
From the 1970s and into the 1980s the requirement remains for valves that allow even faster filling of the aerosol container. This has come about partly because of changes in propellant and requirements for lower discharge valves, and partly because of a need to increase output speeds of production lines. A number of valve companies, notably the Metal Box Company, UK, with their CL-F valve, have developed systems whereby propellant may also by-pass the discharge path during the filling operation and thus ensure fast gassing no matter what the size of the internal orifices within the valve. Such valves may be used also for direct injection of compressed gases into standard cans using standard equipment.

Transfer Valves
These valves are used to provide a connection with smooth flow between the stems of two standard valves with their buttons removed. They are used for recharging the smaller daily use pack of, for example, 'mother and daughter' pack hair lacquers, for refilling cigarette lighters with liquid butane and in the laboratory for pressurizing aerosol cans with propellant from other cans.

Speciality Valves
An innovation in 1968 was Risdon's valve which was designed to work effectively in any position, from upright to inverted, without loss of vapour gas.[10] The valve has, in addition to the usual assembly, a double dip tube with a transfer connection opened and closed by a ball valve to allow the product to flow between the tubes as the container is tilted. It is claimed that the valve can be used with all types of propellants and with any actuator.

PROPELLANTS

The distinguishing and essential feature of an aerosol is the propellant, which may be a liquefied gas (which will vaporize at atmospheric pressure), a compressible gas or a mixture of the two.

Liquefied Gas Propellants

The majority of aerosols employ liquefied gas propellants, that is propellants which are in the gaseous state at atmospheric pressure and room temperature, but which liquefy on compression. When liquefied, their vapour pressure will vary with temperature. The high vapour pressures of some of the lower boiling compounds must be reduced, by the use of pressure depressants, so as to achieve pressure compatibility with common aerosol containers and also to impart consumer-acceptable spray characteristics.

The important feature of a liquefied gas propellant, distinguishing it from compressible gases, is that as long as any propellant is still present in the pack in its liquefied form the internal pressure within the container will, at any given temperature, be constant irrespective of how much product or propellant has been discharged. As the head space in the pressure pack increases, more of the liquid propellant evaporates to maintain the internal pressure at an approximately constant value. This ensures that the spray characteristics of a pack are virtually unchanged, irrespective of whether the package is full, half-full or almost empty, and hence ensures consistent performance throughout the life of the aerosol. Of course the temperature of usage affects spray performance.

Spray characteristics differ at extremes of temperature, because they are directly affected by pressure which is temperature-dependent. The normal leakage experienced through valves and seams of sound aerosols is increased at higher temperatures. At very high temperatures deformation and rupture may occur under the influence of the attendant high internal pressures.

Liquefied gas propellants include chlorofluorocarbons, hydrocarbons and dimethyl ether.

Chlorofluorocarbon Propellants

The most important chlorofluorocarbon propellants are trichlorofluoromethane, dichlorodifluoromethane, dichlorotetrafluoroethane, and their mixtures.

In addition to the 'Freons' of E. I. du Pont de Nemours Co., USA, who were the first to produce them on the commercial scale, trade names under which these products are sold include:

Algofrene (Montecatini Societa, Italy)
Arcton (Imperial Chemical Industries Limited, UK)
Forane (Elektrochimie Ugine, France)
Frigen (Hoechst Chemicals, Germany)
Isceon (ISC Chemicals Limited, UK)

An international code is used for the various chlorofluorocarbon compounds, and is interpreted as follows:

(i) The right-hand number represents the number of fluorine atoms.
(ii) The second number from the right represents the number of hydrogen atoms +1.
(iii) The third number from the right represents the number of carbon atoms −1.
(iv) Spare valencies are filled by chlorine atoms.
(v) The fourth number, if present, is the number of double bonds, and a prefix C indicates that the compound is cyclic.
(vi) Conversely, if 90 is added to the code number, the last three digits, starting from the right, give the number of fluorine, hydrogen and carbon atoms respectively, with chlorine atoms to make up the valency requirements.

Details of molecular weights, vapour pressures and boiling points for the three most commonly used chlorofluorocarbon propellants are given in Table 40.1.

Propellant 12 may be used on its own, without auxiliary pressure depressants, as for feminine hygiene sprays. It is more usual to associate this propellant with

Table 40.1 Properties of Chlorofluorocarbon Propellants

Propellant	International Code No.	Chemical formula	Molecular weight	Boiling point	Vapour pressure at 21°C (psig)	(kPa)
Dichlorodifluoromethane	12	CCl_2F_2	120·9	−29·8°C	70·2	484
Trichlorofluoromethane	11	CCl_3F	137·4	−23·8°C	13·4	92·5
Dichlorotetrafluoroethane	114	$C_2Cl_2F_4$	170·9	+ 3·6°C	12·9	88·9

pressure depressants deliberately incorporated into the product formulation or occurring there naturally to achieve lower internal pressures compatible with commonly available containers.

The traditional pressure depressant propellant 11 may, under certain conditions, cause corrosion of metal containers in the presence of water or the lower alcohols. In these circumstances or where there is an adverse effect upon certain perfumes, it is advisable to replace it by propellant 114. Propellant 114 is extremely stable; it exerts a low pressure of about 13 psig (90 kPa) at 21°C, and this makes it generally unsuitable for use on its own in temperate climates.

By mixing propellants 11, 12 and 114, all practically required pressures may be obtained. Since these mixtures obey Raoult's law, their pressure characteristics can be predicted. If, however, they are mixed with other types of propellants or with solvents (for example ethanol) the resulting mixtures will deviate from Raoult's law.

The most commonly used propellant mixture contains equal parts by weight of propellants 11 and 12 and exerts a pressure of about 37 psig (255 kPa) at 21°C; it is used in such applications as solvent-based space insecticides and air fresheners. Mixtures in different ratios will give rise to different pressures more suitable for hair sprays, deodorants and antiperspirants. Mixtures of propellants 12/114 in weight ratios ranging between 10/90 and 40/60 are used to pressurize products such as shaving creams, colognes and perfumes.

Vapour pressures exerted by various mixtures of propellants 11 and 12 and propellants 12 and 114 at 21°C are given in Table 40.2.

Chlorofluorocarbons 11, 12 and 114 are characterized by a high degree of stability and a low order of toxicity; studies on the possibility that chlorofluorocarbons may be toxic if inhaled have been reviewed in Chapter 25 under the heading of Hair Sprays. They do not form explosive mixtures with air. They are also nonflammable and furthermore they depress the flammability of aerosol formulations which include flammable solvents.

The Chlorofluorocarbon Ozone Controversy. The hypothesis that emissions of chlorofluorocarbons to the atmosphere would cause depletion of ozone, and thus increase the amount of harmful UV radiation reaching the surface of the earth, was first postulated by Rowland and Molina in 1974.[11] Since then a great deal of research effort has been applied in attempts to validate the theory. While this has increased our knowledge, it has shown that our understanding of the

Table 40.2 Vapour Pressures of Mixtures of Propellants

Propellant mixture 11/12 ratio by weight	**Vapour pressure at 21°C** (psig)	(kPa)	**Propellant mixture 114/12 ratio by weight**	**Vapour pressure at 21°C** (psig)	(kPa)
70/30	23	158	90/10	20	138
65/35	27	186	85/15	24	165
60/40	30	207	80/20	27	186
50/50	37	255	70/30	34	234
40/60	44	303	60/40	40	276
35/65	47	324	50/50	46	317
30/70	51	352	40/60	51	352

atmosphere is very much less complete than had been assumed. The atmosphere is being revealed as a very complex interlinked system and the consideration of a single effect or reaction sequence in isolation can give very misleading results.

Numerous reports, for example that by Brasseur,[12] have been written reviewing the scientific developments together with calculations of potential ozone depletion. Such calculations, which are based on many assumptions, may be regarded as doubtful forecasts as they are a compound of uncertain science and uncertain scenarios projected many decades ahead. Research into the problem continues and new assessments of the science are prepared as more information becomes available.

Statistical treatment of actual ozone measurement data has so far failed to reveal a depletion trend.

Notwithstanding the lack of scientific validation of the Rowland–Molina theory, some countries have regulated the use of chlorofluorocarbons as aerosol propellants.

Hydrocarbon Propellants

Hydrocarbons such as propane, n-butane and isobutane, which have a low order of toxicity,[13] are used in aerosol packs as cheap, stable and non-corrosive propellants which are handled safely in transport, storage and filling operations.[14–16] They are used with water-based products to give low-cost non-flammable end products that present no special corrosive hazards. They are blended with other propellants and solvents to lower the overall costs and in some cases actually improve the product.

These hydrocarbons, which must be relatively free of impurities, may be blended with each other or with the chlorofluorocarbons to give the required vapour pressures for the propellant charge content to dispense aerosol products satisfactorily.

Propane, n-butane and isobutane are the only naturally occurring hydrocarbons which have the stability and availability that make them suitable for use on a commercial scale. These hydrocarbons are readily separated from naturally occurring mixtures to give stable odour-free propellants in large volume at low cost. Their physical properties are given in Table 40.3. Hydrocarbons are used alone in water-based aerosol products such as shaving creams, window cleaners,

Table 40.3 Physical Properties of Hydrocarbon Propellants

Propellant	Specific gravity of liquid	Formula	M.W.	Boiling point at 1 atm (°C)	Vapour pressure at 100°F (37·8°C) (psig)	(kPa)
Propane	0·508	C_3H_8	44·09	−42·1	189·5	1306
n-Butane	0·584	C_4H_{10}	58·12	−40·5	52·0	358
Isobutane	0·563	C_4H_{10}	58·12	−11·7	73·5	507

air fresheners and spray starches. They may also be used alone or combined with other propellants to dispense many water-free products.

Hydrocarbons with their low specific gravity are lighter than most products so the liquefied hydrocarbons will float on top of the product when they are immiscible.

Herzka[17] has indicated that there are advantages in using propellants that are lighter than the product; in particular there is no danger of discharging pure propellant with three-phase systems, at least until all the product has been dispensed.

Besse, Haase and Johnsen[18] have presented graphs showing that up to 30 mole per cent of propane can be blended with propellant 12 before a pressure of 100 psig at 70°F is reached (689 kPa at 21·1°C).

Smaller amounts (about one-third by weight) of hydrocarbon propellants, because of their lower specific gravity, are required compared with chlorofluorocarbon propellants in order to give the same pressure in an aerosol container and to give the same volume content in a formulation.

The hydrocarbon propellants are soluble in chlorofluorocarbons, alcohol, chloroform, methylene chloride, ether and the higher hydrocarbons such as n-pentane and n-hexane. They are non-polar and are not hydrolysed by water. There is little chance of the formation of corrosive substances from the hydrocarbons. This non-corrosive characteristic is an important advantage in the use of hydrocarbons in aerosol propellants. They are also essentially immiscible with water. Butane–water–alcohol systems, designed to produce a single liquid phase, are of increasing importance for personal care products such as hair sprays. Care needs to be exercised with respect to corrosion and therefore product/container compatibility.

Hydrocarbon propellants have a low toxicity rating, and they do not form toxic decomposition products at high temperatures. Normal care should be exercised, however, to avoid contact of the liquid with the skin or eyes.

Hydrocarbons have low flash points and low explosion limits.

	Flash point (°C)
Propane	−104
Butane	−74
Isobutane	−83

However, from the point of view of flammability the properties of the complete formulation are most important. Many formulations using hydrocarbons as the sole propellant and many containing a combination of halocarbons and hydrocarbons are also non-flammable. Components of an aerosol mixture other than the propellant, such as solvents, resins, etc., may increase flammability risks.

Many examples of non-flammable products containing hydrocarbon propellants have been reported in the literature. Other instances have been reported in which the presence of hydrocarbon in the propellant actually reduced flammability.

Reed[19] described a propellant comprised of:

	per cent w/w
Propellant 11	45
Propellant 12	45
Isobutane	10

Similar blends are being brought into use increasingly for alcohol-based hair sprays and deodorants.

Fowks[20] cited an example in which the length of the flame in flame extension tests was notably reduced when a chlorofluorocarbon propellant in a hair spray was replaced by a chlorofluorocarbon–hydrocarbon mixture.

Calor Gas (Limited) UK supplies propane–butane mixtures now used as propellants in a large number of important aerosol products. They offer 'destenched gas' as opposed to 'unstenched gas'—the latter being collected at the refinery and containing a small quantity of residual sulphur compounds.

Destenched gas is obtained by further processing through molecular sieves or other absorbents. Calor aerosol propellant is available as a standard at pressures of 30, 40, and 48 psig (207, 276 and 331 kPa) at 21°C. The different pressures are obtained by varying the percentages of butanes and propane present, since each has a different vapour pressure.

Dimethyl Ether

Rotheim, in Norway, first suggested the use of dimethyl ether for aerosols.[21] Since 1966, aerosols using dimethyl ether as a propellant have been produced, with the major contribution from Holland.

Highly purified dimethyl ether is an attractive propellant.[22,23] It is virtually odourless, has specific dissolving powers and, as it is miscible with water, is capable of producing single liquid phase product systems (with ethanol, for example) thus commending its application to various personal care products.

Current toxicological studies,[24] wide in their scope, provide encouraging data to date. The toxicity studies continue and further data are awaited with interest.

Flammability characteristics of dimethyl ether are such that when it is handled with appropriate caution and care, both in production and in product formulations, commercial experience reveals that no unacceptable hazard is posed.

Increasingly dimethyl ether is being used both in Europe and now in America for many types of aerosol product, including hair sprays, setting lotions, deodorants, deo-colognes, perfumes and toilet waters.

Compressed Gas Propellants

The use of compressed gases as propellants for pressurized packs was first proposed in the second half of the nineteenth century—well before the introduction of chlorofluorocarbons.[25,26] The term 'compressed gas aerosol propellants', as far as the aerosol field is concerned, is applied to gases that can be liquefied only at very low temperatures, or under extremely high pressures. The compressed gases mostly used as propellants are nitrous oxide, carbon dioxide and nitrogen.

Before 1958 the use of carbon dioxide and nitrous oxide was introduced for whipped cream aerosols, while carbon dioxide alone is used for de-icers and fire extinguishers. It was only after special valves became available that it was possible to use nitrogen as the propellant for dispensing toothpaste, thick emulsions in stream form, and various liquids in spray form, for example, perfumes in glass bottles.

There are a number of differences in behaviour, connected chiefly with volume and pressure changes, between compressed gases on the one hand and liquefied propellants on the other, and even in some respects between compressed gases that are soluble in the product and those that are not. These differences produce different performance characteristics in aerosols. Some of the differences are listed in Table 40.4.

In practice, nitrogen is used for dispensing products in an unchanged form. An extension of this principle is illustrated by the free piston aluminium aerosol can developed by the Bradley Sun Division of the American Can Company.[27] This provided a true solid-stream dispensing of the product which was separated from the propellant by means of an internal plastic diaphragm made from medium density polyethylene. The propellant was nitrogen, introduced at 90 psig (620 kPa) through an aperture in the concave base of the can, which was then immediately closed with a rubber plug. The pressure of nitrogen caused further compression of the walls of the piston against the can, thus providing a seal which prevented mixing of the product and propellant. The piston was free to move up and down within the container as the pressure on either side changed. An almost complete (99 per cent) discharge of product was achieved.

Carbon dioxide used alone gives a wet product and a large fall-off in pressure leading to changed discharge characteristics and possible unexpelled product. Large head spaces, which invite adverse consumer reaction, tend to reduce these problems slightly. With aqueous products there is a possibility of product reaction and also of corrosion, which necessitates the use of special valves and cups. A system developed in Germany, the Carbosol system,[28] relied on the fact that at a given temperature and pressure a liquid absorbs a given quantity of carbon dioxide. Facilities for saturation of solvents to be used in a special Carbomix or Carbomat reactor were provided. Although this system on the face of it offered an alternative to the conventional propellant system, no significant commercial success was achieved.

Mixed Compressed Gases and Solvents

Mixtures of carbon dioxide with chlorofluorocarbon have been proposed, and although pressures are higher (60–70 psig or 414–483 kPa) they can usually be

Table 40.4 Comparison of Compressed Gas Propellants and Liquefied Propellants

Characteristics/Properties	Compressed gas propellant		Liquefied propellant
Solubility in product	Insoluble	Partly soluble	Partly soluble
Example	Nitrogen	Carbon dioxide	Chlorofluorocarbon or hydrocarbon
Pressure effects	Decreasing pressure as product is used, with decreasing discharge rate, changing spray pattern, and product retention, especially if volume fill of container high.		Constant pressure throughout use of pack, with constant discharge characteristics and low product retention.
Product residues	Tend to be high although less with soluble propellants.		Generally low.
Volume changes on discharge to atmospheric pressure	Gas not discharged with product, so no break-up effect, except that imparted by special break-up actuators.	Gas dissolved in product under pressure escapes with a 10-fold volume increase. Break-up effect small but significant. Correct choice of valve specification very important.	Propellant in formulation vaporizes with a 250-fold volume increase. Consequently large break-up effect which can be enhanced still further by correct choice of valve and, especially, the actuator.
Pressure change with temperature rise	Propellant largely obeys gas laws and shows small pressure changes with temperature.		Propellant is not an 'ideal' gas and shows considerable pressure change under influence of temperature.
Effect when product sprayed on skin	No chilling.	Slight chilling.	Marked chilling effect due to heat required for vaporization.

safely used with tinplate and aluminium cans. Filling presents difficulties, and has to be carried out as a two-stage process which slows down production, or else by special methods such as pre-saturation of the product with carbon dioxide. However, new generation valves may be employed which avoid all of these difficulties.

Both nitrous oxide and carbon dioxide have been used in conjunction with chlorofluorocarbon for dispensing space sprays.

Another mixed propellant system, developed by the Dow Chemical Co., is based on a mixture of a soluble compressed gas such as nitrous oxide with chlorinated hydrocarbons such as methylene chloride or 1,1,1-trichloroethane. This system is claimed to be cheaper than the chlorofluorocarbons which are replaced either in part or totally.

Depending on the solvency of the mixture and nature of spray required, the chlorinated hydrocarbon may be methylene chloride or 1,1,1-trichloroethane or blends of the two in the range 1:2 to 2:1. Grades of 1,1,1-trichloroethane developed for cold degreasing or vapour degreasing are generally applicable to aerosol use, for example:

Genklene N } ICI Ltd
Genklene LV }

Chlorothene NU } Dow Chemical Co.
Chlorothene VG }

The following examples illustrate hair spray formulations based on this system:

	(1) *per cent*	(2) *per cent*	(3) *per cent*
Methylene chloride	15	15	16
1,1,1-trichloroethane	10	31	16
Propellant 12	25	—	16
Nitrous oxide (60 psig, 414 kPa)	—	—	2
Nitrous oxide (80 psig, 551 kPa)	—	4	—
Hair spray concentrate	50	50	50
Perfume	*q.s.*	*q.s.*	*q.s.*

Methylene chloride and 1,1,1-trichloroethane are exceptionally good solvents for compressed gases, thus ensuring good filling characteristics and spray break-up while minimizing product residues.

With such solvent–compressed gas systems aerosols are filled to about 80 psig (551 kPa), and it is claimed that a pressure drop of 30 psig does not result in any appreciable change in spray quality if the valve system is suitably chosen to give the required spray characteristics. It has been found that such aerosols will deliver the product at a rate 60 per cent faster than a container using liquefied propellant at 40 psig (276 kPa).

Use of Methylene Chloride and 1,1,1-Trichloroethane

Methylene chloride and 1,1,1-trichloroethane are being used as solvents in aerosol hair sprays, insecticides, air fresheners, paints, oven cleaners and lubricants.

These solvents function as co-solvents for active ingredients and propellants and vapour pressure depressants. They are non-flammable under normal conditions and so help to control flammability and flammable contents within prescribed legal limits. The high densities of the two solvents provide mass compensation when used with low density propellants.

They are available in stabilized forms to give good performance in the presence of moisture, heat, light, metals and air.

Chlorinated solvents are usually non-corrosive when used in anhydrous aerosol formulations. Some hydrolysis may occur, however, in the presence of water, similar to that experienced with some chlorofluorocarbons.

The use of special inhibitors to protect against corrosion has been suggested. The results have been published[29] of tests showing superior properties in respect of corrosion of tin, steel and solder. However, it is recommended that water-based aerosols should be thoroughly shelf tested before marketing—this is sound advice for any new or modified aerosol product.

Once the balance of solvents to produce the required spray characteristics from the total formulation has been established, the choice of gasket should be carefully made, as with any new product formulation.

Gelled Propellants

When a product is dispensed from a conventional aerosol container, some of the liquid propellant is included in the spray. The additional droplets of liquid propellant can cause non-uniform deposition of product, a more rapid change in the characteristics of the spray during long bursts owing to cooling caused by evaporation of propellant, and increased cooling effect on skin with such products as antiperspirants and deodorants. It has been claimed[30] that these undesirable effects are minimized by the use of gelled versions of any liquefied propellants. The propellant mixes with the solution of the product in its dry vaporized form, and additional liquid droplets are not introduced into the spray.

The initial spraying times of gelled propellants were claimed to be approximately one-third greater than those of conventional propellants, and the overall dispensing capacity was said to be approximately 40 per cent greater than that provided by the conventional types of propellant.

FILLING OF AEROSOLS

As the industry began, the most important method for filling aerosols with liquefied propellants was cold filling. This method is still used today for some specialized filling, particularly in the perfume and pharmaceutical industries.

A second important method—under-cup filling—became a dominant method of production filling aerosols in the late 1960s and the 1970s in the United States, using machines from Kartridg Pak, USA. It was also adopted, particularly for large filling operations, in other countries.

With ever increasing importance, pressure filling is the most important production method employed world-wide today. Eminently suitable for liquefied propellants of all types, it is now also accepted as a proven method for compressed gases provided that suitable aerosol valve designs, originating in the 1970s and improved in the 1980s, are employed.

Cold Filling

In cold filling, the propellant is cooled to a low temperature at which it will handle as a liquid, and is run through the open aperture of the aerosol container to join chilled product concentrate already there. The valve assembly is then sealed on to the container.

This method facilitates the rapid filling of propellant while at the same time expelling air from the aerosol container. It cannot be used where the chilling effect produced by the addition of the cold propellant would adversely affect the product concentrate as, for example, with some emulsions and water-based products.

The temperature of the contents in the sealed aerosol container soon returns to normal, assisted by passage through the essential hot water test bath.

In spite of the simplicity of the cold filling method, both capital investment costs and refrigeration running costs are high.

Under-cup Filling

All operations in under-cup filling are conducted at ambient temperatures. Product is filled into the open-aperture container. The valve, complete with actuator, where dimensions allow, is placed loosely in the one inch (25·4 mm) aperture and the filling head is lowered to seal over and around the container top.

The valve is lifted slightly, a vacuum is drawn in the container head space, and propellant is injected between the valve and the can curl. Then the valve is sealed into the container aperture and the filling head recedes.

Pressure Filling

With pressure filling, the propellant at room temperature is injected under pressure through the aerosol valve itself (see Figure 40.4).

The sequence of operations entails the addition of the concentrate to the empty container, displacement of air from the head space, sealing on the valve assembly, metering the required quantity of propellant and injecting it under pressure through the aerosol valve.

With cold filling, the introduction of propellant into the open container displaces almost all the air in the head space. It is a self-purging operation. However, when pressure filling is used, there is a seal between the charging head and the aerosol valve; air remaining inside the container could not escape and would be compressed as the propellant was added. This air would contribute to undesirable excess pressures in containers and the oxygen in the air would affect corrosion performance. Therefore it is necessary to evacuate the head space by vacuum closing the valve on to the container, or by using an appropriate method of purging.

Pressure filling machines employ pneumatic or hydraulic systems to operate the charging cylinder. The quantity of propellant is varied by the adjustment of

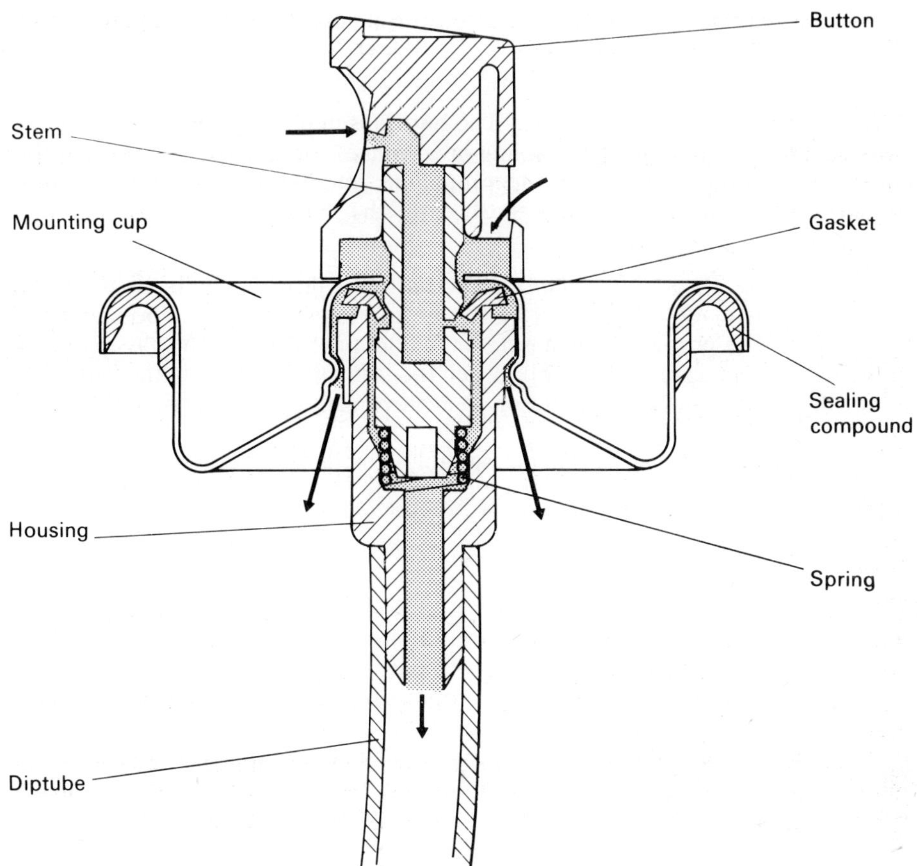

Figure 40.4 The CLF valve with 2000 series button
The valve is mounted in a metal mounting cup which can be swaged (clinched) into the one-inch aperture of the container. It is shown in the open position with the spring compressed and the gasket depressed, and with an indication of the propellant injection flow paths during pressure filling (Courtesy Metal Box Ltd)

the stroke of the charging piston. The filling head on the valve is opened mechanically or by propellant pressure, or both, to permit propellant injection to take place.

Rapid injection even when there is a high propellant charge is possible with current valves, irrespective of specification.

Pressure Filling Variants

There are several variants of the actual process of pressure filling the propellant. These variants include:

Pressure Filling with Button off. After the valve has been sealed into the container aperture, propellant is injected through and around the valve stem. Then the actuator button is fitted.

Pressure Filling Around the Button. The button is in place on the valve which is closed on to the container. The filling head makes a seal concentrically on the cup boss, and the injection of the propellant into the container takes place primarily between the stem and the hole in the valve cup boss.

Pressure Filling Through the Button. The button is provided with gassing holes and the seal is made by the skirt on the button between the cup boss and the filling head. The propellant then enters the container through the gassing holes in the button and again primarily between the stem and the hole in the valve cup boss. Cups and stems may be chosen to achieve very fast production filling and line speeds.

Again, pressure filling is equally suitable for all liquefied and compressed gases.

Solvent Saturation

Solvent saturation is an alternative procedure for compressed gases which entails the pre-saturation, in a pressure tower, of part of the solvent used in the liquid product concentrate with a soluble compressed gas, to the desired pressure, before combining it with the balance of the concentrate in the container. This procedure offers an alternative to expensive gasser shakers and allows improved filling rates, making it attractive compared with other methods of filling compressed gases. This type of aerosol filling has been adopted by a number of aerosol fillers. However, the most important development in aerosol filling of compressed gases is their metered introduction into containers through the valves, using pressures and equipment similar to those employed for liquefied propellants—that is, impact gassing with the latest aerosol valves.

Free Space in an Aerosol

In any closed container completely filled with liquid, a rise in temperature will result in the development of hydraulic pressure which may deform and even rupture the package. Chlorofluorocarbon propellants have a high coefficient of expansion. However, in all aerosol containers allowance must be made for the expansion of the contents by provision of adequate ullage or head space. Also it is wise and common practice to make provision for testing all filled aerosols in a hot water bath, for example at 50°–55°C for metal aerosol containers, as a safety check and to permit removal of 'leakers'.

The consumer's requirement of maximum fill of product conflicts with the need for safety, that is, the avoidance of hazards to the consumer from excessive fills. In Europe, a general rule is becoming accepted of a 75 per cent product fill, and thus 25 per cent head space (or ullage) at 20°C.

Check Weighing

In production it is customary to weigh a proportion of the filled aerosol containers, in order to check the accuracy of the filling operation, and to ensure that filled units correspond with requirements.

Hot Water Bath Testing

After production filling and closing, metal aerosol containers are tested for leaks and checked for safety and integrity by immersing in a water bath at 50°–55°C. Similar arrangements are required for glass and other non-metal containers. In small-scale operations, a simple tank with a thermostat and immersion heater may be used, in which the containers are manually immersed using a wire basket. In large-scale filling, the containers are usually carried on a conveyor through the water bath. Safety protection is provided for operators. The testing area should be well illuminated to allow minute bubbles from a leak to be readily detected. After leak and safety testing, the residual water must be removed from the valve cups before the filled containers are packed into cartons. Air blast tunnel driers and similar devices are used for this purpose.

TYPES OF DISPENSED AEROSOL PRODUCT

There are two sides to an aerosol—the inside (the contents) and the outside (dispensed contents).

Inside contents of aerosols are either two-phase (gas and liquid) or three-phase (gas, liquid gas and liquid concentrate); the latter occurs when liquid propellant and liquid product lack miscibility.

Products can be dispensed in various forms (sprays consist of liquid and/or solid particles):

1. Space sprays (true aerosols) composed of minute particles which remain suspended in the air for long periods of time. Examples are insecticides and air fresheners.
2. Surface (wet sprays) with larger particles. Examples are hair sprays and deodorants.
3. Surface sprays as jets. Examples are de-icers and lubricants.
4. Foams in which the liquefied gas propellant is partially emulsified with the active components of the product; they may be stable or collapse readily and present various consistencies. Examples are soaps, shaving creams and suntan foams.
5. Original unchanged physical form; the product is dispensed with the same physical form as that existing inside the container. This is generally the conventional form in which it would be obtained from a non-pressurized container but in a more convenient form, for example a liquid cream or a ribbon of solid cream or paste. Toothpaste is an example.

TWO-PHASE SYSTEMS

A large proportion of all aerosols are two-phase systems containing a liquid and a vapour phase (Figure 40.5). When the actuator, or button, of the aerosol is

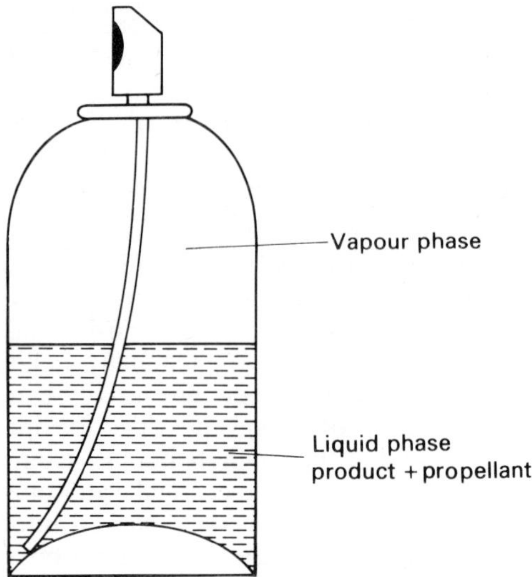

Figure 40.5 Two-phase aerosol system

pressed, the pressure inside the container causes the liquid to rise up the dip tube, through the open valve, and out of the button into the atmosphere where the lower pressure allows the immediate expansion and vaporization of the propellant and transforms the product into a spray mist, jet or foam. The solvent will also evaporate rapidly leaving behind the active constituents in a finely concentrated and dispersed form.

The majority of two-phase packs consist of a homogeneous solution of active matter, solvent and liquid propellant as the liquid phase, with propellant as the vapour phase. Compressed gases which may or may not be soluble in the liquid phase may also be used. The size of the emerging product particles is largely determined by the propellant/product ratio, and is influenced by the design of the valve. Examples of products based on two-phase systems are hair sprays, deodorants and colognes. Space sprays, air fresheners and insecticides of the solvent-based type are also two-phase systems.

The solvents employed in two-phase systems must effect a complete solution of all the active ingredients, for example hair spray resin in ethanol, so that the liquid phase of the filled pack is completely homogeneous. In addition to dissolving the active substance, solvents in two-phase systems must also be miscible with the propellant. Common solvents with these properties, and which show sufficiently low irritancy and toxicity when used topically or inhaled, include water, ethanol, isopropanol, propylene glycol and hexylene glycol. Other solvents include methylene chloride and 1,1,1-trichloroethane.

Space Sprays

Space sprays of the two-phase variety require high propellant/product ratios to achieve optimum particle size. Usually they contain not less than 80 per cent of liquefied gas propellant and consequently the amount of product concentrate is relatively small. The sprays emerging from such packs are true aerosols, with particle sizes of the order of 50 μm or less.

A good example of the functional use of space sprays is the aerosol room deodorant, which may be considered as bordering on a toilet preparation, and which must be atomized (as in the case of insecticides) in order to be effective. Such products may function by a number of mechanisms including the physical removal of malodours, by literally washing the air, by chemical neutralization or by an odour-masking or reodorant effect.

The formula given in example 4 functions primarily by the reodorant principle, but the glycols assist in the physical removal of malodours.

	(4) *per cent w/w*
Triethylene glycol	4·5
Propylene glycol	4·6
Ethanol	6·0
Perfume	0·2
Propellant 11	42·5
Propellant 12	42·5

Internal pressure at 20°C: approximately 40 psig (276 kPa).
Suitable container: tinplate.

Inexpensive hydrocarbon-propelled solvent-based insecticides based on two-phase systems are important worldwide. However, three-phase hydrocarbon-propelled water-based emulsions employing vapour-phase tap valves are important for insecticides and dominate the air freshener market for reasons of cost.

Surface Sprays

When the product constitutes 20–75 per cent by weight of the pack (with appropriate propellant loadings to complement), the resulting particle size of the sprayed product may be of the order of 50–200 μm. This type of spray, exemplified by hair sprays, is referred to as a surface spray because the particles are of such a size that they cannot remain suspended in the air but deposit and coalesce on the available surfaces to which they are applied. They are sometimes referred to as wet sprays, and include body deodorants, colognes, perfumes and suntan sprays.

These sprays are commonly of the two-phase type. Although alcohol-based products can be produced with a liquefied propellant concentration as low as 25 per cent, or even 5 per cent in the case of compressed gases, the resulting sprays are relatively coarse.

Other personal products may contain as much as 95–99 per cent propellant, for example feminine hygiene sprays, so chosen, in combination with a vapour-base tap valve, that the chilling/freezing effect is avoided.

Some surface spray products using two-phase systems are described below.

Hair Spray
Example 5 is a typical hair spray formula.

	(5) *per cent*
Resin	1·5
Soluble lanolin derivative	0·2
Isopropyl myristate	0·2
Perfume	0·2
Ethanol	32·9
Propellant 11	39·0
Propellant 12	26·0

Internal pressure at 20°C: approx. 30 psig (207 kPa).
Suitable container: tinplate.

Tinted hair sprays can be obtained by incorporating coloured aluminium powder or soluble dyes into the formula given in example 5. Colours and concentrations must be carefully selected to obtain the desired colour effect and also to avoid valve blockage in the case of solid colorants. Up to 4 per cent of solid material can normally be tolerated but it is advisable to include agitator pellets to assist dispersion.

Hair Dressings
Oil-based hair dressings can be formulated relatively easily. Products can range from simple solutions in mineral oil to more complex preparations containing, in addition, fatty esters, hair conditioning ingredients or aqueous alcohol.

Body Deodorants
Giacomo[31] has given various formulations. The bacteriostatic and bactericidal compounds discussed in Chapter 35 can be incorporated in aerosol formulations, for example deodorant colognes.

	(6) *per cent*
Perfume	0·7
Bactericide	0·3
Isopropyl myristate	1·0
Dipropylene glycol	3·0
Ethanol	65·0
Propellant 12	30·0

Internal pressure at 20°C: approx. 45 psig (310 kPa).
Suitable container: lacquered tinplate or lacquered aluminium.

Foams

The two-phase system can be used to produce transient coarse foams, using liquid propellants at low concentrations, particularly in product systems that are relatively viscous. Stable foams can be obtained using compressed gases that are soluble in the product phase. These systems may yet be shown to have application for toilet preparations.

Unchanged Products

Two-phase systems may also include certain unchanged products, such as pastes and creams dispensed by the use of an insoluble compressed gas such as nitrogen as the propellant. Large product retentions have been experienced, especially with viscous products. This particular difficulty may be overcome by employing a device which prevents contact between the liquefied propellant used and the product. This has been covered by a number of patents.[32–35]

The two-phase system using nitrogen as the propellant provides the simplest and cheapest method of dispensing products in an unchanged form. It is essential that the product is viscous enough to be extruded in a convenient form, yet sufficiently fluid to flow readily in the container. A rheopectic product can be ideal for this form of dispensing.

The most important product of this type to date is toothpaste, but many other products including liquid soaps, hand and hair creams, suntan creams and antiperspirant creams can be dispensed. Formulae can be adapted readily from standard conventional products.

THREE-PHASE SYSTEMS

In the treatment of two-phase systems it is assumed that the liquid phase is a completely homogeneous solution. However, many of today's products are based on emulsions, in accord with a trend towards water-based systems, for economic reasons, to avoid difficult solvents and partly for ease of formulation. If a product is in the form of a stable emulsion it is not difficult to consider it as a single phase in the context of a two-phase aerosol pack. But it is not always easy to preform stable emulsions to include liquid propellants and this leads to the concept of three-phase packs with two liquid phases which are emulsified, or mixed, by shaking at the time of use.

In a three-phase system the liquid propellant and the product no longer form a homogeneous liquid phase, as they do in the two-phase system, but are present as two distinct liquid phases. This system is used primarily for dispensing water-based formulations, including shaving creams and shampoos. The three phases of such a system are the aqueous phase, the non-aqueous liquid phase and the vapour phase—see Figure 40.6 a and b. The non-aqueous liquid phase, consisting largely or entirely of propellant, may be continuous, or it may be wholly or partly emulsified in the aqueous phase.

The three-phase system can be used to dispense space and surface sprays, foams and unchanged products.

The amount of propellant used in most three-phase systems is relatively small. Just sufficient of the chosen propellant is added to the aerosol to ensure the correct and complete discharge of the contents.

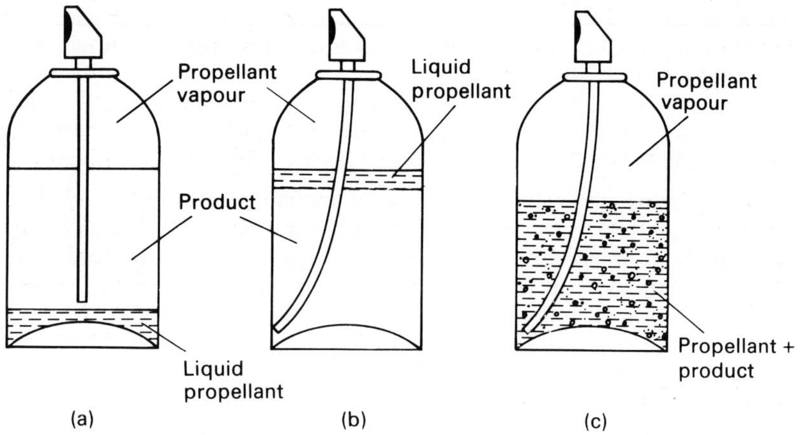

Figure 40.6 Three-phase aerosol systems

Another type of three-phase system is also possible in which a liquid phase, a solid phase and a vapour phase are present, as exemplified by antiperspirant and powder sprays—see Figure 40.6 c.

Liquid Sprays

Sprays can be obtained even though the liquid propellant and the liquid product are present as two distinct layers. Sprays can also be obtained by emulsifying the liquid product in a lipophilic external phase consisting entirely or partially, of liquefied propellant. These systems have been described in Du Pont publications[36] and also discussed by Root[37] and by Sanders.[38] Both space sprays and wet sprays can be obtained with this type of system which can be used for suntan sprays, room deodorants and others.

Sprays Containing Solids

Antiperspirants

The formulation of aerosol antiperspirants is discussed in detail in Chapter 10.

Suitable containers and valve specifications equal to the requirements of established formulations are readily available. A typical formula is given in example 7.

	(7) *per cent*
Aluminium chlorhydrate	4·0
Emollient, Suspending agent, Perfume	6·0
Propellant 11	35·0
Propellant 12	55·0

Internal pressure at 20°C: approx. 37 psig (255 kPa).
Suitable container: lacquered tinplate.

Powder Spray

Another variant of the three-phase system is the pressurized powder aerosol which gives a powder spray. The formulation of such sprays was discussed by Armstrong.[39] They consist of the product in particulate form constituting the solid phase, while the propellant forms both the liquid and the vapour phases. The liquid phase normally consists of the pure propellant, but sometimes it may also contain a small amount of dissolved lubricant to assist in the dispensing of the product by aiding the passage of the powder through the valve orifices. Sometimes the powder, for example talc, may have been treated with the lubricant. Because of the relatively low concentration of powder which can be incorporated (less than 20 per cent and more usually 10 per cent by weight) this system does not provide a very economic method for dispensing body powders. However, it can be used for dry shampoos[40] and talc and for dispensing antiseptic powders and other speciality products in powder form. Before use, the dispenser is shaken in order to ensure that the powder is uniformly dispersed throughout the liquid phase.

Caking and agglomeration have presented many problems in powder packs. In early sprays, clogging of valves occurred frequently, and dispensing was often poor. There is a limit to the amount of powder that can be incorporated in an aerosol pack, above which severe clogging of the valve and/or dispensing head is experienced, resulting in either intermittent discharge or complete failure. When the powder tends to aggregate, the addition of a suitable surface-active agent may be indicated.

The following formula illustrates this type of preparation.

	(8)
	per cent
Talc	9·0
Lubricant	0·5
Bactericide	0·1
Isopropyl myristate	0·2
Perfume	0·2
Propellant 11	45·0
Propellant 12	45·0

Pressure at 20°C: approx. 35 psig (241 kPa).
Suitable container: lacquered tinplate or lacquered aluminium.

In general, for powder sprays to be effective it is necessary to ensure that the solid particles are sufficiently small to pass screens of about 200 mesh and even finer, and exposure to moisture must be avoided in order to prevent the aggregation of powder particles. Furthermore, the product must be insoluble in the propellant to eliminate the possibility of crystal growth. For best results the densities of product and propellant should be similar.

Electrostatic Charges and Particle Sizes of Solids-containing Sprays

Electrostatic charge and particle size of sprayed solids are two issues of increasing importance which continue to occupy the thoughts and researches of the aerosol industry worldwide.

Electrostatic Charge. In anticipation of, and attendant upon, the virtual ban on chlorofluorocarbons in the USA, alternative propellants have been sought and found for all products, including such personal case products as antiperspirants and hair sprays. In general, hydrocarbons were adopted, with the exception of carbon dioxide in the field of men's hair sprays. Adoption of hydrocarbons for personal care products raised questions of flammability which were resolved by careful formulation and, more importantly, by choice of valve specification. However, one issue remains, that of electrostatic charge created when spraying powders, as with aerosol antiperspirants and talc sprays.

Such phenomena are well known scientifically in industry. The creation of such electrostatic charges on chlorofluorocarbon-propelled antiperspirants, produced in their thousands of millions since 1970, was never an issue. When hydrocarbon was substituted for chlorofluorocarbon as the propellant for powder sprays it was recognized that the electrostatic charge on the discharged contents might spark and cause an explosion.

There has been much activity in relation to this problem. In friendly and competitive vein, two groups, one in America[41] and one in Great Britain,[42] have sought to define and agree test methods. Codes of safe practice have been devised and are generally available. Individual companies[43] have been searching for quenching agents for incorporation into product formulations. Water is good, for example, but is unacceptable in antiperspirant formulations because of adverse effect on the product and problems of corrosion.

Particle Size. Again in relation to antiperspirants, particle size became an issue in America between the Food and Drug Administration and industry. Suppliers of active ingredients agreed to introduce new materials[44] with guaranteed particle size distributions so as to minimize the number of particles below 10 μm. Small particles, if inhaled and retained in the lungs, may constitute a health hazard.

The issue has become much wider and all types of sprayed aerosol products, and other spray forms, are now being examined. Laboratories are utilizing expensive equipment to measure particle size distributions for many products.[45–47] Many individual companies world-wide, as well as various trade associations, are looking at this whole question in order to determine the critical particle sizes, the levels of exposure to which consumers are subjected, and what real hazards are created.[48,49]

The trade associations—the British Aerosol Manufacturers Association (BAMA) in the United Kingdom, the Federation of European Aerosol Associations (FEA) in Europe, and the Chemical Specialties Manufacturers Association (CSMA) in the United States, and others—play a key role, through the collective and cooperative enterprises of committees and working parties, in recognizing such challenges and formulating work programmes to secure the well-being of the industry through the integrity of its products and production.[50]

Foams

Foams are normally obtained by the use of a three-phase rather than a two-phase system. Aerosol foams usually contain up to 15 per cent, usually 6–10 per cent, of liquefied propellant emulsified with an aqueous product in which a fairly high amount of surfactant is present. The container is shaken, the valve is actuated, and the emulsion which is forced up the dip tube emerges through the actuator. As soon as it reaches the atmosphere the dispersed propellant drops will immediately vaporize, and, in so doing, transform the product into a thick foam. Aerosols of this type are often fitted with a special foam head or spout. Such systems are discussed in some of the references quoted earlier in this chapter and in other Du Pont publications, and also under the appropriate product headings elsewhere in this book. The formulation of cosmetic foam preparations has been covered in a number of patents.[51–53]

Foam packs have been used to dispense shaving creams, shampoos and suntan foams.

While the constituents of the aqueous solution or the primary emulsion in the system, including the type and amount of surfactant, will undoubtedly influence the physical and functional properties of the foam, there is no doubt that the consistency of the foam, the rate of its formation and its stability will also be determined to a large extent by the type and the amount of propellant used, and hence the vapour pressure prevailing in the system. For a given concentrate, high vapour pressure propellants and high propellant charges tend to produce stiff, dry and elastic foams with poor wetting and spreading properties. Medium pressure propellants will produce a softer and wetter foam, while low pressure propellants and low propellant concentrations will give rise to soft, wet, less resilient, slowly forming foams. Interest in compressed gases for aerosol foams is developing.

When oil-in-water emulsion systems are to be dispensed as foams, the propellant vapour pressure will usually be between 20 and 50 psig (138–345 kPa) at 20°C and the propellant concentration will be 5–25 per cent of the total formulation, but usually towards the lower end of that range.

This system provides a very valuable method for applying a small quantity of cream to the skin and thus has found application also for hand creams, hand cleansers and suntan creams. It has application for liquid soaps,[54] shampoos and hair conditioners. However, the most important product in this class, until now, has been the aerosol shaving cream.

Shaving Creams

Aerosol shaving cream emerges from the nozzle of the container as a fully developed foam. For this purpose, propellants with vapour pressures of at least 25 psig (172 kPa) at 20°C are employed. A 40/60 mixture of propellants 12 and 114 at a level of 7–10 per cent may be used to produce a vapour pressure of 40 psig (276 kPa) at 20°C. With vapour pressures above 40 psig at 21°C the product will expand rapidly inside the valve actuator and emerge from the container in the form of a compact foam. The texture of the foam alters with increasing level of propellant, becoming more rigid and dry. It is interesting to note that butane foams are not flammable and have virtually replaced other propellants for these formulations.

The formula given in example 9 illustrates the composition of a shaving foam.

	(9)
	per cent
Stearic acid	5·9
Triethanolamine	3·1
Propylene glycol stearate	3·0
Lanolin	1·0
Glycerin	2·0
Lauric diethanolamide	2·0
Perfume	0·5
Water	82·5

92 parts packed with 8 parts of Butane 40.
Internal pressure at 20°C: approx. 40 psig (276 kPa).
Suitable container: lacquered tinplate or lacquered aluminium.

Shampoo

For aerosol shampoos a 'stream' foam, also referred to as a 'lazy' foam, is preferred. This is a soft type of foam which will spread easily but will not break down too readily. For this purpose, low pressure propellants with vapour pressures of 5–20 psig (34–138 kPa) are used, which will cause the product to emerge initially as a liquid or semi-liquid stream, which gradually expands to a foam. Normally about 40 per cent active matter is required. When using non-soap detergents, corrosion problems may be encountered.

The choice of propellants 12 and 114 in foam-type preparations is governed by their stability in the presence of water as well as by the fact that when used in combination they will produce any required vapour pressure. The foam pattern depends on the type of valve and actuator employed as well as on the propellant concentration, propellant type and the actual formulation. If the product used for producing foam is in the form of an emulsion, it should not be too viscous, otherwise it will not mix readily with the propellant and will subsequently dispense unevenly with accompanying spurting and propellant loss. In the case of an emulsion, foaming is also influenced by the type of emulsifying agent that is used in the preparation of the primary emulsion; those of the oil-in-water type are usually selected.

Water-based Systems

In a three-phase system comprising two immiscible liquid layers and containing a water-based product, the product forms one of the liquid layers, and the propellant forms the other as well as the vapour phase. Two possible situations exist, which are illustrated diagrammatically in Figure 40.6 a and b (page 826).

In (a) the propellant is heavier than the product and forms the lower liquid layer. This is generally the case with chlorofluorocarbon propellants. In (b) the propellant is lighter than the product (as in the case when the former is a hydrocarbon) and floats on top of the product. In either case (a) or (b) it is essential to shake the aerosol thoroughly immediately before use. In case (a) this is necessary to allow the two propellant phases to equilibriate so that the pressure in the head space is maintained; this occurs automatically with case (b). Also in case (a) the length of the dip tube requires careful adjustment to avoid dispensing neat liquid propellant which in turn causes product residue problems, through insufficient propellant.

The product is dispensed from the container when the valve is actuated. Special actuators are required to assist break-up of the coarse product particles, even though shaking, which causes entrainment of propellant in the product, materially helps in breaking up the product spray.

Care in the development of water-based products will resolve any problems which may arise with regard to the efficiency of the spray, compatibility of product and container in respect of corrosion, and correct balance of formulation and valve specification to control flammability when hydrocarbon propellants are used. As far as corrosion is concerned, with both chlorofluorocarbons and hydrocarbons full storage tests must be conducted with any development formulations. Mention may also be made of an article by Geary,[55] in which he discussed water–liquefied propellant–co-solvent systems designed to assist in the formulation of low cost water-based spray products.

PERSONAL CARE PRODUCTS WITH ALTERNATIVE PROPELLANTS

The reaction to the Rowland and Molina theory in 1974 imposed significant restriction on the use of chlorofluorocarbon propellants for aerosols in the USA[56] with echoes later in Sweden and Norway, while the EEC recommended a reduction of 30 per cent of the 1976 usage by the end of 1981.[57–59]

The aerosol industry in the USA, as well as in other countries, had already adopted hydrocarbon propellants for non-personal care products such as polishes, starches and paints, so that half of the aerosols produced were already dispensed by a flammable propellant, having an excellent safety record in production and use.[60] Carbon dioxide was used for de-icers, some surface insecticides and fire extinguishers.

The only personal care product, and a very important one in the USA, for which hydrocarbon was used almost exclusively was shaving cream. There were no hazards in this case because the hydrocarbon content was only 5–10 per cent, the product was water-based, it was not dispensed as a spray but as a foam, and this foam could not be ignited. The real problems in adapting to the use of hydrocarbons in the USA related to only two major personal products: antiperspirants, containing 90 per cent chlorofluorocarbon propellant, and hair sprays, which contained less chlorofluorocarbon propellant but a product concentrate rich in ethanol (hair sprays typically contained 50 per cent chlorofluorocarbon and 50 per cent of product based on ethanol).

In Europe, deo-colognes with 35 per cent propellant 12 and 65 per cent ethanol-based product became the third important personal care product requiring consideration of formulation revision with respect to propellant choice.

More recently, dimethyl ether has become of interest as an alternative propellant.[61,62]

Antiperspirants

Antiperspirants [63,64] are an important product category in the UK and the USA but not in continental Europe.

American antiperspirants, after a brief flirtation with hydrocarbon-propelled water-based emulsions, are now based on 80 per cent hydrocarbon and 20 per cent antiperspirant concentrate; such aerosols have also appeared in other markets. The difference in specific gravity—much lower for hydrocarbons than for chlorofluorocarbons—demands careful re-design of concentrate to achieve acceptable suspension characteristics and aesthetics and good spray performance. The absence of propellant 11 from the formulation contributes to the inferior performance of such aerosols.

In spite of the 80 per cent hydrocarbon content, choice of valve (vapour phase tap) and actuator ensures acceptable flammability characteristics when tested by Flame Extension Methods,[65,66] achieved largely by reduced discharge rate. Such aerosols, light in weight because of the reduced specific gravity of the propellant, feature a lower mass declaration on the label, last longer in the hands of the consumer and generally have a poor performance.

In the many countries which still permit the use of chlorofluorocarbons, many fillers and marketers, recognizing the lower costs of hydrocarbon, now use a blend of chlorofluorocarbons and hydrocarbons; they thus avoid the shortcomings mentioned above, while gaining in respect of reduced costs.

Hair Sprays

In the USA[67] basic hair spray formulations, using 25 per cent hydrocarbons and 75 per cent ethanol, are modified to accept (say) 8 per cent water or 10–15 per cent methylene chloride to provide solvency for the resins. Since these products consist of virtually 100 per cent flammables, choice of actuator and valve, even to include vapour phase taps,[68] is designed to produce good sprays[69] but with attenuated discharge rates[70] in order to control flame extension[71,72] and other characteristics. Performance of these products is generally good; they appear to have a longer pack life.

It should be noted that butane–water–alcohol systems will tolerate 8 per cent water in a single liquid phase. Care must be exercised to ensure correct choice of metal container specification in order to avoid corrosion problems.[73]

Carbon dioxide has been adopted in the USA as an alternative propellant for some men's hair sprays.

In Europe[74] the cost advantages of cheaper propellants are of course recognized, but their use is not so extensive because there are several important differences compared with America:

1. A total ban on chlorofluorocarbons has not been imposed.
2. Aerosols containing in excess of 45 per cent flammable materials incur labelling penalties.
3. In many countries ethanol is many times the price of chlorofluorocarbons.
4. The EEC Cosmetics Directive[75] restricts methylene chloride[76,77] contents to less than 35 per cent. The trend, however, is lower still, and in Germany an upper limit of 20 per cent has been imposed.

As for methylene chloride, so 1,1,1-trichloroethane seems to attract the attention of the legislative authorities and real concern is expressed in respect of future formulations, not only for hair sprays, over the possibility of greater restrictions being placed on propellant 11, methylene chloride and 1,1,1-trichloroethane.

One European basic formulation for a hair spray illustrates the approach to the above-mentioned regulations and restrictions:

	(10) *per cent*
Methylene chloride	35
Chlorofluorocarbon	20
Hydrocarbon Ethanol-based product concentrate	45

Hydrocarbon hair sprays similar to the American products are appearing in Europe in limited quantities. In Germany, for a time, specific chlorofluorocarbon hair spray formulations were modified to reduce the propellant content, substituting, say, 3 per cent by carbon dioxide. They were known as 'cocktails' or 'topped-up' systems.

Effective hair sprays based on nitrous oxide and propellant 11 have also been proposed.

Systems propelled by dimethyl ether can be devised to meet current European restrictions and regulations, as for hydrocarbon versions, and this relatively new propellant can accept up to 30 per cent water which also assists with respect to cost, solubility and flammability.

As for antiperspirants, so filler–marketers of chlorofluorocarbon hair sprays are able partially to substitute hydrocarbon in place of chlorofluorocarbons without change of container or valve specifications and without noticeable change of performance.

Deo-colognes

This European product,[78–80] also alcohol-based, suffers the same reformulation problems as hair sprays except that methylene chloride does not enter into the formulation equation. Topped-up (carbon dioxide–chlorofluorocarbon), hydrocarbon, hydrocarbon–water–alcohol, dimethyl ether–water and even straight carbon dioxide propellant systems have been tried, or are currently marketed while the availability of traditionally formulated products continues.

CORROSION IN AEROSOL CONTAINERS

In aerosol containers, as in other metallic containers, corrosion can result in product and/or pack damage. Since the system is under pressure and in the presence of a propellant, the corrosive action of the aerosol product may be accelerated and more severe than it would be if the same system were under atmospheric pressure. Sometimes it culminates in pack failure. Storage tests should take into account the nature of the product, its application and the market for which it is intended.

The study of container corrosion, apart from covering reactions between a product and its environment, also deals with suppression of corrosion by altering the characteristics of metals and their environment, for example by the use of protective lacquer films and the inclusion of corrosion inhibitors in the formulation, or by the modification of the latter.

While it is not proposed to give here a detailed exposition of corrosion theory, it is necessary to refer to the causes of internal corrosion in aerosol containers and to discusss ways of controlling them.

There are three main causes of internal corrosion in aerosols:

(i) Change of the propellant stability in the product environment.
(ii) Attack by the product.
(iii) Electrolytic interaction of dissimilar metals.

Propellant Influences on Corrosion

Some chlorofluorocarbons, although stable by themselves, show varying degrees of instability in contact with the constituents of the product (for example low alcohols) and the materials of the container (for example aluminium). The most stable among the common propellants is propellant 11, trichloromonofluoromethane. Its reaction with alcohol involving free radicals and leading to corrosion has been studied by many workers. It proceeds in the presence of trace amounts of air, but is inhibited if an excess of air is present. Sanders[81] proposed the following reaction chain, resulting in the formation of acetaldehyde, acetal, dichlorofluoromethane, ethyl chloride and symmetric tetrachlorodifluoroethane:

$$C_2H_5OH + CCl_3F \rightarrow CH_3CHO + HCl + CHCl_2F$$

$$CH_3CHO + 2C_2H_5OH \rightarrow CH_3CH(OC_2H_5)_2 + H_2O$$

$$C_2H_5OH + HCl \rightarrow C_2H_5Cl + H_2O$$

$$R\cdot + CCl_3F \rightarrow RCl + \cdot CCl_2F \text{ (free radical)}$$

$$2\ \cdot CCl_2F \rightarrow FCCl_2{-}Cl_2CF$$

R· represents an initiating amount of a free radical derived from a polyhalohydrocarbon by reaction with, for example, a metal.

Nitromethane has been regarded as an effective corrosion inhibitor in reactions involving free radicals.

In the presence of ethyl alcohol, hydrolysis of propellant 11 will occur at a rate depending on the ambient temperature. Its corrosive effect will be even more intense in the presence of large amounts of water.

The corrosion risk inherent in the use of a mixture of propellants 11 and 12 can be minimized by substituting for it a mixture of propellants 12 and 114, though it is by no means certain that these propellants will not react in a similar manner, even if at a slower rate.

In discussing corrosion due to propellants, mention must also be made of methylene chloride and 1,1,1-trichloroethane, both of which are used as auxiliary solvent–propellants or pressure depressants, and to replace the more expensive propellant 11.

Archer[82] correctly anticipated that reactions similar to those between propellant 11 and ethanol would take place between 1,1,1-trichloroethane and ethanol, and showed that appreciable corrosion of tin and iron did occur when metal strips were refluxed with such a mixture. The mixture causing maximum corrosion after 42 hours reflux was:

	per cent
1,1,1-trichloroethane	30
Ethanol	55
Water	15

In practice, the use of 1,1,1-trichloroethane, especially inhibited grades, together with anhydrous ethanol, should not present serious corrosion problems, but the introduction of even small quantities of water will increase the chances of corrosion and thorough testing will be necessary.

Product Influences on Corrosion

Corrosion as a result of direct chemical attack is comparatively rare in aluminium containers. It may result on contact with highly acid or alkaline solutions, and it can also occur in pressurized packs. Anhydrous alcohols (that is those containing less than 0·01 per cent of water) and also fatty acids are liable to attack aluminium. Aluminium containers are thus particularly prone to corrosion in the presence of ethanol or n-propanol, and products containing these alcohols should not be packed in internally unlacquered aluminium containers without adequate storage test experience.

Studies by Du Pont workers[83] on the corrosion of aluminum have indicated that the corrosive action of anhydrous ethanol is much more severe than that of 99 per cent ethanol, although it was pointed out that even with the latter it is still greater than the dispenser will tolerate. A number of reactions appear to be involved when corrosion by anhydrous ethanol occurs, chiefly alcoholysis of the propellant and attack of aluminium by hydrochloric acid produced during this reaction. According to Parmlee and Downing,[84] isopropanol appeared to be generally less corrosive than ethanol, but in many cases the difference was not great.

Products which contain isopropanol, even if they include propellant 11 (trichlorofluoromethane) as a constituent of the propellant mixture, can be packed in internally unlacquered containers, as long as the moisture content within the dispenser is not higher than 0·05 per cent.

Tin, like aluminium, will corrode in contact with solutions of high acidity or alkalinity. Corrosion in acid systems can further be aggravated by the presence of oxygen, following inadequate purging during injection filling, and has resulted, for example, in the discoloration of products by the dissolved tin; this has led to the lacquering of containers. However, Herzka and Pickthall[85] have pointed out that in lacquered containers the corrosive attack of acidic products 'will be concentrated at the pinholes which invariably occur in the lacquer film, and the perforation of the dispensers will be even more rapid than with internally plain dispensers'. These authors have also stated that corrosion of tinplate containers is not straightforward because of such effects as the inhibitory action on mild steel of dissolved tin and the reduction of the tin ion concentration by complex formation. A similar problem has been observed in aerosols containing alkaline products, such as shaving creams, again because of dissolution of the tin in the product, and production of a greyish discoloration. Although there is no danger of container perforation, the discoloration is undesirable and internal lacquering is again resorted to in order to overcome this defect.

Galvanic Action

In metal containers for aerosols more than one metal is often present, so that apart from a purely chemical attack by the product on the metal of the container, there are possibilities of galvanic corrosion, and several sites where it can occur. The two-piece aluminium container may constitute a greater corrosion risk than the one-piece monobloc container. Bimetallic interaction can occur in aluminium containers with tinplate bases or tinplate valve cups, especially with aggressive products and where any protective oxide or lacquer coating has been mechanically damaged. The tinplate base in a two-piece container will usually be cathodic to the aluminium body, but the polarity relationship will depend on the product characteristics. Cathodic metal is not at risk in respect of corrosion but forces metal which is more anodic to dissolve.

Even if all the container components are tinplate, two metals (tin and iron) are still present. The electrical properties of these may depend on the oxygen concentration. With some products when the oxygen concentration is low the tin may be anodic with respect to the iron, but at higher oxygen concentrations the tin may become cathodic and the iron anodic, and the latter becomes corroded. With other products the whole pattern may be reversed. The optimum level of air is a function of the product–package combination, and so vacuum purging of tinplate containers may be effective in reducing the potential corrosion of certain products.

Bimetallic corrosion can also be caused by traces of copper and by salts of heavy metals which must therefore be minimized. The adoption of blackplate and tin-free steel materials requires careful consideration.

Corrosion Inhibition and Prevention

In the preceding paragraphs reference has been made to the parts played by water, by oxygen and by the nature of the metal in contact with the product. Mention has also been made of the fact that some alcohols and some chlorofluorocarbons are more liable to attack containers than others. Attention should be paid, therefore, to control of the water content of the formulation, and to the efficient removal of oxygen from the head space of the aerosol either by vacuum or by purging with a compatible or an inert gas, where this is appropriate. Correct choice of alcohol and propellant will also minimize the possibility of corrosion.

Common corrosion inhibitors added in trace amounts to non-pressurized packs, such as neutral sodium silicate and ethanolamine phosphate, have not found general application in aerosols, because either they have not been found to be effective, or they are incompatible with the product or with its use. However, the use of specific corrosion inhibitors at empirically determined levels of addition has proved to be of enormous value, especially with water-based products.

Lacquers for Internal Protection

An important way of preventing corrosion is to separate the product and propellant from contact with the metal, and this is very frequently achieved by lacquering the interior of the container.

While unlacquered tinplate containers are liable to corrosive attack, particularly at the side seam, they seldom develop pinhole corrosion. Imperfections in a lacquer film, however, may give rise to this defect with acidic products, and perforation of container is likely to result more quickly than in unlacquered containers. Internally plain cans could also experience serious perforation problems if used for products that put steel at risk, for example some water-based products.

Cathodic metal is not at risk in respect of corrosion, but forces metal that is anodic to dissolve. Internally lacquered containers should preferably be used in systems in which the tin is cathodic to steel. In the circumstances of tin being anodic to steel, detinning results. Anodic attack is always more intense when a large cathodic area (tin) is in contact with a small anodic area (steel), a condition which occurs when flaws are present in the tin coating. As already explained, while tin is more noble than iron in the electrochemical series, in many solutions it is anodic to steel. Under those conditions only slow solution of tin will occur because of the small areas available for the equivalent cathodic reaction.

Internal lacquering is used in nearly all forms of metallic container where contact of the contents with the container would be deleterious in any way. This includes canned foods, and, in toiletries, collapsible tubes and aerosols.

Lacquers for metal aerosol containers must form a coherent film which is adherent, impervious, unaffected by the metal, unaffected by the product or propellant, and which does not affect the product. They must be tough but not brittle, having sufficient flexibility to withstand the manufacturing process because, while it would be possible to lacquer the partly made can and cover the base seam and the soldered side seam, it is impossible to lacquer the final seal when the valve assembly has been seamed on. The general practice is to lacquer the tinplate sheet before the three components for the tinplate can are fabricated, likewise for the aerosol cup. The can is formed in such a way that the portion of metal which is to form the side seam is left bare, to assist good welding (or soldering), after which it is later covered separately with a side stripe of lacquer.

Frequently a combination of a phenolic and an epoxy resin is used as a primer coating for the internal lacquering of containers. This may be followed by the application of two or even more coatings of an epoxy resin to prevent pinhole corrosion which, as pointed out above, is much more liable to produce perforation of the container than is the overall etching which occurs in unlacquered aerosols.

Phenolic lacquers are characterized by greater impermeability and chemical resistance than vinyl and epoxy resins, but their flexibility is rather low. Vinyl resins are tough, but have a poor adhesion to bare metal and are normally used over a base coat. Epoxy resins have good fabrication properties and have good adhesion to tinplate; because of this they are used as a base coat in many systems. An internal resin stripe is present in many containers to protect the product from discoloration or changes in odour or flavour following exposure to the metal at the side seam. Side stripes are applied, usually by spraying, immediately after the welding (soldering) operation, and the materials used for this purpose include oleoresins and vinyl resins.

The internal lacquering of aluminium containers is different from and not as easy as that of tinplate containers. While the latter are lacquered on the flat plate before fabrication, aluminium containers must be lacquered by spraying or flushing the finished container—a process which is rather expensive when dealing with monobloc containers. It is preferable, however, not to rely on the protection afforded by the lacquer coating, particularly where aluminium containers are concerned, but rather to use, as far as possible, non-corrosive formulations.

Anodizing

An alternative, or additional, method of protecting the internal surface of an aluminium container is the process known as anodizing, whereby the naturally occurring thin film of aluminium oxide is reinforced by a coherent, relatively thick (4–6 μm) and highly resistant layer of aluminium oxide. This film may be coloured to give, for example, a golden shine effect. Anodizing is expensive but gives a greatly increased corrosion resistance.

Electro-corrosivity[86,87]

Sophisticated and simple electrochemical methods for corrosion prediction have been devised over the years. Their basic purpose is to study the relationship between current and potential applied to all combinations of the basic metals occurring in a particular metal container construction, in the product environment for which the specific container is intended. Account is taken of the areas of metal exposed in the container as specified.

Significant advances in these techniques augur well for the aerosol industry in the 1980s, promising even greater confidence than has already been well established in making predictions about container behaviour after only 24 hours of study. Increasing experience facilitates the elimination of unsatisfactory specifications for the can, while screening out difficult product ingredients, for example unsatisfactory perfumes or surfactants.

Rapid exploration of alternative materials such as corrosion inhibitors ensures that only the product variations most likely to succeed in cans with favourable specifications are put on full storage tests.

In the event of complaint or failure, the services offered by electro-corrosivity facilities are invaluable in the inevitable investigation and especially when marketing continuity demands a temporarily substituted but compatible product.

Storage Tests

Full and adequate storage tests, whose design recognizes the aerosol product, its application and the market for which it is intended, are vital.[88,89] Failure to observe this simple rule is a recipe for ultimate disaster. Many filled aerosols should be stored, upright and inverted, with air levels which reflect production variations including lapses, both at ambient and elevated temperatures, for periods commensurate with the expected shelf life.

Experience should be developed to acquire confidence that results obtained after eight months at 35°–37°C, provided that the results at ambient temperatures are also favourable, are a reasonable guide to performance over two years in temperate climates. More rigorous testing is likely to be required for tropical regions.

ALTERNATIVE SYSTEMS

In the conventional aerosol, the product and propellant are in intimate association with each other. Early attempts were made, however, to develop methods which would enable these two components to be kept separate, particularly when they were found to be incompatible with each other and problems arose on storage and dispensing.

Piston Packs

Early developments led to the piston pack, in which a concentric plastic piston within the aerosol can was moved by the pressure of the propellant. An example of such a system was the American Can Company's container described earlier,[27] which prevented direct contact between the product and propellant and offered the possibility of packing products of relatively high viscosity into pressurized packs without the risk of cavitation and serious product residues. Problems have arisen, however, because of pressure equalization occurring on the two sides of the piston as a result, for example, of container deformation. Some products are still marketed in piston containers, for example pressurized cheese spread. The existence of piston packs has led to the development of different two-compartment packs, such as the Sepro-can, aimed at preventing the propellant from reaching the product.

The Sepro-can (*Continental Can Company, USA*)[90]

In early versions of this type of container the product was introduced into a plastic bag, which was then attached to the orifice of the container and sealed by means of a valve. The propellant was introduced through the bottom of the container, exerting pressure against the bag, so that when the valve was opened the product was dispensed. Further development of this system has led to the Sepro-can. This can contains a bellows-type plastic bag container held rigidly connected to the cone curl of the tinplate aerosol can. The product is filled through the bag aperture and the valve cup is then crimped into position, while the propellant is introduced through the concave bottom of the can by means of an orifice which is then plugged. Permeation through the bag that separates product and propellant may occur; nevertheless, this pack is successfully used in the USA, especially for a shave cream gel.

European Bag-in-can Systems[91]

Permeation through a plastic container was prevented in German developments. The bi-aerosol (Bi-aerosol Verpackungs GmbH, Germany) and the tri-aerosols, which operated on the bag-in-can principle, had thin aluminium foil in place of plastic bags, suspended within a standard aerosol can. The mode of operation was the same.

From Rhen AG, Presspack, and from Comes in Switzerland, commercially successful plastic bag systems are available. Recent innovation permits around-the-valve propellant injection, eliminating the need for propellant injection through an orifice in the can base with subsequent plugging with a rubber component. Claims for 98 per cent delivery are made for such systems, which have considerable potential application[92] in the packaging of food and pharmaceutical products and toothpastes within a wide range of viscosities, where the problem of product–propellant incompatibility has imposed severe limitations on the range of products which could be dispensed in this way. No doubt other applications in the toiletry field are possible to extend the range, and a claim has already been made for depilatories.

Compared with conventional systems, for which large amounts of propellant are used to bring about the dispensing, propellant contents are low for all bag-in-can and piston systems. While conventional valves and spray actuators may be used with these systems, the use of mechanical break-up actuators is necessary to achieve acceptable sprays where appropriate.

Tri-aerosol[91]

The principle of product–propellant separation embodied in the bi-aerosol was extended to the so-called tri-aerosol system which contained two internal containers. This permitted mixing of two products to activate them at the time of dispensing, for example mixing of an alkali base with stabilized hydrogen peroxide for bleaching hair. The outflow was controlled by a two-channel valve and the two products from the inner containers were then mixed in the required ratio, for example by valve nozzles of different diameters in the two-valve channels. The introduction of the tri-aerosol paved the way to co-dispensing, and marketing, of two-component and multiple-component products in a single pack.

Co-dispensing Valves

All major valve manufacturers have worked on the development of valves for co-dispensing. The early work was reviewed in *Aerosol Age* in 1968.[93,94]

The concept of co-dispensing, that is the dispensing of two or more products from a single aerosol container, was originated by Du Pont who secured two patents[95,96] disclosing valve modifications which would allow co-dispensing of two materials from a single aerosol container. Among the first valve manufacturers to obtain a licence under the Du Pont patent was the Clayton Corporation, USA, who developed the 'Clay-Twin' co-dispensing valve allowing the simultaneous dispensing of two different products in specified proportions through a common orifice. The introduction of co-dispensing valves has made it possible to dispense various products, including hot shaving creams and hair dye formulations.

Hot Shave Creams. To dispense a hot shave cream, the product incorporating a reducing agent is stored in a can, while hydrogen peroxide is placed in a laminated plastic bag. When the valve is actuated, the shaving cream and peroxide are dispensed and combined in a balanced proportion. This results in an exothermic reaction between hydrogen peroxide and the reducing agent (for example a mixture of potassium sulphite and potassium thiosulphate) and heating of the shaving cream. The proportions in which the two reactants are provided and combined are important. Serious problems, but different ones, develop with an excess of either component; insufficient peroxide gives only a lukewarm foam, whereas excess peroxide can lead to a liquid rather than a cream product. According to Du Pont, foam temperatures as high as 80°C have been produced with certain combinations of oxidizing and reducing agents.

The self-heating of shaving formulations based on the oxidation reduction principle has also been the subject of a US patent[97] issued to the Gillette Co. in which thiourea and substituted thiobarbituric acid derivatives were mentioned as reducing agents.

Co-dispensing Valves for Upright Containers. Early co-dispensing valves were designed to operate in the inverted position. However, OEL Inc., USA, also licensed under the Du Pont patent, developed a valve designed for use in the upright position.[90] In the package utilizing this valve, the oxidizing agent is enclosed in a polyethylene tube which snaps on the valve. The manufacturers have claimed that is is possible to fill their pack on standard equipment without the need for any special adaptations and that it is foolproof from failure in delivering the correct proportion of shaving creams and peroxide.

A co-dispensing valve system developed by the Valve Corporation of America consists of two distinct valve systems (referred to as a double compartment valve) and is operated with a single actuator.[94] The flow from each component is regulated by the size of the valve body and the depth of the slot in the stem. This valve system can be operated in either the upright or the inverted position.

The Precision Valve Corporation, USA, has fully researched co-dispensing valves operating in the upright position to the stage of final development.

Venturi Spraying

Reference has already been made to the fact that in systems such as the piston container, the Sepro-can and the Presspack system, where product and propellant are separated from each other by a plastic or metallic surface, it is not possible to effect satisfactory atomization. The attempts to combine separation of product and propellant with adequate atomization have led to the development of new systems in which the venturi principle has been utilized. The prototype of this system was a normal aerosol container which was filled with the liquid propellant and connected by means of a plastic bridge with a glass vessel holding the product to be sprayed. In connection with these developments, reference must be made to the Innovair system of Geigy, France, and the Preval Atomizer developed by the Precision Valve Corporation of America. In both these systems propellant and product are separated and only brought together at the moment of spraying. There is no need for the container holding the product to be pressure-resistant, which thus permits the use of materials other than metals.

The Innovair System (ITO). The Innovair system, originally known as ITO and used for insecticides and later air fresheners, was developed by Geigy S.A. of Switzerland and employed a blow-moulded non-pressurized container which held the product to be sprayed. The technical bulletin[98] dealing with this system described it as follows:

> An inner capsule contains the propellant which is maintained under pressure by the high-pressure valve. The latter is fixed in the spray head which comprises a push-button together with the venturi suction-and-spray micro-nozzle. A joint which forms the low-pressure valve controls the out-flow of the liquid and maintains the outer container at atmospheric pressure. When not in use the container is hermetically sealed, the high-pressure valve preventing the escape of the propellant, whilst the joint prevents passage of the liquid and its exposure to the atmosphere.
>
> When the three-way valve controlling the action of the pack is operated, the three exits will be opened, the propellant gas will pass through the spray cone, and will, when flowing through the venturi nozzle, produce a vacuum. The joint will connect the spray chamber of the venturi nozzle with the container holding the product, and will, when opening, compensate for the difference between atmospheric pressure and the vacuum formed in the container by the exit of the liquid. This will result in siphoning of the product from the outer container, and produce a spray. Air will simultaneously enter the container and replace the expelled product.

Several technical advantages were claimed for this system. The outer container which is no longer under pressure can be made from a number of plastic materials which in turn permits the use of containers in a variety of shapes and colours. It also provides chemical and corrosion resistance, thus overcoming many of the problems associated with the possible incompatibility in the pack of product and propellant encountered in tinplate and in aluminium containers. The system allows, for example, for the packing in aerosol containers of water-based products without the danger of producing extensive corrosion and can be used to dispense solutions, emulsions and suspensions. The unit can only be used in the upright position.

As with co-dispensing valves, maintaining the design balance for all production units, recognizing product tolerances, and all conditions and extremes of usage render an excess of one component over the other inevitable.

The Preval Sprayer[90]. The Preval system, developed by the Precision Valve Corporation, USA, also makes use of an aerosol cartridge consisting of a valve and a dip tube assembly, incorporating its own propellant chamber—again making it possible to eliminate the use of pressurized containers for the product and to use metal, glass or plastic. The difference between the Innovair and the Preval systems lies in the fact that in the former the product is supplied to the venturi nozzle around the propellant container, while in the Preval system the product passes through the propellant container into the venturi nozzle. Again, since the propellant joins the product at the point of discharge, the majority of compatibility problems hitherto encountered will not arise. Also, the use of the Preval cartridge will result in a reduction in the amount of propellant used, permitting the use of much lower propellant/product ratios. The Preval system is suitable, for example, for personal care products for salons, touch-up paints in garages, etc., as well as for do-it-yourself and hobby applications.

Aquasol

Aquasol[99–101] was introduced to the aerosol industry at the Chemical Specialties Manufacturers Association meeting in Chicago, in May 1977.

Hydrocarbon-propelled aqueous-based aerosol product systems, featuring vapour-phase tap valves, achieve remarkable results in terms of particle size and efficiencies when used for insecticidal and air freshener applications. Topically, however, such sprays are generally wet, while, in contrast, application for personal care products demands a dry spray.

For reasons of economy, water finds favour in aerosol products especially where the formulator—for example because of legislative restriction on the use of chlorofluorocarbons—has to consider alternative propellants such as hydrocarbon and seeks a ready modification of the flammability characteristics of the sprayed product. Aquasol was designed and offered to achieve dry sprays with such systems consisting—to quote an over-simplified example—of equal parts of hydrocarbon, water and ethanol, using a special valve and actuator button.

The established and simple principle is to introduce propellant gas into the liquid product stream by means of a vapour-phase tap, and this takes place in the housing or body of the valve. In the original Aquasol system the interaction of gas and liquid streams took place in the actuator button. A high velocity vortex of gas was arranged to strike the liquid stream tangentially just prior to release from the actuator button orifice. The violent swirl initiated production of very small uniform particles which resulted in a dry spray. For various design reasons this swirl feature—the interaction of the gas stream tangentially with the liquid stream—was later transferred to the base of the housing, thus permitting the use of simpler valves and a simpler design of actuator button.

Aquasol valve production, with designs at different levels of discharge rates, finds many applications for various types of product formulations beyond, but including, the hydrocarbon, water and ethanol systems originally proposed.

PROPELLANT-FREE DISPENSING PUMPS

The concept of a propellant-free dispensing system[102–109] is attractive for many reasons, chief among which are:

1. Simpler formulation of products.
2. Elimination of concern about propellant toxicity and flammability.
3. Simpler filling operations.
4. Absence of the need for a pressure-resistant container, permitting a wider choice of container material and pack shape.
5. The possibility (without the need to contain propellant material in a sturdy pressure-resistant container) of lighter and more compact packs, leading to improved portability and perhaps greater ease of handling and use.

6. The possibility of persuading the consumer that he is buying 'all product'.
7. Ready availability of refill packs.
8. Disposal the same as for any non-pressurized pack.

The challenge of producing a more efficient, less expensive propellant-free spraying system has existed and been pursued for many years. The objective has been to produce a system that is convenient, simple and safe to use, does not require special preparation before use, leaves no surplus product to be cleaned up and wasted after use and (of paramount importance) disperses the spray product over the application area so that the product is deposited in the required manner.

Hundreds of inventions exist and have been patented as propellant-free spraying devices but very few have been exploited on a commercial scale. The most important devices developed so far are the following.

The Plastic Squeeze Bottle

When the bottle is squeezed in the hand, the spray liquid rises through a dip tube to an injection-moulded spray plug where it is mixed with air entering the plug through vents near the top of the dip tube. The mixture then passes through a constriction in the plug, causing an increase in pressure sufficient to produce a spray break-up when the mixture leaves the plug. This system has been exploited over the years on products as diverse as hair lacquer, de-icer and nasal decongestants. A major disadvantage of the system is that when the pressure is relaxed on the bottle air is drawn through the dip tube and into the product, giving rise to product oxidation in certain cases.

The Rubber Ball Pump (Perfume Atomizer)

This system operates on the venturi principle. Air is blown from a rubber ball over the top of a tube which extends into the spray liquid. The liquid is drawn up the tube into the air-stream and diffused as a spray. This system requires a fairly vigorous pumping of the rubber ball and the rate of spray delivery is very slow. However, these pumps have been associated with the classic quality perfumes and still have a market value based on nostalgia.

The Elastomer Pressure Sprayer

In cases such as the 'Selvac' system, an elastomer bag, sealed to a dispensing valve and housed within an outer casing, is filled through the valve with product under pressure. The pressure energy is stored in the elastomer bag and provides the pressure for spraying. However, the spraying pressure is variable as the product evacuates. Aging and chemical attack may cause deterioration in the strength of the elastomer.

The Mechanical Finger Pump

The first consumer packaging use of mechanical sprayers was in 1946, when units made of PVC acetate were featured on certain household window cleaners. The wider availability of various plastics from 1962 enabled pump components to be moulded to the tolerances necessary to ensure efficient operation and to provide leakproof assemblies. Even then, the pumps that were available produced a coarse wet spray, which was suited to many household products but not at all satisfactory for products requiring fine atomization. The first 'fine-mist' pumps appeared in 1970 and performed well enough to secure by 1975 a significant share of the USA hair spray market. However, the 'first generation' of finc-mist pumps was vulnerable to consumer misuse: if the actuator was not pressed with a positive, firm finger-stroke, dribbling or streaming resulted instead of good break-up.

The difficulty of controlling the pressure at which the pump operates, and consequently of overcoming the problem of dribbling and streaming, is resolved in the 'second generation' of fine-mist pumps. Several excellent systems have been patented and they are usually described using terms such as 'pressure build-up', 'pre-compression' or 'constant pressure'. Characteristically, they feature complicated internal configurations and in manufacturing terms they are intricate, close-tolerance, multi-component assemblies. They have become an economic packaging entity only through the utilization of the most advanced moulding and assembly techniques and the development of sophisticated technical support to ensure consistent product quality and well-executed new product applications.

Particle size in a good spray pattern from a second-generation pump, it is claimed, can range from 10 to 40 μm. However, a careful product formulation is all-important to ensure good spray quality. A formulation with a high surface tension will give poor particle break-up, and high viscosity leads to streaming from the spray orifice. Materials with a broad chemical compatibility are used for the pump components. Pumps are designed essentially to spray solutions, and solids suspended in a liquid can be sprayed only if the particles are very fine and present in quantities of no more than about 10 per cent.

The great majority of conventional spray products can be packaged successfully in pump packs. There are also certain new product applications which have not been possible hitherto that can now be presented in a propellant-free natural spray, for example high concentration aluminium salt antiperspirant solutions and aromatic compounds susceptible to rapid loss of top-note; oxidation may be a problem.

Pumps available today to suit the various product applications fall into four major categories:

Regular sprayers	Sprayers with an output of about 1 ml and a medium-to-coarse spray.
Regular dispensers	Dispensers for lotions with the same output as regular sprayers, for example liquid soaps.
Fine mist sprayers	Sprayers with outputs ranging from 0·05 ml to 0·2 ml of a very fine spray.
Trigger sprayers	Sprayers with an output in excess of 1 ml and a horizontal rather than vertical actuation.

After 30 years of steady technical evolution of pumps the signs are that in the United States, and elsewhere, this particular type of propellant-free spray system—the pump—is achieving consumer acceptance.

REFERENCES

1. US Patent 628 463, Helbing, H. and Pertsch, G., 1899.
2. US Patent 2 321 023, Goodhue, L. D. and Sullivan, W. N., 1943.
3. Ford, G. F., *Aerosol Age*, 1981, **26**(1), 37.
4. Gunn-Smith, R.A. and Simpson, A., *Aerosol Report*, 1980, **19**(6), 214.
5. Special Report, *Aerosol Age*, 1978, **23**(6), 40.
6. Budzilek, E., *Aerosol Age*, 1979, **24**(4), 28.
7. Vinson, N., *Aerosol Age*, 1958, **3**(11), 28.
8. Bespak Industries Ltd, *Chem. Engng News*, 1961, **39**(46), 84.
9. Harris, R. G. and Platt, N. E., *International Encyclopaedia of Aerosol Packaging*, ed. Herzka, A., Oxford, Pergamon, 1965, p. 94.
10. Anon., *Manuf. Chem.*, 1968, **39**(9), 84.
11. Rowland, F. S. and Molina, M. J., *Review of Geophysics and Space Physics*, 1975, **1**(1), 1.
12. Brasseur, G., *Critical Analysis of Recent Reports on the Effect of Chlorofluorocarbons on Atmospheric Ozone*, Eur 7067 EN, Commission of the European Communities, 1980.
13. Sanders, P. A., *Aerosol Age*, 1979, **24**(1), 24.
14. Ayland, J., *Aerosol Age*, 1978, **23**(3), 40.
15. Ford, G. F., *Aerosol Age*, 1978, **23**(3), 41.
16. Shaw, D., *Aerosol Age*, 1978, **23**(8), 30.
17. Herzka, A., *Aerosol Age*, 1960, **5**(5), 72.
18. Besse, J. D., Haase, F. D. and Johnsen, M. A., *Proc. 46th Mid Year Meeting CSMA*, 1960, p. 56.
19. Reed, W. H., *Soap Chem. Spec.*, 1956, **32**(5), 197.
20. Fowks, M., *Aerosol Age*, 1960, **5**(10), 100.
21. US Patent 1 800 156, Rotheim, E., 1931.
22. Bohnenn, L., *Aerosol Rep.*, 1979, **18**(12), 413.
23. Bohnenn, L., *Aerosol Age*, 1981, **26**(1), 26 and **26**(2), 42.
24. Reuzel, P. G. J., Bruyntjes, J. P. and Beems, R. B., *Aerosol Rep.*, 1981, **20**(1), 23.
25. US Patent 34 894, Lynde, J. D., 1862.
26. US Patent 256 129, Decastro, J. W., 1882.
27. Anon., *Modern Packaging*, 1961, **34**(6), 88.
28. Hoffman and Schwerdtel GmbH, Munich, technical literature.
29. Anthony, T., *Aerosol Age*, 1967, **12**(9), 31.
30. US Patent 3 461 079, Goldberg, I. B., 12 August 1969.
31. Giacomo, V. Di, *Am, Perfum. Aromat.*, 1957, **69**(5), 49.
32. British Patent 740 635, Taggart, R., 1955.
33. US Patent 2 689 150, Croce, S. M., 1954.
34. US Patent 2 772 922, Boyd, L. Q., 1956.
35. US Patent 2 794 579, McKernan, E. J., 1957.
36. Du Pont, Publication KTM, 21.
37. Root, M. J., *Am Perfum. Aromat.*, 1958, **71**(6), 63.
38. Sanders, P. A., *J. Soc. cosmet. Chem.*, 1958, **9,** 274.
39. Armstrong, G. L., *Soap Chem. Spec.*, 1958, **34**(12), 127.
40. Hauser, N., *Aerosol Rep.*, 1978, **17**(5), 130.
41. Reusser, R. E., O'Shaughnessy, M. T. and Williams, R. P., *Aerosol Age*, 1979, **24**(3), 17.
42. Greaves, J. R. and Makin, B., *Aerosol Age*, 1980, **25**(2), 18.
43. Johnson, S. C., *Aerosol Age*, 1979, **24**(10), 28.
44. Rubino, A., Siciliano, A. A. and Margres, J. J., *Aerosol Age*, 1978, **23**(11), 22.

45. Greaves, J. R., *Manuf. Chem.*, 1980, **51**(12), 3.
46. Pengilly, R. W., *Manuf. Chem.*, 1980, **51**(7), 49.
47. Turner, K., *Aerosol Rep.*, 1981, **20**(4), 114.
48. Berres, C. R., *Aerosol Age*, 1979, **24**(7), 32.
49. Whyte, D. E., *Aerosol Age*, 1979, **24**(5), 29, and **24**(6), 30.
50. Dixon, K., *Aerosol Age*, 1979, **24**(4), 20.
51. British Patent 719 647, Colgate-Palmolive-Peet Co. Inc., 1954.
52. British Patent 748 411, Spitzer, J. G., 1956.
53. British Patent 780 885, Innoxa (England) Ltd, 1957.
54. Coupland, K. and Chester, J. F. L., *Manuf. Chem.*, 1980, **51**(10), 39.
55. Geary, D. C., *Soap Chem. Spec.*, 1960, **36**(3), 135.
56. Von Schweinichen, J. G., *Aerosol Rep.*, 1981, **20**(3), 77.
57. MacMillan, D. and Simpson, A., *Manuf. Chem.*, 1980, **51**(6), 45.
58. Hyland, J. G., *Manuf. Chem.*, 1981, **52**(5), 51.
59. Kelly, S. W., *Manuf. Chem.*, 1981, **52**(1), 33.
60. Special Report: *Aerosol Age*, 1978, **23**(8), 16.
61. Braune, B. V., *Aerosol Rep.*, 1980, **19**(9), 294.
62. Blakeway, J. and Salerno, M. S., *Aerosol Rep.*, 1980, **19**(10), 330.
63. Anon., *Manuf. Chem.*, 1978, **49**(6), 37.
64. Anon., *Manuf. Chem.*, 1980, **51**(7), 31.
65. *BAMA Standard Test Methods*, UK, 1981.
66. *CSMA Aerosol Guide*, USA, 1981.
67. Murphy, E. J., *Aerosol Age*, 1981, **26**(3), 20.
68. Sanders, P. A., *Aerosol Age*, 1978, **23**(10), 38.
69. Kopenetz, A., *Aerosol Rep.*, 1978, **17**(10), 335.
70. Kinglake, V., *Aerosol Age*, 1978, **23**(8), 24.
71. Tauscher, W., *Aerosol Rep.*, 1980, **19**(12), 412.
72. Tauscher, W., *Aerosol Rep.*, 1979, **18**(2), 60.
73. Nowak, F. A., Koehler, F. T. and Micchelli, A. L., *Aerosol Age*, 1978, **23**(9), 24.
74. Klenliewski, A., *Aerosol Rep.*, 1981, **20**(1), 8.
75. Eisberg, N., *Manuf. Chem.*, 1980, **51**(2), 30.
76. Johnsen, M. A., *Aerosol Age*, 1979, **24**(6), 20.
77. Special Report, *Aerosol Rep.*, 1978, **17**(11), 410.
78. Schonfeld, H. W., *Aerosol Age*, 1979, **24**(5), 36.
79. Schonfeld, H. W., *Aerosol Rep.*, 1979, **18**(1), 5.
80. Schonfeld, H. W., *Aerosol Rep.*, 1981, **20**(3), 94.
81. Sanders, P. A., *Soap Chem. Spec.*, 1960, **36**(7), 95.
82. Archer, W. L., *Aerosol Age*, 1967, **12**(8), 16.
83. Downing, R. C. and Young, E. G., *Proc. sci. Sect. Toilet Goods Assoc.*, 1953, (19), 19.
84. Parmlee, H. M. and Downing, R. C., *Soap Sanit. Chem., (Special issue, Official Proceedings CSMA)*, Vol. XXVI-CSMA (2).
85. Herzka, A. and Pickthall, J., *Pressurized Packaging (Aerosols)*, London, Butterworth, 1958.
86. Murphy, T. P. and Walpole, J. F., *Aerosol Rep.*, 1972, **11**(11), 525.
87. Ziegler, H. K., *Aerosol Age*, 1980, **25**(9), 23 and **25**(10), 26.
88. British Aerosol Manufacturers Association Test Methods, *Aerosol Age*, 1981, **26**(2), 26.
89. Kleniewski, A., *Aerosol Rep.*, 1979, **18**(7/8), 235.
90. Meuresch, H., *Aerosol Age*, 1967, **12**(10), 32.
91. Anon., *Aerosol Rep.*, 1968, **7**(6), 265.
92. Braune, B. V., *Aerosol Rep.*, 1981, **20**(5), 171.
93. Anon., *Aerosol Age*, 1968, **13**(1), 19.

94. Anon., *Aerosol Age*, 1968, **13**(2), 17.
95. US Patent 3 325 056, Du Pont, 23 February 1966.
96. US Patent 3 326 416, Du Pont, 14 January 1966.
97. US Patent 3 341 418, Gillette Co., 3 March 1965.
98. Geigy, Technical Bulletin: *Innovair*.
99. Abplanalp, R. H., *Aerosol Age*, 1977, **22**(6), 35.
100. Kubler, H., *Aerosol Rep.*, 1979, **18**(1), 27.
101. Bronnsack, A. H., *Aerosol Rep.*, 1979, **18**(2), 39.
102. Davies, P., *Soap Perfum. Cosmet.*, 1978, **54**(6), 241.
103. Anon., *Aerosol Age*, 1975, **20**(9), 36.
104. Davies, P. W., *Manuf. Chem.*, 1980, **51**(4), 52.
105. Anon., *Aerosol Age*, 1979, **24**(9), 22.
106. De Vera, A. T., *Aerosol Age*, 1975, **20**(6), 46.
107. Nash, R. J., Rus, R. R. and Kleppe, Jr., P. H., *Aerosol Age*, 1975, **20**(5), 49.
108. Murphy, E. J. and Bronnsack, A. H., *Aerosol Rep.*, 1978, **17**(7), 232.
109. Anon., *Aerosol Age*, 1979, **24**(7), 25.

Chapter Forty-one

Packaging

Introduction

The packaging of cosmetics and toiletries is in principle no different from the packaging of any other product, but aspects of package design and development are of such prime importance in the successful marketing of cosmetic products that it has a more important role in this industry than in almost any other.

Packaging is very diverse, and utilizes a wide variety of materials such as plastics, glass, paper, board, metal and wood combined with a wide range of technologies including printing, machinery design and tool making. There is in fact no clearly defined packaging industry, since many companies in packaging also manufacture other products. The purpose of this chapter is to give a broad spread of information in this wide field; further details can be obtained from textbooks such as those of MacChesney,[1] Paine[2] and Park,[3] and from the *Modern Packaging Encyclopaedia*.[4]

Packaging has been defined as the means of ensuring the safe delivery of a product to the ultimate consumer, in sound condition at the minimum overall cost. Other definitions are:

> Packaging is the art or science of, and the operations involved in, the preparation of articles or commodities for carriage, storage and delivery to the customer (BSI *Glossary of Packaging Terms*[5]).
>
> Packaging sells what it protects and protects what it sells.

Principles of Packaging

Packaging must:

- Contain the product
- Restrain the product
- Protect the product
- Identify the product
- Sell the product
- Give information about the product

and do this within a cost related to the marketing, profit margin, selling price and image of the product.

Marketing and Packaging

The package projects the style and image, not only of the product, but often of the company which markets the brand. The pack must therefore project the image that it has been designed for, not only to the customer through advertising and point of sale but also to the retailer and wholesaler chain.

Packaging is particularly important in the self-service retail trade. The package designer has a responsibility to ensure not only that the pack has the type of appeal that will make the customer pick it up and be encouraged to purchase on impulse, but also to ensure that the pack will stack on self-service shelves and give the retailer the maximum profit per linear unit of shelf space.

Advertising has ensured that the package is now more widely seen than ever before. With the predominance of colour in advertising—in television, cinema and press advertisements and on posters—the package must be made of materials that have good aesthetic appeal and which will take and hold colour.

TECHNOLOGY AND COMPONENTS

Plastics

The use of plastics for producing primary components and point-of-sale material now dominates packaging technology. Two main groups are used—thermoplastic resins and thermosetting resins. Thermoplastics can be extruded at their melt temperature and then blow moulded or injection moulded. After cooling, the resin can be remelted by heating to the limits of thermal fatigue and oxidation. Thermosetting resins, by contrast, are moulded using an irreversible chemical reaction and the resins tend to be rigid, hard, insoluble and unaffected by heat up to decomposition temperature.

Thermoplastic Resins

Polyvinyl chloride (PVC) is the most familiar of all plastics, certainly as far as the general public is concerned. The basic polymer varies from transparent to opaque. In its unplasticized state, known as UPVC, the product is rigid and is used chiefly for transparent bottles and blow mouldings of various kinds. Plasticized, or flexible, PVC is widely used in sheet form, either by itself or reinforced and supported by other materials, for flexible laminates.

Polyethylene is of the class known as polyolefins, which includes two types of polyethylene and an allied material, polypropylene. Low density polyethylene (LDPE) is the more flexible form. It has high cold flow characteristics with no indicated break point under flexing or impact; when formed into film, it can be stretched to cause an increase of up to 600 per cent in tensile strength, resulting from the realignment of molecules. It is chiefly used as film. About 70 per cent of output is in this form which can be used for packaging, and in building and horticulture. Its injection moulding characteristics are excellent and it is used for closures and fittings.

High density polyethylene (HDPE) is the stiffer, more rigid form of polyethylene with generally greater mechanical strength. Its chief use (about 40 per cent of output) is as a blow moulding material for bottles and small-to-medium size containers. HDPE does not mould as well as LDPE but finds outlets as polymer for injection moulded milk crates or general industrial pallets.

Polypropylene (PP) could be described as the best of LDPE and HDPE—at a competitive price. Its chief use is for blow and injection moulding in all kinds of packaging. One particular quality is its high fatigue resistance which makes it popular for the manufacture of durable closures. To some extent PP is also used for making packaging film and extrusions.

Polystyrene (PS) is a rigid transparent material with excellent flow properties which allow it be formed into intricate shapes, so that it is principally processed by injection moulding. It is widely used for jars, bottles and mascara cases. It is somewhat brittle, but this can be remedied by mixing it with synthetic rubber to form toughened polystyrene (TPS) which is widely used for all types of packaging components where solvent attack is not a problem.

Expanded polystyrene (XPS) is used in sheet form for building and industrial insulation because of its low thermal conductivity. It is also moulded to provide close-fitting packs for fragile products and used as inert filler in secondary packaging.

Thermosetting Resins

The generic term 'aminoplastics' is used for plastics produced by reacting formaldehyde with amino compounds. These have the advantage of not being dependent on crude oil supplies for a feedstock and, because of this, in recent years have been in greater demand. Their applications range from electrical equipment such as switch plates, sockets or circuit breakers, to toilet seats, bathroom cabinets and work surface laminates. In packaging, caps and closures are the main uses. Generally aminoplastics are processed by compression moulding.

Phenolics (PF) are related to aminoplastics in that they are formed by a reaction between formaldehyde and phenol. Their general characteristics are similar but they are usually only available in black or brown, although they will accept paints without difficulty. Some grades can be injection moulded and compression moulded.

Plastics Technology

There are five main methods of converting plastic resin into packaging components:

(i) *Injection moulding* is used in thermoplastic conversion, where molten plastic is injected under pressure into a mould and allowed to cool. The mould is then split, the component is removed and the cycle repeated. This type of moulding is used mainly for caps, closures, fittings and small trays or boxes.
(ii) *Extrusion blow moulding*, again used in thermoplastic conversion, is a process in which a tube or 'parison' of molten plastic is extruded from a die. The tube is cut to length while still hot, and transferred to a blow mould where air pressure is applied through the tube, forcing it to expand into the shape of the mould. The mould is split, the component removed and the cycle repeated. Extrusion blow moulding is used mainly for bottles and jars.
(iii) *Compression moulding* is used in the conversion of thermosetting resins. The resin and catalyst are placed in a mould and held under pressure of up to 6000 psi (41 N mm^{-2}). The mould is then heated by steam, electricity or induction heaters until the reaction is complete (generally only a few

seconds). The mould is parted and the component is removed. The main uses in packaging are in high quality caps and closures.

(iv) In the process of *thermoforming*, a pre-extruded sheet of thermoplastic is placed over a table mould and the sheet is heated from above by infrared heaters. When the softening temperature of the sheet is reached, vacuum suction is applied from below the mould to pull the sheet tightly onto the mould table. The heat is switched off, the vacuum is released and the component is removed from the mould table. The process is used mainly for low quality trays and point-of-sale display items.

(v) *Injection blow moulding* is a combination of injection and extrusion blow moulding, and is used for bottles in which tight-tolerance neck and shoulder measurements are required. The parison, including the neck of the bottle, is first made in an injection mould and then transferred while still hot to another mould where the final bottle shape is blown.

It should be noted that all containers made from thermoplastics are permeable to air, perfume and water vapour to some degree, and furthermore the product may react chemically with the plastic. Full compatibility studies are essential, therefore, before packaging cosmetics in plastic.

Metals

The particular applications for which metals are most suited in cosmetics packaging are aerosol containers, powder dispensers, shallow tins and collapsible tubes. Tinplate is the most commonly used metal for the rigid packs, although aluminium also finds widespread use. As aerosols are by far the largest of these applications, the use of rigid tinplate and aluminium has already been discussed in the chapter on aerosol packaging (Chapter 40). The metal required for collapsible tubes has, on the other hand, to be readily deformable but must not fatigue or crack under stress. Suitable metals for this purpose are aluminium, tin and lead, when of the correct gauge and purity. The use of collapsible aluminium tubes in particular is extremely widespread, and almost all varieties of semi-solid, cosmetic and toiletry products, including emulsions, pastes and gels, can be purchased in collapsible tubes, but toothpaste remains the most popular application. The impervious nature of the metal gives the collapsible tube the great advantages of reduced risk of contamination of the product and reduced losses of volatile materials from the contents.

Lead. The use of lead for the manufacture of tubes varies to some extent with the availability and cost of the different metals in a particular area, but its use, particularly in Europe, is not widespread. Lead is more resistant to corrosion than aluminium, being much lower in the electrochemical series, and has been used for the packaging of products such as fluoride toothpaste that are acid and attack aluminium very rapidly. As lead can cause discoloration in some products, and as it might be a source of contamination, particularly undesirable in oral products such as toothpastes, the interior of lead tubes is usually lined with wax to reduce contact between the products and the metal. Apart from the high cost of lead tubes in many areas, their weight is an obvious disadvantage.

Tin. Tin (99·5% pure) has a great many properties which commend it for collapsible tubes and it held sway for many years in the first half of the century. However, by about 1950 the technology of aluminium had so improved that satisfactory tubes could be made from aluminium which was available in a purity of 99·7% at a considerably lower cost than tin.

In some places there has been an attempt to obtain some of the benefits of tin by using it in conjunction with lead, either as alloy or as an internal tin lining. Neither of these approaches has gained widespread acceptance, due largely to the large proportion of tin which has to be used to prevent contamination of the product by lead.

Aluminium. Structurally, high purity aluminium is a very suitable material for collapsible tubes as it is readily deformable, it is light, does not fracture under normal use conditions, and is totally impermeable to water, oils, solvents, and gases such as oxygen. Aluminium is, however, a reactive metal and one that is easily corroded. Compared with plastic tubes, aluminium tubes have the advantages of being impermeable and of being permanently deformable rather than flexible, so that there is no 'suck-back' on release of the tube, but they suffer from the disadvantages of having poor resistance to corrosion and of being rather unattractive in appearance when they have been squeezed or crumpled, particularly if there is a complex or colourful design on the pack.

When packing a product in an aluminium tube, careful attention must be paid to the problem of corrosion. Aluminium can be corroded by a galvanic action or a direct chemical reaction. Chemical attack is common at extremes of pH and corrosion is generally rapid and accompanied by hydrogen evolution. Very alkaline products such as depilatories and hair straighteners and acid products such as hydroquinone skin bleach creams cannot be packed in plain aluminium tubes. Internal coatings of lacquer, usually two, are required and in many cases, where the product is liable to strip the lacquer, a coat of wax over the lacquer is required. Galvanic corrosion of aluminium is very common and takes place under slightly acid or alkaline or even neutral conditions if electrolytes are present. Migration of ionic dyestuffs is often a very good indication that galvanic action is taking place. It is accelerated by entrapped air, corrosion often taking place at the air–metal–product boundary, and also depends to a considerable extent on the purity of the aluminium. Sodium silicate is a very effective anodic corrosion inhibitor and can be used in mildly alkaline products such as chalk-based toothpastes and soap-based lather shaving creams. Corrosion in slightly acid conditions is rather more difficult to control chemically. Both types of corrosion can lead to hydrogen evolution and eventual blowing of tubes. The usual method of reducing or preventing galvanic corrosion is by internally lacquering the tubes so that the metal is insulated from the product. Vinyl, phenolic and epoxy resins are used as lacquers.

Laminates

The various requirements of packages for cosmetics and toiletries (such as attractive appearance; impermeability to water and volatile oils) are not always available from a single material. This problem can sometimes be solved by the use of composite materials in laminar form. Laminates have found particular application in the production of sachets and of collapsible tubes as alternatives to pure metal tubes for toothpastes.

Laminates are used for *flat sachets* that are heat sealed around the periphery (as distinct from the fatter *pillow sachets* which are made from PVC tube and sealed ultrasonically). The laminate must be able to withstand the pressure of the contained product and provide leakproof seals. It must also prevent loss of water and other volatile substances such as perfume; these barrier properties can be provided by aluminium foil or by combinations of plastics with complementary properties, such as polyethylene and cellulose acetate films. Polyester films can be used to impart strength, and a paper layer will give both strength and stiffness. A typical three-ply laminate suitable for sachets is made up of cellulose acetate, aluminium foil and polyethylene, and a tougher four-ply laminate has a layer of paper between the aluminium and the polyethylene. Dweck[6] has described the manufacture of filled sachets from laminated materials.

Laminates for *collapsible tubes* aim to combine the appearance of plastics with the impermeability and collapsibility of aluminium. Tubes made from plastics material only—usually polyethylene, PVC or (to a lesser extent) polypropylene—do not collapse when squeezed: air is sucked back into the tube when the product has been dispensed. This is not acceptable for products such as toothpastes which are expected to be discharged as a continuous ribbon without air bubbles. A further disadvantage is that a plastics tube which retains its shape gives no indication of the amount of product remaining. The basic materials used for laminates to overcome these shortcomings are polyethylene film and aluminium foil, but complex combinations have been developed in order to achieve the desired properties. For example, a suitable laminate consists of polyethylene (or polypropylene) in contact with the product, then aluminium foil, polyethylene, paper and polyethylene.

A comparative review of the various metals, plastics and laminates available for the fabrication of tubes, together with a short description of tube filling and cartoning operations, has been given by Guise.[7]

Glass

Glass containers are still used widely in the toiletries industry by virtue of the basic packaging characteristics of glass. Glass is chemically very inert and generally will not react with or contaminate high quality cosmetic and perfume products; it has the approval of the US Food and Drug Administration (FDA) for a wide range of products. With a properly designed closure, glass is a 100 per cent barrier material and provides protection from oxidation, moisture loss or gain, perfume loss, etc. Glass is transparent, sparkles and provides an excellent point-of-sale display. Alternatively, for a product that is sensitive to light, amber glass or cartoning can be used. Finally, glass can be moulded into very attractive designs and provides an excellent brand or product image, especially at the high quality end of the market.

Glass is manufactured in many different formulations but the most common in packaging is soda lime glass composed as follows:

	per cent
Silica obtained from sand or quartz	72
Calcium carbonate (limestone)	11
Sodium carbonate (soda ash)	14
Aluminium oxide	2
Trace oxides	1

It is the trace oxides that provide colour to the glass, and green and amber containers are readily available. Trace selenium compounds are sometimes added as a decolorizer. An important 'raw material' in glass manufacture is the oil or gas required to melt the materials, which constitutes a very significant element in the cost of glass containers.

Glass Technology

The technology of glass making is thousands of years old but it is only in recent years that fully automatic methods have been developed for the manufacture of glass components. Molten glass is made in a furnace in a continuous flow, that is, the raw material input matches the rate at which the molten glass is drawn off. A furnace will run continuously for about eight years and during this time a temperature of around 1500°C must be maintained. The molten glass is fed to the conversion machines where the containers are made. Details of glass moulding processes have been described by Moody;[8] the basic techniques are the following.

The Suction Process. Molten glass is sucked into an initial or parison mould where the neck is formed. This parison shape is transferred to a blow mould where the final shape is made using air pressure.

Press and Blow Flow Process. A molten gob of glass of predetermined weight drops into the parison mould, a plunger presses the parison into shape. This is then transferred to a blow mould where the final shape is blown.

Blow and Blow Flow Process. The gob drops into the mould where the neck is formed, assisted by air pressure. A parison shape is then blown, which is then transferred to the final mould and the finished container is blown.

Glass, even with its inherent disadvantages of fragility and weight which generally cause transport and secondary packaging costs to be high, continues to provide the cosmetic industry with aesthetically pleasing, stable and high quality containers and bottles.

Paper and Board

Practically every cosmetic and toiletry product uses paper or paperboard in some form. Many grades of paper and board are available; the main types used in packaging are listed in Tables 41.1 and 41.2. Uses of paper and board in cosmetics packaging include labels, leaflets, corrugated cases, printed cartons and soap wraps.

Table 41.1 Main Types of Packaging Board

Type	Made from	Properties	Uses
Plain chipboard	100% low-grade waste, e.g. old newspapers	Cheap Prints poorly Light grey/tan colour Folds poorly	Rigid boxes Packing pieces
Cream-lined chipboard	Two layers: (a) cream liner from new pulp (b) as plain chipboard	Poor quality Folds satisfactorily	Low-quality cartons
Duplex board	Two layers, both from new pulp	Prints well Folds well Smooth surface	Quality cartons Point-of-sale displays
White-lined chipboard	Two layers: (a) white 100% chemical pulp (b) as plain chipboard	Prints well Folds well Smooth surface	Quality cartons Point-of-sale displays
Clay-coated boards	As duplex or white-lined + clay coat	Excellent print, fold and gloss	Top-quality cartons for high-price cosmetics
Solid bleached sulphate or sulphite boards	100% sulphate or sulphite pulps	Good strength Excellent print and whiteness Odourless for food contact	Frozen food, ice cream, etc.

Table 41.2 Main Types of Packaging Paper

Type	Made from	Weight ($g\ m^{-2}$)	Properties	Uses
Kraft	Sulphate pulp	65–300	Heavy duty paper	Corrugated case liners Multiwall sacks
Sulphite	Mixture of hardwood and softwood, i.e. mechanical and chemical pulp Usually bleached	34–300	Bright paper Prints well	Envelopes Foil lamination Labels Leaflets
Greaseproof	Heavily beaten pulp	65–150	Translucent Grease-resistant	Fatty products
Glassine	Greaseproof, calendered	39–150	Resistant to oil and grease Good odour barrier	Soap wraps
Vegetable parchment	Unsized paper treated with concentrated sulphuric acid	59–370	Non-toxic High wet strength Resistant to grease and oil	Mainly food products
Tissue	Any virgin pulp	20–50	Low strength Light weight	Wrapping of goods

Printing and Decoration

All packaging components can be printed to give a wide range of decorative effects. Different processes are used depending on the application. The five main processes used in the printing of packaging components are described briefly below.

Screen Printing
Method: porous stencil printing, in which a rubber blade forces ink through unblocked print areas of a screen. The screen is generally made from nylon sheet or, sometimes, silk.

Main packaging uses: printing of plastics and glass containers, point-of-sale display items.

Letterpress
Method: the face of the printed image is a raised surface above a metal blade—a relief print process. Ink is applied direct to the face and is transferred to paper or other printing surface directly.

Main packaging uses: all types of labels.

Flexography
Method: as in letterpress, but the raised image is made from a flexible rubber or composition plate. Similar in operation to a rubber stamp.

Main packaging uses: corrugated cartons and transit shippers, some films and labels.

Offset Lithography
Method: print and non-print areas are hydrophobic and hydrophilic areas in the same plane on the printing plate. The plate is wetted and the inked hydrophobic areas reject water and accept ink, whereas hydrophilic areas accept the water and thus reject ink. The image formed is transferred to a rubber-covered cylinder, which then transfers the ink to the printing substrate.

Main packaging uses: all types of cartons and high-quality metal containers.

Gravure Printing
Method: the reverse of letterpress. The image is made up from subsurface 'cells' etched into a metal cylinder. The depth of the cells varies according to the depth of ink to be transferred. The cylinder is inked, and excess ink is wiped from the surface by a wiper or 'doctor' blade. The ink in the subsurface cells is transferred to the printing surface.

Main packaging uses: long-run flexible package printing and label printing—a very high quality of work is achieved.

PACKAGE DEVELOPMENT AND DESIGN

Package development has the aim of increased sales and profit through the correct design of the package. Packaging must be considered as early as possible in the development of a new product to allow time to ensure that pack and product are compatible. The development process begins with a detailed analysis of the product so that a pack can be designed to give protection. Marketing factors must also be considered—the pack must be suitable for the product and its market. Easy opening, convenience factors and ease of handling are of prime importance to the ultimate user. Graphics and aesthetic design should also be considered at this stage, any important related marketing criteria again being taken into account, and the interaction between the product and the primary container and other packaging should be investigated. Finally, in the development of the package the environmental aspects of the pack should be considered, not only in terms of disposability and litter, but also from the point of view of the re-use of scarce raw materials.

Technical Aspects of Design

The technical aspects of packaging design are rather diverse and too involved for any detailed discussion in this text, but, generally speaking, given a particular packaging material, the final pack must be sufficiently strong to survive any treatment that it is liable to receive from the time of first delivery at a factory, through to filling, distribution, sale and actual use. There is bound to be a certain failure of packs, largely due to accidents, but it is essential that this failure rate is very low if the product is to be a commercial success, and hence very strict testing and quality control of packs is essential in a factory producing toiletries or cosmetics. Typical problems that must be watched for in package design are thin areas in bottles, which are quite common in flat glass or plastic bottles, highly stressed areas, of particular importance when polyolefins are used, and very small radius convex areas on the outside of clear packs that can act as effective lenses for focusing ultraviolet radiation, and so concentrate its harmful effects on to small areas of a product. Bottles which are unstable on their bases can give rise to troubles on a mechanical filling line if they are unstable when empty, and to spillage and breakage in use if they are unstable when filled.

Not only must a pack be attractive to the consumer and contain the product in an efficient manner, but it must also render the product available as soon as the consumer desires to use it. There is nothing more infuriating to a consumer than a hand cream that will not come out of a bottle because the hole is too small, or a squeeze pack that squirts a direct jet of liquid rather than the expected fine spray. The simple example of a badly-judged orifice size is an extremely important one, as it is absolutely essential that the orifice size of a bottle or tube is correct so that the product is dispensed at the desired rate. Similarly, in the second example the orifice dimensions and design must be correct for the desired effect, that is a fine spray, but also the pack must be sufficiently flexible to allow the consumer to dispense easily the required volume of product. These points of orifice size and design, wall flexibility, and also general product accessibility in open containers such as jars, would seem to be somewhat obvious, but are factors that can be very easily overlooked and hence merit considerable emphasis.

Applicators are widely used for products such as antiperspirant–deodorants and the various forms of decorative make-up that require application to specific areas of the body. In the field of deodorants in particular, there has been considerable activity in applicator design. Products in deodorant ranges can be obtained in the form of pressurized packs, finger-operated pumps, squeeze packs, roll-balls and sticks.

The roll-ball (or roll-on) type of pack is in widespread use and is particularly ingenious as it dispenses a convenient quantity of product only when the pack is actually applied to the skin. It usually consists of a glass or plastic bottle with a snap-on polyethylene housing containing a polystyrene or glass ball. On inversion of the pack the product flows through a hole in the centre of the housing on to the ball and can then be rolled on to the skin. The closure can be obtained in two ways, both of which are patented;[9,10] either a raised portion in the cap can be made to engage on the ball and push it on to a seating in the housing, so restricting flow, or the interior of the cap can be so shaped that, when it is replaced, the lip of the housing is compressed on to the ball, so giving the seal round the ball rather than below it. The development of the roll-ball applicator has been described by Hanlon.[11] Stick containers are generally polystyrene cylinders containing a polyethylene godet. The container containing the godet is used as a mould for the stick which can then be extended or retracted manually, or in more sophisticated packs by means of a screw device operated by revolving the base of the pack. It is of prime importance that the interior of the cylinder is true so that there is no possibility of evaporation of the water or alcohol in the stick past the godet. Lipstick containers can also be of the screw type and are generally made in metal. As evaporation is not a problem, lipsticks can be moulded separately.

Another interesting area, where there has been considerable activity, is in the design of collapsible tubes to contain two incompatible materials which are mixed on extrusion. A large number of patents have been filed for such a tube, which would have particularly wide application in the field of hair dyes and bleaches, but as yet no completely successful pack is available. A related design is that of the collapsible tube which dispenses product (usually a toothpaste) in a striped form. The separate parts of the extruded product may differ in colour but may also be different in other ways, for example clear gel and opaque toothpastes may be co-extruded. This is achieved by placing one product in a specially designed insert in the tube nozzle, so that, when the other product is extruded past the insert, the first product is drawn into the ribbon to form a stripe.[12]

Closures

No container, however perfect, is of any value unless it has an efficient closure. Ideally the closure should be easy to remove and replace, but should give a seal that prevents the diffusion of gases and vapours and the seepage of liquids. Many bottle and tube closures consist of a screw cap containing a compressible wad, but wadless closures—achieved by new designs of thermoplastic screw and snap-on caps—are now extensively used for liquid products such as shampoos and hair conditioners.

Caps containing wads are usually manufactured from a thermosetting plastics material such as urea- or phenol-formaldehyde, although metal caps are still sometimes used. Wads usually consist of a stiff base material such as wood pulp, which can be treated in various ways. Wood pulp used alone gives only a poor seal but this can be very much improved by a coating of wax. This type of wad, however, has a rather poor resistance to moisture and oils and is inclined to become soggy on long-term storage. A more satisfactory wad for cosmetic packs is the vinyl-coated paper-faced wood pulp wad which is very resistant to moisture and oil, although the seal obtained on glass and metal rims is not particularly good. If products are to be packed that are sensitive to moisture loss, for example oil-in-water emulsions, the vinyl coating should be further coated with a layer of wax.

Where the neck of a container is small, as in a bottle, there is more choice of wadding material, as cork-backed wads can be used. Aluminium-foil-faced cork wads are quite common, but give a poor seal for oily materials. The vinyl-coated wads are rather superior in this respect, but are inclined to stick to the container and split when the cap is removed, if the product is resinous. Waxing should, however, overcome this problem. Solid, shaped flexible plugs or wads have also been used for bottles, particularly those with a sprinkler neck, where a central spike can be designed to engage in the sprinkler orifice to give a very effective seal.

The closures used for collapsible metal tubes are basically similar to those used for bottles. The caps are made of a thermosetting resin and the wads are usually of vinyl-coated paper-faced wood pulp. It is normally only necessary to use special wads where the product is air-sensitive or very sensitive to moisture loss, and in such cases solid polyethylene or polyethylene-coated paper-faced wood pulp wads are particularly suitable. An alternative is to use a solid moulded high-density polyethylene or polypropylene cap containing a flat area that seals on the neck of the tube. These caps give a good seal but are inclined to expand and loosen slightly at high temperatures. For extreme cases, to give a perfect seal, an aluminium tube with a pierceable membrane across the nozzle can be obtained. At the crimp end of an aluminium tube a latex end seal, an internal band of latex rubber about 6 mm in width, can be used to give a good seal, but care must be taken in filling to ensure that no product comes into contact with the latex. An end seal is not necessary, however, if the interior of the tube is waxed. In the case of plastic (polyethylene or PVC) tubes, a solid polypropylene cap is usual as the seal obtained between two fairly flexible plastics is usually quite adequate. Attractive designs for this type of package can be achieved by the use of caps of the same diameter as the body of the tube, so allowing stand-up storage on a shelf.

For some products with appropriate rheological characteristics, for example skin creams, the closure may incorporate a dispenser pump, so designed that a downward pressure causes the discharge of a fixed amount of product. Such dispensers are usually formed from a polyolefin material.

PACKAGE TESTING AND COMPATIBILITY

Testing

There are three main reasons for testing packaging materials and finished packs: to provide information vital to the designer to enable him to make effective material selection; to assess the performance of a material in relation to the duty that it has to perform; and to provide an continuing check on quality.

Comparison of results is only valid if the same standard test method is used each time the test is performed, and the quotation of results without the relevant standard must be suspect. Therefore it is important to use an accepted standard method for testing whenever possible. Standard test methods are drawn up and published by national organizations such as the British Standards Institution (BSI), the American Society for Testing and Materials (ASTM), Deutsches Institut für Normung (DIN), etc., using panels of experts from industry and trade bodies, and are coordinated by the International Standards Organization (ISO). As with all testing procedures, a basic knowledge of statistical methods is essential, both in making sure that the sample selected for testing are representative, and in the interpretation of the results.

The main testing methods used in the packaging industry, both on materials and finished packs, investigate the following:

Mechanical properties, for example, compression, tensile, flexural and impact strengths.

Physical properties, for example, water absorption, moisture vapour transmission rates, accelerated aging, flammability and thermal conductivity.

Chemical properties, for example, resistance to the product or the chemical environment, and corrosion testing.

Compatibility

Compatibility testing is performed when the final product formulation and packaging system have been decided. Samples of the product should ideally be taken from trial batches, and the complete packaging system should be assembled using actual samples or pilot tooling samples representing the final component.

The general compatibility of the pack and product needs to be checked by storage testing which will enable an assessment to be made of the effect of the pack on the product as well as that of the product on the pack. It is important to remember that the effect of spillage on the outside of the pack is important, and should certainly be included in any testing schedule. For example, it is possible to formulate alcoholic products containing a small amount of non-volatile ester which have no effect on, say, polystyrene if the polystyrene is immersed in the product, but if the product is allowed to dry out on the surface of a polystyrene container, the evaporation of the alcohol allows a high concentration of ester to occur and this can dissolve in the polystyrene to make the pack sticky.

Shelf life testing is also necessary in order to determine the rate at which volatiles may be lost and this includes not only water or alcohol but the perfume. The assessment of loss of perfume can only sensibly be done by nose, although weighing will obviously allow determination to be made of the loss of solvents.

Finally, the convenience of any cosmetics pack must be tested by in-use tests as well as by laboratory tests and for this purpose it is necessary to take into account both the local customs and the climate of the country in which the product is to be marketed: for instance, bathrooms in the UK tend to be cold, and tests for pourability should reflect this; similarly, products for tropical countries should be tested in conditions of high temperature and humidity.

REFERENCES

1. MacChesney, J. C., *Packaging of Cosmetics and Toiletries*, London, Butterworth, 1974.
2. Paine, F. A., *Packaging Evaluation: The Testing of Filled Transport Packages*, London, Butterworth, 1974.
3. Park, W. R. R. (Ed.), *Plastics Film Technology*, New York, Van Nostrand Reinhold, 1970.
4. *Modern Packaging*, 1979, **52**(12).
5. BS 3130: 1973, *Glossary of Packaging Terms*.
6. Dweck, A. C., *Cosmet. Toiletries*, 1981, **96**(6), 17.
7. Guise, W., *Manuf. Chem.*, 1981, **52**(8), 24.
8. Moody, B. E., *Packaging in Glass*, London, Hutchinson Benham, 1977.
9. British Patent 740 220, Bristol-Myers Co., 17 March 1954.
10. British Patent 843 315, Owens Illinois Glass Co., 25 April 1958.
11. Hanlon, J. F., *Soap Cosm. Chem. Spec.*, 1981, **57**(6), 67.
12. British Patent 813 514, Marraffino, L. L., 11 June 1956.

Chapter Forty-two

The Use of Water in the Cosmetics Industry

Of all the raw materials used in the formulation and manufacture of cosmetics, water is almost certainly the most widely used. Without water, our range of cosmetic products would be drastically reduced, yet because it is relatively cheap and abundant, water is often taken for granted. We should not be so complacent: of all the fresh water on this planet (and there is not much of that) only 0·03 per cent is readily available to the world's population, which is growing at a staggering rate.

Properties and Cosmetic Uses of Water

Water is an extremely reactive substance, much more so than most of the raw materials of cosmetics. This is made manifest by its extreme corrosiveness—water rusts metals and rots animal and vegetable matter. It is surprising, therefore, that it should be physiologically innocuous—it rots dead, but not living, material.

Water takes part in four types of chemical reaction, namely oxidation, reduction, condensation and hydrolysis. All four are represented in the various biochemical processes in which water is involved. For this reason, water is an essential requisite of all living organisms—without water, life itself cannot exist. Once water is present, however, life of some kind will almost certainly be found. Moreover, water is distributed very heterogeneously among living organisms. For instance, the jellyfish is 97 per cent water, adult human beings 70 per cent and bacterial spores only 50 per cent, which seems to be about the lower limit at which life can continue.

In the manufacture of cosmetics, use is made of water as a solvent and a relatively innocuous raw material rather than as an essential biochemical ingredient. Among other applications, water forms a significant part of shampoos, bath products, spirituous preparations, soaps and emulsions. It is because of its easy availability and its cheapness that water plays an important part in these cosmetic products; it is, however, through neglect of the quality of the water used that many otherwise satisfactory cosmetic products can be made useless.

Composition of Mains Water

In many cosmetic-producing countries, water is available from the mains supply and may be obtained by the turn of a tap. Since pure water is an extremely aggressive solvent, mains water will inevitably contain traces of contaminants and it is the presence of these that should be the concern of the cosmetic chemist. The identities of the contaminants that reach our manufacturing plants in mains water depend upon the original source of the water and upon the nature and the extent of any purification processes to which it may have been subjected by national or municipal water authorities. Generally, water from rural areas contains the following inorganic ions to varying extents: calcium, magnesium, sodium, potassium, bicarbonate, sulphate, chloride and silicate. In addition, the water may have a recognizable organic content, particularly humic and fulvic acids (polycarboxylic materials derived from the breakdown of natural vegetation), amino acids, carbohydrates and proteins (from decaying leaves), high molecular weight alkanes and alkenes (from algal growth) and possibly traces of organic sulphur compounds (from sewage effluent or contact with animal life).

From urban areas, more polluted water has a wider range of contaminants. Among inorganic traces to be found in polluted water are ammonia, phosphates, arsenates, borates, chromium, zinc, beryllium, cadmium, copper, cobalt, nickel, iron and manganese. Organic contaminants in polluted water include petrol, chlorinated solvents and traces of surface-active agents such as alkyl benzene sulphonic compounds (although the levels of these have been greatly reduced by the introduction of biodegradable surfactants). Whatever the origin of the water, it is almost certain to contain bacteria, viruses, pyrogens, moulds and yeasts.

Even relatively uncontaminated raw water from rural areas would no longer be deemed fit for municipal supply in most industrial countries today. Consequently, such water is purified before distribution. The object of such purification is not to produce completely pure water, but to produce potable water for general consumption that is pleasant to see, taste and drink and which contains nothing that is dangerous to human health. To reach this standard, the water is freed from most of the suspended solids, humic acids and living organisms but it still contains enough dissolved salts and gases to be pleasant to drink.

Water Purity Requirements for Cosmetics

Cosmetics were once made exclusively from mains water which had not been further purified. To meet today's high standards of product stability, however, two aspects of mains water contamination need to be investigated thoroughly.

The first of these is the inorganic ion concentration. Mains water, even after its initial purification, still contains the majority of its sodium, calcium, magnesium and potassium salts; it will also contain 50 per cent of its original concentration of heavy metals, particularly mercury, cadmium, zinc and chromium, as well as traces of iron and other materials that may be picked up from supply pipes.

In the manufacture of colognes and aftershaves (which usually contain between 15 and 40 per cent water), trace quantities of calcium, magnesium, iron and aluminium can give rise to the slow formation of unsightly insoluble residues—often made worse by the co-precipitation of the least soluble components of the perfume compound. In addition, the presence of phenolic organic compounds such as antioxidants and UV-stabilizers can be the cause of discoloration in such cosmetics by reacting with trace metals to form coloured products.

In the field of emulsion technology, it is well known that the presence of large inorganic ions such as magnesium and zinc can, by interfering with the balance of static charges responsible for the proper functioning of certain surfactants, bring about the separation of otherwise stable emulsions. Even where complete separation is not caused, the presence of such ions in the water phase can give rise to extremely unpredictable viscosity characteristics in the cream or lotion, and similar viscosity effects can result in shampoos and other products containing surfactants.

The second aspect of unpurified mains water which should be of concern to cosmetic chemists is the presence in such water of micro-organisms. If micro-organisms are introduced into cosmetics and are allowed to thrive, the result will be spoilage of the product by the development of unpleasant odours, visible colonies of bacteria, moulds or fungi and eventually (in the case of emulsions) product separation—and all this in addition to potential harm to the consumer. Any cosmetic product containing even a surface coating of water is at microbiological risk and the most common source of such contamination is the water itself. Modern cosmetics manufacturing plants must therefore be supplied with water that is as free from microbiological contamination as is technically and economically feasible. (It is ironic, as will be seen, that a potential source of contamination with bacteria is the very apparatus that may be used to remove stray ion contamination from the water).

In general, the level of microbiological contamination of mains water from the tap is very variable; since microbes will multiply best in stagnant or quiescent water, the level of contamination by the time the water reaches the consumer depends not only on its purity on leaving the water authority's plant, but also upon the layout and frequency of use of the distributive system.

Further Purification of Mains Water

Deionization

The further deionization of mains water may be effected by means of ion-exchange systems, by reverse osmosis or by distillation. Of these, the most popular method is ion exchange.

In order to remove all ions from water completely, deionization systems comprise at least two types of resin—one to remove cations (and therefore called the cation resin) and a second, the anion resin, to remove anions. Figure 42.1 illustrates the principle diagrammatically. As the active sites on both types of resin become occupied with inorganic ions, the deionizer becomes increasingly ineffective until the feed water passes through the columns virtually unchanged. The resins must now be 'regenerated' in order to continue to function. During regeneration, the active sites are swamped with free hydrogen and hydroxyl ions by lengthy contact with strong mineral acid and base respectively, when the sorbed inorganic cations and anions are once more replaced by hydrogen and hydroxide. After thorough washing in deionized water, the resins are fit for use again.

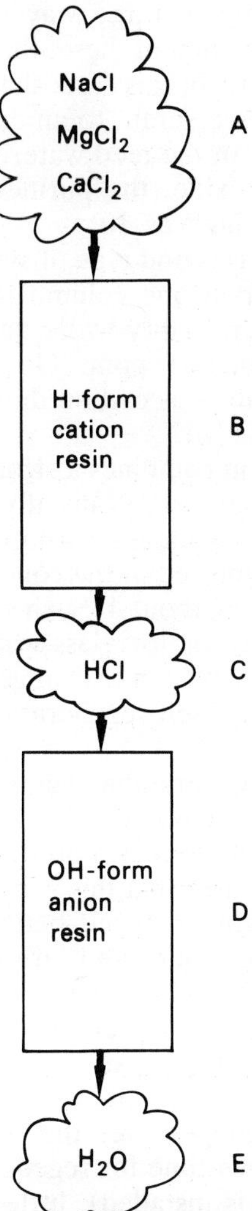

Figure 42.1 Principle of ion-exchange columns
The mains water (A) contains various inorganic ions to be removed by the process. The cation resin (B) through which this water passes is in the acid or hydrogen form. As water percolates through the resin beds, each cation becomes bound to an active site on the resin surface, replacing a hydrogen ion. Given a sufficient number of such sites and a long enough contact time between water and resin, quantitative replacement by hydrogen of all metal ions in the original solution is possible; the water then has the composition shown at C. This acidic water is now passed through the anion column (D) where a similar exchange of anions for hydroxyl ions takes place, so that the eluent at E contains a further concentration of hydrogen and hydroxyl ions equivalent to the quantities of ionizable inorganic impurities initially present.

Ion-exchange resins are used in three ways, each giving rise to a slightly different quality of water. In the first of these, the two resins are kept in separate columns (the 'twin bed' system). In this case, there is invariably a slight leakage of sodium salts from the cation column, to an extent depending on the ratio of sodium to total cation content in the feed water. Since these sodium salts leave the deionizer as sodium hydroxide, the purified water obtained from such a deionizer may have a pH as high as 10.

To overcome this problem, a second type of system exists in which both resins are mixed intimately together in one column (the 'mixed bed' system). This arrangement produces a higher quality water having a neutral reaction and an ionic concentration of less than 1 ppm. On exposure to the atmosphere, however, such water rapidly absorbs carbon dioxide to form carbonic acid and will therefore exhibit an acidic pH.

The third type of resin system combines a strong cationic with a weak anionic resin in either twin or mixed bed form. Naturally, such a system will not remove such weakly acidic materials as silica and carbon dioxide. The resulting purified water will therefore be undiminished in the concentration of these ionic species and frequently has a pH value of about 4. Such systems are used to purify water for the topping up of batteries or for glass-washing.

From a practical point of view, ion-exchange systems can be purchased or hired in three different forms: self-regeneratable, cartridge exchange or as throw-away resins.

As the name implies, self-regeneratable deionizers may be regenerated by the user either manually (by the manipulation of valves) or automatically at a pre-set time or when the effluent water reaches a predetermined quality. The convenience of the automatic type ensures that this is by far the most popular and such deionizers are available in twin or mixed bed form.

Cartridge deionizers, on the other hand, are usually of the mixed bed type, this being packed into a replaceable cartridge. When the mixed bed requires regeneration this cartridge is simply removed and replaced by a spare one containing fresh resin, the exhausted cartridge being returned to the manufacturer for regeneration. Depending on the uses to which they are put, cartridge deionizers have several advantages over the self-regeneratable variety. They produce purer water, shut-down time for regeneration is negligible (or nil if an automatic changeover device is installed), little skill is required, there are no chemicals to handle or effluent to dispose of and they are quick and easy to install. The main disadvantage is their comparatively high running cost, especially for larger throughputs of water. Throw-away cartridges are, as the name implies, designed to be discarded when they are exhausted. This is particularly useful in areas where regeneration facilities are not readily available.

Another type of ion-exchange technique in common use in the cosmetics industry is known as 'base-exchange' or 'water-softener'. In this case the feed water is passed over a cation resin in the sodium form. In this way, calcium and magnesium ions in the hard water are replaced by an equivalent number of sodium ions, so producing soft water. The equivalent ionic concentration of water treated this way therefore remains the same—only the nature of the cation changes.

Distillation

Ion-exchange resins do not provide the means for removing nonionic or weakly ionic contaminants from the water. In particular, deionized water may still contain pyrogens and, for this reason, distillation is sometimes used as an alternative or an adjunct to ion-exchange.[1] Although simple baffle stills are sometimes to be found in laboratories, the two most commonly used types of still for large volumes of water are the 'thermo-compression' and 'turn-back' types. The former, however, is difficult and complex to operate, while the latter utilizes comparatively large amounts of cooling water.

The latest development is to use multi-stage pressure column stills which require little or no cooling water. Generally, distillation is more commonly used in the pharmaceutical industry where the use of sterile water is a necessity. As a means of obtaining large quantities of pure water for cosmetic use, distillation is far too costly. The associated nature of the water molecule lends it a heat capacity and latent heat of evaporation which is disproportionately large for such a small molecule. If no heat is recovered from the distillation process, the minimum energy requirement for distillation is of the order of 0·8 kW of power for every kg of water distilled.

Ultrafiltration

Ultrafiltration is a simple and rapid method for separating dissolved molecules on the basis of size by pumping water through a filter of molecular size. Rates of throughput (up to 10 litres per hour) are far too low for this technique to be of much use for production purposes.

Reverse Osmosis

Reverse osmosis[2–5] is the most widely applicable of all the purification techniques involving membranes. The principle which gives rise to its name involves the forcing of water through a semi-permeable membrane from a concentrated solution to a weak one against osmotic pressure. Thus the concentrate becomes increasingly more concentrated in solutes. The value of this technique is that the dilute solution is exceedingly dilute: the membrane is able to prevent the passage of 95 per cent of inorganic ions, 100 per cent of bacteria and viruses and a very high percentage of other organic species, depending on their molecular weight.

As with the more familiar osmosis phenomenon, the membrane does not play a passive part, its nature and structure being of great importance to the efficiency of the process. Several types of membrane are available, but still the most widely used types are the anisotropic cellulose acetate and the hollow polyamide fibre. The latter suffers the disadvantage of being vulnerable to bacteriological attack, high temperatures and pH changes outside a narrow range, and this has led to the search for other anisotropic membranes, which are unfortunately more expensive.

Typical cellulose acetate membranes consist of a relatively dense non-porous layer (the 'active layer') with a thickness of 1500–2500 Å supported on a highly porous substructure which comprises the bulk of the membrane. This substructure contains about 55 per cent void space with pores of average diameter 20 Å.

Although the mechanism of rejection is not fully understood, it is apparently concerned with the formation of hydrogen bonds between the feed water and the membrane polymer, making the water less easily available to solvate the solutes.

The membranes are constructed in the form of cylinders capable of withstanding working pressures of 400–600 psi (2·7–4·1 N mm^{-2}), and the whole reverse osmosis device works on a continuous basis. Feed water passes across and through the membrane (through the 'active' layer first) and is collected as purified water on the other side of the membrane. Approximately 75 per cent of the feed water is collected as purified and 25 per cent continuously discarded as concentrate.

Reverse osmosis plants are available to suit almost any usage rate of water and, apart from the high initial capital cost, are an ideal way to provide pure water for cosmetics processing. The membranes need replacing typically at intervals of 5–10 years for continuous operation.

Microbiological Purification

In the UK, mains water from the tap is far from sterile. Plate counts of between 10^2 and 10^3 organisms per ml are often obtainable and if the supply is from storage tanks (commonly used in the cosmetics industry) these counts can easily reach 10^5 or 10^6 organisms per ml. The types of organisms that will actually thrive in tap water are limited to those of meagre nutritional requirements—largely Gram-negative types of which *Pseudomonas*, *Achromobacter* and *Alcaligenes* are representative. These, however, are precisely the organisms that will proliferate most readily in aqueous-based products such as emulsions.

Other types of organism may, once they have found their way into the water supply, survive the chlorination procedures and adapt to a non-proliferating form until suitable substrates become available—even spore-forming varieties can behave in this way. On leaving the treatment plant, however, the municipal water supply is likely to be free from pyrogens, algae and viruses and therefore the factory supply should also be free of these contaminants unless they are picked up in the supply pipework.

The raw material, before further purification, may therefore already be contaminated with micro-organisms to an unacceptable degree. Passage over ion-exchange resins can cause even higher contamination levels: these resins form an ideal breeding ground for microbes since they contain large areas of thin films of stagnant water. Even the manufacturing plant itself may not be free from contamination—every pump, metering device, joint, pipeline, gauge and valve can provide a stagnant pool of water in which microbial growth can take place.

There are five practical methods for the reduction or elimination of microbiological contamination of water in the cosmetics plant: chemical treatment, heat treatment, filtration, UV treatment and reverse osmosis. These may be used separately or in combination.

Chemical Treatment. Contaminated resin beds and distribution systems can be sterilized and cleaned by the use of dilute solutions of formaldehyde or of chlorine (usually in the form of hypochlorite solution). Before the resins are treated they must be completely exhausted by prolonged contact with brine; if this is not done, formaldehyde is likely to be converted into paraformaldehyde (a polymer) and hypochlorite may give rise to free chlorine gas. The usual method of treatment is to leave the beds in contact with a 1 per cent aqueous solution of either chemical overnight.

Once water has passed through the deionizer, one method of ensuring that micro-organisms cannot thrive in the storage tank or supply system is to dose it with a very low concentration of either disinfectant. Cosmetics plants can be maintained at levels of contamination of less than 100 colony-forming units by dosing the post-deionized storage tank with between 1 and 4 ppm available chlorine (5 ppm is just detectable by smell without any apparent detrimental effect on the vast majority of cosmetic products). In order to achieve this, however, constant monitoring of the levels of chlorine must be carried out and re-dosing effected once the chlorine level falls below 1 ppm.

A less usual means of achieving sterile or near sterile water involves treatment with preservative and heat. For example, a 0·1–0·5 per cent solution of methyl parabens when heated to 70°C gives almost sterile water which can be used for cleaning out plant.

Heat Treatment. The decontamination of process water by heat treatment in the cosmetics industry is carried out most frequently in the process vessel itself. The vessel is charged with the appropriate quantity of water which may be heated to 85°–90°C and held at this temperature for at least 20 minutes. Such treatment is sufficient to eliminate all the common water-borne bacteria but will not destroy spore-formers—in the unlikely event of any being present. (In fact heat treatment such as this is likely to cause any spores to germinate. If spores are suspected to be present, the same heating procedure should be repeated a second time after the lapse of two hours and, for absolute safety, for a third time two hours later still.)

An alternative arrangement is available which heats the water to 120°C in a thin film and then instantaneously cools it again. This is an in-line device known as a UHST treatment unit ('ultra high short term'). It is claimed that such units are able to destroy all bacteria.

In some pharmaceutical manufacturing plants where sterilized water has to be stored, the storage tanks are maintained at a constant heat of 70°C to prevent the growth of any stray contaminants which may have escaped the sterilization procedures.

UV Radiation. Ultraviolet radiation of wavelengths below 300 nm has been shown to destroy most of the micro-organisms that commonly contaminate water—including viruses, bacteria and most moulds. The mechanism by which the organisms are killed appears to be the photochemical effect of such UV radiation on the DNA and RNA content of their thinly protected nuclei. Since light of this wavelength does not penetrate very far through water, however, the supply has to be brought into very close contact with the UV sources to prove effective; in effect, this means that the water has to be spread into a thin film, thus providing a restriction in the supply system and cutting down the flow rate.

Although UV sterilization is a useful means of microbiological control in some types of installation, care must be taken to ensure that the efficiency of the source does not become impaired either by the build-up of slime around the source or the inevitable deterioration of the source itself. As with all cold methods of sterilization, UV radiation is never completely effective—a few organisms usually manage to escape even the most efficient systems and these will thrive and multiply if they are allowed to.

Filtration. In theory, all known bacteria can be removed from water by passing it through a filter having a pore size of 0·2 μm or less. Practically, such filters exist in cartridge form and are sometimes recommended as an in-line method of sterilizing water in cosmetics plants either alone or in conjunction with other means (although 0·45 μm filters are sometimes suggested as a more practical alternative).

While it can be shown that membrane filters can very effectively remove microbiological contamination from water, the method has a number of drawbacks. These filters create a very high resistance to flow and are extremely expensive to replace—running costs can be disproportionately large compared with other methods. A more fundamental objection, however, is that eventually the build-up of organisms caught in the filter matrix increases the resistance to flow of water until eventually the pressure reaches a point at which some organisms break through the membrane or the water ceases to flow at all. Moreover some organisms—particularly moulds—are able to multiply in the membrane matrix and literally grow through to the other side. The rate at which this happens depends upon the volume of water passing through and its level of contamination. Installations employing the constant recirculation principle are especially susceptible, since the constant passage of water through pumps and filters warms the circulating water and speeds up the rate of growth of organisms trapped in the membranes. For these reasons, many people believe that it is fundamentally wrong to use membrane filters to hold back living organisms in this way.

Finally, it should be noted that only distillation, ultra-filtration and reverse osmosis give water which is free from pyrogens.

Distribution Systems

The quality of the water obtainable from a practical piped distribution system depends substantially on the quality of water entering the system (if this is poor, there is no hope of obtaining good quality of water at the point of use), the nature of the materials which come into contact with the water, the design of the system and the maintenance of the system. Of these factors, the first has already been discussed in some detail while the second is usually assumed, for cosmetics purposes, to be relatively unimportant. This is far from true, however.[6–8]

There are two good reasons for concern about the nature of the materials from which the distribution system is constructed: the likelihood of contamination by substances leaching out into or reacting with the water, and the ease of cleaning the system. The ideal material from which to construct pipework for water distribution is probably stainless steel, but the prohibitively high cost normally precludes its use in most cosmetics plants. It is theoretically possible to manufacture pipework from other non-corroding metals, but such materials are also relatively expensive and difficult to join together in such a way as to exclude bacteria and air. Many distribution systems employ plastic pipework, especially pipes fabricated from unplasticized polyvinyl chloride, polypropylene and acrylonitrile–butadiene–styrene (ABS). All three materials suffer the disadvantage of not being able to withstand sterilization by live steam and each contains a variety of additives that could leach out and contaminate the water. Such additives include catalysts, pigments, plasticizers, antioxidants, lubricants, stabilizers, antistatics, high impact modifiers, and monomers or low molecular weight polymers.

In practice, many cosmetics installations have been constructed of plastic with no obvious detriment to the products although it would be foolish to assume that interaction between product and contaminant could never occur.

Physical Layout of the Distribution System

The choice of material from which the pipework is to be made is only one design element of a distribution system which has to be decided upon. The choosing of the physical layout of the system is a very important part of the design process; even the best equipment will create serious problems in use if it is not correctly chosen for the purpose and connected together in a sensible and appropriate manner.

Probably the first step in this process is to decide which type of deionizer is to be used, bearing in mind the quality and quantity of water required, the level of expertise for use and maintenance that is available and the running and capital cost limitations that have to be met. This will to some extent help in the choice of equipment for controlling the microbiological quality of the water: filters, UV, chemical treatment or perhaps a combination. The distance over which the system is to stretch and the minimum number of outlets needed also has to be fixed before the first draft drawings can be made. The design engineer should then be in a position to decide on the fundamental features of the layout and, in particular, whether to choose a centralized or a peripheral purification system and whether to choose a 'ring main' or a 'dead leg' distributive system. These features are illustrated in Figures 42.2–4.

Figure 42.2 represents a conventional ring main, constant circulation system in which the water is continuously circulated through the pipework, the deionizer, the filters and/or UV sterilizer and back to its start point again. Thus circulation takes place 24 hours per day whether or not water is being drawn off, so that the system, if properly designed, has no dead spots where stagnant water can accumulate. By being continuously pumped through the purification system, the water improves in quality with every pass.

Provided that the pipework is properly laid out with no sagging or tight bends and that the take-off valves have a minimum of dead space within them, the constant circulation system has been shown to be very effective in supplying water of good quality. Its chief disadvantage is that the filters clog rapidly if the quality of the raw water coming from the mains is not good to start with. Once bacteriological growth begins in the filter system the increased pressure so generated and the constant passage of water through other restrictions in the layout can lead to a considerable rise in temperature, so encouraging further unwanted growth.

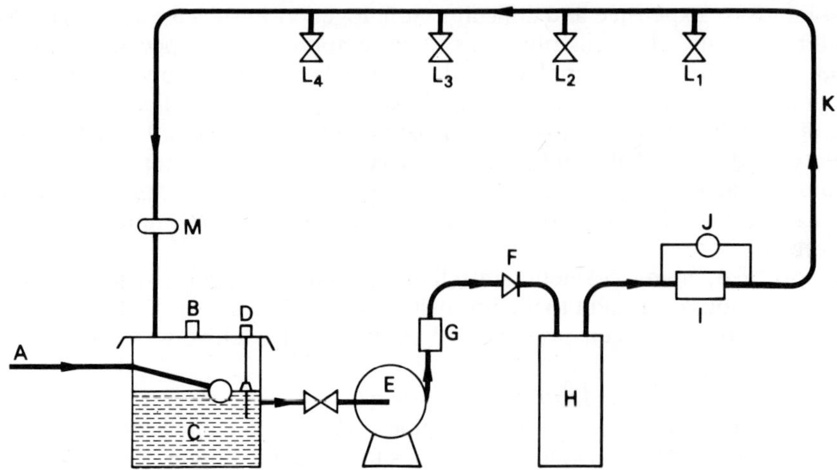

Figure 42.2 Ring main circulation system
The raw water supply (A) feeds directly into a break tank (C) fitted with a ball valve, a sealed top, a bacteriological air-filtered breather tube (B) and low-level float switch (D). The pump (E) pushes the water via a non-return valve (F) through a coarse pre-filter whose purpose is to protect the deionizer (H) from any suspended particulate matter. The deionized water emerges from H via microbiological control (I) (filter system or UV sterilizer) and on to take-off points L_1–L_4, finally returning via a constant-flow tank (M) to the break tank.

The popularity of the ring main system (especially with the suppliers of water purification equipment), has led to the underrating of the 'dead leg' system (Figure 42.3). The obvious disadvantage of such a system is that water lies stagnant in the pipes when it is not actually being drawn. Nevertheless, 'dead leg' layouts have proved their worth in certain circumstances, particularly in small plants where the distribution pipework can be kept short and all the take-off points are frequently used. Figure 42.3 shows two such deionizers installed in parallel. With the proper microbiological and chemical monitoring that is essential for any well-run purified water system, whatever its design, such a 'dead leg' layout can be effective, comparatively trouble-free and inexpensive to run.

For larger plants and where especially pure water is required, a decentralized system such as that shown in Figure 42.4 may be appropriate. In such a system, each take-off point is fitted individually with a deionizer and a microbiological control device. It will be obvious that high installation cost is a serious disadvantage of a layout of this kind; nevertheless in appropriate circumstances it represents the safest and most trouble-free system of all.

The range of possible designs is not limited to the three described here: combinations of all these illustrated layouts can be used as well as additional features such as cooling columns and UHST treatment units. Before installing any system it is wise to contact a reputable manufacturer of water-treatment equipment for advice on the most appropriate design.

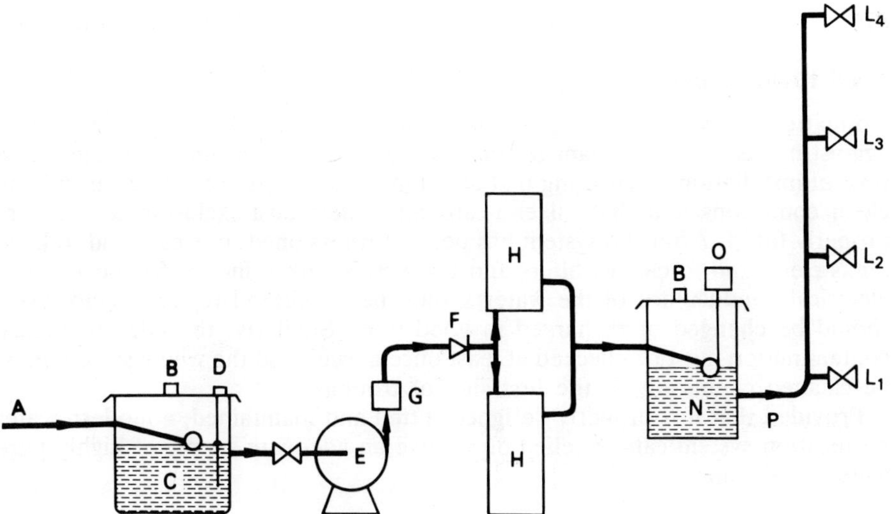

Figure 42.3 'Dead leg' distribution system
Water is pumped into a storage tank (N) fitted with a tight lid, a microbiologically filtered breather (B) and a chlorine dosing unit (O) which keeps the chlorine content of the water constant at 1 ppm. The distribution is gravity fed via a very short run of pipework (P). Before use, sufficient water is run to waste from each take-off point to empty P, thus ensuring that water that has remained stagnant in the part of the system after N is not used.

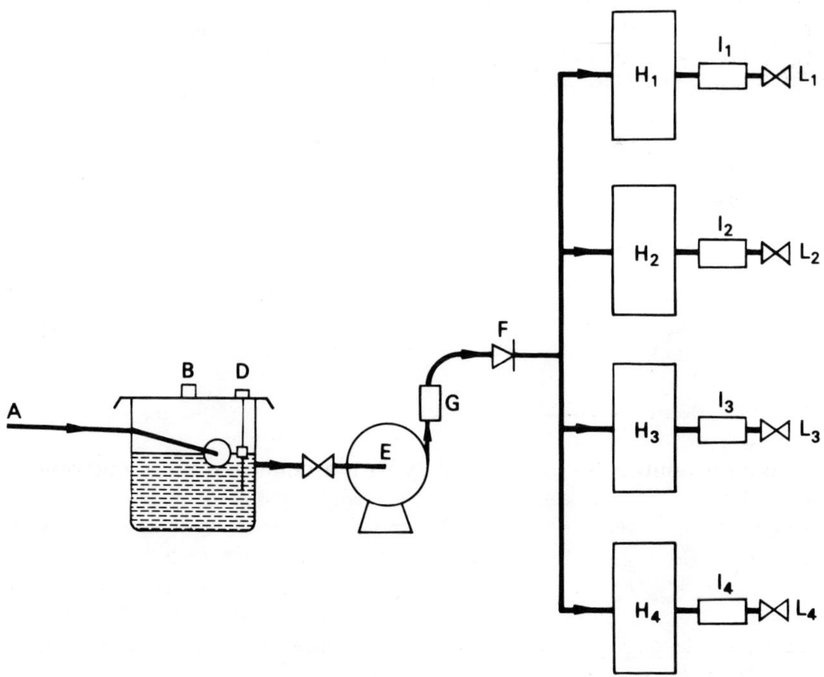

Figure 42.4 Decentralized distribution system (key as Figures 42.2 and 3)

Good Houskeeping

No matter how well it is designed and made, no purified water system is proof against neglect and bad management. Good housekeeping should begin at the time of installation by ensuring that all piping and fittings are stored carefully in clean conditions and that all end-caps and other dust-exclusion devices are properly fitted. After the system has been commissioned, it is essential to keep tanks clean and to change filters and UV lamps with sufficient frequency. The electrical conductance of the water should be monitored regularly and resins should be changed or recharged in good time. Similarly, the microbiological contamination must be checked at least once a week and the whole system must be cleaned chemically at the first sign of trouble.

Provided that it is properly designed, fitted and maintained, a modern water purification system can be relied on to give an adequate supply of highly pure water at all times.

REFERENCES

1. Shvedov, A., *Med. Tek.*, 1974, 36.
2. Reid, C. E. and Lonsdale, H. K., *Industrial Processing with Membranes*, ed. Lacy, R. E. and Loeb, S., London and New York, Wiley-Interscience, 1972.
3. Loeb, S. and Sourirajan, S., *Adv. Chem. Ser.*, 1963, **38,** 117.
4. Carter, J. W., Psaras, G. and Price, M. T., *Desalination*, 1973, **12,** 117.
5. Carter, J. W. and De, S. C., *Trans. Inst. Chem. Eng.*, 1975, **53,** 16.
6. Packham, R. F., *Water Treat. Exam.*, 1971, **20,** Parts 2 and 3.
7. Goodhall, J. B., *Manuf. Chem. Aerosol News*, 1973, **44**(9), 58.
8. Conacher, J. G., *Filtr. Sep.* 1976, **13,** 251.

Chapter Forty-three

Cleanliness, Hygiene and Microbiological Control in Manufacture

Introduction

The control of quality begins before any material is purchased, continues throughout manufacture, assembly and distribution and cannot be 'inspected into' a product at the end of the manufacturing process. Products of high microbiological quality do not just happen; they are so designed from the earliest stages of manufacture. Frequent occurrence of low-level microbial contamination of the finished product is a warning that the equipment on which it was made may not have been adequately sterilized. Finished product testing is a measure (and only a measure) of good manufacturing practice. Appropriate precautions should be taken against product contamination risks of all kinds. Insanitary processing and filling equipment can also be detrimental to the finished item. Dust, dirt or irregular air flow in the manufacturing area can easily compromise an otherwise acceptable product[1] and there should be a cleaning routine for all equipment and manufacturing areas. Toilet and washing facilities must be appropriately located, designed and equipped.

The following are some of the essential requirements for manufacturing a quality cosmetic:

- The maintenance of clean premises
- Attention to personal hygiene of operatives
- Development of an effective cleaning and sterilization programme
- Continuous monitoring of the water supply
- Adherence to rigid microbiological criteria for raw materials
- Incorporation of an adequate preservative system
- Continuous microbiological monitoring of all stages of the cosmetic in production

The presence of micro-organisms in large numbers in cosmetic preparations is undesirable because spoilage of the product may result. The product can change in colour, odour or consistency, or manifest visible growth. Furthermore, the presence of microbial contaminants constitutes a potential hazard to public health, although reports of infections traced to contaminated cosmetics are rare. These problems were emphasized in the early 1970s by several surveys of cosmetic products on the US market; one showed a high proportion to be contaminated, some with bacteria counts as high as 12×10^6 per ml. A survey by the US Food and Drug Administration also showed a fairly high rate of heavy microbial contamination.[2]

SOURCES OF CONTAMINATION

Environment

Control of the manufacturing environment significantly reduces the risk of product contamination. Floors and wall surfaces should be impervious and resistant to antimicrobial agents. Walls should be washed periodically and floors should be washed each night. Puddles, dirt and debris should be removed as soon as practical. Drains on the floor should be kept covered and should receive particular attention as they can easily become sources of contamination. Air flow through production areas should be minimized. Air ducts, light fittings and other piping can create problems unless they are routinely cleaned. In general, an acceptable and effectively controlled production environment requires all exposed surfaces to be kept clean. Other potential hazards of contamination are cleaning utensils—buckets, mops, brushes, etc., should themselves be kept in a clinically clean condition. Sinks and areas where equipment is washed should be kept clean at all times. Traffic through production areas must be restricted to essential personnel and necessary equipment.

The factors mentioned above must be carefully considered as all can influence the microbial content of the environment. Although the product itself may be protected by the use of preservatives, the load of micro-organisms present in the environment of manufacture may affect the long-term activity of the preservative system. The presence of large numbers of micro-organisms in the manufacturing atmosphere will shorten the period of preservative efficacy. A product manufactured in a contaminated environment will have a reduced preservative activity and as a result, when some of the contents are withdrawn from the pack, further contamination may occur and present a hazard to the consumer.

Factory Buildings

Buildings should be constructed to protect as far as possible against the entrance and harbouring of vermin, birds and other pests. The buildings should be effectively lit and ventilated with air control facilities appropriate both to the operations undertaken within them and the external environment. Certain areas should not be used as a general right of way for materials or personnel passing through to other parts of the factory. Laboratories, processing areas, stores and the factory in general should be maintained in a clean, neat and tidy condition, and free from accumulated waste. Waste material should be collected into suitable receptacles for removal to collection points outside the buildings. It should be then disposed of at regular and frequent intervals. The operations carried out in any particular area of the premises should be such as to minimize the contamination of one product by another.

Floors and Walls

Floors should be made of impervious materials, laid to an even surface, and free from cracks and open joints. They should be maintained in a good state of repair. Dust, dirt and other contaminants that may be carried into manufacturing and filling areas on the soles of shoes and on trolley wheels can be collected on special plastic screens that pick up such material on their surface, have good non-slip properties and are easy to maintain.

Walls and ceilings should be finished with a smooth, impervious and washable surface and maintained in a good state of repair. Pipework should not contain uncleanable recesses and should be effectively sealed into walls and partitions through which it passes. Extraction fans should be sited to avoid cross-contamination hazards caused by intake or exhaust.

Equipment

Frequently, equipment is purchased with little regard for ease of clean-up or sanitization after breakdown. Before new equipment is bought, not only should rate of output be considered but also the susceptibility of the product to contamination. A piece of equipment may be installed in a manufacturing or filling area where there is no one who really knows how to dismantle, clean or sanitize it properly.

Hygienic considerations are often neglected in equipment design. Cooperation between production managers and microbiologists is essential and this applies even to minor modifications to existing equipment. Equipment should be made of materials capable of withstanding conventional cleaning methods, such as the use of steam, detergents and chemical antimicrobial agents; the most efficient and widely used material is stainless steel. Other essential components of production machinery such as gaskets, fittings, tubing and rails should also receive proper attention. Hoses that are old, cracked and decaying and equipment with crevices and corners are nearly impossible to clean and sanitize properly. Pumps that are old or frequently taken apart can be a major source of contamination. Such hazards can be avoided by establishing routine cleaning and maintenance procedures. Discretion and a general awareness of all aspects of sanitary hygiene on the part of the microbiologist and the engineering personnel enable appropriate equipment choices to be made and routine maintenance procedures to be established, so that unnecessary clean-up and prolonged stoppages in processing and filling are avoided.

The Manufacturing Process

Potential sources of microbial contamination during the manufacturing process can be summarized as follows:[3]

	Cause of contamination
Storage	Airborne micro-organisms
	Inadequate cleaning of vessel/tank
Filling	Airborne micro-organisms
	Transfer via filling machine
	Transfer via container
	Transfer via operator

Unsuspected Sources of Contamination

Where all reasonable precautions have been taken but sporadic and serious outbreaks of infection still occur, the source is likely to be an unsuspected reservoir of contamination. The underlying cause is probably among the following:

Poor communication between management and staff
Poor supervision, especially early in the morning
Poor hygienic design of equipment or layout
Changes in cleaning/sterilizing procedure, introduced to reduce costs
Rapid staff turnover
Assumptions made without verification by laboratory test

CLEANING AND DISINFECTION

The utmost care taken in formulating a quality cosmetic, in purchasing raw materials to high specification and marketing elegant cosmetics can be defeated by failure to exercise the same strict control over the cleaning and sanitation of buildings, plant and equipment and over the conditions of bulk manufacture, filling and storage of products. Effective cleaning, when viewed as part of the normal production cycle, requires not only a detailed knowledge of the manufacturing operations and plant but also an appreciation of all aspects of cleaning. The decisions to be made relate to the labour requirement, the cleaning equipment, the detergents and the standard of cleanliness required. The following important considerations must be taken into account:

1. The quality and therefore the value of the end product depends essentially on the cleanliness of the production plant.
2. The sterilization capacity of the product is a function of the initial microbial count.
3. The shelf life of creams, lotions and powders that are made from sterilized ingredients is affected by reinfection during manufacture and filling.
4. Production of pathogenic micro-organisms such as staphylococci, *Pseudomonas*, coliforms, clostridia and *Candida* must be prevented.

Cleaning schedules can only be established by careful study of the problem, an appreciation of what is required, and a knowledge of how cleaning and sterilization are best carried out.

The Cleaning Logbook
Despite the exercise of every care in the cleaning and sterilization of equipment and plant, unexpected problems and unusual situations may occur. These can be detected easily by the senior staff of the production area if the senior cleaner or the operatives in manufacturing areas are responsible for keeping a logbook of daily occurrences reported. The supervisor or manager responsible can quickly make himself familiar with any variation from the expected pattern and take remedial action. When poor microbiological results are obtained from manufactured bulk or filled finished product, reference to the cleaning logbook can help to identify the cause if it lies in a variation in cleaning procedure.

Cleaning Staff

Because of the failure to appreciate that efficient cleaning can materially contribute to the total profitability of the business it has too often been a custom in the past to employ for this purpose labour considered unsuitable for other activities. This is a grave error, for failure by these employees in the intelligent performance of their duties can do greater harm to a company's products and reputation than a mistake by the more highly regarded production line employees. Staff of suitable quality should be engaged, their role in the business should be explained to them and they should be instructed in the type of plant employed in the factory, and the cleaning equipment, procedures and materials to be used.

Cleaning staff should be supplied with overalls, rubber gloves, rubber boots and hats. Arrangements should be made to replace this equipment where necessary with adequately clean items in order to minimize the possibility of cross-infection from dirty clothing. The cleaner should be provided with a composite cleaning kit comprising brushes, abrasives, detergents and sterilants. Cleaning procedures will not be carried out correctly and within reasonable time if the cleaner has to interrupt his work to look for cleaning materials, and a trolley to carry this composite cleaning kit is strongly recommended. Also advantageous would be an equipment cupboard with glass sliding doors for stocking cleaned equipment; cleaning personnel should be given the responsibility for maintaining it in a scrupulously clean condition.

Many of the cleaning and disinfecting materials described later in this chapter are potentially hazardous, so it is essential to instruct and train cleaning personnel to handle them in a completely safe manner. Supervisors must know about the toxic properties and dangers of the chemicals used and be aware of suitable treatments for injuries arising from mishandling.

Equipment Cleaning

However well designed the equipment may be, unless the method of cleaning is effective all the design efforts will be nullified. Establishment of an effective cleaning method requires:

(a) An appreciation of the type of plant and equipment used, with special reference to the type of contamination likely to occur and the consequences of failure to remove it.
(b) Instruction in the need for a planned and rational approach to cleaning.
(c) The selection and use of appropriate cleaning equipment.
(d) The selection and use of correct detergent concentrations.
(e) The selection and use of correct sterilant and sterilant concentrations.

The method to be employed to clean a specific item of equipment will vary according to the nature of the product being processed and the quality of the surfaces, but certain general principles can be established (Figure 43.1). The equipment should first be dismantled and all product residues removed. Any part of the equipment that comes in contact with the product should be removed and cleaned: all processing vessels, pumps, hoses, valves, storage cans, various hand utensils, and the filling equipment should undergo this cleaning process. Inaccessible parts of this equipment should be given extra attention. This preparation may take a long time, but it is well worth the effort if a serious contamination problem is thereby avoided.

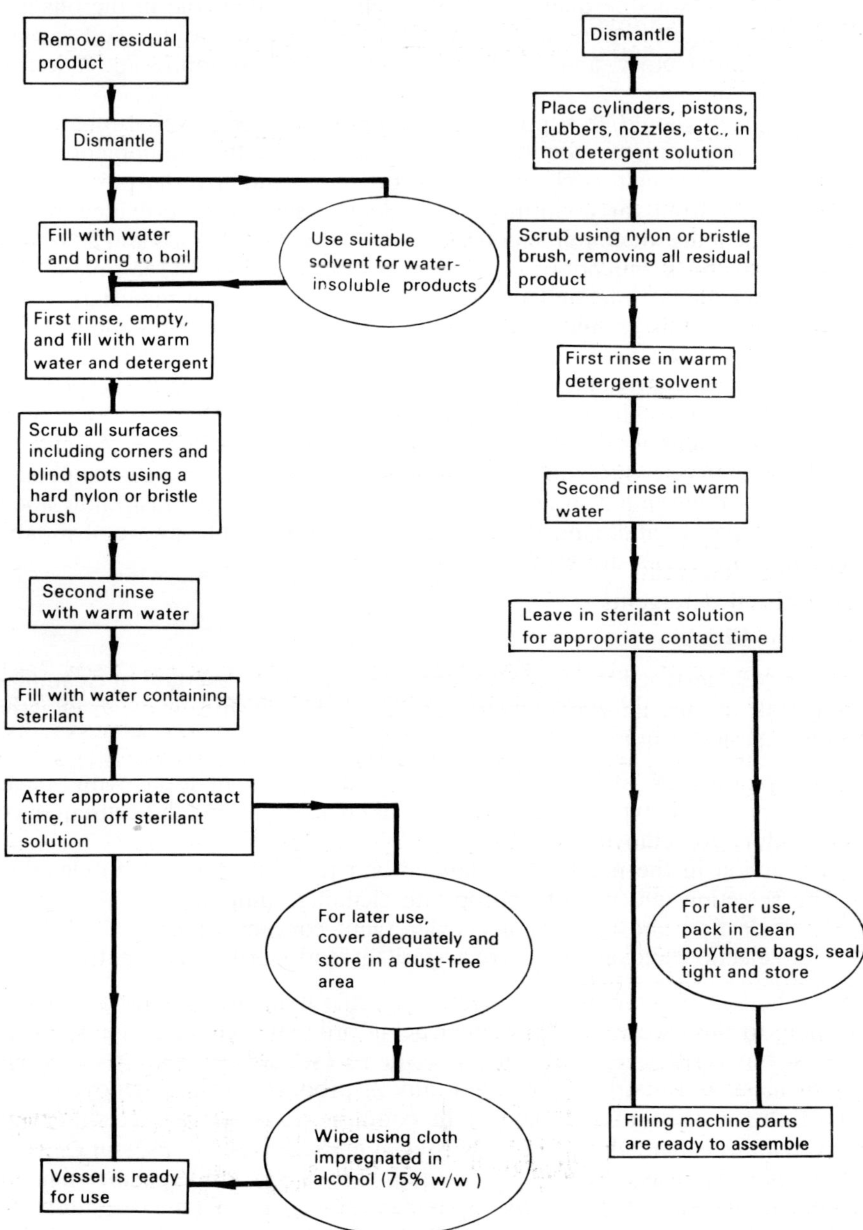

Figure 43.1 Sequence of equipment cleaning and sterilization operations

Detergents

Detergents employed in cleaning plant and equipment must be able:

(i) thoroughly to wet the surface to be cleaned;
(ii) to remove the residual product from the surface;
(iii) to hold the removed product in suspension;
(iv) to resist deposition of the residual product on dilution, that is, have good rinsability.

In addition, other factors need to be considered in choosing a suitable detergent, such as:

Prevention of corrosion
Dissolving and emulsifying action on cosmetic solids, that is, emulsification of dried cosmetic product.
Prevention of scale formation
Quick and complete solubility
Economy in use

No single detergent or class of detergents can meet all these requirements, and hence a vast range of detergents is offered for both domestic and industrial purposes. They fall into four main categories:

Soaps
Synthetic soapless detergents
Alkaline detergents
Specialized detergents

Soaps. Soaps are produced in flake, bar, tablet, liquid and powder form. Flake and powder soaps are used for laundry processes, tablet soap for personal hygiene, liquid soap for cleaning equipment, plant, etc. Soaps, however, lack the powerful wetting action of synthetic detergents and the powerful dissolving properties of alkalis.

Synthetic Soapless Detergents. Synthetic detergents are divided into four classes:

Cationic detergents
Anionic detergents
Nonionic detergents
Amphoteric surface-active agents

Cationic detergents are not normally used for detergent purposes but some are employed as sterilizers, since they possess bactericidal properties. The quaternary ammonium compounds belong to this group.

Anionic detergents, to which class belong the sulphated fatty alcohols and alkyl aryl sulphonates, are the most widely manufactured synthetic detergents.

Nonionic detergents have certain unique properties. In general they are unaffected by any ionic reactions, for example they do not react with the hardness of water. Foaming is generally less than with the anionics and their ability to remove oily product residues is excellent.

Amphoteric surface-active agents: amphoteric compounds exhibit the property of being able to function either as acids or as bases—the electrochemistry of the molecule depends on the pH of the medium in which it is present. They exhibit a very marked degree of detergent activity. In working solutions amphoteric compounds produce good wetting and penetrating effects and efficient grease-dissolving and cleaning action. Solutions at 0·5–1·0 per cent in water show adequate microbiocidal activity and therefore these are widely used as sterilant–disinfectants.

Synthetic detergents lack the powerful dissolving properties of alkalis.

Alkaline Detergents. Alkaline detergents (for example, solutions of sodium hydroxide or alkaline phosphates used with treatment times of at least 30 minutes) are very effective in removing hard product residues and therefore are invariably used in machine and plant cleaning where foam is a disadvantage. They do not, however, possess the same wetting properties as synthetic detergents or the same emulsifying properties as soap.

Although the first detergents ever to be used were alkaline, their continued use in the cleaning field is a clear indication of their adaptability and efficiency.

Specialized Detergents. These include the detergent sanitizers which are designed to incorporate bactericidal properties with the other desirable detergent features.

Bactericidal Action of Detergents. Many detergents have marked germicidal properties although they are used primarily as detergents. Hot water at 60°–80°C will kill most or all vegetative cells but few spores. A detergent will always enhance the killing effect of heat; probably the best example is sodium hydroxide: a treatment at 63°C for 30 minutes in water will kill all bacteria except thermoduric ones and spores, but a 1–3 per cent NaOH solution under these conditions will kill all thermoduric cells and a considerable proportion of spores. Detergents are nearly always used hot and so act by enhancing the bactericidal effects of heat. This effect is especially valuable against spores in those industrial applications, for example bottle-washing followed by cold filling, in which excessive temperatures have to be avoided. Detergent–sterilants are particularly useful when high temperatures cannot be used, as in manual dishwashing, or because of delicacy of materials. Even apparently innocuous detergents such as anionics, nonionics and trisodium phosphate exert a powerful killing effect against most pathogens but not against tubercle bacilli or spores.

Equipment Disinfection

Once equipment is thoroughly cleaned, the actual disinfection process can be initiated, using steam or chemical sterilants.

Disinfection by Steam

Steam is the most effective and reliable means of sterilizing equipment but the steam tolerance of equipment material should be checked before steam is used as a sterilizing agent. As with other sterilizing agents, contact time is important.

The minimum temperature at the exit end of the steam generating system should be in the range 72°–80°C. Contact times for open vessels are normally 30 minutes or more, since the steam is at zero pressure. For closed systems, contact times vary according to steam pressure; for example, at pressures of about 1–5 psi (5–100 kPa), a 20 minute contact time is necessary. At higher pressures the contact time may be as little as 5 minutes. In addition, the steam used must be free from particulate and other extraneous matter. This will eliminate the problem of residues remaining after the sterilization process.

Disinfection by Chemical Sterilants

At sufficiently high concentrations many chemicals, including nutrients such as oxygen and fatty acids, are bacteriostatic and even bactericidal. The term 'disinfectant' is restricted to substances that are rapidly bactericidal at low concentrations. Most sterilizing agents act either by dissolving lipids from the cell membrane (detergents, lipid solvents) or by damaging proteins (denaturants, oxidants, alkylating agents and sulphydryl reagents). The rate of killing by disinfectants increases with concentration and with temperature. Anionic compounds are more active at low pH and cationic compounds at high pH. This effect results from the greater penetration of the undissociated form of the inhibitor, and possibly also from the increase in opposite charges in cell constituents. Strongly acid and alkaline solutions are actively bactericidal. Weak acids exert a greater effect than can be accounted for by pH: the presence of highly permeable undissociated molecules promotes penetration of the acid into the cells, and increasing activity with chain-length suggests that direct action of the organic compound itself plays a part. Lactic acid is the natural preservative of many fermentation products, and salts of propionic acid are now frequently added to foodstuffs such as bread to retard mould growth. Halogens such as iodine and chlorine combine irreversibly with proteins and they are oxidizing agents. Chlorine was the antiseptic introduced as chlorinated lime by O. W. Holmes in Boston in 1835, and by Semmelweis in Vienna in 1847, to prevent transmission of puerperal sepsis by the physician's hands. Chlorine is a reliable, rapidly acting sterilant–disinfectant for 'cleaning' materials but it is less satisfactory for materials that are subject to attacked by chlorine.

Alkylating Agents. Suitable dilutions of formaldehyde and of ethylene oxide in carbon dioxide (in appropriate gastight enclosures) replace the labile H atoms on—NH_2 and—OH groups which abound in proteins and nucleic acids, and also on —COOH and—SH groups of proteins (Figure 43.2). The reactions of formaldehyde are in part reversible, but the high-energy epoxide bridge of ethylene oxide leads to irreversible reactions. These alkylating agents, in contrast to other disinfectants, are nearly as active against spores as against vegetative bacterial cells, presumably because they can penetrate easily (being small and uncharged) and do not require water for their action.

Phenols. Phenol itself is both an effective denaturant of proteins and a detergent. Its bactericidal action involves cell lysis. The antibacterial activity of phenol is increased by halogen or alkyl substitutes on the ring, which increase polarity of the phenolic OH group and also make the rest of the molecule more

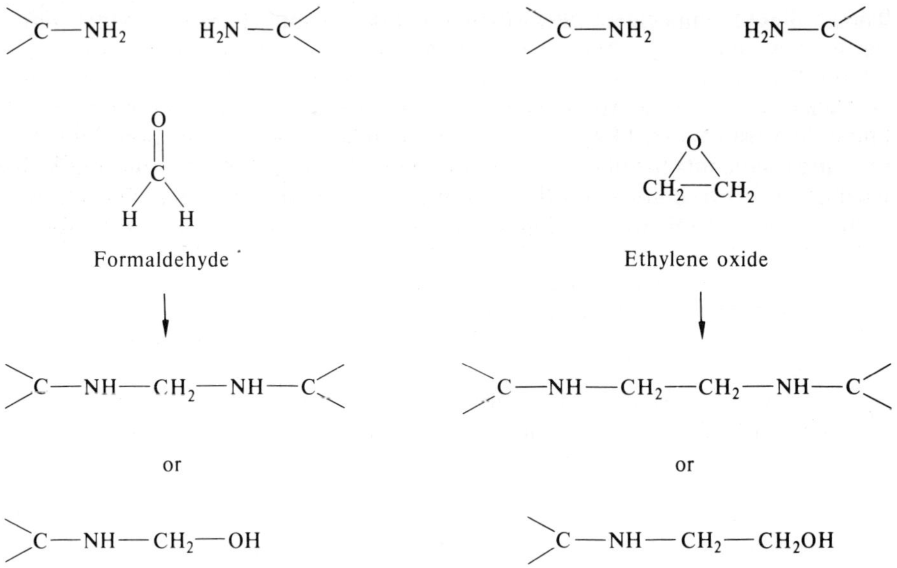

$>C-NH_2 \quad H_2N-C<$ + $HCHO$ (Formaldehyde) $\rightarrow$ $>C-NH-CH_2-NH-C<$ or $>C-NH-CH_2-OH$

$>C-NH_2 \quad H_2N-C<$ + CH_2-CH_2 (epoxide, O bridging) (Ethylene oxide) $\rightarrow$ $>C-NH-CH_2-CH_2-NH-C<$ or $>C-NH-CH_2-CH_2OH$

Figure 43.2 Reactions of formaldehyde and ethylene oxide with amino groups. Bridges may be formed between groups on the same molecule or on different molecules

hydrophobic; the molecule becomes more surface-active and its antibacterial potency may be increased a hundred-fold or more. Phenols are more active when mixed with soaps, which increase their solubility and promote penetration. However, too high a proportion of soap impairs activity, presumably by dissolving the disinfectant too completely in soap micelles. With increasing chain-length the potency of phenols first increases and then decreases, presumably because of low solubility; with Gram-negative organisms the maximum is reached at a relative short chain-length.

Alcohols. The sterilant–disinfectant action of the aliphatic alcohols increases with chain-length up to 8–10 carbon atoms, above which the water solubility becomes too low. Although ethyl alcohol has received the widest use, isopropyl alcohol has the advantages of being less volatile and slightly more potent. The sterilant–disinfectant action of alcohols, like the denaturing effect on proteins, involves the participation of water. Ethyl alcohol is most effective in 50–70 per cent aqueous solution; at 100 per cent it is a poor sterilant, in which anthrax spores have been reported to survive for as long as 50 days, and its bactericidal action is negligible at concentrations below 10–20 per cent. Some organic disinfectants such as formaldehyde and phenol are less effective in alcohol than in water because of the lower affinity of the disinfectant for the bacteria relative to the solvent. On the other hand, alcohol removes lipid layers that may protect skin organisms from some other disinfectants.

Other Chemical Sterilants. Organic solvents such as ether, benzene, acetone and chloroform also kill bacteria but are not reliable disinfectants. Glycerol is bacteriostatic at concentrations exceeding 50 per cent and is used as a preservative for vaccines and other biologicals, since it is not irritating to tissues.

Propylene glycol and diethylene glycol reduce the bacterial count in air when dispersed in fine droplets at concentrations non-toxic to man, but their activity is unfortunately highly sensitive to humidity; at high humidities the glycol droplets take up water and become too dilute, while at low humidities the desiccated bacteria no longer attract glycols. Moreover, the glycols do not disinfect surfaces such as floors, walls or working surfaces, from which aerial contamination is renewed. The chemical sterilants mentioned above are widely used in the cosmetics industry. They are generally used on equipment that cannot tolerate steam or to which steam cannot be applied.

Chlorine is commonly used in the food and cosmetics industries, usually by employment of hypochlorites or chloramines, and has a relatively broad spectrum of activity. It is fairly inexpensive but it is corrosive and is not a cleaning agent. Equipment must be thoroughly cleaned or the chlorine will be inactivated.

Iodine-containing sterilants are useful and can be formulated with detergents in order both to clean and to sterilize. The disadvantages of materials containing iodine are their poor rinsability and their less than satisfactory effect against certain micro-organisms. Staining of equipment components may also be a problem.

Quaternary ammonium compounds are odourless and less corrosive than some of the other chemical agents. Their antimicrobial effectiveness is slightly less than that of alternative materials, but when used at appropriate concentrations and elevated temperatures they can be satisfactory. Quaternary ammonium compounds have the advantage of being non-toxic and are therefore more easily handled than other sterilizing agents.

Detergent sterilizers work more efficiently at higher temperatures, requiring a lower concentration of sterilant. At lower temperatures, on the other hand, not only may a higher concentration be needed but also increased contact time. Different types of surface may require different contact times with the sterilant; smooth, non-porous surfaces require less time than moving parts. Processing systems containing moving parts that are not always in contact with the sterilant will require frequent cycling of the parts during the exposure time to give adequate contact time.

Table 43.1 Properties of Some Major Classes of Antimicrobial Chemicals

Property	**Inorganic hypochlorites**	**Detergent quaternaries**	**Detergent iodophors**
Broad microbial spectrum	Yes	No	Yes
Activity against pseudomonads	Yes	No	Yes
Product stability	Limited	Yes	Yes
Interference by water hardness	No	Yes	Yes
Interference by alkalinity	Yes	No	Yes
Odour (use dilution)	Yes	No	Slight
Visual indicator of activity	No	No	Yes

Selection of the appropriate sterilant for sterilizing equipment is a difficult problem. The first step in the evaluation of a sterilant is to review the technical literature and set out its advantages and disadvantages. In the microbiological laboratory basic cleaning procedures, such as the determination of minimum inhibitory concentrations, should be conducted using a wide variety of possible contaminants and these should preferably be the types encountered in cosmetic products. The results will give information on both the spectrum of effectiveness and the potential range of concentration required for the sterilant. Once an effective concentration has been established, additional testing using equipment components can be conducted. From further investigations into temperatures and contact times, a degree of sterilant effectiveness may be established. Comparative testing of two or more sterilants or conditions of sterilization can also be conducted easily.

Once the new sterilant has demonstrated acceptable ability, it should be experimentally evaluated on full-scale equipment. Since all conditions of use have been previously established, this final evaluation should not create substantial difficulties.

Regular use of a single type of sterilant can result in the development of strains of organisms resistant to the antimicrobial agent in use. Over long periods of time organisms that were once eliminated are found to have adapted and to survive; this adaptation of equipment-borne micro-organisms can be prevented by the rotation of sterilants. Regular use of a different sterilant will eliminate the occurrence of resistant organisms. For instance, a quaternary sterilant may be used for a given period of time, followed by a chlorine-based sterilant for a further period of time. However, if steam is used as the sterilant, rotation is not normally necessary. In order to avoid serious equipment-borne contamination particular attention should be given to the phenomenon of microbial adaptation.

Parameters of Cleaning, Disinfection and Rinsing

Cleaning. In this procedure the dirt or the product residue is separated from the surfaces through the combined action of cleaning agent and mechanical energy. Investigations show that the removal of product residue or dirt proceeds formally as a chemical reaction of the first order. If m_A is the quantity of residue per unit surface area and t is the time taken to remove it, the relationship between them can be described as $dm_A/dt = -k_R m_A$, where k_R is a rate constant describing the velocity of cleaning. Thus a plot of t versus log m_A yields a straight line (Figure 43.3). This behaviour could be due to the removal of the residue in layers, there being only weak cohesive forces between the layers, but in the end the stronger adhesion force of the final residual layer to the walls of the vessel must be overcome.

The value of the constant k_R depends on several parameters:

1. Type and concentration of cleaning agent
2. Material and state of surface
3. Type and state of dirt or residue
4. Temperature of cleaning solution
5. Mechanical energy supplied.

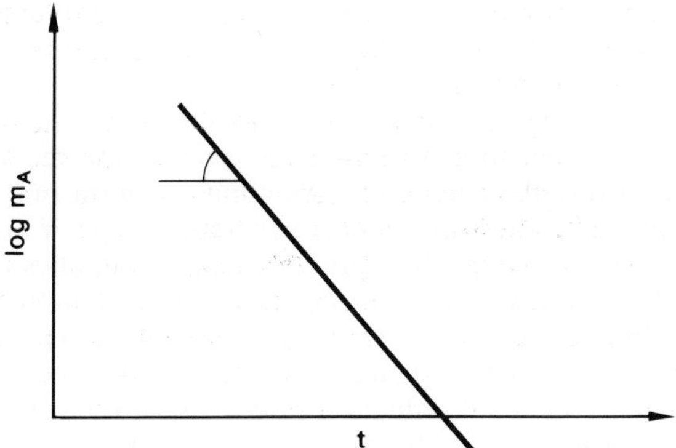

Figure 43.3 Kinetics of product residue removal: $dm_A/dt = -k_R m_A$

Schlussler[4] found that for many cleaning agents there is an optimum concentration at which the cleaning speed (defined as the amount of dirt or residual product removed in unit time) is maximal. This means that an overdose can hinder the cleaning action just as much as an underdose. It is well known that smooth surfaces can be cleaned more easily than rough surfaces. It is not only the depth of roughness, but also the structure of the roughness which influences the cleaning speed.

There are essentially two types of cleaning procedure: open cleaning and closed cleaning. In these two methods the same sequence of preparation is followed—firstly, removal of residual product and, secondly, disinfection followed by finally rinsing. When there is heavy residue to be removed there may be rinsing before disinfection; this sequence will avoid residues and detergents affecting the disinfectant. If the residue to be cleaned is light and the bacterial count is low, cleaning and disinfection can be carried out in one step.

(a) Open cleaning is used on open surfaces, such as working areas, open vessels and manufacturing equipment such as spatulas and stirrer blades. Strong foaming detergents can be used. This has the advantage that the cleaning foam can work longer on the surface than a quickly disappearing liquid film. The final outcome of cleaning operations in open cleaning, which forms the bulk of the cleaning in any industry, depends on the understanding of the procedure and the care taken by the cleaning personnel.
(b) Closed cleaning or CIP (cleaning in place) is used for plant and storage vessels that are connected together and consist of pipes and tanks. Here the cleaning or disinfectant solutions are held in storage tanks and supplied through the plant to be cleaned. Cleaning or disinfectant solutions are applied with a turbulence, often at high pressures in the form of a spray. CIP avoids operator errors but, because of the more complex cleaning machinery, there is an increased tendency for machine wear and, therefore, failure of some parts of the plant.

Disinfection. For disinfection of solid surfaces two processes are in general use: (i) the application of heat and (ii) the use of chemicals, when an aqueous disinfecting solution is pumped through the plant or straight on to the surface. In order to kill the micro-organisms either the lethal temperature or the lethal dose of the disinfecting agent must be exceeded.

Kinetics of disinfection. Investigations by Prado[5] and Han[6] on the killing of micro-organisms on aluminium foil have shown that for the killing of micro-organisms on solid surfaces the same laws apply as those that have long been known for the killing of micro-organisms in suspension. If N_A is the number of micro-organisms per unit area, then the following law applies both for thermal and for chemical death: $dN_A/dt = -kN_A$. If log N_A is plotted against time, t, a straight line is obtained (provided that the temperature and concentration of disinfectant remain constant) (Figure 43.4). Under these circumstances, it is possible to describe a 'decimal reduction value', D, this being the time taken to reduce the population of micro-organisms to one-tenth of its initial concentration. It is possible to show that $D = 2{\cdot}3/k$. When the temperature is allowed to vary, D varies inversely with it. The temperature rise, Z, at which the value of D is reduced to one-tenth of the initial value, is also a useful practical parameter. For most micro-organisms Z is about 5°C, and for spores it is about 10°C.[7]

The effect of disinfecting agents can increase for a given surface through increasing concentration, but for disinfectants with active chlorine there is a corrosion-limited upper concentration. Increase in temperature enhances the action of the disinfectant. For a given temperature, a separation into thermal and chemical action is no longer possible. The effect of disinfectants also depends on the pH value. Active chlorine compounds develop their greatest effect when neutral, and iodophors are most effective at an acid pH.

The condition of the surface has a strong influence on disinfection. The rougher the surface, the greater is the danger of formation of nests of micro-organisms in the pores, and the more difficult is the disinfection. The presence of organic materials will greatly reduce the action of a disinfectant, and so pre-cleaning is necessary for efficient disinfection. Apart from removing undesirable organic material, pre-cleaning also reduces the bacterial population by washing out the micro-organisms.

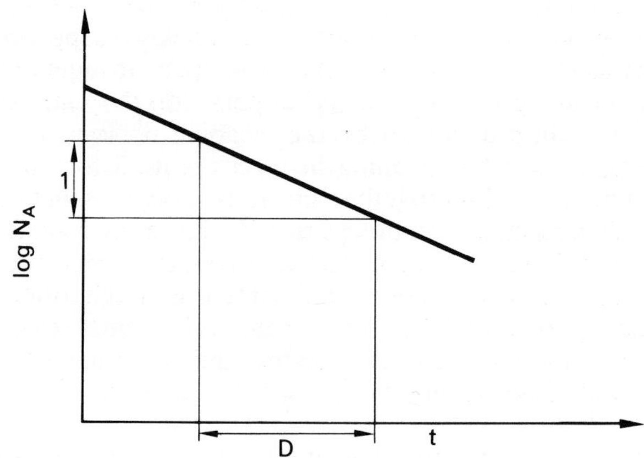

Figure 43.4 Kinetics of disinfection: $dN_A/dt = -kN_A$; $D = 2{\cdot}3/k$

CONTROL OF CONTAMINATION

Hazards from Personnel

Eating, drinking and smoking should be forbidden in all production areas, and high standards of personal hygiene should be observed by all persons concerned in production processes; direct contact between materials and operators' hands should be avoided by the use of correct and suitable equipment. All operatives should wear protective garments appropriate to the process being carried out. The garments should be regularly and frequently laundered. Persons not regularly employed in a production area, whether employees of the firm or not, should wear protective garments where appropriate and necessary. No person known to be suffering from a disease in communicable form, or to be the carrier of such disease, and no person with open lesions or skin infection on the exposed surface of the body, should be employed on production processes.

The bacteria that live on human skin are adapted to the skin habitat and use skin secretions such as sebum and sweat as food; furthermore, pores and cavities function as shelter. Skin bacteria such as *Staphylococcus epidermidis* and *Corynebacterium acnes* adhere to epithelial layers that form the cornified skin surface, and extend between the squames and down the mouths of the hair follicles and glands opening onto the skin surface. These bacteria can only be reduced in numbers, but never eliminated, by scrubbing and washing. They produce odoriferous substances by metabolizing secretions of the apocrine sweat glands, giving the body a smell that modern man, at least, finds offensive.

The ability of micro-organisms to spread from one host to another is of great importance for bacterial distribution and their eventual survival. Clearly if micro-organisms do not spread from individual to individual they will die with the host and will be unable to persist in nature. A classical example of this distribution is the spread of respiratory disease from one individual to a score of others in the course of an innocent hour in a crowded room. Thus, a successful parasitic micro-organism is one that lives on or in the individual host, multiplies, spreads to fresh individuals, leave descendants, and from an evolutionary point of view avoids its extinction and that of its host. On the other hand, if an infection is too lethal or crippling, there will obviously be a reduction in numbers of the host species and thus in the numbers of the micro-organisms. Although a few micro-organisms cause disease in the majority of those infected, it is to be expected that most are comparatively harmless, causing even no disease, or disease in only a small proportion of those infected. Successful parasites cannot afford to become too pathogenic; some degree of tissue damage may be necessary for the effective shedding of micro-organisms to the exterior, as for instance in the flow of infected fluids from the nose in the common cold or from the alimentary canal in infectious diarrhoea, but otherwise there is ideally very little tissue damage. A few microbial parasites achieve the supreme success of causing zero damage, thus failing to be recognized as parasites by the host. It is the virulence and pathogenicity of micro-organisms, their ability to kill and damage the host, that makes them important to the medical scientist and to the microbiologist.

Skin bacteria are mostly shed attached to desquamated skin scales, and an average of about 5×10^8 scales, 1×10^7 of them carrying bacteria, are shed per person per day. The fine white dust that collects on surfaces in hospital wards and in bedrooms consists to a large extent of skin scales. Shedding also takes place from the nose and notably from the perineal area. Potentially pathogenic *Staphylococcus aureus* colonizes the nose, fingers and the perineum. The effect of shedding can be reduced by wearing suitable clothing. A good staphylococcal shedder can raise the staphylococcal count in the air from less than 36 per m^3 to 360 per m^3. It is not known why only some individuals are profuse shedders, but the phenomenon is significant in cross-infection in hospitals and other areas where cross-contamination is critical. Many micro-organisms are effectively transmitted from faeces to mouth after contamination of water used for drinking. In the bowel *E. coli*, *Cl. welchii*, and *Cl. tetani* are nearly always present. All these organisms are potentially pathogenic, and there can be little doubt that they can very easily be transferred and cause contamination and infection. Microbial transmission by the respiratory route depends on the production of aerosols containing micro-organisms. These are air-borne particles produced to some extent in the larynx, mouth, and throat during speech and normal breathing. Harmless commensal bacteria are shed and more pathogenic streptococci, meningococci and other micro-organisms are also spread in this way, especially when people are crowded in small rooms. Micro-organisms in the mouth, throat, larynx and lungs are expelled to the exterior with much great efficiency during coughing; a cough will project bacteria into the air.

In factory areas such as filling lines, where there is a high density of people, cough and sneeze spray is probably the most important vehicle or likely cause of contamination. In a sneeze up to 20 000 droplets carrying micro-organisms are produced and these droplets, depending on their size, may travel up to 5 metres. They evaporate and release micro-organisms into the environment.[8] A streptococcal sore throat can be acquired by subjects at a distance of 6 metres from a carrier. Dust, dandruff, hair, skin particles, dried sputum and droplets from coughing and sneezing have all been shown to harbour all types of bacteria associated with the human body, and these can obviously be transferred to products.

Washrooms and Toilets

Stringent procedures should be set up and adhered to for the cleaning and disinfection of sinks, toilets, showers, tools, lockers and floors. The use of tablet soap and roll towels for handwashing should be discouraged and discontinued. Instead, antiseptic liquid soap and disposable paper towel dispensers should be installed. Work clothing should be provided to employees and freshly laundered garments should be made available on a regular and frequent basis. Dirty clothing cannot be overlooked as a source of contamination and therefore every effort should be made to eliminate it.

Despite the fact that washrooms and toilets may be cleaned daily, bacteria originating from faeces are often found in large numbers on all surfaces. These areas readily come into contact with people. One third by weight of faeces is bacteria, many of which are alive.[9] The areas likely to cause cross-infection are the toilet seat, wash basin, overflow tap, handles, inside handle of the entrance door. Splashing of the water in the toilet pedestal during defecation is another potential source of cross-infection; this varies with cistern height and bowl design. Contamination originating from these areas is unlikely, unless the handles or toilet seats are particularly soiled. This may happen in mental hospitals or children's wards, and outbreaks of bowel infections have occurred in such places, but it is unlikely elsewhere.[10] Salmonella are potentially dangerous organisms, because they resist drying and small-dose infection is thought to occur, but no evidence is available to incriminate the toilet in such infections. Water that 'looks clean' in pedestals is not necessarily bacteriologically clean. Bacterial populations of 1×10^6 organisms per ml have been found in fresh water in pedestals.[11] In the UK drought of 1976 when 'economy flush' systems were used, large numbers of bacteria remained unflushed, whereas with full flushing the bacteria were removed from the pedestals effectively. Thus simple cleaning and the maintenance of a good flushing cistern are the main principles of toilet care. Flush handles can be a source of transmission of contamination from one individual to another and foot-operated pedals for flushing are recommended.

Toilet care should be mainly mechanical: the ball valve should be set to the correct height and the flush mechanism made to work properly. The surfaces and handles must be kept clean. In the UK the best equipment is the British Standard wash-down bowl with a low cistern, but the latter must be capable of a good cleaning action. The care of the bowl should be restricted to cleaning. A standard scouring powder is adequate for this if used in conjunction with a brush and flushed away. It should not be just poured into the bowl and left, for it might choke up the toilet. The more conventional descaling agents may be required in hard-water areas. There is no reason to pour large quantities of disinfectant into the bowl; this is uneconomic and aimed at a non-existent danger. Finally, as always, the care of the hands is of paramount importance in preventing transmission of contamination; operatives can be encouraged to use paper tissues when handling flushing handles, water taps and door knobs in the toilet area. Also necessary is the provision of a wash-basin close to the toilet area.

Raw Materials

Ingredients used in cosmetic formulae should be quarantined until their quality is determined; those found to be acceptable should be protected from contamination during storage. Raw materials are likely sources of microbial contamination and should be examined microbiologically on a routine basis. A major raw material in cosmetic products is water. If contaminated, the entire water system should be shut down, emptied, thoroughly cleaned, and disinfected by disinfectant solution circulating throughout the entire system, including all feed lines, tanks, and resin beds. After disinfection the system should be drained and flushed with clean water until all traces of the disinfectant solution have been rinsed away. On the other hand if a reduced bacterial count is detected on routine testing, and found to be increasing, it may be sufficient to inject a small quantity of disinfectant solution (say, on Fridays) as a prophylactic measure in order to control the bacterial count. The water supply may be a major source of *Pseudomonas* contamination, and where ambient temperatures exceed 18°C few bacterial cells should be allowed to reach over 10^6 bacteria per gramme. Raw materials of natural origin, such as the natural gums tragacanth and acacia, may be very heavily contaminated and so may kaolin, chalk and starch. Significant microbial contamination has been reported in bentonite, Quaternium-18, hectorite, sodium magnesium silicate and aluminium powder.

Among other raw materials, fats and waxes contain relatively few organisms but Hall (cited by Boehm and Maddox[2]) found counts of up to 16×10^6 bacteria per gramme in dried natural products such as gums and herbs. Also encountered were a variety of fungi and yeasts. Many synthetic materials may also be contaminated.

For microbiological examination of raw materials there is a number of methods for taking samples aseptically. The aim of aseptic sampling is to limit the possibility of adding contamination to the product, so that the test results indicate the microbial status of the material on its arrival at the plant. A sample is scooped or pipetted out from the raw material container, placed in a sterilized sample container and returned to the laboratory for immediate testing. The raw material container should immediately be sealed.

If, after completion of all microbiological tests, the material is accepted, it is then removed from the holding quarantine area and placed in stock. It should be borne in mind that excessive humidity and large fluctuations in temperature from ambient may effect the physical, chemical and microbial properties of raw materials. Raw materials that are retained for long periods of time should be retested at specific intervals (at least every six months). Raw materials that are purchased pre-sterilized to assure the microbiological quality should also be tested to verify (i) that the sterilization procedure was effective and (ii) that post-sterilization contamination has not occurred.

Storage Areas

All raw materials should be stored in such a way that the tested degree of microbial purity is maintained. Storage areas for raw materials should be maintained in as clean a condition as bulk product manufacturing and product filling areas.

When it is necessary to store raw materials, packing materials, intermediate products or finished products in a special environment, these goods should be stored off the floor where possible, to allow their maintenance in a clean dry and orderly condition. This does not, however, prevent outdoor storage of materials whose condition is not adversely affected thereby.

Product Packaging

The protection of the product, once it is filled into an adequate container, depends purely on the efficiency of the closure. The seal should be secured and tight, thus protecting the product from microbes for an indefinite period. This protection may be described as a criterion of the physical preservation of the product. Packaging components such as jars, bottles, tubes, container closures and closure liners should not be ignored as sources of contamination.[12] They should be stored in clean and dry areas and suitably packed or covered to prevent dust-borne microbes from settling on them. Before use if possible they should be air-blown with dry, clean air to rid them of any extraneous particles. It is recommended strongly that air-blowing is carried out under an extraction hood with vacuum facilities so that the particles blown off are immediately extracted from the filling area. If possible, the supplier should be persuaded to supply the packaging components in sealed plastics bags which should be opened just prior to use. This will obviously avoid the complications of air rinsing.

Microbiological Standards

Microbiological standards serve several quite distinct purposes:

1. Control of the danger from pathogenic organisms
2. Assurance that the cosmetic product has never been grossly contaminated
3. Confirmation of a reasonable expected life in storage—that is, an estimate of perishability

Freedom from pathogens is the requirement that is usually paramount in microbiological standards. The usual guidance given to the industry is to exclude 'named pathogens' from cosmetics. With few exceptions, it is doubtful whether a single bacterial cell ever did anyone any harm. It appears to be necessary for growth that at least a few organisms should establish themselves and adapt to their environment. This is clearly illustrated by figures for the 'minimum effective dose' for well recognized diseases (Table 43.2).[13]

The need to ensure that the cosmetic product has never been seriously contaminated emphasizes the importance of quantitative tests in microbiological control. An attempt should always be made to assess the number of organisms present, even though the error may be large. The reason for this is that the onset of an infection, or the development of a microbiological defect in a product, is dependent on the number of organisms originally present.

The third consideration, that of shelf life or perishability, relates to the retail and consumer end of the cosmetics chain. In general, it is always better to prevent microbiological growth by formulation rather than to rely on preservatives. Preservatives, like antioxidants, are rarely completely satisfactory for a long period, especially where ambient temperatures are warm, whereas control by formulation lasts indefinitely.

Microbial specifications within the cosmetics industry are essential aids in maintaining the sanitary quality and stability of the end product, when used to control raw materials, processes and factory hygiene.

Table 43.2 Minimum Effective Doses (approx.) from Pathogens in Human Beings

Disease	Number of cells
Typhoid fever	3
Tuberculosis	100
Cutaneous moniliasis	100 000
Salmonelloses (other than typhoid fever)	100 000–1 000 000

If equipment has been properly cleaned and sterilized the number of bacteria left will not exceed one per cm^2 by a swab test or one per ml by a rinse test. These tests are therefore quite adequate to assess the efficiency of cleaning in a general sense. It can be assumed under ordinary working conditions that if the results are satisfactory (that is, less than 1 colony per cm^2 or ml) then all pathogens have been killed or removed. It is also unlikely that micro-organisms will have survived in sufficient numbers to cause trouble.

Conclusion

The cleanest factory conditions can easily be invalidated if unhygienic methods are practised; microbial contamination will only be avoided by careful attention to all aspects of production and control. Disregard for any of the following points results in weaknesses and, eventually, costly complications.

1. Check the quality of raw materials
2. Unpack raw materials in a separate building, especially if embedded in sawdust, cotton waste, straw, etc.
3. Apply a biocidal treatment where necessary if this is practicable
4. Control the quality of water used in the factory
5. Check the bacterial purity of the air near fillers, etc.
6. Check the hygienic condition of all containers
7. Remove all residue, broken or split containers, etc. as soon as possible
8. Do not use cloths for mopping up spillages, unless these are maintained in a clean condition; paper towels are much better
9. Thoroughly clean equipment immediately after use
10. Sterilize or cleanse all equipment, as may be necessary, immediately before use

Common Fallacies in Hygiene
It is not true that:

(a) splashing disinfectant over floors, etc., 'solves the hygiene problem';
(b) forcing steam round a circuit, with the production of great clouds of 'steam' and considerable noise, necessarily sterilizes the equipment;
(c) if equipment looks clean, then it must be clean—this is not true microbiologically.

'Window-dressing' devices, such as making people wear white coats and caps, provision of glaring UV lamps, use of disinfectant aerosols, may serve some useful purpose and undoubtedly exert a psychological influence, but in real terms their value is not great.

A cosmetic product contaminated by micro-organisms is generally held to be a potential danger to the user, especially if harmful pathogens are present. Paetzoid examined 129 products and found 17·1 per cent of them to be contaminated with bacteria that were 43·9 per cent pathogens and 56·1 per cent non-pathogens.[14] It has been pointed out by Knothe[15] that the skin possesses self-disinfection properties with a greater bactericidal effect on pathogens than on non-pathogens. The conclusion from his work is that cosmetic products need not necessarily be sterile in order to protect the normal healthy user. So why does the cosmetics industry attempt to limit the micro-organism content of products? The answer is that if micro-organisms are introduced into a cosmetic product the general properties of the cosmetic can be altered (a) by the micro-organism, (b) by multiplication of the micro-organism and (c) by microbial metabolites.

The following phenomena are common in insufficiently preserved cosmetics: creams and lotions can exhibit visible mould colonies; products can separate with or without the production of a rancid or putrid odour; discoloration can occur; clear preparations can become turbid owing to precipitation; fermentation can produce gas, sometimes causing tubes to swell or glass bottles to burst. The result is that a product manufactured at high cost and believed to have received all possible care has become unsaleable.

The sanitary quality of a cosmetic product is normally the summation of the level of sanitation in the production of raw material ingredients, the sanitation quality of the finished raw materials, and level of sanitation in the cosmetic product manufacturing plant. The EEC definition of the term cosmetic is 'any substance or preparation intended to be placed in contact with various superficial parts of the human body (epidermis, hair system, nails, lips and external genital organs) or with the teeth and mucous membrane of the oral cavity with a view exclusively or principally to perfume them, clean them, protect them and keep them in good condition, to change their appearance or to correct body odours'. As such, cosmetics demonstrating poor microbiological quality have a potential for being 'injurious to health' with unpredictable frequencies and at unpredictable times.

REFERENCES

1. Yablonski, J. L., *Cosmet. Toiletries*, 1978, **93**(9), 37.
2. Boehm, E. E. and Maddox, D. N., *Manuf. Chem. Aerosol News*, 1971, **42**(4), 41.
3. Kano, C., Nakata, O., Kurosaki, S. and Yanagi, M., *J. Soc. cosmet. Chem.*, 1976, **27**, 73.
4. Schlüssler, H. J., *Milchwissenschaft*, 1970, **25**(3), 133.
5. Prado Fihlo, L. G., *Lebensm.-Wiss. Technol.*, 1975, **8**(1), 29.
6. Han, B-H., Dissertation, University of Karlsruhe, 1977.
7. Thor, W. and Loncin, M., *Chemie Ingenieur Technik*, 1978, **50**(3), 118.
8. Mims, C. A., *Pathogenesis of Infectious Disease*, London, Academic Press, 1977.
9. Mendes, M., *New Scientist*, 1977, **76**(1079), 507.
10. Newsom, S. W. B., *Lancet*, 1972, 30 September, 700.
11. *Guide to Good Pharmaceutical Manufacturing Practice*, London, HMSO, 1971.
12. Most, S. and Katz, A., *Am. Perfum. Cosmet.*, 1970, **85**(3), 67.
13. Davis, J. G., *Soap Perfum. Cosmet.*, 1973, **46**(1), 37.
14. Paetzoid, H., *Vortrag anl. der D. G. F.-Vortragstagung*, Mainz, 1967.
15. Knothe, H., *Referat anl. d. Vortrags- und Diskussionstagung der Gesellschaft deutscher Kosmetika-Chemiker e.V.*, Hamburg, 1967.

Appendix

Proprietary Materials Cited in this Book

Entries in italics are the CTFA adopted names as listed in the *CTFA Cosmetic Ingredient Dictionary*; they are reprinted with the permission of the Cosmetic, Toiletry and Fragrance Association, Inc., Washington DC, USA.

It should be noted that CTFA does not 'approve', 'certify' or 'endorse' particular ingredients for use in cosmetic products. CTFA has a committee that assigns 'adopted names' to certain cosmetic ingredients for inclusion in the *CTFA Cosmetic Ingredient Dictionary*, which is recognized by regulations of the United States Food and Drug Administration (Title 21, Code of Federal Regulations, Section 701.3) as an authoritative reference of proper nomenclature of an ingredient for purposes of cosmetic ingredient labelling. Assignment of an 'adopted name' and inclusion in the *Dictionary* establishes only proper *nomenclature* for an ingredient for purposes of ingredient labelling on cosmetic products sold in the United States; it does *not* signify CTFA (or FDA) 'approval' of the ingredient.

In the interests of achieving uniform nomenclature for ingredients available for use in the cosmetic industry, the CTFA adopted names are becoming increasingly used in countries other than the United States of America.

Proprietary name	**Chemical description**	**Supplier**
Acetol	*Acetylated lanolin alcohol*	Emery
Actamer	2,2-Thiobis-(4,6-dichlorophenol)	Monsanto
Aduvex 2211	2-Hydroxy-4-methoxy-4′-methyl-benzophenone	Ward Blenkinsop
Aerosil	Finely divided silica	Degussa
Aerosol OT	Dioctyl sodium sulphosuccinate	American Cyanamid
Aethoxal	Ethoxylated propoxylated fatty alcohol *PPG-10-Ceteareth-20*	Henkel
Alcloxa	Aluminium chlorhydroxy allantoinate *Alcloxa*	Hoechst
Alrosal	Fatty acid amide concentrate	Ciba-Geigy
Amerchol CAB	Cholesterol absorption base *Petrolatum* and *Lanolin Alcohol*	Amerchol
Amerchol L-101	Mineral oil and lanolin alcohols	Amerchol
Aminoxid WS35	Cocamidopropylamine oxide	Goldschmidt

Proprietary name	Chemical description	Supplier
Ammonyx 4002	Benzyldimethylstearylammonium chloride *Stearalkonium Chloride*	Onyx
Amphomer	*Octylacrylamide/Acrylates/Butyl-aminoethyl Methacrylate Polymer*	National Starch
ANM starch powders	Starch ethers	Neckar-Chemie
Anobial	3′,4′,5-Trichlorosalicylanilide	Firmenich
Anobial TFC	3-Trifluoromethyl-4, 4′-dichlorocarbanilide	Firmenich
Antara opacifiers		GAF
Antifoam AF	Mixture of dimethylpolysiloxane (*dimethicone*) and silica gel *Simethicone*	Dow Corning
Antiviray	Homomenthyl salicylate (*Homosalate*), benzyl salicylate and methyl eugenol	Bush Boake Allen
Arlacel A	Mannide mono-oleate	ICI
Arlacel C	*Sorbitan Sesquioleate*	ICI
Arlacel 83	*Sorbitan Sesquioleate*	ICI
Arlacel 165	Glycerylmonostearate and polyoxyethylene stearate *Glyceryl Stearate* and *PEG-100 Stearate*	ICI
Arlamol E	Fatty acid propoxylate *PPG-15 Stearyl Ether*	ICI
Arlatone T	Sorbitan polyoxyethylene fatty acid ester *PEG-40 Sorbitan Peroleate*	ICI
Arlex	*Sorbitol*	ICI
Arochlor 5460		Monsanto
Arquad 2HT	Dimethyldi (hydrogenated tallow) ammonium chloride *Quaternium-18*	Armak
Arquad 16	Alkyltrimethylammonium chloride *Cetrimonium Chloride*	Armak
Arquad S50	Trimethylsoyammonium chloride *Soyatrimonium Chloride*	Armak
Atlas G-271	N-soya-N-ethylmorpholinium ethosulphate *Quaternium-2*	ICI
Atlas G-1086	Polyethylene glycol sorbitol hexaoleate	ICI
Atlas G-1425	Polyoxyethylene sorbitol lanolin derivative	ICI

Proprietary name	Chemical description	Supplier
Atlas G-1441	Polyoxyethylene sorbital lanolin derivative *PEG-40 Sorbitan Lanolate*	ICI
Atlas G-1690	Polyoxyethylene ether of alkyl phenol	ICI
Atlas G-1790	Polyoxyethylene lanolin condensate *PEG-20 Lanolin*	ICI
Atlas G-2132	Polyoxyethylene lauryl ether	ICI
Atlas G-2240	Polyoxyethylene sorbitol	ICI
Atlas G-2320	Polyoxyethylene sorbitol	ICI
Atlas G-2859	Polyoxyethylene sorbitol 4, 5-oleate	ICI
Atlas G-3721	Polyoxyethylene-2-butyl octanol	ICI
Atlas G-7596J	Polyoxyethylene sorbitan monolaurate *PEG-10 Sorbitan Laurate*	ICI
Avicel	Microcrystalline cellulose	FMC
Avitex ML	Cationic emulsifying agent	DuPont
Bentones	Quaternary hectorites	NL Industries
Bentonite	Clay mineral, montmorillonite *Bentonite*	Berk
Betadine	lodophor (i.e. mixture of iodine with surface-active agent)	Berk
BHT	Butylated hydroxytoluene (i.e. 2,6-di-*tert*-butyl-4-methylphenol) *BHT*	Kodak Chemical
Bithionol	2,2′-thio-*bis*-(4,6-dichlorophenol)	Hilton Davis Chemicals
Bradosol	β-Phenoxyethyldimethyldodecyl-ammonium bromide	CIBA
Brij 30	Polyoxyethylene lauryl ether *Laureth-4*	ICI
Brij 35	Polyoxyethylene lauryl ether *Laureth-23*	ICI
Brij 93	Polyoxyethylene (2) oleyl ether *Oleth-2*	ICI
Bronopol	*2-Bromo-2-Nitropropane-1,3-Diol*	Boots
BTC	Lauryldimethylbenzylammonium chloride	Onyx
BTC 2125M	Tetradecyldimethylbenzylammonium and dodecyldimethyl-*p*-ethylbenzylammonium chlorides *Myristalkonium Chloride* and *Quaternium-14*	Onyx
Cab-o-sil	Fumed silica	Cabot

Proprietary name	**Chemical description**	**Supplier**
Calflo E	Calcium silicate	Johns-Manville
Carbopol 934	Polymer of acrylic acid cross-linked with a polyfunctional agent *Carbomer-934*	Goodrich
Carbopol 940	Polymer of acrylic acid cross-linked with a polyfunctional agent *Carbomer-940*	Goodrich
Carbopol 941	Polymer of acrylic acid cross-linked with a polyfunctional agent *Carbomer-941*	Goodrich
Carbopol 960	Ammonium salt of Carbopol 934 *Carbopol-960*	Goodrich
Carbowax 400	Ethylene oxide polymer *PEG-8*	Union Carbide
Carbowax 1000	Ethylene oxide polymer *PEG-20*	Union Carbide
Carbowax 1500	Ethylene oxide polymer *PEG-6-32*	Union Carbide
Carbowax 1540	Ethylene oxide polymer *PEG-32*	Union Carbide
Carbowax 4000	Ethylene oxide polymer *PEG-75*	Union Carbide
Carbowax 6000	Ethylene oxide polymer *PEG-150*	Union Carbide
Catrex	Interpolymer of aminoethylacrylate phosphate and acrylic acid *Aminoethylacrylate Phosphate/Acrylate*	National Starch
Ceepryn	Cetylpyridinium bromide or chloride	Merrell
Celacol	Sodium carboxymethylcellulose	British Celanese
Cellofas	Sodium carboxymethylcellulose	ICI
Cellosize	*Hydroxyethylcellulose*	Union Carbide
Ceraphyl 60	γ-Gluconamidopropyldimethyl-2-hydroxyethylammonium chloride *Quaternium-22*	Van Dyk
Ceraphyl 65	Mink aminopropyldimethyl-2-hydroxyethylammonium chloride *Quaternium-26*	Van Dyk
Cetavlon	Cetyltrimethylammonium bromide	ICI
Cetiol HE	Polyol fatty acid ester *PEG-7 Glyceryl Cocoate*	Henkel
Chlorhexidine	1,6-di-(N-*p*-chlorophenylguanidino)hexane *Chlorhexidine*	ICI

Proprietary name	**Chemical description**	**Supplier**
Chlorhydrol	Aluminium chlorhydrate, 50% solution *Aluminium Chlorohydrate*	Reheis
Comperlan HS	Stearic monoethanolamide	Henkel
Comperlan KD	Cocamide diethanolamide *Cocamide DEA*	Henkel
Courlose	Sodium carboxymethylcellulose	British Celanese
Crillet 3	Mixture of stearate esters of sorbitol and sorbitol anhydrides condensed with approx. 20 moles of ethylene oxide *Polysorbate 60*	Croda
Crodafos N3 acid	Phosphated oleyl ether (3EO) *Oleth-3 Phosphate*	Croda
Crodafos N3 neutral	Diethanolamine salt of a complex mixture of esters of phosphoric acid and *Oleth-3* (which is the polyethylene glycol ether of oleyl alcohol) *DEA-Oleth-3 Phosphate*	
Crodalan IPL	*Isopropyl lanolate*	Croda
Crodamol CSP	Cetostearyl palmitate	Croda
Crodamol IPP	*Isopropyl Palmitate*	Croda
Crodamol ML	*Myristyl Lactate*	Croda
Crodamol OP	*Octyl Palmitate*	Croda
Crodaterge LS (now named Crodasinic LS)	*Lauroyl Sarcosine*	Croda
Crodaterge OS (now named Crodasinic OS)	*Oleoyl Sarcosine*	Croda
Crodesta F70, F160 and F110	Sucrose esters of palmitic acid and stearic acids	Croda
Cromeen	Substituted alkyl amine derivative of various lanolin acids	Croda
Crotein Q	Quaternized hydrolysed animal protein	Croda
DC 200	Dimethylpolysiloxane	Dow Corning
Dehyquart SP	Ethoxylated quaternary ammonium phosphate *Quaternium-52*	Henkel
Deriphat 170C	Lauryl aminopropionic acid *Lauraminopropionic Acid*	General Mills
Detergent 1011	Secondary amide of lauric acid	
Dichlorophene	2,2-methylene-*bis*-(4-chlorophenol)	Givaudan
Dicrylan 325	Acrylate/acrylamide polymer	Ciba-Geigy

Proprietary name	Chemical description	Supplier
Diometam	Di-(n-octyl)-dimethylammonium bromide	British Hydrological
Dioxin	6-Acetoxy-2, 4-dimethyl-*m*-dioxane	Givaudan
Dowfax 2A		Dow
Dowicil 200	1-(3-chloroallyl)-3,5,7-triaza-l-azoniaadamantane chloride *Quaternium-15*	Dow
Duponol C	Sodium lauryl sulphate	DuPont
Duponol WA	Sodium lauryl sulphate	DuPont
Duponol WAT	Triethanolamine salt of lauryl sulphate *TEA-Lauryl Sulfate and TEA-Oleyl Sulfate*	DuPont
Edifas	Sodium carboxymethylcellulose	ICI
Emcol CD-18	Propoxylated polyol	Witco
Emcol E-607	*N*-(Acylcolaminoformylmethyl) pyridinium chloride	Witco
Emcol E-6075	*N*-Stearoylcolaminoformyl-methylpyridinium chloride *Quaternium-7*	Witco
Emerest 2400	*Glyceryl Stearate*	Malmstrom
Emersol 132	Stearic acid	Malmstrom
Empicol LZ	Sodium lauryl sulphate	Albright & Wilson
Empigen BB	Alkyl dimethyl betaine	Albright & Wilson
Empigen BT	Alkyl amido betaine	Albright & Wilson
Empigen CDR10	Coconut imidazoline betaine	Albright & Wilson
Empigen CDR30	Modified coconut imidazoline betaine	Albright & Wilson
Emsorb 6915	*Polysorbate 20*	Malmstrom
Escalol 506	Amyldimethyl-*p*-aminobenzoic acid *Amyl Dimethyl PABA*	Van Dyk
Estol 1461	Glycerol monostearate nse	Unichema
Ethomeen C/25	*PEG-15 Cocamine*	Armak
Eutanol G	2-Octyldodecanol *Octyl Dodecanol*	Henkel
Eutanol LST		Henkel
Evanol	Proprietary cream base	Evans Chemetics
Extrapones	Herbal extracts	Dragoco
Fentichlor	*Bis*-(2-hydroxy-5-chlorophenyl) sulphide	Cocker Chemical
F.H.P.	Sodium carboxymethylcellulose	Hercules

Proprietary name	Chemical description	Supplier
Filtrosol A-1000 and B	UV screens—proprietary mixtures	Norda
Fixanol C	Cetylpyridinium bromide or chloride	ICI
Fixanol VR	Tetradecylpyridinium bromide	ICI
Fluilanol	Lanolin oil and *Oleth-3* (which is the polyethylene glycol ether of oleyl alcohol)	Croda
Fluorophene	3,5-dibromo-3′-trifluoromethylsalicylanilide *Fluorosalan*	Pfister
F.M.P.	Sodium carboxymethylcellulose	Hercules
Foromycen F10	Antifungal agent	Petrosin Laboratorium
Fungicide DA	Undecylenamide diethanolamide *Undecylenamide DEA*	Dragoco
Fungicide UMA	Undecylenamide monoethanolamide *Undecylenamide MEA*	Dragoco
Gafquat 734 and 755	Quaternary ammonium polymers formed by the reaction of dimethyl sulphate and a copolymer of vinyl pyrrolidone and dimethylaminoethylmethacrylate *Quaternium-23*	GAF
Gantrez ES-225	Monoethyl ester of methylvinylether/maleic acid copolymer (50% in ethanol) *Ethyl ester of PVM/MA copolymer*	GAF
Gantrez ES-425	Monobutyl ester of methylvinylether/maleic acid copolymer (50% in ethanol) *Butyl ester of PVM/MA copolymer*	GAF
Genapol S200		Hoechst
Germall 115	Imidazonidyl urea	Sutton Laboratories
Giv-tan F	2-Ethoxy-*p*-methoxycinnamate *Cinoxate*	Givaudan
Glucam-P20	*PPG-20 methylglucose ether*	Amerchol
HD Eutanol	Oleyl alcohol	Henkel
Hexachlorophene	2,2′-methylene-*bis*-(3,4,6-trichlorophenol)	Givaudan
Hibitane	1,6-di-(N-*p*-chlorophenylguanidino) hexane	ICI
Hostaphat KL340N	Triester of the polyethylene glycol ether of lauryl alcohol and phosphoric acid *Trilaureth-4 Phosphate*	Hoechst

Proprietary name	Chemical description	Supplier
Hostapur SAS	Secondary alkane sulphonate	Hoechst
Hyamine 10X	Methylbenzethonium chloride	Rohm & Haas
Hyamine 1622	*p*-Diisobutylphenoxyethoxyethyl-dimethylbenzylammonium chloride *Benzethonium Chloride*	Rohm & Haas
Hyamine 2389	Alkyltolylmethyltrimethylammonium chloride *Quaternium-28* and *Quaternium-29*	Rohm & Haas
Ionol	Butylated hydroxytoluene, i.e. 2,6-di-*tert*-butyl-4-methylphenol *BHT*	Shell
Ionol CP	Purified grade of Ionol *BHT*	Shell
Irgasan BS200	3,3′,4,5′-Tetrachlorosalicylanilide	Ciba-Geigy
Irgasan CF-3	3-Trifluoromethyl-4,4′-dichlorocarbanilide *Cloflucarban*	Ciba-Geigy
Irgasan DP-300	2,4,4′-Trichloro-2″-hydroxydiphenylether *Triclosan*	Ciba-Geigy
Isopar E	Isoparaffinic solvent	Esso
Isopropylan 33	*Isopropyl Lanolate* and *Lanolin oil*	Robinson-Wagner
Isothan Q	Alkylisoquinolinium bromide	Onyx
Isothan Q-15	*Lauryl Isoquinolinium Bromide*	Onyx
Kelzan	*Xanthan gum*	Kelco
Klucel	*Hydroxypropylcellulose*	Hercules
Klucel HA	Hydroxyalkylcellulose	Hercules
L-43 Silicone		Union Carbide
Laneto 50, 100	Polyethylene glycol-50 lanolin *PEG-50 Lanolin*	R.I.T.A.
Lanoquats	Quaternized fatty acid amides derived from lanolin acid	Malmstrom
Lanoquat DES	Lanolin quaternary	Malmstrom
Lantrol	Liquid lanolin *Lanolin oil*	Malmstrom
Laponite	Hectorite-type synthetic clay	Laporte
Lathanol LAL	*Sodium Lauryl Sulphoacetate*	Stepan
Lexate TA	Glyceryl stearate, isopropyl myristate and stearyl stearate	Inolex
Lexein X-250	Hydrolysed animal protein	Inolex

Proprietary name	Chemical description	Supplier
Lexemul AR	Glycerol sterate	Inolex
Lexemul AS	Glyceryl stearate and sodium lauryl sulphate	Inolex
Liquid Base CB. 3929	Mineral oil and lanolin alcohols	Croda
Loramine DU 185	Undecylenamide diethanolamide	Rewo
Loramine OM 101	Monoalkylolamide of mixed fatty acids	Rewo
Loramine SBU 185	Disodium monoundecylenamido monoethanolamide sulphosuccinate	Rewo
Loramine U 185	Undecylenamide monoethanolamide	Rewo
Luviset CE 5055	Vinyl acetate/crotonic acid copolymer	BASF
Marinol	Alkyldimethylbenzylammonium chloride	Berk
Maypon 4C	Potassium cocylhydrolysed animal protein	Stepan
Merquat-550	Polymeric quaternary ammonium salt consisting of acrylamide and dimethyldiallyl ammonium chloride monomers *Quaternium-41*	Merck
Merquat resins	Dialkyldimethyl ammonium chloride cyclopolymers	Merck
Methocel	*Hydroxypropylmethylcellulose*	Dow
Methofas	Methylcellulose	ICI
Microdry	Aluminium chlorhydrate fine powder *Aluminum Chlorohydrate*	Reheis
Miranol C2M, C2M-SF	A long-chain imidazoline type of zwitterion *Amphoteric-2*	Miranol
Miranol SM	Capryl imidazoline derivative	Miranol
Mirapol A15	Polyquaternary ammonium compound	Miranol
Modulan	*Acetylated Lanolin*	Amerchol
Morpan CHSA	Cetyltrimethylammonium bromide	Glovers
Myacide SP	2,4-Dichlorobenzyl alcohol	Boots
Myrj 45	Polyoxyethylene monostearate *PEG-8 Stearate*	ICI
Myrj 52	Polyoxyethylene monostearate *PEG-40 Stearate*	ICI
Myverol 18-17	Molecularly distilled monoglyceride	Eastman

Proprietary name	Chemical description	Supplier
Nacconal NRSF	Alkylaryl sulphonate	Allied Chemical
Natrosol	*Hydroxyethylcellulose*	Hercules
Neo-fat 18-55	*Stearic Acid*	Armak
Neo-PCL, water-soluble	Ethoxylated alkyl phenol and polyethylene glycol ester of 2-ethylhexanoic acid *Nonoxynol-14* and *PEG-4 Octanoate*	Dragoco
Neosyl	Finely divided silica	Crosfield
Neutronyx 600	Ethoxylated alkyl phenol *Nonoxynol-9*	Onyx
Nimlesterol D	*Mineral Oil* and *Lanolin Alcohol*	Malmstrom
Ninol 2012	Fatty acid alkanolamine concentrate	Stepan
Nipagin M	Methyl ester of *p*-hydroxybenzoic acid *Methylparaben*	Nipa
Nipagin P		Nipa
Nipasol M	Propyl-*p*-hydroxybenzoate *Propylparaben*	Nipa
Nonic 218	Polyethylene glycol *tert*-dodecyl thioether	Sharples Chem.
Novol	*Oleyl Alcohol*	Croda
Oat-Pro	*OatFlour*	Quaker Oats
Octaphen	*p-tert*-Octylphenoxyethoxyethyl dimethylbenzylammonium chloride	Ward Blenkinsop
Omadine	1-Hydroxypyridine-2-thione	Olin Mathieson
Omamids	Polyamide resins	Olin Mathieson
Onamer	Poly(dimethylbutenylammonium chloride)-α, ω-*bis*(triethanolamine chloride)	Onyx
Oracid	Urea-formaldehyde foam	Chemische Fabrik Frankenthal
Orvus WA	*Sodium Lauryl Sulfate*	Procter & Gamble
Ottasept extra	*Chloroxylenol*	Ottawa Chemical
PCL Liquid	*Cetearyl Octanoate*	Dragoco
Phemerol	*p-tert*-Octylphenoxyethoxyethyl dimethylbenzylammonium chloride	Parke-Davis
Phenonip	Combination of parabens and phenoxyethanol	Nipa
Pluronic F-127	A polyoxyethylene–polyoxypropylene block polymer *Poloxamer-407*	BASF-Wyandotte

Proprietary name	Chemical description	Supplier
Pluronic L64D	A polyoxyethylene–polyoxypropylene block polymer *Poloxamer-184*	BASF-Wyandotte
Polawax	Nonionic emulsifying wax	Croda
Polawax A.31	Blend of cetyl stearyl alcohol and EO condensate products	Croda
Polychol 5	Polyethylene glycol condensate of wool wax alcohols *Laneth-5*	Croda
Polychol 15	Polyethylene glycol ester of lanolin alcohol with an average ethoxylation value of 15 *Laneth-15*	Croda
Polyglycol 400	Polyethylene glycol 400	Hoechst
Polymer JR	Cationic cellulose ether derivative *Quaternium-19*	Union Carbide
Povidone-Iodine	*PVP-iodine*	Berk
Procetyl AWS	Polyoxypropylene, polyoxyethylene ether of cetyl alcohol *PPG-5-Ceteth-20*	Croda
Promulgen D	Cetylstearyl alcohol and its polyethoxylated (20) derivative *Cetearyl Alcohol* and *Ceteareth-20*	Robinson-Wagner
Prosolal S8	Octyl cinnamate sunscreen	Dragoco
Prosolal 58	Mixture of phenylacrylic esters and oxybenzoic acid esters	Dragoco
Protolate WS	PEG-75-lanolin and hydrolysed animal protein *PEG-75 Lanolin oil*	Malmstrom
PVP-VA E-735	*PVP/VA Copolymer*	GAF
Quadramer		American Cyanamid
Renex	Polyoxyethylene esters of mixed resin and fatty acids	ICI
Resyn 28-1310	*Vinyl Acetate/Crotonic Acid Copolymer*	National Starch
Resyn 28-2930	*Vinyl Acetate/Crotonic Acid/Vinyl Neodecanoate Polymer*	National Starch
Rewopol SBFA 30	Disodium lauryl alcohol polyglycol ether sulphosuccinate	Rewo
Roccal	Alkyldimethylbenzylammonium chloride *Benzalkonium Chloride*	Bayer

Proprietary name	Chemical description	Supplier
Sandopan DTC acid	α-(Carboxymethyl)-ω-(tridecyloxy) poly (oxy-1,2-ethanediyl) *Trideceth-7-Carboxylic Acid*	Sandoz
Sandopan TFL	Sulphoamidobetaine *Amphoteric-7*	Sandoz
Santicizer 8	Sulphonamide plasticizer	Monsanto
Santicizer 160	*Butyl Benzyl Phthalate*	Monsanto
Santocel 54	Hydrated silica	Monsanto
Santolite MHP	Toluenesulphonamide/formaldehyde resin	Monsanto
Santolite MS 80%	Toluenesulphonamide/formaldehyde resin	Monsanto
Schercoquat	Quaternized fatty acid amides derived from isostearic acid	Scher Chemicals
Silicone fluid DC-556	Polyphenylmethyl siloxane *Phenyl Dimethicone*	Dow Corning
Silicone fluid L-45	Dimethyl silicone *Dimethicone*	Union Carbide
Sodium silicate 'O'	*Sodium Silicate*	Philadelphia Quartz
Softigen 767	Ethoxylated partial glyceride fatty acid ester *PEG-6 Caprylic/Capric Glycerides*	Dynamit-Nobel
Solprotex	Digalloyl trioleate	Firmenich
Solulan 98	Acetylated polyoxyethylene (10) lanolin alcohol *Laneth-10 Acetate*	Amerchol
Sorbo	*Sorbitol*	ICI
Span 20	Sorbitan monolaurate *Sorbitan Laurate*	ICI
Span 60	Sorbitan monostearate *Sorbitan Stearate*	ICI
Span 80	Sorbitan monooleate *Sorbitan Oleate*	ICI
Span 85	*Sorbitan Trioleate*	ICI
Standapol OLP	Oleyl betaine	Henkel
Steinapon AM-B13	Alkylamido betaine	Goldschmidt
Stepanhold R-1	*PVP/Ethyl Methacrylate/Methacrylic Acid Polymer*	Stepan Chemical
Sunscreen 3573	Oil-soluble suncreen	Merck
Surfynol 82	*Dimethyl Octynediol*	Air Products
Syloid 72	*Hydrated Silica*	Grace
Syncrowax ERLC	*C18-C36 Acid Glycol Esters*	Croda
Syncrowax HGLC	*C18-C36 Acid Triglyceride*	Croda

Proprietary name	Chemical description	Supplier
Syncrowax HRC	*Glyceryl Tribehenate*	Croda
Syncrowax PRLC	Propylene glycol ester of mixed fatty/wax acids	Croda
TBS	3,4′,5-Tribromosalicylanilide	Theodore St. Just
TCC	3,4,4′-Trichlorocarbanilide *Triclocarban*	Monsanto
TCS	2,3,3′, 5-Tetrachlorosalicylanilide	Ciba-Geigy
Tegin 515	Glyceryl monostearate *Glyceryl Stearate*	Goldschmidt
Tego 103S		Goldschmidt
Tegobetaine L7	Cocamidopropyl betaine	Goldschmidt
Temasept IV	3,4′, 5-Tribromosalicylanilide	Fine Organics
Tergitol NPX	Alkylaryl polyethylene glycol ether *Nonoxynol-10*	Union Carbide
Texapon Extract N25	Sodium lauryl ether sulphate *Sodium Laureth Sulfate*	Henkel
Timica	Mica coated with titanium dioxide	Mearl
Tinuvin P	UV-absorber	Ciba-Geigy
Tiona G	Titanium dioxide, oil-dispersible	Laporte
Topanol O	Butylated hydroxytoluene, i.e. 2,6-di-*tert*-butyl-4-methylphenol *BHT*	ICI
Topanol OC	Purified grade of Topanol O	ICI
Triton X-100	Alkylated aryl polyether alcohol *Octoxynol-9*	Rohm and Haas
Triton X-200	Sodium salt of alkylated aryl polyether sulphate *Sodium Octoxynol-3 Sulfonate*	Rohm and Haas
Triton X-400	Stearyldimethylbenzylammonium chloride *Stearalkonium Chloride*	Rohm and Haas
Tuasal 100	3,4′, 5-Tribromosalicylanilide	Dow
Tween 20	Polyoxyethylene sorbitan monolaurate *Polysorbate-20*	ICI
Tween 40	Polyoxyethylene sorbitan monopalmitate *Polysorbate-40*	ICI
Tween 60	Polyoxyethylene sorbitan monostearate *Polysorbate-60*	ICI
Tween 65	Polyoxyethylene sorbitan tristearate *Polysorbate-65*	ICI

Proprietary name	Chemical description	Supplier
Tween 80	Polyoxyethylene sorbitan monooleate *Polysorbate-80*	ICI
Tylose	Sodium carboxymethylcellulose	Hoechst
Ucon LB-1715	Polypropylene glycol butyl ether *PPG-40 Butyl Ether*	Union Carbide
Ucon 50-HB-660	Polyoxypropylene, polyoxyethylene monobutyl ether *PPG-12-Buteth-16*	Union Carbide
Ultrawet 60L	Triethanolamine salt of alkylaryl sulphonate *TEA-Dodecylbenzene Sulfonate*	ARCO
Uvinul D-50	2,2′,4,4′-Tetrahydroxybenzophenone *Benzophenone-2*	GAF
Uvistat	2-Hydroxy-4-methoxy-4′-methyl benzophenone	Ward Blenkinsop
Vancide 89RE	N-Trichloromethylthio-4-cyclohexene-1,2-dicarboximide *Captan*	Vanderbilt
Vancide BL	2,2′-Thiobis-(4,6-dichlorophenol)	Vanderbilt
Vantoc AL	Alkyltrimethylammonium bromide	ICI
Vantoc B	Tetradecylpyridinium bromide	ICI
Vantoc CL	Alkyldimethylbenzylammonium chloride	ICI
Veegum	*Magnesium Aluminum Silicate*	Vanderbilt
Veegum HV	*Magnesium Aluminum Silicate*	Vanderbilt
Veegum K	*Magnesium Aluminum Silicate*	Vanderbilt
VEM resin	*PVP/Ethyl Methacrylate/Methacrylic Acid Polymer*	Barr-Stalfort
Versamids	Polyamide resins	General Mills
Versene	Ethylenediamine tetraacetic acid *EDTA*	Dow
Virac	Iodophor (i.e. mixture of iodine with surface-active agent)	Ruson Laboratories
Volatile Silicone 7158, 7207	Cyclic dimethyl polysiloxane compound *Cyclomethicone*	Union Carbide
Volpo N3	Oleyl alcohol/ethylene oxide condensate	Croda
Volpo N5	Polyoxyethylene oleyl ether	Croda

Proprietary name	Chemical description	Supplier
Volpo S10	Polyoxyethylene (10) stearyl ether *Steareth-10*	Croda
Volpo S20	Polyoxyethylene (20) stearyl ether *Steareth-20*	Croda
Wescodyne	Iodophor (i.e. mixture of iodine with surface-active agent)	Bebgue
Zephiran	Alkyldimethylbenzylammonium chloride	Bayer
Zephirol	Alkyldimethylbenzylammonium chloride	Bayer
Zetesol 856T	Alkyl ether sulphate	Zschimmer & Schwarz
78-4329 (now named Celquat)	Cationic polymer grafted on a cellulosic chain	National Starch

Index

Abrasive action of toothpastes 619
Abrasives in toothpastes 609
Absorbency of face powders 289
Absorption of radiation by sunscreens 242, 250
Acacia gum in face masks 278
Acetals as insect repellants 212
Acetates as astringents 74
Acetic acid as astringent 75
Acetone as nail lacquer remover 386
Acetone sodium metabisulphite as antioxidant 718
Acetophenones in sunscreens 233, 235, 237
Acetylcholine 13
Acid dyes 526
Acid theory of dental decay 596
Acids, dissociation constants 689
Acids, weak, as preservatives 687, 689
Acne 17, 46, 119
  incidence 119
  products 121
Acquired pellicle 590, 591
Acrylic resins
  in nail elongators 389
  in nail lacquers 377, 384
Acyl lactylates in shampoos 437
Acyl peptides 637
  in shampoos 437
Acylamino acids 637
Acylaminopropionates 637
Acylsarcosinates in shampoos 436
Adamantane chloride, chloroalkyltriazonia-, as
    preservative 687
Adhesion of face powder 293
Adipates in moisturizing creams 65
Adrenalin 13
Adrenocorticotrophic hormone 47
Adsorption, role in surface activity 632
Aerosol products
  antiperspirants 134, 826, 831
  coloured setting lotion 528
  deo-colognes 833
  deodorants 140, 824
  depilatories 149, 840
  foams 824, 828
  foot powders 197
  hair creams 825
  hair sprays 474, 824, 831, 832
  hairdressings 490, 824
  hand creams 825
  men's hairdressings 483
  mouth fresheners 629
  powder sprays 827
  shampoos 454, 830
  shaving foams 161, 829, 840
  sun products 255, 825
  toothpaste 825
Aerosols, corrosion 833
  storage testing 838
Aesculin in sunscreens 256
Affinity, molecular 729, 734
After-bath products 108
After shave products
  crackling foam 185
  cream and balm 186
  foam 184
  gel 185
  lotions 75, 77, 181
  powder 187
Aggregation in emulsions 732
Air Spun process for face powders 300
Alanine derivatives as insect repellants 212
Albumin in moisturizing creams 63
Alcohols
  as astringents 75, 76
  as preservatives 687, 688, 691
  fatty, in lipsticks 319
  for disinfection of plant 886
Aldehydes
  oxidation 713
  in rancid fats 720
Alkali preparations for hair straightening 581
Alkaline agents in hair waving 556, 568
Alkanolamide sulphates 634
Alkanolamides 636
  as opacifiers in shampoos 446
  as thickeners in shampoos 446
  fatty acid, in shampoos 437
  in foam baths 95
Alkanolamine lauryl sulphates in shampoos 434
Alkyl aryl polyglycol ethers 636
Alkyl aryl sulphonates 634
Alkyl benzene sulphonates
  in foam baths 93, 94
  in shampoos 433
Alkyl ether carboxylates 634
Alkyl ether sulphates 634
  in shampoos 435
Alkyl imidazolines 637
Alkyl imidazolinium salts 635
Alkyl morpholinium salts 635
Alkyl phosphates 634
Alkyl polyethylene glycol sulphates in shampoos 435
Alkyl polyethyleneimine 636
Alkyl polyglycol ethers 636
Alkyl propionates 637
Alkyl pyridinium salts 635
Alkyl sulphates 634
  in shampoos 433
Alkyl sulphonates 634
Alkylamino acids 637
Alkyldimethylbenzylammonium chloride in baby
    products 112
Alkyldimethylbenzylammonium salts 635
Alkyltrimethylammonium salts 635
Alginates
  in protective creams 83, 85
  in rouge 340
Alginic acid, triethanolamine salt, in shampoos 445
Allantoin
  in hand creams 69
  in skin tonics 76, 77
  in sunscreens 243
Allergens 33, 39
Allergic contact urticaria 37
Allergy 33
Alloxan for skin colouring 333
All-purpose skin creams 51, 70
Almond oil
  in hair conditioners 508
  in hairdressings 484
Alopecia
  androgenetica 402, 416
  areata 416, 417
  diffuse 46, 418
  male pattern 15, 402, 416, 418

Alopecia (*cont.*)
  post-febrile 417
  post-partum 416, 417
Alpha olefin sulphonates
  in foam baths 94
  in shampoos 433
Alternaria solani 701
Alumina in protective creams 83
Alumina, hydrated, in toothpastes 611
Aluminium
  cans for aerosols 801
  corrosion 835, 853
  foil in laminates 854
  metal in eye cosmetics 342, 347
  powder, contamination 893
  soaps, in shampoos 445
  tubes for packaging 853
Aluminium bromohydrate as antiperspirant 129
Aluminium chlorhydrate
  as antiperspirant 127, 128, 129, 134, 136, 137, 138
  as astringent 74
Aluminium chloride as antiperspirant 126, 127, 129, 131
Aluminium lactate in face masks 282
Aluminium oxide in setting lotions 473
Aluminium salts
  as astringents 74, 75
  in toothpastes 604
Aluminium silicate in baby powder 113
Aluminium stearate in mascara 346
Aluminium sulphate
  as antiperspirant 129
  in nail strengtheners 373
Aluminium zirconium chlorhydrate as antiperspirant 129, 137
Alums
  as astringents 74, 76
  in mouthwashes '628
  in nail strengtheners 373
Amidoamine oxides in shampoos 446
Amine fluorides in toothpastes 603
Amine oxides
  in foam baths 95
  in hair conditioners 516
  in shampoos 440
Amines in antioxidants 724
Amino acids
  effect on hair growth 403
  in hair keratin 403, 412
  in hair tonics 504
  in moisturizing creams 63
  in skin 43
Amino acids, N-alkyl, in shampoos 440
Aminoalkanethiol derivatives in shampoos 446
Aminoanthraquinone dyes 530
Aminobenzaldehydes in sunscreens 235
Aminobenzenesulphonic acid in skin lighteners 272
Aminobenzoates in sunscreens 231, 232, 234, 235, 236, 238, 239, 243, 244
Aminobenzoic acid
  in hair pigmentation 416
  in sunscreens 232, 235, 243, 244, 256
  phototoxicity 37
Aminoplastics in packaging 851
Aminopropionic acid esters as insect repellants 212
Ammoniated mercury for skin lightening 267
Ammonium lauryl ether sulphate in shampoos 435
Ammonium lauryl sulphate
  in foam baths 93
  in shampoos 434
Ammonium salts in toothpastes 601
Amniotic fluid in hair tonics 504
Amphipathic molecules 632
Ampholytic surfactants 633
Amphoteric surfactants
  as germicides 668
  in foam baths 95
  in shampoos 440
Amyl acetate as nail lacquer remover 386
Anabasine sulphate as insect repellant 212
Anagen 14, 398
Anagen effluvium 416, 417
Anaphylactic sensitivity 36
Androgens
  absorption by skin 45
  in acne 18
  in hair growth 15, 402
  in sebaceous gland activity 16
  in skin pigmentation 9
Animal oils in hairdressings 484
Anionic surfactants 633
  as preservatives 693
  effects on hair 427
  in foam baths 93
  in shampoos 432
  interaction with preservatives 697
Anodizing of aluminium aerosol cans 838
Anonychia 365
Anthranilates in sunscreens 233, 236, 243, 244
Anthraquinone dyes 526, 527
Antiandrogens
  in hair growth 15, 402, 418
  in sebaceous gland activity 16, 18
  in skin 46
  in treatment of hair loss 502
Antibacterial agents
  as deodorants 132
  substantivity 658
Antibiotics
  in acne products 123
  in oral products 598
Antibodies 33, 34
Anticholinergic substances 12, 126
Antifungal agents for tinea pedis 204
Antigens 33
Antimony trisulphide in eye make-up 341
Antioxidants
  in hair dyes 542
  in shampoos 432, 446
Antiperspirant products
  aerosols 134, 826, 831
  creams 137
  foot cream 199
  foot powder 197
  foot spray 197
  OTC panel 128, 130
  roll-ons 138
  sticks 136
Antiperspirants
  action 126
  efficacy 130
  evaluation 130
  ingredient classification 128
  purpose 124
Antiseptics in sunburn products 257
Anti-wrinkle products 282
Apatite, synthetic 610
Apricot oil in hairdressings 484
Apocrine glands 16, 125
Aquasol aerosol 843
Arachidonic acid in skin 44
Aragonite 610
Argillaceous face masks 280
Arnica in hair tonics 503
Arquad antibacterial agent 665
Arrector pili muscles 12, 165
Arsenical pyrites as depilatory agent 142
Aryl sulphonamide–formaldehyde resins
  in nail lacquers 377
  in nail mending compositions 391
Ascorbic acid
  as antioxidant 718, 721, 724
  deficiency 44
  in skin creams 61
  in skin lighteners 271, 272
Ascorbyl oleate in skin lighteners 272
Ascorbyl palmitate
  as antioxidant 718, 721, 724
  in skin lighteners 271
Aspergillus niger 701
Astringents in face masks 281

Athlete's foot 192
 products 202
Atmospheric oxidation in hair waving 572
Atomizer spray 844
Attrition mills 300, 334
Autoxidation 707, 724
Avocado oil
 in hair conditioners 508
 in hairdressings 484
Azine dyes 527
Azo dyes 526, 527, 532
Azoles in sunscreens 233
Azulene in skin tonics 76

Baby products
 creams 112, 114
 lotions 113, 114
 oils 113, 116
 ointment 113, 116
 powder 113, 116
Bacillus subtilis 701
Bacteria
 role in dental caries 598
 in shampoos 447
 on the skin 675, 891
Bacteriostats
 in shaving foams 165
 in toothpaste 600
Bag-in-the-can aerosols 839
Baking soda in deodorants 132
Baldness 15, 418, 501
Ball mill 780
Bamboo extract in moisturizing creams 63
Bandrowski bases 538
Barcroft manometer 720
Barium sulphide as depilatory 151
Basal lamina 5
Bases in shaving creams 163
Basic dyes 526, 527
Bath products
 cubes 103
 oils 103
 salts 101, 193
 satins 109
 tablets 103
Beard hair 156, 165
 softening 156
 softening cream 157
Beau's lines 366
Beeswax
 in all-purpose creams 71
 in baby products 113, 116
 in cold creams 55
 in depilatories 142
 in eyebrow pencil 353
 in eyeshadow 348, 349
 in face powder 299
 in hairdressings 485
 in lip salves 331
 in lipsticks 322, 323, 324, 325, 327
 in mascara 343, 345, 346
 in rouge 335, 336, 337, 338
 in stick make-up 311
Bentonite
 contamination 893
 in face masks 277, 280
 in products for oily skin 121
 in protective creams 83
Bentonite clays
 in nail lacquers 383
 in nail lacquer removers 388
Benzalkonium chloride 664, 665, 666, 678, 687, 691, 698, 699
Benzethonium chloride 665, 666, 687
 in baby products 112
 in deodorants 132
 in mouthwashes 628
 in toothpaste 600
Benzoates
 as antioxidants 722
 as astringents 74
Benzocaine
 in depilatories 143
 in sunburn products 258
Benzoic acid as preservative 687, 688, 689
Benzophenones in sunscreens 225, 243, 244, 256
Benzoyl peroxide in acne products 121
Benzyl alcohol in shampoos 445
Benzyl benzoate as insect repellant 207, 208, 211
Bergamot oil
 as insect repellant 207
 in Berlock dermatitis 224
Berlock dermatitis 224
Betaines 637
 in shampoos 440, 441
Binding agents
 for compact face powders 302
 for toothpastes 612
Biodegradation of surfactants 639
Biodeterioration 675
Biphenyldisulphonates, hydroxy-, in sunscreens 233
Birch tar oil
 as insect repellant 207
 in hair tonics 504
Bismuth oxychloride
 in eyeshadow 347
 in lipsticks 320
 in rouge 336
Bismuth subnitrate in face powder 308
Bisphenols 659, 672
Bisulphite in hair straightening 585
Bithionol 657, 659, 660, 661, 672
 phototoxicity 37
Black skin, depilation 150
Blackheads 18, 119
Bleaches
 for hair 547
 in face masks 281
 in toothpaste 615
Blood vessels of skin 12
Bloom of face powders 294
Blooming bath oils 106
Blow waves 474
Blusher 334
Body odour 124
Body powders 108
Body talcs 108
Boiling points of propellants 809, 812
Borates in baby powder 113
Borax
 in bath salts 102
 in cold creams 55
 in hair waving 555, 568
Boric acid
 as preservative 687, 689
 in baby powder 113
Bovine growth hormone 16
Bovine serum albumin in anti-wrinkle products 282
Bradosol 665
Bradykinin in inflammation 31
Brassidic acid in hair tonics 503
Brevibacterium ammoniagenes in nappy rash 112
Brilliantines 484
Bromo acid dyes in lipsticks 315
Bronopol 668, 690
Bronze powder in eye shadow 347
Brushite 592
Brushless shaving cream 171
Brushless shaving stick 173
Bubble bath 92
Buccal epithelial test 626
Buildings, cleanliness 878
Bunions 191
Butane as propellant 479, 811
Butanediol, interaction with preservatives 695
Butoxypolypropyleneglycol as insect repellant 211
Butoxypyranoxyl as insect repellant 207, 209, 213
Butyl acetate as nail lacquer remover 387, 388

Butylated hydroxyanisole as antioxidant 718, 720, 721, 724
Butylated hydroxytoluene as antioxidant 718, 720, 721, 724
Butylene glycol as humectant 644
Butylethylpropanediol as insect repellant 207, 211

Cactus extract in moisturizing creams 63
Cade tar oil in hair tonics 504
Cadmium sulphide as antidandruff agent 421
Cake make-up 304, 306
Calamine in sunburn products 257
Calciferol in skin creams 61
Calcite 610
Calcium carbonate
in baby powder 113, 117
in face powders 291
in sunscreens 232
in toothpaste 610
Calcium chloride as humectant 642
Calcium hydroxide in hairdressings 491
Calcium phosphates in toothpaste 610
Calcium pyrophosphate 610
Calcium stearate in shampoos 446
Calcium thioglycollate as depilatory 147, 151, 153
Calculus, dental 590, 592, 600
Calluses 191, 200
Camphor
as insect repellant 206
in astringents 76, 78
in depilatories 143
in hair tonics 503
Camphor derivatives in sunscreens 243, 244
Cancer of skin 224
Candelilla wax
in eyebrow pencil 353
in eyeshadow 349
in lipsticks 321, 324, 325, 326, 327
in rouge 338
Candida albicans 701
in the mouth 626
in nappy rash 112
Candida parapsilosis in products 679
Cantharides tinctures as hair tonics 503
Cantharidin in hair tonics 503
Caprylates in athlete's foot products 203
Capsaicin in hair tonics 503
Capsicum tinctures in hair tonics 503
Captan 669
Carbanilides 657, 662, 663
Carbohydrates, effect on skin 42
Carbon black in mascara 343, 344
Carbon dioxide as propellant 479, 814, 815, 832, 833
Carboxymethylcellulose 613
Carboxymethylmercaptosuccinic acid, esters of, as antioxidants 725
Carboxyvinyl polymers
in hairdressings 490
in shampoos 446
Carcinogenicity 522
Caries, dental 593, 594
Carmine
for skin colouring 333
in eye cosmetics 342
Carnauba wax
in eyebrow pencil 353
in eyeshadow 350
in hairdressings 485
in lipsticks 321, 323, 324, 325, 326, 327
in mascara 343, 345, 346
in rouge 336, 337, 338
in stick make-up 311
Carotene in skin 264
Carragheen gum 612
in face masks 278, 281
Carthamine for skin colouring 333
Casein in face masks 278
Casein hydrolysate in hair conditioners 507
Cassia oil as insect repellant 207
Castor oil
in baby products 113
in eyebrow pencil 352
in eyeshadow 350
in hair conditioners 508
in hairdressings 487
in lipsticks 319, 324, 325, 326, 327, 330
in mascara 347
in rouge 338, 341
Catagen 14, 398
Catechol in skin lighteners 272
Cationic antibacterials 664
Cationic polymers
in hair conditioners 507
in hair rinses 517
in hair thickeners 512
in protective creams 86
in setting lotions 473
in shampoos 444
in skin tonics 76
Cationic surfactants 633
as antidandruff agents 421
as preservatives 690, 692
in hair conditioners 506
in hair rinses 513
in hairdressings 484
in hand creams 69, 70
in moisturizing creams 63
in shampoos 442
Cedar tar oil in hair tonics 504
Cedarleaf oil as insect repellant 207
Cedrus atlantica oil as insect repellant 207
Ceepryn 665, 666
Cell-mediated response 33, 34
Cellulose acetate
in nail elongators 389
in laminates 854
Cellulose derivatives
in protective creams 83
in shampoos 443, 446
in toothpaste 613
interaction with preservatives 697
Cellulose ethers in toothpastes 613
Cellulose, microcrystalline
in antiperspirant sticks 137
in face powders 290
Cellulose nitrate in nail lacquers 376
Ceresin wax
in eyeshadow 350
in hairdressings 485
in lip salves 331
in lipsticks 323
Cervical erosion 619
Cetavlon 665
as antidandruff agent 421
Cetomacrogol, interaction with preservatives 694
Cetrimide 654, 664, 665, 666
as preservative 690
Cetyl alcohol in eyebrow pencil 353
in shampoos 446
Cetylpyridinium chloride
as preservative 687, 691, 694, 698
in baby products 112
Cetyltrimethylammonium bromide 665
as preservative 687
in baby products 112
Chalones in skin 47
Chalk
contamination 682
in cosmetic stockings 309
in face powders 286, 291, 297, 298, 299, 303, 306, 308
in liquid make-up 310
in rouge 334
in toothpaste 610
Chamomile hair dyes 546
Chelating agents, role in oxidation 717, 725
Chemical disinfection of plant 885
Chilblain products 201
China clay in face masks 280
Chloramine T 670

Chloramines for plant disinfection 887
Chlorbutanol as preservative 687
Chlorhexidine 654, 658, 664, 666, 667, 670
as preservative 691, 699
in deodorants 132
in mouthwashes 628
in toothpaste 593, 600, 604, 614
Chlorhydroxides as astringents 74
Chlorides as astringents 74
Chlorinated phenols
in mouthwashes 627
in shampoos 458
Chlorine dioxide, effect on hair 412
Chlorine for disinfection of plant 887
Chlorobenzoic acid as preservative 687, 689
Chlorobutadiene as depilatory 151
Chlorocresol 659, 660
Chlorodiethylbenzamide as insect repellant 207
Chlorofluorocarbon propellants 809, 815
effect on ozone layer 474, 810
Chlorophenols as antidandruff agents 499
Chlorophyll in toothpaste 600
Chloroquine, effect on nails 367
Chloroxylenol 654, 658, 659
Cholesterol derivatives in hair rinses 517
Cholesterol in hairdressings 492
Cholic acid in hair tonics 504
Chromium oxide in eye cosmetics 342, 349, 351
Chromium salts as astringents 74
Cinnabar for skin colouring 333
Cinnamates in sunscreens 234, 235, 236, 243, 244
Cinnamic acid derivatives in sunscreens 233
Citrate esters in aerosol antiperspirants 135
Citrates as astringents 74
Citric acid
as astringent 75, 77, 78
role in oxidation 717, 718, 721, 722, 723
Citronella oil as insect repellant 207
Citronellol in hair tonics 503
Cladosporium resinae in products 678
Clarifying agents in shampoos 446
Clays in protective creams 83
Cleaning of manufacturing plant 880, 881, 888
Cleansing creams 53
Clinical trials for toothpastes 602, 603
Cloflucarban in deodorant soaps 139
Closures for packages 860
Clove oil as insect repellant 206
CMC 633
Coal tar 654
as antidandruff agent 499
Cobalt salts in eye cosmetics 342
Cochineal for skin colouring 333
Cocoa butter
in eyebrow pencil 352
in eyeshadow 348
in lipsticks 322, 323
in mascara 346
in rouge 338
Coconut oil in hairdressings 485
Co-dispensing aerosol valves 840, 841
Cold creams 51, 55
Collagen 10
Collagen hydrolysate in hair conditioners 507
Colloid mill 778
Colognes 75
Colour lightening 534
Colours
for bath salts 102
for eye cosmetics 342
for face powders 295
for foam baths 97
for lipsticks 315, 328
for nail lacquers 381
for rouge 335
for shampoos 432
for shaving foams 164
for toothpastes 615
Combing, ease of 430, 431
Comedones 18, 119
Compact face powder 301
Complement 29, 31
Conditioning agents
in hair dyes 544
in shampoos 431, 443
Contact urticaria 31
Contamination
clinical significance 678
of products by bacteria 677
Cooling agents in shaving foams 164
Copper
in formation of melanin 415
in greying of hair 416
Copper salts as astringents 74
Corn preparations 200
Corns 191
Corrosion
by humectants 642
of aerosol cans 835
of aluminium 835, 853
Corrosion inhibitors
in shaving foams 165
in toothpastes 615
Corticosteroids
absorption by skin 45
in dermatology 47
in treatment of vitiligo 17
Cortisone in dermatology 47
Corynebacterium acnes 18, 120
Cottonseed oil in eyebrow pencil 352
Coulter counter 301
Coumarin derivatives in sunscreens 233, 235, 247
Covering power of face powders 285
Crackling foam 185
Crayons for hair colouring 528
Cream, definition 50
Creams, *see* After-shave, All-purpose, Antiperspirant, Baby, Beard-softening, Cleansing, Cold, Foot, Foundation, Hand, Hand and body, Insect repellant, Make-up, Massage, Moisturizing, Nail, Night, Nutritive, Protective, Shaving, Sports, Vanishing
Cresols
as antiseptics 659
as preservatives 687, 691
Critical Micelle Concentration 633
Curlers, heated 473
Cuticle of the nail 363
removers 369
softeners 370
Cyanurates in nail bleaches 371
Cyclohexylacetoacetate as insect repellant 207
Cyproterone acetate
in hair growth 419
in sebaceous gland activity 16, 18
in skin 46
Cysteine in shampoos 446
Cysteine hydrochloride as antioxidant 718

Damage to hair by waving solutions 575
Dandruff 19, 419, 498
Decenoic acid as insect repellant 206
Decyl alcohol as insect repellant 212
Dehydroacetic acid as preservative 687, 689, 699, 700
Delayed hypersensitivity 34, 39
Demineralization of enamel 595
Density of toothpaste 617
Dental calculus 590, 592, 600
Dental caries 593, 594
Dental plaque 590, 591
Dentifrice, solid 618
Dentine 589
Denture cleansers 622
Deo-colognes, aerosol 833
Deodorant products foot cream 198
foot spray 198
soaps 139
sticks 139
Deodorants,
action 132

Deodorants (*cont.*)
 assessment 133
 purpose 124
Dermatan sulphate 11
Dermis 10
Dermo-epidermal junction 5, 13
Desmosomes 5, 6
Detergency 429, 637
Detergents
 evaluation as shampoo bases 429
 for cleaning plant 883
 for shampoos 432
 in toothpaste 611
Dexamethasone in skin lighteners 270
Dialkyl adipates in hairdressings 484
Dialkyl sebacates in hairdressings 484
Dialkyl sulphosuccinates 634
Dialkyldimethylammonium salts 635
Diaper rash 111
Diapers, cleansing 117
Dibenzalacetone in sunscreens 233
Dibromosalicylanilide 663
Dibutyl phthalate as insect repellant 207, 211
Dibutyl succinate as insect repellant 211
Dicalcium phosphate, anhydrous 610, 611
Dicalcium phosphate dihydrate 610
Dichlorophene 659, 661
 as preservative 687, 691
Dicyanamide in depilatories 148
Diethylene glycol as humectant 647
Diethylolthiourea in nail strengtheners 373
Diethyltoluamide as insect repellant 207, 208, 209, 210
Digalloyl trioleate
 in sunscreens 243, 244, 247
 phototoxicity 37
Diglycol stearate in rouge 339
Diguanidohexane, bischlorophenyl-, as preservative 687
Dihydrotestosterone
 in hair growth 418
 in skin 47
Dihydroxyacetone
 in sunscreens 243
 in suntan products 259, 260
Dihydroxynaphthoic acid in sunscreens 233
Dihydroxyphenylalanine in formation of melanin 9, 268, 413
Diisopropyl tartrate as insect repellant 207
Dilatency 769
Dilaurylthiodipropionates as antioxidants 721
Diluents in nail lacquers 380
Dimethyl carbate as insect repellant 207, 208
Dimethyl ether as propellant 479, 813, 831, 833
Dimethyl phthalate as insect repellant 207, 208
Dimethylol thiourea
 for strengthening hair 575
 in nail strengtheners 373
Dimethylol urea for strengthening hair 575
Dimethylsulphoxide, effect on skin penetration 45
Diometam 665
Dioxan, acetoxydimethyl-, as preservative 687
Dioxan, bromonitro-, as preservative 687
Dioxin 669
Diphenylamines in hair dyes 536
Diphosphonate in toothpastes 604
Diphtheroids in products 678, 679, 681
Dipropylene glycol as humectant 647
Disinfectants 653
Disinfection of manufacturing plant 880, 890
Dispenser pumps 844, 861
Disperse dyes 526, 527
Dispersible bath oils 106
Dissociation constants of acids used as preservatives 689
Distillation of water 869
Disulphide links
 in hair 406
 in keratin 406, 412
 in weakened hair 575
Dodecyl alcohol as insect repellant 212
Domiphen bromide 665, 666
Dopa in formation of melanin 9, 268, 413
Dopa quinone in formation of melanin 9
Double layer in emulsions 742
Dough mixer 775
Dowicil 667
Draize test for shampoos 460
Drying of hair 430
Drying out of products 641
Drying time of shampoos 445
Dusting powders 108

Earth-based face masks 280
Eccrine sweat glands 10, 12, 13, 125
Eczema, infantile 112
EDTA in shampoos 447
Egg yolk in hair tonics 504
Elastic fluids 769
Elastin 11
Elastomer pressure sprayer 844
Electrical charge in emulsions 741
Electro-corrosivity 838
Electrolysis for depilation 144
Electrolytes in shampoos 446
Electrostatic charge in aerosols 827
Embden-Meyerhof pathway in skin cells 42
Emollience 65
Emollients
 after-bath 109
 in bath oils 106
 in foam baths 96
 in hand cleansers 89
Emulsification 638, 794
Emulsifying agents 734
Emulsions
 analysis 754
 appearance 750
 application properties 754
 as hairdressings 491
 breakdown 731
 coalescence 732
 creaming 732
 determination of type 754
 electrical charge 741
 manufacture 747
 pH 744
 quality control 754
 rheological properties 752
 stability 650, 731, 733, 738, 749
 water loss 646, 650
Enamel, dental 588, 589
Endorsement of toothpastes 601
Enterobacter aerogenes 701
Enzyme inhibitors in toothpaste 600
Enzymes
 as depilatories 150
 in dental caries 596, 598
 in denture cleansers 624
 in hair dyes 543
 in hair waving 561
 in inflammation 29, 31
 in toothpastes 601
Eosin
 phototoxicity 37
 in lipsticks 315
 in rouge 333
 solubility 320
Eosinophils in inflammation 31
Ephilides 17
Epidermis 5
Epidermophyton floccusum in athlete's foot 192
Epidermophyton inguinale in athlete's foot 192
Epilation 142
Epoxy resins
 as aerosol can lacquers 837
 as tube lacquers 853
Equipment
 contamination 682
 design for cleanliness 879
Erythema 30
 caused by sunlight 222, 223, 225, 227
Erythemogenic radiation 222, 225

Erythrulose in suntan products 260
Escherichia coli in products 679, 701
Essential fatty acids
  in skin 44
  in skin creams 61
Essential oils in deodorants 133
Esters
  as plasticizers in setting lotions 472
  in moisturizing creams 65
  in vanishing creams 62
Ethanol
  as astringent 75
  in hair sprays 477
Ethoxylated alkyldimethylammonium salts 635
Ethoxylated alkylphenols in shampoos 439
Ethoxylated fatty acid amides in shampoos 439
Ethoxylated fatty acid diesters in shampoos 446
Ethoxylated fatty acid esters in shampoos 439
Ethoxylated fatty alcohols in shampoos 439
Ethoxylated fatty amines in shampoos 439
Ethyl acetate as nail lacquer remover 386
Ethyl alcohol
  as deodorant 133
  interaction with preservatives 695
Ethyl hexanediol as insect repellant 207, 208
Ethylene glycol
  as humectant 643, 645, 647, 651
  in moisturizing creams 64
Ethylene oxide for disinfection of plant 885
Ethylenediamine tetraacetic acid
  in oxidation 717, 721, 725
  in shampoos 447
Ethylhexyl palmitate in sunscreens 251
Ethylhexyl salicylate in sunscreens 243, 244
Ethylhexylcyanodiphenyl acrylate in sunscreens 243
Ethylhexyl palmitate in sunscreens 251
Ethylhexyl salicylate in sunscreens 243, 244
Eucalyptus oil as insect repellant 207
Eumelanin 9, 265, 413
  in skin 43
Evaporation rates of solvents 379
E-viton concept 227
Extinction coefficients of sunscreens 236
Extruder 776
Eye cosmetics, contamination 678
Eye irritation by surfactants 461
Eye make-up 341
  application 358
Eyebrow pencil 352
  application 358
Eyeliner 351
  application 359
Eyeshadow 347
  application 359
  cream 348
  liquid 351
  stick 350

Face masks 78, 79
Fast-filling valves for aerosols 808
Fats,
  contamination 681
  oxidation 707, 719
Fatty acid alkanolamides in shampoos 437
Fatty acid amides in hairdressings 484
Fatty acid esters in hair conditioners 508
Fatty acid soaps 634
Fatty acids
  in hair conditioners 508
  in shaving creams 160
  in shaving foams 162
  oxidation 707, 713
Fatty alcohol ether sulphates in foam baths 93
Fatty alcohol lactates in hairdressings 484
Fatty alcohol sulphates in foam baths 93
Fatty alcohols
  in hair conditioners 508
  in hairdressings 484
Fennel oil as insect repellant 207
Fentichlor 662
Ferric fluoride in toothpaste 603
Ferulic acid in hair tonics 504
Fibroblasts 10, 11
Fibrocyte 11
Fillers in foam baths 99
Filling equipment, contamination 683
Film thickness of sunscreen products 245
Filtration of water 872
Finger pumps 844
Finsen unit of erythemal flux 227
Fixanol 665
Flammability of aerosols 479, 481, 832
Flash points of hydrocarbon propellants 812
Flavours
  for mouthwashes 628
  for toothpastes 614
Floating bath oils 104
Flocculation 790
  in emulsions 732
Floors, cleanliness 878
Fluid energy mill 766
Fluorapatite 602
Fluoride,
  effect on dental enamel 598
  in dentine and enamel 589
  in tooth powder 618
  in toothpaste 601, 616
  in water supplies 589, 598
Fluorinated polymers
  in hair sprays 477
  in setting lotions 473
  in shampoos 446
Fluorocarbon resins in hair conditioners 508
Fluorocarbons, effect on ozone layer 474, 810, 831
Fluorometholone in skin lighteners 271
Fluorophene 663
Fluorozirconates in toothpaste 604
Foam of shampoo 429
Foam baths 92
  assessment 100
  dry 99
  foaming agents 93
  gels 99
  irritation 92, 93, 94, 99
Foam boosters
  for foam baths 95
  for shampoos 431
Foam stabilizers for shampoos 431
Foam valves for aerosols 806
Foaming 637
Foaming bath oils 107
Follicular keratosis 44
Food debris 593
  in dental caries 597
Foot bath preparations 193
Foot products
  creams 198
  massage emulsion 200
  powder 196
  sprays 197
Footwear, influence on foot health 190
Formaldehyde
  as antiperspirant 126
  as preservative 687, 688, 701
  for disinfection of plant 682, 885
  for disinfection of water 870
  in shampoos 447
  in toothpastes 614
  release during hair strengthening treatments 576
Formaldehyde resins in nail strengtheners 373
Formalin in toothpastes 614
Formates as astringents 74
Formic acid
  as astringent 75
  as preservative 687, 689
Foundation
  application 357
  creams 51, 62
  pigmented 67
Freckles 17

Free radicals in oxidation 709, 714, 716
Friction during shaving 157
Fruit extracts in moisturizing creams 63
Fuller's earth in face masks 280
Fungi
  in products 677
  in raw materials 681
  on skin 675

Gallates, alkyl, as antioxidants 718, 720, 721, 722
Gallic acid as antioxidant 722, 723
Galvanic action in corrosion 836
Gamma-valerolactone as nail lacquer remover 387
Gaskets for aerosol cans 804, 805
Gel hairdressings 494
Gelatin,
  effect on nail growth 366
  in face masks 278, 279
  in moisturizing creams 63
Gelling agents
  in hairdressings 486, 494
  in toothpastes 612
Germall 670
Germicides
  as dandruff treatments 421
  in baby products 112
  in shaving foams 165
  in skin products 120, 121
Gingivitis 599
Ginseng in hair tonics 504
Glass containers for aerosols 802
Glass in packaging 854
Gleamer 336
Glomerae 13
Gloss of teeth 620
Gluconates as astringents 74
Glucose
  absorption by skin 45
  as humectant 647
  in skin 42
Glutathione in shampoos 446
Glycerides in hairdressings 484
Glycerol
  as humectant 612, 643
  in antiwrinkle preparations 283
  in face masks 278, 279, 280, 281
  in hair sprays 477
  in moisturizing creams 64, 65
  inhibition of bacteria by 685
  interaction with preservatives 695
  metabolization 684, 697
Glycerol palmitate in shampoos 446
Glycerol stearate in shampoos 446
Glyceryl ether sulphonates in shampoos 436
Glyceryl monostearate
  in face powder 299, 302, 306, 307
  in liquid make-up 310
Glycol stearates in shampoos 446
Glycollates as astringents 74
Glycols in hair sprays 477
Glycolytic sequence in skin cells 42
Glyoxylates in sunscreens 243, 244
Godet 301
Gold metal in eyeshadow 347
Golgi-Mazzoni corpuscles 12
Grease
  reappearance on hair 446
  removal from hair 427, 428
Greasy hair 500
Greying of hair 415
Ground substance 11
Guaiacum resin as antioxidant 721, 722, 724
Guanidine in acne products 121
Guanine
  in nail lacquers 382
  in sunscreens 244
Guar gum in face masks 278
Gum arabic
  in mascara 344
  in setting lotions 471
Gum tragacanth
  interaction with preservatives 695
  in face masks 278, 279
  in hairdressings 489
  in mascara 347
  in rouge 340
  in setting lotions 471
  in toothpastes 612
Gums
  contamination 681
  degradation 684
  interaction with preservatives 697

Hair
  chemistry 403, 411
  colour 413
  condition 427, 428, 438
  cortex 396
  cross-section 555, 581
  cuticle 396
  damage 413
  density 397
  disorders 416
  extraction with solvents 428
  follicle 10, 12, 13, 14, 396, 397, 398, 400
  greying 415
  growth cycle 398, 401
  growth rate 400
  hydration 156
  lanugo 397
  loss 416, 501
  lustre 430
  mechanical properties 564
  medulla 396
  mineral constituents 409
  surface area 428, 526
  terminal 398
  vellus 398
  yield point 562
Hair colour rinses 528
Hair dye removers 547
Hair dyeing, mechanism 524
Hair dyes
  aromatic polyhydroxy 544
  henna 545
  oxidation 533
  para 533
  permanent 522, 533
  semi-permanent 522, 528
  staining of scalp 525
  temporary 521, 526
  uptake by scalp 525
  vegetable 545
Hair rinses 513
Hair sprays 474
  evaluation 480
  for men 490
  toxicity 481
Hair thickeners 512
Hair waving
  chemical heating 569
  chemistry 556
  cold processes 569
  evaluation 562
  foam products 572
  hot processes 567
  warm air processes 574
Hammer mills 300, 334, 762
Hand and body creams 69
Hand creams 51, 69
  aerosol 825
Handwashing test for antibacterials 657, 671
Haptens 33
Heat treatment of water 871
Heated curlers 473
Hectorite clays
  contamination 893
  in nail lacquers 383
  in toothpaste 614

Henna hair dyes 545
Herbal extracts in foam baths 97
Hexachlorophene 654
as dandruff treatment 421
as preservative 687, 691
in deodorants 132, 139
in mouthwashes 628
in toothpaste 600
phototoxicity 38
Hexadecyl alcohol
in lipsticks 319, 324
in mascara 345
Hexahydrophthalic acid, diethyl ester, as insect repellant 207
Hexamethylene tetramine
as preservative 690
in deodorants 132
Hexamethylenecarbamide as insect repellant 212
Hexamine as preservative 687
Hexokinase 638
Hexose-monophosphate shunt in skin cells 42
Hexylene glycol, interaction with preservatives 695
Hibitane 667
in toothpaste 600, 604
High shear mixer 776
Hirsutism 15, 46, 402, 418
Histamine 11, 13
in anaphylactic sensitivity 36
in inflammation 29, 31, 32
Histidine in skin 43
Histiocytes 10
HLB value 105, 737
determination 745
limitations 746
Holocrine glands 15
Homomenthyl salicylate in sunscreens 236, 243, 244
Honey in shampoos 445
Hormones
effect on hair growth 401
effect on sebum excretion 46
effect on skin 42, 45
Hot comb for hair straightening 581
Hot shaving foams 167, 840
Hotroom tests 130
Humectants
in cake make-up 305
in face masks 278
in moisturizing creams 63
in shaving creams 160
in shaving foams 164
in toothpastes 612
Humidity, controlled 643
Hyaluronic acid 11
Hyamine 665
Hydantoin, monomethyloldimethyl-, as preservative 687
Hydrocarbons
as propellants 479, 811, 815
in sunscreen products 233
Hydrocolloid-based face masks 278
Hydrogen bonds in keratin 404
Hydrogen peroxide
in hair bleach 547
in hair dyes 533
in mouthwashes 628
in nail bleaches 371
in toothpaste 615
Hydrolysed protein in moisturizing creams 63
Hydroperoxide in oxidation 709, 710, 721
Hydroquinone
as antioxidant 718, 724
in skin lightening 265, 267
in sunscreen products 233
Hydroquinone, butyl-, as antioxidant 721
Hydroquinone, monobenzyl ether, in skin lighteners 266, 271
Hydroquinone, monoethyl ether, in skin lighteners 265, 271
Hydroquinone, monomethyl ether, in skin lighteners 265, 271
Hydroxyapatite 588, 602
Hydroxybenzoates
as antioxidants 722
as preservatives 687, 688, 691, 698
Hydroxybenzoic acid as preservative 687, 689
Hydroxyethylcellulose 613
Hydroxyl values 717
Hydroxypropylcellulose in hairdressings 490
Hydroxyquinoline as preservative 687
Hydroxyquinoline, dichloro-, in shampoos 458
Hygroscopicity,
equilibrium, of humectants 644, 648
of products 64, 643
Hyperhidrosis 125
of the feet 193
Hyperkeratosis 30
Hyperpigmentation of skin 270
Hyperplasia 30
Hypersensitivity 33
Hypertrichosis lanuginosa 398
Hypochlorite
for plant disinfection 682, 887
for water disinfection 870
in nail bleaches 371

Ichthyocolla hydrolysate in hair conditioners 507
Imidazole sulphates 634
Imidazolidinyl urea as preservative 687, 691, 699
Imidazolines, alkyl- 637
Imidazolines in shampoos 440, 442
Immunoglobulins 34, 36
Immunological responses 33
Impedance of skin 128
Impetigo neonatorum 112
Inactivation of preservatives 692, 693
Indalone 207, 209
Indamine dyes 527
Indium fluorozirconate in toothpastes 604
Indium salts in toothpastes 604
Indoaniline dyes 527
Indole-5, 6-quinone in formation of melanin 6
Indophenol dyes 527
Infection of baby skin 111
Inflammation 11, 28
Inhalation toxicity of hair sprays 482
Innovair aerosol 842
Inoculation tests for preservatives 702
Inorganic fluorides in toothpastes 602
Insect repellant products
aerosol sprays 213
creams 215
gels 217
liquid creams 215
lotions 213
oils 214
pump sprays 214
sticks 218
towelettes 219
Insoluble sodium metaphosphate 611
Instant perms 575
Iodine as antiseptic 670
Iodine-containing sterilants for plant 887
Iodine values 717
Iodophors 654, 658, 670
Ion exchange for water purification 866
Ion exchange resins in foot products 204
Irgasan 654, 657, 659, 660, 661, 664
Irish moss 612
Iron
in hair 410, 413
in product deterioration 725
Iron oxides
in cosmetic stockings 310
in eyebrow pencil 352
in eyeshadow 351
in face powder 299, 305, 306
in liquid make-up 311
in mascara 342, 346, 349, 350
in nail lacquers 381
in rouge 335, 336
in stick make-up 311, 312

Iron salts as astringents 74
Irritation of baby skin 111
Isethionates 634
  in shampoos 436
Isoascorbic acid as antioxidant 718
Isopropanol
  as astringent 75
  in sunscreens 250
Isopropyl myristate
  in aerosol antiperspirants 134, 135
  in bath oils 105, 106
  in cuticle softeners 371
  in eyeshadow 349
  in foam baths 96
  in hair sprays 476, 477
  in hairdressings 486, 488
  in mascara 346
  in protective creams 84, 86
Isopropyl palmitate in aerosol antiperspirants 135
Isopropylcatechol in skin depigmentation 265, 272
Isothan 665
Ivory black in mascara 344

Jaborandi tinctures as hair tonics 503
Juglone in suntan products 260

Kaolin,
  contamination 682
  in baby powder 113, 117
  in face masks 278, 279, 280, 281
  in face powders 289, 291, 292, 298, 299, 303, 306, 308
  in liquid make-up 310
  in mascara 346
  in nail polishes 374
  in nail white 373
  in protective creams 83
  in rouge 334, 335, 337
  in stick make-up 312
  in sunscreen products 232
  interaction with preservatives 695
Karaya gum
  in setting lotions 471
  in shampoos 446
Keratin 403
  affinity of hair dyes 523
  disulphide links 406
  hydrogen bonds 404
  salt links 406
  structure 407
Keratin hydrolysate in hair conditioners 507
Keratinase as depilatory 150
Keratinization
  of hair 14
  of skin 5
Keratinocytes 5, 6, 8, 10, 264, 265
Keratohyalin 6
Kerosene, deodorized
  in hairdressings 484, 486
  in hand cleansers 89, 90, 91
Ketones, oxidation 714
Kieselguhr in face powders 291
Kinins 29
Klebsiella pneumoniae in hand creams 678
Kohl for eye make-up 341
Koilonychia 366
Krause end bulbs 12
Krebs cycle in skin cells 42
Kreis test for rancidity 719
Kritchevsky amides in shampoos 438
Kwashiorkor 44, 402

Lacquers
  for aerosol cans 837
  for collapsible tubes 853
Lactalbumin in anti-wrinkle products 283
Lactates
  as astringents 74
  in moisturizing creams 64
Lactic acid as astringent 75, 77
Lactoglobulin in anti-wrinkle products 283
Lactylates, acyl-, in shampoos 437
Lamina lucida 5
Laminar flow 767
Laminates
  for collapsible tubes 853
  for sachets 854
Lampblack in mascara 343, 344, 345, 347
Langerhans cells 5, 9
  in allergy 34
Lanolin
  in all-purpose creams 71
  in baby products 113, 115, 116
  in cleansing creams 56, 58, 59
  in eyebrow pencil 353
  in eyeshadow 348, 349
  in hand cleansers 89
  in hand creams 70
  in lipsticks 322, 323, 325, 327
  in mascara 343, 344, 345
  in massage and night creams 62
  in moisturizing creams 63, 65
  in protective creams 83, 86
  in rouge 335, 336, 338
  interaction with preservatives 695
Lanolin absorption bases
  in eyebrow pencil 352
  in eyeshadow 349
  in lipsticks 322
Lanolin derivatives
  in hair conditioners 508
  in hair sprays 476
  in hairdressings 484, 488
  in setting lotions 472
  in shampoos 445
Lanthionine in hair waving 556, 558, 562
Lanugo hair 397
Lather of shampoo 429
Laurel leaf oil as insect repellant 207
Lauryl sulphates in shampoos 433, 438
Lauryl sulphoacetates in foam baths 94
Lavender oil as insect repellant 207
Lawsone
  in sunscreen products 243
  in suntan products 260
Lead
  in hair 410
  in hair dyes 546
Lead salts as astringents 74
Lead tubes for packaging 852
Leak testing of aerosols 821
Lecithin
  as antioxidant 718, 721, 722, 723, 724, 725
  in hair conditioners 508
  in hairdressings 484
  in lipsticks 322
Legislation
  on antimicrobial agents 139, 654
  on antioxidants 721
  on antiperspirant ingredients 127
  on chlorofluorocarbons 474
  on colour additives 315, 381
  on fluorides 616
  on hair colorants 538
  on methylene chloride 832
  on preservatives 704
  on skin lightening agents 267
  on sunscreen ingredients 242
Lemon juice in cleansing creams 58
Lentigens 17
Leucocytes 10
  in inflammation 28
Leukonychia 366
Lighteners, hair 547
Lime water in hairdressings 491
Linear alkyl benzene sulphonates in foam baths 94, 100
Linoleates in moisturizing creams 65
Linoleic acid in skin 44
Lip salves 330
Lipids, ethoxylated, in moisturizing creams 65
Lipids in skin 43, 64

Lipoproteins in cell membranes 44
Lipstick, application 360
Liquid crystals in shaving creams 160
Liquid face powders 307
Liquid lipsticks 332
Liquid make-up 310
Lithium bromide, effect on hair 412, 413
Lithium mercaptopropionate in depilatories 148
Lithium thioglycollate in depilatories 148
Lotion
  definition 50
  for hair dyes 532
Lubricants in shaving foams 164
Lubrication of skin 157, 174
Lunula of the nail 363
Lustre
  of hair 430
  of teeth 620
Lymphocytes 10
  in allergy 34

Macromolecules, interaction with preservatives 695, 698
Macrophages 10
  in allergy 34
  in delayed hypersensitivity 35
  in inflammation 31
Magnesium aluminium silicate in shampoos 446
Magnesium carbonate
  in baby powder 113
  in face powders 291, 292, 309
  in liquid make-up 310
  in rouge 335
Magnesium lauryl ether sulphate in shampoos 435
Magnesium lauryl sulphate in shampoos 435, 453
Magnesium oxide
  in face powders 291, 292
  in sunscreen products 232
Magnesium peroxide
  in face masks 282
  in toothpastes 615
Magnesium stearate
  in face powders 288, 298
  in shampoos 446
Make-up cream 307
Malachite for eye make-up 341
Maleic alkyd resins in nail lacquers 377
Manganese borate in bath salts 194
Mannitol as humectant 647
Manufacture, hygienic 676, 877
Marasmus 403
Marinol 665
Mascara 341
  application 359
  cake or block 343
  contamination 678
  cream 344
Massage creams 51, 60
Massage, facial 356
Mast cells 10, 11
  in inflammation 31
Materia alba 593
MECSA antioxidant 725
Medicines act 600
Meissner corpuscles 12
Melamine in depilatories 148
Melanin 7, 261
  formation 413
  in hair 413
  in solar protection 9
  in tanning 222, 230
Melanocytes 5, 7, 10, 264
  in hair 413
  in skin 43
Melanocyte-stimulating hormone 9, 16, 47, 264
Melanogenesis 222
Melanosomes 8, 264, 413
Melissa oil as insect repellant 207
Membrane-coating granules 6
Men's hairdressings 483
Menthol
  in astringents 75, 77, 78
  in skin products 120
Mepacrine, effect on nails 367
Mercaptans as depilatories 146
Mercapto-amines in skin lighteners 272
Mercaptoethanol, effect on hair 412
Mercaptopropanediol as depilatory 150
Mercaptopropionic acid as depilatory 150, 152
Mercaptoquinoline oxide in shampoos 459
Mercaptoquinoxaline oxide in shampoos 459
Mercurithiosalicylate, sodium ethyl, as preservative 687
Mercury compounds
  as antiseptics 670
  for skin lightening 266
Mercury salts as astringents 74
Merkel cells 5
Metabolism, microbiol 676
Metallized dyes 526, 527, 532
Metals
  in hair 410
  in packaging 852
Metering valves for aerosols 807
Methyl eleostearate, oxidation 712
Methyl ethyl ketone as nail lacquer
  remover 388
Methyl linoleate 713, 723
Methyl linolenate 713, 723
Methyl oleate, oxidation 707, 710, 713, 723
Methyl salicylate in sunscreens 256
Methyl stearate, oxidation 713
Methyl taurides in shampoos 436
Methyl-B-nortestosterone in skin 46
Methylcellulose 613
  in eyeshadow 351
  in face masks 278
  in rouge 340
  interaction with preservatives 695, 698
Methylene chloride in propellants 816, 833, 834
Methylglycerin as humectant 647
Methylols for strengthening hair 575
Methylpentanediol, interaction with preservatives 695
METSA antioxidant 725
MGK insect repellants 210, 213, 214, 217, 218
Mica in face powder 299
Mica, titanium dioxide-coated
  in eyeshadow 349, 351
  in lipsticks 320
  in rouge 335
Micelles 634, 735
Microbial flora of the body 654, 657
Microbial growth in products 683
Microbial metabolism 676
Microbial standards 895
Micrococci in products 678, 680, 681
Microcrystalline waxes in hairdressings 485
Micro-organisms
  in mains water 866, 870
  in products 877
Micropulverizers 300
Mildness of shampoos 427, 457
Miliaria rubra 18
Mineral oil
  in aerosol antiperspirants 135
  in all-purpose creams 71
  in bath oils 105, 106
  in cleansing creams 57, 58, 59
  in depilatories 143
  in eyeliner 351
  in eyeshadow 350
  in foundation creams 68
  in hairdressings 484
  in hand cleansers 89
  in hand creams 70
  in lip salves 331
  in lipsticks 323, 327
  in mascara 345, 346
  in moisturizing creams 63, 65
  in protective creams 85, 86
  in rouge 335, 336, 337, 338

Mineral oil (*cont.*)
 in shampoos 445
 in sunscreen products 250
Mineral oil, gelled 486
Minimum erythema dose 227, 240, 248
Minimum infective doses of pathogens 895
Mink oil
 in hair conditioners 508
 in shampoos 445
Mixing 757
 dispersive 767
 of liquids with liquids 793
 of solids with liquids 786
 of solids with solids 760
 time 783
Moisturization 62
Moisturizers 109
Moisturizing creams 51, 60, 62
Moisturizing of skin 651
Moles 17
Molybdenum in water supplies 604
Monilethrix 416
Monilia albicans 701
Monobloc aerosol cans 801
Monoethanolamine lauryl ether sulphates in shampoos 435
Monoethanolamine lauryl sulphate in shampoos 434
Monofluorophosphate
 in toothpastes 601, 602, 616
 reaction with enamel 603
Monoglyceride sulphates 634
 in shampoos 435
Monoisopropyl citrate as antioxidant 721
Morpan 665
Moulding of lipsticks 329
Moulds in products 677
Moulting 15
Mucopolysaccharides
 in moisturizing creams 63
 in skin 42
Mud packs 280
Mutagenicity 522
Myristates in moisturizing creams 65
Myristyl myristate in hand cleansers 89, 91

Nacreous pigments in nail lacquers 382
Nail bleaches 371
Nail creams 372
Nail driers 389
Nail elongators 389
Nail lacquer 375
 base coat 385
 manufacture 385
 removers 386
 top coat 385
Nail mending compositions 391
Nail patella syndrome 365
Nail polishes 374
Nail strengtheners 372
Nail white 373
Naphthol in hair tonics 503
Naphthosulphonates in sunscreen products 233
Naphthosulphonic acids in sunscreen products 237
Naphthylthiourea, effect on hair pigmentation 415
Nappies, cleansing 117
Nappy rash 111
Natural moisturizing factor in skin 64
Nerves of skin 11
Neutralizers
 in cold waving 572
 in hair straightening 584
Neutrophils
 in delayed hypersensitivity 35
 in inflammation 31
Niacin in skin lighteners 273
Nicotinates in hair tonics 503
Nicotinic acid deficiency 44
Night creams 51, 60
Niosomes in skin 64
Nitroaminophenol dyes 530
Nitrocellulose
 in nail lacquers 376
 in nail mending compositions 391
Nitrogen as propellant 814, 815, 825
Nitromethane as corrosion inhibitor 834
Nitrophenylenediamine dyes 529
Nitrosamine formation 438
Nitrous oxide as propellant 814, 816
Nonionic surfactants 633
 as preservatives 693
 in foam baths 94
 in shampoos 437, 447
 interaction with preservatives 692, 693
Norbornylidenepentenone, dimethyl-, in sunscreen products 243
Nordihydroguaiaretic acid as antioxidant 718, 722, 723, 724
Nutrition, effect on hair growth 402
Nutritive creams 60
Nylon
 in nail lacquers 377
 interaction with preservatives 696
 titanium dioxide-coated, in face masks 278
Nylon fibres in mascara 346

Octacalcium phosphate 592
Octaphen 665
Octopirox 422
Odland bodies 6
Odour
 axillary 125, 132
 body 124
 foot 191
Oedema 30
 caused by sunlight 223, 228
Oestrogens
 absorption by skin 46
 in hair growth
 in sebaceous gland activity 16, 18
Oil absorption
 by skin 103
 of face powder materials 291
Oils, contamination 681
Oily skin, products 120
Oleates
 in all-purpose creams 71
 in baby products 113, 115
 in cleansing creams 56, 57, 58
 in hand creams 70
 in moisturizing creams 65
Oleyl alcohol in hairdressings 488
Olive oil
 in cold creams 55
 in hairdressings 484
Omadine 671
Onycholysis 365
Opacifying agents
 in foam baths 98
 in shampoos 446
Opacity of face powder materials 287
Optical density of sunscreen films 245, 247
Organic fluorides in toothpastes 603
Organosilicon compounds in sunscreen products 235
Orpiment as depilatory 142
Orthocortex 581
Osmotic pressure, effect on microbial growth 685
OTC antiperspirant panel 128, 130
OTC sunscreen panel 240, 242, 248, 249
Oxidation dyes 533
 bases 534
 couplers 535
 mechanism 535
 modifiers 535
Oxidation of products 707
Oxidizing agents for skin lightening 266
Oxybenzone in sunscreen products 231, 243
Oxygen tension, effect on microbial growth 685
Oxygenated face mask 282
Ozokerite wax
 in eyebrow pencil 352, 353

Ozokerite wax (*cont.*)
in eyeshadow 350
in hairdressings 485
in lipsticks 323, 325, 327
in mascara 346
in rouge 336, 337, 338
in stick make-up 312
Ozone layer, depletion by fluorocarbons 474, 810, 831

Pacinian corpuscles 12
Packaging
as source of contamination 894
for antiperspirants and deodorants 124
to prevent photodeterioration 725
Paddle mixers 771
Pangamic acid in skin creams 61
Pantethine in skin creams 61
Panthenol
in hair tonics 504
in setting lotions 473
in skin creams 61
Pantothenic acid 416
in hair tonics 504
in skin creams 61
Paper
in laminates 854
in packaging 855
Parabens 614, 678, 688
Paracortex 581
Paraffin sulphonates
in foam baths 94
in shampoos 433
Paraffin wax
in cleansing creams 57
in eyeshadow 349
in hairdressings 485
in mascara 346
in stick make-up 311
Parahydroxybenzoates 614
Parakeratosis 30
Particle size
of aerosol antiperspirants 128
of aerosols 828, 845
of face powders 286, 300
of hair sprays 483
Partition coefficient of preservatives 691, 697
Patch test 35
Peach kernel oil in hairdressings 484
Peanut oil in hairdressings 484
Pearlescent agents
in nail lacquers 382
in shampoos 446
Peeling agents in acne products 121
Pellagra 44
Pellicle 590, 591
Penicillium in products 679
Penicillium chrysogenum 701
Penicillium notatum 684
Pennyroyal oil as insect repellant 207
Peppermint oil as insect repellant 207
Peptides, acyl, in shampoos 437
Peracetic acid, effect on hair 412
Percutaneous absorption 45
Perfume atomizer 844
Perfume solubilization 96
Perfumes
as sensitizers 36
in bath oils 106
in bath salts 102
in deodorants 133
in face powders 296, 300
in foam baths 96
in hair sprays 477
in hair waving lotions 575
in lipsticks 322
in shampoos 432, 448
in shaving foams 164
in sunscreens 256
phototoxicity 38
Periodontal disease 592, 595, 599
Permeability of skin 45
Peroxide
determination 718
in oxidation 708, 713, 717, 721
values 717
Personnel as source of contamination 683, 891
Perspiration 124
Perspiration-resistance of sunscreen products 249
Persulphate in hair bleach 548
Petrolatum
in all-purpose creams 71
in baby products 113,114
in cleansing creams 57, 58, 59
in massage and night creams 62
in protective creams 83
in sunscreens 243
Petroleum jelly
in eyebrow pencil 352, 353
in eyeshadow 348, 349
in face powder 299
in lip salves 331
in lipsticks 322, 327
in mascara 345, 346
in rouge 336, 337, 338
in sunscreens 233
pH
correlation with preservative activity 686, 697
effect on microbial growth 685, 686
Phaeomelanin 9, 265, 413
in skin 43
Phase inversion 56, 746, 747
Phase separation 731
Phemerol 665
Phenol ether sulphates 634
Phenolic resins
as lacquers for aerosol cans 837
as lacquers for tubes 853
in packaging 851
Phenols
as antiseptics 654, 659
as preservatives 687, 688, 689, 698
for disinfection of plant 885
Phenolsulphonates as astringents 74, 76, 78
Phenoxyethyl alcohol as preservative 687, 691, 699
Phenoxypropyl alcohol as preservative 687
Phenyl salicylate in sunscreen products 244
Phenylenediamines in hair dyes 534
Phenylethyl alcohol in shampoos 445
Phenylmercuric salts
as antiseptics 670
as preservatives 687, 688, 691, 700
in shampoos 447
Phenylnaphthylamine as antioxidant 718
Phenylphenols as insect repellants 212
Pheromones 126
Phosphate esters in shampoos 446
Phosphates in bath salts 101
Phosphonium salts 635
Phosphoric acid as antioxidant 718, 722, 725
Photo-allergic reactions 37
Photodeterioration 725
Photosensitization by germicides 663
Photo-toxic reactions 37
Pigmentation of skin 7, 17
Pigments
in face powders 286
in lipsticks 318, 328
interaction with preservatives 695
ultraviolet transmission 287
Pili annulati 416
Pili torti 416
Pilocarpine in hair tonics 503
Pilomotor agents in shaving products 165, 177
Pilosebaceous apparatus 119
Pimento oil as insect repellant 207
Pimples 119
Pin disc mills 300
Pin mill 766
Pin perms 574

Pine oil as insect repellant 207
Pine tar oil in hair tonics 504
Piperonyl ether butoxide as insect repellant 207
Pipework for water systems 872
Piston pack aerosols 839
Pituitary hormones in skin 47
Pityriasis capitis 19
Pityrosporum orbiculare 19, 419
Pityrosporum ovale 19, 419, 421, 458
Placenta extract in hair tonics 504
Plant, contamination 682
Plaque, dental 590, 591
Plastic containers for aerosols 802
Plastic fingernails 389
Plastic fluids 769
Plasticizers
  in hair sprays 476, 477
  in nail elongators 389
  in nail lacquers 377
  in setting lotions 471
Plastics
  in face powders 291
  in packaging 850
  interaction with preservatives 696
Plough–shear mixer 766
Polish on teeth 620
Pollen in skin creams 61
Polyalkoxylated derivatives in shampoos 438
Polyalkoxylated ether glycollates in shampoos 437
Polyalkylene glycols in aerosol antiperspirants 135
Polyester films in laminates 854
Polyethylene glycol alkylethers in shampoos 445
Polyethylene glycol as humectant 643, 647
Polyethylene glycol derivatives 636
Polyethylene glycol ester sulphates 634
Polyethylene glycols
  in hair conditioners 508
  in hairdressings 484, 488, 494
  in setting lotions 472
  interaction with preservatives 694, 695, 698
Polyethylene in packaging 850
Polyethylene wax in baby products 113
Polyethyleneimine amides 636
Polyglyceryl ethers in shampoos 439
Polyglycol esters 636
Polyisobutylene in nail elongators 389
Polymeric sunscreen materials 239
Polymers, adsorbing, in antiperspirants 130
Polymers in setting lotions 471
Polymethacrylic acid in shampoos 445
Polyolefins in packaging 850, 861
Polyoxyethylene glycol as humectant 647
Polyoxyethylene sorbitol as humectant 647
Polyphenols in hair tonics 504
Polypropylene glycol alkylethers in shampoos 445
Polypropylene glycols in hairdressings 484, 488
Polypropylene in packaging 850, 854
Polysiloxanes
  in hair sprays 477
  in setting lotions 472
Polystyrene
  in face powders 292
  in nail elongators 389
  in packaging 851, 862
Polythene
  in face powders 292
  in laminates 854
  in packaging 850
  in products for oily skin 121
Polyunsaturated compounds, oxidation 711
Polyurethane, interaction with preservatives 696
Polyvinyl alcohol in face masks 277, 278
Polyvinyl alcohols in shampoos 446
Polyvinyl chloride in packaging 850, 854
Polyvinylpyrrolidone
  in coloured setting lotions 528
  in face masks 278
  in hair sprays 476
  in hairdressings 488
  in setting lotions 471
  in shampoos 446
  interaction with preservatives 695
Polyvinylpyrrolidone–iodine complex
  as antidandruff agent 499
  in shampoos 458
Polyvinylpyrrolidone–vinyl acetate copolymers
  in hair sprays 476
  in setting lotions 472
Pomades 485
Porositones in foundation creams 67
Potash alum in nail strengtheners 373
Potassium aluminium sulphate as antiperspirant agent 129
Potassium cyanide, effect on hair 412
Powder
  application 358
  pigmented 762, 765
Powder sprays, aerosol 827
Powder valves for aerosols 806
Powders
  cohesive 760
  segregating 760
  storage 766
Power number 783
Preservatives
  in foam baths 97
  in shampoos 431, 447
  in shaving foams 164
  in toothpaste 614
Pressure depressants for aerosols 810
Preval aerosol 842
Prick test for soluble antigens 37
Prickly heat 18
Primary irritation 38
Printing on packages 858
Procollagen 10
Proline in antioxidants 724
Propane as propellant 479, 811
Propanediol, bromonitro-, as preservative 687, 690, 693
Propanediol, chlorphenyl-, as preservative 687
Propanediol, interaction with preservatives 695
Propellants
  effect on ozone layer 474, 810, 831
  in aerosol shaving foams 164, 166
  toxicity 483
Propeller mixers 773
Propionates in athlete's foot products 202, 203
Propionic acid as preservatives 687, 689
Propyl gallate as antioxidant 718, 721, 722
Propylene glycol
  as humectant 643, 645, 646, 647, 650
  as preservative 692
  in anti-wrinkle products 283
  in face masks 278
  in hand cleansers 89, 90
  in moisturizing creams 64, 65
  interaction with preservatives 695
Propylene glycol glucoside as humectant 647
Propylene glycol palmitate in shampoos 446
Propylene glycol stearate in shampoos 446
Prostaglandins in irritation 31
Protective creams 51
Protein deficiency 44
  effect on hair growth 44
Protein hydrolysates
  in hair conditioners 507
  in hair sprays 477
Protein, hydrolysed, in moisturizing creams 63
Proteins
  in hairdressings 484
  in setting lotions 473
  in shampoos 445
  interaction with preservatives 697
  synthesis in skin 43
Proteolytic theory of dental decay 596
Pseudofolliculitis barbae 151
Pseudomonas aeruginosa 701
  in nail infections 367
Pseudomonas sp. in products 678, 679, 684, 685
Pseudoplasticity 768

Psoralens
  for treatment of vitiligo 17
  in suntan products 224, 258
Psoriasis 19
  effect on nails 365, 366
Pumice in nail cleaners 371
Pump applicators 844, 860, 861
  for hair sprays 478
Pumping capacity 786
Purcellin oil in vanishing creams 62
PVP–iodine complex
  as antidandruff agent 499
  in shampoos 458
Pyrethrum as insect repellant 206, 212
Pyridine-N-oxides as antiseptics 671
Pyridinethiol-N-oxide derivatives as antidandruff agents 421, 458, 499
Pyridinethiones as antiseptics 671
Pyrogallol as antioxidant 723

Quaternary ammonium compounds 654
  as antidandruff agents 499
  as antiseptics 665
  as preservatives 687, 688, 690, 694, 695, 700
  for disinfection of plant 682, 887
  in baby products 112
  in cuticle softeners 371
  in hair conditioners 506, 513, 516
  in mouthwashes 628
  in setting lotions 473
Quaternary ammonium surfactants in shampoos 458
Quaternium-18, contamination 893
Quaternized amides
  of ethylenediamine 635
  of polyethyleneimine 635
Quaternized diamine salts 635
Quince seeds, mucilage, in mascara 344
Quinine salts
  in hair tonics 503
  in sunscreens 233, 237
Quinoline derivatives in sunscreens 233
Quinquina in hair tonics 503

Radiotracer method for abrasivity of toothpaste 619
Rancidity 717
Rate of shear 768
Raw materials, contamination 681, 893
Rayon fibres
  in mascara 346
  in nail mending products 391
Razor blades, corrosion 171, 174
Reactive dyes 531
Reducer in cold waving 570
Reducing agents in hair waving 558
Refractive index of face powder materials 286
Regeneration time of cells 5
Relaxers in hair straightening 583
Remineralization of enamel 604
Replacement time of cells 5
Resins in hair sprays 475
Resorcinol
  as antidandruff agent 421, 499
  as antiseptic 654
  in acne products 121
Reticulin 11
Retinoic acid
  in acne products 122
  in skin lighteners 270, 271
Reverse osmosis for purification of water 869
Reynolds number 767, 783, 786, 788
Rhodotorula rubra in products 679
Rhubarb extract in hair tonics 503
Rhusma as depilatory 142
Ribbon blender 763
Riboflavine deficiency 44
Ribosomes 43
Ricinoleates in deodorants 133
Ringed hair 416
Ringworm of the feet 192
Rinses for hair 513
Rinsing of shampoo 430
Robertson-Berger meter 248
Roccal 665
Roll ball applicators 860
  for antiperspirants 138
  for pre-shave lotions 177
Roll mill 779
Roller perms 574
Rolling skin preparation 201
Roll-on antiperspirants 138
Rose water in astringents 75, 76
Rosin
  in depilatories 142
  in mascara 347
Rosins, wood, in shampoos 445
Rouge
  application 358
  compact 334
  dry 334
  liquid 340
  wax-based 336
Royal jelly in skin creams 61
Rubber
  as propellant reservoir 167
  in depilatories 143
  interaction with preservatives 696
Rubber ball pump 844
Rubber-based face masks 277
Ruffini end organs 12
Rutgers 612 207, 208

Salicylanilides 657, 662
  halogenated, in shampoos 458
Salicylates
  as astringents 74
  in sunscreens 233, 234, 236, 243, 247
Salicylic acid
  as antidandruff agent 19, 421, 499
  as preservative 687, 689
  effect on nails 366
  in acne products 122
  in face masks 282
  in ointments 45
Saliva 590
Salol in sunscreen products 256
Salt links in keratin 406
Sand mill 780
Sandalwood oil as insect repellant 207
Sarcosinates 634, 638
Sarcosinates, acyl-, in shampoos 436
Sassafrass oil as insect repellant 207
Saturated compounds, oxidation 713
Scale-up 788
Scalp, area 525
Schiff test for aldehydes 719
Scum 168
Sea salt foot baths 194
Sebaceous glands 15, 119, 124
Seborrhea 500
Seborrheic dermatitis 420
Sebum 16, 43, 124
  composition 16
  effect on hair 427
  excretion 500
  in acne 119, 120
Secondary alkyl sulphates 634
Segregation of mixtures 760
Selenium disulphide 654
  as antidandruff agent 19, 421, 499
Sensitization by preservatives 700
Sensitizers 35, 39
Sepro can aerosol 839
Sequestering agents in shampoos 431, 447
Sericine in hair tonics 504
Sesame oil
  in cleansing creams 55
  in foundation creams 69
  in hairdressings 484
  in shampoos 445

Sesamol as antioxidant 724
Setting lotions 470
  coloured 528
Shampoos
  acid balanced 459
  aerosol 454, 830
  antidandruff 458
  baby 435, 457
  basic requirements 427
  clear liquid 449
  conditioning 455
  detergency 428
  dry 454
  for hair dyes 533
  gel 451
  instrumental evaluation 431
  liquid cream 450
  lotion 450
  mild 434, 457
  oil 453
  powder 453
  raw materials 431
  safety 431, 460
  solid cream 451
Shaving cream
  brushless 171
  cream, lather 159
  foam, aerosol 161, 829, 840
  foam, heated 167, 840
  foam, pre-electric 178
  gel stick, pre-electric 173
  lotion, pre-electric 176
  lotion, pre-shave 75
  powder, pre-electric 174
  stick 161
  stick, brushless 173
  talc stick, pre-electric 173
Shear stress 768, 778
Shellac
  in eyeliner 351
  in hair sprays 475
  in setting lotions 471
Silica
  in baby powders 113, 116
  in face powders 291, 294
  in nail polishes 374
  in nail white 373
  in products for oily skin 121
  in setting lotions 473
Silicates
  in face powders 294
  in setting lotions 473
Silicon compounds, organo-, in sunscreens 235
Silicones
  in aerosol antiperspirants 135
  in antiperspirant creams 137
  in antiperspirant roll-ons 138
  in antiperspirant sticks 137
  in baby products 113, 115
  in hair conditioners 508
  in hair rinses 517
  in hair straightening 581
  in hand creams 70
  in lip salves 331
  in lipsticks 323
  in mascara 344, 345
  in moisturizing creams 63, 65
  in protective creams 83, 84, 85, 86
  in setting lotions 472
  in shampoos 445
  in shaving products 157, 164, 172, 174, 175
Silicone waxes in lipsticks 322
Silk, powdered, in face powders 294
Silver metal in eye cosmetics 342
Silver salts as astringents 74
Skin
  amino acids 43
  area in human beings 3
  blood vessels 12
  colour 264
  disorders 16
  friction 157
  innervation 11
  lipids 43
  lubrication 157
  permeability 3
  pigmentation 7, 43, 47
  racial differences 264
  respiration 43
  temperature control 13
  weight in human beings 3
  wounding 47
Skin bacteria of plant operators 891
Skin fresheners 75
Skin test for hair dyes 539
Skin toners 75, 76
Slip of face powder materials 292
Slip point of creams 51
Soap-based mouthwashes 627
Soaps
  antibacterial 656
  as emulsifiers in hand cleansers 89
  as preservatives 693
  in shaving creams 159, 171
  interaction with preservatives 697
Sodium aluminium chlorohydroxylactate as antiperspirant agent 129, 136
Sodium benzoate in toothpastes 614
Sodium bicarbonate in deodorants 132
Sodium bisulphite as antioxidant 718
Sodium carbonate in bath salts 101
Sodium carboxymethylcellulose
  in face masks 278
  in hairdressings 490
Sodium chloride in bath salts 102
Sodium fluoride in toothpaste 601, 602, 616
Sodium formaldehyde sulphoxylate as antioxidant 718
Sodium hypochlorite
  as antiseptic 670
  in denture cleansers 624
  in mouthwashes 628
Sodium lactate as humectant 643, 645, 647
Sodium lauryl ether sulphate
  in foam baths 93
  in shampoos 435
  in toothpastes 612
Sodium lauryl sulphate
  in foam baths 93
  in shampoos 434
  in toothpastes 611
Sodium metabisulphite as antioxidant 718
Sodium metaphosphate, insoluble 611
Sodium metasilicate in depilatories 149
Sodium monofluorophosphate
  in toothpastes 602, 616
  reaction with enamel 603
Sodium N-lauroyl sarcosinate in toothpastes 598, 600
Sodium perborate
  in denture cleansers 622
  in mouthwashes 628
  in nail bleaches 371
  in toothpastes 615
Sodium percarbonate in denture cleansers 622
Sodium pyrrolidone carboxylate in skin 64
Sodium ricinoleate in toothpastes 612
Sodium sesquicarbonate in bath salts 101
Sodium stearate in deodorant sticks 139
Sodium sulphide
  as depilatory 145
  effect on hair 412
Sodium sulphite as antioxidant 718
Sodium sulphoricinoleate in toothpastes 612
Sodium thiosulphate as antioxidant 718
Sodium xylene sulphonate in shampoos 446
Soft permanent waves 563, 575
Solder for aerosol cans 801
Solehorn of the nail 363
Solid dentifrice 618
Solids, interaction with preservatives 695

Solubilization 638
Soluble bath oils 107
Solvent-assisted hair dyeing 524, 532
Solvent extraction of hair 428
Solvents
  for nail lacquers 378
  in hair sprays 477
  in shampoos 446
Sorbic acid as preservative 687, 689, 696, 698, 700
Sorbitan sesquioleate in hairdressings 492
Sorbitol
  as humectant 612, 643, 644, 645, 646, 647, 650, 651
  in all-purpose creams 71
  in face masks 277
  in moisturizing creams 64, 65
  inhibition of bacteria 685
  metabolization 684, 697
Sorbitol esters, polyoxyethylated, in shampoos 439
Spermaceti
  in cleansing creams 55, 57, 58
  in eyeshadow 348, 349
  in foundation creams 68
  in hairdressings 485
  in mascara 346
Split ends of hair 508
Spoilage of products 675, 677
Spoon nails 366
Sports creams 71
Spots, incidence 119
Spreading bath oils 104
Spreading coefficient 105
Spreading of shampoo 430
Squalane
  in cleansing creams 58
  in hand creams 70
Squeeze packs 844, 859
Staining
  of clothes by aerosol antiperspirants 134, 135
  of scalp by hair dyes 525
Staining dyes in lipsticks 315
Stains for artificial skin tanning 258
Stannic oxide in nail polishes 374
Stannites as depilatories 146
Stannous fluoride in toothpastes 601, 602, 616
Stannous fluorozirconate in toothpastes 604
Staphylococci in products 678, 680
Staphylococcus albus in acne 120
Staphylococcus aureus 678, 701
  in acne 120
  on plant operatives 892
Staphylococcus epidermidis 18
  in foot odour 191
  on plant operatives 891
Starch
  contamination 682
  degradation 684
  in antiperspirants 129, 137
  in baby powders 113, 117
  in face powders 290, 291, 292, 294, 297, 298, 299, 302, 303, 308
Starch ethers in toothpastes 613
Static electricity on hair 431
Static mixers 776
Steam disinfection of plant 884
Stearates
  in all-purpose creams 71
  in baby products 113, 115
  in body powders 108
  in cleansing creams 56, 57
  in face powders 286, 291, 297, 298, 299, 302, 304, 307, 310
  in foundation creams 68, 69
  in hand creams 70
  in moisturizing creams 63
Stearic acid
  in all-purpose creams 71
  in hand creams 70
  in rouge 339
  in skin creams 83, 84, 86, 89, 90
  in vanishing creams 66
Stearin
  in face powder 299, 302
  in rouge 339
Stearyl alcohol
  in antiperspirant sticks 136
  in shampoos 446
Stick applicators 860
  for antiperspirants 136
  for deodorants 139
  for depilatories 149
  for make-up 311
Stinging 32
Stockings, cosmetic 309
Stone mills 778
Storage testing of aerosols 838
Stratum basale 5
Stratum corneum 6
  hydration 45
Stratum germinativum 5
Stratum granulosum 6
Stratum intermedium 6
Stratum lucidum 6
Stratum spinosum 6
Streptococcus mitis 701
Strontium sulphide as depilatory 145
Strontium sulphydrate as depilatory 145
Styptic pencils 75, 80
Styrax, tincture, in rouge 341
Sucrose esters in baby products 114
Sugar, role in dental caries 596
Sugar syrup in mascara 344
Sulphates as astringents 74
Sulphides as depilatories 145
Sulphites
  in hair straightening 585
  in hair waving 558
  in nail bleaches 371
Sulphonated oils in hand cleansers 88
Sulphonated polystyrenes in setting lotions 473
Sulphonium salts 635
Sulphonyl fatty acids 634
Sulphosuccinates
  in foam baths 94
  in shampoos 435
Sulphosuccinates, dialkyl- 634
Sulphur
  as antidandruff agent 421, 499
  as antiseptic 654
  in acne products 121
  in face masks 281
  in hair tonics 503
Sulphurous acid as preservative 687, 689
Sun products, aerosol 825
Sun Protection Factor 240, 248
Sunburn 223, 228, 230
Sunflower seed oil 44
  in hairdressings 484
Sunlight, effects on the body 222
Sunscreen index 234, 245, 247
Superamides in shampoos 438
Surface energy 632
Surface tension 731, 735, 738
  effect on microbial growth 685
Surfactants
  for shampoos 432
  in shaving products 157, 163, 169
  in toothpastes 609, 611
  metabolization 684, 697
  physical properties 464
Suspending agents
  for nail lacquers 383
  for shampoos 432
Suspension of solids in liquids 792
Sweat 14
  glands 13, 125
Sweating 13, 124
Sweetening agents in toothpastes 614
Swelling agents in hair straightening 585
Swift stability test for rancidity 720

Synergism
  in antioxidants 716
  in antiseptics 672
  in preservatives 691

Tablet for toothcleaning 597
Talc
  in baby powder 113, 117
  in eyeliner 351
  in face powders 286, 288, 291, 292, 293, 297, 298, 299, 303, 304, 305, 306, 308, 309
  in liquid make-up 310
  in make-up cream 307
  in nail polishes 374
  in products for oily skin 121
  in protective creams 83
  in rouge 334, 335, 337
  in sunscreens 232
  interaction with preservatives 695
  pre-electric shave stick 173
  sterilization 113
Talcum powder 108
Tannic acid
  in mouthwashes 628
  in sunscreens 233, 256
Tanning 222, 225
Tar
  as dandruff treatment 421
  in hair tonics 504
Tartaric acid
  interaction with preservatives 695
  role in oxidation 717, 725
Tartrates as astringents 74
Taurides, methyl-, in shampoos 436
Taurines 634
Tea-tree oil as insect repellant 207
Tellurium dioxide as dandruff treatment 421
Telogen 14, 398
Telogen effluvium 415, 416, 417
Temperature control in skin 13
Temperature, effect on microbial growth 686
Teratogenicity 522
Terminal hair 15, 398
Terpenes in deodorants 133
Testosterone
  in hair growth 15, 418
  in sebaceous gland activity 16
  in skin 46, 47
  in skin pigmentation 9
Tetrachlorophene 672
Tetrachlorosalicylanilide 663
  as preservative 687
  phototoxicity 38
Tetracycline
  effect on nails 367
  in acne therapy 18, 123
Tetramethylthiuram disulphide 657, 669, 671
  as preservative 687
Thermoplastic resins in packaging 850
Thermosetting resins in packaging 851
Thesaurosis caused by hair sprays 482
Thiamine hydrochloride as insect repellant 206
Thickening agents for shampoos 432
Thickening of hair 512
Thinning agents for shampoos 432
Thiodiglycol as depilatory 152
Thiodipropionates as antioxidants 725
Thiodipropionic acid as antioxidant 721
Thiodisuccinic acid, esters, as antioxidants 725
Thioglycerol
  as antioxidant 718
  as depilatory 150, 152
Thioglycollates
  as depilatories 146, 148, 149
  effect on hair 412, 558, 559, 560, 570
  in hair straighteners 583
Thioglycollic acid
  as antioxidant 718
  in hair dyes 532
  in hair waving 558, 570
Thiolactates as depilatories 148
Thiolactic acid as depilatory 148, 150, 153
Thiols in hair waving 558, 560
Thiomalic acid as depilatory 152
Thiomersal preservative 691
Thiosorbitol as antioxidant 718
Thiouracil, effect on hair pigmentation 415
Thiourea in depilatories 149
Thixotropy 769
Thymol
  in mouthwashes 627
  in shampoos 458
Thymol derivatives as preservatives 687
Thyroxine 401
Tin compounds
  as astringents 74
  in toothpaste 604
Tin, corrosion 835
Tin tubes for packaging 853
Tinea pedis 192
Tinea unguium 192
Tinplate, corrosion 836
Tinplate cans for aerosols 801
Titanium compounds in toothpaste 604
Titanium dioxide
  in cosmetic stockings 310
  in eye cosmetics 342
  in eyeliner 351
  in eyeshadow 348, 350
  in face powders 285, 292, 298, 304, 305, 306, 308
  in lipsticks 318
  in liquid make-up 310, 311
  in make-up cream 307
  in nail lacquers 381
  in nail white 373
  in rouge 334, 335, 336, 337
  in shampoos 446
  in stick make-up 311, 312
  in sunscreens 231, 232, 233, 243
  interaction with preservatives 695
Tocopherols as antioxidants 718, 721, 722, 724, 725
Toe-nails, in-growing 192
Toilets in manufacturing plant 892
Toluenediamines in hair dyes 535
Tonofibrils 6
Tonofilaments 6
Toothbrush 621
  electric 622
Toothbrushing 621
Toothcleaning tablet 597
Toothpaste
  aerosol 825
  formulation 615
  manufacture 616
  striped 615, 860
  transparent 611
Toothpowders 617
  manufacture 618
Toxicity
  of antioxidants 723
  of hair dyes 522, 525, 538
  of hair sprays 481
  of hair waving lotions 575
  of humectants 642, 651
  of preservatives 699
  of propellants 483
  of surfactants 639
Tragacanth gum in shampoos 446
Transepidermal water loss 62
Transfer valves for aerosols 808
Transit time of cell cycle 5
Transition cells 6
Transparent lipsticks 330
Transparent toothpaste 611
Tretinoin in acne products 122
Tri-aerosol 840
Tribromosalicylanilide 654, 657, 663
Tribromsalan in deodorant soaps 139
Tricalcium phosphate 610
Tricarboxylic acid cycle in skin cells 42

Trichlorocarbon in deodorant soaps 139
Trichlorocarbanilide 654, 657, 658, 663, 672
  as preservative 687
  in shampoos 458
Trichloroethane in propellants 816, 832, 834
Trichlorophenylacetic acid as preservative 687
Trichlorosalicylanilide 657, 663
  as preservative 687
Trichophyton floccusum in foot odour 191
Trichophyton interdigitale in athlete's foot 192
Trichophyton mentagrophytes in athlete's foot 192
Trichophyton rubrum in athlete's foot 192
Trichophytosis pedis 192
Trichorrhexis nodosa 416
Trichorrhexis invaginata 416
Trichosiderins in hair 413
Triclosan
  in deodorants 132, 139, 140
  in skin products 120
  in sunburn products 258
Triethanolamine as humectant 647
Triethanolamine lactate as humectant 647
Triethanolamine lauryl ether sulphate in shampoos 435
Triethanolamine lauryl sulphate in shampoos 435
Triethanolamine soaps in shampoos 446
Triethanolamine stearate in sunburn products 257
Triethylene glycol as humectant 647
Trigger sprayers 845
Trihydroxybutyrophenone as antioxidant 721, 724
Triphenylmethane dyes 527
Triple roll mill 779
Tropocollagen 10
Tubes for packaging 852, 854, 860
Turbine mixers 771
Turbulent flow 767
Turkey red oil in toothpastes 612
Turnover time of cell cycle 5
Turpentine in hair tonics 503
Tyrosinase in melanin formation 8, 265, 267, 413, 415
Tyrosine in melanin formation 9, 43, 265, 268, 413

Ultrafiltration of water 869
Ultrasonic homogenizer 780
Ultraviolet absorbers for shampoos 432
Ultraviolet radiation—A, B and C ranges 226
Ultraviolet sterilization of water 871
Umbelliferone in sunscreen products 235, 236, 247
Undecenoic acid as insect repellant 211
Undecenyl alcohol as insect repellant 212
Undecylenates in athlete's foot products 203
Undecylenic acid ethanolamides in shampoos 458
Ungulina officinalis extract in deodorants 133
Urea
  absorption by skin 45
  as humectant 647
  in cuticle softeners 371
  in toothpastes 601
  interaction with preservatives 697
Urea-bisulphite, effect on hair 412
Urea-formaldehyde foam in face powders 291
Urginea maritima in hair tonics 503
Uric acid in sunscreen products 233
Urocanic acid in skin 230
  in skin lighteners 273
Urticaria 31

Vaccine against dental caries 598
Valerolactone as nail lacquer remover 387
Valve homogenizer 779
Vancide 669
  as dandruff treatment 421
Vanillic acid as preservative 687
Vanillin
  as preservative 687
  in athlete's foot products 204
Vanishing creams 51, 62
Vantoc 665
Vapour phase taps for aerosols 807
Vapour pressure of propellants 810, 811, 812
Vaseline in face powder 299
Vegetable extracts in moisturizing creams 63
Vegetable oils
  in depilatories 143
  in hairdressings 484
  in moisturizing creams 63, 65
Vellus hair 15
Velocity head 786
Venturi aerosols 841, 844
Verrucae 191
Vertical vortex mixer 766
Vinyl acetate–crotonic acid copolymers
  in hairsprays 476
  in setting lotions 472
Vinyl-based face masks 277
Vinyl emulsions in shampoos 446
Vinyl latexes in shampoos 446
Vinyl resins as tube lacquers 853
Violuric acid in sunscreen products 233, 237
Viscosity 768
  of foam baths 97
  of humectants 642, 647
Viscosity modifiers in shampoos 432, 446
Vitamin A
  deficiency in skin 44
  in acne products 122
  in skin lighteners 270
Vitamin B_2 deficiency 44
Vitamin C
  deficiency 44
  in prickly heat therapy 19
Vitamin D in skin 223, 265
Vitamins
  effect on hair growth 402
  in hair tonics 504
  in hairdressings 484
  in skin creams 61
Vitiligo 17

Wads for closures 860
Walls, cleanliness 878
Warburg-Dickens shunt in skin cells 42
Warburg respirometer 720
Washability of skin creams 59
Washrooms in manufacturing plant 892
Water absorption of face powder materials 292
Water-based aerosols 830
Water bath testing of aerosols 821
Water, contamination 681, 865
Water fluoridation 598
Water jets for toothcleaning 622
Water loss from skin 62, 111, 651
Water resistance of sunscreen products 249
Wax, depilatory 142
Wax-based face masks 276
Waxes
  contamination 681
  in cleansing creams 57, 58, 59
  in foundation creams 68
  in hair conditioners 508
  in hairdressings 485
  in lipsticks 321, 328
  in massage and night creams 62
  in moisturizing creams 65
  in rouge 336
Wet white face powder 307
Wetting 637
Wheat germ oil in hair conditioners 508
White spot lesion 595
Whitlockite 592
Witch hazel distillate as astringent 75
Women's hairdressings 470
Wool
  composition 404, 409
  properties 412
Wool wax alcohols in hairdressings 492
Wormwood oil as insect repellant 207

Xanthene dyes 527
Xylenol, metadichloro- 659

Xylenols as preservatives 687, 692
Xylitol as humectant 647

Yeasts
  in products 678, 679, 680, 681, 701
  in raw materials 681
Yellow nail syndrome 367

Zephiran 666
Zeta potential 742
Zinc acetate in nail hardeners 373
Zinc and castor oil ointment 116
Zinc hydroxymethylsulphinate as antidandruff agent 499
Zinc in hair 410
Zinc omadine as antidandruff agent 421
Zinc oxide
  in baby products 113, 116, 117
  in cosmetic stockings 309
  in eye cosmetics 342
  in eyeshadow 348
  in face powders 286, 292, 297, 298, 299, 303, 305, 306, 308
  in foundation creams 69
  in liquid make-up 310
  in nail white 373
  in protective creams 83
  in rouge 334, 335
  in shampoos 446
  in sunburn products 257
  in sunscreen products 231, 232, 233
  interaction with preservatives 695
Zinc peroxide in nail bleaches 371
Zinc phenolsulphonate 654
Zinc pyridinethiol-N-oxide 654, 671
  as antidandruff agent 421
  in shampoos 458
Zinc pyrithione as antidandruff agent 19, 421
Zinc ricinoleate in deodorants 133
Zinc salts
  as astringents 74, 75, 76, 77
  in mouthwashes 628
  in toothpastes 604
Zinc stearate
  in cosmetic stockings 310
  in eyeshadow 351
  in face powders 288, 291, 292, 297, 298, 299, 304
  in protective creams 83
  in rouge 334, 335
  in shampoos 446
Zinc sulphate in acne therapy 123
Zinc undecylenate
  in athlete's foot products 203
  in shampoos 458
Zirconium chloride in nail strengtheners 373
Zirconium compounds as antiperspirants 127, 137
Zirconium fluoride in toothpastes 604
Zirconium salts as astringents 74, 75
Zirconium silicate in toothpastes 611